Solid Mechanics and Its A

Volume 213

Series editor

G. M. L. Gladwell, Waterloo, Canada

For further volumes:
http://www.springer.com/series/6557

Aims and Scope

The fundamental questions arising in mechanics are: *Why*? *How*? and *How much*? The aim of this series is to provide lucid accounts written by authoritative researchers giving vision and insight in answering these questions on the subject of mechanics as it relates to solids.

The scope of the series covers the entire spectrum of solid mechanics. Thus it includes the foundation of mechanics; variational formulations; computational mechanics; statics, kinematics and dynamics of rigid and elastic bodies: vibrations of solids and structures; dynamical systems and chaos; the theories of elasticity, plasticity and viscoelasticity; composite materials; rods, beams, shells and membranes; structural control and stability; soils, rocks and geomechanics; fracture; tribology; experimental mechanics; biomechanics and machine design.

The median level of presentation is the first year graduate student. Some texts are monographs defining the current state of the field; others are accessible to final year undergraduates; but essentially the emphasis is on readability and clarity.

Joshua Pelleg

Mechanical Properties of Ceramics

Joshua Pelleg
Materials Engineering
Ben Gurion University of the Negev
Beer Sheva
Israel

ISSN 0925-0042 ISSN 2214-7764 (electronic)
ISBN 978-3-319-37699-8 ISBN 978-3-319-04492-7 (eBook)
DOI 10.1007/978-3-319-04492-7
Springer Cham Heidelberg New York Dordrecht London

Softcover reprint of the hardcover 1st edition 2014

Printed on acid-free paper

Springer is part of Springer Science+Business Media (www.springer.com)

To my wife Ada, children Deenah, Ruth and Asher and their families

In memoriam—*my parents*

Gizella Chaya Golda and Zsigmond Asher Zelig Bruck

Preface

This textbook presents a general review of the *Mechanical Properties of Ceramics* and aims to provide an overall understanding of the subject. It surveys the various behaviors characteristic of ceramics in response to applied forces. The present approach emphasizes common denominators in the responses of all these materials to certain applied forces, while delineating the differences found between various classes of the large families of materials. By discussing the general mechanical behaviors of ceramics as a whole, rather than the specific behaviors of each type of ceramic separately, it is hoped that the readers, students, and engineers alike, will understand that these mechanical properties are governed by physical laws common to them all and relevant to all their applications.

The book has been arranged in a manner different from that found in many, if not most, of the textbooks dealing with mechanical behavior and follows, to some extent, the same framework established in my earlier book on the Mechanical Properties of Materials. As such, Chap. 1 presents all the basic tests and equations useful for a student entering a materials testing laboratory for the first time. Chapter 2 shows that there are ductile ceramics also at ambient temperature and discusses ductility at elevated temperatures, superplasticity, and other features directly influencing ductility or related to their strength properties. Chapter 3 establishes the theoretical foundations of mechanical properties and considers imperfections—point defects and dislocation-related concepts, setting the background for experimental observations. In Chap. 4, deformations, elastic and plastic, are discussed both at low and elevated temperatures, as are slip and twinning. Here, high-temperature deformation is emphasized, because even ceramics that are brittle at room temperature show some ductility at high temperatures. Ceramic strength and strengthening mechanisms are dealt with in Chap. 5. Time-dependent deformation (creep) is the subject of Chap. 6 and ceramics that are brittle or ductile at room temperature and superplastic ceramics are discussed. This chapter also considers the phenomenon of rupture and the design of materials to prevent creep. Cyclic deformation (fatigue) in ceramics is broadly discussed in Chap. 7, whereas static, time-dependent fracture, dynamic deformation and the theoretical strength of materials are the subjects of Chap. 8. In the ninth and final chapter, small grain-sized ceramics, in the nanosize range, are considered.

Actual problems are not presented for solution, so that each lecturer may devise his/her own problems to challenge the students. There is no need to repeat problems that appear in other textbooks. Suffice it to say that those interested in conceiving of practical problems that may arise in the field are encouraged to seek them (and their probable solutions) for their own benefit.

I would like to express my gratitude to all publishers and authors for permission to use and reproduce some of their illustrations and microstructures.

Finally, without the tireless devotion, help, understanding, and unlimited patience of my wife Ada, I could never have completed this book, despite my decades of teaching in this field; her encouragement was essential and her helpful attitude was instrumental in inspiring to write this book. Thanks to Ethelea Katzenell of Ben-Gurion University for improving the English.

Contents

Abbreviations

2D	Two-dimensional
3D	Three-dimensional
A-AT	Alumina-aluminum titanate
AFM	Atomic force microscopy
AT	Aluminum titanate
BDT	Brittle-to-ductile transition
BF	Bright field
BHN	Brinell hardness number
BMG	Bulk metallic glass
BNNTs	Boron nitride nanotubes
BT	Barium titanate, $BaTiO_3$
CAS	Calcium aluminosilicate
CDF	Central dark field
CG	Coarse-grained
CH	Crack healing
CIP	Cold isostatic pressing
CIT	Charpy impact test
CMC	Ceramic-matrix composite
CRSS	Critical resolved shear stress
CSL	Coincident site lattice
CVD	Chemical-vapor deposition
CVI	Chemical-vapor infiltration
DPH	Diamond pyramid hardness
DPN	Diamond pyramid hardness number
DSE	Directionally solidified eutectic
EDS	Energy dispersive analysis system, EDAX
FCC	Face-centered cubic
FG	Fine-grain
FIM	Field-ion microscopy
FWHM	Full width at half maximum
GBS	Grain-boundary sliding
HCP	Hexagonal closed packed
HEBM	High-energy ball milling
HIP	Hot isostatic pressure

HK	Knoop hardness
H-P	Hall-Petch
HREM	High-resolution electron microscopy
HRTEM	High-resolution transmission electron microscopy
HVEM	High-voltage electron microscopy
ISE	Indentation size effect
KHN	Knoop hardness number
KHT	Knoop hardness test
LD	Lomer dislocation
LH	Left-hand
LMP	Larson-Miller parameter
LPCVD	Low-pressure chemical vapor deposition
MAS	Magnesium alumino-silicate
MD	Molecular dynamic
MG	Monkman-Grant
Mg-PSZ	MgO-partially stabilized ZrO_2
MOE	Modulus of elasticity
MOR	Modulus of rupture
NI-FEM	Nano-indentation finite element method
NIST	National Institute of Standards and Technology
nt-cBN	Nano-twinned cubic boron nitride
NW	Nanowire
PBS	Phosphate buffered saline
PLC	Portevin-Le Chatelier
PMN	Lead magnesium niobate
PS	Plasma-sprayed
PSD	Particle-size distribution
PSR	Proportional specimen resistance
PSZ	Partially stabilized zirconia
PT	Lead titanate
PTC	Positive temperature coefficient
PZT	Lead-zirconate-titanate
RE	Rare earth
RH	Right-hand
RT	Room temperature
SANS	Small-angle neutron scattering
SCG	Subcritical crack growth
SEM	Scanning electron microscope
SENB	Single-edge notched beam
SEVNB	Single-edge, V-notched beam
SF	Stacking fault
SP	Shot peening
SPS	Spark plasma sintering
SPT	Strength parameter time
ST	Strontium titanate, $SrTiO_3$

STEM	Scanning-transmission electron microscopy
TBC	Thermal-barrier coating
TEM	Transmission electron microscopy
TGG	Templated grain growth
TZP	Tetragonal zirconia polycrystals
UH-HIP	Ultrahigh hot isostatic pressure
UHT	Ultra-high temperature
UHTC	Ultra-high temperature ceramics
VHT	Vickers hardness test
XRD	X-ray diffraction
Y-PSZ	Yttria partially stabilized zirconia
YSZ	Y_2O_3-stabilized zirconia
Y-TZP	Yttria-stabilized tetragonal zirconia polycrystals
ZTA	Zirconia toughened alumina

About the Author

Joshua Pelleg received his B.S. in Chemical Engineering Technion Institute of Technology, Haifa, Israel, M. S. in Metallurgy, Illinois Institute of Technology, Chicago, IL, USA and Ph.D. in Metallurgy, University of Wisconsin, Madison, WI. He is with Ben Gurion University of the Negev, Materials Engineering Department, Beer Sheva, Israel since 1970, was among the founders of the department, and its second chairman. Professor Pelleg was the recipient of the Sam Ayrton Chair in Metallurgy. He has taught ever since the subjects of Mechanical Properties of Materials, Diffusion in Solids and Defects in Solids. He has chaired several University committees and served four terms as the Chairman of Advanced Studies in Ben Gurion University. Prior to arriving at BGU, Pelleg acted as Assistant Professor and then Associate Professor in the Department of Materials and Metallurgy, University of Kansas, Lawrence, KS, USA. Professor Pelleg was Visiting Professor in: Department of Metallurgy, Iowa State University, Institute for Atomic Research, US Atomic Energy Commission, Ames, IA, USA, McGill University, Montreal, QC, Canada, Tokyo Institute of Technology, Applied Electronics Department Nagatsuta Campus, Yokohama, Japan and in Curtin University, Department of Physics, Perth, Australia. Among his non-academic research and industrial experience one can note: Chief Metallurgist in Urdan Netallurgical Works LTD., Netanya, Israel, Research Engineer in International Harvester, Manufacturing Research, Chicago IL., Associate Research Officer, National Research Council of Canada, Structures and Materials, National Aeronautical Establishment, Ottawa, ON, Physics Senior Research Scientist, Nuclear Research Center, Beer Sheva, Israel. Materials Science Division, Argonne National Labs, Argonne, IL, USA., Atomic Energy of Canada, Chalk River, Ont.

Canada, Visiting Scientist, CSIR, National Accelerator Centre, Van de Graaf Group Faure, South Africa, Bell Laboratories, Murray Hill, NJ, USA, GTE Laboratories, Waltham, MA, USA. His current research interests are diffusion in solids, thin film deposition and properties (mostly by sputtering) and characterization of thin films, among them various silicides.

Chapter 1
Mechanical Testing of Ceramics

Abstract This chapter considers the most common mechanical testing methods which are usually expected to be performed by students entering the first time into a lab. Tensile test-related parameters are evaluated. Very popular tests of ceramics are the various hardness tests (for example Vickers hardness test), which is not only a cost saving test, but also requires shorter times, since no specific specimen preparation, except of a smooth (often polished) surface is required. On small size specimens, Knoop hardness test is the general approach to obtain hardness data. Another accepted method of evaluating the mechanical properties of a ceramic is by a bending (flexural) test. The tests can be performed by three or four point bending tests. Compression tests are more popular than tension tests, since they tend to close pores, cracks and other flaws resulting in higher test results than by those obtained by tension, which tends to open rather than close cracks and microcracks. Toughness is an important criterion in ceramic properties (mechanical) evaluation. Because of the brittle nature of ceramics, special instrumented Charpy Impact Test machines were developed, primarily to evaluate the dynamic toughness of such materials. Creep and Fatigue tests are not included in this chapter and they will be evaluated in separate chapters. Because of the large scatter in the experimental results, Weibull statistical distribution is applied to obtain a mean value of the experimental results.

1.1 Introduction

Ceramic materials have been used by Mankind on Earth since the dawn of time. Early ceramics were made of clays hardened into various desirable and practical shapes. Such objects are called 'ceramics clay' and served both as useful tools and things of beauty. Ceramic materials are the end products of clays fired at high temperatures. Generally, ceramics clay articles are made by moistening a mixture of clays, casting it into desired shapes and then firing it to a high temperature, a process known as 'vitrification'. It would be hard to imagine human progress

J. Pelleg, *Mechanical Properties of Ceramics*, Solid Mechanics and Its Applications 213, DOI: 10.1007/978-3-319-04492-7_1,

without those very first steps when humans began to make utensils to meet daily needs. Tools and weapons were necessary for the struggle to survive the perils to which humans were exposed. Ceramics antedated most other materials and one cannot imagine metal smelting and casting without the necessary ceramic crucibles. In fact, the relatively late development of metallurgy in the history of Mankind was contingent on the availability of ceramics and the know-how to mold them into the appropriate forms. Without that primordial knowledge of ceramics fabrication today's expertise would not exist.

Moreover, one of the hottest scientific topics currently being debated is the role played by ceramics, or more specifically by clay, in the origin of life. Cath Harris discusses this in a recent publication entitled: "Did clay mould life's origins?" [54]. She reports that clay was suggested by crystallographer John Bernal as a means of concentrating primitive biomolecules onto its surface, making them available for further reactions [51]. Clays were spotlighted again more recently when James Ferris showed that they can act as catalysts for the formation of long strands of RNA (ribonucleic acid), crucial elements of the DNA proteins essential for the origin of life. Professor Don Fraser from the Department of Earth Sciences at the University of Oxford has carried out neutron scattering experiments to try and find out more about the role of geochemistry in determining the origin of our amino acids—key building blocks of life on Earth—and specifically why the DNA-coded amino acids that make up our proteins are all left-handed [henceforth: LH]—which has led many researchers to believe that clay served as a template for life [53]. The modern identification of various ceramics by their chemical composition and structure follows.

The atoms of the elements comprising the structures of various ceramic materials are attached to each other by: (a) ionic bonding, (b) covalent bonding or (c) a mixture of ionic and covalent bonding. Most ceramics, however, have mixed bonding. The ratio of ionic to covalent bonding determines the properties of the ceramics. This structural feature is the easiest way to refer to a particular ceramic. The elements comprising ceramics are either metallic or non-metallic (such as O, N, C); accordingly, ceramics are often classified as being 'oxide ceramics', 'nitride ceramics' or 'carbide ceramics'.

Some of the general features of ceramics may be summarized as follows:

(i) Contrary to metals, ceramics show low electric and thermal conductivities, due to the absence of free electrons which produce the bonds between atoms.
(ii) Ceramic materials, in general, are harder and stronger than metals, since their ionic and covalent bonds are much stronger than the relatively weak metallic bonds.
(iii) Ceramic structures are characterized by high melting points, high moduli of elasticity and high temperature and chemical stabilities. These qualities are strength related, namely, a consequence of the inter-atomic potential function versus the inter-atomic distance of covalent and ionic bonds. The minimum potential energy, pictured graphically as the potential well, is deeper for materials held together by these bonds.

(iv) Usually, ceramics are brittle, since free dislocation glide through the planes of the structure is impeded.
(v) As in metallic structures, ceramics are either crystalline or amorphous.

Because of the aforementioned characteristics of ceramics, they offer great advantages over metals in specific applications in which hardness, wear resistance and chemical stability at high temperatures are essential. However, the wide 'all-purpose' use of ceramics remains hampered by room (and lower) temperature brittleness, namely, the lack of sufficient ductility. If this disadvantage, which is an integral property of ceramic materials, could be overcome, then ceramics could be more widely applied. Thus, much research has been done to remove this limitation by improving ceramic brittleness by various methods with appreciable success (see, for example, [27, 45, 29]). Moreover, it has been observed that brittle ceramics can be made ductile in special cases, permitting large plastic deformation, even up to ~100 % at low temperatures, if a polycrystalline ceramic is produced with a low crystal size, of the order of a few nm, which apparently allows for the diffusional flow of atoms along the grain boundaries [29].

In the following sections, the commonly used tests applied to ceramic materials are discussed, mostly with respect to the monolithic ceramics, such as oxides, nitrides, carbides and borides. The vast number and types of existing ceramics preclude covering all of them here. However, consideration and emphasis will be focused on their structural feasibility for various applications and on their relevant, practical mechanical properties.

1.2 Tension Test

Tension tests provide information on the strength and deformation of materials under uniaxial tensile stresses. To evaluate the strength of ceramics, typically a brittle material, one must test a statistically significant number of specimens in order to obtain a reliable average value. Note that specimen size affects the strength values. This test provides information on strength and deformation, but uniform stress states are required to obtain a meaningful value characterizing the ceramics under uniaxial tensile stress. The test conditions, subcritical cracks (which grow relatively slowly, i.e., fast test rates are recommended), other flaws and environmental effects resulting in stress corrosion all influence the outcome of such tests. Therefore, specimens from particular ceramics should be produced in standard dimensions, although one must still bear in mind that the test results for specimens do not necessarily totally represent the strength properties of an entire, full-sized item. Thus, even the results of tests done on standardized specimens only represent the overall strength properties of a certain ceramic material processed in a particular manner and are less indicative of the same or similar ceramics processed and treated differently. Furthermore, a uniaxial tension test is meaningful if it is applied primarily to ceramics that macroscopically exhibit isotropic, homogeneous and continuous behavior. This way, the probabilistic strength distribution of the brittle ceramic material may be converged into a more characteristic value for the tested material. In addition, the method of specimen fabrication, the testing technique, the strain rate used, etc., are important factors to be considered when the results are expected to be characteristic of a particular ceramic.

Figure 1.1a is a typical stress–strain curve for a brittle ceramic having only elastic deformation up to the point of fracture. As indicated in the introduction, ceramics fail in a typically brittle manner, due to the ionic nature of the bonds, which prevent slip via dislocation motion. The fact that brittle catastrophic failure in ceramics is likely is an indication that very little energy is absorbed in the process of fracture. Pure aluminum oxide behaves as indicated in Fig. 1.1b. The fracture strain of a ceramic is $\sim$0.0008–0.001. One can state that ceramics at room temperature are Hookean until fracture. In general, ceramic materials experience very little or no plastic deformation prior to fracture. Slip is difficult due to the structure and the strong local electrostatic potentials (a consequence of the ionic or covalent bonds).

In Fig. 1.1b, a glass-like ceramic, which is usually amorphous, is also shown indicating the same Hookean behavior without plastic deformation, but at a much lower fracture stress than in a crystalline alumina, for example.

The brittle fracture of ceramics is predominantly the result of unavoidable microscopic flaws (micro-cracks, internal pores and atmospheric contaminants), the presence of which are production outcomes and occur during cooling from the melt. It is difficult to thoroughly control the formation of these flaws during manufacturing with the consequent large scatter in the experimental results of the test specimens. The growth of micro-cracks into a crack formation and its propagation occur perpendicular to the applied stress transgranularly along specific cleavage planes until fracture sets in.

The tensile testing of brittle materials is difficult to perform satisfactorily by straight uniaxial testing in the conventional tensile testing machines. Specimen gripping is also a common problem in uniaxial testing. The disagreement between various tests is high, sometimes approaching 100 %, due to the lack of sufficient ductility to allow the relief of misalignment stresses. Efforts to overcome technical problems in the uniaxial tensile testing of brittle materials are described in the literature for round and flat specimens. Seshadri and Chia [43] developed a special test fixture for the uniaxial testing of flat ceramic specimens which eliminates premature gripping failures and stress-concentration-related problems arising from misalignment in uniaxial tension. Some tests have been performed on SiC specimens using this fixture. Figure 1.2 is reproduced from their work. A tension test was carried out at 23° C in air using a crosshead rate of $8.4 7 \times 10^{-6}$ m/s (0.02 in./min). This test resulted in a Young's modulus of 420.9 ± 9.8 GPa, which is very consistent with the published data. Their specimens are essentially plate specimens of uniform thickness (3.2 mm) with wedge-shaped ends for improved gripping. The fraction of specimens failing in the gage section may be used as a qualitative measure of the alignment. The fractions for the sintered alpha SiC were $\sim$55 %.

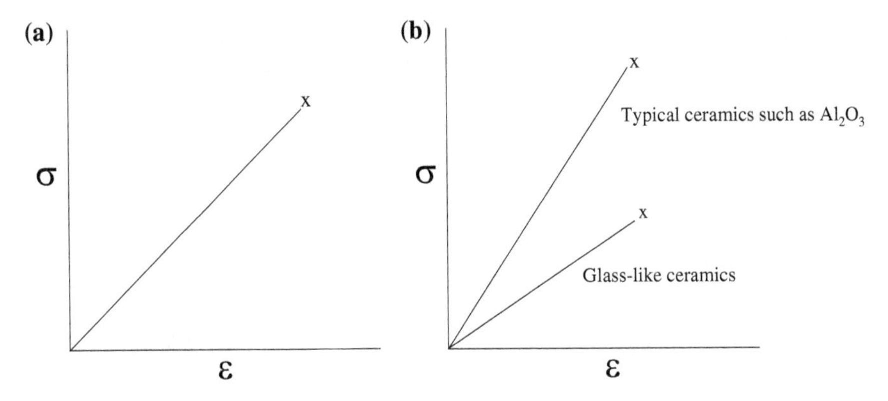

Fig. 1.1 Stress–strain curve of a ceramic material: **a** only elastic deformation; **b** typical ceramics, such as Al_2O_3 and glass

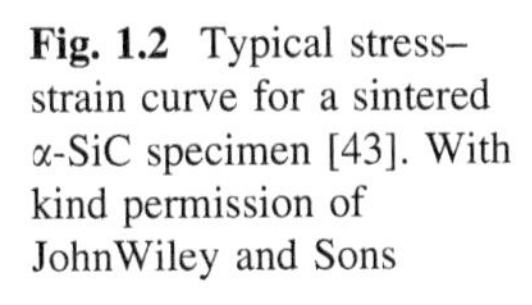

Fig. 1.2 Typical stress–strain curve for a sintered α-SiC specimen [43]. With kind permission of JohnWiley and Sons

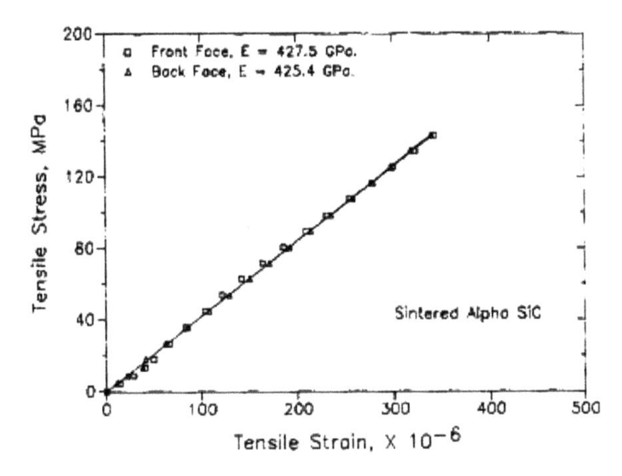

Seshadri and Chia [43] claim that these values are quite comparable to similar test configurations used in sintered copper-steel specimens (i.e., about the same as in metallic specimens).

The reproducibility and applicability of this test procedure has been established as being appropriate for advanced structural ceramic specimens. Figure 1.3 shows the microstructure of the tensile fracture surface for the sintered alpha SiC.

As indicated, the inability to slip makes ceramics more difficult to deform. However, since ceramics behave like a Hookean body until fracture, the known stress–strain relations in elastic deformation can be applied. Assuming that the force, P, is acting normally on a small area, ΔA, of a ceramic test specimen:

$$\sigma = \lim_{A \to 0} \frac{dP}{dA} = \frac{dP}{dA} \tag{1.1}$$

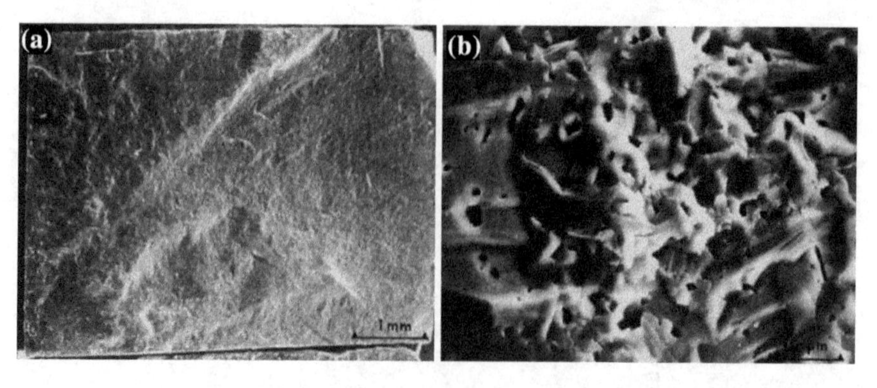

Fig. 1.3 **a** Typical tensile fracture surface of sintered α-SiC ($\sigma = 163$ MPa). **b** Higher magnification shows the transgranular fracture mode [43]. With kind permission of John Wiley and Sons

or

$$\mathrm{dP} = \sigma \mathrm{dA} \tag{1.2}$$

Integrate (1.2)

$$P = \int \sigma dA \tag{1.3}$$

Although ceramics are often not completely uniform in their structure (as mentioned above) and have various flaws, it is assumed that the stress is distributed uniformly over the cross-section of the test specimen and, thus, acts at a constant level, rewriting (1.3) as:

$$P = \sigma \int dA = \sigma A \tag{1.4}$$

As such, the stress is absorbed by the fracture as:

$$\sigma = \frac{P}{A} \tag{1.4a}$$

Figure 1.2 for α-SiC was obtained by using thin-plate specimens (3.2 mm) of uniform thickness with tapered regions for gripping, as illustrated schematically in Fig. 1.4. The purpose of the tapered parts at the ends of the specimens was to accommodate the gripping blocks and to provide good alignment in the plane parallel to the specimen (suggested by Seshadri and Chia [43]). The size of the region between these tapered edges is 50 mm. Tensile specimen dimensions are given in the ASTM standards for the testing of ambient advanced ceramics with solid rectangular cross-sections. The Hookean behavior of the stress–strain relation gives:

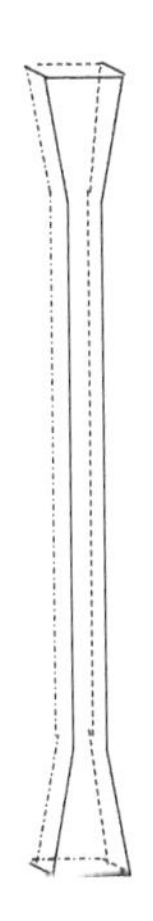

Fig. 1.4 Schematic shape of the plate-like specimen used in tensile testing resulting in uniform and reproducible values

$$\sigma = \text{Ee} \tag{1.5}$$

In (1.5), e is the average linear strain, which correlates the change in specimen dimension with its original length and which may be expressed as:

$$e = \frac{l - l_0}{l_0} = \frac{\Delta l}{l_0} = \frac{1}{l_0} - 1 \tag{1.6}$$

l_0 is the original length within the gage length of the specimen, while Δl is the axial change resulting from the elastic deformation. Thus, Δl is often referred to as the 'deformation'. Equation (1.6), linear strain, may also be expressed as:

$$e = \int_{l_0}^{l} \frac{dl}{l_0} = \frac{l - l_0}{l_0} \tag{1.7}$$

Important amorphous ceramics are the glass-like materials. A schematic stress–strain curve is indicated in Fig. 1.1b and shows a lower overall stress–strain curve. However, it would be difficult to enumerate all the many types of glass with a wide variety of compositions, ranging from the most common window glass to the various metallic glasses. In general however, internal and external factors influence the performance of ceramic materials even to the point at which ductility can be induced. Here are some of these factors, especially those that have critical effects on ceramic (and glass) behavior:

Internal factors of major influence are:

(a) Grain size
(b) Pores
(c) Other flaws, such as micro-cracks.

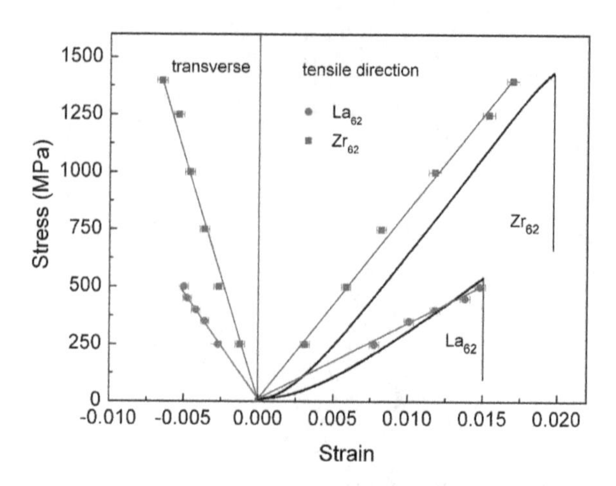

Fig. 1.5 (*Color online*) Strains determined from the diffraction data of tensile/transverse directions. In addition, the tensile stress–strain curves of $Zr_{62}Al_8Ni_{13}Cu_{17}$ and $La_{62}Al_{14}(Cu_{5/6}Ag_{1/6})Co_5Ni_5$ bulk metallic glasses (BMG) are also included for comparison [46]. Reprinted with kind permission of the American Institute of Physics

External factors:

(a) Composition
(b) Specimen size
(c) Specimen shape.

For instance, the tensile stress–strain properties of a bulk metallic glass (BMG) is shown in Fig. 1.5.

Metallic glasses are recognized as being true glasses and behave as any of the aforementioned brittle materials. BMGs are of current interest, because of their wide spectrum of applications. Similarly to conventional oxide glasses (such as silica glasses), BMGs are brittle at room temperature when tested under tension. This is shown in Fig. 1.5 for two types of BMG, where the test was performed in both directions, i.e., also in the transverse direction. It is interesting that the colored lines shown in Fig. 1.5 were obtained in situ by using synchrotron X-ray diffraction [henceforth: XRD]. The tensile stress–strain curves of these two BMGs at room temperature with a strain rate of $1 \times 10^{-4}\ s^{-1}$ are also illustrated Fig. 1.5. Although some BMGs exhibit pronounced plasticity under uniaxial compression or bending conditions, they are generally destroyed with catastrophic failure under tension at room temperature and at a slow strain rate. No tensile plasticity is observed in the specimens illustrated in Fig. 1.5. By linearly fitting the points and calculating the ratio of the strains between the transverse and tensile directions for each alloy, the tensile elastic modulus and Poisson's ratio were obtained as ~83 GPa and 0.37 for $Zr_{62}Al_8Ni_{13}Cu_{17}$ BMG and 34 GPa and 0.36 for $La_{62}Al_{14}(Cu_{5/6}Ag_{1/6})Co_5Ni_5$ BMG, respectively. The strain was calculated by using the relation:

$$\varepsilon = \frac{d - d_0}{d_0} = \frac{q_0 - q}{q} \tag{1.8}$$

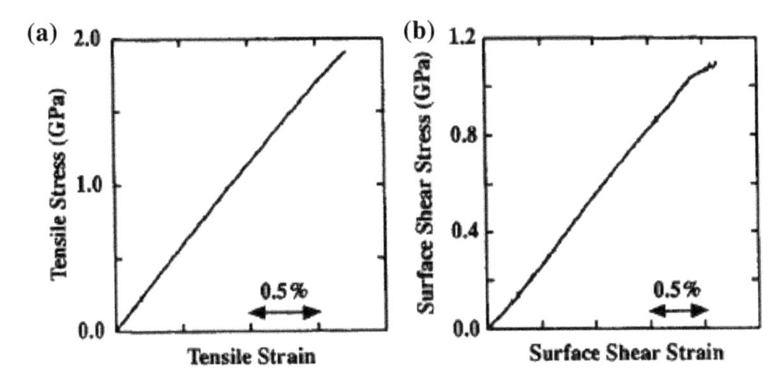

Fig. 1.6 **a** Tension stress–strain curve; **b** torsion stress–strain curve [23]. With kind permission of Elsevier

where d_0 and d indicate the peak positions in real space under zero stress and σ stress while the difference between q_0 and q (the first peak positions under the aforementioned stress conditions) were derived from the use of structural measurements from in situ synchrotron XRD.

To indicate the different deformation-mode-dependent behaviors in amorphous BMGs, Figs. 1.6 and 1.7 indicate the stress–strain relations under tension and compression of $Zr_{41.25}Ti_{13.75}Ni_{10}Cu_{12.5}Be_{22.5}$ bulk amorphous alloys. The strength of amorphous BMGs often approach theoretical limits, while the plastic deformation depends predominantly on temperature, but also on strain rate and mode of deformation.

In Fig. 1.7, the $Zr_{41.25}Ti_{13.75}Ni_{10}Cu_{12.5}Be_{22.5}$ alloy indicates yielding under compression and the plastic flow proceeds at a constant stress, contrary to the tension test shown in Fig. 1.6a. Under tension, the yield stress coincides with the fracture stress, which is usually observed in brittle materials without ductility.

Specimen size, in general, has an effect on mechanical tests, and ceramics or glasses are no exception. The shapes of ceramic and glass test specimens are often a matter of convenience, but, in most cases, the specimen's form dictates the test conditions. Various other techniques have also been suggested to improve the reliability of the strength result. One such technique, used for the tensile testing of ceramic fibers, is video extensometry [25]. Figure 1.8 illustrates a schematic video set up for the evaluation of the results of SiC monofilaments.

The video-extensometer eliminates the use of an external extensometer, which can cause micro-cracking during handling. Furthermore, small-sized specimens in the micron range may be tested using this method. Here, the video-extensometer was used to evaluate the tensile properties of silicon carbide monofilaments 100 μm in diameter and to determine the change in distance, Δl between the marked targets, caused by the mechanical strain to the specimen. The strain is then calculated in the conventional manner, as indicated in Eq. (1.6), rewritten here as:

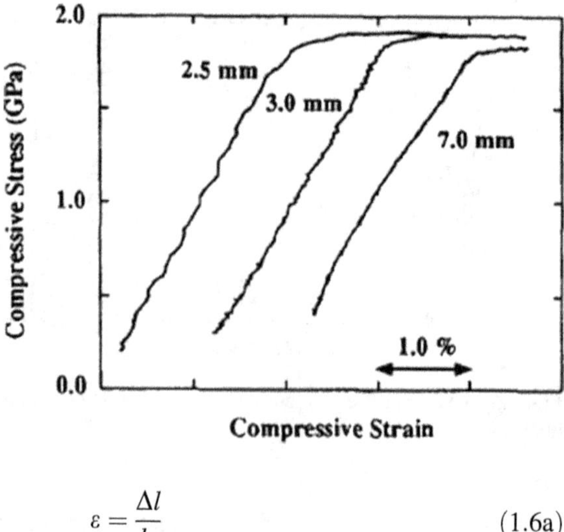

Fig. 1.7 Compressive stress–strain curves [23]. With kind permission of Elsevier

$$\varepsilon = \frac{\Delta l}{l_0} \tag{1.6a}$$

The camera (shown in Fig. 1.8) is focused on the specimen. For details of the measuring technique, the original work of Coimbra et al. [25] may be consulted. The sensitivity of the video-extensometer was tested by varying the crosshead speed, using the rates of 0.1, 0.2 and 0.6 mm/min. The crosshead speed had no effect on the mechanical properties of SiC monofilaments between 0.1 and 0.6 mm/min. Very good agreement was found between the Young's modulus experimental data and the expected values. The tensile strength values (2100–2600 MPa), however, were lower than 3500 MPa, due to surface micro-cracking caused by handling. Different fiber lengths were tested (80, 120 and 160 mm) at a crosshead speed of 0.2 mm/min and the initial distance between the targets was kept between 40 and 60 mm in all cases. No influence of the specimen length was noticed, which means that testing fibers with the video-extensometer does not require specific specimen lengths to get reproducible results. Figure 1.9 is an illustration of the stress-strain curve obtained by Coimbra et al. [25].

Polycrystalline alumina fibers were also tested in this way with very good agreement between the experimental mean values and the expected data (less than 2 %). It was shown that video-extensometry could be a successful method for recording the elastic deformation and brittle failure of ceramic fibers.

1.2.1 Tension of Ductile Ceramics

Most ceramic materials are brittle at room temperature. However, in the case of ductile materials, namely when the elastic limit and the fracture stress do not coincide due to some plasticity observed in the deformation process, the relation can be given, similarly to Eq. 1.7, as:

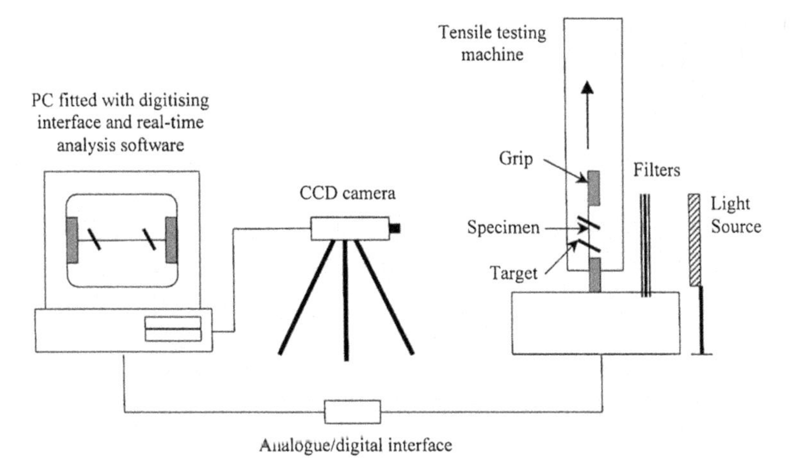

Fig. 1.8 General diagram of the video-extensometer connected to the tensile testing system. PC is a fast processor which allows real-time acquisition and analysis of the data [25]. With kind permission of Springer

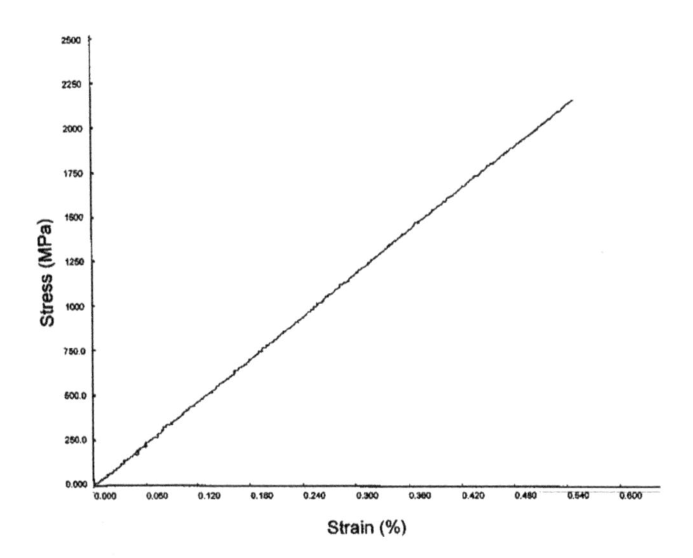

Fig. 1.9 Typical experimental stress-strain curve of SiC monofilament [25]. With kind permission of Springer

$$\varepsilon = \int_{l_0}^{l_i} \frac{dl}{l} = \ln \frac{l_i}{l_0} \tag{1.9}$$

Here clearly, l_i is the instantaneous length of the test specimen. Equation 1.9 is the summation of all the small changes between the two values and, thus, may be expressed as:

$$\varepsilon = \Sigma\left(\frac{l_1 - l_0}{l_0} + \frac{l_2 - l_1}{l_1} + \frac{l_3 - l_2}{l_2} + \ldots\right) \tag{1.10}$$

Sometimes, it is necessary to alternate between these two definitions of strain, namely e and ε, which can be easily done, as shown below, by rendering Eq. (1.6) as:

$$(e + 1) = \frac{l}{l_0} \tag{1.6b}$$

Then, it is possible to apply Eq. (1.9) as follows:

$$\varepsilon = \ln\frac{l}{l_0} = \ln(e + 1) \tag{1.9a}$$

expanding Eq. (1.9a):

$$\varepsilon = \ln(e + 1) = e - \frac{e^2}{2} + \frac{e^3}{3} - \ldots \tag{1.9b}$$

and, for small strains, as is the case in most (brittle) ceramic materials showing limited ductility, Eq. (1.9b) actually becomes:

$$\varepsilon \cong e \tag{1.9c}$$

When, for example, a load is acting on an elastic body in x direction, it elongates not only in the direction of the acting load, but contracts laterally, as well. Thus, contraction must occur in the transverse y and z directions. Empirically, it was observed that transverse strains are constant fractions of longitudinal extension. The ratio of the lateral contractive strain to the axial strain is called 'Poisson's ratio', denoted by ν and expressed as:

$$\nu = -\frac{\textbf{\textit{lateral strain}}}{\textbf{\textit{axial strain}}} = -\frac{\varepsilon_y}{\varepsilon_x} = -\frac{\varepsilon_z}{\varepsilon_x}. \tag{1.11}$$

This ratio denotes a reduction in cross-section elongation. In brittle materials, there is a small change in the cross-section with elongation, so ν is low. Thus, when a sample of material is stretched on one axis, it tends to get thinner also on the other two axes. If, during the uniaxial tension, no lateral contraction occurs, then $\nu = 0$.

Another expression for the Poisson's ratio connects the elastic modulus, E, with the shear modulus and is given by:

$$G = \frac{E}{2(1 + \nu)} \tag{1.12}$$

or

$$\nu = \frac{E}{2G} - 1 \tag{1.12a}$$

1.2.2 Stress Tensor

This section introduces the concept of the 'stress tensor'. Stress tensors are especially significant for brittle materials, such as ceramics and essential for defining stress at a certain point. In addition, they help to construct a 'Mohr's circle of stress' (discussed later). The following discusses Eq. (1.12), explaining what happens when three normal stresses act simultaneously on a test specimen. The simplest way to understand this is by visualizing a cube of unit dimensions under σ_x, σ_y and σ_z stresses, creating a three-dimensional state-of-stress problem. Consider an elementary cube with edges dx, dy and dz removed from a structural body in equilibrium upon which external forces are applied. The stress acting on each plane (shown in Fig. 1.10) can be resolved into normal stress components (σ_x, σ_y and σ_z) and two shear-stress components.

The stress defined at a specific point (described by the elementary cube, i.e., upon which the cube is converging) is a tensor:

$$\sigma_{ij} = \begin{matrix} \sigma_{11} & \sigma_{12} & \sigma_{13} \\ \sigma_{21} & \sigma_{22} & \sigma_{23} \\ \sigma_{31} & \sigma_{32} & \sigma_{33} \end{matrix} \tag{1.13}$$

All the stress symbols in Eq. (1.13) are denoted by subscripts, though the designations from Fig. 1.10 are very often used. Thus, equivalent designations for the normal and the shear stresses are shown respectively in (1.13a).

$$\begin{aligned} &\sigma_{11}, \sigma_{22}, \sigma_{33} \text{ are equivalent to } \sigma_x, \sigma_y, \sigma_z \\ &\sigma_{12}, \sigma_{13}, \sigma_{21}, \sigma_{23}, \sigma_{31}, \sigma_{32} \text{ are equivalent to} \\ &\qquad\qquad \tau_{xy}, \tau_{xz}, \tau_{yx}, \tau_{yz}, \tau_{zx}, \tau_{zy}. \end{aligned} \tag{1.13a}$$

Expression (1.13) may also be written as:

$$\sigma_{ij} = \begin{matrix} \sigma_x & \tau_{xy} & \tau_{xz} \\ \tau_{yx} & \sigma_y & \tau_{yz} \\ \tau_{zx} & \tau_{zy} & \sigma_z \end{matrix} \tag{1.13b}$$

In Eq. (1.13) or (1.13b), the diagonal symbols are the normal stress components, while the off-diagonal elements are the shear-stress components acting tangentially on the shown faces of the elemental cube. Note that, when the normal

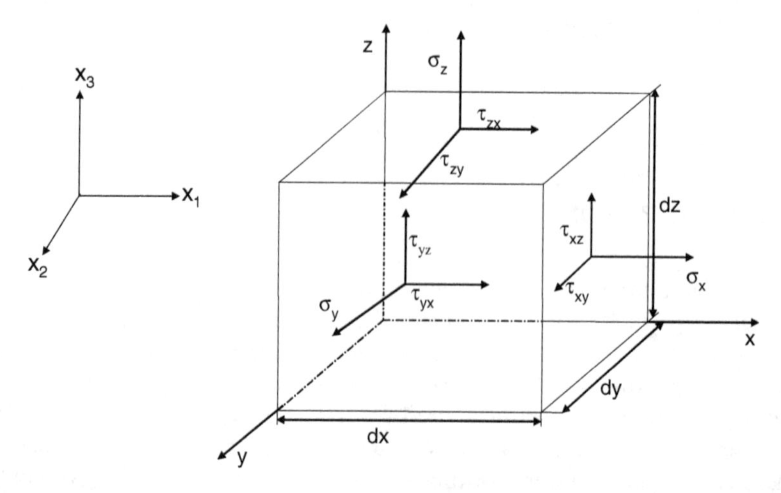

Fig. 1.10 An elementary cube, removed from a structural material, is shown under stress-like forces. These are the components of a stress tensor. Often, the coordinate system is shown as indicated on the left of this figure

stress components are directed outward from the plane, they are defined as positive (i.e., tensile stress). Furthermore, if the positive, normal-stress component is directed in the positive coordinate direction (as shown in Fig. 1.10), then the shear-stress components are also positive in the positive coordinate direction.

1.2.3 Stress on an Inclined Plane

The 'Cauchy stress tensor', $\boldsymbol{\sigma}$, at any point of a body (assumed to behave as a continuum) is completely defined by nine component stresses–three orthogonal, normal stresses and six orthogonal, shear stresses. It is used for the stress analysis of materials undergoing small deformations, in which the differences in stress distribution, in most cases, can be neglected.

Figure 1.10 shows the nine components of the stress tensor required to determine the stress state around an arbitrary point. Assuming that a point, P, is at the origin of a Cartesian coordinate system, Fig. 1.11 shows the stress components acting at an inclined plane labeled A, B and C. The directional cosines of the inclined plane are α_i (i.e., α_1, α_2, α_3).

Recall that the 'directional cosines' of a vector are often defined as being the cosines of the angles that the vector makes with the x, y and z axes, respectively. These angles are labeled: α (the x axis' angle), β (the y axis' angle) and γ (the z axis' angle), while defining: $l = \cos\alpha$, $m = \cos\beta$ and $n = \cos\gamma$.

Let the stress on the inclined plane in Fig. 1.11 be $\sigma_i = (\sigma_1, \sigma_2, \sigma_3)$ and using the labels for the notation from Eq. (1.13):

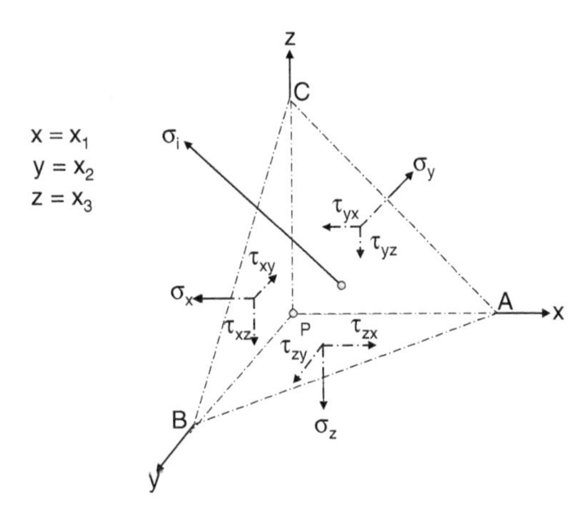

Fig. 1.11 A stress acting on an inclined surface

Let the area of ABC (Fig. 1.11) be s, giving the following for the areas in the tetrahedron:

$$\begin{aligned} \text{Area BPC} &= s_1 = s\alpha_1 \\ \text{Area APC} &= s_2 = s\alpha_2 \\ \text{Area APB} &= s_3 = s\alpha_3 \end{aligned} \tag{1.14}$$

The sum of the forces acting on tetrahedron ABCP should be zero. The force components, in terms of their x, y and z directions, may be expressed as:

$$\sum F_1 = \sum F_2 = \sum F_3 = 0 \tag{1.15}$$

The forces in the x, y, z directions are obtained by multiplying the stresses by the areas upon which they act, giving these forces in their respective directions as:

in the x direction:

$$\sum F_1 = 0 = \sigma_1 s - \sigma_{11} s\alpha_1 - \sigma_{21} s\alpha_2 - \sigma_{31} s\alpha_3 = 0 \tag{1.16a}$$

When eliminating s and expressing the relation for σ_1:

$$\sigma_1 = \sigma_{11}\alpha_1 + \sigma_{21}\alpha_2 + \sigma_{31}\alpha_3 \equiv \sigma_{i1}\alpha_i \tag{1.16b}$$

In line with the notation in Figs. 1.10 and 1.11 and as indicated in Eqs. (1.13a) and (1.16b), this equation may also be rewritten as:

$$\sigma_1 = \sigma_x\alpha_1 + \tau_{yx}\alpha_2 + \tau_{zx}\alpha_3 \equiv \sigma_{i1}\alpha_i \tag{1.16c}$$

Note that σ_i is not a normal stress and is acting on a surface at a point indicated in Fig. 1.11. In Eqs. (1.16b) and (1.16c), the sum of all the components acting in the x direction are indicated.

In the same way, a similar relation in the y direction may be given as:

$$\sum F_2 = 0: \qquad \sigma_2 = \sigma_{12}\alpha_1 + \sigma_{22}\alpha_2 + \sigma_{32}\alpha_3 \equiv \sigma_{i2}\alpha_i \tag{1.16d}$$

and in the notation of (1.13a) as:

$$\sigma_2 = \tau_{xy}\alpha_1 + \sigma_y\alpha_2 + \tau_{zy}\alpha_3 \tag{1.16e}$$

This may also be repeated in the z direction as:

$$\sum F_3 = 0: \qquad \sigma_3 = \sigma_{13}\alpha_1 + \sigma_{23}\alpha_2 + \sigma_{33}\alpha_3 \equiv \sigma_{i3}\alpha_i \tag{1.16f}$$

or, as before, in the notation of (1.13a) as:

$$\sigma_3 = \tau_{xz}\alpha_1 + \tau_{yz}\alpha_2 + \sigma_z\alpha_3 \tag{1.16g}$$

Thus, the Eqs. (1.16b), (1.16d) and (1.16f) may be rewritten as:

$$\begin{aligned}
\sigma_1 &= \sigma_{11}\alpha_1 + \sigma_{21}\alpha_2 + \sigma_{31}\alpha_3 \equiv \sigma_{i1}\alpha_i \\
\sigma_2 &= \sigma_{12}\alpha_1 + \sigma_{22}\alpha_2 + \sigma_{32}\alpha_3 \equiv \sigma_{i2}\alpha_i \\
\sigma_3 &= \sigma_{13}\alpha_1 + \sigma_{23}\alpha_2 + \sigma_{33}\alpha_3 \equiv \sigma_{i3}\alpha_i
\end{aligned} \tag{1.17}$$

One can briefly summarize the above Eqs. (1.16b), (1.16d) and (1.16f) for the stresses at any point on the inclined plane by writing the stress tensor in any Cartesian coordinate system as:

$$\sigma_i = \sigma_{ij}\alpha_j \tag{1.17a}$$

Adopting the notation in Eqs. (1.16c), (1.16d) and (1.16g) for the notation of Figs. 1.10 and 1.11, they may also be rewritten as:

$$\begin{aligned}
\sigma_1 &= \sigma_x\alpha_1 + \tau_{yx}\alpha_2 + \tau_{zx}\alpha_3 \\
\sigma_2 &= \tau_{xy}\alpha_1 + \sigma_y\alpha_2 + \tau_{zy}\alpha_3 \\
\sigma_1 &= \tau_{xz}\alpha_1 + \tau_{yz}\alpha_2 + \sigma_z\alpha_3
\end{aligned} \tag{1.17b}$$

In the above, α_1, α_2 and α_3 are called the 'directional cosines' and are the components of the normal unit vector, i.e., the line of unit length perpendicular to a surface at point P. They are usually written as:

$$\begin{aligned}
\alpha_1 &= \cos(n, x_1) \\
\alpha_2 &= \cos(n, x_2) \\
\alpha_3 &= \cos(n, x_3)
\end{aligned} \tag{1.18}$$

or:

$$\alpha_i = \cos(n, x_i) \tag{1.18a}$$

where i is 1, 2, 3.

In Eq. (1.18), the angles between the normal and the x_i axes are the respective angles. (Clearly, x_1, x_2 and x_3 are the x, y and z coordinates, as indicated in Fig. 1.11.) The stress vector is dependent on the position in space and on the inclination, as determined by the directional cosines of the normal to the surface upon which it is acting.

Depending on the orientation of the plane under consideration, a stress vector may not necessarily be perpendicular to that plane. In Fig. 1.11, σ_i is not normal to the inclined plane. It is often advantageous to resolve this stress both for its normal component vis-a-vis the surface, σ_N (also indicated as N) and for its shear component parallel to the surface, T (τ is a very common designation). The components of the normal σ_N are:

$$\sigma_N = \sigma_1\alpha_1 + \sigma_2\alpha_2 + \sigma_3\alpha_3 \tag{1.19}$$

and those of the shear components are:

$$T = (\sigma^2 - \sigma_N^2)^{1/2} = (\sigma_1^2 + \sigma_2^2 + \sigma_3^2 - N^2)^{1/2} = (\sigma_i\sigma_i N^2)^{1/2} \tag{1.19a}$$

T is often referred to it as 'traction' and clearly it represents a stress vector acting parallel to the surface. From the Pythagorean relation shown in Fig. 1.12, the sum of the squares of the normal and tangential stresses on any face of an elementary cube under arbitrary stress is equal to their sum. Furthermore, the shear component, T, is usually not parallel to any of the axes of a chosen coordinate system, as indicated in Fig. 1.12. It is, however, common to resolve this shear stress into two components, each of which is parallel to a chosen reference coordinate system. In Fig. 1.13, the stresses are indicated on the z plane.

Substituting the respective components of (1.17) into (1.19), one obtains:

$$\sigma_N = \sigma_{11}\alpha_1^2 + \sigma_{22}\alpha_2^2 + \sigma_{33}a_3^2 + 2\sigma_{12}\alpha_1\alpha_2 + 2\sigma_{23}\alpha_2 a_3 + \sigma_{31}\alpha_3\alpha_1 \tag{1.20}$$

If instead, (1.17b) is substituted into (1.19), the resulting notation of Fig. 1.11 becomes:

$$\sigma_N = \sigma_x\alpha_1^2 + \sigma_y\alpha_2^2 + \sigma_z\alpha_3^2 + 2\tau_{xy}\alpha_1\alpha_2 + 2\tau_{yz}\alpha_2\alpha_3 + 2\tau_{zx}\alpha_3\alpha_1 \tag{1.20b}$$

Equation (1.20b) is often found written differently. The directional cosines from Sect. 1.2.3, paragraph 2 (above) were in brackets, in terms of l, m and n; thus, one can rewrite (1.20b) accordingly as:

$$\sigma_N = \sigma_x l^2 + \sigma_y m^2 + \sigma_z n^2 + 2\tau_{xy}lm + 2\tau_{yz}mn + 2\tau_{zx}nl \tag{1.20c}$$

Each of the shear components in (1.20), (1.20b) and (1.20c) are then multiplied by 2 as a consequence of the following. Consider the elementary cube in Fig. 1.10, with dimensions dx, dy and dz, and imagine that this cube is shrinking to a very small size, almost to the point about which the cube was drawn. In the absence of gravitational or other forces to maintain equilibrium, the moments about the axes on the opposite faces must be equal. Thus, the moment produced by τ_{xz} on the z axis acting on area (dydz) at a distance, dx, may be written as: τ_{xy}(dydz)dx.

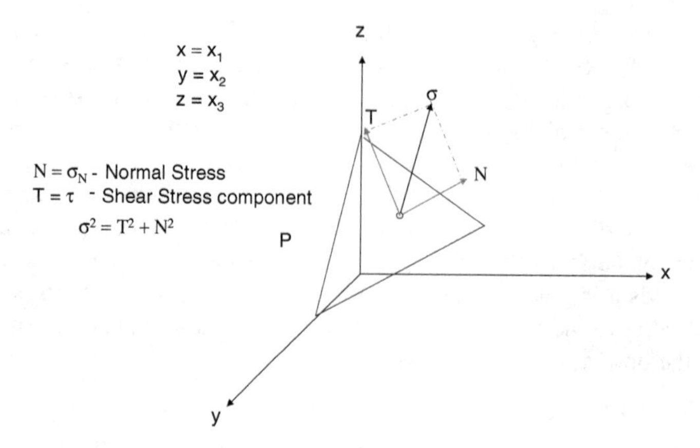

Fig. 1.12 An arbitrary stress acting on a point is resolved into normal and shear stress components

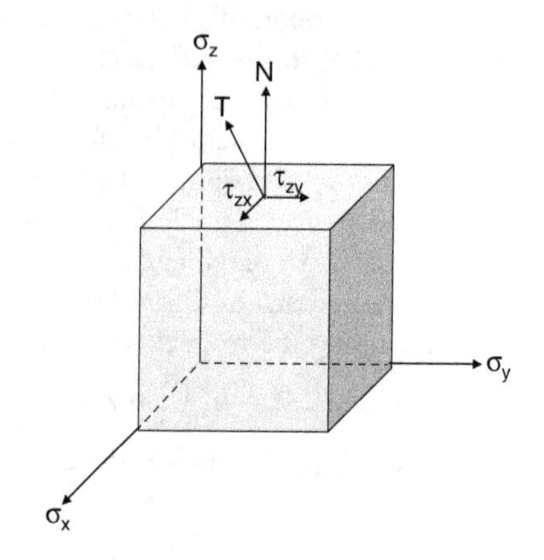

Fig. 1.13 The shear stress, T, in Fig. 1.12 is further resolved into two components parallel to the respective coordinate axes. Both the normal and shear stresses are acting on the *z* plane

Similarly, the moment produced by τ_{yx} acting on the (dxdz) plane at a distance, dy, is τ_{yx}(dzdz)dy. Therefore:

$$\tau_{xy}(\mathrm{dydz})\mathrm{dx} = \tau_{yx}(\mathrm{dzdz})\mathrm{dy} \tag{1.21}$$

or

$$\tau_{xy} = \tau_{yx} \tag{1.21b}$$

Similar relations are obtained at equilibrium of the cube about axes x and y, resulting in:

$$\tau_{yz} = \tau_{zy} \quad \text{and} \quad \tau_{zx} = \tau_{xz} \tag{1.21c}$$

These relations explain the factor 2 in the above relations for σ_N. Since $\tau_{xy} = \tau_{yx}$, $\tau_{yz} = \tau_{zy}$ and $\tau_{zx} = \tau_{xz}$ for arbitrarily chosen orthogonal axes, only six stress components are needed to define the stress at any point (instead of nine).

1.2.4 Principal Stresses

When a stress perpendicular to a surface acts at a point on that surface in the absence of shear stresses, that stress is called 'principal stress'. In other words, it is possible to find an orientation such that, for any point on a surface represented by the elementary cube, the shear stresses on the cube's surfaces vanish. Thus, $\tau_{xy} = \tau_{yz} = \tau_{zx} = 0$ and only three normal stresses (of the nine) remain. The three perpendicular planes and the three coordinates are the principal planes and the principal axes, respectively. Directions along the principal axes are known as 'principal directions'. The stress components vary as the orientation of the orthogonal, coordinate axes changes. Figure 1.14 illustrates the principal stresses and their directions for a point inside a body, compared to the initial system of coordinates and the stress tensor after rotation.

1.2.4.1 Calculation of Principal Stresses

In Fig. 1.15, an inclined plane is indicated with a normal stress, σ ($\equiv \sigma_n \equiv \sigma_N$), which, when multiplied by the directional cosines, may be resolved into its components on the basis of Eq. (1.17) by means of the designations indicated in Fig. 1.15 (on its left side) as:

In Eq. (1.17), the indicated stress is acting at a point on an inclined plane, reproduced here as Eq. (1.22). An inclined plane is defined by its directional cosines; in the absence of shear stress and, if the stress is normal (as shown in Fig. 1.15), it is a 'principal stress'. It follows that:

$$\begin{aligned}
\sigma_1 &= \sigma\alpha_1 = \sigma_{11}\alpha_1 + \sigma_{21}\alpha_2 + \sigma_{31}\alpha_3 \equiv \sigma_{i1}\alpha_i \\
\sigma_2 &= \sigma\alpha_2 = \sigma_{12}\alpha_1 + \sigma_{22}\alpha_2 + \sigma_{32}\alpha_3 \equiv \sigma_{i2}\alpha_i \\
\sigma_3 &= \sigma\alpha_3 = \sigma_{13}\alpha_1 + \sigma_{23}\alpha_2 + \sigma_{33}\alpha_3 \equiv \sigma_{i3}\alpha
\end{aligned} \tag{1.22}$$

Expression (1.22) may be written as:

$$\begin{aligned}
(\sigma - \sigma_{11})\alpha_1 - \sigma_{21}\alpha_2 - \sigma_{31}\alpha_3 &= 0 \\
-\sigma_{12}\alpha_1 + (\sigma - \sigma_{22})\alpha_2 - \sigma_{32}\alpha_3 &= 0 \\
-\sigma_{13}\alpha_1 - \sigma_{23}\alpha_2 + (\sigma - \sigma_{33})\alpha_3 &= 0
\end{aligned} \tag{1.22b}$$

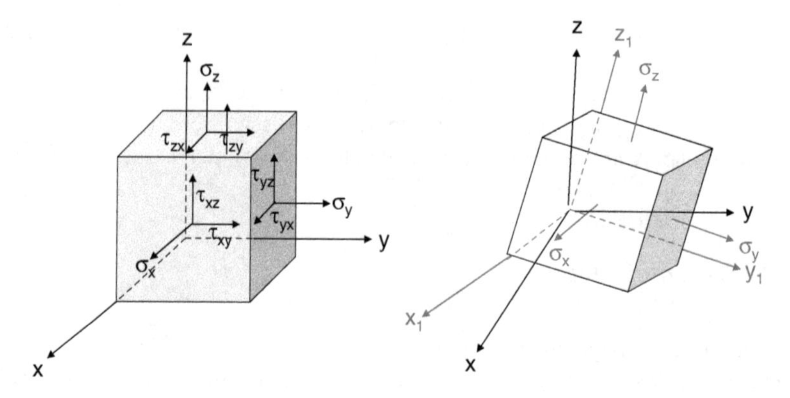

Fig. 1.14 Rotation of the elementary cube to a position where the shear stresses vanish is indicated. **a** Before rotation; **b** after rotation only the normal stresses (in *red*), known as 'principal stresses', remain. The new axes are the 'principal axes' (*red*)

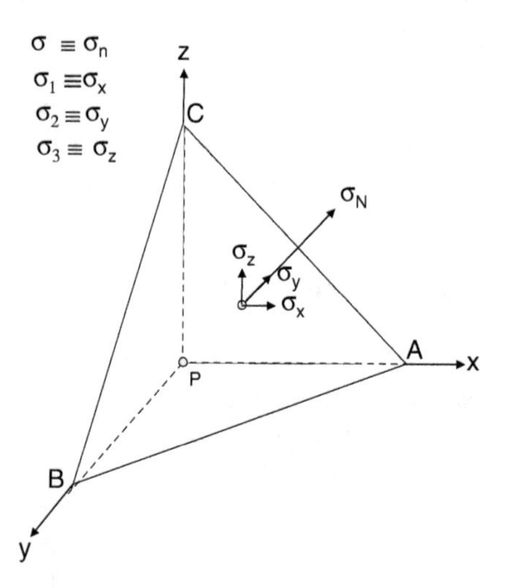

Fig. 1.15 The normal stress on an inclined plane is resolved into its components, acting along the directions of the indicated coordinate system

These three homogeneous linear equations for the directional cosines can only be solved for a non-zero solution, when their determinant equals zero. Thus:

$$\begin{vmatrix} (\sigma-\sigma_{11}) & -\sigma_{21} & -\sigma_{31} \\ -\sigma_{12} & (\sigma-\sigma_{22}) & -\sigma_{32} \\ -\sigma_{13} & -\sigma_{23} & (\sigma-\sigma_{33}) \end{vmatrix} = 0 \tag{1.22c}$$

The solution is:

$$\begin{aligned} &\sigma^3 - (\sigma_{11}+\sigma_{22}+\sigma_{33})\sigma^2 + (\sigma_{11}\sigma_{22}+\sigma_{22}\sigma_{33}+\sigma_{11}\sigma_{33}-\sigma_{12}^2-\sigma_{23}^2-\sigma_{31}^2)\sigma \\ &\quad - (\sigma_{11}\sigma_{22}\sigma_{33}+2\sigma_{12}\sigma_{23}\sigma_{31}-\sigma_{11}\sigma_{23}^2-\sigma_{22}\sigma_{13}^2-\sigma_{33}\sigma_{12}^2) = 0 \end{aligned} \tag{1.22d}$$

The components of the stress tensor usually change with the reference coordinates, but there are functions of these components that do not change. These functions are known as 'stress invariants'. The terms in the parentheses of Eq. (1.22d) may be rewritten in terms of such invariants as:

$$\begin{aligned} J_1 &= (\sigma_{11} + \sigma_{22} + \sigma_{33}) \\ J_2 &= (\sigma_{11}\sigma_{22} + \sigma_{22}\sigma_{33} + \sigma_{11}\sigma_{33} - \sigma_{12}^2 - \sigma_{23}^2 - \sigma_{31}^2) \\ J_3 &= (\sigma_{11}\sigma_{22}\sigma_{33} + 2\sigma_{12}\sigma_{23}\sigma_{31} - \sigma_{11}\sigma_{23}^2 - \sigma_{22}\sigma_{13}^2 - \sigma_{33}\sigma_{12}^2) \end{aligned} \quad (1.22e)$$

By using the terms from (1.22e) in (1.22d), it can be shortened to:

$$\sigma^3 - J_1\sigma^2 - J_2\sigma - J_3 = 0 \quad (1.23)$$

J_1, J_2 and J_3 are invariant coefficients of stress and, therefore, independent of the choice of coordinate system. Equation (1.23) is a third-order equation for stress, σ, with three roots, σ_I, σ_{II}, σ_{III}, also termed 'principal stresses'. Such solutions may be inserted into Eq. (1.22c) by utilizing the relation:

$$\alpha_1^2 + \alpha_2^2 + \alpha_3^2 = 1 \quad (1.24)$$

to obtain three sets of directional cosines. Each set is associated with one principal stress and its direction is called a 'principal axis of stress'. These three principal axes are mutually perpendicular. It is customary to designate σ_I as the largest principal stress and σ_{III} as the smallest one; thus, $\sigma_I \geq \sigma_{II} \geq \sigma_{III}$.

It is possible to determine the values of the directional cosines that give the maximum (or minimum) shear stress by using the principal stress axes, the normal stresses σ_I, σ_{II}, σ_{III} and Eq. (1.17), while all the shear components are zero. Equation (1.17) reduces to:

$$\begin{aligned} \sigma_1 &= \sigma_I\alpha_1 \\ \sigma_2 &= \sigma_{II}\alpha_2 \\ \sigma_3 &= \sigma_{III}\alpha_3 \end{aligned} \quad (1.17c)$$

Note that, in Fig. 1.14, the principal stresses are σ_x ($\equiv\sigma_I$), σ_y ($\equiv\sigma_{II}$) and σ_z ($\equiv\sigma_{III}$), indicated in red with the new coordinate system after rotation (also in red). Now, by inserting Eq. (1.17c) into Eq. (1.19), Eq. (1.19b) is obtained:

$$\sigma_N = \sigma_1\alpha_1 + \sigma_2\alpha_2 + \sigma_3\alpha_3 \quad (1.19)$$

$$\sigma_N = \sigma_I\alpha_1^2 + \sigma_{II}\alpha_2^2 + \sigma_{III}\alpha_3^2 \quad (1.19b)$$

In Fig. 1.15, stress is resolved into three components parallel to the axes (designated by symbols often found in the literature) of the reference coordinate system. Often, it is useful to resolve stress into its normal, N, and shear, T, components (Fig. 1.12), i.e., normal and parallel to the surface. The components of σ_N (Fig. 1.15) and T (Fig. 1.13) are respectively:

$$\sigma_N = \sigma_1\alpha_1 + \sigma_2\alpha_2 + \sigma_3\alpha_3 \tag{1.19}$$

Those of the shear components are:

$$T = (\sigma^2 - \sigma_N^2)^{1/2} = (\sigma_1^2 + \sigma_2^2 + \sigma_3^2 - \sigma_N^2)^{1/2} = (\sigma_i\sigma_i - N^2)^{1/2} \tag{1.19a}$$

The second term in the parentheses above is an expression of the Pythagorean relation between σ and σ_N (N) and T, such as $T^2 = \sigma^2 - \sigma_N^2$. By taking T^2 and replacing σ_N in Eq. (1.19), one obtains a relation in terms of the directional cosines, given as:

$$T^2 = \sigma_1^2 + \sigma_2^2 + \sigma_3^2 - (\sigma_1\alpha_1 + \sigma_2\alpha_2 + \sigma_3\alpha_3)^2 \tag{1.25}$$

with the principal stresses from Eq. (1.17c) introduced into Eq. (1.25) to give:

$$T^2 = \sigma_I^2\alpha_1^2 + \sigma_{II}^2 a_2{}^2 + \sigma_{III}^2\alpha_3^2 - \left(\sigma_I\alpha_1^2 + \sigma_{II}\alpha_2^2 + \sigma_{III}\alpha_3^2\right)^2 \tag{1.26}$$

Using Eq. (1.24), a directional cosine, such as α_3^2, may be written as:

$$\alpha_3^2 = 1 - \alpha_1^2 - \alpha_2^2 \tag{1.24a}$$

By inserting this value into Eq. (1.26) and taking partial derivatives with respect to α_1 and α_2 as indicted below, the positions of planes upon which T reaches its extreme values (maximum or minimum) may be defined by:

$$\frac{\partial T^2}{\partial \alpha_1} = \frac{\partial T^2}{\partial \alpha_2} = 0 \tag{1.27}$$

Similarly, one can take the partial derivative with respect to α_3 $\frac{\partial T^2}{\partial \alpha_3} = 0$.

Equation (1.26) may be written, after partial differentiation according to Eq. (1.27) and with the substitution for α_3^2 from Eq. (1.24a), as:

$$\begin{aligned}
\frac{\partial T^2}{\partial \alpha_1} &= 0 = \alpha_1\left[(\sigma_I - \sigma_{III})a_1^2 + (\sigma_{II} - \sigma_{III})\alpha_2^2 - \frac{1}{2}(\sigma_I - \sigma_{III})\right] \\
\frac{\partial T^2}{\partial \alpha_2} &= 0 = \alpha_2\left[(\sigma_I - \sigma_{III})\alpha_1^2 + (\sigma_{II} - \sigma_{III})\alpha_2^2 - \frac{1}{2}(\sigma_{II} - \sigma_{III})\right] \\
\frac{2}{\partial \alpha_3} &= 0 = \alpha_3\left[(\sigma_I - \sigma_{II})\alpha_2^2 + (\sigma_1 - \sigma_{III})\alpha_3^2 - \frac{1}{2}(\sigma_I - \sigma_{III})\right]
\end{aligned} \tag{1.28}$$

By equating the partial differentials of Eq. (1.26) to zero, the positions of the planes are defined by where the shear stress reaches its extreme values, namely its maximum and minimum. One can find solutions to these equations when α_1 or α_2 are set to zero. When α_1 is zero, the second relation in Eq. (1.28) for α_2 yields:

$$\alpha_2 = \pm\sqrt{\frac{1}{2}}.$$

From the first relation, for $\alpha_2 = 0$ one gets:

$$\alpha_1 = \pm\sqrt{\frac{1}{2}}.$$

From the third relation in Eq. (1.28), for $\alpha_2 = 0$ one obtains:

$$\alpha_3 = \pm\sqrt{\frac{1}{2}}.$$

Another set of three equations may be derived to yield values for the directional cosines: $\alpha_1 = \pm\sqrt{\frac{1}{2}}$, $\alpha_2 = \pm\sqrt{\frac{1}{2}}$ and $\alpha_3 = \pm\sqrt{\frac{1}{2}}$; thus, altogether six values of $\alpha_2 = \pm\sqrt{\frac{1}{2}}$ can be obtained by using equations similar to Eq. (1.24a) and expressed as:

$$\alpha_1^2 = 1 - \alpha_2^2 - \alpha_3^2 \tag{1.24b}$$

$$\alpha_2^2 = 1 - \alpha_1^2 - \alpha_3^2 \tag{1.24c}$$

Inserting these values of α_1, α_2 and α_3 successively into Eq. (1.26), the proper shear-stress components are:

$$\begin{aligned} \tau_{12} &= \pm\frac{1}{2}(\sigma_I - \sigma_{II}) \\ \tau_{23} &= \pm\frac{1}{2}(\sigma_{II} - \sigma_{III}) \\ \tau_{13} &= \pm\frac{1}{2}(\sigma_I - \sigma_{III}) \end{aligned} \tag{1.29}$$

These are principal shear stresses. Conventionally, σ_{I} and σ_{III} refer to the largest and smallest principal normal stresses, respectively. τ_{13} represents the largest acting shear stress and is, therefore, referred to as the 'maximum shear stress', τ_{max}. In Fig. 1.16, an example of a plane with one of the principal shear stresses is shown for an elementary cube whose faces are principal faces on which principal stresses are acting. Note that each pair of principal stresses has two planes of principal shear stress bisecting the planes on which the principal stresses are acting. In Fig. 1.16, only one of these planes is shown, while the second should be perpendicular to it (90°). Maximum shear stress acts on planes inclined at 45° to two principal planes and perpendicular to the remaining plane.

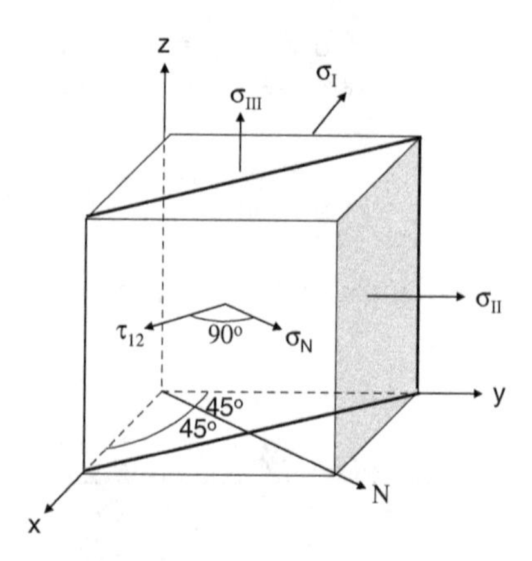

Fig. 1.16 Plane of principal shear stress (45° shear plane)

Table 1.1 Directional cosines of planes that give maximum shear stress in terms of principal axes

Directional cosines	Plane I	Plane II	Plane III
α_1	0	$\pm\sqrt{\frac{1}{2}}$	$\pm\sqrt{\frac{1}{2}}$
α_2	$\pm\sqrt{\frac{1}{2}}$	0	$\pm\sqrt{\frac{1}{2}}$
α_3	$\pm\sqrt{\frac{1}{2}}$	$\pm\sqrt{\frac{1}{2}}$	0
Shear stress	$\tau_{23} = \pm\frac{1}{2}(\sigma_{II} - \sigma_{III})$	$\tau_{13} = \pm\frac{1}{2}(\sigma_I - \sigma_{III})$	$\tau_{12} = \pm\frac{1}{2}(\sigma_I - \sigma_{II})$

Here, the shear stress shown is $\tau_{12} = \pm\frac{1}{2}(\sigma_I - \sigma_{II})$. Altogether, six such planes exist in an elementary cube. For the principal planes in Fig. 1.16, for example, the normal, N, to the 45° plane shown has directions $a_1 = a_2 = \cos 45° = \pm\frac{1}{\sqrt{2}}$ and $a_3 = \cos 90° = 0$.

When $\sigma_{\mathrm{I}} = \sigma_{\mathrm{II}} = \sigma_{\mathrm{III}}$, no principal shear stresses exist on any inclined plane according to Eq. (1.29) and stress state is 'hydrostatic'. Table 1.1 provides information on the directional cosine values giving maximum shear stresses.

1.2.5 Plain Stress: Two-Dimensional Stress

In modern technology, thin films play an important role in many applications. Thin films provide an example in which one of the dimensions is very small compared to the other two; consequently, the stress acting in the microscopic direction will also be much smaller than in the other two and, as such, may be disregarded. Sheets, thin films or thin-walled bodies are other examples in which one of the stresses may be ignored. Thus, a state of plane stress exists when one of the principal stresses is assumed to be zero and the flat body or the thin film may be

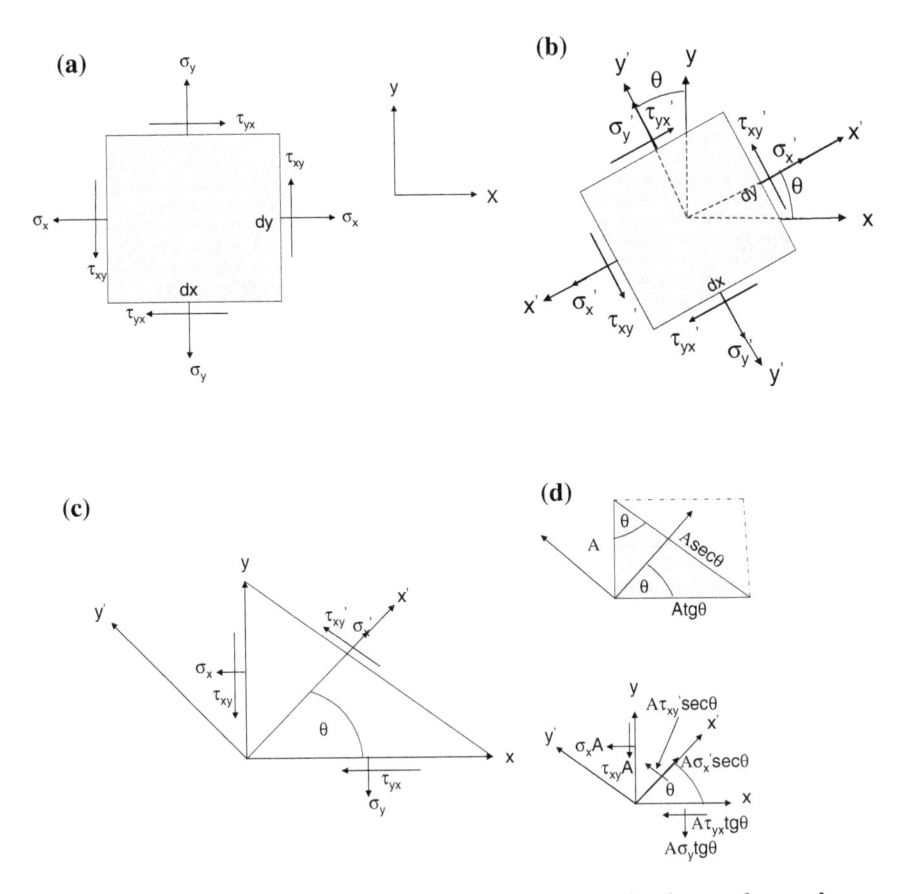

Fig. 1.17 Three stresses, σ_x, σ_y and τ_{xy}, **a** act in the two-dimensional state of stress; **b** stress transformed into another coordinate system; **c** the triangle is a portion of (**b**). Surface areas associated with the sides of the triangle (marked in *blue*) are indicated. A is the area shown on one side of the triangle. **d** The stress distribution after rotation on an inclined plane

analyzed two-dimensionally with the absence of one stress component. A stress condition, in which the stress in one dimension (a primary direction) is zero, is termed a 'plane stress'. To show a two-dimensional state of stress, the elementary cube in Fig. 1.10 has been redrawn in two dimensions, as illustrated in Fig. 1.17a. Rotation of the square transforms this figure into a different coordinate system. In the following, θ is used for the angle of rotation instead of α, previously used for the directional cosines.

Multiplying the stress by the respective areas gives the force acting on the element, as indicated to the right of Fig. 1.17c (the bottom figure). The surface areas associated with the sides of the triangle are indicated to the right of Fig. 1.17c at the top (above d)).

The sum of the forces acting in the x′ direction is:

$$A\sigma_x' \sec\theta - A\sigma_x \cos\theta - A\tau_{xy} \sin\theta - A\sigma_y \tan\theta \sin\theta - A\tau_{yx} \tan\theta \cos\theta = 0 \quad (1.30)$$

and the sum of the forces acting in the y′ direction is:

$$A\tau_{xy}' \sec\theta + A\sigma_x \sin\theta - A\tau_{xy} \cos\theta - A\sigma_y \tan\theta \cos\theta - A\tau_{yx} \tan\theta \sin\theta = 0 \quad (1.30a)$$

Recall that:

$$\tau_{xy} = \tau_{yx} \text{ and } \sec\theta = 1/\cos\theta \text{ and } \tan\theta = \frac{\sin\theta}{\cos\theta}.$$

Thus Eq. (1.30) may be rewritten as:

$$\sigma_x' = \sigma_x \cos^2\theta + \sigma_y \sin^2\theta + 2\tau_{xy} \sin\theta \cos\theta \quad (1.30b)$$

and Eq. (1.30a) as

$$\tau_{xy}' = -(\sigma_x - \sigma_y)\sin\theta\cos\theta + \tau_{xy}(\cos^2\theta - \sin^2\theta) \quad (1.30c)$$

Notice the different signs designating force, $A\sigma_x$ in Eq. (1.30) and $A\tau_{xy}$ in Eq. (1.30a), due to the different directions of the force. It is possible to express Eqs. (1.30b) and (1.30c) by substituting the known relations from Eq. (1.31), resulting in Eqs. (1.32)–(1.32d):

$$\cos^2\theta = \frac{1 + \cos 2\theta}{2} \quad \sin^2\theta = \frac{1 - \cos 2\theta}{2} \quad \sin\theta\cos\theta = \frac{\sin 2\theta}{2} \quad (1.31)$$

$$\sigma_x' = \sigma_x\left(\frac{1 + \cos 2\theta}{2}\right) + \sigma_y\left(\frac{1 - \cos 2\theta}{2}\right) + 2\tau_{xy}\frac{\sin 2\theta}{2} \quad (1.32)$$

rearranging Eq. (1.32) to give:

$$\sigma_x' = \frac{\sigma_x + \sigma_y}{2} + \frac{\sigma_x - \sigma_y}{2}\cos 2\theta + \tau_{xy}\sin 2\theta, \quad (1.32a)$$

and, similarly, for τ_{xy}':

$$\tau_{xy}' = -\frac{\sigma_x - \sigma_y}{2}\sin 2\theta + \tau_{xy}\cos 2\theta \quad (1.32b)$$

As seen from Fig. 1.17c, y′ is 90° + θ away from x. For the evaluation of σ_y', θ should be replaced by (90° + θ), namely by cos2 (90° + θ) or sin2 (90° + θ) using the relations:

$$\cos(180 + \alpha) = -\cos\alpha \quad \text{and} \quad \sin(180 + \alpha) = -\sin\alpha \quad (1.32c)$$

Thus, Eq. (1.32a) may be written for $\sigma_y{}'$ as:

$$\sigma_y' = \frac{\sigma_x + \sigma_y}{2} - \frac{\sigma_x - \sigma_y}{2} \cos 2\theta - \tau_{xy} \sin 2\theta \tag{1.32d}$$

An important concept is gleaned from the results of adding Eqs. (1.32a) and (1.32d):

$$\sigma_x' + \sigma_y' = \sigma_x + \sigma_y = I \tag{1.33}$$

It means that the sum of the normal stresses on two perpendicular planes is an invariant.

1.2.5.1 Principal Stresses

In Sects. 1.2.4 and 1.2.4.1, the concept of principal stresses was discussed regarding the three-dimensional case of stress acting at a point. As previously mentioned, in problems involving structural materials, like thin plates, one of the principal stresses is very small compared to the other two. Often in problems involving such structural elements, the small principal stress is assumed to be zero and the three-dimensional stress state can be reduced to two dimensions; the remaining two principal stresses lay and act in a plane. Figure 1.17a illustrates a two-dimensional case in the z direction, in which the 'ignored' principle stress is actually absent. The stress matrix in Eq. (1.13b) above is presented for the two- dimensional case in which all the components having z subscripts have been removed:

$$\sigma_{ij} = \begin{vmatrix} \sigma_x & \tau_{xy} \\ \tau_{yx} & \sigma_y \end{vmatrix} \tag{1.13c}$$

Clearly, in static equilibrium, $\tau_{xy} = \tau_{yx}$. The principal stresses may be obtained by differentiating Eq. (1.32a) with respect to θ, equating it to zero and discovering where the shear stress vanishes:

$$\frac{d\sigma_x'}{d\theta} = -\frac{\sigma_x - \sigma_y}{2} 2 \sin 2\theta + \tau_{xy} 2 \cos 2\theta = 0 \tag{1.34}$$

$$\frac{d\sigma_x'}{d\theta} = -\frac{\sigma_x - \sigma_y}{2} \sin 2\theta + \tau_{xy} \cos 2\theta = 0 \tag{1.34a}$$

$$\tan 2\theta = \frac{2\tau_{xy}}{\sigma_x - \sigma_y} \tag{1.35}$$

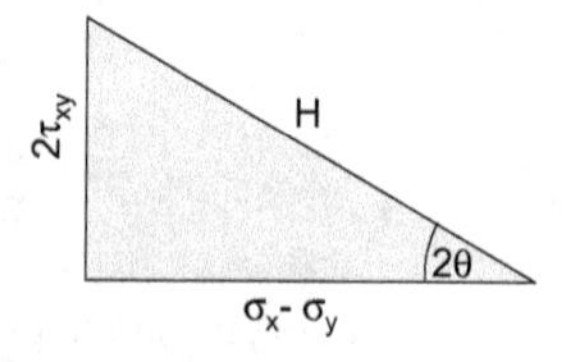

Fig. 1.18 Equation (1.35) can be derived from this triangle

Let us denote this angle by θ_p and, thus, Eq. (1.35) can be rewritten as:

$$\tan 2\theta_p = \frac{2\tau_{xy}}{\sigma_x - \sigma_y} \tag{1.35a}$$

Equation (1.35) may be expressed in a right-handed triangle by an angle, 2θ, between the hypotenuse and its adjacent side, as shown in Fig. 1.18.

This triangle may be also used to evaluate the cosine and sine functions in Eq. (1.32a). The significant data are as follows:

$$\sin 2\theta = \frac{2\tau_{xy}}{H} \quad \cos 2\theta = \frac{\sigma_x - \sigma_y}{H} \quad H^2 = (2\tau_{xy})^2 + (\sigma_x - \sigma_y)^2$$
$$H = \sqrt{(\sigma_x - \sigma_y)^2 + (2\tau_{xy})^2} \tag{1.35b}$$

Inserting the respective values into Eq. (1.32a) and rearranging the equation, one obtains the principal stress:

$$\sigma_x'(principal) = \sigma_I = \frac{\sigma_x + \sigma_y}{2} + \sqrt{\left(\frac{\sigma_x - \sigma_y}{2}\right)^2 + \tau_{xy}^2} \tag{1.36}$$

The second principal axis can be obtained from Eq. (1.33), rewritten as:

$$\sigma_{\mathrm{I}} + \sigma_{\mathrm{II}} = \sigma_1 + \sigma_2 = \sigma_{\mathrm{x}} + \sigma_{\mathrm{y}} \tag{1.33}$$

$$\tan 2\theta = \frac{2\tau_{xy}}{\sigma_x - \sigma_y}$$

Thus:

$$\sigma_{II} = \sigma_x + \sigma_y - \sigma_I = \frac{\sigma_x + \sigma_y}{2} - \sqrt{\left(\frac{\sigma_x - \sigma_y}{2}\right)^2 + \tau_{xy}^2} \tag{1.36a}$$

Furthermore, it is possible to take a derivative of Eq. (1.32a) to find the maximum shear stress:

$$\frac{d\tau_{xy}'}{d\theta} = -\frac{\sigma_x - \sigma_y}{2} 2\cos 2\theta - 2\tau_{xy}\sin 2\theta = 0 \tag{1.37}$$

$$\tan 2\theta = -\frac{\sigma_x - \sigma_y}{2\tau_{xy}} \tag{1.37a}$$

By reassigning this angle as θ_s (Eq. 1.37a), it can be rewritten as:

$$\tan 2\theta_s = -\frac{\sigma_x - \sigma_y}{2\tau_{xy}} \tag{1.37b}$$

As in the technique applied to principal stresses, an equation for the maximum shear stress can be obtained from the triangle in Fig. 1.18 with the geometrical relations in Eqs. (1.35a) or (1.35b) after inserting the proper data into Eq. (1.32b). The results are:

$$\tau_{\max} = \sqrt{\left(\frac{\sigma_x - \sigma_y}{2}\right)^2 + \tau_{xy}^2} \tag{1.38}$$

Note that Eq. (1.37a) is the negative reciprocal of Eq. (1.35), which means that both the angle associated with the maximum shear stress and the angle in which the shear stress vanishes are orthogonal. This is based on the known fact that two curves are orthogonal if, at each point of intersection, their tangent lines are perpendicular, i.e., the slopes of their tangent lines at their point of intersection are negative reciprocals of each other. It is readily shown that the planes of maximum shear stress, θ_s, occur at 45° to the principal planes, θ_p. Subtract the principal stresses, σ_I and σ_{II}, Eqs. (1.36) and (1.36a) and write:

$$\sigma_I - \sigma_{II} = 2\sqrt{\left(\frac{\sigma_x - \sigma_y}{2}\right)^2 + \tau_{xy}^2} \tag{1.39}$$

$$\sigma_I - \sigma_{II} = 2\tau_{\max} \tag{1.40}$$

$$\tau_{\max} = \frac{\sigma_I - \sigma_{II}}{2} \tag{1.40a}$$

Furthermore, Eq. (1.37a) is equal to:

$$\tan 2\theta_s = -\frac{\sigma_x - \sigma_y}{2\tau_{xy}} = -\cot 2\theta_p = -\frac{2\tau_{xy}}{\sigma_x - \sigma_y} \tag{1.41}$$

or

$$\tan 2\theta_s + \cot 2\theta_p = 0 \tag{1.41a}$$

$$\frac{\sin 2\theta_s}{\cos 2\theta_s} + \frac{\cos 2\theta_p}{\sin 2\theta_p} = 0 \tag{1.41b}$$

$$\sin 2\theta_s \sin 2\theta_p + \cos 2\theta_s \cos 2\theta_p = 0 \tag{1.41c}$$

Recalling the relation:

$$\cos(\alpha - \beta) = \cos\alpha\cos\beta + \sin\alpha\sin\beta \tag{1.41d}$$

Equation (1.41c) may be rewritten as:

$$\cos(2\theta_s - 2\theta_p) = 0 \tag{1.41e}$$

or

$$2\theta_s - 2\theta_p = \pm 90^\circ \tag{1.41f}$$

$$\theta_s - \theta_p = \pm 45^\circ \tag{1.41g}$$

$$\theta_s = \pm 45^\circ + \theta_p \tag{1.41h}$$

Thus, shear stress reaches a maximum on planes which are oriented at ± 45° to the principal planes. Recall that the stress system at a given point depends on the inclination of the plane. The principal stresses are often referred to as being 'major' and 'minor' principal stresses, referring algebraically to the largest and smallest normal stresses.

1.2.6 Mohr's Circle in Two Dimensions

There is a graphical way to evaluate principal stress using the well-known concept of 'Mohr's circle'. Basically, this is a graphic method of representing the plane stress state at a given point. In essence, a circle is drawn on the abscissa and ordinate axes, representing σ and τ. All the stress states obtained, as the angle θ varies, fall on the circle (as will be discussed below). It is simpler to describe the two-dimensional case. In essence, the most common, graphic illustration presented in almost every book and the associated methodology is as follows. First, rewrite the squared Eqs. (1.32a) and (1.32b) to get:

$$\left(\sigma_x' - \frac{\sigma_x + \sigma_y}{2}\right)^2 = \left[\left(\frac{\sigma_x - \sigma_y}{2}\right)\cos 2\theta + \tau_{xy}\sin 2\theta\right]^2 \tag{1.42}$$

$$(\tau_{xy}')^2 = \left[-\left(\frac{\sigma_x - \sigma_y}{2}\right)\sin 2\theta + \tau_{xy}\cos 2\theta\right]^2 \tag{1.43}$$

Then, combine the two equations after squaring them, to get:

$$\left(\sigma_x' - \frac{\sigma_x + \sigma_y}{2}\right)^2 + \left(\tau_{xy}'\right)^2 = \left(\frac{\sigma_x - \sigma_y}{2}\right)^2\left[\cos^2 2\theta + \sin^2 2\theta\right] + \tau_{xy}^2\left[\sin^2 2\theta + \cos^2 2\theta\right] \tag{1.44}$$

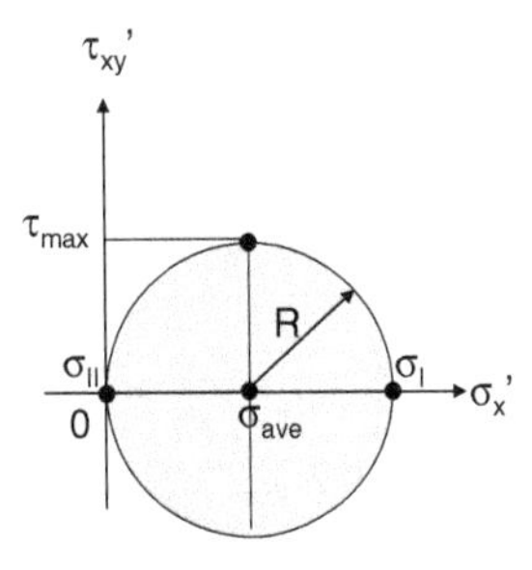

Fig. 1.19 A circle according to Eq. (1.47)

This reduces to:

$$\left(\sigma_x^{'} - \frac{\sigma_x + \sigma_y}{2}\right)^2 + \left(\tau_{xy}^{'}\right)^2 = \left(\frac{\sigma_x - \sigma_y}{2}\right)^2 + \tau_{xy}^2 \tag{1.45}$$

Equation (1.45) is further modified for the purpose of the graphical presentation as follows:

$$\sigma_{ave} = \frac{\sigma_x + \sigma_y}{2} \text{ and } R = \sqrt{\left(\frac{\sigma_x - \sigma_y}{2}\right)^2 + \tau_{xy}^2} \tag{1.46}$$

and

$$\left(\sigma_x^{'} - \sigma_{ave}\right)^2 + \left(\tau_{xy}^{'}\right)^2 = \mathrm{R}^2 \tag{1.47}$$

Equation (1.47) has the form of the equation for a circle, such as:

$$(x - a)^2 + (y - b)^2 = r^2.$$

This circle is in a Cartesian coordinate system with center coordinates a, b and radius r. This circle equation applies to any point on the circle where the radius is the hypotenuse of a right-angled triangle whose other sides are of length (x – a) and (y – b). If the circle is centered at the origin (0, 0), then the equation simplifies to:

$$x^2 + y^2 = r^2$$

A circle with center coordinates (a, b) and radius, r, is the set of all points (x, y). In our case, the circle is such that $(\sigma_{\mathrm{ave}}, 0)$ are the coordinates and the radius is R. The parameters of Eq. (1.47) are defined as indicated in Eq. (1.47a):

$$\sigma_{\mathrm{x}}' \equiv \mathrm{x};\, \tau_{\mathrm{xy}}' \equiv \mathrm{y}; \quad \sigma_{\mathrm{ave}} \equiv \mathrm{a} = \frac{\sigma_x + \sigma_y}{2}; \quad \mathrm{b} = 0; \quad \mathrm{r} = \sqrt{\left(\frac{\sigma_x - \sigma_y}{2}\right)^2 + \tau_{xy}^2} = \tau_{\mathrm{max}} \tag{1.47a}$$

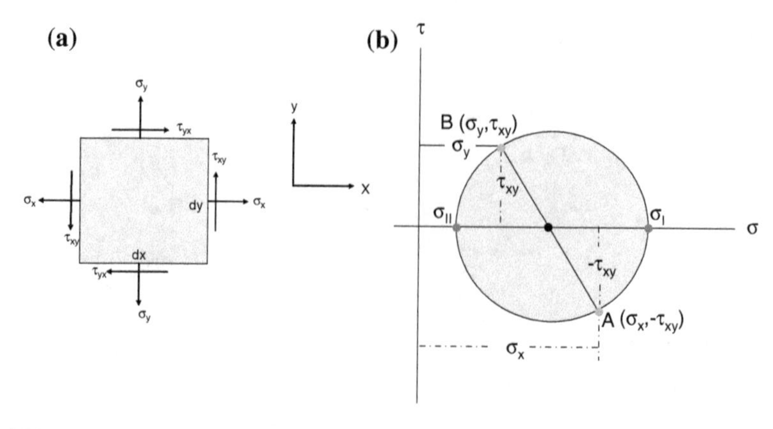

Fig. 1.20 Mohr's circle construction

Equation (1.47) of the circle can be redrawn by replacing coordinates a, b, and r with σ_{ave}, $\tau_{xy}{'}$ and R, as defined in Eq. (1.47a). In this circle, the points along the abscissa (at zero shear $\tau_{xy}{'}$) are the principal stresses, as indicated in Eq. (1.46), in which $\sigma_x{'}$ is the principal stress designated as σ_I. Any point on the circle can be obtained by the Pythagorean theorem. Such constructions serve as the basis for Mohr's circle, yielding the particular stress at each point. Bear in mind that the normal and shear stress components in the z direction are zero or negligible.

Figure 1.19 demonstrates that the relation found in Eq. (1.33): $\sigma_I + \sigma_{II} = \sigma_x + \sigma_y = I$ (an invariant) was applied for the sake of simplicity. However, to draw Mohr's circle, the accepted procedure is as follows. Consider Fig. 1.17a, redrawn in Fig. 1.20a. The plot is in Cartesian coordinates; the abscissa is for normal stresses and the ordinate for shear stresses. Two points on Fig. 1.20a are the coordinates on the diameter of the circle as indicated: A (σ_x, $-\tau_{xy}$) and B (σ_y, τ_{xy}). (Recall that $\tau_{xy} = \tau_{yx}$). Tensile stresses are considered positive, while compressive ones are negative. However, when constructing Mohr's circle, the shear stresses are considered positive if they make a counter-clockwise moment around the stress element. In Fig. 1.20a or 1.21a, τ_{xy} is counter-clockwise (negative), while τ_{yx} is clockwise (positive), as seen in the rotation in Fig. 1.21b and in the Mohr's circle in Fig. 1.21c. This convention is important for determining the proper orientation of the principal stress relative to the x, y coordinate system.

Figure 1.21 shows the construction of a Mohr's circle with a counter-clockwise rotation of an element. It intersects the axis at two points, **C** and **D**. The stresses at these two end points of the horizontal diameter are σ_I and σ_{II}, the principal stresses. In Fig. 1.21, the equation is basically that of Eq. (1.39), defined like Eq. (1.46) for 2R.

$$\sigma_I - \sigma_{II} = 2\sqrt{\left(\frac{\sigma_x - \sigma_y}{2}\right)^2 + \tau_{xy}^2} = 2R$$

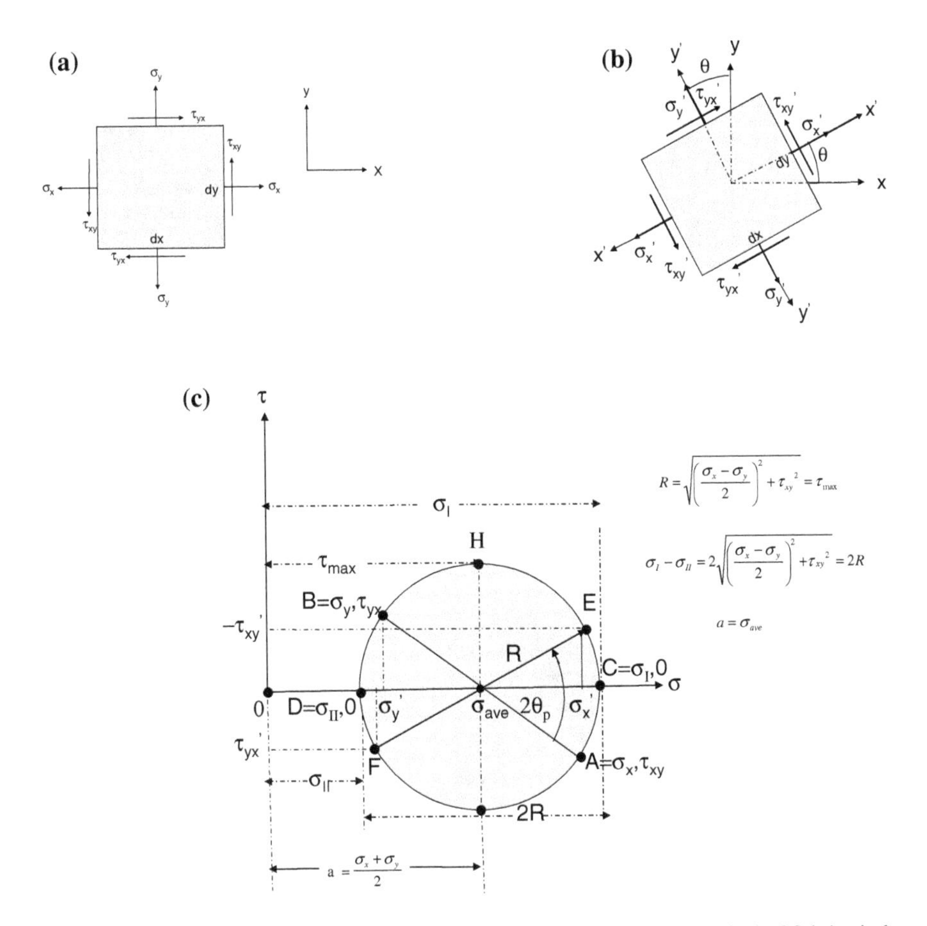

Fig. 1.21 Construction of a Mohr's circle. **a** and **b** from Fig. 1.17a, 1.17b; **c** is the Mohr's circle construction for the two-dimensional case

To construct a Mohr's circle for the stress state shown in Fig. 1.21, first identify the x and y axes of a Cartesian coordinate system as the σ_x and σ_y axes, respectively. Figure 1.21a, redrawn from Fig. 1.17a, serves this purpose. Points A and B, representing normal stresses, have coordinates for A σ_x and τ_{xy}, while point B is defined by σ_y and τ_{yx}; these are then plotted in Fig. 1.21c. Now draw a line through points A and B. The point of intersection of this line with the abscissa serves as the center about which a circle is drawn with diameter AB. The circle passing through points A and B intersects the abscissa at points C and D. These define the distances $\overline{OC}$ and $\overline{OD}$, which represent the principal stresses, σ_I and σ_{II}. At points C and D. the shear stresses are zero and, therefore, they correspond to the principal stresses. The circle's radius, normal to the abscissa, $\overline{C_{ave}H}$, is associated with the maximum shear stress. Next, a rotation is performed, as indicated in Fig. 1.21b) to angle θ_p, which is the same as given in Eq. (1.35), newly positioning

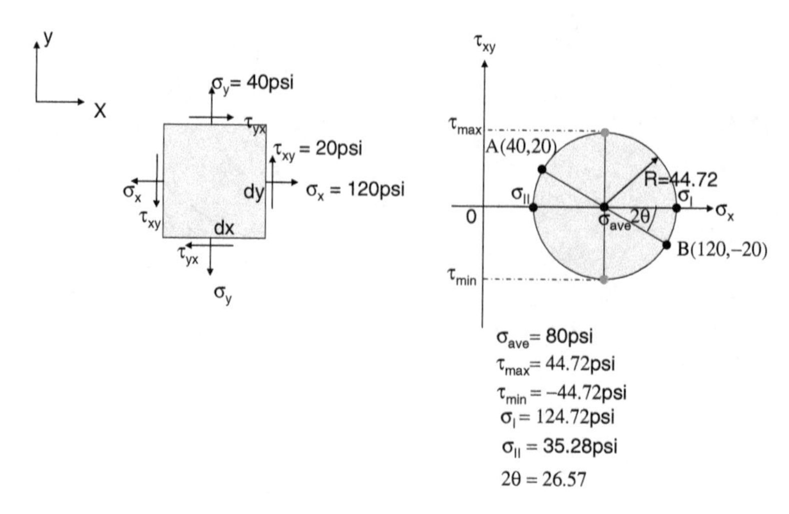

Fig. 1.22 Example of plain stress without rotation

the coordinate system between the abscissas x and x′. However, in the Mohr illustration, the angle shown is $2\theta_p$, since this angle is measured counter-clockwise on the Mohr's circle (see Fig. 1.21b or 1.17b for the angle) and it is between the rotated and non-rotated systems. Thus, it is evident from its construction that the coordinates of the points of line $\overline{EF}$ may be used to obtain the values of stresses in the rotated system, namely those of σ_x' and τ_{xy}'. One may note that, of the axes, the ordinate in Fig. 1.21c is a shear stress and the abscissa is a normal stress, as indicated earlier.

Mohr's circle may be used in the transformation of stresses from one coordinate system to another. Figure 1.21 may also be used for this purpose. Consider Fig. 1.21a or 1.17a representing the normal and shear stresses, σ_x, σ_y and τ_{xy} acting on the respective planes in the body characterized by the coordinate system, x and y. The stresses acting in the new coordinate system, x′ and y′, after rotation to an angle θ, from x towards x′, are indicated in Fig. 1.21b. The previous Mohr's circle shows the stress state of Fig. 1.21a at points A and B with coordinates σ_x, τ_{xy} and σ_y, τ_{yx}, respectively. Now, a line may be drawn between these two points, and then rotated to angle 2θ, which is twice the angle θ between x and x′ and in the opposite direction of θ. A line drawn after the rotation between the two new points, E and F, provides the new stresses, σ_x', σ_y' and τ_{xy}', in the new coordinate system.

A short exercise can illustrate how to use a Mohr's circle to get the principal stresses. A priori the angles are not needed for this. Figure 1.22 indicates the method when no rotation of the coordinate system has occurred. The magnitudes of σ_x, σ_y and τ_{yx} (= τ_{xy}) are indicated in Fig. 1.22.

Equation (1.46) may be used to evaluate σ_{ave} and R as:

$$\sigma_{ave} = \frac{\sigma_x + \sigma_y}{2} = \frac{120 + 40}{2} = 80\,\text{psi}$$

and

$$R = \sqrt{\frac{(\sigma_x - \sigma_y)^2}{2} + \tau_{xy}^2} = \sqrt{40^2 + 20^2} = 44.72.$$

According to Eq. (1.47a):

$$\tau_{max} = R = 44.72$$

From Eqs. (1.36) and (1.36a):

$$\sigma_I = \sigma_{ave} + R = 80 + 44.72 = 124.72$$
$$\sigma_{II} = \sigma_{ave} - R = 80 - 44.72 = 35.28.$$

In this case, a Mohr's circle is constructed as follows. A horizontal axis is drawn for the normal stresses, like σ_x in the figure, while the vertical axis, τ_{xy}, represents shear stresses. Two points, A and B, are indicated by the coordinates (σ_y, τ_{xy}) and $(\sigma_x, -\tau_{yx})$, namely (40, 20) and B (120, –20). Join these points by a line; the midpoint of this line is σ_{ave}, which is the center of a (Mohr's) circle having the coordinates ($\sigma_{ave} = \frac{\sigma_x + \sigma_y}{2}$, 0) or a value of (80, 0). Now, a circle is drawn from the midpoint, with a radius, R, of 44.72. This circle cuts the abscissa (the σ axis) at two points, providing the principal stresses, σ_I and σ_{II}, for the indicated values.

Earlier, the concept of 'principal stresses' was discussed and it was mentioned that at some stressed positions, when shear stress vanishes, the position depends upon the angle θ. When Eq. (1.32b) is differentiated with respect to θ and equated to zero, Eq. (1.35) is obtained:

$$\tan 2\theta_p = \frac{2\tau_{xy}}{\sigma_x - \sigma_y} \tag{1.35}$$

$$2\theta = \frac{2\tau_{xy}}{\sigma_x - \sigma_y} \tag{1.35a}$$

There are two solutions for θ in the range of $-180° \leq \theta° \leq 180°$ that may be attained by inserting the appropriate values shown earlier, $2\theta_1 = 26.57$ or $\theta_1 = 13.29$ and $2\theta_2 = 26.57 + 180°$. It was previously stated that the angle θ, located between x and x′ in Fig. 1.21b on the Mohr's circle, is 2θ. In Fig. 1.22b, the angle following the rotation of the square of Fig. 1.21b is also shown.

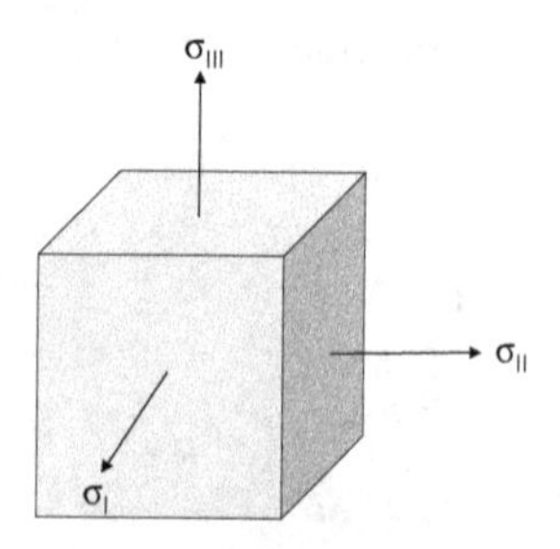

Fig. 1.23 The resultant normal stress system on principal axes without shear stresses

Summarizing this section on the Mohr's circle, the following features have been described:

(1) The principal axes are indicated as σ_I and σ_{II}.
(2) The center of the circle, σ_{ave}, is always on the normal stress axis, σ_x.
(3) The center of a Mohr's circle and its radius are determined by Eq. (1.46).
(4) A Mohr's circle represents all stress states, namely normal and shear that can exist on the surface of an elementary cube as it is being rotated.
(5) The maximum shear stress equals R.
(6) The center of a circle in terms of the principal stresses is $\tau_{max} = \frac{\sigma_I - \sigma_{II}}{2}$.

1.2.7 Three-Dimensional Mohr's Circles

The stress state in Fig. 1.23 is associated with the principal stresses located at the position where the shear stresses vanish. The stress tensor for such a condition may be written as:

$$\sigma_{ij} = \begin{matrix} \sigma_x & \tau_{xy} & \tau_{xz} \\ \tau_{yx} & \sigma_y & \tau_{yz} \\ \tau_{zx} & \tau_{zy} & \sigma_z \end{matrix} \tag{1.13b}$$

or replaced by its reduced form, as represented by Fig. 1.23.

$$\sigma_{ij} = \begin{matrix} \sigma_x & 0 & 0 \\ 0 & \sigma_y & 0 \\ 0 & 0 & \sigma_z \end{matrix}$$

As in the two-dimensional case, the direct stresses are on the horizontal axis and the shear stresses are on the vertical axis. For the construction of the Mohr's circle, three circles are required. The stresses on any plane at any rotation, when plotted in the three-dimensional Mohr's circle diagram, are represented by a point located either on one of the three circles or within the area between the largest and the two smaller circles. The maximum shear stress is given by the radius of the largest circle. When constructing the Mohr's circle, the angle of rotation is double that of the real stress system. Shear stresses are positive if they cause clockwise rotation,

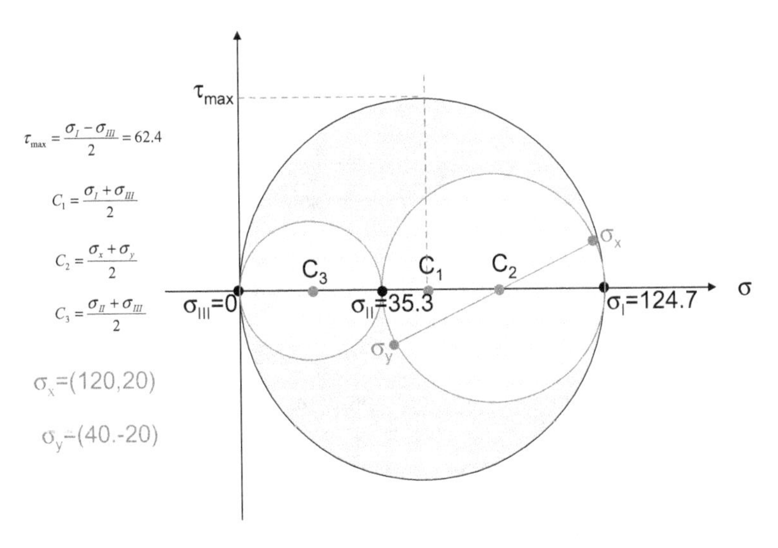

Fig. 1.24 Construction of a three-dimensional Mohr's circle

while those inducing counter-clockwise rotation are negative, as indicated above; thus, the conventional sign for tensile stress is positive and for compressive stress is negative. The practical steps are as follows:

(1) Draw these axes, namely the direct and shear stresses;
(2) Mark the three principal stresses on the abscissa and designate them as σ_{I}, σ_{II} and σ_{III};
(3) Draw the three circles as indicated in Fig. 1.24.

The center of each Mohr's circle which lies on the σ axis is given by:

$$C_1 = \frac{\sigma_I + \sigma_{III}}{2} \quad C_2 = \frac{\sigma_x + \sigma_y}{2} \quad C_3 = \frac{\sigma_{II} + \sigma_{III}}{2}$$

The radii of the circles are:

$$R_1 = \tfrac{1}{2}\sqrt{(\sigma_x + \sigma_y)^2 + (\tau_{xy} + \tau_{yx})^2} \quad R_2 = \tfrac{1}{2}\sqrt{(\sigma_x + \sigma_z)^2 + (\tau_{xz} + \tau_{zx})^2}$$

$$R_3 = \tfrac{1}{2}\sqrt{(\sigma_y + \sigma_z)^2 + (\tau_{yz} + \tau_{zy})^2}$$

Note that the center, C_2, may be evaluated from the values of σ_{x} and σ_{y}, which is equivalent to the center obtained from the principal stresses, σ_{I} and σ_{II}. These principal stresses are obtained at a certain orientation of the stress system of an element in space, in which all the shear-stress components equal zero. In this special orientation, the normals of the faces correspond to the principal directions and the normal stresses associated with these faces are the principal stresses.

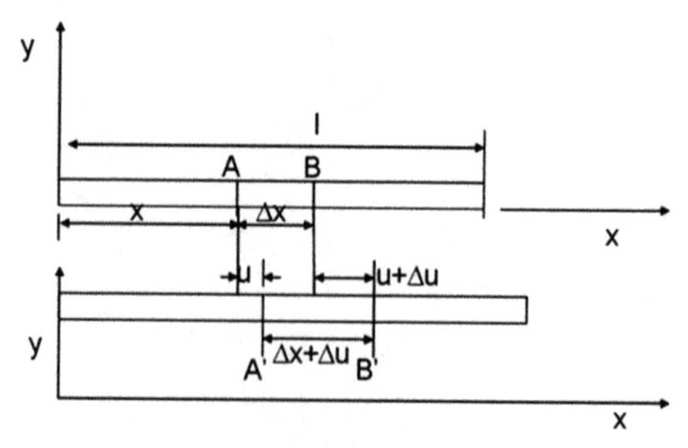

Fig. 1.25 One-dimensional deformation of a bar [14]. With permission of the author Joshua Pelleg

In Fig. 1.24, the solution of a sample problem solved is also indicated by the data below. The following measurements were performed on a body under stress, $\sigma_x = 120$ MPa, $\sigma_y = 40$ MPa and $\tau_{xy} = 20$ MPa. The stresses σ_z, τ_{yz} and τ_{zx} are zero. Here, a Mohr's circle is used to evaluate the principal stresses and the largest shear stress for this stress state. The relations in Eqs. (1.36) and (1.36a) are for the principal stresses, resulting in:

$$\sigma'_x(principal) = \sigma_I = \frac{\sigma_x + \sigma_y}{2} + \sqrt{\left(\frac{\sigma_x - \sigma_y}{2}\right)^2 + \tau_{xy}^2} = 124.7 \text{ MPa} \tag{1.36}$$

$$\sigma_{II} = \sigma_x + \sigma_y - \sigma_I = \frac{\sigma_x + \sigma_y}{2} - \sqrt{\left(\frac{\sigma_x - \sigma_y}{2}\right)^2 + \tau_{xy}^2} = 62.36 \text{ MPa} \tag{1.36a}$$

$$\tau_{\max} = \frac{\sigma_I - \sigma_{III}}{2} = 62.36 \,\text{MPa}$$

The other shear stresses in the respective circles, defined by the principal stresses, $\sigma_I - \sigma_{II}$ and $\sigma_{II} - \sigma_{III}$, are:

$$\tau_{23} = \sigma_2 - \sigma_3/2$$
$$\tau_{12} = \sigma_1 - \sigma_2/2.$$

These relations were given earlier as Eq. (1.27), in terms of the principal stresses:

$$\begin{aligned} \tau_{12} &= \pm\frac{1}{2}(\sigma_I - \sigma_{II}) \\ \tau_{23} &= \pm\frac{1}{2}(\sigma_{II} - \sigma_{III}) \\ \tau_{13} &= \pm\frac{1}{2}(\sigma_I - \sigma_{III}). \end{aligned} \tag{1.27}$$

The stress state at the point upon which the Mohr's circle is based was discussed earlier in Sects. 1.2–1.2.5.

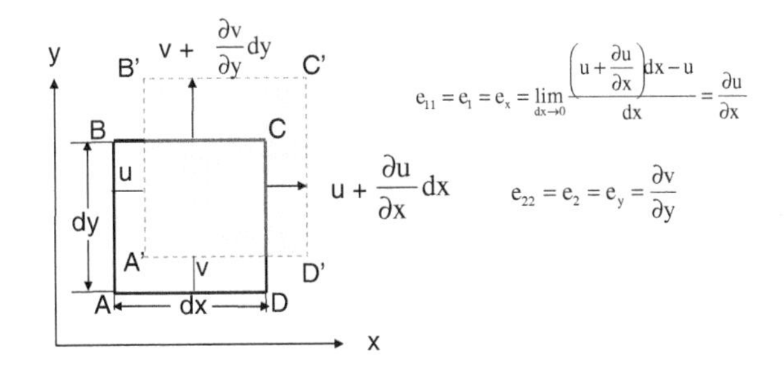

Fig. 1.26 Geometric interpretation of normal strains

1.2.8 Strain Tensor

In Sect. 1.2 above, the simple equations relating to strain were discussed. 'Strain' is defined as the relative change in the position of points within a body that has undergone deformation. In deformed materials, the points have undergone positional changes and, even in isotropic materials, these changes are not uniform, except in the one-dimensional case. Thus, for example, a polycrystalline material with a random orientation of the grains (crystals) is isotropic; otherwise, it might be anisotropic, developing a preferred orientation. To better explain the concept of strain, here is a description of the one-dimensional case. Figure 1.25 shows a one-dimensional change occurring in a bar under uniaxial deformation, e.g., by tension.

Figure 1.25 shows the condition before and after the application of force. A consequent displacement of the points occurs in the body, as exemplified by the displacement of point A to point A′, designated by u. Point B, Δx apart from A, is also displaced to point B′, moving to u + Δu. Due to the assumption that the bar is a rigid body, the elongation is Δu. This case basically considers a small normal strain, i.e., in the elastic region in one direction, shown as:

$$e_1 = \frac{A'B' - AB}{AB} = \frac{dx + \frac{\partial u}{\partial x} dx - dx}{dx} = \frac{\partial u}{\partial x} \tag{1.48}$$

In this Eq. (1.48), the Δx and Δu in Fig. 1.25 are replaced by dx and du or in terms of their partials as ∂x and ∂u, thus indicating the instantaneous change of u with respect to x. Recall that $e_1 \equiv e_{11} \equiv e_x \equiv e_{xx}$, depending on the type of notation. Nomenclatures may vary in accordance with the original research applications. For instance, the Voigt notation is useful in calculations involving constitutive models, such as the generalized Hooke's Law, as well as for finite element analysis.

One may also consider the accepted example of the two-dimensional case of a square which has been deformed into a parallelepiped as an additional step in understanding strain. Various shapes may result from deformation, depending on the types of forces applied and the directions in which they act. Figure 1.26 illustrates the displacement of a square without a change in shape and without shear deformation. Figure 1.26 is a geometric representation of the normal strains

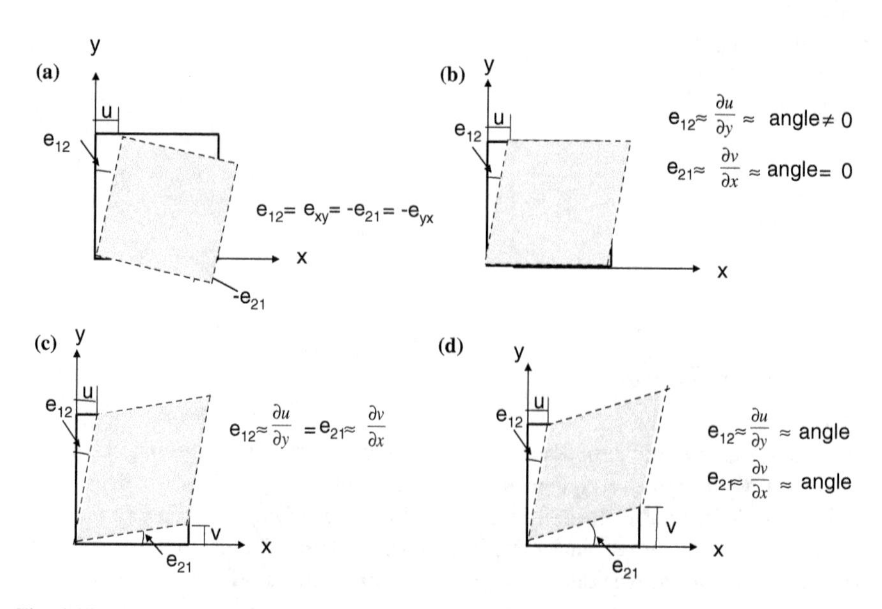

Fig. 1.27 Examples of deformation when shear and/or shear and rotation of the square element occur. **a** Pure rotation no shear. **b** Simple shear. **c** Pure shear without rotation. **d** Angular deformation

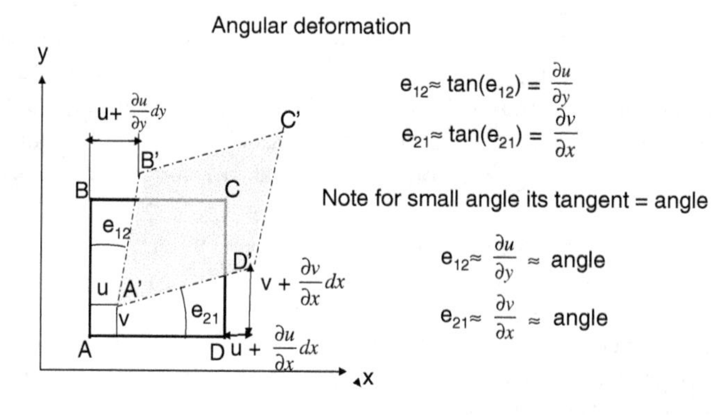

Fig. 1.28 For clarity, Fig. 1.27d is redrawn and more detailed

e_1 ($\equiv e_{11} \equiv e_x \equiv e_{xx}$) and e_2, in which the displacements, u_1 ($\equiv$u) and u_2 ($\equiv$v) and the respective equations, are also shown.

Figure 1.27 also shows various changes occurring in a square under active forces, when $e_{12} = -e_{21}$, the elemental square, is rotated by pure shear, without any change in its dimensions. If e_{12} has a small value but $e_{21} = 0$, it is a simple shear of the element. Figure 1.27c indicates a pure shear deformation, without rotation of the element. However, if $e_{12} \neq e_{21}$, there is angular deformation during the rotation, as shown in Fig. 1.27d.

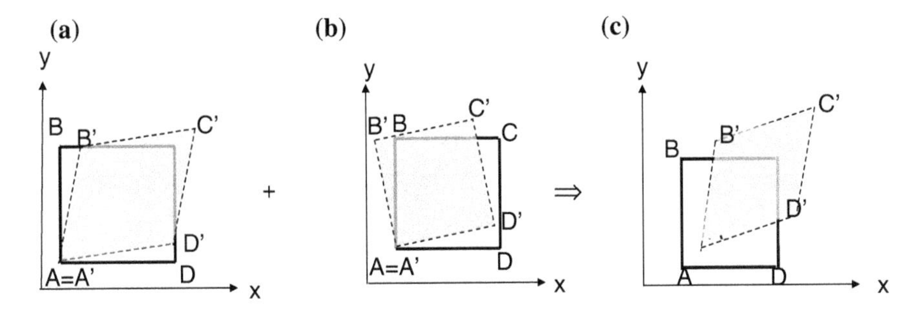

Fig. 1.29 An arbitrary shear strain (**c**) may be decomposed into a pure shear (**a**) and a pure rotation (**b**). **a** Pure shear without rotation **b** Pure rotation no shear **c** Arbitary shear

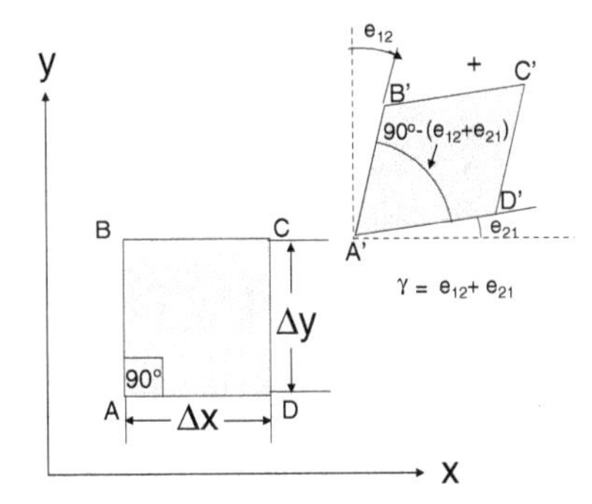

Fig. 1.30 Deformation of a two-dimensional square by shear. Note that the overall shear deformation is composed of two shear strains associated with the two angles indicated

To show more details of the angular deformation shown in Fig. 1.27d, it has been redrawn and illustrated in Fig. 1.28.

In Figs. 1.26, 1.27 and 1.28, e_{11} and e_{22} are normal strains and provide the relative changes in the length of the lines parallel to the coordinate axes following some deformation. The shear stresses are e_{12} and e_{21}. Figure 1.28 or 1.27d are basically the results of pure shear and rotation of the body (represented by a square for two dimensions). This is shown schematically in Fig. 1.29.

Thus, one can state that 'arbitrary shear strain' is the result of pure shear and some rotation occurring simultaneously during a deformation process.

At this stage, further clarification of 'shear' is required in terms of engineering strain, commonly designated as γ. $\gamma_{ave} = e_{12} + e_{21}$ is the overall shear strain, as seen from Fig. 1.30. Here, the sides, Δx and Δy, of a rectangle, ABCD, along axes x and y, respectively, are shear-deformed into a rhombus-like shape, A′B′C′D′. The change from the 90° angle of the undeformed square is associated (as seen in Fig. 1.30) with a shear producing two angles; the right angle becomes $90° - (e_{12} + e_{21})$. Hence, the shear is expressed as $\gamma = e_{12} + e_{21}$ for the two-dimensional case in an x–y coordinate system. (Total shear deformation is also written as γ_{12}, indicating the two angles involved).

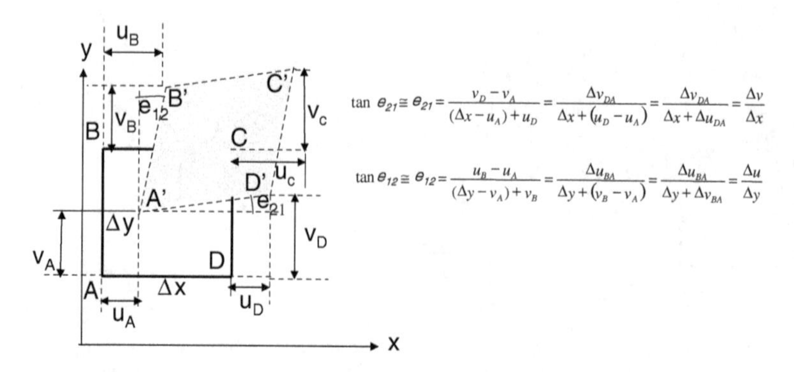

Fig. 1.31 Evaluation of shear deformation in terms of the displacements

As seen in Fig. 1.27c or 1.28, strains are associated with displacements. In the general, three-dimensional case, the displacement of a point in an elemental body is denoted by u, v, and w, which are functions of position in the coordinate system x, y and z. In other words:

$$u = u(x, y, z); \quad v = v(x, y, z); \quad w = w(x, y, z) \tag{1.49}$$

Returning to the two-dimensional case indicated for the shear strain in Fig. 1.30, the deformation may be expressed in terms of the displacements. In Fig. 1.26, the normal strain components were indicated in terms of u and v, which are displacement components in the x, y directions. (Other symbols are also used, such as u, v and w in the three-dimensional case.) In fact, displacement may be described in terms of its components by projecting the displacement vector onto the coordinate axes. In this manner, the average, normal-strain components, in terms of u and v after deformation, may be given as in Eq. (1.50), and in the limit, the displacement vector as in Eq. (1.50a):

$$e_{11,ave} = \frac{u + \Delta u - u}{\Delta x} = \frac{\Delta u}{\Delta x} \quad e_{22,ave} = \frac{v - \Delta v - v}{\Delta y} = \frac{\Delta v}{\Delta y} \tag{1.50}$$

$$e_{11} = \lim_{\Delta x \to 0} \frac{\Delta u}{\Delta x} = \frac{\partial u}{\partial x} \quad e_{22} = \lim_{\Delta y \to 0} \frac{\Delta v}{\Delta y} = \frac{\partial v}{\partial y} \tag{1.50a}$$

Here, partial differentiation must be used to express the shear strains, because u is also a function of y (and, in the three-dimensional case, also of z), as indicated in Fig. 1.28, showing the displacements involved. Figure 1.30 shows that the four corners of the square are displaced. e_{21} and e_{12}, taken from Fig. 1.28, may be used to express displacement in Fig. 1.31.

When evaluating $\gamma_{ave} = e_{21} + e_{12}$ in the x–y system in Eqs. (1.51) and (1.51a), some approximations are made as indicated:

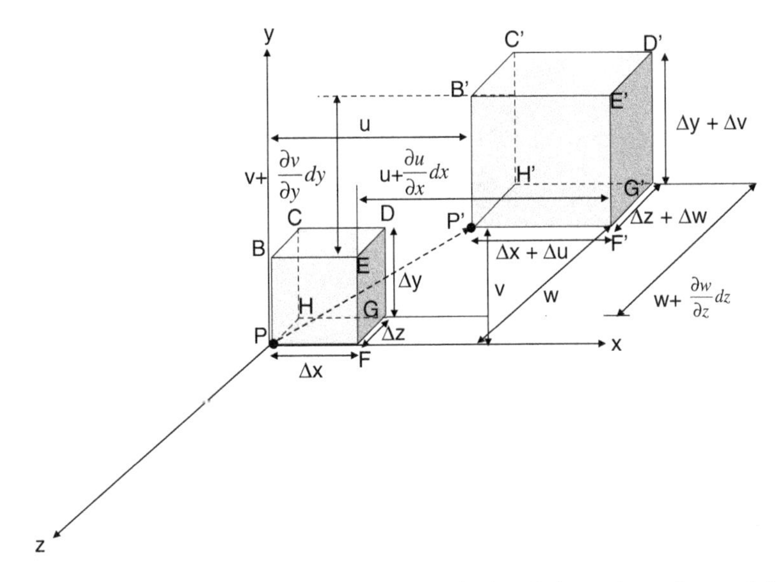

Fig. 1.32 The change in the dimensions of a deformed cube relative to those of a non-deformed one is indicated (the dimensions of the deformed cube are exaggerated). Only normal strains are present. PP′ is the displacement vector with u, v and w components

$$\tan e_{21} \cong e_{21} = \frac{v_D - v_A}{(\Delta x - u_A) + u_D} = \frac{\Delta v_{DA}}{\Delta x + (u_D - u_A)} = \frac{\Delta v_{DA}}{\Delta x + \Delta u_{DA}} = \frac{\Delta v}{\Delta x} \tag{1.51}$$

$$\tan e_{12} \cong e_{12} = \frac{u_B - u_A}{(\Delta y + v_B) - v_A} = \frac{\Delta u_{BA}}{\Delta y + (v_B - v_A)} = \frac{\Delta u_{BA}}{\Delta y + \Delta v_{BA}} = \frac{\Delta u}{\Delta y} \tag{1.51a}$$

$$\gamma_{ave} \equiv \gamma_{12} = e_{21} + e_{12} = \lim_{\Delta x \to 0, \lim \Delta y \to 0} \frac{\Delta v}{\Delta x} + \frac{\Delta u}{\Delta y} = \frac{\partial v}{\partial x} + \frac{\partial u}{\partial y} \tag{1.51b}$$

Strains are assumed to be infinitesimal in the elastic region. Hence, the angles are small and their tangents are about the same as the angles themselves. Furthermore, it may be assumed that Δv_{DA} and Δu_{BA} are smaller than Δx and Δy, respectively. Therefore, one can write $\Delta x + \Delta u_{DA} \approx \Delta x$ and $\Delta y + \Delta v_{BA} \approx \Delta y$. These simplifications produced the last terms in Eqs. (1.51) and (1.51a), also indicated by the sum in Eq. (1.51b).

To summarize, the strain state at a point in a two-dimensional case may be given in terms of the normal and shear strains

$$e_{ij} = \begin{vmatrix} e_{11} & e_{12} \\ e_{21} & e_{22} \end{vmatrix} = \begin{vmatrix} \frac{\partial u}{\partial x} & \frac{\partial u}{\partial y} \\ \frac{\partial v}{\partial x} & \frac{\partial v}{\partial y} \end{vmatrix} \tag{1.51c}$$

Often, the above is given in the literature as:

$$e_{ij} = \begin{vmatrix} e_x & \gamma_{xy} \\ \gamma_{yx} & e_y \end{vmatrix} \tag{1.51d}$$

Analogous to stress, normal strain is obtained when $i = j$ and, for shear strain, $i \neq j$. Finally, in the three-dimensional case, an additional term exists, and Eq. (1.50a) becomes by the addition of a term in z:

$$e_{11} = \lim_{\Delta x \to 0} \tfrac{\Delta u}{\Delta x} = \tfrac{\partial u}{\partial x} \quad e_{22} = \lim_{\Delta y \to 0} \tfrac{\Delta v}{\Delta y} = \tfrac{\partial v}{\partial y} \quad e_{33} = \lim_{\Delta z \to 0} \tfrac{\Delta w}{\Delta z} = \tfrac{\partial w}{\partial z} \tag{1.51b}$$

The normal strains in three dimensions may be derived from the deformed cube shown in Fig. 1.32, where normal strains are operating and all shear strains are zero. Thus, no angular deformation has occurred, but there is a change in the dimensions of the cube. These changes in the sides of the deformed cube, resulting from the displacement, are $\Delta x + \Delta u$, $\Delta y + \Delta v$ and $\Delta z + \Delta w$. The following equations may be given for the averaged normal strain components of such a three-dimensional deformation:

$$\begin{aligned} e_{11} &= \frac{u + \Delta u - u}{\Delta x} = \frac{\Delta u}{\Delta x} \Rightarrow e_{11} = \lim_{\Delta x \to 0} \frac{\partial u}{\partial x} \\ e_{22} &= \frac{v + \Delta v - v}{\Delta y} = \frac{\Delta v}{\Delta y} \Rightarrow e_{22} = \lim_{\Delta y \to 0} \frac{\partial v}{\partial y} \\ e_{33} &= \frac{w + \Delta w - w}{\Delta z} = \frac{\Delta w}{\Delta z} \Rightarrow e_{33} = \lim_{\Delta z \to 0} \frac{\partial w}{\partial z} \end{aligned} \tag{1.52}$$

For each case of Eq. (1.52):

$$u = e_{11}x; \quad v = e_{22}y; \quad w = e_{33}z \tag{1.52a}$$

As a consequence of displacements u, v and z, the three-dimensional components are given as:

$$\begin{aligned} u &= e_{11}x + e_{12}y + e_{13}z \\ v &= e_{21}x + e_{22}y + e_{23}z \\ w &= e_{31}x + e_{32}y + e_{33}z \end{aligned} \tag{1.53}$$

In Fig. 1.32, the displacements are shown indicating how Eqs. (1.52) may be obtained. As mentioned in the two-dimensional case, there is an assumption that the strains are infinitesimal.

The two-dimensional shear components of a deformed body are given in Eqs. (1.51), (1.52a) and their sum, giving the shear strain, γ_{12}. The graphic presentation of the angular distortion of an elementary cube into a rhomboid is more complicated and will not be presented here. However, the following results can be

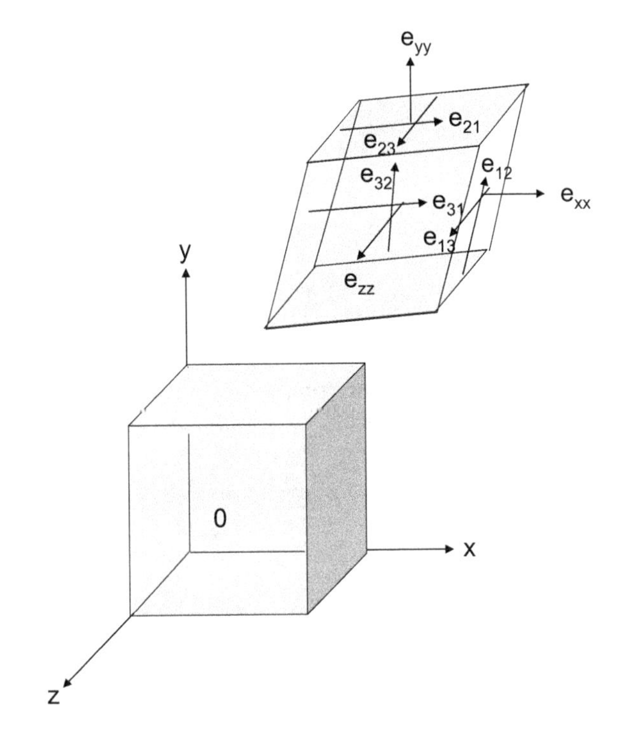

Fig. 1.33 The normal and shear components of a distorted cube are shown

obtained. A shear component of an angle $e_{13} = e_{31}$ (in a three-dimensional illustration) is:

$$\tan e_{13} \cong e_{13} = \frac{\Delta u}{\Delta z} \tag{1.54}$$

Additional tangents of appropriate angles are:

$$\gamma_4 = \frac{A'B' - AB}{\text{AB}} = \frac{w_A - w_B}{\Delta y} = \frac{\Delta w_{AB}}{\Delta y} = \lim_{\Delta y \to 0} \frac{\partial w}{\partial y}$$
$$\gamma_5 = \frac{B'E' - BE}{BE} = \frac{u_E - u_A}{} = \frac{\Delta U_{EU}}{\Delta z} = \lim_{\Delta z \to 0} \frac{\partial u}{\partial z}$$
$$\gamma_6 = \frac{A'D' - AD}{AD} = \frac{w_D - w_A}{\Delta x} = \frac{\Delta w_{DA}}{\Delta x} = \lim_{\Delta x \to 0} \frac{\partial w}{\partial x}.$$

Note again that the 'engineering shear strain' is defined as the total change in the right angle of an elementary cube, being the sum of the appropriate angles, given as (for γ_{12} in Eq. (1.51b)) and rewritten in Eq. (1.55) as:

$$\gamma_{12} \equiv \gamma_{xy} = e_{12} + e_{21} = \frac{\partial u}{\partial v} + \frac{\partial v}{\partial x} = \gamma_{12} \tag{1.55}$$

$$\gamma_{13} \equiv \gamma_{xz} = \frac{\partial u}{\partial z} + \frac{\partial w}{\partial x} = \gamma_{31} \tag{1.56}$$

$$\gamma_{23} \equiv \gamma_{yz} = \frac{\partial v}{\partial z} + \frac{\partial w}{\partial y} = \gamma_{32} \tag{1.57}$$

One can switch the subscripts to obtain:

$$\gamma_{21} = \frac{\partial v}{\partial x} + \frac{\partial u}{\partial y}; \quad \gamma_{32} = \frac{\partial w}{\partial y} + \frac{\partial v}{\partial z}; \quad \gamma_{31} = \frac{\partial w}{\partial x} + \frac{\partial u}{\partial z} \tag{1.58}$$

The strain at a point in a three-dimensional system can be arranged in a 3 × 3 matrix comprising nine components. Of these components, three are normal and six are shear components, as indicated in Eq. (1.59):

$$e_{ij} = \begin{vmatrix} e_{11} & e_{12} & e_{13} \\ e_{21} & e_{22} & e_{23} \\ e_{31} & e_{32} & e_{33} \end{vmatrix} = \begin{vmatrix} \frac{\partial u}{\partial x} & \frac{\partial u}{\partial y} & \frac{\partial u}{\partial z} \\ \frac{\partial v}{\partial x} & \frac{\partial v}{\partial y} & \frac{\partial v}{\partial z} \\ \frac{\partial w}{\partial x} & \frac{\partial w}{\partial y} & \frac{\partial w}{\partial z} \end{vmatrix} \tag{1.59}$$

Actually, this matrix is symmetrical and can, therefore, be reduced to only six independent components.

Equations (1.57) and (1.58) indicate the reciprocity concept–expressions that do not change if the subscripts are reversed. Equation (1.59) may also be expressed in terms of γ, as follows:

$$\begin{matrix} e_{xx} & \gamma_{xy} & \gamma_{xz} \\ \gamma_{yx} & e_{yy} & \gamma_{yz} \\ \gamma_{zx} & \gamma_{zy} & e_{zz} \end{matrix} \tag{1.60}$$

In terms of these strain components, Fig. 1.33 shows the distorted elementary cube around a point for small displacements. Recall that the cube and the rhomboid exaggerate the space around the point considered.

It was indicated earlier that the shear strain $\gamma = e_{12} + e_{21}$, as indicated in Fig. 1.31. Therefore, it is possible to write $\gamma = 2e_{12}$, which is the engineering strain. Also note that:

$$\frac{\gamma_{ij}}{2} = e_{ij}$$

Equation (1.59) is a second-rank tensor and, as such, it can be decomposed into symmetric and anti-symmetric parts, also known as 'skew tensors' (see, for example [6]). Any second-rank tensor, T_{ij}, may be expressed as their sum or as:

$$\mathrm{T_{ij}} = \mathrm{S_{ij}} + \mathrm{A_{ij}} \tag{1.61}$$

where:

$$S_{ij} = \tfrac{1}{2}(T_{ij} + T_{ji}) \quad A_{ij} = \tfrac{1}{2}(T_{ij} - T_{ji}) \tag{1.62}$$

Based on this concept, deformation may be written as a strain tensor, ε_{ij}, and a rigid body rotation, w_{ij}, as follows:

$$e_{ij} = \frac{1}{2}(e_{ij} + e_{ji}) + \frac{1}{2}(e_{ij} - e_{ji}) \tag{1.63}$$

This expression is often written in the literature as:

$$e_{ij} = \varepsilon_{ij} + w_{ij} \tag{1.63a}$$

Applying these concepts, it is possible to write the following for ε_{ij} and w_{ij}:

$$\begin{vmatrix} \varepsilon_{11} & \varepsilon_{12} & \varepsilon_{13} \\ \varepsilon_{21} & \varepsilon_{22} & \varepsilon_{23} \\ \varepsilon_{31} & \varepsilon_{32} & \varepsilon_{33} \end{vmatrix} = \begin{vmatrix} \frac{1}{2}\left(\frac{\partial u}{\partial x} + \frac{\partial u}{\partial x}\right) & \frac{1}{2}\left(\frac{\partial u}{\partial y} + \frac{\partial v}{\partial x}\right) & \frac{1}{2}\left(\frac{\partial u}{\partial z} + \frac{\partial w}{\partial x}\right) \\ \frac{1}{2}\left(\frac{\partial v}{\partial x} + \frac{\partial u}{\partial y}\right) & \frac{1}{2}\left(\frac{\partial v}{\partial y} + \frac{\partial v}{\partial y}\right) & \frac{1}{2}\left(\frac{\partial v}{\partial z} + \frac{\partial w}{\partial y}\right) \\ \frac{1}{2}\left(\frac{\partial w}{\partial x} + \frac{\partial u}{\partial z}\right) & \frac{1}{2}\left(\frac{\partial w}{\partial y} + \frac{\partial v}{\partial z}\right) & \frac{1}{2}\left(\frac{\partial w}{\partial z} + \frac{\partial w}{\partial z}\right) \end{vmatrix}$$
$$= \begin{vmatrix} \frac{\partial u}{\partial x} & \frac{1}{2}\left(\frac{\partial u}{\partial y} + \frac{\partial v}{\partial x}\right) & \frac{1}{2}\left(\frac{\partial u}{\partial z} + \frac{\partial w}{\partial x}\right) \\ \frac{1}{2}\left(\frac{\partial v}{\partial x} + \frac{\partial u}{\partial y}\right) & \frac{\partial v}{\partial y} & \frac{1}{2}\left(\frac{\partial v}{\partial z} + \frac{\partial w}{\partial y}\right) \\ \frac{1}{2}\left(\frac{\partial w}{\partial x} + \frac{\partial u}{\partial z}\right) & \frac{1}{2}\left(\frac{\partial w}{\partial y} + \frac{\partial v}{\partial z}\right) & \frac{\partial w}{\partial z} \end{vmatrix} \tag{1.64}$$

$$\begin{vmatrix} w_{11} & w_{12} & w_{13} \\ w_{21} & w_{22} & w_{23} \\ w_{31} & w_{32} & w_{33} \end{vmatrix} = \begin{vmatrix} 0 & \frac{1}{2}\left(\frac{\partial u}{\partial y} - \frac{\partial v}{\partial x}\right) & \frac{1}{2}\left(\frac{\partial u}{\partial z} - \frac{\partial w}{\partial x}\right) \\ \frac{1}{2}\left(\frac{\partial v}{\partial x} - \frac{\partial u}{\partial y}\right) & 0 & \frac{1}{2}\left(\frac{\partial v}{\partial z} - \frac{\partial w}{\partial y}\right) \\ \frac{1}{2}\left(\frac{\partial w}{\partial x} - \frac{\partial u}{\partial z}\right) & \frac{1}{2}\left(\frac{\partial w}{\partial y} - \frac{\partial v}{\partial z}\right) & 0 \end{vmatrix} \tag{1.65}$$

The shear strain, ε_{ij}, is seen to be symmetric from the following:

$$\varepsilon_{12} = \frac{1}{2}\left(\frac{\partial u}{\partial v} + \frac{\partial v}{\partial x}\right) \equiv \varepsilon_{21} = \frac{1}{2}\left(\frac{\partial v}{\partial x} + \frac{\partial u}{\partial v}\right) \tag{1.65a}$$

as are the other shear strain components, i.e., $\varepsilon_{13} \equiv \varepsilon_{31}$ and $\varepsilon_{23} \equiv \varepsilon_{32}$, as indicated in Eq. (1.64).

The normal and shear strains may be abbreviated as:

$$\varepsilon_{ij} = \frac{1}{2}\left(\frac{\partial u_i}{\partial x_j} + \frac{\partial u_j}{\partial x_i}\right) \tag{1.65b}$$

Keeping in mind the above notation in terms of u, v, w with coordinates x, y, z, when using another nomenclature, i, and j have the values of 1, 2 or 3 in the above formula. Clearly, u_2 and x_2 stand for v and y.

1.2.9 Generalized Hooke's Law

In Sect. 1.2 above, the stress–strain relation in uniaxial tension tests was given in Eq. (1.5), indicating a Hookean behavior. This section now considers linear elastic solids, as described by Hooke, according to which σ_{ij} is linearly proportional to the strain, ε_{ij}. Each stress component is expected to depend linearly on each strain component. For example, the σ_{11} may be expressed as follows:

$$\begin{array}{lll} & c_{1111}\varepsilon_{11}+ & c_{1211}\varepsilon_{12}+ & c_{1311}\varepsilon_{13}+ \\ \sigma_{11} = & c_{2111}\varepsilon_{21}+ & c_{2211}\varepsilon_{22}+ & c_{2311}\varepsilon_{23}+ \\ & c_{3111}\varepsilon_{31}+ & c_{3211}\varepsilon_{32}+ & c_{3311}\varepsilon_{33} \end{array} \tag{1.66}$$

Clearly, similar relations may be written for other stress components. Succinctly, Eq. (1.66) may be presented as:

$$\sigma_{ij} = \sum_{k=1}^{3}\sum_{l=1}^{3} c_{ijkl}\varepsilon_{kl} \tag{1.67}$$

or in terms of the strain as:

$$\varepsilon_{ij} = \sum_{k=1}^{3}\sum_{l=1}^{3} S_{ijkl}\sigma_{kl} \tag{1.68}$$

c_{ijkl}'s and the S_{ijkl}'s are proportionality constants and are called the 'stiffness and compliance constants', where i and j can have any value from 1 to 3 [12]. The number of constants is $3^4 = 81$. However, in the most general case, only 21 independent constants are sufficient, since the following relations hold:

$$\sigma_{ij} = \sigma_{ji} \quad \varepsilon_{ij} = \varepsilon_{ji} \quad \text{and it can be argued that} \quad c_{ijkl} = c_{klij}$$

Note that: $c_{ijkl} = c_{jikl} = c_{ijlk} = c_{klij}$ ... etc. for the other possible combinations.
Thus, for the most general elastic expression of a material, 21 constants must be specified. Yet, considering the various symmetries of crystals, this number can be further reduced. For instance, in cubic crystals having coordinate systems along the cube edges, many of the constants, such as $c_{1111} = c_{2222}$ ... etc. or $c_{1212} = c_{1313} = c_{2323}$ ..., are equivalent. The notation shown in Eq. (1.66) may be simplified here as it has been in many publications:

$$\begin{array}{lll} \sigma_{11} \rightarrow \sigma_1 & \sigma_{22} \rightarrow \sigma_2 & \sigma_{33} \rightarrow \sigma_3 \\ \sigma_{23} \rightarrow \sigma_4 & \sigma_{31} \rightarrow \sigma_5 & \sigma_{12} \rightarrow \sigma_6 \\ \varepsilon_{11} \rightarrow \varepsilon_1 & \varepsilon_{22} \rightarrow \varepsilon_2 & \varepsilon_{33} \rightarrow \varepsilon_3 \\ 2\varepsilon_{23} \rightarrow \gamma_4 & 2\varepsilon_{31} \rightarrow \gamma_5 & 2\varepsilon_{12} \rightarrow \gamma_6 \end{array} \tag{1.69}$$

making Eq. (1.66):

$$\sigma_1 = \begin{array}{c} c_{11}\varepsilon_1 + c_{12}\varepsilon_2 + c_{13}\varepsilon_3 + \\ c_{14}\gamma_4 + c_{15}\gamma_5 + c_{16}\lambda_6 \end{array} \tag{1.70}$$

Furthermore, instead of the γ's, the shear components are often given as ε_4, ε_5 and ε_6. Thus, one can write Eq. (1.70) compactly by replacing the γ's with ε's:

$$\sigma_i = c_{ij}\varepsilon_j \tag{1.71}$$

Notice that, in Eq. (1.70), the definition of the shear components, such as ε_{12}, ε_{21} etc., are indicated in Eq. (1.69) as $2\varepsilon_{ij} = \gamma_k$ (where k = 4, 5 or 6); therefore, it is sufficient to write Eq. (1.70) with six components.

Crystals with cubic symmetry have three independent elastic constants and the following equivalences apply:

$$\begin{array}{l} c_{11} = c_{22} = c_{33} \\ c_{12} = c_{23} = c_{31} \\ c_{44} = c_{55} = c_{66} \end{array} \tag{1.72}$$

and all the other elastic constants are zeros. Consequently, the tensor for the stiffness components for cubic symmetry is given as:

$$\left| \begin{array}{cccccc} c_{11} & c_{12} & c_{12} & 0 & 0 & 0 \\ c_{12} & c_{11} & c_{12} & 0 & 0 & 0 \\ c_{12} & c_{12} & c_{11} & 0 & 0 & 0 \\ 0 & 0 & 0 & c_{44} & 0 & 0 \\ 0 & 0 & 0 & 0 & c_{44} & 0 \\ 0 & 0 & 0 & 0 & 0 & c_{44} \end{array} \right| \tag{1.73}$$

Thus, there are only three elastic constants in cubic crystals: c_{11}, c_{12} and c_{44}. Similarly, the compliance components of materials with cubic symmetry are given:

$$\left| \begin{array}{cccccc} S_{11} & S_{12} & S_{12} & 0 & 0 & 0 \\ S_{12} & S_{11} & S_{12} & 0 & 0 & 0 \\ S_{12} & S_{12} & S_{11} & 0 & 0 & 0 \\ 0 & 0 & 0 & S_{44} & 0 & 0 \\ 0 & 0 & 0 & 0 & S_{44} & 0 \\ 0 & 0 & 0 & 0 & 0 & S_{44} \end{array} \right| \tag{1.74}$$

For cubic crystals, the elastic compliance constants are related to those of stiffness [12]:

$$\begin{aligned} S_{11} &= \frac{c_{11}+c_{12}}{(c_{11}-c_{12})(c_{11}+2c_{12})} \\ S_{12} &= \frac{-c_{12}}{(c_{11}-c_{12})(c_{11}+2c_{12})} \\ S_{44} &= \frac{2}{(c_{11}-c_{12})} \end{aligned} \tag{1.75}$$

The different forms of compliance and stiffness, expressed as tensors for the various crystal classes and Eq. (1.75) are found in the literature. The Eq. (1.75), expressing the stiffness components and related to the c's, are given as:

$$\begin{aligned} c_{11} &= \frac{S_{11}+S_{12}}{(S_{11}-S_{12})(S_{11}+2S_{12})} \\ c_{12} &= \frac{-S_{12}}{(S_{11}-S_{12})(S_{11}+2S_{12})} \\ c_{44} &= \frac{1}{S_{44}} \end{aligned} \tag{1.75b}$$

In isotropic crystals, the number of constants may be further reduced; instead of three constants (cubic crystals), only two independent constants are required to describe a state of stress at a point, given as matrix (1.76):

$$\begin{vmatrix} c_{11} & c_{12} & c_{12} & 0 & 0 & 0 \\ c_{12} & c_{11} & c_{12} & 0 & 0 & 0 \\ c_{12} & c_{12} & c_{11} & 0 & 0 & 0 \\ 0 & 0 & 0 & (c_{11}-c_{12})/2 & 0 & 0 \\ 0 & 0 & 0 & 0 & (c_{11}-c_{12})/2 & 0 \\ 0 & 0 & 0 & 0 & 0 & (c_{11}-c_{12})/2 \end{vmatrix} \tag{1.76}$$

The tensor for compliance is similar and the $(c_{11} - c_{12})/2$ is replaced by $2(S_{11} - S_{12})$:

$$\begin{vmatrix} S_{11} & S_{12} & S_{12} & 0 & 0 & 0 \\ S_{12} & S_{11} & S_{12} & 0 & 0 & 0 \\ S_{12} & S_{12} & S_{11} & 0 & 0 & 0 \\ 0 & 0 & 0 & 2(S_{11}-S_{12}) & 0 & 0 \\ 0 & 0 & 0 & 0 & 2(S_{11}-S_{12}) & 0 \\ 0 & 0 & 0 & 0 & 0 & 2(S_{11}-S_{12}) \end{vmatrix} \tag{1.77}$$

Recall that isotropy is uniformity in all orientations; therefore, relations must be independent of the coordinate system chosen to represent isotropy.

Table 1.2 Elastic constants, bulk modulus, Young's modulus, shear modulus, and Poisson's ratio for cubic ceramic crystals with both LDA/USPP and GGA/USPP exchange–correlation potentials [50] (with kind permission of Wiley)

System	Exchange correlation	# of k-points	$a(\dot{A})$	$V_0(\dot{A}^3)$	C_{11} (GPa)	C_{12}	C_{44}	K	G	E	η
MgO	LDA/USPP		4.2309	75.74	291	92	156	158.3	130.3	306.7	0.1772
	GGA/USPP	180	4.2317	75.78	276	86	149	149.3	124.4	292.1	0.1740
	Experiment[32, 33]		4.2130	74.78	294	93	155	160.0	130.3	–	0.18
CaO	LDA/USPP		4.5677	95.30	240.5	48.2	93.2	112.3	94.4	221.2	0.1718
	GGA/USPP	150	4.7668	108.30	207.8	49.9	79.5	102.5	79.3	189.1	0.1926
	Experiment[32–35]		4.8071	111.08	224	60	80.6	114.7	81.2	–	–
Y_2O_3	LDA/USPP		10.4942	1155.64	241.9	128.0	85.1	166.0	72.4	189.7	0.3095
	GGA/USPP	2	10.6822	1218.86	213.6	112.9	72.6	155.0	66.6	174.8	0.3121
	Experiment[36, 37]		10.6073	1193.48	224	112	74.6	150.	66.3	–	–
In_2O_3	LDA/USPP		9.9839	995.10	269.9	138.8	70.5	181.8	68.1	181.5	0.3336
	GGA/USPP	2	10.2374	1072.91	234.3	107.2	62.7	149.6	63	165.8	0.3152
	Experiment[38]		10.1170	1035.51	–	–	–	–	–	–	–
$MgAl_2O_4$	LDA/USPP		8.0200	515.82	273.6	149.6	150.7	191.0	105.6	267.4	0.2666
	GGA/USPP	4	8.1610	543.50	256.5	133.2	142.4	174.3	101.8	255.6	0.2556
	Experiment[32, 39]		8.0860	528.69	282.9	155.4	154.8	197.9	108.5	–	–
γ-Mg_2SiO_4	LDA/USPP		7.9557	503.53	338.3	111.8	130.0	187.3	128.6	314.0	0.2206
	GGA/USPP	4	8.1139	534.17	299.6	103.3	128.2	168.7	115.2	281.5	0.2219
	Experiment[40, 39]		8.1700	545.34	327	112	126	184	119	–	–
$SrTiO_3$	LDA/USPP		3.8572	57.38	397.2	100.8	114.1	199.6	126.7	313.7	0.2380
	GGA/USPP	80	3.9403	61.17	324.7	89.8	106.3	168.1	110.6	272.2	0.2301
	Experiment[32, 39]		3.9040	59.50	335.0	105	127				
YAG	LDA/USPP		11.8465	1662.37	356.7	122.6	114.3	200.6	115.4	290.5	0.2587
	Experiment[32, 41]	1	12.0000	1728.00	333	113	115	–	–	–	–
BSO	LDA/USPP		9.9152	974.74	150.1	36.5	31.4	74.4	39.9	101.5	0.2724

(continued)

Table 1.2 (continued)

System	Exchange correlation	# of k-points	$a(\dot{A})$	$V_0(\dot{A}^3)$	C_{11} (GPa)	C_{12}	C_{44}	K	G	E	η
	GGA/USPP	2	10.2227	1068.27	116.5	30.1	23.9	58.9	30.4	77.7	0.2801
	Experiment[42, 43]		10.1043	1031.62	129.8	30.2	24.7	–	–	–	–
c-BN	LDA/USPP		3.5725	45.6	825	193	475	404	403	908	0.125
	GGA/USPP	260	3.6653	49.2	783	172	444	376	382	856	0.120
	Experiment[32, 44]		3.6150	47.2	820	190	480	400	405	–	–
γ-Si_3N_4	LDA/USPP		7.7861	472.0	529	169	334	289	261	601	0.153
	GGA/USPP	4	7.7868	472.1	504	177	327	286	248	576	0.164
	Experiment[45, 46]		7.8369	481.3	–	–	–	300	–	–	–
γ-Sn_3N_4	LDA/USPP		8.9521	717.38	282.1	139.5	140.4	187	107	269.5	0.2598
	GGA/USPP	12	9.1171	757.82	246.1	116.9	128.0	160.0	97.3	242.6	0.2472
	Experiment[47]		9.037	738.03	–	–	–	–	–	–	–
TiN	LDA/USPP		4.1865	73.4	688	124	171	312	209	513	0.226
	GGA/USPP	180	4.2606	77.3	680	130	171	313	207	509	0.229
	Experiment[32, 48]		4.2387	76.2	625	165	163	325	192	–	–
HfN	LDA/USPP		4.4446	87.80	734.8	105.8	129.4	315.5	186.3	467.1	0.2532
	GGA/USPP	150	4.5219	92.46	628.9	103.6	119.5	278.7	164.8	413	0.2530
	Experiment[32, 49]		4.5400	93.58	679	119	150	306	202	387	0.15
Diamond	LDA/USPP		3.5268	43.86	1107.2	144.7	598.1	465.5	548.3	1181.1	0.0771
	GGA/USPP	80	3.5674	45.40	1055.0	120.4	559.0	431.9	520.3	1113.7	0.0702
	Experiment[32, 50]		3.5670	45.38	1079	124	578	443.0	535.7	–	–
SiC-3C	LDA/USPP		4.3194	80.58	401.9	136.4	255.7	224.9	196.5	456.6	0.1616
	GGA/USPP	80	4.3694	83.42	384.5	121.5	243.3	209.2	190.1	437.6	0.1513
	Experiment[51, 52]		4.3597	82.86	390	142	256	225	–	–	–
c-ZnS	LDA/USPP		5.3087	149.60	116.6	71.8	50.9	86.7	36.6	96.3	0.3150
	GGA/USPP	80	5.4512	161.98	96.5	56.5	44.9	69.8	32.5	84.3	0.2988
	Experiment[53, 54]		5.4102	158.36	102	64.6	44.6	77.1	31.5	–	–

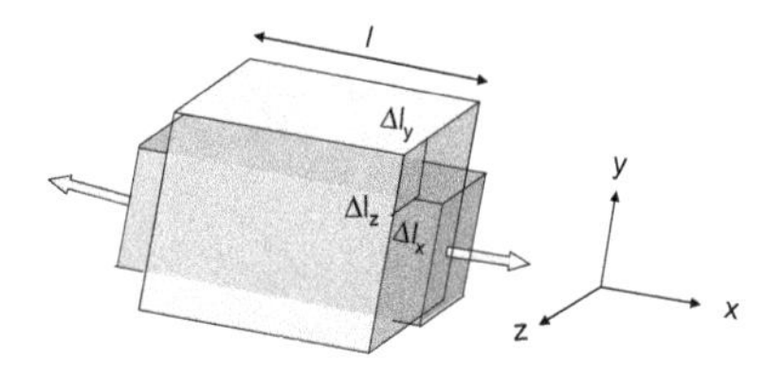

Fig. 1.34 The application of a tensile stress on a cube in the x direction results in a Δl contraction in both the y and z directions

The equations $\sigma_{ij} = c_{ijkl}\varepsilon_{kl}$ and $\varepsilon_{ij} = S_{ijkl}\sigma_{kl}$, in which i, j, k, l = 1, 2 or 3, may be meaningfully expressed in matrix form for the general case as:

$$\begin{vmatrix} \sigma_{xx} \\ \sigma_{yy} \\ \sigma_{zz} \\ \sigma_{yz} \\ \sigma_{zx} \\ \sigma_{xy} \end{vmatrix} = \begin{vmatrix} c_{11} & c_{12} & c_{13} & c_{14} & c_{15} & c_{16} \\ c_{21} & c_{22} & c_{23} & c_{24} & c_{25} & c_{26} \\ c_{31} & c_{32} & c_{33} & c_{34} & c_{35} & c_{36} \\ c_{41} & c_{42} & c_{43} & c_{44} & c_{45} & c_{46} \\ c_{51} & c_{52} & c_{53} & c_{54} & c_{55} & c_{56} \\ c_{61} & c_{62} & c_{63} & c_{64} & c_{65} & c_{66} \end{vmatrix} \begin{vmatrix} \varepsilon_{xx} \\ \varepsilon_{yy} \\ \varepsilon_{zz} \\ \varepsilon_{yz} \\ \varepsilon_{zx} \\ \varepsilon_{xy} \end{vmatrix} \tag{1.78}$$

and

$$\begin{vmatrix} \varepsilon_{xx} \\ \varepsilon_{yy} \\ \varepsilon_{zz} \\ \varepsilon_{yz} \\ \varepsilon_{zx} \\ \varepsilon_{xy} \end{vmatrix} = \begin{vmatrix} S_{11} & S_{12} & S_{13} & S_{14} & S_{15} & S_{16} \\ S_{21} & S_{22} & S_{23} & S_{24} & S_{25} & S_{26} \\ S_{31} & S_{32} & S_{33} & S_{34} & S_{35} & S_{36} \\ S_{41} & S_{42} & S_{43} & S_{44} & S_{45} & S_{46} \\ S_{51} & S_{52} & S_{53} & S_{54} & S_{55} & S_{56} \\ S_{61} & S_{62} & S_{63} & S_{64} & S_{65} & S_{66} \end{vmatrix} \begin{vmatrix} \sigma_{xx} \\ \sigma_{yy} \\ \sigma_{zz} \\ \sigma_{yz} \\ \sigma_{zx} \\ \sigma_{xy} \end{vmatrix} \tag{1.79}$$

The above are known as 'constitutive relations', due to Cauchy, who generalized Hooke's Law by assuming that, in elastic bodies, the 6 components of stress are linearly related to the 6 components of strain.

In Table 1.2, calculated elastic constants and other data compared with experimental data are presented for various cubic ceramic crystals taken from Yao et al. [50].

1.3 Poisson Ratio

Figure 1.34 illustrates a cube with l side of an isotropic, linearly elastic material, subjected to tension along the x axis by a small Δl_x extension. After the cube underwent strain, contraction occurred along the y and z directions at magnitudes Δl_y and Δl_z.

Basically, Fig. 1.33 illustrates the Poisson effect, previously considered in the section on tension and expressed by Eqs. (1.11)—(1.12a) as rewritten below:

$$\nu = -\frac{\textit{lateral strain}}{\textit{axial strain}} = -\frac{\varepsilon_y}{\varepsilon_x}\left(= -\frac{\varepsilon_{22}}{\varepsilon_{11}}\right) = -\frac{\varepsilon_z}{\varepsilon_x}\left(-\frac{\varepsilon_{33}}{\varepsilon_{xx}}\right) \tag{1.11}$$

$$G = \frac{E}{2(1+\nu)} \qquad \nu = \frac{E}{2G} - 1 \tag{1.12a}$$

Given that this is in the elastic region and that the acting stresses are small, the assumption can be made that the normal stress (e.g., σ_x) does not produce shear strains on the appropriate planes. Using the principle of superposition while recalling that the stress components act in their respective directions and using appropriate strains expressed in terms of Poisson's ratio, the effects of the three individual, uniaxial loadings may be summed up as:

$$\begin{aligned}
\varepsilon_x &= \frac{\sigma_x}{E} - \frac{\nu\sigma_y}{E} - \frac{\nu\sigma_z}{E} = \frac{\sigma_x - \nu(\sigma_y + \sigma_z)}{E} \\
\varepsilon_y &= \frac{\sigma_y}{E} - \frac{\nu\sigma_z}{E} - \frac{\nu\sigma_x}{E} = \frac{\sigma_y - \nu(\sigma_z + \sigma_x)}{E} \\
\varepsilon_z &= \frac{\sigma_z}{E} - \frac{\nu\sigma_x}{E} - \frac{\nu\sigma_y}{E} = \frac{\sigma_z - \nu(\sigma_x + \sigma_y)}{E}
\end{aligned} \tag{1.80}$$

The strain components in the x, y and z directions used to obtain Eq. (1.80) are:

$$\varepsilon'_x = \frac{\sigma_x}{E} \qquad \varepsilon'_y = -\frac{\nu}{E}\sigma_x \qquad \varepsilon'_z = -\frac{\nu}{E}\sigma_x \tag{1.81}$$

$$\varepsilon''_x = -\frac{\nu}{E}\sigma_y \qquad \varepsilon''_y = \frac{\sigma_y}{E} \qquad \varepsilon''_z = -\frac{\varepsilon_z}{E}\sigma_y \tag{1.82}$$

$$\varepsilon'''_x = -\frac{\nu}{E}\sigma_z \qquad \varepsilon'''_y = -\frac{\nu}{E}\sigma_z \qquad \varepsilon'''_z = \frac{\sigma_z}{E} \tag{1.83}$$

Eliminating E from the two relevant equations in (1.80) yields:

$$\frac{\varepsilon_x}{\sigma_x - \nu(\sigma_y + \sigma_z)} = \frac{\varepsilon_y}{\sigma_y - \nu(\sigma_z + \sigma_x)} = \frac{\varepsilon_z}{\sigma_z - \nu(\sigma_x + \sigma_y)} \tag{1.80a}$$

expressing Hooke's Law for isotropic materials. It is easy to remember the relations shown in Eq. (1.80), since they are in succession, x, y and z.

If one of the normal strains in one of the directions, e.g., the z direction, is zero, then strain ε_z may be reduced as follows:

$$\varepsilon_z = 0 = \frac{\sigma_z - \nu(\sigma_x + \sigma_y)}{E} \tag{1.80b}$$

making the stress component σ_z:

$$\sigma_z = \nu(\sigma_x + \sigma_y) \tag{1.80c}$$

Similar relations can be obtained for strains in the x or y directions, when the appropriate normal strain in the respective directions is zero. Thus, it is possible to write the following relations for ε_x and ε_y:

$$\varepsilon_x = 0 = \frac{\sigma_x - \nu(\sigma_y + \sigma_z)}{E} \tag{1.80d}$$

$$\sigma_x = \nu(\sigma_y + \sigma_z) \tag{1.80e}$$

and:

$$\varepsilon_y = 0 = \frac{\sigma_y - \nu(\sigma_z + \sigma_x)}{E} \tag{1.80f}$$

$$\sigma_y = \nu(\sigma_z + \sigma_x) \tag{1.80g}$$

When an applied load produces a normal stress in one of the three directions (assumed to be zero), a plain-stress condition prevails. In thin films, for example, the z dimension is small compared to the other directions; the approximation that the normal stress is $\sigma_z = 0$ is a good one. In this case, the equations for plain stress may be obtained from Eqs. (1.80). These strains are:

$$\begin{aligned} \boldsymbol{\varepsilon_x} &= \frac{\boldsymbol{\sigma_x - \nu\sigma_y}}{\boldsymbol{E}} \\ \varepsilon_y &= \frac{\sigma_y - \nu\sigma_x}{E} \\ \varepsilon_z &= -\frac{\nu(\sigma_x - \sigma_y)}{E} \end{aligned} \tag{1.80h}$$

In terms of planar stresses, one can easily infer the following expressions from Eq. (1.80h):

$$\begin{aligned} \boldsymbol{\sigma_x} &= \frac{\mathbf{E}(\boldsymbol{\varepsilon_x + \nu\varepsilon_y})}{\mathbf{1 - \nu^2}} \\ \boldsymbol{\sigma_y} &= \frac{\mathbf{E}(\boldsymbol{\varepsilon_y + \nu\varepsilon_x})}{\mathbf{1 - \nu^2}} \end{aligned} \tag{1.80i}$$

For instance, to show how σ_x is obtained from Eq. (1.80h), the following steps are taken:

(a) Express σ_x from the first relation as:

$$\sigma_x = E\varepsilon_x + \nu\sigma_y;$$

(b) Express σ_y from the second relation in Eq. (1.80h) and multiply by ν to obtain:

$$\nu\sigma_y = \nu E\varepsilon_y + \nu^2\sigma_x$$

(c) Replace $\nu\sigma_y$ in (a) from (b) and rearrange the relation as:

$$\sigma_x = \frac{E\left(\varepsilon_x + \nu\varepsilon_y\right)}{(1-\nu^2)}.$$

In a similar manner, σ_y may be obtained.

Despite the assumption that in a thin film $\sigma_z = 0$, $\varepsilon_z \neq 0$, as seen from Eq. (1.80b). It is also easy to get the plain strain conditions when $\varepsilon_z = 0$.

Useful relations can be obtained when Eq. (1.80c) is inserted into Eq. (1.80) giving:

$$\begin{aligned}\varepsilon_x &= \frac{\sigma_x(1-\nu^2) - \nu\sigma_y(1+\nu)}{E}\\ \varepsilon_y &= \frac{\sigma_y(1-\nu^2) - \nu\sigma_x(1+\nu)}{E}\\ \varepsilon_z &= 0.\end{aligned} \tag{1.80j}$$

Even though strain $\varepsilon_z = 0$, the stress is not as found in Eq. (1.80c). Now, returning to the three-dimensional case, one gets the following relations from Eq. (1.80) by adding and subtracting the appropriate terms:

$$\begin{aligned}\varepsilon_x &= \frac{1}{E}\left[\sigma_x(1+\nu) - \nu\left(\sigma_x + \sigma_y + \sigma_z\right)\right]\\ \varepsilon_y &= \frac{1}{E}\left[\sigma_y(1+\nu) - \nu\left(\sigma_x + \sigma_y + \sigma_z\right)\right]\\ \varepsilon_z &= \frac{1}{E}\left[\sigma_z(1+\nu) - \nu\left(\sigma_x + \sigma_y + \sigma_z\right)\right]\end{aligned} \tag{1.84}$$

Thus, for ε_x the steps are:

(a) From Eq. (1.80):

$$\varepsilon_x = \frac{\sigma_x - \nu(\sigma_y + \sigma_z)}{E}$$

(b) Add and subtract the appropriate term of $\frac{\nu\sigma_x}{E}$ to get:

$$\varepsilon_x = \frac{\sigma_x - \nu(\sigma_y + \sigma_z)}{E} + \frac{\nu\sigma_x}{E} - \frac{\nu\sigma_x}{E}$$

(c) Rearrange the above relation to obtain:

$$\varepsilon_x = \frac{1}{E}\left[\sigma_x(1+\nu) - \nu\left(\sigma_x + \sigma_y + \sigma_z\right)\right]$$

Similarly, one can obtain the relations for ε_y and ε_z.

Another useful relation may be obtained by adding the Eq. (1.80) as:

$$\varepsilon_x + \varepsilon_y + \varepsilon_z = \frac{(1-2v)}{E}\left(\sigma_x + \sigma_y + \sigma_z\right) \tag{1.85}$$

Equation (1.85) is often expressed as:

$$\left(\sigma_x + \sigma_y + \sigma_z\right) = \frac{E}{(1-2v)}\left(\varepsilon_x + \varepsilon_y + \varepsilon_z\right) \tag{1.85a}$$

Additional relations may be derived from Eqs. (1.84) and (1.85) to give Eq. (1.86) as follows:

(a) Take σ_{x} from Eq. (1.84) as:

$$\sigma_x = \frac{E\varepsilon_x}{(1+v)} + \frac{v}{(1+v)}\left(\sigma_x + \sigma_y + \sigma_z\right)$$

(b) Replace (σ_{x}, + σ_{y} + σ_{y}) in Eq. (1.85) to get Eq. (1.86):

$$\sigma_x = \frac{E\varepsilon_x}{(1+v))} + \frac{Ev}{(1+v)(1-2v)}\left(\varepsilon_x + \varepsilon_y + \varepsilon_z\right) \tag{1.86}$$

Similar relations may be obtained in the same manner for σ_{y} and σ_{z}. Thus, Eq. (1.86) may be rewritten in terms of Lame's parameters μ and λ as:

$$\sigma_x = 2\mu\varepsilon_x + \lambda(\varepsilon_x + \varepsilon_y + \varepsilon_z) \tag{1.86a}$$

λ is known as 'Lame's constant' and from Eq. (1.86) it is:

$$\lambda = \frac{Ev}{(1+v)(1-2v)} \tag{1.86b}$$

Hooke's Law for isotropic materials, in terms of Poisson's ratio, is given in matrix form as:

$$\begin{vmatrix} \varepsilon_{11} \\ \varepsilon_{22} \\ \varepsilon_{33} \\ 2\varepsilon_{23} \\ 2\varepsilon_{31} \\ 2\varepsilon_{12} \end{vmatrix} = \begin{vmatrix} \varepsilon_{11} \\ \varepsilon_{22} \\ \varepsilon_{33} \\ \gamma_{23} \\ \gamma_{31} \\ \lambda_{21} \end{vmatrix} = \frac{1}{E}\begin{vmatrix} 1 & -v & -v & 0 & 0 & 0 \\ -v & 1 & -v & 0 & 0 & 0 \\ -v & -v & 1 & 0 & 0 & 0 \\ 0 & 0 & 0 & 2(1+v) & 0 & 0 \\ 0 & 0 & 0 & 0 & 2(1+v) & 0 \\ 0 & 0 & 0 & 0 & 0 & 2(1+v) \end{vmatrix} = \begin{vmatrix} \sigma_{11} \\ \sigma_{22} \\ \sigma_{33} \\ \sigma_{23} \\ \sigma_{31} \\ \sigma_{12} \end{vmatrix} \tag{1.87}$$

Recall that the engineering shear strain was previously given as $\gamma_{ij} = 2\varepsilon_{ij}$. The above stress may also be expressed as being a reverse relation:

$$\begin{vmatrix} \sigma_{11} \\ \sigma_{22} \\ \sigma_{33} \\ \sigma_{23} \\ \sigma_{31} \\ \sigma_{12} \end{vmatrix} = \frac{E}{(1+\nu)(1-\nu)} \begin{vmatrix} (1-\nu) & \nu & \nu & 0 & 0 & 0 \\ \nu & (1-\nu) & \nu & 0 & 0 & 0 \\ \nu & \nu & (1-\nu) & 0 & 0 & 0 \\ 0 & 0 & 0 & (1-2\nu)/2 & 0 & 0 \\ 0 & 0 & 0 & 0 & (1-2\nu)/2 & 0 \\ 0 & 0 & 0 & 0 & 0 & (1-2\nu)/2 \end{vmatrix} \begin{vmatrix} \varepsilon_{11} \\ \varepsilon_{22} \\ \varepsilon_{33} \\ 2\varepsilon_{23} \\ 2\varepsilon_{31} \\ 2\varepsilon_{12} \end{vmatrix} \tag{1.88}$$

Furthermore, considering Eq. (1.86) or (1.86a), matrix (1.88) may be also given in terms of Lame's constants as (1.89):

Clearly, one could rewrite any of the tensors in Eqs. (1.87)–(1.89) for plane stress conditions. In this case, Hooke's Law can be rewritten with the understanding that $\sigma_{31} = \sigma_{13} = \sigma_{32} = \sigma_{23} = 0$. The only shear stress operating in the case of planar stress conditions is σ_{12}.

$$\begin{vmatrix} \sigma_{11} \\ \sigma_{22} \\ \sigma_{33} \\ \sigma_{23} \\ \sigma_{31} \\ \sigma_{12} \end{vmatrix} = \begin{vmatrix} (2\mu+\lambda) & \lambda & \lambda & 0 & 0 & 0 \\ \lambda & (2\mu+\lambda) & \lambda & 0 & 0 & 0 \\ \lambda & \lambda & (2\mu+\lambda) & 0 & 0 & 0 \\ 0 & 0 & 0 & \mu & 0 & 0 \\ 0 & 0 & 0 & 0 & \mu & 0 \\ 0 & 0 & 0 & 0 & 0 & \mu \end{vmatrix} \begin{vmatrix} \varepsilon_{11} \\ \varepsilon_{22} \\ \varepsilon_{33} \\ 2\varepsilon_{23} \\ 2\varepsilon_{31} \\ 2\varepsilon_{12} \end{vmatrix} \tag{1.89}$$

For ceramics, Poisson's ratio is smaller than it is for metals, namely ≤ 0.3. Poisson's ratio is an indication of the ability of a material to undergo deformation and, except for ductile or high-temperature ceramics, regular ceramics indeed have quite low Poisson ratios in accordance with their brittle behavior.

Structural ceramic materials are required to withstand high-temperature use and, therefore, their elastic properties are of great interest. As such, the focus will now shift to the Poisson's ratio. A common method for evaluating the Young's modulus at room temperature (found in all above relations) is either by a resonance technique [44] or by an ultrasonic pulse method [38]. A more appropriate technique for such high-temperature measurements is by a laser ultrasonic method [16]. Matsumoto et al. [9] used a laser ultrasonic method coupled with a Fabry–Perot interferometer (LUFP) to eliminate certain drawbacks encountered during high temperature measurements. Two techniques were compared when measuring samples of sintered SiC in the range of 20–1600 °C, as illustrated in Fig. 1.35. The results were also compared with those of the standard resonance technique. Using this technique Matsumoto et al. [9] determined the Young's modulus and Poisson's ratio for SiC ceramics, with and without additions, as a function of temperature, as shown in Fig. 1.36 for the Poisson's ratio.

These measurements indicated that, as the temperature increased, the Poisson's ratio remained relatively constant. However, even for the same material, there may

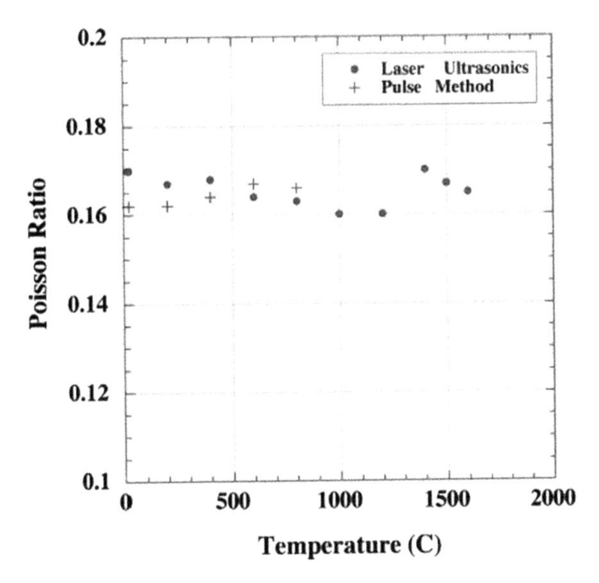

Fig. 1.35 Poisson's ratio measured by the (*filled circle*) LUFP (laser ultrasonics coupled with a Fabry–Perot interferometer) method and (*plus sign*) laser ultrasonic pulse technique using SiC as a standard [9]. With kind permission of Wiley and Sons

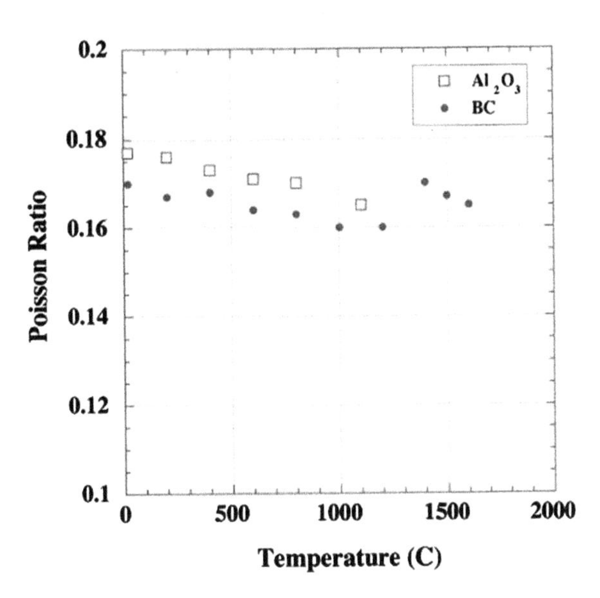

Fig. 1.36 Temperature dependence of Poisson's ratio of (*filled circle*) SiC ceramics with boron and carbon additions and (*open square*) SiC ceramics with Al_2O_3 additions [9]. With kind permission of Wiley and Sons

be differences in the test results from different laboratories as a consequence of the sensitivity of ceramics to production, minor chemical changes, testing techniques and also due to the effects of temperature on the composition of structures, especially the potential loss of O.

Thus, the information on SiC may be cited as an example, as illustrated in Fig. 1.37, where the Poisson' ratio seems to increase above 1000 °C, unlike the case shown in Fig. 1.36. However, this value was below 0.2 and at about the same

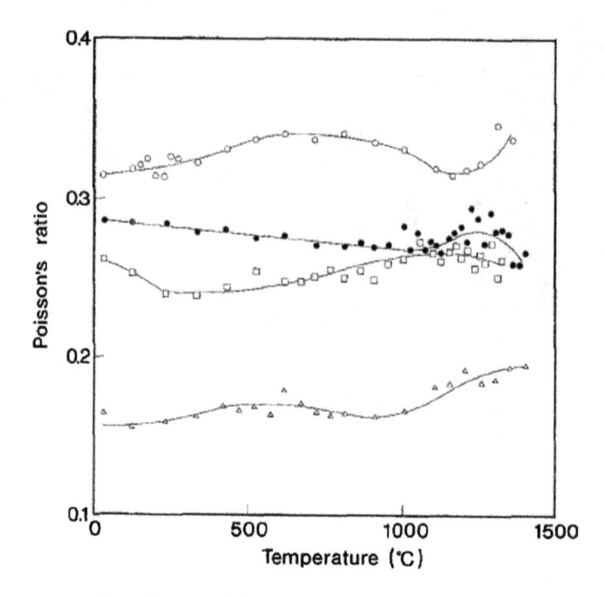

Fig. 1.37 Temperature dependence of the Poisson's ratio for the materials indicated: (*open circle*) TZP (tetragonal zirconia polycrystal), (*filled circle*)Si_3N_4, (*open square*) alumina and (*open traingle*) silicon carbide. Sakaguchi et al. [42] with kind permission of Springer

Table 1.3 Comparison of the elastic constants from the resonance and ultrasonic sound velocities [42] (with kind permission of Springer)

		Ultrasonic	Resonance	Error (%)
Si_3N_4	E (GPa)	310.2	314.4	+1.4
	G (GPa)	121.4	122.2	+0.7
	v	0.278	0.286	+2.9
SiC	*E* (GPa)	368.7	388.6	+5.4
	G (GPa)	158.0	166.8	+5.6
	v	0.167	0.156	−6.6
Al_2O_3	*E* (GPa)	372.2	375.6	+0.9
	G (GPa)	151.1	148.9	−1.5
	v	0.232	0.261	+12.5
TZP	*E* (GPa)	212.2	212.2	±0.0
	G (GPa)	80.7	80.8	±0.1
	v	0.314	0.313	−0.3

level as in Fig. 1.36. The increase in the Poisson's ratio may indicate some deformability, but, according to the authors, no macroscopic deformation was observed below 1400 °C. In Fig. 1.37, the variation in the Poisson's ratios of some ceramics at various temperatures is also shown. Note that TZP (tetragonal zirconia polycrystal) has a Poisson's ratio at a level comparable to metals. It is likely that this reflects the non-elastic behavior of TZP at room temperature. Another reason for the high Poisson value in TZP is the lower Young's modulus, as indicated in Table 1.3.

A linear degradation of the Poisson's ratio with temperature increase was observed in Si_3N_4 (as indicated in Fig. 1.37), which may be a consequence of the macroscopic non-elastic behavior of Si_3N_4.

1.4 Volume Strain (Dilatation)

The concept of 'volume strain' is defined as the volume change per unit volume of a deformed body, $\frac{\Delta V}{V}$. Here, ΔV are the normal strain components and Eq. 1.85a is reintroduced, as above:

$$\varepsilon_x + \varepsilon_y + \varepsilon_z = \frac{(1-2\nu)}{E}\left(\sigma_x + \sigma_y + \sigma_z\right) \tag{1.85a}$$

The term on the left, often denoted as Δ, represents the volume strain, ε_V. ΔV, the normal strain components, refer to cases in which the strains are small. Basically, when a rectangular parallelepiped of initial volume V with sides a, b, c is deformed to a value of V′, as shown in Fig. 1.38, the following relations hold:

$$\begin{aligned} V' &= a'b'c' = a(1+\varepsilon_x)b(1+\varepsilon_y)c(1+\varepsilon_z) \\ &= abc(1+\varepsilon_x+\varepsilon_y+\varepsilon_z+\varepsilon_x\varepsilon_y+\varepsilon_x\varepsilon_z+\varepsilon_y\varepsilon_z+\varepsilon_x\varepsilon_y\varepsilon_z) \end{aligned} \tag{1.90}$$

When linear strains are small, as they are in the elastic region, their products can be ignored, rendering Eq. (1.90) into:

$$\varepsilon_V = \frac{\Delta V}{V} = \frac{V'-V}{abc} = \frac{abc(1+\varepsilon_x+\varepsilon_y+\varepsilon_z) - abc}{abc} = \left(\varepsilon_x+\varepsilon_y+\varepsilon_z\right) \tag{1.91}$$

In this case, ε_V is equal to the sum of the strains on the left side of Eq. 1.85a. In terms of the mean strain, ε_α, it is expressed as:

$$\varepsilon_V = \varepsilon_x + \varepsilon_y + \varepsilon_z = 3\varepsilon_\alpha \tag{1.92}$$

The volume strain may be expressed in terms of Eq. (1.85a) as:

$$\varepsilon_V = \varepsilon_x + \varepsilon_y + \varepsilon_z = \frac{(1-2\nu)}{E}\left(\sigma_x+\sigma_y+\sigma_z\right) = \frac{(1-2\nu)}{E}3\sigma_a \tag{1.93}$$

where:

$$\sigma_a = \frac{1}{3}\left(\sigma_x+\sigma_y+\sigma_z\right) \tag{1.92a}$$

and expressing σ_a, from Eq. (1.93) as:

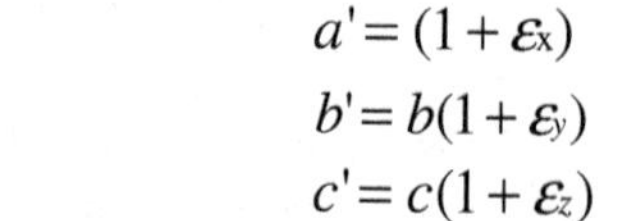

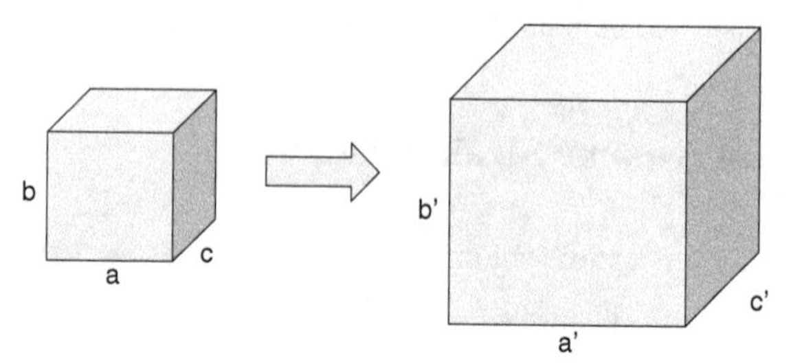

Fig. 1.38 Changing volume V, with sides a, b, c, into volume V′, with the indicated change in length

$$\sigma_a = \varepsilon_v \frac{E}{3(1 - 2v)} = K\varepsilon_v = 3K\varepsilon_a \tag{1.94}$$

K is known as the 'bulk modulus' and also as the 'volumetric modulus of elasticity' and is given by:

$$K = \frac{E}{3(1 - 2v)} \tag{1.94a}$$

K is often expressed as:

$$K = \frac{\sigma_a}{\varepsilon_v} = \frac{-p}{\varepsilon_v} = \frac{1}{\beta} \tag{1.95}$$

−p is the hydrostatic pressure, which is the negative of Eq. (1.92a) and β is the compressibility factor. In this case (hydrostatic pressure), the volume change of an elemental cube is the result of the acting pressure, p, on all six faces of the cube.

1.5 Principal Strain

Figure 1.39 is reproduced from Fig. 1.21 in terms of strain in the two-dimensional case to show the principal strains. A transformation to the principal direction is performed by rotating the x, y axes to x′, y′, the principal directions of those axes. The principal strains are ε_{I} and ε_{II}. Due to the similarity between the plane-stress and plane-strain transformation equations, the orientation of the principal axes and the principal strains are given below. First, there is an angle, θ_p, at which the shear strain, ε_{xy}, vanishes. In analogy to Eq. (1.35a), this is now given as:

$$\tan 2\theta_p = \frac{2\tau_{xy}}{\sigma_x - \sigma_y} \tag{1.35a}$$

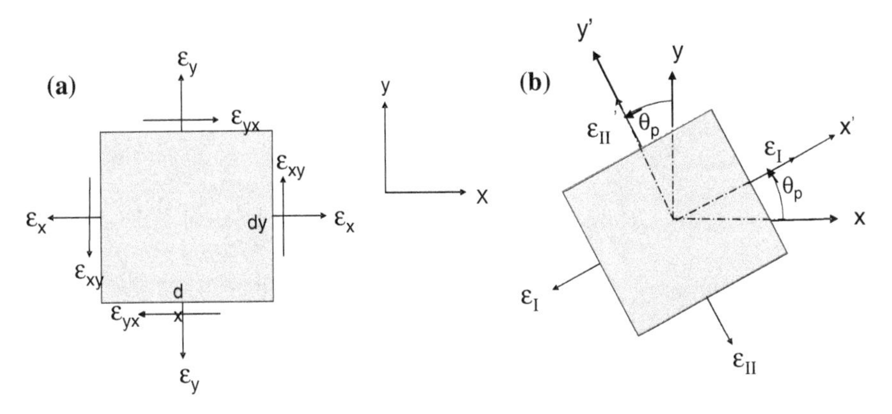

Fig. 1.39 The strains of an elementary square of sides dx and dy, before and after rotation: **a** the strains in a given coordinate system, **b** the rotation performed by θ_p is to a position at which no shear strains are present. ε_I and ε_{II} are the principal strains acting in the principal directions

A similar relation may be given for the strain defining the principal directions:

$$\tan 2\theta_p = \frac{2\varepsilon_{xy}}{\varepsilon_x - \varepsilon_y} \tag{1.96}$$

These principal strains also take the form of Eqs. (1.36) and (1.36a), previously considered in Sect. 1.2.5.1 and reproduced here for comparison. Equation (1.97) shows the principal strains at which, in the rotated square, ε'_{xy} vanishes. This coordinate system is marked as x′ and y′.

$$\sigma'_x(principal) = \sigma_I = \frac{\sigma_x + \sigma_y}{2} + \sqrt{\left(\frac{\sigma_x - \sigma_y}{2}\right)^2 + \tau_{xy}^2} \tag{1.36}$$

$$\sigma_{II} = \sigma_x + \sigma_y - \sigma_I = \frac{\sigma_x + \sigma_y}{2} - \sqrt{\left(\frac{\sigma_x - \sigma_y}{2}\right)^2 + \tau_{xy}^2} \tag{1.36a}$$

$$\varepsilon_I, \varepsilon_{II}(principal) = \frac{\varepsilon_x + \varepsilon_y}{2} \pm \sqrt{\left(\frac{\varepsilon_x - \varepsilon_y}{2}\right)^2 + \varepsilon_{xy}^2} \tag{1.97}$$

Analogous to principal stresses, there are principal strains acting on the principal planes of the strains. Also as in the principal stresses, the shear strains on the principal planes of those strains equal zero, i.e., the normal strains on these planes are actually the principal strains. Following convention, the maximum principal strain of the three is called the 'major principal strain', while the smallest strain is known as the 'minor principal strain'. In an isotropic elastic material, the principal planes of strain coincide with the principal planes of stress. In a manner similar to that in Eqs. (1.22e) and (1.23), it is possible to write the following for the strain:

$$\varepsilon_i^3 = I_1\varepsilon_i^2 + I_2\varepsilon_i + I_3 = 0 \tag{1.98}$$

where:

$$I_1 = \varepsilon_x + \varepsilon_y + \varepsilon_z \tag{1.98a}$$

$$I_2 = \varepsilon_x\varepsilon_y + \varepsilon_y\varepsilon_z + \varepsilon_z\varepsilon_x - \varepsilon_{xy}^2 - \varepsilon_{yz}^2 - \varepsilon_{zx}^2 = \frac{1}{2}\left(\varepsilon_{ii}\varepsilon_{jj} - \varepsilon_{ij}\varepsilon_{ji}\right) \tag{1.98b}$$

$$\begin{aligned} I_3 &= \varepsilon_x\varepsilon_y\varepsilon_z + 2\varepsilon_{xy}\varepsilon_{yz}\varepsilon_{zx} - \varepsilon_x\varepsilon_{yz}^2 - \varepsilon_y\varepsilon_{zx}^2 - \varepsilon_z\varepsilon_{xy}^2 \\ &= \varepsilon_x\varepsilon_y\varepsilon_x + \frac{1}{4}\gamma_{xy}\gamma_{yz}\gamma_{zx} - \frac{1}{4}\varepsilon_x\gamma_{yz}^2 - \frac{1}{4}\varepsilon_y\gamma_{zx}^2 - \frac{1}{4}\varepsilon_z\gamma_{xy}^2 = \det(\varepsilon_{ij}) \end{aligned} \tag{1.98c}$$

The symbol I represents the strain invariants analogous to the stress invariants given as J in Eqs. (1.22e) and (1.23). The coefficients in Eq. (1.98c) are the results of the engineering shear strain being:

$$\gamma_{\mathrm{ij}} = 2\varepsilon_{\mathrm{ij}},$$

as given earlier.

The principal strains are determined as being the roots of Eq. (1.98), quite similarly to the evaluation of the principal stresses.

1.6 Thermal Strains

In many engineering applications, uniaxial thermal strain is useful when defined as:

$$\varepsilon = \frac{\Delta l}{l_0} = \alpha\Delta T \tag{1.99}$$

α is the coefficient of thermal expansion, assumed to be temperature-independent and, thus, a material constant. However, one must be aware that chemical changes may occur with increasing temperature, especially in transition-metal oxides [21], which can induce strain change and, thereby, the linear relation between thermal strain and temperature will not necessarily be maintained. Such changes may occur, for example, when the oxygen content of certain ceramics changes, thus modifying the metal–oxygen bond length responsible for the strain deviation from linearity. These effects are observed in $La_{1\text{-}x}Sr_xCoO_{3\text{-}\delta}$ (LSCF) and other La-Sr-based ceramics. An illustration of this effect is shown in Fig. 1.40 for $La_{0.6}Sr_{0.4}$-$Co_{0.2}Fe_{0.8}O_3$ (LSCF 6428).

In general, however, when no chemical changes occur in materials, Eq. 1.99 is helpful for describing thermal strain. It expresses the change when a uniform temperature is applied to an unconstrained three-dimensional element experiencing thermal expansion or contraction. Free, unhindered thermal expansion produces normal strains. The values of α (when no chemical effects are involved, as

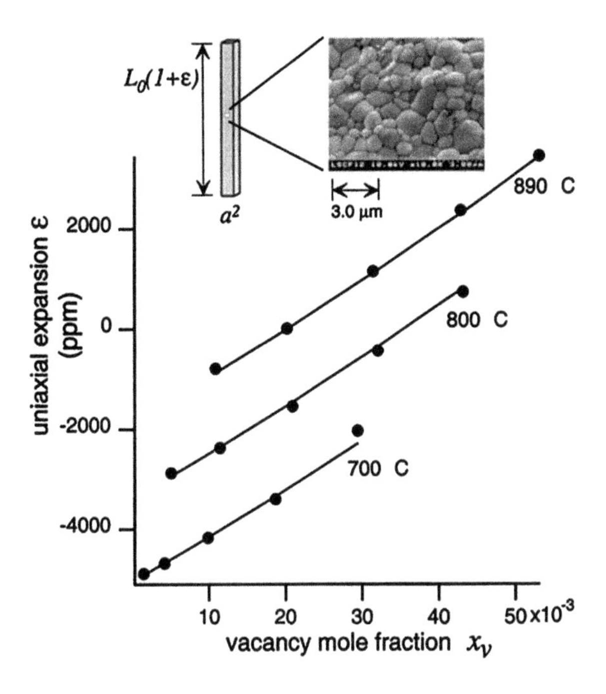

Fig. 1.40 Equilibrium uniform expansion of a dense, square prismatic bar of LSCF 6428 as a function of oxygen content and temperature [21]. With kind permission of Wiley

stated above) may be found in the literature on structural materials. The most practical way to take the thermal strain occurring, when structural ceramics are used at high temperatures, into account is by directly adding the thermal strain to the respective stress–strain relations. Thus, Eq. (1.80) may be expressed as:

$$\begin{aligned} \varepsilon_x &= \frac{\sigma_x - \nu(\sigma_y + \sigma_z)}{E} + \alpha\Delta T \\ \varepsilon_y &= \frac{\sigma_y - \nu(\sigma_z + \sigma_x)}{E} + \alpha\Delta T \\ \varepsilon_z &= \frac{\sigma_z - \nu(\sigma_x + \sigma_y)}{E} + \alpha\Delta T \end{aligned} \tag{1.100}$$

It should be emphasized that thermal expansion does not induce angular deformation, thus no shear stresses are involved. Also note the expressions for stresses Eq. (1.86) modified by thermal strain:

$$\begin{aligned} \sigma_x &= \frac{E}{(1+\nu)}\varepsilon_x + \frac{E\nu}{(1+\nu)(1-2\nu)}\left(\varepsilon_x + \varepsilon_y + \varepsilon_z\right) - \frac{E\alpha\Delta T}{(1-2V)} \\ \sigma_y &= \frac{E}{(1+\nu)}\varepsilon_y + \frac{E\nu}{(1+\nu)(1-2\nu)}\left(\varepsilon_x + \varepsilon_y + \varepsilon_z\right) - \frac{E\alpha\Delta T}{(1-2V)} \\ \sigma_z &= \frac{E}{(1+\nu)}\varepsilon_z + \frac{E\nu}{(1+\nu)(1-2\nu)}\left(\varepsilon_x + \varepsilon_y + \varepsilon_z\right) - \frac{E\alpha\Delta T}{(1-2V)} \end{aligned} \tag{1.101}$$

Equation (1.86) may be rearranged and the thermal strain factor added, giving:

$$\begin{aligned}\sigma_x &= \frac{E}{(1+\nu)(1-2\nu)}\left[\varepsilon_x(1-\nu)+\nu(\varepsilon_y+\varepsilon_z)\right]-\frac{E\alpha\Delta T}{(1-2V)}\\ \sigma_y &= \frac{E}{(1+\nu)(1-2\nu)}\left[\varepsilon_y(1-\nu)+\nu(\varepsilon_z+\varepsilon_x)\right]-\frac{E\alpha\Delta T}{(1-2V)}\\ \sigma_z &= \frac{E}{(1+\nu)(1-2\nu)}\left[\varepsilon_z(1-\nu)+\nu(\varepsilon_x+\varepsilon_y)\right]-\frac{E\alpha\Delta T}{(1-2V)}\end{aligned} \tag{1.102}$$

Equation (1.102) is equivalent to Eq. (1.101) and may, for instance, be obtained by rearranging Eq. (1.86), as presented below:

$$\sigma_x = \frac{E}{(1+\nu)}\varepsilon_x + \frac{E\nu}{(1+\nu)(1-2\nu)}\left(\varepsilon_x+\varepsilon_y+\varepsilon_z\right) \tag{1.86}$$

(a) Express the two terms of Eq. (1.86) by a common denominator as:

$$\sigma_x = E\left[\frac{\varepsilon_x(1-2\nu)\nu+\nu(\varepsilon_x+\varepsilon_y+\varepsilon_z)}{(1+\nu)(1-2\nu)}\right] \tag{1.102a}$$

(b) Multiply the two terms by their respective coefficients to get:

$$\sigma_x = E\left[\frac{\varepsilon_x-2\varepsilon_x\nu+\varepsilon_x\nu+\nu(\varepsilon_y+\varepsilon_z)}{(1+\nu)(1-2\nu)}\right] \tag{1.102b}$$

(c) Rearrange (b) to obtain:

$$\frac{E}{(1+\nu)(1-2\nu)}\left[\varepsilon_x(1-\nu)\right]+\nu\left(\varepsilon_y+\varepsilon_z\right) \tag{1.102c}$$

By adding the thermal factors, Eqs. (1.102) are obtained.

1.7 Relations Among Some Elastic Constants

In the previous sections, some elastic constants were mentioned, namely, E, G, ν, λ and K. The constants E and G, the Young's and the shear modules, respectively, relate to stress–strain relations represented as:

$$\sigma = \mathrm{Ee}$$

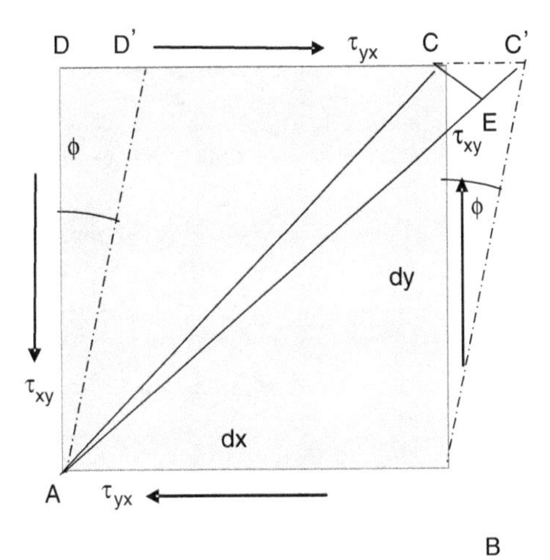

Fig. 1.41 A square distorted by pure shear. The result of the shear is indicated by ABC′D′

and shear-stress strain:

$$\tau = G\gamma.$$

Again, Poisson's ratio, Lame's constant and the bulk modulus were given earlier, respectively, as:

$$v = -\frac{lateral\,strain}{axial\,strain} = -\frac{\varepsilon_y}{\varepsilon_x}\left(= -\frac{\varepsilon_{22}}{\varepsilon_{11}}\right) = -\frac{\varepsilon_z}{\varepsilon_x}\left(-\frac{\varepsilon_{33}}{\varepsilon_{xx}}\right) \tag{1.11}$$

$$\lambda = \frac{Ev}{(1+v)(1-2v)} \tag{1.86b}$$

$$K = \frac{E}{3(1-2v)} \tag{1.94a}$$

These three constants are not independent, but if a solid is isotropic, these constants are sufficient for describing the mode of deformation, as given below.

(a) In order to study the relation between E, G and v, the following section considers pure shear deformation. In the illustration, take, for example, a body comprising a two-dimensional cube element around a point (see the square in Fig. 1.41). This square, ABCD, is distorted by shear to ABC′D′; the diagonal AC bisects the square, ABCD, thus, the angle, ABC, is 45°. Due to this deformation, the diagonal elongates to AC′; simultaneously, the diagonal, BD, of the square gets shorter to a value of BD′. Now draw a perpendicular to the new diagonal, i.e., AC′ of the distorted parallelepiped. Since this deformation is a very small, angle, AC′C, may be taken as 45°. The resultant shear strain from the triangle, CC′B, is:

$$\text{shear strain} = \tan \gamma = \gamma = \frac{CC'}{BC} \tag{1.103}$$

This shear strain is also given as stress per the shear modulus, namely:

$$\text{shear strain} = \frac{\tau}{G} \tag{1.103a}$$

By equating Eqs. (1.103) and (1.103a), one gets:

$$\frac{\tau}{G} = \frac{CC'}{BC} \tag{1.103b}$$

There is a right-angle triangle, CC′E, where CE is perpendicular to AC′. Two angles in this right-angle triangle are 45°.

$$\cos 45 = \frac{C'E}{CC'} \tag{1, 104}$$

or:

$$CC' = \frac{C'E}{\cos 45} = \sqrt{2}C'E \tag{1.104a}$$

Moreover, in triangle ABC, which is also a right-angle triangle

$$\cos 45 = \frac{BC}{AC} \tag{1.105}$$

and, thus:

$$BC = AC \cos 45 = \frac{AC}{\sqrt{2}} \tag{1.105a}$$

Equations (1.105a) of BC and (1.104a) of CC′ may be inserted into Eq. (1.103b) to obtain

$$\frac{\tau}{G} = \frac{\sqrt{2}C'E}{AC}\sqrt{2} = 2\frac{C'E}{AC} \tag{1.106}$$

Since AC is almost equal to AE, C′E can be considered to represent the increase in length of the diagonal, AC. The strain in the diagonal length is given as the ratio of the increase in the length and the original length. Thus:

$$\textit{strain in diagonal} = \frac{\textit{increase in diagonal length}}{\textit{original length}} = \frac{C'E}{AC} \tag{1.107}$$

From Eqs. (1.106) and (1.107):

$$\frac{\tau}{2G} = \frac{C'E}{AC} = \frac{\gamma}{2} \tag{1.108}$$

One can state that, due to the increase in the diagonal length, AC is subjected to a tensile stress, but BD must be subjected to a compressive stress, since it shortens as a consequence of shear deformation. Under these conditions, the tensile shear stress is $\sigma_1 = \tau$ and the compressive one is $\sigma_2 = -\tau$. Remember that:

$$\nu = -\frac{\varepsilon_2}{\varepsilon_1} = -\frac{\varepsilon_3}{\varepsilon_1}.$$

Now, the direct state of the stress system along the diagonals may be written as:

$$\textit{strain on diagonal} = \tfrac{\sigma_1}{E} - \frac{\nu\sigma_2}{E} = \frac{\tau}{E} - \frac{(-\tau)}{E} = \frac{\tau}{E}(1+\nu) \tag{1.109}$$

Equating both the expressions of strain found in Eq. (1.107), i.e., Eqs. (1.108) and (1.109) give:

$$\frac{\tau}{2G} = \frac{\tau}{E}(1+\nu) \tag{1.110}$$

thus, the relation between E and G, in terms of Poisson's ratio, is given by:

$$E = 2G(1+\nu) \tag{1.110a}$$

or as indicated more often in the literature and previously shown in Eq. (1.12):

$$G = \frac{E}{2(1+\nu)} \tag{1.12}$$

(b) The relation between E, K and ν.

Equation (1.85) is rewritten below as:

$$\varepsilon_x + \varepsilon_y + \varepsilon_z = \frac{(1-2\nu)}{E}\left(\sigma_x + \sigma_y + \sigma_z\right) = 3\frac{(1-2\nu)}{E}\sigma_a \tag{1.85}$$

where σ_a is the average (or mean) stress. It is also given (for isotropic materials) as Eq. (1.92), which, when combined with Eq. (1.85), gives:

$$\varepsilon_v = \varepsilon_x + \varepsilon_y + \varepsilon_z = 3\varepsilon_\alpha = 3\frac{(1-2\nu)}{E}\sigma_a \tag{1.93}$$

Furthermore, in Eq. (1.94), K is related to the average stress, σ_a, and average strain, ε_a:

$$\sigma_a = \varepsilon_v \frac{E}{3(1-2v)} = K\varepsilon_v = 3K\varepsilon_a \tag{1.94}$$

Using Eq. (1.94):

$$K = \frac{\sigma_a}{3\varepsilon_a} \tag{1.94b}$$

Taking σ_a from Eq. (1.94) gives:

$$\sigma_a = \varepsilon_v \frac{E}{3(1-2v)} \tag{1.94c}$$

Then, equate this with σ_a from Eq. (1.94b) to express this equality as:

$$K3\varepsilon_a = \frac{\varepsilon_v E}{3(1-2v)} \tag{1.94d}$$

or:

$$E = \frac{3K\varepsilon_a 3(1-2v)}{\varepsilon_v} \tag{1.94e}$$

Take $\varepsilon_v = 3\varepsilon_\alpha$ from Eq. (1.94) and substitute this value for ε_v into Eq. (1.94e) to get:

$$E = 3K(1-2v) \tag{1.111}$$

which is basically the expression for the bulk modulus usually given as:

$$K = \frac{E}{3(1-2v)} \tag{1.93}$$

thus expressing the relation between K, E and v.

(c) The relation between G, K, E

In Eq. (1.12), G is given as:

$$G = \frac{E}{2(1+v)} \tag{1.12}$$

or:

$$E = 2G(1+v) \tag{1.112}$$

while in Eq. (1.111), E is given in terms of K as:

$$E = 3K(1 - 2\nu) \tag{1.111}$$

When equating Eqs. (1.111) and (1.112), one gets:

$$3K(1 - 2\nu) = 2G(1 + \nu) \tag{1.113}$$

$$G = \frac{3K(1 - 2\nu)}{2(1 + \nu)} \tag{1.114}$$

In order to get the relation between G, K and E, first express the following from Eq. (1.112):

$$\frac{E}{2G} - 1 = \nu \tag{1.112a}$$

and express ν from Eq. (1.111) as follows:

$$\frac{\mathrm{E}}{3\mathrm{K}} = 1 - 2\nu \tag{1.111a}$$

I. $\nu = -\dfrac{E - 3K}{6K}$

Now, equate Eqs. (1.111a) and (1.112a) to get:

II. $\dfrac{E}{2G} - 1 = -\dfrac{E - 3K}{6K}$

Cross-multiplying II, gives:

III. $6EK - 12GK = -2GE + 6KG$

IV. $E(6K + 2G) = 18KG$

$$E = \frac{18KG}{(6K + 2G)} = \frac{9KG}{3K + G} \tag{1.115}$$

(d) The relation between G, K and ν.

Express E from both Eqs. (1.112) and (1.111) and equate them to obtain:

$$2G(1 + \nu) = 3K(1 - 2\nu) \tag{1.116}$$

Get G as:

$$G = \frac{3K(1 - 2\nu)}{2(1 + \nu)} \tag{1.117}$$

Table 1.4 Relations between some elastic constants for isotropic materials

The relation between E, G and v:	$G = \frac{E}{2(1+v)}$
The relation between E, K and v:	$E = 3\,K(1-2v)$
The relation between G, K and E:	$E = \frac{9KG}{3K+G}$
The relation between G, K and v:	$G = \frac{3K(1-2v)}{2(1+v)}$
The relation between λ, E and v:	$\lambda = \frac{Ev}{(1+v)(1-2v)}$
The relation between λ, G and v	$\lambda = \frac{2Gv}{(1-2v)}$

Equation (1.117) is one of the required relations between G, K and v.

(e) The relation between λ, E and v.

Is given in Eq. (1.86b), shown above:

$$\lambda = \frac{Ev}{(1+v)(1-2v)} \tag{1.86b}$$

(f) The relation between λ, G and v.

Is found by inserting E from Eq. (1.112) into Eq. (1.86b):

$$E = 2G(1+v) \tag{1.112}$$

to obtain:

$$\lambda = \frac{2Gv}{(1-2v)} \tag{1.118}$$

Additional relations between λ and elastic constants that may be derived are presented below without proofs, such as those between: λ, E and K; λ, K and G; λ, K and v, respectively.

It is convenient for users of these elastic constants to get them from tables summarizing all these relations, such as Table 1.4. For isotropic materials, two independent elastic constants are sufficient (as indicated in the Table 1.4) for describing a stress–strain relation. There are different stress–strain constants for various other deformation conditions.

1.8 Compression of Ceramics

Experimental observations indicate that, in general, the true stress–strain curves of ductile materials coincide. Brittle materials, among them brittle ceramics, do not show a similar behavior. Usually, experimental observations indicate that brittle materials are stronger under compression than under tension.

Almost all materials are candidates for compression tests. Thus, ductile metals used for various applications, when shaped by forging, drawing, extrusion, etc., experience pressure components during their commercial fabrication and, therefore, are tested by loading under compression. Very soft materials, such as plastics or soft metals, are often exposed to compression experiments. However, ceramics, like most of the brittle materials intended for structural applications are invariably tested under compression as the prime mechanical evaluation test. Furthermore, compressive strength plays an important role in the performance of machine components including ceramic armor, machine tool bits and bioceramic body parts.

True stress–strain curves of tension and compression have been experimentally observed to coincide. Yet, this observation is not true for brittle materials (that behave like glass), in which fracture and yield stress coincide. Thus, in brittle materials, e.g., ceramics, the strain seldom exceeds ~ 1 %, depending to a large degree on the type of deformation involved. Classical Hookean behavior is observed in such brittle materials. The presence of imperfections of various kinds, including porosities, has a profound effect on the mechanical behavior of ceramics.

Uniaxial compression tests may be done either on unconfined or confined specimens. Unconfined uniaxial tests of porous calcium phosphate ceramics (intended for use as a bioceramic body part) are shown in Fig. 1.43.

These specimens were cleaned by soaking in a phosphate-buffered saline (PBS) after deaeration, which is a recommended biomimetic process ('biomimetic' refers to man-made processes, substances, devices or systems that imitate nature). The designations in Fig. 1.42 indicate the conditions of the porous specimens. Specimens A, B and D show a stepwise collapse, due to the stepwise collapse of the weaker part in each pore wall. Specimens E, F and G reveal a dense body-like stress–strain curve, i.e., their pore wall has a similar structure to dense ceramics. These results suggest that, to enhance the compressive strength of bioactive ceramics in the initial stage of their implantation, they should be kept in air after briefly removing the surface liquid remaining after a 24 h soaking in PBS at 25 °C, followed by the addition of PBS after the de-aeration of the specimens.

Uniaxial and triaxial compression tests of silicon-carbide ceramics under quasi-static loading conditions were performed by Brannon et al. [33]. Their SiC-N specimens were prepared in the form of a right circular cylinder, as is indicated schematically in Fig. 1.43a next to the experimental set-up.

These specimens were porous, as illustrated in Fig. 1.44. Triaxial compression tests were also performed under unconfined and confined conditions. The confined tests employed lateral confining pressure. The confinement of specimens during compression tests relates to the accompanying reduction of transverse tensile stress, which induces fracture. This confinement may be lateral or created by producing true hydrostatic pressure by submerging the specimens (or the entire test assembly) in a pressurized liquid. For more on the testing of brittle specimens by uniaxial and triaxial compression, the reader is referred to Chapter 11 of the textbook by Polakowski and Ripling [15]. Table 1.5 is a summary of the compression tests performed on SiC-N and includes: specimen size, fracture stress, Poisson's ratios and elastic module confinement levels. In Fig. 1.46, the axial stress, σ_a, is plotted against ε_a (axial) and ε_l (lateral) strains, respectively.

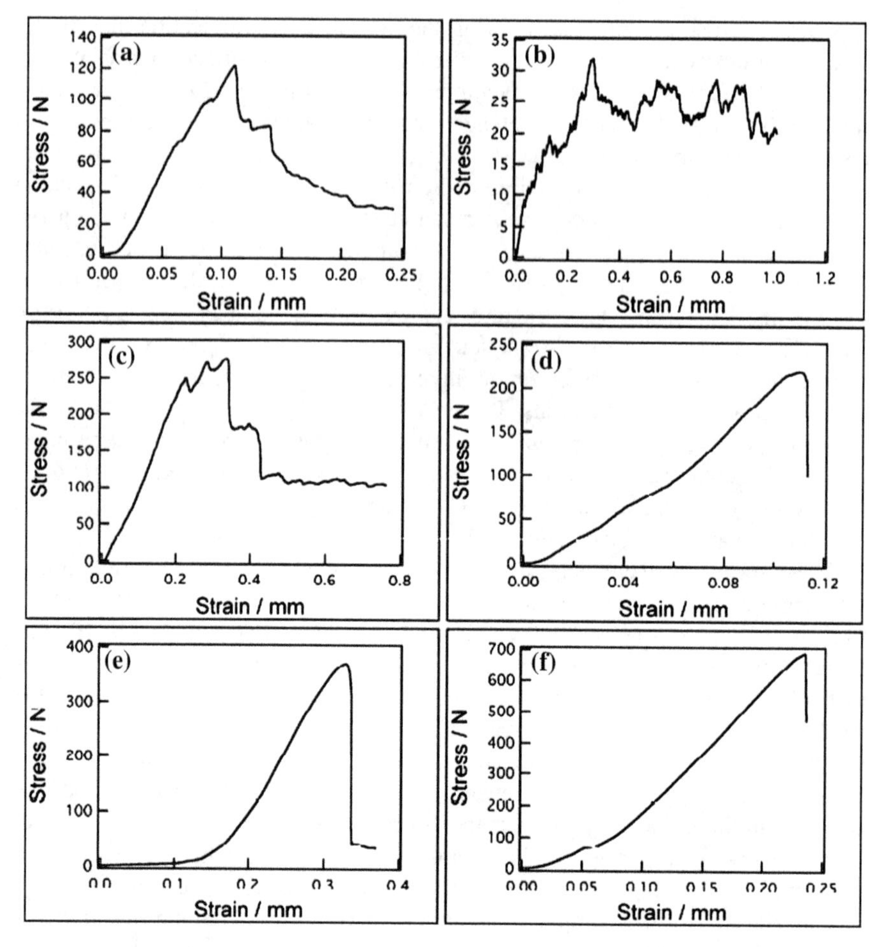

Fig. 1.42 Typical uniaxial stress–strain curves obtained by compression of porous calcium phosphate ceramics for biomedical purposes as bone fillers [30]. With kind permission of Ashdin Publishing

During the uniaxial compression, the specimens were loaded at a constant axial strain rate of 2×10^{-5}/s, until the peak stress was reached and the specimen failed in an explosive manner, as shown in Fig. 1.45.

Figure 1.46 illustrates the stress–strain curves of an unconfined, uniaxial, compressive deformation of a SiC-N specimen. For uniaxial compression, the axial load was applied without confining pressure (P = 0). The confining pressure is indicated in Fig. 1.47, where stress–strain plots obtained from the uniaxial/triaxial compression tests of SiC-N are illustrated. In general, the application of

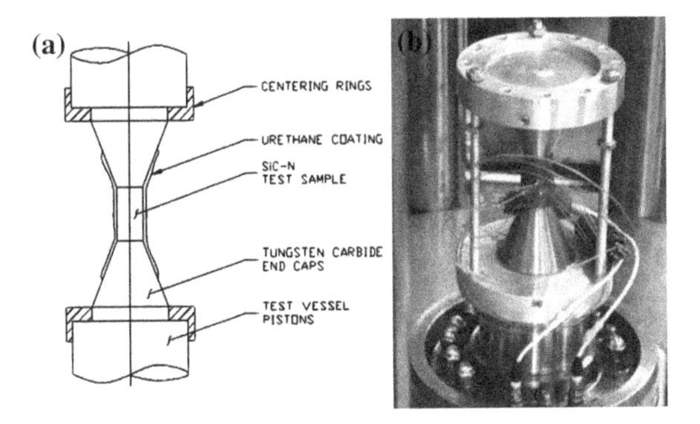

Fig. 1.43 Sample alignment jig designed for coaxial alignment of different components of the test set-up. The strain gaged SiC-N specimen coated with flexible polyurethane membrane is also shown. The strain gage signal was transmitted to the data acquisition system through the high-pressure coaxial feed-through connectors. **a** Schematic; **b** actual experimental set-up [33]. With kind permission of Professor Brannon

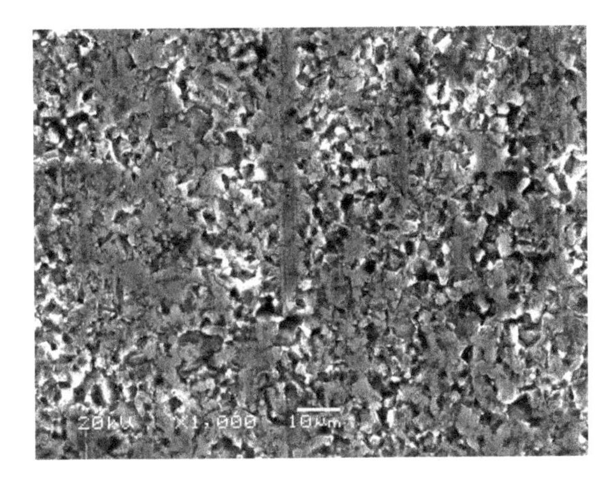

Fig. 1.44 SEM micrograph of the surface of a SiC-N specimen prepared for mechanical testing. Grain and pore sizes are distributed uniformly [33]. With kind permission of Professor Brannon

confinement effects stress and strain and is observed to increase with the increase of P for various specimens tested [33].

The unconfined compressive strength used in the above equation is 3,872 ± 126 MPa, as seen in Fig. 1.46. A discussion about the effects of temperature and strain rates on the compression stress–strain curves is found in later sections dealing with ductile ceramics and the influence of impact on the strength properties of ceramics.

Table 1.5 Summary of uniaxial and triaxial compression tests for SiC-N specimens [33] (with kind permission of Professor Brannon)

Specimen no.	Diameter (mm)	Length (mm)	P (MPa)	σ_f (MPa)	E (GPa)	V	I_1(MPa)	$\frac{I_1}{3}$ (MPa)	$\sqrt{J_2}$ (MPa)
SiCN-UC01	12.7	25.4	0	3738	464	0.156	3738	1246	2158
SiCN-UC02	12.7	25.4	0	3988	463	0.153	3988	1329	2302
SiCN-UC03	12.7	25.4	0	3890	467	0.154	3890	1297	2246
SiCN-TA01[a]	12.7	25.4	200	6326	NA	NA	6726	2242	3537
SiCN-TA02	12.7	25.4	350	5948	466	0.161	6648	2216	3232
SiCN-TA03[b]	12.7	25.4	200	NA[b]	442[c]	NA	NA[b]	NA[b]	NA[b]
SiCN-TA04	12.7	25.4	100	5508	480	0.167	5708	1903	3122
SiCN-TA05	12.7	25.4	200	6120	480	0.169	6520	2173	3418
SiCN-TA06	12.7	25.4	350	6422	484	0.172	7122	2374	3506
SiCN-TA07	12.7	25.4	350	6515	482	0.173	7214	2405	3559
SiCN-TA08[b]	12.7	25.4	100	NA[b]	474[c]	0.159[c]	NA[b]	NA[b]	NA[b]
SiCN-TA09	12.7	25.4	100	5283	478	0.166	5483	1828	2992

$P(= \sigma_2 = \sigma_3)$ = lateral confining pressure
σ_f = failure stress (maximum σ_1)
E = Young's modulus
V = Poisson's ratio
$I_1 = \sigma_1 + \sigma_2 + \sigma_3$ at failure $= \sigma_f + 2P$
$\frac{I_1}{3}$ = mean stress
$\sqrt{J_2} = \frac{\sigma_f - P}{\sqrt{3}}$
[a] Strains were not measured
[b] Premature failure of the tungsten carbide end-caps at 2284 MPa for SiCN-TA03 and 3477 MPa for SiCN-TA08
[c] Uncertain value due to premature of the WC end-caps

1.9 Bend (Flexural) Tests of Ceramics

It has been stated that brittle materials, such as ceramics, are preferentially tested by means of compressive and bending deformations. The main reason for this is that specimens of this nature tend to fail at relatively low stresses, not only because of the flaws and cracks commonly found in various sized ceramics, but for other reasons as well, as listed below:

(a) The preparation of tensile specimens of ceramics (invariably brittle substances) to the proper size and dimensions is quite difficult, because it is problematic to machine them to the desired shape.
(b) Once a specimen is prepared, difficulties may arise in their proper alignment, as required in tensile testing, resulting in a non-uniform transfer of load across the specimen's area.

Fig. 1.45 Explosive failure of the SiC-N-UC02 specimen (12.7 mm in diameter and 25.4 mm in length) subjected to the unconfined uniaxial compressive stress condition ($\sigma_1 = 3988$ MPa at failure and $\sigma_2 = \sigma_3 = 0$) [33]. With kind permission of Professor Brannon

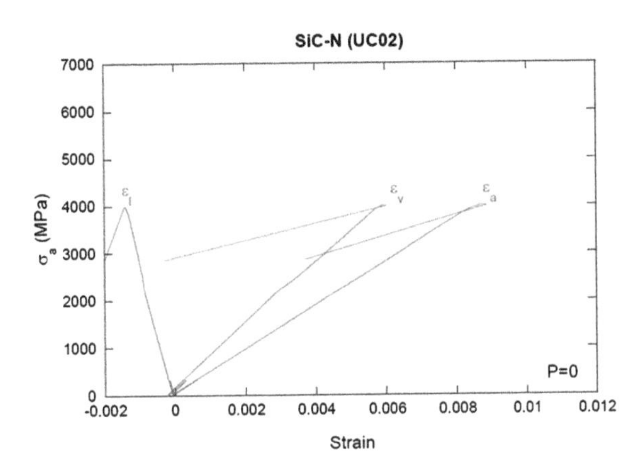

Fig. 1.46 Stress–strain plot for the uniaxial compression test of specimen SiCN-UC01. ε_a, ε_l, and ε_v are axial, lateral and volumetric strains, respectively. P is the confining pressure [33]. With kind permission of Professor Brannon

(c) The presence of various imperfections among the cracks of various sizes act not only as weak regions, but also as stress raisers. Compression tests, for example, tend to close pores and cracks, whereas tension opens them and increases their size.
(d) Surface defects, such as crazes, are common in brittle materials, like ceramics (and more so in glass), and these act as notches, namely stress raisers.
(e) There is often also difficulty gripping or clamping ceramic tension-test specimens.
(f) Finally, due to the above problems, a large number of specimens must be tested in order to get a representative strength value via statistical evaluation, which is both time consuming and costly.

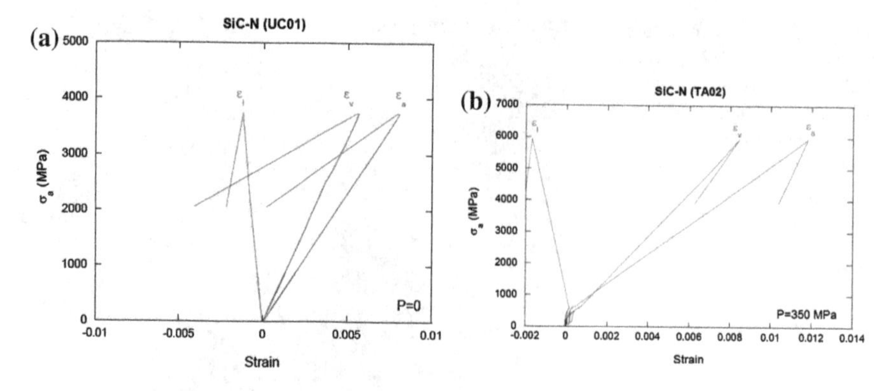

Fig. 1.47 Stress–strain plots obtained from the uniaxial/triaxial compression tests of SiC-N specimens (σ_a—axial stress, ε_a—axial strain, ε_l—lateral strain, ε_v—volumetric strain). **a** without confining pressure; **b** confining pressure of 350 MPa [33]. With kind permission of Professor Brannon

Two methods are used for bend testing—three- and four-point bending tests. Here, the specimens are rectangular, without notches. It is obvious from Fig. 1.48 that the applied force (downward arrows) is compressive by nature, resisted by the tensional force (upward arrows). Thus, the longitudinal stresses at the lower surfaces (convex) in the specimens are tensile and compressive at their upper surfaces (concave). As a consequence, a calculable bending moment develops. The 'modulus of rupture' is the stress of the specimen at its failure and represents the flexural strength of the specimen.

The arrow pointing downwards in **a**, for example, is the center point for the load application. A large variety of machines are available for flexural tests, such as MTS, Instron, Universal Testing Machine, etc. The basic assumption in flexural tests may be summarized briefly as follows:

(i) The beam material is isotropic and homogeneous.
(ii) Perpendicular planes to the longitudinal axis of the specimen are assumed to remain plane after applying the load for bending.
(iii) Since bending is associated with tension (the lower longitudinal plane is under tension during load application and upper plane is in compression), the elastic modulus is considered to be about equal both under compression and tension.
(iv) This test is based on a small deflection compared to the beam depth.
(v) It is reasonable to believe that stress and strain are proportional to the distance from the neutral axis. The neutral axis is shown in the schematic specimen of Fig. 1.49 (at half of h).
(vi) Shear stress and its consequences are not taken into account in the structural stress of the rectangular bar under consideration.

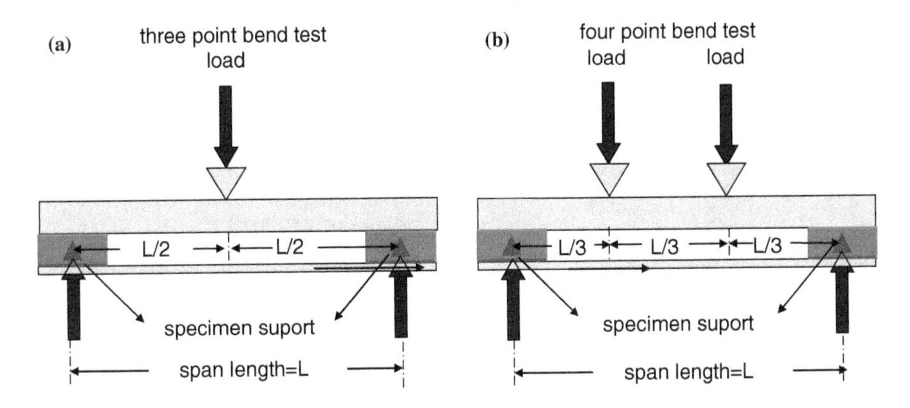

Fig. 1.48 Schematic bend-test configurations: **a** three point; **b** four point

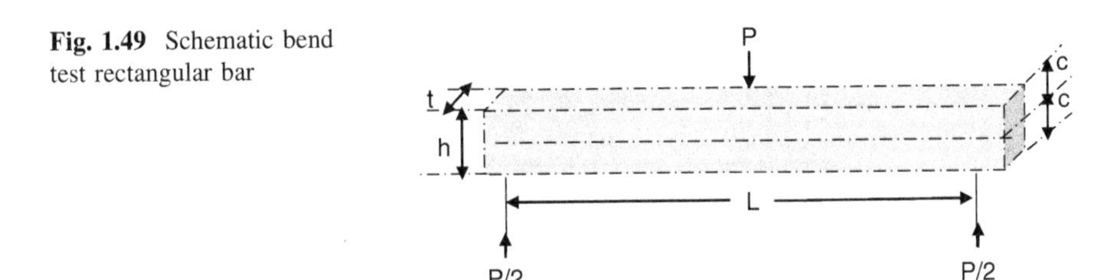

Fig. 1.49 Schematic bend test rectangular bar

A circular cross-section is not used as frequently as the rectangular beam shown above.

The fracture stress, σ_f, is determined by:

$$\sigma_f = \frac{Mc}{I} \tag{1.120}$$

$$I = \frac{2tc^3}{3} \tag{1.121}$$

Replacing I in Eq. 1.120 produces the following for σ_f:

$$\sigma_f = \frac{3M}{2tc^2} \tag{1.120a}$$

M is the bending moment, c is half the specimen width, t is the thickness and I is the moment of inertia of the cross-sectional area. Lists of the moments of inertia of plane figures and areas are found in the literature and also in the Appendix of Timoshenko's book [19]. Basically, the plane under consideration is divided into small pieces and the contribution of each individual piece to the moment of inertia is evaluated by integration:

$$I = \frac{bh^3}{12} \equiv \frac{t(2c)^3}{12} = \frac{2tc^3}{3} \tag{1.121a}$$

Since I is the moment of inertia of the cross-sectional area, expressing the moment as the force times the lever, allows Eq. (1.120) to be modified as:

$$\sigma_f = \frac{2P\frac{L}{2}c}{\frac{2tc^3}{3}} = \frac{3PL}{2tc^2} \tag{1.122}$$

Below, a method for evaluating the inertia is presented.

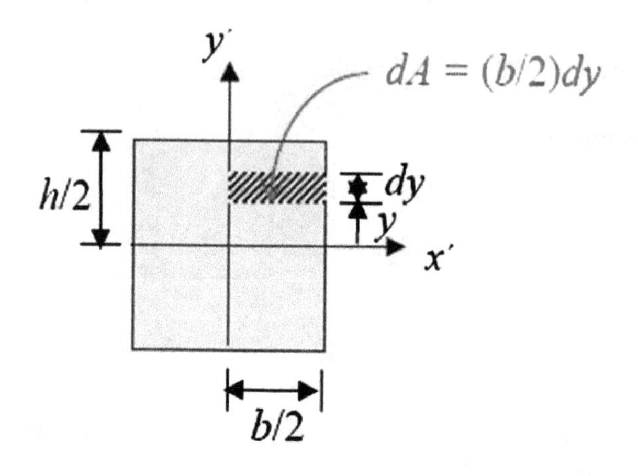

$$\bar{I}_x = I_{x'} = \int_A y^2 dA$$

$$= 4\int_0^h y^2\left(\frac{b}{2}dy\right)$$

$$= 4\left(\frac{b}{2}\right)\frac{y^3}{3}\Bigg|_0^{h/2}$$

$$= \frac{bh^3}{12} \Leftarrow$$

Note the final relation $\frac{bh^3}{12}$ for the inertia obtained above and shown in Eq. (1.121a). When considering the notation in Fig. 1.49, Eq. (1.121) may be obtained. Both values for inertia: $\frac{bh^3}{12}$ and $I = \frac{2tc^3}{3}$ from Eq. (1.121) are indicated in Eq. (1.121a).

Equation (1.122) is the flexural strength for the three-point test of a rectangular bar. In the above relation, a force is acting on a lever of size L/2 (1/2 of the bar at the support) and this force is supported or balanced at the two supporting points marked by the arrows close to the ends of the rectangular bar (i.e., M = PL (force × arm), which yields the same answer as given in Eq. (1.122)).

The four-point bend setup is illustrated in Fig. 1.50 for two cases: for the loading span at L/2 (illustration **a**) and for loading span of L/3 (b). The following relations apply to the L/2 span, using Eq. (1.120a) with the appropriate substitution for M as.

$$\sigma_f = \frac{3M}{2tc^2} = \frac{3P\frac{L}{2}}{2tc^2} = \frac{3PL}{4tc^2} \tag{1.123}$$

By using Eq. (1.120a) again, the L/3 span shown in Fig. 1.50b may be written as:

$$\sigma_f = \frac{3M}{2tc^2} = \frac{3P\frac{L}{3}}{2tc^2} = \frac{PL}{2tc^2} \tag{1.124}$$

In the general case, when the loading span is different from L/2 or L/3 in a four point bend test, the stress is given as:

$$\sigma_f = \frac{3P(L - L_i)}{2tc^2} \tag{1.125}$$

Equation (1.125) is obtained in a manner similar to other bend test relations, namely:

$$\sigma_f = \frac{3M}{2tc^2} = \frac{3P(L - L_i)}{2tc^2} \tag{1.125a}$$

Figure 1.51 illustrates the results of a bend test of five specimens of three-dimensional carbon-silicon carbide compared with tension tests (Fig. 1.52) on the same material. The results of the bend test are similar to stress–strain curves; however, the stress is plotted versus deflection, rather than versus strain. The results of these bend and tension tests appear in Tables 1.6 and 1.7.

The relations expressing the flexural strength, σ_f, actually represent the highest stress of the ceramics at the time of rupture. While tension or compression tests of metals are commonly used to characterize and development new materials for design purposes, bend tests of ceramics are the preferred test method. The flexural strength of a ceramic is dependent on its inherent properties, especially flaws and crack sizes (common features in ceramics). Variations in size, distribution and the nature of such cracks cause a natural scatter in test-sample results, requiring the testing of several test specimens in order to get a statistical value for the inherent flexural strength.

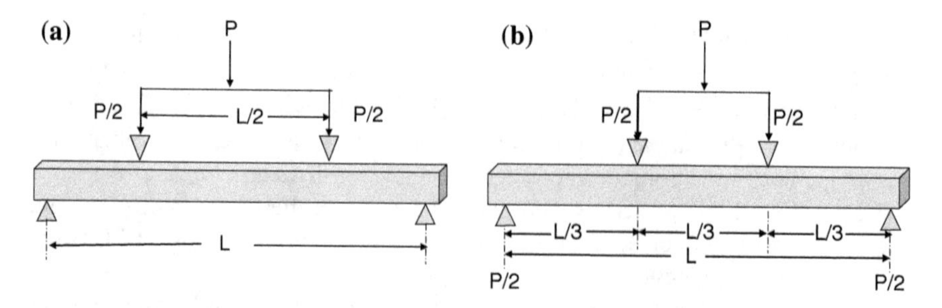

Fig. 1.50 Rectangular beams in a four-point bending test; **a** loading span L/2; **b** loading span L/3. Note that the loading span may be different from L/2 or L/3. In this case, it is customary to denote the load span as L_i

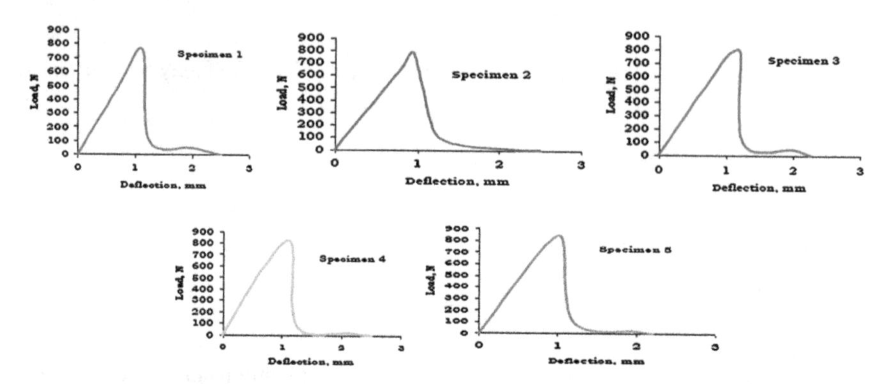

Fig. 1.51 Load-deflection curves of 3D carbon-reinforced SiC obtained by flexural tests [7]. With kind permission of Professor Chetan Sharma, Editor in Chief

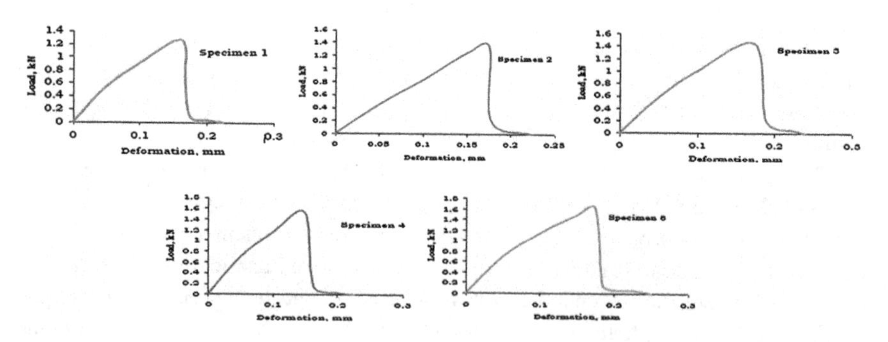

Fig. 1.52 Load-deformation curves of 3D Carbon reinforced SiC obtained by tensile tests [7]. With kind permission of Professor Chetan Sharma Editor in Chief

Table 1.6 Experimental results of flexural strengths of 3D C-SiC specimens [7] (with kind permission of Professor Chetan Sharma, Editor in Chief)

Specimen No.	Max. load (P_U) recorded during flexural test (N)	Flexural strength, σ_f (MPa) from Experiment
1	757	210.2
2	780	216.6
3	789	219.2
4	805	223.6
5	829	230.3

Table 1.7 Experimental results of tensile strengths of 3D C-SiC specimens [7] (with kind permission of Professor Chetan Sharma Editor in Chief)

Specimen No.	Max. load (P_U) during tensile test (kN)	Tensile strength, σ_t (MPa) From Experiment
1	1.264	70.2
2	1.346	74.8
3	1.437	79.8
4	1.525	84.7
5	1.619	89.9

If the loading span is neither 1/3 or 1/2 the support span for the 4 pt bend setup is used.

Note the relation between three- and four-point bend tests and test-specimen size. It is expected that the larger volume specimens will show a lower modulus of rupture than smaller sized specimens, since there is a higher probability that more defects (micro-cracks, for example) will exist in larger specimens. Therefore, test specimens have to be standardized. The lower modulus of rupture in four-point bend tests than in three-point bend tests is a consequence of the size effect.

1.10 Hardness Tests in Ceramics

1.10.1 Introduction

As is known, 'hardness', H, is defined as the ratio of applied load to the projected area of indentation and is generally expressed as:

$$H = \alpha \frac{P}{d^2} \tag{1.126}$$

where P is the load, d is the size of the measured impression with α, the indenter constant, taking the indenter geometry into consideration. Any of the known hardness tests are possible candidates for evaluating the hardness of ceramics with the proper choice of test conditions, e.g., the load, the indenter, the time of load application, etc. The most common hardness testing methods are those of Vickers and Knoop, the latter being used for ceramic coatings. Here, general hardness testing methods will not be discussed (details may be found in [13]), rather the following section focuses on Vickers' and Knoop's techniques. In fact, the scatter in ceramic hardness test results is larger than that found in metals and, therefore, the reproducibility of hardness is reduced. Therefore, the expected indentation response of ceramic materials must be considered before selecting the most reproducible and reliable hardness measurement method and the appropriate test force. It is recommended to use standard ceramics reference blocks as measurement guides to enhance test reliability. Clearly, the choice of technique and applied load must be such that no cracking of the ceramics in the vicinity of indentation occurs. These tests are indentation tests, mainly intended to provide information on the resistance of these materials to deformation. Resistance to deformation provides significant information for engineers, stemming from the relations of the hardness values to other parameters obtained by different testing methods, mainly to the stress data obtained under tension or compression. The hardness of a material correlates directly with its strength, wear resistance and other properties. It is hoped that similar relations also exist for ceramics or can be developed. Thus, hardness testing is widely used for material evaluation, because of its simplicity and low cost relative to direct measurements.

The indenters and their indentation shapes vary widely, leaving different impressions. The appropriate dimensions of these impressions on the surfaces are measured to obtain interesting research data. Conversion charts, from hardness to other properties of interest, appear in the literature. Basically, glass and ceramic hardness tests are carried out using static methods, usually by means of a diamond indenter and low test loads, because brittle materials tend to propagate cracks.

1.10.2 Vickers Hardness Test (VHT)

The Vickers Hardness Test (henceforth: VHT) uses a diamond pyramid with a square base for indentation, so it is also known as the Diamond Pyramid Hardness Test [henceforth: DPH]. The indenter has an angle of $\phi = 136°$ between the two opposite faces. The results of this test are also known as the 'diamond pyramid number' [henceforth: DPN], defined as the load divided by the surface area of the pyramid-shaped indentation (impression). This area is simple to evaluate from the geometry of the shape of the indentation, which requires measuring the diagonals and using the known angle between the two opposite faces. These two diagonals are measured on the screen of the Vickers Tester and their average is used in the DPH formula. The area of the sloping surface of the indentation is calculated as indicated in Fig. 1.53 and the steps are also shown for deriving the expression for DPH measurements:

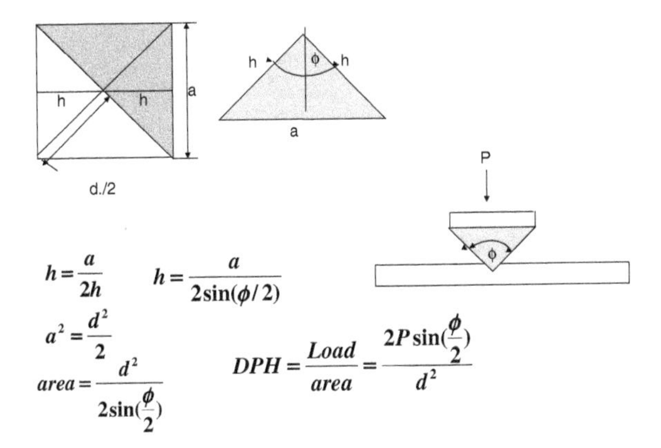

Fig. 1.53 Vicker's indentation [13]

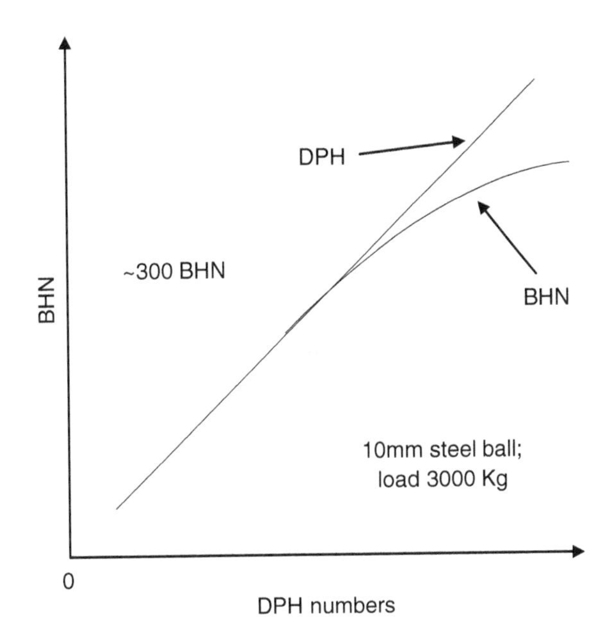

Fig. 1.54 Beyond 300 BHN, the lines diverge [13]

$$DPH = \frac{2P\sin(\phi/2)}{d^2} = \frac{1.854P}{d^2} \tag{1.127}$$

The 136° angle between the opposite faces of the indenter was chosen because of the similarity between the Brinell hardness number (henceforth: BHN) and the DPN. Due to this similarity, the DPN is on the same hardness scale as the BHN and their values are about the same up to ~300 BHN. Only beyond this BHN value do the two curves shown in Fig. 1.54 deviate one from the other. Significant deviation between these two types of measurement occurs at high hardness values, particularly above ~600 BHN, due to the deformation of the indenter. The DPH

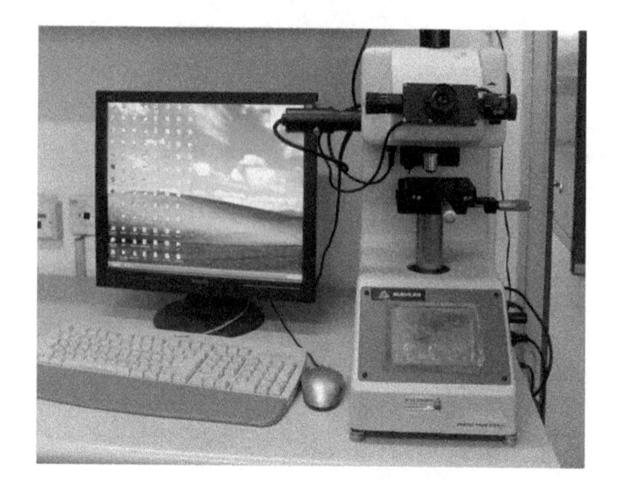

Fig. 1.55 Buehler made a combined hardness tester for Vickers, Knoop and Rockwell. Model #1600-6400. Load range 1–2000 g

curve is linear, as seen in the schematic Fig. 1.54, because of the insignificant deformation of the diamond indenter compared to that of the BHN.

The load of a conventional Vickers machine varies from 1 to 120 kg and may be easily reset to the desired load at the push of a button. Thus, a VHT may be applied to different materials having a very wide range of hardness values. However, at the Ben-Gurion University of the Negev's ceramics testing laboratory, the Vickers tester has a load range of 1–2000 g. A photograph of the Buehler testing unit is seen in Fig. 1.55. This tester may be used to make Vickers, Knoop and Rockwell hardness measurements. The load is applied via the square-based pyramid indenter against the smooth, firmly-supported, flat surface of the test specimen. It is recommended to polish the specimen's surface, because a small impression is usually obtained in such a test.

The advantages of the VHT are:

(I) Despite several different loading settings (the application and removal of loads is controlled automatically) almost identical hardness numbers are obtained on the same ceramics if the distribution and size of the flaws are uniform. However, a scatter in the results may occur when the flaws are not uniformly distributed or when there are variations in their sizes within the test piece.
(II) The VHT yields more accurate diagonal readings than other testing methods.
(III) Only one type of indenter is used for all types of ceramics and the load applied may be changed by push button. The VHT covers a wide range of hardness and, as a result, a continuous scale can be established.
(IV) Because of the wide load range, this test is adaptable for the testing of almost any ceramic material, the softest and hardest ones.

In Fig. 1.56, graphs show the Vickers hardness variations for several refractory carbides under loads. These experiments were performed for a wide range of test forces, between 0.49 N and 196 N. All three graphs below show the same pattern

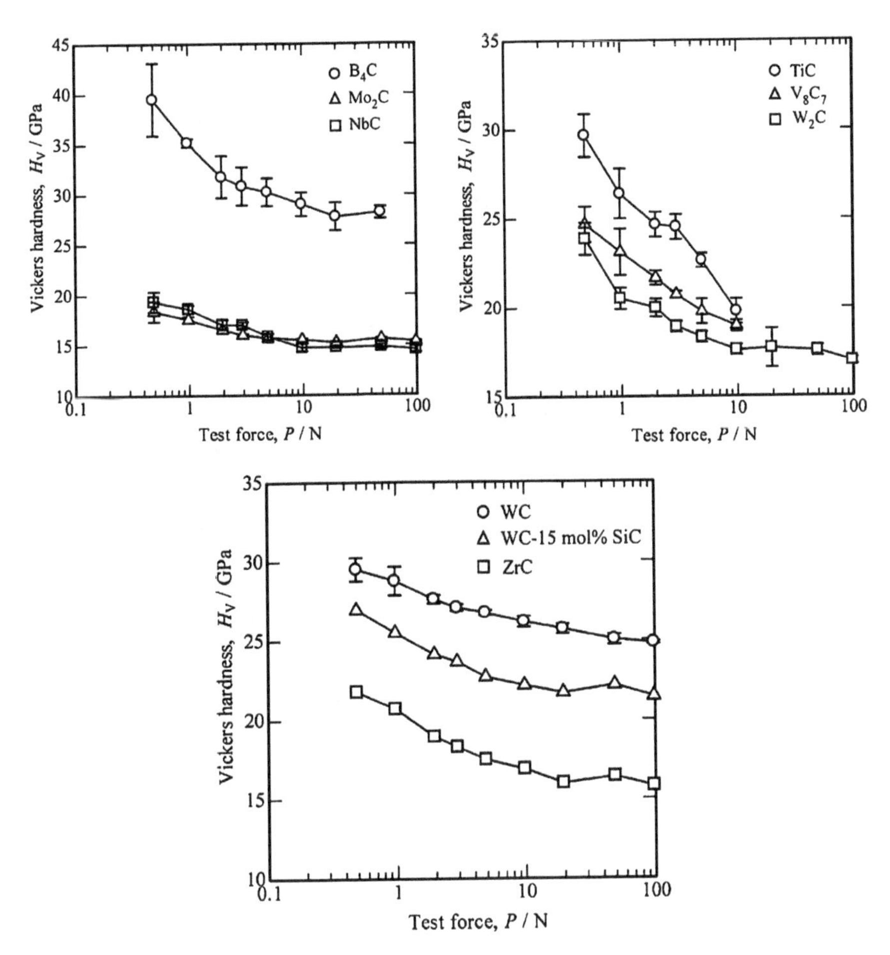

Fig. 1.56 Variation of Vickers hardness as a function of the test force **a** for B4C, Mo2C, and NbC; **b** for TiC, V_8C_7 and W_2C; **c** for WC, WC-15 mol% SiC and ZrC [40]. With kind permission of Professor Akihiro Nino and the Japan Institute of Metals

of decreasing hardness with increased load. The only high hardness found in this carbide series was observed in B_4C, reaching a value of 39.5 GPa at 0.49 N. One of the microstructures, that of tungsten carbide, is shown in Fig. 1.57 (a back-scattered electron micrograph), which indicates their grain sizes.

There is no preferential orientation and the grain sizes are 0.43 mm for (a), 1.4 mm for (b) and 2.5 mm for (c). Furthermore, as expected, an indentation size effect with hardness values should be observed for all carbides tested, because the size of the impressions are related to the load applied, being smaller for smaller loads or for harder materials. This, indeed, is the case, as shown in Fig. 1.58. The shapes of the curves in the illustrated materials are similar to those given by Li [34] and associates as:

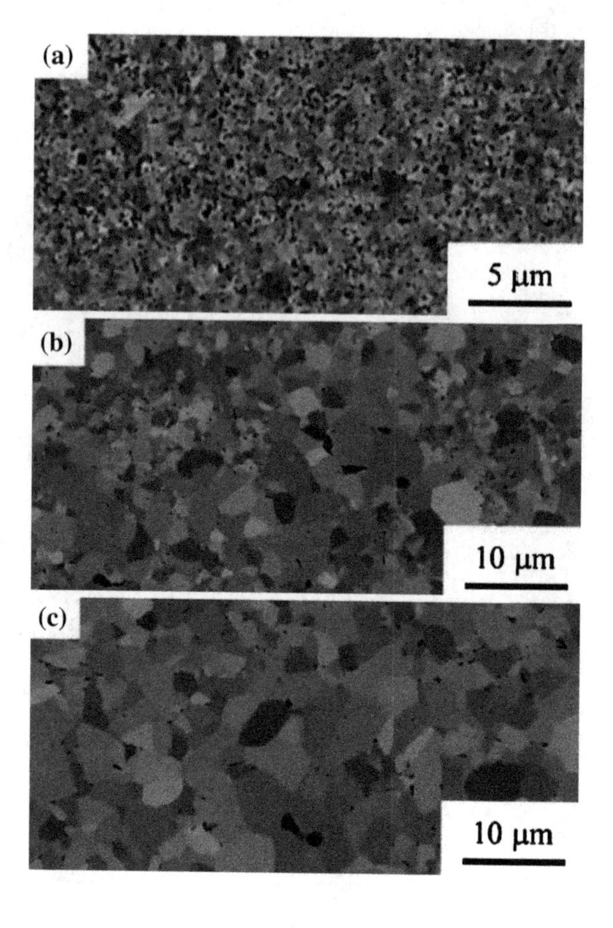

Fig. 1.57 Backscattered electron micrographs of tungsten carbides [40]. With kind permission of Professor Akihiro Nino and the Japan Institute of Metals. **a** 5 μm, **b** 10 μm. **c**, 10 μm

$$P = a_1 d + a_2 d^2 \tag{1.128}$$

and may be expressed as:

$$\frac{P}{d^2} = \frac{a_1}{d} + a_2^2 \tag{1.128a}$$

In the above relations, d is the indentation size; a_1 and a_2 are constants, which can be determined from the lines in Fig. 1.58. Thus, a_1 and a_2 are represented by the slope and the intercept, respectively. Recalling the definition of hardness in Eq. (1.126) and combining it with Eq. (1.128a), one obtains:

$$H = \alpha \frac{a_1}{d} + \alpha a_2 = \frac{a_1'}{d} + a_2' \tag{1.129}$$

A few more examples of the versatility of the VHT may be cited that were performed on polycrystalline ceramics, as shown in the following graphs. Note

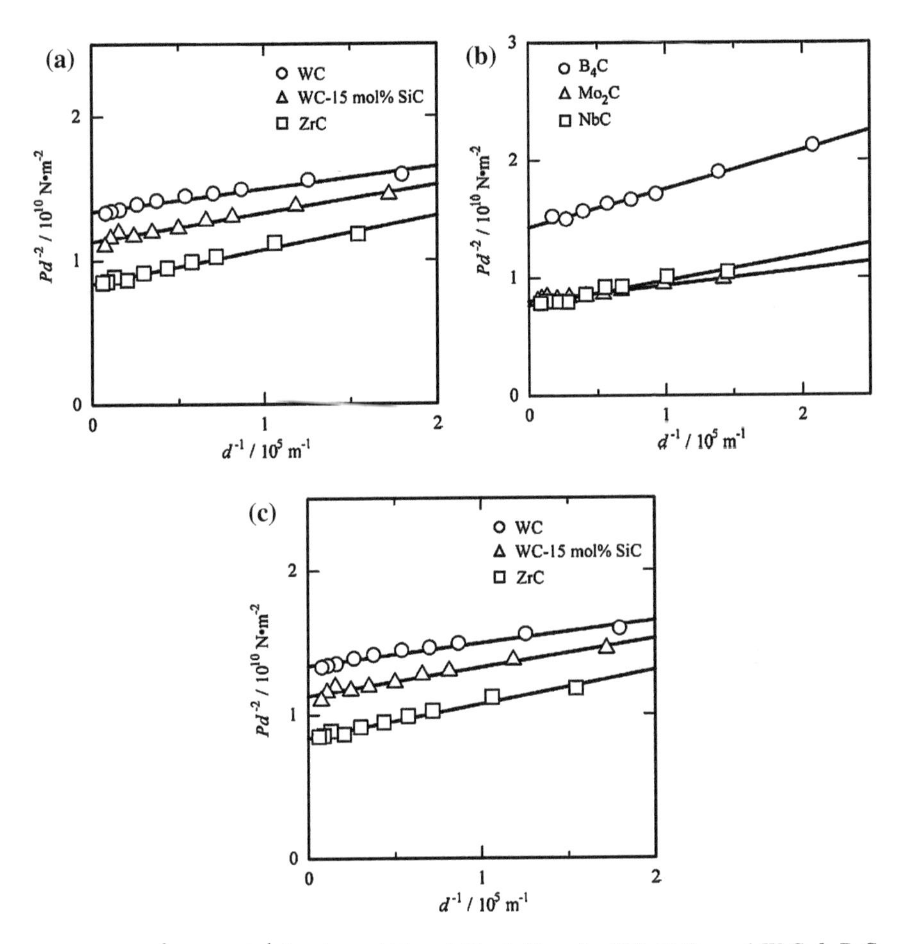

Fig. 1.58 Pd^{-2} versus d^{-1} for the carbides of Fig. 1.55: **a** for TiC, V_8C_7, and W_2C; **b** B_4C, Mo_2C, and NbC; **c** for WC, WC–15 mol% SiC, and ZrC [40]. With kind permission of Professor Akihiro Nino and the Japan Institute of Metals

that the pattern of these graphs is quite similar to those in Fig. 1.56. In all the lines in Fig. 1.59, a well-defined plateau is observed, usually at higher loads. Thus, it seems likely that there is a region on the hardness curves where hardness is load-dependent until the seemingly constant plateau has been reached, defining a transition point. Such a transition point appears to be associated with the onset of extensive cracking around and underneath the indentation. Cracking is an integral response of ceramics to indentation, often observed in structural ceramics, sometimes even at small loads. At higher loads, the cracking of ceramics usually occurs as observed in Fig. 1.60. The hardness measurement of ceramic materials is unsatisfactory and not reproducible when cracking, chipping or other flaws are observed in the test-piece, caused by the indentation. Cracks formed during

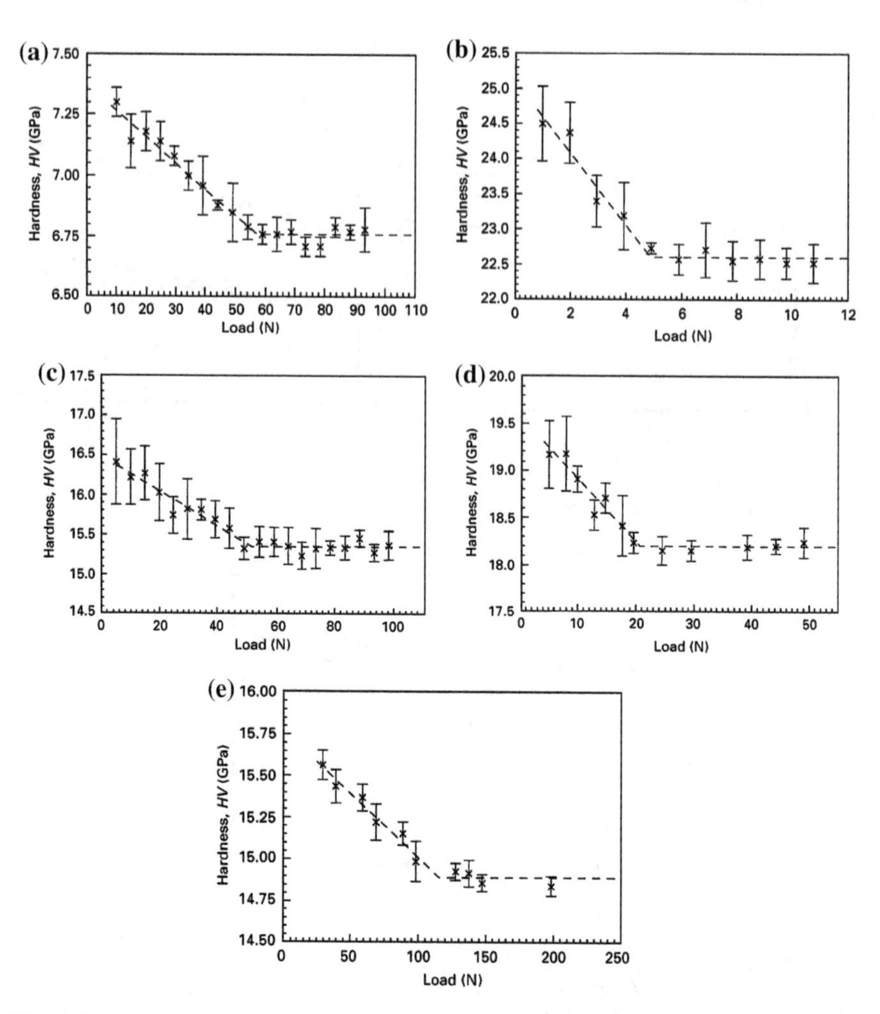

Fig. 1.59 Vickers hardness as a function of load for: **a** Pyroceram 9603; **b** sintered α-SiC; **c** NC132 silicon nitride; **d** AD 999 alumina; **e** NBD200 silicon nitride [17]. With kind permission of Springer

indentation can be radial or lateral and, in extreme cases, chipping may accompany crack formation or the test specimen may even be crushed. The actual indentation response of a tested ceramic material must be considered before selecting the appropriate hardness technique and test force level, in order to secure the most crack-free measurement possible and to enable reproducibility.

In Fig. 1.60b, the radial crack is visible and its occurrence occurs at a threshold load. It has been observed experimentally that, for most ceramic materials indented by Vickers tester, the threshold load is at an indentation of ~250 N.

It is of practical interest to specify a load for the testing of ceramics; however, this has not yet been realized by VHT measurements (or any other testing

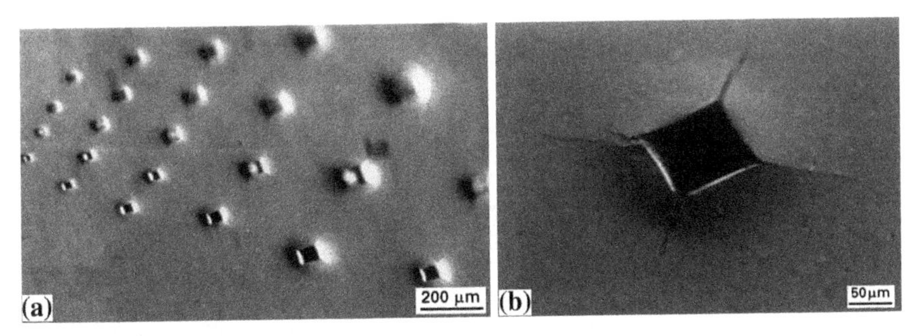

Fig. 1.60 Indentations in NC 132 (Si_3N_4, hot-pressed silicon nitride) with a stereo optical microscope with low angle incident lighting and a severe specimen tilt so as to accentuate surface detail: **a** shows rows of indents at different loads, from left to right: 9.8, 19.6, 29.4, 49, 98 and 73.5 N. **b** Shows a close-up of a 98 N indentation. [17]. With kind permission of Springer

methods). Obviously, the main difficulty is the propensity for cracking; as such, it is not practical to expect that two ceramics, one with a crack and another intact, will respond similarly to an applied load. Cracking may influence the hardness of a crack-free material, making indentation unreliable.

Figure 1.59 showed the load-dependence of hardness; increasing load results in decreasing hardness. This is equivalent to the size effect on the hardness value, since increasing a load means increasing the indentation, which is size-related. Clearly, the effect of the degree of brittleness must be considered when discussing crack formation, size and resistance to indentation in ceramics.

The VHT is a versatile hardness test, since it may be adapted for micro-hardness measurements and for a variety of materials, such as metals, ceramics, composites and plastics. It is useful for applications in which macro-hardness measurements are unsuitable, e.g., testing very thin materials, such as coatings, and for measuring individual phases within a larger matrix composed of more than a single phase.

Another micro-hardness test is the Knoop Hardness Test (henceforth: KHT, discussed below). It is worth mentioning that superficial Rockwell tests are also used for ceramics measurements. Hardness tests have been upgraded by the application of instrumented Knoop and Vickers hardness measurements.

1.10.3 The Knoop Hardness Test: Microhardness

The KHT was devised in 1939 by F. Knoop. A Knoop diamond indenter may be used in a Tukon Hardness Tester (or in a Buehler Instrument Model 1600-6400). The indentation produced by the KHT resembles that made by a pyramid-shaped diamond indenter (developed by the U.S. National Bureau of Standards). The apical angles are 130° and 172°. Thus, a narrow, rhombus-shaped indenter is used to produce a rhomboid-shaped impression, as seen in Fig. 1.61.

$$\frac{\textit{Projected area}}{\textit{longitudinal diagonal}^2} = \textit{Conversion factor } C$$

$$HK = KHN = \frac{P}{Cl^2} = \frac{14.286P}{l_1{}^2}$$

The long diagonal is seven times longer than the short diagonal, i.e., d_{long}: $d_{short} = 7{:}1$. As a result, the length of the impression is approximately seven times the width and the depth is 1/30 of the length of the longer diameter. The loads are often less than one kilogram-force and even a value of 25 g may be used. This test has practical applications for the testing of ceramics, thin films, coatings, and for phase identification in microstructures. The latter use is a consequence of the KHT indenter, used to sample minute grains of interest in microstructures. In modern technology, where thin-films play an important role, micro-hardness testing has become crucial and, thus, micro-indentation, such as the KHT test, plays a major role. During this test, the pyramid-shaped diamond indenter is pressed against a ceramic (or other material), making a rhombohedral impression with one diagonal seven times longer than the other. Then, the Knoop hardness number (henceforth: KHN) is determined by the depth to which the Knoop indenter penetrates. The KHT is useful for the hardness testing of brittle materials, such as glass and ceramics, because the indentation pressures are lower than in the VHT. A Knoop indenter leaves an impression of ca. 0.01 to 0.1 mm in size. Because of the small size of the impression under load, hardness may only be calculated after measuring the length of the longest diameter with the aid of a calibrated microscope. The KHN is usually given by the load (in kgf) per projected area (in mm^2), as:

$$\boldsymbol{HK = KHN = \frac{P}{Cl_1^2} = \frac{14.286P}{l_1^2}} \tag{1.130}$$

l_1 is the long diagonal and C a conversion factor, which ideally is 7.028×10^{-2}, but this depends on the load. The derivation of the formula is given in Fig. 1.61. Again, the accepted way to express a KHN is as follows. For instance, 356HK0.5, where the first number and the letter are the measured hardness value, the second letter indicates the KHT and 0.5 is the load in kgf.

The measuring apparatus is preset to apply a 25 g load. The duration of the contact between the indenter and the specimen should be 10–15 s. The length of the long diagonal of the impression is measured with a high-powered microscope. This procedure is repeated until at least five impressions have been made at widely spaced locations. The KHN is then calculated. The test loads are in the range of 10–1000 g. The samples are normally mounted and polished. One scale covers the entire hardness range.

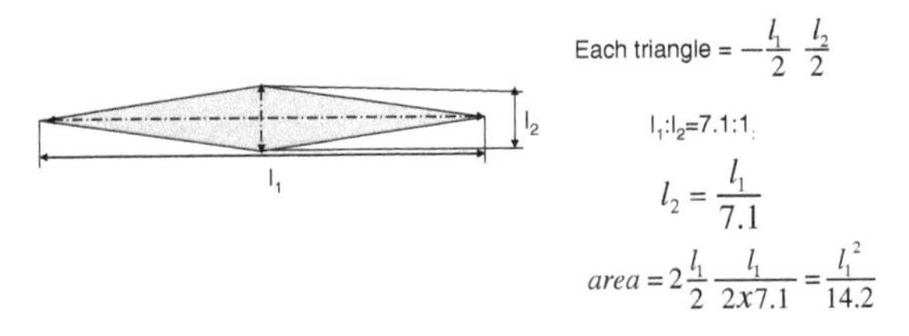

Fig. 1.61 Knoop Hardness Test (KHT) [13]

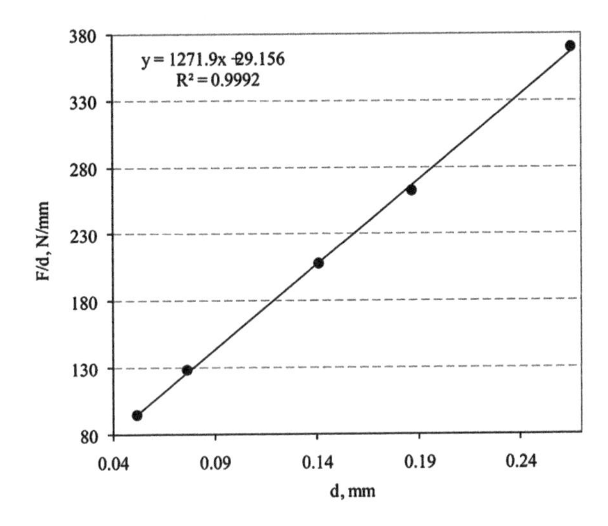

Fig. 1.62 Knoop hardness plotted according to relation (1.128) for SiC ceramics. Note that in the graph P of relation (1.128) is replaced by F/d. [8]. With kind permission of John Wiley and Sons

Li and associates [34] suggested a model known as the 'proportional specimen resistance' (henceforth: PSR) model for Eq. (1.128), which expresses the relation between the applied load and the diagonal. This has also been used to get the KHN, as indicated for SiC in Fig. 1.62. Thus, according to the above authors of the PSR model, Eq. (1.128) adequately describes the results of the Knoop hardness measurements for SiC. The actual Knoop hardness variation with load is shown in Fig. 1.63. However, according to Gong and Guan [28] the PSR model holds only at the lower load range, as seen in Fig. 1.64. A similar phenomenon was also observed by analyzing the experimental data for other materials investigated. These analyses indicate that the existing PSR model does not provide a satisfactory description for the observed indentation size effect (henceforth: ISE) in ceramics. In Eq. (1.128), the constants, a_1 and a_2, are related to the specimen's proportional resistance and load-independent hardness. Thus, a modified PSR model was suggested, known as the MPSR model, basically similar to Eq. (1.128) with an additional term, P_0, given as:

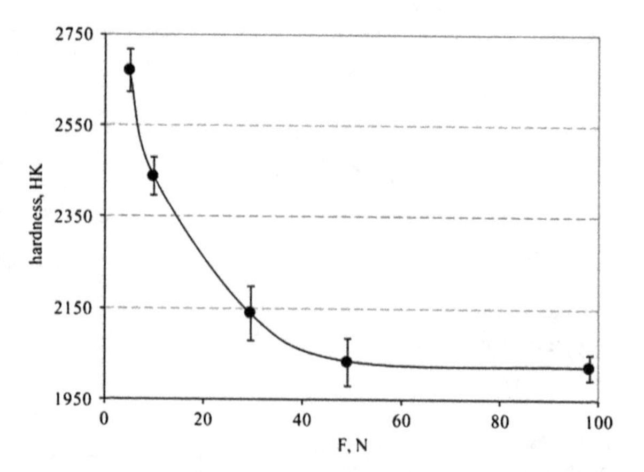

Fig. 1.63 Knoop hardness, HK, as a function of applied load for SiC [8]. With kind permission of John Wiley and Sons

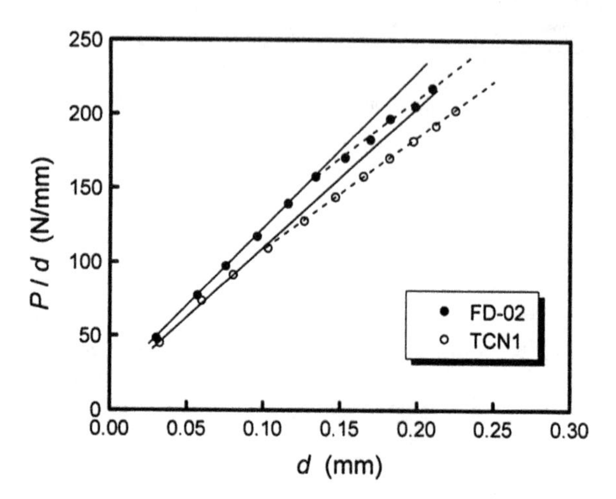

Fig. 1.64 A plot according to relation (1.128) for FD-02 (hot pressed Si nitride) and TCN1 (Ti based cermet) by Knoop hardness measurements [28]. With kind permission of Elsevier

$$P = P_0 + a_1 d + a_2 d^2 \tag{1.131}$$

P_0 is related to residual surface stresses in the test specimen. Analyses indicate that the existing PSR model does not provide a satisfactory description for the observed ISE in ceramics and warrants further modification.

The Knoop hardness for several ceramics, as a function of load, is indicated in Fig. 1.65. The pattern of the lines in Fig. 1.65 is similar to those in Figs. 1.56 and 1.59 in the VHT. The plateaus are not well defined, though there is such a tendency. The load variation with d is illustrated for all the materials shown in Figs. 1.65 and 1.66. Experience with a wide range of ceramics has proven that Knoop indentations are far less likely to crack than Vickers indentations. This explains the development of KHTs as an alternative to VHTs. Indeed, for a wide range of ceramics and other brittle materials, this is justified.

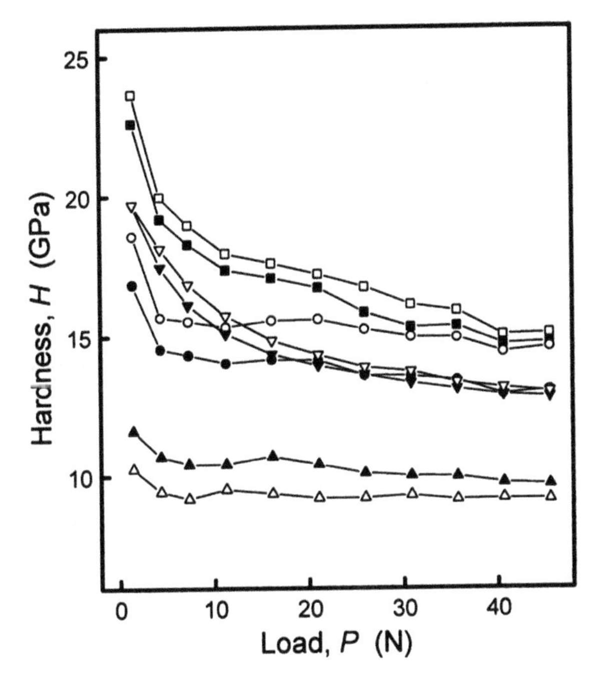

Fig. 1.65 Knoop hardness as functions of the applied test load for materials tested. FD-02 (sintered Si_3N_4) (*filled square*); FD-03 (hot pressed Si_3N_4) (*open square*); SN-W (SiC whisker reinforced Si_3N_4 (*filled circle*); Al_2O_3 (*opened circle*); TZP (Tetragonal zirconia polycrystal) (*filled triangle*); mullite; TCN1 (Ti(C,N)- based cermet) (*filled inverted triangle*); TCN2 (Ti(C,N)-based cermet) (*open inverted triangle*). [28]. With kind permission of Elsevier

This section has emphasized the most widely used techniques for hardness determinations. In fact, all the other known hardness measurement techniques (see [13]) may be adopted under appropriate conditions for the determination of the resistance of ceramics to indentation. Specifically, the superficial Rockwell method is useful for testing hard tiles, but is beyond the scope of this section.

1.11 Impact Testing of Ceramics

One of the important features of impact testing is the evaluation of the ductile-to-brittle transition temperature. What, then, is the purpose of discussing the impact testing of ceramics, since most are brittle at ambient temperature (and clearly at low temperatures)? Yet, impact tests are also performed on classic, brittle materials in order to evaluate the energy absorbed during the fracturing process. Furthermore, some brittle ceramics are ductile at sufficiently elevated temperatures, so the brittle-ductile transition may still be of interest. Ductile and superplastic ceramics will be discussed in depth in Chap. 2 (on ductile ceramics), while the present section deals with the actual process of performing impact tests.

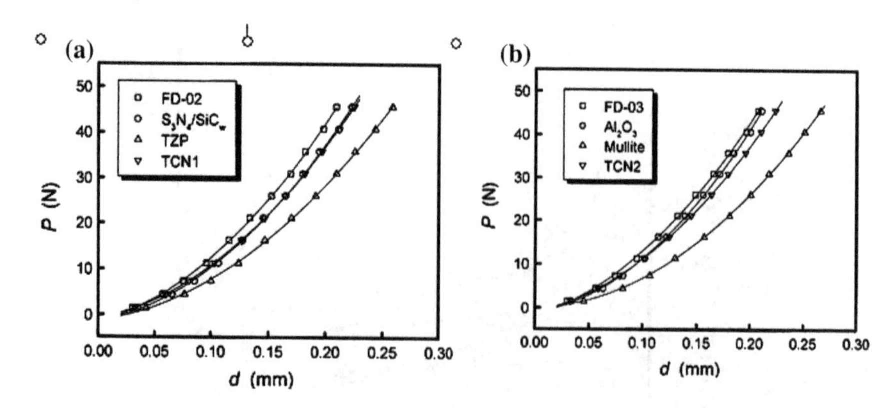

Fig. 1.66 Indentation size versus the applied test load for the materials of Fig. 1.65 [28]. With kind permission of Elsevier

The primary significance of impact tests is to determine the amount of energy absorbed by a material when it is struck by a moving object, such as a pendulum or a falling steel ball. Impact tests expose ceramic materials and other brittle products to dynamic events, forcing these materials to absorb energy quickly. Thus, an impact tests measures the energy absorbed when the material fractures or breaks due to a high-speed collision with a pendulum or a falling load, such as a ball. The amount of energy absorbed during an impact test indicates the resistance of the tested material to impact (known as its 'toughness'), measured as J m^{-3}. The data collected from such tests are usually expressed in terms of 'load' versus 'deflection', and the area under the curve represents the 'impact energy'. Modern impact testing is sophisticated, as will be discussed below. One of the common methods for getting impact information is via instrumented impact-testing machines, by means of which a load may be increased steadily and the results recorded simultaneously. A load-extension curve is shown schematically in Fig. 1.67.

The data provided by this type of testing is useful for understanding how a material will function in actual applications. The main goal of such tests is to characterize the ceramics under dynamic loading and to get information on the dynamic failure process. However, impact tests are rarely done to evaluate the impact toughness of brittle ceramics, because there is a major difficulty in evaluating the real energy of a given ceramic material, which may be masked by/or incorporated into the energy of the impact itself. Ceramics that are candidates for such studies are the engineering ceramics, such as Si_3N_4 [36] and others.

Modern impact testing is generally performed by instrumented machines and with other sophisticated impact testers. Ultra-high temperature ceramics (UHTC) are tested by using unique testing apparatus, as illustrated in Fig. 1.68 [37]. These tests were performed using a compressed-gas gun to fire ~0.5 mm diameter steel and tungsten carbide balls at velocities ranging 100–300 m s^{-1}. In addition, a 2MV Van de Graaff particle accelerator was used to fire micron-scale iron particles in the 1–3 km s^{-1} range. The results were documented by means of optical and

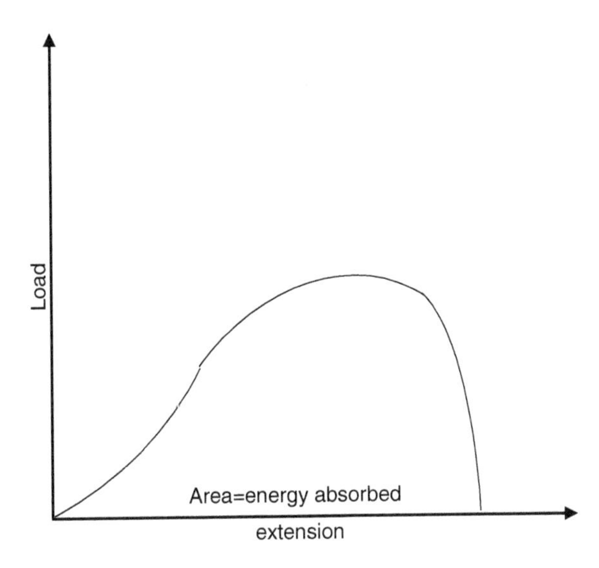

Fig. 1.67 Schematic impact load-extension curve. Area under the curve = energy absorbed

scanning-electron microscopy. The UHTC, such as monolithic ZrB_2/SiC and HfB_2/SiC, were high-velocity impact tested by the above apparatus. These materials are relatively hard, having micro-hardness values of ~15–20 GPa. The above impact tester operates via stainless steel or tungsten carbide spheres, with diameters in the 500–800 micron range, producing impact velocities of 200–300 m s^{-1}. This impact produced minimal plastic deformation, but significant radial and ring cracking at the impact sites occurred. NASA, while developing ceramics for use in hypersonic flight tests, discovered that the addition of ~20 % SiC to these ceramics enhances their oxidation resistance. Impact damage is a major concern, since it directly influences vehicle flight characteristics.

Note that, unlike conventional Charpy or Izod impact tests, during which the test pieces are usually broken (or bent to different degrees, depending on the ductility of the tested material), here the shape of the crack and its propagation are the measure of impact resistance. Figure 1.70 illustrates the type of ring crack generated by impact on an HfB_2/SiC surface.

Figure 1.69 illustrates some of the test results in terms of the crack diameter formed following the impact tests.

Furthermore, a radial-crack formation, resulting from the impact on the ZrB_2/SiC surface, is shown in Fig. 1.71. The authors note that their experiments, using this technique, are of an investigative nature. Further investigations should be made to systematically quantify the effects of impact damage on the strength and reliability of ultra-high temperature (henceforth: UHT) components.

Because of the brittle nature of ceramics, special instrumented Charpy Impact Test (henceforth: CIT) machines were developed, primarily to evaluate the

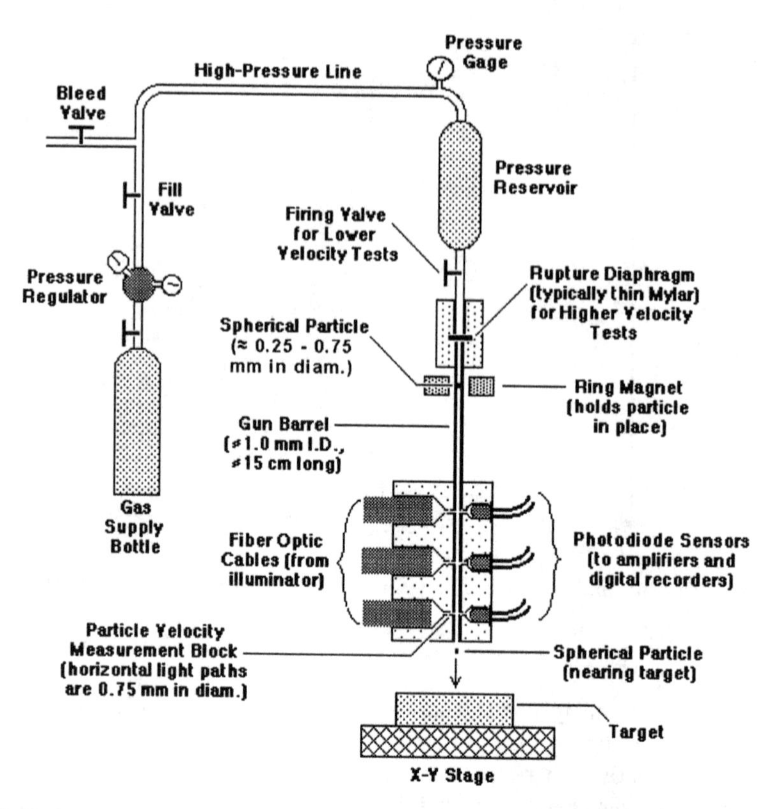

Fig. 1.68 Gas gun impact apparatus used for studying resistance to impact of ZrB_2/SiC and HfB_2/SiC [37]. With kind permission of Springer

dynamic toughness of such materials. The resulting load–deflection curves obtained by this instrumented CIT and the absorbed energy were analyzed to evaluate the characteristics of brittle ceramics exposed to dynamic loading. As mentioned above, the ductile-to-brittle transition is not relevant at ambient temperatures for such materials, but the purpose of the test is rather to evaluate toughness in service. Kobayashi et al. [32] was using an instrumented CIT to study zirconia (PSZ: ZrO_2-3 mol%Y_2O_3) and SiC. The resulting load–deflection curves obtained for these ceramics are illustrated in Fig. 1.72. Except for PSZ(a), all the lines are linear up to the maximum load, indicating that no macroscopic plastic deformation exists. This is in line with what was said earlier regarding brittle-ductile transition. No such transition exists in these ceramic materials, attested by the fact that the load in Fig. 1.71 drops vertically from the maximum to zero load. The area underneath the load–deflection curves represents the absorbed energy, E_t, which may be expressed as:

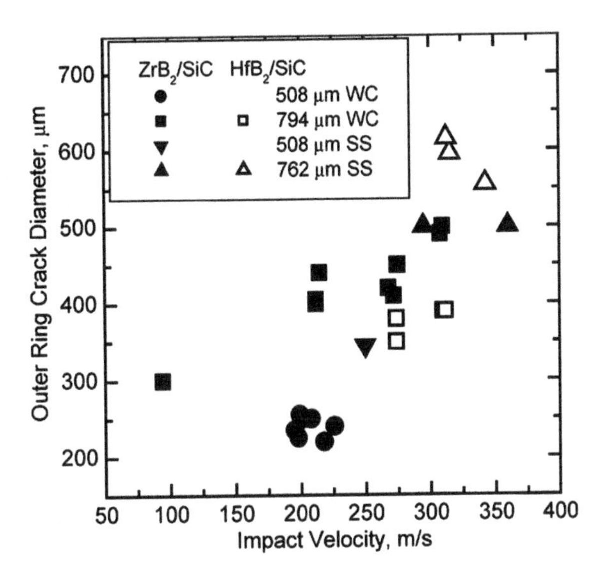

Fig. 1.69 Summary of outer ring crack diameters observed on ZrB_2/SiC and HfB_2/SiC surfaces as a function of size, composition, and velocity of the impacting ball. SS refers to steel and WC to carbide balls, respectively [37]. With kind permission of Springer

$$E_t = E_i + E_p \tag{1.132}$$

E_i is usually associated with crack initiation and, on a load–deflection curve, spans the area up to the maximum load, while E_p is believed to represent the apparent crack-propagation energy, appearing after the maximum load. Considering the curves of these ceramic materials shown in Fig. 1.72, it is clear that the load drops vertically and there is no additional area to represent E_p. This observation indicates that PSZ and SiC show typically elastic brittle fracture without plastic deformation and that:

$$E_t \cong E_i \tag{1.132a}$$

To further analyze this absorbed energy, the elastic deformation of the machine, E_m, and the kinetic energy (which tossed the broken specimen), E_k, must be taken into account when calculating the total energy, E_t. Thus, during the fracturing process, where E_f represents fracturing, the total energy, E_t, absorbed is:

$$E_t = E_m + E_k + E_f \tag{1.132b}$$

As such, the absorbed energy included in E_t consists partially of the elastic deformation of the testing machine, E_m, and partially of the stored energy in the specimen, E_s, which is composed of:

$$E_s = E_k + E_f \tag{1.132c}$$

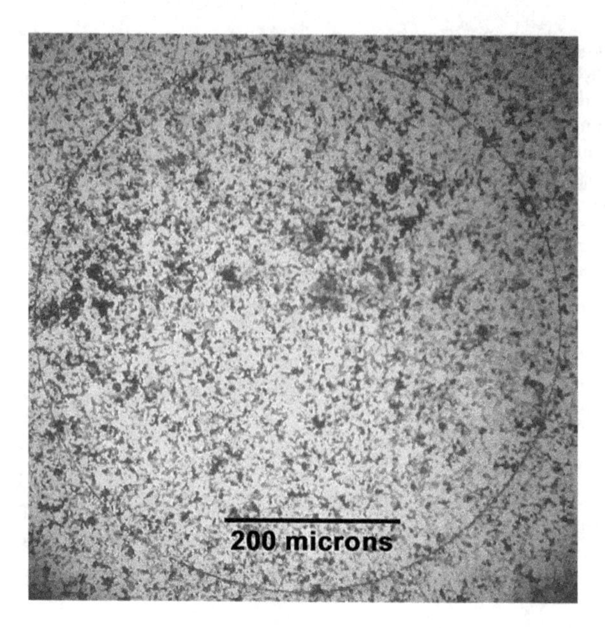

Fig. 1.70 Optical microscope image of the ring crack generated by the 316 m s^{-1} impact of a 762 μm diameter stainless steel ball on a HfB_2/SiC surface [37]. With kind permission of Springer

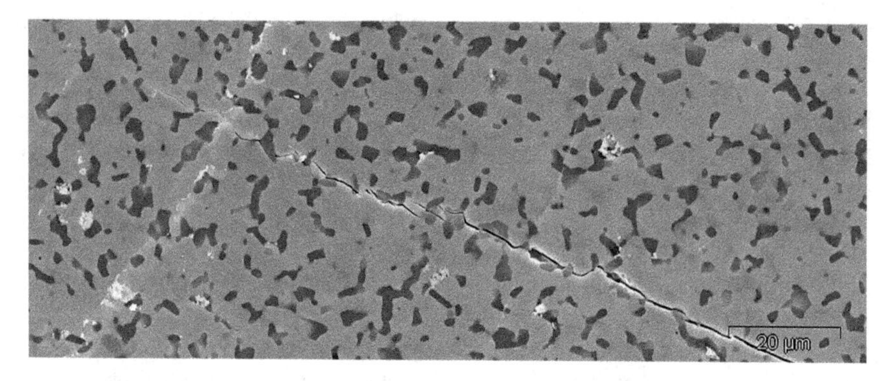

Fig. 1.71 SEM image of a radial crack at boundary of ring crack pattern on ZrB_2/SiC surface [37]. With kind permission of Springer

This kind of test enables an evaluation based on the reciprocals of the curves, the compliances of the machine, C_m, and those of the specimen, C_s, to provide total compliance, C_t, as given by Eq. (1.133). In other words:

$$C_t = C_s + C_m \tag{1.133}$$

No correction is needed for C_s, since no plastic deformation has occurred in these specimens.

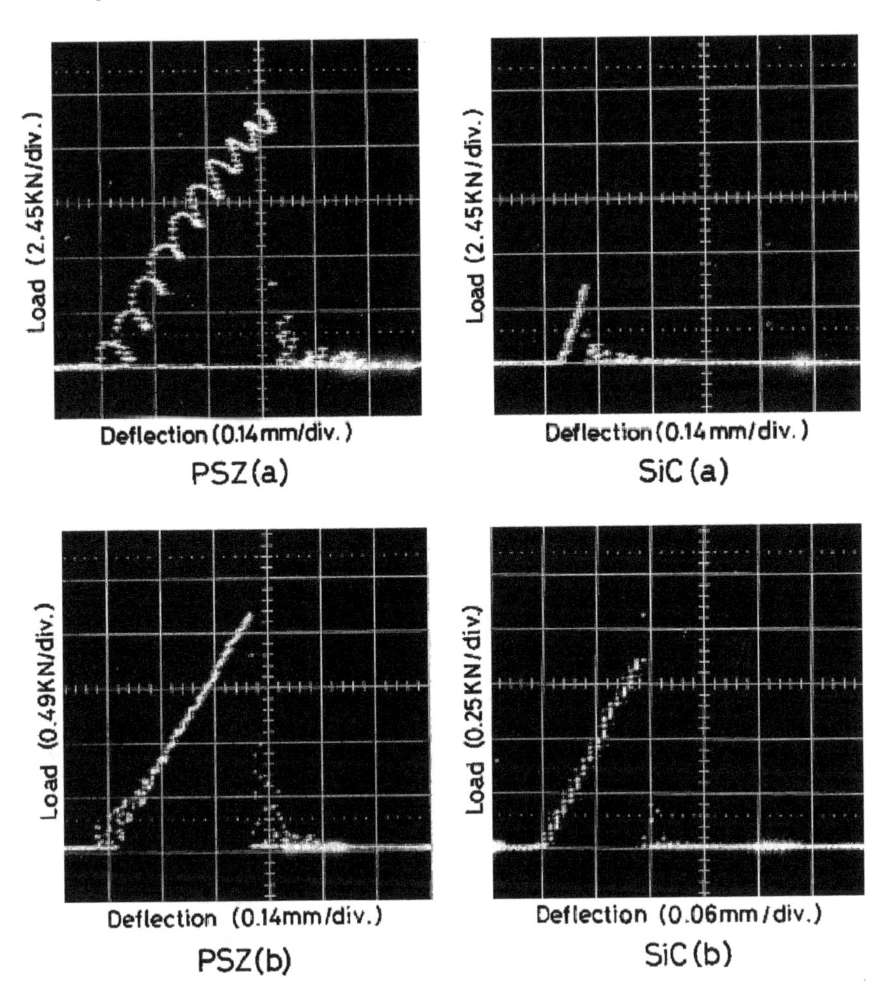

Fig. 1.72 Typical load–deflection curves for these materials [31]. With kind permission of Professor Toshiro Kobayashi

Figure 1.73 illustrates the instrumented CIT system in a block diagram and the impact-response curve method, as applied to the above tests. In addition to the aforementioned partially stabilized zirconia (PSZ: ZrO_2-3 mol%Y_2O_3), samples of S_3N_4 were also investigated by Kobayashi et al. [32]. Typical hammer load-times and strain-gage signal-time curves of PSZ and Si_3N_4 are found in Fig. 1.74. The impact-response curves of PSZ and S_3N_4, at several impact velocities are shown in Fig. 1.75. The impact curves of these specimens are impact-velocity-dependent. This technique enables the determination of the dynamic fracture toughness of these ceramics by means of the impact-response curve method. Impact-response curves quantitatively relate the response of the specimen to the impact, which depends solely on the elastic reaction between the specimen and the actual impact.

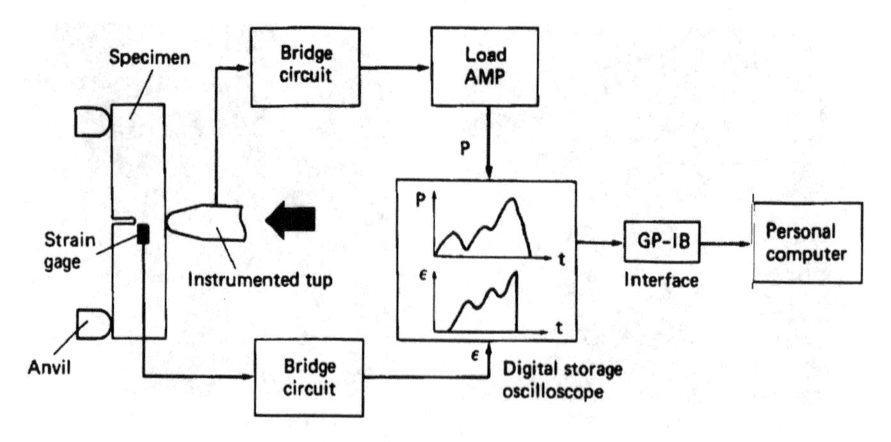

Fig. 1.73 Block diagram of instrumented Charpy impact testing system [32]. With kind permission of Professor Toshiro Kobayashi

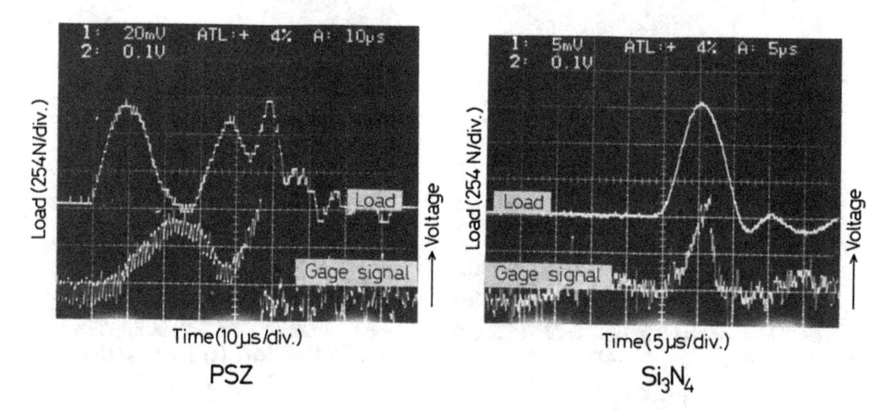

Fig. 1.74 Typical load–time curves and strain gage signal–time curves recorded by a instrumented Charpy impact testing machine for PSZ and Si_3N_4 [32]. With kind permission of Professor Toshiro Kobayashi

The impact-response curve is previously determined under several conditions and dynamic fracture toughness may be obtained from the measured time-to-fracture, t_f. An optical method is generally used for this purpose, but, in the case of for PSZ and Si_3N_4, a strain-gage was directly attached to the specimens. This procedure may also be applied to obtain the linear elastic brittle fracture range.

Dynamic fracture toughness, K_d, evaluated from the impact-response curve and the time-to-fracture was $6.2 MN/m^{3/2}$ for PSZ and $1.7 MN/m^{3/2}$ for Si_3N_4. The effects of the notch-root radius on static and dynamic fracture toughness were also evaluated. The specimens were blunt-notched with radii $\rho = 25$, 50, 100 and 150 µm. They were tested by instrumented CIT, applying the impact-response

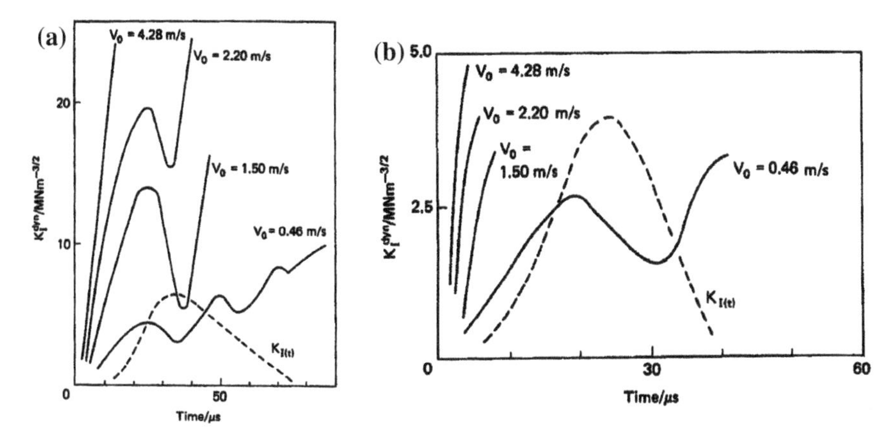

Fig. 1.75 Impact response curves in several impact velocities: **a** for PSZ; **b** for Si_3N_4 [32]. With kind permission of Professor Toshiro Kobayashi

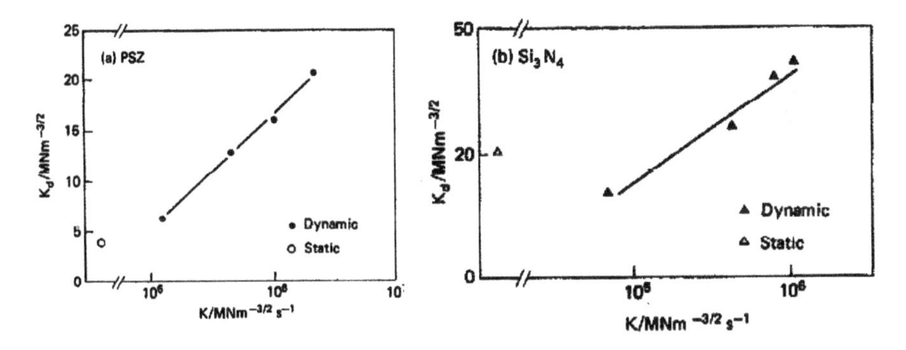

Fig. 1.76 Dynamic fracture toughness, K_d, against stress intensity rate, $\dot{K}$, for: **a** PSZ; **b** Si_3N_4 [32]. With kind permission of Professor Toshiro Kobayashi

curves in order to investigate the effects of the notch-root radius on the apparent fracture toughness.

It is generally considered that dynamic fracture toughness is lower than static toughness in many metallic materials; however, in the case of PSZ, the dynamic fracture toughness appears to be higher than the static. This is apparently related to the stress-induced phase transformation which occurs in PSZ.

The effect of impact velocity, or rather the stress-intensity rate, $\dot{K}(K_d/t_f)$, on K_d, is shown in Fig. 1.76. Dynamic fracture toughness, K_d, in partially stabilized zirconia has been observed to increase simply with stress-intensity rate, $\dot{K}$, but K_d in Si_3N_4 initially decreases and then increases with increased $\dot{K}$. These specimens were pre-cracked and the same observation that was made for Si_3N_4 was also made

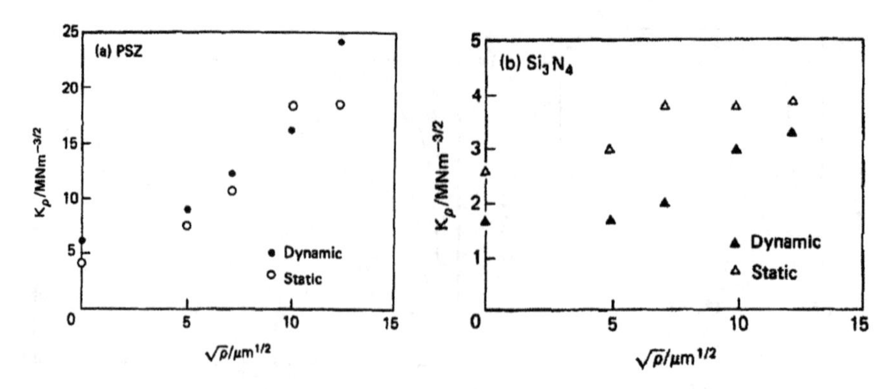

Fig. 1.77 Apparent dynamic and static fracture toughness, K_P, as a function of the notch root radius ρ: **a** for PSZ; **b** for Si_3N_4 [32]. With kind permission of Professor Toshiro Kobayashi

for metals (e.g., steel) and is associated with the crack becoming unstable when the intensity of the dynamic stress at the crack tip exceeds a critical value.

The effect of the notch-root radius on the apparent dynamic fracture toughness, K_d, is shown in Fig. 1.77. K_d increases with the radius of the notch for both static and dynamic fracture toughness. A similar notch-radius effect is observed for Al_2O_3. Note that the static fracture toughness of Si_3N_4 increases at first, but then becomes constant (Fig. 1.77b).

The critical strain-energy release rate, G_ρ, in the specimen with notch-root radius, ρ, is indicated in Fig. 1.78 for both static and dynamic ceramic fracture. The linear relation of the lines in Fig. 1.78 are based on the Williams' relation [49], given as:

$$G_\rho = G_C \frac{\left(1 + \frac{\rho}{2l_0}\right)^3}{\left(1 + \frac{\rho}{l_0}\right)^2} \tag{1.134}$$

Here, G_ρ is the critical strain-energy release rate in a notched specimen with radius, ρ, at the root, l_0 (equal to the characteristic distance). When $\rho \gg l_0$, Eq. (1.134) reduces to:

$$G_\rho = G_0\left(\frac{1}{2} + \frac{\rho}{8l_0}\right) \tag{1.134a}$$

The ceramic specimens mentioned above, tested by dynamic loading, namely impact testing, are notched as is usual during the performance of such tests. During conventional impact testing, the radius at the notch-root is significant, since it affects the outcomes. For metals, the dimensions of the specimens and the notches are standard. A standard CIT specimen consists of a bar of metal or other material (ceramics included), 55 × 10 × 10 mm having a notch machined across one of

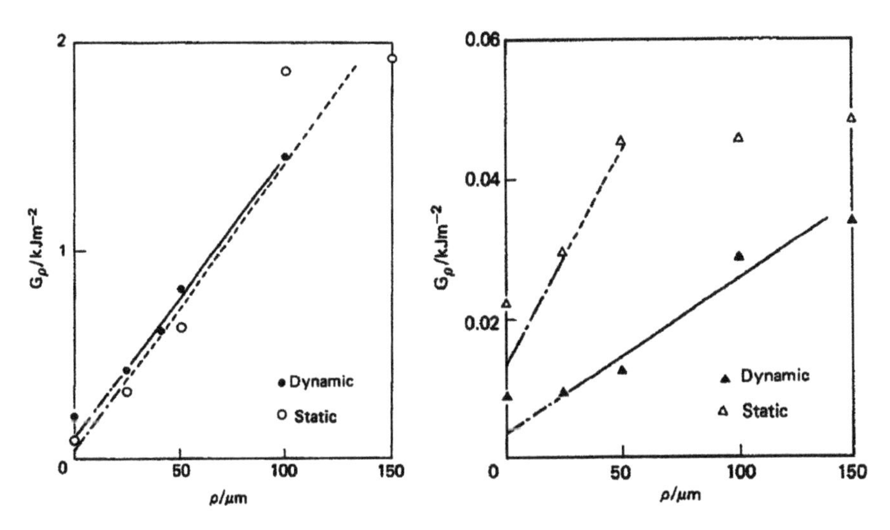

Fig. 1.78 Relationship between critical elastic strain energy release rate, G, and notch root radius, ρ: **a** for PSZ; **b** for Si_3N_4 [32]. With kind permission of Professor Toshiro Kobayashi

the larger dimensions. The most common is a V-notch, 2 mm deep, with a 45° angle and a 0.25 mm radius along the base.

If U-notch and keyhole notch specimens are used, their characteristics are a 5 mm deep notch with a 1 mm radius. Furthermore, specimens should be selected that represent the lot being sampled. In general, impact tests should be based on a minimum of ten specimens that must be inspected for soundness and obvious physical defects before testing. Aigbodion et al. [22] conducted CITs on notched samples of ceramic-based composites. Their standard, square impact-test specimens measured 50 × 10 × 10 mm with notch depths of 2 mm and notch-tip radii of 0.02 mm at 45° angles. Their testing machine could provide a range of impact energies from 0 to 300 J. The mass of the hammer was 22.7 kg and the striking velocity was 3.5 m/s.

1.12 Weibull Statistical Distribution

Despite the use of 'identical' specimens, the mechanical properties of ceramics show considerable scatter in the measured results. The main reason for the scatter in the values measured is a consequence of the presence, size and distribution of cracks in ceramics. A mean value must be determined via statistical evaluation. The most commonly used statistical approach for describing experimental data is Gaussian normal distribution. In ceramics, however, the use of the Weibull distribution is preferable, reviewed below.

Weibull distribution assumes that each elemental part of a bulk material has an individual property, i.e., a local strength. The fracture or failure probability, P_f, of each such element is integrated over the entire test piece, giving:

$$P_f = 1 - \exp\left[-\int_V \left(\frac{\sigma - \sigma_u}{\sigma_0}\right)^m dV\right] \tag{1.135}$$

σ is the stress at a point and V is the volume of the specimen; σ_u, σ and m are the Weibull parameters representing location, scale and shape-modulus parameters, respectively, also known as 'threshold strength', 'characteristic strength' and the 'Weibull modulus'. This function is known as the 'Weibull three-parameter strength distribution'. The threshold stress parameter, σ_u, represents a minimum stress, below which a test specimen will not break. The probability of failure, P_f, increases with the fracture-stress variable, σ. Also, σ_0 is dependent on the stress configuration and the size of the test piece. Clearly, the second term is the survival probability and may be expressed as:

$$P_s = \exp\left[-\int_V \left(\frac{\sigma - \sigma_u}{\sigma_0}\right)^m dV\right] \tag{1.135a}$$

In cases of fracture with a fracture strength (or stress), σ_f, the first term, under the integral in the nominator in the above relations, may be replaced to obtain:

$$P_f = 1 - \exp\left[-V_e\left(\frac{\sigma_f - \sigma_u}{\sigma_0}\right)^m\right] \tag{1.136}$$

The higher is m and the lower represents strength variability. V_e is the effective volume of the specimen and may be expressed as:

$$V_e = \int_V \left(\frac{\sigma - \sigma_u}{\sigma_f - \sigma_0}\right)^m dV \tag{1.137}$$

In uniaxial tension, for example, in brittle materials (such as ceramics), the stress $\sigma = \sigma_f$ and Eq. (1.137) yields $V_e = V$. Thus, Eq. (1.136) may be written as:

$$P_f = 1 - \exp\left[-\left(\frac{\sigma_f - \sigma_u}{\sigma_0 V_e^{-1/m}}\right)^m\right] \tag{1.138}$$

where $\sigma_0 V_e^{-1/m}$ is a constant.

Equation (1.138) may be obtained as follows. Multiply and divide Eq. (1.135) by $\left(\frac{1}{\sigma_f - \sigma_u}\right)^m$ and rewrite it either as:

$$P_f = 1 - \exp\left[-\int_V \left(\frac{\sigma - \sigma_u}{\sigma_0}\right)^m \frac{1}{(\sigma_f - \sigma_u)^m}(\sigma_f - \sigma_u)^m dV\right] \quad (1.135a)$$

or:

$$P_f = 1 - \exp\left[-\int_v \left(\frac{\sigma - \sigma_u}{(\sigma_f - \sigma_u)}\right)^m dV\left(\frac{1}{\sigma_0}\right)^m (\sigma_f - \sigma_u)^m\right] \quad (1.135b)$$

In Eq. (1.137), V_e is the first term of the integral being equal to V, since, in the nominator, $\sigma = \sigma_f$ for brittle materials:

$$P_f = 1 - \exp\left[-V_e(\sigma_f - \sigma_u)^m\left(\frac{1}{\sigma_0}\right)^m\right] = 1 - \exp\left[-V_e\left(\frac{\sigma_f - \sigma_u}{\sigma_0}\right)^m\right] \quad (1.135c)$$

Rearranging Eq. (1.135c) results in Eq. (1.138) being:

$$P_f = 1 - \exp\left[-\left(\frac{\sigma_f - \sigma_u}{\sigma_0 V_e^{-1/m}}\right)^m\right] \quad (1.138)$$

Equation (1.138) may be expressed after rearrangement as:

$$\begin{aligned}\ln\left[\ln\left(\frac{1}{1 - P_f}\right)\right] &= m\ln(\sigma_f - \sigma_u) - m\ln\sigma_0 V_e^{-\frac{1}{m}}\\ &= m\ln(\sigma_f - \sigma_u) - m\ln\sigma_0 + \ln V_e \end{aligned} \quad (1.139)$$

Equation (1.139) represents a straight line when $\ln\left[\ln\left(\frac{1}{1-P_f}\right)\right]$ is plotted versus $\ln(\sigma_f - \sigma_u)$ with a slope of m. The intercept of the curve is either $-m\ln\sigma_0 V^{-1/m}$, or $(-m \ln \sigma_0 + \ln V_e)$, or

$$\text{intercept} = -m\ln\sigma_0 + \ln V_e.$$

For the unit volume one calculates σ_0 by:

$$\sigma_0 = [\exp(-\text{intercept})]^{1/m} \quad (1.140)$$

In the above relations, σ_u is the stress level below which the probability of failure is zero, in other words, the probability of survival is 1.0. The Weibull modulus, m, principally has values in the range 0–∞. In metals, the value of m is ~100 and for ceramics m < 3, but this value depends on the soundness of the ceramics. Well-controlled engineering ceramics with fewer flaws may even have an m value in the range of 5–10.

Again, for brittle materials, such as ceramics, σ_u represents the minimum stress below which a test specimen will not break and it is assumed to be equal to zero. Thus, omitting σ_u from the above relations, the survival probability Eq. (1.135a) may be rendered as:

$$P_s = \exp\left[-\int_V \left(\frac{\sigma}{\sigma_0}\right)^m dV\right] \tag{1.141}$$

and the failure probability as:

$$P_f = 1 - P_s = 1 - \exp\left[-\int_V \left(\frac{\sigma}{\sigma_0}\right)^m dV\right] \tag{1.142}$$

$$P_f = 1 - \exp\left[-V_e\left(\frac{\sigma_f}{\sigma_0}\right)^m\right] \tag{1.143}$$

As above in Eq. (1.137) (and with $\sigma_u = 0$), $V_e = V$; thus, for the integral from Eq. (1.142) one can obtain Eq. (1.143). Then, by taking logarithms twice, Eq. (1.143) becomes:

$$\ln\left[\ln\left(\frac{1}{1-P_f}\right)\right] = m\ln\sigma_f - m\ln\sigma_0 V_e^{-\frac{1}{m}} = m\ln\sigma_f - m\ln\sigma_0 + \ln V_e \tag{1.143}$$

As in the three-parameter case, plotting $\ln\left[\ln\left(\frac{1}{1-P_f}\right)\right]$ versus $\ln\sigma_f$ results in a linear relation with slope m and intercept ($\ln V_e - m \ln \sigma_o$), or :

$$\text{intercept} = \ln V_e - m\ln\sigma_0 \tag{1.144}$$

and for unit volume σ_0:

$$\sigma_0 = [\exp(-\text{intercept})]^{1/m} \tag{1.144a}$$

Note that σ in Eq. (1.137) has been replaced by σ_f, the fracture strength, which, in brittle materials, is almost identical to σ. This value of σ_f was carried through in all following equations.

Using the Weibull distribution, plots are made from measured strength data. These data are arranged in ascending order and assigned numbers beginning with 1 and ending with n. The survival probability (or expected life-time) is usually assigned to the ith strength value, expressed as:

$$P_s = 1 - \frac{i - 0.5}{n} \tag{1.145}$$

In case of fracture (failure), for example, Eq. (1.145) is written for failure probability as:

$$P_f = \frac{i - 0.5}{n} \tag{1.145a}$$

Thus, the lowest stress under tension for each configuration represents the first value ($i = 1$), the next lowest stress value is the second measured value ($i^{th} = 2$), etc., and the highest stress is represented by the nth value measured. This enables ranking the probability of failure, P_f, assigned to each datum according to Eq. (1.145a). Now that the fracture stress and the associated P_f have been evaluated according to Eq. (1.145a), a graph may be constructed of $\ln\left[\ln\left(\frac{1}{1-P_f}\right)\right]$ versus $\ln\sigma_0$ according to Eq. (1.143). The fitting of the straight line is often performed by linear regression. An evaluation may be made using Eq. (1.145a) by expressing the first, second, etc. (up to n) percent of each value measured for P_f. In this manner, if there are many broken specimens (failures under test conditions) the percent value of each failed specimen of the total indicates the weakness (in percentage or fraction) of the specimen compared to the measured stress value. It is commonly accepted to make a table recording the values of i in the first column, indicating the measured strength in the second column, the measured ln (strength) in the third column, the fourth column should give the value of P_f according to Eq. (1.145a) (in terms of percent or fraction) and the last column should give the value of $\ln\left[\ln\left(\frac{1}{1-P_f}\right)\right]$. This facilitates making the plot of $\ln\left[\ln\left(\frac{1}{1-P_f}\right)\right]$ versus ln(strength).

A large number of test specimens are necessary to determine the Weibull parameters, about 30, that must be broken to obtain reasonable accuracy. Figure 1.79 shows Weibull plots for four values of m obtained on the basis of Eq. (1.139), obtained under tension. In Eq. (1.139), $\sigma_0 V^{-1/m} = MOR_0$.

Often in ceramics, fracture tests are replaced by a bending test (a flexural test), discussed earlier in Sect. 1.9, rather than by a tension test, for the aforementioned reasons. Figure 1.80 is a Weibull plot. In Fig. 1.80, Weibull plots of "measured" bending strength, $\sigma_{f,B}$, are shown for different m values.

The lines are not completely linear; one of the reasons may be the probability of the different distribution, amount and size of cracks. Specimens containing larger flaws will break or fail at lower stresses than those predicted. As indicated above, the higher the m, the lower the strength variability.

Concluding this section, materials susceptible to brittle fracture, such a ceramics, are known to behave in this fashion and they usually fail without advanced warning. No visible plastic deformation is observed, in general, and,

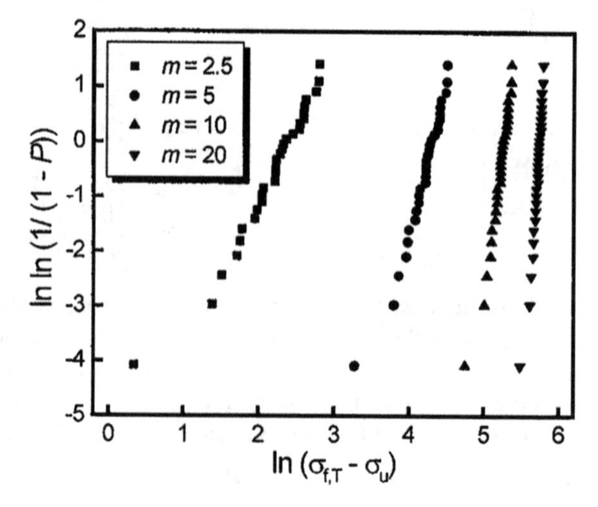

Fig. 1.79 Weibull plots of the "measured" tensile strength, $\sigma_{f,T}$. The strength data were generated using $\sigma_u = 250$ MPa and $MOR_0 = 500$ MPa. ($MOR_0 = \sigma_0 V^{-1/m}$) [2]. With kind permission of Professor Gong

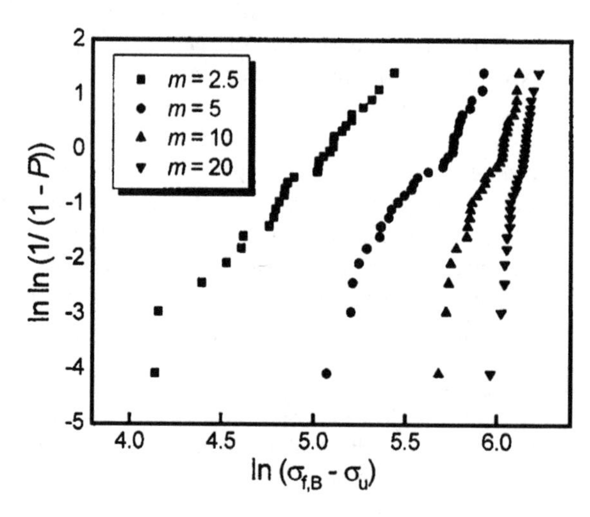

Fig. 1.80 Weibull plots of the "measured" bending strength, $\sigma_{f,B}$. The strength data were generated using $\sigma_u = 250$ MPa and $MOR_0 = 500$ MPa [2]. With kind permission of Professor Gong

thus, fracture sets in quite suddenly. Stress concentrations at various locations, due to intrinsic flaws, are responsible for (often premature) failure, since they cannot be relieved by plastic deformation. Therefore, the design methods and tests applied to ductile materials are not suitable for brittle materials. The smallest strength of such a material is associated with the largest flaw size. For purposes of design and failure prediction, the preferred statistical method for obtaining the probable strength of brittle materials is generally not the Gaussian, but rather the Weibull distribution approach (described above) for two- and three-parameter methods. However, for specific tests, these equations may be changed according to the characteristics of the measurements. Extensive discussion on the use of the Weibull distribution for the various tests and the required modifications needed for analysis are found in the literature.

References

1. Davies JR (ed) (2004) Tensile testing, 2nd edn. ASM International, Materials Park, Ohio (#05106G)
2. Gong J (2003) J Mater Sci 38:2541
3. Green DJ (1998) An introduction to the mechanical properties of ceramics. Cambridge University Press, Cambridge
4. Hague B (1959) An introduction to vector analysis. Methuen & Co. LTD, London
5. Hodgkinson JM (2000) Mechanical Testing of Advanced Fibre Composites. Woodhead Publishing, Abingdon Hall
6. Hunt RE (2002) Lecture notes and handouts. University of Cambridge, Cambridge
7. Khadar Vali S, Ravinder Reddy P, Ram Reddy P (2012) International Journal of Computational Engineering Research, 2, 1165
8. Majic FM, Curkovic L, Coric D (2011) Mat-wiss u Werkstofftech 42(3):S313–S319
9. Matsumoto T, Nose T, Nagata Y, Kawashima K, Yamada T, Nakano H, Nagai S (2001) J Am Cer Soc 84:1521
10. Muntz D, Fett T (2001) Ceramics: mechanical properties, failure behavior, materials selection. Springer, New York
11. NIST/SEMATECH e-Handbook of Statistical Methods (2012) http://www.itl.nist.gov/div898/handbook/, April 2012, 1.3.6.6.8. Weibull Distribution
12. Nye JF (1957) Physical properties of crystals. Oxford University Press, New York
13. Pelleg J (2012) Mehanical properties of materials. Springer, Dordrecht
14. Pelleg J (2013) Mechanical properties of materials. Springer, Dordrecht, pp 36–50
15. Polakowski NH, Ripling EJ (1966) Strength and structure of engineering materials. Prentice-Hall, Englewood
16. Scruby CB, Drain LE (1990). Laser Ultrasonic Technologies and Applications, CRC Press
17. Quinn JB, Quinn GD (1997) Indentation brittleness of ceramics: a fresh approach. J Mater Sci 32:4331
18. Rice RW (2000) Mechanical properties of ceramics and composites: grain and particle effects. CRC Press, New York
19. Timoshenko S (1940) Strength of materials. D. Van Nostrand Company, New York
20. Wachtman JB, Cannon WR, Matthewson MJ (2009) Mechanical properties of ceramics. Wiley, Hoboken

Further References

21. Adler SB (2001) J Am Ceram Soc 84:2117
22. Aigbodion VS, Agunsoye JO, Kalu V, Asuke F, Ola S (2010) J Miner Mater Charact Eng 9:527
23. Bruck HA, Christman T, Rosakis AJ, Johnson WL (1994) Scr Metall Mater 30:429
24. Buresch FE (1978) ASTM STP 678:151
25. Coimbra D, Greenwood R, Kendall K (2000) J Mater Sci 35:3341
26. Davis JR (ed) (2004) Tensile testing ASM international. Materials Park, Ohio (ASTM C 1273)
27. Domínguez-Rodríguez A, Gómez-García D, Zapata-Solvas E, Shen JZ, Chaim R (2007) Scripta Materialia 56, 89
28. Gong J, Guan Z (2001) Mater Lett 47:140
29. Karch J, Birringer R, Gleiter H (1987) Nature 330:556
30. Kikuchi M (2011) Bioceramics Dev Appl 1:1
31. Kobayashi T, Niinomi M, Koide Y, Matsunuma K (1986) Trans Jpn Inst Met 27:775
32. Kobayashi T, Matsunuma K, Ikawa H, Motoyoshi K (1988) Eng Fract Mech 31:873

33. Lee MY, Brannon RM, Bronowski DR (2005) SAND 2004–6005 Unlimited Release February 2005
34. Li H, Bradt RC (1992) Diam Relat Mater 1:1161
35. Li H, Bradt RC (1993) J Mater Sci 28:917
36. Maria Berkes Maros (2009) Nikoletta Kaulics Helmeczi, Péter Araté and Csaba Balázsi. Key Eng Mater 409:338
37. Marschall J, Erlich DC, Manning H, Duppler W, Ellerby D, Gasch M, Mmat J (2004) Science 39:5959
38. Nakano H, Nagai S (1988) Ultrasonics 26:256
39. Naplocha K, Janus A, Kaczmar JW, Samsonowicz Z (2000) J Mater Process Technol 106:119
40. Nino A, Tanaka A, Sugiyama S, Taimatsu H (2010) Materialstransactions. Jpn Inst Met 51:1621
41. Robinson JN (1971) J Phys E Sci Instrum 5:171
42. Sakaguchi S, Murayama N, Kodama Y, Wakai F (1991) J Mater Sci Lett 10:282
43. Seshadri SG, Chia K-Y (1987) J Am Ceram Soc 70, C-242
44. Shimada M, Matsushita K, Kuratani S, Okamoto T, Koizumi M, Tsukuma K, Tsukidate T (1984) J Am Ceram Soc 67:C-23
45. Shimazu T, Miura M, Isu N, Ogawa T, Ota K, Maeda H, Ishida EH (2008) Mater Sci Eng A 487:340
46. Wang XD, Bednarcik J, Saksl K, Franz H, Cao QP, Jiang JZ (2007) Appl Phys Lett 91:081913
47. Weibull W (1939) A statistical theory of the strength of materials. Royal Institute for Engineering Research, Stockholm, p 151
48. Weibull W (1951) J Appl Mech 18:293
49. Williams JG (1980) Met Sci 345
50. Yao H, Ouyang L, Ching W-Y (2007) J Am Ceram Soc 90:3194

Special References

51. Bernal JD (1949) The physical basis of life. Proc R Soc Lond 357A:537
52. Ferris JP (2005) Catalysis and prebiotic synthesis. Mol Geomicrobiol 59:187
53. Fraser DG, Fitz D, Jakschitz T, Rode BM (2011) Phys Chem Chem Phys 13:83
54. University of Oxford (2011) Did clay mould life's origins by Cath Harris, 01 Apr 11

Chapter 2
Ductile Ceramics

Abstract Not all ceramics are brittle at room temperature. There are some ceramics which are ductile at ambient temperatures. Such ceramics, for example are single crystals MgO, $SrTiO_3$, etc. They undergo plastic deformation and by dislocation motion slip lines are observed on the deformed specimens. In pure MgO at room temperature, dislocations are very mobile at comparatively low stresses. Changing the microstructure, possibly by alloying, the mobility of dislocations may be reduced and an increase in strength may be achieved. As usually observed, material undergoing plastic deformation tend to strain harden, a feature observed also in ductile ceramics. Of the several factors influencing the strength properties of ductile ceramics, grain size is outstanding. Fine grained ceramics are desirable. Originally brittle ceramics show elongation at high temperatures which is a usual observation. There is a transition temperature from brittle to ductile behavior which depends on the ceramics. One of the common methods to determine the brittle to transition temperature is by impact testing, and for this purpose various sophisticated machines have been developed. An extraordinary phenomenon related to ductility is superplasticity, where very high values of strains can be achieved before fracture. Superplastic ceramics are oxide (zirconia) or non-oxide ceramics. Well-known superplastic ceramics are SiC and FeC. The common feature of superplastic materials is the requirement of very fine grains, namely, in the nanosize range.

2.1 Introduction

This chapter considers the mechanical properties of ductile ceramics, which can be grouped into three categories as classified below:

(a) Ductility at elevated temperatures. As is commonly known, some ceramics are ductile at high temperatures and it is meaningful to consider brittle–ductile transition. This was mentioned in Chap. 1, Sect. 1.11 on the Impact Testing of Ceramics.

J. Pelleg, *Mechanical Properties of Ceramics*, Solid Mechanics and Its Applications 213, DOI: 10.1007/978-3-319-04492-7_2,

(b) Ductile ceramics that show plasticity at ambient temperatures. Their features and, in some cases, modifications in their compositions are significant. Furthermore, certain additions to base ceramics, those which promote ductility (such as metals), are considered.
(c) Superplasticity in ceramics. Some ceramics show plastic behavior with elongations of $\sim$100 % and more. This is an important feature of this class of ceramics, because it is of structural interest for technical or industrial applications.

2.1.1 Ceramics at Elevated Temperatures

The most outstanding feature of ceramics in this category is the various degrees of plasticity that occur following the transition from brittle to ductile condition. This is the first transition to be discussed here, since all the other features are consequences of the brittle-to-ductile transition (henceforth: BDT).

2.1.1.1 Transition Temperature

The classic method for evaluating the transition temperature from a ductile to a brittle state is by impact testing. The basic reasons for using such a test are the high strain rate that can be achieved by impact and its simplicity. Though there are currently many other ways to vary strain rate, those who choose to perform impact tests can enjoy the use of modern, instrumented impact machines. For most ceramics which are brittle at room temperature (henceforth: RT), ductility is a high-temperature feature; thus, it is more meaningful to discuss BDT, rather than ductile-to-brittle transition (DBT), the more common nomenclature.

Relatively few impact strength data are available in the literature on ceramics and there are even fewer recorded experimental reports. The major limitations of performing such impact tests are the brittleness and low impact strength of ceramics at low and ambient temperatures, especially when the focus is on their applications at elevated temperatures, in light of their high strength properties.

For an early work on the determination of the transition temperature by impact, one may consult the paper by Kingery and Pappis [29]. Above a critical transition temperature, ductility increases markedly and ductile fractures are observed in ceramics. An illustration of the transition temperatures for a few ceramics may be seen in Fig. 2.1.

The experimental set-up for impact loading is shown in Fig. 2.2. The samples are cylindrical, 6 in. long and $^1/_2$ in. in diameter, supported on dense, sintered alumina knife edges across a $4^1/_2$ in. span.

The furnace is heated with silicon carbide resistant elements to 1600 °C. The samples were impacted in the furnace by a pendulum hammer having a 23.65 in. arm length and a 0.411 lbs wedge-shaped, sintered alumina head, as shown in Fig. 2.2.

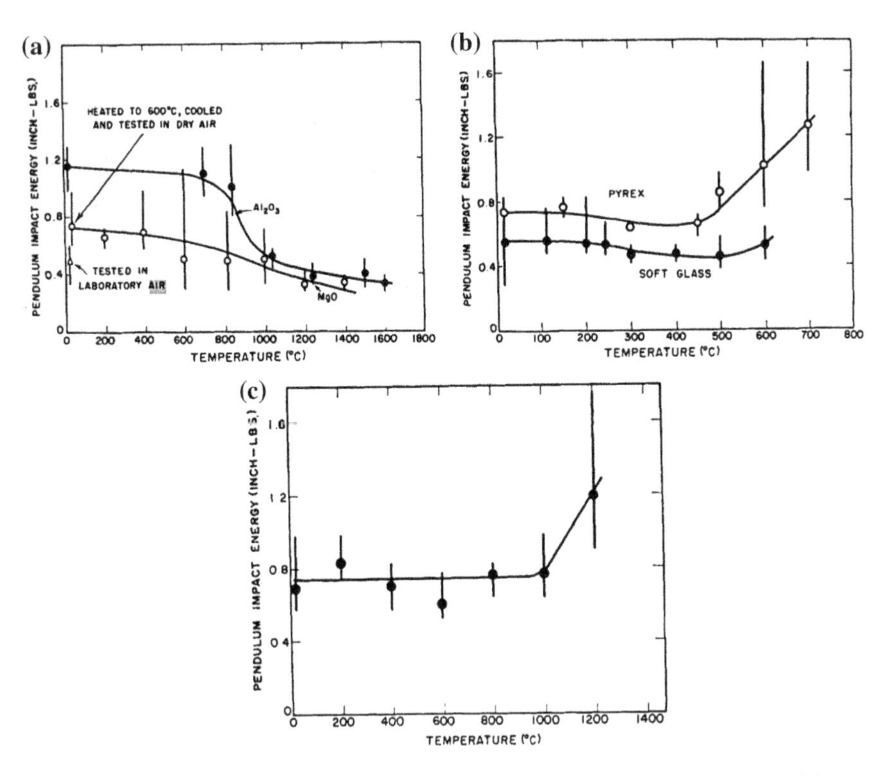

Fig. 2.1 Impact strength of: **a** Al_2O_3 and MgO; **b** Pyrex and soft glass; **c** semivitreous whiteware body [29]. With kind permission of Wiley and Sons

Tests were performed until fracture occurred after successive blows of the hammer. About 12 specimens were tested at each temperature. The pendulum velocity at impact was approximately 41 in./s.

These impact test outcomes are different from the familiar impact results for metals, where impact strength increases above the transition temperature; this is not observed in Al_2O_3 and MgO. Furthermore, in all these cases, an examination of the fracture surfaces indicates that brittle fracture has occurred, whereas the fracture in metals is ductile. The transition temperature in Al_2O_3 is about 900 °C. Both the soft glass and Pyrex-brand glass samples showed little variation of impact strength at temperatures up to 500 °C, which can be considered their transition temperatures. However, the results for Pyrex are questionable, because the specimen slumped at ~600 °C. The semivitreous, white-ware body tested showed no variation in impact strength at temperatures up to 1000 °C, its probable transition temperature. The 900 °C temperature, in the case of Al_2O_3, at which the impact strength decreases, is about the temperature at which plastic flow and creep commence. The expectation that, by analogy with metals, an increase in impact strength and ductile failure should also be observed at sufficiently high temperatures for ceramics was not observed in these impact tests.

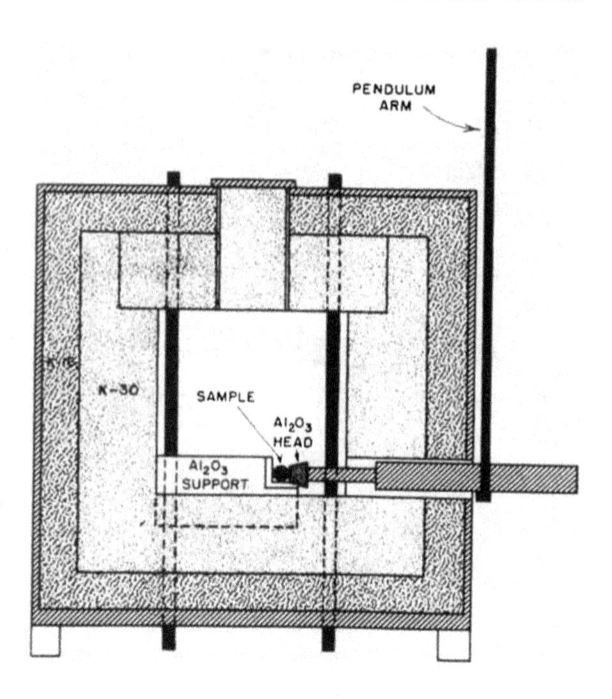

Fig. 2.2 The apparatus for impact testing at elevated temperatures [29]. With kind permission of Wiley and Sons

Recent experiments, however, determine transition temperature by varying the strain rates, whereas, in the impact tests, the information is obtained at the single strain rate characteristic of the given machine. Strain rate influences transition temperature both in single crystals and in polycrystalline materials. Moreover, in single crystals all the effects should be taken into account. In Fig. 2.3, the influence of orientation is indicated. Here, the BDT of precracked sapphire in four-point bending is shown as a function of orientation and strain rate from RT to 1500 °C. As can be seen in the figures below, T_c specimens fracture at the stresses indicated in the temperature range ~20–1000 °C. Above T_c, specimens exhibit general yield with a yield stress that falls with decreasing temperature. Transition occurs at the temperature at which the yield stress in bending is equal to the fracture stress for the specimen geometry used. T_c is the BDT of the specimens shown.

These tests were performed using an Instron Model 8561 (single screw) machine in air and the furnace was adapted to perform four-point bend tests. The rates indicated in Fig. 2.3 relate to crosshead displacement. Figure 2.4 shows the resolved shear stress at yield for the specimens tested at $\dot{e} = 4.2 \times 10^{-7}\ \text{s}^{-1}$ above T_c at the indicated orientations. The mechanism for slip is dislocation glide, which explains the orientation dependence of yield, as seen in Fig. 2.4. Thus, the BDT temperature, T_c, of the sapphire (Al_2O_3) varies not only with the strain rate, but also with the crystallographic orientation of the fracture plane.

The activation energy of the process controlling the BDT in sapphires, derived from the strain rate variation of T, is approximately 3.2 eV, close to that for dislocation glide. This was obtained by the Eq. (2.1):

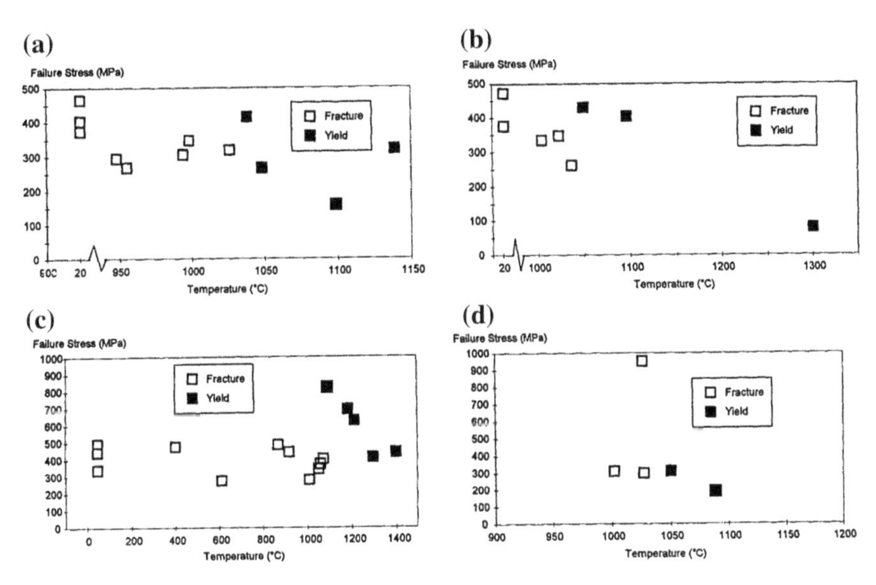

Fig. 2.3 Fracture stress versus temperature data: **a** specimens of orientation **A** (1012) (transition temperature, T = 1035 ± 5 °C), **b** specimens of orientation **B** (1105) (T = 1055 ± 5 °C), **c** specimens of orientation **C** (11$\bar{2}$0) (T = 1090 ± 10 °C), **d** specimens of orientation **D** (0001) (T = 1025 – 1050 °C). $\dot{e}_l = 4.2 \times 10^{-7} s^{-1}$ for (**a**), (**b**) and (**c**), and 1.3×10^{-6} s^{-1} for (**d**) [26]. With kind permission of Wiley and Sons

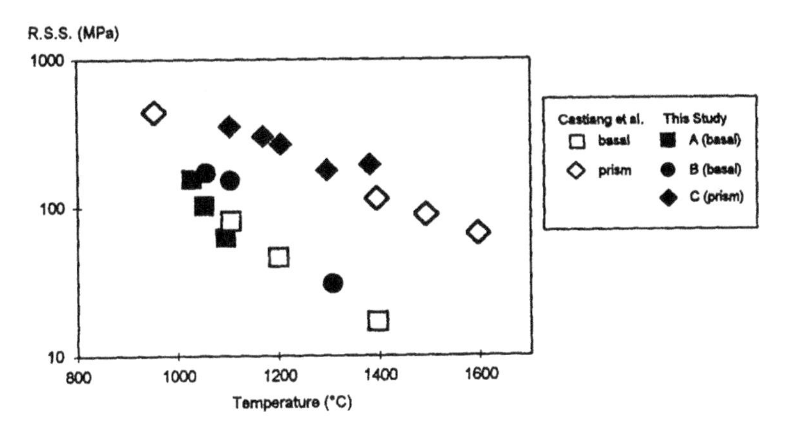

Fig. 2.4 Critical resolved shear stress for yield (above the transition temperature) in bending for all specimens tested at $\dot{e} = 1.3 \times 10^{-6}$ s^{-1} compared with data obtained by compression tests of Castaing et al. **C** specimens show prismatic slip and those of **A** and **B** show basal slip [26]. With kind permission of Wiley and Sons

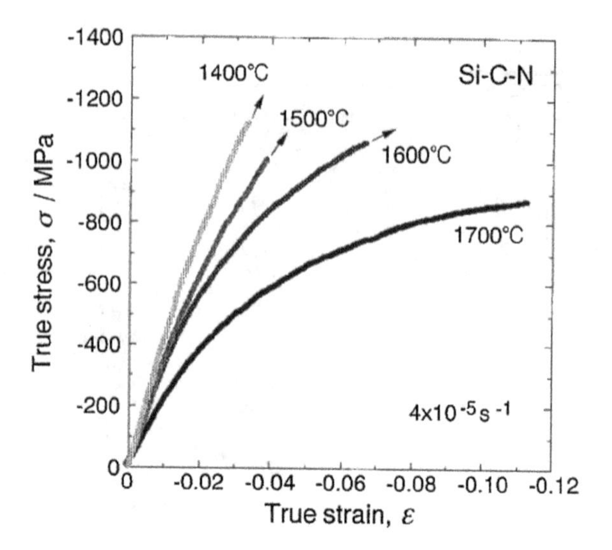

Fig. 2.5 Stress–strain curves of the Si–C–N ceramics tested in compression at high temperatures in the 1400–1700 °C range at a strain rate of $4 \times 10^{-5}\ s^{-1}$ [25]. With kind permission of Etsuko Hasebe of the editorial staff of the Japan Institute of Metals (JIM)

$$\dot{e} = \exp\left(-\frac{E_a}{kT}\right) \tag{2.1}$$

The plastic deformation of the sapphire occurred due to basal and prismatic slip during loading above T. Basal slip was found in **A**- and **B**-oriented specimens and prismatic slip in **C**-oriented specimens. The resolved stresses at yield (according to the author) are comparable to those measured by other researchers under compression in the appropriate slip system.

Note that the impact BDT in Al_2O_3, as indicated above, is ~900 °C, whereas for sapphires it was ~1000 °C. The reason for this may be that, in polycrystalline Al_2O_3, grain-boundary sliding contributed to the onset of plastic flow, which is absent in single crystal Al_2O_3. The effect of strain rate on BDT is observed also in superplastic deformation, discussed below in Sect. 2.2c.

2.1.1.2 Ductility and Strength

Not unlike the case of superplastic ceramics, ductility and strength relations are influenced by strain rate. The conditions of the experiment must be above the DBT to observe plastic flow, which is different for various ceramics. An illustration of the effect of strain rate and temperature on the strain (ductility) at some stress level can be seen in monolithic Si–C–N. Silicon–nitride-based ceramics are quite promising candidates for mechanical applications at elevated temperatures. Specimens were prepared by hot isostatic pressure (henceforth: HIP) of pyrolyzed powder compact at 1500 °C and 950 MPa, without any sintering additives. These compression tests were conducted at temperatures from 1400 to 1700 °C in a nitrogen atmosphere with a servo-hydraulic-type testing machine at constant crosshead speed in an induction heating furnace. In Fig. 2.5, stress–strain curves

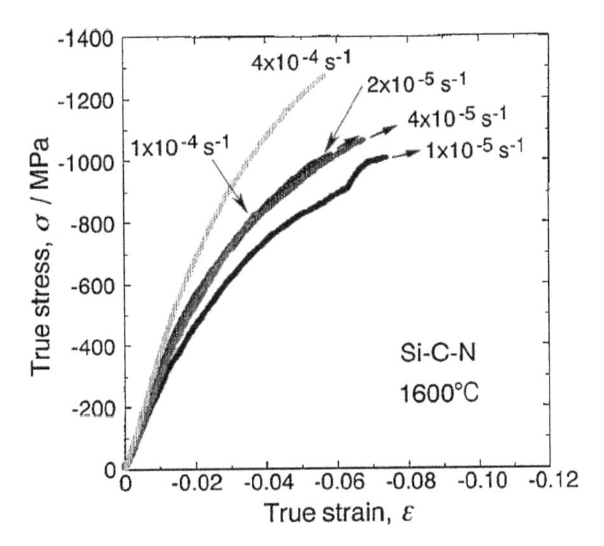

Fig. 2.6 Stress–strain curves of the Si–C–N ceramics tested in compression at 1600 °C at various strain rates [25]. With kind permission of Etsuko Hasebe of the editorial staff of the Japan Institute of Metals (JIM)

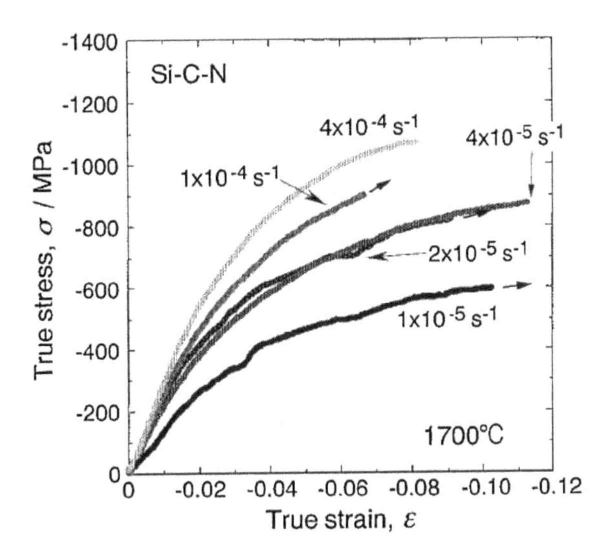

Fig. 2.7 Stress–strain curves of the Si–C–N ceramics tested in compression at 1700 °C at the strain rates indicated [25]. With kind permission of Etsuko Hasebe of the editorial staff of the Japan Institute of Metals (JIM)

obtained by compression tests at various temperatures and at an initial strain rate of $4 \times 10^{-5}\ s^{-1}$ are shown.

The arrows in the curves indicate the locations where the tests were interrupted. In curves when no arrows are shown, the specimen has failed at the end of the curve. Deviation from linearity may be seen even at 1400 °C, but at this temperature and at 1500 °C, only slight plastic deformation was observed, in spite of the high compressive stress over 1000 MPa. At the same stress, a compressive strain of ~7 % was obtained at 1600 °C. Yet, a compressive strain of about 11 % was achieved at 1700 °C at a lower compressive stress. Figures 2.6 and 2.7 illustrate compressive stress–strain curves at various strain rates at 1600 and 1700 °C.

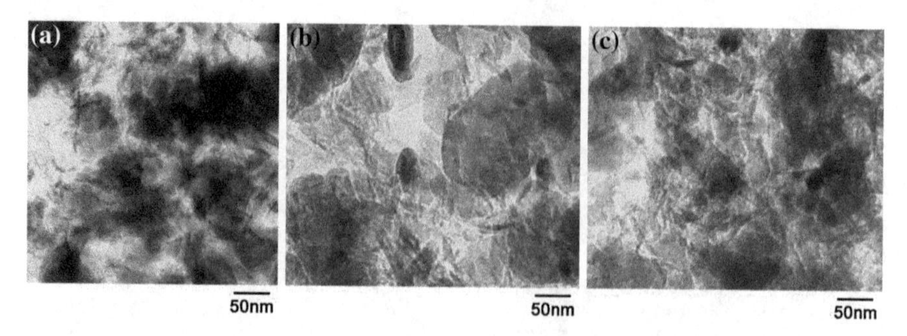

Fig. 2.8 TEM micrographs of the Si–C–N ceramics, **a** before test (as-HIP-treated), and after compression tests at **b** 1700 °C, $2 \times 10^{-5}\ s^{-1}$, and **c** 1700 °C, $4 \times 10^{-5}\ s^{-1}$ [25]. With kind permission of Etsuko Hasebe of the editorial staff of the Japan Institute of Metals (JIM)

The strain-rate exponent of the compressive stress in the strain-rate range above $4 \times 10^{-5}\ s^{-1}$ at 1600 and 1700 °C is about 0.1. Recall that the strain rate is related to stress as:

$$\sigma = C\dot{\varepsilon}^{m}|_{\varepsilon,T} \tag{2.2}$$

where m is strain-rate sensitivity or the strain-rate exponent and $\dot{\varepsilon}$ is the strain rate. Usually, m can be evaluated from the slope of a plot of log σ versus log $\dot{\varepsilon}$. In metals, m is low, <0.1, at RT, but may increase with T. Ceramics fail in pure compression by the coalescence of axially oriented microcracks. Lankford expresses the compressive stress for low rates of loading (10^{-5}–$15^{-1}\ s^{-1}$) as:

$$\sigma_c \propto \dot{\varepsilon}^{1/(1+n)} \tag{2.3}$$

Here, σ_c is the compressive strength, n is the stress-intensity exponent in the macroscopic tensile crack velocity relationship and n is in the range 50–200. Equation (2.3) expresses a relatively strain-rate insensitive process controlled by the thermally-activated growth of microcracks. This process is said to dominate the compressive failure of some ceramics, such as Al_2O_3, SiC and Si_3N_4. An additional equation (Grady and Lipkin) predicts a transition from the dependence expressed by Eq. (2.3) to one given as:

$$\sigma_c \propto \dot{\varepsilon}^{-1/3} \tag{2.4}$$

which occurs at a material-dependent characteristic strain rate that, for most ceramics, lies within the range of 10^3–$10^4\ s^{-1}$.

TEM micrographs of these specimens after the compression tests at 1700 °C are shown in Fig. 2.8, together with the results of the pretest sample.

The microstructure before the test (Fig. 2.8a) is a fine, two-phase structure, consisting of roundish Si_3N_4 crystalline grains, 20–80 nm in size, in a turbostratic, graphite-like phase. These phases are homogeneously distributed. The microstructures of the specimens after the compression tests, at both the initial strain

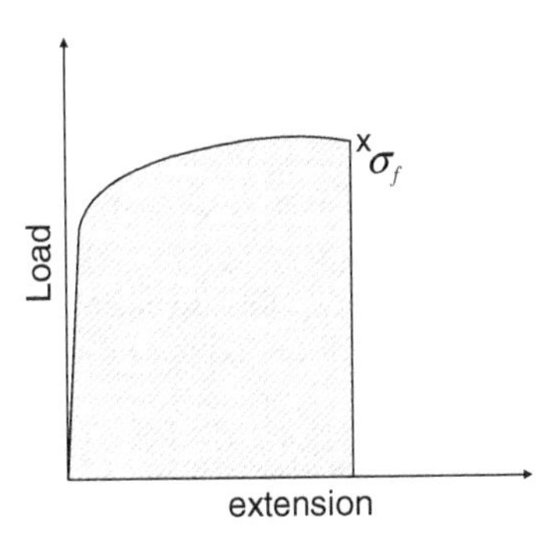

Fig. 2.9 Schematic curve for toughness

rates of $2 \times 10^{-5}\ s^{-1}$ (Fig. 2.8b) and $4 \times 10^{-5}\ s^{-1}$ (Fig. 2.8c), also consist of Si_3N_4 and graphite-like phases, similar to the structure before the test. However, the Si_3N_4 grains in the structures at both the initial strain rates were considerably coarsened during testing. The sizes of the Si_3N_4 grains after the compression tests were about 200 and 150 nm for the 2×10^{-5} and $4 \times 10^{-5}\ s^{-1}$, respectively.

Si–C–N ceramics maintained their mechanical strength up to 1500 °C and plastic flow was observed during the compressive stress tests at 1600 and 1700 °C, making Si–N-based ceramics one of the most promising candidates for mechanical applications at elevated temperatures.

2.1.1.3 Toughness

In defect-free solid materials, the ability to absorb energy, expressed by the area under the stress–strain curve, is known as 'toughness'. This represents the resistance of a material to fracture and is expressed as the amount of energy per volume having the units $\left[\frac{kg}{m^2} \times \frac{m}{m} = \frac{kg.m}{m^3}\right]$, given in SI system notation as joules per cubic meter (J/m^3). For such defect-free solids (as in the case of metals), the relation is:

$$U_T = area\,under\,the\,curve = \int_0^{e_f} \sigma\, de \tag{2.5}$$

Graphically, the area under the curve and up to the fracture is shown in Fig. 2.9.

In Eq. (2.5), the elastic contribution is neglected. Toughness and strength are related, but not necessarily the same, especially when brittle and ductile materials are considered. Whereas strength indicates how much force a material can support before breaking, toughness shows how much energy a material can absorb before fracture. A material may be strong and tough, if it fractures by withstanding a high force and exhibits high strain, but brittle materials with high strength (~equal to its yield stress) may be strong, but not tough, since their strain values are limited. Variables that greatly influence the toughness of a material are: strain rate (rate of loading), temperature and notch effect (for more details see [7]).

The fracture toughness [7] resulting from crack propagation is related to a critical stress intensity factor, K_c, which is a measure of fracture toughness representing the resistance of the material to failure from fracture initiated by a preexisting crack. One may add to the aforementioned variables that influence toughness: loading rate, composition, environment, microstructure and the tip geometry. A subscript is used to denote the crack opening mode to K_c and the equation can be expressed as:

$$K_{Ic} = \sigma\sqrt{\pi a B} \tag{2.6}$$

Here, a is the crack length and B is a dimensionless parameter. From Eq. (2.6), the critical stress, σ_f, is:

$$\sigma_f = \frac{K_{Ic}}{\sqrt{\pi a B}} = \frac{K_{Ic}}{\alpha\sqrt{\pi a}} \tag{2.6a}$$

and $\alpha = B^{1/2}$. B is a crack length and component geometry factor that is different for each specimen and is a dimensionless parameter. Expressions for α are tabulated for a wide variety of specimen and crack geometries, and specialty finite-element methods are available to compute it for new situations. K_{Ic} values are also used to calculate the critical-stress value, when a crack of a given length is found in a component. The critical-crack length is given from Eq. (2.6) as:

$$a_c = \frac{1}{\pi}\left(\frac{K_{Ic}}{\sigma B}\right)^2 \tag{2.6b}$$

For edge crack, a is the crack length or one-half crack length for internal cracks. In the literature, the above relations are often given in terms of Y, rather than B, with the same meaning. One immediately realizes the connection between this equation and the one derived by Griffith in his theory on brittle fracture.

The best quality ceramics contain imperfections, all of which have remarkable, but detrimental influences on the mechanical properties. Therefore, toughness or rather 'fracture toughness' is of critical importance for design purposes. Like impact tests, a very common method for testing toughness in ceramics is by the introduction of a notch, usually a V notch. Ceramics and ceramic-based composites, having high strength but low crack resistance, are considered for application due to their high strength. However, wide-scale application is still hindered by the presence of cracks. One of the accepted testing methods for the evaluation of critical stress intensity factor, K_{Ic}, is the single-edge, V-notched beam (henceforth: SEVNB) method. Several ceramics (of zirconia, alumina and silicon–nitride ceramics, zirconia and alumina single crystals, silicon carbide, etc.) were tested for fracture toughness by the SEVNB method [17]. V-notched specimens were tested using flexural tests and K_{Ic} values were calculated by means of three-point and four-point flexure tests. The load–deflection diagrams for V-notched specimens contributed to better understanding of the deformation behavior of ceramics at RT and 1300–1400 °C. Figure 2.10 illustrates the effect of V-notch

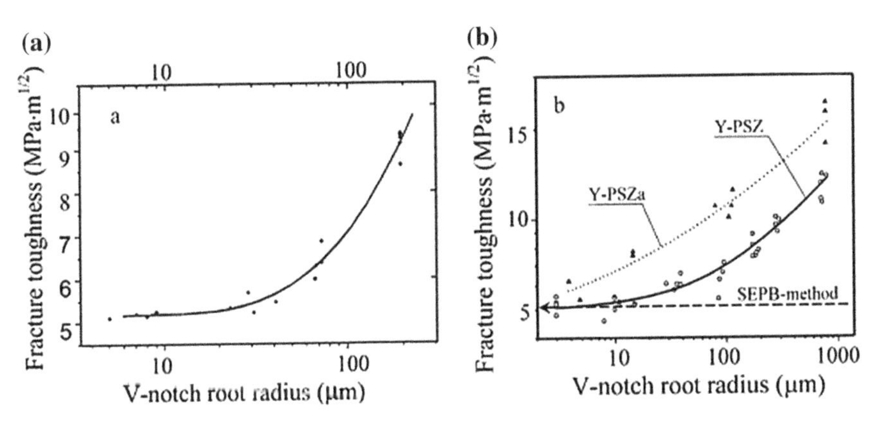

Fig. 2.10 Effect of V-notch root radii on the K_{Ic} values: **a** for Si_3N_4, and **b** for Y-PSZ ceramics [17]. With kind permission of Elsevier

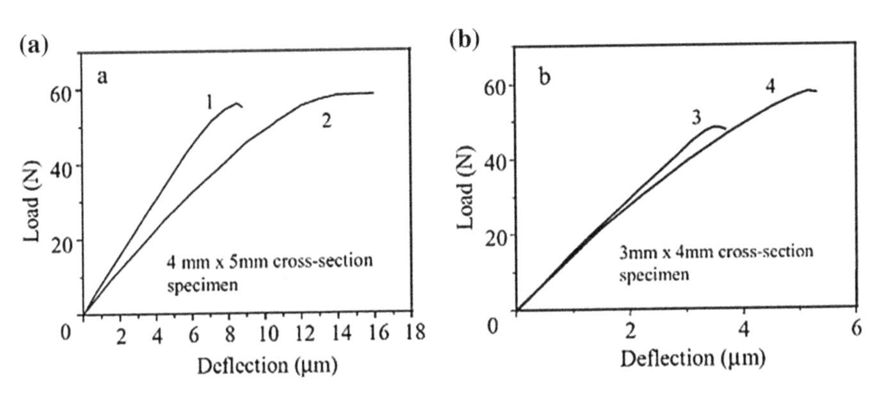

Fig. 2.11 Load-deflection of notched test: **a** Si_3N_4 + 30 % SiC + 3 % MgO; **b** SiC + 50 % ZrB_2 + 10 % B_4C. Specimens 1 and 3 tested at room temperature and specimens 2 and 4 tested at 1400 °C [17]. With kind permission of Elsevier

radii on K_{Ic} values for Si_3N_4 and yttria partially-stabilized zirconia (henceforth: Y-PSZ) ceramics, while the load–deflection relation is illustrated in Fig. 2.11.

Analysis of the data of the fractured specimens reveals that a fracture crack propagated from the points where additional stress concentrations were present. This confirms the assumption that the fracture of a loaded ceramic specimen starts from a small crack ahead of a machined notch root. It is believed that K_{Ic} is influenced more by the sharpness of the notch root, rather than by its shape. The data in Table 2.1 are from three to four-point flexure tests performed on several monolithic ceramics. Table 2.2 shows the K_{Ic} values at RT and high temperatures attained by the SEVNB method for these notched specimens.

Single crystals of zirconia and alumina were tested by SEVNB and by the single-edge notched beam (henceforth: SENB) method, and the results are shown

Table 2.1 Comparative fracture toughness tests (MPa $m^{1/2}$) of several ceramics [17] (with kind permission of Elsevier)

Test method	Three-point flexure (a/W ≈ 0.5)	Four-point flexure (a/W ≈ 0.2...0.3)	
		Our results	RRFT'97 results
Si_3N_4	5.5 ± 0.07(5)[a]	5.35 ± 0.16(5)	–
GPSSN	5.3 ± 0.04(5)	5.2 ± 0.18(5)	5.36 ± 0.34(129)
Si_3N_4 + 30 % SiC + 3 % MgO	2.27 ± 0.14(4)	2.40 ± 0.16(5)	–
SSiC	2.45(1)	2.66 ± 0.20(4)	2.61 ± 0.18(56)
SiC + 50 % ZrB_2 + 10 % B_4C	3.59 ± 0.12(3)	3.51 ± 0.15(3)	–
Al_2O_3-998	3.5 ± 0.05(5)	3.6 ± 0.06(5)	3.57 ± 0.22(135)
Y-PSZ	5.7 ± 0.17(5)	5.9 ± 0.19(5)	–

±Standard deviation
[a] The number of specimen tested (in parentheses)

Table 2.2 High temperature fracture toughness test results (SEVNB method) [17] (with kind permission of Elsevier)

Materials	K_{Ic} (MPa $m^{1/2}$)		
	20 °C	1300 °C	1400 °C
Si_3N_4	5.5 ± 0.1	4.2 ± 0.3	–
Si_3N_4 + 30 % SiC + 3 % MgO	2.27 ± 0.1	–	2.68 ± 0.1
SiC + 50 % ZrB_2 + 10 % B_4C	3.52 ± 0.1	3.63 ± 0.3	3.70 ± 0.1
Si_3N_4 [22][a]	5.6 ± 0.5	5.0 ± 0.4	–

[a] The notches were produced by diamond saw with V-shaped tip

Table 2.3 K_{Ic} values for single crystals obtained by SEVNB and SENB methods (MPa $m^{1/2}$) [17] (with kind permission of Elsevier)

Single crystals	Peculiarity	Elastic modulus (GPa)	Test method		Brittleness measure, χ	Index φ
			SENB	SEVNB		
Zirconia	Partially stabilized (3 % Y_2O_3)	245	9.33 ± 0.95	10.33 ± 2.17	1	0.9
Alumina	Specimen axis 45° to optical axis of crystal	403	2.31 ± 0.34	2.45 ± 0.29	1	0.94
	Specimen axis 90° to optical axis of crystal	410	3.19 ± 0.53	2.85 ± 0.50	1	1.12

in Table 2.3. In Table 2.3, χ is a measure of brittleness and is defined by the ratio of the specific elastic energy accumulated in the ceramics by the moment of fracture to the total energy spent on its deformation. χ was evaluated from stress–strain curves obtained during a four-point bend test.

The index of sensitivity, φ, to stress concentrations, which is equal to the ratio of the K_{Ic} values, obtained by the SEVNB and SENB methods is shown in Table 2.3. In the tests of elastic materials, φ ($\chi = 1$) is about 0.6, and in those of inelastic materials, ($\varphi < 1$), exceeding 0.9. Almost all the studies on the deformation behavior of V-notched ceramic specimens with a φ value of about 0.6 produced linear load–deflection diagrams or diagrams with small nonlinearity, as seen in Fig. 2.11 of lines 1 and 3, which are the results of comparatively slow crack growth.

A practical and relatively easy method for obtaining mechanical properties is by indentation tests. For ceramics, the most common methods of performing such tests are the Vickers and the Knoop hardness tests, which are very attractive, practical and relatively less expensive than the other tests discussed in Chap. 1. Assessing the toughness of ceramic materials by means of indentation testing is often done, also due to the ease of performance and low cost of conducting the measurements. Vickers hardness can be expressed [7] as:

$$DPH = HV = 1.854\frac{P}{d^2} = \alpha\frac{P}{d^2} \tag{2.7}$$

P is the load and d is the diagonal. α is a numerical factor (1.854) that depends on shape and it is quoted as ~2 for ceramics, as a consequence of using the projected area of the indenter contact with the surface plane. Vickers or Knoop indentations introduce cracks into the ceramics the sizes of which may be measured. The sizes of these artificial surface cracks are related to K_{Ic}. In particular, the lengths of these impression cracks are related to K_{Ic} and the connection between them has been evaluated, for example, by Anstis et al. [11]. A Vickers indentation is performed on a flat ceramic surface so that cracks develop around the indentation. By measuring the crack lengths, it is possible to estimate K_{Ic}, which are in inverse proportion to the toughness of the material. The crack-length method for evaluating toughness, according to Anstis et al. [11] is given by:

$$K_{Ic} = 0.016\left(\frac{E}{H}\right)^{1/2} \times \frac{P}{c^{3/2}} \tag{2.8}$$

Thus, toughness, measured and expressed by K_{Ic}, is dependent on the elastic modulus, E, of the material, its hardness, H, (microindentation is often preferable for the proper evaluation of the indentation crack), crack length, c, and the applied load. Anstis et al. [11] employed a two-dimensional fracture mechanics analysis. The crack length, c, is measured from the center of the impression to the crack tip in meters; E is in GPa and H is the Vickers hardness in GPa. The height of the opposite triangular faces is h. It is clear that under small indentation loads, only small cracks form, as indicated schematically in Fig. 2.12. Actual Vickers indentation cracks are shown in Fig. 2.13. Equation (2.8) is often also expressed as:

$$K_{Ic} = \alpha\left(\frac{E}{H}\right)^{1/2} \times \frac{a^2}{c^{3/2}} \tag{2.8a}$$

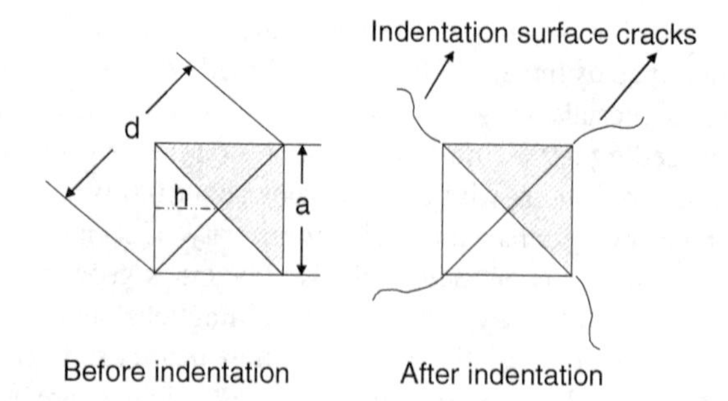

Fig. 2.12 Schematic indentation of a Vickers test before and after indentation. Crack resulting from the indentation are shown

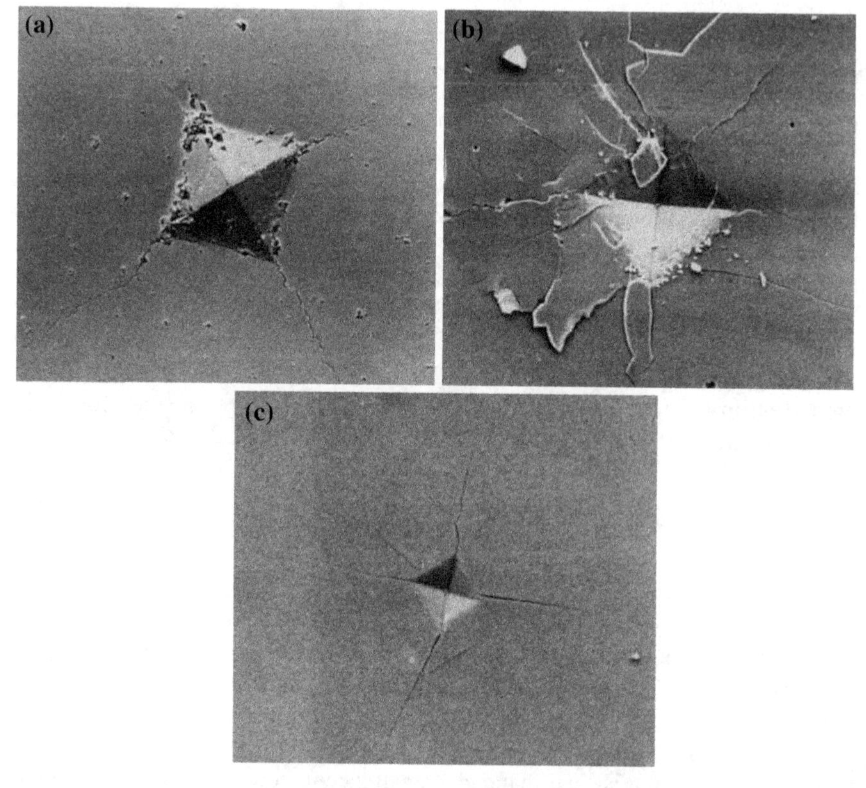

Fig. 2.13 Scanning electron micrographs of radial crack systems of Al_2O_3 in 3 modifications **a** AD999 (P = 50 N), **b** Vi (P = 50 N), and **c** sapphire (P = 10 N), showing the effect of increasing grain size on pattern definition; width of field 200 μm [11]. With kind permission of John Wiley and Sons

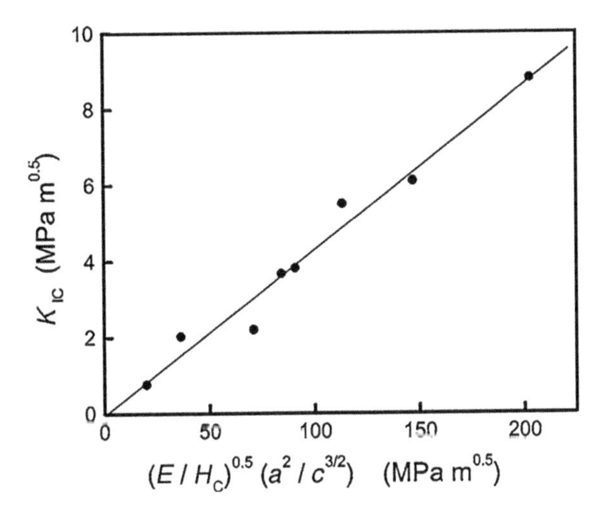

Fig. 2.14 Plot of K_{IC} determined with conventional methods versus the quantity $(E/H_C)^{0.5}(a^2/c^{3/2})$ [19]. With kind permission of Springer

since Vickers hardness may be given in terms of the half-diagonal, a, as:

$$H = \frac{P}{2a^2} \tag{2.7a}$$

by substituting the value of P from Eq. (2.7a) into Eq. (2.8), one obtains Eq. (2.8a).

Equation (2.8a) is expressed in Fig. 2.14 in terms of K_{Ic} versus $\left(\frac{E}{H_c}\right)^{12}\left(\frac{a^2}{c^{3/2}}\right)$.

More about toughness and fracture toughness, expressed in terms of K_{Ic}, will be discussed in the chapter on Fracture in Ceramics.

2.1.1.4 Grain Size Effect

Various strength properties of materials (especially metals) are related to the grain size effect. The well-known, empirical Hall–Petch (henceforth: H–P) relation addresses the grain size effect of these properties [7], expressed as:

$$\sigma_y = \sigma_0 + \frac{k_y}{\sqrt{d}} \tag{2.9}$$

σ_y is the yield stress, σ_0 represents resistance to dislocation glide, k_y is a measure of dislocation pile-up behind an obstacle (a grain boundary, for example) and d is the size of the grain. One may ask how this relates to ceramics. Indeed, various mechanical tests have indicated that the H–P relation also applies to ceramics in many cases [10, 9] and, thus, this relation has been successfully extended to the study of ceramics, as well. Various mechanical properties have been applied to test this relation, but a very common property, hardness, is very often used to indicate the H–P concept. Figure 2.15 shows H–P variation with the grain size of

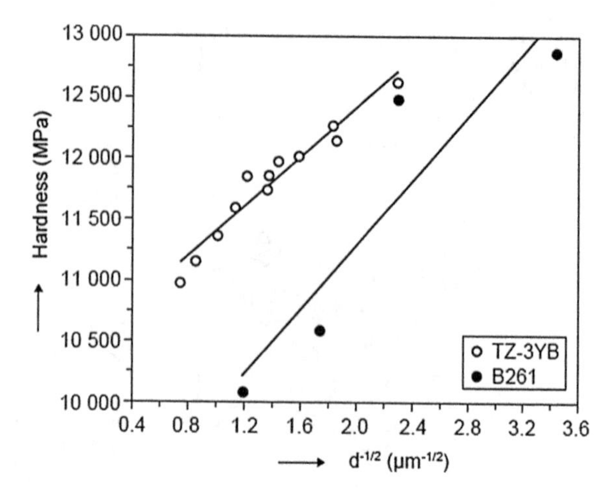

Fig. 2.15 The dependence of the hardness of TZ-3YB and B261 ceramics on the inverse square root of grain size [10]. With kind permission of Dr. Trunec

Fig. 2.16 SEM micrograph showing the microstructure of the TZ-3YB ceramics sintered at 1400 °C for 2 h [10]. With kind permission of Dr. Trunec

yttria-stabilized tetragonal zirconia polycrystals (TZ-3YB) and zirconia nano-powder (B261).

The hardness of TZ-3YB ceramics clearly decreases with increasing grain size from HV = 12620 MPa at a grain size of 0.19 μm to HV = 10971 MPa at a grain size of 1.79 μm. The hardness of B261 ceramics shows a higher dispersion, but the linear fit seems to be quite reasonable. These hardness values were determined by Eq. (2.7). The values in the graph may be reasonably expressed by a line ($r^2 = 0.94$), which means that this dependence follows the H–P relation. Figure 2.16 illustrates the microstructure of TZ-3YB.

Miyoshi [36] has confirmed the H–P relation by measuring Wickers hardness as a function of $d^{-1/2}$ for almost the entire grain-size range investigated. Furthermore, using bending strength tests, Rothman et al. [9] reported the adherence to the H–P relation even for spinel, such as magnesium aluminate ($MgAl_2O_4$) (Fig. 2.17).

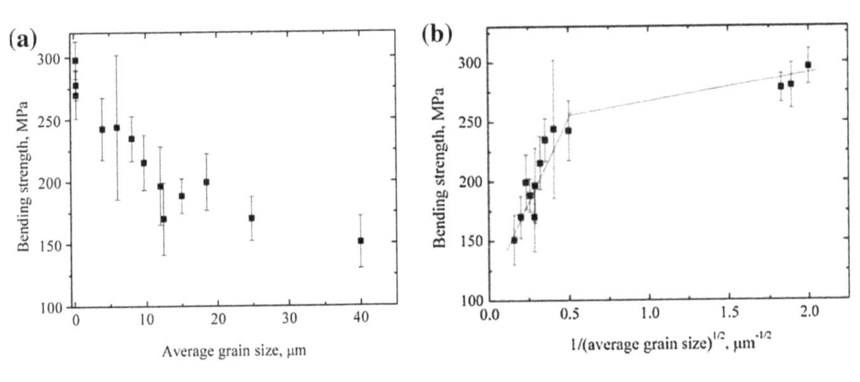

Fig. 2.17 **a** Bending strength as a function of average grain size; **b** shows agreement between the experimental data and the Hall–Petch relation [9]. With kind permission of Mr. Rothman for the authors

The grain-size dependence of mechanical properties is also observed in ceramics. Thus, one can state that the H–P relation is a general observation of most materials, metals and ceramics alike. Deviations from the linear H–P plots, known as the inverse H–P trend, observed in metals having small dimensions, will be discussed later on in the chapter dedicated to that topic.

2.1.1.5 Addition of a Second Phase

A second phase may be either in solution in the matrix or in a state of dispersion. Often, a soluble second phase is in an undissolved stage, since the processing, mainly the thermal treatment, occurs at a low temperature or for an insufficient amount of time for the induction of complete solubility. A second phase may be another ceramic, polymer or metal in various shapes and sizes. Fibers and whiskers are also often used for strengthening ceramic base materials. Here, the focus is on the mechanical properties and not on the influence of a second phase on other physical characteristics (e.g., electrical, magnetic or optical). Furthermore, the method of fabrication of ceramics necessary to densify compacted powder samples (green bodies) in order to form a continuous three-dimensional (henceforth = 3D) structure and, thus, to get ceramic pieces appropriate for the selected application is of critical importance. Moreover the second phase may be crystalline or in an amorphous stage. Although the present objective of adding a second phase is to enhance the mechanical performance of ceramics, it often occurs that the addition of a second phase weakens some mechanical properties, probably due for its promotion of pore formation. An insoluble second phase is a discontinuity in the matrix and, from this standpoint, pores and cracks, though they are not a genuine second phase, may be considered as such.

Usually, the consolidation of the constituents of a ceramic is done by means of a sintering process, which is a densification of the granular (or powder) compact

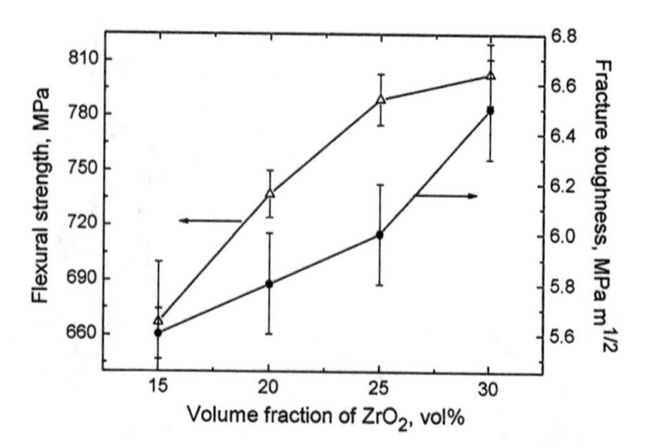

Fig. 2.18 Plots of flexural strength and fracture toughness (by SENB) of hot pressed ZrB_2–ZrO_2 ceramics [34]. With kind permission of Elsevier

by heat treatment. During this stage, the microstructures of the samples evolve into the initial stage of basic ceramics. Note that for maximum densification, HIP is a common practice.

Several examples will be presented to show the influence of a second phase, for example in the ZrB_2 case. There is a growing interest in ZrB_2-based ceramics for their outstanding properties: high melting point, high electrical and thermal conductivities, chemical inertness and good oxidation resistance. These properties make them attractive candidates for high-temperature applications, in which corrosion-wear-oxidation resistance is of interest, for example, for use as ballistic armor, coatings on cutting tools, electrical devices, nozzles etc [16]. Furthermore, refractory diborides exhibit partial or complete solid solution with other transition-metal diborides, which allows compositional tailoring of properties such as the thermal expansion coefficient and hardness.

The fracture toughness of ZrB_2, with and without additives, is generally in the range of 3.5–4.5 MPa $m^{1/2}$. For most applications, however, the value of toughness is unsatisfactory, which hinders its wider use and, therefore, the incorporation of various additives is expected to remedy this problem. ZrO_2 additive was found to improve the mechanical properties of ZrB_2 and to enhance its toughness. Figure 2.18 is an illustration of the effect of ZrO_2 additive in ZrB_2 on flexural strength and fracture toughness.

The densification of ZrB_2–ZrO_2 is improved with increasing amounts of ZrO_2, which is attributed to the smaller grain size of ZrO_2. Denser aggregates are expected to provide better mechanical properties, as indeed observed in Fig. 2.18. Figure 2.19b illustrates the effect of increasing the volume fraction of ZrO_2. Note that the increase of the relative density follows the trend of the plots shown in Fig. 2.18.

Considering the microstructure, energy dispersive spectroscopy (henceforth EDS) patterns reveal that it is characterized by the presence of a coarser and elongated ZrB_2 matrix and relatively finer and equiaxed ZrO_2 grains. This appears in Fig. 2.20. With the increase in the amount of ZrO_2, a denser microstructure is obtained. The fracture surface indicates that ZrB_2 grains fracture predominantly

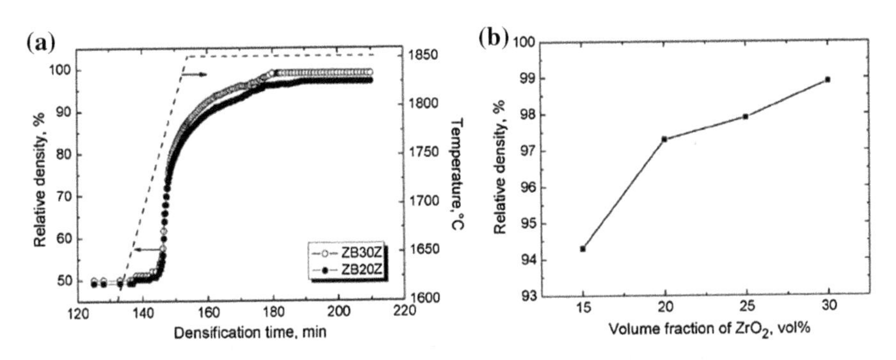

Fig. 2.19 **a** Densification curves of hot-pressed ZB20Z and ZB30Z; **b** relative density (%) of hot-pressed ceramics with increased ZrO_2 content from 15 to 30 vol% [34]. With kind permission of Elsevier

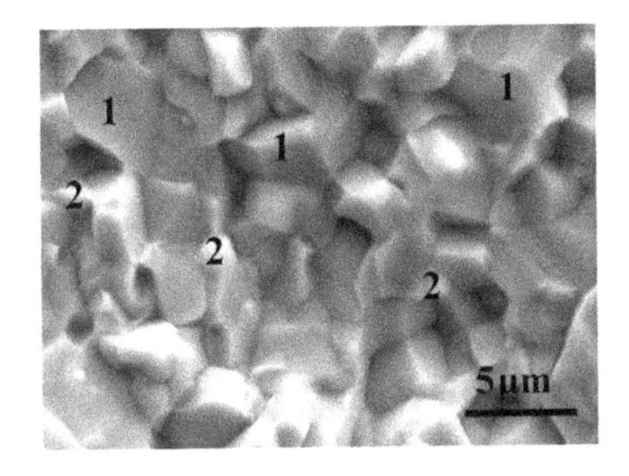

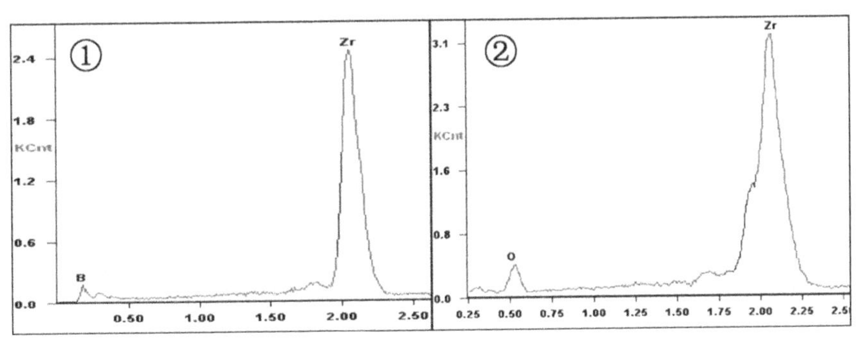

Fig. 2.20 Fracture surface of ZB20Z (ZrB_2-20 vol% ZrO_2), EDS patterns show that the bigger and coarser grains are ZrB_2, the finer and equiaxed grains are ZrO_2 [34]. With kind permission of Elsevier

transgranularly and ZrO_2 grains, which are dispersed among ZrB_2 grain boundaries, fracture intergranularly. The introduction of the smaller second phase of ZrO_2 effectively restrained the growth of grains during hot pressing, becoming more significant with the higher content of ZrO_2.

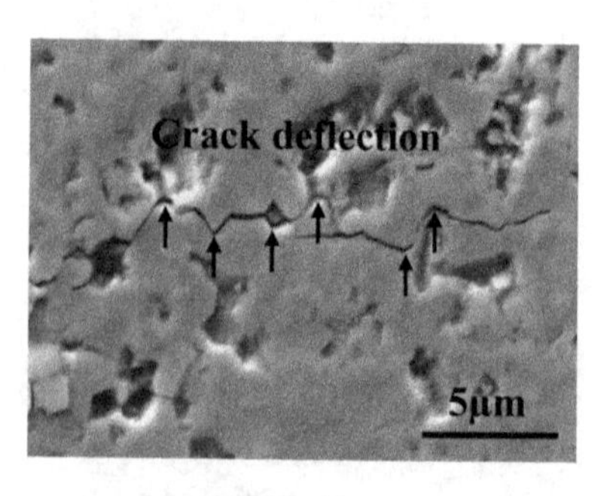

Fig. 2.21 SEM image of microcrack from Vickers indentation on the polished surface of ZB25Z [34]. With kind permission of Elsevier

Flexural strength improved with the increase in the amount of ZrO_2 and the strength increased from 667 MPa for the ZB15Z (ZrB_2-15 vol% ZrO_2) to 803 MPa for the ZB30Z (ZrB_2-30 vol% ZrO_2) ceramics, as indicated in Fig. 2.18. The finer grains are responsible for the improved strength, which can be associated with the H–P relation. Similarly, the increased ZrO_2 content plays an active role in the toughness of the ZrB_2-based ceramics. An increase from 15 to 30 vol% ZrO_2 increases toughness from 5.6 to 6.5 MPa $m^{1/2}$. There are two reasons for this increased toughness: deflection of the crack and stress-induced phase transformation. In the case of crack-deflection toughening, ZrO_2 grains hinder crack growth or its propagation, as indicated in Fig. 2.21. The increase in toughness due to phase transformation is associated with the transformation of tetragonal ZrO_2 into the monoclinic phase. The more tetragonal ZrO_2 is present in the ceramic, the more monoclinic ZrO_2 transformation will occur during the process. Both kinds of toughening, namely, from phase transformation and crack deflection, were largest in ZB30Z (ZrB_2-30 vol% ZrO_2). It is common to express toughness in terms of hardness measurements, involving the dimension of the crack formed on the surface in the vicinity of the indentation, as follows:

$$K_{Ic} = \eta \left(\frac{E}{H}\right)^{2/5} \frac{P}{(al^{1/2})} \tag{2.10}$$

In Eq. (2.10), η is a dimensionless constant for a given indenter geometry, provided the volume is conserved within the 'plastic zone' (adjacent to the indentation). E is the elastic modulus; H is the Vickers hardness; P is the indent load; 2a is the average indentation diagonal length; 2c is the crack length; and $l = c - a$. This relation is applied for toughness where the samples exhibit Palmqvist-type cracks ($0.25 < l/a < 2.5$). The hardness of the ceramic also increases with increasing ZrO_2 and can reach a value of 22.7 GPa under a load of 9.8 N having the composition of ZrB_2-30 vol% ZrO_2. The load dependence of hardness is quite pronounced and the nature of decrease in hardness with increased load has the same form in all ceramics.

A second phase might weaken the ceramics by reducing some of the mechanical properties, probably because, in some way, it promotes pore formation. Above, pores were described as a "discontinuity in the material" and, thus, as having an undesirable effect. In the following, the effect of ceria on the mechanical

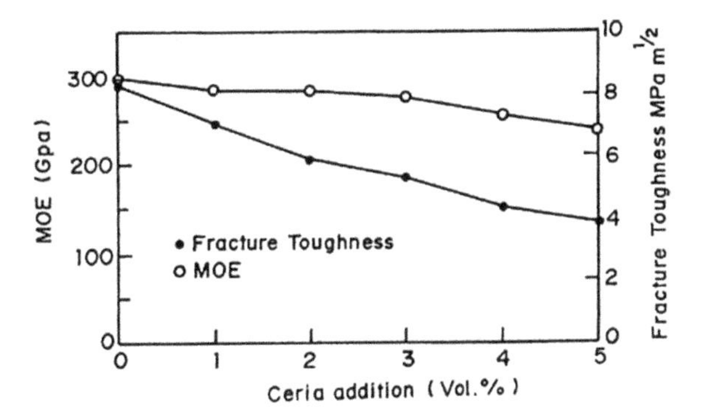

Fig. 2.22 Effect of ceria addition on fracture toughness and modulus of elasticity of ZTA (zirconia toughened alumina) [35]. With kind permission of Elsevier

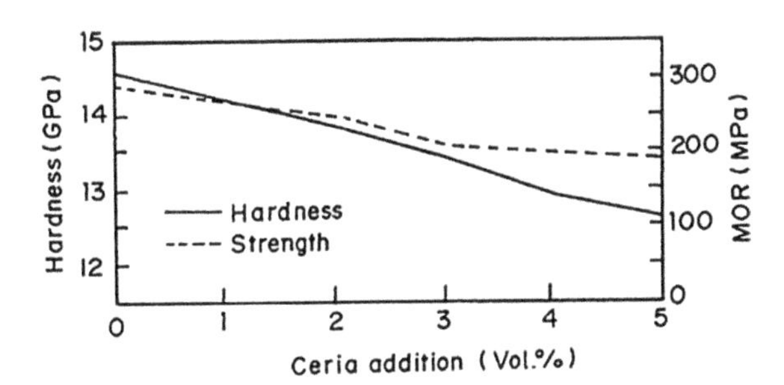

Fig. 2.23 Effect of ceria on hardness and strength of ZTA [35]. With kind permission of Elsevier

properties of yttria-stabilized zirconia-toughened alumina (henceforth: ZTA) will be considered to show how pores can, indeed, degrade mechanical properties. Figures 2.22 and 2.23 clearly show the degradation of the modulus of elasticity (henceforth: MOE), the fracture toughness, the hardness and the modulus of rupture (henceforth: MOR) with the addition of ceria, respectively.

The authors (Mangalaraja et al. [35]) claim that the possible reason for this reduction in strength is due to the high density of the microcracks. As the distances between the microcracks decrease, they near each other and coalescence occurs spontaneously, which substantially reduces the strength of the materials. Moreover, the decreased mechanical properties are found to be due to the higher apparent porosity, possibly resulting from the addition of ceria. The higher degree of apparent porosity is a result of the solid-state mixing of powders.

It was observed that the addition of ceria deteriorates the mechanical properties, including the fracture toughness of yttria-stabilized ZTA, although reports exist to the contrary [46] for polycrystalline yttria-stabilized ZTA, in which fracture toughness is increased by the addition of ceria.

2.1.1.6 Particle Size Effect

The mechanical properties of ceramics are influenced by the particle size of both the base ceramic and of the added phases. Thus, to obtain a fine-grained

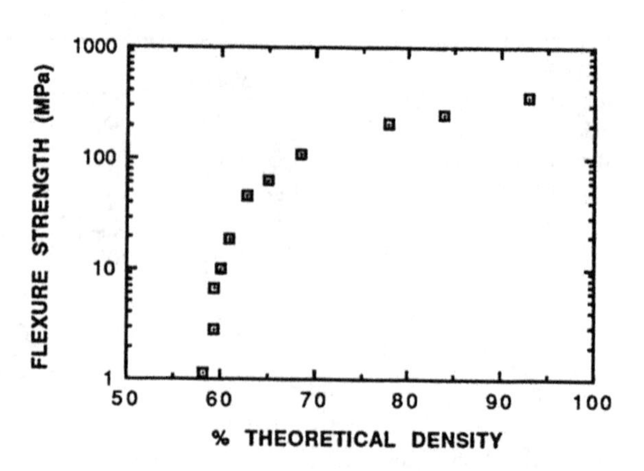

Fig. 2.24 Variation of flexural strength of alumina measured at room temperature after sintering to various temperatures (800–1600 °C). The increase in strength occurs at low theoretical density after which a leveling in strength occurs [5]. With kind permission of John Wiley and Sons

(henceforth: FG), uniform ceramic microstructure after sintering, the distribution of the particle sizes in the slurries of the components (while in the green stage) is of major importance for the production of a high viscosity substance. A proper particle-size distribution facilitates particle arrangement and the development of a relatively dense structural packing. Particle size affects densification, which is a prerequisite for the reduction of pores, thus enhancing the mechanical properties. The range of particle sizes is quite broad—from less than 100 nm to greater than 100 μm. However, the 4-sieve particle-size distribution (henceforth: PSD), which was borrowed from soil mechanics, is of great importance. Sieving is one of the oldest techniques for powder separation (based on size or some other physical characteristic) still in use today. It is among the most widely used and least expensive methods due to its relative simplicity, low capital investment, high reliability and the low level of technical expertise required for the determination of the PSD for a broad range of sizes. There are both dry and wet sieving processes. Typically, wet sieving is used for the analysis of particles finer than ~200 mesh (75 μm). For particle-size analysis, one can consult Special Publication 960-1 of the National Institute of Standards and Technology (henceforth: NIST).

Products used in ceramic or abrasive applications are generally manufactured from powders. The PSD has profound effects on the processing and functioning of these products, which include most oxides and minerals ranging from aluminum oxide to zirconium oxide. Laser diffraction, dynamic light scattering and acoustic spectroscopy have all been successfully utilized to characterize ceramic materials.

It has been mentioned that particle size affects densification, which is a prerequisite for the reduction of pores, thus enhancing mechanical properties. The effect of pores in alumina serves as an example of the direct influence of densification which is influenced by particle size. Figure 2.24 shows flexural strengths following subjection to varying degrees of densification. In addition, changes in Young's modulus may be seen in Fig. 2.25. Changes occurring with different degrees of densification are expressed by the two relations given below: Eq. (2.11), suggested by Lam et al. [30] and Eq. (2.12), its modification given by Mangalaraja et al. [35]:

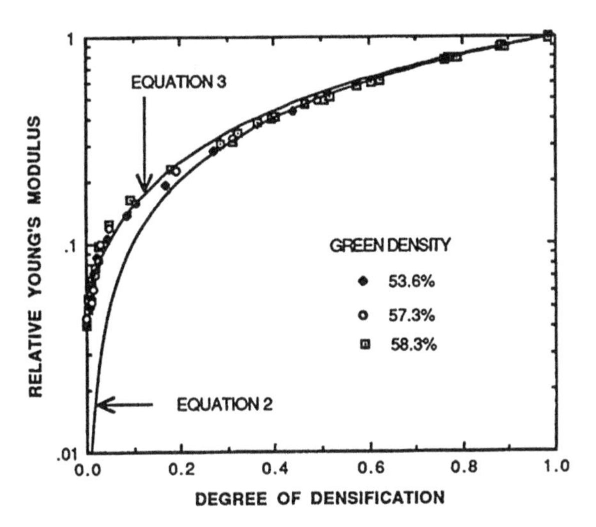

Fig. 2.25 Experimental Young's modulus data of high purity alumina for three different green densities expressed by the theoretical expressions mentioned above [5]. With kind permission of John Wiley and Sons

$$\frac{E}{E_0} = \left(1 - \frac{P}{P_g}\right) \tag{2.11}$$

and

$$\frac{E - E'}{E_0 - E'} = \left(1 - \frac{P}{P_g}\right) \tag{2.12}$$

As in the case of flexural strength, variation of the Young's modulus with the degree of densification increases during the initial stages of sintering.

In the above equations, E and E' are Young's modulus values of the porous and the theoretically dense materials, respectively, and P is the fractional porosity (not load). The subscript g refers to the green body. The right-hand side of the equation represents the degree of densification. In Eq. (2.12), E' is the Young's modulus at the onset of densification. The increase in strength properties, even with minimum densification, is an indication that strength property improvements may be made by controlling the sintering mechanism and the geometry of the particle structure.

Figure 2.26 shows the apparent porosity versus the average grain size of ceramic specimens sintered at ~1340 °C. These specimens were prepared from magnesite and bauxite by means of an in situ pore-forming technique.

In general the size of the pores depends, among other factors, on the particle size of the aggregate. Particle size is the most important parameter in the production of ceramic products; it must be optimized to ensure that the desired mechanical and physical properties are achieved. The majority of ceramic products are manufactured by the process of slip casting in a mold. Maintenance of the desired PSD requires control of the dispersion stability of the ceramic slip. Like all ceramic materials and castables, in alumina refractories a proper PSD is of importance,

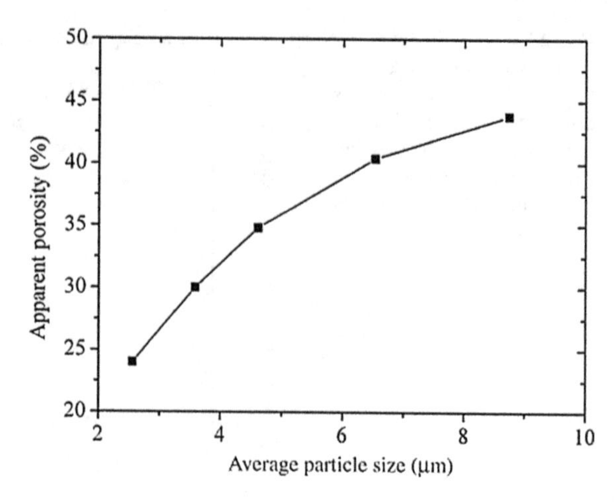

Fig. 2.26 Apparent porosity of specimens sintered at 1340 °C [50]. With kind permission of Professor Wen Yan

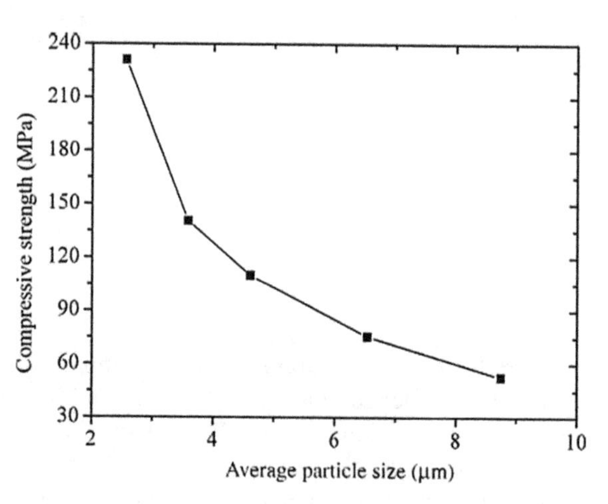

Fig. 2.27 Compressive strength of sintered specimes versus average particle size sintered at 1340 °C [50]. With kind permission of Professor Wen Yan

rather than using a single-sized component. This helps to create better packing, compared to the use of mono-sized particles. Besides influencing packing, PSD effects flow, apparent porosity and, hence, the strength of castables. Some research has been done to discover the relationship between particle size and physical properties. Figure 2.27 shows compressive strength versus average particle size.

Having established that the size of the pores depends on particle size, some other examples may now be considered, in which the effect of grain size is expressed in terms of pore density. In shell casting, ceramic molds are often used. Previous studies have indicated that zirconia is one of the least reactive materials and apparently holds promise as a mold refractory material, viable for metallurgical processing and the investment casting of TiAl alloys [15]. An important factor in zirconia mold properties is the role played by PSD on the packing. Information on this experimental procedure and the effect of PSD on zirconia mold properties may be found in the original paper of Chen Yan-fei et al. [15] FG,

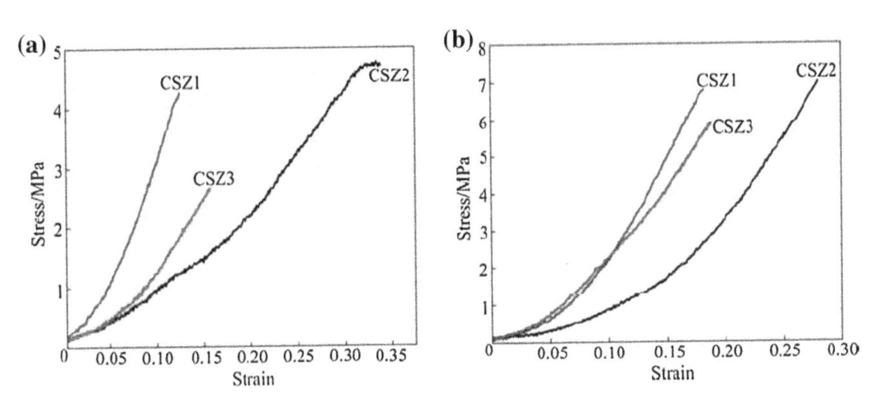

Fig. 2.28 Stress–strain relations for the CSZ1, CSZ2 and CSZ3 specimens; **a** green stage, **b** sintered [15]. With kind permission of Elsevier

uniform, sintered microstructure is obtained from slurries with median PSD and optimal viscosity, which facilitate particle arrangement and a structure with good packing. The mechanical properties of these molds were evaluated at RT by three-point bend tests of green and sintered (950 °C for 2 h) aggregates. Instron was used for the constant load test on five samples. The bending is given as:

$$\sigma_w = \frac{3FL}{2ah^2} \tag{2.13}$$

where F is the fracture load, L is the span length, and a and h are the width and thickness of the sample fracture area, respectively. Equation (2.13) is basically the same relation given in Eq. (1.122) from Chap. 1, Sect. 1.9, which was then given as:

$$\sigma_f = \frac{2P\frac{L}{2}c}{\frac{2tc^3}{3}} = \frac{3PL}{2tc^2} \tag{1.122}$$

In 1.122 P = F, t = a and $c^2 = h^2$.

The load–deflection relation in Eq. (2.13) may be expressed in terms of the stress–strain relation as:

$$\varepsilon = \frac{6h\delta}{L^2} \tag{2.14}$$

where ε and δ are the strain and deflection, respectively. Figures 2.28a and 2.28b are the stress–strain curves, in accordance with Eqs. (2.13) and (2.14) at the green and sintered stages, respectively, for the CSZ1, CSZ2 and CSZ3 specimens having different PSDs. It is interesting to compare the aggregates of these samples in the green stage and post-sintering (see Figs. 2.29 and 2.30). The PSDs of the powders used for these zirconia mold preparations appear in Fig. 2.31, showing that the median particle diameters are 20, 30 and 40 μm, respectively.

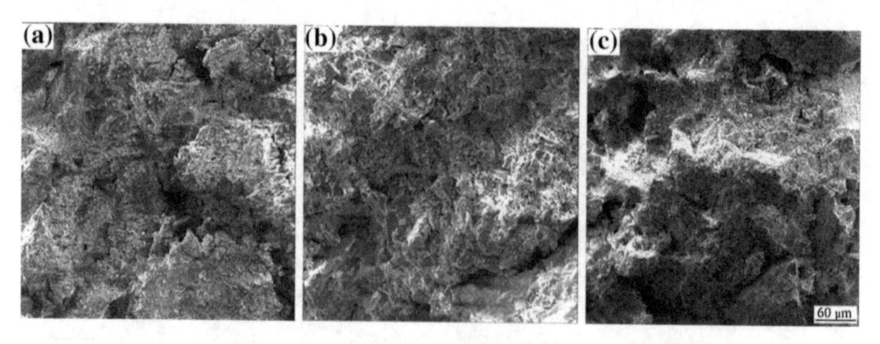

Fig. 2.29 Morphologies of fracture surfaces of green zirconia ceramic moulds with different PSDs: **a** CSZ1; **b** CSZ2; **c** CSZ3 [15]. With kind permission of Elsevier

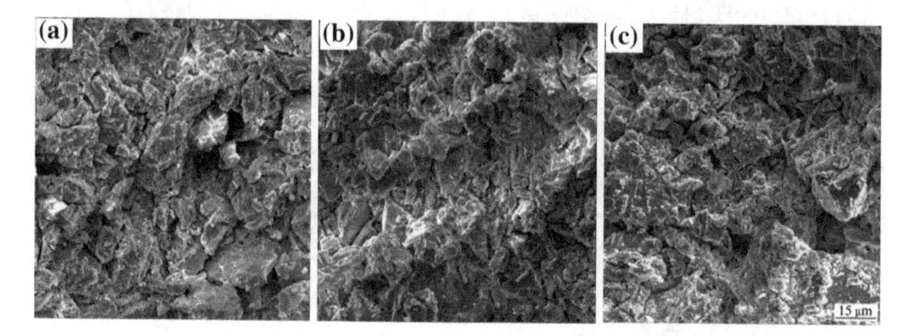

Fig. 2.30 Morphologies of fracture surfaces of sintered zirconia ceramic moulds with different PSDs: **a** CSZ1; **b** CSZ2; **c** CSZ3 [15]. With kind permission of Elsevier

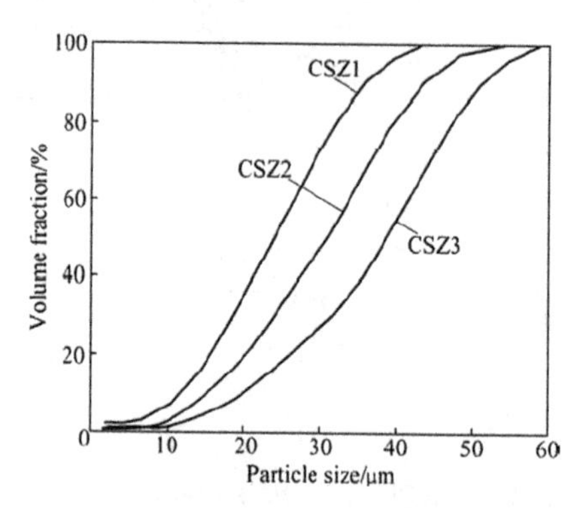

Fig. 2.31 Particle size distributions of zirconia powders [15]. With kind permission of Elsevier

The particle morphologies of the zirconia powders observed by scanning electron microscope (henceforth: SEM) are shown in Fig. 2.32.

Observations indicate that particle morphology and PSD greatly influence the quality of slurries in the reproduction of details and thin sections of ceramic molds. Consequently, good cast-metal quality may be obtained. It is clear from the results

Fig. 2.32 SEM micrographs of zirconia powder: **a** CSZ1; **b** CSZ2; **c** CSZ3 [15]. With kind permission of Elsevier

that the surface roughness of castings is directly related to the fineness of the investment powders. Decreasing the size of powders causes the formation of extremely small pores in the zirconia ceramic mold which prevents the deeper penetration of molten TiAl into these cavities under the same hydrostatic or centrifugal pressure and, thus, improving surface quality. Thus, pore size, in particle-size-dependent zirconia molds, greatly influences casting quality; the finer the particle size, the smaller the pore diameter in the ceramic mold. The relation between the zirconia powder characteristics determines the resulting mechanical properties of the ceramic molds. The bend strength of zirconia ceramic molds is directly related to the PSDs in the green and sintered ceramic (bar-shaped) specimens.

2.1.2 Ductile Ceramics at Low or Ambient Temperatures

This section deals with the features of ductile ceramics and, in some cases, the modifications in composition that induce ductility at low temperatures, as well as additives to base ceramics, such as metals, which also promote ductility.

Polycrystalline ceramics are of great interest for specific industrial applications, but the primary drawback of using ceramic materials in structural applications is their inherent brittleness, which results from the strong bonding between the metallic and non-metallic components. In general, most ceramics are brittle and

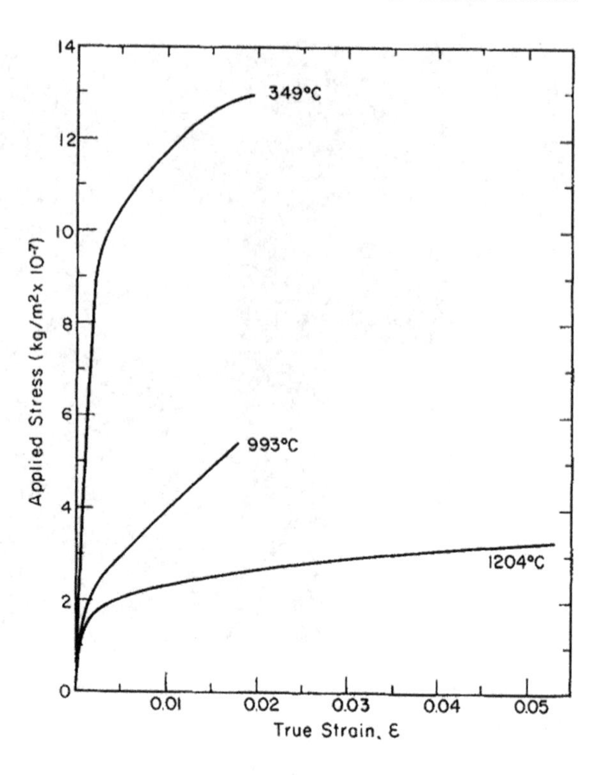

Fig. 2.33 Stress–strain curves for single crystals of MgO with a ⟨111⟩ loading axis [23]. With kind permission of Elsevier

various internal imperfections, such as porosity, reduce both strength and ductility. Due to the fact that most engineering ceramics are compacted from powders, the presence of some porosity is inevitable, which makes most ceramics very brittle. Furthermore, ceramics suffer from the presence of microcracks, which act as stress raisers. Generally, tensile stresses must be kept low, if sudden failure is to be avoided. Although the production of ductile polycrystalline ceramics (or other ionic solids) at RT has not yet been achieved in a satisfactory manner, it has been the objective of many research studies. Material purity is a contributing factor, though not necessarily a controlling one in governing the RT ductility of solids, in general, and of ceramics, in particular. Moreover, complications exist in polycrystalline ceramics, involving the presence of particles in the grain structure that may induce grain-boundary sliding, thus masking the possibility of real ceramic ductility in the absence of the contribution of grain-boundary sliding. Therefore, attempts have been made to study ductility in single crystals, in which a contribution from grain boundaries is ruled out.

A typical and much studied example of ductility in single crystals is MgO. Parker et al. [6] were the first to suggest that single crystals of MgO could be deformed at RT. Ever since the probability of RT ductility was confirmed, much attention has been given to evaluate the factors that affect such ductility and the mechanical properties of this structure. In addition to the other effects, crystal orientation also affects the deformation of single crystals. Figure 2.33 relates

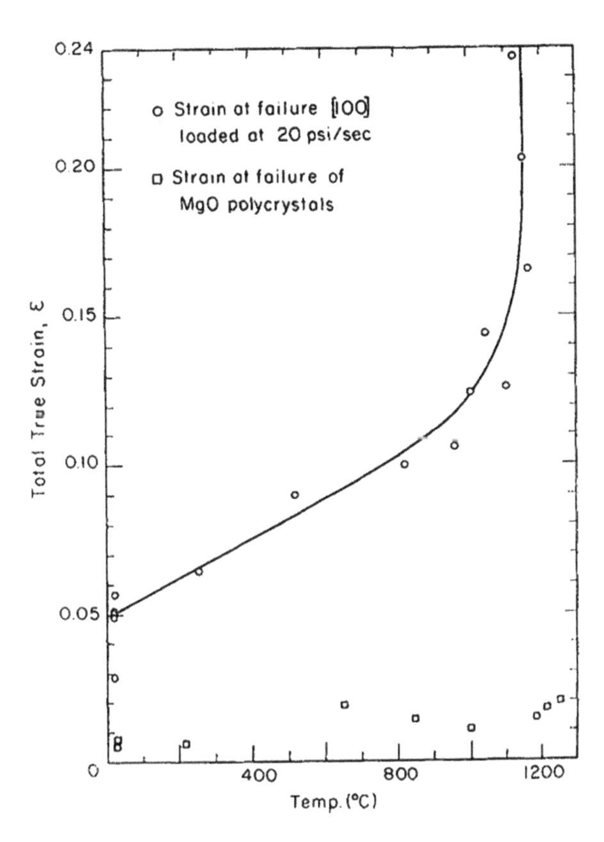

Fig. 2.34 Total strain versus temperature for single-crystal MgO with a ⟨100⟩ loading axis and for polycrystalline MgO [23]. With kind permission of Elsevier

applied stress to true strain at three temperatures. The lowest temperature in this figure is 349 °C, not strictly RT. However Fig. 2.34 also shows RT ductility (strain). Figure 2.35 shows stress–strain curves for small-grained polycrystalline material. When this polycrystalline material was yielded at RT using the Instron machine, yielding was followed by a decrease in stress at an increasing rate until final fracture occurred.

The temperature dependence of the bulk yield stress of the small-grained polycrystalline specimens is included in Fig. 2.34 for comparison with the single-crystal results. Note that the RT strain of the small-grained polycrystalline MgO is 0.005. Also notice that the loading axis orientation is ⟨100⟩ and not ⟨111⟩ as in Fig. 2.33. Figure 2.36 shows the variation with temperature of yield stresses for MgO single crystals with various loading axes. The yield-stress variation is also indicated for polycrystalline MgO. Typical behavior for single crystals loaded with a ⟨111⟩ axis at RT failed without plastic deformation.

A slip-band structure may be seen in Fig. 2.37, showing specimens deformed at RT and at 1240 °C. This band structure is similar to the dislocation band structure revealed by hot etching {100} faces of a ⟨111⟩ specimen just yielded at about 650 °C and then air-quenched (Fig. 2.38). In Fig. 2.34, the RT true strain at failure

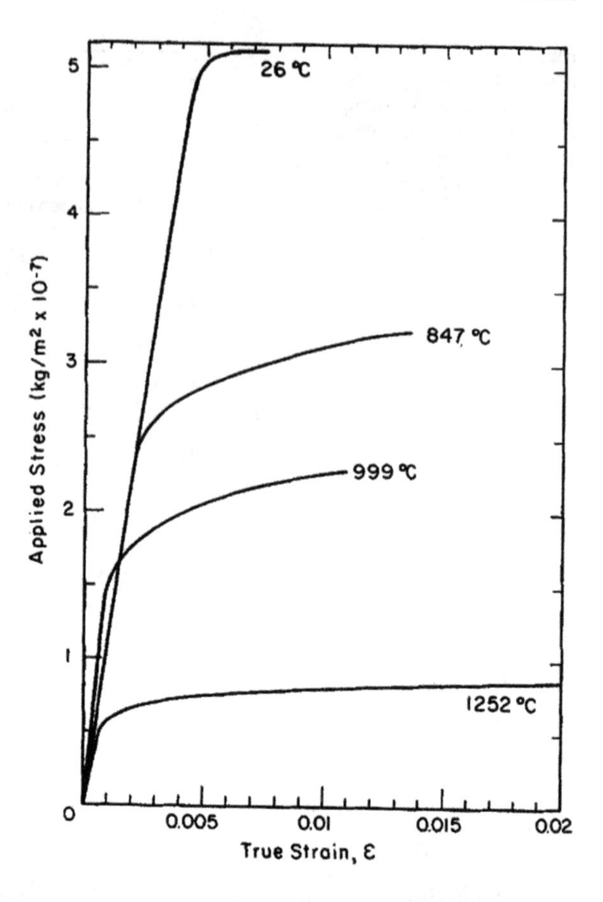

Fig. 2.35 Stress–strain curves far small-grained polycrystalline MgO [23]. With kind permission of Elsevier

of [100] loaded specimen at 20 psi/s was about 0.05. The polycrystalline MgO at RT had a strain at fracture of about 0.6 %, as seen in Fig. 2.34 and was about 2 % above 600 °C. The observed poor ductility of the polycrystalline MgO is attributed to cleavage fracture, slip non-uniformity and a lack of five independent slip systems, which is a requirement for polycrystalline ductility according to Taylor [45]. Above 600 °C, slip can occur on the {100} ⟨110⟩ slip system. At higher temperatures, stress-induced climb and high dislocation mobility inhibit cleavage fracture. Surface effects are extremely important and good ductility can be obtained only with specimens having carefully prepared, chemically polished surfaces. Thus, the ⟨110⟩ slip systems provide the additional slip systems necessary to satisfy Taylor's criterion. It was also found that MgO single crystals are ductile at RT and elongation values in excess of 10 % were obtained regularly [6]. Elongations of as much as 20 % on the tension side of a single-crystal bend-test specimen have also been reported. High purity is essential. Thus, purity and environmental effects play major roles in brittleness [41].

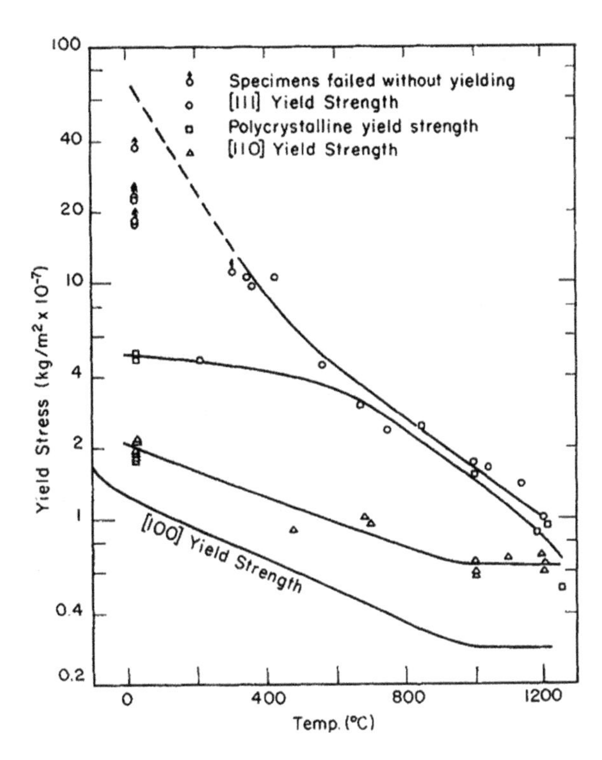

Fig. 2.36 Yield stress versus temperature for single-crystal and polycrystalline MgO [23]. With kind permission of Elsevier

Studies on single crystals, bicrystals and polycrystalline materials of MgO have shown that the strength of magnesium oxide depends on the availability of mobile sources. Research on single crystals has shown that the mechanical properties of magnesium oxide fall into two categories, namely, they are either extremely strong and elastic in the complete absence of mobile dislocation sources or relatively weak and ductile in their presence [43]. The mobility of a dislocation depends on a number of factors, such as crystal structure, bond character, temperature and microstructure. In pure magnesium oxide at RT, dislocations are very mobile at comparatively low stresses [42]. Changing the microstructure, possibly by alloying, the mobility of dislocations may be reduced and an increase in strength may be achieved.

As indicated above, the effects of impurities and surface reactions with components of the air exert control over ductility. It was predicted by researchers that a class of materials, normally considered brittle (i.e., ionic solids having cubic crystal structures) would possess a degree of ductility. Experiments performed on ionic materials indicate that face-centered cubic and body-centered cubic ionic materials can exhibit a considerable amount of ductility under controlled conditions, such as induced by impurities and surface effects. For instance, the ductility of MgO is shown in Fig. 2.39.

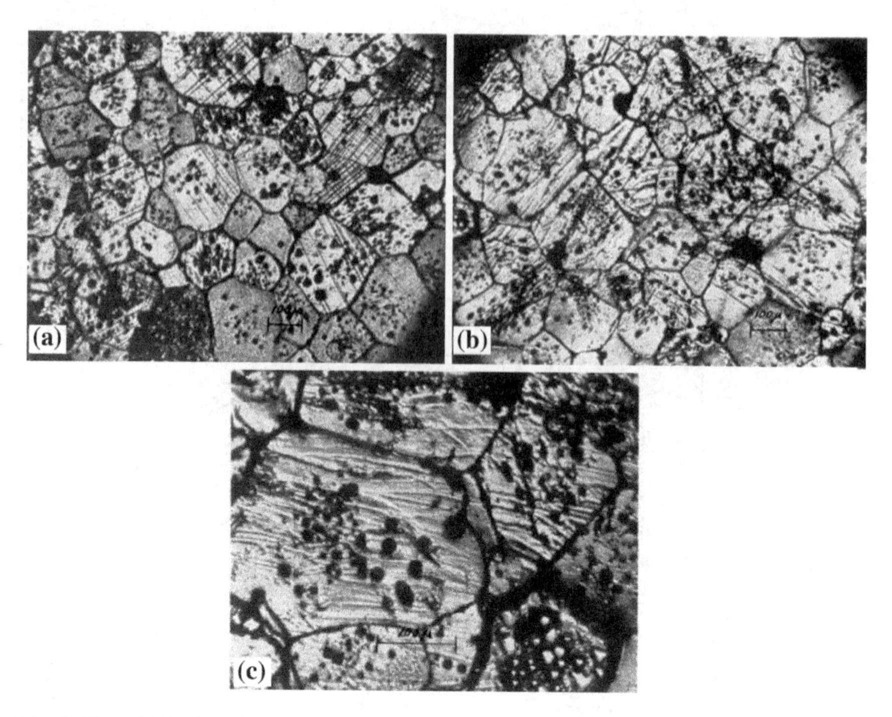

Fig. 2.37 **a** Etched grains of polycrystalline MgO deformed at room temperature, X95; **b** deformed at 1240 °C, X95; **c** deformed at 1240 °C, X190 [23]. With kind permission of Elsevier

Fig. 2.38 Etched {100} ⟨110⟩ slip bands (X75) [23]. With kind permission of Elsevier

The stress–deflection curve for a MgO single crystal is indicated in Fig. 2.40. To eliminate the environmental effect, these specimens were cleaved and tested under oil. The base material was of commercial grade, thus the crystals were impure,

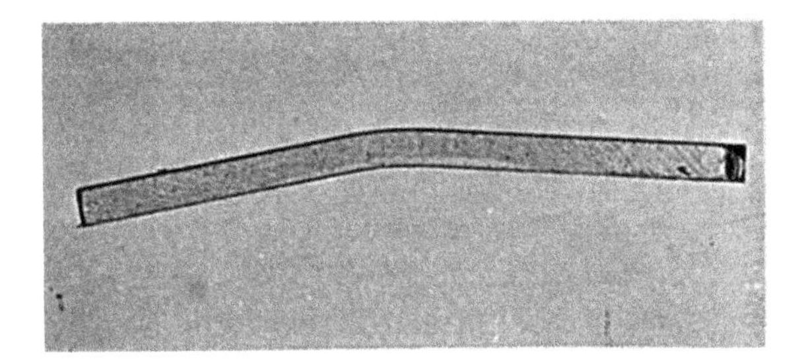

Fig. 2.39 MgO single crystal which was bent in air at room temperature [20]. With kind permission of John Wiley and Sons

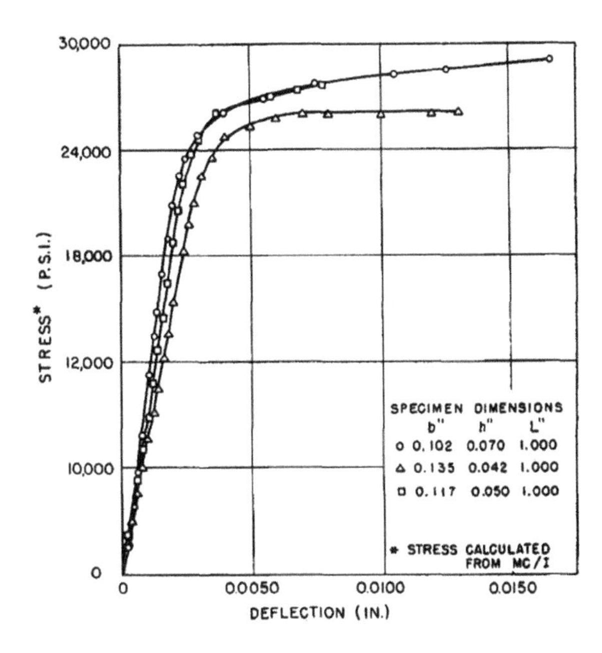

Fig. 2.40 MgO cleaved under oil, stored 48 h, and then tested under oil [20]. With kind permission of John Wiley and Sons

containing ~ 1/2 % or even more of foreign material (mostly silica). Nevertheless, the amount of elongation on the tension side of the specimens varied from 0 to 20 %, with an average of perhaps 5 % for the batch tested. In Fig. 2.40, three typical stress–deflection curves for MgO are seen. Undoubtedly, the variation in ductility was due, in part, to variation in purity. The effect of impurities and ambient gases, such as O or N, are assumed to be associated with the dislocation-impurity interaction. O or N can diffuse from the surface into the interior of the ceramic, acting like inherent impurities. It is well known that the reaction of the impurities with the dislocations and their pinning form Cottrell-like atmospheres. Dragging dislocations with their atmospheres and their immobilization are dependent on the amount of impurities and the amount of locks formed by impurity-dislocations interactions. Dislocation sources may have been activated by the high local stresses required to activate the motion of the dislocations having impure atmospheres. Eventually, when sufficient atmospheres form, the dislocation is immobilized and cleavage sets in. Thus, the free motion of dislocations gradually becomes more

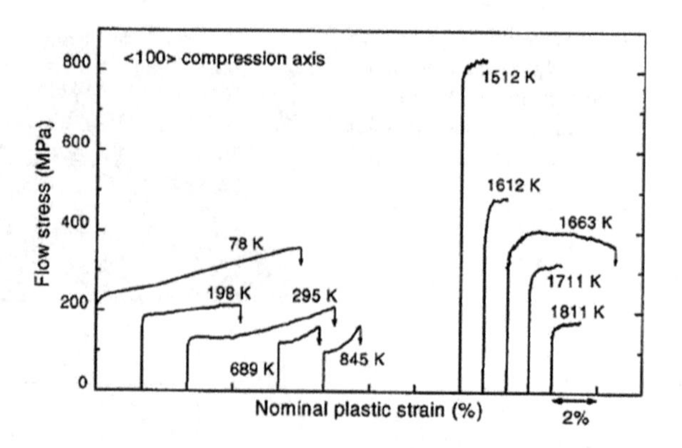

Fig. 2.41 Stress–strain curves of $SrTiO_3$ at different temperatures (arrows indicate load release) [13]. With kind permission of John Wiley and Sons

difficult and ductility becomes restricted, requiring higher stresses. The effect of embrittlement by impurities or by ambient gases depends on their amount.

One interesting case is that of the ceramic $SrTiO_3$, which belongs to the class of ceramics having a perovskite structure. This ceramic is expected to fail in a brittle manner at low temperatures, but, surprisingly, it was found to exhibit plastic behavior when deformed by compression in the range of 78–1050 K. Above this temperature, in the 1500–1800 K range, it behaved as a brittle material [13]. Such materials are often used as substrates for electronic devices for high-temperature superconductors. Therefore, there is a need to obtain information about the mechanical properties of $SrTiO_3$. Single crystals of $SrTiO_3$ were tested by compression in the ⟨001⟩ orientation at 78–1811 K. Figure 2.41 displays representative curves of the true stress versus the nominal plastic strain in $SrTiO_3$ specimens that were deformed at 78–1811 K.

Several features may be observed in Fig. 2.41 at several temperatures:

(1) The stress at the beginning of plastic deformation is weakly dependent on temperature;
(2) The stress plateau of plastic deformation, being the most pronounced at 296 K, decreases with increasing temperature and;
(3) The work-hardening rate after the short plateau increases with increasing temperature.

The most striking feature of the stress–strain curve at RT is the extended plastic deformation, reaching a plastic compressive strain of up to 8 % before fracture. At 78 K, a specimen can be plastically strained to 9 % before fracture. Figure 2.42 shows the two side faces of a specimen that has been deformed to 3 % at RT in transmitted polarized light. The bands of birefringence lie at an angle of 45° to the [001] compression axis. These structures are typical of plastically-deformed specimens at temperatures below 900 K; above 1500 K, no such structures can be detected within the deformed, transparent samples, since no plastic deformation has taken place. Figure 2.43 suggests that the temperature range under investigation may be subdivided into four regimes: (i) regime I ($T \leq 300$ K), the low-temperature regime, where σ_c (critical flow stress) decreases as the temperature

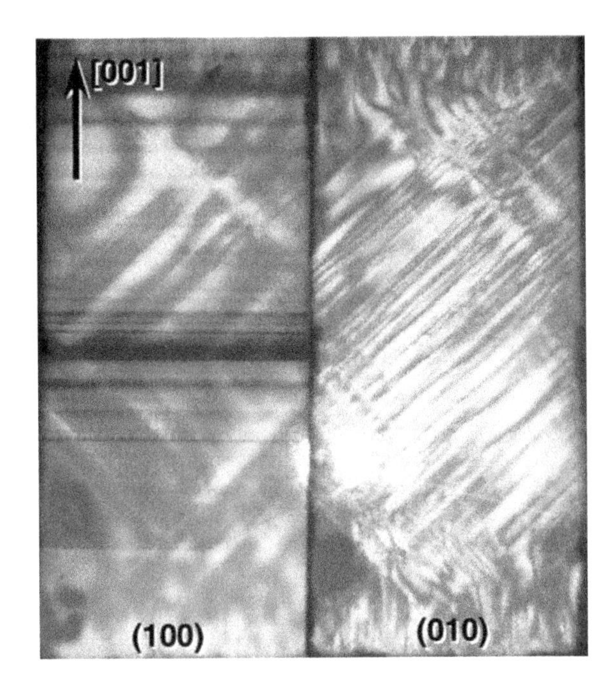

Fig. 2.42 Polarized-light micrograph showing single-crystal $SrTiO_3$ after 3 % plastic deformation at RT [13]. With kind permission of John Wiley and Sons

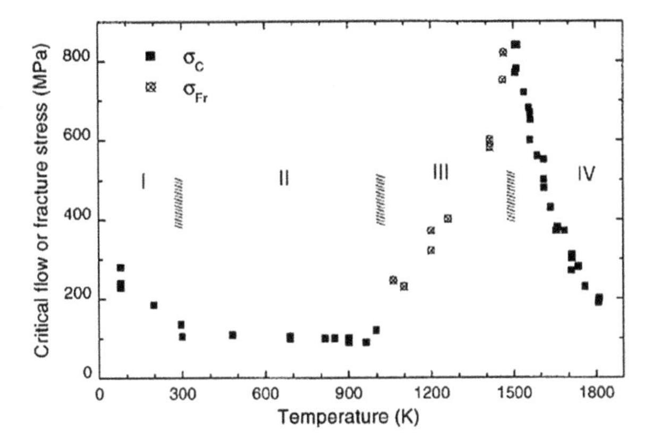

Fig. 2.43 Critical flow stress (σ_c) and fracture stress (σ_{Fr}), as a function of temperature [13]. With kind permission of John Wiley and Sons

increases and high ductility is observed; (ii) regime II (T = 300–1050 K), which is characterized by an almost-constant σ_c value and ductile behavior decreases as the temperature increases; (iii) regime III (T = 1050–1500 K), which is characterized by complete brittleness of the specimens, but an increase in σ_{Fr} (fracture stress) is observed with increasing temperature; (iv) regime IV (T = 1500–1811 K), where ductility occurs again, but σ_c rapidly decreases if the temperature is increased. Thus, a 'two-directional' transition phenomenon-a ductile–brittle-ductile transition—is observed in these ceramics.

It can be summarized that a class of ceramics materials, normally considered brittle could posses a degree of ductility depending on the production technique,

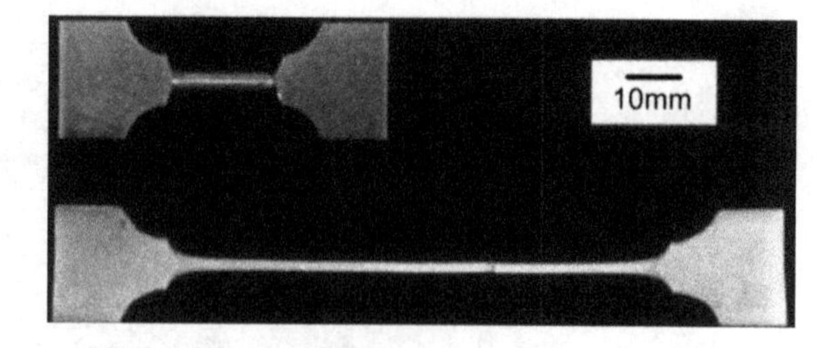

Fig. 2.44 Undeformed and superplastically deformed Si_3N_4. An elongation over 470 % is noted [47]. With kind permission of Elsevier

and among the many factors the impurity concentration exerts control over the degree of ductility. The restriction to use ceramic materials at low temperature by the transition from ductile behavior to brittle fracture can thus be overcome in cases when production occurs under strict controlled conditions.

2.1.3 Superplasticity in Ceramics

2.1.3.1 Introduction

A superplastic phenomenon occurs in solid crystalline materials, including ceramics, and is a state in which the material may be deformed before fracture and may reach large strains, well above 100 %, often even in the range of 200–500 %. Figure 2.44 shows superplastic behavior in Si_3N_4 with a 470 % elongation.

This usually occurs at high, homologous temperatures of about 0.5 T_m, where T_m is the absolute melting temperature. However, often a superplastic state is found in metals and alloys even at RT. An essential feature of materials exhibiting superplastic behavior is their fine grains. Superplastic materials may be thinned down, usually in a uniform manner, before breaking, without neck formation, unlike ductile metals, where necking is a common feature before fracture sets in. Two-phase ceramics seem to be desirable for superplasticity, since the second-phase particles are finely dispersed to pin the grain boundaries, thus maintaining the FG structure. The particles in superplastic materials are thermally stable. In addition, these ceramics must be strain-rate sensitive, with a value >0.3. Recently, superplastic behavior was also observed in iron aluminides with coarse grain structures. It is believed that this is due to recovery and dynamic recrystallization. Some relate superplasticity to grain boundary sliding [31]. Most of the reports consider ZrO_2 (zirconia) as a typical superplastic ceramic. New developments have also been achieved in the superplasticity of Si_3N_4 and SiC.

2.1.3.2 Oxide Superplastic Ceramics

As previously indicated, zirconia is a typical superplastic ceramic and was among the first oxide ceramics to be studied. As early as 1986, Wakai et al. studied

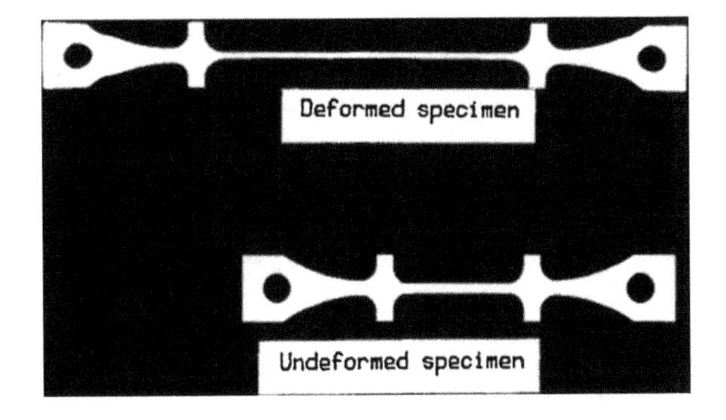

Fig. 2.45 Superplastically elongated specimen of Y-TZP at 1450 °C [48]. With kind permission of Professor Wakai

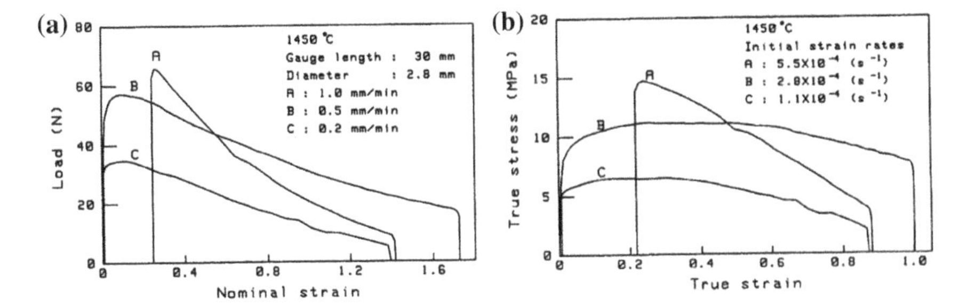

Fig. 2.46 **a** Load-nominal strain under constant displacement rate; **b** estimated true stress–true strain curves assuming a uniform deformation without necking [48]. With kind permission of Professor Wakai

yttria-stabilized tetragonal zirconia FG polycrystals (henceforth: Y-TZP) in the temperature range of 1000–1500 °C. The grain size was $\leq$0.3 and the tensile experiments were performed at strain rates of 1.1×10^{-4}–5.5×10^{-4} s^{-1}. Over 120 % strain was obtained at 1450 °C. Furthermore, the interest in this ceramic is a consequence of its excellent bending strength and toughness. The yttria (3 mol percent) was in solid solution. Figure 2.45 illustrates the superplasticity observed in this alloy.

The Y-TZP specimen showed an elongation >120 %. It is compared with an undeformed specimen. The elongation in the gage length was uniform and no local necking was observed. These tests were performed at constant displacement in a universal tensile testing machine in air and at 1450 °C. The results of these tests are shown in Fig. 2.46. As usual, the true strain, ε_t, was obtained (see Chap. 1, Sect. 1.2.1) by using:

$$\varepsilon_t = \ln\frac{l}{l_0} = \ln\left(1 - \frac{\Delta l}{l_0}\right) \tag{1.9}$$

In the above, l and l_0 are the elongated and original gage lengths, respectively. The true stress is related to the true strain by:

$$\sigma_t = \sigma \exp(\varepsilon_t) \tag{2.15}$$

where the nominal stress is the ratio of the load, P, to the original cross-section of the specimen, A_0, i.e., $\sigma = (P/A_0)$.

2.1.3.3 Other Non-oxide Superplastic Ceramics

Superplasticity is not limited to zirconia-type ceramics. Other ceramics have also been found to exhibit superplasticity, such as nitrides or carbides. Representative examples are Si_3N_4 and SiC. In these cases, superplasticity occurs in single-phase ceramics. Section 2.1.3.1 (i) A superplasticity of $\sim$ 470 % in Si_3N_4 (Fig. 2.44) has been mentioned above.

Superplasticity is one of the common properties of FG ceramics at elevated temperatures. Superplastic forming and strengthening by superplastic forging are applicable to a wide range of ceramics, including oxides and non-oxides. Zhan et al. [51] have studied the superplastic behavior of FG β-silicon nitrides (with 5 wt% Y_2O_3) under compression in the temperature range 1450–1650 °C at various strain rates. It was found that β-Si_3N_4 can be deformed at high strain rates ($\sim 10^{-4}$–10^{-3} s^{-1}) in a range of temperatures and at pressures of 5–100 MPa. No strain hardening occurs even during slow deformation. Beside Eqs. (1.9) and (2.15), the initial strain rate may be expressed as:

$$\dot{\varepsilon} = \frac{\dot{l}}{l_0} \tag{2.16}$$

Here, $\dot{l}$ is a constant. The immediate strain rate is expressed as:

$$\dot{\varepsilon} = \dot{\varepsilon}_0 \exp(-\varepsilon) \tag{2.17}$$

The corrected flow stress is given by:

$$\sigma_c = \sigma_0 [\exp(\varepsilon)]^{\frac{1}{n}} \tag{2.18}$$

For the corrected flow stress the stress exponent, n, must be known and is given as:

$$\dot{\varepsilon} = A\sigma^n \tag{2.19}$$

A typical, corrected true stress-true strain curve is shown in Fig. 2.47, together with the uncorrected curve. The effect of the strain rate at 1550 °C during a compression test is seen in Fig. 2.48. After the initial transient state, a steady state is reached for all the strain rates. The true strain rates are based on the corrected data. As can be seen, no strain hardening occurred in these tests, even at low strain rates, unlike other cases in which pronounced strain hardening has been observed. In those cases, the starting powder was α-Si_3N_4, rather than β-Si_3N_4. The strain hardening was attributed to microstructural changes during deformation, such as dynamic grain growth and α-to-β phase transformation. However, no shape change

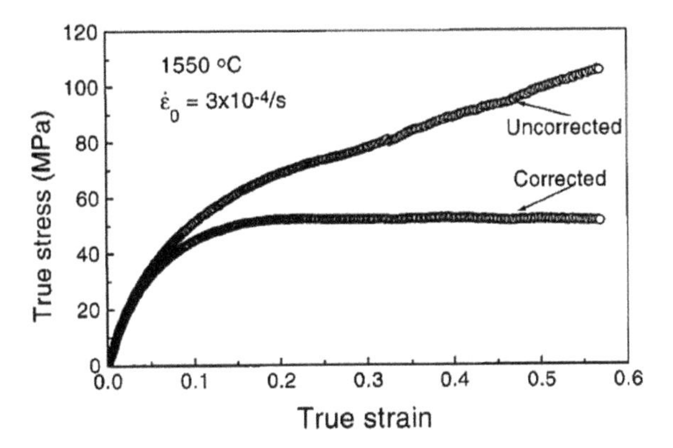

Fig. 2.47 Typical correction curve for a compression test at 1550 °C and an initial strain rate of 3×10^4/s, in the as-hot-pressed β-Si_3N_4 [51]. With kind permission of John Wiley and Sons

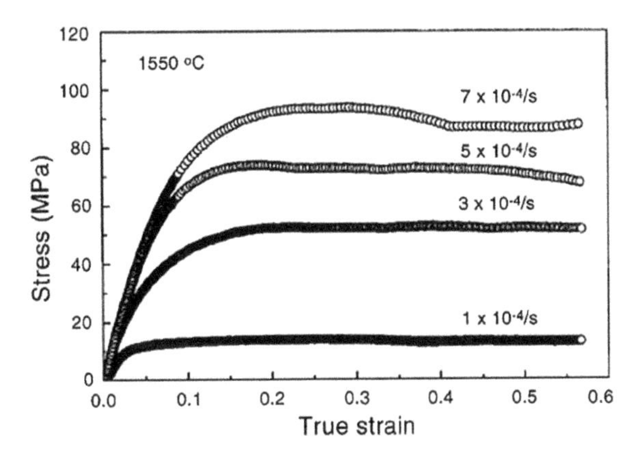

Fig. 2.48 Compressive stress–strain curves for various strain rates of the as-hot-pressed Si_3N_4 at 1550 °C [51]. With kind permission of John Wiley and Sons

occurred in the present material, due to the uniform PSD of the starting powder and the absence of an α-to-β phase transformation, as illustrated in Fig. 2.49.

The mechanism that controls the deformation rate at high temperatures, i.e., the plastic flow, may be expressed (constitutive equation) by:

$$\dot{\varepsilon} = \frac{A\sigma^n}{d^p} \exp\left(-\frac{Q}{RT}\right) \tag{2.20}$$

where $\dot{\varepsilon}$ is the strain rate, σ the flow stress, A is a temperature-dependent constant, d the grain size, n and p the stress and grain-size exponents, respectively, and Q the activation energy for flow.

In order to use Eq. (2.20), the stress exponent, n, must be determined. A plot, according to Eq. (2.19), expressed on a logarithmic scale at various temperatures provides the values of n, as shown in Fig. 2.50. The slopes of these curves give the values of n at the temperatures indicated. The initial flow stress regions were ignored and only the quasi-steady-state part of the flow stress is plotted in Fig. 2.50. Note that the values of n that barely change with temperature are ~1–1.4.

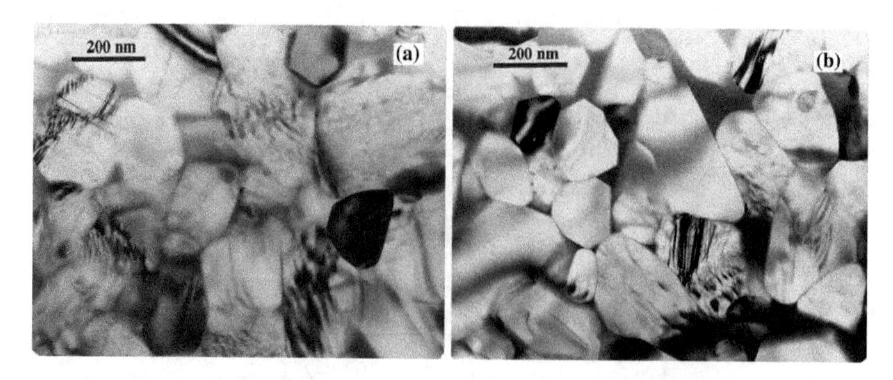

Fig. 2.49 TEM photographs of: **a** an undeformed sample and **b** a deformed sample at 1600 °C, with a true strain of −1.1, showing no dynamic grain growth [51]. With kind permission of John Wiley and Sons

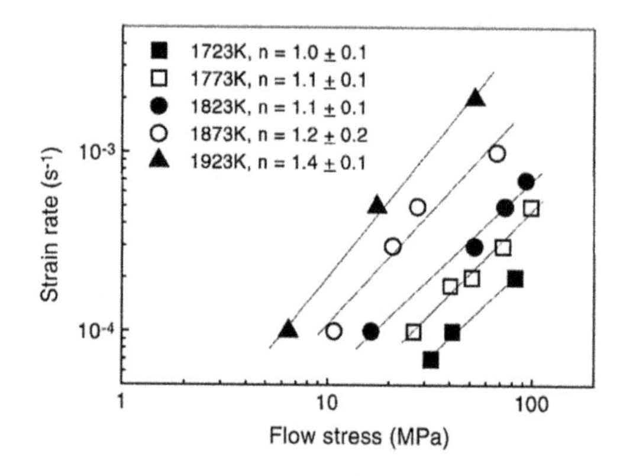

Fig. 2.50 Strain rate versus stress at various temperatures, under compression (n = slope), in the as-hot-pressed β-Si_3N_4 [51]. With kind permission of John Wiley and Sons

The higher values of the stress exponent seem to occur at temperatures above 1823 K. The temperature dependence of the strain rate is shown in Fig. 2.51, where the strain rate is plotted against the reciprocal, absolute temperature. The activation energy, Q, calculated from the slopes of the lines in Fig. 2.51, are 344 ± 26 kJ/mol at 20 MPa and 410 ± kJ/mol at 100 MPa. High-resolution transmission-electron microscopy (henceforth: HRTEM) observations of materials, both before and after deformation, are shown in Fig. 2.52. Observe that most of the grain boundaries have a glass film, although some grain boundaries were free of such film. This indicates that the formation of glass film is dependent on grain-boundary orientation and whether they were perpendicular or parallel to the direction of the applied force and on grain orientation. Those boundaries oriented in parallel show wide films (Fig. 2.52a), whereas the film thickness on grain boundaries perpendicular to the applied load direction were smaller (Fig. 2.52b).

It was also mentioned above that some relate superplasticity to grain-boundary sliding (see, for example [31]). Guo-Dong Zhan et al. [51] report that grain-boundary sliding may also be the mechanism of superplasticity in Si_3N_4, in

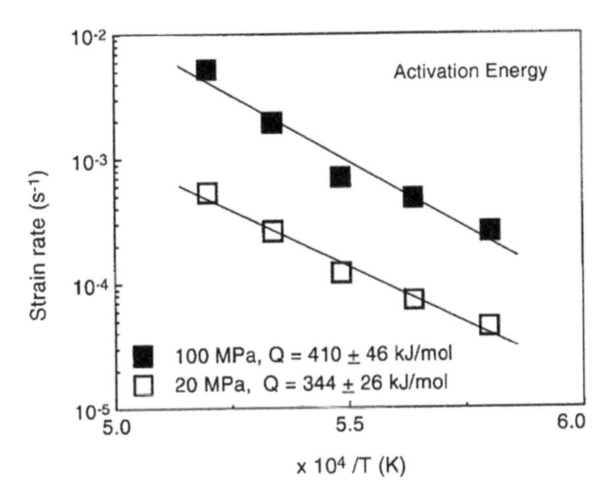

Fig. 2.51 Determination of activation energy for flow equation in the as-hot-pressed β-Si_3N_4 [51]. With kind permission of John Wiley and Sons

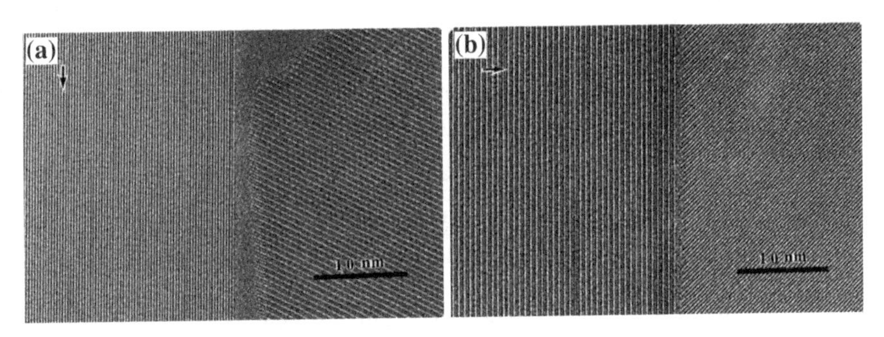

Fig. 2.52 Representative HRTEM photographs of boundaries oriented **a** parallel and **b** perpendicular to the applied load direction, indicating that the grain-boundary film thickness decreased after superplastic deformation, under compression ((→) applied stress direction during deformation) [51]. With kind permission of John Wiley and Sons

addition to grain rotation, accommodated by viscous flow. Furthermore, FG β-Si_3N_4 exhibits high grain-size stability against dynamic grain growth during sintering and deformation, a characteristic that satisfies the microstructural requirement for classic superplasticity. This kind of Si_3N_4 does not work-harden as do the other silicon nitrides discussed earlier.

2.1.3.4 Superplasticity in Carbides

'Superplasticity' is basically defined as the ability of a material to exhibit exceptionally large tensile elongation during stretching. In addition to oxides and nitride-like materials, some carbides also show large elongation and frequent

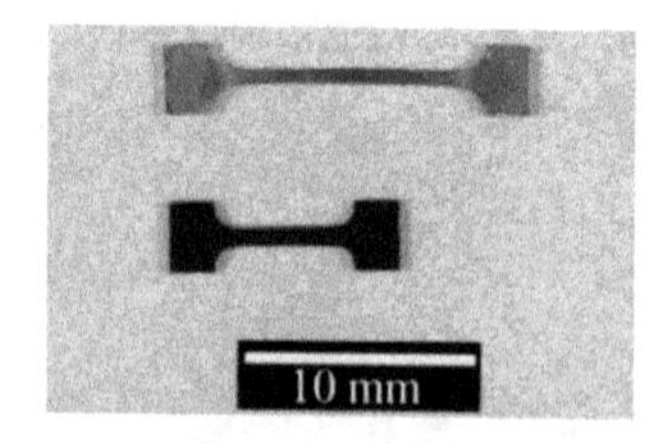

Fig. 2.53 HIPed B, C-SiC specimens before and after tensile deformation. The tensile test was conducted at 1800 °C, and at an initial strain rate of 3×10^{-5} s^{-1} in an argon atmosphere. The specimen deformed uniformly, and a superplastic elongation of 140 % was achieved [40]. With kind permission of John Wiley and Sons

superplastic behavior. For better familiarity with carbide behavior, this section will discuss SiC and FeC as exemplars.

(1) **SiC**

It is somewhat difficult to produce SiC without additives, because it is hard to obtain a dense material due to its low self-diffusivity and covalent nature. Thus, the role of the additives is to loosen or modify this bond structure. The most common additives are B and C [8]. B additives, for instance, provide superior mechanical strength at elevated temperatures. High superplasticity at a level of ~140 % has been observed in β-SiC doped with B or C and having a small grain size of 0.2 µm, fabricated by ultra-high hot isostatic pressure (henceforth: UH-HIP). The B segregated at the grain boundaries and apparently promoted grain-boundary sliding, one mechanism of superplasticity [4]. However, when liquid-phase sintering is the fabrication method and there is an amorphous phase at the grain boundary, rather than solid-phase sintering with no amorphous phase, it is easier to deform the product.

Basically, the degree of elongation depends on the additive. Thus, when 1 % B and 3.5 % free C are added to SiC fabricated by HIP at 980 MPa at a temperature of 1660 °C with an average grain size of 200 nm, a 140 % superplastic elongation is obtained at 1800 °C [40]. When β-SiC was produced by liquid-phase sintering, prepared with different oxynitride glasses in an N_2 atmosphere [4], the elongation was either 74 or 153 % at the initial strain rate of 1×10^{-5} s^{-1} at 2023 K under tension, depending on its composition (the additives forming the oxynitride glasses). Figure 2.53 compares specimens before and after deformation. A superplastic elongation of 140 % was achieved.

This specimen deformed uniformly. Stress–strain curves following HIP and hot-pressed B, C-SiC are shown in Fig. 2.54. The hot-pressed SiC was sintered under a pressure of 30 MPa at 2000 °C for 1 h and the average grain size was 2 µm. The B, C-SiC, after HIP, exhibited a superplastic elongation of >100 %, whereas the hot-pressed B, C-SiC fractured without significant plastic deformation. Thus, grain refinement was effective for obtaining superplasticity in SiC. HRTEM observation and electron energy-loss spectroscopy analysis revealed that there was no glassy phase at the grain boundaries, but boron segregation and carbon excess

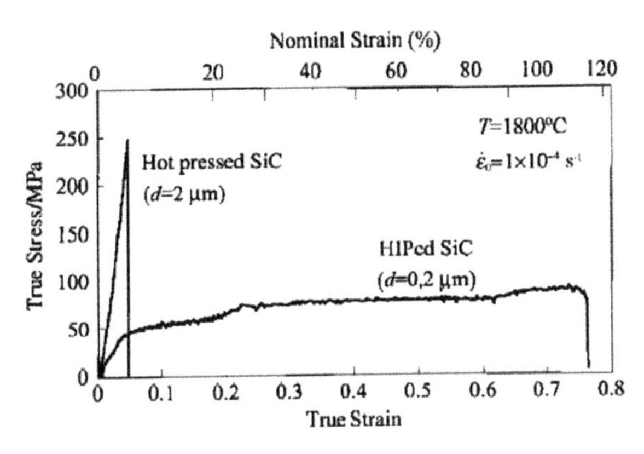

Fig. 2.54 Stress–strain curves of HIPed B, C-SiC and hot-pressed B, C-SiC. The tensile tests were conducted at 1800 °C and an initial strain rate of $1 \times 10^{-4}\ s^{-1}$ in an argon atmosphere. HIPed, B, C-SiC exhibited superplastic elongation of 114 %, because of grain refinement. On the other hand, hot-pressed B, C-SiC fractured without plastic deformation [40]. With kind permission of John Wiley and Sons

Table 2.4 Chemical compositions and some properties of as-sintered materials [4] (with kind permission of Elsevier)

Material	Additives (wt%)					Grain size (nm)	Density (g/cm^3)	Sintering condition
	SiO_2	MgO	Al_2O_3	Y_2O_3	AlN			
SiC (G1)	3.78	0.981	1.17	2.13	0.936	260	3.15	2073 K 30 MPa 20 min in N_2
SiC (G2)			5.022	3.321	0.657	230	3.21	2073 K 30 MPa 15 min in N_2

were observed there. A small amount of oxygen segregation was also detected at the grain boundaries. However, the amount of segregated atoms was not enough to form an intergranular glassy phase, such as had formed in the experiments of Nagano et al. [4]. Yet, Wang et al. [49] indicate that the microstructure of the material, both before and after the superplastic deformation, retains the microstructural features of that material before its deformation.

In the work of Nagano et al. [4], the starting material was ultra-fine β-SiC powder with a particle size of ~90 nm. The mixtures of SiO_2, MgO, Al_2O_3, Y_2O_3 and AlN were then rendered into oxynitride compositions by SiC ball milling in n-hexane. The SiC was mixed with 9 wt% oxynitride powders by SiC ball milling in n-hexane. The mixed powder was hot-pressed at 2073 K under a stress of 30 MPa in N_2. The chemical compositions and other properties of the sintered materials are shown in Table 2.4. Compression and tension tests at constant crosshead speeds were performed using a universal testing machine with a furnace at initial strain rates from 1×10^{-4} to $5 \times 10^{-6}\ s^{-1}$ at temperatures ranging from 1973 to 2048 K in N_2. The degree of specimen deformation was evaluated from the displacement of the crosshead. Compressive and tensile directions were

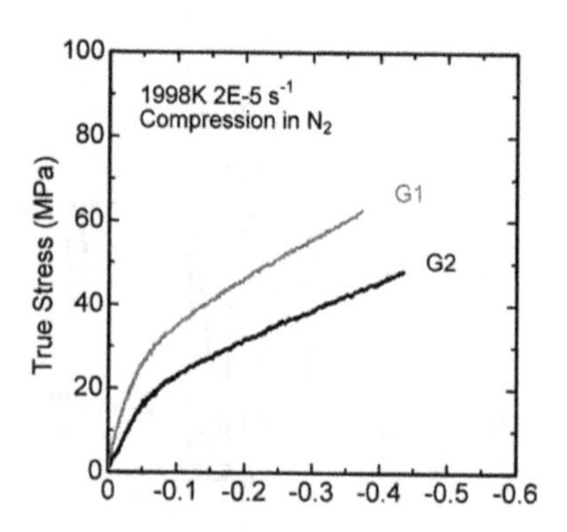

Fig. 2.55 True stress–true strain curves at 1998 K in compression tests [4]. With kind permission of Elsevier

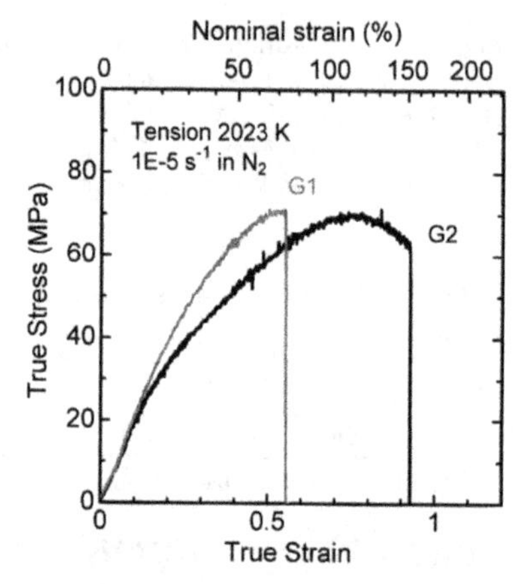

Fig. 2.56 True stress–true strain curves at 2023 K in tension tests [4]. With kind permission of Elsevier

perpendicular to the hot-pressing direction. The true stress–true strain curves at 1998 K are shown in Fig. 2.55. No cracks were observed in both compressed specimens. SiC (G1) showed higher flow stress and higher strain hardening in comparison with SiC (G2). The true stress–true strain relation is seen in Fig. 2.57. These tests were performed under tension. Strain hardening was observed in both specimens, i.e., SiC (G1) and SiC (G2). The SiC (G1) showed higher strain hardening and fractured at a 74 % elongation, while the SiC (G2) showed strain hardening to a 110 % elongation and then showed strain softening. The final elongation of SiC (G2) achieved was 153 % (Fig. 2.56).

These figures are based on Eqs. (2.16)–(2.19) and the strain-rate variation with temperature for the activation energy evaluation is based on Eq. (2.20). To use Eq. (2.20), the stress exponent, n, is required for various temperatures, which may be evaluated by using Eq. (2.19). Plots of this relation are shown in Fig. 2.57. HRTEM images at grain boundaries are shown in Fig. 2.58. An amorphous phase, from 1 to 2 nm, is evident in SiC (G1) and SiC (G2). However, some grain boundaries with no amorphous phases may be seen in SiC (G1). SEM photographs of the gauge portions of the elongated specimens are shown in Fig. 2.59. The cavitation damage of SiC (G1) after 74 % elongation was higher than that of SiC

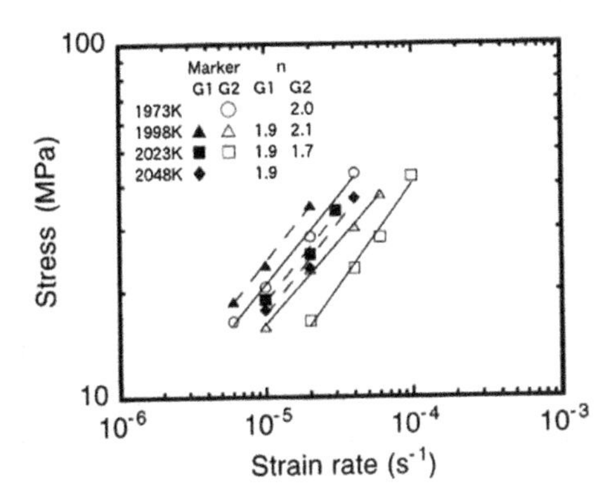

Fig. 2.57 Relationship between flow stress and strain rate [4]. With kind permission of Elsevier

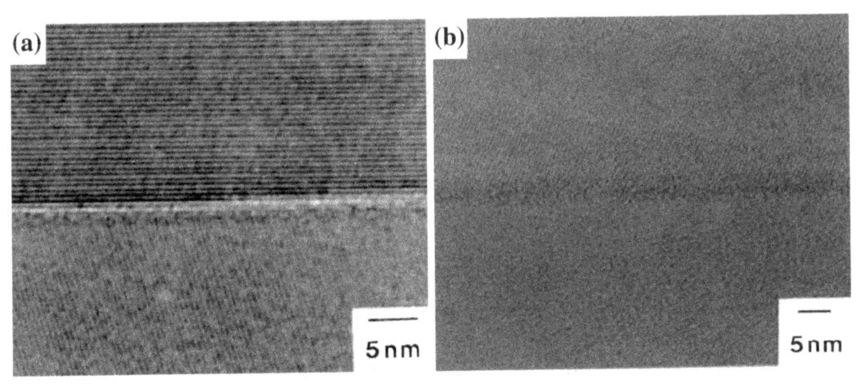

Fig. 2.58 HRTEM images at grain boundaries of **a** SiC(G1) and **b** SiC(G2) in as-sintered materials [4]. With kind permission of Elsevier

(G2) after 153 % elongation. On the one hand, the average grain sizes of SiC (G1) after 74 % elongation were 350 and 500 nm in the vertical and horizontal directions of the tensile axis, respectively; on the other hand, the average grain sizes of SiC (G2) after 153 % elongation were 300 and 430 nm in the vertical and horizontal directions of tensile axis, respectively. Most of the residual grain-boundary phase after 153 % elongation was at the triple points. The contribution of grain-boundary sliding to the total strain was calculated to be in the order of ~76 % in elongated specimens of SiC (G1) and SiC (G2). Therefore, the critical deformation mechanism was thought to be grain-boundary sliding in both the SiC (G1) and SiC (G2) specimens. This being the case, the initial grain size of the as-sintered material, the grain-growth rate, the cavitation damage during deformation, the vaporization of the grain-boundary phase and the formation of crystalline phases at triple points are all significant factors for the improvement of superplastic deformation behavior in liquid-phase sintered SiC with an amorphous phase.

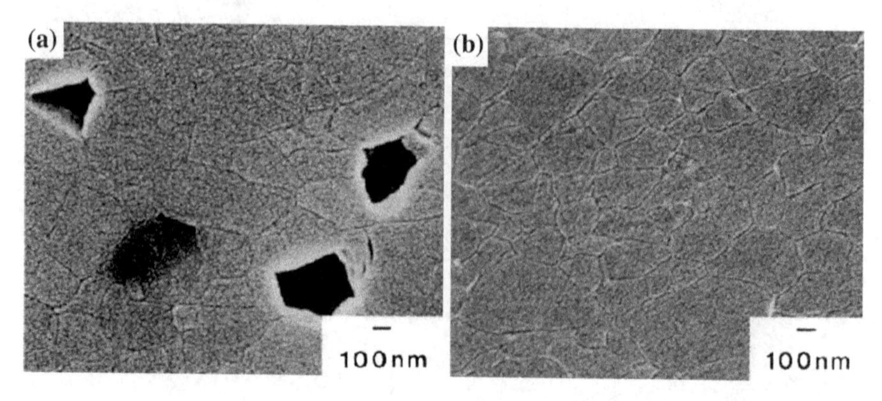

Fig. 2.59 SEM photographs of **a** 74 % elongated SiC(G1) and **b** 153 % elongated SiC(G2) [4]. With kind permission of Elsevier

(2) **FeC**

By now, it is clear that superplasticity is not restricted to a special group of materials. Even other carbides than the aforementioned SiC show superplasticity under certain conditions. Iron carbide, FeC, is such a carbide.

As indicated earlier, grain size, in general, is an important factor in superplastic phenomena and, in this regard, FeC is no exception. Thus, the processing of FeC for superplasticity studies is usually done on FG structures. As in other superplastic ceramic materials, a second phase is present. In the work of Kim et al. [28], for example, an iron-based second phase was added to the carbide. Strain-rate sensitivity is usually evaluated by testing for strain-rate changes. In such tests, a certain strain rate is applied to provide a strain creating isostructural conditions (similar structures), in order to obtain a stable grain size. While the strain rate is changed by specific strain-rate values, the stress must be recorded. A plot is made of flow stress versus strain rate on a logarithmic scale and from the slope of such curves, m, the strain-rate sensitivity is determined. It is necessary to perform such tests at various temperatures to determine the activation energy. The relation used is

$$\sigma = K\left[\dot{\varepsilon}\exp\left(\frac{Q_c}{RT}\right)\right]^m \tag{2.21}$$

The term:

$$\dot{\varepsilon}\exp\left(\frac{Q}{Rt}\right) \tag{2.21a}$$

is the Zener-Hollomon parameter (Dieter).

In Fig. 2.60, the flow stress is shown as a function of the strain rate-temperature parameter, $\dot{\varepsilon}\exp\frac{Q_c}{RT}$. The activation energy was evaluated as 200 kJ/mol. The strain-rate sensitivity exponent, m, is 0.5 as derived from the slope. K is a material

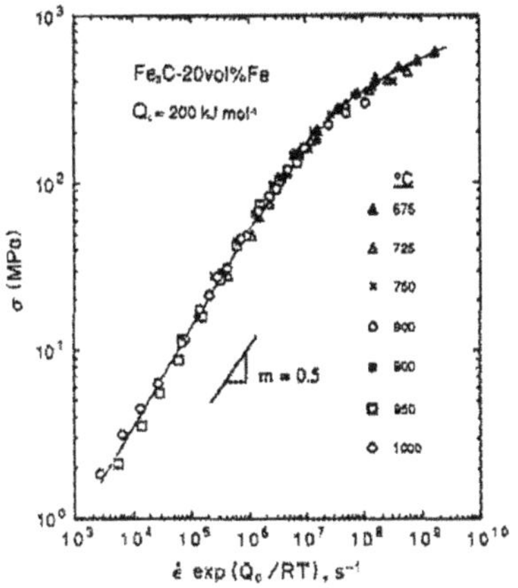

Fig. 2.60 The flow stress as a function of the strain rate-temperature parameter $\dot{\varepsilon}\exp\frac{Q_c}{RT}$ for a superplastic fine grained iron carbide (Fe_2C–20 % Fe) [28]. With kind permission of Elsevier

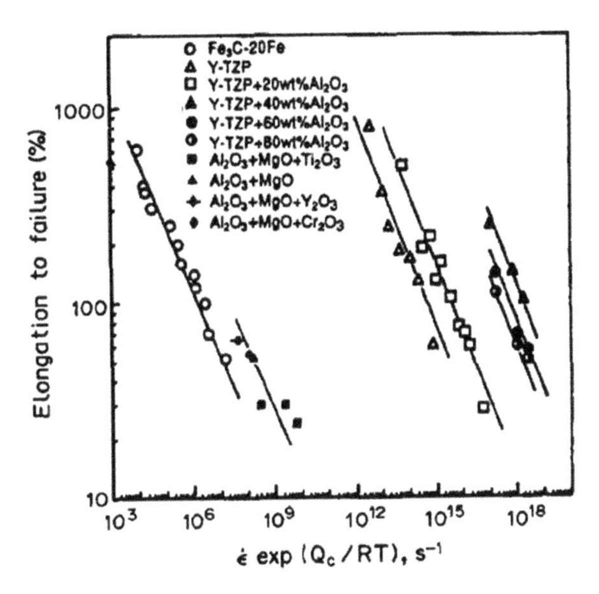

Fig. 2.61 Tensile ductility of fine-grained iron carbide as function of the Zener-Hollomon parameter is compared with some superplastic ceramics doped with various impurities. The strain rate sensitivity parameter is in the range m = 0.5–0.6 [28]. With kind permission of Elsevier

constant, which is a structural factor and a function of the modulus; $\dot{\varepsilon}$ is the steady-state strain rate and the other parameters are familiar.

The tensile ductility of various superplastic ceramics are compared with that of iron carbide in Fig. 2.61. All the curves show the same tendency, namely that tensile ductility decreases with increased strain rate-temperature, $\dot{\varepsilon}\exp\left(\frac{Q_c}{RT}\right)$. This decrease has been explained by grain growth. It is possible to superimpose all the superplastic ceramics data shown in Fig. 2.51 on a common curve when $\dot{\varepsilon}\exp\left(\frac{Q_c}{RT}\right)$ is multiplied by A, which is unique for each ceramic. The results are shown in Fig. 2.62.

Kim et al. [28], in their extensive work on superplasticity, classified materials on the basis of their elongations, defining 'superplasticity' as being ductility beyond 200 %. In accordance with their classification system: superplastic-like materials are those with elongations in the 50–200 % range; ductile ceramics have

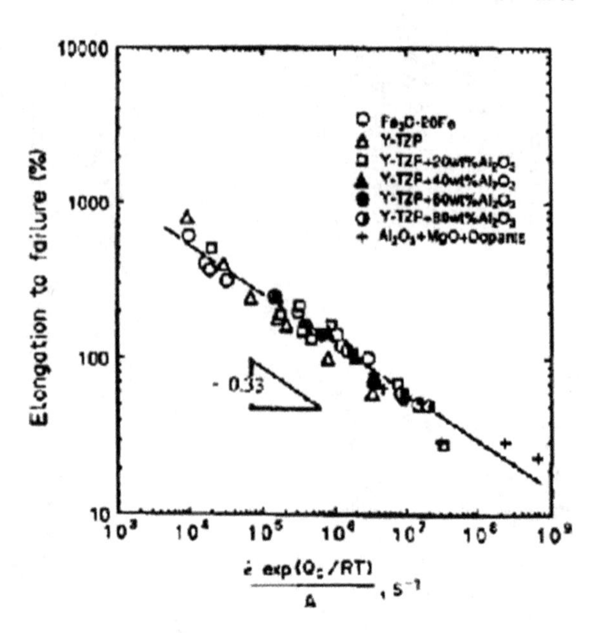

Fig. 2.62 Tensile ductility for fine-grained ceramics as a function of $\dot{\varepsilon}\exp\left(\frac{Q_c}{RT}\right)$ normalized by the material parameter [28]. With kind permission of Elsevier

elongations ranging from 3 to 50 %; and in brittle ceramics, elongations are below 3 %. They obtained tensile elongations as high as 600 % in FG iron carbide (Fe_3C–20 % Fe). The strain-rate sensitivity is an important parameter.

In an additional work, Kim [27] indicated a tensile elongation to fracture in the 200–400 % range, as shown Fig. 2.63. Here, strain hardening may be observed in the curve. The deformation mechanism in his tests was grain-boundary sliding. To calculate the grain-size exponent, p, one can use Eq. (2.20), as follows:

Rewrite relation (2.20) as

$$\dot{\varepsilon} = \frac{A\sigma^n}{d^p}\exp\left(-\frac{Q}{RT}\right) \tag{2.20}$$

This relation is often expressed as:

$$\dot{\varepsilon} = K\left(\frac{b}{L}\right)^p \sigma^n \exp\left(-\frac{Q_c}{RT}\right) \tag{2.22}$$

Clearly, these relations are equivalent when $A \equiv Kb^p$. In Eq. (2.22), $L \equiv d$ and b is the Burgers vector. Expressing Eq. (2.22) on a logarithmic scale and taking the derivative for the constant strain rate and temperature, one obtains the grain size exponent:

$$p = n\frac{\partial \ln \sigma}{\partial \ln L}\bigg|_{\dot{\varepsilon},T} \tag{2.23}$$

The final grain size, evaluated from fracture test specimens, is related to the maximum flow stress of the tensile test. Table 2.5 lists the final grain sizes, the

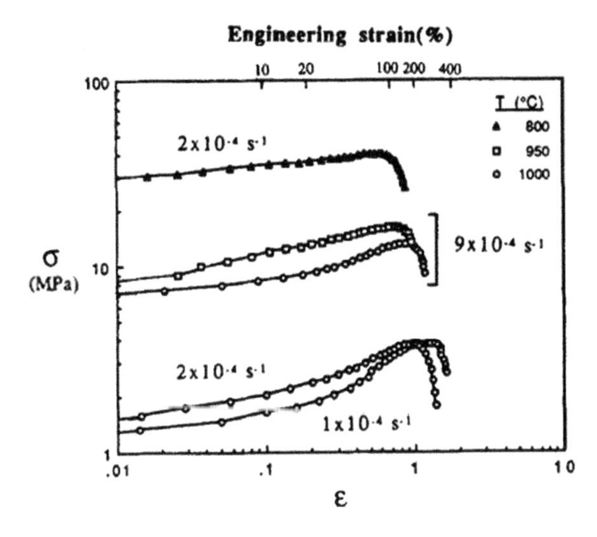

Fig. 2.63 True stress–true strain curves obtained by tension of extruded and pressed iron carbide at the temperatures indicated and at various constant true strain rates [27]. With kind permission from Springer Science+Business Media B.V.

Table 2.5 Values of grain size exponent, p for the extruded and pressed 5.25 % C carbide [27] (with kind permission from Springer Science+Business Media B.V.)

T (°C)	$\dot{\varepsilon}$ (s^{-1})	$L_{initial}$ (μm)	L_{final} (μm)	σ_{max} (MPa)	σ_{min} (MPa)	n	p
1000	1×10^{-4}	3.4	6.70	3.75	1.25	1.66	2.69
1000	2×10^{-4}	3.4	5.36	3.70	1.5	1.66	3.28
1000	9×10^{-4}	3.4	5.10	13.2	7.0	1.66	2.73
950	9×10^{-4}	3.4	4.94	16.0	8.7	1.66	2.71
800	2×10^{-4}	3.4	3.98	40.05	30.05	1.66	3.03
						Average	2.9

stress exponents, the grain-size exponents and other relevant parameters of this test.

In addition to the extruded and pressed carbides, also fabrication by HIP and pressure were performed. A comparison of the strain rate versus flow stress of these two kinds fabrication methods in carbides may be seen in Fig. 2.64 at three temperatures. Figure 2.65 shows the variations of strain rate versus flow stress on a logarithmic scale for extruded and pressed iron carbide at several temperatures. Depending on the temperatures, two values of the stress exponent were calculated from the slopes, as shown in Fig. 2.65: the value of 2 represents testing at the

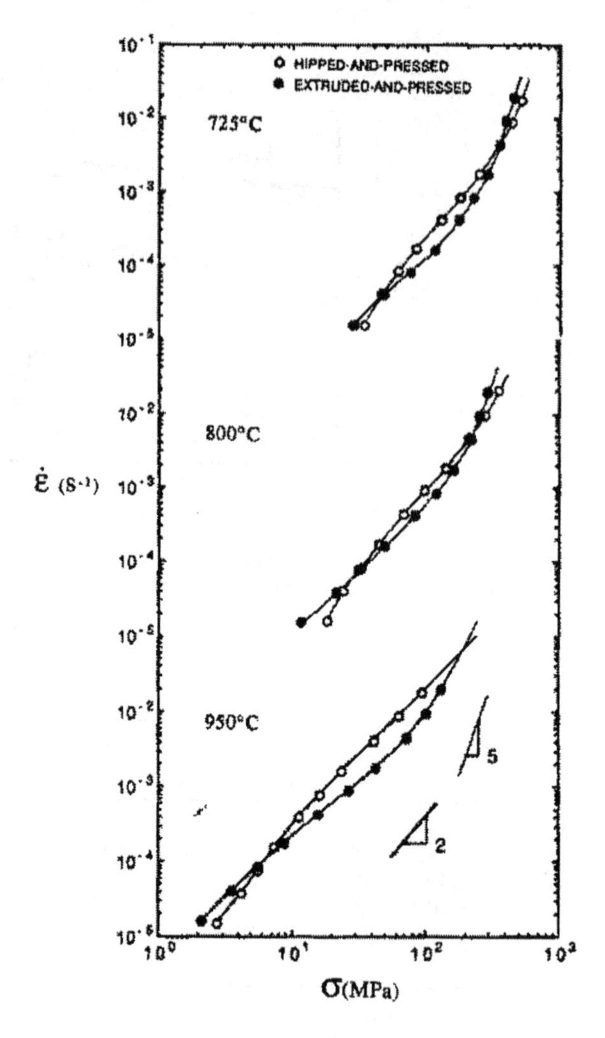

Fig. 2.64 Comparison of strain rate versus flow stress for the carbide fabricated by the methods indicated in the figure at three temperatures [27]. With kind permission from Springer Science+Business Media B.V.

725–1050 °C temperature range, whereas, above this temperature, the value is 1. The change in slope value is likely indicative of a different plastic flow mechanism. It is possible that at these low stresses and higher temperatures creep occurs.

The strain rate change with flow stress relation determined by tension is compared with the data obtained under compression, shown in Fig. 2.66 at two temperatures. Note the microstructure of the hipped and pressed iron carbide after large compressive deformation at 950 °C and at an initial strain rate of $6.67 \times 10^{-3}\ s^{-1}$, as shown in Fig. 2.67. Observe the grains that remain equiaxed after the deformation, suggesting that the deformation mechanism was, indeed, grain-boundary sliding. Also note that no cracks developed (see Fig. 2.67). The activation energy may be expressed either from Eq. (2.20) or (2.22). Consider

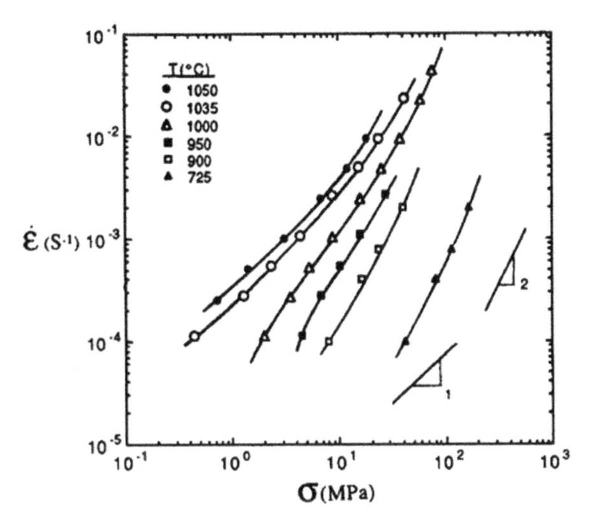

Fig. 2.65 Strain rate versus flow stress by tension test is indicated for iron carbide [27]. With kind permission from Springer Science+Business Media B.V.

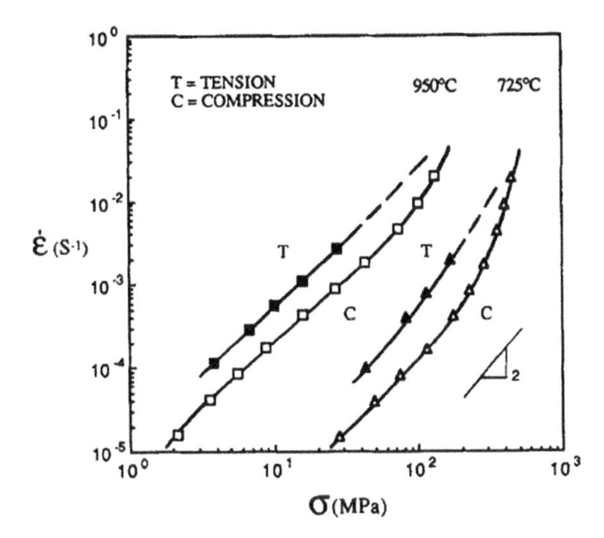

Fig. 2.66 Comparison of the tension and compression of the strain rate change versus flow stress at two temperatures [27]. With kind permission from Springer Science+Business Media B.V.

Eq. (2.22); express it on a logarithmic scale and take the derivative of strain rate with 1/T at constant stress and grain size to obtain the expression:

$$Q_c = -R\frac{\partial \ln \dot{\varepsilon}}{\partial 1/T}\bigg|_{\sigma,L} \tag{2.24}$$

A plot expressing this relation for deformation under tension and compression is shown in Fig. 2.68 for several stress exponents. The activation-energy values are indicated on the plot and they are in the range of 200–420 kJ/mol; the lower value is for n = 2 and the higher activation energy is for n = ~1.

Fig. 2.67 Microstructure of the hipped and pressed iron carbide after large compressive deformation ($\varepsilon = -2.81$) [27]. With kind permission from Springer Science+Business Media B.V.

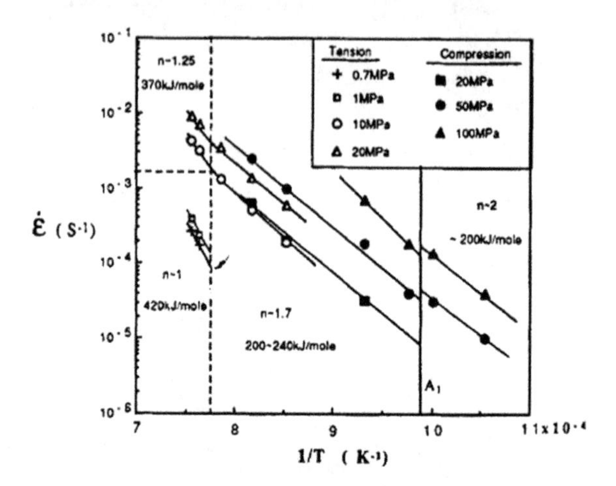

Fig. 2.68 Activation energies from tension and compression strain rate change [27]

2.2 Ductility in Single Crystal Ceramics

Figure 2.69 shows additional stress–strain curves of single crystals of MgO, deformed under compression in the [100] direction and temperature range of −196 to 1200 °C. In Fig. 2.33, the loading axis was ⟨111⟩ and in Fig. 2.36, the deformation was performed on a polycrystalline material. Note that the temperature range in Fig. 2.33 is 349–1204 °C, different than in Fig. 2.69. These crystals show considerable elongation at all temperatures and, at RT, they may be deformed plastically to about 6 %, as seen in Fig. 2.69a, before fracture sets in. This value of strain is considerably higher than the one obtained for polycrystalline MgO (as shown in Fig. 2.36), where only a strain of ~0.0075 is indicated. Strain hardening is observed in these figures at all temperatures. It is known that the resistance of a material to deformation increases with the number of slip systems activated and the dislocation bands, which act as barriers to dislocation movement, generally occurring at slip-band intersections. Figure 2.70 shows the dislocation

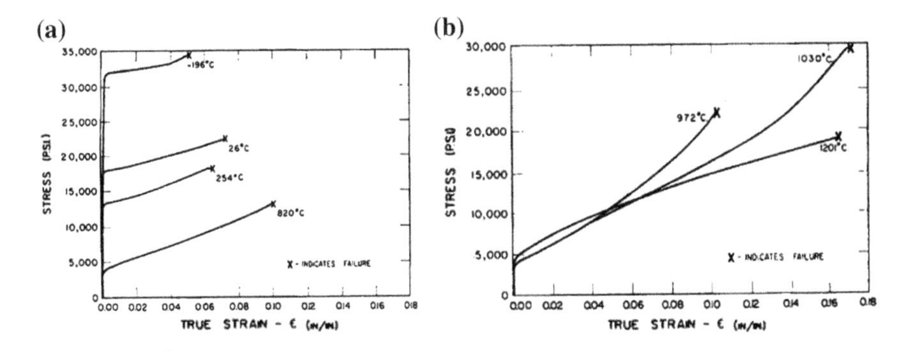

Fig. 2.69 Stress–strain curves of MgO single crystals deformed by compression in the [100] direction at various temperatures. **a** Temperature range −196 to 820 °C; **b** temperature range 972–1200 °C [2]. With kind permission of Wiley and Sons

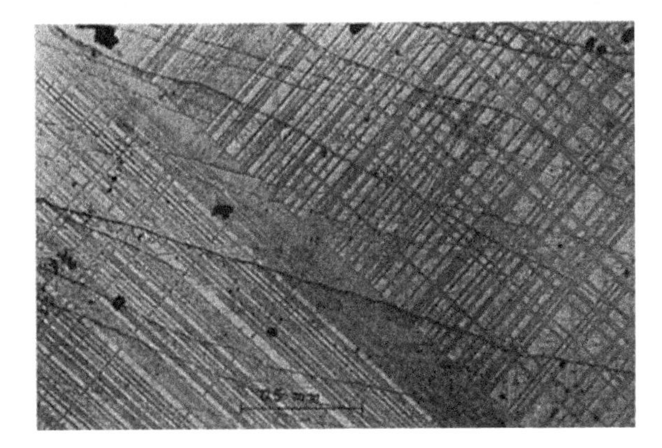

Fig. 2.70 Dislocation band structure. The single crystal was deformed in [100] direction [2]. With kind permission of Wiley and Sons

band structure deformed in the [100] direction. In this figure, the slip on one set of (110) planes encountered difficulty in passing through the slip bands formed on conjugate planes, a set of (110) planes. The thickness of the bands increases the resistance of the dislocations passing through them.

The stress–strain data were obtained at a constant rate of loading. The effect of the loading rate at three different rates is shown in Fig. 2.71. As may be seen from the curve, the slower loading rate resulted in (a) a lowering of the yield stress, (b) an increase in the plastic strain and (c) a decrease in the work-hardening rate. In other words, when testing these MgO crystals at a loading rate of 0.382 lbs/sq. in./s, the elongation was above 9 %.

Bear in mind that tests providing mechanical property data, such as ultimate strength, etc., may be considered extremely sensitive to the experimental conditions and the data are usually more scattered than those in similar tests of metallic materials.

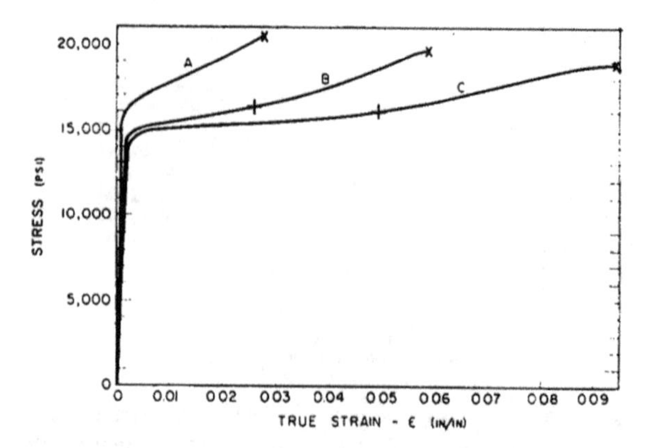

Fig. 2.71 Stress–strain curves of MgO obtained by compression in the [100] direction loaded at the strain rates of A 80.7, B 3.48 and C 0.382 lbs/sq. in./s. The + sign indicates the appearance of first visible crack [2]. With kind permission of Wiley and Sons

Another good example of single-crystal data regarding ductility or stress–strain relations is the case of zirconia (ZrO_2). ZrO_2 is considered to be a promising engineering material in comparison with other ceramics, because it reveals higher serviceability under different loading conditions, even up to 1400 °C.

Figure 2.72 presents an Y_2O_3-stabilized single crystal of ZrO_2 with other ceramics. Here, the crosshead speed, V_{ch}, is in the 0.005–5.0 mm/min range. These specimens were loaded in four-point bending under isothermal conditions. In this figure, the power parameter in the relation $V = M.K_I^N$ is on the ordinate N ($\equiv$ n in other terminology), where M is a parameter and K_I is the stress-intensity factor at the crack tip. Here, the deformation and strength of zirconia crystals at different temperatures and deformation rates are compared with the most widely used engineering ceramics, such as silicon nitride and alumina, under similar loading conditions. Thus, a general, comparative picture of the mechanical behavior of different types of ceramics and zirconia single crystals, over a wide range of temperatures and deformation rates, is obtained.

Compare the stress values and the deflection of the single-crystal zirconia (in particular, Y-PSZ-3) with those of the other ceramics presented. Table 2.6 shows some of the properties investigated, among them that of Y-PSZ-3. Note its ultimate strain reaching 44.0×10^{-4} m/m, obtained at a crosshead rate of $V_{ch} = 0.5$ mm/min. Furthermore, Table 2.7 may be also consulted for some of the properties of the various zirconia at 1400 °C. The deviation from linearity (elastic deformation) for zirconia was in the 1000–1100 °C range (not shown in Fig. 2.72). The strain at the elastic limit was in the range of $31.2–35.8 \times 10^{-4}$ m/m, depending on the rate of deformation (see Table 2.7). Also note that the ultimate strain of Y-PSZ-3 at an orientation of $\langle 111 \rangle$ is the highest of the various zirconia. In Fig. 2.73, the load versus the deflection is indicated for different strain rates and temperatures.

Load versus deflection diagrams for Y-FSZC-10 and Y-PSZC-3 are compared in Fig. 2.74.

Additional experiments on stress and strain, providing information on ductility, may be seen in Fig. 2.75. This figure illustrates the brittle and ductile failure modes and the strain-rate dependence of yield stress. These different failure

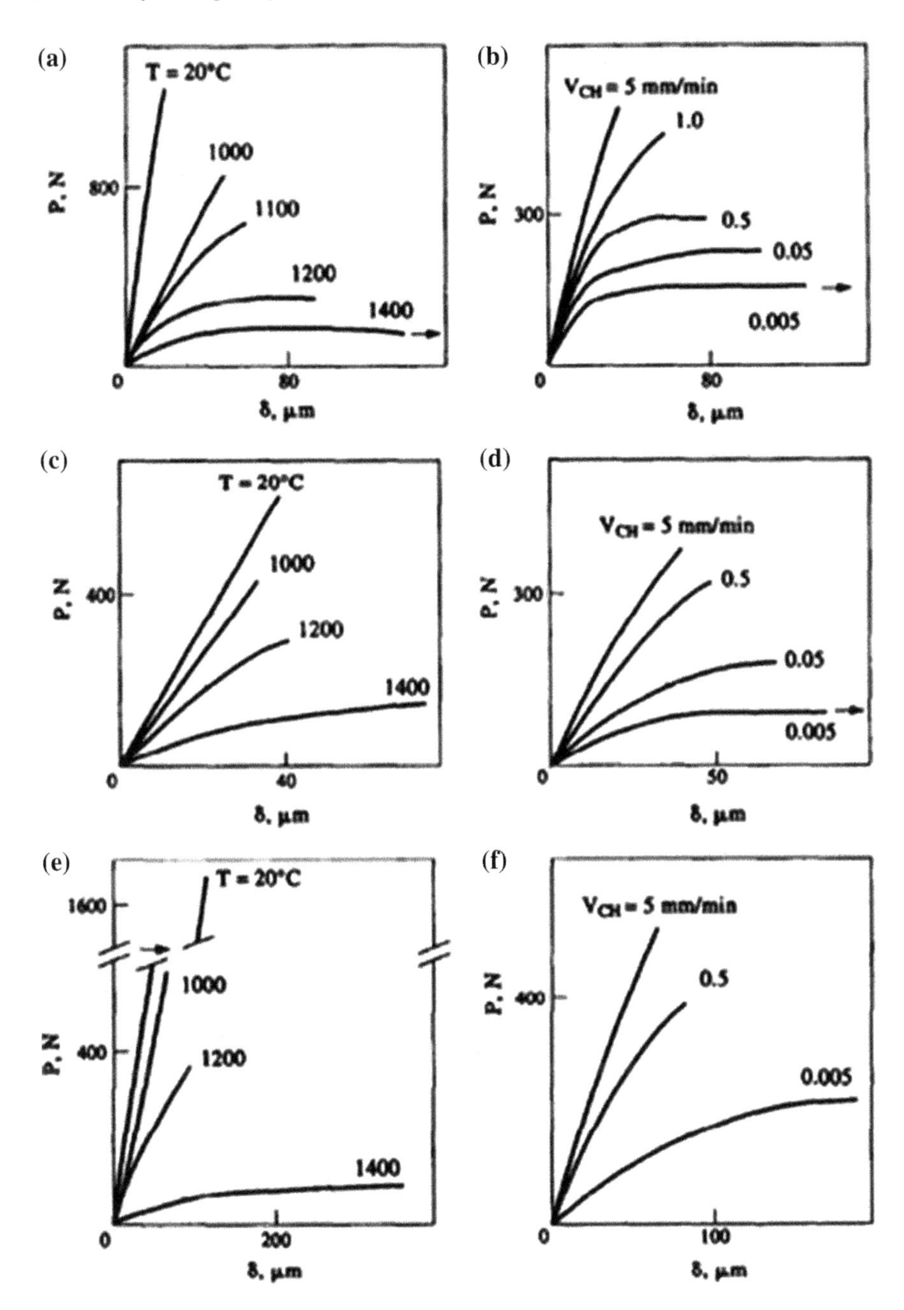

Fig. 2.72 The influence of the temperature, T (diagrams **a**, **c**, **e**) at V_{ch} = 0.5 mm/min and of the speed V_{ch}, (diagrams **b**, **d**, **f**) at T = 1200 °C on the appearance of the load (P) versus deflection δ: **a**, **b** SN-1; **c**, **d** A-l; **e**, **f** Y-PSZ-3. SN-1 is a silicon nitride-based ceramic with additions of Y_2O_3 and Al_2O_3; A-1 is an alumina-based ceramic with an addition of MgO; and Y-PSZ-3 is a Y_2O_3 stabilized zirconia based single crystal [18]. With kind permission of Elsevier

Table 2.6 Physico-mechanical characteristics of investigated ceramics and crystals at ambient temperature (average values) [18] (with kind permission of Elsevier)

Material (orientation of the specimen axis)	Density (g/cm^3)	Ultrasonic velocity (m/s)	Elastic modulus (GPa)		Bending strength (MOR)[a] (MPa)	Ultimate strain[a] ($\times 10^{-4}$ m/m)
			Dynamic	Static[a]		
Ceramics						
SN-1	3.21	9289	277	280	465	16.6
SN-2a	3.27	9690	307	305	510	16.7
A-1	3.70	9347	323	320	300	9.4
Y-PSZ-3	5.94	5951	210	207	908	44.0
Crystals						
Y-FSZC-10 (⟨101⟩)	5.87	5933	207	180	134	7.4
Y-FSZC-20 (⟨101⟩)	5.76	6116	215	185	140	7.6
Y-PSZC-3 (⟨111⟩)	6.04	5394	176	149	642	43.0

Table 2.7 Average values of mechanical characteristics of investigated crystals at 1400 °C [18] (with kind permission of Elsevier)

Material	V_{ch} (mm/min)	Elastic limit (MPa)	Upper yield point[a] (MPa)	Lower yield point (MPa)	Strain at elastic limit ($\times 10^{-4}$ m/m)	Strain[a] at upper yield point (MPa)	Elastic modulus (GPa)
Y-FSZC-10	5.0	274	–	–	17.5	–	157
	0.5	327	445	445	23.2	51.5	141
	0.05	269	338	318	18.7	32.2	144
	0.005	210	250	234	16.9	28.4	124
Y-FSZC-20	5.0	213	–	–	–15.3	–	140
	0.5	300	–	–	21.6	–	139
	0.005	276	324	295	23.2	35.7	119
Y-PSZC-3	5.0	393	444	–	31.2	37.5	126
	0.5	403	475	–	33.0	42.5	122
	0.05	400	550	–	32.8	75.2	122
	0.005	390	556	–	35.8	74.4	109

[a] For Y-PSZC-3 crystals—ultimate stress and strain values

modes, which relate to the presence or absence of plastic deformation, presumably result from the various orientations of the easy slip planes in the specimens, relative to the applied stress. Notice the inset indicating a 7.8 % permanent strain. In Fig. 2.76, strength is plotted versus test temperature for partially-stabilized zirconia (henceforth: PSZ), cubic ZrO_2 crystals, polycrystalline PSZ and hot-pressed Si_3N_4. At 1500 °C, the cubic ZrO_2 specimens either failed in a brittle manner or exhibited significant plasticity, depending on their crystallographic orientation. The strengths of those cubic ZrO_2 specimens exhibiting plasticity depended significantly on strain rate, with their flexural strengths decreasing with decreasing strain rate.

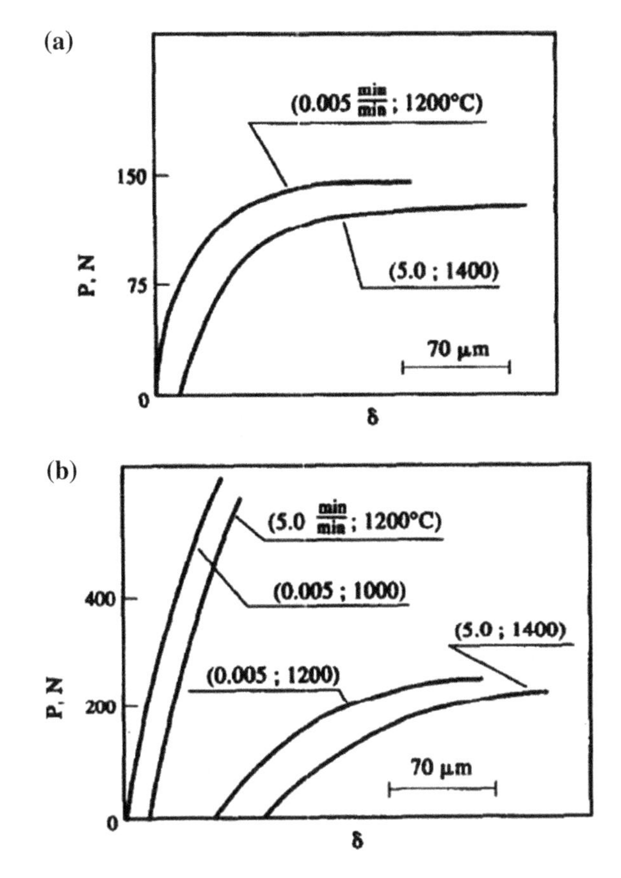

Fig. 2.73 Load versus deflection diagrams for **a** SN-1 and **b** Y-PSZ-3 under different conditions of deformation [18]. With kind permission of Elsevier

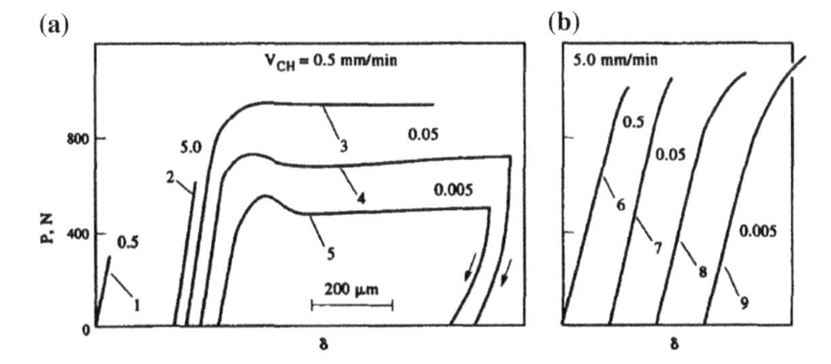

Fig. 2.74 Load versus deflection diagrams for **a** Y-FSZC-10 and **b** Y-PSZC-3 at various V_{ch} and **T** = 20 °C (1) and 1400 °C (2–9). *Arrows* show specimen unloading [18]. With kind permission of Elsevier

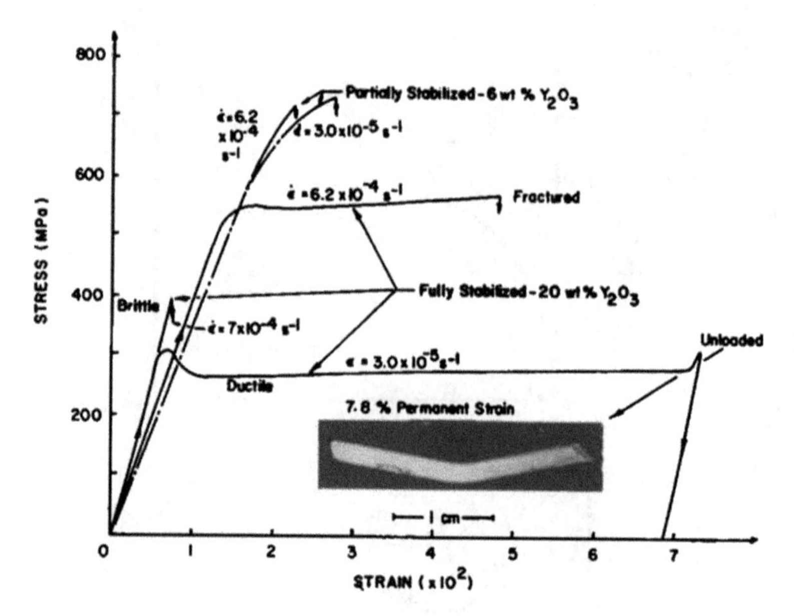

Fig. 2.75 Stress–strain behavior at 1500 °C of partially and fully stabilized zirconia single crystals vis. strain rate. Inset shows permanent deformation achieved in fully stabilized specimen [24]. With kind permission of Wiley and Sons

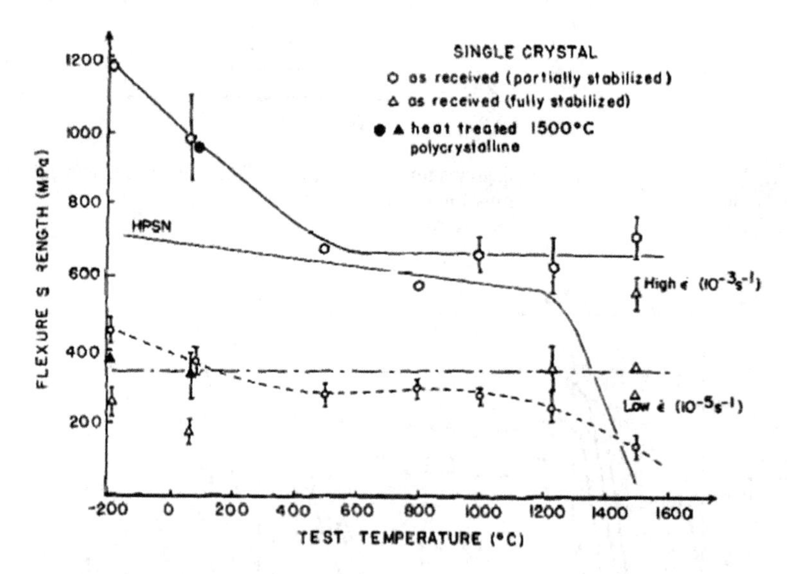

Fig. 2.76 Temperature dependence of flexural strength of partially and fully stabilized zirconia single crystals, polycrystalline PSZ, and hot-pressed Si_3N_4 (HPSN). PSZ is partially stabilized ZrO_2 with Y_2O_3 [24]. With kind permission of Wiley and Sons

The results for strength and toughness of PSZ single crystals indicate a significant potential for the development of high-strength, high-temperature structural ceramics, in which excellent retention of mechanical properties up to ~ 1500 °C is achieved. Thus, in contrast to many ceramics, the strength of zirconia crystals remains practically unchanged in partially-stabilized crystals and even increases in fully-stabilized ones.

2.3 Summary

The challenge of ceramists and materials scientists is to make ceramics ductile at RT and low temperatures, in order to extend the use of ceramic materials to more applications. This is not a trivial goal, since the nature of ionic and covalent bonding must be overcome by introducing dislocations and making their mobility possible. Dislocations are primarily responsible for plastic deformation. In the last decade, much effort has been made to develop ductile materials exhibiting proper slip planes that enable dislocation movement. Dislocations are present in ceramics and their role, in most ceramics, is best observed at high temperatures. However, at low and moderate temperatures, their motion is limited, since the stress required for their movement is high, usually above their fracture stress, so that fracture sets in before yielding. Different mechanisms may be responsible for the BDT, such as: grain size, an additional second phase, structural modification (preferably to nanostructure), etc. Single crystals are likely to be the best candidates for the development of appreciable ductility, since they have no inherent grain boundaries (or sliding) to interfere with dislocation motion.

References

1. Dieter GE (1986) Mechanical metallurgy, 2nd edn. McGraw-Hill Inc., New York
2. Hulse CO, Pask JA (2006) J Am Ceramic Soc 43, 373
3. Jillavenkatesa A, Dapkunas SJ, Lum, L-SH (2001) Special publication 960-1, Particle size characterization
4. Nagano T, Kaneko K, Zhan G-D, Mitomo M, Kim Y-W (2002) J Eur Ceram Soc 22:263
5. Nanjangud SC, Brezny R, Green DJ (1995) J Am Ceram Soc 78:266
6. Parker ER, Pask JA Washburn J (1962) Ductile ceramics research, California University, Berkley Institute of Engineering Research, Defense Technical Information Center
7. Pelleg J (2013) Mechanical properties of materials. Springer, Dordrecht
8. Prochazka S (1975) The role of boron and carbon in the sintering of silicon carbide. In: Popper P (ed.) Special ceramics, 6th edn. British Ceramic Research Association, Stoke-on-Trent, p 171
9. Rothman A, Kalabukhov S, Sverdlov N, Dariel MP, Frage N (2012) Int. J Appl Ceram Technol 1:11
10. Trunec M (2008) Ceramics—Silikáty 52:165

Further References

11. Anstis GR, Chantikul P, Lawn BR, Marshall DB (1981) J Am Ceram Soc 64:533
12. Basu B, Venkateswaran T, Kim D-Y (2006) J Am Ceram Soc 89:2405
13. Brunner D, Baghbadrani ST, Sigle W, Rühle M (2001) J Am Ceram Soc 84:1161
14. Castaing J, Cadoz J, Kirby SH (1981) J Phys (Paris), Colloq C3 42:43–47
15. Chen Y-f., Xiao S-l, Tian J, Xu L-j, Chen Y-y (2011) Trans Nonferrous Met Soc China 21:342
16. Fahrenholtz WG, Hilmas GE, Talmy IG, James A, Zaykoski J (2007) Am Ceram Soc 90:1347
17. Gogotsi GA (2003) Ceram Int 29:777
18. Gogotsi GA, Ostrovoy DY (1995) J Eur Cer Soc 15:271. (With kind permission of Elsevier)
19. Gong J, Wang J, Guan Z (2002) J Mater Sci 32:865
20. Gorun AE, Parker ER, Pask JA (1958) J Am Ceram Soc 41:151
21. Grady DE, Lipkin J (1980) Geophys Res Lett 7:255
22. Hall EO (1951) Proc Phys Soc London 643:747
23. Hulse CO, Copley SM, Pask JA (1963) J Am Ceram Soc 46:317
24. Ingel RP, Lewis D, Bender BA, Rice RW (1982) Commun Am Ceram Soc C-150:65
25. Ishihara S, Bill J, Aldinger F, Wakai F (2003) Mater Trans 44:794
26. Kim H-S, Roberts S (1994) J Am Ceram Soc 77:3099
27. Kim W-J, Wolfenstine J, Ruano OA, Frommeyer G, Sherby O (1992) Met Trans A 23A:527
28. Kim W-J, Wolfenstine J, Sherby O (1991) Acta Metal Mater 39:199
29. Kingery WD, Pappis J (1956) J Am Ceram Soc 39:64
30. Lam DCC, Lange FF, Evans AG (1994) J Am Ceram Soc 77:2113
31. Langdon TG (1994) Acta Metall At Mater 42:2437
32. Lankford T (1996) J Mater Sci Letters 15:745
33. Lankford T Jr (1995) Office of naval research, Arlington, VA 22217, prepared by Southwest Research Institute, San Antonio
34. Li W, Zhang X, Hong C, Han W, Han J (2009) J Eur Ceram Soc 29:779
35. Mangalaraja RV, Chandrasekhar BK, Manohar P (2003) J Mater Sci Eng A343:71
36. Miyoshi T, Funakubo H (2010) Jpn J Appl Phys 49, 09MD13
37. Palmqvist S (1962) Arch Eisenhuettenwes 33:629
38. Parker ER, Pask JA, Washburn J, Gorum AE, Luhman W (1958) J Metals 10:351
39. Petch NJ (1953) J Iron Steel Inst London 173:25
40. Shinoda Y, Nagano T, Gu H (1999) J Am Ceram Soc 82:2916
41. Srinivasan M, Stoebe TG (1970) J Appl Phys 41:3726
42. Stokes RJ, Johnston TL, Li CH (1959) Trans AIME 215:43744
43. Stokes RJ, Li CH (1963) J Am Ceram Soc 46:423
44. Suzuki H (1983) Ceramics 3
45. Taylor GI (1938) Plastic strain in metals. J Inst Metals 62:307
46. Tsukuma K, Vasahiko (1985) J Mat Sci 20:1178
47. Wakai F, Kondo N, Shinoda Y (1999) Curr Opin Solid State Mater Sci 4:461
48. Wakai F, Sakaguchi S, Matsuno Y (1986) Adv Ceram Mater 1:259
49. Wang C-M, Mitomo M, Emoto H (1997) J Mater Res 12:3266
50. Yan W, Li N, Li Y, Liu G, Han B, Xu J (2011) Bull Mater Sci 34(5):1109 (Indian Academy of Sciences)
51. Zhan G-D, Mitomo M, Nishimura T, Xie R-J, Sakuma T, Ikuhara Y (2000) J Am Ceram Soc 83:841

Chapter 3
Imperfections (Defects) in Ceramics

Abstract The periodic nature of crystalline materials can be interrupted by imperfections. The relevant imperfection determining the mechanical properties of ceramics are point defects, or dislocations, or both. The major point defects considered in the chapter are vacancies and interstitials, which are responsible for some observed phenomena via diffusional exchange with atoms in their vicinity. One such process relates to climb which is an essential process in creep phenomena. Edge dislocations are involved in the climb process which occurs by leaving the glide plane, either in the positive, or the negative direction. Point defect-atom exchange by diffusion is the basic mechanism. Although one can talk about point defect hardening, the important defects that determine the mechanical properties of materials are line defects, commonly known as dislocations (edge or screw character). Their presence in crystals is essential, because of the orders of difference between the theoretical and actual strength of materials. The presence of dislocations makes deformation easier by the application of smaller stress than would be required in their absence. Conservative motion of dislocations occurs by slip, whereas non-conservative motion is associated with climb. The strengthening of material is a consequence of retarding the motion of dislocations, either by their intersection, or by particles of a second phase or by grain boundaries. Closely associated with dislocations are partial dislocations which usually produce stacking faults when they form. Basically stacking faults are surface defects. The association of partial dislocations and stacking faults define the extended dislocation, which makes cross slip more difficult, thus strengthening the material against deformation. Various properties of dislocations are one of the subjects of this chapter.

3.1 Introduction

When the periodic arrangement of a crystal is interrupted, a deviation from a perfect and orderly arrangement of the array of the lattice points generally occurs. Such a deviation from the periodic arrangement may be localized, in the

J. Pelleg, *Mechanical Properties of Ceramics*, Solid Mechanics and Its Applications 213, DOI: 10.1007/978-3-319-04492-7_3,

immediate vicinity of an atom or a few atoms, or it may occur across microscopic regions of a crystal. On this basis, it is customary to classify imperfections as being either:

(a) point defects;
(b) line defects;
(c) planar defects; and
(d) volume defects.

Although a 'point defect' may be considered to be a volume defect of atomic dimensions, in this section the common term 'point defect' will be used. Impurity atoms are often also considered to be point defects. Perhaps a solute atom at a substitutional or interstitial site should also be considered as an imperfection, namely as a point defect, since there is a deviation from the original periodicity of the pure crystal. However, an impurity atom is present in a crystal unintentionally, whereas solute atoms are purposely added to a pure material. Generally, atoms are added to pure materials to enhance certain properties, mechanical or physical. Regarding mechanical properties, bear in mind that some impurities may strengthen a material, but others may be detrimental to its mechanical properties. Clearly, the common interest is to enhance the mechanical properties of a given material.

The basic, structural point defects in very pure crystals are the vacancies and the interstitials, the former representing a vacant lattice site, while the latter is an extra atom at a non-lattice site. Either one of them is highly localized and characterized, as mentioned above, by the disturbance around a single atomic site. A perfect crystal is thermodynamically stable only at absolute zero temperature. At any higher temperature, the crystal must contain a certain number of point defects. For example, it is probable that an atomic site is vacant at low temperature, i.e., a vacancy is only $\sim 10^{-6}$, whereas, at the melting point, this probability is $\sim 10^{-3}$. Thus, point defects are a thermodynamic feature, unlike other defects such as line defects.

'Line defect' is the term used for the various configurations of dislocations. Since this kind of defect, as mentioned above, extends through a microscopic region, a common term used is 'lattice imperfection'. Such dislocations have either of edge or screw orientations or are of a mixed character. Their connotations are derived from the way in which they propagate in crystals.

Various internal and external surfaces, such as grain boundaries and stacking faults [henceforth: SF], are considered as two-dimensional defects and they comprise the class known as planar defects.

Finally, voids, pores and precipitates are also defects that interrupt the periodicity of a crystal and are known as 'volume defects'. This chapter begins with various point defects followed by lattice defects; no extensive coverage of surface or volume defects is included here.

3.2 Point Defects

3.2.1 Schematic Illustration of Point Defects

3.2.1.1 Vacancy

The first materials under consideration are ionic crystals, rather than metals that have a simpler structure. Figure 3.1 illustrates a vacancy schematically; note where a lattice site is missing.

3.2.1.2 Interstitial

An atom of a lattice site may become displaced to an interstitial position, as shown in Fig. 3.2.

The displaced atom may be an atom of the crystal and, in that case, the term 'self- interstitialcy' is used, or it may be a foreign atom in the structure, located at an interstitial site. In Fig. 3.2, a self-interstitial or 'self-interstitialcy' is shown.

Point defects in crystals can have mixed configurations and are best illustrated by ionic systems, in which ions with positive and negative charges occur. Ceramics provide good examples of mixed point configurations. However, for the sake of clarity, no specific substance is discussed here. Examples of multiple point defects are shown below. The main requirement is to maintain charge neutrality, namely charges should be balanced. Figure 3.3 is an illustration of an ionic crystal representing AB compounds, where A is a metal, such as Na or Mg, and B is a non-metallic constituent, such as Cl or O.

Note that in Fig. 3.3 charge neutrality is maintained, since one positive and one negative ion are missing. The larger atom is a non-metallic element, an anion (carrying a negative charge) and the smaller one represents a metallic atom and is known as a 'cation' when charged. This kind of defect is known as a 'Schottkey defect'. Figure 3.4 shows a defect known as a 'Frenkel defect', in which a positively charged cation (marked red) is displaced from its normal site into an interstitial, leaving a positively charged vacancy behind. Charge neutrality is maintained, since no charged atom was removed from the structure, only displaced to a different location.

To summarize, Fig. 3.5 presents all the above point defects in one figure, showing the possible configurations in ceramics.

Both these kinds of vacancies may occur in ceramics, but, only cation interstitials are likely to form, because anion interstitials are too large and the very heavy distortion prevents such a defect formation.

As mentioned in the introduction, impurities are considered as point defects. There are no perfect crystals; in addition to point defects, they also contain a large number of various impurities.

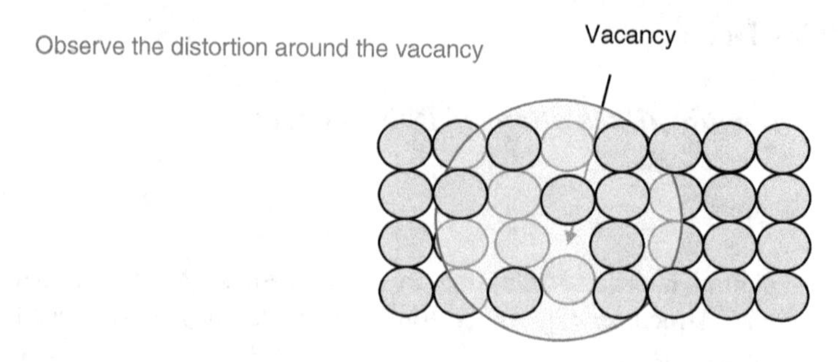

Fig. 3.1 Schematic illustration of a vacancy

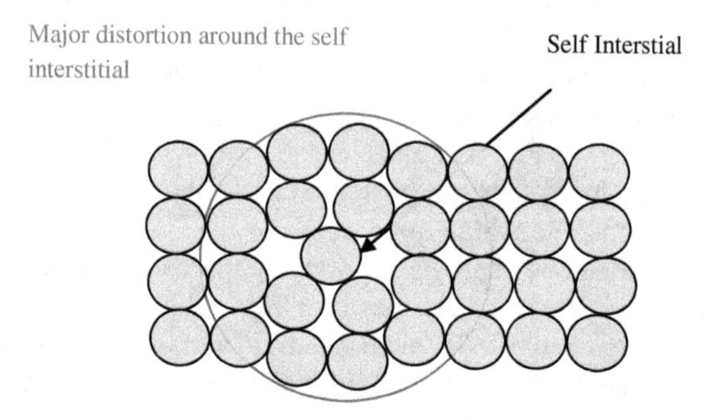

Fig. 3.2 Self-interstitialcy

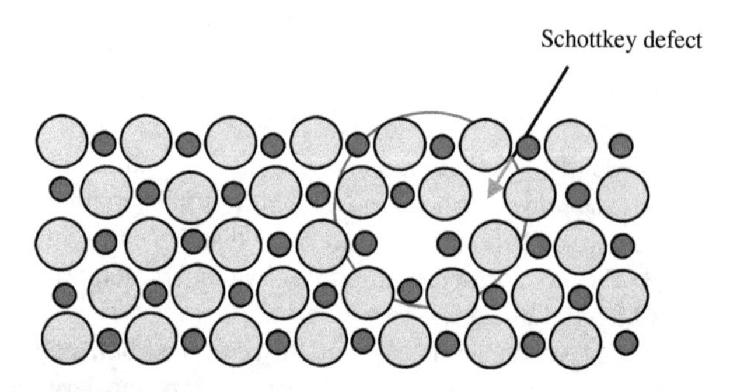

Fig. 3.3 Two vacancies with opposite signs; a 'Schottkey defect'

3.2.2 Thermodynamics of Vacancy Formation

As stated above, point defects can be characterized by thermodynamics. Though processes, like the formation of point defects, are associated with positive entropy,

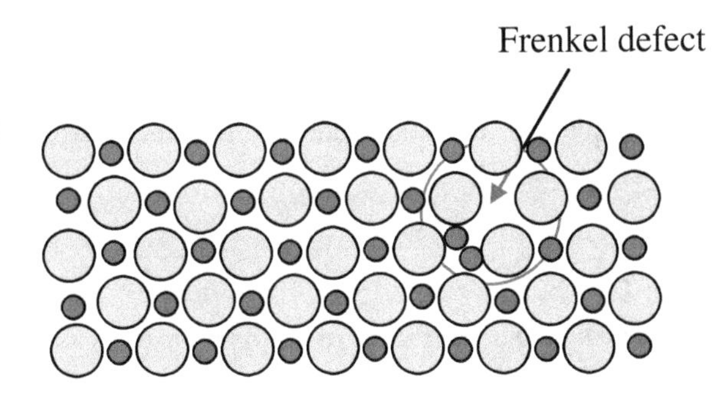

Fig. 3.4 A 'Frenkel defect' is shown; a positive ion has been displaced to an interstitial site leaving behind a vacancy. Charge neutrality is maintained

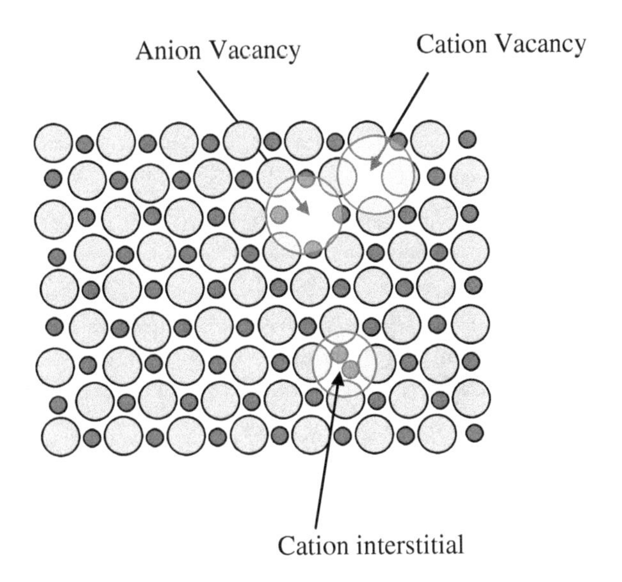

Fig. 3.5 An illustration of point defects in ceramics

they are usually expressed by their free energy. Forming a point defect requires a certain amount of positive work, which increases the internal energy of a crystal. Entropy contains a term known as 'configurational entropy' (or 'the entropy of mixing'), which expresses the way of distributing point defects in a material on the lattice sites. There are many ways to arrange point defects at a certain temperature. At any temperature above absolute zero, the free energy will be minimal for a certain amount of point defects, as determined by both terms of the free energy, F, given as:

$$F = nE_F - TS \tag{3.1}$$

where E_F is the energy to form a single defect and S is the entropy. As indicated, entropy has a configurational entropy component, which determines the number of ways that point defects may be arranged on the lattice sites of a crystal. Denote the number of ways to arrange defects on N lattice sites as W and express it as:

$$W = \frac{N(N-1)(N-2)(N-3)\ldots(N-n+3)(N-n+2(N-n+1)}{n!} \quad (3.2)$$

Multiply the numerator and denominator by (N–n)! and rearrange Eq. (3.2) giving:

$$W = \frac{N!}{(N-n)!n!} \quad (3.3)$$

The component of the aforementioned entropy is the 'configurational entropy', which is related to W on the basis of statistical mechanics as:

$$S = k \ln W = k \ln \frac{N!}{(N-n)!n!} \quad (3.4)$$

For large numbers, the logarithm of the factorial may be approximated by Stirling's relation:

$$\ln x! \cong x \ln x - x \quad (3.5)$$

Using Stirling's approximation, Eq. (3.4) may be expressed as:

$$S = k[N \ \ln N - (N-n)\ln(N-n) - n \ \ln n] \quad (3.6)$$

Equation (3.1) gives the free energy to form n defects where E_F refers to one defect. Substituting for S in the free energy Eqs. (3.1) from (3.6) and taking the derivative, with respect to the number of defects to get the minimum, results in:

$$\frac{\partial F}{\partial n} = 0 = E_F - kT\left(\ln \frac{N-n}{n}\right) \quad (3.7)$$

Equation (3.7) may be rearranged as:

$$\frac{n}{(N-n)} = \exp\left(-\frac{E_F}{kT}\right) \quad (3.8)$$

For n << N, Eq. (3.8) may be expressed as:

$$\frac{n}{N} = \exp\left(-\frac{E_F}{kT}\right) \quad (3.9)$$

This relation indicates that at absolute zero the concentration of defects is zero and with increasing temperature, the number of defects increases rapidly. The radiation of specimens largely increases its defect concentration. In the derivation of Eq. (3.9), only the configurational entropy was used and all the other entropy terms were neglected. Therefore, Eq. (3.9) is usually given as:

$$\frac{n}{N} = A \exp\left(-\frac{E_F}{kT}\right) \quad (3.10)$$

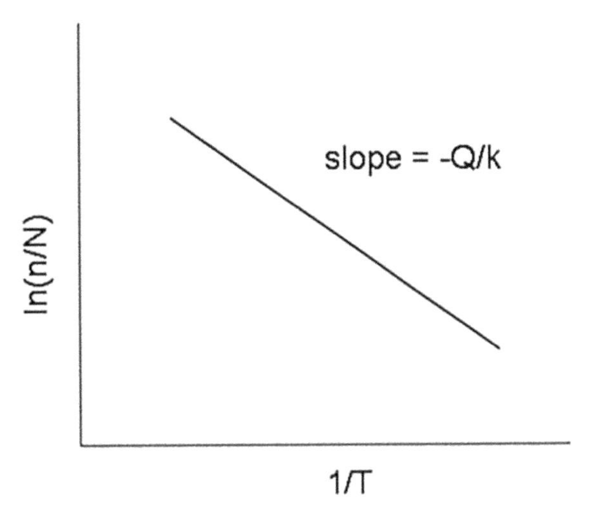

Fig. 3.6 The plot for determining the activation energy needed to form a defect

Recall from thermodynamics that in a process at constant volume ΔE = Q (see, for example, Cahn [4] or any book on thermodynamics). Equation (3.10) may also be expressed in terms of activation energy, Q, to form a vacancy:

$$\frac{n}{N} = A' \exp\left(-\frac{Q}{kT}\right) \tag{3.11}$$

A′ is known as the 'pre-exponential factor'. Taking a semi-logarithm on both sides of Eq. (3.11) enables the plotting of n/N versus 1/T and the derivation of Q by proper experimentation (see Fig. 3.6).

Of the experimental methods used to determine Q, an often-used method is the resistivity technique; resistivity is measured, since electrical conductivity in most ceramics is determined by the number and type of defects present. This technique involves quenching a specimen and then measuring and evaluating its resistivity, as given by:

$$\Delta\rho = A \exp(-\frac{E_F}{kT}) \tag{3.12}$$

A schematic illustration on a semi-logarithmic scale of Eq. (3.12) is shown in Fig. 3.7.

Water is the best quenching medium and is used in such experiments. Quenched specimens should not be oxidized and no reaction with the quenching medium should occur. Electrical conductivity (or its inverse) is directly related to the concentration of mobile electronic defects.

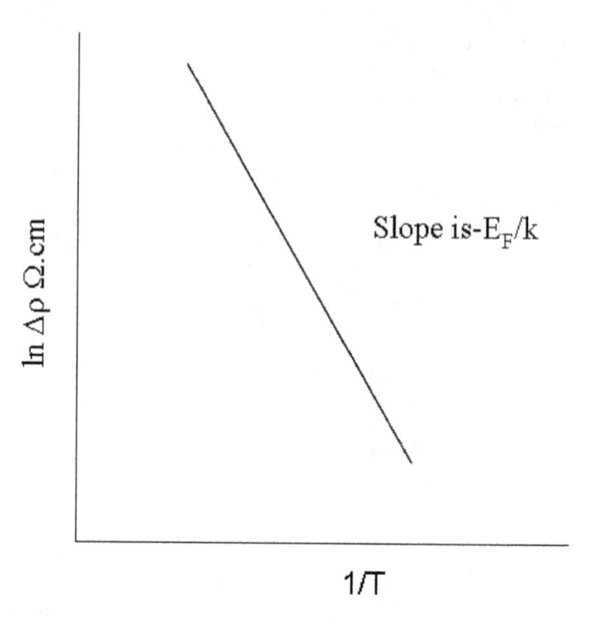

Fig. 3.7 Schematic, semi-logarithmic plot of quenched-in resistivity versus reciprocal absolute quench temperature

3.2.3 Strengthening (Hardening) in Ceramics by Point Defects

Radiation more effectively increases the concentration of point defects than an increase in temperature. To study the effect of point defects on mechanical properties, such as strength or hardness-related features, large amounts of point defects are preferable. Therefore, radiation is useful for studying the effects of point defects in crystals and studies on the effects of point defects are done on irradiated materials.

MgO.3A$1_2 0_3$ spinels provide a good example of the hardening effect of point defects. Neutron irradiation results in an increase in compressive strength and toughness [51]. In order to eliminate dislocation strengthening, irradiation must be performed at low temperatures, as in the case of the MgO.3Al_2O_3 spinel. Irradiation-induced hardening was observed only in a sample irradiated at 100 °C, with no clear dislocation loops observed. Thus, hardening is due mainly to point defects, i.e., interstitials and vacancies. Although small defect clusters are probably present in any sample, the assumption is, for the sake of simplicity, that the point defects are present in a dispersed state. During the irradiation process, interstitial and vacancy formation occurs. Swelling is an additional effect of irradiation and is associated with point defect formation. Annealing induces softening in ceramic spinels and the properties of the pre-radiated state are restored. During annealing, the point defects become mobile, like in metals. Spinel interstitials probably have higher mobility than vacancies and when they encounter vacancies, interstitial-vacancy recombination occurs. Recombination is the predominant mechanism in irradiated ceramics

Table 3.1 Irradiation conditions and the consequent swelling [51] (with kind permission of John Wiley and Sons and Professor Suematsu)

Irradiation temperature (°C)	Fluence (n/m^2)	Swelling (%)
100	8.3×10^{22}	0.083
470	2.4×10^{24}	0.008

*$E > 1$ MeV

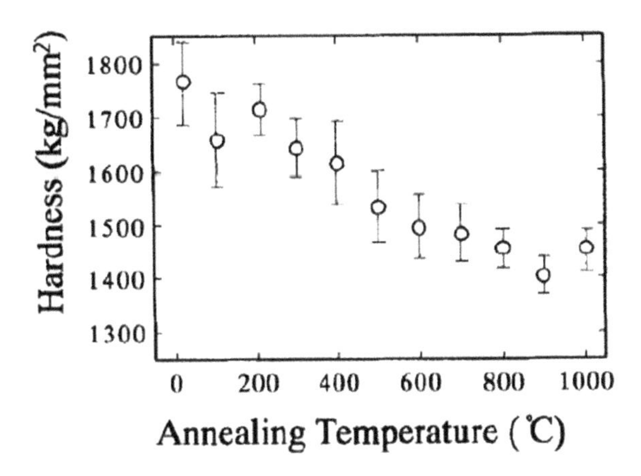

Fig. 3.8 Hardness irradiated at 100 °C and annealed in the temperature range shown [51]. With kind permission of John Wiley and Sons and Professor Suematsu

during annealing and has also been observed in SiC [39]. Irradiation at relatively low temperatures induces point defects with consequent strengthening. However, at high-temperature irradiation, softening occurs with a reduction of point defects (vacancies), because the more mobile interstitials recombine with the vacancies and, by doing so, the remaining number of point defects is reduced with the consequent reduction in strengthening. As stated above, irradiated samples softened after being annealed. Table 3.1 shows the irradiation conditions and the swelling after the irradiation. The smaller swelling after irradiation at 470 °C is the result of interstitial and vacancy recombination.

Figure 3.8 is a plot of hardness versus temperature of spinel irradiated at 100 °C with a fluence shown in Table 3.1 and then isochronally annealed.

The hardness decreases with increasing annealing temperature above the irradiation temperature and eventually reaches the value of the unirradiated spinel, when annealed at 600 °C and above. The most rapid decrease in hardness occurs between 200 and 600 °C, the latter is just above the temperature of 470 °C, at which the irradiation by 2.4×10^{24} n/m^2 induced practically no hardening. Compare the hardness change versus the temperature shown in Fig. 3.9 irradiated at 470 °C at a fluence of 2.4×10^{24} after isochronal annealing; almost no hardness change is observed following annealing, because most of the point defects have been annealed out. The length change in the same sample as that for Fig. 3.8 follows the same pattern of hardness versus temperature. The length change also decreases gradually at the irradiation temperature of 100 °C. Almost no change is

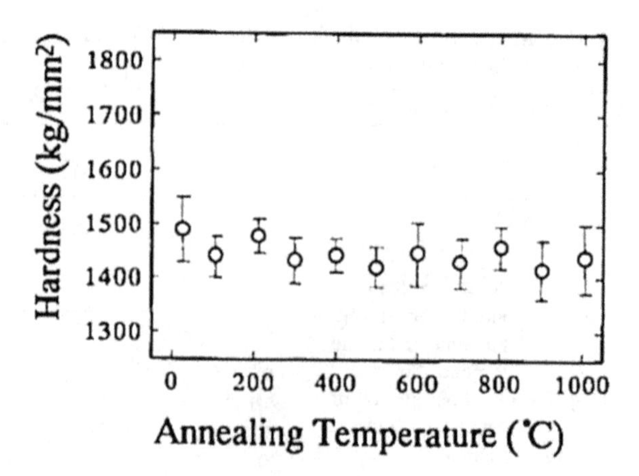

Fig. 3.9 Hardness of spinel irradiated at 470 °C at a fluence of 2.4×10^{24} n/m^2 after isochronal annealing [51]. With kind permission of John Wiley and Sons and Professor Suematsu

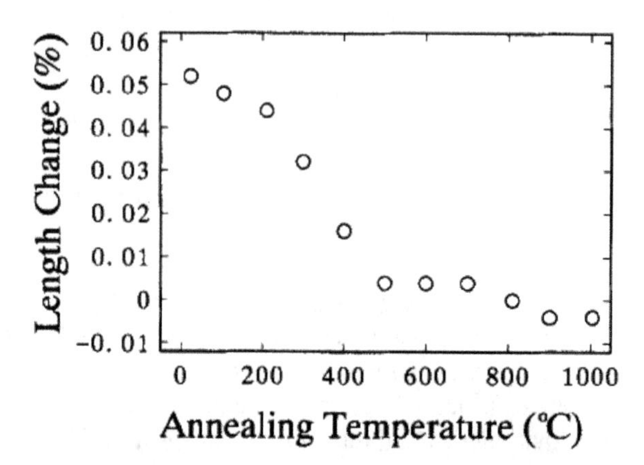

Fig. 3.10 Length change of spinel irradiated at 100 °C to a fluence of 8.3×10^{22} n/m^2 during isochronal annealing [51]. With kind permission of John Wiley and Sons and Professor Suematsu

observed at of 500 °C and higher, and the sample length becomes equal to that before irradiation (i.e., zero change). This is shown in Fig. 3.10, which should be compared to Fig. 3.11 for specimens irradiated at 470 °C to a fluence of 2.4×10^{24} n/m^2 after isochronal annealing. The length change is ~zero. This sample showed almost no swelling. As expected, practically no change in length was detected, which explains the absence of hardness change in Fig. 3.9.

TEM images have revealed the following. Irradiation at 100 °C (fluence 8.3×10^{22} n/m^2) did not induce observable dislocation loops and the defects produced are isolated point defects or small clusters. This means that irradiation-induced hardening in a sample irradiated at 100 °C, which had no clear dislocation loops, is due mainly to point defects, i.e., interstitials and vacancies. Probably Frenkel pairs are formed in this manner. At 470 °C with a fluence of 2.4×10^{24}, dislocation loops were also found by TEM observations

In general, an increase in hardness (or strength) is the result of retarded dislocation movement due to point defects introduced by the irradiation. This is

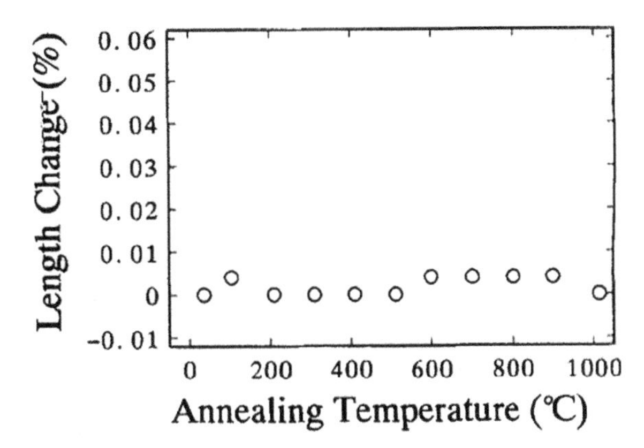

Fig. 3.11 Length change of spinel irradiated at 470 °C to a fluence of 2.4×10^{22} n/m^2 after isochronal annealing [51]. With kind permission of John Wiley and Sons and Professor Suematsu

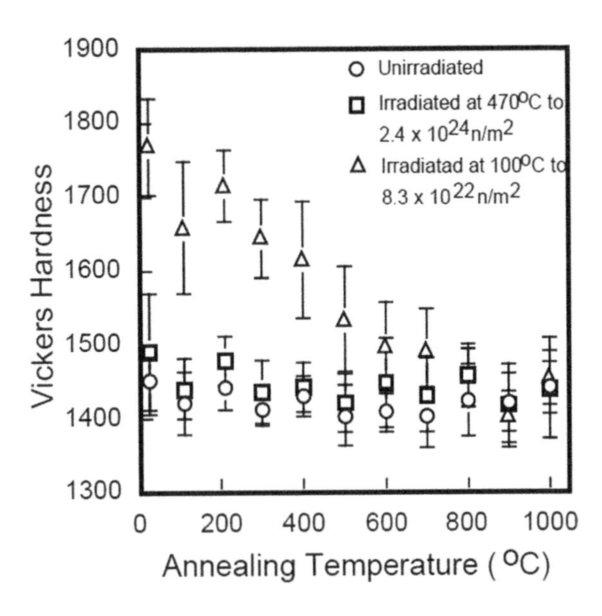

Fig. 3.12 Comparison of unirradiated, 100 °C and 470 °C irradiated samples of single crystal $MgO.3Al_2O_3$ after isochronal annealing [52]. With kind permission of John Wiley and Sons and Professor Suematsu

similar to strengthening by impurities, which is commonly known as 'dislocation pinning'.

Similar hardening results were also observed in single crystal $MgO.3Al_2O_3$, as can be seen in Fig. 3.12. Note that also in the single crystal the 100 °C irradiation gave a highest hardness result comparable to the irradiated specimen at 470 °C with the higher flux and to the unirradiated one. This trend is similar to the one observed in polycrystalline $MgO.3Al_2O_3$. Here, the mechanism of hardening is dislocation pinning.

Radiation-induced point defects are usually preferred over thermal point defects (obtained by quench-in from some higher temperature and freeze-in these point defects) for studying their effects on the physical and mechanical behavior of ceramics. Radiation affects mechanical properties by way of changes in strength, ductility, etc. These changes stem from changes in micro-structural characteristics. In general, the types of radiation applied can be neutron, ionic or X-ray and their corresponding effects on materials are:

(a) Impurity production usually occurs during neutron radiation. The resultant changes are related to fission and capture. The creation of hydrogen or helium when a proton or alpha particle becomes neutralized in a material may be involved in impurity formation. Atoms displaced from their normal positions due to radiation may leave behind vacancies and, simultaneously, interstitials are created;
(b) During ionic radiation, electrons may be removed from atoms to form ion pairs in the path of the charged particles;
(c) Thermal heating may be involved when a large amount of energy is released in a small volume.

In general, nuclear radiation tends to destroy the crystalline structure of a material and may eventually lead to amorphization. Radiation effects (damage) are mainly due to the formation of point defects. These imperfections alter the original properties of the material. As indicated earlier, point defects can interact with dislocations, inhibiting their slip process. This means that more energy will be required for dislocation slip initiation and that the resistance to penetration by hardness indentation and the stress required to initiate failure will increase, due to the resistance of the material. At the same time, radiation is accompanied by a decrease in energy to failure, namely fracture strength, (because of lower toughness) and ductility. These observations may be of special interest and importance for material exposed to radiation, such as in nuclear reactors.

Now, materials based on alumina will be considered. Figure 3.13 shows the effect of radiation fluence on hardness in Al_2O_3 at several temperatures. Observe that, in this and following graphs, the fluence is shown on the abscissa, which is proportional to the point defects induced in the material during radiation. In the above figure, ΔH is the difference between unirradiated and irradiated samples. During RT irradiation, three stages can be seen in the ΔH vis. fluence relation. ΔH increases with fluence and reaches a maximum at 5×10^{19} He^+/m^2 (stage I). At slightly higher fluences, ΔH decreases sharply (stage II). At fluences higher than 7×10^{19} He^+/m^2, ΔH increases again gradually with fluence (stage III). These stages are also observed at higher temperatures, but ΔH at stages I and III decreases with increasing temperature.

In Fig. 3.14, the changes in several ceramics are shown at the same fluence and at an irradiation temperature of 300 K. The radiation hardening at stage I in Al_2O_3 is attributed to both plastic and elastic hardening. This interpretation by Izumi et al. [11] is based on the dissipation of the elastic and plastic energies, W_e and W_p, respectively during the indentation process.

Figure 3.14 indicates the variation of W_e and W_p as a function of the He fluence for the materials investigated. The W_p in α-Al_2O_3 decreases at the beginning of irradiation and keeps a constant value up to 5×10^{19} He^+/m^2 (stage I), but then recovers to the same value as in an unirradiated sample (stage II). However, W_e in

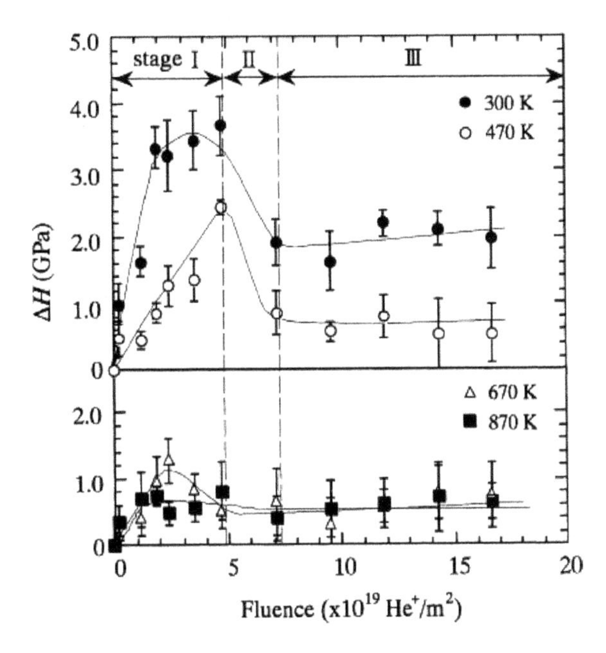

Fig. 3.13 Change in apparent hardness ΔH in α-Al_2O_3 as a function of 100 keV He-ion fluence at irradiation temperatures 300, 470, 670 and 870 K [11]. With kind permission of Elsevier

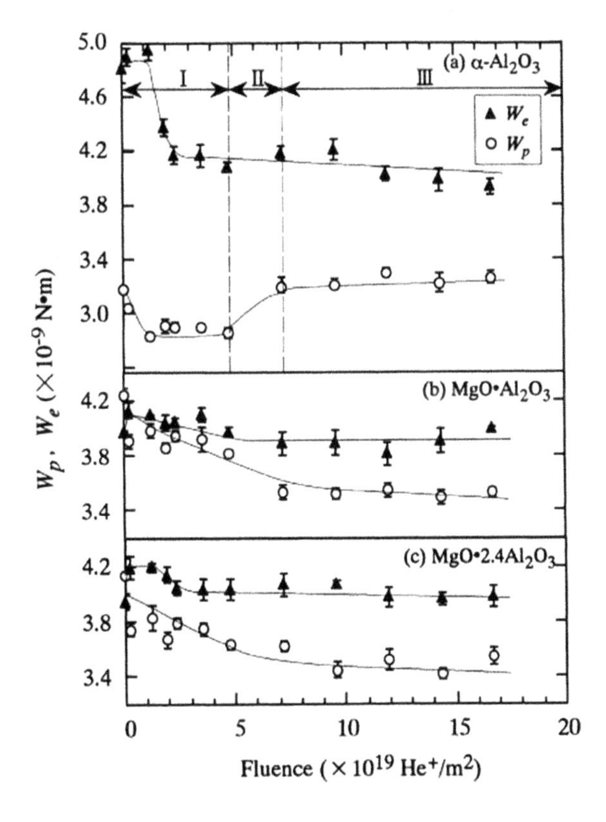

Fig. 3.14 Variations of W_p and W_e of **a** α-Al_2O_3, **b** MgO.Al_2O_3 and **c** MgO. 2.4Al_2O_3 against 100 keV He-ion fluence at an irradiation temperature of 300 K [11]. With kind permission of Elsevier

α-Al_2O_3 decreases sharply around 2×10^{19} He^+/m^2 (stage I) and gradually decreases with increasing fluence (stage II and III). Analogous variations of W_p and W_e are seen at higher temperatures of 470, 670 and 870 K, as indicated in Fig. 3.13. The amount of change in W_p and W_e, however, decreases with increasing temperature. Yet in $MgO.Al_2O_3$ and $MgO.2.4Al_2O_3$, W_p decreases monotonically with increasing fluence (though We increases slightly with fluence in the lower fluence range and almost recovers to the level of the unirradiated sample). From a comparison of Figs. 3.13 and 3.14, the increase in ΔH of α-Al_2O_3 in stage I is due to the decrease in W_p and W_e. Radiation hardening at stage I is, therefore, attributed to both plastic and elastic hardening. The decrease in ΔH of α-Al_2O_3 at stage II is attributed to the recovery of W_p to the same level of the unirradiated sample, indicating that plastic softening is the main reason for the decrease in ΔH at stages II and III. No significant changes are observed in W_e against fluence for $MgO.Al_2O_3$ and MgO $2.4Al_2O_3$ (Fig. 3.14b and c), indicating that the variation of ΔH is mainly due to the decrease of W_p with fluence. Thus, plastic and elastic hardenings are responsible for the variation of ΔH in α-Al_2O_3, whereas plastic hardening is the main hardening mechanism in $MgO.Al_2O_3$ and $MgO.2.4Al_2O_3$. In the case of α-Al_2O_3, ΔH has three stages, whereas in $MgO.Al_2O_3$ and $MgO.2.4Al_2O_3$ it increases monotonically with fluence. As such, the difference in the recombination rates of point defects in α-Al_2O_3, $MgO.Al_2O_3$ and $MgO.2.4Al_2O_3$ is probably the reason for the difference in their hardening mechanisms. Note that point defect formation and mobility in ceramics vary vastly from material to material.

In summary, TEM observations suggest that point defects and/or 'invisible' defect clusters are the main cause of the radiation hardening in α-Al_2O_3, $MgO.Al_2O_3$ and $MgO.2.4Al_2O_3$. Point defects and/or 'invisible' defect clusters are more effective in radiation hardening than 'visible' dislocation loops (also observed in microstructures). The decrease of ΔH in α-Al_2O_3 at stage II may be explained by the decrease in point defect concentration, due to aggregation with dislocation loops.

3.2.4 Point Defects in Amorphous Ceramics and Their Strengthening (Effect)

The most representative amorphous ceramics are the various glasses, among them the well-known silica glass. There are various silicate ceramics, such as oxide and halide glasses. Amorphous ceramics can be obtained by several means:

(a) Radiation damage;
(b) Polymer-derived ceramics (PDCs);
(c) Ion implantation, etc.;
(d) Mechanochemical activation (amorphous powders are hot-pressed and sintered).

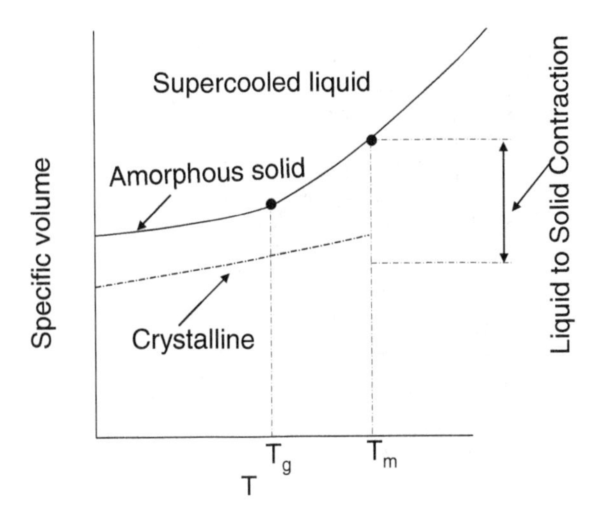

Fig. 3.15 Schematic illustration comparing crystalline and glassy amorphous materials (supercooled liquids)

A major characteristic of amorphous ceramics, such as glasses, is their glass temperature, T_g, which is usually observed by the change in the slope of a plot of specific volume against temperature, as shown schematically in Fig. 3.15 (other properties may be used instead of the volume, for example, enthalpy vs. temperature); however, some amorphous materials or ceramics first form as glassy materials and then crystallize. Point defects are present in various stages of formation in amorphous ceramics and have an effect on the mechanical behavior of those amorphous ceramics, including softening at some stage.

For instance, Zr-based, bulk amorphous metallic glasses, obtained by arc melting and drop casting (Zr55Cu30Al10Ni5), induction melting and injection casting (Zr52.5Al10Ti5Cu17.9Ni14.6), have been studied by both tensile and compressive tests at RT in various test environments [32]. That these materials were indeed amorphous may be seen in Fig. 3.16, which presents the TEM and XRD results.

TEM electron diffraction and XRD show no evidence of crystalline phases in the cast alloy bars. Tensile tests were performed on both alloys at RT in various test environments, including air, water, vacuum and dry oxygen. Figure 3.17 shows a photo of BAA-11 at the moment of tensile fracture.

Tables 3.2 and 3.3 show the test results for the Zr55Cu30Al10Ni5 (BAA 10) and BAA 11 specimens, respectively.

Shear deformation was observed in the tension test indicating some ductility of these glasses before fracture. Ductile fracture of the specimen is generally characterized by slip bands and necking down before fracture.

These ductile features of these glasses may be observed microscopically in Fig. 3.18a and b, respectively. However, the stress–strain curves of these Zr-based amorphous glasses did not display appreciable macroscopic plastic deformation prior to catastrophic fracture, rather they mainly deformed elastically, followed by catastrophic failure along their shear bands. Examination of the fracture regions

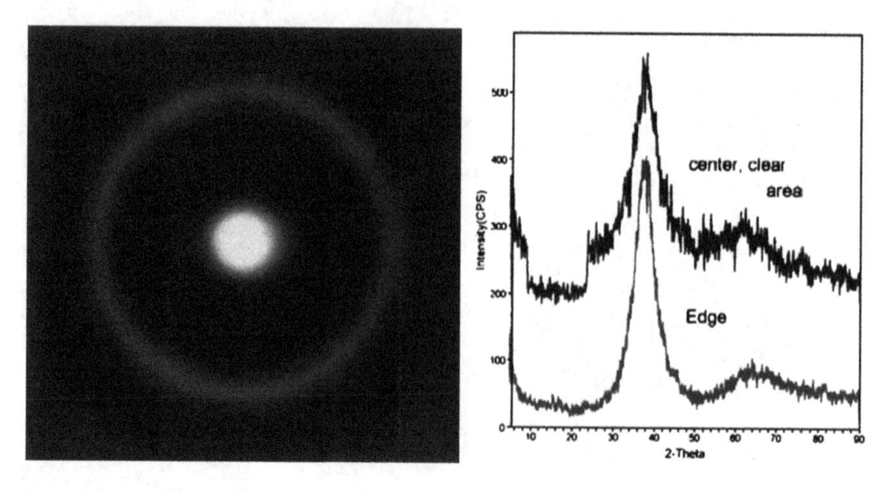

Fig. 3.16 Electron diffraction and XRD showing the formation of amorphous structures in Zr52.5Al10Ti5Cu17.9Ni14.6 (BAA-11) ingot [32]. With kind permission from Springer Science+Business Media B.V.

Fig. 3.17 Photo showing the moment of fracturing a BAA 11 specimen tested at room temperature in air [32]. With kind permission from Springer Science+Business Media B.V.

revealed ductile fracture features, as indicated above, resulting from the substantial increase in temperature, attributed to the conversion of the stored elastic strain energy into heat.

As mentioned above, one of the methods to amorphisize ceramics is by ion irradiation. Changes in strength properties occur following irradiation due to irradiation-induced structural modifications. Hardness measurement is an accepted method for evaluating the properties of ceramics after irradiation in the wake of atomic displacements. Information on ionization-induced changes in materials is particularly important in nuclear applications, where it is essential to evaluate reactor lifetimes. It is important to know the mechanical properties of materials

Table 3.2 Effect of Test Environment on Room-Temperature Tensile and Compressive Properties of BAA 11 [32] (with kind permission from Springer Science+Business Media B.V.)

Alloy number	Test environment	Elastic limit (MPa)	Fracture strength (MPa)	Fracture strain (Pct)
Tensile properties				
BAA-11-23	air	–	1650	1.8
BAA-11-16	air	–	1650	2.0
BAA-11-23	vacum	–	1750	~2.0
BAA-11-28	vacum	–	1720	~2.0
Compressive properties				
BAA-11-25	air	1670	1850	2.6
BAA-11-25	air	1770	1880	2.5

*Prepared from drop-cast 7-mm ingots

Table 3.3 Effect of Test Environment on Room-Temperature Tensile Properties of Zr55Cu30Al10Ni5 (BAA 10) [32] (with kind permission from Springer Science+Business Media B.V.)

Alloy preparation* (and Ingot diameter)	Test environment	Fracture strength (MPa)	Fracture strain (Pct)
Alloy Ingots made at Tohoku University			
IC, 5 mm	water	1210	1.40
IC, 5 mm	air	1310	–
IC, 5 mm	vacum	1410	–
IC, 5 mm	dry oxygen	1320	–
Alloy Ingots made at ORNL			
IC, 5 mm	water	1640	1.63
IC, 5 mm	air	1340	1.33
DC, 7 mm	air	1450	1.60

*IC = injection casting. DC = drop casting

and, specifically, the resistance of structural materials during long exposure to energetic particle irradiation. Defects, such as lattice and point defects, are induced during irradiation and influence the elastic and plastic properties of structural materials.

Figure 3.19 shows the changes in hardness as a function of the irradiation dose estimated by the nano-indentation finite element method [henceforth: NI-FEM]. In Fig. 3.19, it is seen for all the crystalline samples tested, that hardness increased during the initial stage of ion irradiation and then gradually decreased with the dose. The irradiation doses at which hardness starts to decrease, in the case of SiC and α-quartz, approximately correspond to their reported critical doses for amorphization. Equation (3.13), for the line fitting curves, is given by:

$$H = C_1H_1 + C_2H_2 + C_3H_3 \tag{3.13}$$

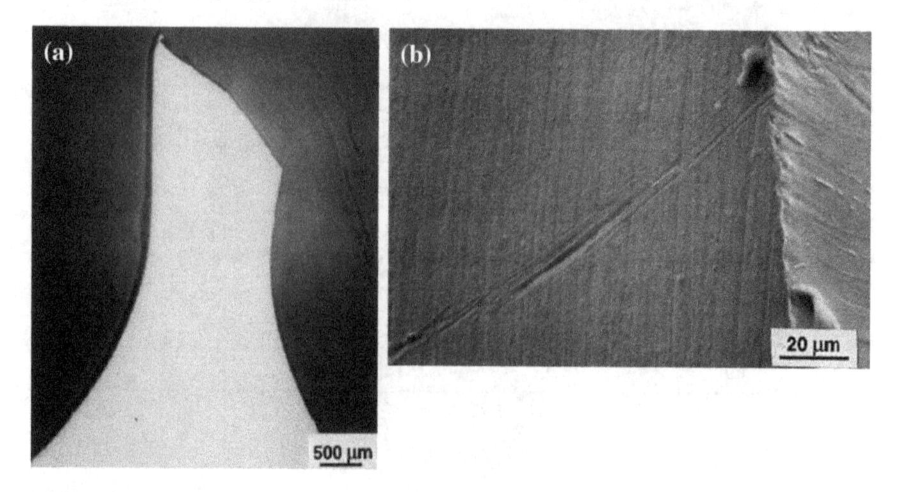

Fig. 3.18 Shear deformation indicating ductile behavior **a** neck formation in tension; **b** shear-deformation bands [32]. With kind permission from Springer Science+Business Media B.V.

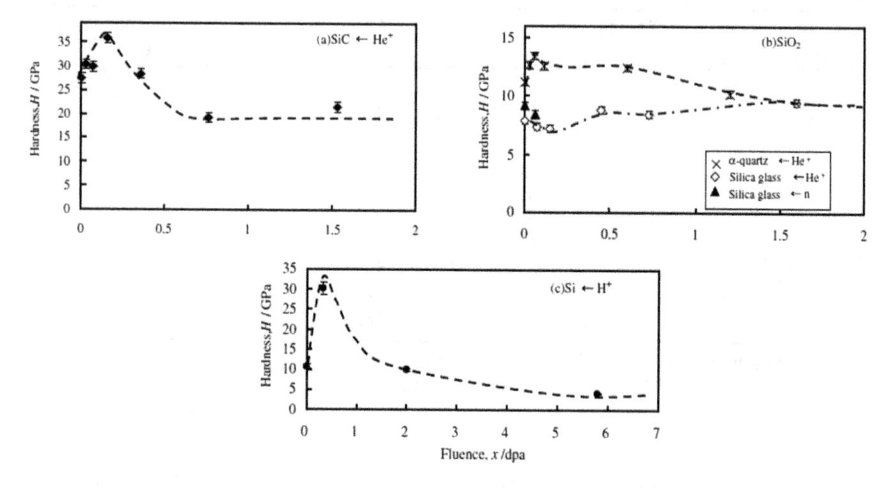

Fig. 3.19 Variation in hardness with irradiation dose in He^+ irradiated materials: **a** SiC; **b** SiO_2 (α-quartz and silica glass); **c** estimated by the present NI-FEM method. Broken lines are fitting curves based on Eq. (3.1). Solid triangles in (**b**) represent the experimental data of neutron-irradiated silica glass [42]. With kind permission of Ms. Etsuko Hasebe JIM staff member, The Japan Institute of Metals and Materials

and

$$C_1 + C_2 + C_3 = 1 \tag{3.14}$$

Ci and Hi (i = 1, 2, 3) correspond, respectively, to the concentration and hardness of the crystalline (non-irradiated) phase, hardened phase and amorphous (softened) phase.

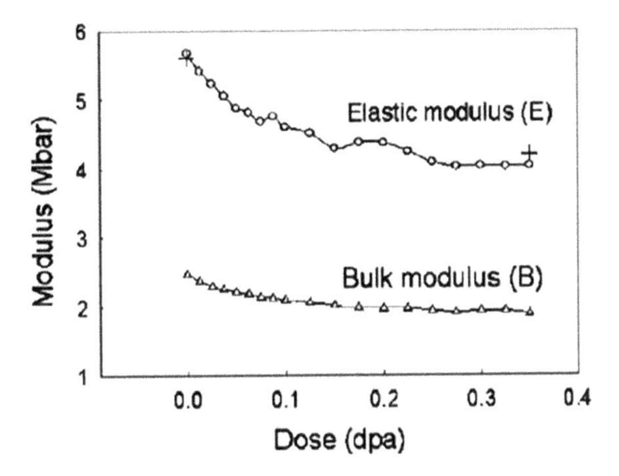

Fig. 3.20 Variation of the elastic modulus, E, and the bulk modulus, B, with dose. (dpa is displacement per atom) [18]. Free to copy

According to the work of Nakano on the mechanical properties of ion-irradiated ceramics, the increase in hardness of crystalline materials (Fig. 3.19) at the initial stage of irradiation suggests that covalent and ionic materials can be plastically deformed through local dislocation motion, which is hindered by the defects generated. These defects are likely to be point defects. The increase is seen to occur up to 0.1–0.2 dose per atom. This suggests that the deformation of crystalline ceramics is induced by dislocation motion. Upon prolonged irradiation, the hardness of crystalline materials decreases and, as a result of amorphization, asymptotically approaches a certain value. This behavior is basically related to a change in interatomic bonds, since irradiation induces bond breakage. The plastic deformation of amorphous structures proceeds slowly, like creep deformation.

Ion implantation was previously mentioned as being one of the methods for obtaining amorphous ceramics. Actually, the structure of ion-implanted ceramics may either be crystalline, having large concentrations of point defects, point defect clusters and dislocations, or it may be amorphous [35]. The implanted microstructure depends upon the implantation parameters, including ion species, fluence and substrate temperature. The chemical bonding present in a ceramic also plays a significant role. The amorphous state may contain different short-range order for different implanted ion species. Covalent-bonded SiC is amorphized at deposited damage-energy densities of 0.02 keV/atom at RT, but remains crystalline to values as high as 1.6 keV/atom for implantation at 1050 K. Therefore, effective amorphization is performed at low temperatures [35]. There is a critical, temperature-dependent dose for amorphization. Some researchers claim that Frenkel pairs and 'antisite defects' (a type of point defect distinct from a vacancy, interstitial or impurity) play significant roles in the amorphization process [33]. Molecular dynamic [henceforth: MD] calculation results suggest that a large number of irradiation-induced Frenkel pairs are formed in metastable configurations and majority of close pairs would recombine during annealing at 200 and 300 K. Figure 3.20 shows typical results for the variations of elastic modulus E, and bulk modulus, B.

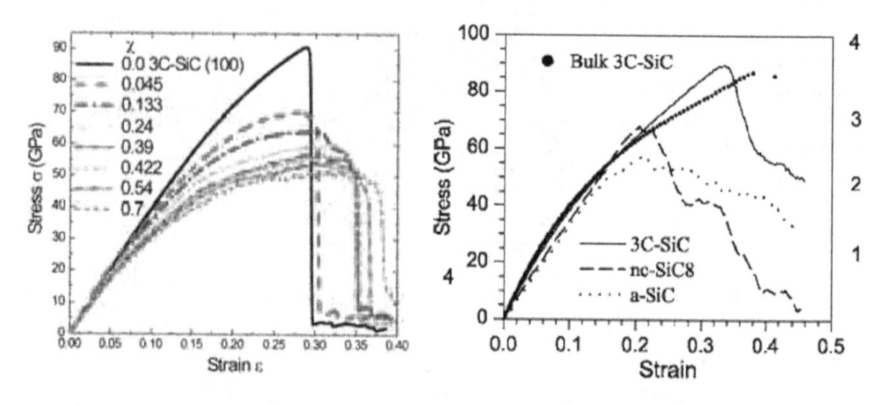

Fig. 3.21 Stress versus strain curves: **a** for *a*-SiC samples with varying chemical disorder at a strain rate of $10^8\ s^{-1}$. **b** Stress versus strain curves for 3C-SiC, nc-SiC and melt-quenched *a*-SiC with an extension rate of 100 m/s [18, 29]. (dpa is displacement per atom)

As seen in Fig. 3.20, these elastic properties decrease rapidly at doses less than 0.1 MD-dpa and the decrease becomes smaller at high dose levels. Due to the dose dependence of the formation and coalescence of point defects and small clusters, it was concluded [18] that point defects and small clusters contribute much more than topological disorder to the degradation of the elastic properties of *a*-SiC. The elastic constants and elastic modulus showed the expected softening behavior under irradiation for the dose range of interest.

Stress–strain curves for amorphous SiC samples (*a* -SiC) are indicated in Fig. 3.21 for various values of the chemical disorder, χ. Experimental simulation and MD calculations indicate a strong correlation between chemical and topological disorders. To maintain topological perfection for $\chi \geq 0.54$ seems impossible and so a stable amorphous structure is achieved.

Point defects, small clusters and topological disorder, as forms of defect accumulation, may somehow indicate the dependence of the mechanical properties of a-SiC on the disordered microstructure. For example, by the irradiation-induced amorphization of SiC, it was shown via MD calculations that a fully amorphous state is reached at a dose of about 0.28 MD-dpa and it was suggested that point defects and small clusters may contribute more significantly to the changes of elastic constants than the topological disorder associated with amorphization dose. These changes involve the degradation of the elastic properties of a-SiC, attributed to the dose-dependent formation of point defects and small clusters. As stated for $\chi \geq 0.54$, topological perfection seems impossible to maintain though a stable amorphous structure is achieved. In Fig. 3.21, simulated axial tensile testing is carried out on a set of SiC assemblies with varying chemical disorder, χ, representing a range of disordered structures from crystalline to completely amorphous. The full stress–strain dependencies for different SiC assemblies with varying χ are shown. An appreciable softening of SiC, after the stress reaches σ_{max}, is evidenced by the decrease of the stress before fracture. This ductile-like behavior increases with increasing χ and a pronounced plastic-flow plateau is observed with increased strain. Such plastic-like behavior is reproduced in the tension of simulated melt-quenched a-SiC, with a drastic reduction in the number of tetrahedrally-coordinated atoms [29]. Note that in Fig. 3.21b the stress–strain dependencies of the SiC structures were obtained from rod extension tests.

In single-crystal Al_2O_3 (i.e., sapphire), optical absorption measurements provide ample evidence for charged-point defect formation [35]. The presence of an 'F-center' (i.e., an oxygen vacancy containing two electrons) and an 'F^+- center' (i.e., an oxygen vacancy containing one electron) has been definitively established in both doped crystals and irradiated crystals. The ion implantation of ceramics, such as Al_2O_3 and SiC, may produce either a highly damaged, but crystalline, surface layer or an amorphous surface [36]. The specific structure depends upon the implantation parameters. Studies using microindentation show that a crystalline-implanted surface has a higher hardness (by 10–50 %) than a corresponding unimplanted crystal, but the MOE is essentially unchanged. The hardness and elastic modulus of amorphous implanted surfaces are less than those of crystalline materials (See Fig. 3.21b). Estimates of the residual stress may be obtained from microindentation tests.

3.3 Introduction to Dislocations

3.3.1 Introduction

Important defects that determine the mechanical properties of materials are line defects, commonly known as 'dislocations'. The original postulate for the existence of such defects was put forward in 1934 by Taylor, Orowan and Polanyi, the fathers of modern dislocation theory. Their postulate was intended to explain the large discrepancies in strength between the theory and actual observations of deformation. Their postulate for solids has been confirmed by TEM, field ion microscopy [henceforth: FIM] and atom probe techniques, which permit direct observations of dislocations at high magnification on an atomic scale. Etch-pit techniques may also be used as indirect techniques to detect the presence of dislocations in solids. An exemplary use of the etch pit technique in a grain of polycrystalline Nb is illustrated in Fig. 3.22. The etch pits outline the substructure inside the grain, as do the dislocation arrays.

This technique involves etching a surface of the material with an etching solution appropriate to the material under study. When a dislocation line emerges from the surface of a metallic material and intersects it, the local strain field existing around it increases the relative susceptibility of that material to etching and an etch pit of some geometrical shape forms. These etch pits can be counted,

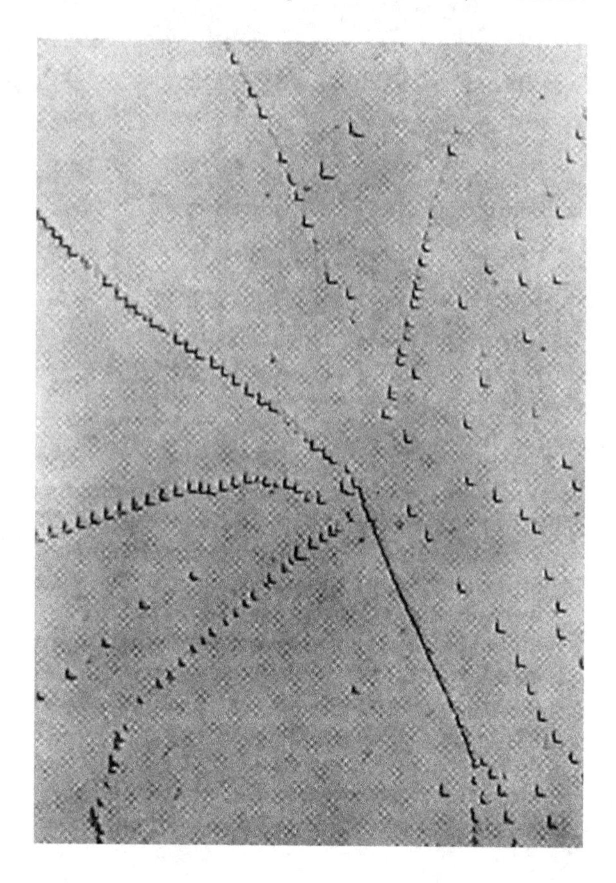

Fig. 3.22 Etch pits in one grain of a polycrystalline Nb, outlining the substructure in the grain [14]

giving some idea of the density of the dislocations. The shapes of these etch pits are orientation dependent [14]. Such dislocation structures in the TEM bright field [henceforth: BF] in Y_2O_3-stabilized zirconia [henceforth: YSZ] are shown in Fig. 3.23. Cubic zirconia has a fluoric-type structure and $\{001\}\ \langle 110 \rangle$ and $\{111\}\ \langle 110 \rangle$ slip systems [28]. In Fig. 3.23, the main slip system is the $\{001\}\ \langle 110 \rangle$ system. FIM micrographs, before and after field evaporation, appear in Fig. 3.24 [46]. The location of the dislocation is outlined by a closed red loop. FIM (developed by Muller [12]) allows for the study of structures on an atomic scale, at a resolution of 2–3 Å. By means of field evaporation, a layer of atoms can be peeled off in order to observe the successive layer of the remaining structure and to study point-defect concentration and dislocations. Sharp tungsten tips, electro-polished to a hemispherical shape of ~300 Å or less and positively charged at high voltages in the kV range, are used. After reaching a high vacuum, He atoms are bled into the system, which is then ionized and produces an ionic image on a screen equipped with a channel plate. The FIM micrograph in Fig. 3.24 shows that the field-evaporated zirconiated W tip at 8.75 kV, obtained at a He pressure of 2.93×10^{-3} Pa and 7.48 kV, is oriented at $\langle 100 \rangle$.

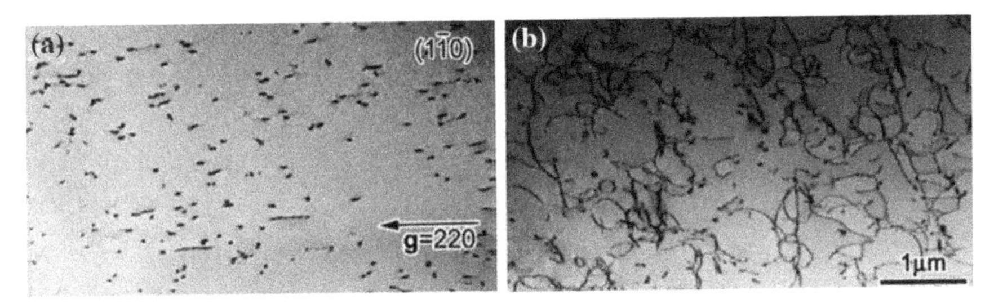

Fig. 3.23 Bright field images of the dislocation structures in (YSZ) for the specimens deformed up to **a** 1 and **b** 10 % strain with the incident beam parallel to the $[1\bar{1}0]$ direction, which is parallel to the dislocation lines introduced by primary slip [28]. With kind permission of Professor Yuichi Ikuhara

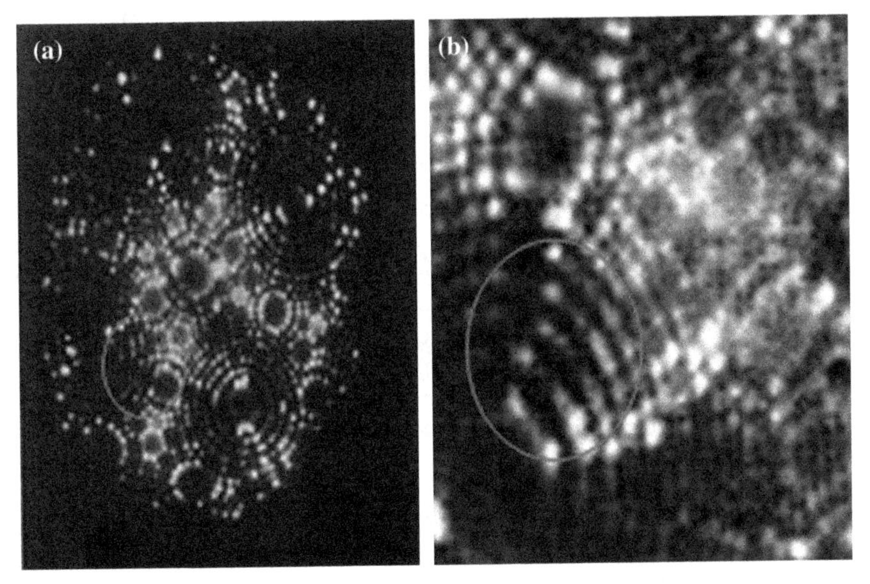

Fig. 3.24 **a** Field ion micrograph of clean W tip outlining the location of a dislocation; $\langle 100 \rangle$ oriented tip; 7.48 kV and He pressure of 2.93×10^{-3} Pa. **b**. Pattern after field evaporation at 9.76 kV. Dislocation is outlined in the micrograph after stripping Zr from the tip. 8.75 kV [46]

The line defects discussed in the next part of this chapter determine not only the role played by dislocations in the deformation characteristics of materials, but in determining their mechanical properties. Understanding dislocation theory makes it possible to better grasp the properties of materials, especially the mechanical properties of engineering materials, such as ceramics. However, before discussing the theoretical strength of crystals, the characteristics of dislocation lines should be considered.

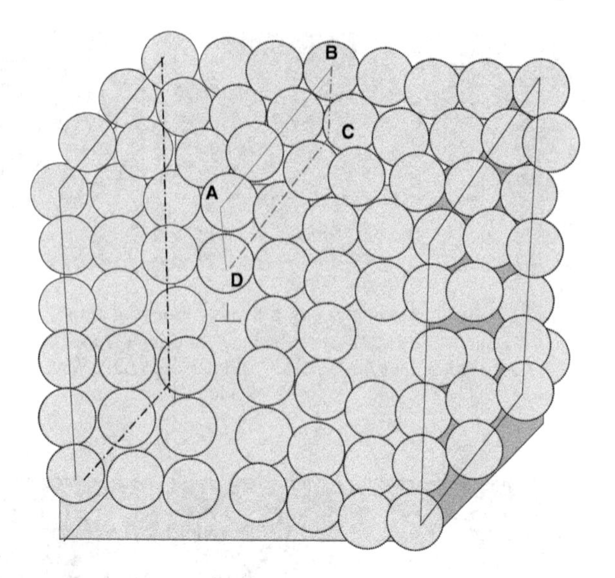

Fig. 3.25 Line DC defines an edge dislocation, which is the termination of an extra plane in a cubic crystal. A missing atom is also shown which is a vacancy [14]

3.3.2 Deformation or No Deformation in Crystals

Assuming familiarity with the concept of a line defect, this section deals first with what makes a material brittle or deformable without detailed consideration of various aspects of dislocations. A schematic figure of a dislocation (edge) is shown in Fig. 3.25 (more on its formation and motion will be discussed below). The edge of the red line defines an 'edge dislocation'. The parameter of 'width' is used as a measure of the disruption of a perfect crystal, as a consequence of the presence of a line dislocation. One may also define a 'dislocation' as the area between slipped and unslipped locations following deformation, in which case 'width' basically indicates the number of neighboring atoms pushed aside while maintaining bond continuity across the plane (slip plane) below the edge where the dislocation ends. Thus, 'width' means the width of the area on a slip plane where the atoms are out of register by a certain amount, conventionally given as one-half the maximum shear strain. Dislocated individual atoms move less than the Burgers vector when a material is deformed along its slip plane.

The 'width' of a dislocation determines whether deformation will occur or not and whether a material will show brittleness. Figure 3.26 is a schematic illustration of the width of a dislocation and the disregistry of the atoms in the dislocation surroundings.

At the center of a perfect edge dislocation, the disregistry is always b/2. The edge dislocation width, w, is defined as the distance over which the disregistry is greater than one quarter of the magnitude of the Burgers vector, b. The disregistry is the magnitude of the displacement of the atoms from their perfect crystal positions. When the displacement of the upper half-plane is b, the crystal is again in complete registry. The magnitude of the 'width' is defined according to the

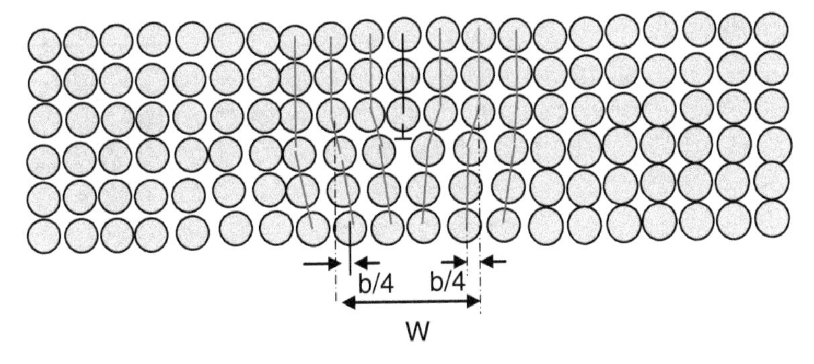

Fig. 3.26 A schematic illustration showing the width, w, of a dislocation. Actual distortion of the atoms is not indicated. The magnitude, b, is the slip vector (Burgers vector)

number of atomic spacings (an atomic spacing equals the distance between the atomic nuclei), where the disregistry is pronounced. If w is of the order of *one or two atomic spacings*, the width of the dislocation is considered narrow; contrasting cases have several atomic spacings and are considered to have a wide dimension; the dislocation is 'wide'. Glide in a material readily occurs when dislocation is wide; this occurs in materials classified as ductile. The opposite is true in cases with narrow dislocation width, such as most ceramics at RT, in which no glide generally occurs; they are brittle and most feature high strength. The tendency to fail suddenly, with little or no *plastic deformation*, limits the wider use of ceramics than exists today. When the dislocation width is between narrow and wide dimensions, materials with low ductility can be expected.

Several researchers have observed that the halides and some sulphides and carbonates are among the ceramics showing plasticity at ambient temperatures. Of the oxide ceramics, MgO is one that shows ductility at RT, a behavior quite exceptional in the family of oxide ceramics. Usually, most plastic deformation experiments are conducted at high temperatures under compression.

3.3.3 The Theoretical Strength of Crystals

Although Orowan [13] has calculated the strength under tension of a perfect crystal, the following section will discuss theoretical strength under shear, since plastic deformation involves shear stress acting on some preferential slip plane. Frenkel [26] has calculated the critical amount of shear stress required to move adjacent atomic planes past one another, i.e., the energy-per-unit area involved in shearing an atomic layer from its equilibrium configuration past the one below it. Figure 3.27 schematically depicts two rows of atoms in a closely-packed structure, one of which is sheared over the other one inter-atomic distance.

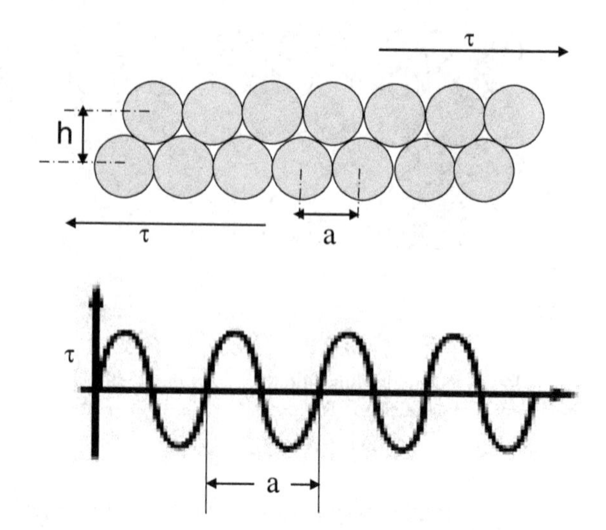

Fig. 3.27 The slip of one row of atoms over another by a distance, a, of the lattice spacing. A sinusoidal motion is assumed for the application of shear stress [14]

The top row had to move a distance, a, from one atomic equilibrium position to another. An approximation, made using a sinusoidal function, was employed to evaluate the theoretical shear stress, τ, given below as:

$$\tau = k \sin \frac{2\pi x}{a} \tag{3.15}$$

The applied shear stress, τ, produces a displacement, x, of “a”, which is basically the inter-atomic distance. This shear brings the atoms in the two rows of atoms in the crystal into registry again. For small displacements, the value of the sine in Eq. (3.15) equals: $\frac{2\pi x}{a}$ and, thus, Eq. (3.15) becomes

$$\tau = k \frac{2\pi x}{a} \tag{3.15a}$$

Expressing Hooke’s Law in terms of the shear modulus and shear stress one obtains:

$$\tau = \frac{Gx}{h} \tag{3.16}$$

where h is the height between the two rows (or planes of atoms) and G is the shear modulus. x/h is obviously the shear strain, since x represents the displacement of the atoms from their equilibrium position. (recalling that the shear strain is $\tan\alpha \cong \sin\alpha$ and for small values of tan or sin it is $\cong$ the value itself, namely, $\tan\alpha = x/h$ and thus the shear strain $= x/h$): By equating Eqs. (3.16) and (3.15a), a value can be obtained for k as:

$$k = \frac{Ga}{2\pi h} \tag{3.15a}$$

Substituting this value of k into Eq. (3.15), we can write:

$$\tau = \frac{Ga}{2\pi h}\sin\left(\frac{2\pi x}{a}\right) \tag{3.15b}$$

When $\sin\frac{2\pi x}{a} = 1$, the shear stress is maximal; this occurs at x = a/4 (since $\sin(\pi/2) = 1$) and Eq. (3.15b) becomes:

$$\tau_{\max} = \tau_0 = \frac{Ga}{2\pi h} \tag{3.17}$$

At a/4, the lattice becomes unstable, i.e., it will yield at a value of τ_0, which is the critical shear stress. Since h ≅ a, the maximum shear stress for the instability of the lattice is about:

$$\tau_{\max} = \tau_0 \sim \frac{G}{2\pi} \sim \frac{G}{6} \tag{3.18}$$

For metals, G is ~27–77 GPa (for W ~ 161). Thus, according to Eq. (3.18), τ_0 ~ 4.5–13 GPa. The observed critical shear-stress value is ~0.0069 GPa and, as such are two to three orders of magnitude smaller than the theoretical value. In a more realistic, refined approach, without using a sinusoidal function, $\tau_{\max}$ is ≈ G/10–G/30. Even these refined values are ~2 orders of magnitude greater than the experimental values.

The inevitable conclusion is that real crystals must contain defects, such as dislocations suggested by Taylor, Orowan and Polanyi, which reduce their mechanical strength or, more specifically, their resistance to slip when the applied stress reaches a critical value. The 1934 postulate showed that shear is possible at much lower stresses than in a perfect crystal.

3.3.4 The Geometric Characterization of Dislocations

As indicated in Sect. 3.3.1, dislocations are line defects. The two basic types of dislocations are the edge and screw dislocations. A schematic three-dimensional (3D) illustration of an edge dislocation appears in Fig. 3.25. A (100) plane of Fig. 3.25 in a simple cubic crystal is illustrated schematically in Fig. 3.28. This illustration will help to define the Burgers vector later on. Figure 3.29 is a schematic view of edge and screw dislocations.

Again, a dislocation is characterized by its 'Burgers vector', which is defined by what is known as the 'Burgers circuit'. A dislocation has two properties–a line direction and its Burgers vector. In an edge dislocation, the Burgers vector is perpendicular to the line direction, whereas, in screw dislocations, it is parallel. Figure 3.30 illustrates how to describe a 'Burgers circuit' and a 'Burgers vector'. There are two ways to proceed with the formation of the stepwise circuits. One

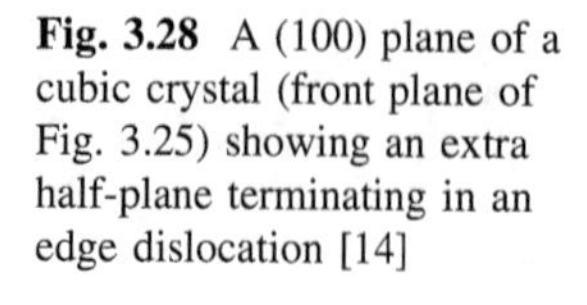

Fig. 3.28 A (100) plane of a cubic crystal (front plane of Fig. 3.25) showing an extra half-plane terminating in an edge dislocation [14]

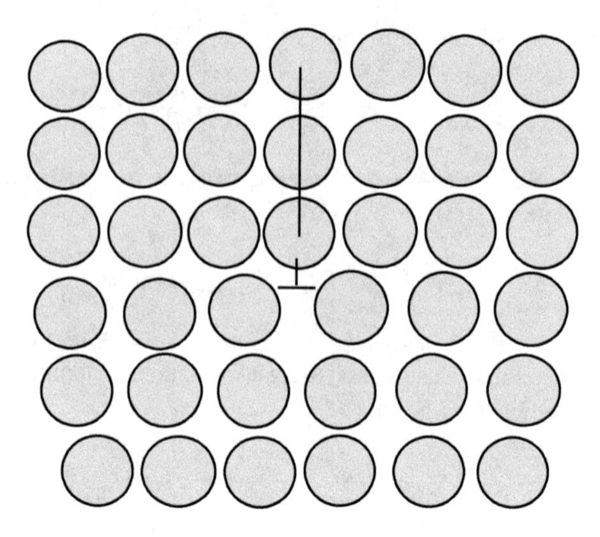

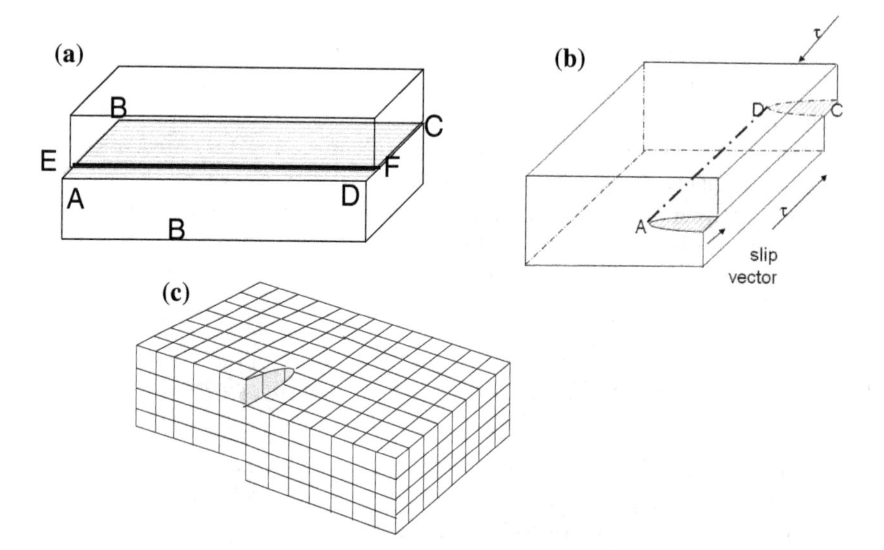

Fig. 3.29 Dislocations outlined: **a** an edge dislocation, EF; **b** a screw dislocation, AD, formed by slip by a pair of shear stresses; **c** the same as **b**, but more detailed [14]

way is to start a circuit from S (for start) and end it at F (for finish), as shown in Fig. 3.30.

This circuit closes perfectly on itself (see Fig. 3.30a). In a chosen region of the crystal, a loop is made by a stepwise procedure, moving from one atomic position to the next, starting at some point, S. The conventional direction is clockwise and, while performing this procedure, the number of lattice steps in each direction is

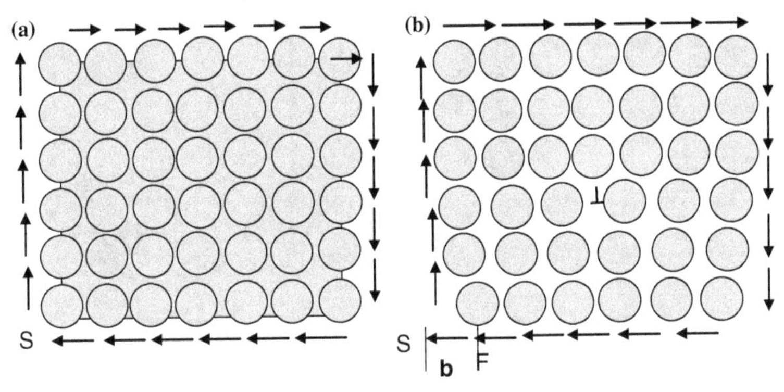

Fig. 3.30 The Burgers circuits in a perfect crystal and in a faulted one. The steps of the circuit from one atomic position to the next must be the same in both lattices: **a** the circuit in a dislocation-free crystal; **b** the circuit around a dislocation. **b** is the Burgers vector [14]

counted until the loop is closed at F. The circuit finishes at the starting point S (not shown for clarity). In Fig. 3.30a, this represents a perfect region of a dislocation free crystal. In the faulted region of a crystal, a dislocation is shown. Making the same circuit, as done in the perfect crystal, no closure of the circuit occurs and the excess step indicates the presence of a dislocation as seen in Fig. 3.30b. The excess step is the Burgers vector. The number of excess steps is an indication of the number of dislocations in the faulty crystal.

The Burgers vector, which is a unit-slip vector, is supposed to be the same as an interatomic distance of the lattice; however, this is not so, as seen in Fig. 3.30b. In the vicinity of this dislocation, the lattice is strained elastically, as indicated. Performing a Burgers circuit by the second method, the Burgers vector indeed has the interatomic distance shown in Fig. 3.31b. Initially, a closed circuit is made in a crystal containing a dislocation, starting at point S and finishing at F; the same lattice point is chosen for this step-by-step procedure as was done for Fig. 3.30 (shown in Fig. 3.32). Then, exactly the same circuit is done in the reference crystal, counting the same number of steps in all directions. In this method for making a Burgers circuit, the position of F is separated from S by one lattice unit, which is the special vector needed to close the circuit in the reference crystal–by definition, this is the true Burgers vector, **b**. Again, in an edge dislocation, the Burgers vector is perpendicular to the line direction, whereas, in screw dislocations, it is parallel. The true Burgers vector, **b,** appears in the reference crystal, rather than in the lattice containing the dislocation (Fig. 3.31), as indicated in the first version (Fig. 3.30). Note that if, in the first method (see Fig. 3.30), the circuit is sufficiently farther away from the 'bad region' (near the dislocation), then the vector **b** obtained would be practically the same as the one obtained by the second

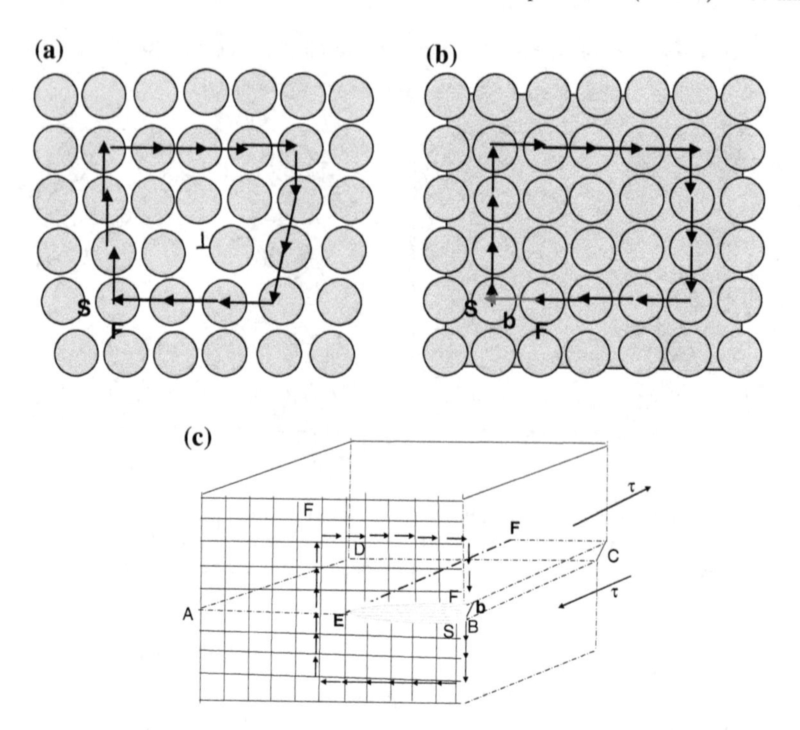

Fig. 3.31 Burgers circuits to show the strength of vector **b** as having the magnitude of the distance between the atoms of the lattice; **a** around a dislocation, **b** in the reference crystal (dislocation free) [14] and **c** a Burgers circuit around a screw dislocation (schematic). The Burgers vector **b** is required to complete the circuit

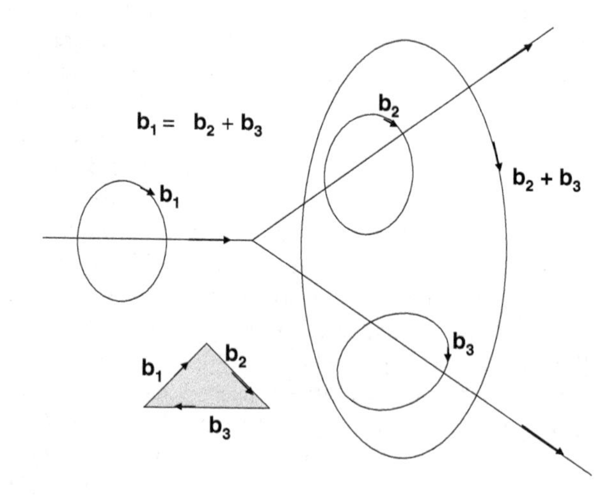

Fig. 3.32 Schematic illustration of a dislocation with Burgers vector $\mathbf{b_1}$ is branching off into two dislocations with vectors $\mathbf{b_2}$ and $\mathbf{b_3}$ [14]

method (see Fig. 3.31), since distortion decreases with distance. It follows from the above that the Burgers vector of a (perfect) dislocation is, by necessity, a lattice vector.

The Burgers vector has both magnitude and orientation showing the direction of slip during deformation. There is a significant controversy among researchers in the literature regarding the sign of the dislocation and a question still remains regarding the notation of the direction of slip in the Burgers vector. One school favors the method indicated in Figs. 3.30 and 3.31, assuming that the direction of the dislocation line is into the drawing and that the circuit is made clockwise. In this case, the dislocation is considered to be positive, as the dislocation symbol appears in the figure. Another group of researchers in the dislocation field (including Burgers himself) takes an opposite approach. Thus, one must clearly define whether the circuit is drawn clockwise or counter-clockwise around the dislocation. The vector is always the same, but its sign will be different and, therefore, consistency in describing a dislocation is important. The presentation of a Burgers circuit for the evaluation of **b** is called the right-handed [henceforth: RH] convention' (i.e., the right-handed convention, clockwise and positive). The opposite approach defines what is called the 'LH convention' (i.e., the left-handed convention). In Fig. 3.31c, a Burgers circuit is drawn around a screw dislocation on the front plane of a parallelepiped. Deformation (slip) has occurred on the plane ABCD by applying a pair of shear stresses, as indicated. Deformation has resulted as a screw dislocation marked as **EF,** demarcating the slipped from the unslipped regions. The step required to close the circuit is the Burgers vector **b**.

A dislocation can change from an edge orientation to a screw one. In real crystals, a dislocation is seldom pure edge or pure screw. Furthermore, dislocations are rarely found only in one plane. In general, dislocations are curved or form a loop.

3.3.4.1 A Few Comments on the Burgers Vector

A dislocation line cannot end inside a crystal, because the Burgers vector cannot become zero (it is constant) at some point along its line. It can terminate at a free surface, internal surface or interface (e.g., a grain boundary), form a closed loop inside a crystal or branch into other dislocations (forming a node). At a node (where a dislocation branches out), for all dislocations *pointing away from it*, Burgers-vector conservation must exist, analogous to Kirchoff''s Law for the flow of electric current. This is Frank's Rule of the conservation of the Burgers vector. Thus:

$$\mathbf{b}_1 + \mathbf{b}_2 + \mathbf{b}_3 = 0 \tag{3.19}$$

and more generally:

$$\sum_{i}^{n} \mathbf{b}_\mathrm{i} = \mathbf{0} \tag{3.20}$$

So, for three dislocations, or rather for a dislocation that branches into two other dislocations (not all dislocation lines leave the node), one may write:

$$\mathbf{b_1} = \mathbf{b_2} + \mathbf{b_3} \tag{3.21}$$

This is shown graphically in Fig. 3.32. In Fig. 3.32, Burgers circuits are drawn around the dislocation line entering the node and around the two lines leaving the node in the positive direction. After the node, the dislocations can be enclosed by one single extended Burgers circuit, like the one drawn around the dislocation before the branching. The sum of the Burgers vectors given in Eq. (3.21) clearly applies in this case, as is indicated above, because the dislocation line branches out and not all of them are pointing away from the node. In summary (and since no agreement on the convention for **b** is available), in order to define a dislocation line, a unit vector, **t** (which is the same as a translation vector in crystals), is chosen, such that it is tangent to the dislocation line and the positive direction of the dislocation is taken in the positive direction of **t**. This helps to establish the sign of the **b** vector. Thus, Fig. 3.30b or 3.31c, view the dislocation line as running into the sheet of paper and consider it to be positive; therefore, the Burgers vector, **b**, is defined by the RH convention. When reversing the direction of **t** (i.e., the dislocation line is running out of the drawing), the sign of **b** is also reversed. On the basis of the above convention, a dislocation may be characterized by **t** and **b** as: a) edge dislocation $\mathbf{b} \cdot \mathbf{t} = 0$ (i.e., the direction and Burgers vector are perpendicular); RH screw dislocation $\mathbf{b} \cdot \mathbf{t} = \mathrm{b}$ and LH screw dislocation $\mathbf{b} \cdot \mathbf{t} = -\mathrm{b}$ (in both directions the dislocation and Burgers vector are parallel).

Thus, one can sum up this section by stating that the most important feature of a dislocation is its Burgers vector.

3.3.5 Producing Dislocations in an Elastic Body (Schematic)

Hypothetically, one might imagine two ways of producing a dislocation artificially:

(i) Look at Figs. 3.25 and 3.26 or the simple two-dimensional illustration in Fig. 3.28. Assume that a slot has been cut in any of the materials represented by these figures by means of a knife of atomic scale; now insert a partial plane (not necessarily a half-plane) of atoms into that slot and then glue them together. The termination of the plane at its edge, as defined by the row of atoms, is a line defect in the crystal, i.e., an 'edge dislocation' (Fig. 3.25). This imaginative method of producing a dislocation distorts the region around the line defect, as shown in Fig. 3.26. This procedure can be further visualized by an analogy to a stack of cards. If half a card is inserted in a deck of cards, a defect is formed in the deck at the termination line of the half-card. *In real crystals, due to the straining caused by the extra half-plane, elastic distortion occurs around it and in its immediate vicinity extending out for several planes. The dislocation line itself has a high*

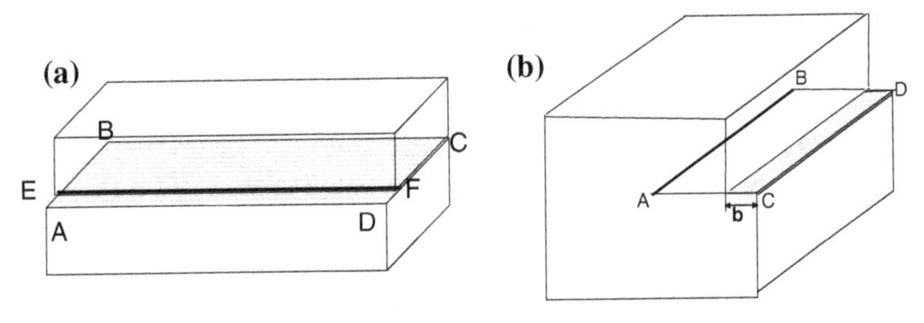

Fig. 3.33 Dislocation formation by cutting a slot in the body and gluing the faces formed together: **a** ABCD is the slip plane and EF is the dislocation line at the start of deformation; **b** edge dislocation **AB**, formed by shear displacement in the slip plane ABCD following a cut in the plane and the gluing of the faces formed by the cut [14]

energy, while strain decreases over distance. Clearly, this method of producing dislocations is useful for understanding their nature.

(ii) In addition, one may produce a dislocation as is indicated in Fig. 3.29a, redrawn in Fig. 3.33a. Visualize Fig. 3.33a as being the result of cutting a slot part way into a plane of an elastic body (a parallelepiped), applying shear stress at the start of the deformation and gluing the faces together to form a solid body. A dislocation line formed in this way is clearly the demarcation line or the boundary between the slipped and unslipped parts indicated in Fig. 3.33a (line EF). Here, the upper part illustrated slipped over the shaded slip plane, marked as ABCD, forming one atomic displacement **b.** In Fig. 3.33b, this slip has progressed over the shaded slip plane and a dislocation line has formed at AB. In a similar manner, a screw dislocation may be formed (see Fig. 3.29b), but the slip vector, in that case, will be parallel to the dislocation line.

Dislocations in a real crystal are present in sufficient amounts, depending on the material and its production history. The growth of a single crystal by itself is pictured as occurring via a process of screw dislocation (which is an integral necessity for growth per se). However, the dislocation concept of strain hardening, for example, requires the formation of a large number of dislocations and their multiplication. Plastic deformation can produce large numbers of dislocations (discussed below). During plastic deformation, the density of the dislocations increases greatly, even to a level of $10^{10}/mm^2$, depending on the severity of the deformation. Annealed material has a density of $\sim 10^6$–$10^7/mm^2$. Silicon wafers usually have a very low dislocation density, $<100\ cm^{-2}$, while semi-insulating GaAs wafers have a density on the order of $\sim 10^5\ cm^{-2}$.

Three mechanisms for dislocation formation in polycrystalline materials are considered here: (a) by homogeneous nucleation; (b) by induction at grain boundaries and interfaces in the lattice; and (c) by surface stress. In regard to homogeneous nucleation, it is very unlikely, since it requires high stresses of about the same level as the theoretical stress in a dislocation-free crystal. When considering grain boundaries in materials, they can produce dislocations which propagate into the grain; in particular, sites of irregularities and steps or ledges at the boundaries are dislocation sources in the early stages of plastic deformation. Lastly, stresses on surfaces are usually larger than the average stress within the lattice; the presence of irregularities, such as steps, can induce dislocation propagation in crystals. Thus, dislocations nucleate from the surface.

In single crystals, dislocations are formed at the free surfaces. Suffice it to say that sites, such as precipitates, dispersed phases, oxides, etc., can all serve as sources for dislocation initiation and generation. Note that an important concept for dislocation generation is the collapse of vacancies to form disc-shaped dislocations, as suggested by Nabarro [41].

3.3.6 The Motion of Dislocations

The motion of dislocations is directly linked to the fact that the plastic deformation of crystalline materials is mostly carried out by either their (a) conservative or (b) non-conservative advance. Non-conservative motion refers to 'climb'. When enough force is applied to a crystal structure (but by orders-of-magnitude less than the theoretical strength in a dislocation-free crystal), the atoms at the edge of the extra plane (the dislocation line) pass through the planes of atoms, breaking and forming bonds with them, until they leave from the opposite side of the crystal, forming a step, as shown in Fig. 3.34. This procedure occurs in many discrete steps, until the dislocation has moved through the entire lattice, leaving plastic deformation in its wake. One question immediately arises: Why does this deformation require so much force? The atoms around these dislocations are symmetrically located on opposite sides of the extra half-plane. In Fig. 3.34, this is indicated by short dashes at the dislocation site, such that the forces acting on the dislocation, on its both sides, are equal and opposite in sign.

The force opposing the motion of a dislocation is balanced by the force encouraging its motion. Thus, in a first approximation, there is no net force on the dislocation and the motive stress required is ~zero. Nevertheless, the small force actually required to move a dislocation has been explained by Peierls-Nabarro. 'Peierls-Nabarro stress', as it became known, is given as:

$$\tau_{PN} = \frac{2G}{(1-\nu)} \exp\left(-\frac{2\pi w}{b}\right) \tag{3.22}$$

Here, w is the width of the dislocation defined in Fig. 3.26 (Sect. 3.3.2), written as:

$$w = \frac{a}{2(1-\nu)} \tag{3.23}$$

In Eq. (3.17), h was used for the distance between the interatomic planes, "a". Equation (3.22) may be substituted for the width in Eq. (3.23) to obtain:

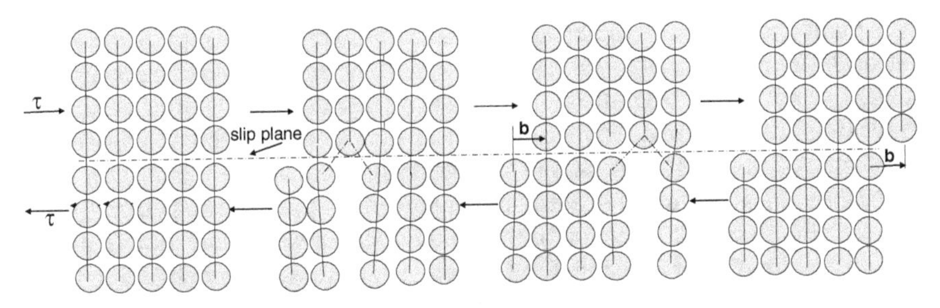

Fig. 3.34 The motion of a dislocation involves the breaking and reforming of a bond (short dashes) between an edge atom of the dislocation and its two neighbors, only one bond at a time, resulting in plastic deformation. A step is formed when the dislocation leaves the crystal [14]

$$\tau_{PN} = \frac{2G}{(1-\nu)}\exp\left[-\frac{\pi a}{(1-\nu)b}\right] \tag{3.24}$$

The width of the dislocation, Eq. (3.23), defines the magnitude of the Peierls-Nabarro stress and is a measure of the degree of distortion that has occurred due to dislocation. It is basically the distance over which the dislocation causes disregistry; thus, it is the magnitude of the displacement of the atoms from their perfect-crystal positions. As indicated in Sect. 3.3.2, when w is several atomic spacings, a dislocation is considered to be wide; if it is on the order of one or two atomic spacings, it is narrow.

Wide dislocations glide easily, such as in closely-packed crystal structures, where the atomic spacing is large. Hence, these materials are *ductile* and only small stresses are needed to produce large strains. Ceramics, for example, tend to be *narrow* in width, so that high stresses are required to move dislocations and the ceramics become hard and *brittle* as a result. Therefore, w is defined in terms of a, the atomic spacing in Eq. (3.23). The Peirels-Nabaro shear stress, τ_{PN}, varies exponentially with $\left(\frac{a}{b}\right)$, i.e., for large, inter-planar spacing, which corresponds to the distance between closely-packed planes, the shear stress is minimal; inter-atomic bonds across such slip planes are weak and the stress is small. Having a small **b** makes the ratio in the exponent larger, so it is clear why dislocations tend to move in the closest-packed direction; shear stress is much smaller than theoretical strength in closely-packed planes. Actual calculations for an FCC lattice gives $\tau_{PN} \sim 10^{-3}$G, which can be obtained on an FCC (111) plane with $a = b\sqrt{\frac{2}{3}}$. Thus, we understand why the stress to move a dislocation is not zero. Though the Peirels-Nabaro equation indicates the low stress required for dislocation motion, it is not accurate, since the structure at the core of the dislocation and the changes in energy during slip are not known.

To briefly review, the dislocation motion described above largely explains, for example, why FCC metals can be ductile, as long as some obstacle does not retard their motion. Contrary to the breaking of all the atomic bonds on a plane simultaneously (requiring high stress, as calculated in the section above on theoretical strength), in order to break the bonds between individual neighboring atoms at a dislocation site, the stress required is lower by orders. The motion of a dislocation line (depicted in Fig. 3.34) relates to an edge dislocation, but the motion of a screw dislocation is also the result of shear stress; the difference between these two kinds of dislocations is in the direction of their motion. Whereas the direction of motion of an edge dislocation is parallel to the applied stress, in screw dislocation, it is perpendicular. Nevertheless, the net plastic deformation of both the edge and screw dislocations is the same. It has been indicated above that, in general, most dislocations are of a mixed type, exhibiting both characteristics, i.e., at one part of the line there is an edge, while at the other, there is a screw orientation.

3.3.6.1 Conservative Motion

In conservative motion, only shear stresses are of interest, since a shear component acting in the glide plane is effectively involved in the dislocation motion. A normal stress component acting on the same glide plane will not contribute to this motion. Glide occurs on a slip plane without any change in the extra half-plane atomic configuration, namely no atoms are added or removed while the dislocation advances under the influence of stress. Glide does occur in preferential planes and directions, depending on structure type. In conservative motion, whether caused by edge or screw dislocations, the resultant step formed is the same, despite the difference in movement direction, as shown in Fig. 3.35 for both positive and negative edge dislocations under the same shear stresses. These figures schematically depict dislocation motion in three dimensions in a simple cubic structure.

Regarding screw dislocations–the directions of the motion of positive and negative screw dislocations are opposite, yet the end results will be similar, as long as the screw dislocation is moving according to the same pair of shear stresses. Both the RH and LH screw dislocations produce the same steps under the same stress as an edge dislocation (Fig. 3.35). LH screw dislocations move to the rear, while RH screw dislocations progress to the front. Students should practice schematically drawing the motions of RH and LH screw dislocations. Figure 3.36 is a schematic presentation of the motion of a positive-edge dislocation on a (100) plane.

Figure 3.36a shows the (100) plane before deformation. Crystal deformation is indicated in Fig. 3.36b–d. Schematic d depicts the step after the dislocation has covered the entire slip plane.

At a first glance, one might suggest how to make a perfect, defect-free crystal. One would simply have to apply an appropriate shear and squeeze the dislocation out of the crystal, leaving behind a dislocation-free crystal, thus obtaining a high-strength material (as calculated in Sect. 3.3.3). This is a naive thought, since during the deformation process so many dislocations are formed, multiply and interact as stress is applied that their movement is often blocked by a host of

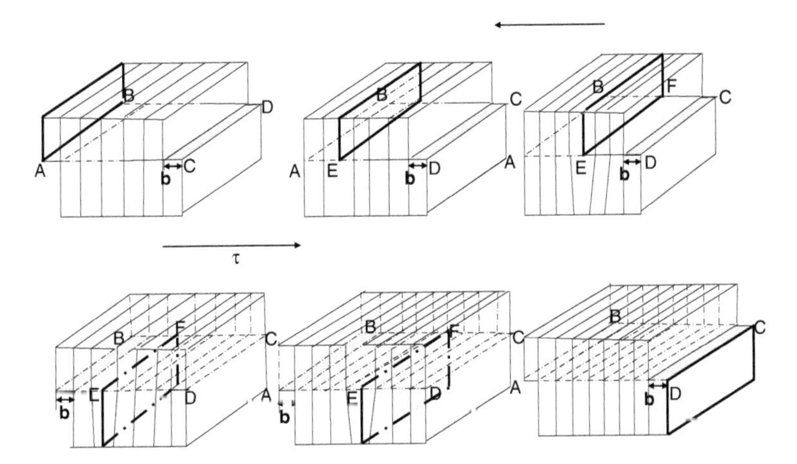

Fig. 3.35 Movements of positive and negative dislocations acting under the same shear stresses: top figures are positive dislocations; those at the bottom are negative dislocations. The movement directions of the dislocations are opposite, but the results of the shear are the same–a step is formed on the same crystal plane (slip plane). The step **b** is the Burgers vector [14]

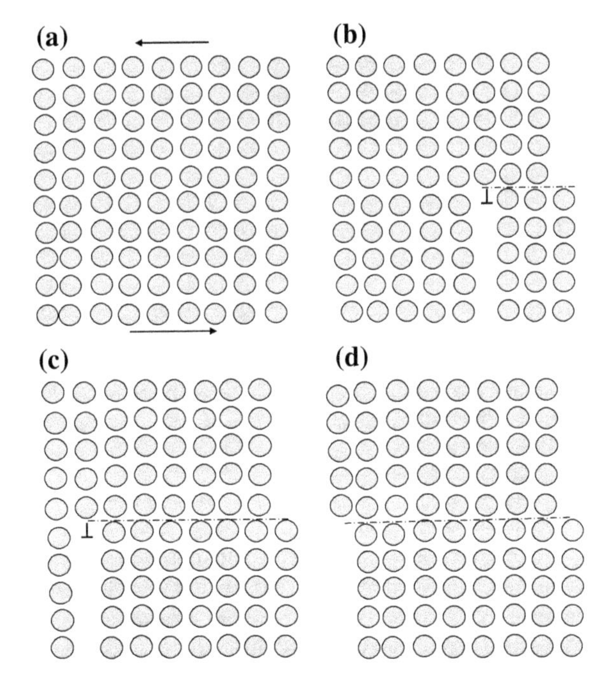

Fig. 3.36 Edge-dislocation motion on its slip plane under the influence of shear stress. The crystal deforms as seen in illustrations **a–d**. When the dislocation has left the crystal (**d**), a step is formed having the slip vector's dimension [14]

possible obstacles. In fact, materials actually become stronger, not because the dislocations have been squeezed out of them, but rather due to the application of stress and the resulting interactions between those dislocations formed.

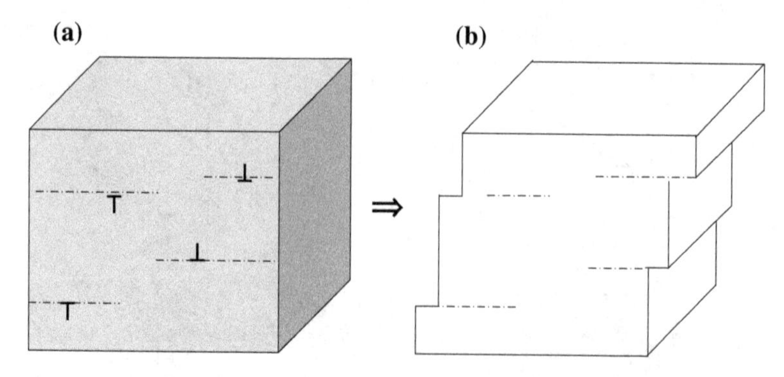

Fig. 3.37 Steps formed by shear stress caused by dislocations leaving their respective slip planes [14]

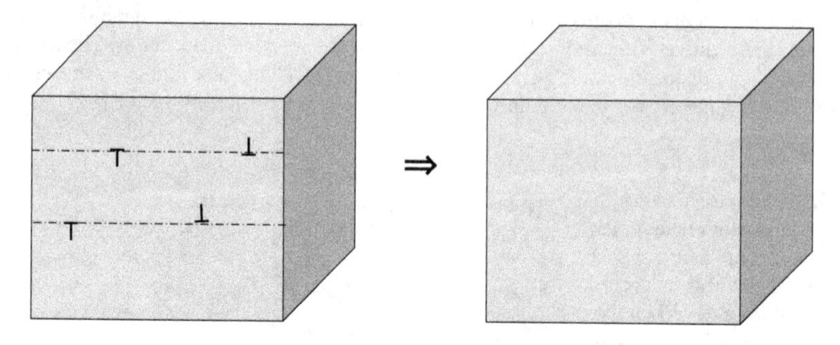

Fig. 3.38 Two sets of dislocations having opposite signs on the same slip plane. They annihilate each other, resulting in a perfect crystal [14]

When more than one set of dislocations leave a crystal, more steps are formed. To illustrate this effect, take the following Fig. 3.37, having two positive and two negative edge dislocations in their respective slip planes and apply a pair of shear stresses. The slip planes, shown (partially) in Fig. 3.37b, reflect the planes shown in Fig. 3.37a, where the dislocations have moved forming the steps upon leaving the crystal. If the positive and negative dislocations were on the same planes, as shown in Fig. 3.38, they would annihilate each other, resulting in a perfect lattice. One must realize that, above an edge dislocation, the atomic planes experience compression (because an extra half-plane has been inserted), while, below it, the atoms are being pulled apart by tension. In other words, the strain fields above and below a dislocation line are compressive and tensile, respectively. The energy of a crystal is reduced by the reaction of the compressive and tensile strain fields, resulting in the cancellation of the pair of dislocations.

An edge dislocation is able to glide across a plane, containing its edge dislocation line and its Burgers vector. Thus, an edge dislocation is confined to movement within one plane, unlike the motion of screw dislocation. A glide plane is defined by its normal, given as $\mathbf{b} \times \mathbf{t}$. A screw dislocation does not have a

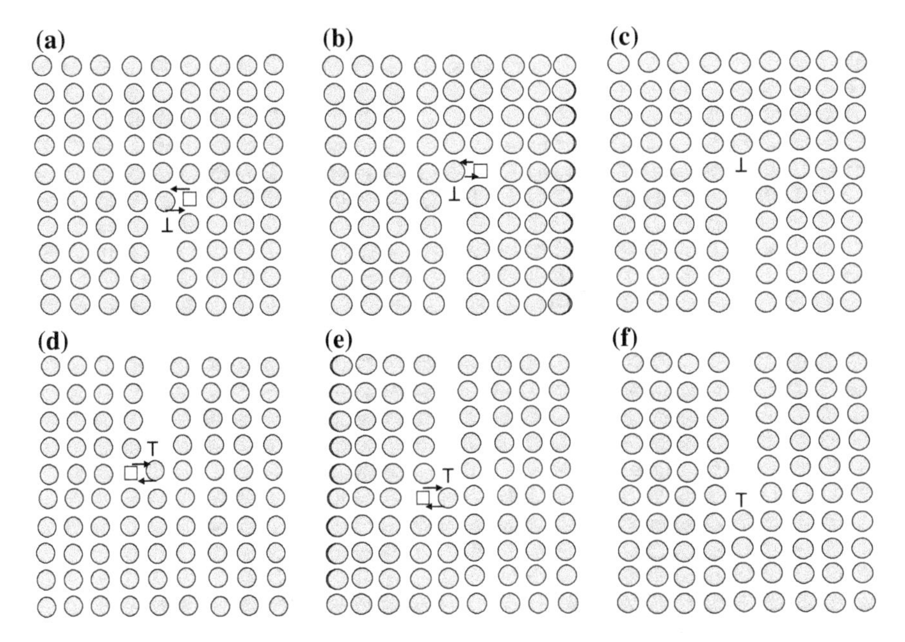

Fig. 3.39 Climb in positive and negative dislocations. When a point defect, such as a vacancy, reaches the vicinity of a dislocation, as shown in (a) it changes places with the indicated atom of the dislocation and climb occurs, resulting in (b) and (c). Dislocation climb means that a part of the dislocation has left its slip plane. Illustrations **d–f** indicate the same for a negative dislocation [14]

defined glide plane, since $\mathbf{b} \times \mathbf{t} = 0$ and it has a cylindrical symmetry about its axis. Every plane passing through that axis (known as the 'zone axis') may become a glide plane through which a screw dislocation may move. Therefore, screw dislocation is not restricted to only one plane. Often, the step-by-step motion of edge dislocation, breaking and reforming each individual bond, is more easily explained by an analogy to the motion of a caterpillar or to the removal of a wrinkle in a large carpet. The localized force employed by the special motion of a caterpillar or the force needed to move each carpet wrinkle into its correct place is much smaller than that required to pull an entire carpet straight.

3.3.6.2 Non-conservative Motion ('Climb')

The non-conservative motion (or 'climb') of a dislocation is indicated in a series of illustrations in Fig. 3.39 for positive and negative edge dislocations.

The motion of an edge dislocation is restricted to one plane only–the glide (or slip) plane. Both a positive and a negative edge dislocation can only exit a glide plane by means of the climb process, as illustrated in Fig. 3.39. For climb to happen in a positive *dislocation* (the upper illustrations in Fig. 3.39), vacancies, arriving by diffusion, must occur adjacent to the dislocation line by replacing one

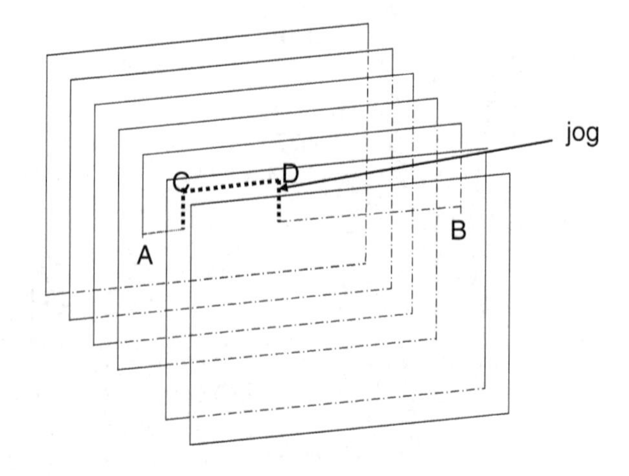

Fig. 3.40 Jog formation by climb [14]

or more of its atoms. This is a thermally activated process, causing part of the extra plane defining the dislocation become shorter. Since these vacancies never appear at the same time in sufficient amounts, only part of the dislocation climbs. The climb of a segment of a dislocation forms a 'jog' (i.e., a segment of a dislocation line having a vector normal to the glide plane). Jog formation is shown in Fig. 3.40. Here, the dislocation segment that climbed, **CD,** and the jogs formed are heavily outlined by dashed lines. Thus, rather than an entire dislocation line shifting all at once, only segments of a dislocation climb. In addition to the jogs, 'kinks' are also formed during the process of climb; these are the segments of the dislocation line that have remained within the glide plane that did not climb. Jogs are immobile, unlike kinks, and cannot glide together with the dislocation segments that did not climb (i.e., kinks), since their direction of motion is in a plane normal to the glide plane and not within the glide plane itself. This is shown in Fig. 3.41. A kink is basically an edge dislocation and, if it is lying in the slip plane, it does not provide resistance to motion and may be considered a step in the dislocation line. Diffusion of vacancies (or interstitials) is required for the motion of jogs. Jogs and kinks can also be formed by dislocation interactions (discussed later on).

A screw dislocation cannot climb, since it does not have a specific, well-defined glide plane. Every plane passing through the zone axis may be considered a glide plane; when in the motion, a screw dislocation can pass from one glide plane to another, within the family of planes sharing a common zone axis. 'Cross-slip' is the term used for the process of passing a screw dislocation from one plane onto another with an equivalent glide plane to that of the original one. Thus, cross-slip involves the initial slip plane and the cross-slip plane, where the glide continues. Usually, when a screw dislocation meets some obstacle during its glide, it can circumvent that obstacle by the process of cross-slip. Many obstacles can impede dislocation movement, such as: other dislocations, precipitates, grain boundaries, etc.

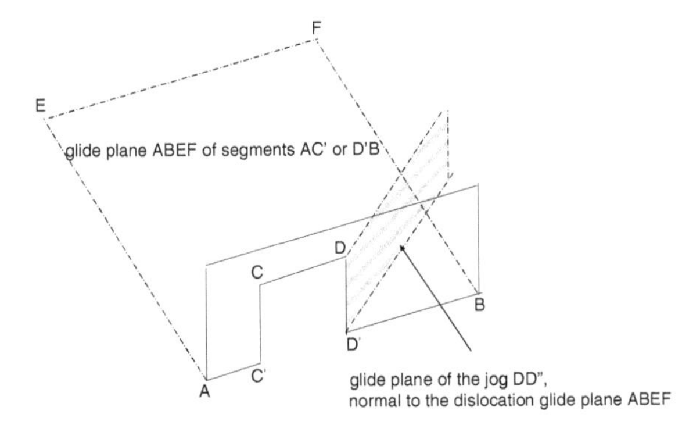

Fig. 3.41 One of the jogs, DD, is shown with its glide plane shaded; it is normal to the glide plane of the dislocation before its climb [14]

To conclude Sect. 3.3.6 on dislocation motion, the following may be added. Dislocations that are able to move by pure slip are called 'glissiles' and dislocations which cannot glide without some thermal process, such as point defect diffusion, are called 'sessiles'. Glissile dislocations play a role in plastic deformation, in which good elongation is of importance, since large strains can be achieved by the motion of many dislocations without considerable impedance to their motion. Sessile dislocations, on the other hand, are among the factors responsible for high strength, but lower strains. Thus far, nothing has been said about a very important factor that influences the motion of dislocations, namely the structure of materials, which will be discussed in a later section in regard to FCC metals.

3.3.7 The Energy of Dislocations

Dislocation formation requires work and so the resultant energy becomes stored in the material. The application of stress to produce strain introduces elastic strain energy caused by atoms displaced from their equilibrium position. Whenever atoms are displaced from their lattice equilibrium position, strain energy is introduced into the lattice. A dislocation must have some strain energy, since it distorts the lattice and displaces atoms from their lattice equilibrium position; therefore, a strain is introduced into the surroundings. A strain field, thus, exists at the site of a dislocation, affecting its motion. Bonds outside the radius of the dislocation line, say at a distance r_0, are elastically strained, but some of the energy lies within the radius, r_0, stored in what is known as the 'core of the dislocation'. Thus, the total energy-per-unit length of the dislocation line, E_{TE}, is composed of the core energy and the energy outside, r_0, given by:

$$E_{TE} = E + E_{core} \tag{3.25}$$

For the present, consider only E (discussion on the contribution of the core follows later on).

Tension applied to a rod of length, l, and cross-section, A, extends the rod by dl and a strain energy, dE, is introduced. This may be expressed as:

$$\mathrm{dE} = \mathrm{Fdl} \tag{3.26}$$

Force is given by:

$$\mathrm{F} = \sigma \mathrm{A} \tag{3.27}$$

Also, based on Chap. 1, one can write:

$$d\varepsilon = \frac{dl}{l} \tag{3.28}$$

Volume V = Al remains constant during deformation and, by simple substitution, one can write:

$$dE = \sigma V \frac{dl}{l} = \sigma V d\varepsilon \tag{3.29}$$

This is obtained as follows: Rewrite Eq. (3.26), substitute F from Eq. (3.27); then express A in terms of volume and express dl/l by dε from Eq. (3.28). This produces Eq. (3.30):

$$\mathrm{dE} = \mathrm{Fdl} = \sigma \mathrm{Adl} = \sigma V \frac{dl}{l} = \sigma \mathrm{V} \mathrm{d}\varepsilon \tag{3.30}$$

Due to the constancy of the volume (V = Al), Eq. (3.30) can now be replaced by the energy-per-unit volume, dE′:

$$\frac{dE}{V} = dE' = \int_0^{\varepsilon} \sigma d\varepsilon \tag{3.31}$$

After integrating Eq. (3.31) and applying Hooke's Law, Eq. (3.32) is obtained for the energy-per-unit volume:

$$E' = \frac{1}{2}\sigma\varepsilon = \frac{1}{2}E\varepsilon^2 \tag{3.32}$$

In terms of shear stress and shear modulus, Eq. (3.32) should be written as:

$$\mathrm{E}' = 1/2\tau\gamma = 1/2\mathrm{G}\gamma^2 \tag{3.33}$$

The general relation from the principle of superposition is:

$$E' = \frac{1}{2}\left(\sigma_{xx}\varepsilon_{xx} + \sigma_{yy}\varepsilon_{yy} + \sigma_{zz}\varepsilon_{zz} + \tau_{xy}\gamma_{xy} + \tau_{xz}\gamma_{xz} + \tau_{yz}\gamma_{yz}\right) \tag{3.34}$$

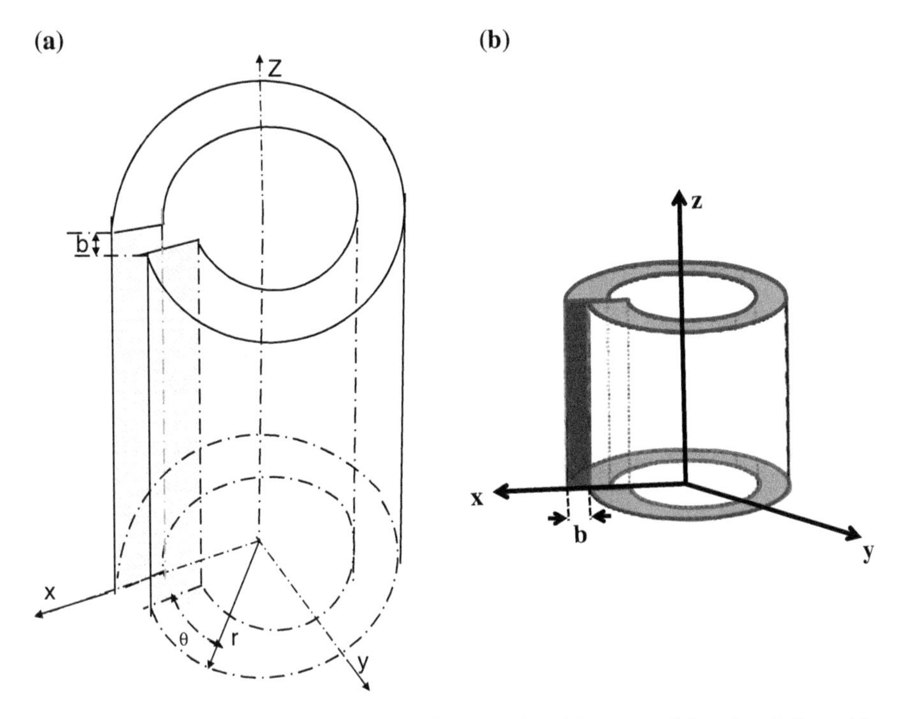

Fig. 3.42 Elastic distortion of a cylindrical ring **a** produced by screw dislocation; **b** formed by edge dislocation [14]

However, when a single strain component exists, the strain energy per volume is either $1/2E\varepsilon^2$ or $1/2G\gamma^2$. In Eq. (3.32), E is Young's modulus and not the energy.

3.3.7.1 Screw Dislocation

It is simpler to consider a case of screw dislocation in which only a shear stress is active, resulting in shear strain. Consider a segment taken from a cylinder with length, l, and radius, r, as shown in Fig. 3.42a (in polar coordinates) and which behaves like an elastic continuum. The strain in the shell of radius dr is presented below. A radial slit has been cut in the cylinder (not shown) parallel to the z axis and then the cut surfaces were displaced one over the other by shear deformation at amount **b** and glued together. This displacement by the shear is **b** pointing in the direction of z. Now, unfold this shell and spread it flat, for the sake of understanding the derivation. The unfolded cylindrical shell from Fig. 3.42a is shown in Fig. 3.43, indicating the distortion that has occurred due to shear stress. The circumference is $2\pi r$ and the angle of the shear strain is $\gamma = b/2\pi r$. The displacement is in the z direction and the stored energy-per-unit volume, between r and dr in this

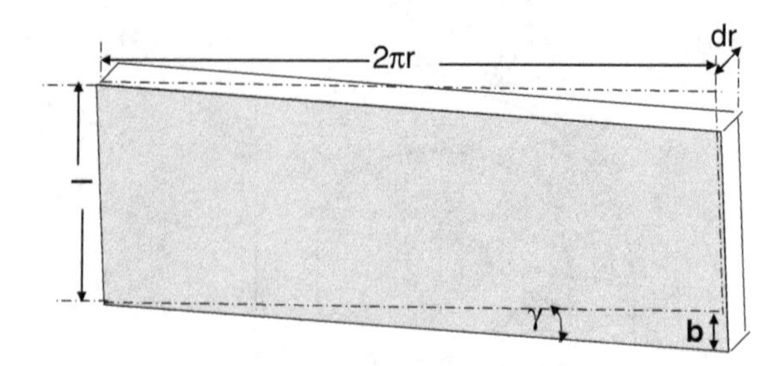

Fig. 3.43 The segment of the cylinder seen in Fig. 3.42a in unfolded form [14]

case may be written as $Gb^2ldr/4\pi r$. Then after integrating, Eq. (3.35) may be obtained for r, between the limits r_0 and r:

$$E' = \int_{r_0}^{r} \frac{Gb^2 l}{4\pi}\frac{dr}{r} = \frac{Gb^2 l}{4\pi}\ln\frac{r}{r_0} \tag{3.35}$$

For a unit dislocation line, $E'/l = E$, one can give the energy of the screw dislocation in Eq. (3.35) as:

$$E = \frac{Gb^2}{4\pi}\ln\frac{r}{r_0} \tag{3.36}$$

Equation (3.36) is the result of the following steps:

(a) the area under a stress–strain curve is the strain energy-per-unit volume (see Eqs. 3.31 and 3.32); in our case, for shear, it is: $E' = \frac{1}{2}\tau\gamma$;
(b) $\tau = G\gamma$;
(c) substituting for τ leads to $E' = \frac{1}{2}G\gamma^2$;
(d) the volume of the thin shell (Fig. 3.42 or rather 3.43) is $dV = 2\pi rl$; therefore, the energy in the elemental volume is:
(e)

$$E' = \frac{1}{2}\frac{Gb^2}{(2\pi r)^2}2\pi rldr = \frac{Gb^2}{4\pi r}ldr;$$

(f) integrating between the limits r_0 and r and taking the energy-per-unit line gives Eq. (3.36).

Since the above displacement is in the z direction, there are no displacements in the x and y directions. Thus, all the hydrostatic and shear components are inactive except for the shear stress in the z direction and its respective strain, as derived above. This strain is γ and, thus, the stress is $\tau = Gb/2\pi r$ (see the shear strain in Fig. 3.43). Note that, in the above equation, the strain should actually be written as $\gamma_{z\theta} = \gamma_{\theta z}$ and the shear stress, accordingly, as $\tau_{z\theta} = \tau_{\theta z}$; but for simplicity, these indexes were dropped. The above is a consequence of the fact that screw dislocation has only two active shear stresses, one on a radial plane in the z direction and the other on a plane perpendicular to the z axis and parallel to the radius, respectively for $\tau_{\theta z}$ and $\tau_{z\theta}$. The stress field has radial symmetry and is independent of θ, as seen in Eq. (3.36). Recall that in the system under consideration (polar coordinates) the shear and, thus, the displacement are in the z direction, such that no strains exist in the x and y directions and the corresponding strains are zero, as are the stresses shown below:

$$\sigma_r = \sigma_\theta = \sigma_z = \tau_{r\theta} = \tau_{\theta r} = \tau_{zr} = \tau_{rz} = 0 \tag{3.37}$$

There is a limit to the validity of Eq. (3.36). For instance, it is unlikely that in the expression $\tau = Gb/2\pi r$, $r \rightarrow 0$ the stress is infinite, since $\tau \rightarrow \infty$. Equation (3.36) is valid only outside a radius of ~5–10 Å. By using reasonable values for r and r_0, Eq. (3.36) may be approximated as:

$$E \approx Gb^2 \tag{3.38}$$

This can be obtained as follows. Assume that $r \sim 10^5 b$ (providing a dislocation density of $\sim 10^{10}$ dislocations/m^2). Since Hooke's Law only applies to small strains and not to the very high values that exist in dislocation cores, a value of b/4 has often been suggested for r_0. By substituting these values into Eqs. (3.36), (3.38) is obtained.

An estimated value of the core energy is then:

$$E_{core} \cong \pi(5b)^2\left(\frac{G}{30}\right)\frac{1}{G} \approx \frac{Gb^2}{10} \tag{3.39}$$

This expression is obtained by the following assumptions:

(a) that the stress level in the vicinity of the core is $\tau = G/6$, as indicated in Eq. (3.18);
(b) that the radius is approximately 5b;
(c) that the area is $\sim \pi\ (5b)^2$;
(d) that the elastic energy-per-unit volume acting on the dislocation core, is $E' = 1/2G\gamma = 1/2\tau^2/G$ (Eq. 3.33);
(e) by using the last term in (d) with the assumption in (a) and multiplying by the area of (c), Eq. (3.39) is obtained:

$$1/2\tau^2/G = \frac{1}{2}\frac{\tau^2}{G} = \pi(5b)^2\frac{1}{2}\left(\frac{G}{30}\right)^2\frac{1}{G} \approx \frac{Gb^2}{10}.$$

The total energy, according to Eq. (3.25), can now be expressed by Eqs. (3.38) and (3.39), yielding:

$$E_{TE} \approx Gb^2 + Gb^2/10 \cong Gb^2(1 + 1/10). \tag{3.40}$$

A core energy of 0.1, according to the estimated expressions, is small or even negligible in its contribution to the overall energy, so it is often neglected.

3.3.7.2 Edge Dislocation

The analysis of an edge dislocation is somewhat more complicated, but it can be done in a manner similar to that of a screw dislocation, again using a ring-shaped segment of a material. An edge dislocation line is also parallel to the z axis, as in the case of a screw dislocation, but the Burgers vector is normal to the z axis, as shown in Fig. 3.42. Thus, the **t** vector of the dislocation line points in the direction of the z axis. Since the line is in the z direction and its displacement is parallel to the x axis, i.e., the Burgers vector is normal to the z axis, there will be no strain (no displacement) in the z direction, though the stress will not be zero.

The energy of an edge dislocation per unit length of the dislocation line (stated without proof) is similar to the energy of a screw dislocation, but it is divided by $(1 - \nu)$, where ν is Poisson's ratio, is given by:

$$E_{edge} = \frac{Gb^2}{(1-\nu)4\pi}\ln\frac{r}{r_0} \tag{3.41}$$

As indicated above, a value of ~0.1 Gb^2 is the contribution of the dislocation core, which is neglected. Equation (3.41) indicates that the energy of an edge dislocation is larger by $1/(1 - \nu)$, i.e., 1.5 times greater than that of a screw dislocation with the same length, l.

It is known that, in materials, the dislocation line's direction and the Burgers vector are neither perpendicular nor parallel, being mixed dislocations, consisting of both screw and edge characteristics. Therefore, it is appropriate to sum up the dislocation energies given in Eqs. (3.36) and (3.41), while neglecting the contribution of the core. This may be expressed as:

$$E_{total} = \frac{Gb^2}{4\pi}\ln\frac{r}{r_0} + \frac{Gb^2}{(1-\nu)4\pi}\ln\frac{r}{r_0} = \frac{Gb^2}{4\pi}\ln\frac{r}{r_0}\left[1 + \frac{1}{(1-\nu)}\right] \tag{3.42}$$

In this summation, done in order to estimate the energies of mixed dislocations, the interactions between the edge and screw components are ignored. Figure 3.44 shows a 'general dislocation line' indicating its mixed character. Note that, here, energy is given per unit length. Several comments are now in order:

(a) Energy is proportional to b^2; thus, a dislocation having a large Burgers vector, e.g. of strength 2b, is not stable and is likely to decompose into two vectors, with lower energies:

$$b^2 + b^2 = 2b^2 < (2b)^2$$

(b) The energy of dislocations, as seen in Eqs. (3.36), (3.41) and (3.42), is proportional to b^2; therefore, the most stable dislocations are those with minimum Burgers vectors, namely those with the closest-packed directions.

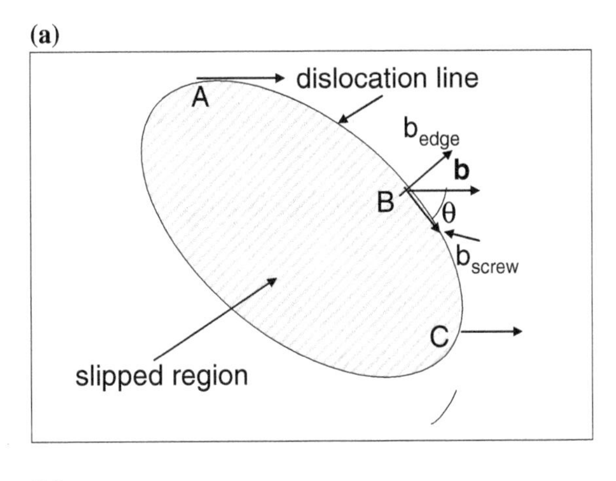

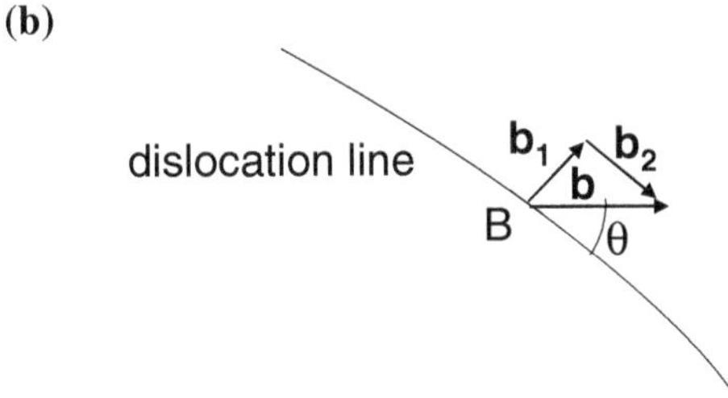

Fig. 3.44 A 'general dislocation line' is the boundary between the slipped and unslipped regions: **a** at point B, the Burgers vector, **b**, can be resolved into $\mathbf{b}_{\mathbf{edge}}$ and $\mathbf{b}_{\mathbf{screw}}$; **b** the dislocation at point B is shown (magnified) and **b** is resolved into edge and screw components [14]

(c) Since $r_0 = 0$ is not reasonable, providing an infinite value for energy, it is customary in the continuum model to avoid this by introducing a hole along the dislocation line. Thus, r_0 is the radius of a hole with a diameter of about 10^{-7}cm.

Equation (3.42) may also be presented in a different form. To this end, one needs a schematic drawing of a mixed dislocation in a slip plane, as shown in Fig. 3.44a. In Fig. 3.44, the dislocation is in the shape of a closed line and, for a given displacement of the vector, **b**, both the edge and screw orientations are shown. Some mixed characteristics may be seen, for example at point B. The Burgers vector is constant, pointing to the displacement of the line, but the dislocation line changes direction. At point A, the line is parallel to **b** and, therefore, it has the character of a screw dislocation, while at point C, the segment of the dislocation line is perpendicular, like a pure edge. At point B, **b** can be resolved into two components with edge and screw orientations, designated as b_{edge} and b_{screw} (Fig. 3.44b). In Fig. 3.44a, the slipped region is shaded and the direction of the dislocation line at point B is shown. For the edge and screw components, one may write:

$$\mathbf{b}_{\mathbf{edge}} = \mathbf{b} \sin \theta$$
$$\mathbf{b}_{\mathbf{screw}} = \mathbf{b} \cos \theta.$$

Substituting these values into Eq. (3.41), one finds the mixed dislocation at point B in Eq. (3.42) to get:

$$E_{total} = \left[\frac{Gb^2 \cos^2 \theta}{4\pi} + \frac{Gb^2 \sin^2 \theta}{(1 - v)4\pi}\right] \ln \frac{r}{r_0} = \frac{Gb^2}{4\pi(1 - v)} \ln \frac{r}{r_0} [1 - v \cos^2 \theta] \quad (3.43)$$

Thus, the value of a mixed dislocation is between that of screw and edge dislocations.

3.3.8 Line Tension

The expressions for the energies above were given per unit length of dislocation. When not normalized for a unit length, this energy is proportional to the length of the dislocation:

$$\mathrm{E} \approx \mathrm{lGb}^2 \quad (3.44)$$

As a consequence, a dislocation will tend to decrease its length to a minimum to decrease its energy. A curved dislocation line has a tension, **T**, which acts along its line and may be expressed as a change in energy with the length:

$$T = \frac{\partial E}{\partial l} \approx Gb^2 \quad (3.45)$$

The units of line tension in Eq. (3.45) are in energy-per-unit length.

A dislocation line can encounter various obstacles to its motion. In such a case, the dislocation becomes pinned by the obstacle and curves, due to the forces exerted by the obstacle resisting its motion. Small forces cannot make such a dislocation move. Figure 3.45 shows the bowing of a dislocation line after it has been pinned at both ends, at points A and B, by obstacles, perhaps foreign particles in the material or some other precipitates. In order to better understand this phenomenon, consider Eq. (3.35), which allows (3.45) to be rewritten as:

$$T = \alpha Gb^2 = \frac{lGb^2}{4\pi} \ln \frac{r}{5b} \quad (3.46)$$

In Eq. (3.35), r_0 was replaced by 5b, assumed to represent the core of the dislocation. The line tension, T, is acting tangentially to shorten the dislocation line and, thus, to reduce the dislocation energy. The curving of the dislocation line segment, due to its normal force, τbl, as a result of the applied shear stress, is shown in Fig. 3.45. For force equilibrium in the y direction, since it is balanced by the line tension, it is possible to write:

$$\tau bl = 2T \sin \frac{\theta}{2} \quad (3.47)$$

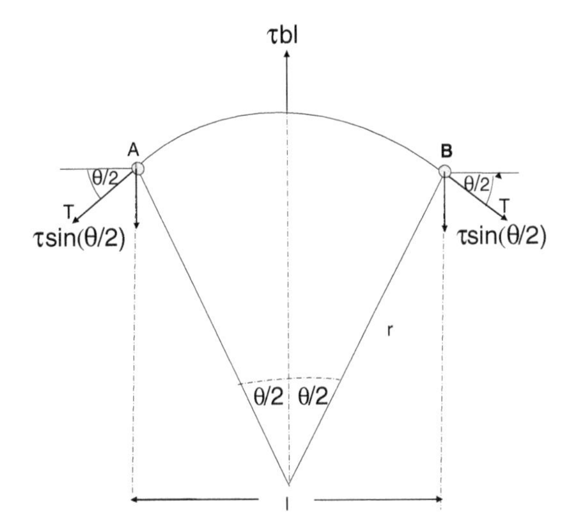

Fig. 3.45 The line tension of a dislocation when blocked by obstacles *A* and *B* [14]

The coefficient 2 arises from the fact that the line tension acts at points A and B of the dislocation line segment. For small angles, sin $(\theta/2) = \theta/2$. Substituting for T from Eq. (3.46), τ can be expressed as:

$$\tau = \frac{\alpha Gb\theta}{l} = \frac{\alpha}{l}Gb\frac{l}{r} = \frac{\alpha Gb}{r} \tag{3.48}$$

The last term in Eq. (3.48) is a consequence of substituting $\theta = $ l/r for $\theta/2$ from Fig. 3.45. Note (from the first term in Eq. 3.48) that the stress required to motivate a dislocation being detained by some obstacle depends on its length, l. For small values of l, the required force is higher.

The line tension of dislocations will be significant for the later discussion of dislocation multiplication sources, according to the ideas of Frank-Read.

3.3.9 *The Stress Field of a Dislocation*

It has been mentioned above that it is simpler to consider screw dislocations, since they possess a cylindrical symmetry and only shear stress acts when displacement occurs in the z direction. Elastic distortion occurs with the introduction of a dislocation into an isotropic material.

3.3.9.1 Screw Dislocations

Here is another case of a cylindrical ring cut from good material surrounding a screw dislocation, as shown in Fig. 3.42a. Recall that dislocations are defects in crystals and, therefore, are accompanied by elastic stress fields. Now, a radial cut is made parallel to the z axis (as shown in Fig. 3.42a), displaced by **b**, and the surfaces are joined together. The shear strain obtained is:

$$\gamma_{\theta z} = \gamma_{z\theta} = \frac{b}{2\pi r} \tag{3.49}$$

The stress field of a screw dislocation is pure shear. As indicated earlier, high strains exist in the core region and, therefore, Hooke's Law of elasticity does not apply and so will not be considered. The dislocation line is parallel to the z axis; there are no displacements in the x and y directions and the other stress components are zero:

$$\sigma_r = \sigma_\theta = \sigma_z = \tau_{r\theta} = \tau_{\theta r} = \tau_{rz} = \tau_{zr} = 0$$

Once again, the magnitude of the elastic stress is directly proportional to the shear modulus, *G*, and the Burgers vector, ***b***, and inversely proportional with increasing distance from the dislocation, given as:

$$\tau_{\theta z} = \tau_{z\vartheta} = \frac{Gb}{2\pi}\frac{1}{r} \tag{3.49a}$$

The stress field of a screw dislocation is manifested by two active stresses: $\tau_{\theta z}$, acting in radial planes parallel to the z axis and $\tau_{z\theta}$, acting in planes normal to the z axis. The stress has a long range, because it is inversely proportional to r and perpendicular to the radius. To get a feeling for this, take a distance of 10^4b; for 1/ r, the stress is $\sim 10^{-5}$G, which is about the yield stress of some crystals. The stress vanishes at infinity.

Generally, these displacements can be expressed as:

$$\begin{aligned} u &= 0 \\ v &= 0 \\ w &= \frac{b}{2\pi}\arctan\frac{y}{x} = \frac{b}{2\pi}\theta \end{aligned} \tag{3.50}$$

The above is obtained from the equilibrium equations of elasticity:

$$\begin{aligned} \gamma_{xy} &= \frac{\partial u}{\partial y} + \frac{\partial v}{\partial x} \\ \gamma_{xz} &= \frac{\partial w}{\partial x} + \frac{\partial u}{\partial z} \\ \gamma_{yz} &= \frac{\partial w}{\partial y} + \frac{\partial v}{\partial z} \end{aligned} \tag{3.51}$$

or generally:

$$\gamma_{ij} = \frac{1}{2}\left(\frac{\partial u_i}{\partial x_j} + \frac{\partial u_j}{\partial x_i}\right) \tag{3.51a}$$

The expression for w in Eq. (3.50) is obtained as follows. While making a circuit around a dislocation line, the arctan changes by 2π and, thus, w changes as π.

For rectangular coordinates and by applying Eq. (3.51), while displacements u and v = 0, one gets:

$$\begin{aligned}\gamma_{xz} &= \frac{\partial w}{\partial x} + \frac{\partial u}{\partial z} = -\frac{b}{2\pi}\frac{y}{(x^2+y^2)} = \frac{b}{4\pi}\sin\frac{\theta}{r}\\ \gamma_{yz} &= \frac{\partial w}{\partial y} + \frac{\partial v}{\partial z} = \frac{b}{2\pi}\frac{x}{(x^2+y^2)} = \frac{b}{4\pi}\cos\frac{\theta}{r}\\ \gamma_{xy} &= \frac{\partial u}{\partial y} + \frac{\partial v}{\partial x} = 0\end{aligned} \tag{3.51b}$$

When deriving displacement via these relations for the strain, the stress is given by multiplying by G. Rectangular coordinates have only two stress components:

$$\begin{aligned}\tau_{xz} &= G\left(\frac{\partial w}{\partial x} + \frac{\partial u}{\partial z}\right) = -\frac{Gb}{2\pi}\frac{y}{(x^2+y^2)}\\ \tau_{yz} &= G\left(\frac{\partial w}{\partial y} + \frac{\partial v}{\partial z}\right) = \frac{Gb}{2\pi}\frac{x}{(x^2+y^2)}\end{aligned} \tag{3.51c}$$

Note that no normal stresses act and, of all the shear stresses, only the two above exist. All the other stresses are zero, as indicated below:

$$\sigma_x = \sigma_y = \sigma_z = \tau_{xy} = \tau_{yx} = 0$$

Also note that when going from Cartesian to cylindrical coordinates (or vice versa), the following relations may be used:

$$\begin{aligned}r^2 &= x^2 + y^2\\ \tan\theta &= y/x\\ z &= z\end{aligned} \tag{3.52}$$

The last term in Eq. (3.52) is a consequence of Fig. 3.42a, schematically representing all the relevant parameters of the cylinder, but not drawn for clarity.

3.3.9.2 Edge Dislocations

As shown in Fig. 3.42b, **b** is parallel to the x axis and vector **t** is in the z direction. The displacement is in the x direction; therefore, no strain in z direction exists. For cylindrical coordinates, one may write:

$$\begin{aligned}\sigma_r &= \sigma_\theta = -D\frac{\sin\theta}{r}\\ \tau_{\theta r} &= \tau_{r\theta} = D\frac{\cos\theta}{r}\\ \sigma_z &= \nu(\sigma_r + \sigma_\theta)\end{aligned} \tag{3.53}$$

where:

$$D = \frac{Gb}{2\pi(1-\nu)} \tag{3.53a}$$

and:

$$\tau_{rz} = \tau_{zr} = \tau_{z\theta} = \tau_{\theta z} = 0.$$

For Cartesian coordinates, the above takes the form of:

$$\begin{aligned}
\sigma_x &= -Dy\frac{(3x^2+y^2)}{(x^2+y^2)^2} \\
\sigma_y &= Dy\frac{(x^2-y^2)}{(x^2+y^2)^2} \\
\tau_{yx} &= \tau_{xy} - Dx\frac{(x^2-y^2)}{(x^2+y^2)^2} \\
\sigma_z &= \nu(\sigma_x+\sigma_y)
\end{aligned} \tag{3.54}$$

and:

$$\sigma_{xz} = \sigma_{zx} = \sigma_{yz} = \sigma_{zy} = \tau_{xz} = \tau_{zx} = \tau_{yz} = \tau_{zy} = 0.$$

Note that, although no displacement and strain evolves in the z direction, a stress is acting, as shown in the above equations. Clearly, dilatational (expanding) stresses are also active, in addition to shear stress. The largest normal stress is σ_x, which acts parallel to the slip vector **b**. The slip plane, according to Fig. 3.42b, is along the plane passing through y = 0. The sign above the edge dislocation is negative; therefore, the stress is compressive and maximal just above the slip plane of the dislocation, whereas the stress is tensile, thus, positive (σ_x) and maximal just below the slip plane. The signs of the stress and the strain reverse if the sign of the Burgers vector is reversed.

In cases of mixed dislocations, it is possible to combine their components, namely the solutions of their screw and edge orientations.

Cases with arbitrary orientations in an isotropic material (i.e., when stress depends on the orientation of the dislocation and the Burgers vector and the crystal axes are significant), stress fields and stress distributions around edge dislocations are shown and discussed in detail in Read's book.

3.3.10 Forces Acting on Dislocations

The presence of forces acting on dislocation lines, balancing line tension, was discussed in Sect. 3.3.8. When an external stress is applied to a dislocation, the dislocation will tend to move. In general, dislocations in stress fields sense a force

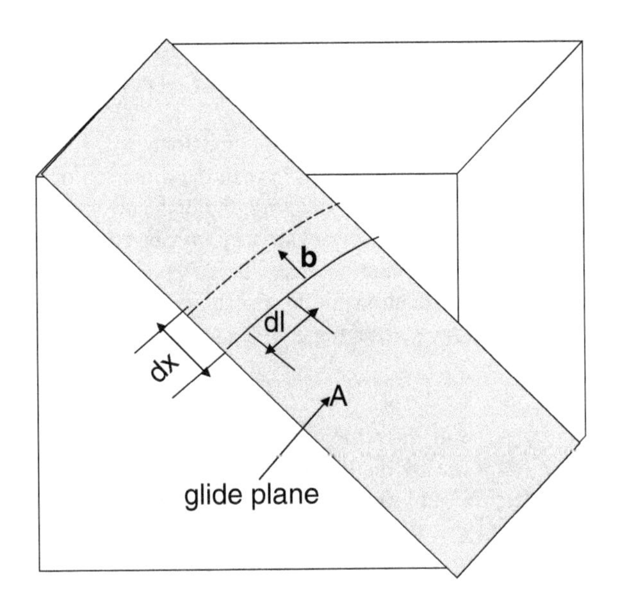

Fig. 3.46 A segment of dislocation, dl, has moved a distance, dx, in the glide plane of area A, as a result of a force acting on it perpendicularly [14]

that tends to move them. The motion of a dislocation (as seen in Sect. 3.3.6) can either be a glide or a climb by edge dislocations or cross-slips made by screw dislocations.

3.3.10.1 Glide Forces

An external force acting on a dislocation may induce its glide; this effect may be seen in Fig. 3.46. A slip plane, A, is shown, where a segment, dl, of a dislocation is depicted after moving a distance, dx. The ratio of the area swept out by the motion of the segment, dA = dldx to the total slip area, multiplied by **b,** gives the displacement, which may be written as:

$$displacement = \frac{dldx}{A}\mathbf{b} \tag{3.55}$$

The applied force is F = τA and, therefore, the work, dw, performed by that force is given by F x distance (i.e., displacement), namely:

$$dw = F\frac{dldx}{A}\mathbf{b} = A\tau\frac{dldx}{A}\mathbf{b} \tag{3.56}$$

If one defines the force on a unit length of dislocation as 'the work done when the unit length of the dislocation segment, dl, moves a unit distance, dx', one gets:

$$F = \frac{dw}{dldx} = \frac{dw}{dA} = \tau\mathbf{b} \tag{3.57}$$

This force is normal to the dislocation line. Apparently, only a shear stress, acting on the dislocation in this plane, is necessary for its motion and only a force component, lying in the glide plane, is effective in this regard. A normal component of force or stress will not contribute to glide motion, since only components of force acting in the glide plane in the direction of the Burgers vector can move a dislocation. Practically speaking, the dislocation line never moves in total, because it may be anchored somewhere by an obstacle (as in Sect. 3.3.8 on line tension, where the dislocation became curved).

The above relation might also have been obtained in the following way. The work, w, done by moving a dislocation a distance, **b**, on slip area, A, is:

$$w = A\tau \mathbf{b} \tag{3.58}$$

where $A\tau = F$. Incremental work, dw, done on an incremental area is dldx (the area swept out by the elemental dislocation line, dl, when moving an incremental distance, dx) and may be related to the work of moving the dislocation. Thus:

$$\frac{dw}{w} = \frac{dldx}{A} \tag{3.59}$$

Following all the prior relations, dw may be expressed as:

$$dw = A\tau b\frac{dldx}{A} = \tau bdldx \tag{3.60}$$

but $dw = Fdx$ and, when acting on the dislocation segment, dl, $dw = Fdxdl$. Redefining the force acting on a unit dislocation as $F' = F/dl$, one obtains for Eq. (3.57):

$$F' = \tau b \tag{3.61}$$

F' is the same as F, given and defined in Eq. (3.57).

3.3.10.2 Climb

Climb was first discussed in Sect. 3.3.6.2. Figure 3.39 illustrated the climb process and Fig. 3.41 showed the jog formation resulting from climb. Climb is the movement of dislocations out of their glide planes as a result of interactions with vacancies (or interstitials, for negative climbs, i.e., the growth of an extra plane or part of it). It is possible that a force acting on an edge dislocation will be directed perpendicularly to the slip plane, normal to its Burgers vector, allowing the edge dislocation to leave its slip plane. The driving force for dislocation climb is the movement of vacancies through a crystal lattice. If a vacancy moves next to the extra half-plane of atoms that defines an edge dislocation, the atom in the half-plane that is closest to the vacancy can trade places with that vacancy, thus causing climbing. Compressive force produces positive climb in the positive y direction, whereas tensile force induces negative climb. During positive climb, a crystal

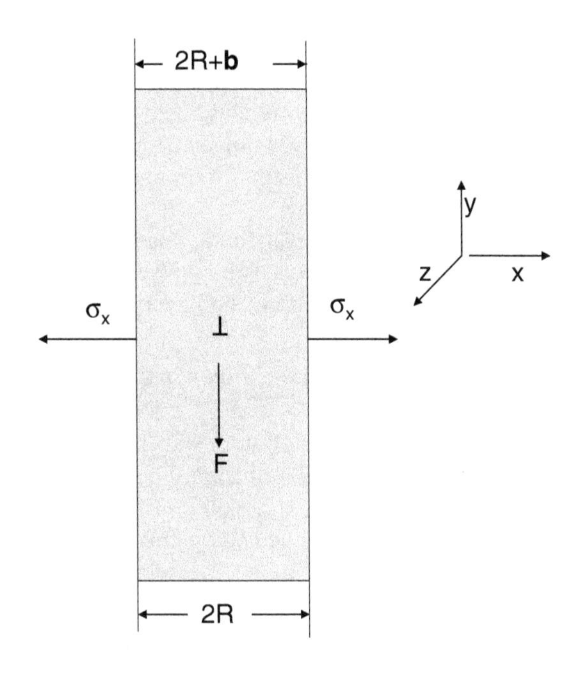

Fig. 3.47 This force is acting in the negative y direction, because tensile stress induces the negative climb of a dislocation. Note that the width at the lower edge of the curve is less by **b** than at the upper edge, since a dislocation is defined by the extra half-plane [14]

shrinks in the direction perpendicular to the extra atomic half-plane, because atoms are being removed from it. Since a negative climb involves the addition of atoms to a half-plane, the crystal grows in the direction perpendicular to that half-plane. Therefore, compressive stress, in the direction perpendicular to the half-plane, promotes positive climb, while tensile stress promotes negative climb. Figure 3.47 illustrates negative climb schematically.

Assume that climb occurred over a distance L in the y direction. The work done is then:

$$\mathrm{w} = \mathrm{FL} = \sigma_{\mathrm{x}} \mathbf{b} \mathrm{L} \tag{3.62}$$

The work per unit distance of the tensile stress, σ_{x}, in the crystal has been given in Eq. (3.61). Thus, the force per unit length is:

$$\mathrm{F}' = \sigma_{\mathrm{x}} \mathbf{b} \tag{3.61a}$$

Jogs are formed by climb. They are favorable sites for the absorption and emission of point defects. In thermal equilibrium, the atoms at jog sites are in dynamic equilibrium and arrive and leave the jog at equal rates. If there is an increase in vacancies, for example, in the vicinity of a dislocation line above the thermal equilibrium value, the probability of atomic exchange at a jog with a vacancy increases, climb occurs and the extra plane (defining the dislocation line) shrinks. Therefore, excess vacancies promote the process of climb. Similarly, an excess of interstitial atoms adds atoms to the existing jog, which causes it to grow. In summary, when atoms are removed from an extra plane, the crystal collapses locally, but when atoms are added, the atomic plane expands locally. As stated above, either compressive or tensile strains are induced during the climb process. These stresses work during climb and the direction of the climb depends on the type of stress. In Fig. 3.47, tensile stress is shown promoting negative climb. If, however, a compressive stress were applied, the process would appear to squeeze out the extra plane.

The formation of jogs by climb may be considered to be a thermally activated process, since point defects are involved. It is proper to talk about nucleation and the motion of jogs. An activation energy exists, expressed by:

$$n_j = n_0 \exp\left(-\frac{U_j}{kT}\right) \tag{3.63}$$

where n_0 is the number of atom sites per unit length of dislocation and U_j is the activation energy of jog formation. The activation energies for vacancy formation (for positive climb), E_f, and motion, E_m, determine the activation energy, E_{SD}, of self-diffusion. Thus, the activation energy of climb, U_c, may be expressed as:

$$U_c = U_j + U_f + U_m = U_j + U_{SD} \tag{3.64}$$

Excess point defects are also a driving force for climb.

3.3.11 Forces Between Dislocations

A dislocation senses a force due to the presence of a stress field around it. The presence of more dislocations, each with its own surrounding stress field, means that all these forces influence one another. Consider the presence of two dislocations without the application of an external stress. The total force is assumed to be the sum of the individual forces. In Sect. 3.3.7, it was indicated that a dislocation must have certain strain energy and, indeed, a strain field exists at the site of each dislocation, affecting its motion.

3.3.11.1 Screw Dislocations

Given two parallel dislocations with the same sign at sites 1 (the origin) and 2 (as shown schematically in Fig. 3.48a), the stress fields at these screw dislocations have a radial symmetry and their Burgers vectors are parallel to the z axis. Due to their radial symmetry, their stress is:

$$\tau_{z\theta} = \frac{Gb_1}{2\pi r} \tag{3.65}$$

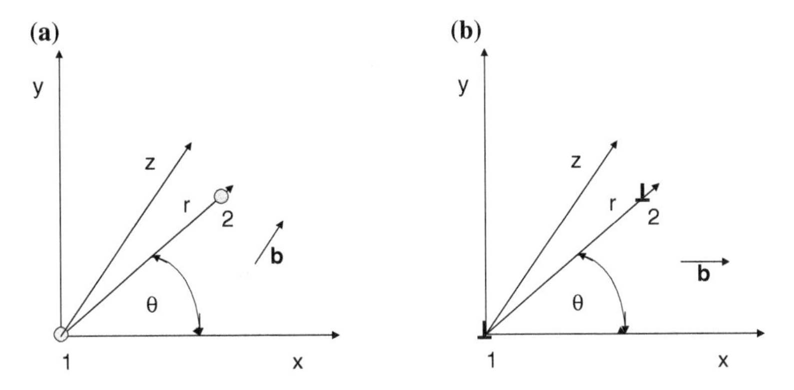

Fig. 3.48 Two same-signed, parallel dislocations at sites *1* and *2*: **a** screw; **b** edge [14]

There is only one force, because of the radial symmetry, F_r, which is a radial component. The first stress field is acting on dislocation 2 at a distance, r, from the first dislocation and, therefore, this force is:

$$F_r = \tau_{z\theta} b_2 = \frac{Gb_1 b_2}{2\pi r} \tag{3.66}$$

These two same-signed, parallel dislocations repel each other, depending inversely on the distance, r, (how far apart they are from each other) and, therefore, they stay relatively distant on the same or on near slip planes.

In Cartesian coordinates, two forces exist, those of F_x and F_y, expressed as:

$$F_x = \frac{Gb_1 b_2 x}{2\pi(x^2 + y^2)} \tag{3.67}$$

$$F_y = \frac{Gb_1 b_2 y}{2\pi(x^2 + y^2)} \tag{3.68}$$

3.3.11.2 Edge Dislocations

Again, consider two parallel edge dislocations, as shown in Fig. 3.48b. They act in plane zx. Here, the dislocation line is parallel to the z axis. The first stress field is acting on dislocation 2, which is a parallel slip plane. The force has two components, F_x and F_y, in the x and y directions, as indicated by the subscripts. For the dislocation at the point of origin, these two components may be expressed as:

$$F_x = \tau_{xy} b_2 = \frac{Gb_1, b_2}{2\pi(1-v)} \frac{x(x^2 - y^2)}{(x^2 + y^2)^2} \tag{3.69}$$

F_x exerts a glide force on the second dislocation in the x direction. The dislocation at the origin (dislocation 1) exerts a climb force component, F_y, on dislocation 2, shown in Fig. 3.48b in the y direction (a force component perpendicular to the glide plane).

$$F_y = -\sigma_1 b_2 = \frac{Gb_1 b_2}{2\pi(1-\nu)} \frac{y(3x^2+y^2)}{(x^2+y^2)^2} \tag{3.70}$$

Note that when dislocation 2 is above dislocation 1, i.e., when its coordinates are (0, y), they do not exert glide force on each other. For polar coordinates, these forces are given as:

$$F_r = F_x \cos\theta + F_y \sin\theta = \frac{Gb_1 b_2}{2\pi(1-\nu)} \frac{1}{r} \tag{3.71}$$

$$F_\theta = F_y \cos\theta - F_x \sin\theta = \frac{Gb_1 b_2}{2\pi(1-\nu)} \frac{\sin 2\theta}{r} \tag{3.72}$$

Equations (3.66–3.72) are based on the expressions in Sect. 3.3.9, specifically Sects. 3.3.9.1 and 3.3.9.2 for screw and edge dislocations, respectively. In Fig. 3.48a and b, the Burgers vectors are parallel to the z and x axes of the screw and edge dislocations, respectively. As mentioned in Sect. 3.3.11.1, there is a repulsive force between same-signed dislocations, an attractive force between unlike dislocations. When $y = 0$, Eq. (3.69) reduces to

$$F_x = \frac{Gb_1 b_2}{2\pi(1-\nu)} \frac{1}{x} \tag{3.73}$$

indicating that the force depends only on the distance between the dislocations and also that $F_y = 0$.

3.3.12 Intersection of Dislocations

In earlier sections, the characteristics of single dislocations were discussed. It was indicated that a dislocation moves quite freely in certain glide planes under applied shear stress. However, even well-annealed crystals contain many dislocations at a level of $\sim 10^4$. Therefore, it is quite probable that moving dislocations will encounter others, known as 'forest dislocations', which will hinder their freedom to glide. Even the most favored planes will contain dislocations and, thus, moving dislocations will interact with those others present in the material. The term 'dislocation intersection' refers to an interaction occurring between a moving dislocation and the others encountered while in motion. For the sake of simplicity, this section will describe the interactions between two dislocations.

When dislocations intersect jogs, 'kinks' may form. 'Kinks' are steps occurring in dislocation lines within the same slip plane, while 'jogs' are steps in another slip

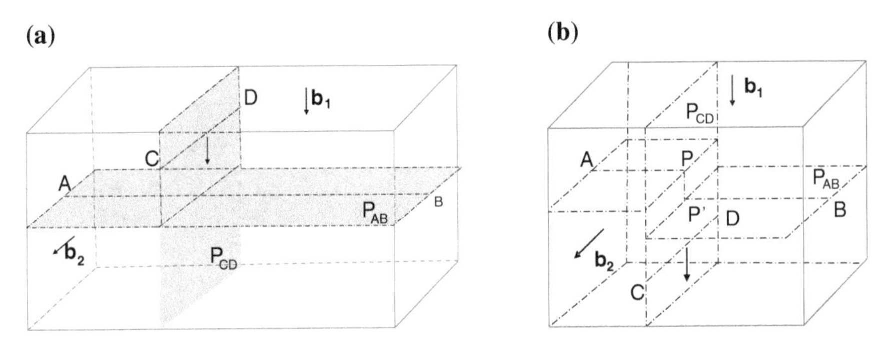

Fig. 3.49 The intersection of two edge dislocations: **a** before intersection; the dislocation in plane P_{CD} is moving in the direction shown; **b** after intersection, jog PP′ is formed on plane P_{AB} [14]

plane. Jogs may form during processes of thermal activation (as mentioned in Sect. 3.3.10.2), at an elevated temperature or at low temperatures by some intersection of dislocations. Of the many possibilities, the simplest case is when two edge dislocations, moving on two planes normal to each other, intersect. In Fig. 3.49, the intersection of two edge dislocations is shown before and after jog formation.

A jog, PP′, is formed after dislocation CD cuts through the AB dislocation line (Fig. 3.49b). The size of the step is that of Burgers vector $\mathbf{b_1}$, but remains a part of dislocation AB. The segment is of edge character, since its Burgers vector is $\mathbf{b_2}$ (that of the overall dislocation AB of which it is a part). Since the Burgers vector and the dislocation line APP'B are in the same glide plane, jog PP′ will not hinder the motion of that dislocation. There is an energy increase in dislocation AB by the length of **b** (i.e., $\mathbf{b_1}$), which may be expressed by $\alpha G b^2 \mathbf{b} = \alpha G b^3$.

The intersection of two orthogonal dislocations with parallel Burgers vectors is illustrated in Fig. 3.50, before and after intersection. In this case, two kinks are formed with the Burgers vectors, namely $\mathbf{b_1}$ and $\mathbf{b_2}$ for the AB and CD dislocations, respectively. Both these kinks have a screw orientation and an increase in energy occurs in each of their dislocation lines.

Other cases are the intersections between edge-screw and screw-screw dislocations, illustrated in Figs. 3.51 and 3.52, respectively. Simply drawn, the intersecting planes in the figures below are orthogonal, but they do not have to be.

Figure 3.52 illustrates that, in the edge-screw case, a jog PP′ in the edge and a kink QQ′ in the screw dislocations, respectively, are formed. The jog is formed, because plane P_{AB}, where dislocation AB is moving, is a spiral surface, not unlike a ramp, so its motion, after cutting the screw dislocation, is not at the same level as it was at the beginning of its motion. The PP′ jog has an edge character and can move with dislocation AB in the P_{AB} plane, whereas QQ′ must climb, since it cannot move up along the screw dislocation. Segments PP′ and QQ′ (in Fig. 3.52)

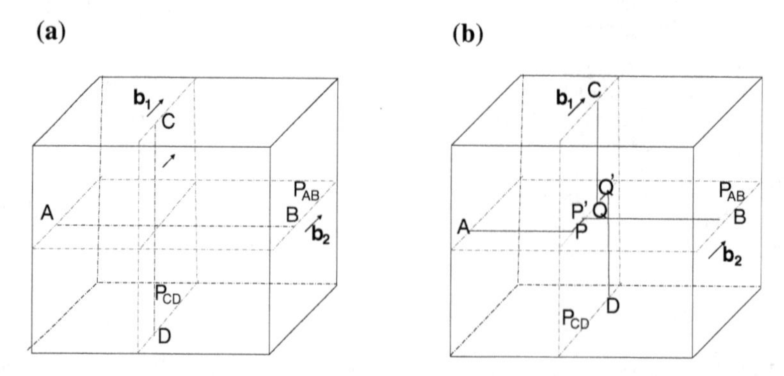

Fig. 3.50 Two orthogonal dislocations with parallel Burgers vectors: **a** *before* intersection; **b** *after* intersection [14]

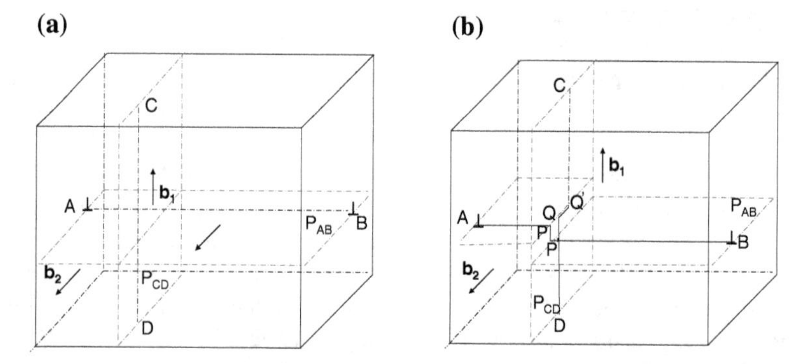

Fig. 3.51 The intersection of edge and screw dislocations: **a** *before* intersection; **b** *after* intersection [14]

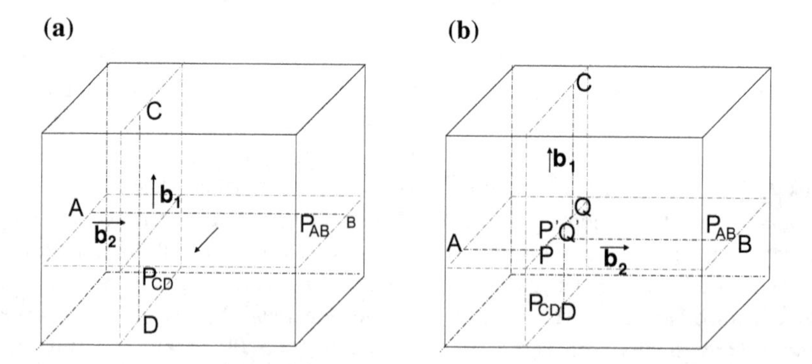

Fig. 3.52 The intersection of screw dislocations: **a** *before* intersection; **b** *after* intersection. Two steps are formed, both having an edge character [14]

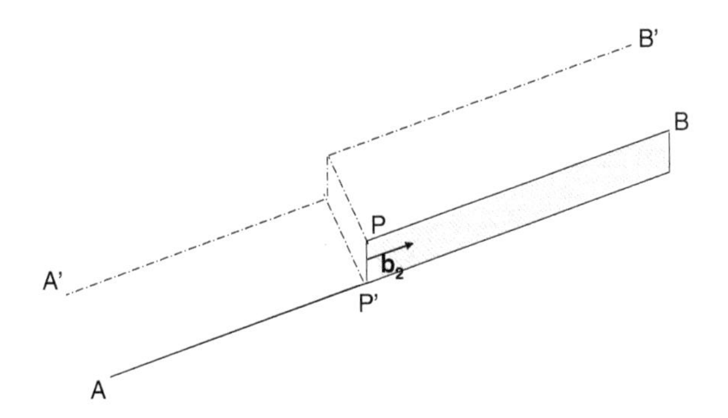

Fig. 3.53 The jog, PP′, and the screw dislocation line, AP′PB, after intersection. This jog has an edge character and its glide plane (i.e., where the jog can glide) is defined by the shaded area. Movement of the screw dislocation to A′B′ can occur only by climb [14]

both have an edge character, since the respective Burgers vectors are perpendicular to them. The AB screw dislocation, after cutting through dislocation CD and moving into P_{XD} again (as described in Fig. 3.51), also concludes its motion at a level higher than where it began, thus forming a jog. The segment, PP′, is actually a jog, but it is not easily drawn. However, by means of Fig. 3.53, the jog formation in Fig. 3.52 and its motion may be better understood. Note that the shaded area defines the glide plane of the jog, which has an edge orientation and, thus, its motion is along the screw axis of AP′PB. The only motion of the screw dislocation to A′B′ is by climb.

Some comments in summation:

(a) the movement of dislocations generates steps as jogs and kinks;
(b) jogs severely influence the glide of dislocations;
(c) jogs and kinks can occur in any dislocation, not just edge dislocations, though they are often difficult to render graphically;
(d) as will be discussed in a next chapter, the role of jogs in strain hardening is of great importance, since no easy glide is possible.

3.3.13 Dislocation Multiplication

As indicated in earlier sections, well-annealed crystals have $\sim 10^4$ dislocations and, in heavily deformed ones, that number may even reach $\sim 10^{10}$. Therefore, a means for producing such a large number of dislocations is required. Several sources of dislocations in materials were mentioned above (Sect. 3.3.5). The formation of a large number of dislocations is a prerequisite for understanding much of the mechanical behavior of materials, such as large plastic strains. One of the well-known concepts for generating a large number of high-velocity dislocations is by means of the 'Frank-Read mechanism', also known as the 'Frank-Read generator'. In essence, according to this method, one single dislocation is sufficient to produce a large number of them at the deformation velocity. Figure 3.54 is the graphic method used by most publications in the field to show how this is done. A

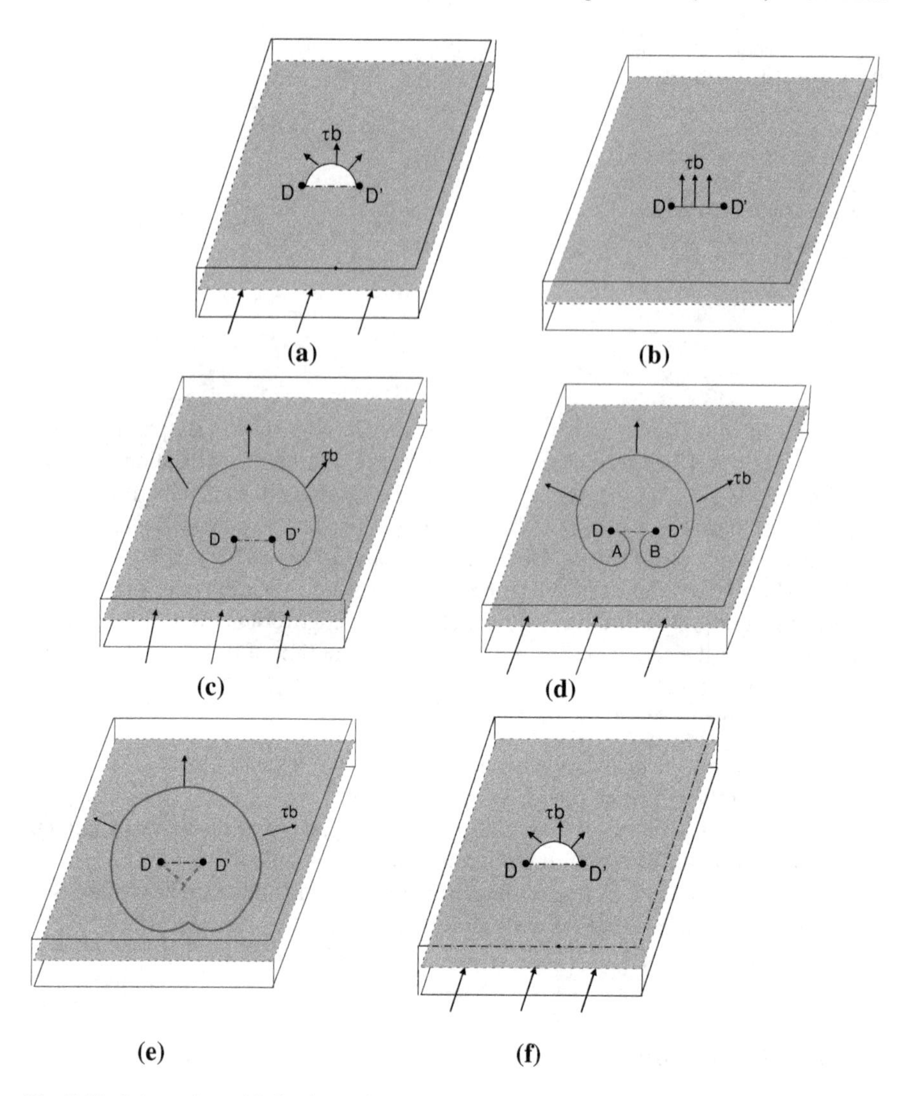

Fig. 3.54 Schematic multiplications of a single dislocation according to the concept of a Frank-Read source [14]

well-documented Frank-Read dislocation source in action in silicon was demonstrated by Dash and is readily available elsewhere in the literature. The stepwise operation of a Frank-Read source on a slip plane is as follows:

(a) assume that a dislocation segment, D-D′, is pinned at both ends, as seen in the illustration;

(b) an applied stress produces a glide force, τb, acting on this segment in its glide plane, causing it to curve maximally into a semicircle shape. The line tension balances the force on the dislocation;
(c) the stress, τ, is increased beyond a critical value (illustration b) and the dislocation becomes unstable, tending to expand indefinitely. The expanding dislocation loop doubles back on itself, forming spirals at each end of the dislocation;
(d) this dislocation continues to expand. In c) and d), slip occurs, as indicated by the area of the expanding loop;
(e) the spirals of the slipped area have joined, forming a closed dislocation loop and that segment of the dislocation is ready to start a new cycle.

The dislocation parts at A and B (illustration d) have opposite signs and attract each other until the loop closes (illustration e). The dislocation segment, D-D′, which was left behind, repeats the same sequence as described above. The force acting on the dislocation, causing it to curve, is balanced by the line tension (discussed in Sect. 3.3.8). The line tension per unit length in that section is equal to the force acting normally on the dislocation segment. In equilibrium with the applied stress, the relation derived as Eq. (3.74) may be rewritten here as:

$$\tau = \frac{\alpha G b}{r} \tag{3.74}$$

which may also be expressed in terms of the radius, r, as:

$$r = \frac{a G b}{\tau} \tag{3.74a}$$

Beyond this radius, at the point of the applied stress, the dislocation becomes unstable, as previously mentioned, and the Frank-Read source, pinned at both ends, expands under a diminishing force. Equation (3.74) represents the stress required to generate dislocations from a Frank-Read source. If the shear stress increases any further and the dislocation passes the semicircular *equilibrium state*, it will spontaneously continue to bend and grow, spiraling around the A and B pinning points until the segments spiraling around the A and B pinning points collide and cancel each other out, as described earlier. This process results in a dislocation loop around D and D′ in the slip plane, which expands under continued shear stress. In addition, the dislocation line between D and D′ may continue to generate dislocation loops in the manner described above. A Frank-Read source can, thus, generate many dislocations in a crystal plane under applied stress. A single Frank-Read source may also operate when the line is pinned only at one end. Intersections with other dislocations, which form jogs that increase the length of the line, may even act as Frank-Read sources.

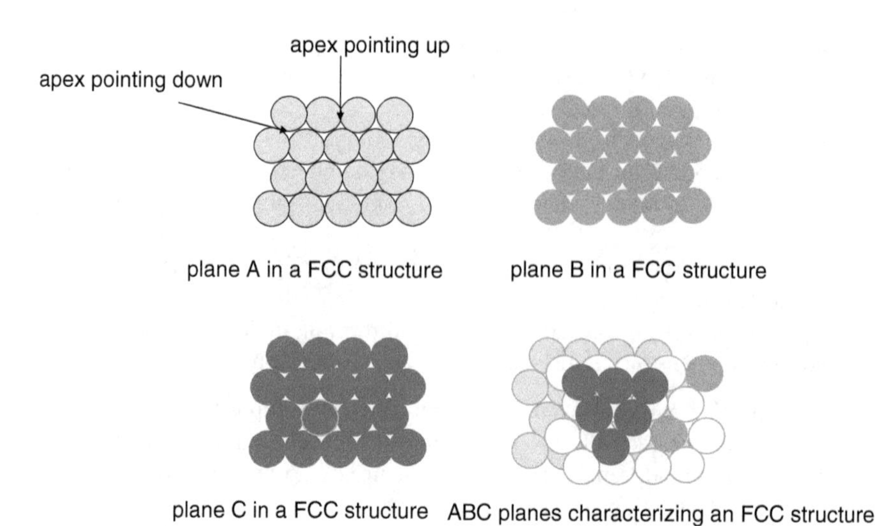

Fig. 3.55 Packed planes in an FCC metal. The sequence of the planes is *A*, *B* and *C* [14]

3.3.14 Partial Dislocations

Usually, there are preferred slip planes and directions in certain crystal systems. The combination of both the slip plane and the direction define a slip system. Slip planes are generally the closest-packed planes in a structure. The slip direction of a slip plane is the one with the highest linear atomic density. Up to this point, no specific consideration has been given to structure; however, when discussing partial dislocations, one cannot avoid the structural aspect. The most common approach is to consider FCC structures. However, before entering into the details, one must define the term 'partial dislocation'. Unlike the dislocation cases discussed thus far, in which the most characteristic feature was the Burgers vector, which is also a 'translation vector' of the lattice, in 'partial dislocation', the Burgers vector is *not* a translation vector of the lattice and partial dislocations are usually bordered by planar defects called 'stacking faults' (SF). For the purpose of illustration, closely-packed planes make a good example. In FCC, slip occurs on $\{111\}$ planes in $\langle 110 \rangle$ directions. In Fig. 3.55, closely-packed planes and their arrangement in an FCC structure are shown. Note that the apexes of the triangular-shaped voids have two orientations, with the apex pointing upward or downward. Plane B can occupy any of these voids. Assume that those with the apex pointing upwards are occupied by this plane. In an FCC structure, in which the sequence of planes is A, B and C, plane C will occupy the voids with the apexes pointing downward (as in the considered case). The entire stacking is shown in the last illustration, such that all three planes may be observed. Thus, by necessity, plane C is only partially shown. An FCC unit cell and one of the $\{111\}$ planes are depicted in Fig. 3.56.

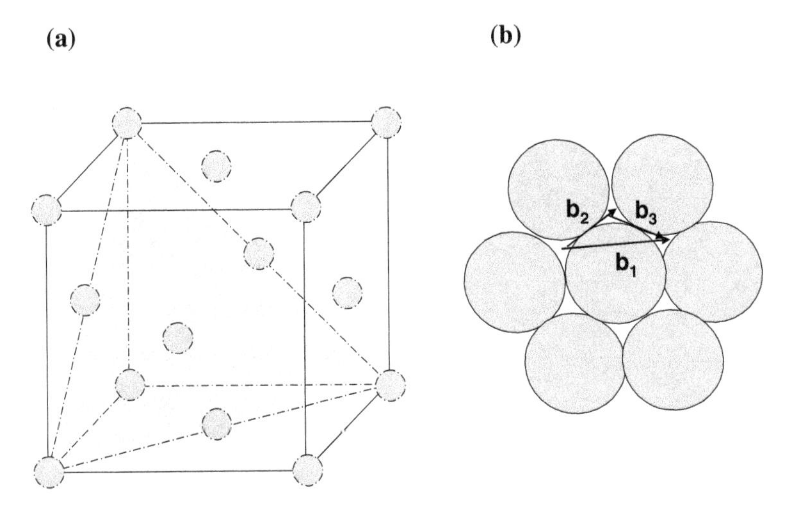

Fig. 3.56 FCC unit cell: **a** shaded area a {111} plane; **b** the shaded {111} plane showing the splitting of Burgers vector $\mathbf{b_1}$ into $\mathbf{b_2}$ and $\mathbf{b_3}$ [14]

3.3.14.1 Shockley Partial Dislocation

As indicated above, the translation vector in FCC structures is of type **a/2** $\langle\mathbf{110}\rangle$; therefore, it should also be the Burgers vector. Yet, for energetic and topographic reasons, this vector splits into what is known as 'partial dislocations'. Two partials are formed and are termed 'Shockley partial dislocations' or simply 'Shockley partials'. First, consider the topological reason why a Burgers vector might split. In Fig. 3.56b, one sees Burger vector $\mathbf{b_1}$, which is the translational vector. To attain motion, it would have to surmount the peak formed by the central atom. Rather than taking the proverbial 'high road', an easier strategy is adopted—it splits into two steps and goes, as it were, "round the mountain", by passing through 'the valleys' on either side, between each of the two neighboring atoms, as represented by vectors $\mathbf{b_2}$ and $\mathbf{b_3}$. The crystallographic orientations of these partials (students should verify this as an exercise) are of the type **a/6** $\langle\mathbf{211}\rangle$.

As indicated in Sect. 3.3.7, the energy of a dislocation is proportional to b^2 in the form of:

$$E = \alpha G b^2 \tag{3.75}$$

where α takes the type of dislocation into account. Note from the following reaction that dislocation splitting is energetically favored. One such reaction is:

$$\frac{\mathbf{a}}{\mathbf{2}}[\mathit{110}] = \frac{\mathbf{a}}{\mathbf{6}}[\mathit{211}] + \frac{\mathbf{a}}{\mathbf{6}}[\mathit{12\bar{1}}] \tag{3.76}$$

Before looking at the energetic profile of this reaction, one must verify its correctness according to the following steps:

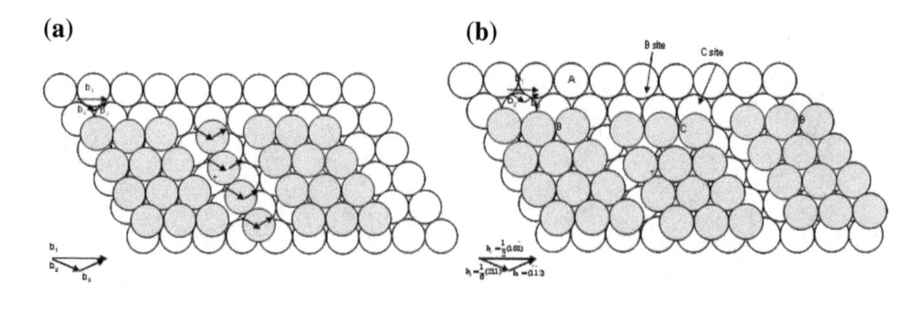

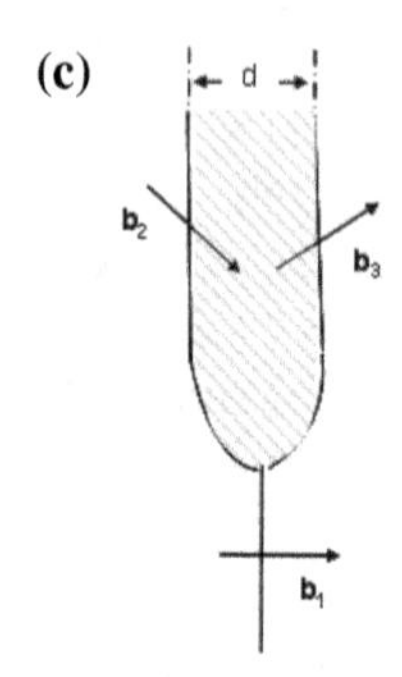

Fig. 3.57 Partial dislocations in an FCC structure: **a** the split motion of atoms from site B to site C; partial dislocation with Burgers vectors $\mathbf{b_2}$ and $\mathbf{b_3}$; **b** extended dislocation composed of the two partials and the SF between them; **c** schematic presentation of a SF [14]

	The un-dissociated vector	The partials
component x	1/2	2/6 + 1/6 = 3/6 = 1/2
component y	1/2	1/6 + 2/6 = 3/6 = 1/2
component z	0	1/6 − 1/6 = 0

Thus, this reaction is proven correct and both sides of Eq. (3.76) are balanced. Next, one must check the energetic aspect. For this purpose, take the absolute values of the indexes of the Burgers vectors. Since their energies are proportional to their squares (as seen in Eq. 3.75), for the left side of Eq. (3.76), one gets $a^2/2$ and, for the sum of the squares on the right-hand side, $a^2/3$. Clearly, $a^2/2 > a^2/3$ and there is a decrease in energy, thus the splitting of the dislocation is also favored energetically.

In Fig. 3.57, the separation of a perfect dislocation in FCC metals is illustrated.

As indicated in Fig. 3.55, the planes are designated as A, B and C and the same notation is also used for the voids between the atoms with up or down apexes. The reaction shown in Fig. 3.57 is:

$$\frac{1}{2}(10\bar{1}) = \frac{1}{6}(21\bar{1}) + \frac{1}{6}(\bar{1}\bar{1}2)$$

This is another form of Eq. (3.76).

In Fig. 3.57a, one row of atoms indicates the formation of the partial dislocations. The dislocation in plane A, with $\mathbf{b_1}$ normal to it, is of edge orientation (it is the total dislocation). The single row of atoms in plane B is zig-zagged on both sides, which is characteristic of partial dislocations. The first movement of the row, by the amount of $\mathbf{b_2}$, shifts it into a zig-zag position, while the second movement, indicated by vector $\mathbf{b_3}$, brings the atoms to the final position, as would have occurred with a single displacement of $\mathbf{b_1}$. These two partial dislocations will repel each other to a point where a balance is reached between the elastic energy decrease, due to the splitting of the dislocation, and the increase in SF energy. The SF energy varies widely from metal to metal, depending on the width of the fault. Thus, the width of a SF in Cu is about 10 atomic spacings, whereas, in Al, it is only ~2 atomic separations. This means that the SF energy of Cu is low (~80) compared to that of Al (~200 mJ/m^2). The combined defect created by partials and a SF is called an 'extended dislocation'. Schematically, the partials and the SF are shown in Fig. 3.57c.

The width of a SF is a consequence of the balance between the repulsive force separating its two partial *dislocations* and its attractive force. When the SF energy is high, the splitting of the perfect dislocation into two partials is unlikely and glide in the material occurs only as a result of perfect dislocation glide. Lower SF energy will promote the formation of wider SFs and cross-slip or climb will be more difficult consequently. As a result, the mobility of the extended dislocation in materials with low SF energy decreases.

The usual schematic presentation of the splitting of a perfect dislocation into Shockley partials with a fault between them is shown in Fig. 3.57c.

In Sect. 3.3.11, the force between two dislocations is given either by Eq. (3.73) or by Eq. (3.67). In analogy to Eq. (3.67), the repulsive force between the two partial dislocations, in terms of the distance, d, may be rewritten as:

$$F = \frac{Gb_2b_3}{2\pi}\frac{1}{d} \tag{3.77}$$

where d, in Eq. (3.77), stands for the thickness of the SF ribbon formed by the partial dislocations with vectors $\mathbf{b_2}$ and $\mathbf{b_3}$. The SF energy-per-unit area is γ (not shear angle!) and, when multiplied by the area, the energy of the fault is obtained. Thus, the force of the SF (i.e., energy per distance) is equivalent to γ:

$$\gamma = \frac{Gb_2b_3}{2\pi}\frac{1}{d} \tag{3.78}$$

By balancing these forces, the equilibrium distance, d, of the separation between the partials may be obtained

$$d = \frac{Gb_2.b_3}{2\pi}\frac{1}{\gamma} \tag{3.78a}$$

The extended dislocation, discussed hitherto, consists of Shockley partials and a SF, which can glide within its own glide plane; therefore, the accepted notation is glissile dislocation.

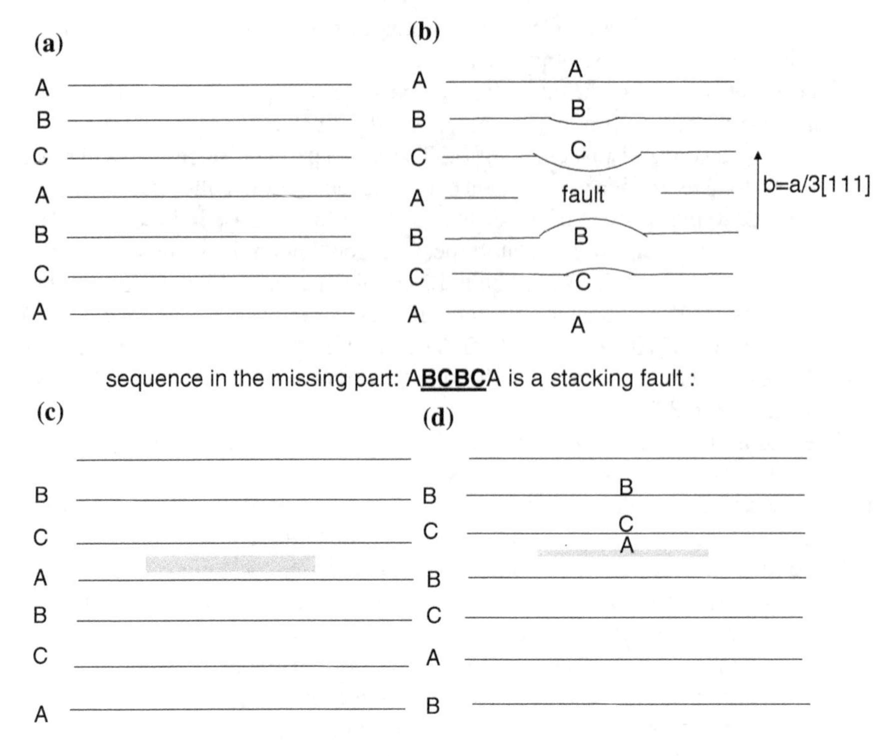

Fig. 3.58 SF formation: **a** the FCC stacking of slip planes before removal of a plane; **b** the stacking sequence of the {111} planes is modified at the region where the fault exists from the FCC to an HCP structure; **c** the sequence before the insertion of an additional plane (indicating the place of insertion); **d** the re-insertion of the plane produces a thin HCP structure

3.3.14.2 Frank Partial Dislocations

Partial dislocations can form not only by splitting a perfect dislocation, but also by inserting or partly removing a {111} plane. In Fig. 3.58a, the FCC stacking of slip planes is illustrated schematically before the removal of a plane. The sequence of the stacking of the {111} planes is modified at the region where the fault exists from the FCC to a hexagonal closely-packed [henceforth: HCP] structure, as seen in Fig. 3.58b.

In the center of this figure, in the region of the fault, the sequence is not …ABCABC…, but …A**BCBC**A…; the underlined bold letters are characteristic of HCP packing. The Burgers vector indicated is in the $\langle 111 \rangle$ direction; its actual value is $\mathbf{b} = a/3\ \langle 111 \rangle$. The missing plane (indicated in Fig. 3.58b) may have been removed by the condensation and collapse of the vacancies, forming a disc-shaped layer. As seen in Fig. 3.58, this fault is bounded by an edge-oriented dislocation at each side. The Burgers vectors are normal to the {111} planes and are not in the

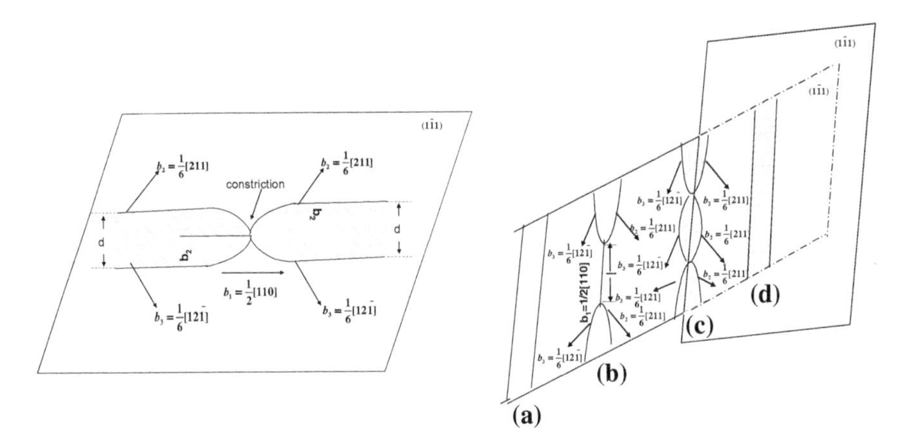

Fig. 3.59 The cross-slip of a screw dislocation: **a** schematic constriction in an extended dislocation, making cross-slip possible; **b** the stages of cross-slip from one octahedral plane to an other in (**a**) a SF before cross-slip; (**b**) the partial dislocations of the extended dislocation combined along distance l; (**c**) the dislocation, after combination of the partials curves under the influence of stress, cross-slipping into the $\bar{1}11$ cross-slip plane; and (**d**) the extended screw dislocation and the SF continue on from one slip plane into another

slip direction in FCC crystals. This is known as 'Frank sessile dislocation'. A Frank dislocation may also be formed by inserting part of a plane, which is also a sessile dislocation, since it cannot glide in any of the ABC slip planes, as shown in Fig. 3.58c. A dislocation formed in this way can only move non-conservatively, which requires the transport of either atoms or vacancies. Note that, when part of a plane is removed, the fault is described as 'intrinsic' while, when a partial plane is inserted, it is called an 'extrinsic-stacking fault'. In practice, both types of Frank partial dislocations may occur, either due to the condensation of excess vacancies during crystal growth or by the condensation of interstitial atoms. Such partial dislocations are also known as 'negative' or 'positive Frank partials', respectively.

3.3.14.3 Cross Slip of Partial Dislocations

As stated earlier, in edge dislocations, the slip direction and the dislocation line define the slip system; however, in screw dislocations, the Burgers vector is parallel to the dislocation line and, thus, it may cross-slip into planes belonging to the same form. The situation is different in cases of extended dislocations, where SFs influence cross-slip. Figure 3.59 illustrates cross-slip in a faulted crystal bounded by two partial dislocations, in accordance with ideas promoted by Seeger [17].

An edge dislocation with partials is able to move within its glide plane along with its faulted region–the extended dislocation–but it will not be able to move into another octahedral plane unless it climbs. A screw dislocation or a screw component will not have such a problem, as long as the direction of slip and the

Burgers vector are common to both $\{111\}$ planes. However, cross-slip can occur only if a 'constriction' (i.e., a joining of the partials) forms. Figure 3.59a shows such a constriction in a fault at the point at which the partials come together, allowing cross-slip to occur. In Fig. 3.59b, the stages of cross-slip are indicated: (a) the faulted region in its glide plane; (b) the partial dislocations reunited by an applied stress over a length of l to form an un-extended dislocation, before it can cross-slip from $(1\bar{1}0)$ to another $\{111\}$ plane (here, an $(\bar{1}11)$ plane is illustrated, which is a cross-slip plane); (c) the dislocation, after the partials have combined under the influence of stress in the $(\bar{1}11)$ cross-slip plane, bows out, also after having first been anchored at two constrictions and then spread as a new extended dislocation; and (d) the extended screw dislocation and the SF continue in another plane (i.e., a cross-slip plane).

Cross-slip depends on the width of the SF. Wide ribbons require considerable force to bring the partials together for cross-slip. Aluminum, for example (as indicated in Sect. 3.3.14.1), has high SF energy and, thus, due to its very thin ribbon, cross-slip will occur very readily as compared to copper. As a matter of fact, it is often stated that, in Al, the dislocations are relatively un-extended and, thus, frequent cross-slip is observed. HCP structures behave quite similarly to FCC metals. In transition BCC crystals or in cubic ionic crystals (NaCl, MgO, etc.) with narrow SFs, cross- slip occurs readily. Note that the process of cross-slip occurs more readily at higher temperatures, since activation energy may be required for the removal of SFs, especially in cases in which the ribbon is broad.

3.3.14.4 Thompson Tetrahedron

It is usually difficult to predict the outcome of dislocation reactions that will occur, especially when partial dislocations are involved. Without having elaborate crystallographic understanding of the subject, it is usual (and practical) to visualize such possible reactions with the help of a Thompson tetrahedron. Figure 3.60 presents a schematic diagram of this commonly used tetrahedron, based on Thompson's concept. Four $\{111\}$ planes form this tetrahedron, labeled ABCD, where the external edges, such as AB, indicate the $\langle 110\rangle$ slip directions common in FCC structures before their dissociation into partials. The Burgers vector of the Shockley partial dislocations, a/6 $\langle 112\rangle$, is obtained by the dissociation of a/2 $\langle 110\rangle$. The Burgers vectors are indicated by lines connecting the corners to the centers of the faces, such as Aγ, Bγ, etc.

Frank partial having a/3 $\langle 111\rangle$ Burgers vectors are represented by lines, such as Aα, Bγ, etc., which connect the corners of the triangles to the opposite triangle centers, labeled by Greek letters. Furthermore, stair-rod dislocations obtained by the intersection of two Shockley partials on two intersecting $\{111\}$ planes are indicated as reactions between the lines connecting the centers of the triangles (i.e., $\{111\}$ planes), for example:

$$\alpha\beta + \beta\gamma = \alpha\gamma.$$

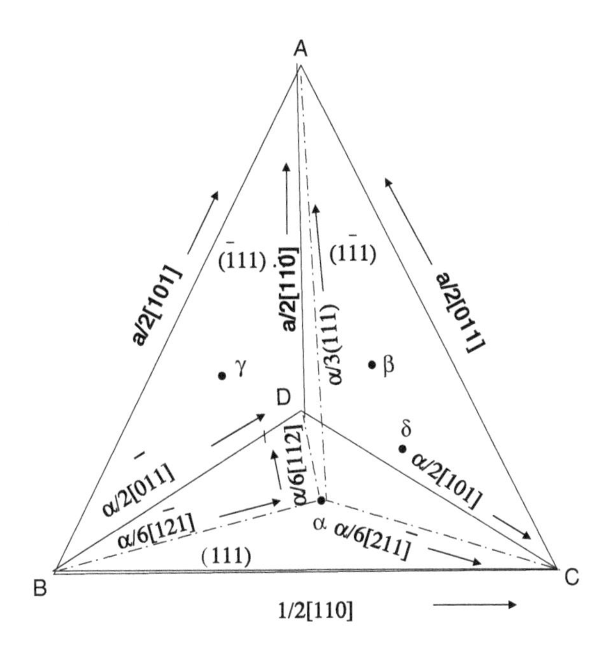

Fig. 3.60 A Thomson's tetrahedron helps in visualizing dislocation reactions in FCC structures on {111} planes. Decomposition into Shockley and Frank partials is shown

3.3.14.5 Lomer-Cottrell Locks

Dislocation gliding on intersecting {111} planes may form a series of obstacles known as 'Lomer-Cottrell barriers', preventing further glide. Since, in FCC crystals (and also in HCP and BCC structures), dislocations generally tend to dissociate into partials, causing extended dislocation faults, it is important to consider the case of Lomer-Cottrell barrier formation. Figure 3.61 illustrates the formation of such a barrier. In Fig. 3.61a, SFs, bounded by partial dislocations, glide on intersecting {111} planes before their interaction. In Fig. 3.61b, the result of the reactions between these partials is shown, forming a 'Lomer-Cottrell lock'. Here, partial leading and tracking dislocations, with a SF between them, glide on their respective planes. The leading partial dislocations on the intersecting planes have formed a new partial dislocation with Burgers vector $a/6[1\bar{1}0]$, according to:

$$a/6[\bar{1}\bar{2}1] + a/6[21\bar{1}] = a/6[1\bar{1}0].$$

This reaction is correct, as may be seen by checking the components of the Burgers vectors and it is also energetically favorable. A consequence of the above reaction is the formation of a sessile dislocation, beyond which the trailing dislocations pile up. The Burgers vector of the newly-formed partial dislocation, i.e.

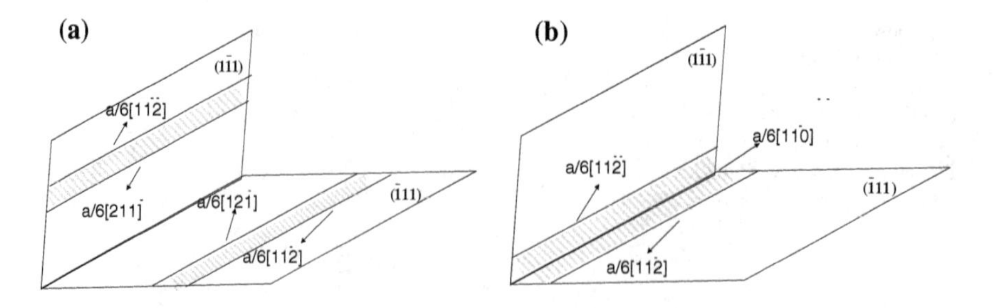

Fig. 3.61 A Lomer-Cottrell lock: **a** two partial dislocations before interaction, **b** after interaction of the leading partials forming a stair-rod dislocation

a/6 $\langle 110 \rangle$, shown in the above reaction, is not the vector of the FCC lattice (but rather a/2 $\langle 110 \rangle$) located in plane $\{001\}$, which is *not* a slip plane in the FCC structure, so it cannot glide. As a matter of fact, all three partial dislocations—both the leading partials which entered into the reaction and the newly-formed partial dislocation–form a very stable lock, which prevents the movement of other dislocations.

3.3.15 Dislocation Pile-Ups

Dislocations generated by a Frank-Read source quite often pile-up at various barriers, such as grain boundaries, precipitates, etc. and cannot glide easily. The later dislocations exert forces on the prior ones. Pushing dislocations into a pile-up configuration requires stresses higher than those which act in the absence of pile-up dislocations. Figure 3.62 schematically illustrates a dislocation pile-up on a $\{111\}$ plane, where movement is blocked by some barrier, e.g., a Lomer-Cottrell lock. High stress concentration is acting on the leading dislocations and, when the pile-up stress is greater than the theoretical shear stress, yielding is induced. A back stress acts on the obstacle, preventing further motion of the dislocations. Near the obstacle, the density of the pile-ups is the greatest, but the distance between the individual dislocations increases as same-signed dislocations repel each other. Since repelling forces are additive, they increase. One could say that a stress concentration exists at or close to the obstacle.

The key factor in the motion of these dislocations is the first or leading dislocation in the vicinity of the obstacle. Assume that the leading dislocation has moved a distance, dx; all the trailing dislocations will move the same distance. The work done per unit length of the dislocation is:

$$work = n\tau b dx \tag{3.79}$$

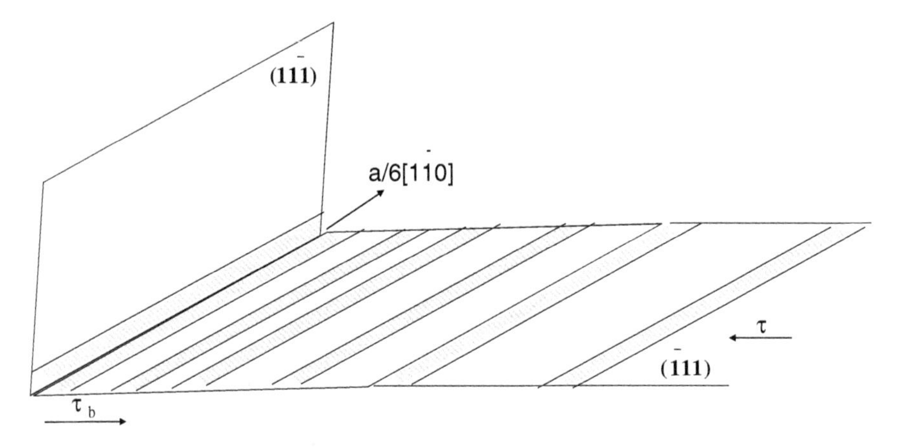

Fig. 3.62 Dislocation pile-ups (partials and their faults are shown) behind a Lomer-Cottrell lock, acting as obstacles to their movement

where n is the number of dislocations and τ is the applied stress. The leading dislocation, τ_i, works against the local stress (internal stress) of the obstacle. Thus, the work of the leading dislocation is:

$$\text{work} = \tau_i \text{bdx} \tag{3.79a}$$

At equilibrium, these equations should be equal and, thus, it is possible to derive the following from Eqs. (3.79) and (3.79a):

$$n\tau = \tau_i b \tag{3.79b}$$

In other words, the internal stress at the head of the pile-up, composed of n dislocations, is n times greater than the applied stress. As seen in later chapters, stress concentration plays an important role both in strain hardening and in brittle-fracture formation. The back stress of the pile-up, τ_b, acts on the source to create new dislocations. As long as:

$$\tau_b - \tau_a = \tau \tag{3.79c}$$

where τ_a is the stress required to operate a source, it will function to produce dislocations.

Eshelby et al. [24] have calculated the number and distribution of dislocations that can pile-up in a slip plane of length, L, with acting shear stress, τ, to be:

$$n = \frac{\pi L \tau k}{Gb} \tag{3.80}$$

Here, k = 1 for screw and $(1 - \nu)$ for edge dislocations.

A pile-up of n dislocations along a distance, L, may be considered to be a giant dislocation with Burgers vector nb. The breakdown of a barrier occurs due to:

(a) slip on a new plane or if the material is polycrystalline in a new grain;
(b) climb;
(c) crack formation due to the high stress.

3.3.16 Low- (Small)-Angle Grain Boundaries

Materials are usually polycrystalline, but in order to study crystal properties without the complexities caused by grain boundaries, single crystals are more suitable subjects. However, if the difference between the grain orientations in a polycrystalline material is sufficiently small, then the problem of studying crystal properties becomes limited to the presentation of an array of dislocations. Figure 3.63a shows two crystals with a tilt misorientation angle of θ before joining. The two grains share a common axis about which both are rotated, defining the angle θ. By joining these two crystals into a bi-crystal, a dislocation wall is formed by an array of dislocations, each dislocation having the same sign and Burgers vector. These dislocations in the array line up one above the other to reduce the system energy, as shown in Fig. 3.63b. Each individual dislocation may be resolved up to a misorientation of $\sim 10^{\circ}$, above which analysis of the grain boundary (in terms of an array of dislocations) is not possible. A simple relation may be obtained from Fig. 3.63b connecting the distance between dislocations, the tilt angle and the Burgers vector (in the simple cubic bi-crystal), given as:

$$2sin\frac{\theta}{2} \approx \theta = \frac{b}{D} \tag{3.81}$$

For $\theta = 1^{\circ}$ and for $\mathbf{b} \sim 2.5 \times 10^{-8}$, the value of D $\sim$ 140 Å. In simple cubic materials, b = $\langle 100 \rangle$. The larger the angle θ is, the closer the spacing between the dislocations will be. Thus, for an angle of $\sim 15^{\circ}$ (when the dislocation model is no longer valid), the dislocations are approximately a few atomic spacings one from another, at $\sim 9.5 \times 10^{-8}$, and it is impossible to distinguish between the individual dislocations.

This kind of presentation, of a bi-crystal as an array of edge dislocations, defines a 'low-angle grain boundary', which is also known as a 'simple tilt boundary'. Low-angle grain boundaries are not usually pure tilt boundaries, since the formation of a bi-crystal may occur via a process of screw dislocation and, thus, may include some characteristics of screw dislocation. Such a low-angle boundary is also known as a 'twist low-angle grain boundary', originating from the fact that the lattices of the grains are twisted when a screw dislocation component is present.

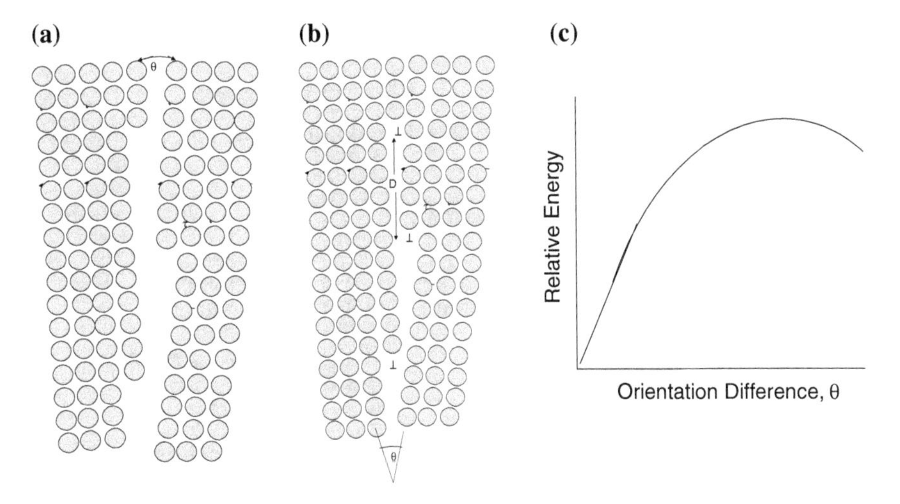

Fig. 3.63 A low-angle grain boundary: **a** two grains are shown with a common cube axis and their angular difference is θ; **b** the two grains are joined to form a bi-crystal. An array of edge dislocations is formed at the joint of the two grains with spacing D between them; and **c** relative, low-angle, grain-boundary energy

The energy of a small-angle grain boundary may be calculated on the basis of the dislocation model. Schematically, the relative energy, as a function of the angle of orientation difference, is shown in Fig. 3.63c. One may express this energy as:

$$E_b = E_0\theta(A - \ln\theta) \tag{3.82}$$

This relation may be obtained as follows:

(a) in Sect. 3.3.7, Eq. (3.25) gives E_{TE}, which includes the core energy:

$$E_{TE} = E + E_{core} \tag{3.25}$$

(b) the core energy in Eq. (3.39) is given by:

$$E_{core} \cong \pi(5b)^2\left(\frac{G}{30}\right)\frac{1}{G} \approx \frac{Gb^2}{10} = CGb^2 \tag{3.39}$$

(c) in Eq. (3.41), the energy-per-unit length of dislocation is given as:

$$E_{edge} = \frac{Gb^2}{(1-v)4\pi}\ln\frac{r}{r_0} \tag{3.41}$$

(d) the energy of the unit-cell area is E_{TE}x1 and E_{TE} is the energy-per-unit length of a dislocation passing or cutting through this area. Therefore, the energy per area of the boundary is E_{TE} x1/1xD; the energy-per-unit area of a boundary of length D is then:

(e)

$$E_b = (E_{TE} \times 1)/(1 \times D) \tag{3.83}$$

(f) substituting from Eqs. (3.25), (3.39) and (3.41) (using the proper subscripts):

$$E_b = \frac{Gb^2}{(1-\nu)4\pi} \ln\left(\frac{r}{r_0}\right)\frac{1}{D} + \frac{CGb^2}{D} \tag{3.84}$$

(g) taking r = D (since, outside this range, the stress fields cancel each other out) and expressing D in terms of θ, from Eq. (3.81), one obtains:

$$E_b = \frac{Gb^2}{(1-\nu)4\pi} \ln\left(\frac{D}{r_0}\right)\frac{1}{D} + \frac{CGb^2}{D} = \frac{Gb^2}{(1-\nu)4\pi} \ln\left(\frac{b}{\theta r_0}\right)\frac{\theta}{b} + \frac{CGb^2\theta}{b} \tag{3.85}$$

(h) arbitrarily taking $r_0 \sim b$, Eq. (3.85) may be rewritten as:

$$E_b = \frac{Gb^2}{(1-\nu)4\pi} \ln\left(\frac{1}{\theta}\right)\frac{\theta}{b} + \frac{C\theta Gb^2}{b} \tag{3.86}$$

$$E_b = \frac{Gb\theta}{(1-\nu)4\pi} [\ln\left(\frac{1}{\theta}\right) + C(1-\nu)4\pi] \tag{3.87}$$

(i) Equation (3.87) expresses the form of the low-angle grain boundary (just as well as does Eq. (3.82), commonly found in the literature) and so:

$$E_0 = \frac{Gb}{(1-\nu)4\pi} \text{ and } A = C4\pi(1-\nu) \text{ and } \ln\frac{1}{\theta} = -\ln\theta \tag{3.88}$$

Note the shortcomings of this derivation: (a) it is valid only for real, small angles, as assumed to be the case when using Eq. (3.81); (b) the core energy is not known. The approximation in Eq. (3.39), at best, only provides an estimation; and (c) it is inherent in the derivation that A and E_0 be independent of θ.

Small-angle grain boundaries are illustrated in Fig. 3.22 by means of the etch-pits technique: Relation (3.82) is rewritten as (3.89);

$$E_b = E_0\theta(A - \ln\theta) \tag{3.89}$$

3.3.17 *Experimental Observations of Dislocations in Ceramics*

After the above conceptual discussion of dislocation and some of its features, the following presents relevant experimental observations regarding dislocation types and characteristics. Since the field of ceramics is huge and includes various structures, a limited number of representative examples are presented below of:

(a) edge dislocations;
(b) screw dislocations;
(c) partial dislocations;
(d) climb;
(e) cross-slip;
(f) Lomer-Cottrell locks;
(g) dislocation pile-ups;
(h) low-angle grain boundaries; and
(i) SFs.

3.3.17.1 Edge Dislocations

As indicated in Chap. 2, MgO single crystals deform plastically at ambient temperature. The dislocation substructures in single crystals of MgO deformed by four-point bending at temperatures from −196 to 1300 °C have been observed by TEM. Although the majority of pairs originated where screw dislocations intersected grown-in dislocations, elongated edge-dislocation pairs were also found at all the deformation temperatures. In Fig. 3.64, pairs of edge dislocations are shown in an MgO specimen deformed at −196 °C. The MgO sample was deformed by four-point bending. In this case, there was an equal shear stress on four of the six possible {110} slip systems. Two of the active systems were on planes intersecting the {100} tension surface at 45° and two intersecting it at 90°. Hereafter, these planes will be referred to as the 45° and 90° planes, respectively. Deformation of as-polished specimens, despite the equal shear stress on the slip planes occurred predominantly on 45° planes, was concentrated under the center knife-edges. The stress–deflection curve for such specimens invariably had a high yield stress followed by a sharp drop and jerky flow, as indicated in Fig. 3.65a. The high yield stress followed by a sharp drop in stress appears to be a result of the higher stress necessary to nucleate dislocations in MgO than that required to move them.

Section 3.3.16 indicated that: low-angle grain boundaries consist of an array of dislocations; each dislocation shares the same sign and Burgers vector; and the dislocations of the array line up, one above the other, to reduce the system energy, forming a dislocation wall. Each individual dislocation may be resolved up to a misorientation of $\sim 10^{\circ}$, above which analysis of the grain boundary in terms of an array of dislocations is not possible. The distance between the dislocations, D, was given as:

$$2\sin\frac{\theta}{2} \approx \theta = \frac{b}{D} \tag{3.81}$$

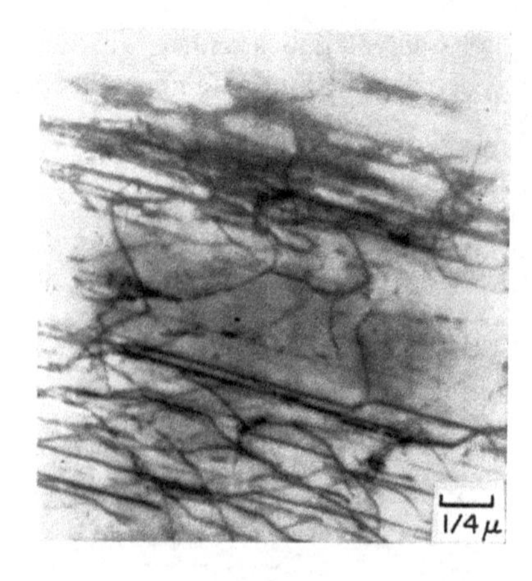

Fig. 3.64 Dislocation pairs in a specimen deformed at −196 °C. Dense dislocation substructure composed largely of close edge. Active slip plane is at 45′ to plane of foil [23]. With kind permission of John Wiley and Sons

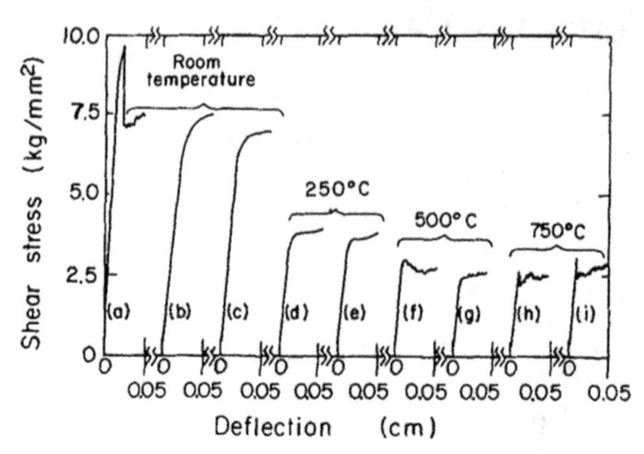

Fig. 3.65 Effect of temperature and surface condition on the stress deflection curve for crystals deformed in four-point bending [23]. With kind permission of John Wiley and Sons

A pure tilt boundary is composed of an array of edge dislocations, while a twist boundary is defined by an array of screw dislocations. In Fig. 3.66, a high-resolution electron microscopy [henceforth: HREM] image of a 5° small- angle tilt boundary around [110] axis is shown for zirconia. Periodic contrasts may be seen along the boundary coming from the array of edge dislocations. This boundary consists of discrete-edge dislocations separated by nearly perfect crystals. The distance between these dislocations is ~4.3 nm. A Burgers circuit around these dislocations is visible and the vector was determined to be $\mathbf{b} = a/2[1\bar{1}0]$.

3.3.17.2 Screw Dislocations

Cubic zirconia single crystals stabilized with 11 mol% yttria were deformed in air at 1400 °C and at ~1200 °C at different strain rates along $[1\bar{1}2]$ and [100] compression directions. The microstructure of the deformed specimens was investigated by high-voltage TEM, including contrast extinction analysis for determining the Burgers vectors, as well as stereo pairs and wide-angle tilting experiments to

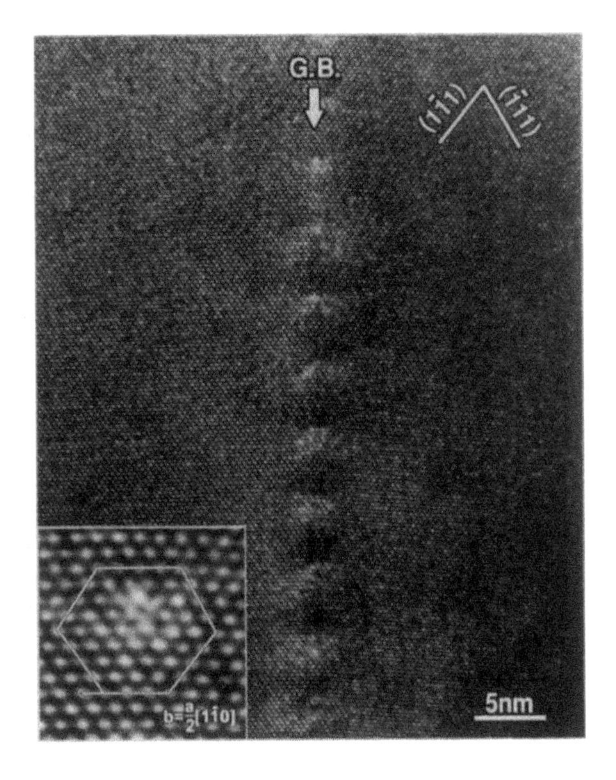

Fig. 3.66 HREM image of $2\theta = 5.0°$ symmetric tilt small angle boundary. Periodical dislocations can be seen along the boundary. Inset is the Burgers circuit around a dislocation, and the Burgers vector is $a/2[1\bar{1}0]$ [50]. With kind permission of Professor Shibata

find the active slip planes. In deformations along $[1\bar{1}2]$, the primary and secondary slip planes are of {100} type. At strains above about 1 %, cubic zirconia deforms by multiple slip. Figure 3.67 shows the dislocation structure of a sample deformed at 1400 °C along $[1\bar{1}2]$ using different g vectors.

The primary dislocations of the $a/2[1\bar{1}0]$ Burgers vector are in contrast only in Fig. 3.67a. These dislocations appear smoothly curved on their inclined (001) slip plane. Stereo pairs prove that dislocations are not always precisely arranged on their slip planes, indicating the action of cross-slip or climb. The only pinning agents acting on primary dislocations seem to be the other two sets of dislocations, which can clearly be discerned in Fig. 3.67b–d. The Burgers vectors of these secondary dislocations are $a/2[01\bar{1}]$ (extinguished in Fig. 3.67c) and 1/2[101] (extinguished in Fig. 3.67d). These secondary dislocations are mainly of screw type, so that they appear as almost straight lines.

3.3.17.3 Partial Dislocations

Wu and Wang [55] analyzed the dissociation of perfect dislocations into partials in $BaTiO_3$ by TEM. Figure 3.68 illustrates the results for this hexagonally-structured ceramic, in which the basal plane is commonly involved in the deformation.

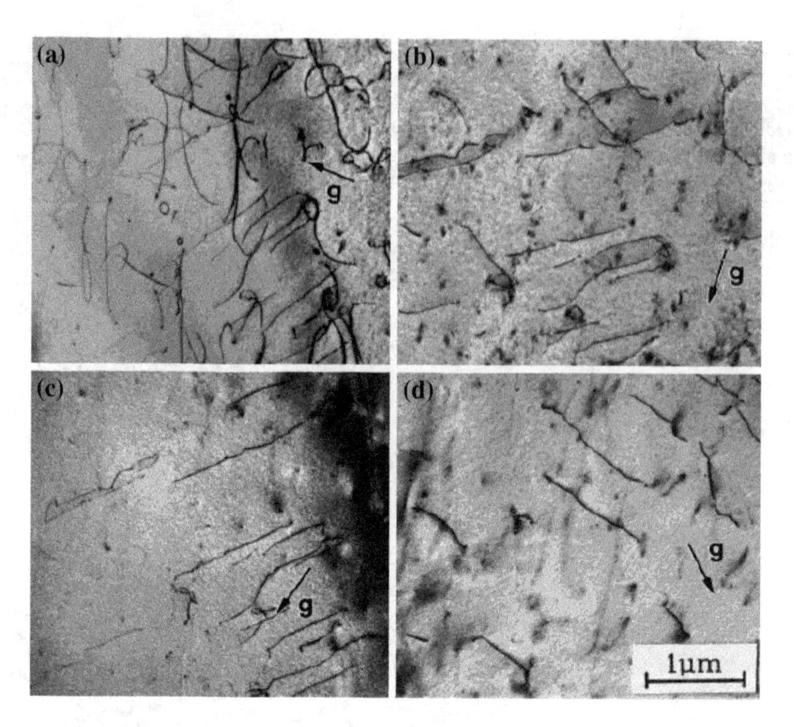

Fig. 3.67 Dislocation structure of a sample deformed along $[1\bar{1}2]$ at 1400 °C up to a plastic strain of 3 % and cooled under full load. **a** g = $[0\bar{2}2]$ near the $[\bar{1}11]$ pole; primary and secondary dislocations visible. **b** g = [220] near the $[\bar{1}11]$ pole; primary dislocations extinguished, secondary ones visible. **c** g = [111] near the $[\bar{2}11]$ pole; primary and secondary dislocations with b = $1/2[01\bar{1}]$ extinguished, secondary ones with b = 1/2[101] visible. **d** g = $[11\bar{1}$ near the $[\bar{1}21]$ pole; primary and secondary dislocations with b = 1/2[101] extinguished, secondary ones with $b = 1/2[01\bar{1}]$ visible [22]. With kind permission of John Wiley and Sons

It was observed that perfect dislocations in the basal plane, embedded in an α-type extended planar SF hexagonal (h-) $BaTiO_3$ with a Burgers vector of $\mathbf{b_B} = 1/3\langle\bar{1}2\bar{1}0\rangle$, were dissociated into prism plane half-partials with Burgers vectors $\mathbf{b_P} = 1/3\langle01\bar{1}0\rangle$.

It was found that the dissociation of a series of basal dislocations occurred by glide in the fault plane (0002). However, the migration of the pair partials, trailing behind in the fault plane, was impeded by the leading pair. Under the stress applied during hot pressing, these partials were gradually piled up with successively decreasing separation between each pair. That consequently led to partial separations ranging from 195 to 56 nm. That dislocations are pair partials is evidenced by the SF fringes in bounded ribbons and by invisibility criteria, when the dislocation pairs were both visible under g = $\bar{1}2\bar{1}0$ (Fig. 3.68b) and, complementarily, one of the two became invisible under g = $11\bar{2}0$ (p_i visible and q_i invisible, where i = 2–9, in Fig. 3.69c) and $\bar{2}110$ (p_i invisible and q_i visible, where i = 2–9, in Fig. 3.68a) from Z_A = [0001]. Only pair partials, p_8 and q_8, are curved

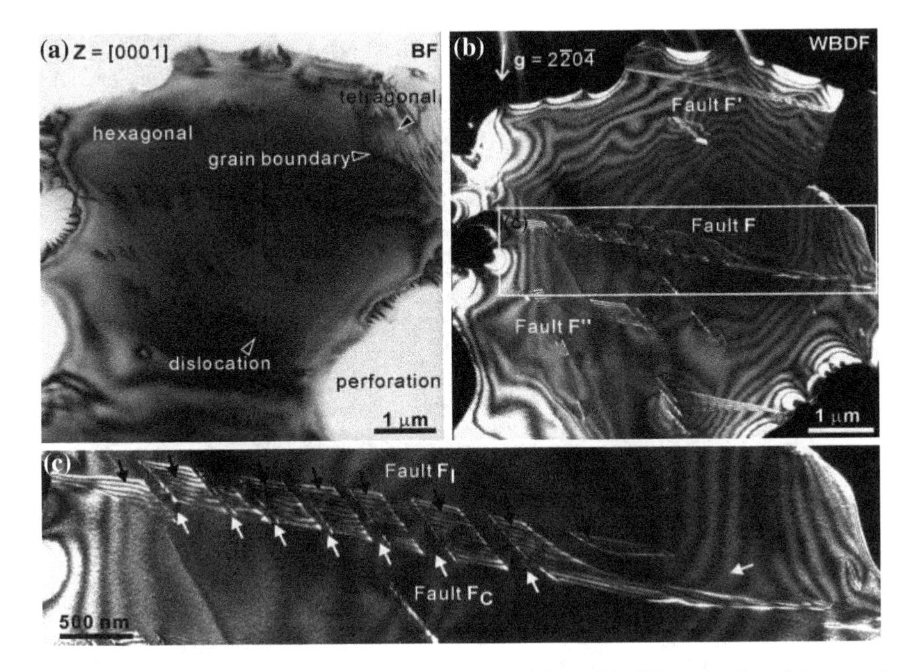

Fig. 3.68 **a** Bright-field (BF) image of the area consisting of both tetragonal and hexagonal grains, **b** faults F, F', and F'' (weak-beam dark-field (WBDF)), and **c** framed area of fault F containing a series of dislocations lying in the fault plane (transmission electron microscopy) [55]. With kind permission of John Wiley and Sons

(Fig. 3.69); other pairs, e.g., p_2, p_3, and p_4, remain mutually in parallel and each partial (p_i) is parallel to its corresponding one in the pair (q_i), i.e., $p_2//q_2$, and $p_3//q_3$. Dislocations p_8 and q_8 are the only pair partials terminated at the same side of fault F, others straddling the fault remain parallel to each other, e.g. (p_2, q_2).

Perfect basal dislocations, dissociated into prism plane half-partials by glide in fault plane F, have been analyzed for their fault vectors, dislocation Burgers vectors and for true line directions. All half-partials traveling behind the leading pair were piled up due to the hindrance of motion caused by one partial of the leading pair that was moving into another fault plane by a mixed mechanism of glide in (0002) and climb down in $(10\bar{1}0)$ by $1/2[000\bar{1}]$. Whether the rate-determining mechanism is climb-controlled dislocation-glide during the hot pressing of h-$BaTiO_3$ requires further systematic, kinetic studies.

3.3.17.4 Dislocation Climb

Previous Sect. 3.3.6 indicated that dislocation motion may be conservative (glide) and nonconservative (climb). For glide, the Burgers vector and dislocation must be in the same plane. Unlike edge dislocations (as indicated above), screw

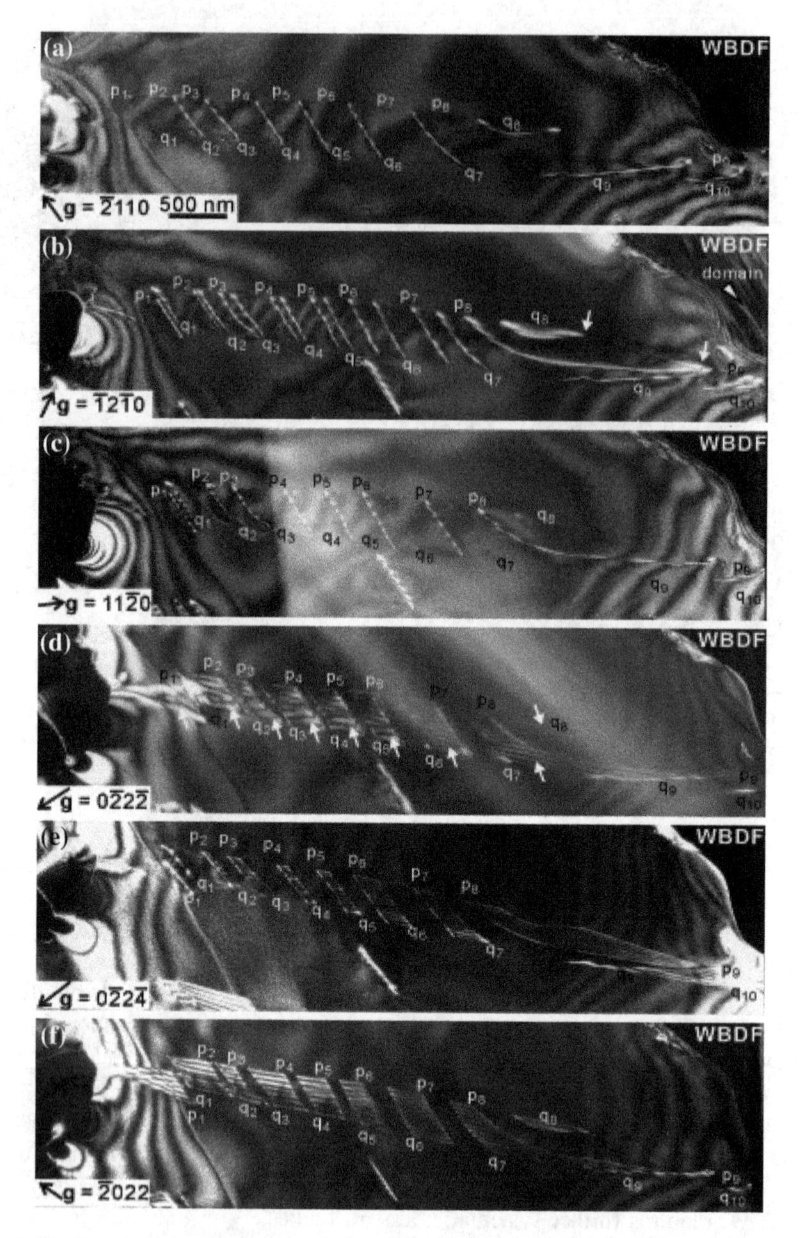

Fig. 3.69 Representative weak-beam dark-field (WBDF) images where dislocations Burgers vectors (**b**) and fault vector (R_F) of fault F were determined (transmission electron microscopy) [55]. With kind permission of John Wiley and Sons

dislocations have Burgers vectors parallel to the dislocation and may continue their deformation in any plane having a common zone axis with the plane in which the screw dislocation is located. An edge dislocation can leave its slip plane only by the climb process, by motion perpendicular to its glide plane. Dislocations can slip in planes containing both the dislocation itself and its Burgers vector. In screw dislocations, the dislocation and the Burgers vector are parallel, so the dislocation may slip into any plane having the same form. In edge dislocations, the dislocation and the Burgers vector are perpendicular, so there is only one plane in which the dislocation can slip. There is an alternative mechanism of dislocation motion, fundamentally different from slip, which allows an edge dislocation to move out of its slip plane, known as 'dislocation climb'. Dislocation climb allows an edge dislocation to move perpendicular to its slip plane. The driving force for dislocation climb is the temperature-dependent movement of vacancies. Vacancies in crystals must diffuse to the vicinity of the half-plane, defining the dislocation, and change places with atoms from the half-plane. This exchange shifts the half-plane or part of it one atomic spacing and is known as 'positive climb'. Contrary to this case, 'negative climb' means addition of atoms (interstial or interstitialcy) to a half-plane, causing the dislocation to grow.

To strengthen two points stated earlier. First of all, during positive climb, crystals shrink in a direction perpendicular to the extra half-plane defining the dislocation, because atoms are being removed from the crystal. Consequently, applying a compressive stress in a direction perpendicular to the half-plane promotes positive climb. Contrariwise, when a tensile stress is applied in a direction perpendicular to a half-plane, negative climb (growth of the half-plane) is promoted. This is contrary to glide, in which shear stress induces slip in its plane. Secondly, climb is temperature-dependent, promoted thermally, since vacancies and their motion increase with temperature. Slip is dependent only to a small degree on temperature.

Further below, time-dependent deformation (creep) initiated by climb will be extensively discussed. In this section, an example of dislocation climb is illustrated. Figure 3.70 shows dislocation climb in an Al_2O_3-YAG specimen. Here, climb was assisted by thermal activation. Such a dislocation network, resulting from the reaction of dislocations from the basal and pyramidal slip systems, involves dislocation climb. It is a diffusion-controlled deformation mode characterizing creep deformation and, in this particular case, the activation energy determined is Q = 670 kJ/mol.

As usual, one of the accepted steady-state creep rates is given as:

$$\dot{\varepsilon} = A\sigma^{n} \exp(-Q/RT) \tag{3.89}$$

which allows for the evaluation of the activation energy for the process. A is a material constant and n is the stress exponent obtained from:

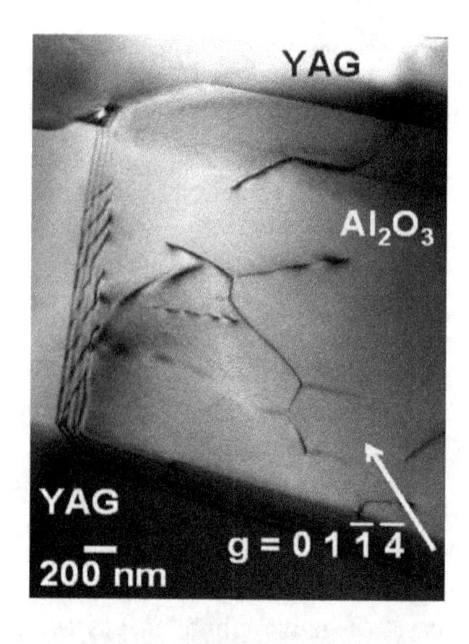

Fig. 3.70 TEM examinations of compression tested specimens. Dislocation reaction involving climb in the Al_2O_3 phase of the Al_2O_3-YAG eutectic composite is shown [44]. With kind permission of Dr. Appriou, Research Director of Aerospace Lab Journal

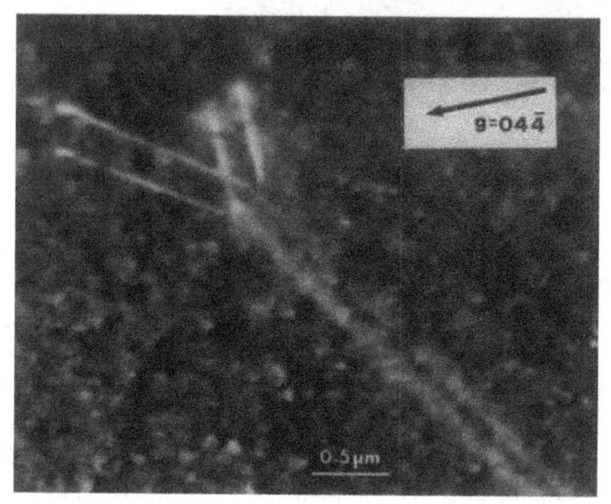

Fig. 3.71 Dissociated dislocation node in nonstoichiometric MgO-3.5 $A1_20_3$ spinel. Node of three 1/2 ⟨110⟩ dislocations has dissociated by climb to produce two partial nodes each consisting of three 1/2 ⟨110⟩ dislocations [38]. With kind permission of John Wiley and Sons

$$n = \left(\frac{\partial \ln \dot{\varepsilon}}{\partial \ln \sigma}\right)_{T,A} \qquad Q = \left(-R\frac{\partial \ln \dot{\varepsilon}}{\partial 1/T}\right)_{\sigma,A} \tag{3.90}$$

Mitchell states that dislocation dissociation is rare in oxides. The only two cases in which it has been clearly observed ($A1_20_3$ and $MgAI_20_4$,) involve dissociation by climb, rather than glide, in situations where the point defects are probably helping the dissociation process. Therefore, it is of interest to study such cases as additional examples in which climb is involved, as one of the mechanisms in the recovery process. Figure 3.71 illustrates dislocation dissociation by climb in MgO-3.5 Al_2O_3 spinel.

Large quantities of secondary slip must have occurred, as well as dislocation climb. Dislocation nodes of the type $1/2\langle 110\rangle + 1/2\langle \bar{1}01\rangle + 1/2\langle 0\bar{1}\bar{1}\rangle$ are

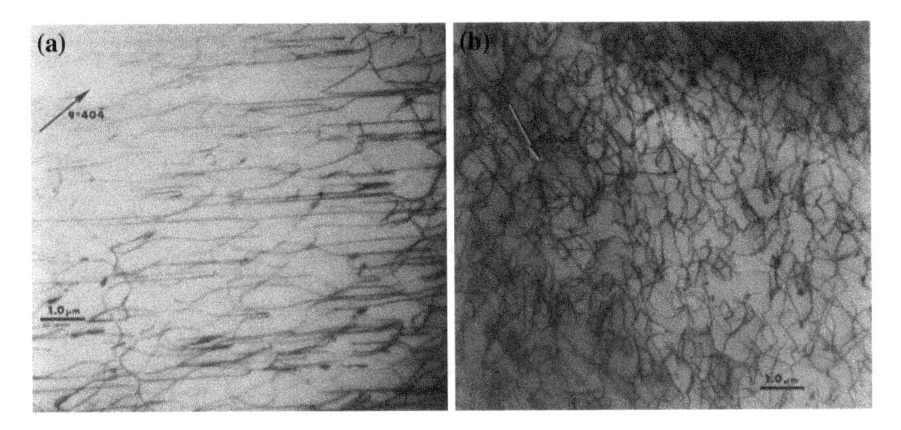

Fig. 3.72 Dislocations in $MgAI_20_4$ spinel deformed at 1800 °C, **a** 2 % strain, showing mostly edge dislocations and dipoles produced by glide, and **b** 3 % strain, showing uniform 3-dimensional network of dislocations produced by glide and climb [38]. With kind permission of John Wiley and Sons

observed. Nodes are frequently extended by climb dissociation, as seen in the above figure. Nonstoichiometric spinel with excess $A1_20_3$ deforms at lower temperatures and lower stresses, with {110} being the preferred glide plane. The dissociation of dislocations by climb is even more apparent in Fig. 3.72, probably because the SF energy decreases with increasing deviation from stoichiometry. It has been suggested that the greater plasticity of nonstoichiometric spinel is due to the diffusion of excess cation vacancies to the moving dislocations. Clearly, the high temperatures required for deformation in spinel are such that dislocation climb is competitive with dislocation glide. Vacancies are essential for climb, thus increasing plasticity (a recovery process).

An additional illustration of climb in Ti-doped sapphire appears in Fig. 3.73. The shortest lattice vector is $1/3\langle 11\bar{2}0\rangle$ and glide is easiest on the basal plane, followed by prismatic and pyramidal slip. The initial dislocation structure consists of edge dipoles, linking glide dislocations and loops. The dipoles form when edge dislocations with opposite signs are trapped on parallel basal planes. As the strain increases, the densities of the glide dislocations and loops also increase, whereas the density of the dipoles stays constant or decreases slightly. As in MgO, loops are formed by self-climb from the edge dipoles by a pinch-off mechanism. The loop-formation mechanism is more complex than in MgO, since dislocations in narrow dipoles preferentially undergo climb dissociation and form faulted dipoles and loops.

The role of diffusion must also be considered, since most oxides require high temperatures for plastic deformation and since measured activation energies for yielding invariably give values close to those for oxygen self-diffusion. This coincidence is not so easily discussed in oxides, in which Burgers vectors are frequently large and where evidence for dislocation climb in the later stages of deformation is evident.

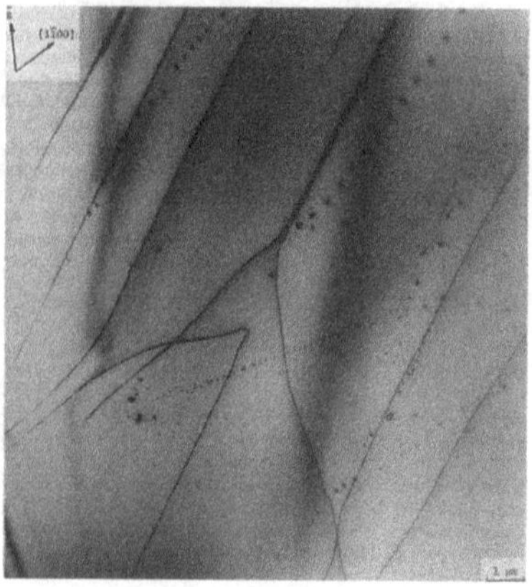

Fig. 3.73 Dislocation substructure in $Ti4^{+}$ -doped sapphire deformed 2.5 % on basal plane at 1520 °C. Strings of loops resulting from breakup of dipoles by self-climb are apparent. Basal foil [38]. With kind permission of John Wiley and Sons

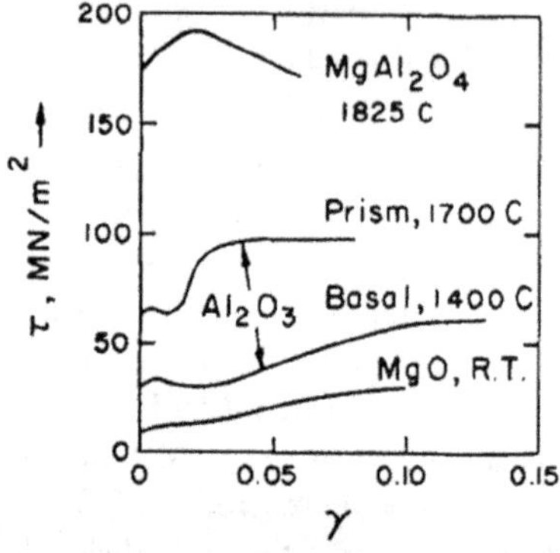

Fig. 3.74 Selected curves of shear stress versus shear strain for stoichiometric spinel, sapphire deformed on prismatic and basal planes and MgO at the temperatures indicated [38]. With kind permission of John Wiley and Sons

Recovery mechanisms quickly become important at higher strains and temperatures, as seen in Fig. 3.74, particularly in prismatic Al_2O_3 and $MgAl_2O_4$, where the initial high work-hardening region is highly transient. Recovery is undoubtedly due to climb, leading to the annihilation and softening of the obstacle network. This figure shows some selected stress–strain curves for: MgO crystals deformed at RT; Al_2O_3 deformed by basal slip at 1400 °C; $A1_2O_3$ deformed by prismatic slip at 1700 °C; and $MgAl_2O_4$, deformed at 1825 °C. For sapphires, yield points are observed, followed by work hardening. Recovery effects set in at higher strains, such that the flow stress reaches a plateau. For spinel, the flow stress actually decreases. Stress–strain curves at other temperature indicate that, for each oxide, there is less work hardening and more recovery with increasing temperature. MgO approximates the three-stage hardening behavior observed in FCC metals, except that stage I is not very clear, presumably because of the inevitable simultaneous operation of orthogonal (110) $[1\bar{1}0]$ and $(1\bar{1}0)$ [110] glide systems. A similar recovery mechanism probably holds for sapphires deformed by prismatic slip, in which the plateau stress is reached when the rate of annihilation of dislocations in the network, due to climb, is balanced by the rate of formation of the network, due to the decomposition of the prismatic dislocations.

These illustrations are an indication of the contribution of transmission electron microscopy [henceforth: TEM] to the understanding of plastic deformation in some oxide ceramics, including softening at the measured temperatures, most probably caused by climb. Clearly, since most of the common ceramics are ductile at high temperatures, Mitchell's [38] above measurements explain the plasticity of oxide ceramics and the recovery by climb, as observed by the softening seen as indicated in Fig. 3.74.

3.3.17.5 Cross-Slip

The observation of processes, such as climb or cross-slip, seems to be quite difficult, since the generation of dislocation motion occurs at high velocities and often one observes only a 'post mortem' figure, as it were, after the termination of the process. These days, fast video frames may be obtained in situ, which is useful for capturing images of this process from beginning to end. Taking into account the instantaneous dislocation generation and the high velocity of dislocation motion during in situ experiments, the changes occurring during the motion of mobile dislocations seem to provide a good interpretation of the observed phenomena, as illustrated by Fig. 3.75. Deformation experiments were carried out in a Deformation-DIA [henceforth: D-DIA] high-pressure apparatus on oriented Mg_2SiO_4 olivine single crystals at pressures ranging from 2.1 to 7.5 GPa in the temperature range 1373–1677 K and in dry conditions.

TEM investigation of the experimental run products reveals that dislocation creep, assisted by dislocation climb and cross-slip, was responsible for the sample's deformation. At $P < 3$ GPa, moderate differential stress ($\sigma \approx$ few hundreds MPa or less) and, in dry conditions, olivine [100](010) and [100](001) slip systems (space group *Pbnm*), also called 'a-slip', dominate the less active [001](010) and [001](100) systems, (referred to as the 'c-slip'). This results in the alignment of the crystal fast-velocity [100] axis with the principal shear direction in deforming aggregates. Yet, recent high-temperature deformation experiments carried out at 11 GPa, as well as a theoretical study based on first-principle calculations, suggest that olivine c-slip may be dominant at high pressures.

Olivine single crystals were specifically studied in order to test the likelihood of an olivine a-slip/c-slip transition at high pressures. To this end, steady-state deformation experiments were done on pure forsterite single crystals using a newly-developed D-DIA apparatus coupled with X-ray synchrotron radiation, which provided an in situ determination of the applied stress (σ) and the sample strain rate ($\dot{\varepsilon}$).

TEM investigation revealed that a-slip was largely dominant in the a-samples and c slip in the c-samples, as shown by the high dislocation densities observed in the samples shown in Fig. 3.75. The a-samples exhibit **a** dislocation loops in

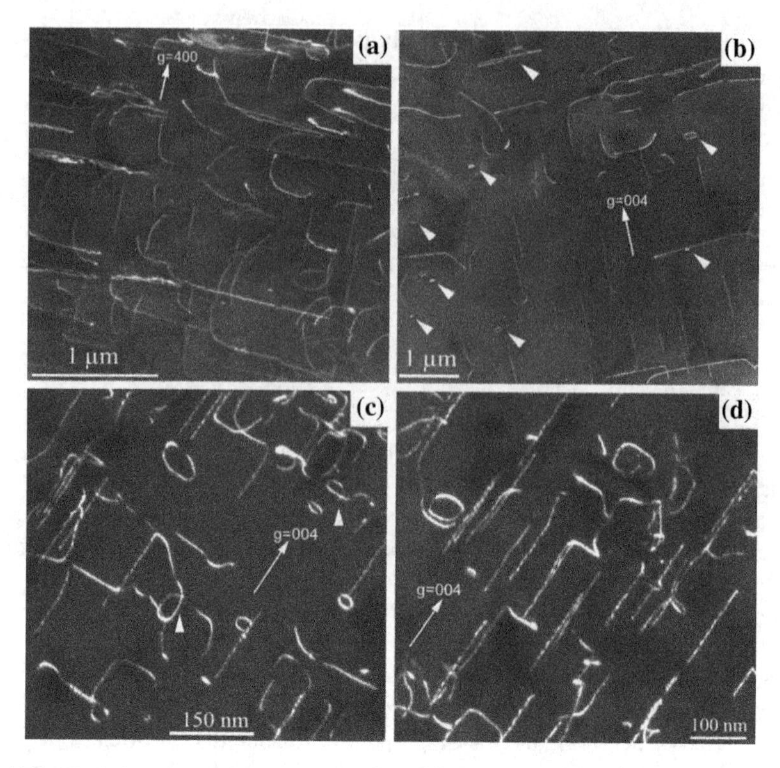

Fig. 3.75 Weak beam dark-field TEM micrographs of deformed specimens quenched from 1377 K (run FOR18) or 1673 K (run FOR20). Diffraction vectors **g** are indicated (long arrows): **a** the a-sample (FOR20) exhibits climb assisted gliding of [100] dislocations with long edge segments, attesting that a-slip was responsible for specimen deformation at high temperature; **b** the c-sample (FOR20) exhibits [001] dislocations showing the activation of c-slip in this sample, with numerous evidences of dislocation cross-slip (short arrows); **c** [001] dislocations in the FOR18 c-sample exhibit numerous direct evidence of cross-slip under the form of open loops (short arrows); **d** details of [001] screw dislocations lines in the FOR18 c-sample exhibiting wavy features with pseudo-period of ~20 nm and amplitudes <7 nm. These features, which are smoother with a larger period ~300 nm at higher temperature (in FOR20 c-sample), may indicate extended or split cores relaxed in several cross-slip planes [47]. With kind permission of J. Alexander Speer, Executive Director, MSA

gliding configuration, i.e., fairly confined in (010) glide plane, with long edge segments. These segments show numerous scallops, revealing (in Fig. 3.75a) that the glide of the **a** dislocations was assisted by climb. Evidence of dislocation cross-slip is also visible in the a-samples. The c-sample exhibits large **c** dislocation loops, mostly confined in their (010) glide planes (Fig. 3.75b), with edge and screw segments of comparable lengths at 1673 K. The **c** dislocation edge segments exhibit some scallops, suggesting climb-assisted gliding, although to a much lesser extend than in the case of the **a** dislocations. The c-samples also exhibit much indirect evidence of dislocation cross-slip in the form of small sessile loops (outside the glide plane) of **c** dislocations. Such sessile loops can form at high

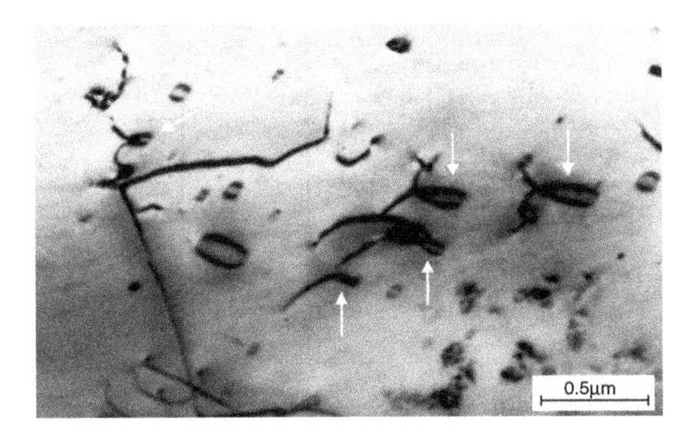

Fig. 3.76 Dislocation structure during in situ deformation of ZrO_2–10 mol% Y_2O_3 at 1150 °C [37]. With kind permission of Elsevier and Professor Messerschmidt

temperatures from equilibrated dislocation dipoles that evolve into dislocation loops by diffusion. However, given the relatively high stress applied to the specimens, it is unlikely that such dipoles were stable during the runs. As such, the observed sessile loops must have formed due to the cross-slip of **c** dislocation screw segments. There is also directly observable evidence of **c** dislocation cross-slip throughout the c-samples in the form of open dislocation loops (e.g., top of Fig. 3.75b and c), confirming that dislocation cross-slip is a very active process during deformation. **c** screw dislocations also exhibit wavy lines at high magnification (Fig. 3.75d), possibly resulting from an extended or split-core structure. Split-screw dislocations are confined to the plane in which they are split and their core must constrict over a certain length in order to change plane (or cross-slip). This process results in observable wavy slip lines. Following observations, it was concluded that **c** and **a** dislocation glides were, respectively, responsible for the deformations of the c- and a-samples. This mechanism, commonly observed in olivine deformed at high temperatures, was assisted by dislocation climb, as well as by dislocation cross-slip.

Another such example of cross-slip in oxide ceramics may be observed during the generation of dislocations in plastic deformation, as illustrated in Fig. 3.76.

A mechanism of the double-cross slip mechanism is suggested for the generation of dislocations during deformation which increases the density in large amounts in addition to the known Frank-Read mechanism.

Double cross-slip is another mechanism suggested as being responsible for the generation of dislocations during deformation; like the known Frank-Read mechanism, it too greatly increases the dislocation density. Figure 3.76 was obtained by the high-voltage electron microscopy (henceforth: HVEM) of plastic deformation during in situ straining experiments. These experiments were carried out either in a quantitative double-tilting straining stage at RT or a high-temperature straining stage for temperatures up to 1150 °C. The actions of the different processes of the double cross-slip mechanism observed in ZrO_2–10 mol% Y_2O_3 at 1150 °C indicated above have been demonstrated in detail by in situ experiments on MgO single crystals by Appel et al. [21]. Cross-slip seems to be easy and appears frequently. A schematic model of double cross-slip is shown in Fig. 3.77.

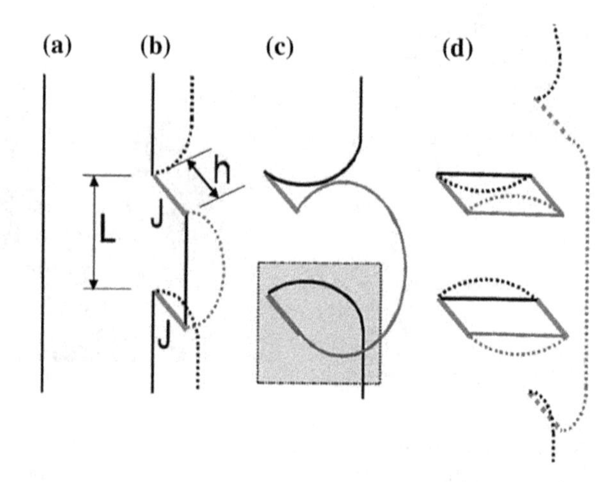

Fig. 3.77 Double cross-slip mechanism for dislocation generation [37]. With kind permission of Elsevier and Professor Messerschmidt

Multiplication events appear instantaneously. Nevertheless, the intermediate α-like configurations of stage (c) in Fig. 3.77 are observed frequently, as marked by the arrows in Fig. 3.76.

In these in situ experiments, with their relatively low foil thickness, a segment may cross-slip while being terminated by the foil surface at one end, so that only a single jog is created. In Fig. 3.76, many dislocation loops are also visible; these may open at a later stage. As described above, the double cross-slip mechanism leads to the sidewise spreading of slip. Since double cross-slip events of smaller height may lead to multiplication at higher stresses, the slip bands become narrower in strong materials and at low temperatures. Figure 3.76 is a typical example of the double cross-slip mechanism. Table 3.4 provides information about dislocation generation in other materials studied by Messerschmidt et al. This summary contains the relevant slip and cross-slip planes.

As discussed in Sect. 3.3.14.5, dislocations become immobilized when they encounter Lomer-Cottrell locks. Figure 3.61 schematically illustrates the formation of a Lomer-Cottrell lock.

As previously discussed, two perfect dislocations along a slip plane, such as the {111} slip planes in an FCC lattice, may split to form two Shockley partial dislocations. One becomes a leading dislocation and the other a trailing dislocation. When the two leading Shockley partials combine, they form a separate dislocation with a Burgers vector that is not in the slip plane–a Lomer-Cottrell dislocation, which is sessile and immobile in the slip plane. This then serves as a barrier against other dislocations that are located in the wake of these partials in their slip plane. The trailing dislocations may pile up behind the Lomer-Cottrell dislocation, thus forming a dislocation configuration known simply as a 'pile-up'. Further pushing of additional dislocations into a pile-up requires increased force.

Experimental observations of a polysilicon Lomer-Cottrell lock is shown in Fig. 3.78.

Table 3.4 Burgers vectors, slip planes as well as type of dislocations generation in different materials [37] (with kind permission of Elsevier and Professor Messerschmidt)

Material load axis	Burgers vector	Slip planes	cross slip planes	Frank-read sources	Double-cross slip mechanism
NaCl ⟨1 0 0⟩	1/2⟨1 1 0⟩	{1 1 0}	{1 0 0}		[10][a]
MgO ⟨1 0 0⟩	1/2⟨1 1 0⟩	{1 1 0}	{1 2 2} {2 1 1}		[7]
ZrO_2 ⟨1 1 2⟩	1/2⟨1 1 0⟩	{1 0 0}	{1 1 0} {1 1 1}		[8]
Si, Ge	1/2⟨1 1 0⟩	{1 1 1}	{1 1 1}	[11]	
Al-Zn-Mg	1/2⟨1 1 0⟩	{1 1 1}	{1 1 1}		[12]
Al-1Ag	1/2⟨1 1 0⟩	{1 1 1}	{1 1 1}		[13]
Al-8Li	1/2⟨1 1 0⟩	{1 1 1}	{1 1 1}		[14]
Duplex steel austenite	1/2⟨1 1 0⟩	{1 1 1}		[9]	
Duplex steel ferrite	1/2⟨1 1 1⟩	{1 1 0} {1 1 2} {1 2 3}	{1 1 0} {1 1 2} {1 2 3}		[9]
Ti-6Al	$a/3\langle 1\,1\,\bar{2}\,0\rangle$	$\{0\,0\,0\,1\}$ $\{1\,\bar{1}\,0\,0\}$	$\{0\,0\,0\,1\}$ $\{1\,\bar{1}\,0\,0\}$	[15]	[15]
γ-Ti-52Al	1/2⟨1 1 0⟩	{1 1 1}	{1 1 1}		[16]
NiAl ⟨1 1 0⟩	⟨1 0 0⟩	{1 0 0} {1 1 0} {2 1 0}	{1 0 0} {1 1 0} {2 1 0}		[17]
$MoSi_2$ ⟨2 0 1⟩	1/2⟨1 1 1⟩	{1 1 0}		[18]	

[a] Cross slip studied by metal surface decoration

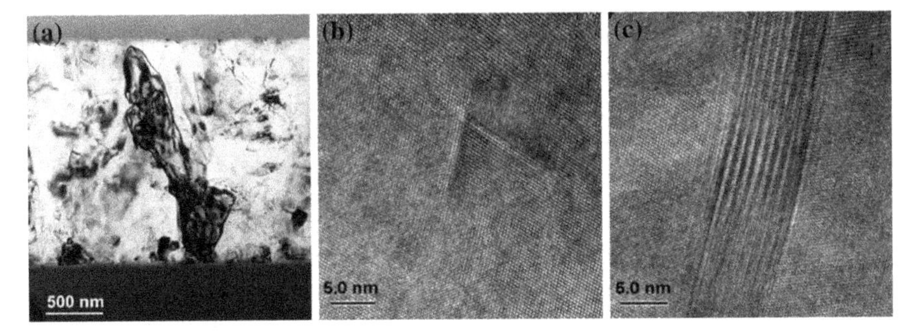

Fig. 3.78 Microstructure of the 2 μm thick polysilicon films: **a** cross-sectional TEM image, **b** Lomer-Cottrell lock, and **c** microtwins [16]. With kind permission of Professor Ritchie

Although Si is not necessarily a ceramic, it fulfills one requirement of the ceramic definition, namely it is covalently bonded. More illuminating illustrations of Lomer-Cottrell locks may be seen in Pt and Fe–Mn–Al–C steel. To observe Lomer-Cottrell locks in Pt, see Fig. 3.79 [54]. Using a newly-developed nanoscale deformation device, atomic scale and time-resolved dislocation dynamics were captured in situ by TEM during the deformation of a Pt ultrathin film with truly nanometer grains (diameter d< ~10 nm). In larger grains (d ~ 10 nm), full dislocations dominate and their evolution sometimes leads to the formation, destruction and reformation of Lomer locks, as seen in Fig. 3.79. The reason for performing experiments on the $<\sim 10$ nm grain size is that, in such small grain-sized polycrystalline materials, there may be little chance of dislocation interaction (they run across the small grains) and post-mortem observation after the applied stress is removed. In larger grains, after large strains (e.g., by cold rolling) Lomer-Cottrell locks are observed. However, in smaller grain dislocation, such activities were rarely observed experimentally, although molecular-dynamic simulations do predict dislocation activities, interactions and the formation and destruction of Lomer and Lomer-Cottrell locks. Here, experimental evidence is only presented for Lomer-Cottrell lock dislocation activities in larger grains. Following the loading of the specimens, movement and the interaction of dislocations were indicated. In Fig. 3.79a and b, HRTEM images were taken 180 s apart; (c) and (d) are the enlarged HRTEM images corresponding to the framed regions in (a) and (b), respectively. Two trapped dislocations are seen in (c) and local Burgers circuits were drawn to identify the Burgers vector. An inverse fast Fourier-filtered [henceforth: IFFT; inverse fast Fourier transform] image is given in (e). Extra half-planes are seen for both the $(\bar{1}11)$ and the $(1\bar{1}1)$ planes. This configuration represents a Lomer dislocation [henceforth: LD] exhibiting the Burgers vector of a a/2 [011] full dislocation.

This LD was formed by the interaction of two nondissociated full dislocations with Burgers vectors $a/2[\bar{1}01]$ and a/2[110], respectively, moving under applied stress on two intersecting slip planes, $(1\bar{1}1)$ and $(\bar{1}11)$. The dislocation reaction may be written as:

$$\mathrm{a}/2[\bar{1}01] + \mathrm{a}/2[110] \rightarrow [011].$$

Atomistic and dislocation dynamic simulations have previously shown that Lomer and Lomer-Cottrell locks have practically the same strength; these stable junctions are dislocation obstacles and, therefore, sources of strain hardening. The LD junction above is expected to be sessile, since it can glide in neither of the slip planes of the reactant dislocations. In (d) and (f), the LD destruction process is observed on the atomic scale. The reformation of Lomer lock in the same region, in the same grain, with continued straining is depicted in (g). Figure 3.80 is a schematic view, illustrating the formation and destruction process of a Lomer lock under applied stress. This figure may be understood in light of the Thompson tetrahedron shown in (a). The case of two interacting a/2[110] {111} dislocations in (b) are considered. Dislocations in the planes ACD (**b** = **CD**) and ABC (**b** = **BG**) interact under applied stress to form **BD**. For Lomer lock formation, one may write:

$$\mathbf{BC} + \mathbf{CD} \rightarrow \mathbf{BD}.$$

With increasing stress, the length of the Lomer segment decreases and then the junction breaks under high stress. In (d) of Fig. 3.80, the unzipping of the LD is depicted schematically. In situ experiments indicate that dislocations are highly

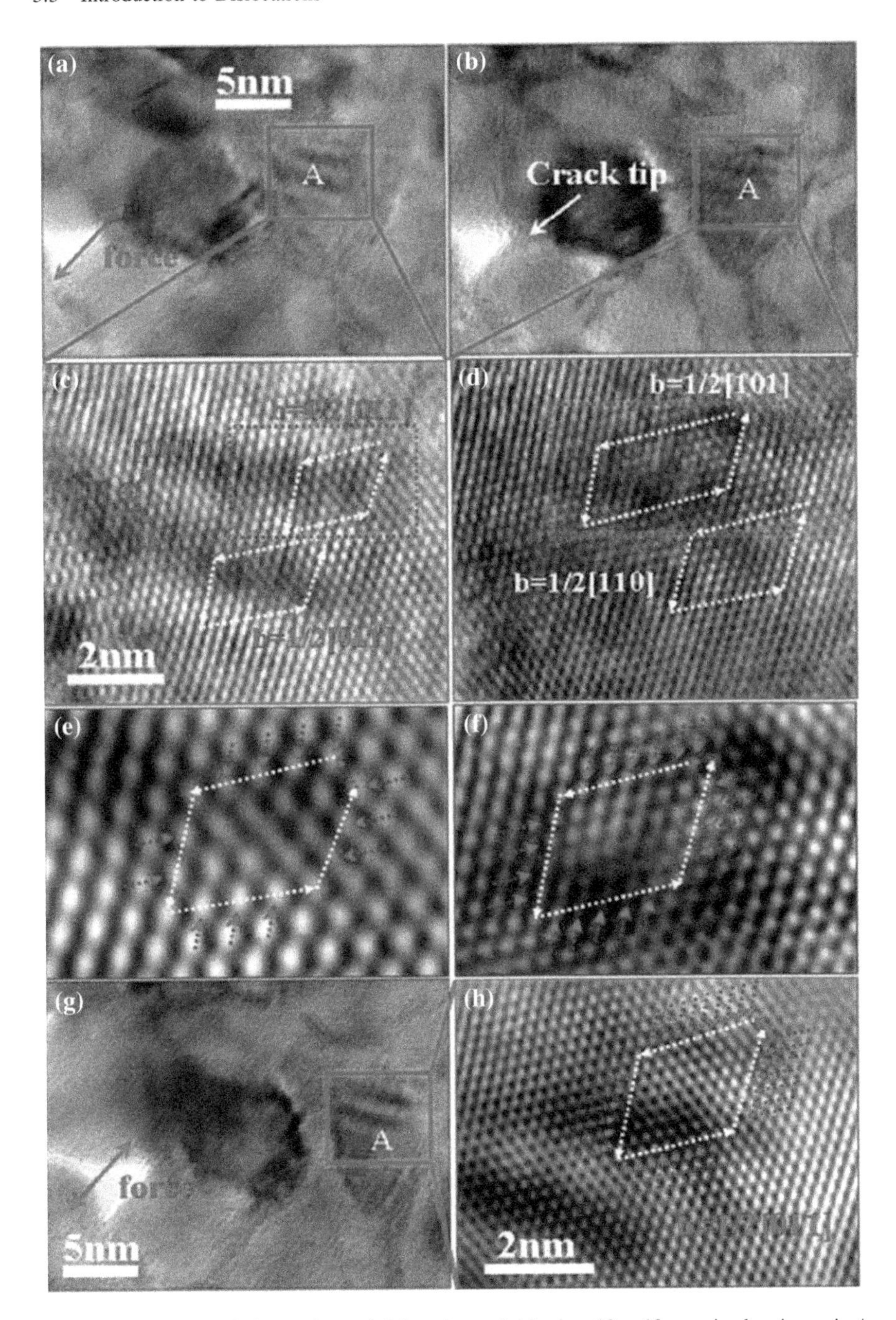

Fig. 3.79 In situ TEM observations of dislocation activities in a 10×13 nm-sized grain, grain A in (**a**). Two Lomer dislocations are found in (**c**); see details in the inverse fast Fourier-filtered (IFFT) shown in (**e**). The corresponding frames taken 180 s later are displayed in (**b**), (**d**), and (**f**), depicting the unzipping of the LD. **g** At a later stage of deformation, LD reforms in this grain, as seen in the IFFT image in (**h**) [54]. With kind permission of Professor Xiaodong Han. (LD means Lomer dislocation)

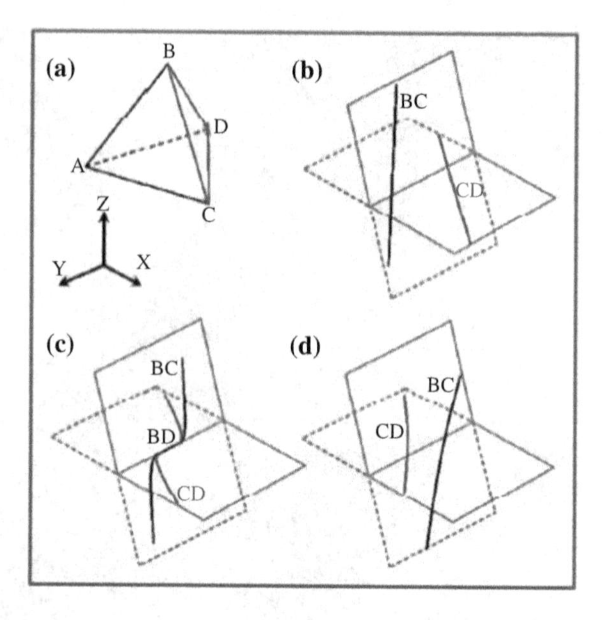

Fig. 3.80 **a** Thompson tetrahedron showing the Burgers vectors of the interacting dislocations. **b**, **c** Schematic view of the Lomer lock formation: the Lomer lock is formed by the interaction of two nondissociated full dislocations. **d** The Lomer dislocation is unlocked under stress [54]. With kind permission of Professor Xiaodong Han

active, even in nanometer grains. Full dislocations and Lomer locks are frequently observed for larger grains (~ 10 nm), whereas partial dislocations generating SFs are prevalent in smaller grains. In smaller grains ($d < 10$ nm), observations of the generation, motion, interaction and storage of full and partial dislocations, including the dynamic process of Lomer lock formation, destruction and reformation, explain the role of dislocation mechanisms in the plastic deformation of nanometer-grained ultrathin films.

The kinetics of deformation structure evolution and its contribution to strain hardening in austenitic steel, with a composition of Fe-30.5Mn-2.1Al-1.2C (wt%), was investigated by means of TEM and electron-channeling contrast imaging combined with electron backscatter diffraction [27]. This alloy exhibits a superior combination of strength and ductility (an ultimate tensile strength of 1.6 GPa and elongation to failure of 55 %), due to its multiple-stage strain hardening. Among the many dislocation configurations and twinnings responsible for its excellent mechanical properties, Lomer-Cottrell locks are also illustrated (Fig. 3.81). Here, various dislocation configurations may be seen, such as nodes (a), hexagonal dislocation networks (b) and Lomer-Cottrell locks (c). The configuration of the Lomer-Cottrell locks is formed by the interaction of dissociated dislocations, provided that the leading $\langle 112 \rangle$ partials attract one another. The product is a $\langle 110 \rangle$ sessile dislocation on a $\{001\}$ plane. Figure (c) reveals Lomer-Cottrell dislocations appearing as straight dislocations along the $[1\bar{1}0]$ crystallographic direction and forming Lomer–Cottrell locks labeled as LC1 and LC2. These locks act as strong barriers to dislocations with slip directions [101] and $[10\bar{1}]$. Lomer–Cottrell locks are known to be among the most important barriers to dislocation glide in the stage II hardening of FCC metals.

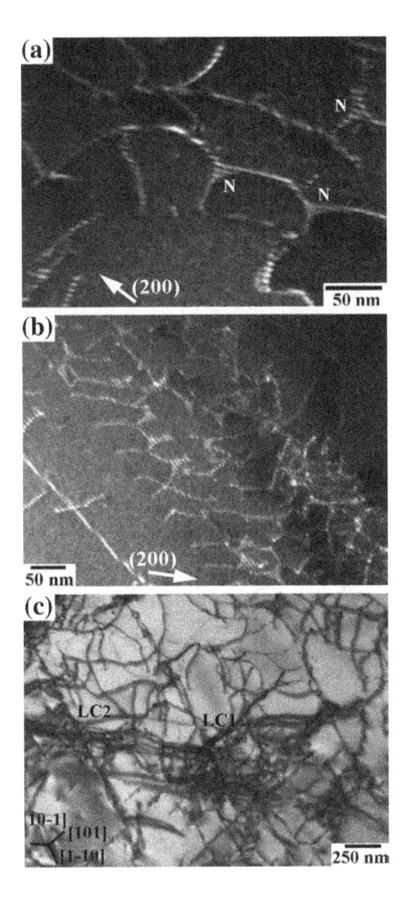

Fig. 3.81 Extended planar dislocation configurations in a sample deformed to 0.05 true strain/450 MPa: **a** dislocation nodes (N), **b** hexagonal dislocation networks. The stacking faults associated to the partial dislocations are visible. Diffraction vectors are indicated by an arrow. **c** Lomer–Cottrell locks (LC1 and LC2). Lomer–Cottrell dislocations appear as straight dislocations lying along the $[1\bar{1}0]$ direction [27]. With kind permission of Elsevier

3.3.17.6 Dislocation Pile-Ups

In Sect. 3.3.15, the dislocation of pile-ups was discussed and a schematic illustration presented. Figure 3.82 shows another informative, schematic view of a pile-up. Dislocation pile-ups at grain boundaries act as a barriers preventing cross-over into the next grain. This grain is favorably oriented for slip, but deformation requires dislocation mobility in many grains. For slip to occur in a neighboring grain, the stress field of a pile-up dislocation must either (a) induce sufficient stress into the second grain at the grain boundary or (b) initiate a new source in the neighboring grain. In the first case, the boundaries emit dislocations for further deformation by slip; in the second case, the source to be activated requires high stress.

Experimental pile ups in a ceramic can be seen in Fig. 3.83 in an Al_2O_3-$GdAlO_3$ binary eutectic which piles up in the $GdAlO_3$ phase. The deformation was by compressive creep test in air in the stress range of 50–200 MPa and temperature range 1450–1600 °C.

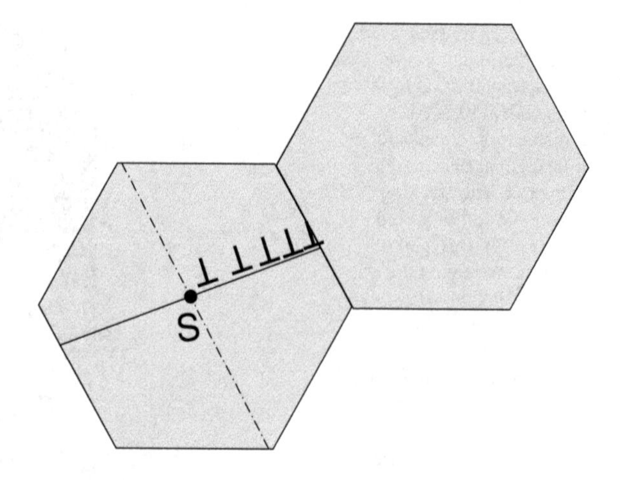

Fig. 3.82 A schematic view of a dislocation pile up at a grain boundary

In another example, amorphization by dislocation accumulation in shear bands in a microcrystalline c-$Y_2Si_2O_7$ was investigated by TEM in the vicinity of the resulting microstructural indent obtained by indentation deformation at RT. Figures 3.84 and 3.85 are illustrations of pile-ups in c-$Y_2Si_2O_7$.

The deformation microstructure, as viewed under weak-beam TEM imaging mode, is dominated by individual dislocations. In Fig. 3.84a, the isolated dislocations that expand in their slip plane exhibit favored crystallographic orientations, suggesting a strong stabilizing interaction with the lattice (the so-called 'Peierls valleys'). Such dislocations are often dissociated into two partials with different Burgers vectors about 8 nm apart in projection (white arrow pairs in Fig. 3.84a and c). In this area, it is seen that isolated dislocations may coexist with planar arrays of glissile dislocations, as in Fig. 3.84b and c (the letter H indicates a pile-up spearhead). The fact that dislocations are dissociated encourages planar slip and is consistent with the formation of such arrays and their piling up at obstacles. In Fig. 3.85, a second volume is submitted to stresses larger than those applied to the outer zone (Fig. 3.84). Each slip band comprises dislocations with Burgers vectors all identical in direction and sign (see inset in a). They are grouped in bands under a density far larger than that commonly observed in the peripheral zone (of Fig. 3.84a). For further information on amorphization by dislocation accumulation in shear bands, the work of Lin et al. [31] should be consulted.

3.3.17.7 Low-Angle Boundary

In Sect. 3.3.17.1 (Fig. 3.66), a HREM micrograph of zirconia illustrated a tilt boundary consisting of an array of edge dislocations. Basically, a low-angle grain boundary may be (a) a tilt boundary, (b) a twist boundary, or (c) a mixed boundary, as described in Sect. 3.3.16. Some visual examples follow.

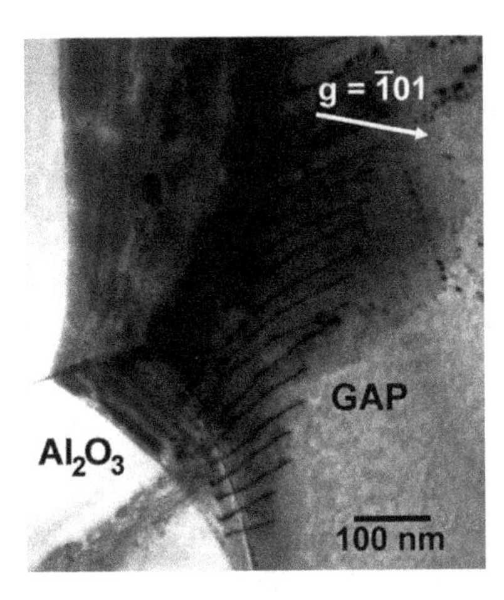

Fig. 3.83 TEM examinations of compression tested specimens. Dislocation pileup in the GAP ($GdAlO_3$) phase of the Al_2O_3-GAP eutectic composite is shown [44]. With kind permission of Dr. Appriou, Research Director of Aerospace Lab Journal

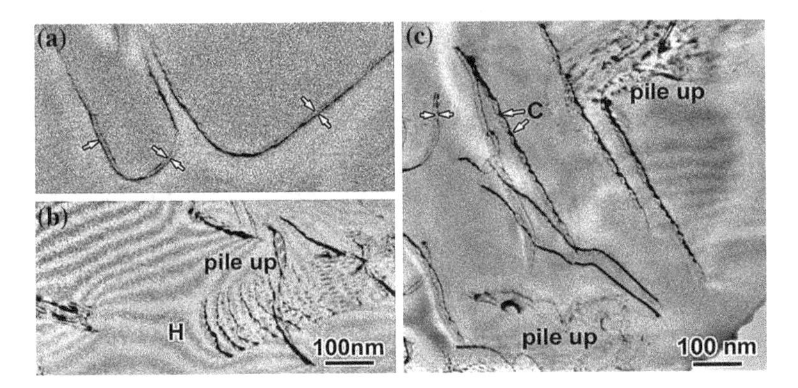

Fig. 3.84 Isolated dislocations populating the outer region of the indented sample, viewed under weak-beam TEM imaging mode. For enhanced visibility, the negative prints of the micrographs are displayed. **a** The isolated dislocations that expand in their easy slip plane. **b** A series of dislocations arranged to form pile-ups. **c** The dislocations show large linear densities of constrictions. The letter H indicates the pile-up spearhead. The images have the same scale bar [31]. With kind permission of Elsevier

(a) Tilt Boundary

Figure 3.86 is a schematic illustration of a tilt boundary, where 2θ is the tilt angle between the **n** direction (i.e., n = $[1\bar{1}00]$ or [1120]) Σ values and grain-boundary planes for the respective boundaries. Note that in coincident site lattice [henceforth: CSL] theory, the degree of fit, Σ, between the structures of the two grains is described by the *reciprocal* of the ratio of coincident sites to the total number of sites. A boundary with high Σ may be expected to have higher energy than one with low Σ.

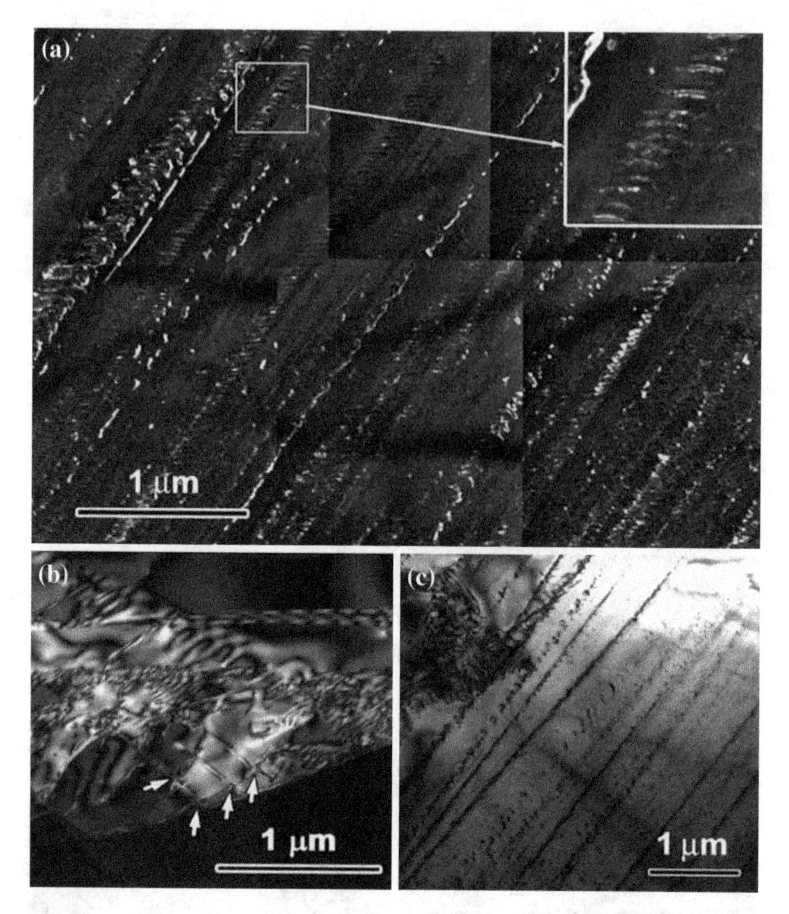

Fig. 3.85 Slip bands in the intermediate region of the indented volume. **a** Slip band planarity and evidence of profuse pile-ups. The dislocations exhibiting paired lines in the boxed area are not dipoles but dissociated dislocations since the distance between partials is constant whether the dipole is imaged with the g or -g reflecting plane. **b** Intersecting slip bands. **c** The slight misalignment and differences in pile-up projected widths indicate that the slip bands are parallel to at least two crystallographically distinct planes [31]. With kind permission of Elsevier

Low-angle boundaries, where the distortion is entirely accommodated by dislocations, are $\Sigma1$. Another example is the coherent twin boundaries ($\Sigma3$). Deviations from the ideal CSL orientation may be accommodated either by local atomic relaxation or by the inclusion of dislocations into the boundary. An actual HRTEM image for three types of $\Sigma7$ symmetrical tilt grain boundaries obtained in the study is illustrated in Fig. 3.87. Cross-sectional specimens for HRTEM observation were prepared using mechanical grinding and dimpling to a thickness of 10 μm, argon-ion-beam milling to electron transparency at $\sim$4 kV, and carbon thin-film coating. Grain boundaries were observed in a cross-sectional view, i.e., $\Sigma7\{11\bar{2}0\}$, $\Sigma7\{1\bar{1}02\}$, for the boundary and [0001] for the $\Sigma7\{2\bar{3}10\}$ and $\Sigma7\{4\bar{5}10\}$

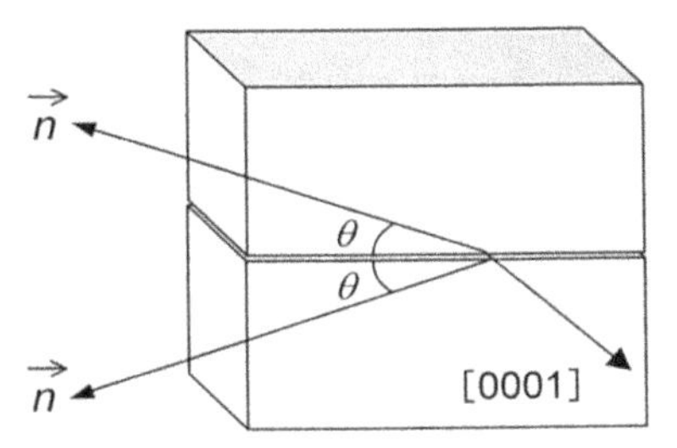

Fig. 3.86 Schematic illustration of [0001] symmetrical tilt grain boundary, indicating the tilt angle is 2θ [43]. With kind permission of John Wiley and Sons

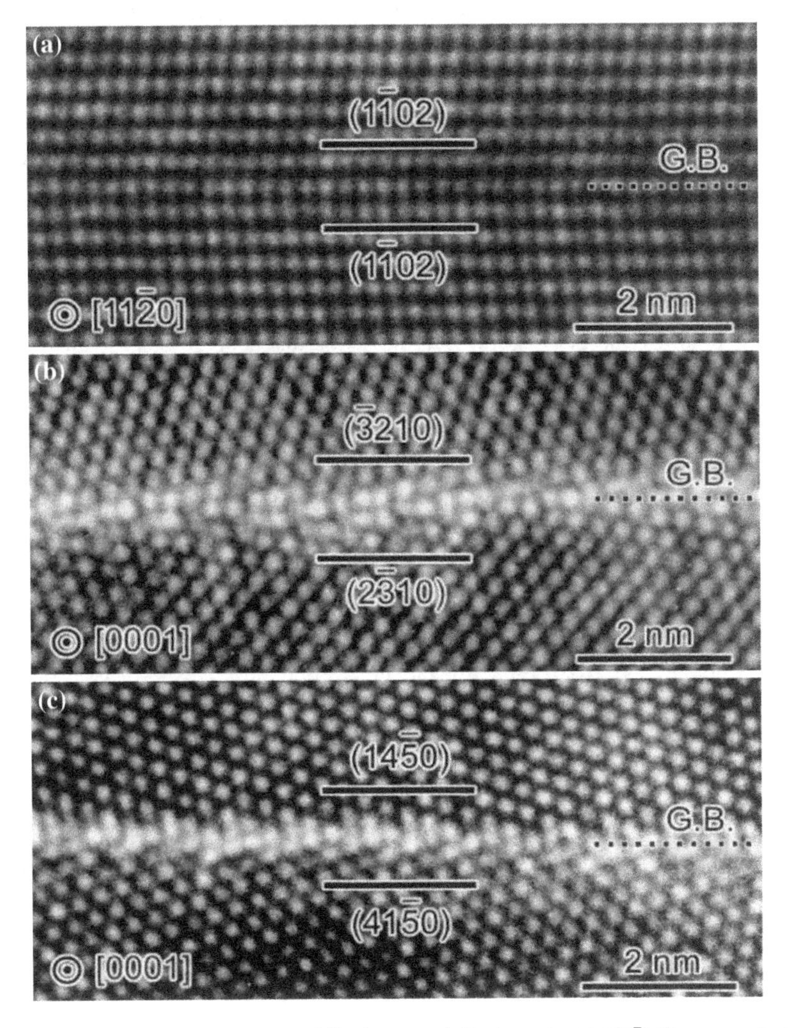

Fig. 3.87 HRTEM images for **a** $\Sigma\, 7\{1\bar{1}02\}$, **b** $\Sigma\, 7\{2\bar{3}10\}$, and **c** $\Sigma\, 7\{4\bar{5}10\}$ symmetrical tilt grain boundaries [43]. With kind permission of John Wiley and Sons

Fig. 3.88 Zeiss interferogram of a thermal groove. Configuration of a pure tilt boundary in aluminum oxide [49]. With kind permission of John Wiley and Sons

boundaries. In this study, the HRTEM images for three types of Σ7 symmetrical tilt-grain boundaries showed no observable amorphous layers or secondary phases at these boundaries. They were systematically taken at defocus values from 0 to ~500 Å at steps of ~50 Å. These HRTEM images match well with the images calculated for defocus and thickness values of $\Delta f \sim 350$ Å, $t = 50$ Å (Fig. 3.87a, b) $\Delta f \sim -130$ Å, $t = 90$ Å (Fig. 3.87b) and $\Delta f = -110$ Å, $t = 90$ Å (Fig. 3.87c).

Another experimental technique for showing grain boundaries is by using thermalgrooving technique, which is basically used for calculating grain-boundary energies. In Fig. 3.88, an illustration of a pure tilt-grain boundary is shown, obtained by this technique.

(b) Twist Boundary

Figures 3.89 and 3.90 show twist boundaries in SiAlON and in sapphire ceramics, respectively. In Fig. 3.89, both tilt and twist boundaries are indicated. Rows of parallel and more complex dislocations are observed. These dislocation structures are periodic. The Burgers vector determined for the dislocations are of type **b = a/3 ⟨110⟩**. The experimental results show that these twist boundaries are stable without an amorphous grain-boundary phase. It appears, according to the experimental results, that boundaries with low Σ misorientation possess relatively low energies and, therefore, are formed favorably during a sintering process.

In Fig. 3.90, a TEM micrograph of a 0.3° [0001] twist boundary, inclined relative to the incident beam and revealed in projection, is shown. This image was recorded in BF mode under strong two-beam diffraction conditions, revealing a regular network of screw dislocations with a large Burgers vector $\mathbf{b} = [10\bar{1}0]$ spaced about 160 nm apart, amid a dense array of secondary grain-boundary dislocations with a smaller strain field and lower contrast.

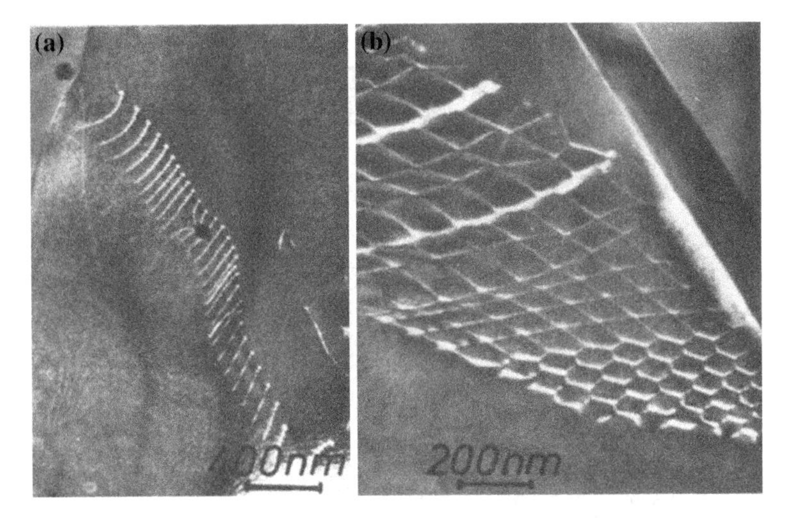

Fig. 3.89 Low angle grain boundaries: Weak beam DF images showing dislocation structure in SIAlON; **a** in a tilt boundary **b** in a twist boundary [48]. With kind permission of Professor Rühle

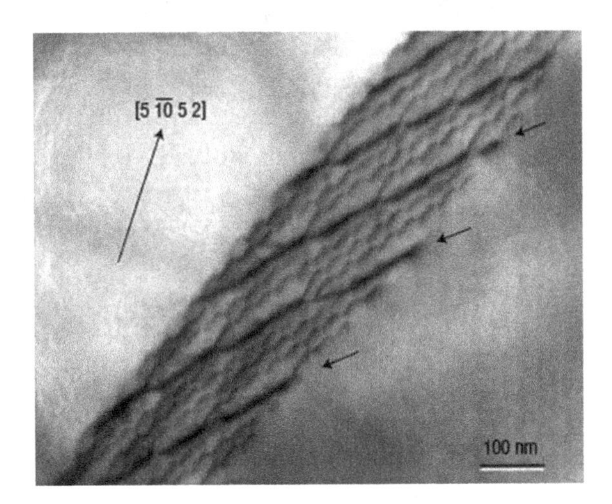

Fig. 3.90 Bright-field TEM image of a 0.3° basal twist boundary, with the primary dislocations arrowed, and secondary grain-boundary dislocations showing lighter contrast [34]. With kind permission of Nature Materials and Dr. Glaeser

3.3.17.8 Stacking Faults (SFs)

SFs in ceramics were shown in Sect. 3.3.17.3, Fig. 3.68, when discussing partial dislocations. However, the SF images below are presented for more clarity. In Fig. 3.91, the microstructures of bulk $Zr_2Al_3C_4$ and $Zr_3Al_3C_5$ ceramics are shown illustrating SF in these materials.

For the sake of those unfamiliar with the term 'Z contrast', this refers to a technique used in scanning-transmission electron microscopy [henceforth: STEM], which provides an incoherent image of crystals at atomic resolution. There are no

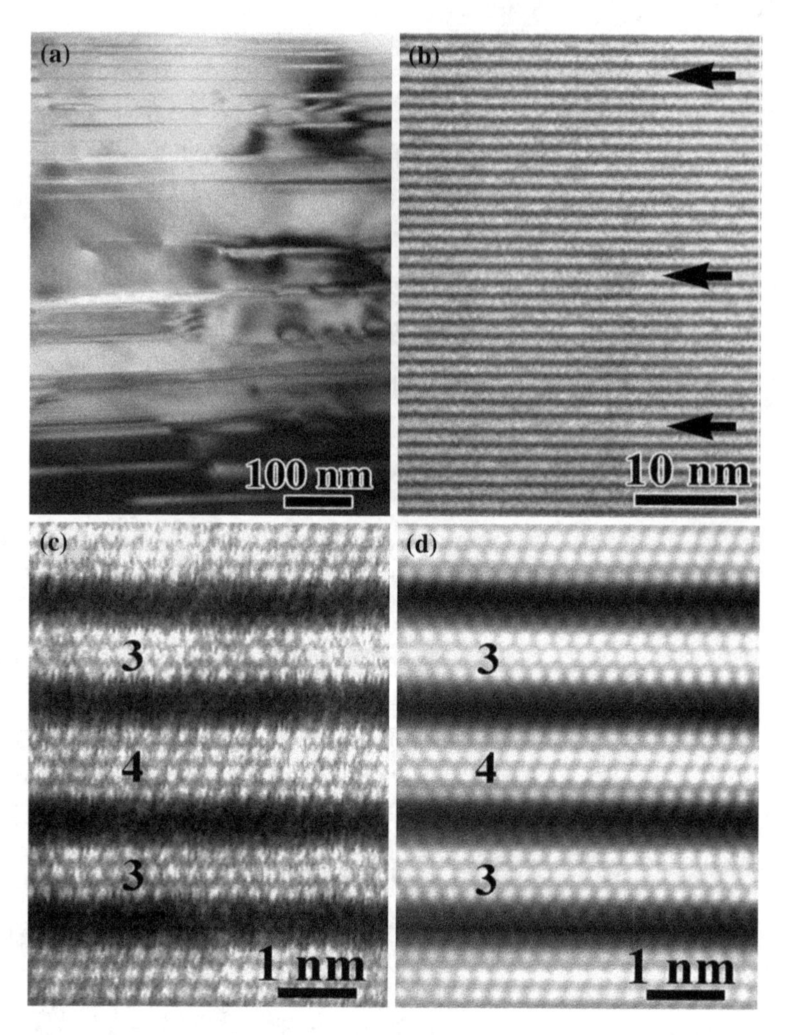

Fig. 3.91 **a** Low-magnification bright-field image of stacking faults of $Zr_3Al_3C_5$. **b** Medium-magnification Z-contrast image showing the periodicity of the stacking faults. **c** Raw high-magnification Z-contrast STEM image of a stacking fault resulting from the insertion of an additional ZrC layer. **d** FFT filtered image of **c**. The arrows in **b** indicate the stacking faults [30]. With kind permission of Elsevier

phases in an incoherent image, so there is no phase problem for structural determination. The location of atom- column positions in an image is greatly simplified. SFs in $Zr_3Al_3C_5$ were found to result from the insertion of an additional Zr-C layer. The interest in these ternary aluminum carbides is due to their attractive properties, such as: easy machinability, damage tolerance, excellent high-temperature oxidation resistance and good electrical and thermal conductivities. Such ceramics have hexagonal symmetry and crystallize in 6/mmm and P63/mmc

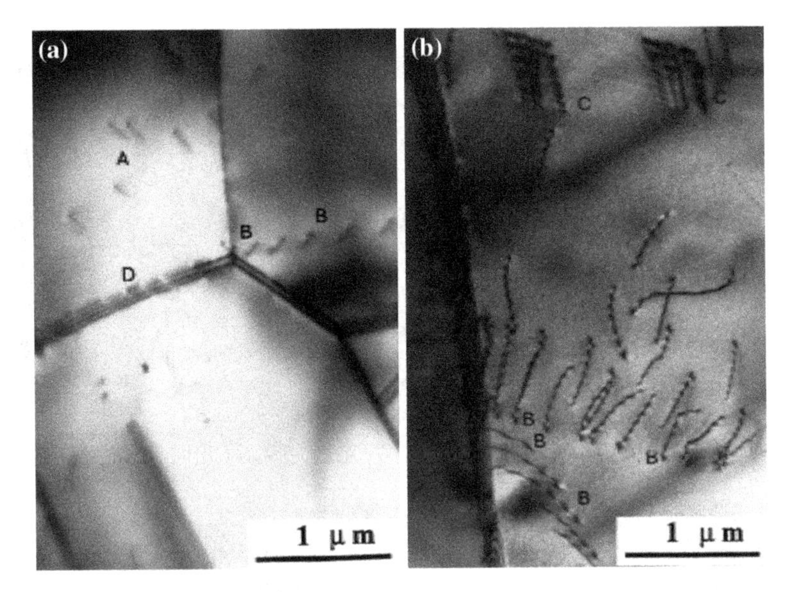

Fig. 3.92 Typical dislocations in Ti_3SiC_2 produced by reactive hot pressing: single dislocations (*A*); arrays of parallel straight dislocations within a grain (*B*); dislocations bounding stacking faults (*C*); dislocations at grain boundaries (*D*) [25]. With kind permission of Wiley and Sons

symmetry for $Zr_2Al_3C_4$ and $Zr_3Al_3C_5$, respectively. Details on the preparation of these ceramics are presented in the work of Lin et al. [31]. Figure 3.90a shows a low-magnification, BF image of SFs in $Zr_3Al_3C_5$. Bright and dark fringes are visible. In order to clarify the microstructure of the SFs, Z-contrast STEM images were obtained. As displayed in Fig. 3.91b, the periodicity of the stacking sequence was not identical. The atomic-scale microstructure of these SFs is further illustrated in Fig. 3.91c and d. The high-resolution Z-contrast image indicates that the SF resulted from the insertion of an additional Zr–C layer.

In an additional example of SF, Fig. 3.92 illustrates their existence in Ti_3SiC_2. Undeformed and deformed samples of Ti_3SiC_2, fabricated by the reactive hot pressing of Ti, SiC, and graphite, were characterized by TEM. Figure 3.92 presents micrographs of undeformed ceramic specimens. The basal plane dislocations are mobile and multiply as a result of deformation at RT. Direct evidence of dislocation mobility and multiplication in Ti_3SiC_2, as a result of deformation, may be seen in Fig. 3.93.

The dislocation density is significantly higher after deformation than before, as may be seen by comparing Figs. 3.93 and 3.92a. A further difference observed is the very limited number of dislocation arrays in the undeformed samples, shown in Fig. 3.92a. In these undeformed samples, the only dislocation arrays observed are adjacent to grain boundaries or cracks; the interiors of the grains are, for the most part, dislocation free. As a result of deformation, the number of arrays is greatly increased. Moreover, in the undeformed samples, the arrays only extend part way

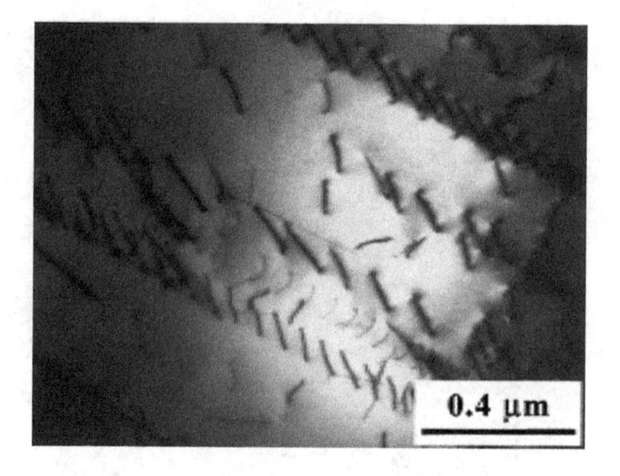

Fig. 3.93 Typical TEM micrograph after deformation at room temperature. Number and spatial extent of dislocation arrays is greatly increased compared to the undeformed samples [25]. With kind permission of John Wiley and Sons

into the grain interior, typically 2–3 μm. After deformation, however, these arrays extend across the whole field of view and, presumably, extend across the whole grain. Note that basal-plane dislocations move and multiply at RT. This is in line with the fact that Ti_3SiC_2 plastically deforms at RT along the basal planes. Furthermore, the absence of slip systems, other than basal, explains the brittleness of randomly oriented polycrystalline Ti_3SiC_2 samples.

Numerous planar defects are also observed, which may be divided into two types, depending on their extent. Defects of the first type extend clear across the grains. Typically, one or two such defects are observed per grain. In Fig. 3.94a and b, such defects, labeled SF, propagate across the grain, G_2. The selected area diffraction patterns for regions on both sides of the defect are identical, confirming that G_2 constitutes a single grain containing this planar defect. Furthermore, the contrast associated with a defect, as well as its absence, is easily observed by tilting the specimens and imaging the defects with various reflections. As shown in Fig. 3.94a and b, the defect is invisible when imaged with $g = \bar{1}2\bar{1}0$, but is visible with $g = 0114$. In the latter case, alternating bright and dark fringes are observed, which are of symmetrical contrast in BF and asymmetrical contrast in dark field (not shown). Thus, the defect must be SF characterized by a displacement vector parallel to [0001]. Defects of the second kind are similar to the first, except that they do not propagate across the whole grain (Fig. 3.94c and d).

3.3.18 Section Summary

This section has discussed point defects, dislocations and experimental observations of the various dislocation configurations observed in ceramics. These imperfections are the basic features of each and every material, including

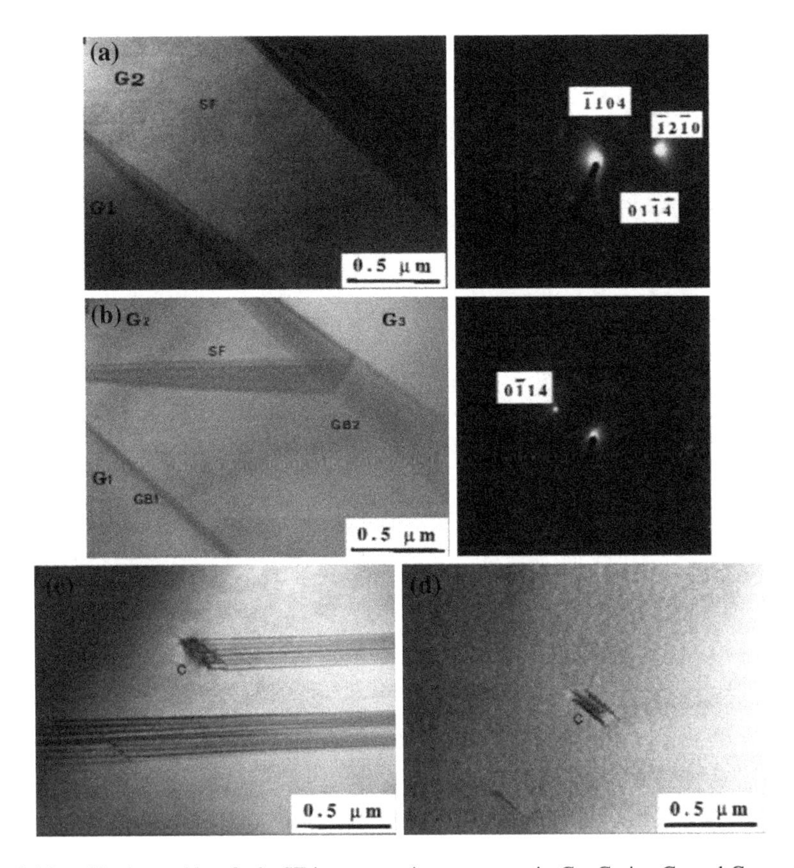

Fig. 3.94 **a** Single stacking fault, SF is propagating across grain G_2. Grains G_1 and G_3 are two adjacent grains separated from G_2 by boundaries GB1 and GB2. Bright field, BF, with $g = \bar{1}2\bar{1}0$. **b** BF with g = 0114. **c** Typical example of second type of stacking faults, in which one side is bounded by partial dislocations and the other (not shown) a grain boundary. **d** Same area as **c**, but imaged such that only the bounding are dislocations are visible [25]. With kind permission of John Wiley and Sons

ceramics, responsible for their various properties. The general brittleness of ceramics at RT, their ductility at high temperature, the ductility exhibited by some ceramics at RT and superplasticity are directly related to the presence of dislocations, their mobility and the influence of their width on the character of a material under stress (load). When the width of a dislocation is such that no room is provided for their accommodation and motion, because the bond character of the specific material (e.g., ceramics) prevents it—as occurs in ionic and covalently-bound materials–inevitably low ductility or brittleness result.

It has also been briefly stated that some of the planar defects, such as low-angle grain boundaries and SFs may be interpreted as being related to specific dislocation configurations.

This section began with point defects, which have either direct or indirect influence on the mechanical properties of materials, discussing point-defect hardening, climb, etc., but a later chapter will show that creep is greatly associated with the mobility of point defects. No discussion was included here on the important aspect of the physical properties of point defects, which is beyond the scope of this book.

References

1. Amelinckx S (1957) Dislocations and mechanical properties of crystals. John Wiley & Sons Inc, New York
2. Amelinckx S (1964) The direct observation of dislocations, solid state physics, supplement 6. Academic Press, New York
3. Barsoum MW (1997) Fundamentals of ceramics. IOP Institute Physics Publishing, Bristol
4. Cahn RW (1970) Physical metallurgy. North Holland Publishing Company, Amsterdam
5. Damask AC, Dienes GJ (1971) Point defects in metals. Gordon and Breach Science Publishers, New York
6. Dash WC (1957) Dislocations and mechanical properties of crystals. John Wiley, New York, p 57
7. Friedel J (1967) Dislocations. Pergamon Press, Oxford
8. Gilman JJ, Johnston WG (1957) Dislocations and mechanical properties of crystals. John Wiley, New York
9. Hirth JP, Lothe J (1968) Theory of dislocations. McGraw-Hill, New York
10. Hull D (1969) Introduction to dislocations. Pergamon Press LTD, Oxford
11. Izumi K, Yasuda K, Kinoshita C, Kutsuwada M (1998) J Nucl Mater 258–263:1856
12. Muller EW, Tsong TT (1970) Field ion microscopy principles and applications. Elsevier Science, New York
13. Orowan E (1949) Fracture and strength of solids. Rep Prog Phys 12:185
14. Pelleg J (2013) Mechanical properties of materials. Springer, Dordrecht
15. Read WT Jr (1953) Dislocations in crystals. McGraw-Hill Book Company Inc, New York
16. Ritchie RO, Kruzic JJ, Muhlstein CL, Nalla RK, Stach EA (2004) Int J Fract 128:1
17. Seeger A (1957) Glide and work hardening in face centered cubic metals, dislocations and mechanical properties of crystals. John Wiley, New York
18. Xue K, Niu L-S, Shi H-J (2011) Silicon carbide—materials, processing and applications in electronic devices, M. Mukherjee (Ed.), www.intechopen.com, 2011,p. 3. Published by InTech

Further References

19. Amelinckx S, Dekeyser W (1958) J Appl Phys 29:1000
20. Amelinckx S, Dekeyser W (1958) J Appl Phys 29:1000
21. Appel F, Bethge H, Messerschmidt U (1977) Phys Stat Solidi (a) 42:61
22. Baufeld B, Baither D, Bartsch M, Messerschmidt U (1998) Phys Stat Sol (a) 166:127
23. Doman RC, Barr JB, McNally RN, Alper AM (1963) J Am Ceram Soc 46:307
24. Eshelby JD, Frank FC, Nabarro FRN (1951) Phil Mag 42:351
25. Farber L, Barsoum MW, Zavaliangos A, El-Raghy T (1998) J Am Ceram Soc 8:1677
26. Frenkel J (1926) Z Phys 37:572
27. Gutierrez-Urrutia I, Raabe D (2012) Acta Mater 60:5791

28. Ikuhara Y (2009) Mater Trans 50:1626
29. Ivashchenko VI, Turchi PEA, Shevchenko VI (2007) Phys Rev B 75:085209
30. Lin ZJ, Zhuo MJ, He LF, Zhou YC, Li MS, Wang JY (2006) Acta Mater 54:3843
31. Lin ZJ, Zhuo MJ, Sun ZQ, Veyssiére P, Zhou YC (2009) Acta Mater 57:2851
32. Liu CT, Heatherly L, Easton DS, Carmichael CA, Schneibel JH, Chen CH, Wright JL, Yoo MH, Horton JA, Inoue A (1998) Met Mater Trans A 29A:1811
33. Malerba L, Perlado JM (2001) J Nucl Mater 289:57
34. Marks RA, Taylor ST, Mammana E, Gronsky R, Glaeser AM (2004) Nat Mater 3:682
35. McHargue CJ, O'Hern ME, Joslin DL (1990) MRS Proc 188:111
36. McHargue CJ, Sklad PS, White CW (1990) Nucl Instr Meth Phys Res B 46:79
37. Messerschmidt U, Bartsch M (2003) Mater Chem Phys 81:518
38. Mitchell TE (1979) J Am Ceram Soc 62:254
39. Miyazaki H, Suzuki T, Yano T, Iseki T (1992) J Nucl Sci Technol 29:656
40. Nabarro FRN (1947) Proc Phys Soc 58:699
41. Nabarro FRN (1952) Proc Phys Soc A 200:279
42. Nakano S, Muto S, Tanabe T (2006) Mater Trans, 47:112. The Japan Institute of Metals
43. Nishimura H, Matsunaga K, Saito T, Yamamoto Y, Ikuhara Y (2003) J Am Ceram Soc 86:574
44. Parlier M, Valle R, Perrière L, Lartigue-Korinek S, Mazerolles L (2011) J Aerosp Lab Issue 3, 1 (November 2011)
45. Peierls R (1949) Proc Phys Soc 52:34
46. Pelleg J, Liu R (1992) Thin Solid Films 221:318
47. Raterron P, Chen J, Li L, Weidner D, Cordier P (2007) Amer Mineral 92:1436
48. Schmid H, Rühle M (1984). J Mater Sci 19:615. With kind permission of Professor Rühle
49. Shackelford JF, Scott WD (1968) J Am Ceram Soc 51:688
50. Shibata N, Yamamoto T, Ikuhara Y, Sakuma T (2002) J Electron Microsc 50:429
51. Suematsu H, Iseki T, Yano T, Saito Y, Suzuki T, Mori T (1992) J Am Ceram Soc 75:1742
52. Suematsu H, Yatsui K, Yano T (2001) Jpn J Appl Phys 40:1097
53. Thompson N (1953) Proc Phys Soc London 366:481
54. Wang L, Han X, Liu P, Yue Y, Zhang Z, Ma E (2010) Phys Rev Lett 105:135501
55. Wu Y-C, Wang S-F (2007) J Am Ceram Soc 90:230

Chapter 4
Deformation in Ceramics

Abstract Deformation can be elastic or plastic. Understanding elastic deformation is very important in ceramics to eliminate instantaneous brittle fracture at some applied stress levels. The fracture stress is usually the same or very close to the elastic limit. Stresses have to be exercised with understanding of the limits of the specific ceramics and the level that it can endure before fracture. No dimensional changes in test pieces occur in elastic deformation. Plastic deformation of ductile ceramics at room temperature, and of low temperature brittle ceramics at elevated temperatures, produce slip marks due to the advance of dislocations. All the characteristic phenomena of plastic deformation are observed either at ambient temperature (of room temperature ductile materials) or at the elevated temperature (of brittle ceramics at low temperature) such as yielding, existence of resolved and critical shear stress and slip. Among the yield phenomena, serrated stress–strain curves and Lüders bands can be noted as existing features. Twinning deformation, -mechanical or annealing twins-, are also observed to operate under proper conditions. Among the many factors influencing the mechanical properties, special consideration should be given to the effect of grain size. Preferred orientation in polycrystalline ceramics are seldom observed in its natural state but it may be induced during processing or fabrication. It might be of interest sometimes to get anisotropy for specific purposes of interest to enhance some directional property.

4.1 Introduction

As previously stated, ceramics are characterized either by the ionic or covalent bonding of their constituents and, consequently, with some exceptions, they exhibit brittle behavior. Also note that the field of ceramics covers a broad range of structures, from completely crystalline to amorphous (mostly glassy structures). Therefore, the main deformation at ambient temperatures is elastic (tending to brittleness); only at elevated temperatures may one speak about plastic deformation, since most ceramics show ductility. Clearly, the temperature level is a

J. Pelleg, *Mechanical Properties of Ceramics*, Solid Mechanics and Its Applications 213, DOI: 10.1007/978-3-319-04492-7_4,

significant parameter that indicates when a particular ceramic will become ductile and exhibit plastic deformation upon the application of stress.

Many factors influence the mechanical behavior of ceramics; among them are impurities and dispersed second phases. In the case of polycrystalline ceramics, the characters of grain boundaries and their vicinities are very significant. Add to the above the inevitable presence of pores, which weaken the overall mechanical strength of ceramics. Deformation, especially plastic deformation, involves dislocation glide, dislocation climb and other related phenomena, while, in time-dependent deformation, grain-boundary sliding (creep) is a major contributor to the process of deformation. Twinning is another factor to consider when considering deformation. Most of these subjects will be considered here; however, later on, a full chapter is devoted to time-dependent deformation.

4.2 Elastic Deformation

This kind of deformation is very important in ceramics, because there is an interest in eliminating instantaneous brittle fracture at some commonly applied stress levels. Also of interest are dimensional stability and the level of elastic strain not to be exceeded. Elastic deformation is recoverable and, in ceramics, it is mostly instantaneous and time-independent when tested under tension, although the presence of micro-cracks may show some time-dependence when the stress level applied induces crack growth (which is clearly time-dependent). Note that such time-dependence is not related to the anelasticity observed in metals, but mainly to that in polymers. Tension testing to observe stress–strain behavior is quite uncommon in ceramics because: (a) it is difficult to machine specimens to the desired shape and dimensions; (b) the gripping of a brittle material, such as ceramics, may induce premature fractures in test specimens; and (c) perfect alignment is necessary in tension tests to eliminate any side effects, such as bending, if a reliable stress–strain relation is required. Therefore, flexural strength is the frequently recommended test involving the transverse bending of the specimens, using a three- or four-point loading technique. Chapter 1 discussed these test methods and a schematic curve for flexural strength was shown in Fig. 1.47. Further experimental results for flexural and tension tests in SiC were compared in Sect. 1.9, Figs. 1.50 and 1.51, respectively.

A method for classifying refractory ceramics by their tendencies to brittleness was suggested by Gogotsi et al. [4], according to index χ. The equation for χ is:

$$\chi = \frac{\sigma_{\text{lim}}^2}{2E \int^{\varepsilon_{\text{lim}}} \sigma dE} \tag{4.1}$$

where σ^{lim} is the strength, E the elastic modulus and ε^{lim} is the strain of the material. As it follows from its definition, 'brittleness' is determined by $0 < \chi < 1$. The parameter, χ, is defined by the ratio of the specific elastic energy, u_c, to the

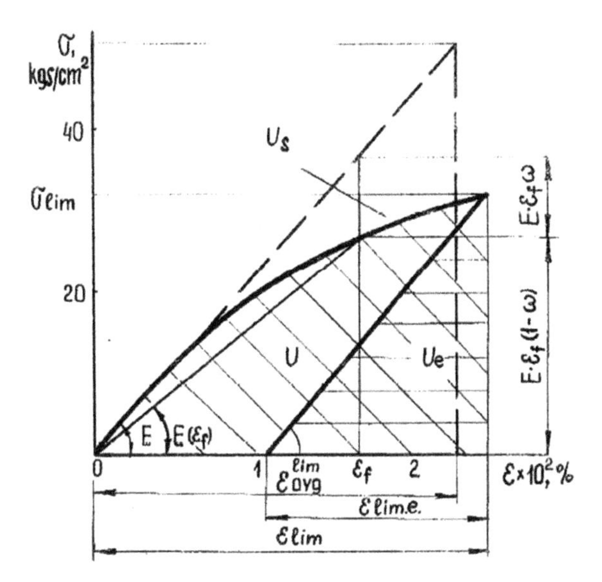

Fig. 4.1 Deformation diagram of zirconium oxide. $\varepsilon_{ave}^{\lim}$ is half sum of limit values ε_{+} and ε_{-} [4]. With kind permission of Elsevier

whole specific energy, u, expended to attain the limiting state. This equation is based on the σ-ε relation in zirconia and the assumption that the elastic moduli are equal under loading and unloading (Fig. 4.1).

Refractory ceramic materials are classified as brittle for $\chi = 1$ and relatively brittle for $\chi < 1$. Additional σ-ε relations for nitride- and oxide-based ceramics are shown in Figs. 4.2 and 4.3. In Fig. 4.2, the relations of two values of χ are shown, one of them for $\chi = 1$.

Brittle materials, characterized by $\chi = 1$ follow Hooke's Law until fracture (see, for example, Fig. 4.3 with $\chi = 1$). Relatively brittle materials, with $\chi < 1$, are not necessarily linear, as seen in Fig. 4.3 for χ. Most brittle materials observed are homogeneous single-phased and FG oxide ceramics or glassy ceramics and porcelains. Such materials are deformed without structural changes to failure. The beginning of failure coincides with the propagation and growth of cracks and their development from micro-cracks. In relatively brittle materials ($\chi < 1$), some amount of energy is dissipated under loading conditions due to non-elastic effects at certain stress levels.

4.3 Plastic Deformation

Bulk plasticity is only partially realized because of the inherent defect distributions in large volume ceramic structures. However, plasticity in some ceramics, such as MgO, has also been observed at ambient temperatures. Thus, in this section, Plastic deformation will be considered at both low and elevated temperatures.

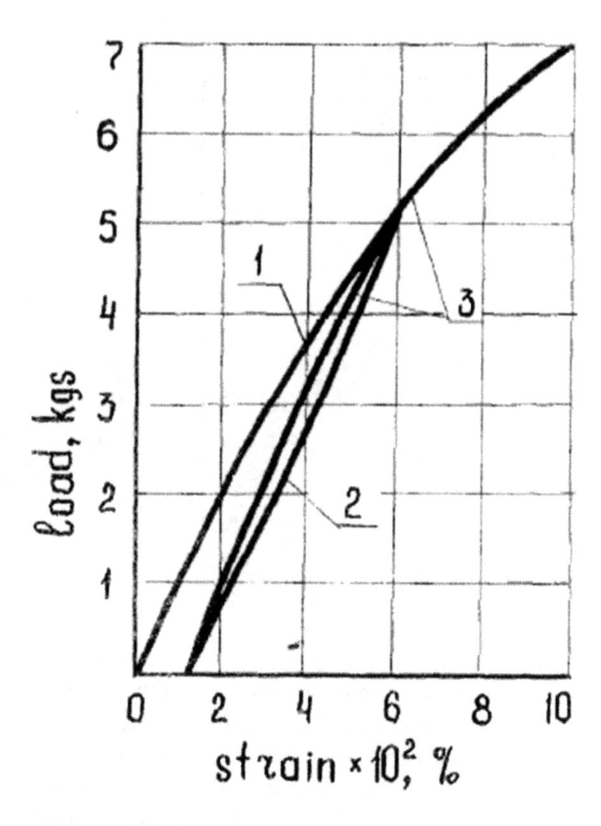

Fig. 4.2 Load-strain dependence for a nitride ceramic specimen *1*—loading; *2*—unloading: *3*—repeated loading [4]. With kind permission of Elsevier

4.3.1 Deformation at Low Temperatures

Low-temperature ductility is rarely observed in ceramics, which are inherently brittle, but some bulk ceramics show plasticity at ambient temperatures. One example of low-temperature plasticity in MgO is considered here. First, consider a single crystal, where i orientation-dependent properties are of interest. Orientation is one of the factors that influence mechanical properties. It was observed (by etch-pit technique) that the flow in MgO occurs on the $\{110\}$ $\langle 110\rangle$ slip system. However, it was also found [28] that the $\{110\}$ $\langle 110\rangle$ slip system contributes to deformation above ~600 °C. Details on Plastic deformation in MgO single crystals were presented in Sect. 2.2, Figs. 2.33 and 2.38. Consequently, some information on deformation in polycrystalline ceramics may be of interest.

As mentioned above, some ceramics may be made plastic even at room temperature. As such, producing them in nano-size form may induce low-temperature plasticity. Figure 4.4 is an illustration of in situ TEM observation of the Plastic deformation features of a SiC nanowire [henceforth: NW]. The in situ SiC NW elastic, elastic–plastic and Plastic Deformations were conducted using an ultra-HRTEM by bending SiC NWs via the mechanical force produced by the TEM specimen-supporting grid under the irradiation of an electron beam. Figure 4.4

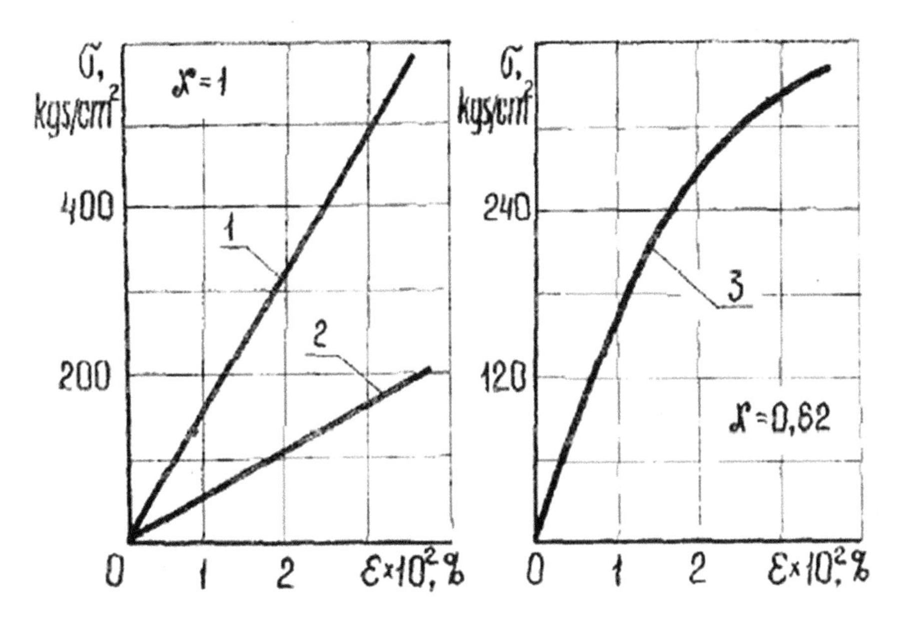

Fig. 4.3 Deformation diagram of yttrium oxide-based materials *1–3*—indexes of materials [4]. With kind permission of Elsevier

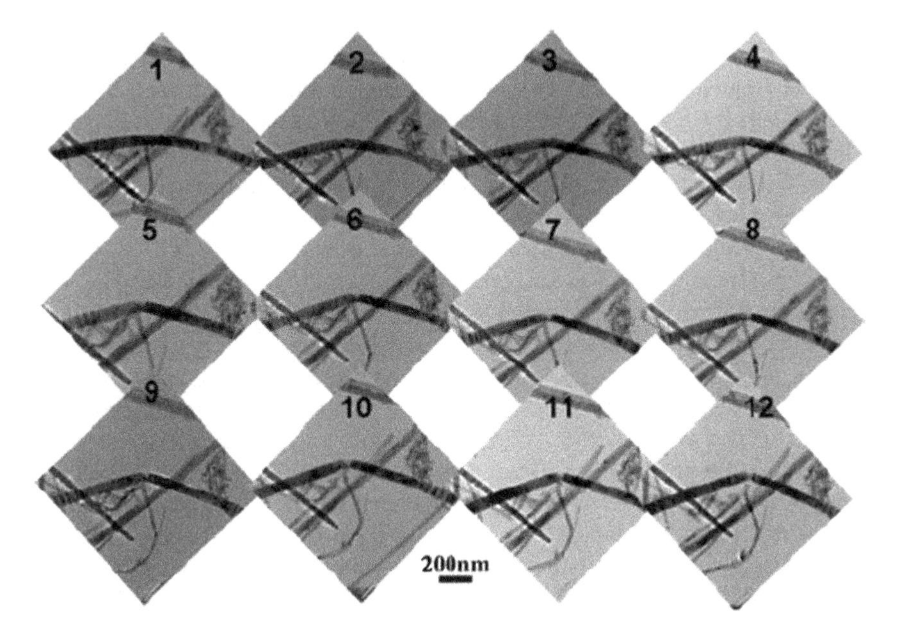

Fig. 4.4 In situ TEM observation of the plastic deformation features of a SiC NW under large angle bending. The plastic deformation was triggered from the second image. The time interval between image 1 and 2 was about 30 min, and the entire procedure lasted about 3 h. Reprinted with permission from Han et al. [5]. With kind permission of the American Chemical Society

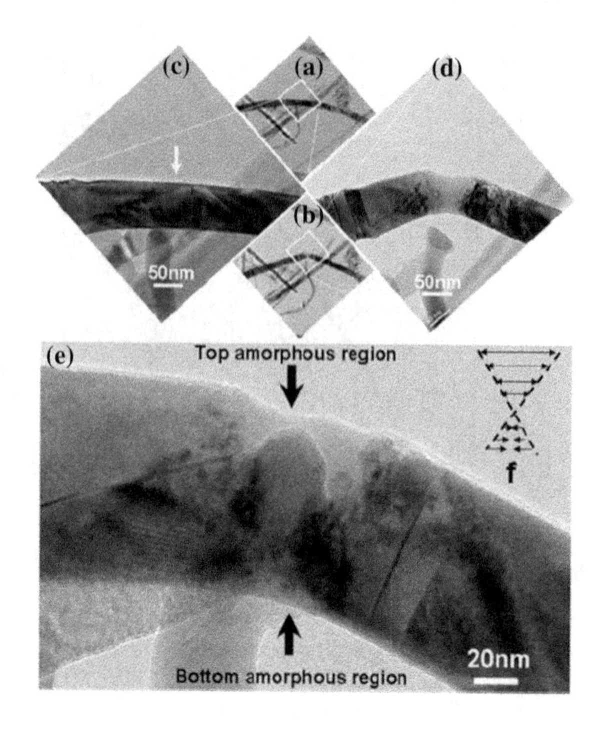

Fig. 4.5 SiC NW morphologies of the elastic bend (**a**, **c**) and the final plastic bend of (**b**) and (**d**). The location indicated by the *white arrow* in (**c**) is the triggering point of plastic deformation. **e** is an enlarged TEM bright field image. The *black arrows* in (**e**) show the *top* and *bottom* deformation-induced amorphous zones. Reprinted with permission from Han et al. [5]. With kind permission of the American Chemical Society

shows a series of images being deformed by the force created by the TEM grid under the irradiation/heating of the electron beam. Plastic deformation is observed at the bending region. Many specimens tested showed large elastic strain (up to 2 %) and four of them revealed plastic deformation with large strain. Figure 4.5 is a more detailed representation of the observed bend deformation in Fig. 4.4. A deformation-induced amorphous zone was observed in these tests. Parts a and c of Fig. 4.5 present examples of the initial elastic deformation in SiC NW, while parts b and d display the final plastically-deformed images of the single bent SiC NW, respectively.

Another example of plastic behavior in ceramics appears in Fig. 4.6. Here, the material is carbide, more specifically Ti_3SiC_2 polycrystalline ceramic. A typical stress–strain curve for samples tested in the z direction is shown in Fig. 4.6. These samples yielded around 200 ± 10 MPa and deformed plastically thereafter, until the test was interrupted. Note that it is possible to deform these samples by compression to strains that exceed 50 pct. As may be seen, when compressed along the x direction, a maximum yield stress is observed, which is followed by a region of strain softening. Two nominally identical samples were tested: one sample yielded at 230 MPa and the second at 290 MPa. In this and other 312- and 211-phase carbides, it was observed in the case of large-grained, oriented polycrystalline samples loaded under compression at RT, plastic deformation might be caused (a) by slip, when the basal planes are oriented to allow for slip deformation or (b) when slip planes are parallel to the applied load making glide impossible.

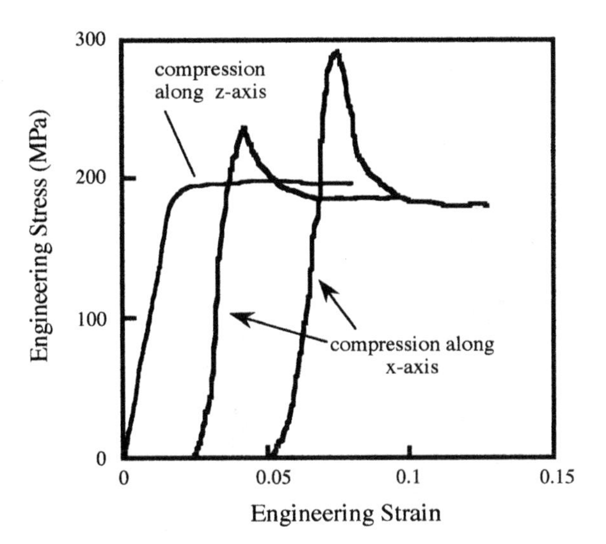

Fig. 4.6 Effect of Ti_3SiC_2 grain orientation on room-temperature engineering stress–strain curves compressed in the *z* directions and the x direction, i.e., parallel to the basal planes. The curves are shifted by 0.025 strain for clarity's sake [14]. With kind permission of Springer and Professor Barsoum

In (a), the minimum critical resolved shear stress [henceforth: CRSS] is 36 MPa. In (b), deformation occurs by a combination of delamination and kink-band formation in individual grains and also by shear-band formation. The multiple modes of deformation allow for plastic behavior in any arbitrary orientation of the compressive load. Notice that the 312- and 211-phases are layered hexagonal carbides and nitrides, having the general formula: $M_{n+1}AX_n$, (MAX), where n = 1 to 3, M is an early transition metal, A is an A-group element (mostly IIIA and IVA, or groups 13 and 14) and X is either *carbon* and/or *nitrogen.*

Some aspects of the dislocation configurations in Ti_3SiC_2 polycrystalline ceramics are indicated in Fig. 4.7.

The arrangement of these dislocations is parallel and they are positioned in different basal planes, one under another, so that the entire arrangement is normal to the basal planes and constitutes a low-angle boundary (a). Contrast analysis (b and c) reveals that the wall is composed of edge and mixed dislocations. For information on the various dislocation configurations observed during RT deformation, refer to the work of Barsoum et al. [14].

4.3.2 Plastic Deformation at Elevated Temperatures

Deformation at elevated temperatures is the commonly observed case in ceramics. Once again using the example of polycrystalline ceramic MgO, the following stress–strain curves are illustrated (Fig. 4.8). One of the possible differences in these stress–strain curves reflects the difference in grain-size; the different grain size before deformation is shown in Fig. 4.9. The composition and porosity were also different in the otherwise nominally pure and dense specimens, as seen in Table 4.1.

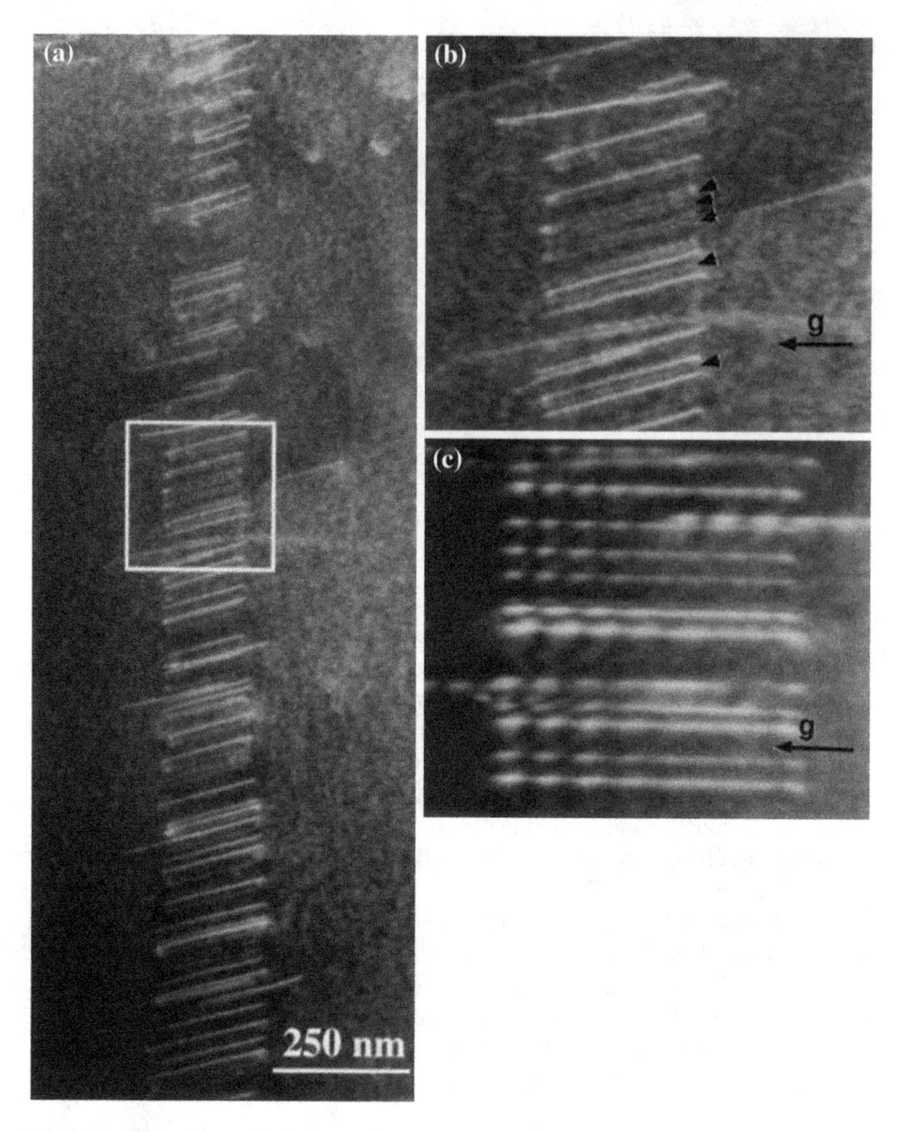

Fig. 4.7 **a** Dislocation wall; dislocations are parallel and positioned in different basal planes one under another. **b** Same area as white square in (**a**), but at higher magnification (imaged in g $\langle 33\bar{6}0\rangle$). **c** Weak beam image of the same area as in (**b**), but tilted and imaged in g of $\langle 3\bar{3}00\rangle$. Dislocations that become invisible in (**c**) are perfect edge dislocation; those that remain visible are mixed dislocations [14]. With kind permission of Springer and Professor Barsoum

The slope of the stress–strain curves can be determined by the

$$\frac{d\sigma}{d\varepsilon} = \frac{d\sigma}{dt}\left(\frac{d\varepsilon}{dt}\right)^{-1}. \tag{4.2}$$

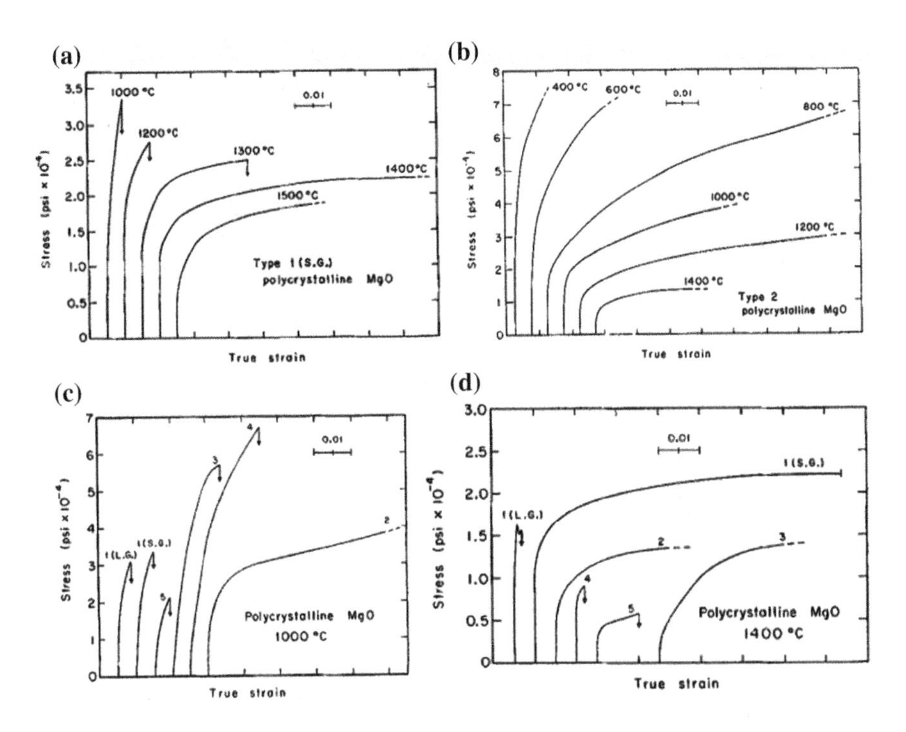

Fig. 4.8 Stress–strain curves for MgO specimens loaded at 20 psi/sec. **a** Type 1 (S.G.). **b** Type 2 MgO; **c** and **d** various types of MgO at 1000 and 1400 °C, respectively [16]. With kind permission of John Wiley and Sons

This equation is related to the strain rate at small strains and the symbols have the usual meaning, namely stress (σ), strain (ε) and time (t). The stress–strain curves of the polycrystalline MgO are indicated as a function of temperature. Above 1200 °C, deformation occurred by grain-boundary shearing accompanied, in some cases, by slip, as seen in Fig. 4.10.

Below 800 °C, specimens fractured primarily due to the grain boundary parting with little permanent strain. Figure 4.11 is an illustration of type 2 MgO deformed at 400 °C. Between 800 and 1200 °C, one type was deformed plastically by slip; the other four types were brittle. The observed behavior is related to the presence of mobile dislocations, resistance to dislocation motion and to the strength of the grain boundaries. The polycrystalline MgO curves at 1000 °C (Fig. 4.8c) are similar in shape, indicating brittleness (types 1, 3, 4 and 5; see Table 4.1), except for the type-2 specimen, whose deformation is characterized by slip. This observation is supported by metallographic examination. At 1400 °C, grain-boundary shearing or separation was evident in all the types and there were greater variations in the shapes of the stress–strain curves and yield stresses (types 1 and 4, while type 5 showed very little strain).

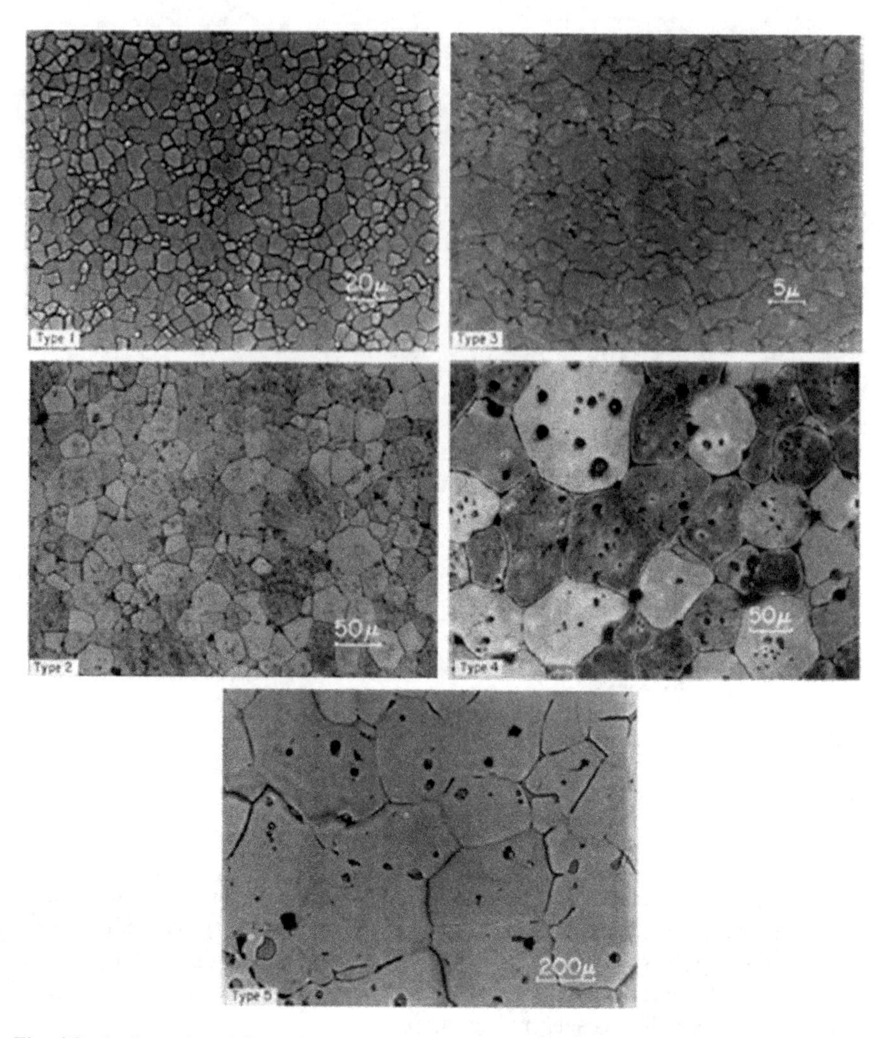

Fig. 4.9 Surface of undeformed specimen polished and etched [16]. With kind permission of John Wiley and Sons

As mentioned above, these different stress–strain curves are probably the results of differences in grain size, porosity and other imperfections. However, to obtain good ceramic, such as MgO, which is known to exhibit appreciable ductility at high temperatures and at RT, it is necessary to reduce the propensity for defect formation in order to eliminate inter-granular and grain fracture, the kind shown in Figs. 4.12 and 4.13. Slip resulting from dislocation motion, as shown in Fig. 4.14, is a general feature of properly prepared MgO and appreciable ductility can be achieved. For example, total axial strains up to 35 % at pressures up to 8 GPa were reported by Uchida et al. [47], indicating that MgO samples yielded at low total strains (<1 %) and the yield strength (~5 GPa) was insensitive to pressure. Beyond the

Table 4.1 Spectroscopic analysis of various types of polycrystalline MgO* [18] (with kind permission of John Wiley and Sons)

Constituents	Types of polycrystalline MgO				
	1 (%)	2 (%)	3 (%)	4 (%)	5 (%)
Mg	Principal constituent of all 5 types				
Fe			0.025	0.03	0.1
Ba				0.004	
Si	0.01	0.015	0.03	0.4	1.0
Mn			0.003	0.003	0.006
Al	<0.005	<0.005	<0.005	<0.005	<0.005
Ca	0.003	0.02	0.25	0.2	0.25
Cu	0.0008	0.0008	0.0008		0.0005
Ti	0.006		0.003		
Li	0.0075†				
Ni			0.002	0.005	0.003
Cr	0.004		0.006	0.001	<0.001
B					0.15
	0.0288	0.0408	0.3248	0.648	1.5155

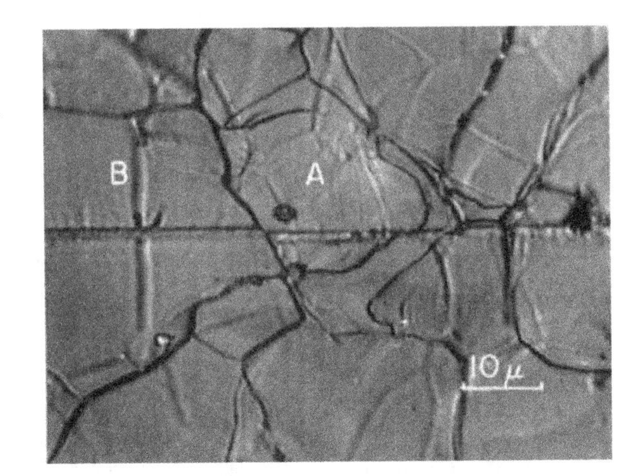

Fig. 4.10 Grain boundary shearing in type 1 (S.G.) MgO strained 3 % at 1400 °C [16]. With kind permission of John Wiley and Sons

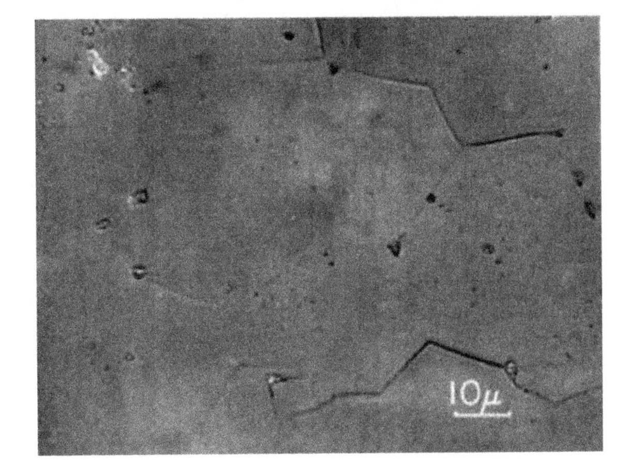

Fig. 4.11 Intergranular fracture in type 2 MgO deformed at 400 °C [16]. With kind permission of John Wiley and Sons

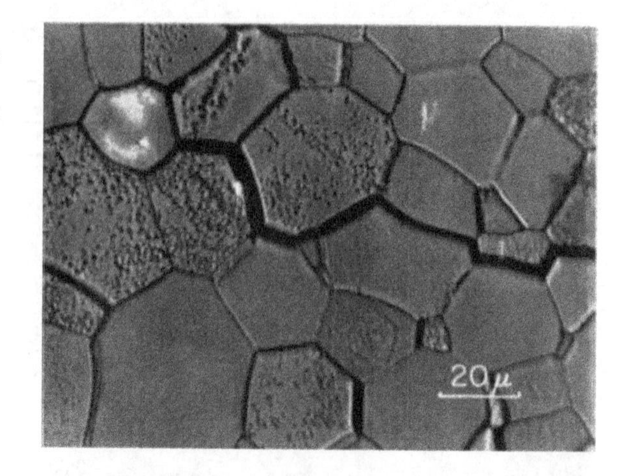

Fig. 4.12 Intergranular fracture in type 1 (L.G.) MgO deformed at 1300 °C [16]. With kind permission of John Wiley and Sons

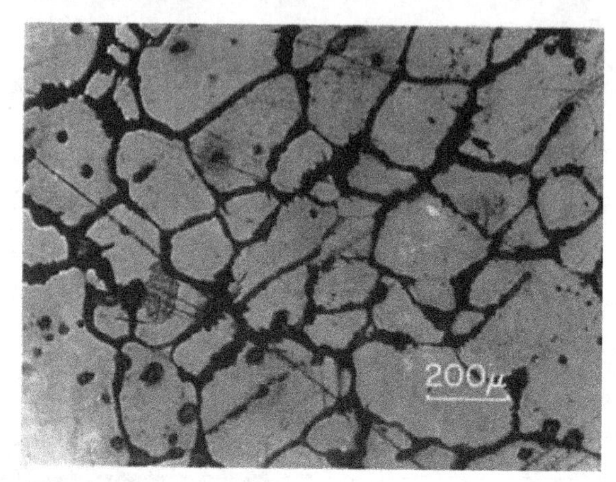

Fig. 4.13 Type 5 MgO deformed at 1400 °C showing fracture of boundary phase [16]. With kind permission of John Wiley and Sons

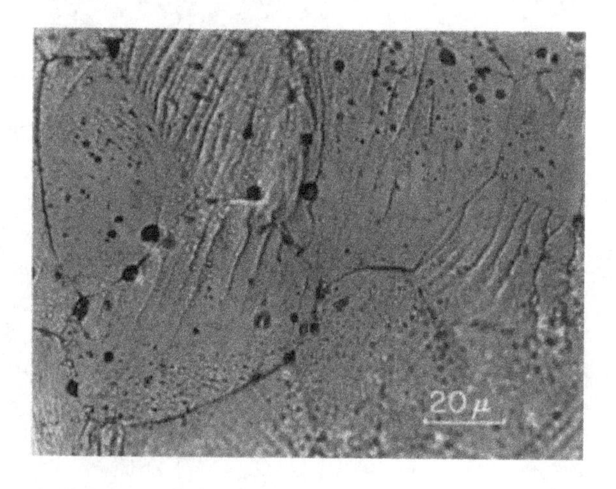

Fig. 4.14 Wavy slip in type 2 MgO strained several percent at 1300 °C [16]. With kind permission of John Wiley and Sons

yield point, MgO showed strong strain-hardening behavior, characterized by an initial rapid increase in flow stress and followed by a linear stage with a constant hardening parameter.

Considerable work was done on MgO, both single crystals and polycrystalline materials alike, toward an understanding of deformation behavior in ceramics. Furthermore, when polycrystalline MgO is ductile, its strain hardening is comparable to that of $\langle 111 \rangle$ oriented single crystals [16].

In summary, ductility at high-temperature deformation is exhibited by most of ceramics, but is best studied in MgO, which, unlike most ceramics, is not brittle at ambient temperatures.

4.4 The Critical Resolved Shear Stress (CRSS) in Ceramics

'Resolved shear stress' is a concept related to plastic deformation and associated with shear stress. Similarly, it is reasonable to talk about ceramics exhibiting ductility as a consequence of acting shear stress. To do so, one must consider stress and strain tensors. The stress tensor in Sects. 1.22 and 1.23 (Eqs. 1.13–1.13b) is rewritten here as:

$$\sigma_{ij} = \begin{matrix} \sigma_x & \tau_{xy} & \tau_{xz} \\ \tau_{yx} & \sigma_y & \tau_{yz} \\ \tau_{zx} & \tau_{zy} & \sigma_z \end{matrix} \tag{1.13b}$$

and the strain tensor from Sect. 1.28 (Eqs. 1.59–1.60) is reproduced as:

$$e_{ij} = \begin{matrix} e_{xx} & \gamma_{xy} & \gamma_{xz} \\ \gamma_{yx} & e_{yy} & \gamma_{yz} \\ \gamma_{zx} & \gamma_{zy} & e_{zz} \end{matrix} \tag{1.60}$$

The off diagonals refer to shear stresses and shear strains, while the diagonals represent normal stresses and strains. When stress exceeds a critical value, materials irreversibly change their dimensions by the process of plastic deformation after the load has been removed from the tested specimen. The stress applied may be tensile or compressive, as indicated by the tensorial expressions. Critical stress is necessary for plastic deformation to occur. Shear stress, acting in a specific plane, is responsible for plastic deformation. Plastic deformation is usually associated with a slip mechanism, but that is not the sole mechanism which may be involved. Twinning may occur simultaneously, if no time element, temperature factor or cyclic stress is involved. However, before considering slip and the systems in which it occurs, it is worth exploring the details regarding the applied stress which initiates slip deformation.

Slip is aided by dislocation movement. Dislocations (discussed in Chap. 3) may have a dual effect on the strength of materials. In the presence of a few dislocations, materials deform at a low-yield stress, τ_0, whereas, when the concentration

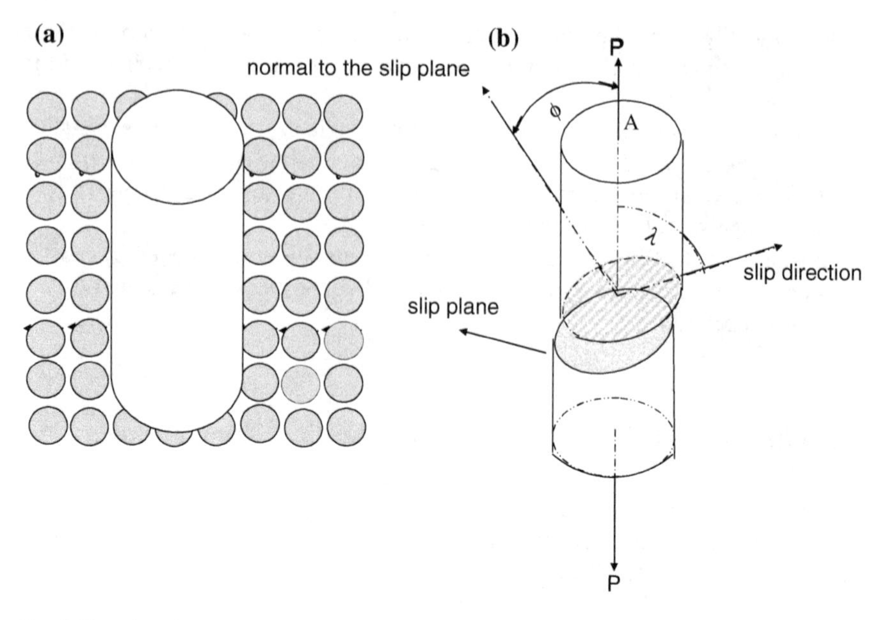

Fig. 4.15 **a** A cube showing the bar cut out of it; **b** the load applied to the bar and the resolved shear stress [7]

of dislocations increases, materials become strong again. In the absence of dislocation (which may be considered a reference point), the resistance to any change in the dimensions of a material is extremely large (close to the theoretical strength). Whiskers tested under tension approach the theoretical strength of materials. Frenkel [20] calculated a value for the yield stress, τ_0:

$$\tau_0 \sim G/6 \tag{4.3}$$

In crystals, slip starts on specific planes and in definite crystallographic directions at some critical value, known as the 'critical resolved shear stress' (CRSS, discussed below). In the following, the minimum stress required for deformation by initiating slip is derived with the aid of Fig. 4.15.

Figure 4.15b is a bar cut from a cubic single crystal, as shown in (a). A tensile load, P, is applied to the bar (as illustrated), resulting in the stress:

$$\sigma = \frac{P}{A} \tag{1.4a}$$

Slip occurs on the slip plane indicated by the dashed area and may be expressed as:

$$\text{slip area} = \frac{A}{\cos \phi} \tag{4.4}$$

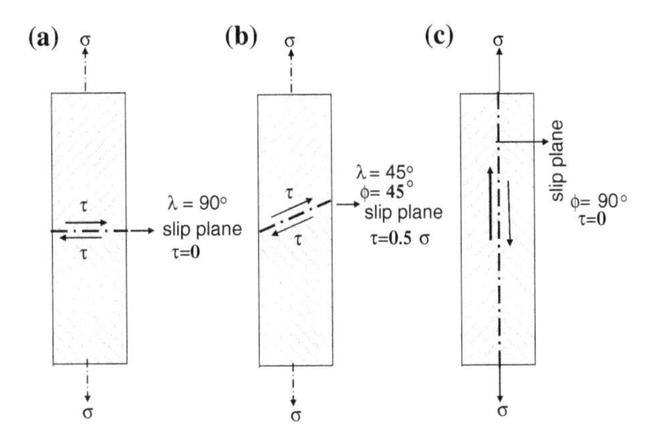

Fig. 4.16 The shear stress for some angles of Eq. (4.6): **a** $\lambda = 90°$, when the tensile axis is parallel to the slip plane, $\tau = 0$; **b** $\lambda = 45°$, $\phi = 45°$, $\tau = 0.5\sigma$, maximum shear stress; and **c** $\phi = 90°$, when the tensile axis is normal to the slip plane, $\tau = 0$ [7]

Figure 4.15b shows that ϕ is the angle between the normal to the slip planes and the tensile axis.

$$\text{The resolved force in the slip direction } = \mathrm{P}\cos\lambda \tag{4.5}$$

In terms of stress, the resolved shear stress in the slip plane in the slip direction obtained from Eqs. (1.4a) and (4.4) is:

$$\tau = \sigma \cos\phi \cos\lambda \tag{4.6}$$

One immediately observes, from Eq. (4.6), that the resolved shear stress is zero, either when $\phi = 90°$ or when $\lambda = 90°$. In the first case, the tensile axis is normal to the slip plane, whereas, in the second case, the tensile axis is parallel to the slip plane. Deformation by slip is not expected when the tensile axis is parallel to the slip direction, because the shear stress is zero. The component of stress, normal to the slip plane, does not influence slip. However, in ceramics, the effect of normal stress on the critical shear stress is not necessarily negligible; but, in the following this is not considered. Maximum shear stress is when $\phi = \lambda = 45°$ (Fig. 4.16).

Equation (4.6) gives the 'resolved shear stress'. The product in the equation is known as the 'Schmid factor' and determines whether the orientation is favorable for slip. The conditions for slip are given by Schmid's Law and the value of Eq. (4.6), often represented in the literature by τ_r, indicating the onset of plastic deformation and called 'critical resolved shear stress'. CRSS is a structure-sensitive property, since it is very dependent on impurities and the way the crystal was grown and handled.

On application of a tensile load, the yield stress, σ_0 varies widely in ceramics, but considering Eq. (4.6) in terms of yield, the yield stress value is minimal when $\phi = \lambda = 45°$, giving a value of 0.5 for the product of the cosines (Schmid's factor). In this case, the maximum value attained for τ is $0.5\sigma_0$. Beyond 0.5, σ_0 increases again.

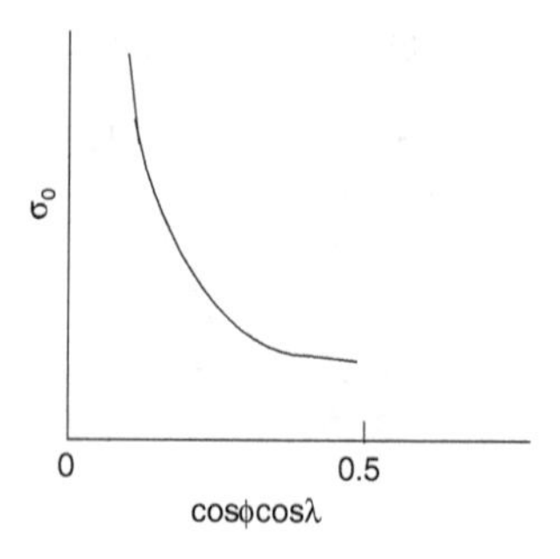

Fig. 4.17 The schematic variation of yield stress with orientation, schematic

The early work of Schmid and Boas [9], who studied HCP structures such as Zn, showed that, despite the wide variation in σ_0 with orientation, the CRSS is a constant of the material:

$$\sigma_0 = \frac{\tau_0}{\cos\varphi\cos\lambda} \tag{4.6a}$$

A schematic illustration of the change in $\sigma_{0,}$ according to Eq. (4.6a) is shown in Fig. 4.17.

Schmid's Law has been experimentally proven for a large number of single crystals, but tests to substantiate this rule are usually more conveniently done on HCP crystals, due to the low number of slip systems; in FCC and BCC, where more slip systems are possible, the active slip plane is usually the one at which CRSS is first reached.

In summary, CRSS is the necessary component of shear stress, resolved in the direction of slip, which initiates slip in the crystal. It is a constant for a given crystal.

Complications may arise in polycrystalline materials, where each grain may be considered as a single crystal having a different orientation than that of its neighbor. Taylor assumed that the strain on the grains (crystallites) and deformation occur with the same value of CRSS on all the available slip systems. According to Schmid's calculations for the yielding behavior of polycrystalline metals, the different initial CRSS values, due to the geometry of the slip systems, have no effect on the macroscopic yield surface of a randomly oriented polycrystalline material.

Among various influential factors, the CRSS generally varies with temperature, as shown in Fig. 4.18 for $TiC_{0.97}$. The CRSS for slip, $\sigma_{r,}$ on the $\{111\}\ \langle 1\bar{1}0\rangle$ slip system in $TiC_{0.97}$ decreases with temperature, T, in the range 800–1280 °C, as shown in Fig. 4.18. The CRSS obeys a relationship shown in (4.7) as:

$$\sigma_T = \sigma_0 \exp\frac{B}{T} \tag{4.7}$$

Equation (4.7), expressed in logarithmic form, is shown in Fig. 4.19, indicating the linear relation of σ_r with 1/T.

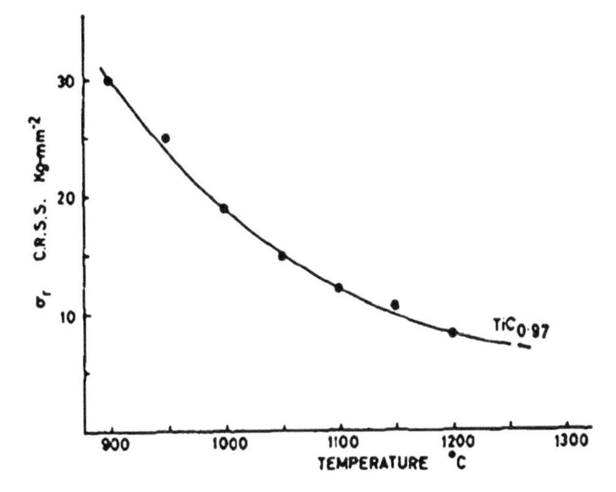

Fig. 4.18 The critical resolved shear stress σ_r as a function of temperature T in $TiC_{0.97}$ [27]. With kind permission of American Institute of Physics

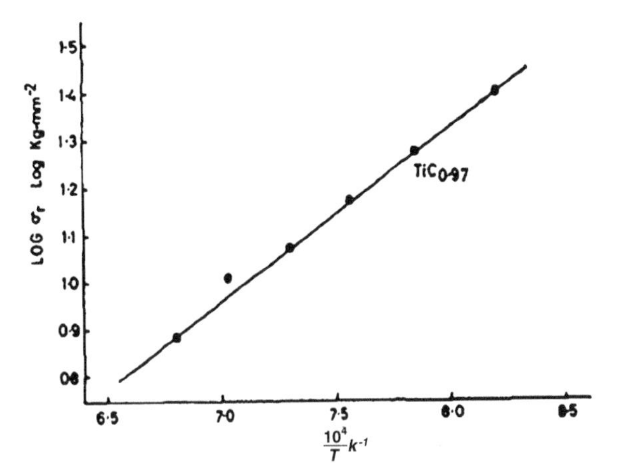

Fig. 4.19 The critical resolved shear stress as a function of temperature in $TiC_{0.97}$ [27]. With kind permission of American Institute of Physics

Titanium carbide has had the rock salt structure assigned to it, e.g. LiF, NaCl, MgO, etc. with slip system assumed to be of {110}. However, the primary slip system turns out to be the one indicated above, namely {111}, rather than {110}, which is the characteristic slip system in FCC structures [27].

Note that no yield-stress drop has been observed in these carbide ceramics (see Fig. 4.20). The curves are parabolic along the entire extension.

In Chap. 2 Sect. 2.2a, the CRSS was shown for Al_2O_3 at $4.2 \times 10^{-7}s^{-1}$ above Tc, indicating basal and prismatic slip. The Schmidt factors are 0.3916 for (0001) slip for specimens with A and B orientation and 0.4330 for $\langle 1\bar{1}00 \rangle$ slip for specimens with C orientation. A and B show basal slip and C has prismatic slip.

In addition, a relatively large number of shear deformation data for ceramics and metals is listed in Table 4.2, indicating the common slip systems. The results

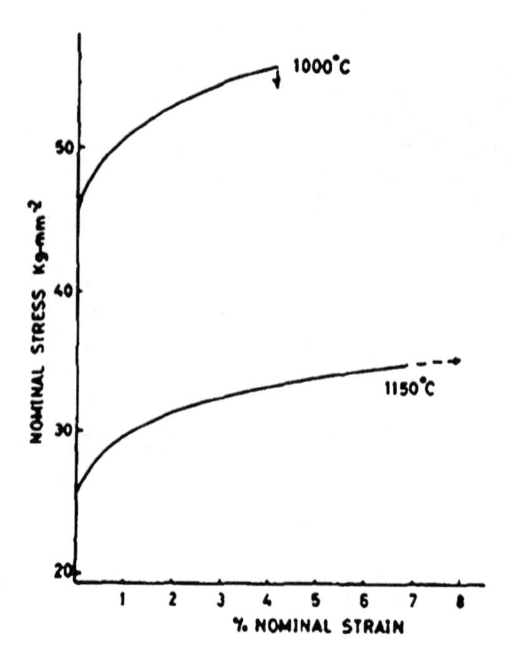

Fig. 4.20 Typical stress–strain curves for $TiC_{0.97}$ compressed in the cube-orientation at temperatures above 900 °C [27]. With kind permission of American Institute of Physics

of tests on 22 materials are presented in Table 4.2 as ideal shear strain. Equation (4.8) was used to renormalize Frenkel's model, as indicated by the heavy dark line in Fig. 4.21. In this relation, G is the shear modulus, s is the shear strain, where their indexes, r and u, refer to relaxed and unrelaxed parameters, respectively. Here, shear stress is indicated for relaxed and unrelaxed parameters by σ_m, rather than by the usual symbol, τ:

$$\sigma_m = \frac{2Gs_m}{\pi} \sin \frac{\pi s}{2s_m}, \quad 0 < s < s_m, \tag{4.8}$$

The authors [6] suggest the term 'shearability', s_m, for the maximum shear strain that a homogeneous crystal can withstand. It is defined by $s_m \equiv \arg\max \sigma(s)$, where $\sigma(s)$, is the resolved shear stress and s is the engineering shear strain in a specified slip system. The relaxed shear stress, σ_m^r in Table 4.2 is normalized by $G_{r.}$ In this table, experimental and calculated values of the relaxed shear vales of G_r are given. For details on these calculations, refer to the work of Ogata et al. [6].

It is, therefore, important to bear in mind the CRSS of plastic deformation in ceramics, because several ceramics show not only high-temperature plasticity, but also low-temperature plasticity.

Table 4.2 Shear moduli G_r (relaxed), G_u (unrelaxed), ideal shear strains s^r_M (relaxed), s^u_M (unrelaxed) and stresses σ^r_m (relaxed), σ^u_m (unrelaxed) in common slip systems [4] (with kind permission of Dr. Yip)

Material	Slip System	Expt.	[GPa][a]	G/B	[GPa][a]	Calc.	[GPa]		Relaxed			UnRelaxed[b]	
		G_r	G_M	G_r/B	G_M/B	G_r	G_M	s^r_M	σ^r_m[Gpa]	σ^r_m/G_r	s^u_m	σ^u_m[GPa]	σ^u_m G_u
C	$\{111\}\langle 110\rangle$	506.8	511.1	1.14	1.15	514.1	519.2	0.325	113.32	0.220	0.374	146.03	0.281
S_i	$\{111\}\langle 110\rangle$	57.9	60.4	0.585	0.610	55.2	58.2	0.275	9.62	0.174	0.262	11.13	0.191
$\beta-SiC$	$\{111\}\langle 110\rangle$	149.7	168.0	0.666	0.748	158.2	173.4	0.350	31.74	0.201	0.348	43.12	0.249
$\alpha-Si_3N_4$	$\{11\overline{2}0\}\langle 0001\rangle$					127.3	128.4	0.259	23.72	0.186	0.295	26.48	0.206
$\beta-Si_3N_4$	$\{10\overline{1}0\}\langle 0001\rangle$	108.0	108.0	0.417	0.417	101.0	102.0	0.232	19.00	0.188	0.244	21.00	0.206
NaCL	$\{110\}\langle 1\overline{1}0\rangle$	22.8	22.8	0.855	0.855	29.4	29.4	0.221	3.69	0.126	0.658	25.55	0.869
KBr	$\{110\}\langle 1\overline{1}0\rangle$	19.1	19.1	1.09	1.09	23.2	23.2	0.211	2.62	0.113	0.610	15.41	0.666
MgO	$\{110\}\langle 1\overline{1}0\rangle$	106.5	106.5	0.649	0.649	109.5	109.5	0.270	17.09	0.156	0.629	74.34	0.679
CaO	$\{110\}\langle 1\overline{1}0\rangle$	71.5	71.5	0.624	0.624	101.3	101.3	0.277	16.18	0.160	0.664	72.35	0.714
Mo	$\{110\}\langle\overline{1}\overline{1}1\rangle$	138.7	142.8	0.525	0.541	126.5	134.5	0.190	15.18	0.120	0.192	16.52	0.123
Mo	$\{211\}\langle\overline{1}\overline{1}1\rangle$	138.7	142.8	0.525	0.541	126.8	134.1	0.175	14.84	0.117	0.177	15.99	0.119
Mo	$\{321\}\langle\overline{1}\overline{1}1\rangle$	138.7	142.8	0.525	0.541	126.8	134.2	0.176	14.87	0.117	0.175	15.93	0.119
W	$\{110\}\langle\overline{1}\overline{1}1\rangle$	164.0	164.0	0.521	0.521	153.7	155.3	0.179	17.52	0.114	0.196	17.63	0.113
W	$\{211\}\langle\overline{1}\overline{1}1\rangle$	164.0	164.0	0.521	0.521	154.0	155.8	0.176	17.37	0.113	0.175	17.28	0.111
W	$\{321\}\langle\overline{1}\overline{1}1\rangle$	164.0	164.0	0.521	0.521	153.9	155.7	0.176	17.33	0.113	0.175	17.27	0.111
Fe[c]	$\{110\}\langle\overline{1}\overline{1}1\rangle$	64.8	75.7	0.375	0.438	76.6	80.6	0.178	8.14	0.106	0.234	11.43	0.142
Fe[c]	$\{211\}\langle\overline{1}\overline{1}1\rangle$	64.8	75.7	0.375	0.438	75.6	79.9	0.184	7.51	0.099	0.236	9.95	0.124
Fe[c]	$\{321\}\langle\overline{1}\overline{1}1\rangle$	64.8	75.7	0.375	0.438	75.7	80.0	0.181	7.57	0.100	0.197	9.43	0.118
Ti[c]	$\{1\overline{1}00\}\langle 11\overline{2}0\rangle$	44.6	44.6	0.406	0.406	47.6	47.8	0.099	2.82	0.059	0.144	4.92	0.103
Mg	$\{0001\}\langle 11\overline{2}\overline{0}\rangle$	18.4	18.4	0.499	0.499	19.2	19.2	0.152	1.84	0.096	0.157	2.04	0.106
Zn	$\{0001\}\langle 11\overline{2}\overline{0}\rangle$	46.0	46.0	0.708	0.708	36.6	36.6	0.132	2.12	0.058	0.136	2.33	0.064
TiAl	$\{111\}\langle 11\overline{2}\rangle$	58.5	61.6	0.524	0.552	50.0	56.4	0.218	5.54	0.111	0.217	6.25	0.111

(continued)

Table 4.2 (continued)

Material	Slip System	Expt.	[GPa][a]	G/B	[GPa][a]	Calc.	[GPa]		Relaxed			UnRelaxed[b]	
		G_r	G_M	G_r/B	G_M/B	G_r	G_M	s^r_M	σ^r_m [Gpa]	σ^r_m/G_r	s^u_m	σ^u_m [GPa]	σ^u_m G_u
Ti_3Al	$\{1\bar{1}00\}\langle 11\bar{2}\bar{0}\rangle$	47.1	47.1	0.416	0.416	50.0	50.8	0.127	5.51	0.110	0.139	5.79	0.114
Al	$\{111\}\langle 11\bar{2}\rangle$	27.4	27.6	0.345	0.348	25.4	25.4	0.200	2.84	0.110	0.210	3.73	0.147
Ni[c]	$\{111\}\langle 11\bar{2}\rangle$	68.8	81.0	0.366	0.431	60.1	79.6	0.140	5.05	0.084	0.160	6.29	0.079
Ni[d]	$\{111\}\langle 11\bar{2}\rangle$					48.8	60.5	0.169	3.17	0.065	0.162	4.70	0.078
Ag	$\{111\}\langle 11\bar{2}\rangle$	22.4	28.7	0.206	0.264	25.0	32.3	0.145	1.65	0.066	0.156	2.57	0.079
Au	$\{111\}\langle 11\bar{2}\rangle$	20.9	26.1	0.116	0.145	17.9	22.9	0.105	0.85	0.048	0.142	1.42	0.062
Cu	$\{111\}\langle 11\bar{2}\rangle$	33.3	44.4	0.235	0.313	31.0	40.9	0.137	2.16	0.070	0.157	3.45	0.084

[a] Computed analytically from expt. elastic constants of Table 4.1
[b] subsidiary stress components are unrelaxed, but internal degrees of freedom are relaxed
[c] Feromagnetic
[d] Paramagnetic
[e] *p*-Valence

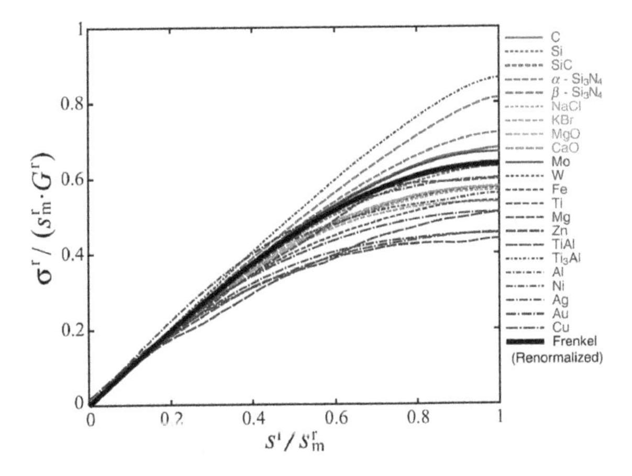

Fig. 4.21 (Color) Relaxed shear stress–strain curves of 22 materials, rescaled such that all have unit slope initially and reach maximum at 1. The renormalized Frenkel model Eq. (4.8) is shown (*in heavy black line*) for comparison [6]. With kind permission of Dr. Yip

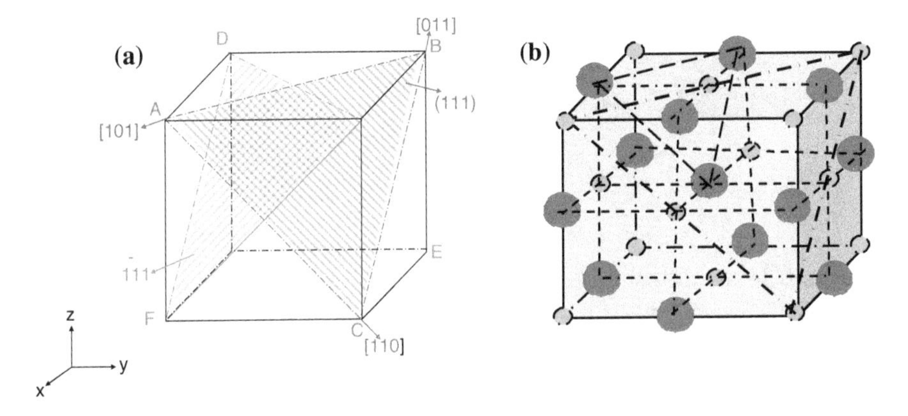

Fig. 4.22 **a** An FCC structure showing two of the four {111} planes and the slip directions of the (111) plane shaded in *red color*; **b** an NaCl structure, also an FCC structure [7]

4.5 Slip in Ceramics

In metals, slip and the common slip systems are usually discussed in terms of the various structures of single crystals: BCC, FCC and HCP. There is no basic difference between the slip systems in ceramics (or other crystalline materials) and metals, since all the aforementioned structures appear in both materials. Thus, NaCl, MgO, CaO, KBr, etc. have an FCC structure with a primary $\{110\}\ \langle 1\bar{1}0\rangle$ slip system. β-Si_3N_4 is HCP with a $\langle 11\bar{2}0\rangle$ (0001) slip system. α-Si_3N_4 with (0001) $\langle 1 1\bar{1} 0\rangle$ and (0001) $\langle 11\bar{2}0\rangle$and β-SiC cubic (zincblende) with a $\{111\}\ \langle 110\rangle$ slip system. However, unlike pure metallic systems (see Fig. 4.22a), ceramic FCC

Fig. 4.23 slip lines on orthogonal planes of **a** unpolished and **b** chemically polished deformed specimens [12]. With kind permission of John Wiley and Sons

structures have two components: a cation and an anion. This is illustrated, for example, in Fig. 4.22b for an MgO ceramic, which has an FCC salt rock structure (NaCl). A (111) plane, heavily outlined in (b), passes through all the sodium atoms (smaller ions), but not through the Cl atoms (larger red ions). The dashed line connects the chlorine atoms. One can see that the unit cell of NaCl contains 8 ions located as follows:

$$4\,\mathrm{Na}+ \text{ at } 0\,0\,0,\ 1/2\,1/2\,0,\ 1/2\,0\,1/2 \text{ and } 0\,1/2\,1/2$$
$$4\,\mathrm{Cl}- \text{ at } 1/2\,1/2\,1/2,\ 0\,0\,1/2,\ 0\,1/2\,0, \text{ and } 1/2\,0\,0.$$

An illustration of slip lines in MgO (having a rocksalt structure is shown in Fig. 4.23. The specimens cleaved along the {100} planes were tested at RT under compression along ⟨100⟩ with an Instron machine using a deformation speed of 0.06 mm/min. The deformation results appear in Fig. 4.24.

In samples annealed after mechanical polishing, the deformation curve has four stages in the plasticity region, showing four-stage work hardening rates (a). After other processing procedures, two stages of work hardening were observed (b).

Molybdenum disilicide ($MoSi_2$), an intermetallic compound, a silicide of molybdenum, is a refractory ceramic primarily used in heating elements. It has

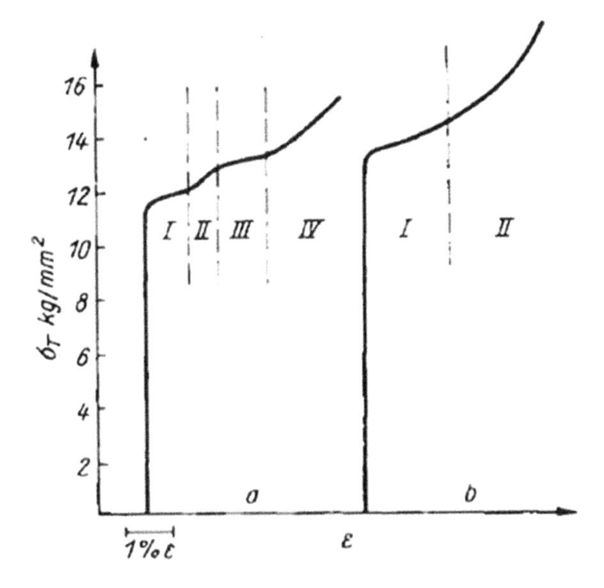

Fig. 4.24 Stress–strain curves for **a** annealed at 1000 °C after mechanical polishing **b** MgO crystals as-cleaved [12]. With kind permission of John Wiley and Sons

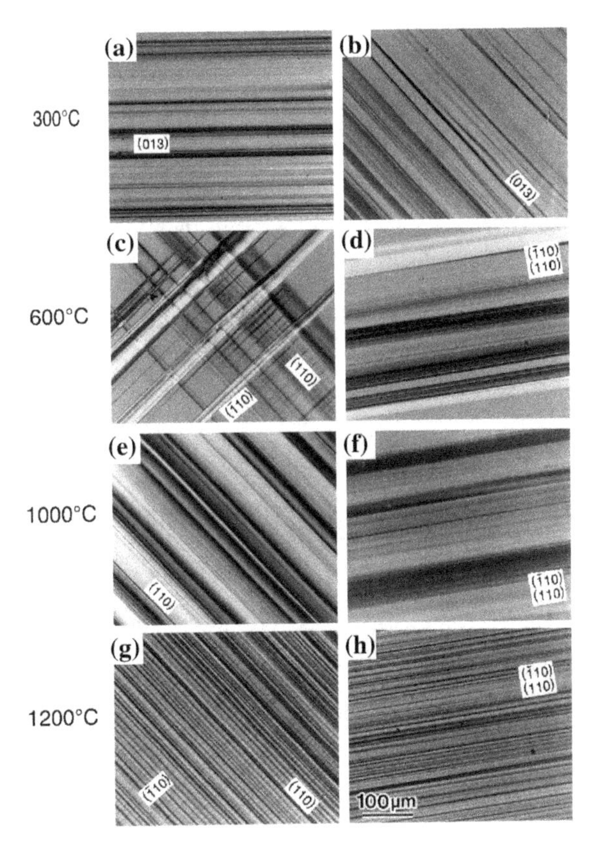

Fig. 4.25 slip lines observed on Nb-bearing $MoSi_2$ single crystals with the [0 15 1] orientation at selected temperatures [29]. With kind permission of Elsevier

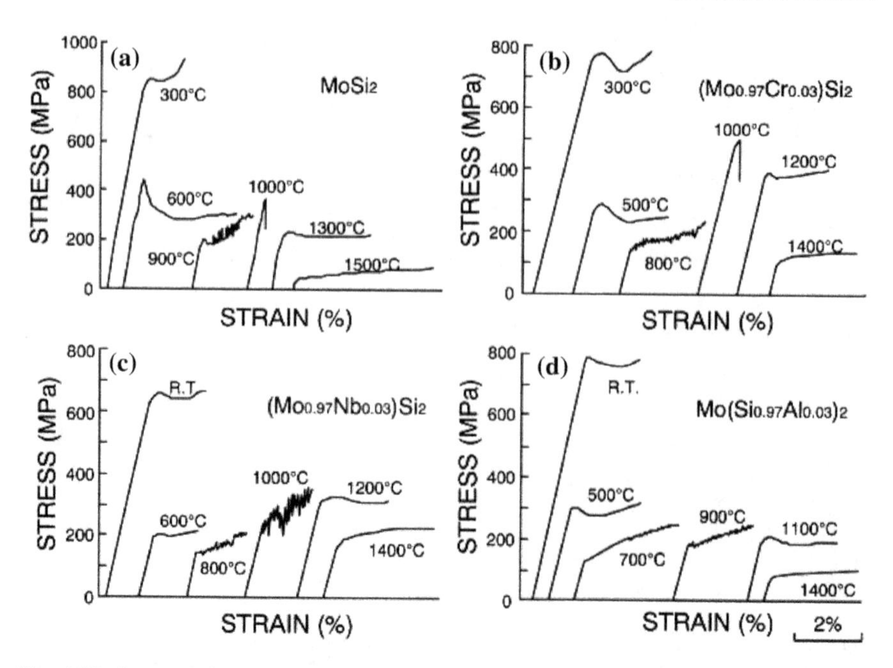

Fig. 4.26 Stress–strain curves of **a** binary, **b** Cr–, **c** Nb– and **d** Al-bearing ternary $MoSi_2$ single crystals with the [0 15 1] orientation at selected temperatures [29]. With kind permission of Elsevier

moderate density, a melting point of 2030 °C and is electrically conductive. Its applications include: the glass industry, ceramic sintering, heat treatment furnaces and semiconductor diffusion furnaces. Slip lines in hexagonal $MoSi_2$ ceramics containing Nb are illustrated in Fig. 4.25. The respective plastic deformation is shown in Fig. 4.26. Note that, in Fig. 4.25, the slip lines at various temperatures refer to $MoSi_2$ single crystals containing 0.03 Nb at the composition indicated in Fig. 4.26c. The deformation curves indicate the effects of additions to the $MoSi_2$ specimens, thus making them basically ternary ceramics. The respective dislocation structures are shown in Fig. 4.27. Further slip lines in $MoSi_2$ containing Re may be seen in Fig. 4.28. The respective deformation curves are indicated in Fig. 4.29. The addition of another constituent to $MoSi_2$ is intended to improve the poor mechanical properties to enable the various aforementioned applications. In Table 4.2, the CRSS values for the listed slip systems are indicated.

Note that the Re and W additions to the $MoSi_2$ form the $C11_b$ tetragonal, whereas the other alloying elements form the C40 hexagonal structures. Thus, Figs. 4.28, 4.29 and 4.30 refer to this tetragonal structure. Five slip systems are reported to operate in $MoSi_2$ [37] slip on $\{110\}\ \langle 111\rangle$ is operative from 500 °C. 1/2 $\langle 111\rangle$ dislocations of this slip system are reported to dissociate into two identical 1/4 $\langle 111\rangle$ partials separated by a stacking fault [30] Table 4.3.

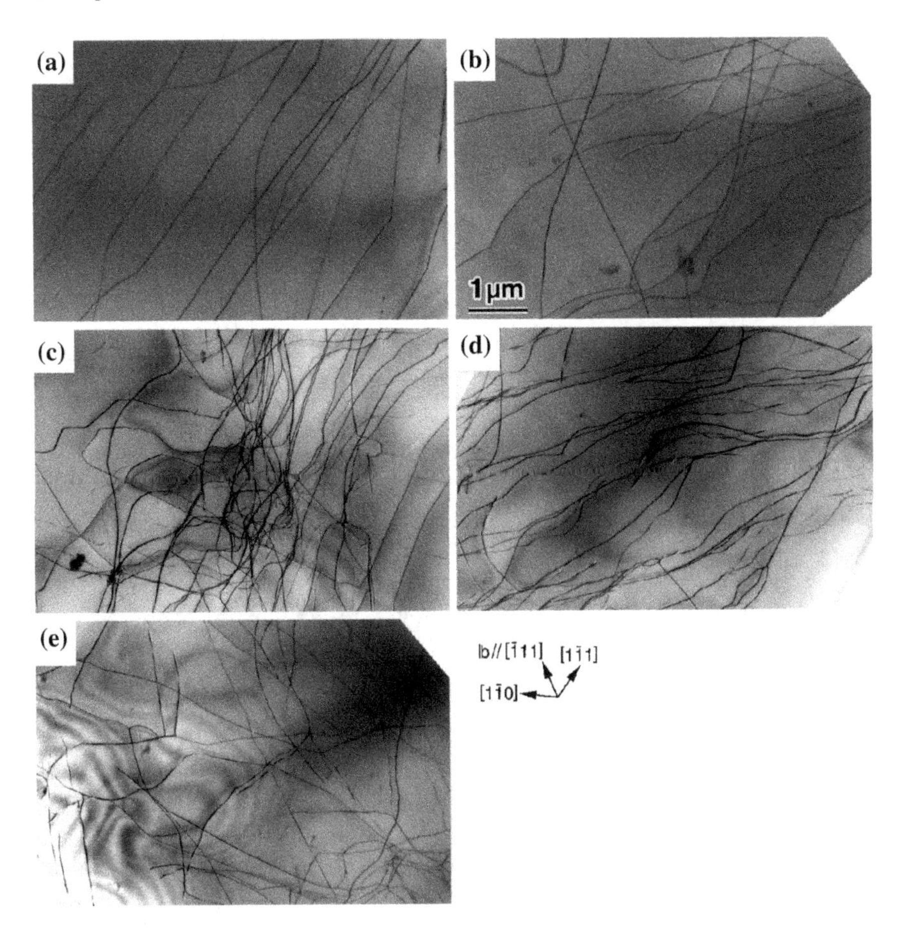

Fig. 4.27 Dislocation structures in **a** binary, **b** V−, **c** Cr−, **d** Nb− and **e** Al-bearing MoSi2 single crystals with the [0 15 1] orientation deformed to 2–3 % plastic strain. Deformation temperatures are 500 °C for (**e**), 600 °C for (**a–c**) and 700 °C for (**d**). All thin foils were cut parallel to (110) slip planes [29]. With kind permission of Elsevier

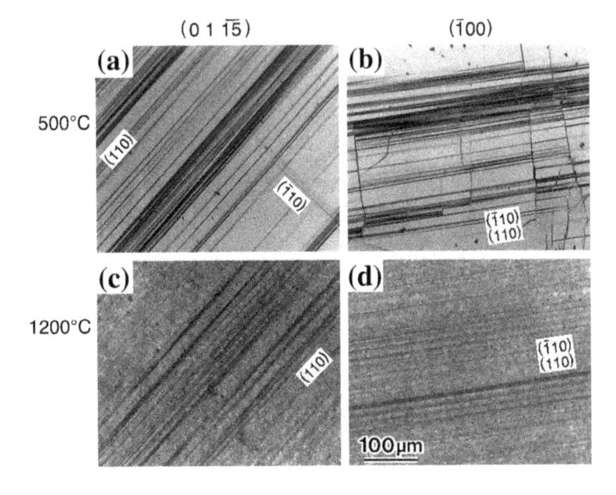

Fig. 4.28 Slip lines observed on Re-bearing $MoSi_2$ single crystals with the [0 15 1] orientation at **a** and **b** 500 and **c** and **d** 1200 °C [29]. With kind permission of Elsevier

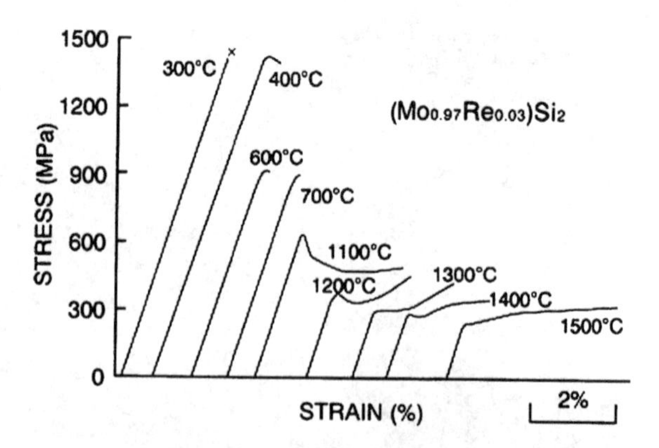

Fig. 4.29 Stress–strain curves of Re-bearing ternary $MoSi_2$ single crystals with the [0 15 1] orientation at selected temperatures [29]. With kind permission of Elsevier

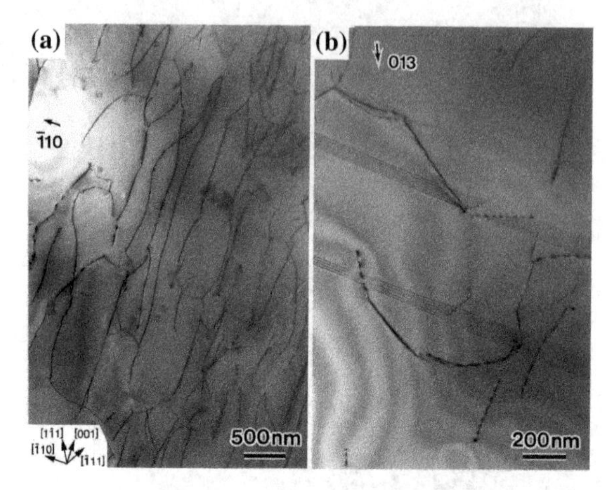

Fig. 4.30 Dislocation structures in a Re-bearing $MoSi_2$ single crystal with the [0 15 1] orientation deformed to 2–3 % plastic strain at 1300 °C. The thin foil was cut parallel to (110) slip planes [29]. With kind permission of Elsevier

There are controversies regarding the slip systems in $MoSi_2$. Mitchel et al. [38] did not find 1/2 (331) dislocations in deformed single crystals and they stated that the presence of such dislocations are either unstable or they occur under special conditions of orientation, temperature and stress state.

4.6 Slip in Polycrystalline Ceramics

In polycrystalline materials, slip planes and directions vary from one crystal (grain) to another; thus, CRSS varies from one crystal to another. The crystal with the largest CRSS yields first. If a grain is oriented unfavorably with respect to the applied-stress direction, its deformation is impeded and vice versa for favorably-oriented grains (Table 4.3).

Grain boundaries are obstacles to slip, since the slip direction of a favorably-oriented crystal may change when it crosses a grain boundary. As a result, the *strength* of polycrystalline materials is higher than that of single-crystal materials. A polycrystalline ceramic can deform plastically (mainly at elevated temperatures,

Table 4.3 The CRSS values for slip on {110} ⟨111⟩ at 500 and 1500 °C and energies of stacking faults on {110} for binary and ternary $MoSi_2$ and the atomic radius, shear modulus and melting temperatures for the corresponding alloying elements [29] (with kind permission of Elsevier)

Cll_b, disilicide	$MoSi_2$	(Mo, Re) Si_2	(Mo, W) Si_2	(Mo, V) Si_2	(Mo, Nb) Si_2	(Mo, Cr) Si_2	Mo $(Si, Al)_2$
CRSS for {110} ⟨111⟩ slip at 500 °C (MPa)	232	405	248	150	95	99	75
CRSS for {110} ⟨111⟩ slip at 1500 °C(MPa)	18	74	25	10	45	27	16
Energy for stacking fault on {110} (mJ/m^2)	365	382	357	321	315	297	231
Alloying element	Mo	Re	W	V	Nb	Cr	Al
Atomic radius (nm)	0.139	0.137	0.139	0.134	0.146	0.127	0.143
Shear modulus (GPa)	122.9	178.6	160.2	47.1	37.7	115.4	26.2
Melting temperature (°C)	2617	3180	3410	1890	2468	1857	660

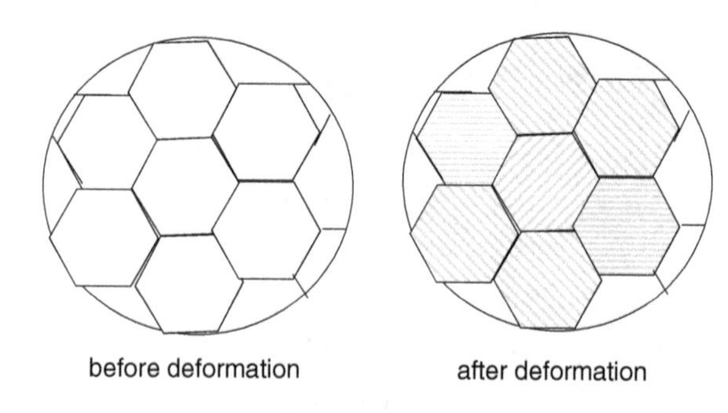

Fig. 4.31 A schematic illustration of a material *before* and *after* deformation. Note the non-realistic assumption that no change in grain shape or size has occurred

except for a few) by dislocation glide, only if each grain conforms to its neighbors in order to maintain the shape changes without disintegration along the grain boundaries. Generally, when straining a polycrystalline ceramic (or any material) five independent strain components of a strain tensor are required. Consequently, 5 independent slip systems, each composed of a slip plane and a slip direction, are necessary for a polycrystalline body to undergo a general strain (von Mises). Figure 4.31 is a schematic illustration of a polycrystalline material as it might be seen under a microscope, before and after deformation, showing slip-line formation in some of its differently oriented grains (crystallites). This figure is hypothetical, since it erroneously assumes that the grains do not change in shape or size. Such a structure (Fig. 4.31) might possibly occur only when macroscopic deformation is very small.

Figure 4.32 provides an illustration of dislocations in a single grain of a polycrystalline Al_2O_3. To eliminate crack formation in the deformation of Al_2O_3, hydrostatic pressure was applied to inhibit the nucleation and growth of cracks during deformation. Deformation at 1150 °C under a hydrostatic confining pressure of 200,000 psi resulted in a total strain of ≈5 %, which was achieved under an axial stress (above the hydrostatic stress) of 280,000 psi. The length of these dislocations suggests that they are probably on the $(10\bar{1}1)$ plane leading to the slip system assignment $(10\bar{1}1)$ $1/3[\bar{1}2\bar{1}0]$. Snow and Heuer [45] indicated that pyramidal slip systems must be activated for the homogeneous deformation of a randomly-oriented Al_2O_3 poly-crystal by slip alone. The electron microscopy of polycrystalline Al_2O_3, deformed at 1150 °C under hydrostatic confining pressure, indicated that several pyramidal systems had been activated. Confinement is used because of the high stresses required to activate pyramidal slip at atmospheric pressure which however induces fracture. By means of the application of hydrostatic pressure, fracture is expected to be eliminated. Note that single crystal measurements proved that CRSS is higher on pyramidal planes than on basal planes.

Slip lines in polycrystalline MgO can be seen in Fig. 4.33. The slip observation was on polished specimens following the deformation on specimens shown as stress–strain curves in Fig. 4.34. A pressure-dependent BDT is observed. At higher pressures, large strains may be achieved without fracture, since the stress required for fracturing increases. The level of the stress–strain curves increases in both the brittle and ductile ranges. Furthermore, the type of jacket influences stress–strain curves. The curves in Fig. 4.34 (and also those in Fig. 4.33) relate to coarse-grained [henceforth: CG] MgO.

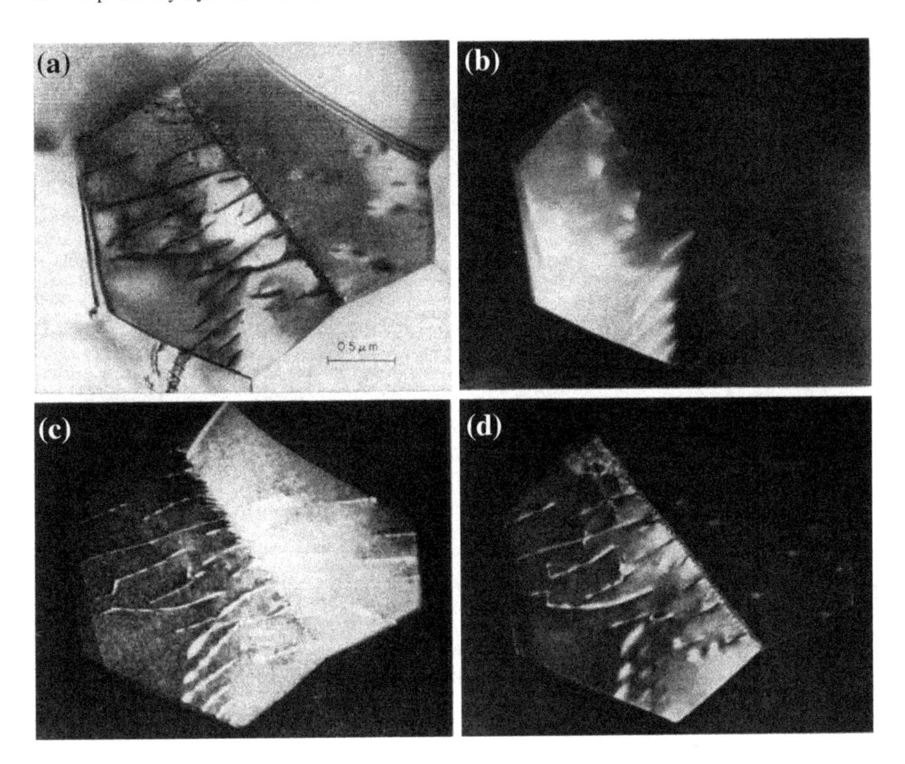

Fig. 4.32 Electron micrographs of dislocations in single grain of Al_2O_3 polycrystal deformed at 1150 °C while under hydrostatic confining pressure and then annealed. **a** Bright-field micrograph; **b**, **c**, and **d** precision dark-field micrographs with operating reflections of $g = (\bar{1}018)$, $(0\bar{1}14)$ and $(\bar{1}104)$, respectively (100 kV) [45]. With kind permission of John Wiley and Sons

Impervious jackets were used in order to seal the specimens, to prevent penetration by the medium used to apply the pressure. The effects of pressure may be seen, in particular, in Fig. 4.34a, where the 0 and 1 kbar curves are also shown.

The temperature dependence of the stress–strain curves at pressures 2 and 5 kbars appear in Fig. 4.35. With increasing temperature, the level of the stress in the stress–strain curve decreases, which is significantly more pronounced in (b), i.e., at the higher confining pressure. The microstructure of the high-temperature specimens is illustrated in Fig. 4.36. Figure 4.34 clearly illustrates the slip lines in the microstructure. Polycrystalline MgO appears to deform in a ductile manner when it is strained under confining pressures greater than **2** kbars at RT. This is shown in Fig. 4.34a. Ceramics having the NaCl structure usually show brittle behavior at RT, when only the $\{110\}$ $\langle 110 \rangle$ slip system is operative, probably due to the lack of the five independent slip systems required by the von Mises concept. Unlike at RT, at high temperatures an additional slip system is observed to be operative, that of $\{100\}$ $\langle 110 \rangle$ and, thus, the von Mises requirement for ductile behavior is met. Polycrystalline MgO, however, can be deformed even at RT under confining pressures greater than 2 kbars [40].

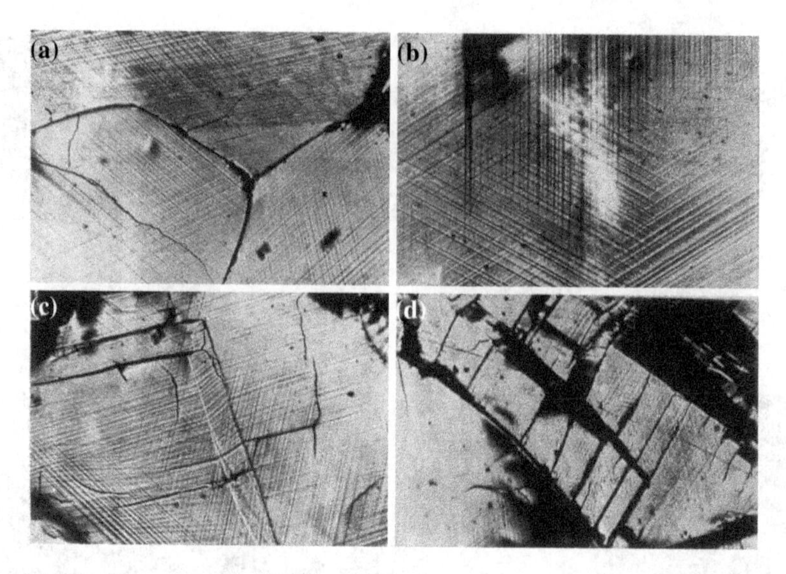

Fig. 4.33 Slip traces in grains of polycrystalline MgO specimen shortened 3.5 % at 8 kbars pressure. Compression axis horizontal. **a** Uniformly distributed slip in group of grains (x150), **b** nonuniform slip in grain (x150), **c** kinked grain (x150), and **d** extension fracturing associated with kinking (x200). All specimens unetched [40]. With kind permission of John Wiley and Son

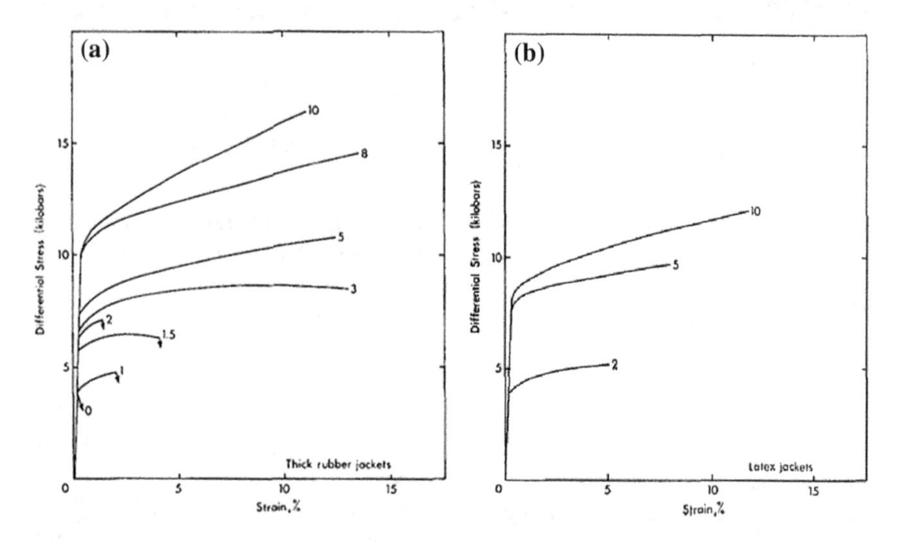

Fig. 4.34 Effect of confining pressure on stress–strain behavior of coarse-grained MgO in **a** thick rubber jackets and **b** latex jackets (Numbers on curves represent confining pressure in kbars and arrows indicate fracture; other tests discontinued where curves end) [40]. With kind permission of John Wiley and Sons

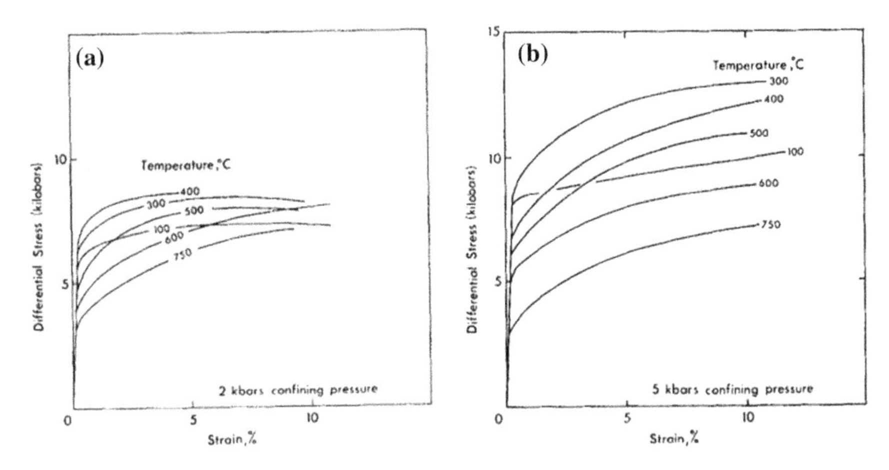

Fig. 4.35 Stress–strain curves of coarse-grained polycrystalline MgO deformed at high temperatures at **a** 2 kbars and **b** 5 kbars confining pressure [40]. With kind permission of John Wiley and Sons

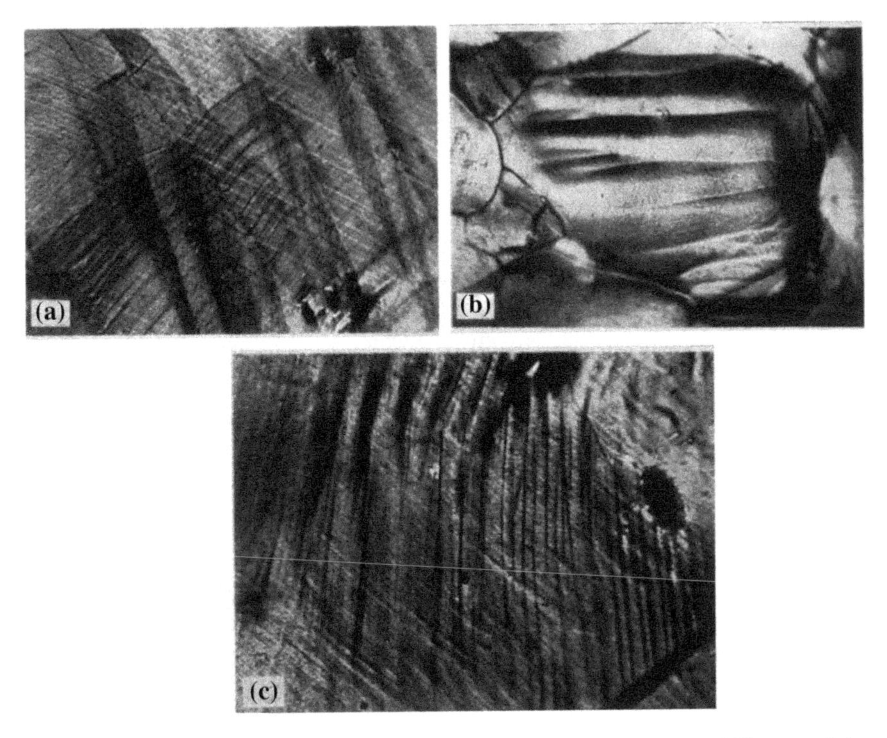

Fig. 4.36 Slip traces on specimens deformed at high temperatures under 5 kbars confining pressure. **a** Kinking in specimen shortened 11 % at 400 °C (×125), **b** wavy noncrystallographic deformation bands in specimen shortened 6 % at 400 °C (×55), and **c** diffuse deformation bands and fine slip in specimen shortened 10 % at 750 °C (× 125) [40]. With kind permission of John Wiley and Sons

To summarize of this section, plastic deformation in ceramics (and other materials) is determined by some compromise between the deformations within the individual grains, which clearly occur by dislocation movement in response to externally applied stress. The movement of crystal dislocations allows grains to deform in certain directions, as dictated by the restrictions of the grain orientations. The grains change shape in order to maintain wholeness, to remain adjoined during the deformation process. Thus, the change in shape is associated with crystallographic-lattice change (or sometimes rotation) of the crystallographic orientation of the grains with respect to each other.

4.7 Twinning in Ceramics

4.7.1 Deformation (Mechanical) Twins

Another way of inducing plastic deformation in materials is via twinning. Thus, plastic deformation may take place not only due to slip, but also by twinning or a combination of both. Whereas, in slip, atoms move a whole number of atomic spacings, in accordance with the Burgers vector of the specific structure, in deformation by twinning, atoms move a fractional atomic spacing, which leads to a rearrangement of the lattice structure itself. There is a difference between the characters of the lines—thin-slip lines are usually observed (visually or under a microscope) on the specimen surfaces of polished test specimens. "Twins" are generally represented as wide bands or broad lines. Slip lines can usually be removed by appropriate etching techniques; 'twins' maintain their microscopic appearance even after relatively deep etching. However, the major difference is related to structural changes in the lattice induced by twinning, i.e., the stacking sequence changes. These changes are associated with the actual motion of the atoms. There is no change in lattice orientation during slip. Figure 4.37 schematically illustrates plastic deformation both by slip and by a combination of slip and twinning. This illustration is hypothetical, since, for the sake of convenience, no changes in the shape or size of the grains has been indicated after deformation. To be more specific, Fig. 4.38 shows images of twinning in barium titanate, $BaTiO_3$ [henceforth: BT], a ferroelectric ceramic obtained by atomic force microscopy [henceforth: AFM]. BT is a semiconductor ceramic and the main constituent of positive temperature coefficient [henceforth: PTC] thermistors.

In particular, Fig. 4.38c shows that the twin patterns on the Y-BT surface are highly ordered. The topographical contrast results from the difference in the etching rate of twins with distinct polar directions and, hence, the interfaces separating the stripes are twin walls.

In FCC metals, there may be a CRSS law for twinning analogous to Schmid's Law, according to Szczerba et al. [46]. Some criteria must be satisfied simultaneously for twinning to occur, namely (a) the ratio of the resolved shear stress,

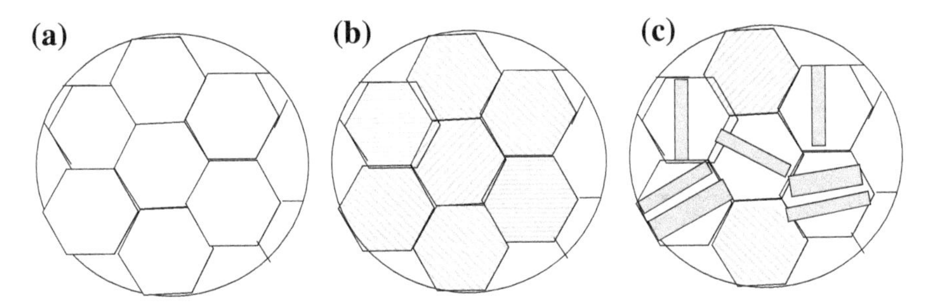

Fig. 4.37 Schematic hypothetical illustration (no change in shape is shown): **a** before deformation; **b** after deformation by slip only; **c** after deformation by slip and twinning. Note the twin bands in some crystallites [7]

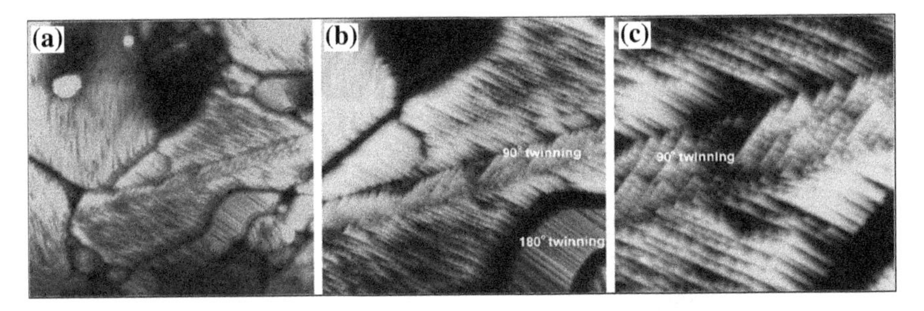

Fig. 4.38 Atomic force microscopy images of Y-BT: **a** 10 × 10 μm; **b** 5 × 5 μm; **c** 2 × 2 μm. (Y-BT stands for yttrium containing Ba titanate to a level of (0.3 at % yttrium) [22]. With kind permission of Elsevier

τ_{RSS}, to the critical stress of a twin system must be greater than that of any slip system, namely $\tau_{RSS}/\tau_C >$ CRSS for slip; (b) the τ_{RSS} should be greater than the threshold (namely greater than some minimum) stress necessary for twinning to occur; and (c) the τ_{RSS} must satisfy the character of a twin shear. It is not currently known if such a concept exists also for ceramics.

In deformation by twinning, the atoms of each slip plane in some part of the lattice move different distances, causing half of the crystal lattice to become a mirror image of the other half. This mirroring is seen schematically in Fig. 4.39. The energy of a twin boundary is very low, compared to grain-boundary energy, about in the range of low-angle grain boundaries. A high-resolution illustration indicating the mirror image of a twinned structure appears in Fig. 4.40. Beautiful twinning in BT ceramics are also illustrated in Figs. 4.41 and 4.42. This material undergoes a ferroelectric structural phase transition as a result of energy minimization associated with stress relief. Its homogeneous elastic energy is reduced at the expense of twin-wall energy. The twin density depends on the grain size, g. When homogeneous stress is applied, the total elastic energy of each grain

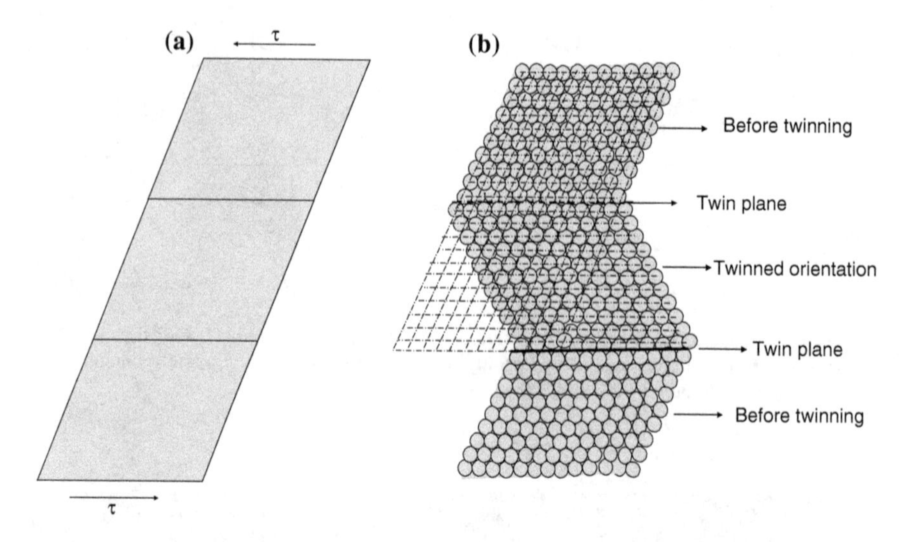

Fig. 4.39 Schematic illustration of a twinned region in the material: **a** before application of shear, τ, and; **b** after twinning deformation. Note that the twinned region is a mirror image of that crystal's part before twinning

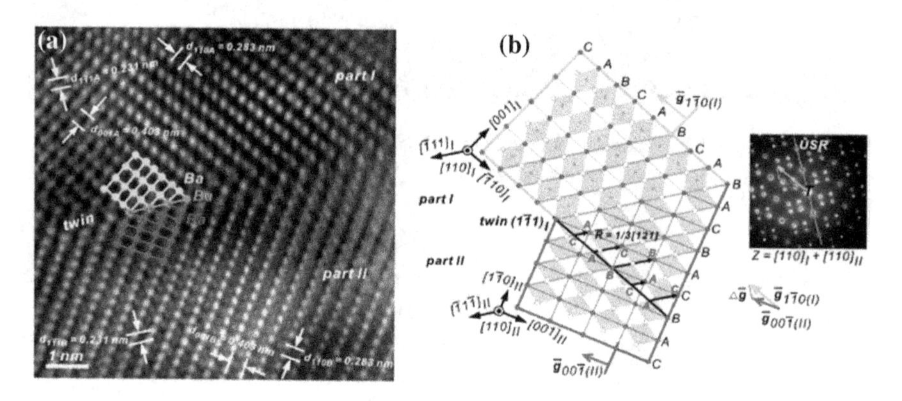

Fig. 4.40 **a** High-resolution image of the $(1\bar{1}1)$ twin boundary and **b** the schematic illustration showing the face-sharing Ti_2O_9 octahedra (TEM) [51]. With kind permission of John Wiley and Sons

increases and becomes proportional, $\propto g^3$. The twin wall, however, increases proportionally to $\propto g^2$.

Graphically, curves pertaining to these two relations intersect and, below their intersection, stress reduction by twinning cannot reduce the total energy of the system. Therefore, there is a grain size below which twinning does not occur. For details on energy minimization, Arlt's [1] review may be consulted.

In a polycrystalline ceramics, such as BT, when structural phase transition occurs, a crystallite (grain) experiences slight deformation. This deformation is

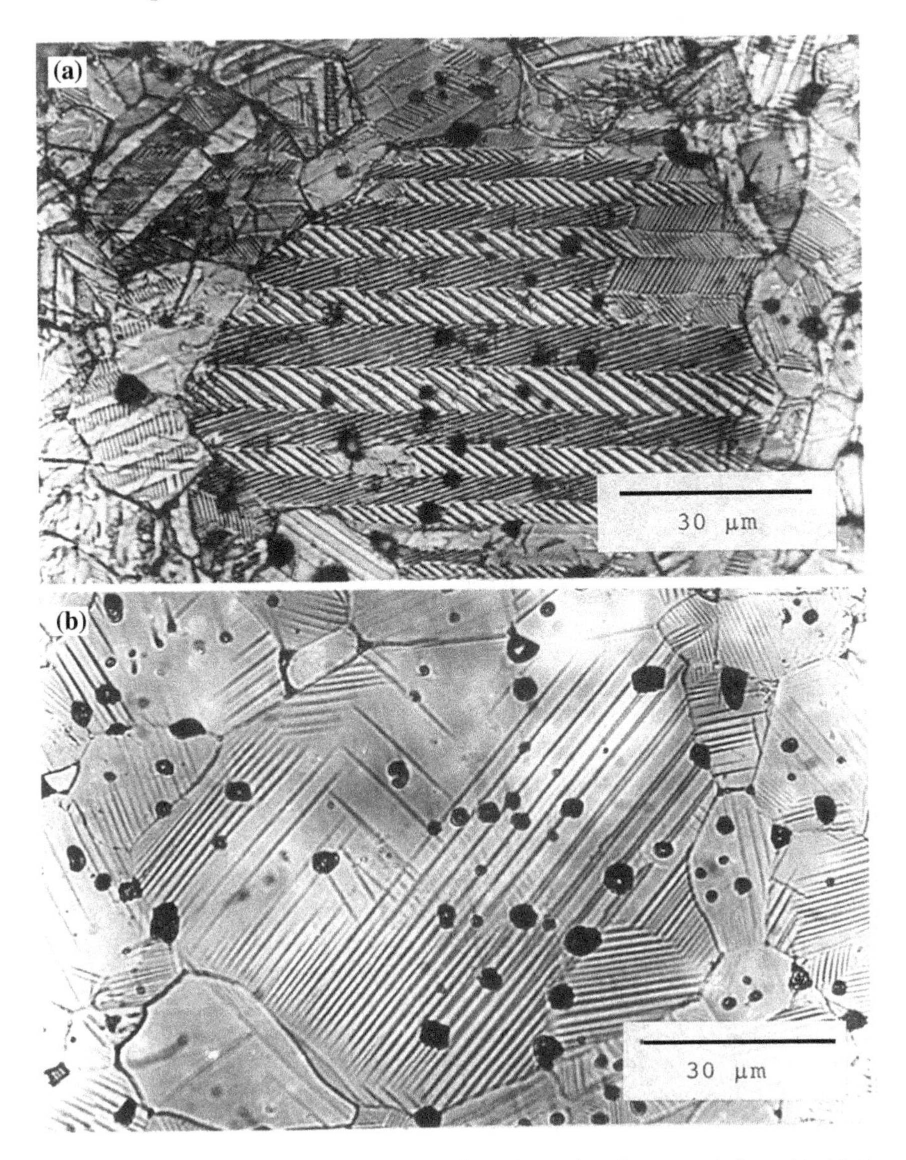

Fig. 4.41 Representative BT domain patterns of a grain: **a** when the pattern is formed inside the ceramic body with three dimensional clamping, **b** the same grain when the pattern is formed under free surface conditions [1]. With kind permission of Professor Arlt

obstructed by surrounding grains. A crystallite is clamped by its neighboring grains and in order to maintain its gross shape high internal stresses or mechanical twinning is required. The illustration in Fig. 4.41a is a representation of twinning when the grain is clamped in three dimensions in the body of the ceramics. The surface grains of a ceramic body do not experience 3D clamping as a crystallite (grain) inside the body. A grain on the surface is clamped only two-dimensionally and will twin differently. In Fig. 4.41b, surface twinning is seen. Two-dimensional grain adjustments can occur in thin films in which 3D clamping is not required, in cubic-tetragonal transition and in tetragonal-orthorhombic transition of the high-T_c superconducting ceramics. Figures 4.43 and 4.44 shows SEM photographs and light micrographs of FG and CG BT and the high-T_c $YBa_2Cu_3O_{7-\delta}$ superconductor. Banded twin structure is shown in Fig. 4.44 in the lead-zirconate-titanate [henceforth: PZT] ceramics, which have a cubic-tetragonal structural phase transition at about 400 °C. The observed domain width, d, and the width of the bands, g_c, are about 3–10 times smaller than those in BT. The grain size is ~10 μm. Twinning in BT, YbaCuO and PZT are seen in Figs. 4.42, 4.43 and 4.44.

Following the above observation on twinning in BT, it may be of interest to visualize the nature and characteristics of twinning in a single crystalline of this ceramic.

4.7.2 *Annealing (Growth) Twins*

The existence of 'annealing twins' (also known as 'growth twins') is another feature of twinning in materials. Examples of this kind of twin formation in $BaTiO_3$ ceramics and in alumina are presented here. Basically, annealing twins are the result of annealing following plastic deformation. Micro-structurally, annealing twins are generally broader and have straighter sides than mechanical twins. Of the many theories attempting to explain the mechanism of annealing-twin formation, one should consider the work of Gleiter, which is based on TEM and FIM observations. In the case of a crystal with an FCC lattice, electron microscopic observations suggest that the steps are formed by the {111}-planes of the grains. During grain growth, atoms may deposit at the steps of the growing grain; therefore, these steps sweep across the grain surface just like the steps on the surface of a crystal grown in vapor. At the points where the grain boundary is parallel to the {111} planes of the growing grain, new planes must be generated.

These TEM observations suggest that the new {111} planes of the growing grain may be generated by two mechanisms: growth spirals and two-dimensional nucleation on close-packed planes of the growing grain. Since a twin has the same lattice structure as the matrix (only in a mirror orientation) and the formation of twins and new grains occurs in a specimen at the same time and under the same conditions (temperature, pressure, etc.), it is assumed that the lattice planes of a twin and of a grain are formed by the same mechanisms. If this assumption is correct, then the mechanisms generating the new lattice planes of a growing grain are also responsible for the generation of twins. Details on such mechanisms may be found in Gleiter's work.

In Fig. 4.45, SEM microstructures of air- and Ar-sintered samples in tetragonal BT, show growth (annealing) twins. Two single twins can be identified by the distinctive contrast across the twin boundary, as indicated in Fig. 4.45a. The twin

Fig. 4.42 **a** Scanning electron micrograph of fine-grained BT. **b** Light micrograph of coarse-grained BT [1]. With kind permission of Professor Arlt

lamellae, 'double twins', indicated in Fig. 4.45b, in contrast to those observed in Fig. 4.45a, were more frequently found in Ar-sintered samples. The twins are lying on the $\{1\bar{1}1\}$ mirror planes and both the single and double twins are growth twins. Those twins lying on the $\{1\bar{1}1\}$ mirror planes, which are not symmetry elements of the (basic) crystal lattice, but rather of the superlattice, are, therefore, 'superlattice twins'. The $\{1\bar{1}1\}$twins, particularly the double twins, were found

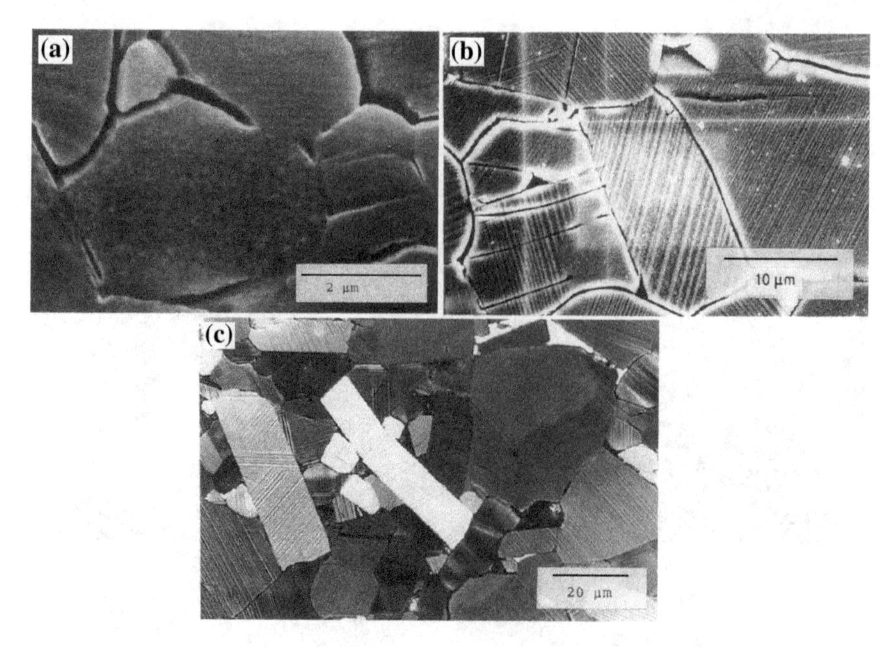

Fig. 4.43 **a** Scanning electron micrograph of fine grained YBaCuO. **b** Scanning electron micrograph of medium grained YBaCuO. **c** Polarized light micrograph of coarse-grained YBaCuO [1]. With kind permission of Professor Arlt

Fig. 4.44 Scanning electron micrograph of $PbZr_xTi_{1-x}O_3$ (PZT) [1]. With kind permission of Professor Arlt

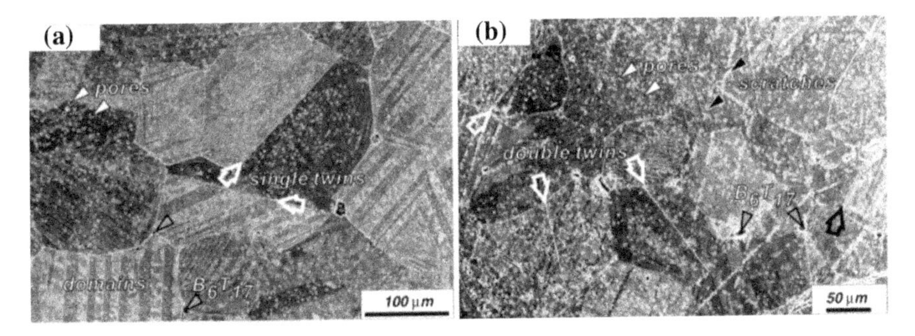

Fig. 4.45 Single- and double-twins in **a** air- and **b** Ar-sintered samples (SEM-SEI) [51]. With kind permission of John Wiley and Sons

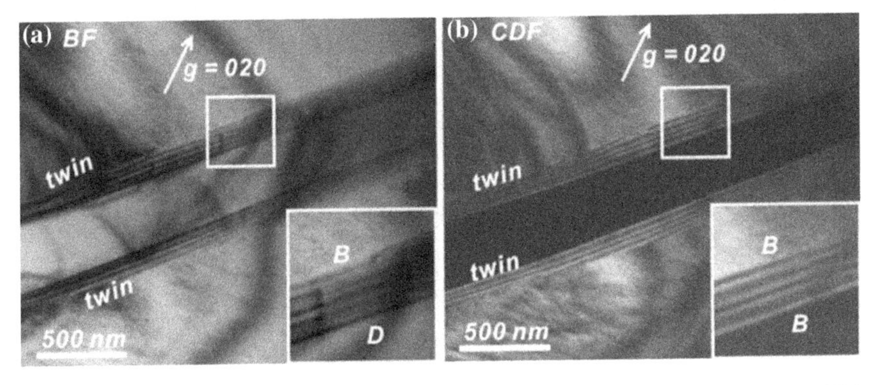

Fig. 4.46 Twin boundaries exhibiting the α-δ fringe patterns in **a** BF and **b** CDF (TEM) [51]. With kind permission of John Wiley and Sons

more frequently in samples sintered in an Ar atmosphere with lower oxygen partial pressure than those annealed in air.

In Fig. 4.46, BF and central dark field [henceforth: CDF] images are showing twin boundaries. The symmetric fringe patterns in the CDF and the asymmetric ones in the BF indicate that the {111} twin boundaries exhibit (α-δ)-fringes. For details on specimen preparation and the procedure for growth twin observation in BT-type samples, consult the original work of Yu-Chuan Wu et al. [51]

4.7.3 Serrated Stress–Strain Curves

Serrated stress–strain curves, similar to those occurring in metals, have also been observed in ceramics. Such stress–strain curves are shown in schematic Fig. 4.47 and experimentally observed serrated curves in alumina are illustrated in Figs. 4.48 and 4.49, formed during deformation at two temperatures and at the

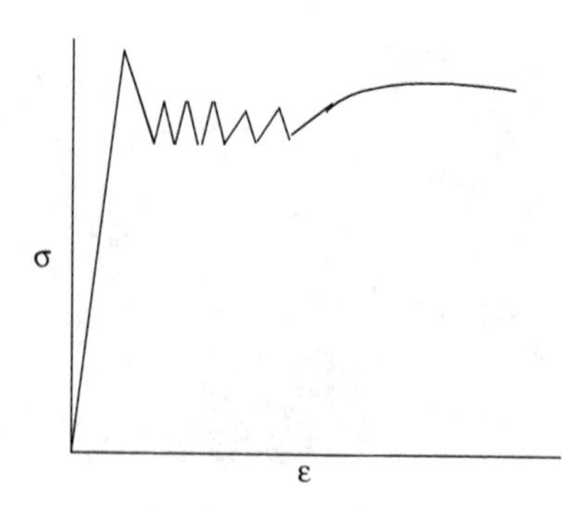

Fig. 4.47 A schematic illustration of the yield drop and serration in the load elongation curve

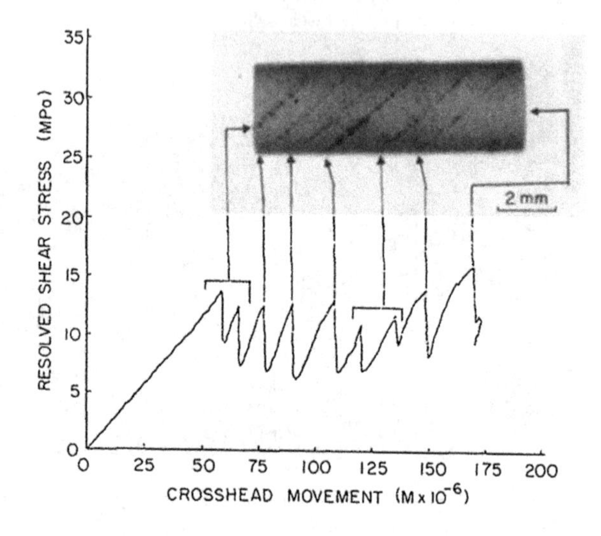

Fig. 4.48 Correlation of stress-displacement curve and twin events in a cylindrical specimen; compression was parallel to cylinder axis, $8.7 \times 10^{-6}\,s^{-1}$ axial strain rate, 1173 K. The occurrence of twins on only one system was unusual [10]. With kind permission of John Wiley and Sons

same strain rate. The load drops in the stress–strain curve are correlated with individual twins formed during the deformation. As is commonly known from twinning in metals, both parameters, those of temperature and strain rate, affect twin formation. Temperature is required in ceramics in addition to strain rate, because they usually deform plastically only at elevated temperatures.

In metals, a yield drop in the tension test of 1020 steel is known to exist. Twinning is very common in iron (deformed by impact or at low temperatures); iron twins are very narrow and are known as 'Neumann bands'. A similar yield drop in tension occurs and a Serrated curve is observed in specimens when twinning deformation occurs as indicated in Figs. 4.48 and 4.49. In FCC materials, such as Cu alloys during low-temperature deformation, serrated ~4.2 K curves occur. Serrated curves are characterized by a sudden fall in stress, followed by a rise in stress and then another fall, repeatedly during mechanical twinning, explaining their observation during tensile stress tests. However, sharp yield drops

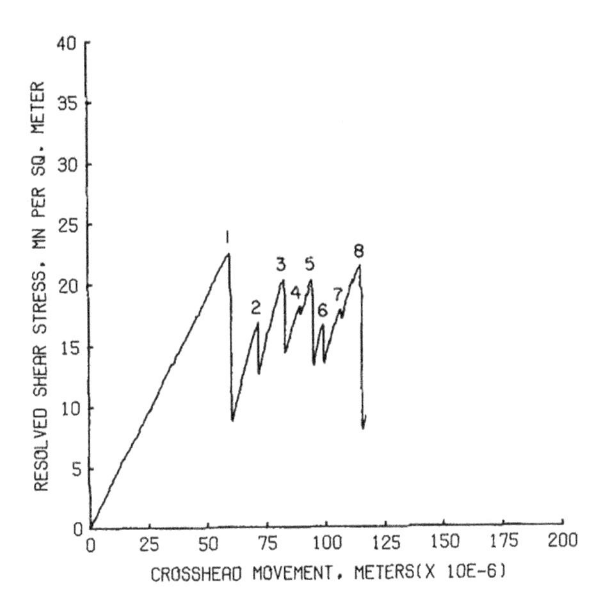

Fig. 4.49 Twinning at 1273 K and axial strain rate of $8.7 \times 10^{-6}\ s^{-1}$. Peak numbers indicate individual twins [10]. With kind permission of John Wiley and Sons

were absent. Thus, a sharp yield drop is not necessarily a feature of twinning in ceramics. In Figs. 4.48 and 4.49, the stress level of the peaks is about the same. This test represents single-crystal alumina specimens loaded under uniaxial compression along the c axis at 623 to 1373 K.

The serration curves in ceramics are analogous with those observed in HCP and BCC metals during low-temperature deformation. Serration is formed by the movement of partial dislocations; this motion converts part of a crystal to a twin orientation. It is believed that dislocations are involved in twinning, but the mechanism is not yet clear.

Twinning is associated with the coordinated deformation of a large number of atoms, possibly leading to serrations in the deformation curves (giving a jagged appearance). Loud clicks are heard during the formation of twins, commonly known as 'tin cry' in metals, but, thus far, such sounds have not been recorded during twinning in ceramics to the author's best knowledge. This occurs because twin formation can be extremely rapid. The serration of a stress–strain ***curve*** is a sign of ***twin*** formation. Many investigators have reported ***twin*** formation and their associated ***serrated*** stress–strain curves during twinning in alumina [49] in Li0.02(K0.45Na0.55) ceramics [15] in BT ceramics [51] etc. For further details on twinning, see the literature on the crystallography of deformation.

In this section, the formation of deformation and annealing twins were discussed with the resulting stress–strain relation characterized by serrated curves. As indicated here, there is no basic difference in twin formation between ceramics and metals.

4.8 Yield Phenomena in Ceramics

4.8.1 Introduction

In metals, several topics are often considered when discussing 'yield phenomena':

(a) sharp yield;
(b) strain aging;
(c) the Portevin–Le Chatelier effect [henceforth: PLC].

Very little or no work on this subject has been reported in the field of ceramics. Therefore, only minor consideration will be given in this section to the above topics.

4.8.2 Sharp Yield

The most commonly known sharp yield was first observed in low carbon-content BCC iron, also known commercially as 'mild steel'. In this case, sharp yield is followed by a sudden drop to a lower value, before further deformation takes place. In Fig. 4.50, such a yield drop in low-carbon steel may be seen from points B to C. Deformation in the C–D region occurs without an increase in the stress level beyond a specific value, an effect known as the 'lower yield point'. The highest stress in the elastic region is known as the 'upper yield point'. The C–D region is not smooth, but jagged (serrated). This kind of yield-point drop may be detected by what is known as a 'hard tensile machine', which is characterized by very little elastic distortion. In Fig. 4.50, other stress–strain curves are also shown for comparison. Although most common ceramics are brittle, sometimes, with minimal deviation from the linear portion (second graph), ductile ceramics at some elevated temperature may be represented by the first graph. Now, focusing on yield phenomena, the last illustration (according to Johnson and Gilman) reveals that sharp yield points also occur in LiF crystals.

Johnson and Gilman reported that, in order to obtain a sharp yield drop, the two necessary criteria are: (a) an increase in the number of moving dislocations and (b) a direct relation between the stress and the velocity of the dislocations. By knowing the strain rate, given as:

$$\dot{\varepsilon} = \mathrm{nvb} \tag{4.9}$$

and the velocity of dislocation motion:

$$\mathrm{v} = \mathrm{k}\tau^{\mathrm{m}} \tag{4.10}$$

they were able to calculate stress-elongation curves for LiF, showing sharp yield drops. By varying the exponent, m, in Eq. (4.10) and the density of the mobile

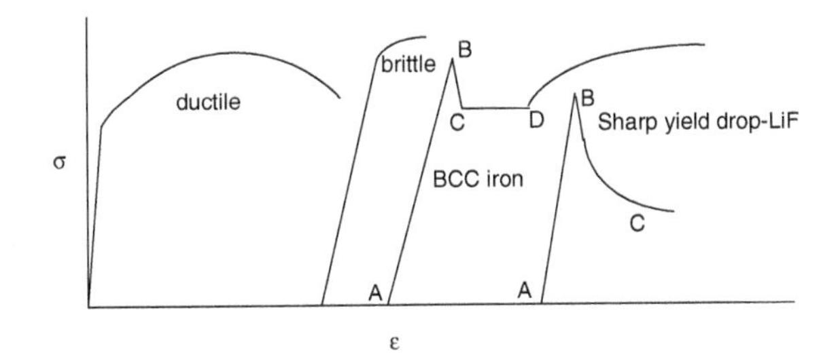

Fig. 4.50 Yielding in ductile materials, brittle materials, BCC iron and LiF. The C–D zone is the 'yield elongation' region

dislocation, ρ, of Eq. (4.11), the magnitude of the yield drop could be changed. In the early stages of deformation, the density of mobile dislocations is given as:

$$\rho = \rho_0 + C\varepsilon^{\alpha} \tag{4.11}$$

The exponent, m, is in the range 1–100. For a certain value of ρ, increasing m decreases the yield drop. The value of m for LiF is ~16.5 and for α-Fe, ~35. Equation (4.10) is an empirical relation with k being a constant. In Eq. (4.9), n is the number of moving dislocations/cm^2. Calculations were performed for $\varepsilon < 0.1$ in the early stages of the deformation. A dislocation-density evaluation was done either using the etch-pits technique or by electron microscopy in the range of $\varepsilon < 0.1$. Equation (4.11) gives ρ_0 as the initial dislocation density, the constant $C = 10^8/\text{cm}^2$ and the constant α is $\sim 1 \pm 0.5$.

Sharp yield and yield drop have been observed in other ceramics, but they were usually not associated with the concept of unpinning impurities from a Cottrell atmosphere, as had been observed in various metals, especially in steel. A popular reason suggested for the yield drop occurring in ceramics was the dislocation multiplication mechanism followed by stress relaxation. An example of the yield drop of the stress–strain curve, as observed in sapphire, is presented in Fig. 4.51. This figure presents typical stress–strain curves for as-received Verneuil- and Czochralski-grown crystals (Fig. 4.51a), demonstrating the marked differences between these types of sapphire. In Fig. 4.51b, the specimen surface finish was improved by grinding both types of crystal to a 'satin finish'. Regrinding raised the upper yield point in the Verneuil crystals and introduced an upper yield point and yield drop in Czochralski materials. The numbers under each curve in Fig. 4.51 refer to the usual yield-drop factor given as:

$$f = (\sigma_u - \sigma_l)/\sigma_l \tag{4.12}$$

where σ_u and σ_l are the upper and lower yield stresses, respectively. Regrinding was essential to provide an improved as-machined surface finish in order to reduce the surface density of surface dislocations, which is the source of mobile dislocations likely to result from the machining. The rod axis of these specimens was parallel to the preferred growth direction for sapphire; the angle between the rod axis and the (0001) direction was ~60° and the angle between the rod axis and the nearest $<11\bar{2}0>$ direction was ~30°. Thus, the Schmid factor for basal slip was

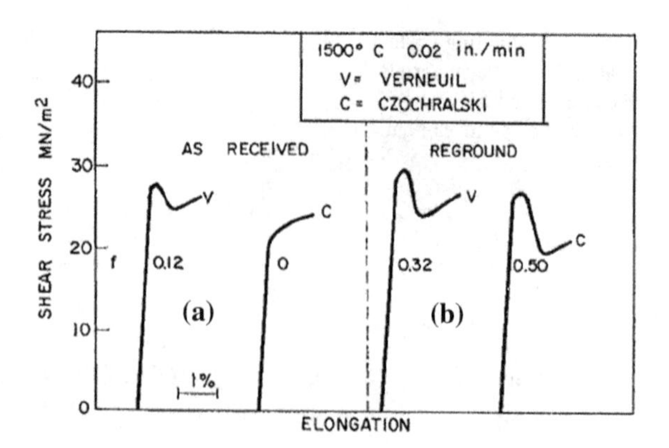

Fig. 4.51 Stress–strain curves showing initial yielding behavior of Verneuil and Czochralski sapphire. **a** As-received specimens and **b** specimens reground to satin finish. Test temperature and crosshead speed appear in inset in this and subsequent figures [19]. With kind permission of John Wiley and Sons

~0.42°. Figure 4.51 indicates that the yielding in sapphire undergoing basal slip is a consequence of dislocation multiplication and is *not due to the unpinning of a Cottrell-type atmosphere, where dislocation pinning results from impurities*. The study of two types of sapphires, with different initial dislocation densities, was meant to point out the difference in their surface dislocation densities and the consequent differences in their yield phenomena.

The modification of surface finish and its effect on yield are not unique to sapphires. It is well known that abrasion will eliminate the yield drop in metals and this effect has been observed in ionic crystals, e.g. LiF, NaCl, AgCl, and MgO. Removing the damaged surface by careful regrinding, reduced the mobile-dislocation-source density to a level low enough that a yield drop appeared in the Czochralski crystals and was enhanced in the Verneuil materials. Various factors influence the appearance and magnitude of yield drop, among them the strain rate and temperature. The effect of temperature on the upper and lower yield points (i.e., the magnitude of the yield drop) is illustrated for sapphire in Fig. 4.52.

Thus, all processes which impede dislocation movement are prone to induce yield point pinning and drop (softening).

4.8.3 Strain Aging

Yield drop is also a feature associated with the observation of strain aging. It has been recorded in two types of ceramic sapphires, known as Verneuil and Czochralski sapphires, that strain aging may be observed during plastic deformation. Figure 4.53 shows well-defined strain aging phenomena.

After initial yielding, the yield drop does not appear for short aging times, but becomes more pronounced as the aging time increases. The lower yield stress increases with increasing strain, indicating work hardening. A smooth transition in the plastic regions is seen without upper yield reoccurrence after a short time has elapsed, as seen in the first two graphs of Fig. 4.53. Thus, a time element is

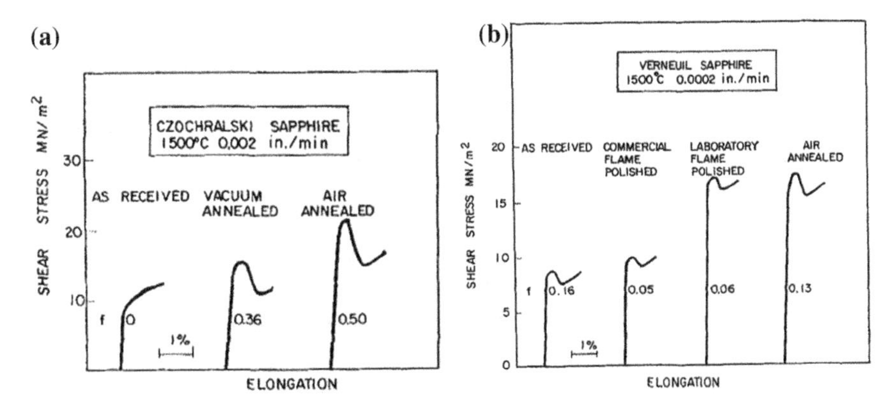

Fig. 4.52 Effect of thermal treatments on initial yielding of **a** Czochralski and **b** Verneuil sapphire crystals [19]. With kind permission of John Wiley and Sons

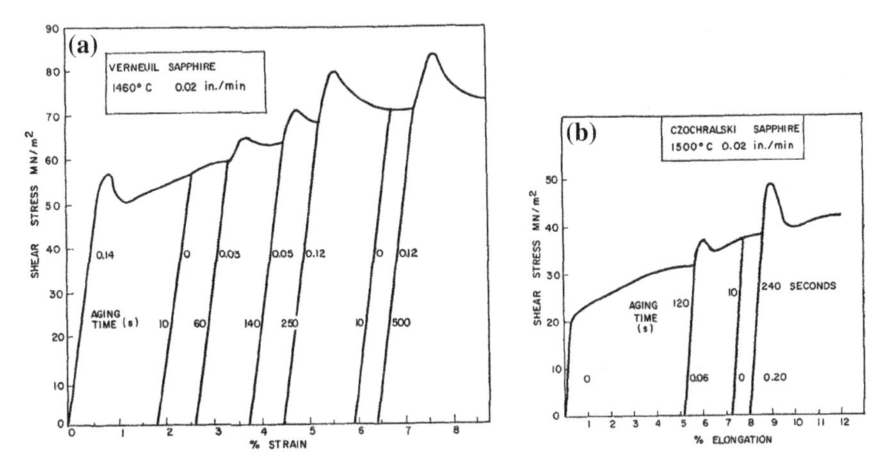

Fig. 4.53 strain aging of **a** Verneuil and **b** Czochralski sapphire crystals [19]. With kind permission of John Wiley and Sons

associated with the reappearance of the upper yield point. The upper yield point that reappears after some aging time is at a higher level than the one that had been observed at an earlier stage. The strain aging observed in sapphire is the result of a different mechanism than the impurity pinning (by O or N) known to occur in steel. In sapphire, it is assumed that a dislocation-multiplication mechanism is associated with strain aging and with the dislocation density, and thus depends on the availability of dislocation sources. The higher aging temperature allows for the rearrangement of dislocation networks at a faster rate. This is possible, since dislocation rearrangement can occur with ease in sapphire at these temperatures and that stable networks can form during strain aging. Such networks would decrease the density of **mobile** dislocations and introduce a yield point. When there are no sufficient dislocations available, then higher stress is required for slip initiation, namely, the yield point will be higher (upper yield point). Removing a damaged surface by careful regrinding reduces the mobile-dislocation-source density to a level low enough that a yield drop appears. Thus, the introduction of a yield drop is attributed to the relief of surface damage (reduction of mobile dislocation sources).

A yield drop may be caused by mechanisms other than the dislocation multiplication of unpinning. In a system of dislocations, which may be or are pinned in position, instability is produced in the stress–strain curve. This instability takes the form of a negative slope, as observed in the yield drop.

The appearance of classic strain aging, as found in metals, has been reported in MnO single crystals. Observations of MnO made by Goretta, et al. showed serrated curves and yield drops in MnO single crystals, which were attributed to solute interactions with dislocations. For sufficient aging time, t_a, serration was observed after reloading the specimen which increased with t_a. These experimental observations are consistent with the PLC model for serrations, but with solute atmospheres causing softening, rather than hardening, as is the case in metals. The saturation of $\Delta\tau_a$ by t_a may be due to solute trapping by immobile dislocations in cell walls ($\Delta\tau_a$ in this work refers to the difference between upper and lower yield points or to the magnitude of the yield drop).

Quite recently Gallardo-Lopez, et al. [21] observed strain aging in Y_2O_3–ZrO_2 [henceforth: YSCZ] single crystals as illustrated in Fig. 4.54. These experiments were carried out by uniaxial compression tests between 1310 and 1450 °C. The experimental results and their comparison with standard models show that these instabilities are caused by the PLC effect. The experiments were carried out in a stable flow domain at 1310 °C. The samples were plastically strained and statically aged for time intervals of 20–45 m in the unloaded state. Upon reloading at the same temperature, a yield point occurred with an approximate amplitude of 20 MPa. This static aging effect suggests the possible pinning of the dislocation lines in the unloaded state by solute atomic diffusion. Note that, also in metals, annealing at some temperatures increases the diffusivity of solutes, which arrive within a shorter time to pin dislocations, essentially causing the upper yield point. Furthermore, it is important to remember that plastic deformation in ceramics usually occurs at some high temperature (depending on the type of ceramic). In metal specimens, left for a short time without reloading, no upper yield is observed, because the diffusion of solutes to re-pin dislocations cannot take place at a sufficient rate, as it does at certain elevated temperatures.

4.8.4 Portevin–Le Chatelier (PLC) Effect

A schematic illustration shows the temperature-dependent PLC effect as observed in metals (Fig. 4.55), which is closely related to yield-point phenomena. Such

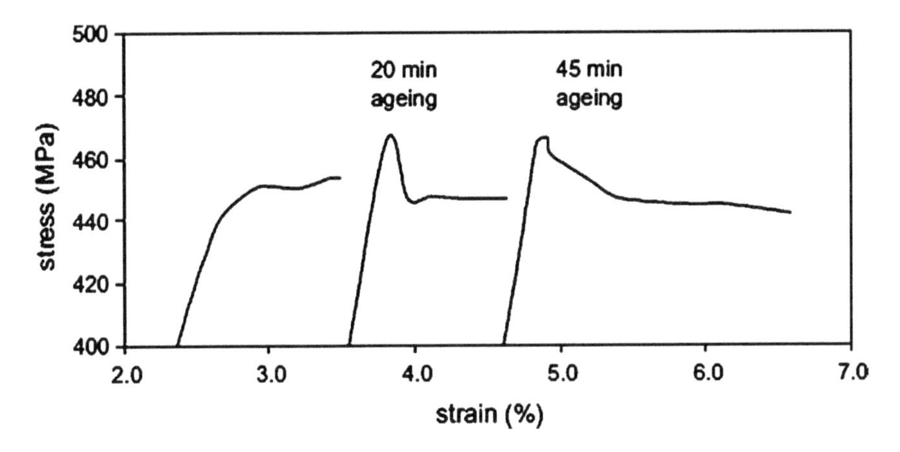

Fig. 4.54 Static ageing of a 24 mol% YCSZ sample at temperatures below the instability range. Conditions: strain-rate $\dot{\varepsilon} = 3 \times 10^{-5}\ s^{-1}$, T = 1310 °C, ageing times 20 and 45 min. Upon reloading after ageing, the specimen exhibits yield points of amplitude 20 MPa [21]. With kind permission of Elsevier

behavior, although observed initially in α Fe, is exhibited by several materials as they undergo plastic deformation, e.g., Al–Cu alloys (see, for example, Liang et al.), substitutional Al-2.5 % Mg alloy [13] etc. For the PLC effect to occur, solute atoms must segregate at the dislocation core. This requires sufficient mobility by diffusion of the segregated atoms. The local site of the dislocation core is energetically favorable, since it has space available to accommodate the solute atom which locks the dislocation, hindering its motion. A larger force (stress) is necessary to move the dislocation, as the cloud of solute atoms is dragged with it. At some stage, the dislocation eventually breaks away from the atmosphere of solute atoms, resulting in reduced drag stress for dislocation movement. Higher temperatures efficiently release the solute from the dislocation core in the process known as 'unpinning'. In Fig. 4.56, for the 24 mol% YCSZ specimen, note the last set of curves indicating the disappearance of serration with increased temperature. Thus, at sufficiently high temperatures and at a specific strain rate, these serrations gradually disappear, due to the relatively higher diffusion of the solute atoms, preventing the recapture of their atmosphere by the dislocations. In YCSZ having a large yttria concentration, the mechanism that governs the yield and flow stresses in easy glide conditions is the interaction of dislocations with yttrium defects, i.e., complex defects formed by the association of two yttrium atoms and one oxygen vacancy.

TEM observations of thin foils parallel to the primary (001) slip plane of the deformed samples show typical dislocation substructures for YCSZ in easy glide. The microstructure consists mostly of straight-edge dislocations and a few dislocation loops, as may be seen in Fig. 4.57. By means of the line-intercept technique, the average dislocation density was estimated to about $10^{13}\ m^{-2}$.

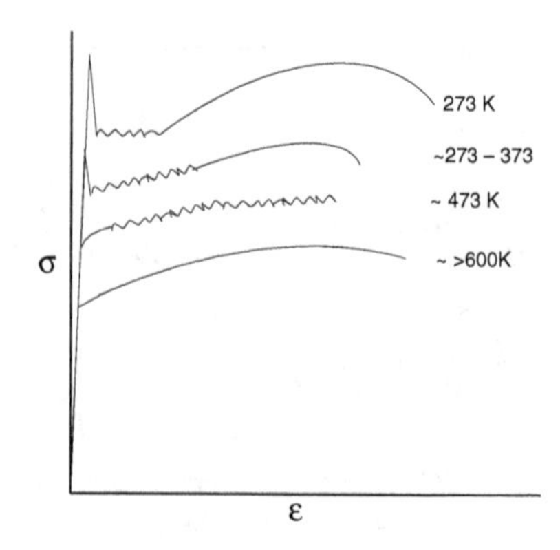

Fig. 4.55 The Portevin–Le Chatelier (PLC) effect. The influence of temperature on the appearance of a stress–strain curve under tension [7]

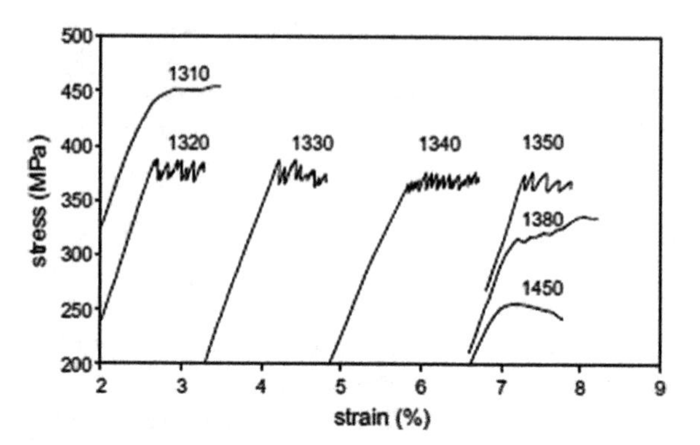

Fig. 4.56 Serrated stress versus strain curves in a 24 mol% Y_2O_3–ZrO_2 specimen, compressed at an imposed strain-rate of $\dot{\varepsilon} = 1.6 \times 10^{-5}\ s^{-1}$ between 1310 and 1450 °C. All the tests are performed in easy glide conditions [21]. With kind permission of Elsevier

Following the approach of Gallardo-Lopez et al., one can evaluate several properties for the PLC. Assume the waiting time for pinning the dislocations, t_w, is:

$$t_w \approx \tau \tag{4.13}$$

Further assume that the flight time to the next obstacle is negligible. Let the concentration of the yttrium defects be c, each defect occupies c/a^2 atomic sites in the slip plane, where the lattice spacing is a. The average distance between the obstacles is:

$$\Lambda = a/\sqrt{c} \tag{4.14}$$

Let the mean dislocation density be $\bar{v}$, expressed as:

$$\bar{v} = \Lambda/t_w = \frac{a}{t_w\sqrt{c}} \tag{4.15}$$

This is the ratio of the mean free path with the waiting time [21].

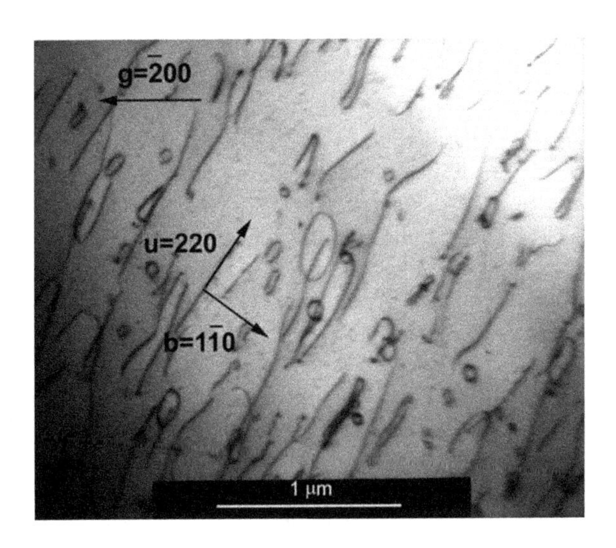

Fig. 4.57 TEM micrograph of the dislocation microstructure in the primary plane of a sample deformed at 1350 °C and with $\dot{\varepsilon} = 1.6 \times 10^{-5}$ up to the yield point, and showing instabilities. The dislocation density is estimated to about 10^{13} m^{-2} [21]. With kind permission of Elsevier

Assume the mobile dislocation density is ρ_m and the Burgers vector is b. The plastic strain, according to Orowan's Law, may be given by:

$$\dot{\varepsilon} = \rho_m b\bar{v} = \frac{\rho_m ba}{t_w\sqrt{c}} \tag{4.16}$$

Taking the following values for YCSZ:

b = 0.36 nm
a = 0.5 nm
c = 0.15 is the concentration of defects in the samples.
$\rho_m = 10^{13}$ m^{-2} is the number of mobile dislocations under single-slip conditions. $\dot{\varepsilon} = 1.6 \times 10^{-5}$ is the imposed strain rate.

Equation (4.16) yields time $t_w = 0.5$ s for the waiting. Being about 1 s, this value is about the same as is typically obtained for PLC in metallic alloys.

For the calculation of the diffusion time, τ, of yttrium atoms, the Friedel approach (his relation 16.3) may be followed (as in Gallardo-Lopez et al. [21]), given as:

$$\tau = \frac{k_B T b^2}{n(n+2)DW_M}\left(\frac{c_1}{ac_0}\right)^{\frac{n+2}{n}} \tag{4.16a}$$

Here, n is an integer which depends on the nature of the solute-dislocation interaction mechanism. For the size effects, n = 1 was used. The diffusion coefficient, D, for yttrium atoms in YCSZ may be expressed, as is usually done, in Eq. (4.17):

$$D = D_0 \exp\left(-\frac{Q}{k_B T}\right) \tag{4.17}$$

in which the parameters have their usual meanings, namely D_0 is the pre-exponential factor, Q the activation energy for diffusion and k_B is the Boltzmann constant. By using relevant values of the parameters from Kilo et al. [31] in Eq. (4.17), the bulk diffusion of yttrium atoms is $D_0 = 4.82 \times 10^{-6}$ ms^{-2}, $Q = 4.86$ eV at T 1640 K (typical for PLC), and, as such, $kT_B \approx 0.144$ eV. The value of D, thus obtained, is $D = 1.11 \times 10^{-20}$ m^2 s^{-1}. W_M is the interaction energy of yttrium solutes with the dislocation core, which is estimated by Friedel as $W_M \geq 1$ eV. c_0 is the atomic concentration of diffusing solutes and is given as $c_0 \approx 0.14$. c_1 is the maximum concentration of solutes on the pinned dislocation lines; since its value is smaller than the saturation value, $c_s = 1$, $c_1 \approx 0{:}5$ was taken, as suggested by Friedel. From all the above data, it was found that $\tau \approx 0.7$ s is comparable to the obtained value of the waiting time, $t_w \approx 0.5$ s.

Although approximate values were used for calculating the waiting time, the value obtained sufficiently confirms that solute atoms of yttrium are responsible for the observations of the PLC effect. Despite the information obtained above for the case of YCSZ, the possibility has not been excluded of observing PLC by dislocation–dislocation interaction leading to pinning, thus producing PLC (or strain aging).

4.9 Deformation in Polycrystalline Ceramics

4.9.1 Introduction

On several occasions throughout this book, no distinction was made whether single or polycrystalline ceramics were used to exemplify certain topics. Single and polycrystalline ceramics were interchangeably used to emphasize the subjects under consideration. Now, this section will focus on several cases of deformation in polycrystalline ceramics not yet discussed in depth. First among the topics to be considered below is that of preferred orientation, a well-known aspect influencing not only the physical properties of ceramics, but also their mechanical behavior.

4.9.2 Preferred Orientation (Texture)

By introducing preferred orientation in polycrystalline ceramics, the advantage of the resulting anisotropies may be used for specific purposes of interest. 'Anisotropy' is a directionally-dependent property of a material, physical or mechanical, in contrast with 'isotropy', i.e., having identical properties in all directions. It is essential to control the orientations of the individual grains of a polycrystalline ceramic in order to basically control its grain-boundary network. Seldom can a textured ceramic be observed in its natural state. Generally, texturing is introduced during processing or fabrication. There are several ways of texturing ceramics, regardless of whether or not improved properties are realized. One of the most popular means for introducing texture is by controlling the microstructures by means of a templated grain growth [henceforth: TGG] technique. This method is discussed below.

4.9.3 Texture by Templating

The TGG process has been used to texture a range of ceramic materials, to provide directional properties and novel microstructures. Some of the requirements of this process are:

(i) Templates serve as nucleation sites for epitaxial growth of the matrix phase and must, therefore, be crystallographically isostructural (though not necessarily chemically identical) with the matrix;
(ii) These templates must also have a high aspect ratio (ratio of height to width) to increase their orientability during forming;
(iii) Also, the templates must be sufficiently larger than the matrix to assure growth, but small enough to have a high ratio of surface area to volume.

Many routes for obtaining template particles are known: using molten salt synthesis, hydrothermal synthesis, sol–gel processing, as well as hybrid, multistep methods. The TGG method has been widely used for the development and texturing of piezoelectric ceramics [44]. One of the reasons for this widespread use is the cost consideration, since TGG enables the relatively inexpensive fabrication of textured ceramics with single crystal-like properties. However, our present discussion pivots around the mechanical properties.

Lead magnesium niobate-lead titanate [henceforth: PMN-PT], 0.675Pb$(Mg_{1/3}Nb_{2/3})O_3$-0.325$PbTiO_3$ (PMN-32.5PT) ceramics were textured (grain-oriented) in the $\langle 001 \rangle$ crystallographic direction by the TGG process. The textured PMN-32.5PT ceramics were produced by orienting {001}-SrTiO3 platelets ($\sim$10 μm in diameter and $\sim$2-μm thickness) in a submicron PMN-32.5PT matrix. The TGG of $\langle 001 \rangle$-oriented PMN-32.5PT grains of the strontium titanate [henceforth: ST] platelets resulted in textured ceramics with $\sim$70 % Lotgering factor and >98 % theoretical density. Note that the '**Lotgering factor**', f, is defined as the fraction of the area textured, i.e., the degree of orientation.

Figure 4.58 is an illustration of tabular {001}-ST particles (5–15-mm width) with an aspect ratio >5, which were mixed in the slurry with a magnetic stir bar.

Fig. 4.58 Molten salt synthesized, tabular SrTiO3 used as templates [33]. With kind permission of John Wiley and Sons

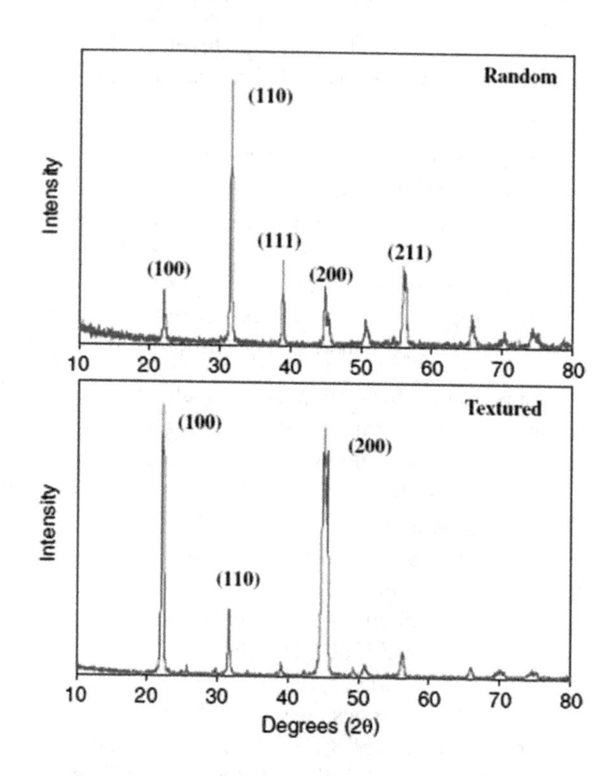

Fig. 4.59 X-ray diffraction patterns of random and textured PMN-32.5PT ceramics [33]. With kind permission of John Wiley and Sons

The tabular ST particles were synthesized by a two-step molten salt process using KCl as a flux. The ST platelets served as templates. Figure 4.59 shows by XRD the random and textured PMN-32.5PT (3 wt% excess PbO, 3 vol% ST) ceramics sintered at 1150 °C for 10 h. The two major peaks in the templated PMN-32.5PT ceramic were (100) and (200), which indicates the high degree of texturing. In the textured XRD, a reflection of the (100) remained, indicating that the texturing did not entirely reach completion. The piezoelectric behavior of the ST and BT6 (i.e.,

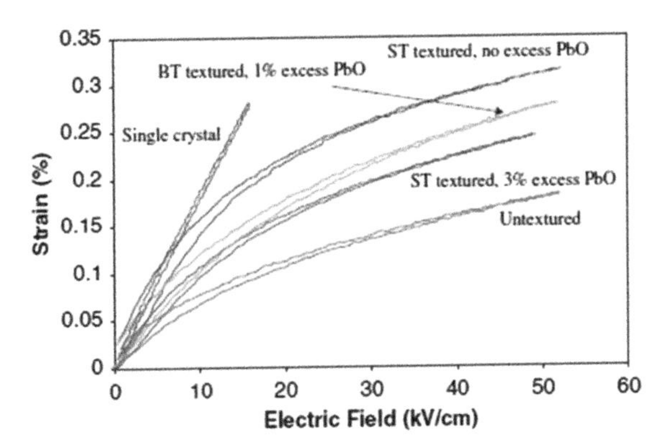

Fig. 4.60 Strain of textured, untextured, and single-crystal PMN-32.5 PT as a function of excess PbO and SrTiO3 templates [33]. With kind permission of John Wiley and Sons

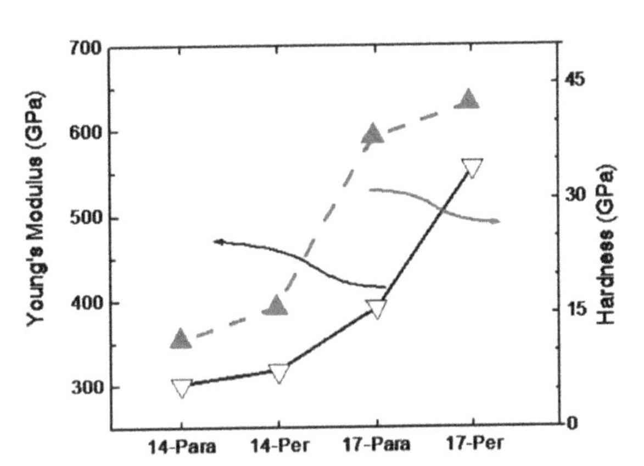

Fig. 4.61 Hardness and Young's modulus measurement for sample sintered at 1400 and 1700 °C MPa for indentation parallel and perpendicular to the direction of uniaxial pressure (Note: per and para stand for indentation perpendicular and parallel to the direction of uniaxial pressure, respectively) [43]. With kind permission of American Institute of Physics

6 % $BaTiO_3$) textured samples and that of a randomly-oriented sample are compared in Fig. 4.60. Textured PMN-32.5PT with 5 vol% ST and no excess PbO showed a maximum strain level of ~0.3 % in an electric field of 40 kV/cm, which is approximately twice the strain response of a random ceramic in the same field. Since this material is used for piezoelectric purposes, Fig. 4.60 shows the strain as a function of the electric field. Due to the fine size of the template particles, it was possible to obtain textured ceramics without the use of excess PbO and without extensive matrix grain growth.

Other techniques for inducing preferred orientations in ceramics are: under strong magnetic field alignment [50]; by means of a high electric field; by the use of a seed layer to control texture [36] etc.

4.9.4 Spark Plasma Sintering (SPS)

In Fig. 4.61, the Young's modulus and hardness are shown for textured alumina obtained by spark plasma sintering [henceforth: SPS] of undoped commercial grade α-Al_2O_3. Of the various parameters measured, these two are shown in Fig. 4.61. The sintering temperatures were 1400 and 1700 °C. The high elastic modulus and hardness justify the advanced use of textured ceramics due to their improved properties.

4.10 Grain Size

Almost all the physical and mechanical properties are strongly microstructure-dependent. The strength of ceramic materials depends, among other parameters, on grain size. The best known concept relating grain size to some strength property is the HP relation, originally applied to metals and alloys, but which holds true within the limits of the theory for ceramics as well. Initially, this relation provided the yield stress, σ_o, and the grain size, d relation according

$$\sigma_o = \sigma_i + kd^{-1/2} \tag{4.18}$$

Here, σ_i represents the resistance of a material to dislocation movement and k is a constant with various interpretations, such as the unpinning of dislocations, the degree of dislocation pile-ups behind some barrier or the measure of the relative grain-boundary strength. k may also be taken to be the grain-boundary resistance to dislocation slip across boundaries. The value of k may be determined from deformation experiments on varying grain sizes. The relation of yield stress to grain size (actually to $d^{-1/2}$) is linear. k is the HP slope (see Fig. 4.62) and σ_i is determined from the intercept of the line with the σ_o axis.

Details of this process may be envisaged as follows. On application of a tensile stress, the resolved shear stress acts on some sources in favorably-oriented grains (note that the boundaries themselves are sources of dislocation) and, on reaching the yield stress, dislocations pile-up behind the boundaries, acting as obstacles. At a sufficiently large stress, other sources in unfavorably-oriented grains start to operate, emitting additional dislocations. Then, general yielding of the entire specimen will commence. Grain boundaries, in general, are considered as barriers limiting the free path of dislocations and, thereby, increasing strain hardening. Thus, the HP model basically indicates the stress required to activate dislocation sources in neighboring grains. At ambient temperatures (when no creep has set in yet), yield strength rises as the grain size decreases. Two main reasons should be noted for this increase in yield stress as a result of smaller grain size: (a) the number of boundaries increases when the grains are smaller, resulting in a smaller number of pile-ups in each small grain (boundaries hinder dislocation movement) and (b) the grain boundaries are also much more disordered than the grain itself, further preventing dislocations from moving from one grain to another on continuous slip planes. Impeding dislocation movement hinders the onset of plasticity, due to the increased yield strength of the material. The higher the yield strength,

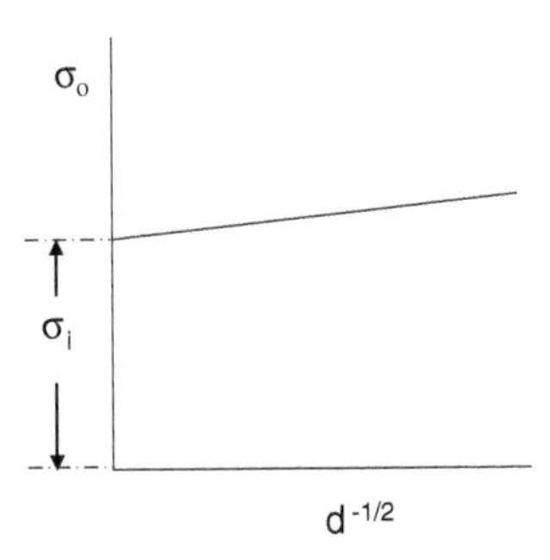

Fig. 4.62 A schematic plot according to Eq. (4.18), showing the HP relation, with intercept σ_i and slope k

the higher the applied stress needed to move a dislocation, in accordance with the inverse relation in the HP Eq. (4.18). Equation (4.18) is often expressed as:

$$\sigma_y = \sigma_0 + \frac{k_y}{\sqrt{d}} \tag{4.18a}$$

The HP relation may be further discussed as follows. A stress concentration exists at or close to the obstacle (a grain boundary in our case). The key factor in the motion of dislocations is the first or leading dislocation in the vicinity of the obstacle. Assume that the leading dislocation has moved a distance, dx; all the trailing dislocations will move the same distance. The work done per unit length of dislocation (in Chap. 3) is:

$$work = n\tau bdx \tag{3.79}$$

where n is the number of dislocations and τ is the applied stress. The leading dislocation, τ_i, works against the local stress (internal stress) of the obstacle. Thus, the work of the leading dislocation is:

$$\mathrm{work} = \tau_i \mathrm{bdx} \tag{3.79a}$$

At equilibrium, these equations should be equal and, thus, from Eqs. (3.79) and (3.79a), one obtains:

$$\mathrm{n}\tau = \tau_i \mathrm{b} \tag{3.79b}$$

The internal stress at the head of the pile-up, composed of n dislocations, is n times greater than the applied stress. The back stress of the pile-up, τ_b, acts on the source to create new dislocations. As long as:

$$\tau_b - \tau_a = \tau \tag{3.79c}$$

where τ_a is the stress required to operate a source, it will function to produce dislocations. Eshelby et al. [18] have calculated the number and distribution of dislocations that can pile-up in a slip plane of length L with acting shear stress τ as:

$$n = \frac{\pi L \tau k}{Gb} \tag{3.80}$$

k = 1 for screw and (1−ν) for edge dislocations.

In Eq. (4.18), stress consists of two terms: the first is independent of grain size and the second is grain-size dependent. The first term at a certain strain, ε, is associated with flow stress in the interior of a grain, while $\frac{k}{\sqrt{d}}$ represents the contribution to the strength, as a consequence of grain-boundary resistance to dislocation movement into another grain (Hansen). The contribution of the grain interior, $\sigma_0(\varepsilon)$, is related to the density of the dislocations accumulated at the grain boundaries. This is detailed in Sect. 3.3.15 on dislocation pile-ups and indicated especially in Eqs. (3.79), (3.79c) and (3.80) above. By applying them in terms of grain boundaries acting as obstacles to dislocation motion and by rewriting Eq. (3.79b), the relation between τ_e and τ_p is obtained:

$$n\tau_e = \tau_p \tag{4.19}$$

τ_e represents an effective stress and τ_p is the stress assumed to exist at the head of the pile-up acting at the boundary. τ_p is n times larger than the effective stress and n is the number of dislocations in the pile-up. The number of dislocations given as Eq. (3.80), in terms of the effective stress (note that τ is given in terms of τ_e), is given by:

$$n = \frac{\pi L \tau_e k}{Gb} \tag{3.80a}$$

A pile-up of n dislocations along a distance, L, may be considered to be a giant dislocation with Burgers vector nb in a pile-up of n dislocations (i.e., it is the length of the pile-up). L cannot be larger than the grain size and its value is usually taken to be d/2. The effective stress (rewriting Eq. (3.79c)) in terms of the applied stress, τ_a, and all the contributions resisting dislocation motion, such as back stress and friction stress, τ_i, may be expressed as:

$$\tau_e = (\tau_a - \tau_i) \tag{4.19a}$$

At the critical shear stress of the applied stress, yielding occurs (=τ_y) and dislocations are nucleated at the head of the pile-up for slip into a neighboring grain (across the grain boundary). At this stage, Eq. (4.19) may be written as:

$$n\tau_e = \tau_p = \tau_c \tag{4.19b}$$

Substituting Eqs. (3.80a) and (4.19a) into Eq. (4.19b) results in:

$$\tau_c = \frac{k\pi d \tau_e^2}{2Gb} \tag{4.20}$$

Expressing τ_e from Eq. (4.20), one obtains:

$$\tau_e = \left(\frac{2Gb\tau_c}{\pi k d}\right)^{1/2} = \left(\frac{2Gb\tau_c}{k\pi}\right)^{1/2} d^{-1/2} \tag{4.21}$$

From Eq. (4.19a), when $\tau_a = \tau_y$, and based on Eq. (4.21), one gets:

$$\tau_e = \tau_y - \tau_i = \left(\frac{2Gb\tau_c}{k\pi}\right)^{1/2} d^{-1.2} \tag{4.22}$$

or:

$$\tau_a = \tau_y = \left(\frac{2Gb\tau_c}{k\pi}\right)^{1/2} d^{-1.2} + \tau_i \tag{4.23}$$

This equation is similar to the HP relation Eq. (4.18a) and, in order to make it equivalent in terms of the yield stress, σ_y, under tension, τ_y is multiplied by the Taylor factor, M, to obtain Eq. (4.18a) as:

$$\sigma_y = \sigma_0 + \frac{k_y}{\sqrt{d}} \tag{4.18a}$$

k_y is:

$$k_y = M\left(\frac{2Gb\tau_c}{k\pi}\right)^{1/2} \tag{4.24}$$

M is ~3.1 in FCC metals. For more such calculations, consult the work of Rollett et al.

The HP model is widely used to the present day and, with the appropriate modification of Eq. (4.18a), its applicability to various materials (particularly metals) has been amply demonstrated. This flow-stress model of deformed metals is applicable as long as the strength contributed by the boundaries is introduced as a variable parameter and not as a constant (as in Eq. (4.18a) of the H–P relation). Hansen has taken this into account. The HP flow-stress model is equivalent to Eq. (4.18a) and is given as:

$$\sigma(\varepsilon) = \sigma_0(\varepsilon) + \frac{k(\varepsilon)}{\sqrt{d}} \tag{4.25}$$

Nonetheless, experimental observations on several polycrystalline materials indicate that σ_0 and k are not always constants under all conditions and are strain-dependent.

Some examples of the grain-size effect in ceramics are illustrated below: Ti_3SiC_2 was chosen as one exemplar, since this ternary compound exhibits a unique combination of properties. It is a layered material that is as machinable as graphite. At the same time, CG (100–300 μm) samples of Ti_3SiC_2 have been observed to be damage-tolerant, not susceptible to thermal shock and oxidation resistant. The specimens are fully dense, bulk, single-phase polycrystalline samples of Ti_3SiC_2. This material exhibits brittle failure characteristics at RT, but is plastic at 1,300 °C with yield points of 300 and 100 MPa under compression and flexure, respectively.

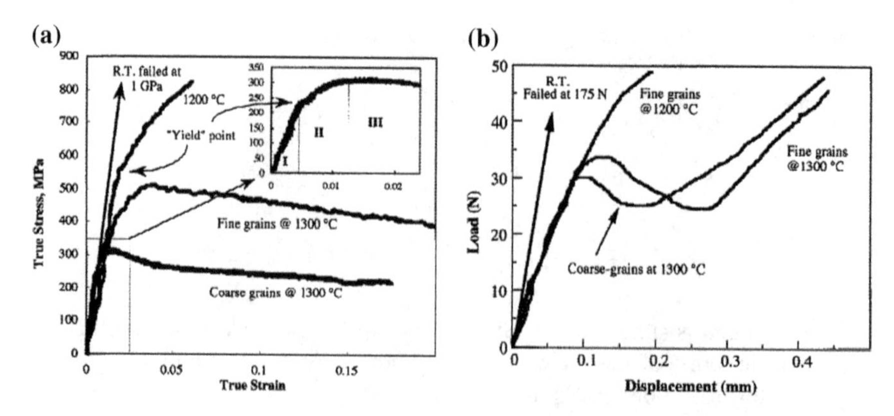

Fig. 4.63 Typical engineering stress–strain curves as a function of temperature and grain size in **a** compression and **b** four-point flexure. Unless otherwise noted, the curves are for the FG microstructure. In both cases, below 1200 °C, the failure is brittle, whereas above 1200 °C the failure is plastic, with significant plasticity [17]. With kind permission of John Wiley and Sons

Furthermore, a temperature/grain size effect is also shown for some mechanical properties. Recall that most ceramics are brittle at RT and become plastic only at elevated temperatures. Figure 4.63a shows the stress–strain relation for FG and CG specimens at RT and at two elevated temperatures, whereas, in Fig. 4.63b, the load–displacement curves are shown at the same temperatures.

The curves of Fig. 4.63a were obtained by compression, whereas, in Fig. 4.63b, the load displacement relation was obtained from flexure tests. As may be seen during compression, failure is brittle, as expected with compressive strengths of 1050 and 720 MPa for the FG and CG ceramics, respectively. The compressive strengths decrease with increasing temperatures above 1200 °C. At and below 1200 °C, the failure is brittle, with strains to failure of <2 %. At 1300 °C, however, both materials exhibit large plastic deformation levels (>20 %) with yield points of 500 and 320 MPa for the FG and CG materials, respectively. The 4-point bend test, i.e., the flexure strength (σ), was calculated using the following relation:

$$\sigma = \frac{3P(l_1 - l_2)}{2BW^2} \tag{4.26}$$

P is the load at fracture, l_1 and l_2 are the outer and the inner spans, respectively. A SiC fixture with an outer span of 20 mm and an inner span of 10 mm was used for the bend test (flexure). B is the specimen width (2 mm) and W the specimen thickness (1.5 mm). The compressive and flexural strengths of the FG and CG specimens (from Fig. 4.63), as a function of temperature, are shown in Fig. 4.64.

At all the temperatures, the FG material is stronger than the CG material and the compressive strength is higher than the flexural strength. A large decrease in strength at ~1200 °C is observed for both FG and CG microstructures, under compression and flexure. Experiments on microhardness by Vickers indenter were also performed on Ti_3SiC_2 ceramics specimens. The results of these tests are

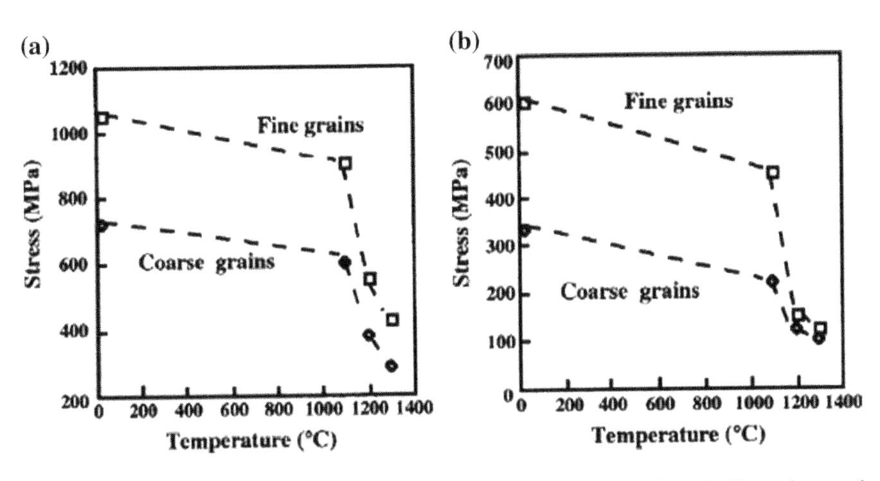

Fig. 4.64 Effect of grain size and temperature on **a** compressive strength and **b** flexural strength of Ti_3SiC_2. The data indicate the behavior of FG and CG specimens [17]. With kind permission of John Wiley and Sons

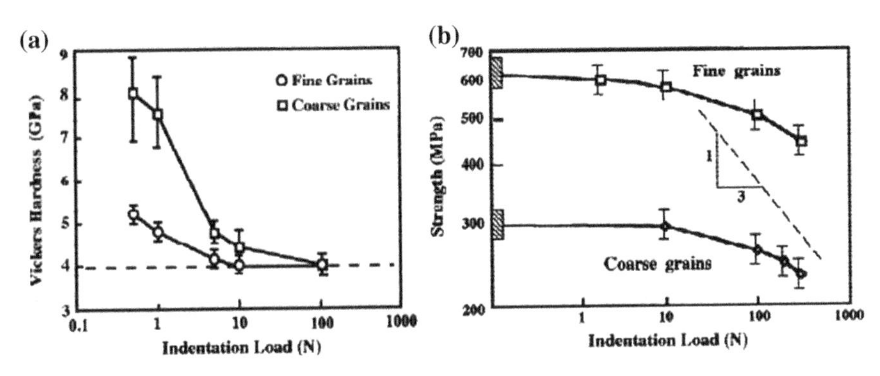

Fig. 4.65 **a** Effect of grain size on Vickers hardness as a function of applied load. For both microstructures, the asymptotic hardness value is 4 GPa. **b** Four-point flexural strength versus indentation load. Inclined *dashed line* has a slope of −1/3, which is the expected behavior for a perfectly brittle material. Hatched area on the left represents the strength of the samples as a result of natural flaws [17]. With kind permission of John Wiley and Sons

presented in Fig. 4.65. Figure 4.65a summarizes the microhardness measurements for the FG and CG microstructures. At higher loads (100 N), the hardness is 4 GPa and is independent of the grain size. The critical indentation load (at which hardness becomes independent of the indentation load) is influenced by grain size and is typically larger for a larger-grain-sized material. The measured values of the retained flexure strength, after the Vickers indentations, are summarized as a log–log plot in for both microstructures. The hatched area on the left represents the strength of the samples plot in Fig. 4.65b as a result of natural flaws. In contrast with the FG material, for which the retained strength decreases immediately with

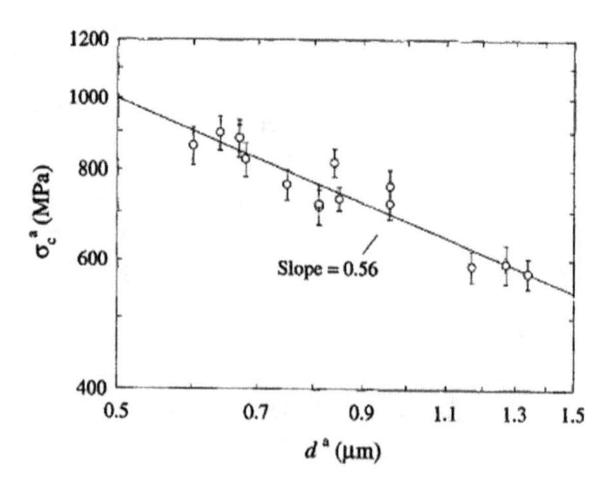

Fig. 4.66 Average strength decreases with grain size [34]. With kind permission of John Wiley and Sons

increasing indentation load, the CG material exhibits a plateau at low loads. Figure 4.65a summarizes the microhardness measurements for the FG and CG microstructures.

It may be seen from these curves that the FG specimens show better strength properties than the CG ones in Ti_3SiC_2 specimens. No consideration has been given to other properties, such as thermal shock resistance, porosity or flaws which accompany almost every ceramic and clearly affect their mechanical behavior.

In another example, the influence of grain size is indicated in the technically important Si_3N_4 polycrystalline ceramics. These ceramics are in situ reinforced [henceforth: ISR] ceramics, accompanied by 'crack bridging' by β-Si_3N_4 grains, i.e., when the silicon nitride has a pronounced acicular microstructure. The strength-flaw size relations are related to the behavior of a bridging zone behind the crack tip. Sintered ISR Si_3N_4 ceramics possess a strong R-curve property due to crack bridging.

Figure 4.66 shows the effect of grain size on strength. In this figure, average strength, σ_c^a, is plotted against average grain size, d^a, on a log–log scale. The slope of the linear-regression line equals 0.56, which indicates that ISR Si_3N_4, basically follows Orowan-type behavior, i.e., strength $\propto$ (grain size)$^{-0.5}$. The microstructures used in evaluating the strength properties of ISR Si_3N_4 are shown in Fig. 4.67 as a SEM micrograph of FG $d^a = 0.64$ μm (a), medium-grained $d^a = 0.84$ μm (b) and CG $d^a = 1.27$ μm (c) samples.

Crack grain interaction in ISR Si_3N_4 is illustrated in Fig. 4.68. In Fig. 4.68a, an intact bridging grain, ~30 μm from the tip of an indentation crack, is shown. This grain is elastically stretched. In Fig. 4.68b a fractured grain of Si_3N_4 is located behind the crack tip. A small bridging grain has been partially pulled out and fractured. It is possible that, after some sliding, the grain became locked up with the matrix and, eventually, the crack opening exceeded the elastic limit of the grain and the grain fractured. In some cases, the strong pullout resistance causes the surrounding matrix to fracture, as seen in Fig. 4.68b. Figure 4.68c shows a long

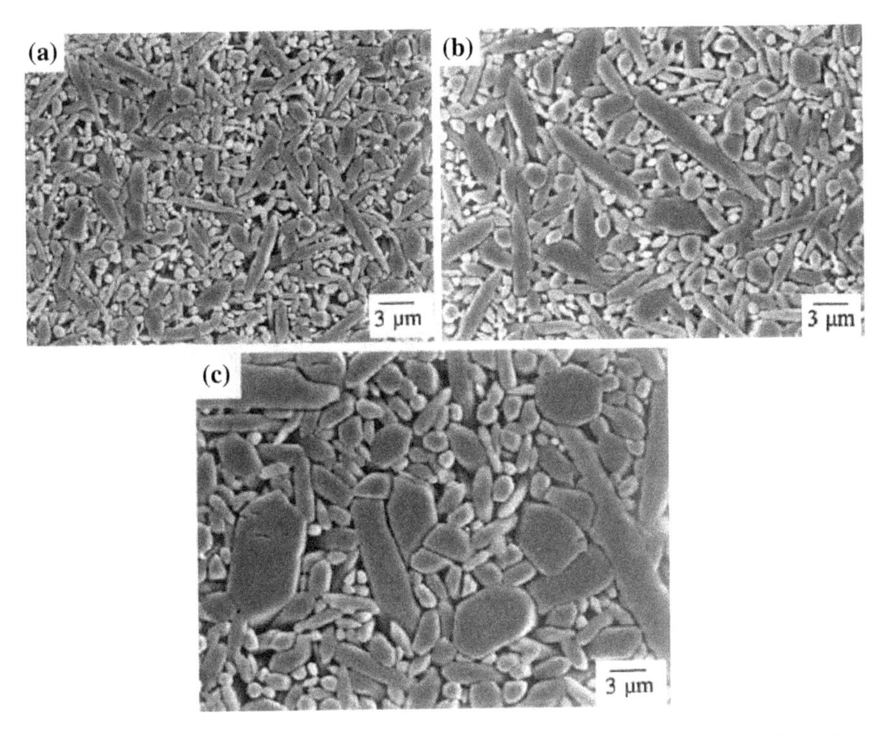

Fig. 4.67 SEM micrographs of polished and etched surfaces for: **a** fine-grained, **b** medium-grained, and **c** coarse-grained ISR Si_3N_4 [34]. With kind permission of John Wiley and Sons

acicular grain, nearly perpendicular to the main crack, that has been fractured and pulled out, a behavior analogous to that of a long fiber. To the left of that long grain, there is an elastic bridging grain with a corresponding crack opening smaller than the one closer to the crack tip. The stress level in ceramics is critically dependent on the flaws (cracks), which must be taken into account when the strength properties are considered. Figure 4.69 is an illustration of the strength-flaw size relation. The best fitting curves for the three types of grain sizes were calculated on the basis of Eq. (4.27), given as:

$$\sigma_c = \frac{\sqrt{\pi}}{2} \frac{K_0}{(c_0 + D_b)^{1/2}} + p^* \frac{\left[(c_0 + D_b)^2 - c_0^2\right]^{1/2}}{(c_0 + D_b)} \tag{4.27}$$

In relation (4.27), c_0 refers to the penny-shaped flaw size at the moment of crack initiation. K_0 is an intensity factor chosen as $K_0 = 2\ MPa.m^{0.5}$. The bridging-zone length, D_b, prior to critical fracture, has a bridging stress, p^*, assumed to be constant. For the derivation of Eq. (4.27), the reader should consult the work of Chien-Wei Li et al. [34]. The estimated p* for the FG, medium and

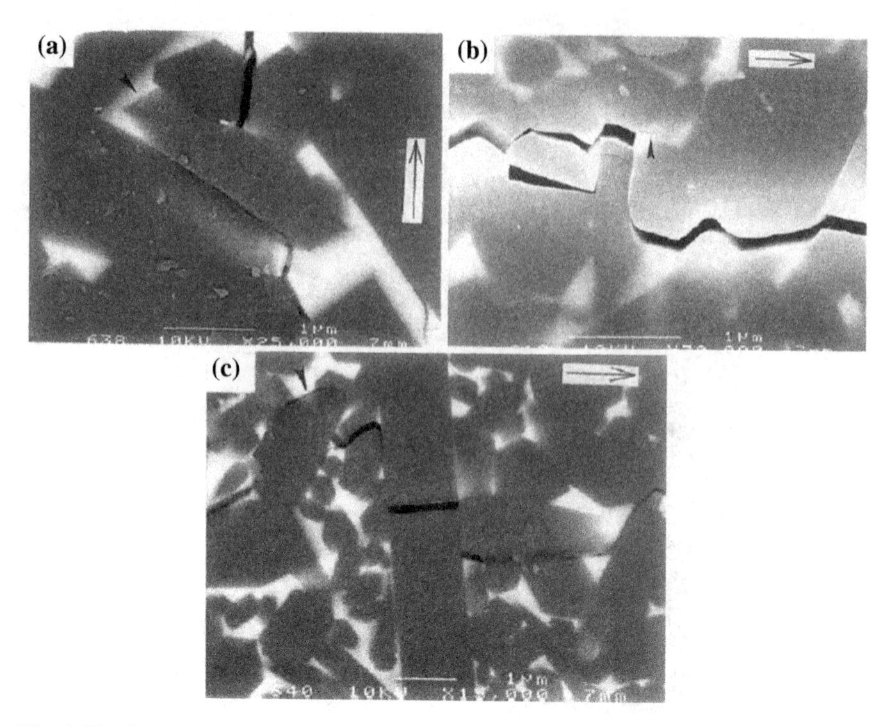

Fig. 4.68 SEM micrographs showing examples of **a** elastic bridging, and **b**, **c** pullout. The arrows indicate crack propagation directions [34]. With kind permission of John Wiley and Sons

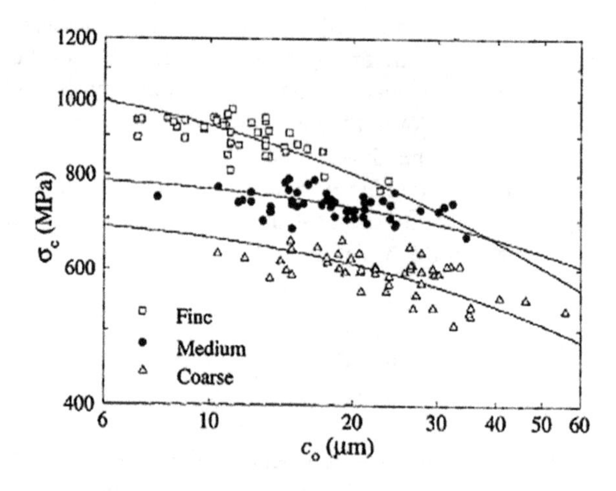

Fig. 4.69 The solid curves are the best fits from Eq. (4.27). Strength-flaw size relations in the three examples of Fig. 4.67 [34]. With kind permission of John Wiley and Sons

CG samples are 628, 556, and 408 MPa, and the corresponding D_b are 20, 73, and 48 μm, respectively. The results of p* and D_b for all the samples are plotted against d^a, the average grain width in Fig. 4.70.

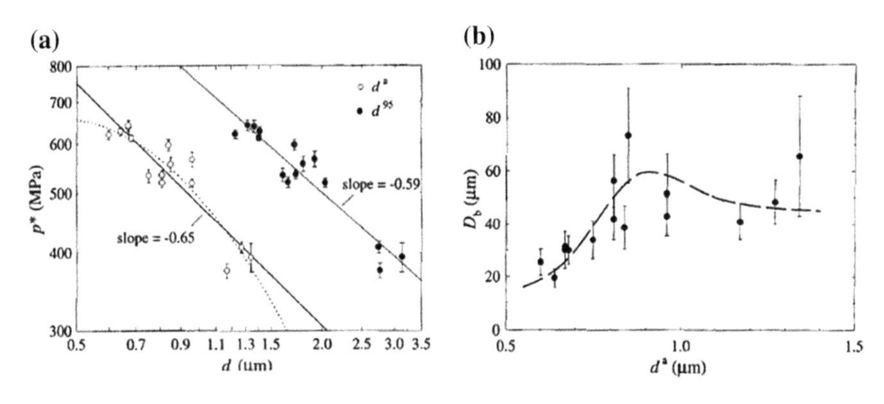

Fig. 4.70 **a** Critical bridging stress decreases with grain width; **b** Variation of bridging zone length with average grain width [34]. With kind permission of John Wiley and Sons

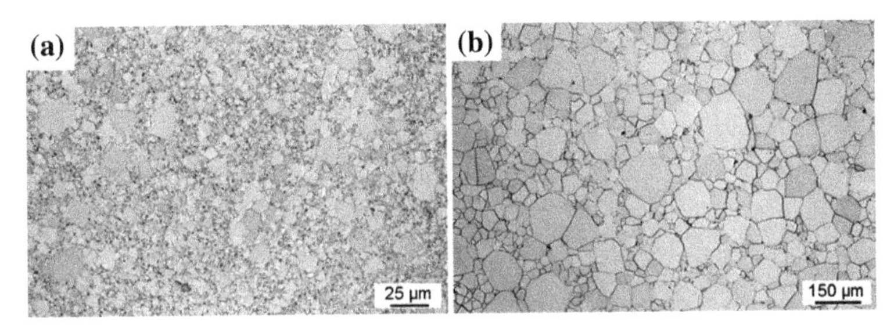

Fig. 4.71 The thermally etched structure of **a** fine- and **b** coarse-grained spinel (light microscopy) [11]. With kind permission of Elsevier

In summary, the average strength of a material decreases with its grain size following the Orowan-type relation, while the average initial flaw size generally increases with grain size. The effect of grain size on the property of strength is demonstrated by a −50 % difference in strength between FG and CG materials having a given initial flaw size. Furthermore, it has been assumed that in a sample the critical bridging stress, p*, is constant and that the critical crack size is the sum of an initial flaw size and a critical bridging-zone size, D_b.

Ceramics are used in various technical applications where either specific functional or structural properties are required. Transparent ceramics have gained importance in technical window applications, one of them is a spinel based on $MgAl_2O_4$. Therefore it was felt to add an additional ceramics of this kind and illustrate the grain size effect on the strength properties. Figure 4.71 illustrates the microstructure of these ceramics. Figure 4.72 shows the changes in the Vickers hardness as a function of applied load for FG and CG spinel, which is less pronounced for the FG. However, the level of hardness with each applied load is higher for the FG ceramics.

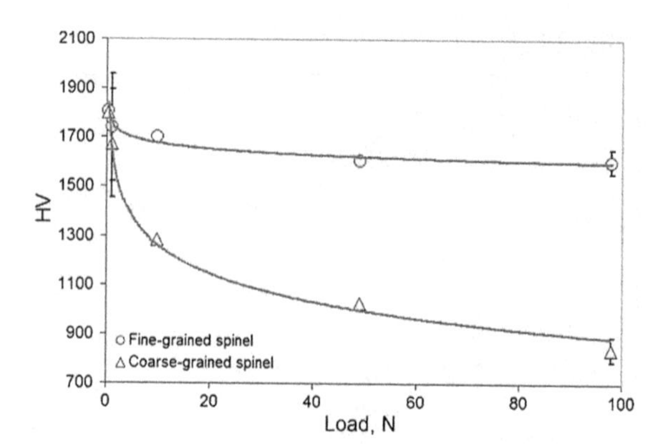

Fig. 4.72 Vickers hardness as a function of the indentation load [11]. With kind permission of Elsevier

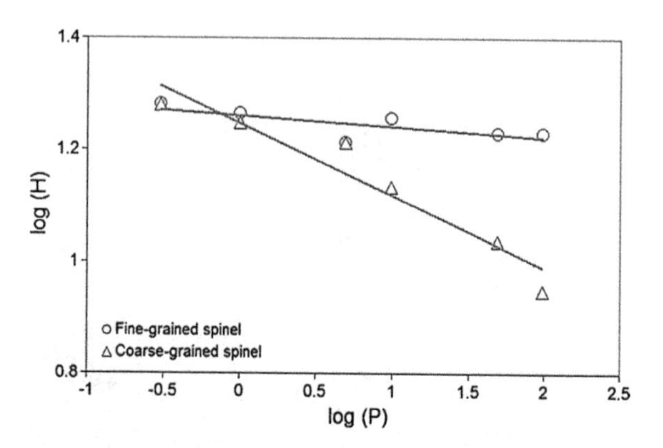

Fig. 4.73 Indentation size effect for fine- and coarse-grained spinel [11]. With kind permission of Elsevier

Hardness versus load may be expressed on a logarithmic scale, as shown in Fig. 4.73, obtaining straight lines for both FG and CG spinels. The line representing the CG material changes with load, whereas that of the FG spinel is almost constant.

Hardness-load curves are usually based on the relation of:

$$HV = kP^c \tag{4.28}$$

and

$$\ln HV = \ln k + c\ln P \tag{4.29}$$

HV (Vickers hardness), P (load) and k are parameters that can be determined from the intercept and slope of the straight lines.

In Fig. 4.74, the microstructure indicates the origin of fracture in FG spinel by bending test. It shows microcracks, before and after the bending, as the origin of

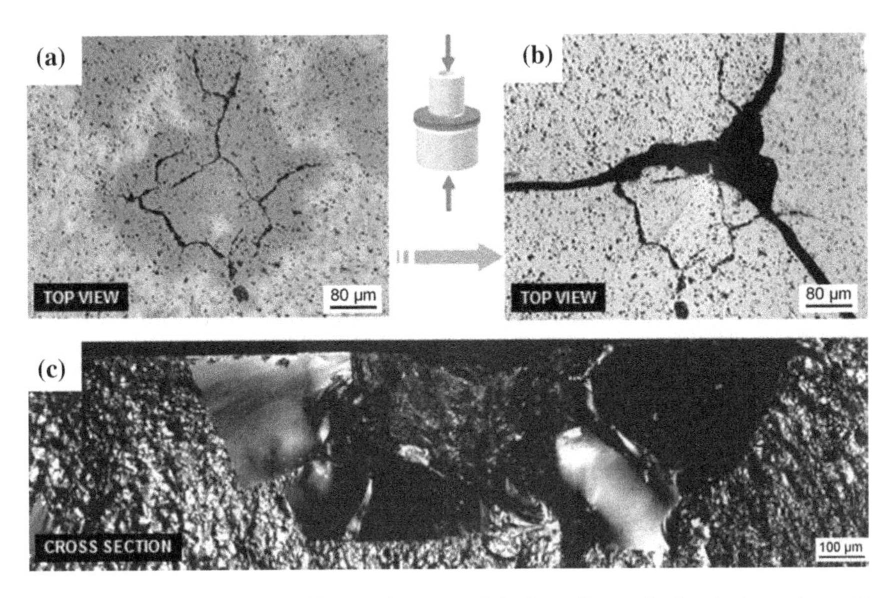

Fig. 4.74 A typical surface defect as fracture origin in a fine-grained spinel specimen. In **a** surface agglomerate and associated microcrack before and **b** after bending test. **c** Fracture surface confirming the extent of the coarse-grained zone [11]. With kind permission of Elsevier

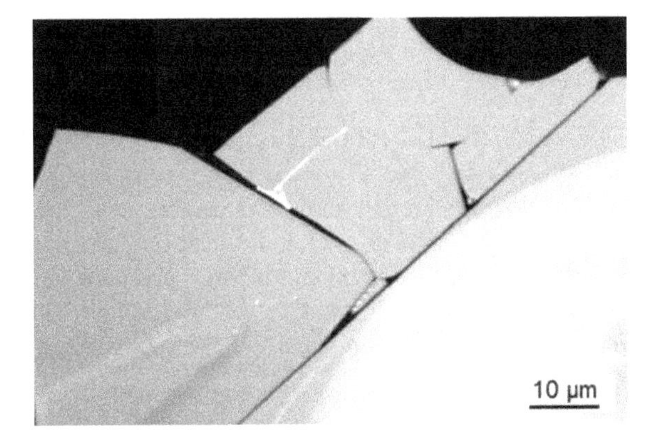

Fig. 4.75 Weak grain boundaries as fracture origins of coarse-grained spinel [11] With kind permission of Elsevier

failure. Compare this microstucture with the origination of a CG fracture, which occurs at weak grain boundaries causing failure (Fig. 4.75). A plot of fracture strength versus the mechanically loaded area is illustrated for both FG and CG in Fig. 4.76. The subcritical crack growth [henceforth: SCG] parameters, n, and fracture stresses determined from the strength parameter time [henceforth: SPT]

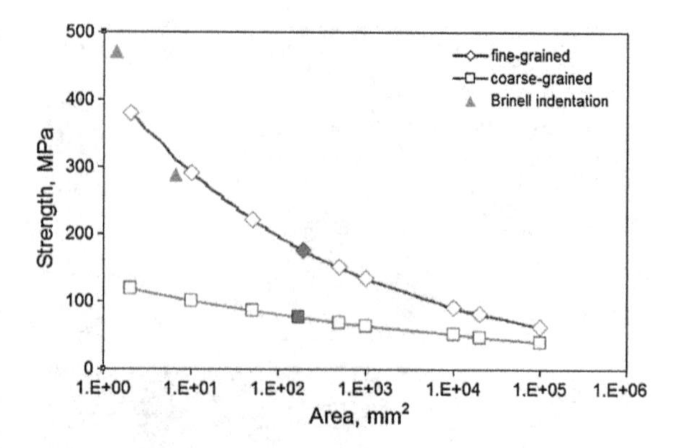

Fig. 4.76 Area effect on fracture strength for fine- and coarse-grained spinels. The experimental strength values are marked by *red color* [11]. With kind permission of Elsevier

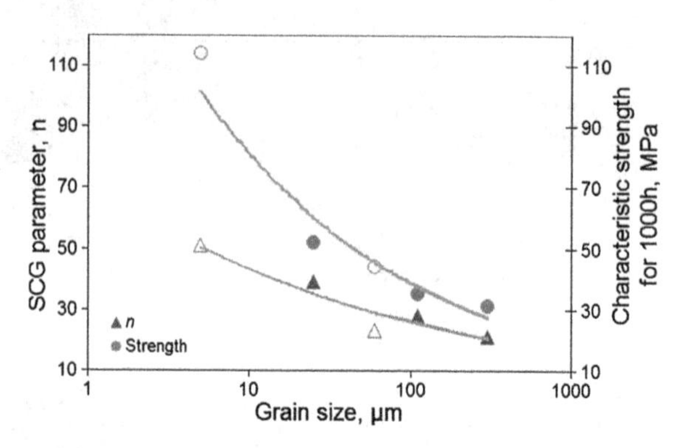

Fig. 4.77 Grain size dependency of SCG parameter and characteristic strength for 1000 h lifetime. *Filled* and *unfilled circles* are literature values and present work, respectively [11]. With kind permission of Elsevier

plot for a lifetime of 1000 h are compared in Fig. 4.77. One can see that both SCG and strength decrease with increasing grain size. The graphic presentation clearly verifies that the strength and SCG parameter decrease with increasing grain size. The grain-size effect appears to be stronger for strength (Fig. 4.77). The SCG was assessed using the strength–loading rate dependence with additional tests at 3.6×10^{-2}, 3.6×10^{-1} and 3.6 MPa/s (around 10 specimens each). The characteristic strength was correlated with the stress rate, $\dot{\sigma}$, for a particular stress rate by means of:

$$\log \sigma_f = \frac{1}{n+1} ogb + \log D \tag{4.30}$$

where n and D (in MPa/s) are SCG parameters. The derived SCG parameters for the FG transparent $MgAl_2O_4$ are $n \sim 51$ and $D \sim 140$ MPa. For the CG spinel $n \sim 23$ and $D \sim 71$ MPa are obtained.

To conclude the above information on spinel, an SCG assessment and strength–probability-time prediction for long-term reliability assessment were based on the loading-rate effect of fracture-strength measurements. The parameter characterizing the slow crack-growth sensitivity suggests that FG materials are less affected. The maximum stress for a lifetime of 40 years (failure probability of 1 %) was

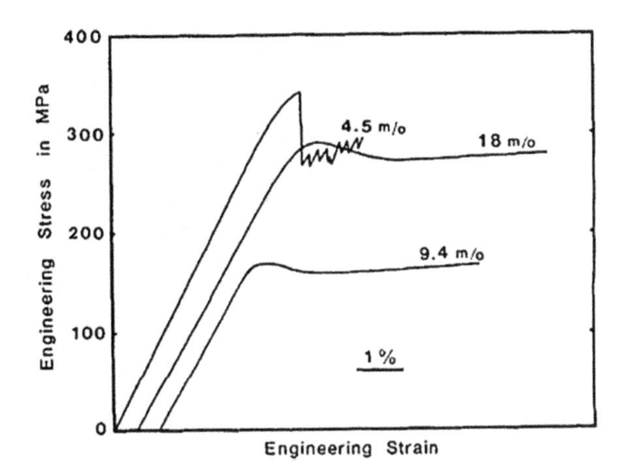

Fig. 4.78 Engineering stress–strain curves for several Y_2O_3-stabilized ZrO_2 single crystals; composition corresponding to each crystal is indicated [2]. With kind permission of John Wiley and Sons

56 MPa for the FG material and 25 MPa for the CG material. Overall, it appears that strength and the slow crack-growth parameter decrease with increasing grain size.

4.11 Closing Remarks

No consideration has been given to two important observations of phenomena known to occur in alloys: (a) that of Lüders bands, which are strongly associated with strain aging, and (b) deformation in reverse directions, known as the 'Bauschinger effect'. Surprisingly, it seems that as yet no reports have been recorded in the literature on ductile ceramics. One cannot help but wonder if these phenomena do not occur in ductile ceramics or have merely been overlooked due to a lack of interest. One may also wonder whether experiments performed on existing ceramics with well-prepared surfaces (polished and possibly etched) after strain aging might enable the observation of the formation of Lüders bands. Similarly, it is of interest to know if well-designed experiments performed on specimens loaded in two directions, namely, loaded, unloaded and reloaded in the opposite direction immediately after the unloading, might provide observations of Bauschinger-like phenomena. For students with an inquisitive nature, these might be valuable topics for exploration.

Appendix

In Sect. 4.11 the question has been raised whether Lüders bands exist in ceramics. A report specifically refers to the observation of Lüders bands in a two-phase ceramics. The material is high temperature precipitation hardening Y_2O_3-partially

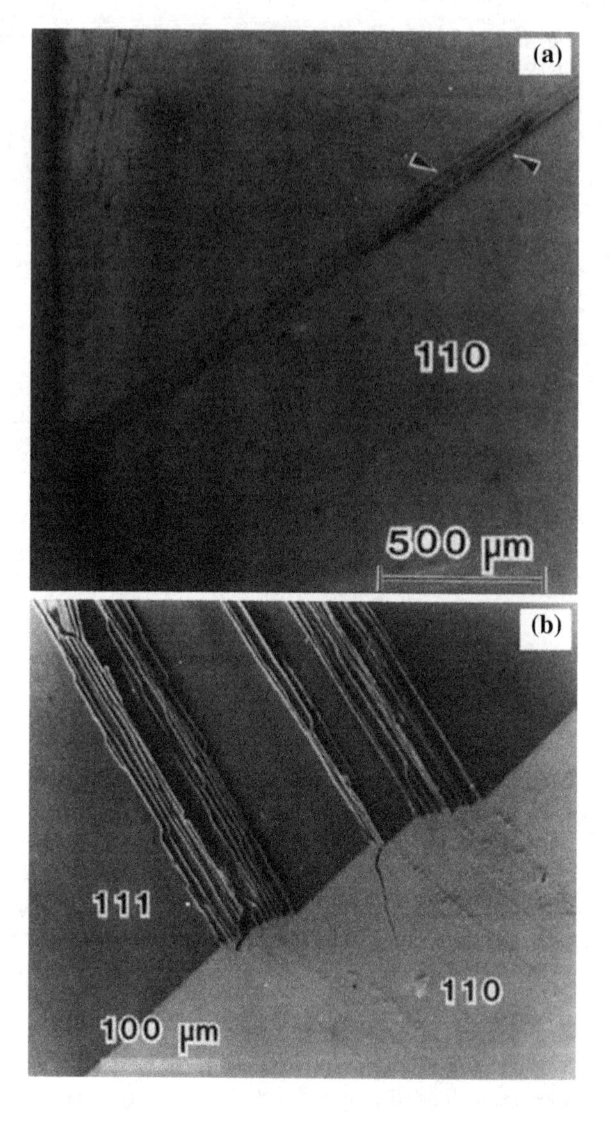

Fig. 4.79 **a** Optical micrograph of Lüders band in deformed specimen ((110) face); arrows show polishing scratch sheared by first Lüders band. **b** Scanning electron micrograph of Lüders band on (111) face [2]. With kind permission of John Wiley and Sons

stabilized-ZrO_2 single crystals. Figure 4.78 shows engineering stress–strain curves of Y_2O_3-stabilized ZrO_2 single crystals. Of the three curves the 4.5 mol% crystal shows a well-developed upper yield point and yield drop, and then serrated flow occurs similar to the one observed in steels. The observed phenomenon is associated with Lüders bands formation as seen in Fig. 4.79.

The specimens were tested by compression on parallel faces to the orientation $\{110\}$ and $\{111\}$ with loading axis parallel to $\langle 112 \rangle$. The Schmid factor evaluated is 0.47 for the slip system $\{001\}$ $\langle 110 \rangle$ which is the easy slip system for the cubic zirconia (c-ZrO_2). The deformation experiments in air at 1400 °C were performed with the crosshead rate of 5 μm/min which provided a strain rate of

$\dot{\varepsilon} = 1.3x10^{-5}/s$. The Lüders bands seen in Fig. 4.79 are parallel to the {001} slip planes. The large yield drop corresponds to the first Lüders band formation while the formation of additional Lüders band is associated with the serration of the stress–strain curve. Many specimens tested gave the same stress–strain curves with Lüders bands formation.

References

1. Arlt G (1990) J Mater Sci 25:2655 (41–44)
2. Dominguez-Rodrigez A, Lanteri V, Heuer AH (1986) J Am Ceram Soc 69:285
3. Friedel F (1964) Dislocations. Pergamon Press, Oxford, Student Edition 1964 (reprinted in 1967). p 405
4. Gogotsi GA, Groushevsky YaL, Strelov KK (1978) Ceramurgia Int 4:113
5. Han XD, Zhang YF, Zheng K, Zhang XN, Zhang Z, Hao YJ, Guo XY, Yuan J, Wang ZL (2007) Nano Lett, 7:452
6. Ogata S, Li J, Hirosaki N, Shibutani Y, Yip S (2004) Phys Rev B 70:104104 (2004)
7. Pelleg J (2013) Mechanical properties of materials. Springer, New York
8. Rollett AD, Garmestani H, Branco G (2005) Polycrystal plasticity-multiple slip, advanced characterization and microstructural analysis, carnegie mellon, department of materials science and engineering
9. Schmid E, Boas W (1935) Kristalplastizitat (trans: Plasticity in Crystals, F. A. Hughes and Co. 1950). Springer Verlag, Berlin
10. Scott WD, Orr KK (1988) J Am Ceram Soc 66:83
11. Tokariev O, Schnetter L, Becka T, Malzbender J (2013) J Eur Ceram Soc 33:749

Further References

12. Alybaeva NR, Berezhkova GV, Govorkov VG (1976) Kristal und technik 11:1315
13. Barat P, Sarkar A, Barat A (2006) Phys Rev B 74:245425
14. Barsoum MW, Farber L, El-raghy T (1999) Met Mater Trans A 30A:363
15. Cho J-H, Lee Y-H, Han K-S, Chun M-P, Nam J-H, Kim B-I (2010) J Korean Phys Soc 57:971
16. Copley SM, Pask JA (1965) J Am Ceram Soc 48:636
17. El-Raghy T, Barsoum MW, Zavaliangos A, Kalidindi SR (1999) J Am Ceram Soc 82:2855
18. Eshelby JD, Frank FC, Nabarro FRN (1951) Philos Mag 42:351
19. Firestone RF, Heuer AH (1973) J Am Ceram Soc 56:136
20. Frenkel J (1926) Z Phys 37:572
21. Gallardo-Lopez A, Gomez-Garcıa D, Domınguez-Rodrıguez A, Kubin L (2004) Scripta Mater 51:203
22. Gheno M, Hasegawa HL, Iris P, Filho P (2006) J Phys Chem Solids 67:2253
23. Gleiter H (1969) Acta Metall 17:1422
24. Goretta KC, Routbort JL, Bloom TA (1986) J Mater Res 1:124
25. Hall EO (1951) Proc Phys Soc B 64:747
26. Hansen N (2005) Adv Eng Mater 7:815
27. Hollox GE, Smallman RE (1966) J. Appl Phys 37:818
28. Hulse CO, Copley SM, Pask JA (1963) J Am Ceram Soc 46:317
29. Inui H, Ishikawa K, Yamaguchi M (2000) Intermetallics 8:1131
30. Ito K, Inui H, Shirai Y, Yamaguchi M (1995) Philos Mag A72:1075

31. Kilo M, Borchar G, Weber, Scherrer S, Tinschert K (1997) Ver Bunsenges Phys Chem 101:1361
32. Kim H-S, Roberts S (1994) J Am Ceram Soc 77:3099
33. Kwon S, Sabolsky EM, Messing GL, McKinstry ST (2005) J Am Ceram Soc 88:312
34. Li C-W, Lui S-C, Goldacker J (1995) J Am Ceram Soc 78:449
35. Liang S, Qingchuan Z, Huifeng J (2007) Front Mater Sci China 1:173
36. Liu L, Zuo R, Liang Q (2013) Ceram Int 39:3865
37. Maloy SA, Mitchell TE, Heuer AH (1995) Acta Metall Mater 43:657
38. Mitchell TE, Castro RG, Petrovic JJ, Maloy SA, Unal O, Chadwick MM (1992) Mater Sci Eng A 155:241
39. Orowan E (1948) Rep Prog Phys 12:186–232
40. Paterson MS, Weaver CW (1970) J Am Ceram Soc 53:463
41. Petch NJ (1953) J Iron Steel Inst Lond 173:25
42. Petch NJ (1964) Acta Met 12:59
43. Pravarthana D, Chateigner D, Lutterotti L, Lacotte M, Marinel S, Dubos PA, Hervas I, Hug E, Salvador P, Prellier W (2013) J Appl Phys 113:153510
44. Seabaugh MM, Cheeney GL, Hasinka K, Azad A-M, Sabolsky EM, Swartz SL, Dawson WJ (2004) J Intell Mater Syste Struct 15:209
45. Snow JD, Heuer AH (1973) J Am Ceram Soc 56:153
46. Szczerba MS, Bajor T, Tokarski T (2004) Philos Mag 84:481
47. Uchida T, Wang Y, Rivers ML, Sutton SR (2004) Earth and Planetary Science Letters 226:117
48. von Mises R (1928) Mechanics of plastic deformation in crystals. Z Angew Math Mech 8:161
49. Wang YG, Bronsveld PM, De Hosson JThM, Djuričić B, McGarry D, Pickering S (1998) J Eur Ceram Soc 18:299
50. Wu W-W, Sakka Y, Suzuki TS (2013) Int J Ceram Technol 1
51. Wu Y-C, Lee C–C, Lu H-Y, McCauley DE, Chu MSH (2006) J Am Ceram Soc 89:1679

Chapter 5
The Strength and Strengthening of Ceramics

Abstract There are several mechanisms by which materials may be strengthened as listed: (i) Strain (or work) hardening in ductile ceramics, (ii) Solid-solution strengthening by pinning dislocations either by interstitial or substitutional atoms, (iii) Second-phase hardening, (iv) Transformation hardening, (v) Strengthening by grain boundaries. Strain hardening is a feature of ductile ceramics, but at high temperatures where brittle materials show plasticity, strain hardening does not necessarily occur. In brittle materials that show plasticity at elevated temperature, strain hardening depends on composition and conditions of the test. It is possible that, due to the recovery process, strain hardening will not be observed. Superplastic materials are characterized by high ductility and no strain hardening occurs. In particular in superplastic materials such as MgO, strain hardening is absent while in β-Si_3N_4 under compression it is not observed. However, it has often been observed that little or no hardening at all occurs. Either interstitial or substitutional atoms can pin dislocations and thus strengthen the material. Second phase particles not in solution can hinder dislocations in their motion with a consequent increase in strength in the ceramics. Ceramics such as those based on zirconia are likely to undergo phase transformation, in particular the yttria stabilized zirconia, which is associated with strengthening of the material. Clearly, grain boundaries are obstacles to dislocation motion and thus harden the ceramic.

5.1 Introduction

The strengthening of ceramics is of great importance for engineering applications, especially at elevated temperatures. Various production processes exist; most are based on powder technology and various sintering techniques (after the preparation of a green structure), but fusion or molten processes are also possible, albeit rarely used. Whatever the production technique may be, the aim is to attain a high-density product with minimal flaws (pores, microcracks, etc.).

J. Pelleg, *Mechanical Properties of Ceramics*, Solid Mechanics and Its Applications 213, DOI: 10.1007/978-3-319-04492-7_5,

Construction parts are designed not only to endure anticipated and intentionally applied forces (those they are meant to withstand during service), but also any sudden, short-duration forces that may cause catastrophic failure, if not taken into account. In order to avoid the probability of such failures, liberal safety factors are generally adopted by designers. Their approach is to strengthen materials beyond the magnitude which is sufficient for the prevention of failure, even if a steady force were to be exerted during the entire period of their use. This extra strength value constitutes the safety factor required for ensuring the safe use of a construction part, even in the event that a sudden force of larger magnitude appears during service. There are several mechanisms by which materials may be strengthened, listed below:

(i) Strain (or work) hardening in ductile ceramics;
(ii) Solid-solution strengthening by pinning dislocations:

  a. interstitial atoms,
  b. substitution atoms;

(iii) Second-phase hardening;
(iv) Transformation hardening;
(v) Grain boundaries and grain size.

These subjects are discussed in this chapter. However, to understand the concept of hardening (strengthening), one must realize that all the aspects of strengthening involve dislocation generation, dislocation movement and the interactions of dislocations with each other or with any of the above-listed entities that might hinder their free motion.

Intentionally in the present discussion, no differentiation is made between single-crystal and polycrystalline ceramics and they are used interchangeably, depending on the topics under consideration.

5.2 Strain Hardening

This section will consider the hardening effects in (a) common ceramics, which are usually brittle up to some elevated temperature; (b) ceramics which show ductility and; (c) superplastic ceramics. But before doing so, it is important to review the general concept of strain hardening (work hardening) in materials. Briefly, dislocations interact with each other and consequently the motion of dislocations is restricted (i.e., the motion of a mobile dislocation may be hindered when it encounters another dislocation). Interacting dislocations may hinder each others' motion either strongly or weakly, depending on their positions, types and other geometric features. Since dislocations are responsible for plastic deformation, they describe strain (work) hardening as the strengthening of the material by plastic

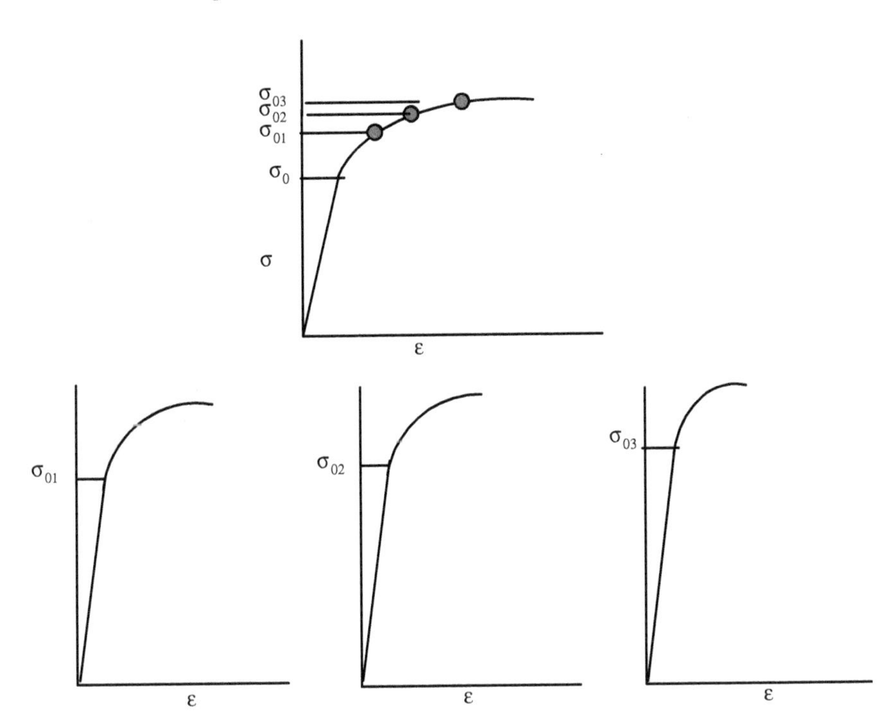

Fig. 5.1 The top schematic figure shows the strain hardening on plotting a tensile stress–strain curve. The bottom curves show that the yield stresses for specimens, reloaded after each unloading, are: σ_{01}, σ_{02} and σ_{03}. Thus, by repeating this procedure, the strength continues to increase, while ductility decreases, until the specimen becomes brittle: $\sigma_{01} < \sigma_{02} < \sigma_{03}$

deformation at low temperatures, where recovery processes do not occur. On a stress–strain plot, a gradual increase in resistance to deformation with plastic strain is observed, requiring higher loads for continued deformation. In terms of shear stress, it is common to relate its variation (increase for further deformation) to dislocation density as:

$$\tau = \tau_0 + \alpha G b \rho^{1/2} \tag{5.1}$$

The figures (Fig. 5.1) show schematically one type of strengthening, namely, work hardening (shown step-by-step); recall that work hardening involves an increase in yield stress (and, hence, of the entire stress–strain curve). Loading a specimen up to the red point (first) and then unloading it, shows a higher yield stress on reloading. The material has been hardened and now its yield stress is $\sigma_{01.}$ As a consequence of hardening, the strength of the specimen increases due to its decreasing ductility. Similarly, if the specimen's loading had been stopped at the second (or third) red point, and it was unloaded and then reloaded, the yield stress would be $\sigma_{02,}$ with a consequent strength increase and decrease in ductility, etc.

5.2.1 Brittle Ceramics

In general, ceramics may contain dislocations, but normally at RT they are brittle to an extent at which no plastic deformation can take place. Thus, ceramics basically cannot strain-harden; consequently, no strengthening is possible at low temperatures. Strain hardening at RT requires the deformation of a material beyond its yield point, but ceramics do not show yield points, being brittle. Under tension, ceramics show only elastic behavior. Ceramic materials, characterized by ionic or covalent bonding, have a limited number of slip systems. (A covalent bond is directional and electrons are shared. Hence, bonds will not reform easily and brittle fracture will occur.) Such materials tend to fracture before any (or very slight) plastic deformation takes place and catastrophic failure may set in if the elastic limit is surpassed. A crack so formed is unstable and propagates rapidly, without any further increase in applied stress. Ceramic materials inherently have cracks, flaws, pores and inclusions. These act as stress raisers that may initiate failure and propagate quickly (because there is no energy-absorbing mechanism, as there is in metals).

5.2.2 Ductile Ceramics

A few ceramics are ductile at RT, but the majority, under common conditions, is ductile only at some elevated temperatures. As such, temperature is a critical factor in making ceramics ductile. When discussing strengthening, in general, and work hardening, in particular, this factor will not be specifically considered unless there are extenuating circumstances.

An accepted way to express strain hardening in materials is by:

$$\sigma = K\varepsilon^n \tag{5.2}$$

where n is the strain-hardening exponent, the value of which is characteristic of the material being deformed. Low values of n mean low work hardening. As previously indicated, ceramics contain dislocations and can be work-hardened to a degree, depending on their temperature-dependent ability to deform.

During the process of making ceramics stronger by means of plastic deformation, dislocations become mobile under the influence of stress and also additional dislocations are generated. The more dislocations that are generated and present (as a result of plastic deformation), the greater the likelihood of encountering other dislocations and interacting with them; this makes their motion more difficult and may result in pinning and tangling. Retarding the mobility of dislocations by pinning and tangling requires additional stress for the continued motion of the dislocations. This is reflected on the stress strain curve, expressing the resistance of ceramics to further deformation (see Fig. 5.1). When ceramics are ductile at RT, the relatively low temperature does not allow rearrangement by atomic diffusion to take place. The term 'recovery' represents a structural rearrangement which commences at some temperature. The strain hardening observed in a stress–strain plot represents the balance between strengthening due to dislocation pinning and recovery. Thus, temperature plays a key role in the strain-hardening process. In such cases, when no strain hardening has been observed (or only to a small degree) a recovery process (involving the re-crystallization of deformed grains) dominates and eliminates the strengthening effect. Thus, often at high temperatures in ductile ceramics, no strain hardening is observed. (Recall that in metals during hot working the dislocations may rearrange and typically little strengthening is achieved). The final stage, i.e., the continued straining of a specimen, will eventually embrittle the material, because increasing the strength also reduces the ductility and leads ultimately to failure.

In ceramics, the extent of plasticity caused by dislocation, when tension tests are performed, is limited and very often accompanied by microcracking. Many of the tests are consequently performed by low-load indentation, such as microindentation, where the indenter may be focused on a grain. Nevertheless, other testing methods are also used. In the figure below, the stress–strain curves represent compressions at a constant strain rate at several temperatures starting with RT and up to 1423 K (Fig. 5.2).

In particular, note the lower temperature curves in the 1218–1323 K range. The strain hardening in this polycrystalline MgO specimen decreases with increasing temperature. The effect of temperature on strain hardening is clear. If the temperature is sufficiently high, the stress–strain plot becomes flat, developing a plateau with zero hardening. The strength of the ceramic drops and its ductility increases, almost as in superplastic ceramics, to be discussed later.

The first signs of macroplasticity are already observed at 1173 K and, with increasing temperature, the plastic strain becomes more extensive. Also, the complete brittle behavior at RT under the stress–strain curve is indicated in Fig. 5.2. The 0.2 % offset yield stress is much lower than that for slip in the $\{001\}\langle 1\bar{1}0\rangle$ systems and, therefore, it is believed that dislocation slip occurs in the $\{110\}\langle 1\bar{1}0\rangle$ system and some grain boundary processes also contribute to plasticity at higher temperatures. The behavior in this very-fine-grained MgO is different than in the coarse-grained, in which, at about 7 % strain, grain-boundary cracking is already well developed, as indicated by microstructural examinations. However, it will be more illuminating to consider MgO single crystals, where no grain and boundary contributions exist. Figure 5.3 shows stress–strain curves of MgO single crystals. The stress axis is in a $\langle 100\rangle$ direction.

All the specimens were loaded under compression at a constant stress rate of 20 psi per second in the temperature range of 1000–1600 °C. Loading the specimens at his orientation results in a maximum resolved shear stress on four of the six possible slip systems belonging to the $\{110\}<1\bar{1}0>$ family. The resolved shear stress vanishes in the other two slip systems, as well as in the $\{001\}<1\bar{1}0>$ family. The four slip systems consist of two pairs of orthogonal slip systems which intersect at 60°. Figure 5.4 shows the results of compression tests performed with the stress axis in the $\langle 111\rangle$ direction.

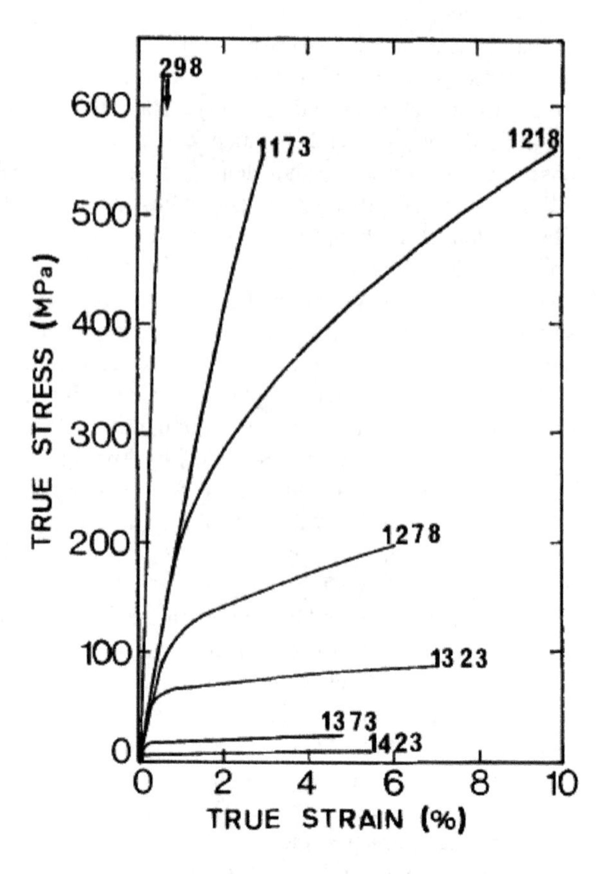

Fig. 5.2 Stress–strain curves in compression at constant strain rate of $\sim 6.7 \times 10^{-6}\ s^{-1}$ as a function of temperature (K). At room temperature, *arrow* indicates fracture of specimen [4]. With kind permission of John Wiley and sons and Professor Escaig

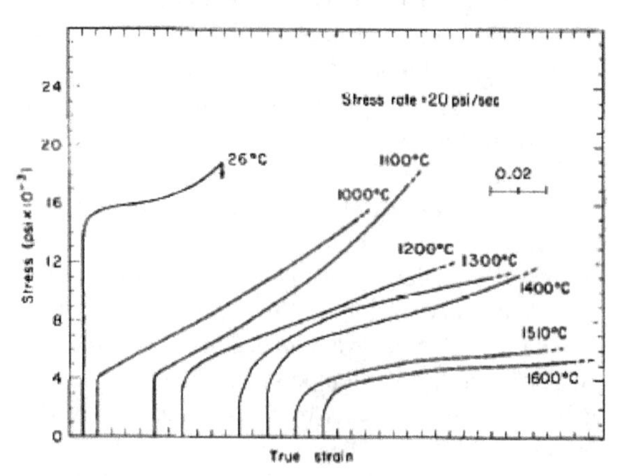

Fig. 5.3 Stress–strain curves for MgO single crystals compressed with a ⟨100⟩ stress axis at various temperatures. All specimens were loaded at 20 psi per second. The specimens deformed at 1300, 1510, and 1600 °C were cleaved from the same crystal block [5]. With kind permission of John Wiley and Sons

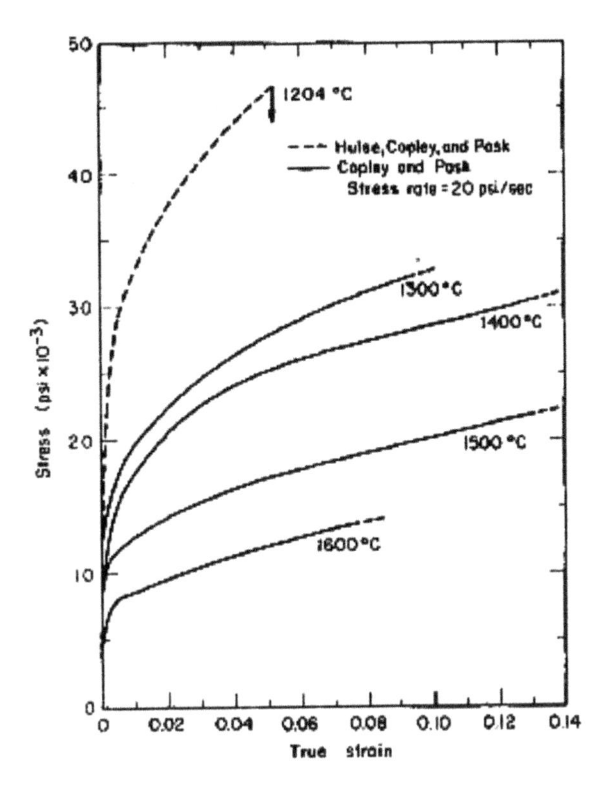

Fig. 5.4 Stress–strain curves for MgO single crystals compressed with a (111) stress axis at various temperatures. All specimens were loaded at 20 psi per second [5]. With kind permission of John Wiley and Sons

In this orientation, equal shear stresses were operating on three of the six slip systems of the $\{001\} < 1\bar{1}0 >$ family. The strain hardening obtained in this case is considerably greater than that observed in curves obtained when the stress axis is in a $\langle 100 \rangle$ direction. This is due to the long-range repulsive stress interactions existing between the intersecting dislocations resulting from the Burgers vectors of the three active slip systems oriented at 60° one to another. Note that the yield stresses for the $\langle 111 \rangle$ oriented specimens were considerably greater than those for the $\langle 100 \rangle$ orientations and continuously decreased up to 1600 °C.

Not only T has an effect on the strain hardening (strain hardening decreases with T), but also the stress rate influences strain hardening, as exemplified for MgO in Fig. 5.5. The effect of stress rate on plastic flow and, thus, on strain hardening is also opposite to that of the temperature. Increasing the stress rate increases the strain hardening of the ceramic MgO. These tests were performed at the same T, at 1321 °C.

It was indicated in the above curves that strain hardening is related to an increase in the yield stress, which can be observed after unloading, followed by an increment of plastic deformation and upon reloading the specimen. The slope of the stress–strain curve, being a measure of the increase in the stress on a stress–strain curve, usually defines the strain hardening rate. The slope, at a constant stress rate, is expressed as:

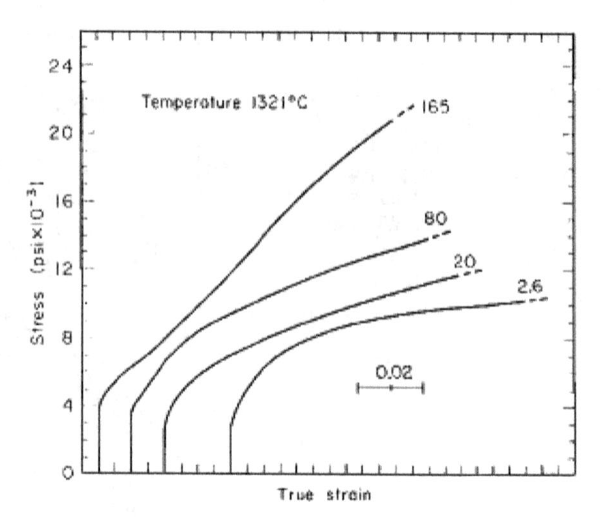

Fig. 5.5 Stress–strain curves for MgO single crystals compressed with a ⟨100⟩ stress axis at various constant stress rates. The specimens deformed at 165, 80, and 20 psi per second were cleaved from the same crystal block [5]. With kind permission of John Wiley and Sons

$$\frac{d\sigma}{d\varepsilon}=\frac{d\sigma}{dt}\left(\frac{d\varepsilon}{dt}\right)^{-1} \tag{5.3}$$

The plastic strain rate is related to the dislocations as:

$$\frac{d\sigma}{dt}=\dot{\varepsilon}=m\rho b\bar{v} \tag{5.4}$$

b is the Burgers vector, $\bar{v}$ is an average dislocation velocity, m is a geometrical constant = 1/2 for single crystals of MgO stressed in the ⟨100⟩ direction and ρ is the total length of the mobile dislocation. LiF and MgO measurements made by Johnston and Gilman [17] and Johnston [18], respectively, have shown that dislocation velocity is related to shear stress by:

$$v=v_0\left(\frac{\tau}{\tau_0}\right)^n \tag{5.5}$$

τ is the resolved shear stress, $v_0 = 1$ cm s^{-1}, and τ_0 and n are constants. They also indicate that the velocity of edge dislocations is much faster (in LiF) than that of the screw dislocation by about:

$$\bar{v}\cong 2v_0 \tag{5.6}$$

The slope of the stress–strain curve from (Eq. 5.3) through (Eq. 5.6) may be written as:

$$\frac{d\sigma}{d\varepsilon}=\frac{1}{\rho b v_0(\tau/\tau_0)^n}\frac{d\sigma}{dt} \tag{5.7}$$

τ_0 is an appropriate constant for screw dislocations. If the constants, τ_0 and n, and the variation of ρ during plastic deformation are known, and m is set to ½, the slope, at a constant stress rate, can be predicted.

As indicated earlier, a very common method for studying the resistance of ceramics to applied loads is by indentation, or rather by micro-indentation, to reduce the propensity of ceramics for premature failure. In Fig. 5.6, an indentation test is presented. The figure was obtained using Hertzian indentation with spherical indenters; plasticity can be attained in ceramics by this test. A glass–ceramic was used, a commercially available Corning product, basically consisting of K_2O–MgF_2–MgO–SiO_2. The specimens were heat treated for 4 h at the temperatures of 1000, 1040, 1060, 1080 and 1120 °C. The microstructure of the glass is illustrated in Fig. 5.7. In Fig. 5.6, the brittle state of the base glass is also shown. The intensity of the stress produced by the spherical indenter is related to the indentation pressure, p_0, by:

$$p_0 = \frac{P}{\pi a^2} \tag{5.8}$$

and the strain is related by the ratio of contact radius, a, to the sphere radius, r, as:

$$\text{strain} = \frac{a}{r} \tag{5.9}$$

In Fig. 5.6 the indentation pressure is plotted versus a/r. The Hertzian relation between the indentation stress, p_0, and the indentation strain, a/r, is linear (see, for example, [2]) and given by:

$$p_0 = \left(\frac{3E}{4\pi k}\right)\frac{a}{r} \tag{5.10}$$

The above considers the elastic region in a perfect Hertzian state of indentation. Details of the analysis of this test in the plastic region may be found in the work of Fischer-Cripps and Lawn [7]. The end result of their analysis may be given as:

$$\tau_F = \left[\frac{1}{2}(\alpha_3 - \alpha_1)\sin 2\Psi\right]p_0 - \tau_c - \mu\left[\frac{1}{2}(\alpha_3 - \alpha_1) - \frac{1}{2}(\alpha_3 - \alpha_1)\cos 2\Psi\right]p_0 \tag{5.11}$$

where τ_F is the frictional shear stress term, Ψ is the angle between the fault plane and σ_3 axis, $\alpha_1 = -\sigma_1/p_0$, $\alpha 3 = -\sigma^3/p_0$, μ is the coefficient of sliding friction and τ_c is the cohesion (adhesion) strength. For more information on the assumptions made to calculate the lines in Fig. 5.6 by theoretical analysis, the reader is referred to the originals work of Fischer-Cripps and Lawn [7].

The Hertzian contact test results described provide basic information on the intrinsic properties of otherwise brittle ceramics, information not generally attainable by more conventional testing procedures. Specifically, it enables one to

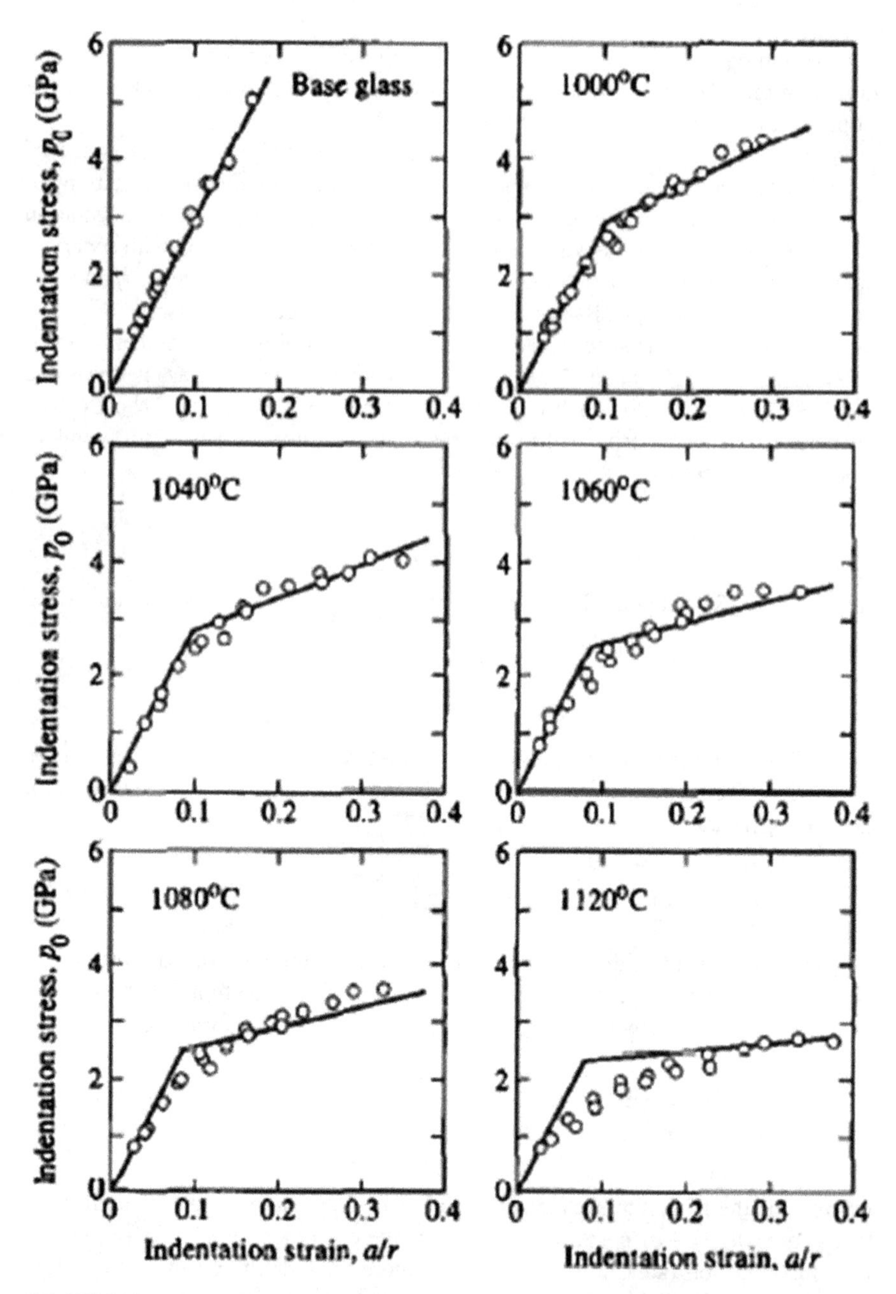

Fig. 5.6 Indentation stress–strain curves for base glass and glass–ceramics heat-treated at specified temperatures for 4 h. Data from experiments line fits from theoretical analysis [7]. With kind permission of Elsevier

Fig. 5.7 Scanning electron micrograph of glass–ceramic heat-treated at 1120 °C, surface-etched to reveal mica platelets in glass matrix [7]. With kind permission of Elsevier

generate (asymptotic) indentation stress–strain curves, such as those shown in Fig. 5.6 for glass–ceramics and thus to quantify any plasticity characteristics.

5.2.3 Superplastic Ceramics

Superplastic structural ceramics were discussed in Sect. 2.2. They are characterized by large strains when deformed. The microstructure of superplastic materials is characterized by fine grains, which resist grain coarsening during sintering and are stable when deformed. Among the best-known superplastic materials are Y-TZP polycrystals, Si_3N_4, Al_2O_3, etc. Second-phase particles are especially effective in suppressing static and dynamic grain growth. Additives that segregate to grain boundaries can be effective in retarding grain coarsening. High strain rates are one of the requirements for inducing superplasticity during deformation. Often, heat treatment can help to obtain the desired fine-grained microstructure required for superplasticity. In Fig. 5.8, the ultrafine-grained structures of several ceramics exhibiting superplasticity are illustrated.

However, it has often been observed that little or no hardening at all occurs.

Superplastic materials are characterized by high ductility, preferably achieved at relatively low temperatures and high deformation rates. The strain rate is expressed in terms of stress and grain size, d, as:

$$\dot{\varepsilon} = \frac{A\sigma^n}{d^p} \tag{5.12}$$

with n and p as the stress and grain size exponents. A is a temperature-dependent coefficient related to diffusion and, thus, may be expressed by an Arrhenius equation. The exponents, n and p, are in the 1–3 range. Figure 5.9 illustrates the stress–strain relation of several zirconia-based ceramics.

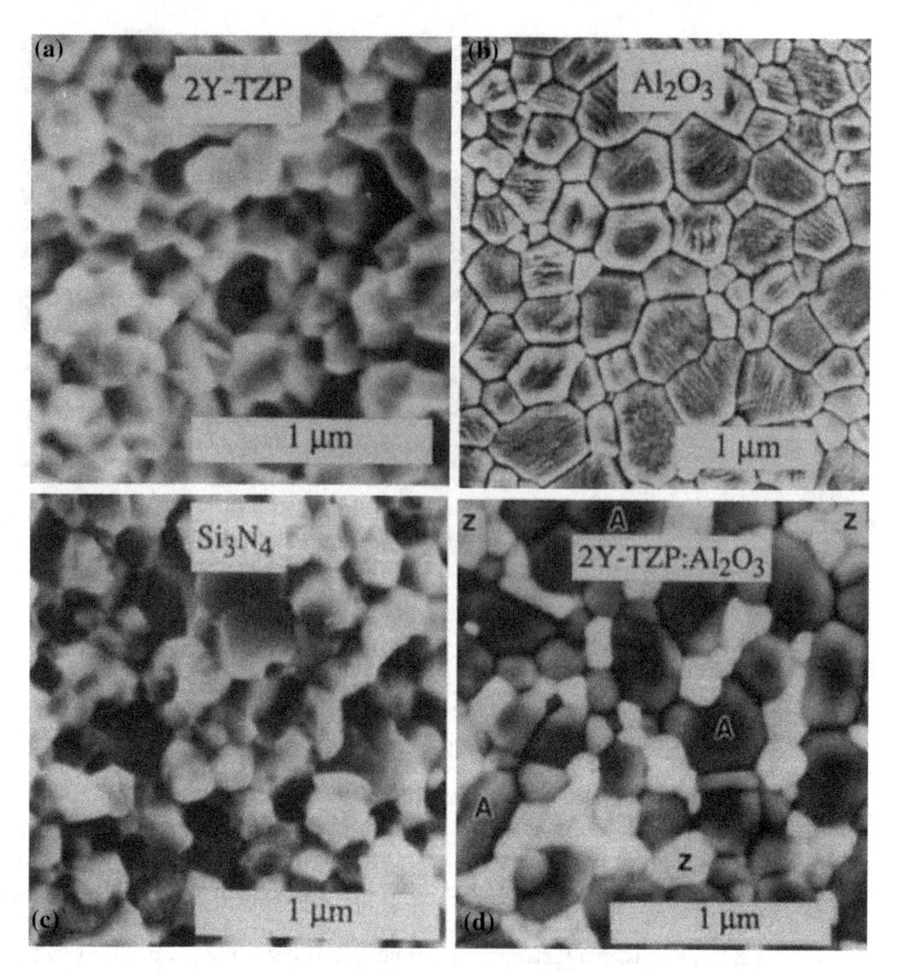

Fig. 5.8 Scanning electron microscopy micrographs of ultrafine grains of superplastic ceramics: **a** 2Y-TZP, **b** alumina, **c** silicon nitride, and **d** 2Y-TZP/alurnina at equal volume fraction [3]. With kind permission of John Wiley and Sons

In the above, the term TZP refers to tetragonal zirconia polycrystals containing 2-mol% yttria, 12 % ceria, while the term CSZ refers to cubic zirconia with the respective additives. Note in the curves that 2Y-TZP and 12Ce-TZP/1%Ca do not show any strain hardening and that these curves are flat. However, it has often been observed, not only in these cases, that little hardening occurs or that there is none at all. The most probable reason may be related to recovery processes which set in immediately after the elastic region is surpassed and equilibrium is maintained by stain hardening. The stress–strain curve of the nanocrystalline (NC) MgO, shown in Fig. 5.10, indicates that no work hardening has occurred and its microstructure is shown in Fig. 5.11.

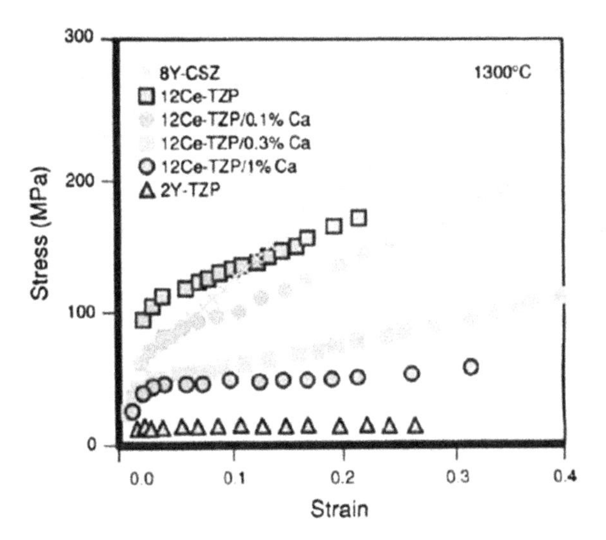

Fig. 5.9 Stress–strain curves of TZP and CSZ. Initial grain sizes are: 0.48 μm (8Y-CSZ), 0.56 μm (12 Ce-TZP with 0, 0.1, and 0.3 % Ca), 0.3 μm (12Ce-TZP with 1 % Ca), and 0.21 μm (2Y-TZP). Tendency for strain hardening directly corresponds to the magnitude of grain-boundary mobility shown in Fig. 5.10 (Strain rate is 10^{-4} s^{-1} except in the case of 8Y-CSZ and 12Ce-TZP with 1 % Ca where 3×10^{-4} s^{-1} is used.) [3]. With kind permission of John Wiley and Sons

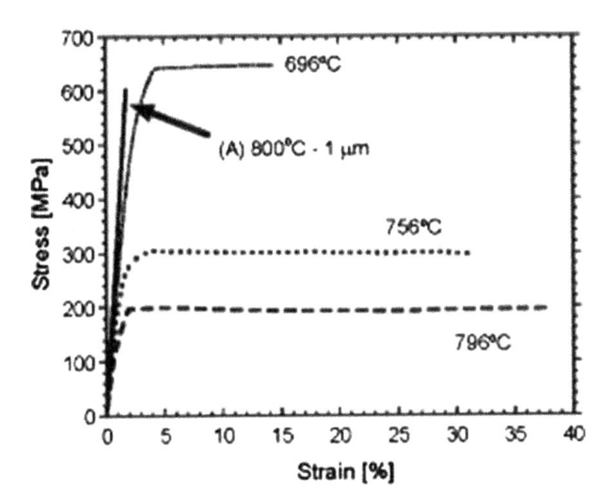

Fig. 5.10 Stress–strain curves recorded by compression of rectangular NC-MgO bars at constant cross-head speed and different temperatures. The curves exhibit elastic and perfectly plastic behavior with no strain hardening. Specimen (A) was annealed to grow the grain size to 1 μm and thus exhibited brittle behavior by compression at 800 °C (*arrowed solid line*) compared with the ductile behavior of its nanocrystalline counterpart specimen (*dashed curve*) at 800 °C [26]

Fig. 5.11 High resolution scanning electron microscope (HRSEM images,) using the secondary electrons, showing the surface microstructure of the NC-MgO composed of equiaxed grains, **a** prior to and **b** after the plastic deformation at 800 °C (40 % strain). No significant changes were visible after the deformation, except some grain boundary cavities and faceting of the surface grains which were in contact with the compressing pad. *Arrows* indicate the applied load direction [26]

Fig. 5.12 Grain-boundary mobility of TZP and CSZ plotted versus reciprocal homologous temperature (T_m is the melting point) [3]. With kind permission of John Wiley and Sons

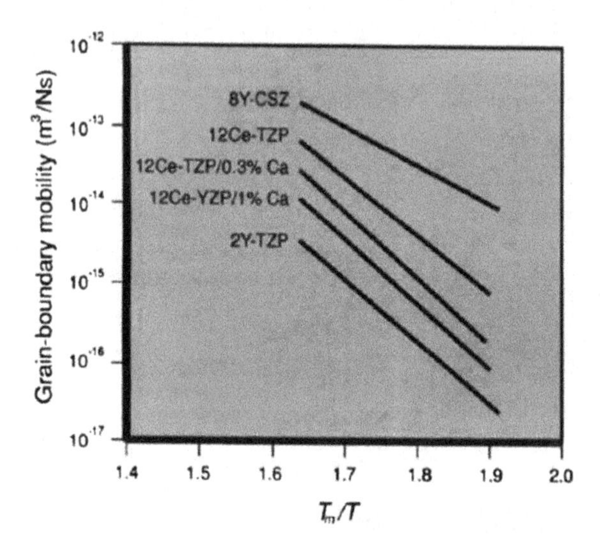

Grain-boundary mobility, which is related to grain growth, may be seen in Fig. 5.12 with the effect of the same additives that was indicated in the stress–strain curves of specimens deformed as illustrated in Fig. 5.9.

In Fig. 5.8, the microstructure of alumina is shown, indicating superplastic behavior (early experiments failed to show superplasticity). Plastic deformation to large strains was achieved, but the window for superplastic fabrication is very narrow, because sintering below 1330 °C is quite difficult and requires much experience for successful production. Above this temperature, grain growth is quite fast. Therefore, proper additives are needed to reduce grain growth above this

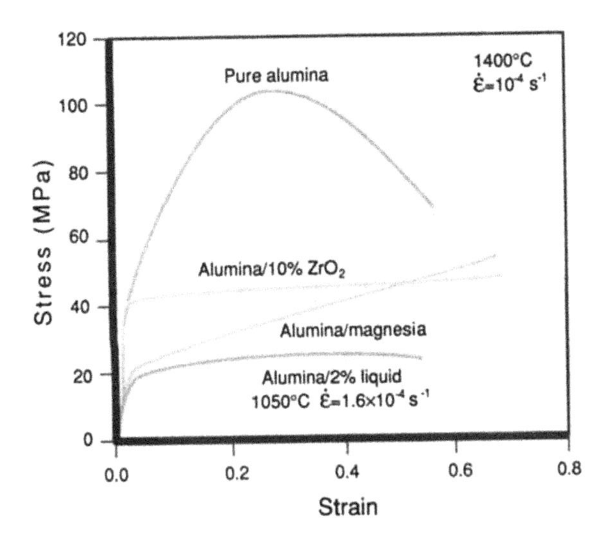

Fig. 5.13 Stress–strain curves for alumina. Strain hardening in pure alumina and magnesia-doped alumina is due to dynamic grain growth. The eventual decrease in stress in pure alumina is a result of cavitation. Note that alumina with 2 % liquid was tested at a lower temperature [3]. With kind permission of John Wiley and Sons

temperature. The deformation of alumina doped with 200 ppm magnesia results in large strain. In Figs. 5.9 and 5.10, compare the stress levels and strain hardening. Except in the magnesia-doped alumina, the strain hardening is negligible or non-existent. The relatively mild strain hardening of the magnesia-doped alumina is a direct consequence of its slower dynamic grain growth. This magnesia-doped alumina may be superplastically deformed to large strains, albeit at the expense of strain hardening. Note that producing superplastic alumina is not that easy, since sintering alumina below 1300 °C requires considerable experience and, when sintering above 1300 °C, grain growth is quite fast; therefore, a very narrow window is available for the processing of superplastic alumina. Grain growth is not desirable for obtaining superplasticity. Adding various additives to ceramics, as have been seen above (Figs. 5.9 and 5.12), is a method often effective for improving superplastic performance while maintaining strain hardening. From the above examples, the effect of an additive on strain hardening is clear.

The microstructure of a sintered pure alumina is shown in Fig. 5.14a. (a) The grain size is a fine 0.5 μm and remains relatively stable during annealing at temperatures below 1300 °C. Pure alumina at such a grain size has a very low initial flow stress, as seen in Fig. 5.13. However, it fails to deform superplastically due to rapid dynamic grain growth, shown in Fig. 5.14. (b) The structure of the deformed specimen, represented by Fig. 5.14a and b, causes strain hardening, as indicated on the stress–strain curve in Fig. 5.13. The grains become elongated when deformed at 1400 °C, as illustrated in Fig. 5.14a. (c) Large grains of this type serve as stress concentrators and potent nucleation sites for cavitation, thus degrading the ductility. The higher deformation temperature only makes the problem worse, as shown in Fig. 5.14c.

Si_3N_4 is another superplastic ceramic. Two polymorphs, α and β silicon nitrides, exist. Both are hexagonal and capable of having a large range of solubility with various constituents. Strain hardening in Si_3N_4 depends on its composition

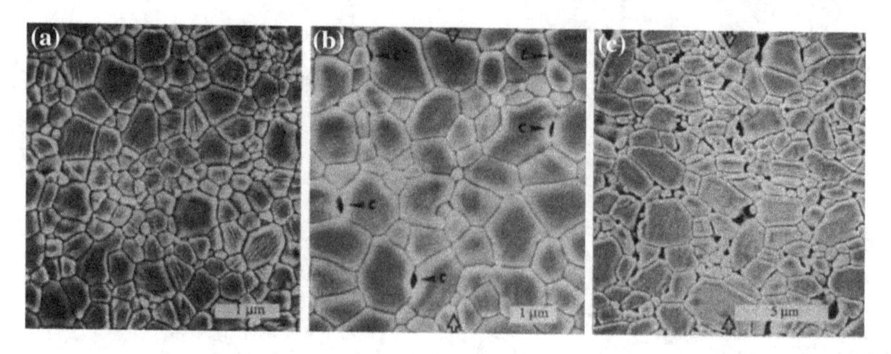

Fig. 5.14 **a** Microstructure of pure alumina: **a** as sintered at 1250 °C; **b** deformed in compression at 1250 °C (strain rate of $1.5 \times 10^{-5}\ \mathrm{s}^{-1}$; total strain of 0.3) with compression axis shown by *hollow arrows* and cavities by *solid arrows*; and **c** same as **b** but deformed at 1400 °C (strain rate of $2.4 \times 10^{-4}\ \mathrm{s}^{-1}$; total strain of 0.68) [3]. With kind permission of John Wiley and Sons

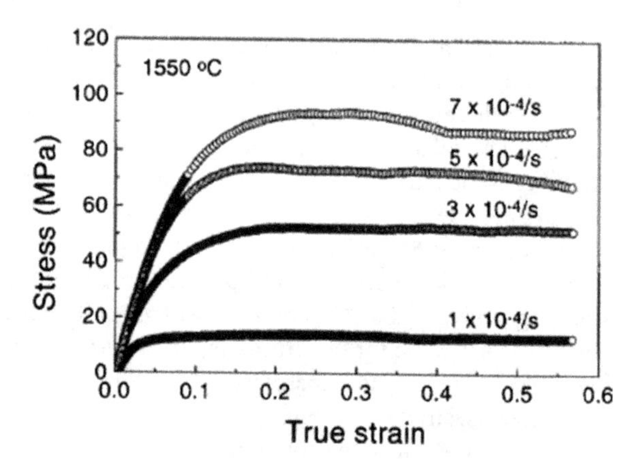

Fig. 5.15 Compressive stress–strain curves for various strain rates of the as-hot-pressed material at 1550 °C [34]. With kind permission of John Wiley and Sons

and its polymorph. In Fig. 5.15, strain hardening in β-Si3N4 is absent at all the strain rates tested under compression at at 1500 °C. The data in Fig. 5.15 are corrected curves that may be obtained by the following correction method for a compression test. Some of the expressions from Chap. 1 may be rewritten as below. Stress may be expressed as:

$$\sigma = \frac{P}{A_0}\exp(\varepsilon) \tag{5.13}$$

with the usual meanings of the parameters (A_0 as known is the initial cross-section). The true strain is:

$$\varepsilon = \ln\frac{l}{l_0} = \ln\left(1 - \frac{\Delta l}{l_0}\right) \tag{5.14}$$

Recall that Δl is the displacement (change in dimensions of the specimen). Denoting the rate of the crosshead displacement, $\dot{l}$, and if it is assumed to be constant during the test, then the initial strain rate is $\dot{\varepsilon}_0$, which is given by:

$$\dot{\varepsilon}_0 = \frac{\dot{l}}{l_0} \tag{5.15}$$

The strain rate at some intermediate stage follows from Eqs. (5.14, 5.15) and (5.16) is obtained. Also assign the symbol $\dot{\varepsilon}$ to $\frac{\dot{l}}{l}$ for the strain rate at the intermediate stage:

$$\dot{\varepsilon}_0 \exp(-\varepsilon) = \frac{\dot{l}}{l} = \dot{\varepsilon} \tag{5.16}$$

Rewrite Eq. (5.16) as:

$$\dot{\varepsilon} = \dot{\varepsilon}_0 \exp(-\varepsilon) \tag{5.16a}$$

Since a constant displacement was used, Eq. (5.16) indicates an increase in strain rate under compression (the sign in the exponent represents compression, since it is negative). Correction for the changing strain rate is possible if the stress exponent is known. The strain rate is related to the stress exponent by:

$$\dot{\varepsilon} = A\sigma^n \tag{5.17}$$

Substituting for $\dot{\varepsilon}$, one obtains:

$$\sigma_{corr.} = \sigma_0[\exp(-\varepsilon)]^{1/n} \tag{5.18}$$

The steps for obtaining Eq. (5.18) are as follows.
From the substitution of Eq. (5.17) into Eq. (5.16), one obtains:

$$\dot{\varepsilon} = A\sigma^n = \dot{\varepsilon}_0 \exp(-\varepsilon) \tag{a}$$

This may also be written as:

$$\sigma^n = \frac{\dot{\varepsilon}_0}{A} \exp(-\varepsilon) \tag{b}$$

$$\sigma = \left(\frac{\dot{\varepsilon}_0}{A}\right)^{1/n} \exp(-\varepsilon) \tag{c}$$

In the initial state, Eq. (5.17) may be expressed as:

$$\dot{\varepsilon}_0 = A\,{\sigma_0}^n \tag{d}$$

By replacing $\left(\frac{\dot{\varepsilon}_0}{A}\right)^{1/n}$ in (c) with (d), one obtains Eq. (5.18), which is the corrected stress:

$$\sigma_{corr.} = \sigma_0[\exp(-\varepsilon)]^{1/n} \tag{5.18}$$

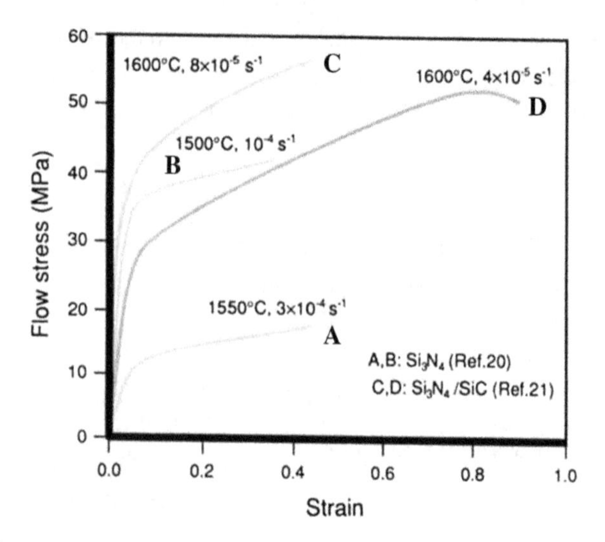

Fig. 5.16 Stress–strain curves of silicon nitrides, where, *A* and *B* are mostly (α'-sialon deformed in compression, and *C* and *D* are mostly β'-sialon with 30 vol% silicon carbide deformed in tension. Note that *A* and *B* have much less strain hardening than *C* and *D* [3]. With kind permission of John Wiley and Sons

The corrected curves of Eq. (5.14) are all at the constant strain rates indicated. No strain hardening was observed during these tests, even at low strain rates. This observation contradicts other observations, in which strain hardening was observed especially at low strain rates. These contradictory reports are probably due to whether α-Si_3N_4 or β-Si_3N_4 was the basic constituent. In Fig. 5.15, the specimens tested are based on β-Si_3N_4. In the following illustration, α-Si_3N_4-based deformation tests were performed, indeed indicating strain hardening of the specimens. In Fig. 5.16, the β-Si_3N_{4-} based specimens indicate considerable strain hardening. Note that the $Si_{6\text{-}x}AI_xO_xN_{8-x}$ is referred *το* as β'-sialon (in solid solution in β-Si_3N_4) and $M_{z/n}Si_{6\text{-}x\text{-}z}$ $AI_{x+z}O_xN_{8-x}$ is known as α'-sialon (in solid solution in the α-phase). Even in this figure, note that specimens A and B work harden much less, being based on an α-Si_3N_4 solid solution (i.e., containing α'-sialon). It is to be expected that, if the α and β Si_3N_4 are mixed to some proportion and tested, strain hardening will occur, depending on the amount of α-Si_3N_4 in the mixture in the material (see Rouxel et al. [27]).

Certain additives, even to β-Si_3N_4, may induce strain hardening, as seen in Fig. 5.17a. The major additive is SiC up to ~30 wt%. Table 5.1 lists the composition of the ceramic Si_3N_4.

Figure 5.17a represents tensile test specimens (compositions and designations in Table 5.1), before and after testing, obtained by optical micrographs, shown in Fig. 5.17b. As may be seen in 5.6a, grades B, C and D exhibit quite large strain, but specimens A, E, F and G fracture at elongations of less than 15 %. The effect of the strain rate on the tensile deformation is illustrated for specimen D (see Table) at 1600 and 1650 °C in Fig. 5.18a and that of the temperature at a constant strain rate is seen in Fig. 5.19b. The generally known fact that the temperature has an opposite effect on the flow curves and on strain hardening may also be seen in Fig. 5.18.

Table 5.1 Nominal compositions and preparation conditions of Si_3N_4/SiC c3 composites [27] (with kind permission of John Wiley and Sons)

Grade	Composition (wt%)		Sintering conditions		
	SiC	Al_2O_3/Y_2O_3	Temperature (°C)	Duration (h)	Density (g/cm^3)
A	28	0/8	1700	1	3.17
B	33	2/6	1800	1	3.26
C	33	2/6	1750	4	3.27
D	33	2/6	1700	1	3.27
E	30	4/4	1650	1	3.26
F	30	4/4	1700	1	3.27
G	30	6/2	1700	1	3.22

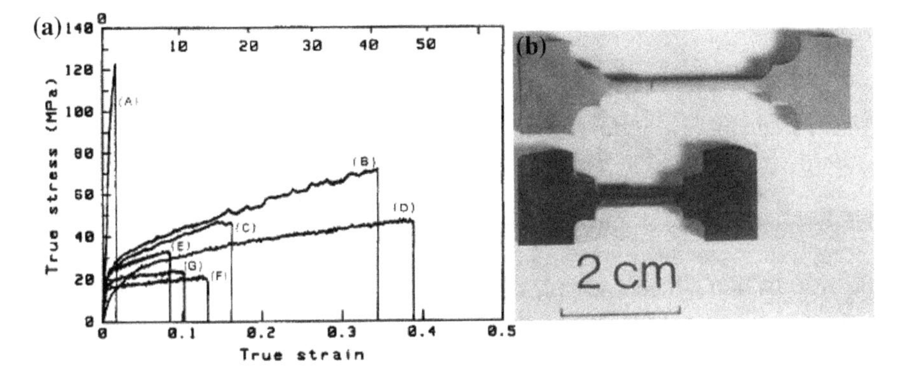

Fig. 5.17 **a** Tensile tests performed on the different grades (letters in brackets represent the grades) in nitrogen atmosphere at 1600 °C with a strain rate of 4×10^{-5}/s. **b** Optical photographs of tensile specimens, before testing and after 114 % elongation [27]. With kind permission of John Wiley and Sons

It is the belief of this author that strain hardening, in the technically important material Si_3N_4 (refer to Figs. 5.16, 5.17, 5.18), is mainly due to its additives, since β-Si_3N_4, as the only phase, did not show work hardening (see Fig. 5.15).

In summary of this section, strengthening by strain hardening was discussed for brittle ceramics (where no strain hardening exists, since there is no plastic deformation), in ductile ceramics and in superplastic ceramics. The absence of strain hardening in some ceramics is a consequence of the recovery process and of dynamic grain growth during deformation. Various recovery processes can set in immediately after elastic deformation. They are continuous processes accompanying strain hardening during deformation. The rate of the two competing processes, i.e., strain hardening and recovery, dictate the outcome of the deformation in the plastic region. When in equilibrium, it appears as illustrated in Fig. 5.15. If the recovery process occurs at a faster rate, the stress–strain curve reaches a maximum early in the deformation, shortly after the elastic region has been

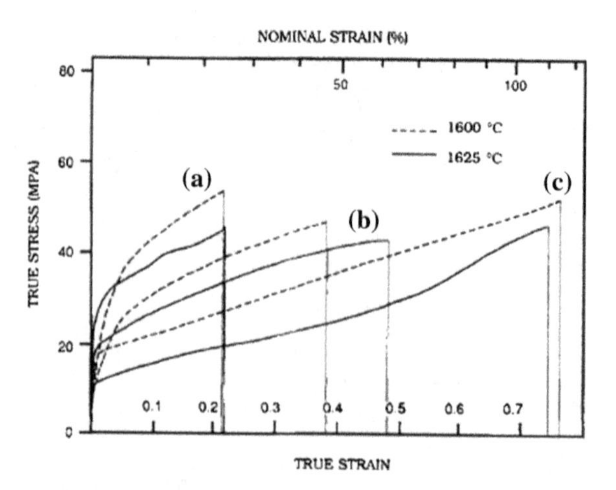

Fig. 5.18 Tensile tests performed in nitrogen atmosphere at 1600 and 1625 °C on grade D with various strain rates: **a** 8×10^{-5}, **b** 4×10^{-5}, and **c** 2×1010^{-5}/s [27]. With kind permission of John Wiley and Sons

surpassed, when the stress for further deformation drops quickly. For meaningful strain hardening, its rate should be greater than that of recovery.

In the present section, strain hardening is discussed as a consequence of dislocation generation and the interactions of mobile dislocations with existing and deformation induced dislocations. However, in general, and in some non-specific exceptional cases, dislocations in ceramics are normally not very mobile, though they can be strain-hardened to a small degree. Polycrystalline ceramics are also porous; as a result, they behave like brittle materials, so significant deformation and strengthening by cold working are not possible. The reason for this was indicated above, namely that ceramics are ionically- or covalently-bonded materials and are too brittle to work harden appreciably. Glasses, in general, comprising a vast variety of ceramics, are amorphous and may contain dislocations (though not like those found in crystalline materials) and, therefore, they cannot be strain-hardened, except in the special cases discussed above.

In the next section, other strengthening mechanisms are considered.

5.3 Solid Solution Strengthening

In metals, solid solution hardening by substitutional and interstitial atoms is related to retarded dislocation motion, which strengthens materials. The effect of carbon in steel, by forming Fe_3C, is the most commonly known form of strengthening by interstitial atoms. In ceramics, solid solution seems to have a smaller effect on hardening and the main reason for the overall strength is dictated by the contribution of flaws. Except for the flawed states, damage modes depend on ceramic types (especially microstructures), loading conditions and geometric factors. An example of flaws in SiC, in the form of pores, may be seen in Fig. 5.20.

Basically, solid solutions have a considerable impact on the physical properties of ceramics, influencing, among others, their magnetic, dielectric and optical

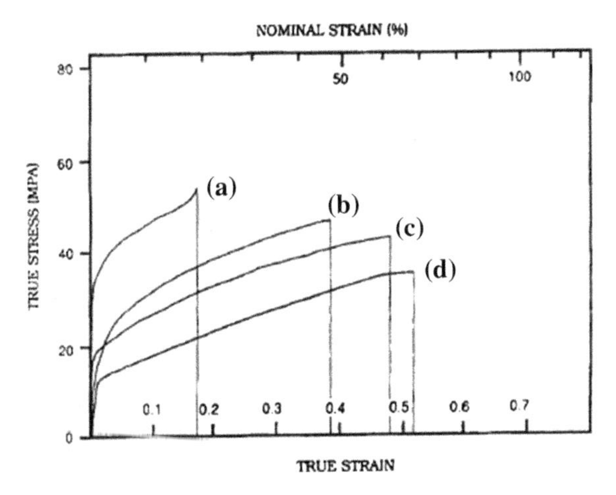

Fig. 5.19 Tensile tests performed in nitrogen atmosphere with a strain rate of 4×10^{-5}/s on samples of grade D at various temperatures: **a** 1575, **b** 1600, **c** 1625, and **d** 1650 °C [27]. With kind permission of John Wiley and Sons

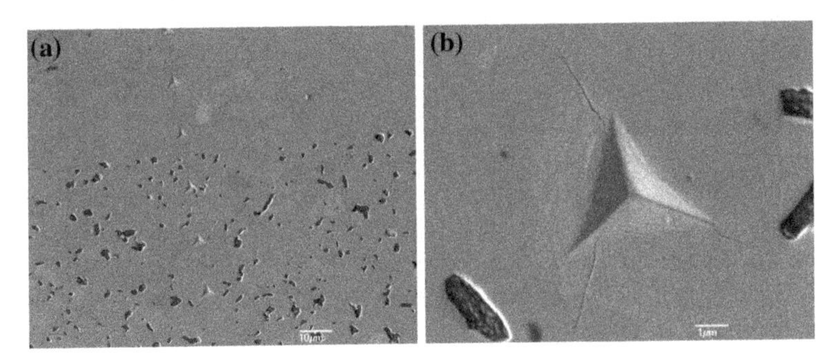

Fig. 5.20 Pores in SiC pipe and a disk joined to it by spark plasma sintering (SPS): **a** pores in the pipe and the interface to pore free the disk is seen, **b** indent by nano indentation and cracks emanation from the corners. Stern et al. (unpublished). Courtesy of Prof. A. Stern

properties. In metals, one usually talks of interstitial atoms and substitutional atoms, as dictated by the phase relations of the constituents (equilibrium phase diagrams); however in ceramics, the concept of solubility often refers to molecular solubility as well. In the following, no distinction will be made between the effects of the soluble entities, either interstitial or substitutional atoms, but rather the term 'solid-solution effects' is considered. Further illustration of these molecular solubility effects will be presented later on for selected ceramics. Here, the influence of atomic solution is discussed.

The addition of Si, in order to strengthen Ti_3AlC_2 by forming a solid solution of $Ti_3Al_{1-x}Si_xC_2$, is presented as an example. Adding Si to Ti_3AlC_2 at a level of $x \geq 0.25$ hardens the ceramics, as illustrated in Fig. 5.21. (The Si is given in wt. percent). As may be seen, hardening (by Vickers) increases with the addition of Si.

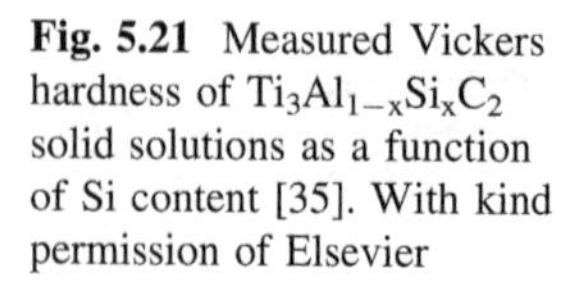

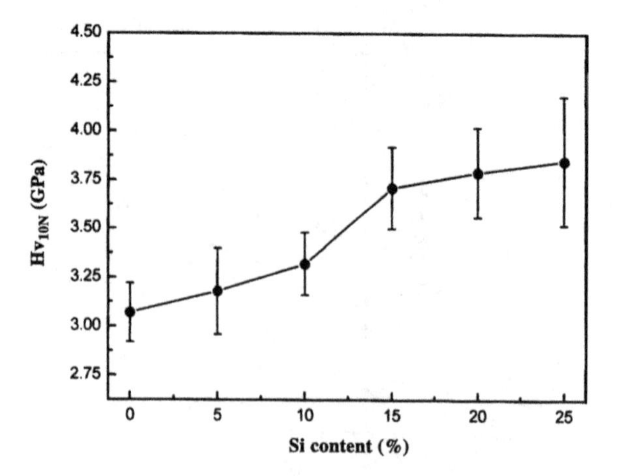

Fig. 5.21 Measured Vickers hardness of $Ti_3Al_{1-x}Si_xC_2$ solid solutions as a function of Si content [35]. With kind permission of Elsevier

The solid-solution specimens in the indicated range were synthesized using an in situ hot-pressing/solid–liquid reaction method. It was observed that the lattice parameter, "c", decreased dramatically, but "a" remained almost unchanged with the increase of Si in the $Ti_3Al_{1-x}Si_xC_2$ solid solutions. A significant strengthening effect was observed when x was greater than 0.15 in the $Ti_3Al_{1-x}Si_xC_2$ solid solutions; the Vickers hardness and other mechanical properties, flexural strength and compressive strength, were enhanced by 26, 12 and 29 %, respectively, for $Ti_3Al_{0.75}Si_{0.25}\ C_2$ solid solution.

Furthermore, it can be seen in Fig. 5.21 that the hardness increases slowly for a small x, but the strengthening effect is significant when x is greater than 0.15. The compressive and flexural strengths variations with Si addition appear in Fig. 5.22. A similar strengthening effect may be seen in the strength versus composition in Fig. 5.22, where the flexural and compressive strengths are plotted as a function of Si content, thus demonstrating a significant strengthening effect, particularly beyond 0.15 wt% Si. The microstructures of the specimens used to evaluate the mechanical properties indicated with and without Si additions are shown in Fig. 5.23. Thus, Ti_3AlC_2 is strengthened by replacing Al with Si to form a $Ti_3Al_{1-x}Si\ _xC_2$ solid solution. Moreover, the addition of Si to Ti_3AlC_2 to form this solid solution has no deleterious effect on the oxidation resistance at 1100 °C due to the formation of a continuous, protective Al_2O_3 layer.

The salient properties of this layered ternary ceramic are: good damage tolerance, good machinability, low density and excellent thermal shock and oxidation resistance. Such unique properties make it possible to use Ti_3AlC_2 in structural components for high-temperature applications and as oxidation-resistant coatings.

Another high-temperature ceramic for which solid-solution strengthening has been evaluated is the solid solution of $MoSi_2$ after Re alloying. Figure 5.24 is a stress–strain curve showing the hardening effect of Re. Re was reported to be a potent solution hardener for $MoSi_2$. Indeed, the addition of 2.5 at% Re increases the hardness of $MoSi_2$ by 30 % at RT and by 100 % at 100 °C.

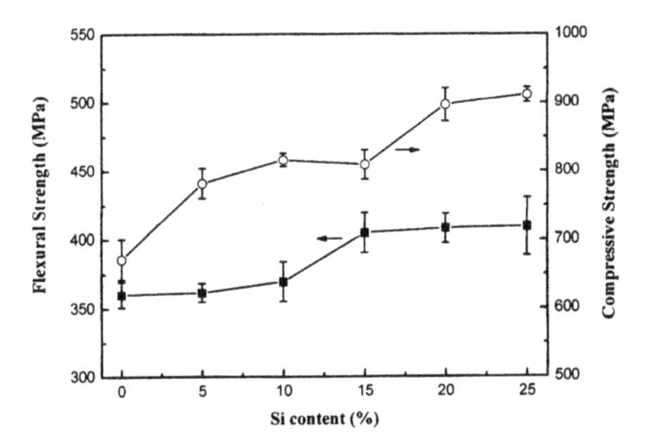

Fig. 5.22 Flexural and compressive strength of $Ti_3Al_{1-x}Si_xC_2$ solid solutions as a function of Si content [35]. With kind permission of Elsevier

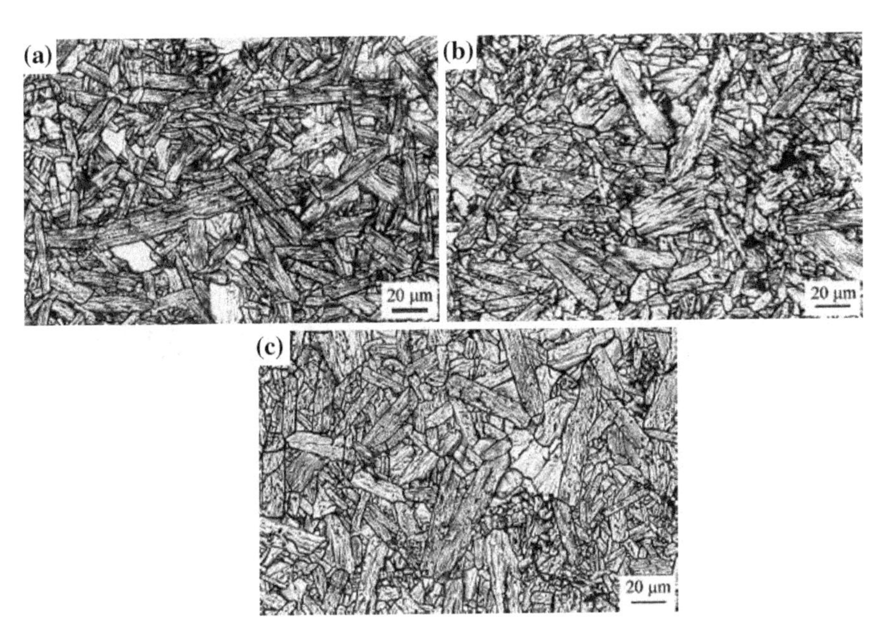

Fig. 5.23 Microstructure of: **a** Ti_3AlC_2, **b** $Ti_3Al_{0.85}Si_{0.15}C_2$ and **c** $Ti_3Al_{0.75}Si_{0.25}C_2$ solid solutions observed using scanning electron microscope [35]. With kind permission of Elsevier

Figure 5.24 compares compression tests of $MoSi_2$ and $MoSi_2$ + 2.5 at% Re alloys. Compressive plasticity in $MoSi_2$ is observed only above 900 °C, whereas in the $MoSi_2$ + 2.5 at% Re alloy only above 1000 °C. This indicates the increase in ductile-to-brittle transition temperature due to the Re alloying. Typical dislocation structures of these two materials are compared in Fig. 5.25. Arrays of subgrains

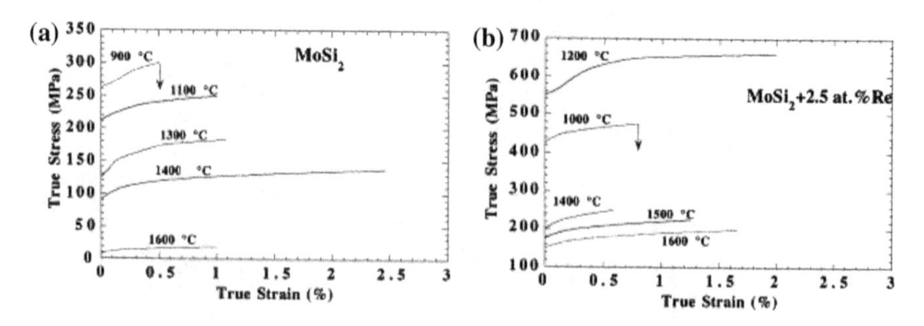

Fig. 5.24 Compressive stress–strain curves at different temperatures for **a** $MoSi_2$ and **b** $MoSi_2$ + 2.5 at% Re alloys. In these graphs, an arrow at the end of the *curve* indicates catastrophic failure of the specimen. The stress–strain curves shown without an arrow at the end represent tests that were stopped prior to fracture [23]. With kind permission of Elsevier

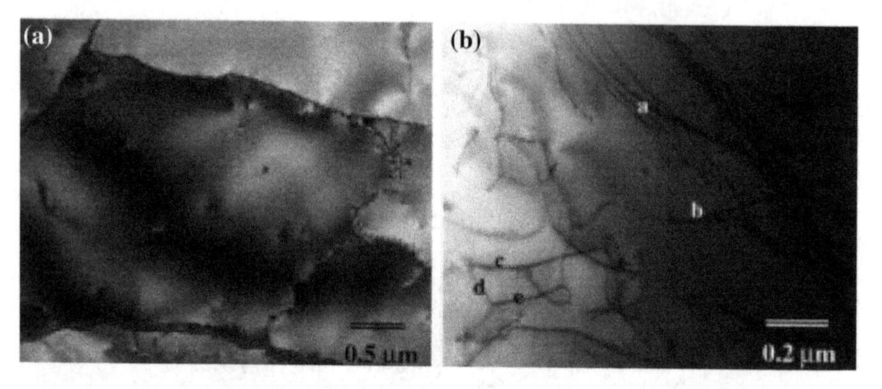

Fig. 5.25 Bright field TEM images showing the dislocation substructures in **a** $MoSi_2$ and **b** $MoSi_2$ + 2.5 at% Re alloys deformed in compression at 1400 °C [23]. With kind permission of Elsevier

and a few isolated dislocations are observed in $MoSi_2$, but the addition of Re changes the dislocation substructure significantly, as may be seen. In $MoSi_2$ + 2.5 at% Re alloys, no tendency toward grain-boundary formation is observed and the dislocations are randomly distributed. Only [100] dislocations were observed in MoS_2. The yield stress was found to be very low at 1400 °C, which suggests that the dislocation glide is very fast and that the climb of dislocations, when encountering obstacles, becomes the rate-controlling factor.

The addition of Re increases the flow stress and changes the high-temperature behavior from climb-controlled to viscous glide-controlled and the glide of the dislocations becomes restricted by the solute Re atoms. The dislocation substructures in Fig. 5.25b mostly have [100] Burgers vectors, but some ½[111] dislocations are also observed. Those labeled Fig. 5.25 as a, b and d have [100] Burgers

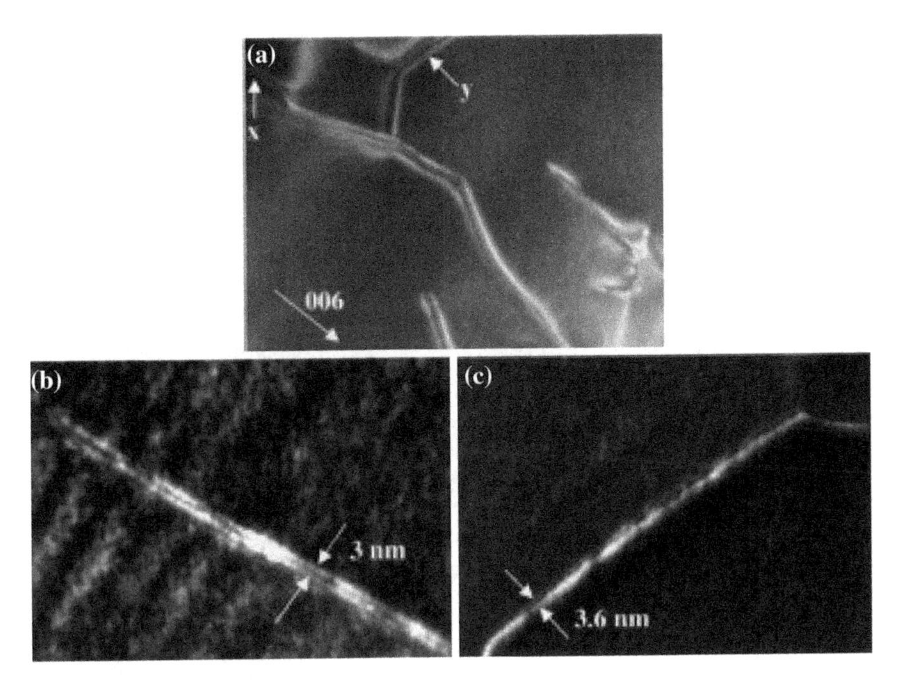

Fig. 5.26 **a** Bright field TEM image of ½ ⟨111⟩ dislocations in $MoSi_2$ + 2.5 at% Re alloys in compression at 1400 °C. Weak beam images of dislocations labeled x in (**a**) are shown in (**b**) and (**c**) respectively. The electron beam direction is ~5–10° from [100] [23]. With kind permission of Elsevier

vectors, while those labeled as c and e are ½[111] dislocations. The ½[111] dislocations in the $MoSi_2$ + 2.5 at% Re alloy, deformed at 1400 °C, dissociate. A BF image of ½ [111] for reflecting vector **g = 006** is shown in Fig. 5.26a. Under these imaging conditions, all the dislocations with [100]-type Burgers vectors are invisible. Weak-beam images (**g/3** g with **g = 002**) of the non-screw dislocations, labeled x and y in Fig. 5.26a are shown in Fig. 5.26b and c, respectively. Split dislocations, with a separation of ~3 and ~3.6 nm, are observed in Fig. 5.26b and c. The electron-beam direction in these images is ~5–10° from [100] towards [110]. Re-induced hardening is illustrated in Fig. 5.27.

The experimental hardness data may be fit into a linear relation when plotted versus the atomic concentration of Re. This relation is plotted versus the square root of the atomic concentration, c. Figure 5.27 indicates a considerable increase in hardness due to the solid solution of Re in the $MoSi_2$. The above figures (Figs. 5.24 and 5.27) show a considerable increase in the strength properties by means of solid-solution strengthening in polycrystalline $MoSi_2$ with small additions of Re (≤2.5 at%) up to a temperature of 1600 °C. This strengthening is the result of the Re atoms acting as obstacles to dislocation glide. When encountering such obstacles, dislocations must climb for continued motion in other respective planes.

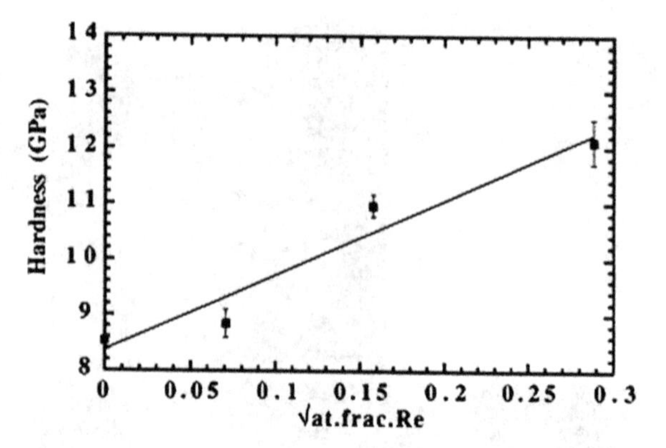

Fig. 5.27 Dependence of room temperature hardness of (Mo, Re)Si_2 alloys on square root of the Re concentration [23]. With kind permission of Elsevier

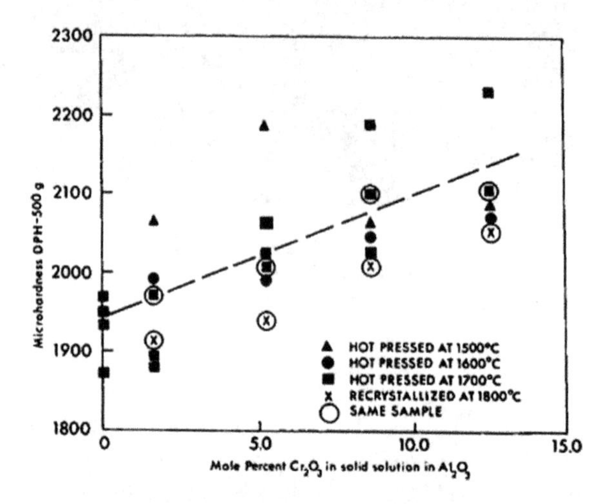

Fig. 5.28 Microhardness of Al_2O_3 as a function of Cr_2O_3 content [1]. With kind permission of John Wiley and Sons

Often, solid solutions in ceramics are expressed in terms of molecular concentrations, rather than atomic ones. An illustration indicating the solid-solution strengthening of Al_2O_3 by Cr_2O_3 is shown in Fig. 5.28. The strengthening effect of increasing the chromia in solid solution in fine-grained alumina is expressed in terms of microhardness. The specimens were prepared by wet ball-milling of the constituents. Grain growth was retarded by the addition of 0.5 wt% magnesia. The magnesia also served for densification during the next step–hot pressing. The microstructures of the hot-pressed and recrystallized specimens are illustrated in Fig. 5.29.

On average, the tests shown in Fig. 5.28 represent ten microhardness indentations per specimen. When cracks occasionally appear to emanate from the corners of the indentations (as also observed in SiC, seen in Fig. 5.29b, these values were not averaged in. The line in Fig. 5.28 is linear and, thus, microhardness values of the solid solutions may be represented as linear relations plotted against mol% chromia. The large scatter around the least-square line is assumed to be associated with the microindentation crossing many grains. Hot-pressed materials contain a

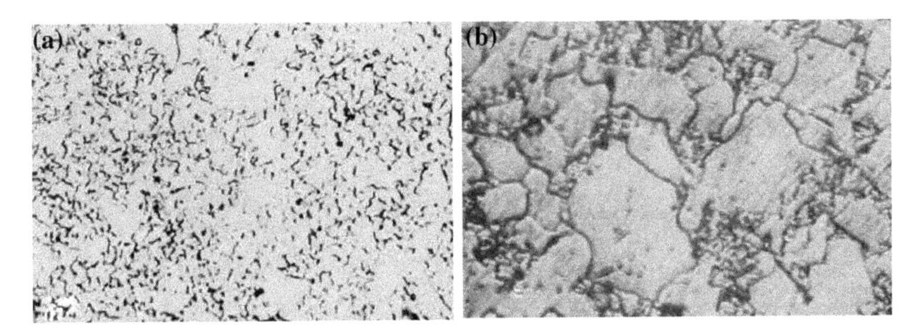

Fig. 5.29 Alumina with 12.5 wt% chromia in solid solution. **a** As-hot-pressed, and **b** recrystallized at 1800 °C for 2 h. (X 1000.) [1]. With kind permission of John Wiley and Sons

large number of fine grains. This assumption was affirmed for recrystallized samples, which provided large grains, as seen in Fig. 5.27b. In Fig. 5.28, the line of the recrystallized samples (not drawn) seems to have a smaller scatter, but the hardness values are lower, as expected. The solid-solution hardening effect of alumina by chromia is clearly visible.

To emphasize the effect of solid-solution strengthening, especially on yield stress and, consequently, on flow stress itself, the case of YSZ serves as a good example. Figure 5.30 is a stress–strain curve illustration with various amounts of Y_2O_3 additives, which are usually used to stabilize ZrO_2. Well-defined upper and lower yield points are observed, somewhat similar to those in steels, but without the serrations characterizing the yield-point phenomena in steels. The yield-point drop is followed by the flow, but with an absence of strain hardening. Additions of Y_2O_2 increase both the yield and the flow stress. The absence of strain hardening is probably associated with the fact that only one slip system, that of $\{001\}$ $\langle 110 \rangle$, is operative. When, however, multiple slips of $\{111\}$ occur, such as in the case of 9.4 mol% YSZ, then more work hardening is observed than in the $\{100\}$ slip. Note that in Fig. 5.30 the solid solution is also expressed in terms of a molecular, rather than an atomic, percentage. The engineering flow stress, plotted against the Y_2O_3 concentration on a logarithmic scale, is illustrated in Fig. 5.31. The solution-hardening rate $d\tau/dc$ is $\mu/200$ ($\mu = G$). The slopes $\alpha = ½$ and 2/3 correspond to the well-known Fleischer and Labusch [8] solution-hardening models, respectively.

The Fleischer's theory was derived for the impurity-controlled dislocation mobility of the flow stress of LiF. It predicts the observed variation in flow-stress values of the slopes of the velocity-stress relations. The critical shear, τ_c, stress needed to move a dislocation through a random array of obstacles on a glide plane is calculated using a statistical theory. The calculated result is an expression of τ_c in terms of the obstacle concentration, the dislocation-line tension and the interaction force between a dislocation and a single obstacle. However, as seen in Fig. 5.31, the experimental slope is $\alpha = 0.87$, which deviates, to a large extent, from the aforementioned theories. This difference between the above theories and the experimental results may be explained by Friedel's model, which assumes a zigzag dislocation line interacting with temporary pinning points (solid atoms in

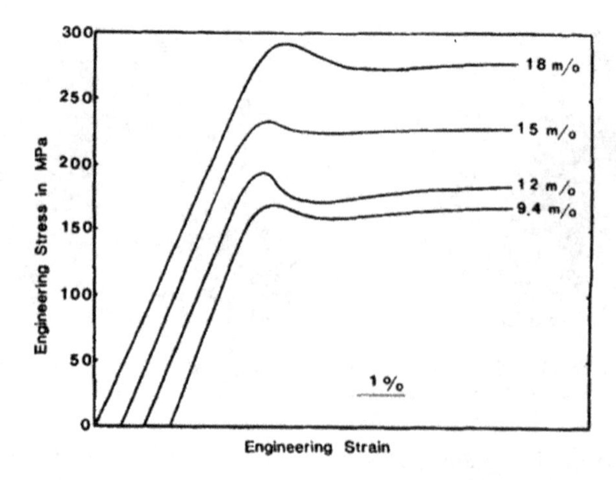

Fig. 5.30 Engineering stress–strain curves of Y_2O_3-stabilized ZrO_2 deformed along ⟨112⟩ at 1400 °C in air; Y_2O_3 content is shown on each *curve* [6]. With kind permission of John Wiley and Sons

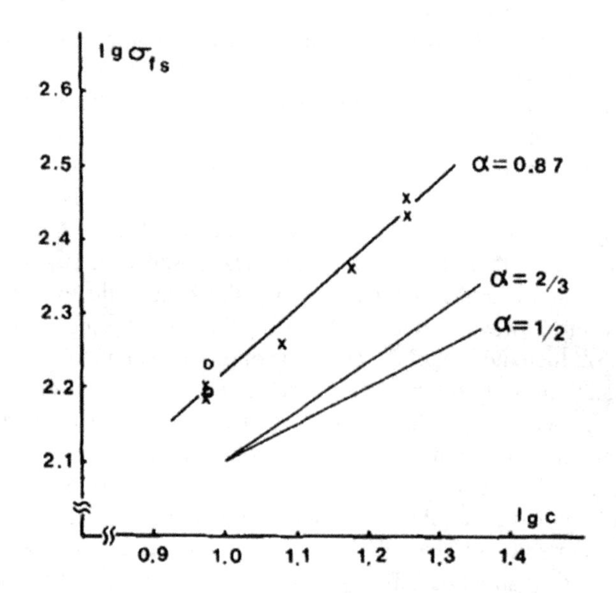

Fig. 5.31 Flow stress vs Y_2O_3 content of samples deformed **(X)** along (112) and (O) a direction 4° from ⟨001⟩ at 1400 °C in air [6]. With kind permission of John Wiley and Sons

structure). In Fig. 14.9 on page 381 of Friedel's book, there is an illustration of a zigzagged dislocation line pinned at several points by impurities. A large enough applied stress is required to tear the dislocation away from its impurity. TEM observations of the dislocation substructure of a 9.4 mol% YSZ crystal, deformed to ≈14 % strain, is shown in Fig. 5.32. Many of these dislocations are curved and a few dislocation loops are also visible. The **g-b** = 0 conditions obtained at two

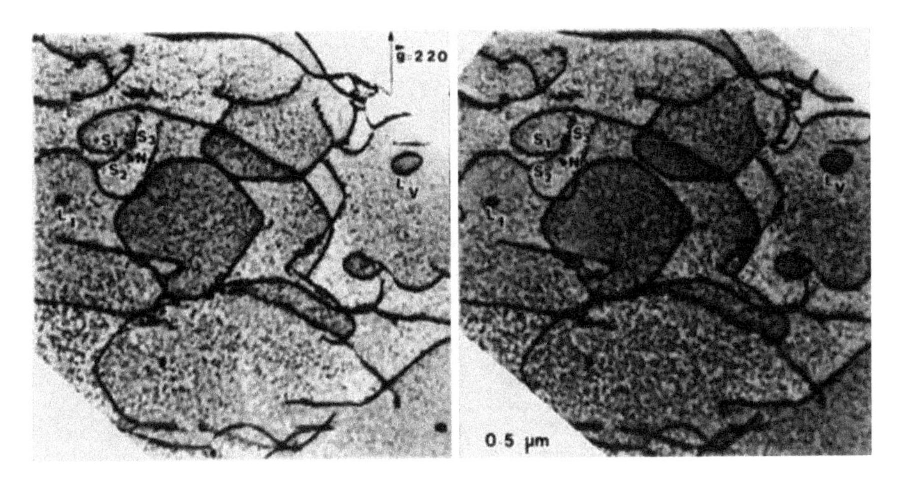

Fig. 5.32 Dislocation substructure in 9.4 mol% Y_2O_3-stabilized ZrO_2 sample deformed along ⟨112⟩ by {001} ⟨110⟩ slip to ≈14 % strain at 1400 °C in air, shown as stereo pair with $\theta = 17^o$, using b = 220 in [00 1] zone. Loops L_y and L_i are of vacancy and interstitial character, respectively. Dislocation reaction at node N gives rise to dislocation segments with different Burgers vectors (S_1, S_2 and S_3) [6]. With kind permission of John Wiley and Sons

different zones, **B** = [001] and [112], confirm that **b** is ½ ⟨110⟩ for the majority of the dislocations. The shapes in Fig. 5.32 also indicate that substantial cross-slip and climb occurred during this deformation. An additional specimen was deformed only to the upper yield point (~0.4 % deformation) and its TEM image was basically the same as that in Fig. 5.32, in regard to density and the form of the substructure. An 18 mol% yttria containing stabilized zirconia, deformed to ≈15 %, is found in Fig. 5.33. Here, the dislocation density is considerably higher, but the other dislocation features are the same as the ones illustrated in Fig. 5.32, with ≈9.4 mol%. More information on dislocation-point defect interactions may be found in the work of Rodrigez et al. [6].

In summary, this section discussed solid-solution strengthening and provided a few illustrations of technically important ceramics. Additives forming solid solutions may enhance the strength properties of ceramics. The applications of some ceramics are often limited by their poor mechanical properties, but the use of certain additives to create solid solutions improve the performance of these enhanced ceramics, enabling their use as structural materials.

5.4 Second Phase Strengthening

Here, the effect of a non-soluble second phase on the strength properties of ceramics will be considered, irrespective of whether the second (or more) phase was originally a part of the process and was mixed with the basic ingredients of a ceramic to be strengthened or whether it was produced as precipitates during

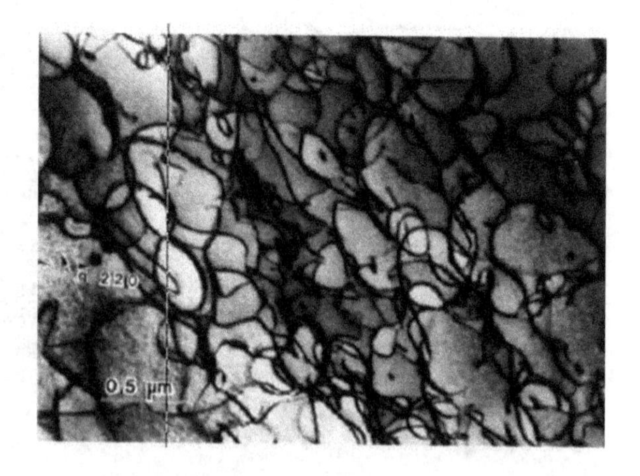

Fig. 5.33 Dislocation substructure in 18 mol% Y_2O_3-stabilized ZrO_2 deformed to ≈15 % under same conditions as sample in Fig. 5.32 [6]. With kind permission of John Wiley and Sons

high-temperature processing. This non-soluble second phase effect will be considered below, regardless of the location of the second phase, whether randomly distributed or segregated at grain boundaries. The focus in this section is on the strengthening effect itself, on the terms 'strengthening' and 'softening'. In fact, the term 'second-phase strengthening' does not exclude the presence of more than one additional constituent. A second phase may take any form, manifesting as: fibers, precipitates, new phases (forming at elevated processing temperatures) and even as single elements. Increased density is a general property involved in any strengthening mechanism and is a good parametric indication of microstructural and, hence, strength modification. Ceramics having a second phase often are referred to as 'composites', since this general term refers to materials composed of two or more constituents with significantly different physical and chemical properties that produce resultant characteristics in the new composite material different from those of the prior individual components. In the following, examples will be given of second-phase hardening attained by means of:

(i) the addition of a ceramic second phase;
(ii) the formation of a second phase during some stage of the processing (precipitation or particle dispersion);
(iii) the addition of fiber;
(iv) the addition of a single element.

Of these, only (i) and (ii) will be discussed in this chapter. Illustrations of these methods of second-phase strengthening are presented below.

5.4.1 Addition of a Ceramic Second Phase

As previously indicated, the methods and techniques used to form such composite materials are not of interest here. Instead, this and the following sections will focus

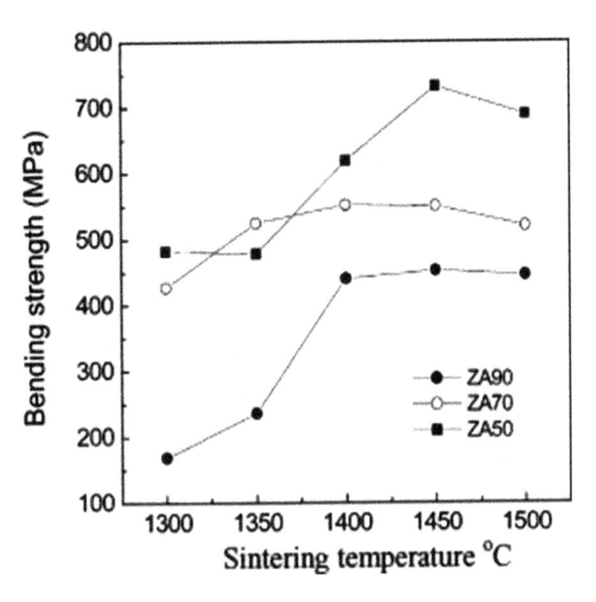

Fig. 5.34 Bending strength as a function of sintering temperature of ZTA samples [15]. With kind permission of Elsevier

on the changes in the mechanical properties of the resultant composites. The first example relates to the effect of a second phase on Al_2O_3, which is one of the most studied composite ceramics, found to have excellent mechanical properties. The properties of alumina (like those of most materials) are microstructure-dependent. To obtain the best properties, fine grain size and homogeneous microstructure are essential. However, fracture toughness is not necessarily the best in fine-grained ceramics. The addition of Zr_2O_3 to Al_2O_3 produces a composite ceramic with very attractive mechanical properties. Zirconia-toughened alumina (ZTA) exhibits excellent wear resistance and good hardness properties. The combination of zirconium oxide and aluminum oxide belongs to a class of composite ceramics called 'AZ composites'. Such composite ceramics are used as components in diverse structural and medical applications. AZ Structures based on alumina–zirconia are characterized by high strength, fracture toughness and elasticity, in addition to the aforementioned hardness and wear resistance and have many structural applications, such as: bearings, joint implants, bushings, cutting tool inserts, wear components, devices for biological use, etc. To this list, one should add the high-temperature stability and corrosion resistance of ZTA. In the following figures, these materials are designated as ZAXX, where ZA represents the ZTA ceramics and XX represents the volume fraction of the alumina present in the specimens. The variation of the bending stress of three Al_2O_3-ZrO_2 composites at sintering temperature is shown in Fig. 5.34.

Note that the specimens containing 90 % Al_2O_3 have the lowest bending stress at all the sintering temperatures, whereas the ceramics containing 50 % Al_2O_3 and 50 % ZrO_2, i.e., ZA50, has the highest value. The microstructure at various stages of the processing may be seen in Fig. 5.35.

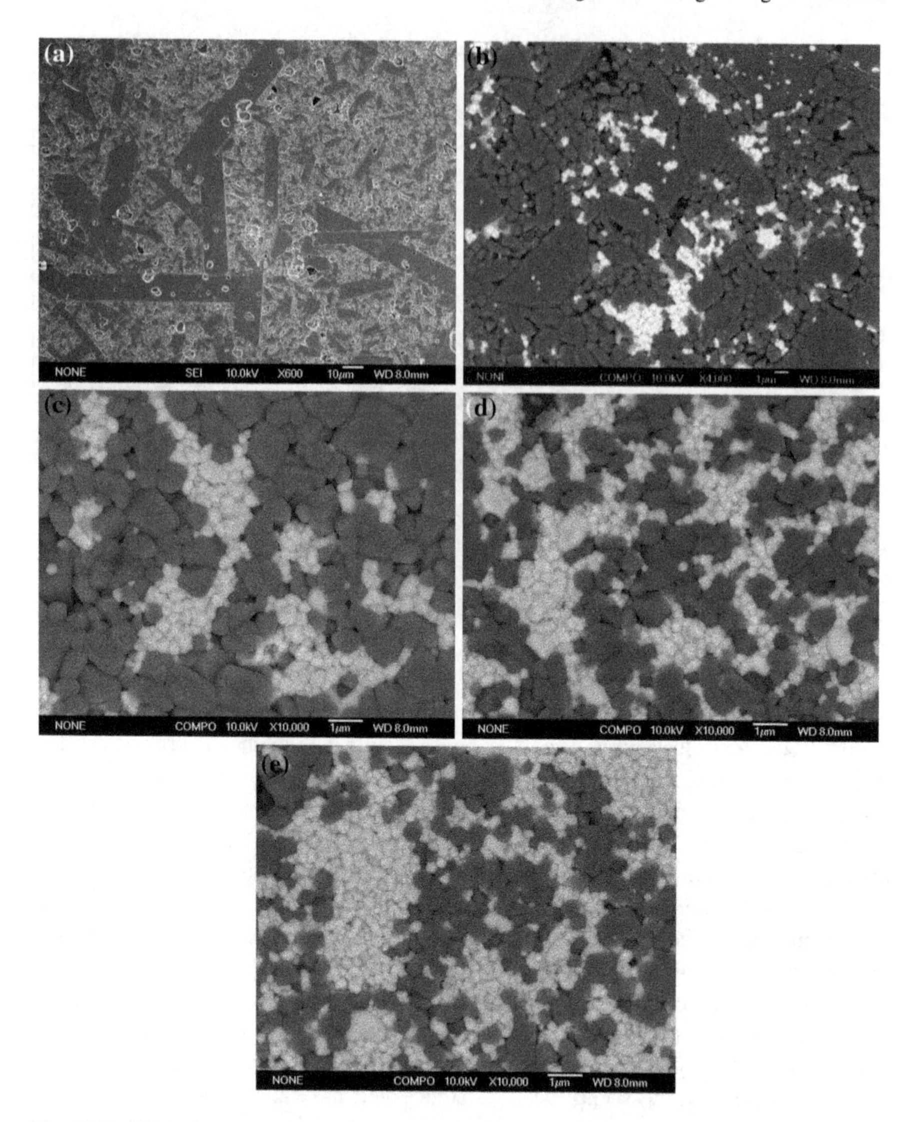

Fig. 5.35 SEM micrographs of polished and thermally etched surfaces for the Al_2O_3/3Y-TZP composites sintered at 1400 °C: **a** Al_2O_3, **b** ZA90, **c** ZA80, **d** ZA60 and **e** ZA50 [15]. With kind permission of Elsevier

Also note that the constituents of the starting powders used to obtain ZTA contain 3Y-TZP as the source for ZrO_2 in the composite ceramics. Observe the change in the bending stress (and fracture toughness) of the ZrO_2 content as expressed in Fig. 5.36 in terms of 3Y-TZP. The microstructure of the composites shown was sintered at 1400 °C. In Fig. 5.35a, plate-like grains formed in the

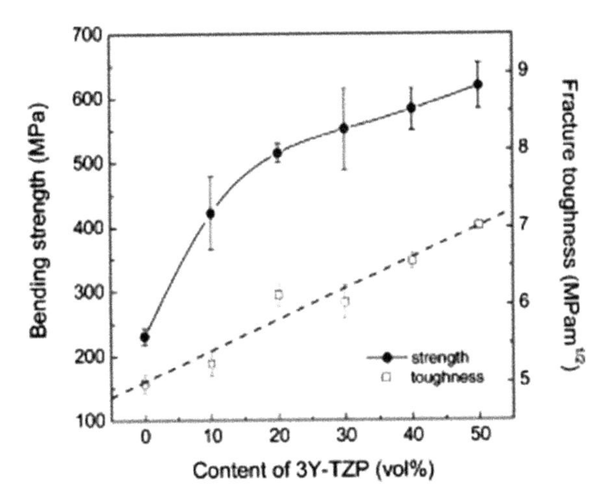

Fig. 5.36 Bending strength and fracture toughness as a function of 3Y-TZP content for the composites sintered at 1400 °C/2 h [15]. With kind permission of Elsevier

sintered Al_2O_3 specimen. It is believed [33] that these plate-like grains of Al_2O_3 are the results of small amounts of impurities, especially of SiO_2, and that their formation requires a higher temperature (≥1500 °C). The pinning effect of ZrO_2 on the grain growth of Al_2O_3 is indicated in Fig. 5.35b–e. Furthermore, with the increase of ZrO_2, the morphology of Al_2O_3 changes to equiaxed grains and their size decreases. At low ZrO_2 concentration, exaggerated grain growth occurs readily (Fig. 5.35b, only 10 vol% 3Y-TZO), whereas at 55 vol%, grain growth in ZTA is prevented. Figure 5.36 shows the average bending strength and fracture toughness of the various specimens sintered at 1400 °C.

This section deals with the effect of a non-soluble second phase, specifically of ZTA. Indeed, the phase diagram shown below in Fig. 5.37 indicates that the system in the temperatures considered above is a two-phase region for both the monoclinic and tetragonal ZrO_2. As may be seen in that phase diagram, 1150 °C alumina and monolithic zirconia coexist, but, at sintering temperatures, the monolithic zirconia transforms into its tetragonal allotrope. During cooling from the fabrication temperature, the dispersed zirconia in the alumina matrix retains a tetragonal phase in the composite, probably as a metastable phase. When strengthening alumina with zirconia, there is the added possibility of tetragonal-zirconia strengthening (especially at temperatures above 1150 °C), but reports also indicate monoclinic zirconia hardening of alumina resulting in ZTA. Moreover, note the concept of transformation toughening (hardening) which is associated with volume expansion and shear-strain change during transformation from the tetragonal to the monoclinic phase.

Alumina may be strengthened by other additives, as well. The most common additives to alumina are: iron oxide, titanium oxide, calcium oxide, magnesium oxide, potassium oxide and sodium oxide.

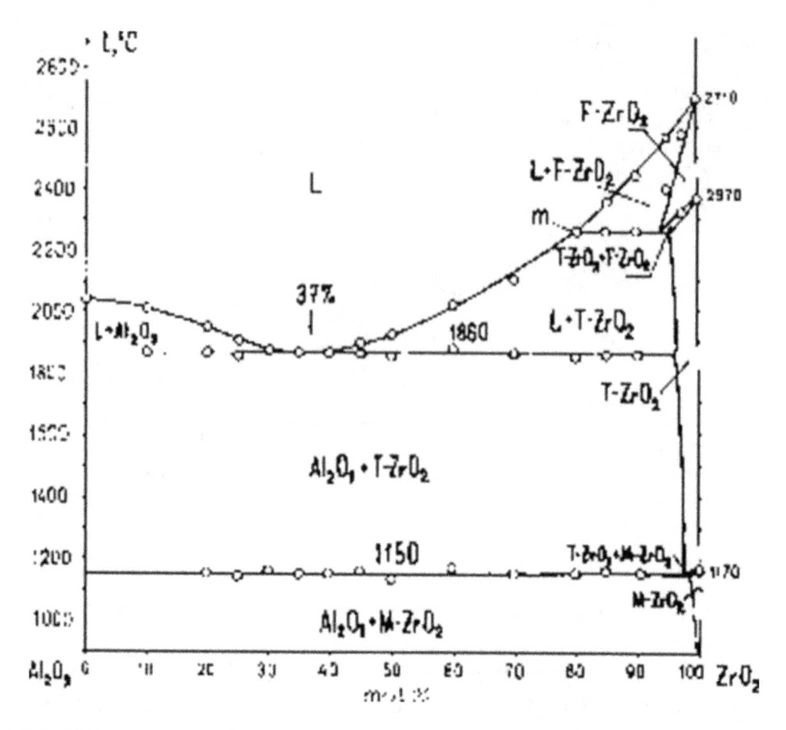

Fig. 5.37 Phase diagram of the system Al_2O_3–ZrO_2 [22]. With kind permission of John Wiley and Sons

5.4.2 *Second Phase Formation During Some Stage of the Processing (Precipitation or Phase Transformation)*

Here, two kinds of strengthening are considered, which are basically process-related effects:

(i) particle dispersion (precipitation) hardening;
(ii) transformation hardening.

Selected examples of each are illustrated below.

(i) As is well known, a solid solution precedes precipitation hardening; the first occurs at some appropriate temperature for an additive solution, while the latter sets in at some lower temperature, rejecting a phase or a precipitate from the solution at favorable sites, which may be grain boundaries, phase boundaries, other surfaces. Later on, at the temperature of the process, this precipitate may become dispersed more randomly within the matrix. In addition to the precipitates obtained during this process, dispersion hardening differs from precipitation hardening in that the small, insoluble particles are added a priori into the ceramics and dispersed in them either randomly or preferentially at various surfaces. To achieve optimal strengthening, small particles (less than 0.25 microns in diameter) are finely distributed in the matrix to impede the movement of dislocations and, thus, the deformation of the material is made more difficult, requiring higher stresses. The overall effect of such strengthening is similar to that of the well known precipitation hardening. However, during precipitation strengthening the particles are dissolved at the high temperature treatment (solution stage) in the matrix, while during dispersion hardening the particles (usually ceramics particles) do not dissolve and are stable. The temperature at which a particle is stable and retained within the matrix is a very important aspect of dispersion hardening, since most ceramics are intended for high-temperature use and applications. The large, second-phase particles dispersed within the matrix have a low strengthening effect. Therefore, the impact of the dispersed particles on ceramic properties depends on the particles' dimensions. Clearly, the properties of the particles, beside their size, have a great influence. A particle can be weaker (softer) or stronger (harder) than the matrix and both types have their own merits. The choice of what kind of particle should be used depends on the purpose of the ceramic application. For instance, soft particles should be added for improved machinability. Particles of iron oxide, titanium oxide, calcium oxide, magnesium oxide, potassium oxide and sodium oxide are some of the common additives in alumina as previously mentioned.

In the following, a matrix reinforced by a dispersed phase in the form of particles will be illustrated. Alumina has been strengthened by the addition of Fe_3Al nanoparticles. Sintering was performed at 1530 and 1600 °C for $A1_2O_3$/5 wt% Fe_3Al and $A1_2O_3$/10 wt% Fe_3Al, respectively. Compared to alumina, ceramics with these additives show strength increase by 132 % and toughness increase by 73 %. Furthermore, their bending and fracture toughness were 832 MPa and 7.96 MPa $m^{1/2}$, respectively. Figure 5.38a shows the variation of hardness versus temperature for $A1_2O_3$ with Fe_3Al additions in the range 5–30 wt%. In (b), the hardness of Al_2O_3/Fe_3Al10 % is shown versus the holding time of the sintering. The change in holding time did affect the hardness values. In Fig. 5.39, the variation in bending strength and fracture toughness are shown as functions of sintering temperature.

The effect of holding time on the bending strength and fracture toughness of an Al_2O_3/Fe_3Al10 wt% sample sintered at 1560 °C is seen in Fig. 5.40. Note that the holding time affects both the bending strength and the fracture toughness. It is likely that the grain size and the particle distribution are also affected by the holding time. This may have occurred after $\sim$30 min., as a decrease in the bending strength occurred. This fracture-toughness improvement is a consequence of the increased Fe_3Al particle size at intergranular locations.

SEM microstructures of the fractured surface are compared in Fig. 5.41 with and without additions to the alumina at 1520–1530 °C. Similar microstructures are shown at 1600 °C for 10 and 20 wt% Fe_3Al. The monolithic Al_2O_3 has platelet-shaped grains with an average size of 51 μm (aspect ratio 4.2). EDS analysis indicated that the alumina phase is dark, whereas the bright phase is Fe_3Al.

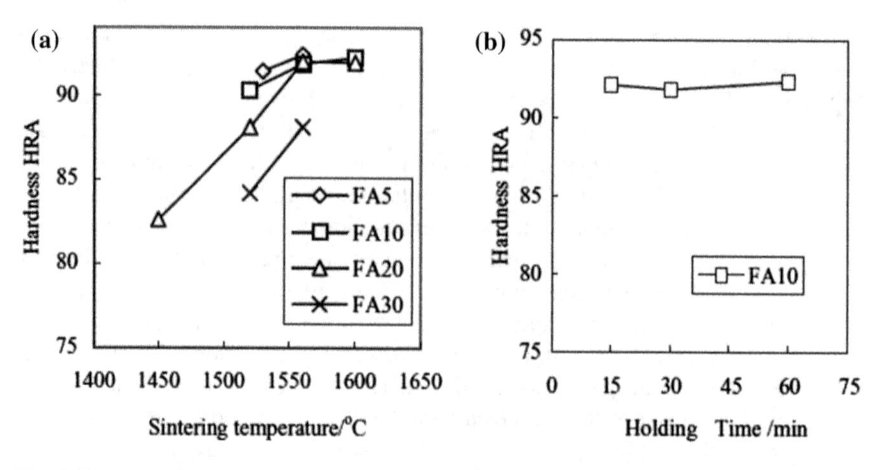

FIg. 5.38 Variation of hardness HRA of Al_2O_3/Fe_3Al nano-composites with **a** sintering temperature and **b** holding time [11]. With kind permission of Elsevier

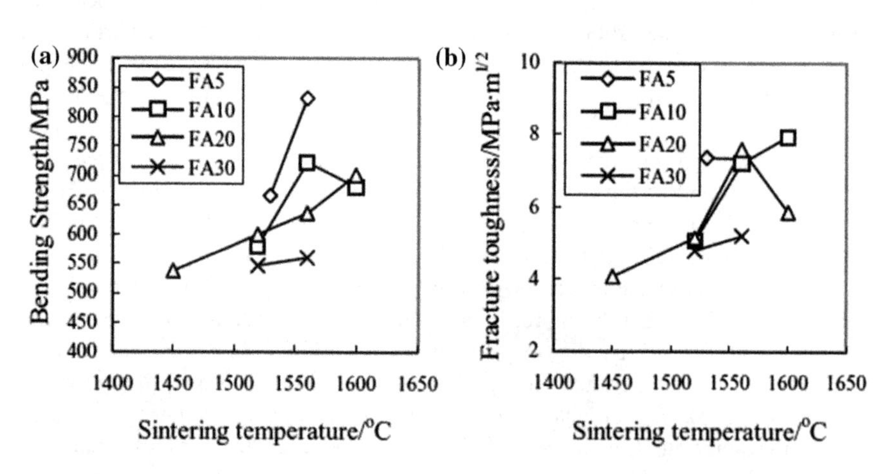

Fig. 5.39 Variation of **a** the bending strength and **b** the fracture toughness (K_{Ic}) of Al_2O_3/Fe_3Al nano-composites with sintering temperature and contents [11]. With kind permission of Elsevier

Abnormal grain growth was observed in the alumina matrix with the consequent deterioration. The addition of Fe_3Al, already at a level of 5 wt% and sintered at 1530 °C, caused the alumina platelets to become smaller and distributed more homogeneously than observed in the alumina without additives (Fig. 5.41b). More additions of Fe_3Al particles further decreased the aspect ratio and the grain size of the matrix. However, these structures show porosity and cracks (Fig. 5.41c and d. Nevertheless, grain growth in the alumina was eliminated. The fracture mode was intergranular and the mechanical properties were low. Figure 5.42 illustrates the microstructure when the hot-pressing temperature is increased to 1600 °C in ceramics containing 10 and 20 wt% Fe_3Al. Now, the fracture is transgranular and

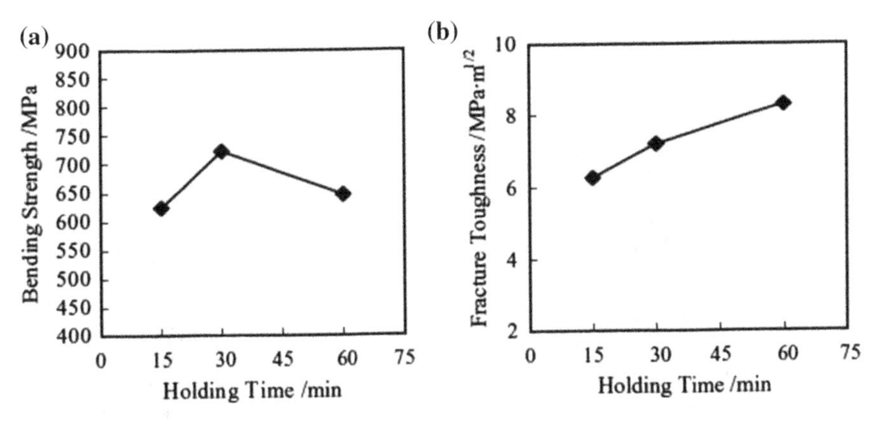

Fig. 5.40 Variation of **a** the bending strength and **b** the fracture toughness (K_{Ic}) of 10 wt% Al_2O_3/Fe_3Al nano-composite sintered at 1560 °C with holding time [11]. With kind permission of Elsevier

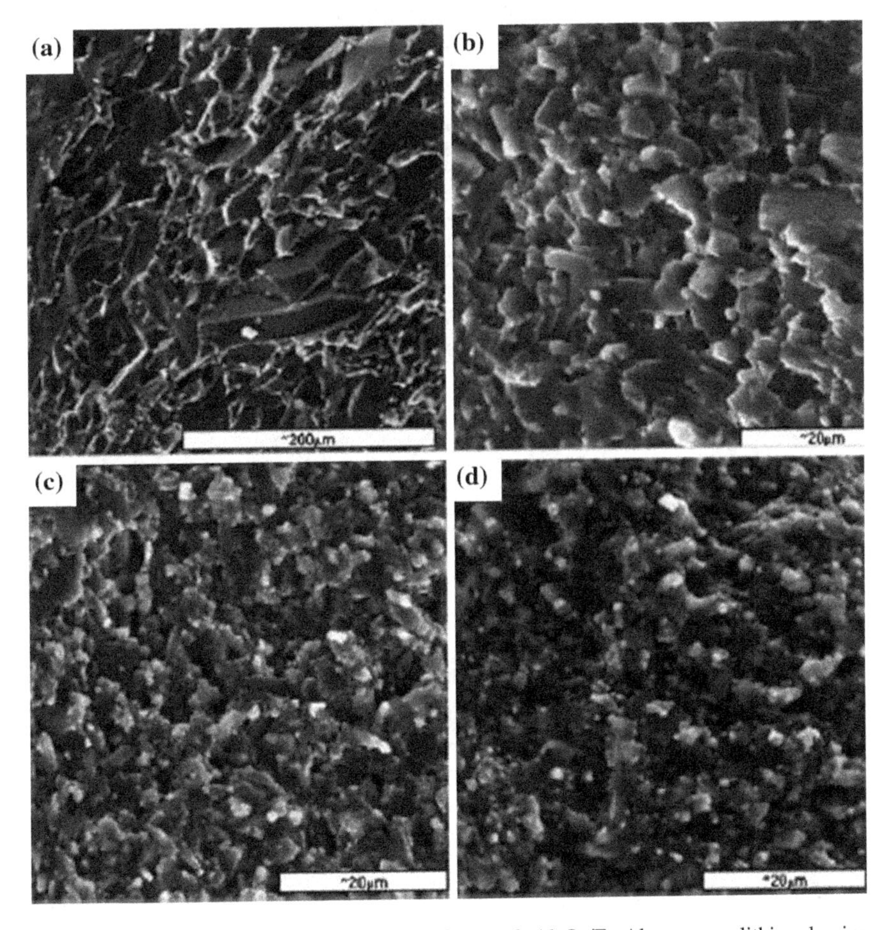

Fig. 5.41 SEM micrographs of fracture surfaces of Al_2O_3/Fe_3Al: **a** monolithic alumina; **b** 1530 °C FA5; **c** 1520 °C FA10; **d** 1520 °C FA20 [11]. With kind permission of Elsevier

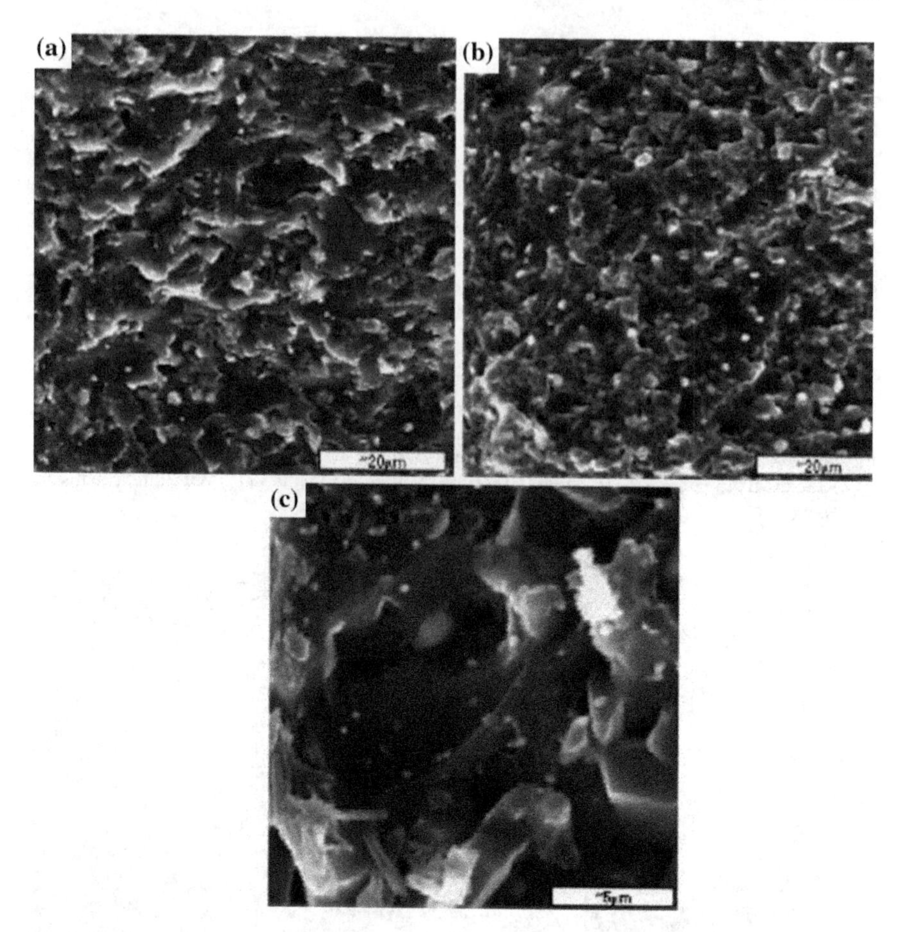

Fig. 5.42 SEM micrographs of fracture surface of Al_2O_3/Fe_3Al: **a** and **c** at 1600 °C, FA10; **b** 1600 °C, FA20 [11]. With kind permission of Elsevier

the intragranular particles (Fe_3Al) are increased. At higher sintering temperatures, some grain growth occurs in the alumina and the Fe_3Al particles coalesce at the alumina grain boundaries.

During the cooling of the specimens from the sintering temperature, residual stress arises in the matrix, due to the mismatch created between the alumina matrix and the Fe_3Al particles during the sintering step. Crack propagation is observed in Al_2O_3/Fe_3Al5 wt% already at the Vickers indentation, 308 N. In some places, the path was deflected by the Al_2O_3 platelet-like grains (Fig. 5.43a). When the crack orientation is parallel to the direction of its propagation, debonding along the Al_2O_3 interface is observed.

Relief of the residual stresses (resulting from the mismatch between the various ceramic constituents) during cooling from the sintering temperature generates dislocations, as illustrated in Fig. 5.44. These dislocations surround the second-

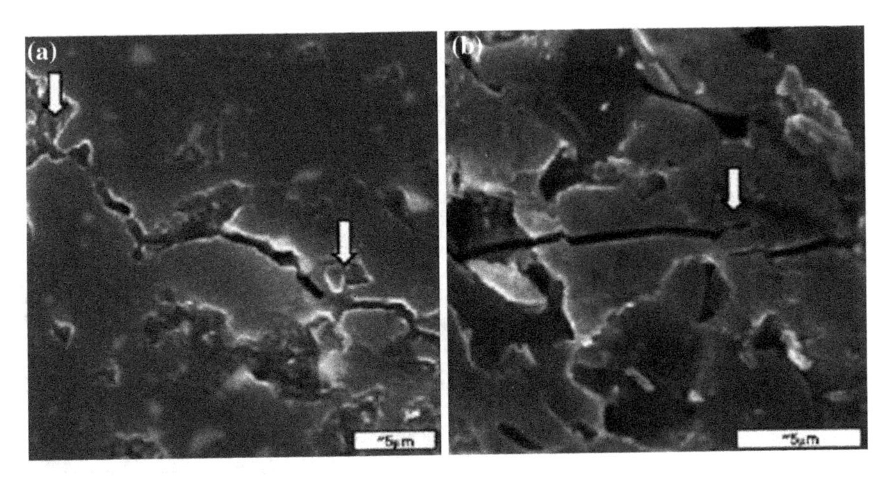

Fig. 5.43 The interaction between crack and Al_2O_3 grains in Al_2O_3/5 wt% Fe_3Al [11]. With kind permission of Elsevier

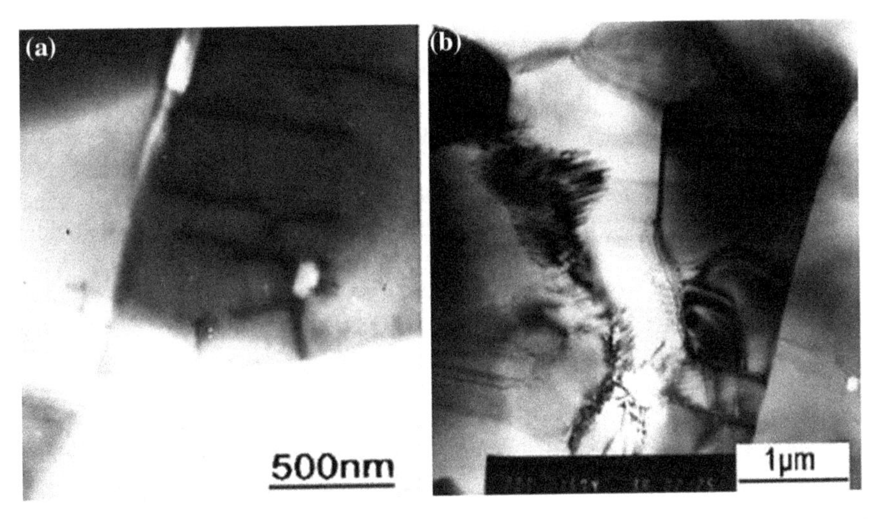

Fig. 5.44 TEM morphology of dislocations in Al_2O_3/Fe_3Al nano-composite [11]. With kind permission of Elsevier

phase particles (Fe_3Al), as seen in this figure at two magnifications. Their location around the particles is consistent with dislocation pinning caused by particles added to a matrix to enhance its mechanical properties.

An interesting microstructure of ZrO_2 particles, toughened by Al_2O_3, is illustrated in Fig. 5.45. Generally, hard particles dispersed in a softer matrix increase wear and abrasion resistance.

To summarize, the effects of the addition of a second phase in the form of particles, as illustrated above, explain why small particles dispersed in a ceramic matrix are an integral part of the matrix strengthening of ceramic materials.

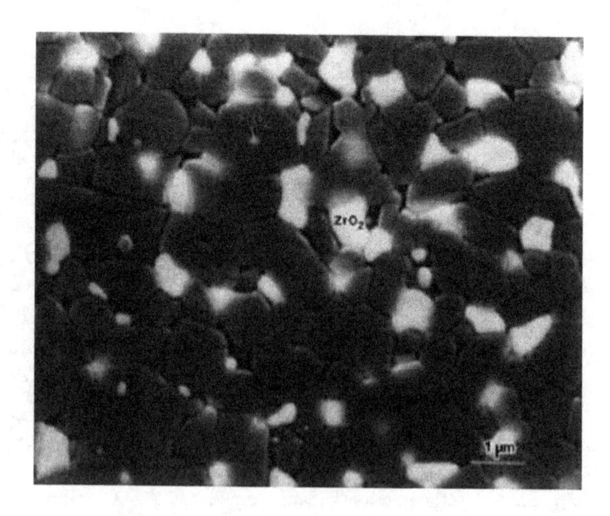

Fig. 5.45 Microstructure of ZrO_2-toughened $A1_20_3$. The ZrO, particles, most of which are intergranular, appear bright in this scanning electron micrograph [14]. With kind permission of John Wiley and Sons

5.5 Transformation Hardening

One of the well-known ceramics undergoing phase transformation is ZrO_2. It has three monoclinic polymorphs at RT and is stable up to 1170 °C. It becomes tetragonal above this temperature and transforms to cubic at 2370 °C, which is stable up to the melting point of 2680 °C. At high temperatures, the major phase is cubic, which may contain tetragonal and/or monoclinic precipitates as minor phases These precipitates may exist at grain boundaries in polycrystalline zirconia. Clearly of interest are the tetragonal and monoclinic phases. Transformation during cooling from tetragonal to monoclinic is somewhat sluggish and takes place in a temperature range of $\sim$100 °C below the transition point of 1170 °C (some reports note the transition temperature on cooling as $\sim$950 °C). Also, the tetragonal to monoclinic phase is considered to be a Martensitic transformation (sometimes reported to set in at 1200 °C), which is not a diffusional transformation. This phase transition is associated with a volume change, $\sim$3–4 %, and, therefore, expansion stresses arise which may cause cracking in pure ZrO_2. Often therefore, various phases are added to stabilize zirconia. Nevertheless, additional strengthening occurs in stabilized zirconia, known as 'transformation toughening'. Thus, a temperature range exists in which both polymorphic ZrO_2s may be present in some proportion in the microstructure, being sintering- and process-temperature-dependent. In pure zirconia, it is essential to be aware of the propensity for crack formation above a certain temperature. However, as mentioned above, various stabilizing agents may be added during fabrication, the most common of which are oxides, such as: CaO, MgO, CeO_2, Y_2O_3, etc. One very effective stabilizer is Y_2O_3 and other potential stabilizers are the PSZ, whose microstructures may contain several phases (a multiphase material). As such, it is possible to produce various matrix phases containing other zirconia- transformed products distributed within them. For example, even a cubic zirconia matrix has been produced with its monoclinic polymorph in the CaO-ZrO_2 system [10]. The high-temperature cubic phase can be stabilized down to RT when cubic oxides, such as MgO, CaO, Y_2O_3, CeO_2 and other rare earth oxides, are added to the zirconia during fabrication. These additives not only strengthen the zirconia matrix, but stabilize the cubic phase by suppressing the transformation temperature during cooling. Tetragonal ZrO_2 may be retained as a metastable phase at RT and this stress induced transformation becomes the basis for transformation toughening. Extensive information on the transformations in the ZrO_2 system appears in several publications [13] which may be consulted.

Below, the exemplary properties of Al_2O_3–ZrO_2 are presented. This system was chosen due to its technological importance for applications at elevated temperatures. As expected, its microstructure dictates the properties of the samples presented below. Figure 5.46 shows the microstructure of this system. Both tetragonal (t-phase) and monoclinic (m-phase) zirconia particles are incorporated into an alumina matrix and their total content is in the range 5–30 vol%. Figure 5.47 shows the variation of the elastic modulus versus the total zirconia content. The best fit line encompassing the points Al_2O_3/t-ZrO_2, Al_2O_3/m-ZrO_2 and Al_2O_3/t-ZrO_2 + m-ZrO_2 was evaluated by the rule of mixtures. One may see that the elastic modulus of pure alumina is higher than those of alumina with zirconia additives, regardless of the zirconia polymorph type. It is assumed that the presence of porosity and microcracks can reduce the elastic modulus associated with zirconia, especially of the m-ZrO_2. The strength of Al_2O_3/ZrO_2, as a function of the total zirconia content, is illustrated for m-ZrO_2, t-ZrO_2 and m-ZrO_2 + t-ZrO_2 toughened alumina in Fig. 5.48. The top curve, associated with Al_2O_3/t-ZrO_2 + m-ZrO_2, shows the highest strength in this system having a value of 940 MPa. The toughness of ZTA is illustrated in Fig. 5.49. Observe that the top curve, representing the highest degree of toughness, occurs in Al_2O_3/m-ZrO_2. The data mentioned above indicate that addition of m-ZrO_2 and t-ZrO_2 improves the mechanical properties of alumina. This alumina, strengthened by both types of zirconia, was sintered at 1600 °C. The t-ZrO_2 also contained some yttria, which had an influence on the transformation. One may see in Fig. 5.49 that the m-ZrO_2 is represented by the top curve with the highest toughness, associated with the effect of the Y_2O_3 originally shown by the t-ZrO_2.

It was observed that ions of yttria are absorbed in the m-ZrO_2, stabilizing it at the expense of the t-ZrO_2. As expected, they affect the microstructure and, consequently, the transformation, as well. Inspecting Figs. 5.49 and 5.50, one finds a decrease in flexure strength and toughness at around 15 vol% of total zirconia. This is likely caused not only by grain-size change, but also by the presence of cracks and pores. Support for this theory may be inferred from Fig. 5.51c, where pores and cracks of various shapes and dimensions may be observed. Nevertheless, the microstructural refinement of alumina by the addition of both types of zirconia may be inferred from Fig. 5.46b, d. In general, adding t-ZrO_2 and m-ZrO_2 to alumina greatly improves its mechanical properties in proportion to the amount of transformable zirconia (a basic precept in transformation hardening).

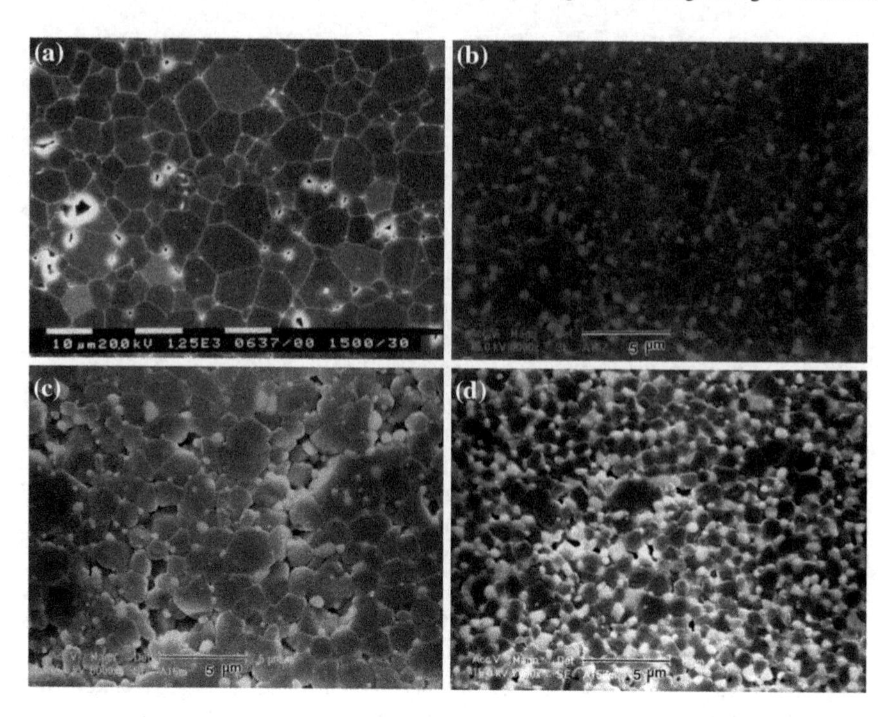

Fig. 5.46 Microstructures of **a** Al_2O_3, **b** Al_2O_3/15 % t-ZrO_2, **c** Al_2O_3/15 % m-ZrO2 and **d** Al_2O_3/(15 % t-ZrO_2 + 15 % m-ZrO_2) composites [31]. With kind permission of Elsevier

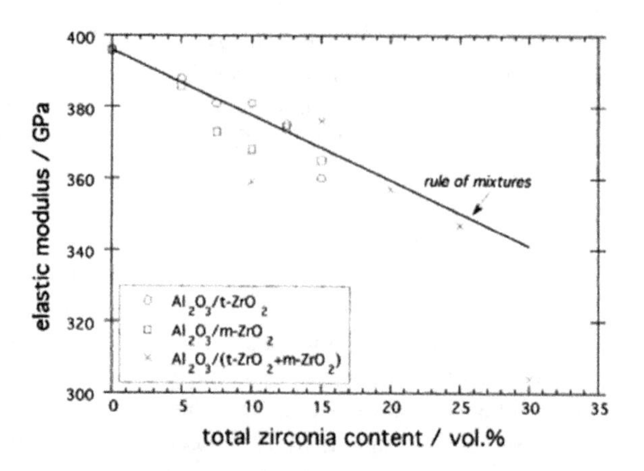

Fig. 5.47 Elastic modulus of composites as function of total zirconia content. The straight line predicted by the rule of mixtures is shown for comparison [31]. With kind permission of Elsevier

The zirconia toughening of various ceramics is of great interest and technological importance. It also undergoes Martensitic transformation. Figure 5.50 shows such a transformation in MgO-partially-stabilized ZrO_2 [henceforth: Mg-PSZ]. The Martensitic transformation upon cooling is a t $\rightarrow$ m transition.

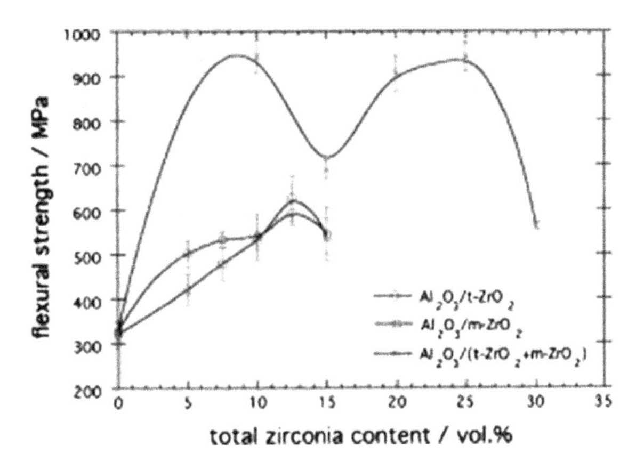

Fig. 5.48 Flexural strength of composites as a function of total zirconia content [31]. With kind permission of Elsevier

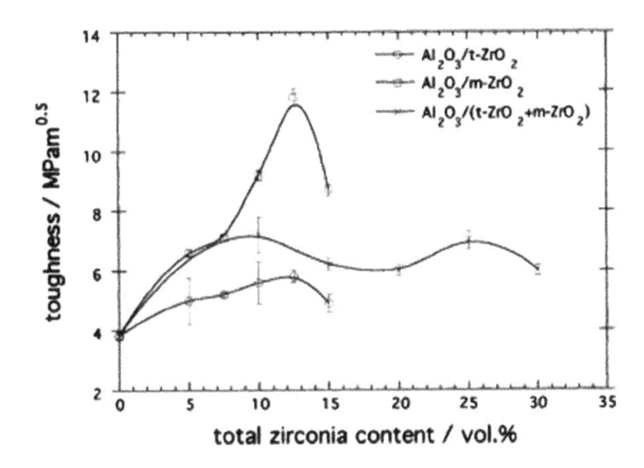

Fig. 5.49 Toughness of composites as a function of total zirconia content [31]. With kind permission of Elsevier

Fig. 5.50 Microstructure of t-ZrO_2, precipitates in c-$Zr0_2$ matrix in Mg-PSZ. The ***c*** axes of two of the three possible precipitate variants, which are parallel to the cube axes, are shown. The third variant cannot be imaged under the diffracting conditions of this example. Bright-field electron micrograph [13]. With kind permission of John Wiley and Sons

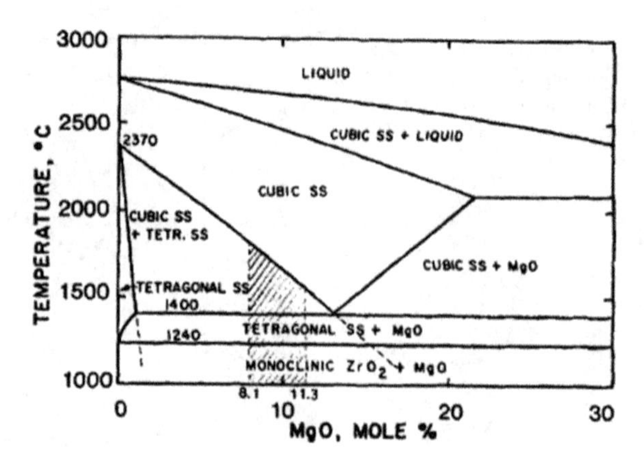

Fig. 5.51 Phase diagram of the high-ZrO, portion of the Zr0$_2$-Mg0 system (grain). The shaded region shows the composition range of commercial alloys. The *dotted lines* are metastable extensions of equilibrium solvuses [13]. With kind permission of John Wiley and Sons

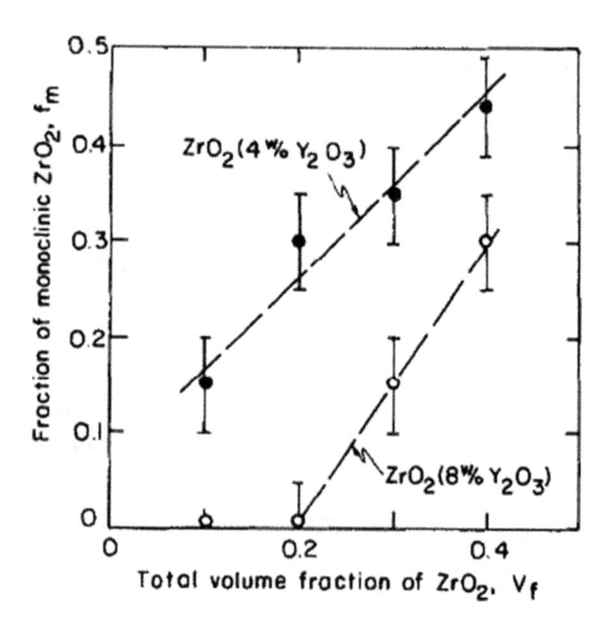

Fig. 5.52 Plot of monoclinic fraction in partially stabilized ZrO,/ZnO composite [28]. With kind permission of John Wiley and Sons

The phase diagram of the ZrO_2–MgO system shows, on the ZrO_2 side, the three polymorphic phases that exist in ZrO_2 (here, the solid solution of MgO in ZrO_2 is shown, indicating the three polymorphs of ZrO_2 at 0 % MgO).

A fraction of the total ZrO_2 which has transformed into a monolithic zirconia (m-ZrO_2) polymorph is shown in Fig. 5.52. Here, two lines, representing 4 and 8 wt% Y_2O_3 in the zirconia, are plotted respectively. A higher monoclinic fraction was obtained for the zirconia containing 4 wt% yttria. The matrix of this ceramic is ZnO. The starting materials, containing pure ZrO_2, indicated the exclusive existence of m-ZrO_2, but the PSZ contained some t-ZrO_2. In Fig. 5.52, the relative amount of m-ZrO_2 obtained for the partially stabilized materials is indicated versus the total volume concentration of ZrO_2.

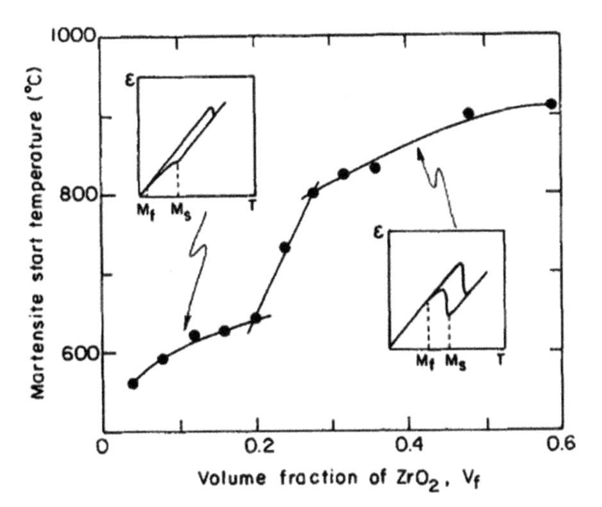

Fig. 5.53 Plot of martensite start temperature, M_s, as a function of ZrO_2 volume concentration (for pure $Zr0_2$), determined from dilatometer studies [28]. With kind permission of John Wiley and Sons

As previously stated, the transformation to m-ZrO_2 is Martensitic and the Martensite starting (M_s) temperature, related to the pure volume fraction (V_f) of the zirconia, is shown in Fig. 5.53. Two regimes are graphically shown in Fig. 5.53. For the volume fraction $\leq$ 0.2, the transformation temperature at the start of the Martensite formation (M_s) varies slowly with the volume fraction of ZrO_2 up to $M_s \approx 600$ °C and this transformation extends over an appreciable temperature range (such that $M_f = 100$ °C). At concentration $\geq$ 0.3, M_s is ~ 850 °C and proceeds to completion over a narrow temperature range ($M_f = 550$ °C). It is believed that transformation at $\geq$0.3 is associated with particle interaction, which supports cooperative transformation in neighboring particles. The TEM micrograph indicates that the m-ZrO_2 particles are twinned, as indicated in Fig. 5.54. The twin spacing seemed to increase with particle size, ranging from 22 to 100 nm for a (zirconia) particle size of 0.16–1.3 μm. Microcracks are observed at the twin terminals, but it was postulated that these are not of sufficient size to affect the mechanical properties and, thus, are unlikely to be a source of stress-induced microcracks.

The elastic modulus (a) and Vickers hardness (b), indented with a 200 N load on polished surfaces, are shown in Fig. 5.55. The lines increase linearly with the increasing fraction of ZrO_2. Such a trend may reflect the absence of spontaneous microcrack formation.

The critical stress-intensity factor, as a function of the total zirconia content, and relative toughness, as a function of the m-ZrO_2 content, when the t-ZrO_2 is excluded, are shown in Fig. 5.56.

SEM micrographs show that fracture in ZnO is mainly transgranular. The cracks are deflected around the ZrO_2 particles along the ZnO/ZrO_2 interface. The crack-surface area increases as the ZrO_2 increases. No zone of microcracks is found (at the tip of an arrested crack). This finding is in line with acoustic emission measurements. Figure 5.57 illustrates the SEM results and the deflection of the

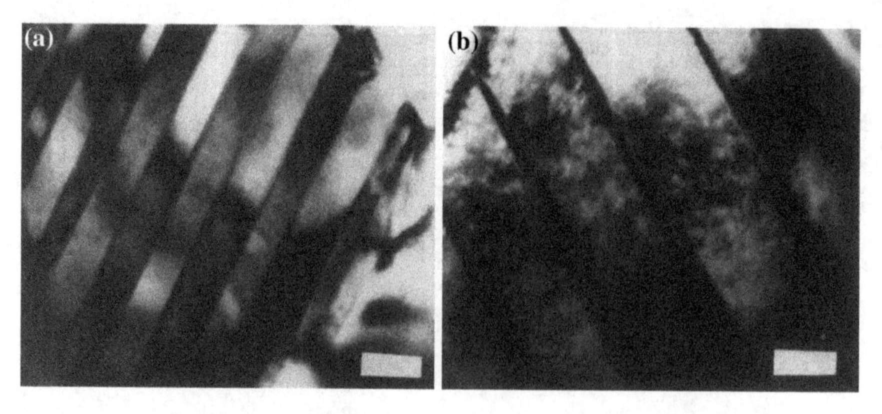

Fig. 5.54 Transmission electron micrographs of twinned ZrO_2 particles in ZnO; bar = (a) 0.02 μm and (b) 0.2 μm [28]. With kind permission of John Wiley and Sons

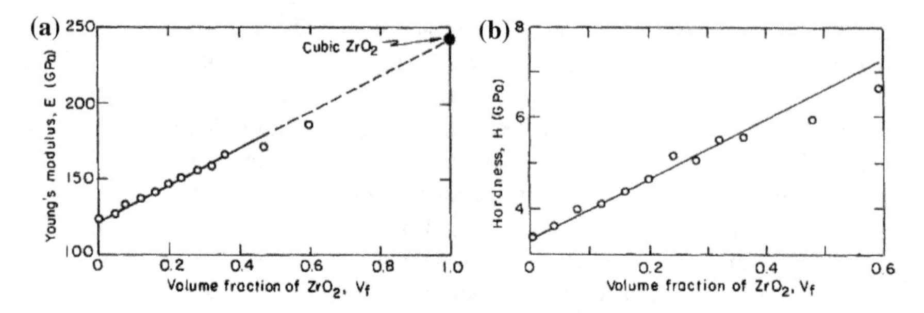

Fig. 5.55 Plots of trends in (a) Young's modulus and (b) hardness versus ZrO_2 volume concentration [28]. With kind permission of John Wiley and Sons

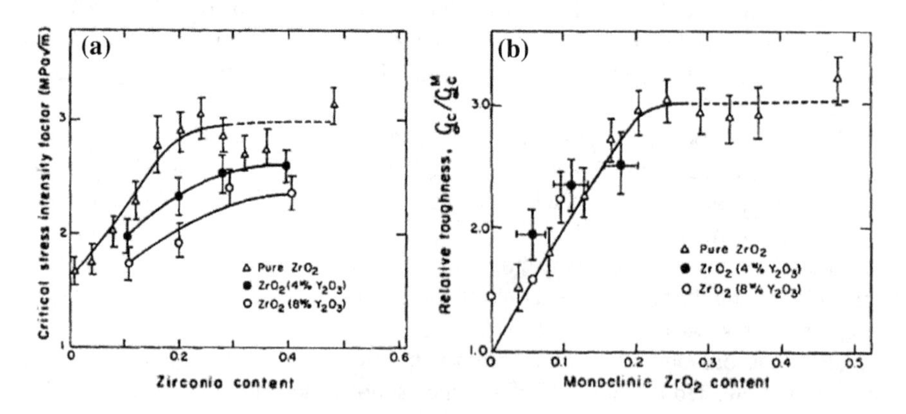

Fig. 5.56 a Critical stress intensity factor as a function of total ZrO_2 content and b relative critical strain energy release rate as a function of monoclinic $Zr0_2$ content [28]. With kind permission of John Wiley and Sons

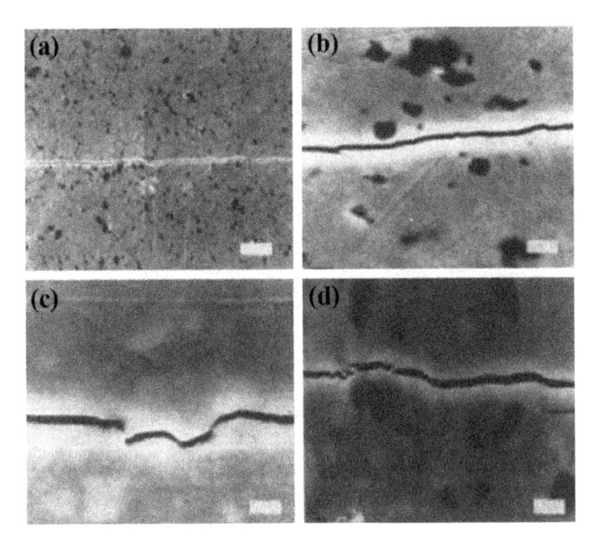

Fig. 5.57 Secondary electron images of indentation cracks indicating the change in crack deflection intensity with ZrO_2 content. **a** $ZnO/0.06ZnAl_2O_4$ (dark particles are $ZnAl_2O_4$ phase, which apparently does not deflect the crack), bar = 10 μm; **b** as in (**a**), bar = 2.5 μm; **c** $ZnO/0.07ZnAl_2O_4/0.02\ ZrO_2$, bar = 2.5 μm; **d** $ZnO/0.07ZnAl_2O_4/0.16\ ZrO_2$, bar = 2.5 μm [28]. With kind permission of John Wiley and Sons

crack. $ZnAl_2O_4$ (detected by XRD) is present in all the samples tested and their size and quantity (4 wt%) are independent of the ZrO_2 content. This phase was incorporated in the material from the Al_2O_3 ball mills during the processing, since fabrication was performed at high temperatures (hot pressing of the powders) and a reaction occurred with the ZnO. Since $ZnAl_2O_4$ is present in all the systems, its contribution to toughness is uniform in all the systems; thus, its influence regarding toughness with ZrO_2 content should not be appreciable.

The relative increase in toughness during transformation toughening is usually expressed [24] as:

$$\Delta K_c^T = 0.22 V_f e^T E \sqrt{w}/(1 - v) \tag{5.19}$$

where ΔK_c^T is the transformation-toughening increment; e^T-the unconstrained, dilatational transformation strain; E is the Young's modulus; w-the width of the transformation zone and v is Poisson's ratio.

Transformation strengthening is associated with its effect on microstructure, particularly on crack conditions and the ways of inhibiting crack propagation. Fracture toughness, based on microstructural considerations, may be achieved by inducing shielding near a crack tip or by locally impeding the motion of a crack front. The stress-induced, Martensitic transformation (Fig. 5.50) and microcracking mechanism represent crack-tip shielding from expansion. However, crack-bowing and crack-deflection effects can also contribute to microstructural toughness enhancement (see Fig. 5.57c and d). Toughening by m-ZrO_2 is a

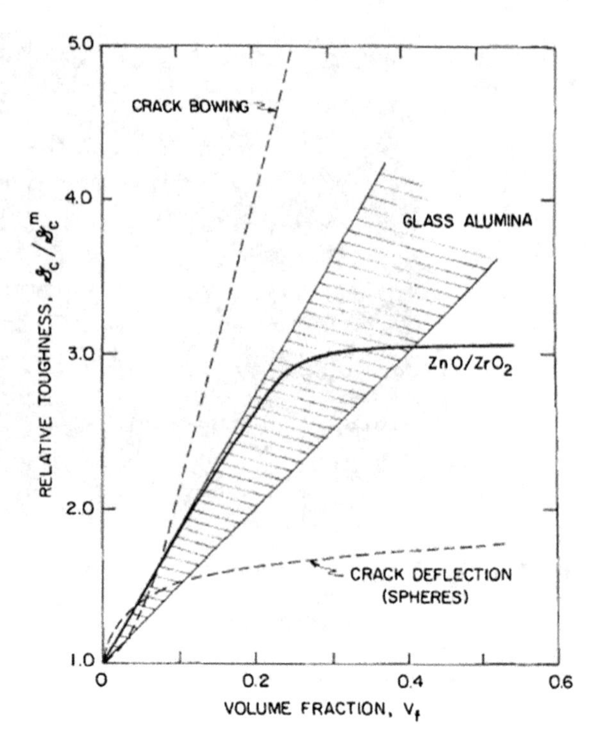

Fig. 5.58 Comparison of toughening in ZnO/ZrO_2 with that for glass/alumina. Also shown are toughness levels predicted for crack-deflection and crack-bowing mechanisms [28]. With kind permission of John Wiley and Sons

consequence of crack retardation by ZrO_2 particles that undergo a stress-induced transformation directly ahead of the crack tip. The monoclinic mechanism of toughening also exists in the presence of t-ZrO_2 particles located in front of the crack tip that undergo a stress-induced transformation into m-ZrO_2 prior to crack extension.

In the case of Zn/ZrO_2 ceramics, the relative peak toughening is achieved at the monoclinic ZrO_2 volume fraction ≥ 0.2. This may be seen in Fig. 5.58, where another comparison is made between the relative toughness values of ZnO/ZrO_2 with glass alumina and a possible stage of crack propagation (whether by crack bowing or crack deflection).

More on the t-ZrO_2 → m-ZrO_2 Martensitic transformation in TZP ceramics is found in the work of Yin in his discussions on the thermodynamics, crystallography and kinetics of this transformation. The solid transformation in pure zirconia is:

$$\text{m-ZrO}_2 \underset{950\,\text{C}}{\overset{1170\,\text{C}}{\rightleftarrows}} \text{t-ZrO}_2 \overset{2370\,\text{C}}{\rightleftarrows} \text{c-ZrO}_2.$$

The free energy of transformation into the Martensitic phase through the monoclinic → tetragonal phases may be presented as:

$$\Delta G_{t\rightarrow m} = V(-\Delta G_{ch} + \Delta G_{str}) + S\Delta G_{sur} \quad (5.20)$$

In the above subscripts, the terms "ch", "str" and "sur" refer, respectively, to the chemical, strain and surface-free energies. As usual, V and S represent the volume and the entropy associated with transformation. Given an equilibrium temperature, T_0, of the t-ZrO_2 → m-ZrO_2 transformation, at which $\Delta G_{ch} = 0$, define M_s as the temperature at which $\Delta G_{t\rightarrow m} = 0$. To describe the free-energy change, divide the RH side of Eq. (5.20) by V and add a term, ΔG_{ext}, for the interaction-energy density due to the external force to obtain:

$$\Delta G_{t\rightarrow m} = -\Delta G_{chem} + \Delta G_{str} + \frac{S}{V}\Delta G_{sur} - \Delta G_{ext} \quad (5.21)$$

This may be rewritten in terms of the barrier, $\Delta G_{barrier}$, which is the sum of the changes in the strain and surface free energies, given as:

$$-\Delta G_{chem} - \Delta G_{ext} + \Delta G_{barrier} \quad (5.21a)$$

Critical transformation stress, σ_c may be defined as:

$$\sigma_c = (-\Delta G_{chem} + \Delta G_{ext})/\varepsilon^t \quad (5.22)$$

where ε^t represents the resulting dilatational transformation strain, which is localized in the vicinity of the crack tip in the transformation zone and is related to transformation toughening.

Again, Martensitic transformations occur at high velocities with shape change and are the basic reason for the toughening which takes place when metastable t-ZrO_2 transforms into a stable m-ZrO_2 phase around a propagating crack. At this same tip location, the transformation introduces a compressive stress, due to the 4–5 % volume change associated with the t-ZrO_2 → m-ZrO_2. As a consequence the local crack-tip, stress intensity is reduced, as well as the driving force for crack propagation.

5.6 Grain Boundaries and Grain-Size Strengthening

It has been indicated on several occasions that plastically deforming a material depends on the ability of dislocations to move within that material. All obstacles, in whatever form they exist in materials, act to retard dislocation motion. Thus, whatever the mechanism may be (work hardening, grain-size reduction, etc.), they all hinder dislocation motion, thus rendering the material stronger than before. In addition to the aforementioned strengthening mechanisms, one should add grain-boundary strengthening. In polycrystalline materials, grain size has a very large influence on overall mechanical properties, not only because small grain-sized materials have a large number of grain boundaries, which act as barriers to dislocation motion, but also due to the various orientations of the grains. Dislocations must change their direction of motion in accordance with the different orientations of the grains; as such, they have to deviate from their unimpeded paths, which would be dictated in a single crystal form. The slip systems, namely the plains and directions dictating dislocation motion, vary from grain to grain and, as such, dislocation motion must accommodate these changes; a compromise must be made and some motion deflected, complicating the resultant route taken by the dislocation. It becomes clear why polycrystalline materials are stronger than their single- crystal counterparts. Thus, by changing the grain size, dislocation movement may be affected. The grain size of a material may be altered by heat treatment after deformation or by changing the rate of solidification, if processing has occurred at the liquid stage.

Recall that, at elevated temperatures, most, if not all, ceramics are ductile. One widely used concept is the H–P relation, showing the inverse variation in yield stress with grain size (in ductile ceramics), given as:

$$\sigma_0 = \sigma_i + kD^{-1/2} = \sigma_i + \alpha Gb\rho^{1/2} \tag{5.23}$$

Here, σ_0 and σ_i represent the yield stress and friction stress, respectively, expressing the resistance of the material to dislocation movement; k is a parameter expressing some contribution to strengthening by the grain boundaries; and D is the grain size. The right side of Eq. (5.23) expresses the dislocation density, ρ, with α being a numerical constant. G and b have their usual meanings—the shear modulus and Burgers vector, respectively. Equation 3.80 (from Sect. 3.3.15) regarding pile-ups is rewritten here as:

$$n = \frac{\pi L\tau k}{Gb} \tag{3.80}$$

n represents the number of dislocations in the pile-up; L is the distance along the pile-up to the obstacle detaining it; τ, as usual, is the applied stress; while the parameter k = 1 is for screw dislocations and $(1-\nu)$ for edge dislocations. Considering that L extends in both directions from the operating source, which is assumed to be located at the center of the D-sized grain, i.e., L/2 in one direction, then the distance, in terms of the grain size at this location, is D/2 and Eq. (3.80) may be rewritten as:

$$n = \frac{\pi D\,\tau k}{2Gb} \tag{3.80a}$$

Considering Eqs. (3.79a–3.79c), it is possible to express the stress needed to overcome the barrier (i.e., the grain boundary) more specifically as τ_s, which is the difference between the applied stress, τ, and the lattice resistance, τ_i, given as:

$$\tau_s = \tau - \tau_i \tag{5.24}$$

(the equivalent of Eq. (3.79c)).

The stress at the end of the pile-up (i.e., at the grain boundary) has to be larger than some critical shear stress, τ_c, in order to overcome the grain-boundary resistance and to enable dislocation motion through it. Thus, Eq. (3.79b) may also be expressed in a different form, with a more specific meaning as:

$$\tau_c = n\tau_s \tag{5.25}$$

Rewrite Eq. (3.80a) in terms of grain size as:

$$n = \frac{\pi D\,\tau_s k}{2Gb} \tag{3.80b}$$

Take n from 3.80b and replace n in Eq. 5.25 to get the following for τ_c:

$$\tau_c = \frac{\pi Dk}{2Gb}\tau_s^2 = \frac{\pi kD}{2Gb}(\tau - \tau_i)^2 \tag{5.26}$$

In Eq. (5.26), the last term is a result of replacing τ_s by its value from Eq. (5.24). This relation may be rewritten as:

$$\tau_c^{1/2} = \left(\frac{\pi kD}{2Gb}\right)^{1/2}(\tau - \tau_i) \tag{5.27}$$

Rearranging this equation, one obtains:

$$\tau - \tau_i = \left(\frac{2Gb}{\pi Dk}\tau_c\right)^{1/2} \tag{5.28}$$

Designate the following expressions in Eq. (5.28) as:

$$\left(\frac{2Gb\tau_c}{\pi}\right)^{1/2} = \alpha \text{ and } \left(\frac{1}{k^{1/2}}\right) = k' \text{ and } \left(\frac{1}{D^{1/2}}\right) \text{x} \frac{D^{1/2}}{D^{12}}.$$

Rewrite Eq. (5.28) with the notation shown as:

$$\tau = \tau_i + \alpha k' D^{-1/2} \tag{5.29}$$

This is equivalent to Eq. (5.23), given in terms of the normal stress, σ.

Experimental observation suggests that dislocation density is an inverse function of grain size, namely $\rho = 1/D$. Equation (5.29), in terms of dislocation density may be given as:

$$\tau = \tau_i + \alpha k' \rho^{1/2} \tag{5.30}$$

Equations (5.23) or (5.30) have been used to calculate various strength properties, for example, one showing bending strength is illustrated in Fig. 5.59.

The data points of the linear relation represent three-point bending test results. Strength increased as the average grain size decreased and the slope was exactly −1/2 in the log plot, in accordance with the H–P relation. Grain size usually increases with temperature and with holding duration at some temperature. Once the characteristic grain size at a specific sintering temperature is achieved, no

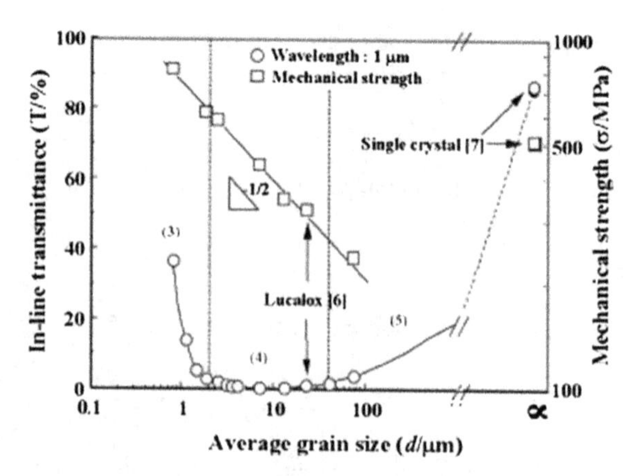

Fig. 5.59 Relationship between transmittance and mechanical strength on average grain size [25]. With kind permission of Elsevier

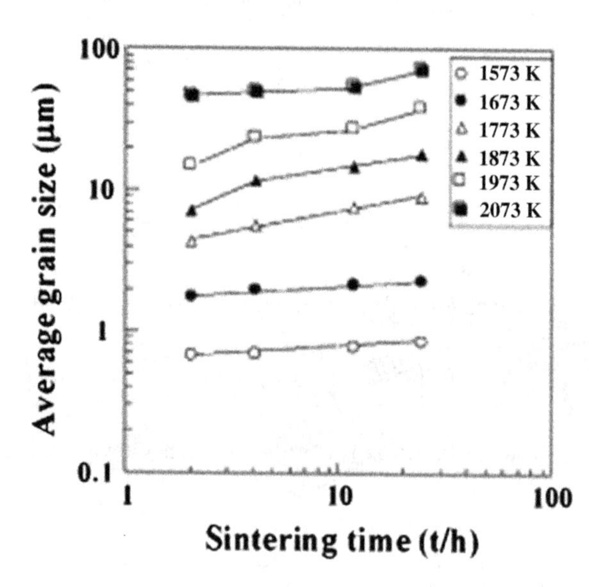

Fig. 5.60 Plots of average grain size versus sintering time at 1300–1800 °C [25]. With kind permission of Elsevier

further increase occurs during additional holding time. The microstructure characterizing the tests reported in Figs. 5.59 and 5.60 is shown in Fig. 5.61.

This microstructure was a result of sintering at 1300 °C for 6 h followed by HIP. Before HIPping, the structure contained micropores at the grain-boundary junction, as seen in Fig. 5.61a; however, when the fabrication process occurred for 12 h, namely sintered at 1300 °C, the specimens did not show micropores and had a relative density of ~ 100 %.

For good mechanical properties of alumina, this ceramic must be largely flaw-free and must have a high density; these properties would make it useful as a structural material and for various industrial applications. The presence of cracks

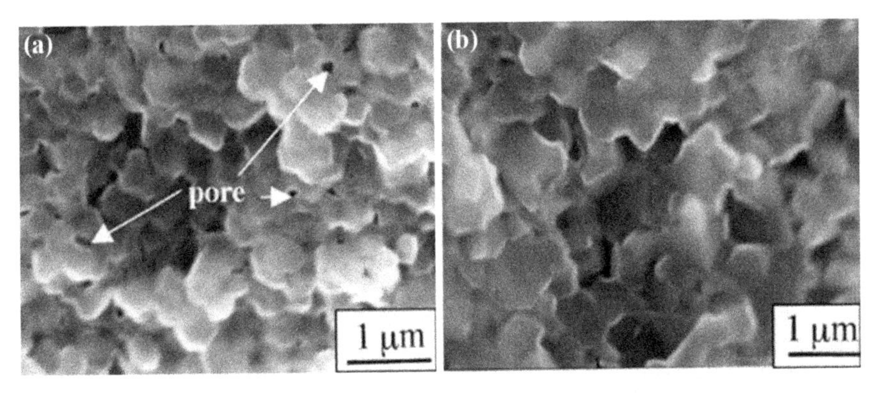

Fig. 5.61 SEM micrographs of sintered alumina (**a**) before and (**b**) after HIP treatment at 1300 °C for 12 h [25]. With kind permission of Elsevier

and various other flaws often override the beneficial effect of small grain size. Considerable scatter may be induced in experimental results as a consequence of such flaws, even in very pure alumina.

One of the important ceramics, of great technological interest, is high-purity alumina, in light of its quality, mechanical properties and many potentially significant applications. Of the many parameters that influence the behavior of alumina, the effect of grain size is of prime importance. However, to take full advantage of the positive impact of grain size, the strengthening of flaws should be kept minimal, otherwise it may nullify the beneficial effect of grain-size strengthening. Flaws can be minimized by proper production techniques to provide dense and low flaw-content ceramics. In the following illustration, the grain-size strengthening in alumina is indicated. Figure 5.62 shows the effect of grain size on bending strength and fracture toughness.

The numbers in Fig. 5.62 refer to the conditions shown in Table 5.2. In this figure, the variation in strength indicates the fabrication effect also on the strength-grain size relation (plots). The lower lines were obtained by dry pressing and cold isostatic pressing [henceforth: CIP]. Beforehand, the CIP samples at 700 MPa were uniaxially pressed at 50 MPa. This procedure increased the green density to 60 % and also reduced the sintering temperature to 1400 °C.

In Fig. 5.63a, the average strength/grain-size relationship is illustrated for selected samples used for subcritical measurements, while in Fig. 5.63b, the measured subcritical growth of the indentation-cracks is indicated. The lines are least-squares approximations. As expected, bending strength decreased with subcritical-crack growth. Figure 5.65 is a plot of subcritical-crack growth versus grain size to obtain information on the shaping of the samples regarding the grain-size effect in reducing subcritical-crack propagation. This figure averages the results of samples produced by different approaches. The numbering is the same as in Fig. 5.63a. High-purity alumina exhibits an increasing potential for higher strengths at decreasing grain sizes. However, this potential depends on the flaw

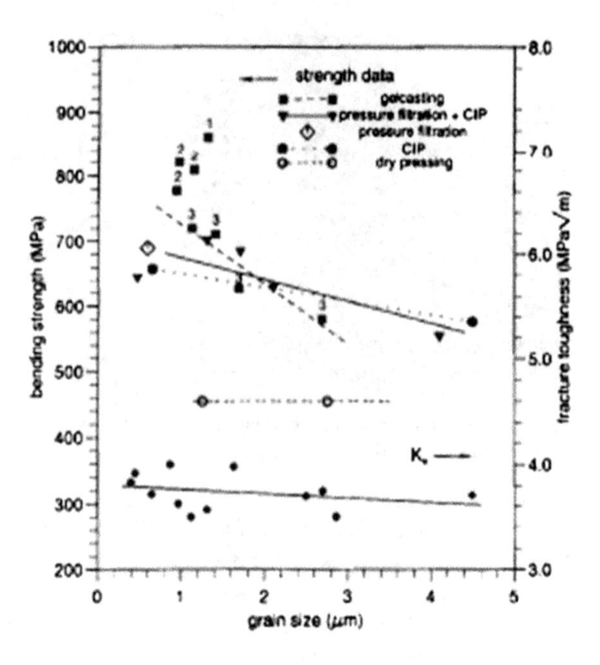

Fig. 5.62 Grain size and technology effects in the strength and toughness of pressureless sintered alumina. Each data point represents the averages of grain size and strength measured for a group of specimens sintered at the same temperature. The numbers that distinguish different gel casting routes refer to Table 5.2. Different data points for one casting route refer to microstructures sintered at different temperatures [20]. With kind permission of Elsevier

Table 5.2 Strength data of gelcast aluminas after pressureless sintering [20] (with kind permission of Elsevier)

Batch number	Monomer content (wt)	Duration of slurry dispersion) min	Grain size (μm)	Average strength (MPa)	Standard deviation (MPa)
1	3.8	130	1.31	862	±102
2	7.4(+CIP)	130	0.97	822	±19
3	3.8	40	1.12	720	±47

population. The great scatter in data in the high purity alumina in Figs. 5.62, 5.63, 5.64 and 5.65 is an indication of the effect of the cracks in the ceramics. This effect has to be eliminated, as much as possible, and, therefore, the most appropriate production method has to be used. It is essential to reduce the number of flaws in order to improve the strength properties while maintaining higher hardness and stability; it is the combination of these properties that make this advanced alumina attractive for wider use and open new fields of application.

Zirconia, in its various forms, is yet another worthy ceramic previously mentioned in this chapter, but not yet discussed in connection with the grain-size effect on strength properties. It is quite difficult to maintain stable, sintered zirconia, because of the large volume change accompanying the $t\text{-}ZrO_2 \rightarrow m\text{-}ZrO_2$ transition; therefore, the stabilization of zirconia is customary. One of the popular stabilizing agents in use is yttria, producing YSZ, a zirconium oxide crystal, stable at RT (and moisture-resistant). Such stability is also important when using zirconia for artificial joint replacements (one of its many applications). Oxide constituents,

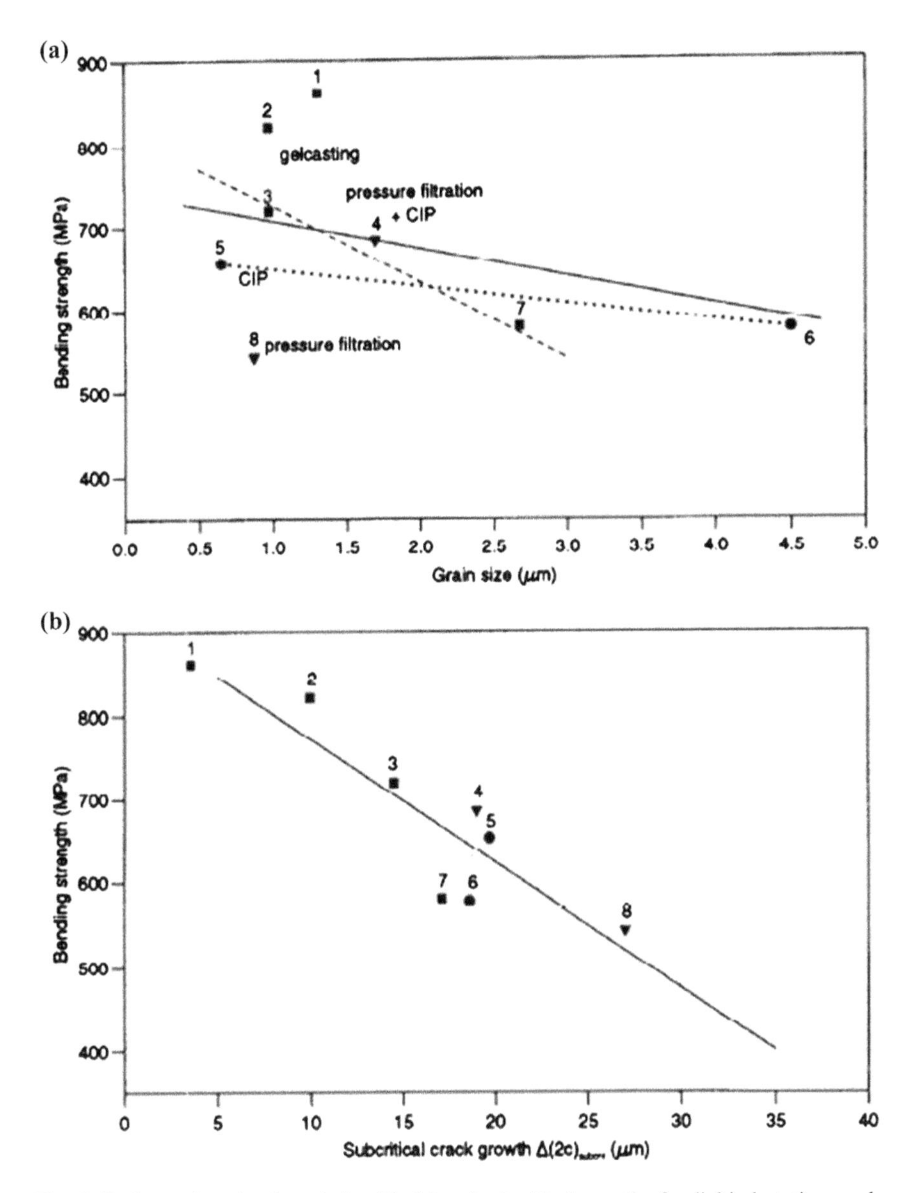

Fig. 5.63 Strength-grain size relationship (**a**) and subcritical growth of radial indentation cracks (**b**) of sintered batches selected from Fig. 5.62. Straight lines are least squares fits. Strength ranking numbers associate specimens in (**b**) to batches in graph (**a**), numbers 1–3 are consistent with Table 5.2 and Fig. 5.62 [20]. With kind permission of Elsevier

such as yttria, are added to further toughen and strengthen the zirconia, which has good mechanical qualities from the outset. The stabilization of zirconia over a wider range of temperatures is accomplished by substituting the Zr^{4+} ions in its crystal lattice with Y^{3+}, in the case of YSZ.

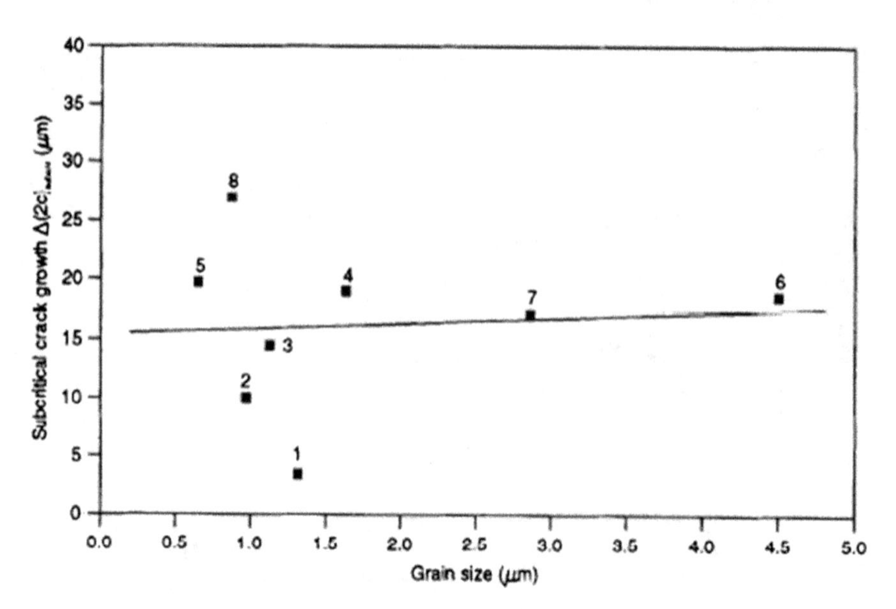

Fig. 5.64 Subcritical growth of radial indentation cracks versus grain size for the time interval between 6 weeks and 6 months after indentation. The *straight line* is a least squares fit. Strength ranking numbers are identical with Fig. 5.63 [20]. With kind permission of Elsevier

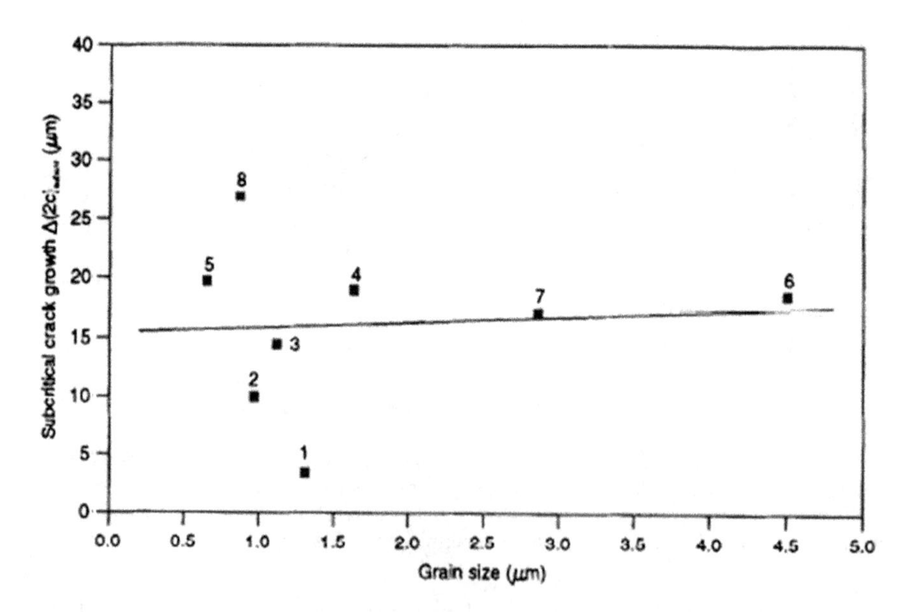

Fig. 5.65 Subcritical growth of radial indentation cracks versus grain size for the time interval between 6 weeks and 6 months after indentation. The *straight line* is a least squares fit. Strength ranking numbers are identical with Fig. 5.63 [20]. With kind permission of Elsevier

Of greatest significance for mechanical properties is the t-ZrO_2 → m-ZrO_2 phase transformation that occurs by a diffusionless shear process at near sonic velocities, similar to those of Martensite formation, as mentioned earlier. Thus, zirconia ceramics will exhibit high strength, provided that the critical-defect size is kept small, which necessitates the optimization of this and other ceramic manufacturing processes. Stress-induced phase transformation involves the transformation of metastable tetragonal crystallites into the monoclinic phase at the crack tip, which, accompanied by volumetric expansion, induces compressive stresses. The extent of strengthening depends on the vol% of transformed ZrO_2, e.g. on the depth of the surface compressive layer, which increases with the thermodynamic instability of the tetragonal ZrO_2 grains. It was shown that zirconia exhibits a transformation-toughening mechanism, acting to resist crack propagation.

Due to their excellent mechanical properties, Y-TZPs are used not only for medical applications, but also for various important technological applications requiring properties such as: high density, high strength, high hardness, good wear resistance and fracture toughness. In its pure form, crystal-structure changes limit mechanical applications, however stabilized zirconias produced by the addition of oxides, such as those of C, Mg or Y, can produce very high strength, hardness and particularly toughness. Various physical properties are also of great interest.

Although the specific subject under discussion is the effect of grain size on strength properties, basically here, the high strength is the result of the transformation occurring in the zirconia; thus, the effect of grain size on the mechanical properties of 3Y-TZP ceramics are considered in this context. Illustration 5.64 shows the variation of the hardness with grain size for two types of stabilized zirconia. Note in the figures TZ-3YB refers to a commercial powder (Japan)

Now, the focus shifts to TZ-3YP which is an yttria-stabilized tetragonal zirconia polycrystalline ceramic (equivalent to 3Y-TZP) containing three moles of yttria. Its hardness was evaluated by an equation given in Sect. 1.10b, as:

$$DPH = \frac{2P\sin(\varphi/2)}{d^2} = \frac{1.854P}{d^2} \tag{1.127}$$

As may be seen in Fig. 5.66, hardness increases with the decrease in grain size. The data in the graph fit reasonably along a line, which means that the dependence follows the H–P relation. The microstructure of TZ-3YP may be seen in Fig. 5.67.

The Fig. 5.67 shows (a) a dense microstructure obtained at 1400 °C sintering. Larger grains were obtained by the action of both higher sintering temperatures and longer sintering times as seen in (b) the structure of TZ-3YB ceramic remains dense, but the theoretical density of these ceramics is decreased due to the spontaneous transformation of the tetragonal into the monoclinic phase. The density change of TZ-3YB as a function of grain size is illustrated in Fig. 5.68.

At higher temperatures, holes and voids are visible in the micrographs, as indicated in Fig. 5.69. However, the presenter of this microstructure claims that they are not pores, but rather the remnants of grains pulled out during processing by grinding and polishing.

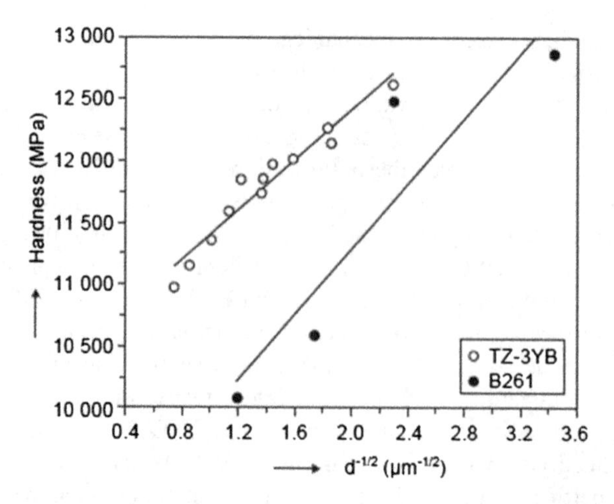

Fig. 5.66 The dependence of the hardness of TZ-3YB and B261 ceramics on the inverse square root of grain size [30]. With kind permission of Dr. Trunec and the author

Fig. 5.67 **a** SEM micrograph showing the microstructure of the TZ-3YB ceramic sintered at 1400 °C for 2 h.; **b** SEM micrograph showing the microstructure of the TZ-3YB ceramic sintered at 1650 °C for 20 h [30]. With kind permission of Dr. Trunec

Fig. 5.68 Density of TZ-3YB and B261 ceramics as a function of grain size [30]. With kind permission of Dr. Trunec

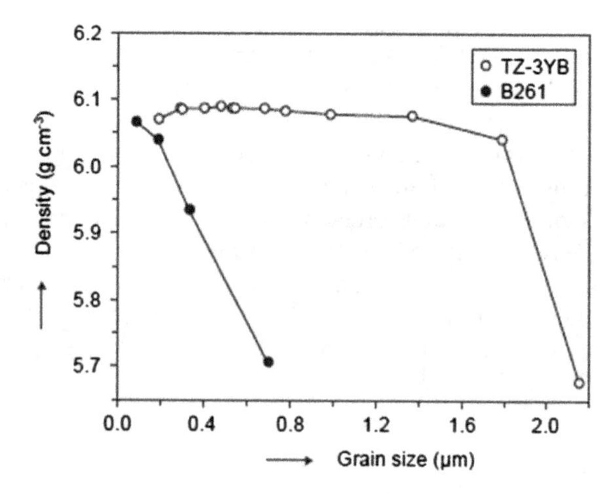

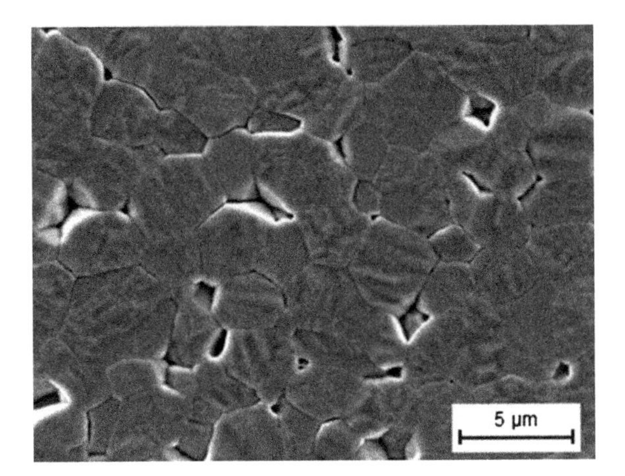

Fig. 5.69 SEM micrograph showing the microstructure of the TZ-3YB ceramic sintered at 1650 °C for 50 h [30]. With kind permission of Dr. Trunec

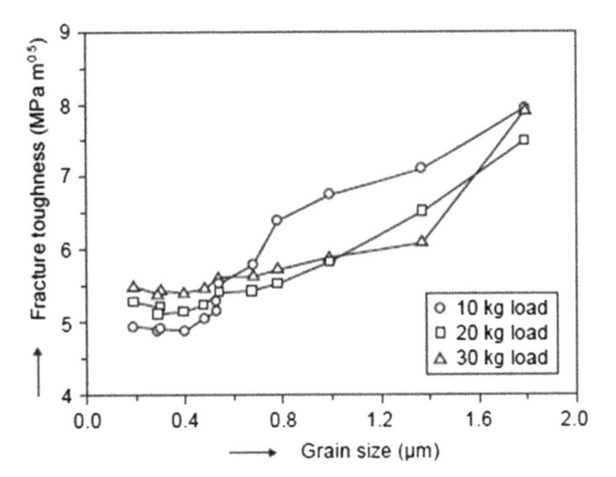

Fig. 5.70 Fracture toughness of TZ-3YB ceramic as a function of grain size, determined at different indentation loads [30]. With kind permission of Dr. Trunec

In Figs. 5.70 and 5.71, the variation of the fracture toughness with grain size is indicated. Figure 5.70 illustrates the effect of the applied indentation load on fracture toughness.

One can see that the fracture toughness is almost constant up to a grain size of ~0.4 μm, above which it increases to ~1.8 μm; however, further grain growth results in spontaneous transformation from the tetragonal to monoclinic phase and damage occurs in the samples, due to cracking.

One should not overlook the exceptional ceramics obtained by the combination of alumina and zirconia. Beyond its many technological applications, ZTA is of increasing interest also for biomedical applications, due to its excellent strength properties, e.g. for improved, long-life, orthopedic implants (current implants are relatively short-lived). In alumina, even in ZTA, age-related degradation is associated with the volume change of the t-ZrO_2 → m-ZrO_2 phase transformation

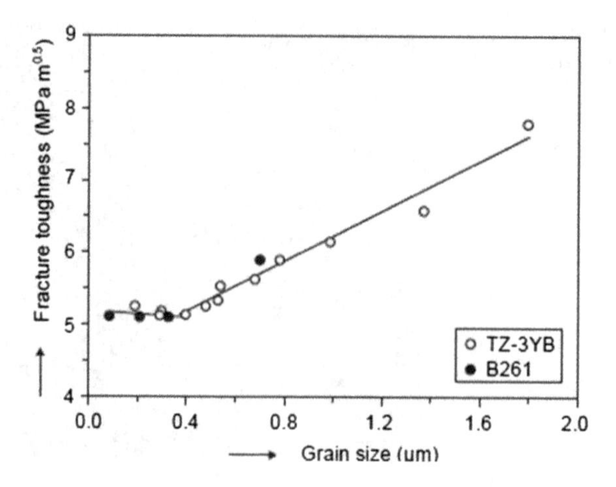

Fig. 5.71 Average fracture toughness of TZ-3YB and B261 ceramics as a function of grain size [30]. With kind permission of Dr. Trunec

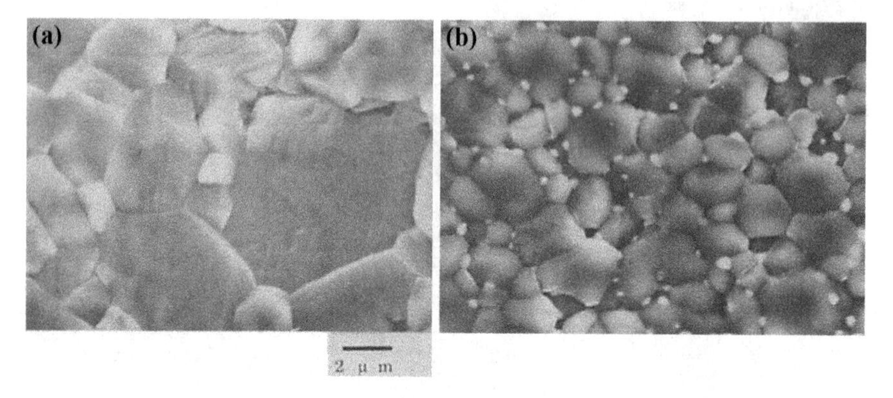

Fig. 5.72 Micrographs of **a** alumina and **b** alumina containing 5 % zirconia, sintered at a rate of 5 °C/min to1700 °C. Particles of zirconia at the triple junctions restrict grain growth during the final stage of sintering (see Fig. 5.73) [32]. With kind permission of John Wiley and Sons

occurring at the surface in a humid atmosphere (or in corrosive surroundings, when implanted in the body), followed by superficial microcracking over time. The addition of one ceramic to another often produces a composite with more desirable properties than those of the individual components themselves, is true for ZTA. A microstructure of alumina, toughened by 5 % zirconia, is illustrated in Fig. 5.72 and compared to unaltered alumina. Note the dispersion of the zirconia in the micrograph comparing the two ceramics, the alumina and zirconia containing alumina. This dispersion, mostly at triple boundary junctions, prevents grain growth, as seen in Fig. 5.73, where grain growth in both ceramics is indicated as a function of density. Small grain size strengthens the material, thus the retardation of grain growth is beneficial for a strong ceramic. In this case, the grain growth is

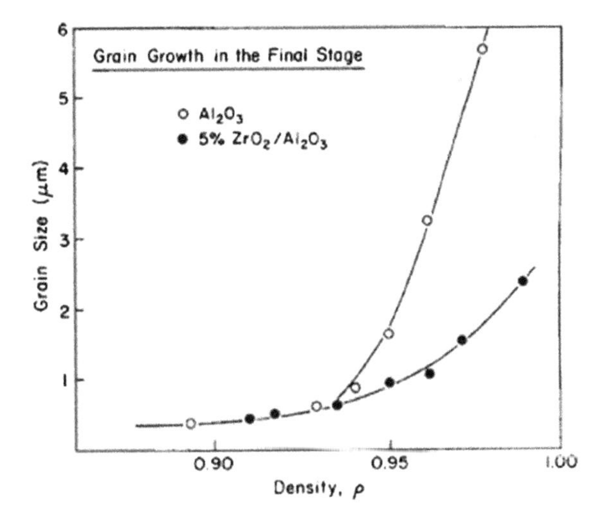

Fig. 5.73 Alumina grain size as a function of density. Growth of the alumina grain is significantly impeded by addition of 5 % zirconia only in the final stage of sintering. In the initial and intermediate stages of sintering, the zirconia particles do not have a measurable effect on grain growth [32]. With kind permission of John Wiley and Sons

retarded almost threefold at a density of ~98 %. The expression for the instantaneous sintering rate, $\dot{\rho}$, is given below as:

$$\dot{\rho} = A\frac{\exp(-Q/RT)}{T}\frac{f(\rho)}{d^n} \tag{5.31}$$

where:

$$A = \frac{C\gamma V^{2/3}}{R} \tag{5.32}$$

$\dot{\rho} = \frac{d\rho}{dt}$ is the rate of instantaneous densification; d is the grain size; f(ρ) is a function only of the density; Q is the activation energy; γ is the surface energy; V is the molar volume; R and T have their usual meanings (the gas constant and the absolute temperature); C is a constant; and A is a material parameter that is insensitive to d. The sintering rate is given by d^n, where n = 3 for controlled lattice diffusion and n = 4 for controlled grain-boundary diffusion. In the case of this ZTA, n = 4, indicating that the sintering of the green compact is grain-boundary diffusion controlled.

For orthopedic applications, alumina–zirconia composites have a higher reliability than single-phase ceramics, due to the combined advantages of both the alumina and the zirconia. With the same pre-existing defects, these composites can work at loads two times higher than monolithic alumina without delayed failure and are not susceptible to the hydrothermal instability (low temperature degradation) observed in the case of stabilized—zirconia bioceramics.

An additional mechanical property of interest in ZTA is its wear resistance. The wear transition load, namely the wear load at which a rapid increase in wear occurs, increases with increased zirconia content. The wear in ZTA depends on the

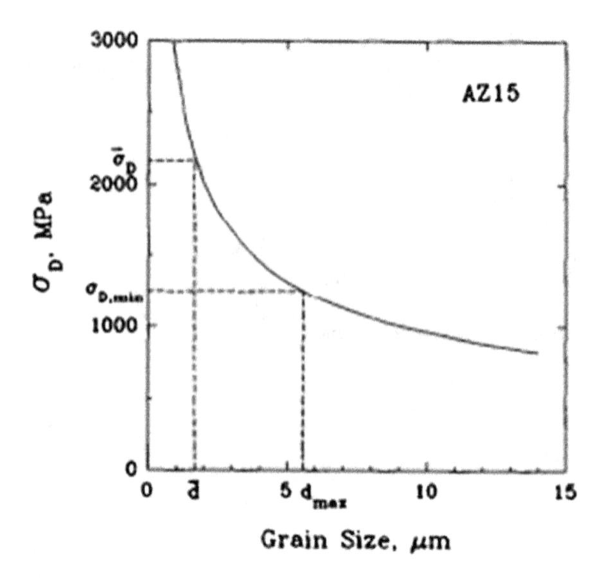

Fig. 5.74 Hall-Petch-type relationship describing the fracture stress between the external sliding damage-induced stresses and grain size in AZ15 material. AZ15 means 15 % zirconia [13]. With kind permission of Elsevier

material properties as influenced by its microstructure. The effect of the microstructure on wear has been demonstrated in various publications and an H–P-type relation between microfracture stresses and grain size was found. Fig. 5.74 shows critical damage stress, σ_D, as a function of grain size.

Figure 5.74 shows the effect of the average grain size on the onset of brittle fracture for AZ15, calculated using Eq. (5.33) given below [33]:

$$\sigma_D = \sigma_1^* \left(\frac{d^*}{d}\right)^{1/2} - \sum \sigma_1 \tag{5.33}$$

where d is the grain size; σ_1 is the thermally-induced internal stress; d^* and σ_1^* are the critical grain size and the critical internal stress, respectively, corresponding to instantaneous microfracture; and $\sigma_D = 0$. If the process parameters remain constant, the following H–P-type relation between grain size and critical damage stress may be obtained from (5.33):

$$\sigma_D = f(d^{-1/2}) \tag{5.34}$$

This relation shows that the smaller grain size results in higher critical damage stress (i.e., a higher transition to brittle fracture). Figure 5.74 also includes data from the literature. In brief, σ_D for an average grain size of 1.7 μm is ~2200 MPa, whereas, for 5.6 μm, a drastic drop in σ_D, the stress required for failure occurs at a level of 1200 MPa, which is significantly lower than the value for the small-grain size (i.e., a mean grain size of 1.7 μm). Figure 5.75 is a SEM microstructure. At loads above the transition load, i.e., above 340 N, the worn surface exhibits massive surface damage, apparently formed by a brittle fracture, as seen in Fig. 5.75b.

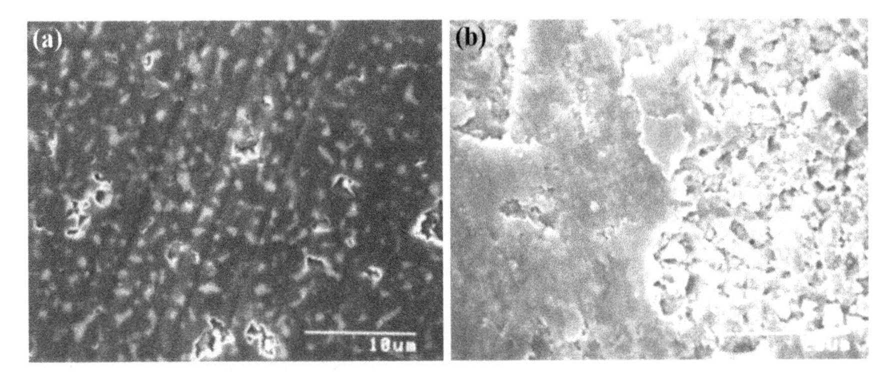

Fig. 5.75 SEM images of worn surface of AZ15: **a** before transition to brittle-fracture-controlled wear (load 260 N; speed 0.23 ms^{-1}; paraffin oil lubricant; sliding distance 2068 m); **b** after transition to brittle-fracture-controlled wear (load 340 N, speed 0.23 ms^{-1}; paraffin oil lubricant; sliding distance 2.55 m) [13]. With kind permission of Elsevier

There is a transition from the plastic deformation to the brittle-fracture-dominated wear process. When the coefficient of the friction level is about 0.1, the maximum shear stress under sliding contact is underneath the sliding surface. This stress induces subsurface cracks, which propagate along the grain boundaries. The rate of inter-granular crack propagation depends on the internal, residual stress at the grain boundaries. Once the grains and particles are detached from the surface, they drastically change the stress distribution at the point of contact. Particle abrasion begins to act, inducing wear on the sliding surface. The lowest wear rate of ZTA was found for a 15 % ZrO addition, namely at AZ15. The reason for this wear reduction is related to: (a) the refinement of the alumina grain size; (b) the reduction of the hardness at the worn region; and (c) the formation of a compressive zone in the wear scar.

Finally, it is worth reemphasizing the microstructural properties of ZTA basically responsible for its advanced performance and key role in biomedical applications, as indicated earlier. In brief:

(i) The microstructure of ZTA reveals that the zirconia grains are quite evenly distributed in the ceramic;
(ii) The zirconia grains are mainly found in the corners, or rather at triple junctions, of the Al_2O_3 matrix;
(iii) There are indications in the relevant literature that zirconia grains remain inter-granular at sintering temperatures;
(iv) Although additions of zirconia to alumina are expected to retard grain growth in an Al_2O_3 matrix, a coarsening of the grains of both components of ZTA may occur at high temperatures; however, despite the change in grain sizes, their ratios remain unchanged;
(v) The fabrication method influences the density of ZTA (as it does in other ceramics). HIP seems to provide a density close to the theoretical one, of $\sim$100 %. Thus, the densities of pores, microcracks and other flows are reduced, which is a prerequisite for the better performance of these ceramics in use.

5.7 Concluding Remarks to this Chapter

Several strengthening mechanisms were discussed in this chapter. The emphasis throughout was on the effects of the flaws and on the contribution of the dislocations. It has been stated several times that defect-free ceramics are much stronger than those usually observed (which are, regretfully, not defect free). To achieve a near-100 % density in ceramics, special production techniques, often in combination, are necessary, and they are both time-consuming and expensive. Nevertheless, the broad applications of ceramics in industry and medicine make these efforts worthwhile, even when cost considerations are critical. The hope is to be able to make common ceramics ductile to various degrees. Doing so will further extend the domain of ceramic applications at a wide range of temperatures while retaining the good strength properties of these ceramics. This would certainly make all the research and experimental efforts worthwhile.

The second point, threaded throughout all the chapters of this book, is particularly relevant to this chapter. When considering strengthening processes in general, the retardation of dislocation motion plays a key role. Regardless of what manner of obstacles dislocations encounter, the end result is material strengthening, because a higher stress is required to maintain dislocation motion. In strain hardening, the dislocations themselves act as obstacles to the motion of other mobile dislocations. A second, insoluble phase or grain boundaries act as barriers to dislocation motion, either pinning or entangling them, stopping or delaying the dislocations. Foreign atoms, when in solution, act no differently than indicated above, since such atoms, even when dissolved in the matrix, disrupt the periodicity of the lattice and, by so doing, the retardation of dislocation motion occurs. Thus, the presence and generation of new dislocations are responsible for the strengthening of ceramics.

References

1. Bradt RC (1967) J Am Ceram Soc 50:54
2. Cai H, Kalceff MAS, Lawn BR (1994) J Mater Res 9:762
3. Chen W-I, Xue LA (1990) J Am Ceram Soc (73):2585
4. Crampon J, Escaig B (1980) J Am Ceram Soc (63):680
5. Copley SM, Pask JA (1965) J Am Ceram Soc 48:139
6. Dominguez-Rodrigez A, Lagerlöf KPD, Heuer AH (1986) J Am Ceram Soc 69:281
7. Fischer-Cripps AC, Lawn BR (1996) Acta Mater (44):519
8. Fleischer RL (1962) J. Appl Phys 33:3504
9. Friedel J (1967) Dislocations. Pergamon Press (student eddition), Oxford
10. Garvie RC, Nicholson PS (1972) J Am Ceram Soc 55:152
11. Gong H, Yin Y, Fan R, Zhang J (2003) Mater Res Bul 38:1509
12. Grain CT (1967) Phase relations in the ZrO_2-MgO system. J Am Ceram Soc 50:288
13. He C, Wang YS, Wallace JS, Hsu SM (1993) Wear 162–164:314
14. Heuer AH (1987) J Am Ceram Soc 70:689
15. Huang XW, Wang SW, Huang XX (2003) Ceram Int 29:765

16. Hulse CO, Copley SM, Pask JA (1963) J Am Ceram Soc (46):317
17. Johnston WG, Gilman JJ (1959) J Appl Phys 30:129
18. Johnston WG (1960) Phil Mag 5:407
19. Jin Xue-Jun (2005) Curr Opin Solid State Mater Sci 9:313
20. Krell A, Blank PJ (1996) Eur Ceram Soc (16):1189
21. Labusch R (1970) (41):659
22. Lakiza SM, Lopato LM (1997) J Am Ceram Soc 80:893
23. Misra A, Sharif AA, Petrovic JJ, Mitchell TE (2000) Acta mater. 48:925
24. McMeeking RM, Evans AG (1982) J Am Ceram Soc 65:242
25. O YT, Koo JB, Hong KJ, Park JS, Shin DC (2004) Mater Sci Eng A (374):191
26. Rodriguez AD, Garcia DG, Solvas EZ, Shen JZ, Chaim R (2007) Scripta mater 56:89
27. Rouxel T, Wakai F, Izaki K (1992) J Am Ceram Soc (75):2363
28. Ruf H, Evans AG (1987) J Am Ceram Soc 70:689
29. Rouxel T, Rossignol F, Besson JL, Goursat P (1997) J Mater Res (12):480
30. Trunec M (2008) Ceram Silikáty (52):165
31. Tuan WH, Chen RZ, Wang TC, Cheng CH, Kuo PS (2002) J Eur Ceram Soc 22:2827
32. Wang J, Raj R (1991) J Am Ceram Soc 74:1959
33. Wang YS, He C, Hockey BJ, Lacey PI, Hsu SM (1995) Wear 156:181
34. Zhan G-D, Mitomo M, Nishimura T, Xie R-J (2000) J Am Ceram Soc (83):841
35. Zhou YC, Chen JX, Wang JY (2006) Acta Mater 54:1317

Chapter 6
Time-Dependent Deformation: Creep

Abstract Creep is time-dependent deformation under constant stress. It may occur at relatively moderate temperatures. Most ceramics are intended for use at high temperatures, where they are ductile and creep deformation might occur. For ceramics with low-temperature ductility, creep may occur at ~0.5 T_m or even at lower temperatures. Creep generally is a function of the stress applied, the time of load duration and the temperature. Many ceramics are characterized by a high melting point even above 2000 °C (MgO, Al_2O_3, SiC, etc.) which makes them natural candidates for high temperature applications without the risk of creep failure during their lifetimes. Single and polycrystals creep, but to eliminate grain boundary sliding single crystals are preferred in certain important applications, despite the cost factor involved. Although small grain size enhances most of the mechanical properties, for creep resistance large grained materials are preferred. Mechanisms of creep that can act individually or simultaneously (depending on conditions) are Nabarro-Herring Creep, Dislocation Creep and Climb, Climb-Controlled Creep, Thermally-Activated Glide via Cross-Slip and Coble Creep, involving Grain-Boundary Diffusion. Creep may terminate in rupture which has to be avoided by choosing the proper ceramics and the safe temperature use for the desired life time. The presence of flaws (cracks) in ceramics intended for high-temperature applications can be controlled by the manufacturing process, which should be reduced to a minimum. It is essential for design purposes to estimate the life time in service of a ceramics to avoid failure, which is evaluated by some parametric method. The most popular methods to predict life time are the Larson-Miller and Monkman–Grant methods.

6.1 Introduction

Most simply and generally defined, creep is 'time-dependent deformation under constant stress'. Even though creep may occur at relatively moderate temperatures, most ceramics are intended for use at high temperatures, where they are ductile. For ceramics with low-temperature ductility, creep may occur at ~0.5 T_m or even at lower temperatures. The term defined below, indicating 'homologous

J. Pelleg, *Mechanical Properties of Ceramics*, Solid Mechanics and Its Applications 213, DOI: 10.1007/978-3-319-04492-7_6,

temperature', may be considered a demarcation point for creep, separating 'low-temperature creep' from 'high-temperature creep'. The homologous temperature of a material is defined in terms of its melting point, T_m, or, more specifically, as the ratio of some absolute temperature, T, at some value to its melting point, namely:

$$\text{homologous temperature} = T/T_{mp} \tag{6.1}$$

Low-temperature creep, at or below $0.5T_m$, is believed to be governed by non-diffusion- controlled mechanisms, whereas high-temperature creep, above $0.5T_m$, is diffusion controlled. Stress, time and temperature, mentioned above, act simultaneously during creep. These three parameters determine the creep rate and may be expressed in terms of strain rate as:

$$\dot{\varepsilon} = f(\sigma, t, T) \tag{6.2}$$

Clearly then, based on the fact that many ceramics are characterized by high temperatures, some even above 2000 °C, there may be safe uses for certain ceramics over long periods of time at high temperature without the risk of creep failure during their lifetimes. Note the very high T_m of some ceramics, such as: MgO (2798 °C), Al_2O_3 (2050 °C), SiC (2500 °C), etc. Some ceramics even have ultrahigh melting points, such as: TiB_2 (3225 °C), ZrC (3400 °C), HfC (3900 °C), etc. Thus, some (high-melting) ceramics are more appropriate than others (low-melting) for use when resistance to creep is an essential prerequisite. Ceramics having high melting points are natural candidates for applications in which resistance to creep is of primary interest. Other requirements for good performance at high temperatures also must not be overlooked, depending on the particular application intended for use. However, common to all these applications is the importance of their resistance to environmental influences, especially corrosion and humidity. For example, ceramics intended for biomedical applications, like implants, are constantly exposed to humid and corrosive surroundings.

In order to study creep without the other additional effects that usually contribute to overall creep deformation, it is essential to investigate single crystals. This eliminates the effects of grain-boundary sliding, which is usually a significant contributor to creep. In polycrystalline materials, grain-boundary sliding generally has a considerable impact on creep strain. In this chapter, the following topics will be discussed (not necessarily in their listed order):

(1) basic concepts;
(2) brittle ceramics (ductile at high temperatures, at which deformation is possible):

(i) single-crystal ceramics,
(ii) polycrystalline ceramics;

(3) RT ductile ceramics:

(i) single-crystal ceramics,
(ii) polycrystalline ceramics;

(4) superplastic ceramics;
(5) creep mechanism;
(6) grain-boundary sliding;
(7) creep rupture;
(8) prediction of service lifetime;
(9) recovery;
(10) design and selection of creep-resistant materials.

As indicated throughout the sections of this book, deformation, including creep deformation, occurs by some form of dislocation motion. The objective of selecting materials or developing new materials is to slow down dislocation motion, as far as possible, in order to ensure their long lifetime of service.

6.2 Basic Concepts

The generally accepted method for recording the results of a creep test is by plotting strain versus time, as shown schematically in Fig. 6.1a. The schematic curve is composed of three stages, known as 'primary', 'secondary' and 'tertiary' creep and an instantaneous elongation upon applying the force. 'Primary creep' is also known as 'transient', or simply 'stage I creep'. For most materials, including metals and alloys, low-temperature creep occurs in a single transient stage, in which the creep rate decreases continuously over time (see schematic in Fig. 6.1b). A decrease in strain rate is associated with increases in dislocation density or changes in the characteristics of the dislocation structure.

Strain hardening occurs during transient creep, which is induced by pure glide. Mobile dislocations, present at the start of creep and continue to move under the influence of an effective stress, which slowly declines as the mobile dislocations are trapped in the network. The total dislocation density, equal to the sum of the mobile and network densities, remains constant. It is clear that strain is a function of stress and increases with stress.

Secondary creep or 'stage II creep' is often referred to as 'steady state' or 'linear creep'. During the tertiary creep or 'stage III creep', the creep rate begins to accelerate as the cross-sectional area of the specimen decreases due to necking, which decreases the effective area of the specimen. If stage III is allowed to proceed, fracture will occur. The instantaneous strain, ε_0, is obtained immediately upon loading; this is not a creep deformation, since it is not dependent on time and is, by its nature, elastic. However, plastic strain also contributes in this case.

The strain rates characterizing these stages are as follows: in stage I, the strain rate, $\dot{\varepsilon}$, is decreasing, while, in stage II, the strain rate is constant, because of the balance between recovery processes and strain hardening. In stage III, creep is accelerated and the strain rate continuously increases until fracture sets in. Figure 6.1b shows the variation of strain rate over time.

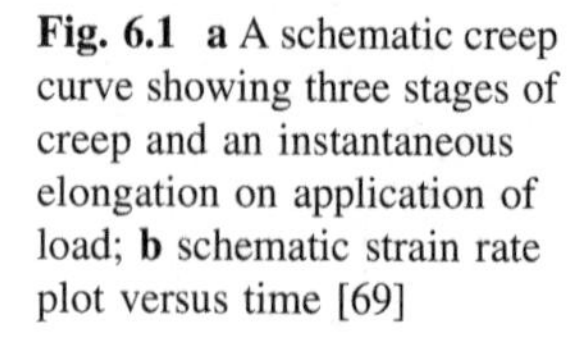

Fig. 6.1 **a** A schematic creep curve showing three stages of creep and an instantaneous elongation on application of load; **b** schematic strain rate plot versus time [69]

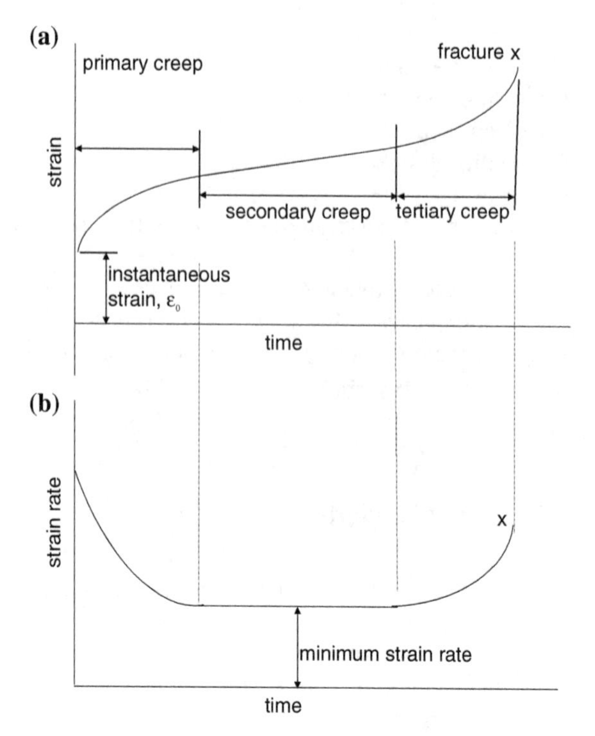

Temperature and stress both affect the shape of a creep curve. Figure 6.2 illustrates these temperature and stress effects on the shape. A schematic representation of this effect appears in Fig. 6.3a for low, medium and high temperatures. At medium temperatures (A in the figure), the commonly observed creep curve resembles the one seen in Fig. 6.1a. Schematic creep curves at a constant temperature, with stress as the varying parameter, are shown in Fig. 6.3b. At the stress indicated by σ_3, the curve is similar to the ideal creep curve shown in Fig. 6.1a, having all three stages. The effects of stress and temperature on the shapes of these creep curves are similar. In Fig. 6.3b, the effect of stress at a constant temperature is shown schematically and similar curves may be obtained when the stress is kept constant and temperature is the varying parameter. The transient creep stage gradually diminishes with increased stress and, at a sufficiently high stress level, it disappears and the steady state dominates the shape of the strain–time relation.

In Fig. 6.1b, a minimum, constant creep rate, which is an important design parameter, is shown. The magnitude of the minimum creep rate on the strain–time relation (see Fig. 6.1a) is associated with steady-state creep and is stress and temperature dependent. Two criteria are commonly applied to alloys: (a) the stress needed to produce a creep rate of 0.1×10^{-3} %/h (or 1 % in 1×10^4 h) and (b) the stress needed to produce a creep rate of 0.1×10^{-4} %/h, namely 1 % in

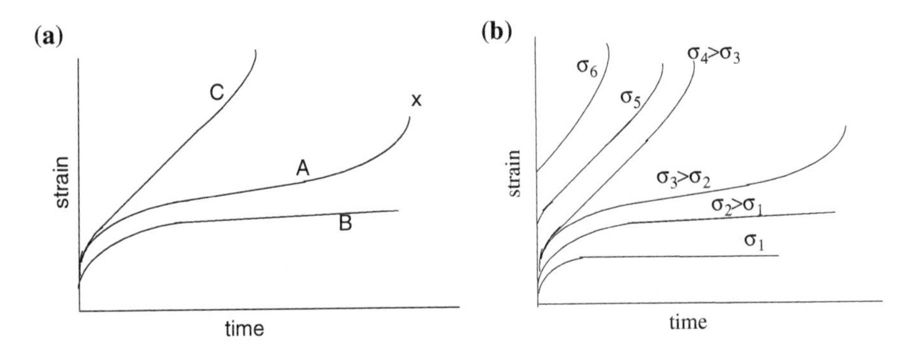

Fig. 6.2 Strain-time creep curves: **a** the shape of creep curves; *A* the standard creep curve (see 6.1a); *B* a creep curve at low temperature and stress and; *C* a high temperature and high stress curve; **b** Schematic creep curves at a constant temperature with variable stress. Note that σ_3 represents the standard creep curve with all three stages

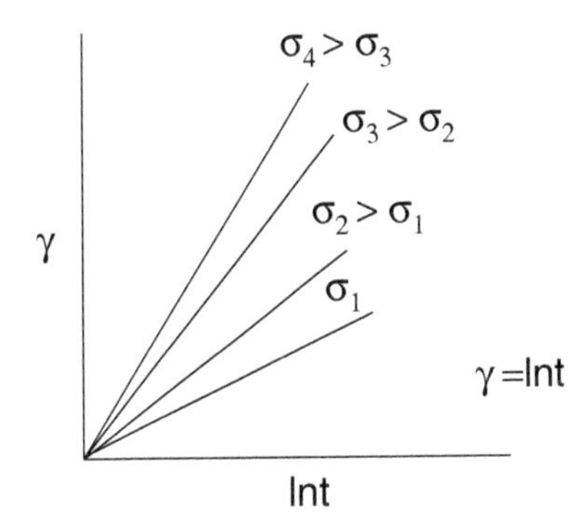

Fig. 6.3 Logarithmic creep. The stress dependence of the *lines* is shown

100×10^3 h (or approximately 11.5 years). Criterion (a) is used for turbine blades, while (b) is commonly applied to steam turbines.

Several empirical models have been suggested for creep. Andrade was the first to consider creep in 1914. He considered creep to be the superposition of transient and viscous creep terms (discussed in the next section dealing with creep in polycrystalline materials). Since creep is a thermally-activated process, the minimum secondary-creep rate may be described by an Arrhenius equation (see McLean [15]) as:

$$\frac{d\varepsilon}{dt} = \dot{\varepsilon} = A \exp - \left(\frac{Q_0 - \alpha\sigma}{kT} \right) \tag{6.3}$$

In Eq. (6.3), A and α are constants and Q_0 is the activation energy for creep at zero stress. A is also known as the 'frequency' or 'pre-exponential factor'.

An additional expression for the creep rate, where the stress and temperature terms are separated, is given as:

$$\dot{\varepsilon} = B\sigma^n \exp -\left(\frac{Q}{kT}\right) \tag{6.4}$$

In Eq. (6.4), the stress affects the frequency factor, B, while Q has the same meaning as Q_0.

Many experimental data indicate that the creep rate, in its early stages, may be expressed by a function suggested by Cottrell [39] as either:

$$\frac{d\gamma}{dt} = \dot{\gamma} = At^{-n} \tag{6.5}$$

or

$$\frac{d\varepsilon}{dt} = \dot{\varepsilon} = \mathrm{B}\mathrm{t}^{-\mathrm{n}} \tag{6.5a}$$

A (B) and the exponent, n, are constants with $0 \leq n \leq 1$.

Equation (6.5) may also be expressed in logarithmic terms and many transient regimes of creep curves may be fitted to a logarithmic law when n = 1. In the extreme case, when n = 1, which is often observed experimentally, one obtains the logarithmic creep law as:

$$\gamma = \alpha \ln t \quad (\mathrm{t} > 1) \tag{6.6}$$

Note that Eq. (6.5) adequately describes the experimental creep data, since the creep rate in the primary stage (transient) decreases over time, as shown in the schematic illustration in Fig. 6.1b. Various values of n, in the range 0–1, may be observed experimentally, but, very frequently, the value of 2/3 is preferred. Thus:

$$\frac{d\gamma}{dt} = \dot{\gamma} = At^{-2/3} \tag{6.5b}$$

and integration gives the equation for strain as:

$$\gamma = \beta t^{1/3} \tag{6.7}$$

Equation (6.7), representing transient creep, is often referred to as 'β-creep' or 'Andrade creep', since Andrade [24] was the first to show that it applies to many materials. The creep behavior obeying Eq. (6.6) is often called 'α' or 'logarithmic creep'.

Often, the instantaneous non-creep strain is also taken into account, suggesting an equation [85] in the form of:

$$\gamma = \gamma_0 + \alpha \ln(\beta \mathrm{t} + 1) \tag{6.8}$$

in which α and β are constants. Figure 6.3 shows schematically logarithmic creep curves at various stresses.

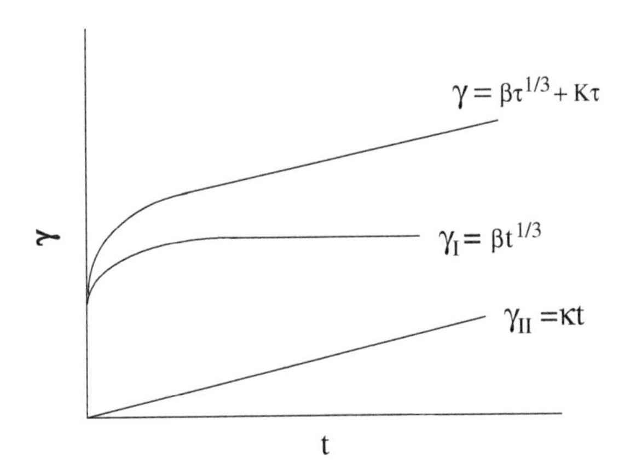

Fig. 6.4 A graphic presentation of Eq. (6.9) without γ_0, obtained from the combination of transient ($\gamma_I = \beta t^{1/3}$) and steady-state ($\gamma_{II} = \kappa t$) creeps

Figure 6.1a shows that stage II creep is linear, such that the function describing this region must also be linear. Much of the creep data is expressed by functions taking this linear contribution into account as:

$$\gamma = \gamma_0 + \beta t^{1/3} + \kappa t \tag{6.9}$$

This relation is a combination of instantaneous strain, γ_0, Eq. (6.7) and the linear contribution of second stage creep, Kt [39], and it well-describes many creep experiments. Usually, especially in experiments performed at high temperatures, transient and steady-state creep occur together. A graphic expression of Eq. (6.9), namely the combination of these stages, is seen in Fig. 6.4 without the instantaneous, non-creep strain, γ_0 obtained upon loading.

In tertiary creep, the strain and strain rate increase until fracture occurs. In ceramics, tertiary creep is usually not recorded, but, if the test is continued long enough, a tertiary creep may develop. In metals, entering stage III occurs when there is a reduction in the cross-sectional area due to necking or internal void formation. In ceramics, it is void formation, in the form of pores or flaws, which effectively causes a reduction in area. Thus, tertiary creep is important in engineering ceramics, because it is often associated with the formation of structural instability, as indicated by void and/or crack formation, leading to failure-by-fracture. The onset of tertiary creep occurs at the end of steady-state creep. It is easier to locate the onset of tertiary creep from the $\dot{\varepsilon}$-t relation than from ε-t, as seen in Fig. 6.1b, since the location of the deviation from the minimum creep rate is well defined. It is clear that the minimum creep-rate parameter must limit allowable stress in practice to prevent the onset of tertiary creep. In light of the minimum creep-rate concept, the attention in experimental creep investigations is focused on steady-state creep, where it is constant over an extensive period of time. Generally in ceramics, tertiary creep is relatively short and sometimes even absent.

Several investigators have shown that the starting time of tertiary creep and rupture life are related for various alloys according to the relation (e.g., Garofalo et al. [11]):

$$t_2 = At_r^{\alpha} \tag{6.10}$$

in which t_r is the rupture life, t_2 is the starting time of the tertiary creep, A and α are constants, often ~1. Equation (6.10) is one of many expressions for creep, in general, and for tertiary creep, in particular, and is widely used for various materials under consideration for high-temperature applications. Other expressions are common in creep studies, such as power, exponential and logarithmic functions. For example, these three functions are shown respectively as:

$$\varepsilon_{III} = \dot{\varepsilon}_{min}t + At^{g} \tag{6.11}$$

$$\varepsilon_{III} = \theta_3(\exp[\theta_4 t] - 1) \tag{6.12}$$

$$\varepsilon_{III} = -(\ln[1 - C\dot{\varepsilon}_{min}t])/C \tag{6.13}$$

In these expressions, for tertiary creep without a primary stage, $\dot{\varepsilon}_{min}$ is the minimum-creep rate, A, g, θ_3, θ_4 and C are parameters. The creep curves with a higher applied stress, which have a pronounced tertiary stage, may be successfully described by all three equations. Dobeš [44] has indicated that the calculated value of g (~7–10) is higher than the one proposed by Graham and Walles [48] ($g = 3$).

6.3 Creep in Brittle Ceramics (Ductile at High Temperature Where Deformation is Possible)

It is important to also consider brittle materials in ceramics, since they deform at high temperatures. Deformation is essential for creep and, since ceramics are potential candidates for high-temperature applications during which deformation occurs, studies of creep and its retardation are important despite their RT brittleness. This discussion of creep will begin with single crystals, since no grain-boundary sliding is involved and, thus, creep occurs only within the lattice.

In Fig. 6.5, experimental transient creep is illustrated. This illustration represents commercially available SiC-fiber/calcium aluminosilicate matrix composites (Nicalon SiCJCAS-11). These composites contain 40 vol% Nicalon SiC fibers and had been hot-pressed.

Table 6.1 summarizes the experimental details together with symbols (designations) including values for: applied load, load time, strain, strain rate and the creep-strain recovery ratio, R_{cr}.

An experimental illustration of the strain-rate variation over time shown for the primary creep rate in Fig. 6.1b is presented in Fig. 6.6 for ceramics having the

Fig. 6.5 **a** Total strain versus time for $[0I]_{16}$ and $[0]_{32}$ Nicalon SiC_f/CAS-II composites crept at 1200 °C in high-purity argon [95]. With kind permission of John Wiley and Sons

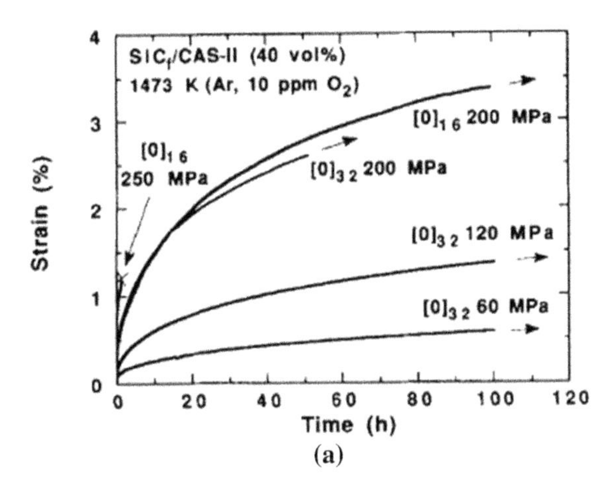

Table 6.1 Summary of loading histories and experimental results [95] (with kind permission of John Wiley and Sons)

Fiber layup	Loading history	ε_{10th} (%)	$\dot{\varepsilon}_{100th}$ (S^{-1})	$R_{cr,100th}$ (%)	$R_{r,100th}$ (%)
$[0]_{16}$	200 MPa/100 h	3.38	2.2×10^{-8}		
$[0]_{32}$	120 MPa/100 h	1.36	1.1×10^{-8}	23	32
$[0]_{32}$	60 MPa/100 h + 2 MPa/100 h	0.58	4.6×10^{-9}	27	33
$[0/90]_{4s}$	60 MPa/100 h + 2 MPa/100 h	0.59	4.0×10^{-9}	49	56
$[0/90]_{4s}$	60 MPa/100 h + 2 MPa/100 h	0.55	2.7×10^{-9}	45	52
$[0/90]_{4s}$	60 MPa/100 h + 2 MPa/100 h	0.62	3.9×10^{-9}	51	56
$[0/90]_{4s}$	60 MPa/40 min + 2 MPa/ 40 min			57/80*	73/70*

*Two 40-min cycles (first cycle/second cycle)

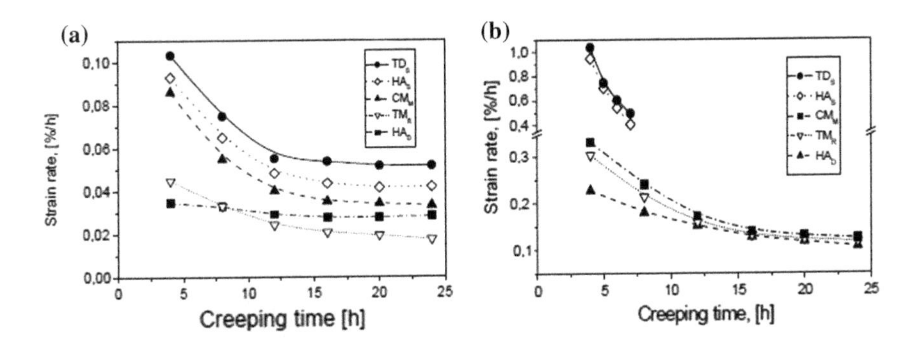

Fig. 6.6 Strain rate versus creeping time, creeping temperature **a** 1400 °C, **b** 1500 °C [84]. With kind permission of Dr. Nina Obradovic, Science of Sintering Editorial Board Secretariat

Table 6.2 Chemical composition of ladle lining bricks [84] (with kind permission of Dr. Nina Obradovic, Science of Sintering Editorial Board Secretariat)

Brick type	Chemical composition, (mass %)							
	SiO_2	Al_2O_3	Fe_2O_3	MgO	CaO	Cr_2O_3	TiO_2	MeO_2
HA_S	25.50	70.20	1.17	1.45	0.41	–	0.56	0.76
HA_D	26.50	72.40	0.54	–	0.21	–	0.11	0.26
TM_R	2.61	–	0.58	95.34	1.37	–	0.05	–
TD_S	1.01	1.40	0.31	43.82	52.74	–	–	–
CM_M	4.38	4.56	4.70	61.00	1.07	21.0	–	–

compositions indicated in Table 6.2 at two temperatures. Note the decrease of the strain rate over time in a similar fashion as shown in the aforementioned schematic Fig. 6.1b.

6.3.1 Creep in Single Crystals

The interest in single-crystal studies is not only a consequence of understanding creep phenomena without the influence of grain-boundary sliding, for example, but also for practical reasons regarding engineering applications. Many components based on single crystals are in use. The common denominator of all these applications is the prerequisite that the materials be stable during high temperature use. At such high temperatures, it is critical to use ceramics that have excellent creep capabilities, which can resist dimensional changes during long-term applications.

The dislocation structures of AI_2O_4Mg single crystals, which have crept (presented in the curves below in Fig. 6.7) may actually be seen in Fig. 6.8. Note that only a few dislocations are visible in the as-grown crystal. There are subgrains of ~500 μm with an inter-boundary misorientation of ~1/2°. Figure 6.8b–d shows a bulk-creep substructure once the damaged, superficial regions have been chemically removed. In Fig. 6.8b, slip lines are visible in the two (011) and (011) directions, but, in some other parts, traces of the two other {110} slip planes are also found. Figure 6.8c and d shows that dislocations at 1 % or 3 % strain are still concentrated in glide bands nearly parallel to the ⟨110⟩ directions. Many bands are seen stopping against each other, so no real cell structure is formed. These structures represent low-strain, transient creep (primary creep). The 1 % creep strain is shown in a more detailed manner in Fig. 6.9. One may observe on the (001) lamella: (a) that the subgrain boundaries were probably already in the as-grown crystals formed by two dislocation families not in any of the {110} planes, seen on the right side of the figure and (b) that the long-edge dislocations are parallel to [100]. More random, steeply-inclined dislocations are also visible; they almost have an edge character and belong to the (110) [110] system. With increased strain, a cell structure develops (see Fig. 6.10) at 7 % creep strain.

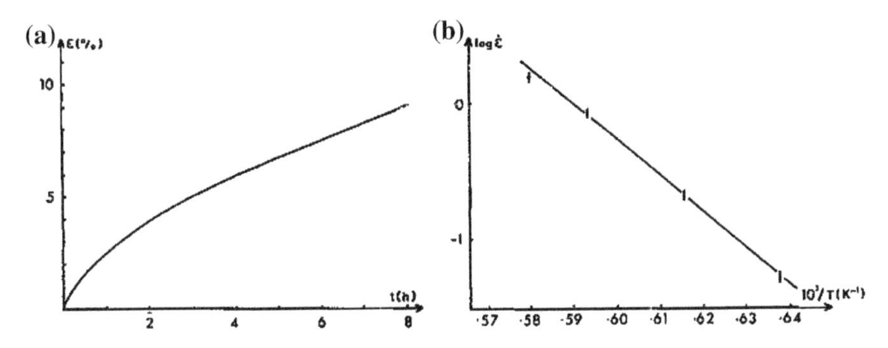

Fig. 6.7 **a** Creep curve at a temperature T = 1412 °C and under a resolved shear stress $\sigma = 5$ kg/mm^2. **b** Creep rate (logarithmic scale) versus 1/kT, for the constant resolved shear stress $\sigma = 5$ kg/mm^2. The slope gives the creep activation energy [45]. With kind permission, Permissions Dept., EDP Sciences by Dr. Corinne Griffon and Professor Escaig

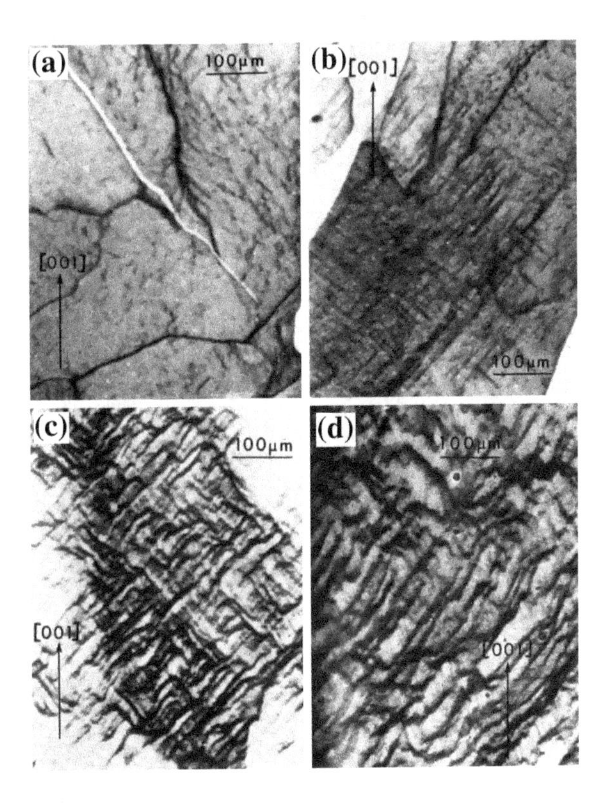

Fig. 6.8 Berg-Barrett topographies of the (100) face of spinel single crystals with the **g** = [4$\bar{4}$0] reflection: **a** as-grown: **b** loaded only a few minutes; **c** one percent creep strain at T = 1450 °C; **d** three percent creep strain at T –1300 °C [45]. With kind permission. Permissions Dept, EDP Sciences by Dr. Corinne Griffon and Professor Escaig

However, it is not yet homogeneous throughout the entire structure (observe Fig. 6.10a and b) until reaching 9 % strain. At 9 % strain, this cell structure is observed throughout the entire sample (Fig. 6.10c and d).

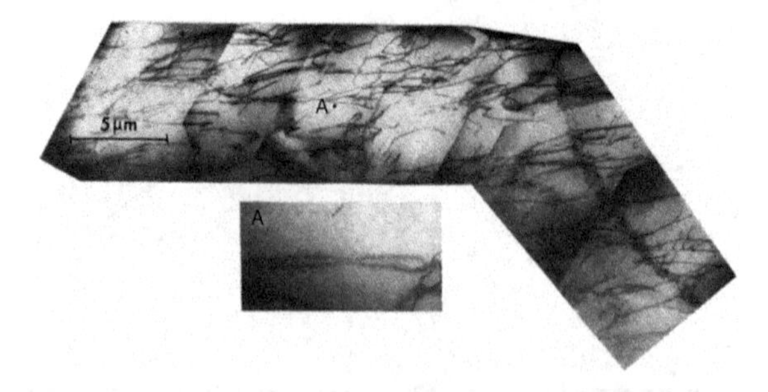

Fig. 6.9 MV electron micrograph of a one percent creep strain specimen: (031) lamella: the edge dipole seen in *A* is enlarged; note an as-grown sub-boundary on the right and numerous long edge dislocations parallel to [100] [45]. With kind permission. Permissions Dept., EDP Sciences by Dr. Corinne Griffon and Professor Escaig

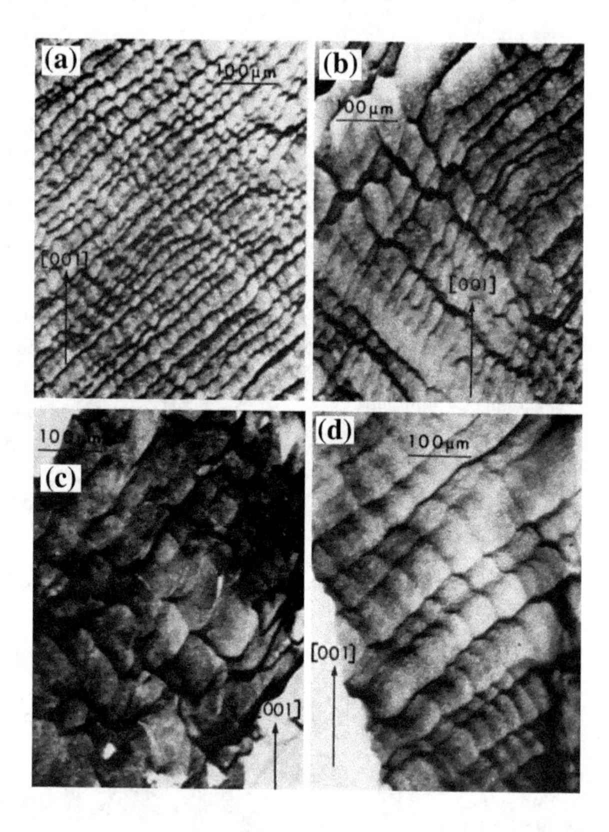

Fig. 6.10 Berg-Barrett topographies of the (100) face of spinel single crystals after creep; **g** = [440]: on these topographics, orientation contrasts would be given by [001] lattice rotation component, and displacement contrasts by [110] components: **a** and **b** 7 % crept specimens at 1350 °C, two regions showing different sizes of cell structure: **c** 9 %, 1412 °C; **d** 12 %, 1450 °C, crept specimens: the cell size is homogeneous everywhere [45]. With kind permission. Permissions Dept., EDP Sciences by Dr. Corinne Griffon and Professor Escaig

Considering Eq. (6.4), the experimental value of the stress exponent, n, obtained for the AI_2O_4Mg single crystal is 3.9 with an activation energy of Q = 5.3 eV. The creep range of this single crystal is 0.65 T_m–0.71 T_m and it follows a dislocation mechanism of the creep law. More specifically, the values of n and Q and the observed dislocation slip indicate that the mechanism is climb-controlled glide creep. The relation given in Eq. (6.4) has been used to calculate the above activation energy, expressed as:

$$\dot{\varepsilon} = A\sigma^{3.9} \exp(-\frac{5.3 \text{ eV}}{kT}) \tag{6.4a}$$

The stress exponent of Eq. (6.4a) may be obtained by the known stress-jump method, giving for n:

$$n = \left(\frac{\partial \ln \dot{\varepsilon}}{\partial \ln \sigma}\right)_{T,str} \cong \frac{\ln \dot{\varepsilon}_2 - \ln \dot{\varepsilon}_1}{\ln \sigma_2 - \ln \sigma_1} \tag{6.14}$$

where $\dot{\varepsilon}_1$and $\dot{\varepsilon}_2$ are the steady-state creep rates before and after the jump. Furthermore, it seems (as understood by Doukhan et al. [45]) that the climb mechanism of creep occurs in the cell walls.

6.3.2 Creep in Polycrystalline Ceramics

Creep in single crystals is not complicated by the presence of grain boundaries, which elsewhere produce a dual effect: (a) strengthening occurs due to retarded dislocation motion through the grains and (b) grain-boundary sliding may occur which contributes significantly to creep at high temperatures. Not surprisingly (and unlike the case of conventional, static plastic deformation), to get the best creep resistance in polycrystalline materials, the contribution of grain-boundary sliding should at least be kept to a minimum (if its complete elimination is unlikely). For this reason, the polycrystalline material chosen should be large-grained. Inducing large grains in a polycrystalline ceramic reduces the number of grain boundaries and, thus, decreases the effects of grain-boundary sliding.

These basic relations and factors also apply to polycrystalline materials with the appropriate variations dictated by experimental observations. Historically, Andrade [24] should be considered 'the father of creep', since it was he who first suggested a unified creep relation. All the many equations describing creep given in the literature follow in the wake of Andrade's concept [24] of this empirical relation. One may express the variation of strain over time as:

$$\varepsilon = \varepsilon_0(1 + \beta t^{1/3}) \exp(\kappa t) \tag{6.15}$$

According to the symbols β and κ in Eq. (6.15), it describes beta or kappa creep. When $\kappa = 0$, the constant β creep is obtained:

$$\varepsilon = \varepsilon_0(1 + \beta t^{1/3}) \tag{6.16}$$

This represents transient creep, since the creep rate is a decreasing function of time. By differentiating Eq. (6.15), with respect to time, one obtains (6.17):

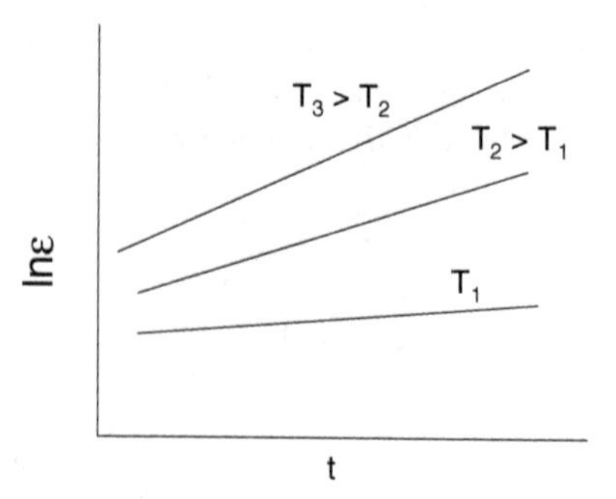

Fig. 6.11 Creep strain versus t for a few temperatures

$$\frac{d\varepsilon}{dt} = \dot{\varepsilon} = \frac{1}{3}\varepsilon_0\beta t^{-2/3} \tag{6.17}$$

However, when $\beta = 0$ in Eq. (6.15):

$$\varepsilon = \varepsilon_0 \exp(\kappa t) \tag{6.18}$$

This is κ creep, describing the stationary stage. Clearly, the strain rate (as obtained from Eq. (6.18)) gives a linear relation, as seen in Eq. (6.19) and, thus, describes steady-state creep:

$$\dot{\varepsilon} = K\varepsilon_0 \exp(\kappa t) = \kappa\varepsilon \tag{6.19}$$

Andrade postulated that β creep is related to dislocation glide within the grain, while κ flow is related to slip along grain boundaries. Ascribing κ flow to grain-boundary sliding is known to be in error. Equation (6.18) may also be expressed as:

$$\ln\varepsilon = \ln\varepsilon_0 + \kappa t \tag{6.18a}$$

A schematic plot of this function at different temperatures is seen in Fig. 6.11. From the intercept of such curves, ε_0, and from the slopes κ for various temperatures can be evaluated. Note that the plots in this figure are similar to those in Fig. 6.12, but this should not be surprising since stress and temperature increase have similar effect of creep as already indicated above.

The term 'viscous creep' is often used for creep at high temperatures with low stresses. Two mechanisms have been proposed to describe such creep in polycrystalline materials. The one known as 'Nabarro-Herring creep' conceives of a stress-directed, diffusional migration of vacancies, while the other, originally suggested by Mott and subsequently elaborated by Weertman, is based on a dislocation-climb model [66]. Extensive experimental evidence also exists to support a 'dislocation-climb model'.

It was suggested by the dislocation-climb model that the activation energy for creep in many materials at high temperatures is equal to their activation energy for self-diffusion. According to both models, the activation energy for high-temperature

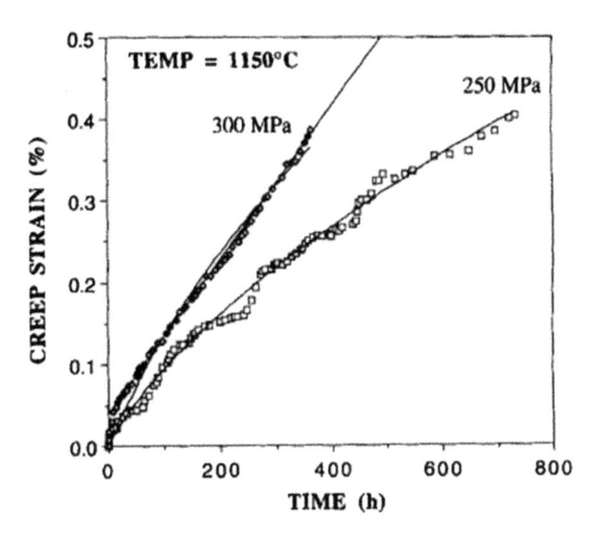

Fig. 6.12 Creep curves of GN-10 Si_3N_4 tested at 1150 °C. (*Symbols* are experimental data; *solid lines* are curve fitting [35]. With kind permission of John Wiley and Sons

low-stress creep should be equal to that of self-diffusion. The Nabarro-Herring model of creep, due to the stress-directed diffusion of vacancies, is given as:

$$Z = \dot{\varepsilon} \exp\left(\frac{\Delta H}{RT}\right) = A\sigma \tag{6.20}$$

where Z = the Zener-Hollomon parameter, A is a constant and ΔH is the activation energy. As seen in Eq. (6.20), Z is related to the strain rate as affected by temperature and it is constant at constant strain. The above relation was given by Sherby [77] as obtained from constant-stress creep tests done on metals at 0.5 T_m and higher:

$$\varepsilon = f\theta \tag{6.19a}$$

with:

$$\theta = \int_0^7 \exp(-\frac{\Delta H_c}{RT})dt \tag{6.19b}$$

Equation (6.20) is obtained by differentiating Eq. (6.19b) with respect to time at a constant stress.

In an empirical relation, as in Eq. (6.20):

$$Z = \dot{\varepsilon}_s \exp\left(\frac{\Delta H}{RT}\right) = f(\sigma) \tag{6.19c}$$

$\dot{\varepsilon}_s$ indicates the secondary-creep rate and f(σ) is a function of stress and the structural changes. By plotting Z as a function of stress, it is revealed that Z

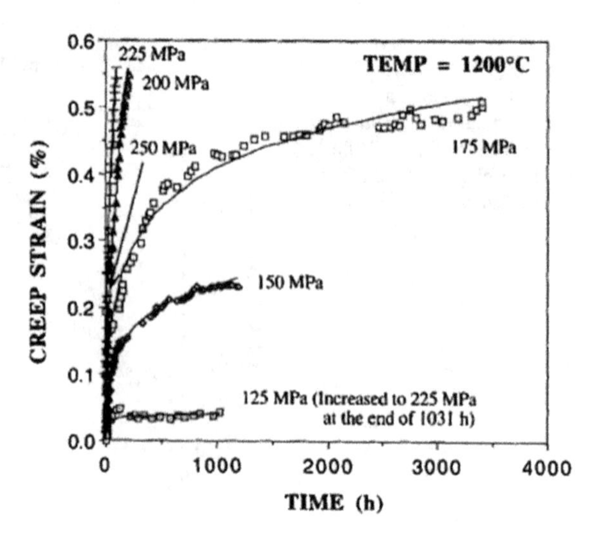

Fig. 6.13 Creep curves of GN-10 Si_3N_4 tested at 1200 °C [35]. With kind permission of John Wiley and Sons

increases linearly up to a certain stress, in agreement with the Nabarro-Herring theory.

Since one, universal creep function can hardly fit all the experimental data, many functions were suggested to provide an empirical description of creep phenomena. As seen above, functions for logarithmic, transient and linear creep have been suggested in various forms and combinations.

One of the high-temperature, high-strength ceramics of technological interest for advanced applications is Si_3N_4. During high-temperature applications, creep must be considered and, since it is usually accelerated at high temperatures, the interest has been focused on its behavior at these temperatures; tests have been performed in this context. Figure 6.12 is a creep curve at 1150 °C of two stresses obtained by uniaxial tension. GN-10 is marketed as a commercial-grade HIPed Si_3N_4, containing Y_2O_3 and SrO as densifiers. Figures 6.13 and 6.14 shows the transition from primary to steady-state creep.

No such transition is indicated in the test at 1150 °C (see Fig. 6.12). Creep failure has occurred in the initial transient stage (see Figs. 6.10–6.12). Note that Figs. 6.14 and 6.15 shows the same character as does Fig. 6.2b), in which the effects of the stresses on the shapes of the creep curves are indicated. Not all the curves show the two stages characterizing transient and steady-state creeps.

The SEM microstructure of this specimen, tested at 1300 °C, is illustrated in Fig. 6.12. No cavity or void formation at the grain boundaries was observed, even after the specimen crept at 1300 °C and 100 MPa for 1721 h.

The fabrication procedure is crucial when a ceramic is intended for high-temperature applications; it must produce good creep resistant performance. Figures 6.17 and 6.18 clearly show that the HIP-fabricated Si_3N_4 is more durable and has a longer service life.

Recall Fig. 6.1b, where the minimum creep rate was indicated schematically. The concept of 'minimum creep rate' is of key importance in the creep-resistant

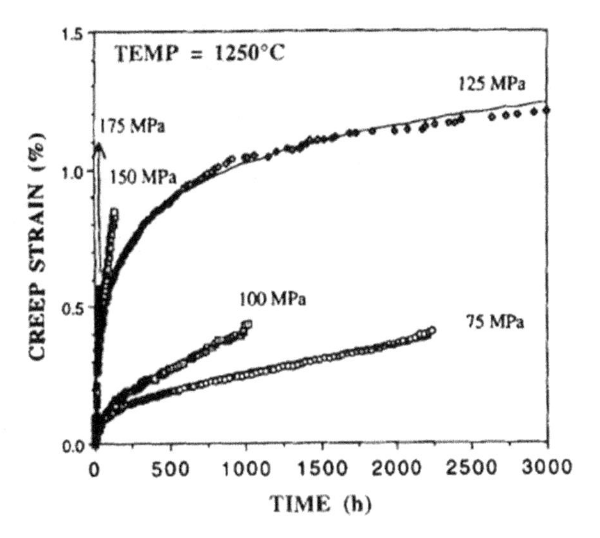

Fig. 6.14 Creep curves of GN-10 Si_3N_4 tested at 1250 °C [35]. With kind permission of John Wiley and Sons

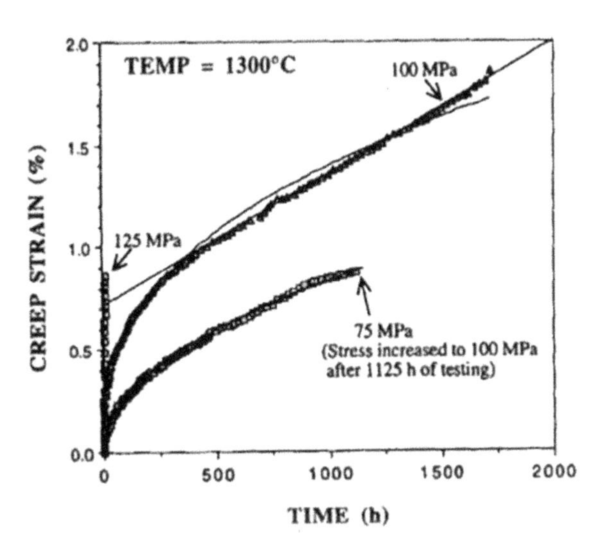

Fig. 6.15 Creep curves of GN-10 $\mathbf{Si_3N_4}$ tested at 1300 °C [35]. With kind permission of John Wiley and Sons

design of ceramics or, as a matter of fact, generally of all materials. Figure 6.19 shows experimental curves by using Norton's power-law relation given below, which is basically Eq. (6.4):

$$\dot{\varepsilon} = A\sigma^n \exp(-\frac{Q}{RT}) \tag{6.4}$$

The data on creep in this important ceramic (Si_3N_4), suitable for high-temperature applications, indicate that no tertiary creep sets in under the experimental conditions.

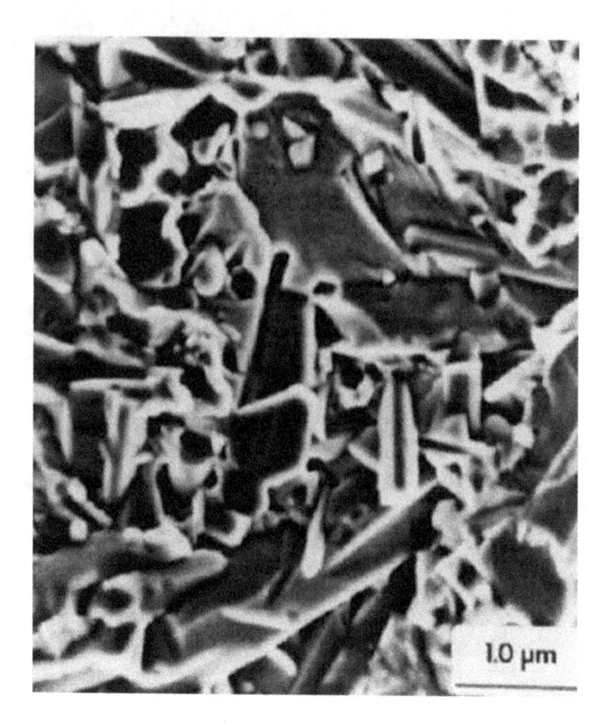

Fig. 6.16 SEM image of the specimen crept at 1300 °C and 100 MPa for 1721 h, as shown in Fig. 6.12. No void formation is visible [35]. With kind permission of John Wiley and Sons

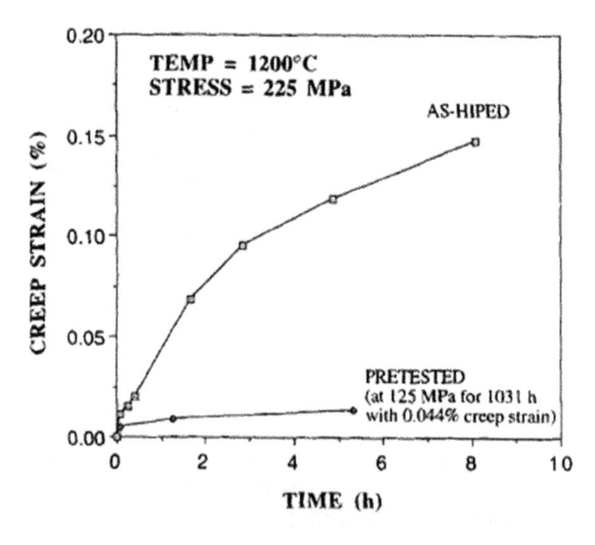

Fig. 6.17 Comparison of the initial transient creep behavior of the as-HIPed specimen tested at 1200 °C with 225 MPa and that of a precrept specimen subjected to the same test condition but preceded by initial testing at 1200 °C with 125 MPa for 1031 h [35]. With kind permission of John Wiley and Sons

Si_3N_4, an exceptional high-strength ceramic at RT, considered for potential use in high-temperature applications, suffers degradation in its strength at high temperatures, which is a function of load and the duration at the specific high temperature. Within the limited ranges of stress and temperature considered above, no tertiary creep or detrimental effects were observed. However, one cannot overlook

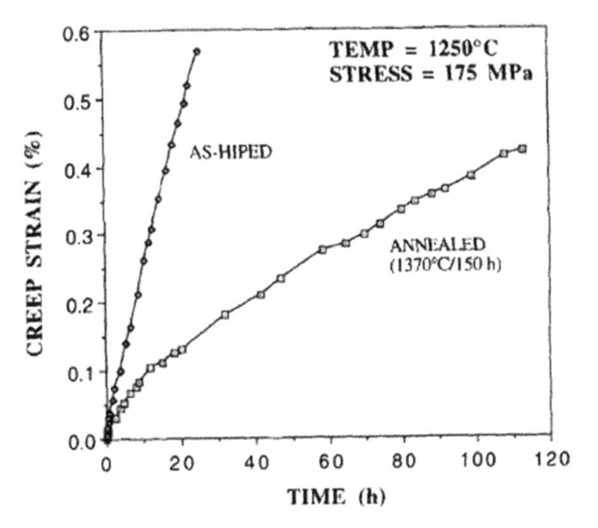

Fig. 6.18 Comparison of the creep behavior of the as-HIPed and that of the annealed specimen both tested at 1250 °C and 175 MPa [35]. With kind permission of John Wiley and Sons

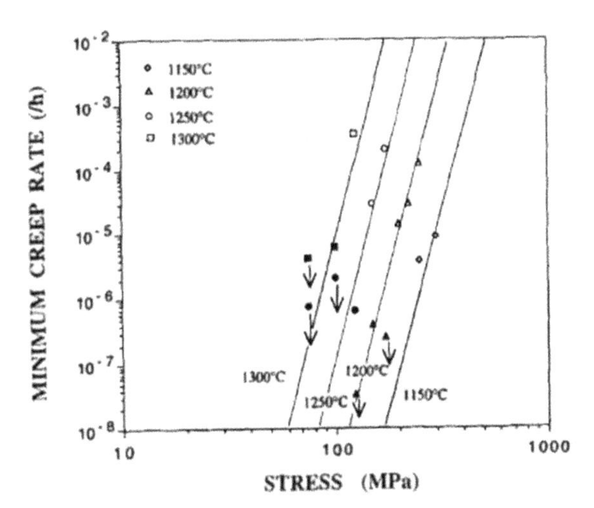

Fig. 6.19 Comparison of the minimum creep rate data and the Norton power-law equation (*straight lines*) with stress exponent and activation energy equal to 12.6 and 1645 kJ/mol, respectively. *Open* and *filled-in symbols* represent the data above and below the experimentally observed transition region, respectively. *Arrowed data* imply the actual creep rate may be lower than the indicated one [35] with kind permission of John Wiley and Sons

the susceptibility of slow crack formation, which limits the reliable, long-range use of Si_3N_4 without appropriate additives. Of the many and various additives used to improve the performance of Si_3N_4, the use of a SiC additive will be illustrated below. SiC is considered an effective additive for stabilizing the structure of Si_3N_4 and for controlling GBS by impeding that process [56]. Furthermore, with this additive, good corrosion resistance is exhibited by Si_3N_4 at high temperatures.

Figure 6.20 illustrates creep plots in the range of 1200–1400 °C under different stresses obtained by a four-point bending test with inner and outer spans of 20 and 40 mm, respectively.

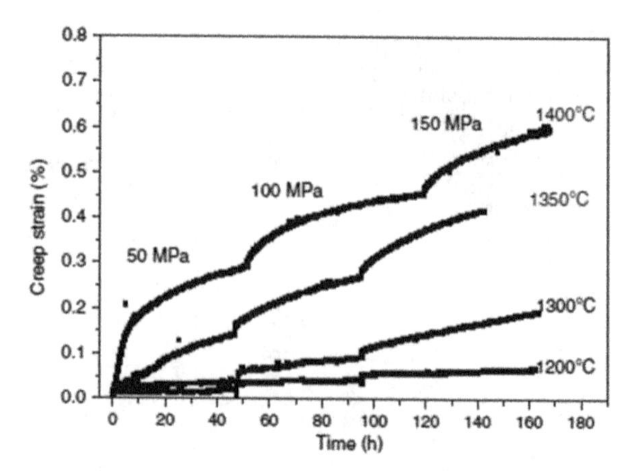

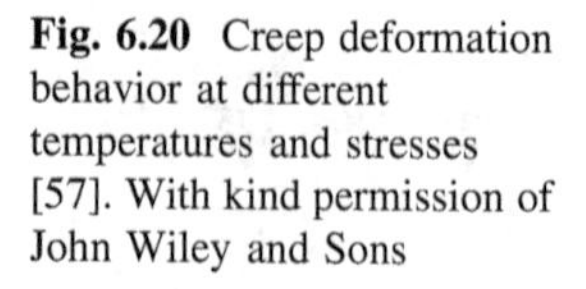

Fig. 6.20 Creep deformation behavior at different temperatures and stresses [57]. With kind permission of John Wiley and Sons

These creep experiments were performed by stepped-load tests to obtain multiple creep-rate data. The crept samples were used for TEM investigations. The tests were carried out on specimens having the sintered microstructure shown in Fig. 6.21. The arrows indicate the globular SiC particles located in Si_3N_4 grains. Their amount is 3.2 vol%. Another 1.9 vol% SiC is distributed intergranularly. BF TEM (Fig. 6.22) revealed that the SiC nanoparticles are located in multigrain pockets, being smaller in size ($\sim$45 nm) than the Si_3N_4 grains, while the intergranularly distributed SiC has larger dimensions ($\sim$150 nm). A tendency was observed for the particles to be trapped at the triple points or the multigrain pockets. The minimum creep rate is shown as a function of stress for each temperature tested (see Fig. 6.23a). An Arrhenius-type relation from Eqs. (6.4) or (6.4a) was used to evaluate the activation energy and the respective plots are shown in Fig. 6.23b.

The creep results in the ranges investigated did not show tertiary creep, because the duration of the tests was not sufficiently long. Compared with Figs. 6.12–6.15 and 6.19, the creep tests, after the addition of SiC, are indeed quite short. The temperatures in the creep tests with and without SiC are about the same (if the 1400 °C test is not considered). Therefore, one cannot safely ascertain that impeding grain-boundary sliding by the addition of SiC–which is one of the claims regarding the effect of SiC–would be true for longer creep tests. Apparently, a 5.1 vol % addition of SiC (3.2 % in the grains and a 1.9 % intergranular distribution) does not have a large impact on improving the high temperature behavior of Si_3N_4. It is likely that more SiC is required in order to attain a significantly improved high-temperature performance. Indeed, numerous reports (e.g., Rendtel et al. [71]) indicate a significant improvement in creep resistance up to 1400 °C and this excellent creep resistance was comparable to or better than other Si_3N_4-base ceramics with other high-temperature strengthening media.

The decision to illustrate a 5.1 vol% SiC addition, rather than the 15–35 wt% additives to the basic Si_3N_4 was done intentionally to indicate the fact that the improvement of a ceramic material is strongly influenced by the amount of the

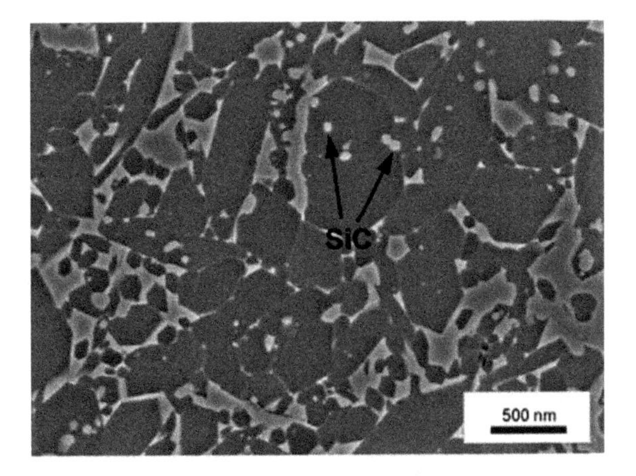

Fig. 6.21 Secondary electron scanning electron microscopy image of the plasma-etched microstructure of a Si_3N_4-5 wt% SiC composite [57]. With kind permission of John Wiley and Sons

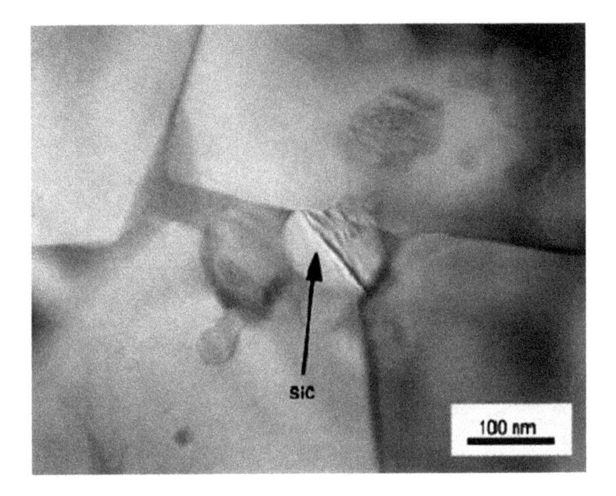

Fig. 6.22 Bright-field transmission electron microscopy micrograph showing the presence of a SiC nanoparticle located in a multigrain pocket (*arrow*) [57]. With kind permission of John Wiley and Sons

additive and certainly by the methods used during all the fabrication stages, including the choice of the constituent powders for the sintering process. The example presented shows that, within the same temperature range and while using the same chemically-composed additives, the outcomes for the base-ceramics may nonetheless be, if not conflicting, then different.

6.4 Creep in RT Ductile Ceramics

More on RT ductile ceramics will be considered in Chap. 9, devoted to ceramics having small dimensions (nano-scale). Many nano-sized ceramics are ductile at ambient temperatures. In this section, those ceramics found to possess ductility at

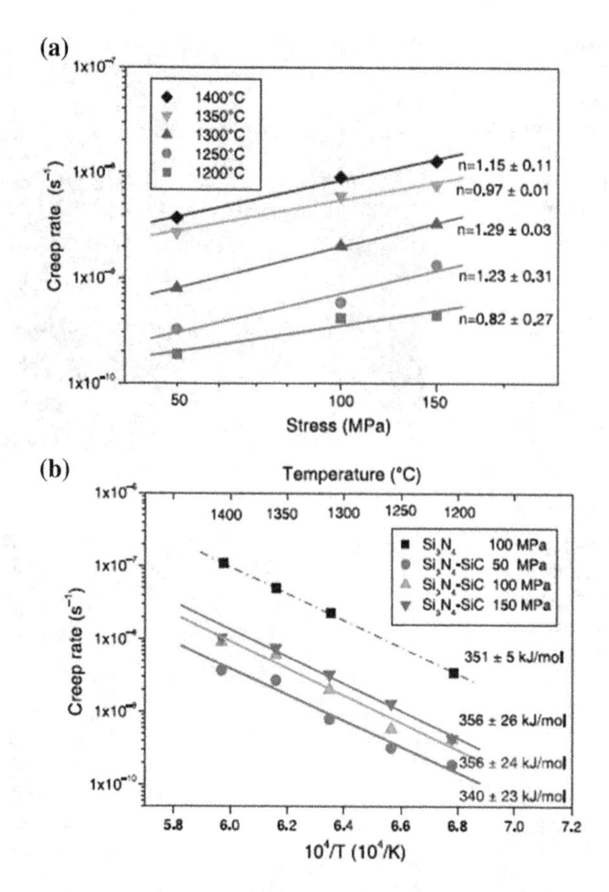

Fig. 6.23 Creep rate as a function of stress for different temperatures—stress exponent n (**a**) and creep rate as a function of temperature for different stresses—activation energy Qc (**b**) [57]. With kind permission of John Wiley and Sons

RT are considered in terms of their high-temperature performances, specifically under creep conditions. In Sect. 5.2.2, MgO was singled out as an example of a low-temperature ductile ceramic (see Figs. 5.3 and 5.4). Of further interest is RT ductile $SrTiO_3$, which is one of many studied ceramics, due to its interesting and unique features (Fig. 6.24).

6.4.1 *Single Crystal Ceramics*

MgO was mentioned in Chap. 5, Sect. 5.2.2 as being ductile at ambient temperature, thus undergoing plastic deformation. Having a high melting point of 2798 °C, it is expected that creep may occur at high homologous temperatures. In

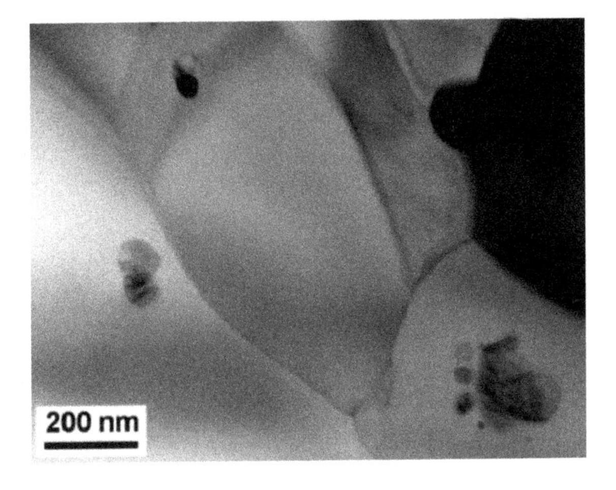

Fig. 6.24 Transmission electron microscopy micrograph of a sample crept at 1350 °C at 150 MPa for 148 h (creep strain 0.53 %) [57]. With kind permission of John Wiley and Sons

Fig. 6.25, a set of creep curves are shown, pre-strained by tension at various rates at 1800 °C to induce dislocation structures. Creep tests were performed at 1400 °C and under a stress of $\sigma = 44.1$ MN/m^2. The focus here is on the curve without pre-strain before the creep test. It represents the classical creep curve (illustrated schematically in Fig. 6.1a) without a tertiary (stage 3) creep region. Not all the test specimens are fractured and, thus, the final stage of creep is missing and is not generally observed in experiments until so desired.

The creep substructures in annealed $\langle 110 \rangle$ MgO single crystals show that no subgrains were formed during tensile creep at 1400 °C. The 1800 °C true-flow stress at 10 % strain increased with the increasing strain rate (i.e., 25.5, 55.3, and 71.6 MNm^2 for strain rates of 0.002, 0.02, and 0.2 min^{-1}, respectively). Note that inducing a substructure, which produces various arrays of dislocations, is equivalent to the effect of stress (as indicated in Fig. 6.2b).

During an additional set of experiments done on single MgO crystals, microstructure evolution and recrystallization during creep were considered [62]. In these experiments, compression was applied to orientation $\langle 100 \rangle$ in the temperature range 1573–1773 K up to 69 % strain. Strain-rate values were obtained from the quasi-steady-state part of the strain versus time curves. As expected, the elastic (instantaneous) strain and the transient creep preceded the quasi-steady-state. The results were expressed in terms of shear strain versus normalized shear stress (τ/G). These curves were fitted to $\dot{\gamma}$ (shear strain rate), σ and T data using a semi-theoretical model of dislocation-climb-controlled creep. The plots are presented in Fig. 6.26. For details on microstructure evolution and recrystallization during creep and on the active slip systems operating during creep, the reader should consult the original work of Mariani et al. [62].

The unique nature (plasticity and creep deformation) of $SrTiO_3$ also merits discussion. Quite surprisingly, this ceramic, unlike most ceramics, can be plastically deformed (under compression) in the temperature ranges 78–1050 K and 1050–1500 K, although it acts in a brittle fashion [31]. Here, the rare phenomenon

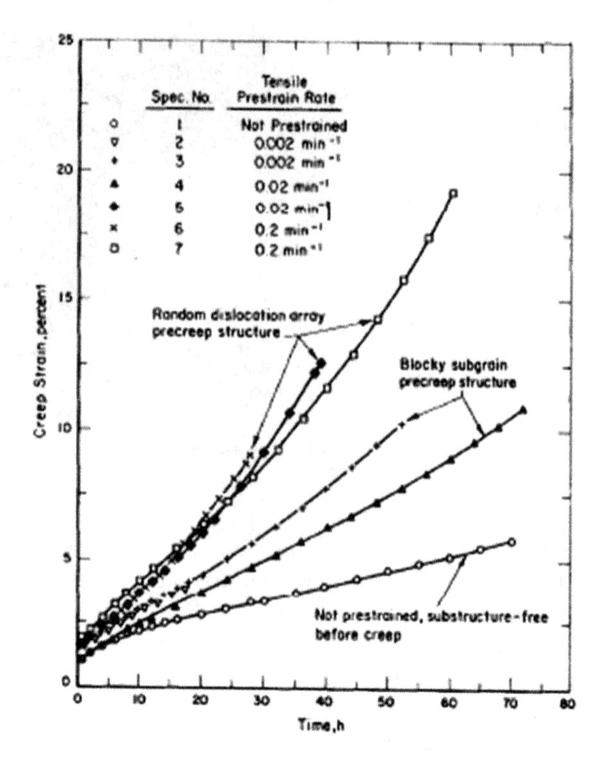

Fig. 6.25 Creep curves of MgO crystals having different pre-creep substructures (T = 1400 °C, $\sigma = 44.1$ MN/m^2) [35]. With kind permission of John Wiley and Sons

of ductile–brittle–ductile transition may be observed. Typical stress–strain creep curves for two orientations are shown in Fig. 6.27 and all the results are summarized in Table 6.3. Compression tests were done at orientations $\langle 100 \rangle$ and $\langle 110 \rangle$ in the temperature range 1473–1793 K. The figures are shown for both orientations. A modification of Eq. (6.4) was applied for the relation of the strain rate, $\dot{\gamma}$, as given below in Eq. (6.21), which expresses the shear stress as normalized by the shear modulus (μ = G):

$$\dot{\gamma} = \frac{\mu b}{kT}\left(\frac{\tau}{\mu}\right) A D_0 \exp\left(-\frac{Q}{RT}\right) \tag{6.21}$$

A short transient stage is observed in the above figures in the 0.3–1 % strain range with work hardening. Following the transient stage, second stage (steady-state) creep sets in as a flat plateau, meaning that the flow stress remains unchanged. Anisotropy in strength is observed in these figures, indicating that the specimens in the $\langle 100 \rangle$ orientation are stronger than those in the $\langle 110 \rangle$ orientation, as seen from their respective stress values. Table 6.3 summarizes the experimental results. Figures 6.28 and 6.29 illustrate the state of creep for both orientations tested.

This analysis was performed as usual for the evaluation of values of interest. Based on Eq. (6.4), the exponent and activation energies are rewritten as:

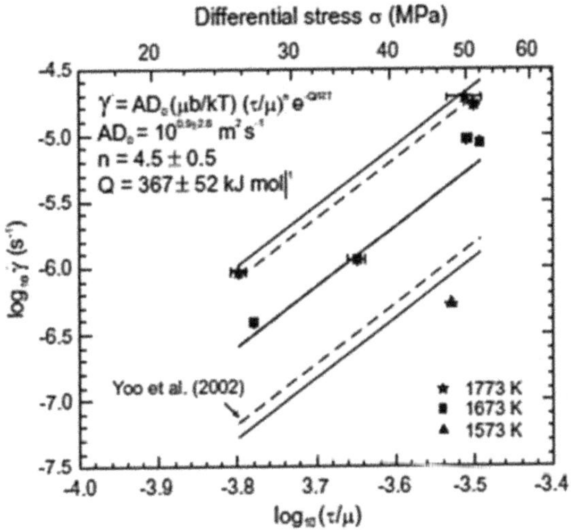

Fig. 6.26 Diagram of log10 shear strain rate versus log10 shear stress/shear modulus curves obtained for temperatures of 1573, 1673 and 1773 K. *Symbols* represent experimental data, *solid lines* are best fit linear regression curves and *dashed lines* are data from Yoo [98] plotted using the value of the stress exponent n = 4.5 obtained in this study. The maximum error on the differential stress is ± 2.5 MPa [63]. With kind permission of Elsevier

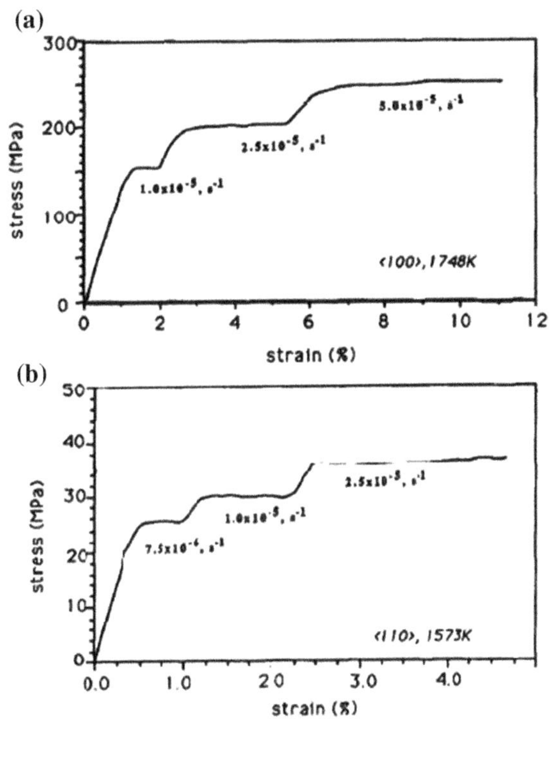

Fig. 6.27 Typical stress–strain curves for samples compressed along **a** (100) and **b** (110) deformed at various strain rates [88]. With kind permission of Elsevier

$$\dot{\varepsilon} = A\sigma^n \exp -(\frac{Q}{kT}) \tag{6.4}$$

where the constant B, has been replaced by A.

Table 6.3 Summary of experimental results [88] (with kind permission of Elsevier)

Run	T(k)	σ(MPa)	(%)	(s^{-1})	$fo_2(p_a)$	Orientation
$SrTiO_34$	1723	140(3)	4.5	5.5×10^{-6}	2.1×10^4	(100)
$SrTiO_36$	1673	216(9)	4.0	7.5×10^{-6}	2.1×10^4	(100)
$SrTiO_3^{J-1}$	1748	149(4)	2.0	1.0×10^{-5}	2.1×10^4	(100)
$SrTiO_3^{J-2}$	1748	193(6)	3.5	2.5×10^{-5}	2.1×10^4	(100)
$SrTiO_3^{J-3}$	1748	236(8)	6.1	5.0×10^{-5}	2.1×10^4	(100)
$SrTiO_3^{9-1}$	1793	76(2)	3.8	5.2×10^{-6}	2.1×10^4	(100)
$SrTiO_3^{9-2}$	1793	125(3)	2.3	2.4×10^{-5}	2.1×10^4	(100)
$SrTiO_3^{9-3}$	1793	144(4)	3.5	4.9×10^{-5}	2.1×10^4	(100)
$SrTiO_3^{10-1}$	1703	155(4)	3.0	4.0×10^{-6}	2.1×10^4	(100)
$SrTiO_3^{10-2}$	1703	185(5)	3.9	7.5×10^{-6}	2.1×10^4	(100)
$SrTiO_3^{10-3}$	1703	233(6)	5.8	1.8×10^{-5}	2.1×10^4	(100)
$SrTiO_3^{11*}$	1723	160(3)	5.0	7.0×10^{-6}	2.1×10^4	(100)
$SrTiO_3^{13-1}$	1623	14(<1)	1.5	7.5×10^{-6}	2.1×10^4	(110)
$SrTiO_3^{13-2}$	1623	16(<1)	2.0	1.0×10^{-5}	2.1×10^4	(110)
$SrTiO_3^{13-3}$	1623	20(<1)	3.0	3.0×10^{-5}	2.1×10^4	(110)
$SrTiO_3^{14-1}$	1573	26(<1)	3.0	7.5×10^{-6}	2.1×10^4	(110)
$SrTiO_3^{14-2}$	1573	30(<1)	1.7	1.0×10^{-5}	2.1×10^4	(110)
$SrTiO_3^{14-3}$	1573	36(1)	2.4	2.5×10^{-5}	2.1×10^4	(110)
$SrTiO_3^{15-1}$	1523	36(1)	3.0	1.0×10^{-5}	2.1×10^4	(110)
$SrTiO_3^{15-2}$	1523	44(1)	2.5	2.5×10^{-5}	2.1×10^4	(110)
$SrTiO_3^{15-3}$	1523	56(1)	2.5	5.0×10^{-5}	2.1×10^4	(110)
$SrTiO_3^{16}$ [a]	1473	49(1)	4.1	8.7×10^{-6}	2.1×10^4	(110)
$SrTiO_3^{17-1}$	1573	20(<1)	1.0	4.8×10^{-6}	1.8×10^4	(110)
$SrTiO_3^{17-2}$	1573	20(<1)	1.2	4.0×10^{-6}	1.3×10^3	(110)
$SrTiO_3^{17-3}$	1573	20(<1)	0.9	3.0×10^{-6}	1.4	(110)
$SrTiO_3^{17-4}$	1573	20(<1)	0.8	1.5×10^{-6}	1.4×10^{-1}	(110)
$SrTiO_3^{17-5}$	1573	20(<1)	0.7	5.0×10^{-6}	2.1×10^{-3}	(110)
$SrTiO_3^{17-6}$	1573	20(<1)	0.8	1.6×10^{-6}	1.2×10^{-1}	(110)
$SrTiO_3^{17-7}$	1573	20(<1)	1.0	2.7×10^{-6}	7.1×10^{-1}	(110)
$SrTiO_3^{17-8}$	1573	20(<1)	1.1	2.9×10^{-6}	2.3	(110)
$SrTiO_3^{17-9}$	1573	20(<1)	1.2	3.1×10^{-6}	15.8	(110)
$SrTiO_3^{17-10}$	1573	20(<1)	1.3	3.5×10^{-6}	4.0×10^3	(110)
$S_rTiO_3^{17-11}$	1573	20(<1)	1.0	4.0×10^{-6}	8.9×10^3	(110)
$S_rTiO_3^{18-1}$	1723	122(5)	1.0	1.6×10^{-6}	2.1×10^4	(100)
$S_rTiO_3^{18-2}$	1723	122(5)	1.0	1.5×10^{-6}	8.2×10^3	(100)
$S_rTiO_3^{18-3}$	1723	122(5)	1.1	1.4×10^{-6}	3.4×10^3	(100)
$S_rTiO_3^{18-4}$	1723	122(5)	0.9	1.3×10^{-6}	1.0×10^2	(100)
$S_rTiO_3^{18-5}$	1723	122(5)	0.8	1.1×10^{-6}	25	(100)
$S_rTiO_3^{18-6}$	1723	122(5)	0.8	1.0×10^{-6}	0.9	(100)
$S_rTiO_3^{18-7}$	1723	122(5)	0.8	1.0×10^{-6}	5.4×10^{-2}	(100)

[a] Data obtained from the initial steady-state creep of stress-dip tests

Only runs reaching steady-state creep have been completed; numberes in parantheses are estimated errors

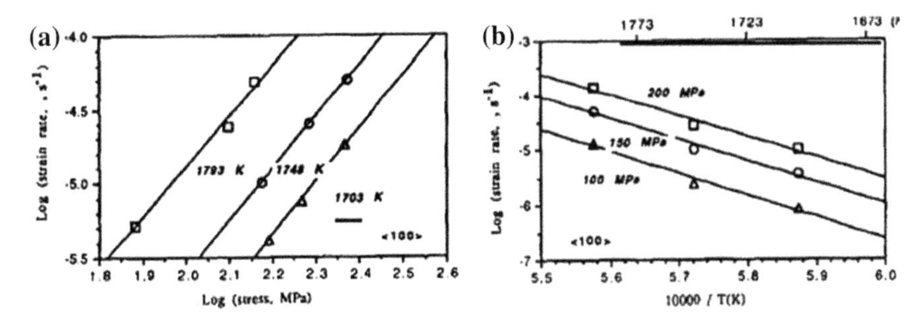

Fig. 6.28 Log–log plots of **a** strain rates versus flow stresses and **b** strain rates versus temperatures for the ⟨100⟩ samples [88]. With kind permission of Elsevier

The stress exponent, n, may be determined from the slopes of the (a) images in Figs. 6.28 and 6.29. Similarly, the activation energy for creep in the steady state may be obtained from the (b) images. The creep parameters are listed in Table 6.4. One observes from the results listed in Table 6.4 that is about the same for both orientations of the $SrTiO_3$ crystals; however, a considerable difference exists in Q_c, the activation energy for creep. TEM observations indicate that gently-curved dislocations with **b** = ⟨**100**⟩ are generated in the crystal under compression along the ⟨110⟩ orientation, whereas very straight dislocations with **b** = ⟨**110**⟩ are generated in the crystals compressed along the ⟨100⟩ orientation.

It was suggested that these different results are due to the creep mechanism itself. Creep, in specimens with $\{1\bar{1}0\}$ ⟨110⟩ orientation slip systems, is controlled by dislocation glide and, due to the high Peierls stress, is essentially determined by the crystal structure. In contrast, creep, in specimens with {100} ⟨010⟩ slip systems, is controlled largely by dislocation climb, which depends on the diffusion of vacancies to the climb site.

6.4.2 Polycrystalline Ceramics

Having described creep in single-crystal MgO, now it is appropriate to illustrate the behavior of its polycrystalline form as well. But first, note that the experimental results (see Fig. 6.30) resemble the shape found in Fig. 6.1b. In addition to the transient curve seen in the Fig. 6.30, this experimental transient curve is joined with the steady-state creep curve. The solid line represents the fitting of both these parts of the creep along the experimental points. The transient creep, $\dot{\varepsilon}_t$, is expressed by an exponential decay relation, given as:

$$\dot{\varepsilon}_t = \dot{\varepsilon}_i \exp(-\frac{t}{\tau}) \tag{6.22}$$

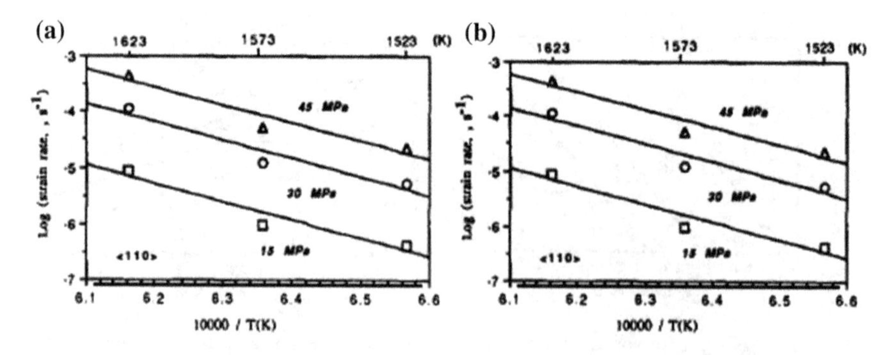

Fig. 6.29 Similar plots as Fig. 6.25 for the ⟨110⟩ samples [88]. With kind permission of Elsevier

Table 6.4 Creep law parameters [86] (with kind permission of Elsevier)

Orientation	T(K)	n	σ(MPa)	Q_c(KJ mol^{-1})	In A (s^{-1} MPa^{-n})
⟨100⟩	1703	3.6 (±0.3)	100	768 (±32)	24.2 (±2.5)
⟨100⟩	1748	3.5 (±0.3)	150	747 (±36)	22.6 (±2.0)
⟨100⟩	1793	3.4 (±0.2)	200	730 (±25)	21.6 (±1.8)
⟨110⟩	1523	3.6 (±0.2)	15	620 (±19)	24.9 (±2.0)
⟨110⟩	1573	3.6 (±0.3)	30	624 (±24)	25.0 (±2.2)
⟨110⟩	1623	3.6 (±0.2)	45	616 (±26)	25.1 (±2.3)

Power law formula, $\dot{\varepsilon} = A\sigma^n \exp(-Q_c/RT)$, is assumed

$\dot{\varepsilon}_i$ is the initial strain rate and τ is a characteristic relaxation time. This exponential relation describes transient creep when time-dependent steady-state creep is assumed to follow a Cobel-type creep-model (described later in this chapter), namely when creep is controlled by grain-boundary diffusion. In the above figures, one may see that the overall creep rate in both stages may be expressed by:

$$\dot{\varepsilon} = \dot{\varepsilon}_t + \dot{\varepsilon}_{ss} \tag{6.23}$$

with $\dot{\varepsilon}_{ss}$ representing the second-stage creep rate.

The overall creep appearance, as indicated in Fig. 6.1a, is illustrated by the experimental results for MgO in Fig. 6.31. Two curves are indicated in Fig. 6.31, both exhibiting a transient creep (decreasing creep rate over time, as seen in Fig. 6.1a) and a constant rate second-stage creep. The larger stress plot also shows a third-stage creep, which corresponds to the accelerated strain rate before fracture. Similar curves were also found in other ceramic materials, indicating that it is quite sound to analyze creep phenomena in terms similar to those used for the creep deformation of metals, in which stage III creep is generally observed if sufficient time is allowed for the measurements and if fracture does not set in earlier.

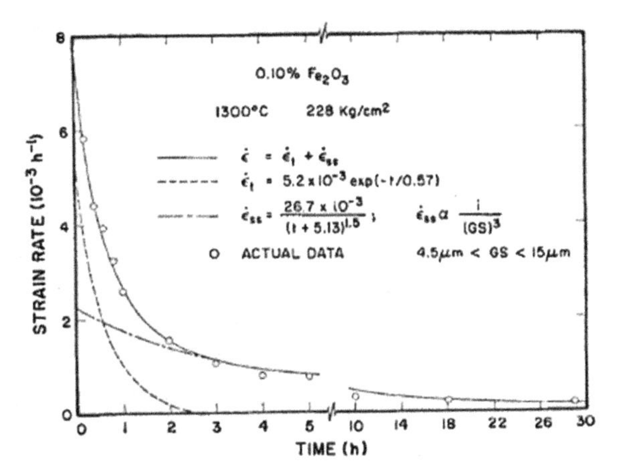

Fig. 6.30 Typical fit to actual data showing necessity for adding steady-state and transient creep rates [49]. With kind permission of John Wiley and Sons

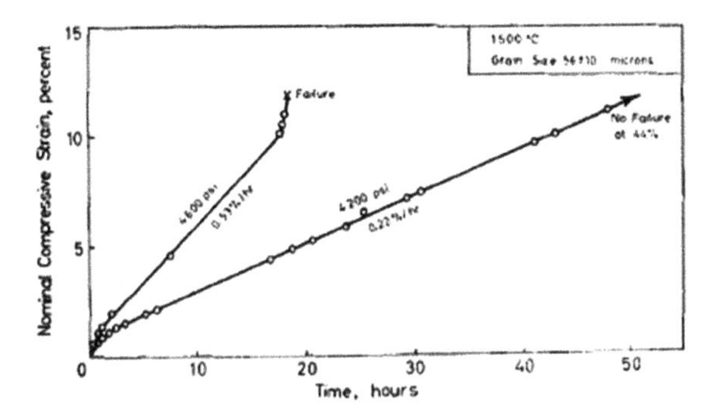

Fig. 6.31 Compressive creep curves for magnesium oxide [52]. With kind permission of John Wiley and Sons

Hot-pressed MgO is deformed for creep studies by compression at stress and temperature ranges of 1000–6500 psi and 1200–1500 °C, respectively, and at steady-state creep rates of 0.001–0.8 %/h. The microstructure has equiaxed grains and porosity mostly in the grain boundaries, as seen in Fig. 6.32. The densities of the specimens were in the range 97.3–98.9 %. Heat treatment produces a stable grain size, but when this stage is omitted during fabrication, grain growth is expected to occur during creep, which is one of the reasons explaining the different results. For good creep resistance, no structural changes should result from creep deformation and no increase in the degree of porosity should occur, in so far as is possible. However, when rupture is approached, changes in the structure are unavoidable. Steady-state creep at several temperatures as a function of stress may be seen in Fig. 6.33. An analysis of the creep data was performed by using two functions respectively, each representing a different creep model. Both are presented below: (a) is the 'power method' and (b) is known as the 'stress-activation model':

Fig. 6.32 Microstructure of MgO before creep deformation (50 % nitric acid etch; X250) [53]. With kind permission of John Wiley and Sons

$$\dot{\varepsilon} = K_1 \sigma^n \tag{6.24}$$

$$\dot{\varepsilon} = K_2 \exp(\alpha\sigma) \tag{6.25}$$

K_1 and K_2 are material-related parameters and α is the stress-activation constant. The values of n and α are indicated on the curves. Clearly, α is given as psi^{-1}. The curves in Fig. 6.33 are isothermal and the lines obtained by least-square regression represent the modes mentioned in Eqs. (6.24) and (6.25), including a thermal activation to calculate the activation energy. Thus, the above relations may be written as:

$$\dot{\varepsilon} = K_3 \sigma^n \exp\left(-\frac{Q_1}{RT}\right) \tag{6.26}$$

and

$$\dot{\varepsilon} = K_4 \exp(\alpha\sigma) \exp\left(-\frac{Q_2}{RT}\right) \tag{6.27}$$

Both models for presenting the experimental data are equally satisfactory, as may be inferred from the values of the activation energies for creep: $Q_1 = 111 \pm 12$ kcal/mole for the power law and $Q_2 = 105 \pm 12$ kcal/mole for the activation law. It has been claimed that creep in MgO appears to be associated with extensive grain-boundary sliding. Certainly, grain-boundary sliding is dependent on the time, applied stress and temperature of creep. Therefore, additives are required to pin-down grain-boundary sliding, as much as possible. Among the constituents in Fig. 6.30, iron is present in different amounts, as indicated in the plots on Fig. 6.34 also (where the strain rate vs. time is shown).

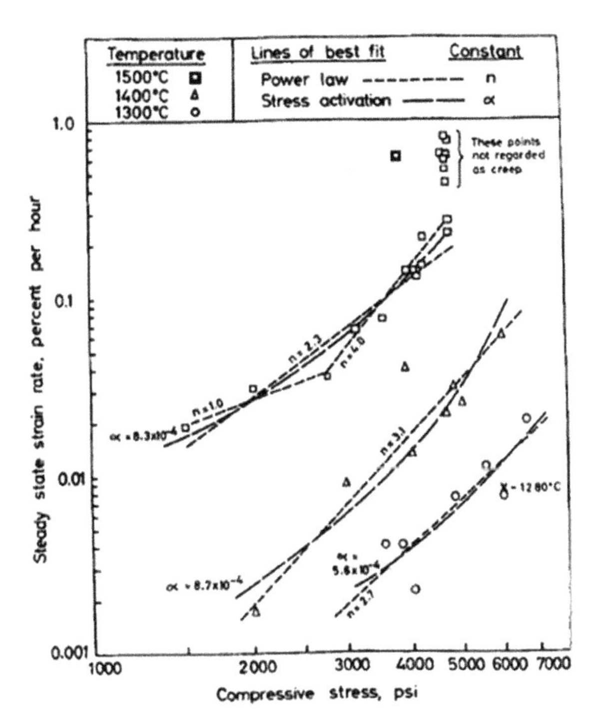

Fig. 6.33 Creep rate measurements and computed lines of best fit [53]. With kind permission of John Wiley and Sons

The curves in Fig. 6.1a and b express the exponential decay of the strain rate during the transient stage, which may be expressed as in Eq. (6.22), which takes into account the initial strain rate, $\dot{\varepsilon}_I$, and a characteristic relaxation time, τ. The stress relaxation process in transient creep may be expressed by incorporating activation energy, ΔH, as given by Eq. (6.28):

$$\frac{1}{\tau} = \frac{1}{\tau_0} \exp - \frac{\Delta H}{RT} \tag{6.28}$$

Plotting this relation provides the activation energy for the relaxation as being $\approx$44 kcal/mol, which may be seen in Fig. 6.35, where $1/\tau$ is plotted versus 1/T. On the basis of the above data, it seems that transient creep in polycrystalline MgO is a diffusion-controlled process.

At this point, it is constructive to observe the dislocation structure in MgO polycrystalline ceramics, as revealed by TEM. In Fig. 6.36, a typical dislocation structure is shown for creep deformation, not unlike that found in metals, with the presence of subgrains, in which a 3D dislocation network may be seen. Note that the long dislocation segments seldom run in a straight line from one node to another, but are bowed out, often in only one plane, though, sometimes, as seen in Fig. 6.36a,

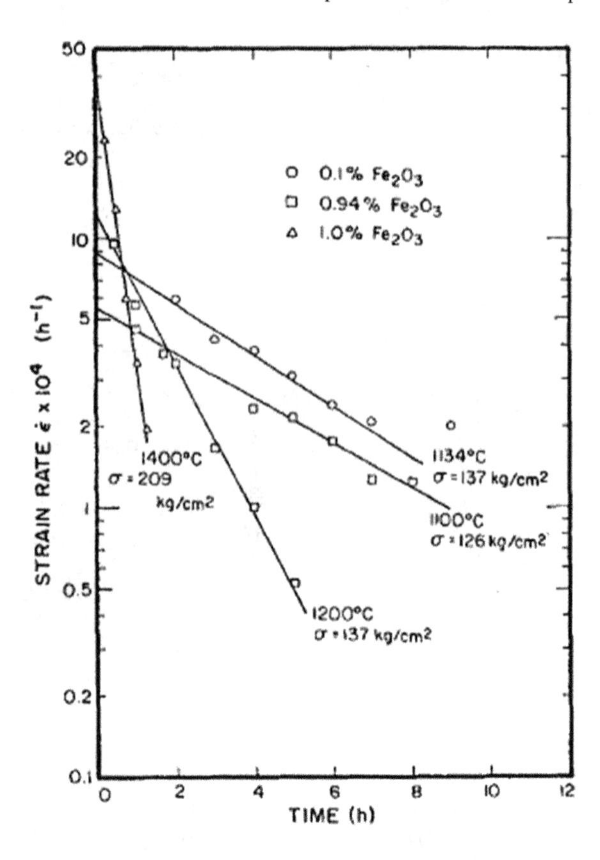

Fig. 6.34 Exponential decay of transient creep in MgO–Fe_2O_3 solid solutions [49]. With kind permission of John Wiley and Sons

apparently in more than one plane. In Fig. 6.36b, the 3D network structure is more obvious, because of the shorter segment lengths. The subgrain boundaries often consist of somewhat irregular hexagonal networks, as seen in Fig. 6.37.

To conclude the subject of creep in MgO polycrystalline ceramics, notice that the vast quantity of experimental data on this ceramic in the literature show marked differences between the various sources. Some of the data discrepancies observed in various investigations on MgO may be attributed to the different methods used for specimen preparation, different densities and grain sizes and, particularly, different loading levels and geometries (torsion, tension, compression and bending). Therefore, the message of this section is that the best and most effective way to study creep is under similar conditions.

As in the previous discussion on single crystals, this section concludes with a consideration of creep in polycrystalline $SrTiO_3$. Experimental data on the high-temperature deformation of $SrTiO_3$ are fairly limited and most of the studies were conducted on single-crystal specimens. The deformation experiments presented below were performed at 1200–1345 °C by compression at strain rates in the range

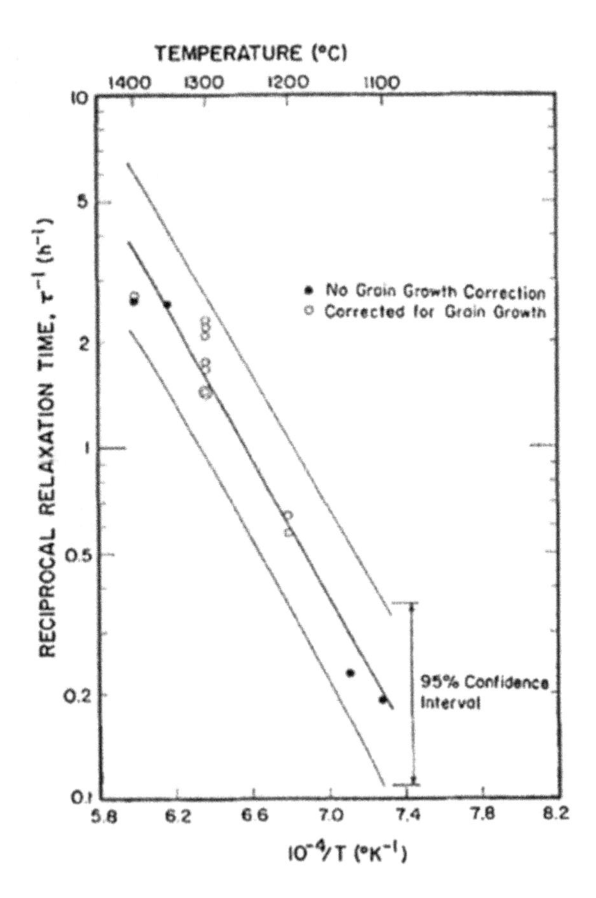

Fig. 6.35 Arrhenius plot of reciprocal relaxation time [49]. With kind permission of John Wiley and Sons

5×10^{-6} to 5×10^{-5} s^{-1}. The steady-state flow stresses were 0.05–30 MPa and increased with increasing strain rates. From the experimental data, one may infer that creep deformation is a diffusion-controlled mechanism.

In Fig. 6.38, the strain-rate dependence on flow stress is shown for the indicated temperatures. The stress exponents at each temperature are also shown on the plots. A steady-state creep equation was used to define the dependence of strain rate on the flow stress, as given in Eq. (6.4), resulting in an activation energy of 628 ± 24 kJ/mol for diffusion-controlled deformation.

The microstructure of an as-fabricated $SrTiO_3$ with a density of $\approx$95 % is shown in Fig. 6.39. The grains are relatively equiaxed with few entrapped pores. The average grain size is $\approx$6 μm. The stress exponents are $\sim$1, except the one related to the 1200 °C plot. The explanation of the higher stress-exponent value in the tests on $SrTiO_3$ at 1200 °C plot is not clear.

This higher value may be due to the onset of microfracture or a change in the deformation mode. A value of $\approx$1 indicates diffusion-controlled flow. The

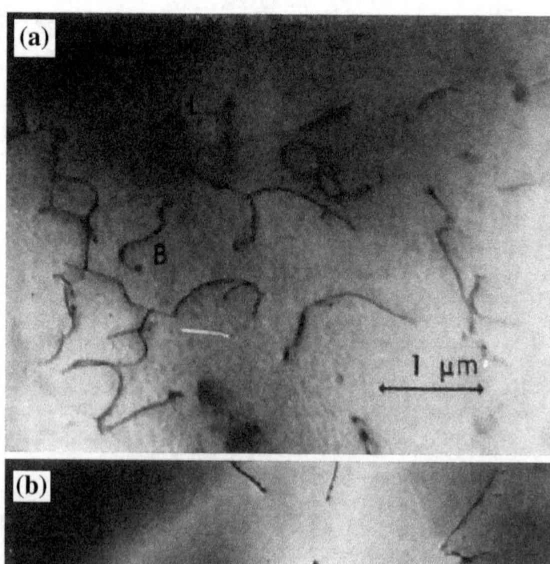

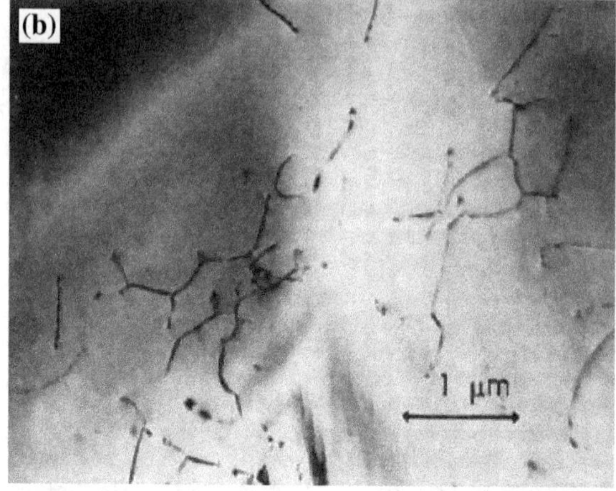

Fig. 6.36 Typical dislocation structures. In **a** loops, L, and bowed-out dislocations, **b** are visible [28]. With kind permission of John Wiley and Sons

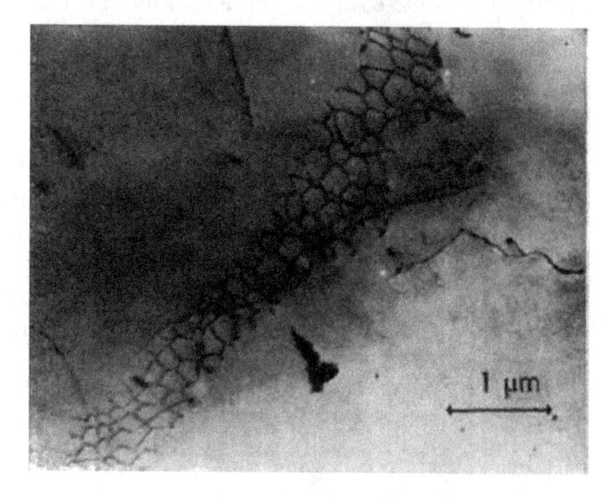

Fig. 6.37 Subgrain boundary consisting of hexagonal network [28]. With kind permission of John Wiley and Sons

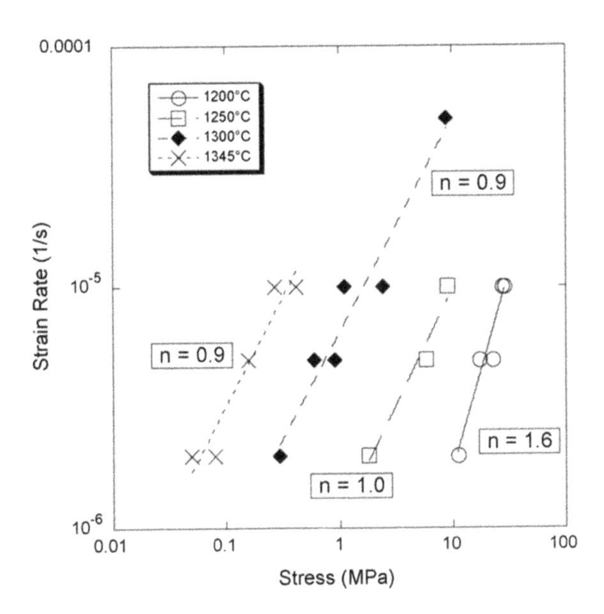

Fig. 6.38 Strain-rate dependence on flow stress for $SrTiO_3$ samples at various test temperatures [21]. With kind permission of Elsevier

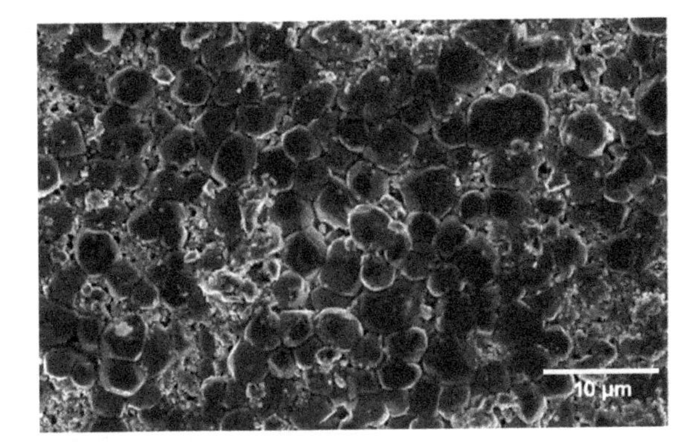

Fig. 6.39 SEM photomicrograph of polished and thermally etched as-fabricated $SrTiO_3$ [21]. With kind permission of Elsevier

activation energy, based on plots such as Eq. (6.4), may be evaluated by expressing the data in an Arrhenius-type relation, as shown Fig. 6.40. SEM and TEM microstructures of the deformed $SrTiO_3$ at 1350 °C are illustrated in Figs. 6.41 and 6.42, respectively. These samples were deformed up to a 5 % strain. Grain-boundary sliding was identified as the principal deformation mechanism. The absence of cavitation and grain-shape changes are consistent with grain-boundary sliding as the principal deformation mechanism.

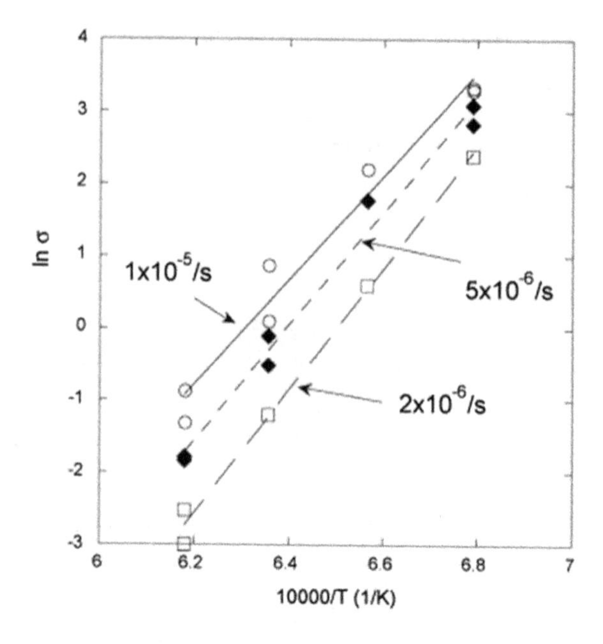

Fig. 6.40 Flow stress in $SrTiO_3$ as a function of test temperature at constant strain rates [21]. With kind permission of Elsevier

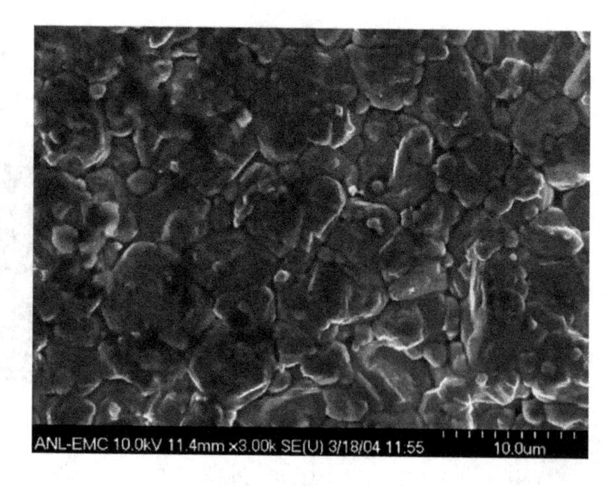

Fig. 6.41 SEM micrograph of $SrTiO_3$ deformed at 1350 °C to $e = 0.05$ [21]. With kind permission of Elsevier

6.5 Superplastic Ceramics

The subject of superplasticity has been extensively discussed in Chaps. 2 (Sect. 2.2) and 5 (Sect. 5.23); therefore, only material relevant to creep will be discussed here. The power law given in Eq. (5.2), rewritten here as:

$$\dot{\varepsilon} = \alpha\sigma^n \tag{6.29}$$

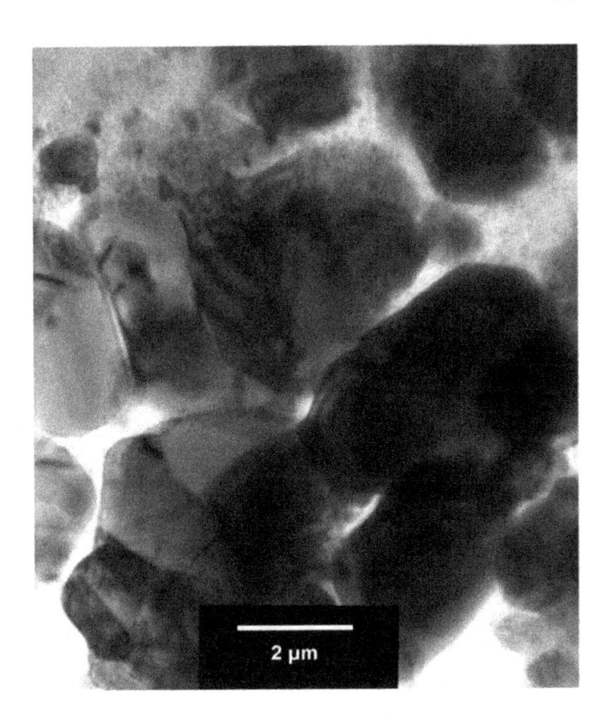

Fig. 6.42 TEM micrograph of $SrTiO_3$ deformed at 1350 °C to $e = 0.05$ [21]. With kind permission of Elsevier

is often used to describe creep and superplasticity, as well. α and n are material constants. Values above one are commonly related to dislocation creep, whereas n = 1 is usually associated with viscous flow in fluids, known as 'Newtonian flow'. A semi-empirical relationship between strain rate and steady-state flow stress is given by:

$$\dot{\varepsilon} = A\sigma^n d^{-p} \exp(-\frac{Q}{RT}) \tag{6.30}$$

This relation determines the activation energy. A is the material constant, σ is the steady-state flow stress, d is the grain size, n and p are the stress and grain exponents, respectively, and Q is the activation energy for superplastic flow. Thus, it may be seen from this equation that the parameters n, p and Q play a role in the deformation mechanism. Superplasticity is considered to be similar to 'diffusion creep'.

'Diffusion creep' refers to the deformation of crystalline solids by the diffusion of vacancies through their crystal lattice and results in plastic deformation, rather than the brittle failure of the material. In fine-grained materials, tested in a high-stress regime, deformation inside the grains is fast and grain-boundary sliding has little effect, leading to stress exponents typical of dislocation creep. Under lower stresses and with sufficiently high temperatures, grain boundaries slide freely and the creep rate is higher than that predicted by only taking into account the effect of dislocation creep. The microstructures of some selected superplastic ceramics are shown in Fig. 6.43.

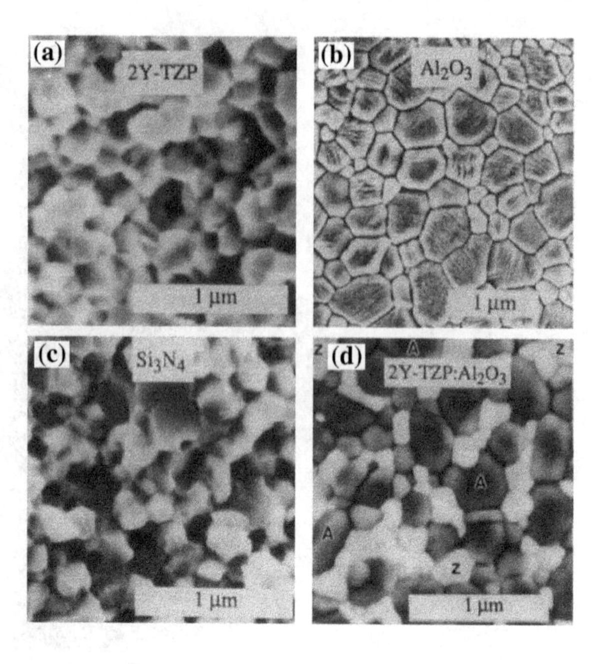

Fig. 6.43 Scanning electron microscopy micrographs of ultrafine grains of superplastic ceramics: **a** 2Y-TZP, **b** alumina, **c** silicon nitride, and **d** 2Y-TAP/alurnina at equal volume fraction [4]. With kind permission of John Wiley and Sons

One ceramic material which may serve to represent superplastic creep behavior is yttria-tetragonal ZrO_2 polycrystals (Y-TZP), illustrated below. Figure 6.44 describes superplastic deformation in this material, though some differences should be noted: (a) the grain-shape change after large deformation is very small; (b) the contribution of grain-boundary sliding is quite large and; (c) the stress and grain-size exponents are often substantially different from those predicted by simple, diffusion creep models.

Although the deformation mechanisms in superplasticity are similar to those in diffusional creep, certain circumstances in interface-controlled deformation may arise when grain boundaries do not act as perfect sources and sinks for vacancies or they do not slide or migrate freely. In such a case, the strain rate may increase nonlinearly with stress and, sometimes, the materials may even exhibit threshold-deformation stress. For example, interface reactions involving dislocation glide or climb result in a quadratic dependence of strain rate on stress, whereas those involving nucleation may show an exponential dependence on stress. The collected strain-rate data comes from various references indicated in the figures, but not listed in the references to this chapter (refer to the work of Chen and Xue [4]). Figure 6.44 shows the stress exponent as influenced by grain size for three ceramics.

Equation 6.29 may be expressed differently for superplastic ceramics, to include the effect of grain size, as seen in Eq. (6.31):

$$\dot{\varepsilon} = A\sigma^n / d^p \tag{6.31}$$

A is a temperature dependent diffusion related coefficient.

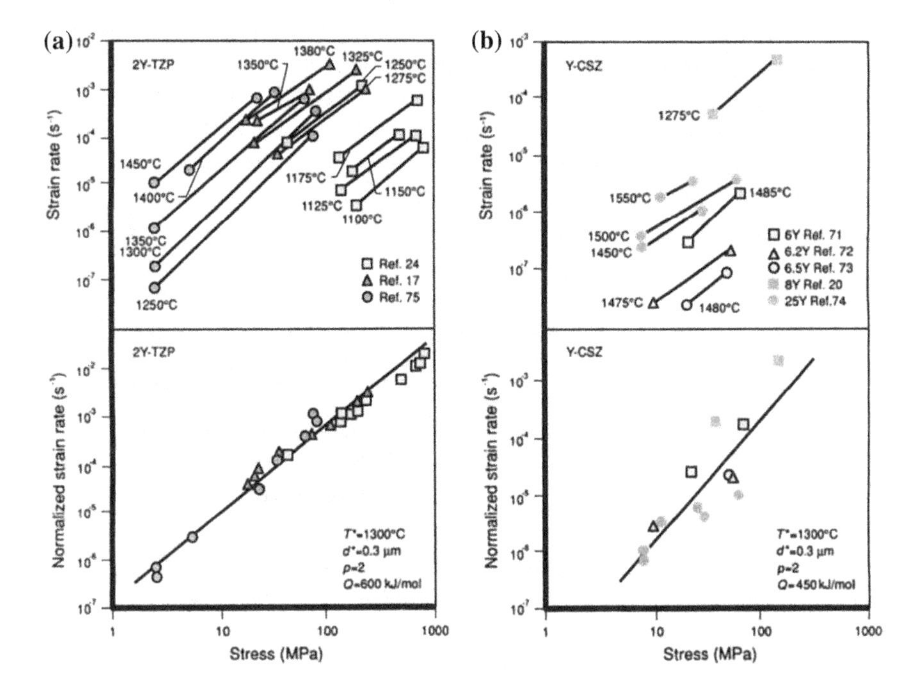

Fig. 6.44 **a** Relationship between strain rate and stress for 2Y-TZP. Grain size is 0.21 μm for Ref. 24, 0.39 μm for Ref. 17, and 0.48 μm for Ref. 75. Data are replotted in the lower figure after normalization to a reference temperature T* and a reference grain size d* using values of grain-size exponent p and activation energy Q, as shown. Very good agreement between different data sets is apparent. **b** Relationship between strain rate and stress for Y-CSZ. Grain size for Ref. 71 is 17 μm, Ref. 72 is 17.5 μm, Ref. 73 is 29 μm, Ref. 20 is 0.48 μm, and Ref. 74 is 2.6, 3.5, and 4.2 μm for 1450, 1500, and 1550 °C, respectively. Compositions of yttria are indicated as well. When the data are normalized, as in Fig. 6.44a, they fall on a single line. Data of the fine-grained Y-CSZ are comparable to those of 2Y-TZP, shown in Fig. 6.44a [4]. With kind permission of John Wiley and Sons

In Fig. 6.44, the grain-size exponent is also shown at a value of $p = 2$. Also note that the additives have an important effect on limiting grain growth. Actually, solute segregation or particle pinning may also hinder diffusional creep by restricting grain-boundary sliding or migration, that are generally necessary for grain-to-grain accommodation. In addition, one may observe (in Fig. 6.45) that the stress exponents tend to decrease with increasing grain size. An increase in grain size is favorable for creep resistance, but reduces the superplasticity of the ceramics and, therefore, reduces the formability of the material into desirable shapes. Therefore, it is a common technological practice to execute the forming operation while the ceramic grain size is very small and may be superplastic, after which it is heat-treated for grain growth to induce creep resistance.

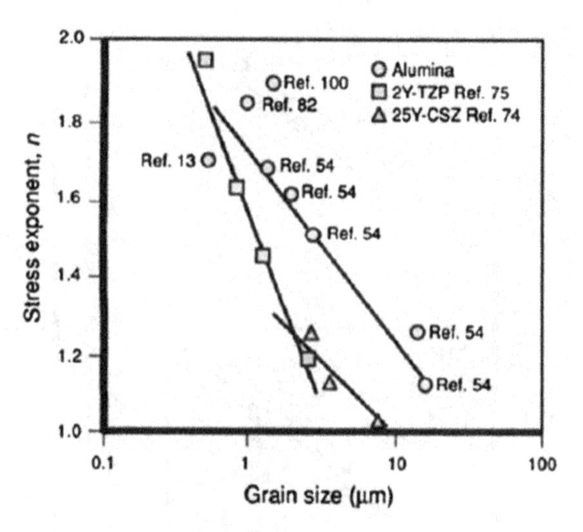

Fig. 6.45 Stress exponent versus grain size for three ceramics [4]. With kind permission of John Wiley and Sons

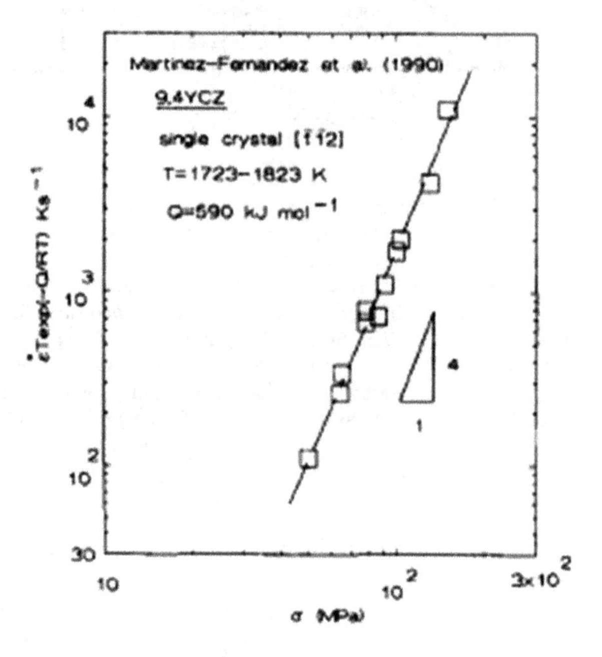

Fig. 6.46 Variation of temperature compensated strain rate with stress for 9.4 mol% yttria stabilized cubic zirconia [6]. With kind permission of Elsevier

Note that many early studies on the subject attributed a stress exponent, n, to some form of intragranular dislocation creep. At 9.4 %, yttria stabilizes cubic zirconia single crystals. The compressive creep characteristics tested in the $[\bar{1}\bar{1}2]$ orientation appear in Fig. 6.46. Here, the strain is normalized for tests in the temperature range 1723–1823 K using an activation energy of 590 kJ mol^{-1}. The data are represented by a straight line with a stress exponent of n $\sim$ 4. In this case, the deformation is attributed to dislocation-climb-controlled intragranular creep (Fig. 6.47).

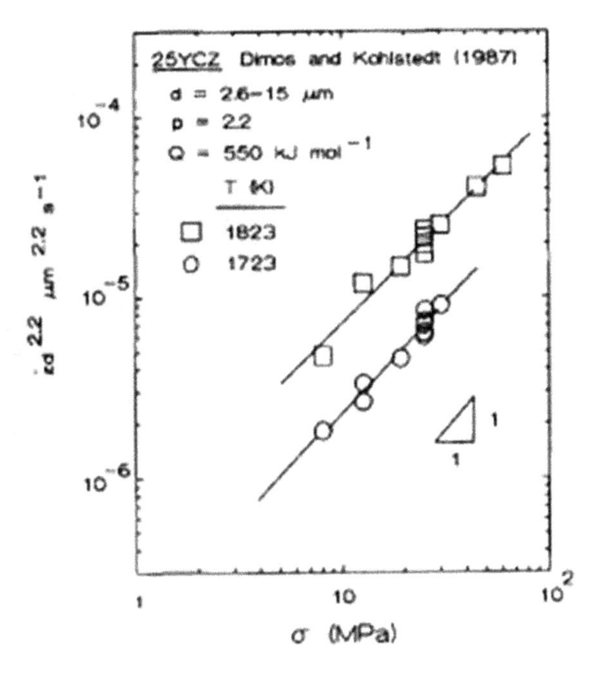

Fig. 6.47 Variation in grain size compensated strain rate with stress for a 25 mol% yttria stabilized cubic zirconia [6]. With kind permission of Elsevier

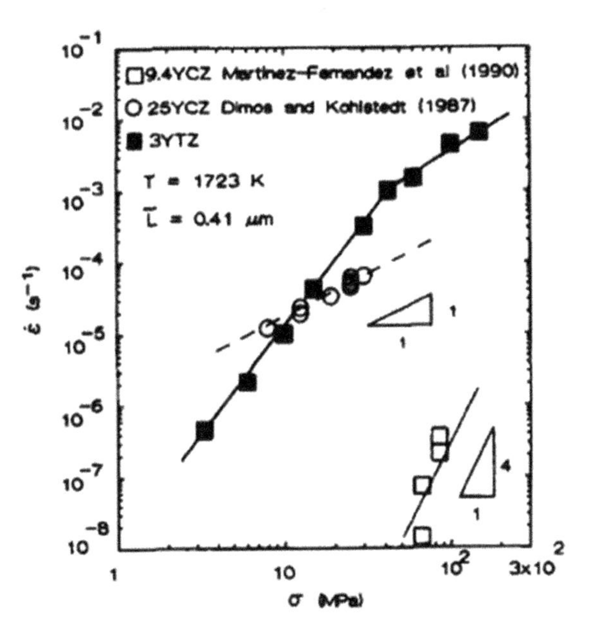

Fig. 6.48 Comparison of experimental data on the superplastic zirconia with the dislocation and diffusion creep data on cubic zirconia [6]. With kind permission of Elsevier

The experimental data on diffusion creep in 25 mol% yttria stabilized cubic zirconia may be seen in Fig. 6.44, where a grain-size, normalized strain rate versus stress is illustrated. The data may be expressed as Nabarro-Herring diffusion creep

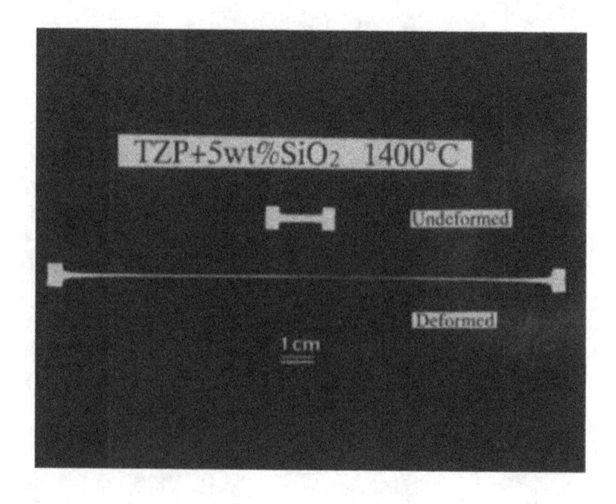

Fig. 6.49 The largest elongation to failure of 1038 % in a superplastic zirconia-5 % silica ceramic [6]. With kind permission of Elsevier

with n ≈ 1, p = 2.2 and Q = 550 kJ mol^{-1}. In Fig. 6.48, the experimental results for dislocation and diffusion creep in cubic zirconia are compared with superplastic 3YTZ having a grain size of 0.4 μm at 1723 K. These data were normalized to a grain size of 0.41 μm (at 1723 K) using p = 2.2.

To conclude this section on superplasticity in yttria stabilized zirconia, it may be of interest to show the amount of strain expressed as elongation in superplastic 5 wt% SiO_2 TZP, in which the extent of elongation is ~1038 % at 1400 °C (Fig. 6.49).

6.6 Mechanisms of Creep

There are several basic mechanisms that may contribute to creep in materials (and ceramics are no exception:

(i) dislocation slip;
(ii) climb, leading to subgrain formation;
(iii) grain-boundary sliding; and
(iv) diffusion flow caused by vacancies.

A short summary of the above contributions to creep follows:

(i) Glide-by-slip strengthens materials as they deform. In primary creep, stress is constant, while strain increases to a certain extent (see Fig. 6.1a) over time, but strain rate decreases (Fig. 6.1b) until a minimum strain rate is achieved. This minimum strain rate, on a strain–time plot, represents steady-state creep.
(ii) During steady-state creep, strain increases over time. The increased strain energy stored in the material, due to deformation, together with the high temperature, provide the driving force for the recovery process. There is, therefore,

a balance between the processes of work hardening and recovery. Recovery involves a reduction in dislocation density and the rearrangement of dislocations into lower energy arrays, such as subgrain boundaries. For this to occur, dislocations must climb, as well as slip, and this, in turn, requires atomic movement or self-diffusion within the lattice. Hence, it is often said that the activation energies for self-diffusion and for creep are almost the same.

Vacancies must be located at a site where climb is supposed to occur, to enable climb by means of a vacancy-atom exchange. As the temperature increases, atoms gain thermal energy and the equilibrium concentrations of these vacancies in the metals increase exponentially. In Chap. 3, Sect. 3.2.1, the number of vacancies, n, was given by Eq. (3.9), which may be rewritten (see, for example, Damak and Dienes [7]) as:

$$n = N \exp(-\frac{E_F}{kT}) \tag{6.32}$$

And again, as given in Chap. 3, N is the number of lattice sites and E_F is the energy of vacancy formation. The activation energy, Q, for the jump rate, J, is given by the sum of the energy of vacancy formation and the vacancy's energy for motion, E_M, ($Q = E_F + E_M$):

$$J = J_0 \exp - \frac{Q}{kT} \tag{6.33}$$

J_0 represents the respective entropies. The diffusion coefficient, D, may be given as:

$$D = D_0 \exp - \frac{Q}{kT} \tag{6.34}$$

D_0, the pre-exponential factor, is equivalent to J_0 and Q is the overall activation energy for self-diffusion. The rate of steady-state creep increases with temperature, as does the essential number of vacancies for effective vacancy-atom exchange for climb.

(iii) Grain-boundary sliding is considered in detail in Sect. 6.7.
(iv) Diffusion flow by vacancies must be considered, since the mechanism of creep depends on both temperature and stress. The various methods detailed below involve some sort of diffusion occurring with vacancy-atom exchange. This may occur either by lattice diffusion or grain-boundary diffusion, or both may be involved. Bulk-diffusion-assisted creep occurs during the processes listed in (a)–(d) below, where the kinetics of atom-vacancy exchange occurs due to lattice diffusion. Afterwards, creep, involving grain-boundary diffusion, will be considered (e).

(a) Nabarro-Herring creep;
(b) climb, in which the strain is actually obtained by climb;

(c) climb-assisted glide, in which climb is a mechanism allowing dislocations to bypass obstacles;
(d) thermally-activated glide via cross-slip;
(e) Coble creep, involving grain-boundary diffusion.

Before entering into a detailed discussion of the above list and based on what has been said thus far on the subject, briefly summarized: (a) creep in materials (including ceramics), namely time-dependent plastic deformation, may occur during mechanical stresses well below the yield stress and (b) in general, two major creep mechanisms characterize the time-dependent plastic-deformation process–dislocation creep and diffusion creep. Now, a detailed discussion of paragraphs (a)–(d) follows.

6.6.1 Nabarro-Herring Creep

One type of creep, in which creep is diffusion-controlled, is Nabarro-Herring creep. In this type of creep, atoms diffuse through a lattice, causing grains to elongate along the stress axis. Mass transport (i.e., the diffusion of atoms) takes place in regions ranging from lower to higher tensile stress. A common illustration may be seen in Fig. 6.50. This schematic figure illustrates the flow of vacancies and atomic movements as induced by tensile stress, σ. During creep deformation, vacancy-atom exchanges take place to and from the grain boundaries. One would expect that, during creep under tension, atoms would tend to diffuse from the sides of the specimen in the direction shown in Fig. 6.50 (a counter-flow of vacancies), causing the sides to lengthen. Assume that local equilibrium of the vacancy concentration exists at the boundaries of the crystal when no stresses are acting on it. Also note that grain boundaries serve as vacancy sources or sinks. In this mechanism, lattice diffusion occurs within the grain and the creep rate (strain rate) is assumed to be proportional to the vacancy flux. See below that the strain rate is inversely proportional to the square of the grain size, i.e., $\dot{\varepsilon} \propto \frac{1}{d^2}$ [11, 20, 72]. In Eq. (6.32), the number of vacancies is given. Equation (6.32), in terms of vacancy concentration at equilibrium, is given as:

$$\frac{n}{N} = C_V^0 = \exp(-\frac{E_F}{kT}) \tag{6.35}$$

The energy to create a vacancy under acting stress is given by:

$$E_F + \sigma V \tag{6.36}$$

V is the atomic volume (here, it is the volume of a vacancy) and E_F is defined by Eq. (6.35). There is a small concentration difference in the vacancies between the faces of AB and BC in the above figure, where tensile and compressive stresses are acting, respectively. Denoting the vacancy concentrations at the respective

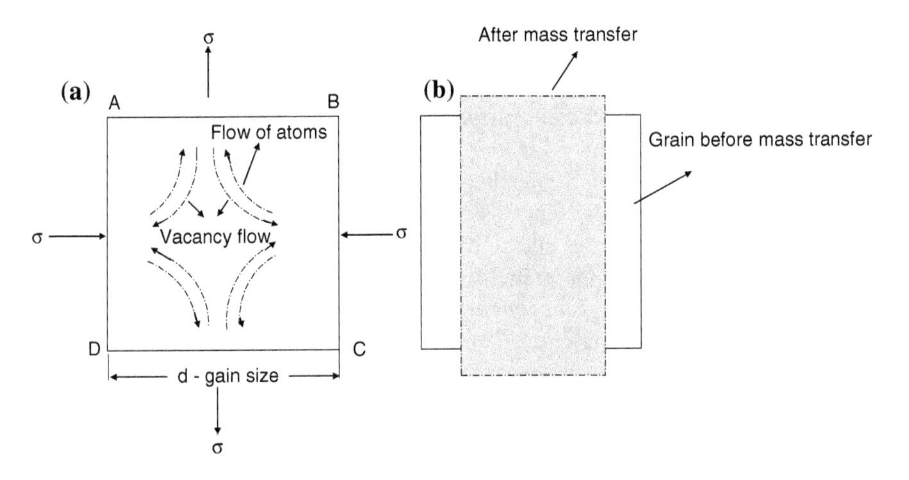

Fig. 6.50 The Nabarro-Herring concept of creep: **a** a schematic of vacancy and mass flow; **b** the elongated grain in the tensile-axis direction after mass flow

faces as C_V^+ and C_V^- and their difference as ΔC, one may write for each of them, by means of Eqs. (6.35) and (6.36), respectively:

$$C_V{}^+ = \exp - \left(\frac{E_F - \sigma V}{kT}\right) = C_V{}^0 \exp\left(\frac{\sigma V}{kT}\right) \tag{6.37}$$

$$C_V{}^- = \exp - \left(\frac{E_F + \sigma V}{kT}\right) = C_V{}^0 \exp\left(-\frac{\sigma V}{kT}\right) \tag{6.38}$$

$$\Delta C = C_V{}^+ - C_V{}^- = \frac{\alpha}{V}\left\{\exp\left(-\frac{E_F}{kT}\right)\left[\exp\left(\frac{\sigma V}{kT}\right) - \exp\left(-\frac{\sigma V}{kT}\right)\right]\right\} \tag{6.39}$$

Clearly, in this relation E_F was replaced by Eq. (6.36). Equations (6.37) and (6.38) represent the local equilibrium concentrations under tension and compression in Fig. 3.50a. Recalling that:

$$\sin h\, x = \frac{1}{2}[\exp(x) - \exp(-x)],$$

Equation 6.39 may be rewritten as:

$$\Delta C = \frac{2\alpha}{V}\exp\left(-\frac{E_F}{kT}\right) \sin h\left(\frac{\sigma V}{kT}\right) = \frac{2\alpha}{V} C_V{}^0 \sin h\left(\frac{\sigma V}{kT}\right) \tag{6.40}$$

where $C_V^0 = \exp(-E_F/kT)$ and E_F is the energy of vacancy formation in the absence of stress.

As indicated, there is a flow of atoms from the tensile to the compressed faces and an opposite flow of vacancies. When a concentration gradient exists, diffusion flux will occur. This flux of vacancies may be expressed as:

$$J = -D_V \nabla C = -\frac{\alpha D_V(\Delta C)}{d} \tag{6.41}$$

D_V is the diffusion coefficient of the vacancies and α is a geometrical factor. The corresponding transport of matter occurs in the opposite direction and produces a creep strain under the applied stress. In a unit time, Jd^2, atoms in the crystal leave the faces under compression and are added to the faces under tension. (Recall that J is the number of atoms in a unit time per unit area; thus, multiplying this value by the square of the grain size, d, one gets the number of atoms per unit time). Consequently, the grain lengthens in the tensile-axis direction and gets thinner in the transverse direction. The change in grain size may be written as:

$$\Delta d = \frac{(Jd^2)V}{d^2} = JV \tag{6.42}$$

where V is the atomic volume (often given as Ω). The strain rate is given as:

$$\dot{\varepsilon} = \frac{\Delta d}{d} = \frac{JV}{d} \tag{6.43}$$

An expression for the strain rate, given by Eq. (6.44), is obtained by substituting the value of Δd from Eq. (6.42) into Eq. (6.43), followed by inserting J from Eq. (6.41) into Eq. (6.43) to get:

$$\dot{\varepsilon} = \frac{\alpha D_V \Delta C}{d}\frac{V}{d} = \frac{\alpha D_V \Delta C V}{d^2} \tag{6.44}$$

With Eq. (6.40) substituted into Eq. (6.44), it is possible to write:

$$\dot{\varepsilon} = \frac{2\beta}{V}\frac{D_V V}{d^2}\exp(-\frac{E_F}{kT})\sinh(\frac{\sigma V}{kT}) \tag{6.45}$$

For small values of stress, and since the nominator is always smaller than the denominator, the quotient is small and $\sinh(\sigma V/kT) = \sigma V/kT$. Substituting this value into Eq. (6.45), one obtains:

$$\dot{\varepsilon} = \frac{2\beta D_V}{d^2} {C_V}^0 \frac{\sigma V}{kT} \tag{6.46}$$

D_V is the diffusion coefficient of the vacancies. $D_V C_V^0$ is, D_S, the self-diffusion coefficient. Thus, Eq. (6.46) may also be expressed as:

$$\dot{\varepsilon} = \frac{2\beta D_S}{d^2}\frac{\sigma V}{kT} \tag{6.47}$$

More exact calculations, in terms of shear strain (i.e., $\gamma = 2b/d$) and macroscopic shear stress, τ (i.e., $\sigma = \beta\tau$ and β is close to unity and recalling that the shear stress at 45° is given by $\tau = \sigma\sqrt{2}$) gives:

$$\dot{\gamma}_S = \frac{32\alpha\beta D_S \tau V}{\pi d^2}\frac{1}{kT} \tag{6.48}$$

This relation defines a simple, ideal, viscous solid. One sees that increasing grain size reduces creep rate. Creep-rate change is proportional to d^{-2}. Nabarro-Herring creep is a low-stress and high-temperature process.

A somewhat alternate method for showing that $\dot{\varepsilon} \propto \frac{1}{d^2}$ is as follows. Based on Eqs. (6.35) through (6.38), the difference in concentration may be expressed as:

$$\Delta C = C_V{}^{+} - C_V{}^{-} = \frac{\alpha}{V}\left\{\exp\left(-\frac{E_F}{kT}\right)\left[\exp(\frac{\sigma V}{kT}) - \exp\left(-\frac{\sigma V}{kT}\right)\right]\right\} \tag{6.39}$$

The flux of the vacancies, going from the tensile to the compressive regions, is:

$$J_V = -D_V\frac{\Delta C}{\Delta x} \tag{6.49}$$

where Δx is the distance in the x direction, so that $\Delta C/\Delta x$ is a gradient. Bear in mind that the atomic flux, J, is in the opposite direction to the vacancy flux, J_V, and, therefore, $D\Delta C = -D_V\Delta C_V$. In our case, the diffusion distance is l.

Stress is not constant along the grain faces, therefore, the diffusion paths are shorter near the corners. Due to stress relaxation, one may assume that $\sigma = \beta\sigma_S$ at distance d/4 from the boundaries (when σ_S is the macroscopic shear stress and β is nearly unity). The length of the diffusion path through this point is $l = \pi/2(d/4)$. The atomic flux across the area of a single atom is given by:

$$J = \alpha D_V\frac{\Delta C}{l} = \alpha D_V\frac{8\Delta C}{\pi d} \tag{6.50}$$

The previous expression is the result of substituting for the value of $l = \pi/2(d/4)$. D_V is the diffusivity of the vacancies. One may rewrite Eqs. (6.42) and (6.43) as:

$$\Delta d = \frac{(Jd^2)V}{d^2} = JV \tag{6.42}$$

$$\dot{\varepsilon} = \frac{\Delta d}{d} = \frac{JV}{d} \tag{6.43}$$

Substituting from Eq. (6.49) for J yields:

$$\dot{\varepsilon} = \alpha D_V\frac{8\Delta C}{\pi d}\frac{V}{d} \tag{6.44}$$

and from Eq. (6.40):

$$\Delta C = \frac{2\alpha}{V}\exp(-\frac{E_F}{kT})\sinh\left(\frac{\sigma V}{kT}\right) = \frac{2\alpha}{V}C_V{}^0\sinh(\frac{\sigma V}{kT}) \tag{6.40}$$

When the argument in the hyperbolic function is small, as mentioned earlier, it is equal to the argument itself; thus, for the strain rate, one may write:

$$\dot{\varepsilon} = 16\alpha\frac{D_V C_V{}^0}{\pi d^2}\frac{\sigma V}{kT} = \frac{16\alpha D_S\sigma V}{\pi d^2 kT} \tag{6.51}$$

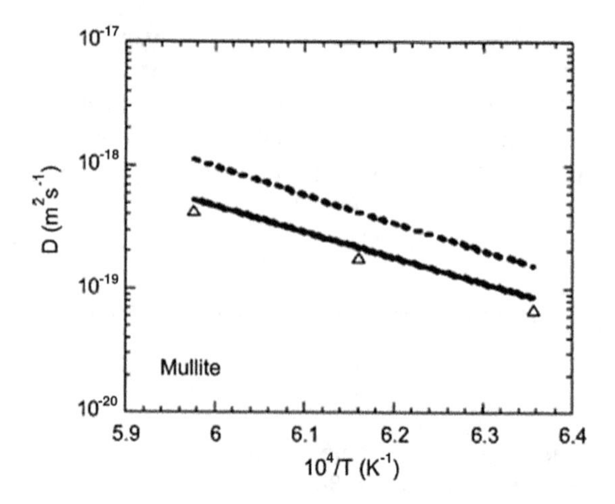

Fig. 6.51 Data for diffusion coefficient for pure mullite calculated from the Nabarro–Herring model (*triangles*) and oxygen diffusion (*bold line*—Ikuma et al., and *dashed line* [40]. With kind permission of Elsevier

D_S is the self-diffusion coefficient and is equal to $C_V^0 D_V$. Again, the strain rate is proportional to d^{-2}.

One sees in Eqs. (6.48) and (6.51) that the strain rate is linearly proportional to the stress and inversely proportional to the grain size. In Eq. (6.48), the expression is given in terms of shear strain and macroscopic shear stress. The above expressions explain why large-grained materials are preferential for creep applications at high temperatures.

As mentioned previously, one of the bulk-diffusion-assisted creeps occurs in the Nabarro-Herring model, though the Coble creep mechanism is also diffusion-assisted. As such, the interpretation of creep results is, to a large extent, chosen by the researchers. In mullite type ceramics, one of the interpretations of the creep results is in line with the Nabarro-Herring model, according to a modified Eq. (6.51) [55, 40]. use:

$$\dot{\varepsilon} = \frac{14\sigma\Omega D_{eff}}{d^2kT} \tag{6.51a}$$

as one of the models, which is consistent with the observed stress dependence and microstructural observations in diffusional creep. In Eq. (6.51a), Ω is equivalent to V in Eq. (6.51). In Fig. 6.51, diffusion data are plotted for mullite calculated from the creep data by the Nabarro–Herring model. The creep data of strain rate versus stress is given in Fig. 6.52.

The designations SP, 5G, and 9G refer, respectively, to pure mullite, 5 and 9 % Y_2O_3 additions to mullite and 5C containing $Y_2Si_2O_7$. Photo micrographs and SEM microstructures are illustrated in Figs. 6.53 and 6.54, respectively.

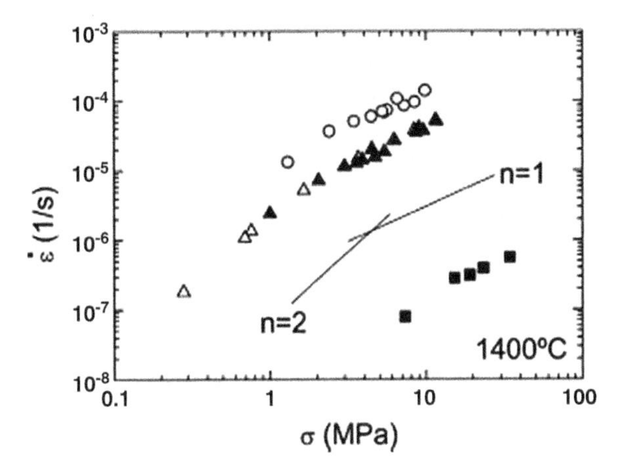

Fig. 6.52 Strain rate versus stress plot. SP Mullite, *filled square*; 5G Mullite, *open triangles*; 5C Mullite, *filled triangles*; and 9G Mullite, *open circles* [40]. With kind permission of Elsevier

Fig. 6.53 **a** TEM photomicrograph of SP Mullite, **b** secondary electrons, and **c** back-scattered electrons SEM photomicrographs of mullite–5 wt% Y_2O_3 sintered for 3 h at 1550 °C [40]. With kind permission of Elsevier

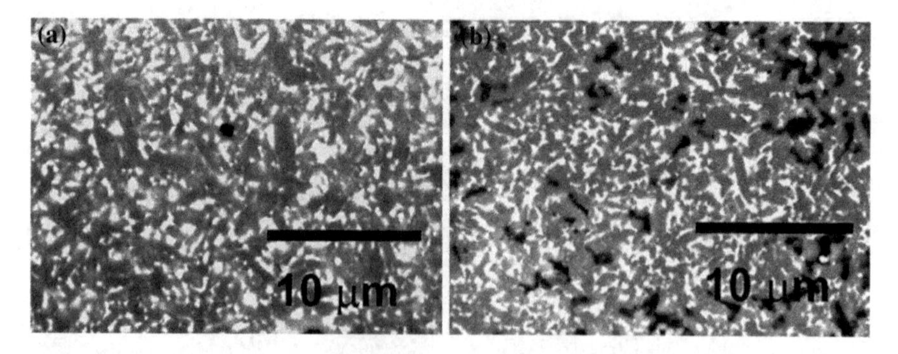

Fig. 6.54 Back-scattered electrons SEM photomicrograph of deformed **a** 5G (54 % strain) and **b** 5C (18.2 % strain) composites. The grain morphology is unchanged. Cavities are observed in 5C composite [40]. With kind permission of Elsevier

6.6.2 Dislocation Creep and Climb

Bulk-diffusion-assisted creep occurs in the processes listed above, namely in (b) climb; (c) climb-assisted glide and; (d) thermally-activated glide via cross-slip. All these are obviously associated with dislocation motion. High stress, below yield stress, causes creep by conservative dislocation motion, namely by dislocation glide within its slip plane. This readily occurs at high temperatures above 0.3 T_M for pure metals and at about 0.4 T_M for alloys, where the dependence on strain rate becomes quite strong. For ceramics, T > 0.4–0.5 T_m (K). A formulation used for such creep is:

$$\dot{\gamma} \sim \left(\frac{\sigma_S}{G}\right)^n \tag{6.52}$$

where n has a value of 3–10 in high-temperature regimes. Since n is in the exponent, this creep is referred to as 'power-law creep'. At high temperatures, obstacle-blocked dislocations can climb, not only glide. If gliding dislocations are blocked by some obstacle, climbing may release them to move on until they meet another obstacle, where the same process is repeated. Climb is performed by the diffusion of vacancies through the lattice or along the dislocation core, diffusing into or out of the dislocation core. By climbing, dislocations change their slip planes, enabling them to bypass their obstacles. Dislocation glide is responsible for most strain, while the average dislocation density is determined by the climb step in the deformation process. This mechanism is known as 'climb-controlled creep'.

6.6.3 Climb-Controlled Creep

At relatively high stresses, beyond the elastic region or the shear moduli, creep is controlled by dislocation-glide movement and by glide in adjacent planes following climb. Real materials contain various internal obstacles (such as dislocations) or external ones (introduced intentionally, such as solutes and particles, or unintentionally by the fabrication process) which block dislocation glide in their respective slip planes. Dislocation motion is also hindered by the crystal structure itself, namely by crystal resistance, an internal obstacle. At high temperatures, obstacle-blocked dislocations can be released by dislocation climb. Creep arises as a consequence of climb, when further deformation by glide is enabled by means of vacancy-atom exchange. The creep rate is a function of several factors, usually given as:

$$\sum \dot{\varepsilon} = f(\sigma, T, S, GS, P)$$

S is the structure, GS is grain size and P represents the material properties, such as the lattice parameter, atomic volume, etc. Vacancies increase with increasing temperature and are likely to diffuse into dislocations, thus, decreasing the overall free energy of the system. By the diffusion of vacancies to locations at which dislocations are blocked by obstacles, climb becomes possible, letting the dislocations bypass those obstacles. Climb allows further glide in an adjacent slip plane to occur and, by such deformation, creep strain arises.

A steady-state-based model for edge-dislocation climb [11, 72] was suggested by Weertman [89]. He assumed that strain hardening occurs whenever dislocations are hindered in their motion by some obstacle and pile up behind it. The dislocations beyond the barrier, such as a Lomer-Cottrell lock, may escape by climbing. However, climb beyond Lomer-Cottrell barriers would lead to the generation of new dislocation loops and to a steady-state creep rate (which is applicable to FCC and BCC structures, but not to HCP ones). Weertman [89] also suggested that edge dislocations with opposite signs, gliding on parallel slip planes, would interact and pile up when a critical distance of 2r between them is not exceeded. In such a case, as in the prior case, dislocations could escape from the piled-up array by means of climb. Dislocation pile-ups lead to work hardening, whereas climb is a recovery process. A steady state is reached when the hardening and recovery rates are equal. The creep rate will, therefore, be controlled by the rate at which dislocations can climb. This climb mechanism requires the creation of vacancies or their destruction at the obstacle-blocked dislocations (in this case, at the pile-up) in order to maintain the equilibrium concentration required to satisfy the climb rate. At the tip of a pile-up dislocation, a non-vanishing, hydrostatic stress $\pm\sigma_i$ may develop, exerting a force on the dislocation in a normal direction to the slip plane and causing a positive (up) or negative (down) climb. Vacancies will be absorbed where the stress is compressive and will be created where the stress is tensile. A change in vacancy concentration develops in the vicinity of the dislocation line and a vacancy flux is established between the segments of the dislocations, acting as sources or segments of sinks.

The vacancy concentration, C_e, in equilibrium with the leading dislocation in the pile-up, is given by:

$$C_e = C_o \exp\left(\frac{\pm 2L\sigma_S^2 b^2}{GkT}\right) \tag{6.53}$$

2L is the length of the dislocation pile-up and C_0 is the equilibrium concentration of the vacancies in a dislocation-free crystal. The vacancy concentration at a distance, r, from each pile-up is assumed to be equal to C_0. The rate of climb, $\dot{X}$, is given (Garofalo) as:

$$\dot{X} = \frac{2C_0 D_V \sigma_S^2 L b^4}{GkT} \tag{6.54}$$

D_V is the vacancy-diffusion coefficient and $2Lb^2\sigma_S^2/GkT < 1$.

When self-diffusion occurs due to the vacancy mechanism, C_0D_V may be replaced by:

$$C_0 D_V = D_S = \frac{v}{b}\exp(\frac{\Delta S}{R})\exp(-\frac{\Delta H}{RT}) \tag{6.55}$$

and $\dot{X}$ is given by:

$$\dot{X} = \frac{2\sigma_S^2 L b^3}{GkT} v \exp(\frac{\Delta S}{R})\exp(-\frac{\Delta H}{RT}) \tag{6.56}$$

ΔH is the activation energy for self-diffusion, v is a frequency factor and S is an entropy term. Equation (6.56) is obtained under the assumption that vacancies are easily destroyed or created and that an equilibrium concentration exists between pile-ups of dislocations. However, the diffusion of flux vacancies may be different in specific climb processes.

In an additional model created by Weertman [89], the rate of dislocation climb is also given by Eqs. (6.54) and (6.56) and the steady-state creep-rate model, in this case, becomes:

$$\dot{\gamma} = NAb\frac{\dot{X}}{2r} \tag{6.57}$$

N is the density of the dislocations participating in the climb process (or the density of the sources), A is the area swept out by a loop in a pile-up and 2r is the separation between those pile-ups. The stress necessary to force two groups of dislocation loops to pass each other on parallel slip planes must be greater than $\frac{Gb}{4\pi\sigma_S}$ (in terms of shear stress it is $\frac{Gb}{4\pi\tau}$). When this relation is satisfied, an estimate for r may be made:

$$r = \frac{Gb}{4\pi\sigma_S} \tag{6.58}$$

The probability, p, of blocking the dislocation loops generated from one source by means of loops emanating from three other sources is given by:

$$p = \frac{8\pi NL^2 r}{3} = \frac{2NL^2 Gb}{3\sigma_S} \tag{6.59}$$

Using Eqs. (6.54) and (6.57)–(6.59) and setting p = 1 and A = $4\pi l^2$, the creep rate at low stresses becomes:

$$\dot{\gamma}_S = \frac{C\pi^2\sigma_S^{4.5}D_S}{\sqrt{bN}G^{3.5}kT} \tag{6.60}$$

C is a numerical constant on the order of 0.25 and D_S is the coefficient of self-diffusion.

Equation (5.54) has been substantiated experimentally for pure metals to a greater extent than other theoretical relations. Exceptions to the exponent 4.5 (Eq. (6.60)) were obtained, but this value is very close to the observed experimental values.

6.6.4 Thermally-Activated Glide via Cross-Slip

Edge dislocations climb when their motion is hindered. The non-conservative motion of screw dislocations is by cross-slip, since they cannot climb. The ease of cross-slip is stacking-fault dependent (see Chap. 3). Materials with high SF energy cross-slip readily, but not so when the SF energy is low. For screw-oriented dislocations, the Burgers vector is parallel to the dislocation line (see Chap. 3) and, therefore, it can move in any plane in which it lies (in isotropic materials). In real crystals (which are in most cases anisotropic), screw dislocations may favor certain planes having the lowest energy. Cross-slip can occur without diffusion, but thermal activation helps cross-slip movement from the original to other slip planes. Climb and cross-slip are recovery processes. Recall that steady-state creep is a deformation process, balanced by work hardening and dynamic recovery. The temperature dependence of creep is:

$$\dot{\varepsilon} \sim \exp -\left(\frac{Q_c}{kT}\right) \tag{6.61}$$

One of the known equations for steady-state creep, indicating stress dependence [10] is:

$$\dot{\varepsilon}_s = A\sigma^n \exp -\left(\frac{Q_c}{kT}\right) \tag{6.62}$$

Here, Q_c is the activation energy for creep and n is the stress exponent. A similar expression may be given for climb-controlled creep:

$$\dot{\varepsilon}_s = A\sigma^n \exp -\left(\frac{Q_c}{kT}\right) \tag{6.63}$$

But in this case, Q_c is independent of applied stress [10]. At lower temperatures, cross-slips made by screw dislocations are the process by which obstacles in the slip plane may be bypassed.

Since the study of cross-slip is more informative in single crystals, many experiments have been performed on single crystals having various structures. For instance, in order to investigate the glide system in FCC metals, Al single crystals were deformed by compression parallel to [00I] at temperatures between 225 and 365 °C and at strain rates between 9×10^{-6} and 9×10^{-4}/s (Le Hazif and Poirer [59]). (Note that Al has high SF energy and readily cross-slips). Their stress–strain curves exhibit three stages, which have been correlated with observations of slip lines and dislocation structures. The unique observation was that, after a small percentage of deformation cross-slips of $\frac{a}{2}\langle 1\bar{1}0\rangle$, screw dislocations from the $\{111\}$ to the $\{110\}$ planes occurred, that might be responsible for the $\{110\}\langle 1\bar{1}0\rangle$ slip. Stage I deformation occurs, as expected in FCC metals, on the $\{111\}$ planes, but, after a small deformation, slip on the $\{110\}$ plane sets in once the stress reaches the critical value, σ_{110}. This stress is thermally activated and decreases with temperature increase. It is not clear why dislocations cross-slip on the $\{110\}$ planes, rather than on the $\{111\}$ planes (as is usually the case), though several explanations have been given. An activation energy for the creep rate, $\dot{\varepsilon}$, of 28 kcal/mol, determined at a constant stress of σ_{110}, is close to the reported cross-slip in Al. It is likely that these observations are compatible with the mechanism of cross-slip by screw dislocations from the $\{111\}$ to the $\{110\}$ planes and that a SF that is stable at high temperatures stabilizes slip in the $\{110\}$ plane. The possibility of a SF in the $\{110\}$ plane is explained on geometrical grounds and the dislocation proposed is expressed as:

$$\frac{a}{2}[110] = \frac{a}{12}[110] + \frac{a}{3}[110] + \frac{a}{12}[110].$$

SF energy, which determines the separation of the partial dislocations, improves creep resistance if it is low. Contrary to the high SF energy observed in Al (in which cross-slip or climb occurs readily), in low-energy SF materials with large separation, cross-slip by creep or climb is suppressed. This was observed by Suzuki et al. [82] in their work on Mg-Y alloys with added zinc. The addition of small amounts of Zn has a beneficial effect on creep resistance, because it widens the separation between the partials by decreasing the SF energy. The average separation of partials in this alloy is given as:

$$d_S = \frac{Gb_1b_2}{8\pi\gamma}\left(\frac{2-\nu}{1-\nu}\right)\left(1 - \frac{2\nu\cos(2\alpha)}{2-\nu}\right) \tag{6.64}$$

where d_S is the separation width between the partials, γ is the SF energy, ν is the Poisson ratio and α is the angle between the total Burgers vector and the dislocation line. A large SF energy drop was calculated, compared with pure magnesium. Mg alloys are being used for more and more applications in which the components are subjected to elevated temperatures. Consequently, research is being focused on the development of alloys able to withstand high stresses at temperatures up to 300 °C, depending on the application. Thus, for example, in other Mg alloys improved creep properties are produced by the addition of rare earth alloys [65]. At low temperatures, a climb mechanism for edge dislocations exists, whereas, at higher temperatures, the cross-slip mechanism of screw dislocations is believed to operate. Opinions about the cross-slip mechanism are not unanimous, but a majority of the researchers support it. Whether the acting mechanism is climb or cross-slip, it is most likely that the beneficial effect of alloying stems from the fact that they both widen the separation between the partials.

During the discussion on cross-slip, metals were considered as examples of its effect on creep. However, it is necessary to emphasize that those examples were provided to clarify the process that might occur at high temperatures, when creep is sometimes unavoidable. However, partial dislocations and SFs occur in structures other than metals (Fig. 6.55). In the following, HREM observations of SFs in β-SiC are illustrated in Figs. 6.56 and 6.57. In Fig. 6.56, a HREM observation was conducted at or near the $\langle 110 \rangle$ zone axis of β-SiC. This direction is perpendicular to that of the stacking sequence $\langle 111 \rangle$ and SFs can be clearly imaged. In this figure, the SFs are observed at 2000 °C along with their structural models. Note that in (A) a twin fault is indicated. This twin fault increases with temperature, but, at the lower sintering temperature of 1750 °C, a heavily-faulted structure with many SFs is observed, as illustrated in Fig. 6.57. The quality of this image is not that good, due to the high fault density, preventing sharp HREM imagery.

The activation energy for cross-slip is rendered by Schoeck and Seeger [76] as:

$$\dot{\varepsilon} = C \exp - \left[\frac{\Delta H_0 - c \ln(\sigma / \sigma_c)}{kT} \right] \tag{6.65}$$

ΔH_0 is the energy for cross-slip, σ_c is the critical resolved shear stress, σ is the applied stress and C and c are constants. A model of creep controlled by cross-slip from the $\{111\}$ to the $\{100\}$ plane in the temperature range of 530–680 °C over the stress range of 360–600 MN m^{-2} was found to be in good agreement with the experimental results. The energy to form a restriction between the partials, namely to recombine the Shockley partials, was evaluated on the basis of Dorn's expression [8]. (See also Hemker et al. [51] for the creep mechanism at intermediate temperatures in Ni_3Al).

The dominant creep mechanism varies from ceramic to ceramic. There is no universal creep behavior which characterizes all high-temperature structural ceramics. In the following figures, the high-temperature creep of YSZ is chosen as an example, because cross-slip and climb mechanisms were seen to act during deformation. This ceramic is a 9.4 mol% YSZ tested in the 1300–1550 °C range. Figure 6.58 represents typical creep experiments.

Both T and σ were changed incrementally during creep. The corresponding strain rates, $\dot{\varepsilon}$, varied between 10^{-7} and 10^{-4} s^{-1}. Creep curves were analyzed according to the usual general constitutive law for high-temperature steady-state creep, as used by Bretheau et al. [30]:

$$\dot{\varepsilon} = A \frac{\mu b}{kT} \left[\frac{\sigma}{\mu} \right]^n \left[\frac{b}{d} \right]^p \left[\frac{p_{O_2}}{p^*_{O_2}} \right]^m \exp \left[- \frac{Q}{kT} \right] \tag{6.66}$$

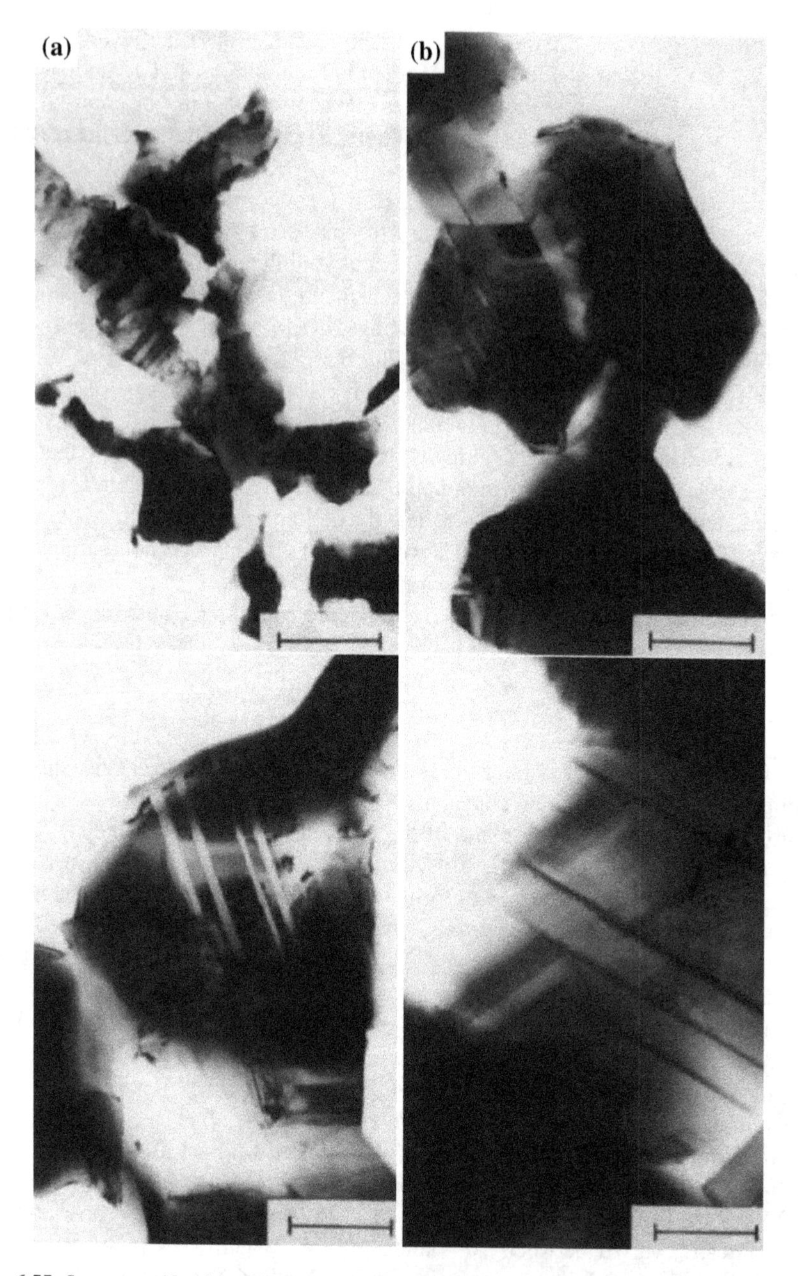

Fig. 6.55 Low-magnification TEM bright-field images of β-SiC ceramics sintered at **a** 1750 °C, **b** 1900 °C, **c** 2000 °C and **d** 2100 °C (bar = 2 μm) [12]. With kind permission of John Wiley and Sons

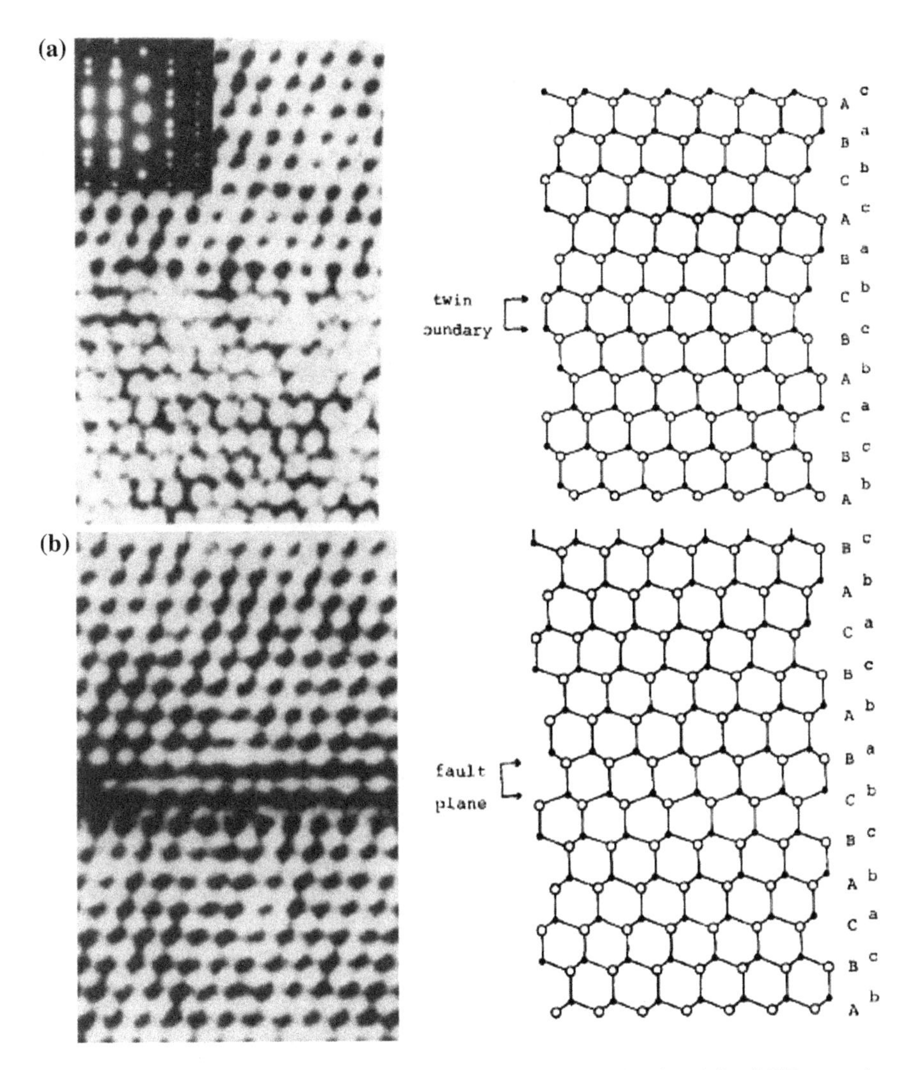

Fig. 6.56 HREM structural images of the typical stacking faults found in β-SiC ceramics sintered at 2000 °C: **a** twin fault and **b** deformation fault. Structural models based on the stacking sequence are shown [12]. With kind permission of John Wiley and Sons

A is a dimensionless constant, b is the Burgers vector of c-ZrO_2 (=3.62×10^{-10} m), d is the grain size (relevant for polycrystals), P^*O_2 is a reference oxygen partial pressure. Deformation and diffusion mechanisms determine the parameters n, p, m and Q. Changes in σ, T or pO_2 (see Fig. 6.58) may be determined experimentally from the variation of $\dot{\varepsilon}$. For single crystals, p = 0.

The dislocation substructures were obtained by TEM. Standard **g.b** analysis was carried out to determine the Burgers vectors of isolated dislocations. The 3D nature of the dislocation arrangement was determined using stereo pairs obtained

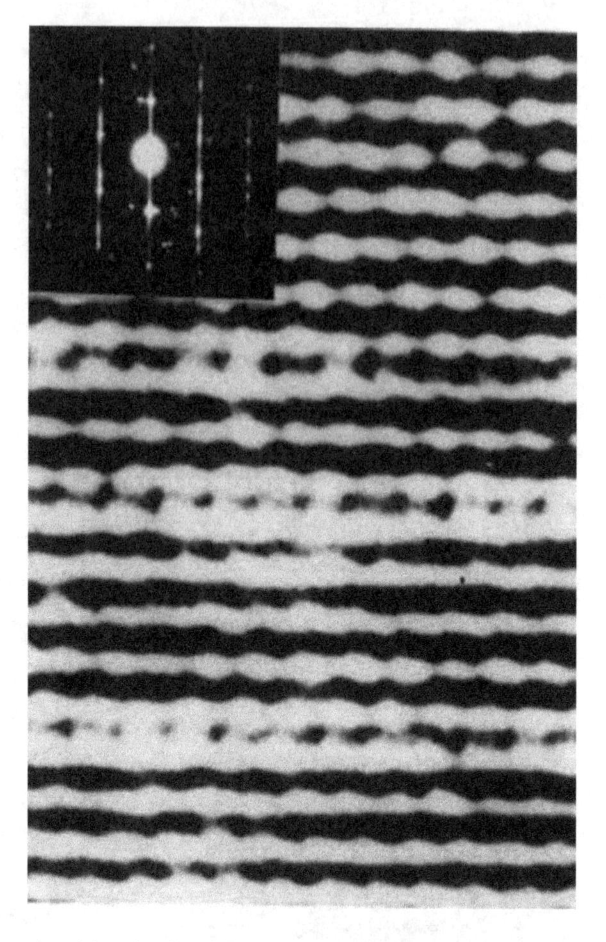

Fig. 6.57 HREM image of faulted structure and streaks in and electron diffraction pattern observed in β-SiC ceramics sintered at 1750 °C [12]. With kind permission of John Wiley and Sons

with **g** = (220). Dislocation densities in the foils were determined by standard means. The steady-state creep rate versus 1/T is shown in Fig. 6.59.

The above figure suggests two deformation regimes, namely two operating mechanisms with a transition in the 1400–1450 °C range. A least square fit for both these regimes gives: $Q = 6.2 \pm 0.4$ eV and $Q = 7.7 \pm 0.4$ eV at $T \leq 1400$ °C. Figures 6.60 and 6.61 show the dislocation substructures. These deformed samples exhibit different dislocation substructures at 'low' and 'high' deformation temperatures. At 1300 °C, the dislocation density, ρ, was high ($\sim 10^{13}$ m^{-2}) and showed substantial dislocation reactions and node formation (Fig. 6.60). **g.b** analysis (not shown in the figure) revealed that most of the dislocations had a Burgers vector **b** = ½[110] and existed along the (001) primary slip plane. However, the stereo pair in Fig. 6.60 indicates that some dislocations, belonging to the primary slip plane, change slip-planes to lie along the $(\bar{1}11)$ and $(1\bar{1}1)$ planes, indicating that a significant amount of cross-slip occurred during that deformation. In other experiments, cross-slip was also observed in samples deformed at temperatures as low as 400 °C under hydrostatic confining pressure.

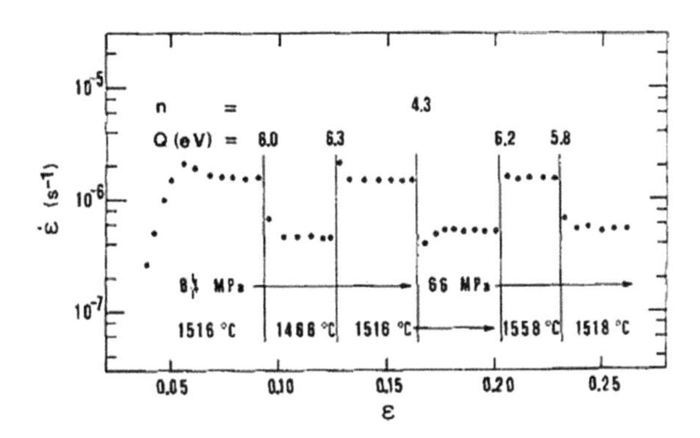

Fig. 6.58 Typical creep experiment in Y_2O_3-stabilized ZrO_2 at stresses of 84 and 66 MPa and three temperatures, 1466, 1516–1518, and 1558 °C [64]. With kind permission of John Wiley and Sons

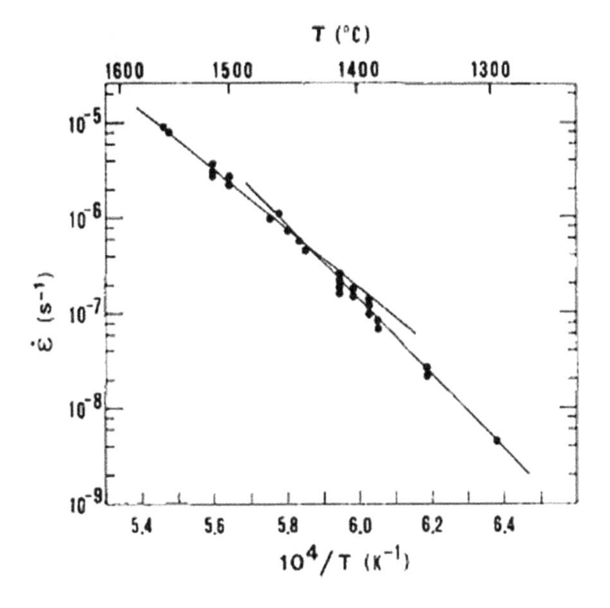

Fig. 6.59 Steady-state creep rate, normalized to a stress of 100 MPa, plotted as a function of reciprocal temperature [64]. With kind permission of John Wiley and Sons

At 1500 °C, the dislocation density is lower ($\rho \sim 5 \times 10^{11}$ m^{-2}). A stereo pair (Fig. 6.61) shows that many dislocation segments are perpendicular to the (001) primary slip plane; these segments lie on the (100) and (010) planes. All six (110) Burgers vectors are present, indicating that numerous slip systems have been activated; however, the low dislocation density indicates that significant recovery has occurred and that diffusion must be reasonably rapid at this temperature.

Figure 6.62 shows the polycrystalline data plotted as a log ($\dot{\varepsilon}$. Td^pexp Q/kT) versus log σ to normalize the temperature and grain-size dependence. Q is taken as 5.7 eV. Data from other experiments are included in the plot, which shows that the creep resistance of single crystals is better than that of ZrO_2 polycrystals having a similar composition under all stresses <100 MPa. From the best fit of the data, a slope of $n = 4.1$ may be obtained. In single crystals, p = 0. At lower temperatures, n rises to 7.5 and Q is ~7.5 eV which is greater than its value at high

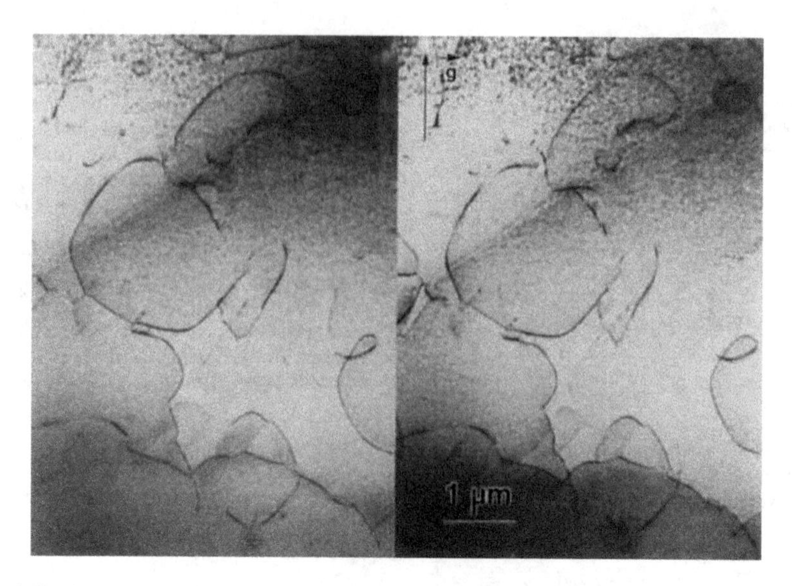

Fig. 6.60 Stereo pair of transmission electron micrographs of dislocation substructure in specimen crept at 1300 °C [64]. With kind permission of John Wiley and Sons

temperatures. This large stress exponent is characteristic of deformation at intermediate homologous temperatures, in which dislocation cross-slip may be as important in controlling creep as is recovery via dislocation climb. In such cases, power-law creep with higher stress exponents is generally found. A high density of dislocations and significant cross-slip is characteristic of such cross-slip-controlled creep, along with stress-dependent activation energy.

Poirier [19] points out that when the cross-slip and climb of dislocations operate at the same time, $\dot{\varepsilon}$ may be written as:

$$\dot{\varepsilon} = \dot{\varepsilon}_{cross-slip} + \dot{\varepsilon}_{clmb} = \dot{\varepsilon}_{0_1}\left(\frac{\sigma}{\mu}\right)^{n1}\exp\left(-\frac{Q_1}{kT}\right) + \dot{\varepsilon}_{0_2}\left(\frac{\sigma}{\mu}\right)^{n2}\exp\left(-\frac{Q_2}{kT}\right) \tag{6.67}$$

The subscripts and superscripts 1 and 2 refer to cross-slip and climb, respectively. Dislocation motion must overcome significant structural barriers or must cross-slip or climb past obstructions. At the lower temperatures, dislocation cross-slip and climb both occur. At the higher temperatures, dislocation climb becomes a rate-controlling mechanism and classic values of the stress exponent (n = 4.5) are obtained. The creep-activation energy is that of cation diffusion.

In general, creep at temperatures below 0.5 T_M is not thought to occur by means of the lattice-diffusion-controlled mechanism.

Seldom does a lone creep mechanism operate at any given time. Creep mechanisms may operate simultaneously (in parallel) or independently. For the two mechanisms, one may write:

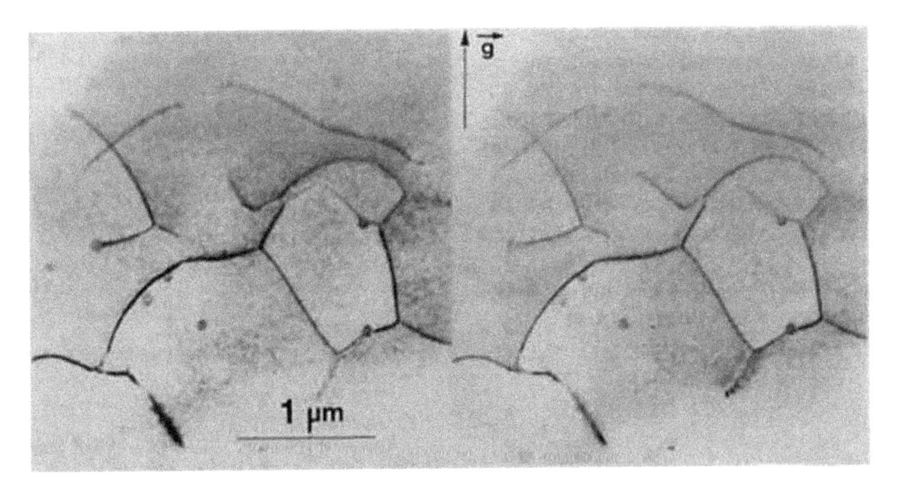

Fig. 6.61 Stereo pair of transmission electron micrographs of dislocation substructure in specimen crept at 1500 °C [64]. With kind permission of John Wiley and Sons

Fig. 6.62 Literature data for creep of polycrystalline ZrO_2, normalized by temperature and grain size, as a function of stress, and compared with the single-crystal data [64]. With kind permission of John Wiley and Sons

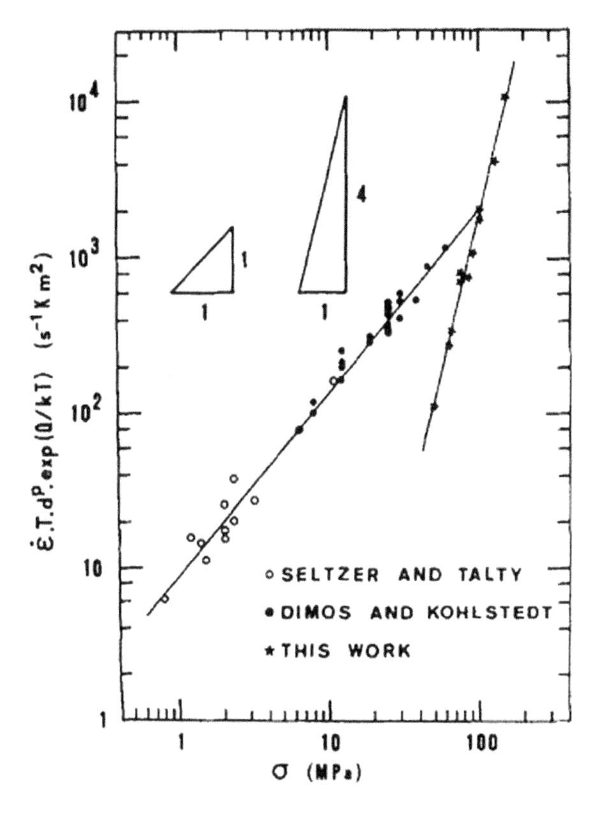

$$\dot{\varepsilon} = \sum_i \dot{\varepsilon}_i \tag{6.68}$$

or

$$\frac{1}{\dot{\varepsilon}} = \sum_i \frac{1}{\dot{\varepsilon}_i} \tag{6.69}$$

In the case of parallel creep mechanisms, the fastest mechanism will dominate the overall creep, whereas, when they operate in sequence, the slowest process controls creep deformation.

6.6.5 Coble Creep, Involving Grain-Boundary Diffusion

Coble creep is also a type of diffusion creep, but involves grain-boundary diffusion. The diffusion of atoms along grain boundaries produces a change in dimensions, due to the flow of the material. Of the two kinds of self-diffusions in polycrystalline materials, the one occurring at low temperatures is grain-boundary dominated, whereas lattice diffusion occurs at high temperatures. Figure 6.63 is an illustration of ideal grain structure, showing the flow of atoms along the boundaries under the influence of a tensile stress. In a polycrystalline matrix, the grain shape is not as indicated in Fig. 6.63 (for an ideal structure), but varies in orientation, making it difficult to analyze.

Coble, in his original paper, used a spherical grain (apparently following the Nabarro-Herring approach for lattice-controlled-diffusion creep). In Coble's [36] analysis of creep, a spherical grain was used once again. Based on the experimental results for Al_2O_3, where it was observed that the Al ion-diffusion coefficient is larger by orders than that of oxygen ions and, since the creep rate in lattice-controlled diffusion is limited by the least mobile species, it was expected that the O^{2-} species would determine the rate of creep. Coble [36] suggested that grain-boundary diffusion, rather than lattice diffusion, might control creep deformation. He proposed that Al diffuses in the lattice and O^{2-} in the grain boundaries, where its diffusion coefficient is enhanced in comparison with the values in the lattice. It was assumed that the spherical grain maintains a constant volume and, thus, the areas of the vacancies at the source and the sink must also be equal (grain boundaries may act as sources or sinks for vacancies). The average gradient of the spherical grain, with a radius, R, is given as $\frac{\Delta C}{(R\pi/2)}$. The problem is to evaluate the concentration gradient at the 60° boundary, which, for equal areas of rotational symmetry, lies at 60° below the pole of a hemisphere.

For steady-state creep, where Fick's law applies, the flux at the 60° boundary is:

$$J_{vac\,\sec^{-1}} = D_V N \left[\frac{\Delta C}{(R\pi/2)}\right](W) 2\pi R \sin 60 \tag{6.70}$$

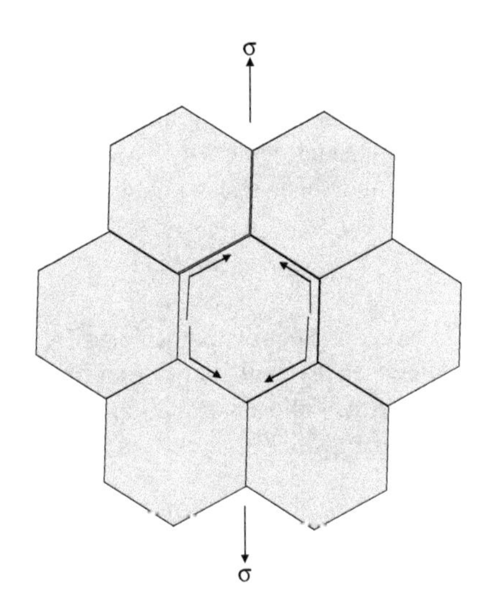

Fig. 6.63 Seven grains are shown in a two-dimensional hexagonal array before creep deformation. Following diffusion, the grains elongate in one direction and decrease perpendicularly to the tensile axis. A void formation develops between the grains, but grain-boundary sliding, which may accompany this process, removes these voids

Here, D_V is the diffusion coefficient of the vacancies in the boundary; N is a proportionality constant relating the average vacancy gradient, $\Delta C/(R\pi/2)$ and the maximum gradient, $1/R(dc/d\theta)_{\theta=60}$; W is the effective boundary width and; ($2\pi R\sin60$) is the length of the zone in which the diffusion flux is at a maximum. Thus, the cross-sectional area for diffusion is $2\pi RW\sin60$. After a detailed and lengthy evaluation of the relevant parameters, Coble [36] arrives at the final equation for creep rate, given as:

$$\dot{e} = \frac{148\sigma(D_b W)a_0^3}{(GS)^3 kT} \tag{6.71}$$

where a_0^3 (=Ω) is the atomic volume of a vacancy. For lattice diffusion, the expression [37] is:

$$\dot{e} = \frac{10\sigma(D_L \Omega)}{(GS)^2 kT} \tag{6.72}$$

Other expressions are given for Coble's creep, the difference being in the coefficient representing the assumptions in each case (in Coble and Guerard [37] it is: $\dot{e} = \frac{150\sigma(D_b W)}{(GS)^3 kT}$).

Thus, by writing the coefficient as a constant, the common expression is:

$$\frac{d\varepsilon_{gb}}{dt} = \dot{\varepsilon} = A\frac{\sigma\Omega D_{gb}\delta}{l^3 kT} \tag{6.73}$$

The subscripts refer to the grain boundaries, Ω is the atomic volume (of a vacancy), δ is the grain-boundary width and l is the grain size. (In Nabarro-Herring, grain size was denoted by d). The D_S in Eqs. (6.47) and (6.48) is replaced, in Coble's equation, by $D_{gb}\delta$. Factor 1/l represents the density of the cross-section of the grain boundaries per unit area. Hence, δ/l is the cross-sectional area of the grain boundaries per unit area. In a realistic structure, A depends on grain structure and how the average grain size is determined. Creep by grain-boundary diffusion has a stronger dependence on grain size than on lattice diffusion. In terms of shear strain and shear stress [73] the expression is:

$$\dot{\gamma} = 42D_S \frac{\pi\delta\tau\Omega}{d^3kT} \tag{6.74}$$

Here, d is equivalent to l and D_S to D_{gb} and V to Ω. When creep deformation is influenced by both lattice- and grain-boundary diffusion, an expression may be derived as follows. Equation (6.51) may be written with the same designations used in Eq. (6.73) as:

$$\dot{\varepsilon} = 16\alpha \frac{D_S\sigma\Omega}{\pi l^2 kt} \tag{6.51}$$

By adding Eqs. (6.51) and (6.73), one gets:

$$2\dot{\varepsilon} = \frac{16\alpha D_S\sigma\Omega}{\pi l^2 kT}\left(1 + \frac{A\pi}{16\alpha}\frac{D_{gb}\delta}{lD_S}\right) \tag{6.75}$$

Designating that $16\alpha/2\pi = B$ and $A\pi/2 \times 16\alpha = C$ gives:

$$\dot{\varepsilon} = \frac{BD_S\sigma\Omega}{l^2kT}\left(1 + \frac{CD_{gb}\delta}{D_S l}\right) \tag{6.75a}$$

An expression for creep may be given in terms of shear-strain rate and shear stress, when both lattice- and grain-boundary diffusion are involved in the deformation. For most polycrystalline materials, diffusion in grain boundaries is more rapid than in the lattice.

To summarize this section, it may be stated that in Coble creep the atoms diffuse along the grain boundaries and elongate the grains along the stress axis. This causes Coble creep to have stronger grain-size dependence than Nabarro-Herring creep. Since the grain boundary is the controlling diffusion mechanism in Coble creep, the process occurs at lower temperatures than Nabarro-Herring creep does. Coble creep is still temperature dependent and, as the temperature increases, so does the grain-boundary diffusion. It also exhibits a linear dependence on stress, as does Nabarro-Herring creep. *Coble creep* and Nabarro-Herring creep can take place in parallel, so that actual *creep* rates may involve both components and both diffusion coefficients.

6.7 Grain-Boundary Sliding

Grains and their sizes are very important variables characterizing the microstructure of polycrystalline materials. Grain-boundary movement plays a significant role in the characteristic behavior of materials for creep application.

Basically, grain-boundary sliding [henceforth: GBS] is a process in which grains slide past each other along their common boundary. It has also been observed that sliding may occur in a zone immediately adjacent to the grain boundary [87]. As seen in Chaps. 2 and 5, the role of grain size in the work-hardening mechanism (strengthening) is crucial, as given by the Hall–Petch relation:

$$\sigma_y = \sigma_0 + \frac{k_y}{\sqrt{d}} \tag{2.9}$$

In primary creep, the required stress increases due to work hardening (which also acts in steady-state creep, but is balanced by various recovery processes). Decreasing grain size should indicate a stronger material, since higher stress is crucial for continued deformation. Thus, one may expect materials with small grain sizes to show better creep resistance, while increasing grain size should cause an increased secondary-creep rate. This is attributed to the decrease in boundary barriers with increasing grain size (less strengthening media exists, because there are less grain-boundary obstacles). However, this is true as long as no undesirable processes occur at the grain boundaries. For example, large-grained materials with a small number of grain boundaries are low sources of vacancies and, therefore, dislocation climb will be reduced compared to small-grained materials. Thus, one can see that grain size in creep has a dual effect, namely small grain size strengthens the ceramics, because the large number of grains act as barrier to dislocation glide. However, in large-grained ceramics, with fewer boundaries, less vacancies are emitted, which is a prerequisite for climb (creep deformation) and, therefore, this process will reduce creep. Note that a suitable choice of grain size in ceramics is critical for achieving the best compromise for good creep resistance.

Major structural changes occur at the start of tertiary creep. Damage is initiated by the formation of multi-shaped cavities (in metals, either wedge-shaped or rounded cavities are seen). Wedge-shaped cavities are primarily seen at grain boundaries and their coalescence is the unmistakable sign that creep rupture will occur. It is believed that GBS is a prerequisite for the nucleation of voids and cavities and that it occurs when a sufficiently high stress concentration develops to create new surfaces. Cavitations increase with increasing strain at high temperatures. The stresses causing GBS are the shear stresses acting on the boundaries. Whether void formation is associated with/or a consequence of GBS has not yet been completely determined, since the experiments found in the literature seem to support both concepts. In Fig. 6.64, cavities at two grain-boundary junctions may be seen in ABC-SiC. The term 'ABC-SiC' refers to SiC, which has been hot-pressed with additions of Al, as well as B and C. This material has been shown to have an ambient-temperature fracture toughness as high as 9 MPa $m^{1/2}$ with strengths of $\sim$650 MPa (mechanical properties that are among the highest reported for SiC).

One of these concepts, regarding GBS, is associated with the presence of an amorphous grain-boundary film along the boundaries between the grains. More specifically, this film has often been termed a 'glassy phase' and considered

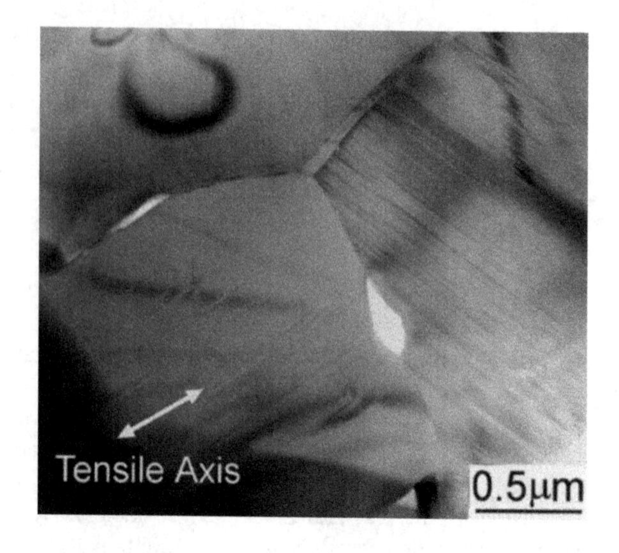

Fig. 6.64 Transmission electron micrograph of ABC-SiC showing grain-boundary cavities at two-grain junctions on the tensile edge of a specimen crept at 1400 °C for 840 h under 200 MPa [41]. With kind permission of Elsevier

responsible for GBS. In Fig. 6.65, such an amorphous film may be seen (before creep deformation) in a hot-pressed material, rich in Al, O and C (common SiC additives).

The term ABC-SiC reflects the mentioned constituents added to SiC. Some believe that the origin of the glassy film has to do with the presence of sintering additives, as is the case in SiC. However, due to the nature of this bonding, as well as vapor pressures and surface and grain-boundary characteristics, densification is difficult in ceramic systems without such additions. HRTEM imaging indicates that glassy film may be fully crystallized after heat treatment above 1100 °C for times in excess of 5 h. Crystallization of the glassy amorphous phase is illustrated in Fig. 6.66.

Clearly, such crystallization of the grain-boundary phase would minimize softening and GBS, which would cause an increase in strength. As stated previously, the microstructure has a major impact on creep properties in ceramics. The microstructural creep properties of ABC-SiC may be observed in Fig. 6.67. The low density of dislocations with no apparent slip bands may occasionally be observed in isolated grains in ABC-SiC. The density of these dislocations increases with increasing temperature. No evidence of cavitation has yet been found after creep testing at 1300 °C (and below this temperature). Creep cavitations appear around 1400 °C, where grain-boundary cavities are observed on the tensile (but not compression) side of the beams (see Fig. 6.53). Cavities form both at two-grain and multiple-grain junctions. Even at higher temperatures, while the dislocation density is higher, slip bands are still not a dominant feature. As noted above, the grain-boundary films become fully crystallized during long durations at creep temperatures (similar to their behavior during pre-annealing). Additionally, the concentration of impurities in the boundary film is found to increase slightly compared with the as-hot-pressed material.

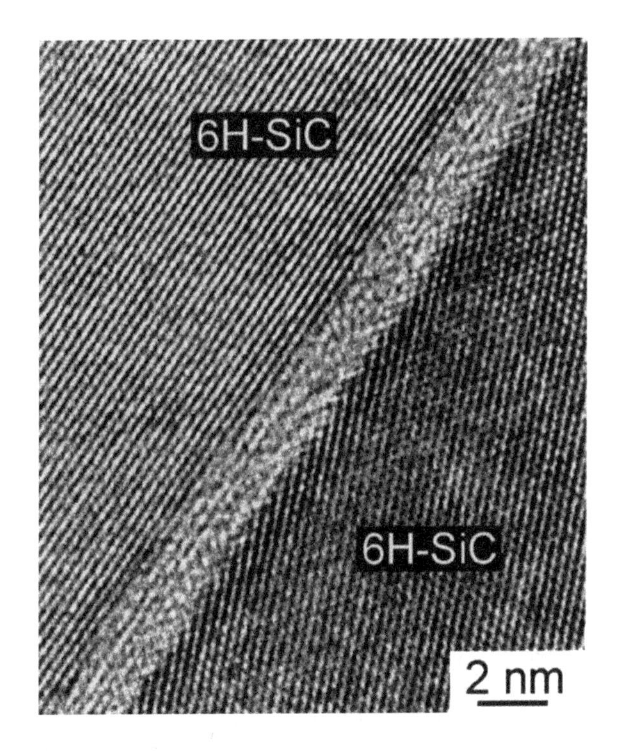

Fig. 6.65 High-resolution transmission electron micrograph of as processed ABC-SiC, showing the amorphous grain-boundary film [41]. With kind permission of Elsevier

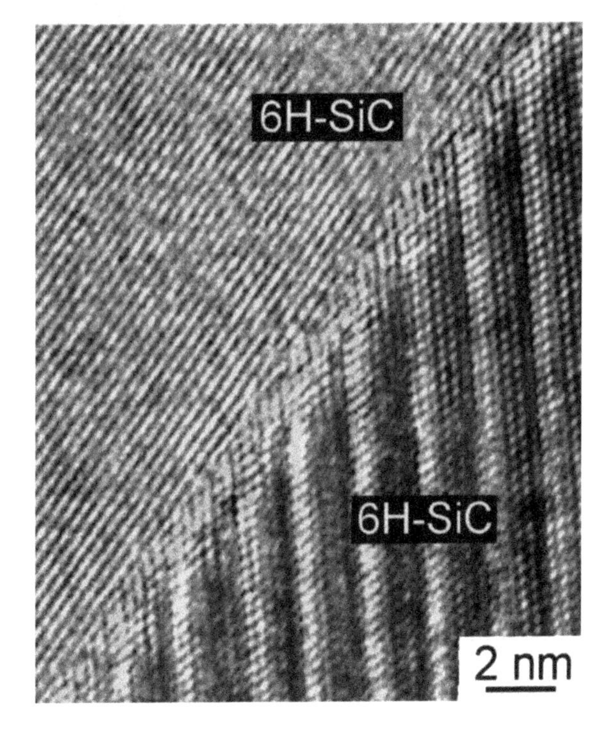

Fig. 6.66 High-resolution transmission electron micrograph of a grain boundary in ABC-SiC after high-temperature annealing (1400 °C/840 h), showing that the amorphous layer has become fully crystallized [41]. With kind permission of Elsevier

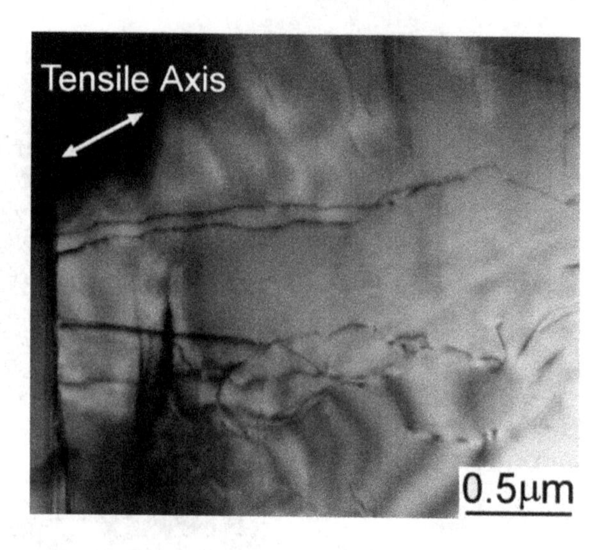

Fig. 6.67 Transmission electron micrograph of ABC-SiC showing dislocations, located in a residual β-SiC grain, on the tensile edge of a specimen crept at 1200 °C for 670 h under a stress of up to 175 MPa [41]. With kind permission of Elsevier

Another ceramics, namely alumina with evidence of grain boundary sliding inducing cavitation when under compression is Al_2O_3. In Fig. 6.68a comparison is shown between crept and uncrept Al_2O_3. Uncrept Al_2O_3 is characterized by the relatively featureless grain boundaries and the presence of residual pores ~ 10 μm in diameter on the grain boundaries. After creep, the typical microstructure, shown in Fig. 6.68b, exhibited creep damage in the form of multiple, individual cavities, that formed primarily on two-grain facets. A relatively smaller volume of cavities is observed at three- and four-grain junctions. Triple-point cavities appeared to be insignificant to the creep damage process in this material, since they were observed only occasionally.

The processes of the nucleation, growth and coalescence of multiple cavities is also visible at three-grain junctions in Lucalox-type alumina, as shown in Fig. 6.69a and b. Figure 6.69a presents both the top (left arrow) and side (right arrow) views of similar cavities formed at three-grain junctions. The two views of the cavities are consistent with the discrete nucleation of individual cavities coalescing together to form a facet crack. Cavities on three- and four-grain junctions generally exhibit the classic, ellipsoidal shape, whereas cavities that form on two-grain facets often appear to exhibit non-ellipsoidal shapes, but ellipsoidal-shaped cavities are also observed. The non-ellipsoidal or irregularly-shaped cavities suggest that their morphologies may be governed by the crystallographic orientation of the grain facet and the corresponding surface energies. An example of this phenomenon is shown in Fig. 6.70a, where the cavities on the three-grain junctions are individual ellipsoidal cavities, while those on the two-grain facets are characterized by their sharp, angular shapes.

Figure 6.71 illustrates completely coalesced facet-sized cavities that formed in the glassy phase of AD99 Al_2O_3, presumably due to the coalescence of individual cavities. No crack-like creep cavities were observed in this material. Small-angle

Fig. 6.68 **a** Baseline fracture surface of Lucalox Al_2O_3 obtained by failing the specimen at 1600 °C before creep. Note the presence of residual pores on the grain boundaries. **b** Typical cavitated fracture surface of Lucalox Al_2O_3, after creep testing. Note the majority of creep cavities nucleated on two-grain facets, while some cavities may be observed at three- and four-grain junctions [29]. With kind permission of John Wiley and Sons

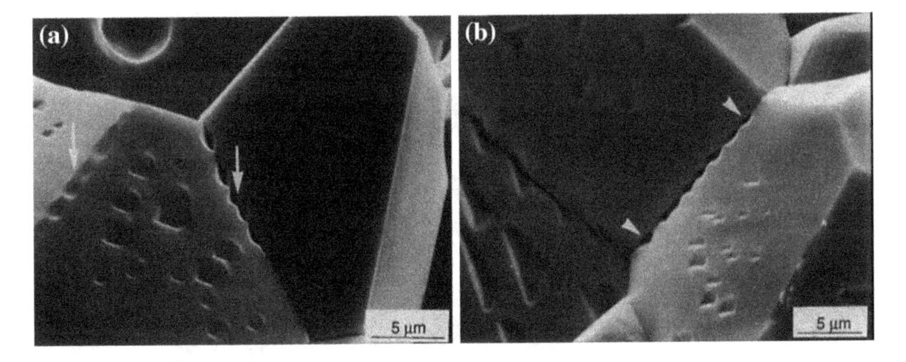

Fig. 6.69 Creep cavities observed at three-grain junctions in Lucalox: **a** the *arrows* point to the *top view* (*left arrow*) and *side view* (*right arrow*) of similar cavities providing evidence for the nucleation and coalescence of multiple cavities; **b** the *arrows* point to the nucleation and coalescence of multiple cavities along the entire length of a three-grain junction. Also, note the various cavity shapes on the two-grain facets [29]. With kind permission of John Wiley and Sons

neutron scattering [henceforth: SANS] measurements of the two kinds of aluminas (illustrated in the above examples of cavitation) are given in Fig. 6.72 in terms of cavity radius as a function of creep time. Though, at present, there is no conclusive data proving that GBS is the driving force for the nucleation and growth of creep cavities, a number of studies have concluded that cavity nucleation is, in fact, induced by GBS.

GBS has been the subject of numerous investigations, in light of the importance of grain boundaries for many aspects of material applications. Understanding the physics of the complex behavior of grain boundaries is of great interest in regard to: grain growth, crystallization and recovery deformation, to mention just a few

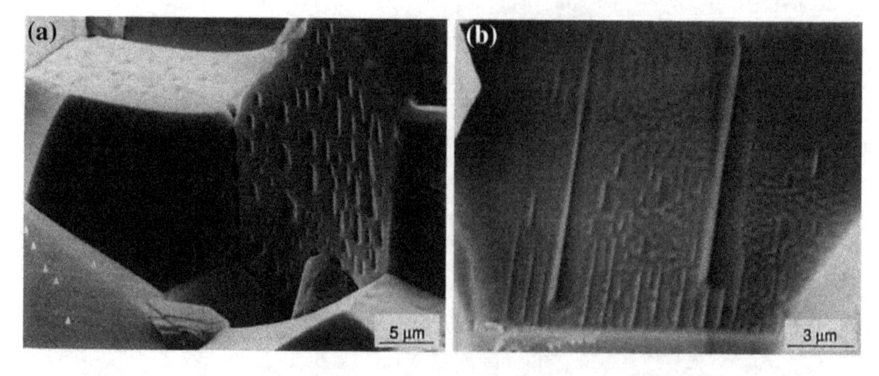

Fig. 6.70 Creep cavities, formed in Lucalox on both two-grain facets and three-grain junctions, exhibit significantly different morphologies: **a** Cavities at three-grain junctions evolve into the more classical ellipsoidal shape often used in creep models, where the shape of cavities on two-grain facets appears to be governed by the crystallographic orientation of the grain facet and surface energy effects. The fracture surface exhibits "arrowhead"-shaped cavities on the right grain facet and "dome"-shaped cavities on the left facet. Also, the *arrows* point to a series of grain boundary slip-plane ledges on an uncavitated two-grain facet. **b** A grain facet exhibiting "needlelike" creep cavities [29]. With kind permission of John Wiley and Sons

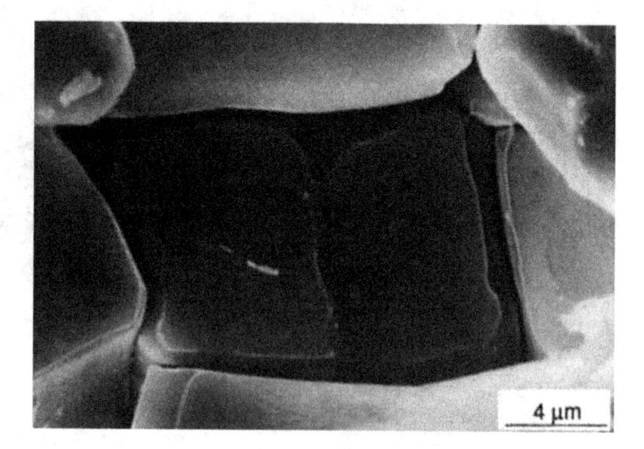

Fig. 6.71 Completely coalesced creep cavity in the glassy phase of AD99 Al_2O_3 [29]. With kind permission of John Wiley and Sons

topics. A general review of the properties of grain boundaries may be found, for example, in the work of Valiev et al. [86]. Here, GBS is of interest in order to gain better practical and theoretical understanding.

In Fig. 6.73, GBS is illustrated in a UO_2 ceramics. Polishing scratches, which cross grain boundaries, have separated, indicating that GBS has occurred during creep. The line, which originally passed the boundaries before creep as a continuous line, has migrated, after creep, as may be seen in the neighboring three grains.

However, some more illuminating research results of GBS may be observed in metals. The instructive photo (Fig. 6.74) was taken of a Mg-0. 78 %Al alloy strained to 2.49 % at a temperature of 473 K and under an applied stress of

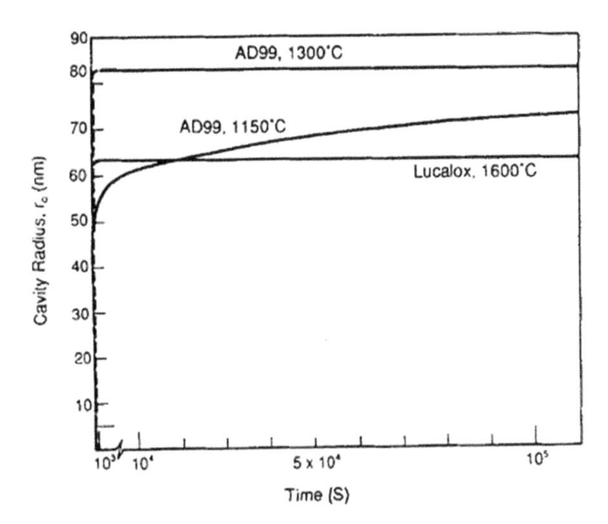

Fig. 6.72 SANS results showing the individual cavity radius versus creep time [29]. With kind permission of John Wiley and Sons

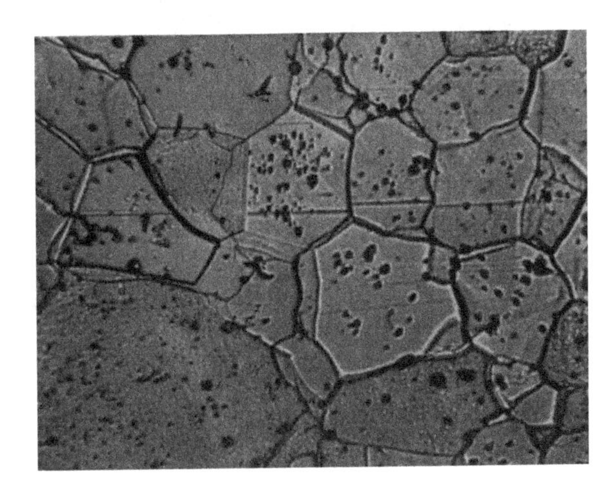

Fig. 6.73 UO_2 sample tested at 1200 °C, 10000 psi; maximum strain = 1.8 %. Sample exhibits grain boundary migration and sliding. 400x [29]. With kind permission of Elsevier

17.2 MPa. The evidence of GBS is the displacement of the scratch lines during creep testing. This figure shows scratch lines displaced across a grain boundary; transverse markings are inscribed perpendicular to the tensile axis. Clear offsets may be seen in the transverse marker line in this Mg-0.78 %Al alloy. The tensile axis in this experiment is horizontal. An alternate method for evaluating GBS is by means of interferometry. An example of the offsets of the same alloy as revealed by interferometry is visible in Fig. 6.75.

Chan and Page [33] have developed a model describing creep-induced transient-cavity growth by assuming cavity growth is governed by the two competing processes, transient creep and sintering. According to this model, the rate of cavity growth is described as:

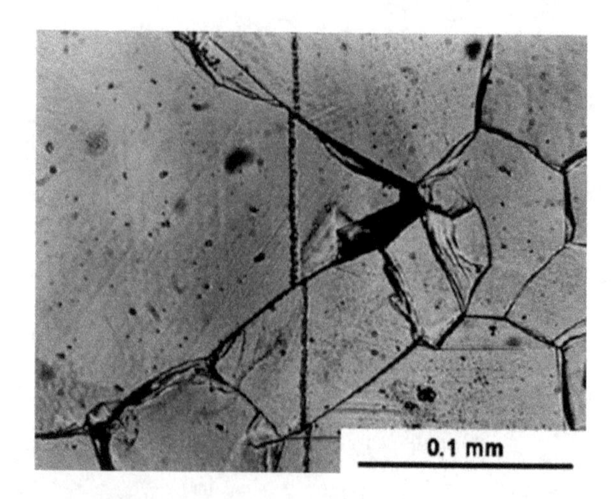

Fig. 6.74 Grain-boundary sliding revealed by the boundary offsets in a transverse marker line in a Mg-0.78 %Al alloy tested under creep conditions at 473 K under a stress of 17.2 MPa. From Bell and Langdon [2], reproduced from Langdon [58]. With kind permission from Springer Science and the author

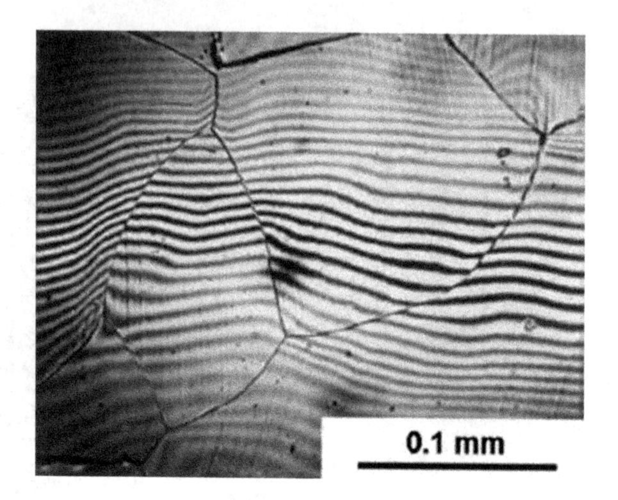

Fig. 6.75 Offset revealed by interferometry in a Mg-0.78 %Al alloy pulled to an elongation of 1.5 % at 473 K under a stress of 27.6 MPa. From Langdon [13], reproduced from Langdon [58]. With kind permission from Springer Science and the author

$$\dot{R} = \frac{33RG(\xi)}{4\pi^2}\left[\dot{\varepsilon}_{ss}(t/t_c)^m - \frac{4\pi}{33}\left(\frac{\gamma_s}{\eta l}\right)(1/\xi - 0.9\xi)\right] \tag{6.75}$$

with

$$\xi = \mathrm{R/I} \tag{6.76}$$

$$G(\xi-) = \frac{2\sqrt{3} - 0.667\pi\xi^2}{0.96\xi^2 - \ln\xi - 0.23\xi^2 - 0.72} \tag{6.77}$$

where R is the cavity radius, $\dot{\varepsilon}_{ss}$ is the steady-state creep rate, t is the creep time, t_c is the characteristic time, m is an exponent ranging from −0.5 to −0.6, γ is the surface energy, η is the viscosity parameter and 2 l is the center-to-center cavity spacing. Note that the first term within the bracket in Eq. (6.75) is the transient

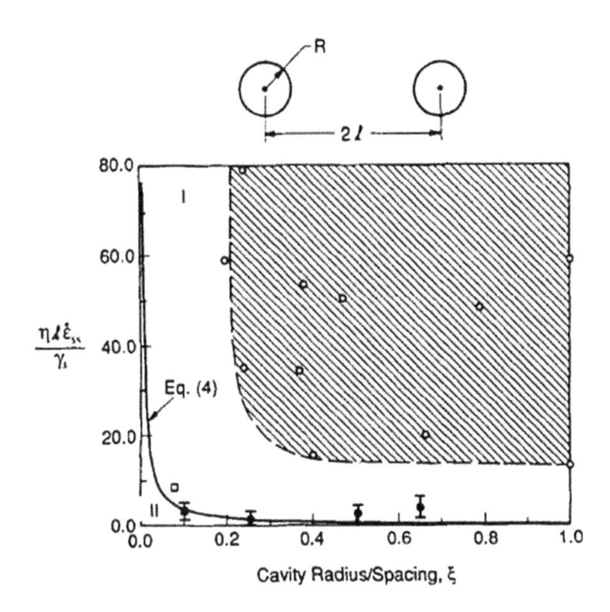

Fig. 6.76 Comparison of the predicted (–) and experimentally observed conditions for zero cavity growth: (*white square*) Lucalox, 1600 °C; (*black square*) AD99, 1300 °C, and for cavity growth; (*white circle*) Lucalox, 1600 °C; (*black circle*) AD99, 1150 °C. Region **I** represents cavity growth and region **II**, cavity shrinkage [29]. With kind permission of John Wiley and Sons

creep rate, $\dot{\varepsilon}_{tr}$, while the second term is the sintering rate, $\dot{s}$. From Eq. (6.75), it is evident that the transient creep rate,$\dot{\varepsilon}_{tr}$, drives cavity growth, whereas the sintering rate term, $\dot{s}$, drives cavity shrinkage. In addition, imposing parameters to reach a state of equilibrium between $\dot{\varepsilon}_{tr}$ and $\dot{s}$ would result in a condition of zero cavity growth. Therefore, a critical value of $\dot{\varepsilon}_{ss}$ ($\dot{\varepsilon}_{cr}$) may be determined by setting $R = 0$ in Eq. (6.75), which defines no-growth behavior as follows:

$$\frac{\eta l \dot{\varepsilon}_{cr}}{\gamma_s} = \frac{4\pi}{33}(1/\xi - 0.9\xi) \tag{6.78}$$

This no-growth boundary is shown in Fig. 6.76 as the solid line. In addition, cavities exhibit continuous growth in region I, where $\dot{s}_{cr} > \dot{s}_{tr}$ and the cavities will shrink when the opposite is true (region **11).**

A quantitative estimate of the contribution of GBS to overall strain, ξ, used by Tan and Tan [82] following Langdon's [13] proposal, is:

$$\xi = \frac{\varepsilon_{GBS}}{\varepsilon_t} \tag{6.79}$$

ε_t, the total strain at high temperatures, is expressed as:

$$\varepsilon_t = \varepsilon_g + \varepsilon_{GBS} + \varepsilon_{dc} \tag{6.80}$$

ε_g is the strain in the grain, due to processes taking place within the grain; ε_{GBS} is the strain due to GBS; and ε_{dc} is the strain due to diffusion creep. In practice, experiments are often performed with a negligible contribution of diffusion creep and, thus, Eq. (6.80) reduces to:

$$\varepsilon_t = \varepsilon_g + \varepsilon_{GBS} \tag{6.81}$$

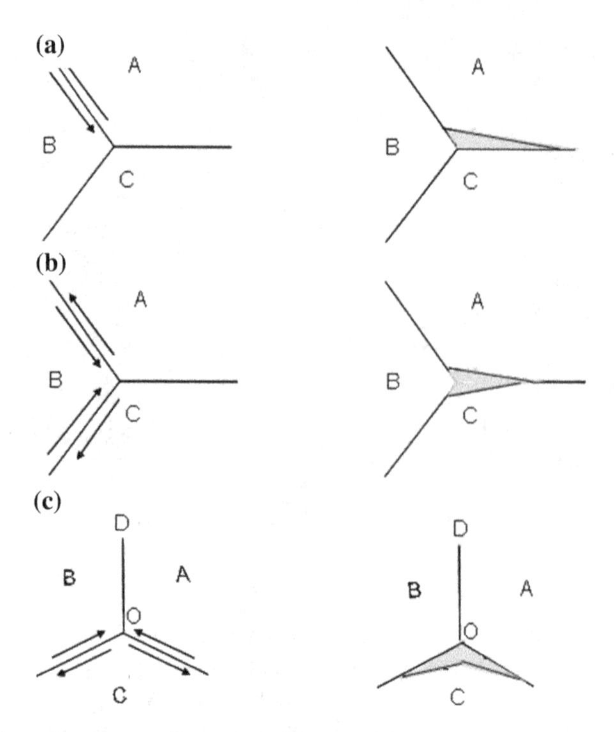

Fig. 6.77 Schematic representation of a w-type crack formation initiated by GBS [3]

Damage leading to failure, in the form of stress rupture, is initiated by void and crack formation. The tertiary creep, per se, is a sign that some sort of structural damage has occurred. Round or wedge-shaped voids, known as 'r-type cavities' and 'w-type cavities', are seen, at first, along grain boundaries and, when they coalesce, creep fracture occurs. As indicated above, the mechanism of void formation is associated with GBS and occurs due to shear stresses acting along the boundaries.

A commonly used illustration of a w-type crack initiation by GBS, its formation and growth (first presented by Chang and Grant [3] and found in almost every publication), is shown in Fig. 6.77. Another configuration for the initiation of intergranular cracks (somewhat more complex) is shown in Fig. 6.78.

A number of w-crack configurations have been experimentally observed at triple points. Figure 6.79 shows crack nucleation at grain boundaries, formed when shear stress acts along the boundaries. Here, a wedge-type cavity was formed under creep in a SiAlON-YAG ceramic in a flexural bar studied at 1170 °C. Figure 6.80 shows the cavitation strain and the total strain as functions of the normalized beam height for the applied moments of $M = 0.165$ N.m and $M = 0.247$ N.m (i.e., at the initial, maximum applied stresses of 80 and 120 MPa). Both tests were carried out at 1170 °C and interrupted before failure. For M = 0.165 N.m (a = 80 MPa), the maximum cavitation strain is ~20 % of the total strain, whereas for M = 0.247 N.m (b = 120 MPa), the cavitation strain

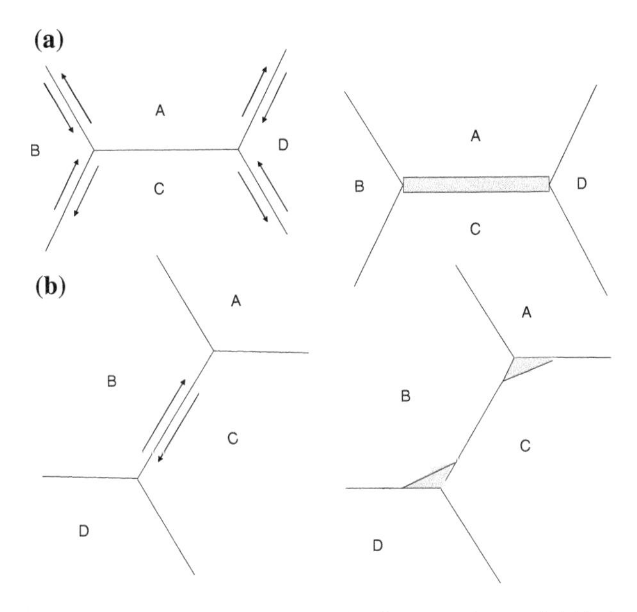

Fig. 6.78 Schematic views showing a more complex intergranular crack initiation by GBS [3]

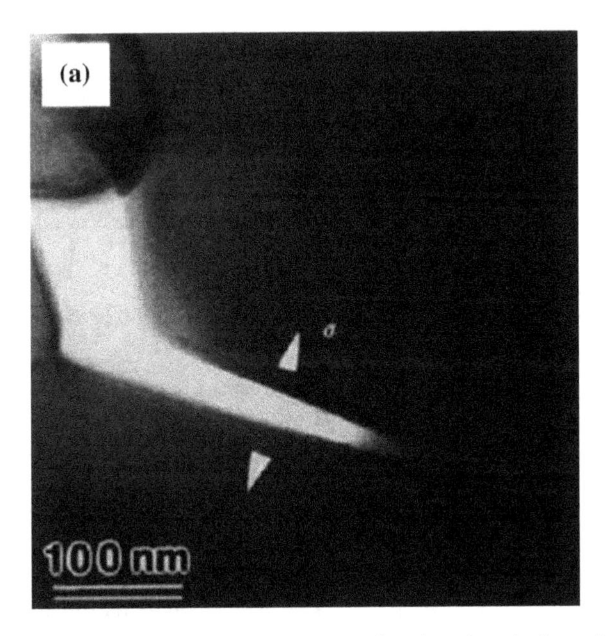

Fig. 6.79 Microstructure of deformed specimens: **a** wedge-shaped cracks formed within a triple junction and propagated between a two-grain interface [5]. With kind permission of John Wiley and Sons

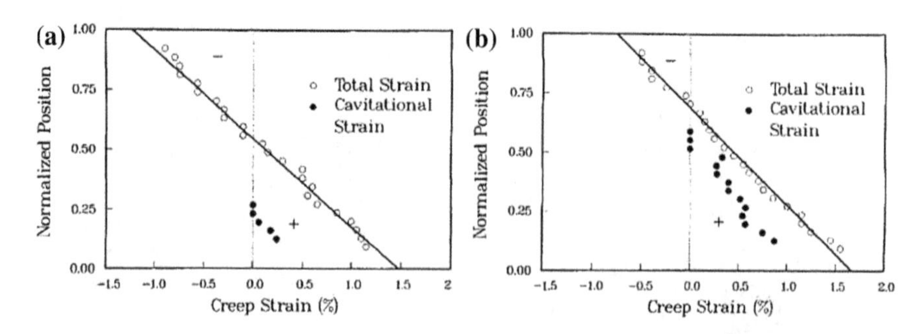

Fig. 6.80 Cavitation strain and total strain as a function of normalized beam height for creep: **a** $M = 0.165$ N-m (80 MPa initial maximum stress), 1170 °C, 300 h; **b** $M = 0.247$ N-m (120 MPa initial maximum stress), 1170 °C, 142 h. The + and − signs in this figure represent tensile and compressive stresses, respectively [5]. With kind permission of John Wiley and Sons

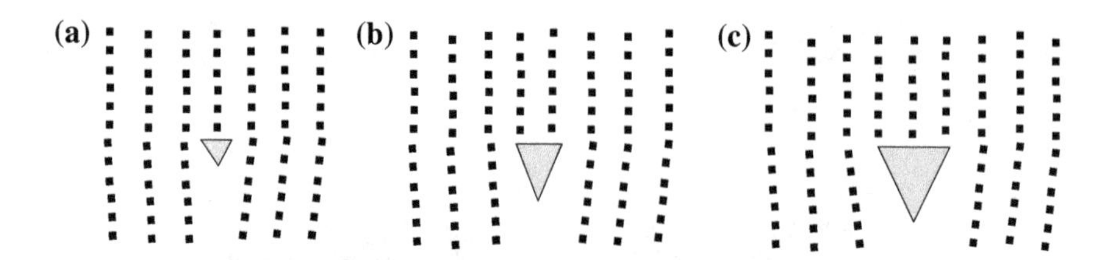

Fig. 6.81 A schematic illustration of Zener's idea, explaining how a crack of atomic dimensions can nucleate at dislocation sites; here, the growth of a crack is initiated by the coalescence of two or three dislocations

is ~60 % of the total strain. These results illustrate that cavitation strain has a strong functional dependence on the applied stress

Wedge-type crack formation at triple points was initially suggested by Zener [23] as early as 1948. According to Zener [23], at sufficiently high temperatures, grain boundaries behave in a viscous manner and, when near triple points under an applied tensile stress, wedge-type cracks develop due to the high stress concentration. Specifically, Zener [23] was among the first to suggest the concept that fracturing is a consequence of plastic deformation, which is required for crack formation. His schematic illustration is shown in Fig. 6.81, where a crack can be nucleated at the site of an edge dislocation (see: Pelleg [69] Fig. 7.9a and c)). In Fig. 6.81b and c, the coalescence of two or three dislocations is illustrated, producing an increase in the size of the crack. The concept of crack origin at dislocation sites has been addressed and modified by various researchers. In essence, Zener suggested that cracks nucleate at pile-ups of dislocations, where sufficient stress develops for the nucleation of cracks.

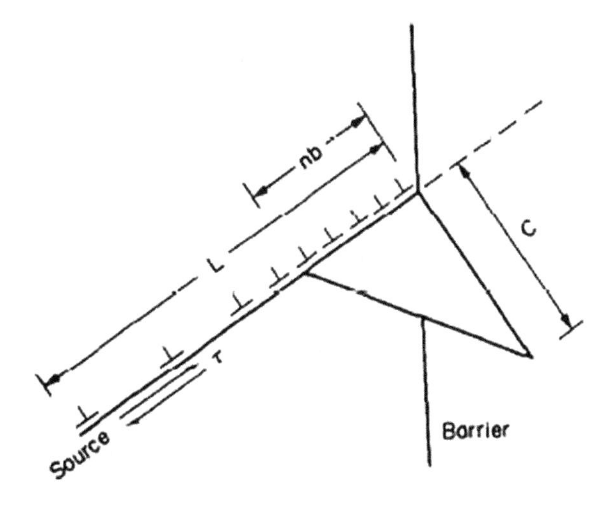

Fig. 6.82 Nucleation of a wedge crack due to pile-up dislocations on a slip plane [75]. With kind permission of Elsevier

A dislocation model for spontaneous microcrack formation was also presented by Stroh [80], by calculating the elastic energy associated with wedge deformation. Stroh also determined that the nucleation of a wedge crack was due to the pile-up of dislocations on a slip plane. In Fig. 6.82, the 2D crack dislocation of a giant Burgers vector, nb, with length, c, extending to a barrier may be seen.

His expression for the elastic energy associated with wedge deformation is:

$$W_e = \frac{Gn^2b^2}{4\pi(1-v)} \ln \frac{4R}{c} \tag{6.82}$$

in which G is the shear modulus (modulus of rigidity), nb is a giant Burgers vector, with n being the number of dislocations comprising the giant vector and R is the bounding radius in the stress field. The surface energy term, $2\gamma_s c$, may be added to obtain the total energy of the system as:

$$W_S = \frac{Gn^2b^2}{4\pi(1-v)} \ln \frac{4R}{c} + 2\gamma_s c \tag{6.83}$$

Differentiating Eq. (6.83), the critical length, c_{min}, may be found:

$$\frac{\partial W_s}{\partial c} = 0 \tag{6.83a}$$

$$c_{\min} = G \frac{n^2b^2}{4\pi(1-v)} \frac{1}{2\gamma_s} \tag{6.84}$$

In polycrystalline solids, the typical values of b, G, v and γ_s are, respectively [75]: $b = 2 \times 10^{-8}$ cm; $G = 10^{12}$ dynes/cm^2; $v = 1/3$; and $\gamma_s = 10^3$ dynes/cm, which gives for c_{min}:

$$c_{min} = 2.4 \times 10^{-8}$$

According to the theoretical presentation of Wu et al. [96], a wedge crack may be formed by the insertion extra material to create the head of a crack. An extra plane, present above a positive-edge dislocation, may serve as the source of a wedge crack. The idea is Stroh's [80], based on Zener's [23] original concept.

GBS may be considered as a deformation mechanism above 0.5 T_m. The strain rate is important to the type of failure caused by GBS. It has been shown that r-type cavities transform into w-types with increased strain rate, leading to transgranular fracture with increasing strain rate [47].

Alloying additions may decrease the tendency for w-type cavity formation. Both cavity types are the results of GBS (Raj). GBS may produce grain-boundary (intergranular) cracking when the grain's interior is stronger than its boundaries. GBS can be reduced by adding intergranular particles or by serrated grain boundaries. These serve as obstacles to GBS, apparently due to an increase in friction between the boundaries. Cavities have been seen to form at grain and phase boundaries preferentially at interfaces or triple points. The process of cavitation, associated with GBS and cavity nucleation, probably occurs at points of stress concentration in the sliding boundaries or interfaces. Creep failure occurs by the nucleation, growth and coalescence of creep cavities at the boundaries predominantly perpendicularly-oriented to the applied stress. An increase in the number of cavitated boundaries over creep-exposure time supports the mechanism of continuous cavity nucleation and growth. Some believe, on the basis of experimental observations, that there are probably pre-existing cavities, voids or pores, previously introduced by the forming processes that are actually responsible for creep cavitations in engineering alloys during long-term service at low stresses and elevated temperatures. Many experiments show that GBS is a necessary condition for cavity nucleation. GBS is a key factor not only in the growth of pre-existing voids, but also in nucleating voids for cavity formation.

An electron micrograph of a sample of 19-μm grain size (Fig. 6.83d) shows r-type cavities, which also preferentially form on boundaries orthogonal to the stress axis. The orientation-dependence of cavity distribution with respect to the stress axis has been investigated by many investigators and by Reynolds et al. [72], in uranium dioxide. It has been proposed that the preferential nucleation of cavities occurs in second-phase particles. Figure 6.83d shows the presence of second-phase particles at grain boundaries and triple points, indicating that the nucleation of r-type cavities preferentially occurs in second-phase particles. In Fig. 6.83, other cavities are also shown. Either type of cavity developed during GBS leads to creep failure.

An additional illustration effectively displays the concept of cavitation and its relation to the condition of creep deformation. This example is of UO_2, shown in Fig. 6.84. Here, the mechanism of creep fracture in uranium dioxide depends on the deformation rate. At high deformation rates, the creep-crack formation starts at triple points having an angular appearance and extends across relatively-large distances. At low rates, rounded cavities grow on boundaries parallel to the compressive stress axis. Figure 6.84 shows a SEM micrograph of specimens fractured at RT after being crept at 1628 K. The microstructures in (a) and (b)

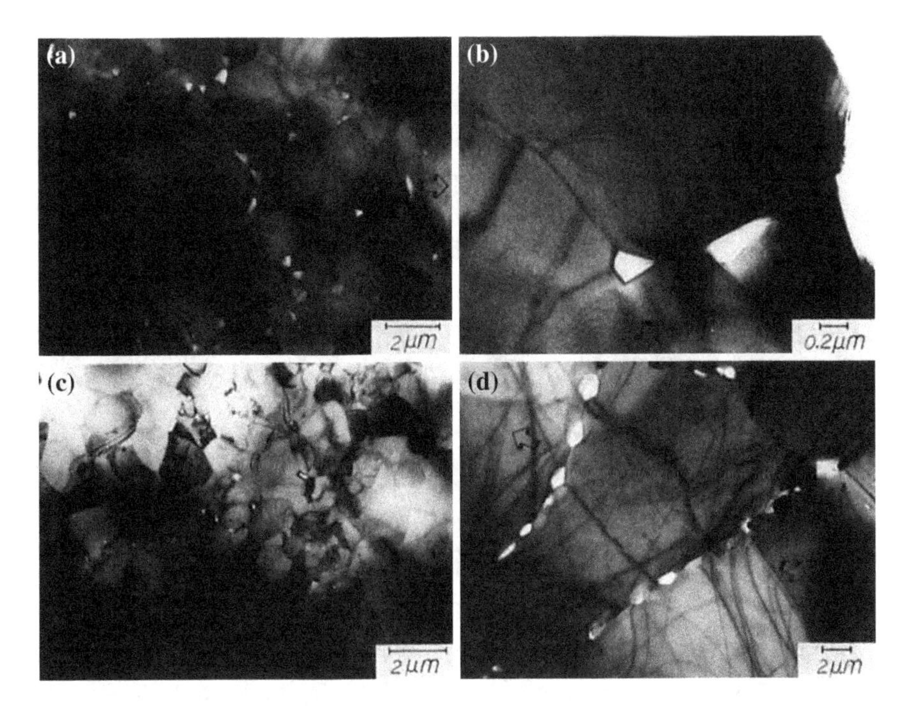

Fig. 6.83 Transmission electron micrographs of samples subjected to creep deformation showing the formation of voids. **a** Tension side of a sample of 1.8-μm grain size with a total of −3.5 % outer fiber strain. Cavities at triple points can be seen. *Arrows* indicate stress axes. **b** A higher magnification micrograph of the sample in (**a**) showing typical elongated cavities formed at triple points. Stress axes are indicated by *arrows*. **c** An electron micrograph of the same sample from the compression side. Note that there are very few cavities in comparison to the tension side. **d** A sample of 19-μm grain size showing numerous cavities (tension side, −3 % strain); note that (1) the cavities form predominantly on boundaries which are nearly orthogonal to the stress axis, (2) cavities form on second-phase particles [32]. With kind permission of John Wiley and Sons

represent high strain rate tests at a stress of 80 MN/m^2, while those in (c) and (d) are low strain rate tests at 10 MN/m^2. Clearly, these creep fracture modes are significantly different. At high rates, typical triple-point cracking is observed with single wedge cracks extending over large distances, whereas, at low rates, numerous rounded cavities are seen at the grain boundaries. The rounded cavities are more clearly seen in Fig. 6.85, where they are located on those boundaries that are parallel to the compressive stress axis. Actually, they may be considered like pores created on the growth cavities by a process involving the diffusion of vacancies into the growing cavities. Cavities are likely to be nucleated in second-phase particles or at kinks in the grain boundary and are found along the grain boundaries, as seen in Fig. 6.85.

In many polycrystalline ceramics at elevated temperatures, GBS contributes significantly to the total strain. GBS can be markedly reduced by introducing additional phases, which form precipitates (such as nitrides, carbides, borides, etc.)

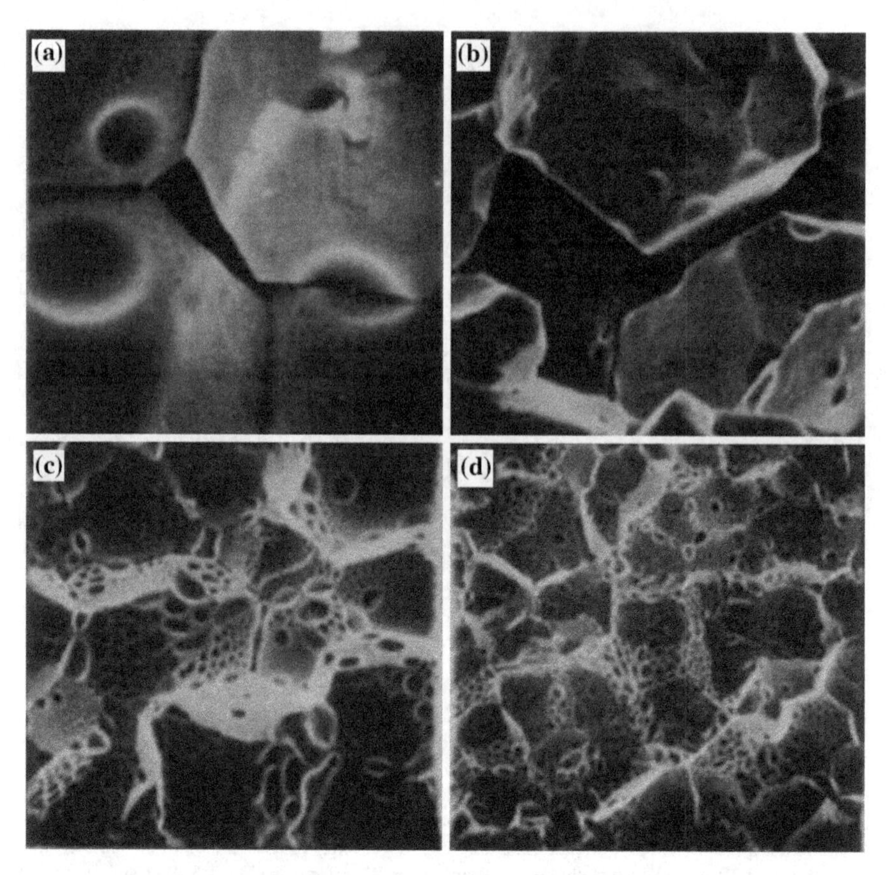

Fig. 6.84 Scanning electron micrographs of specimens tested at 1823 K at high strain rate ($\sim 2 \times 10^{-3}$/s) (**a**, **b**) and low strain rate ($\sim 10^{-6}$/s) (**c**, **d**), showing the differing modes of fracture [72]. With kind permission of Elsevier

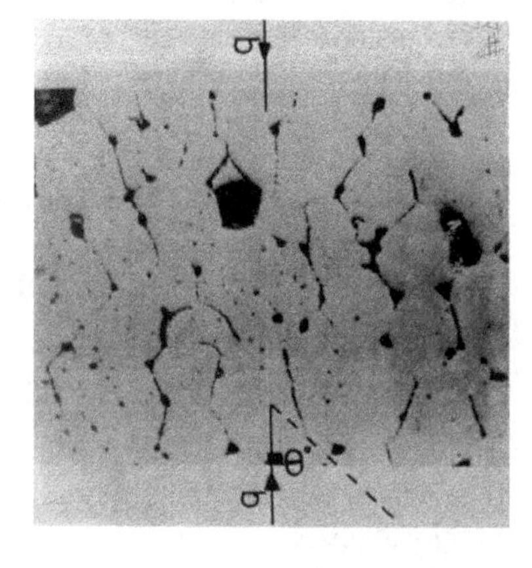

Fig. 6.85 Polished longitudinal section of large grain size (55 μm) creep specimen showing distribution of grain boundary porosity [72]. With kind permission of Elsevier

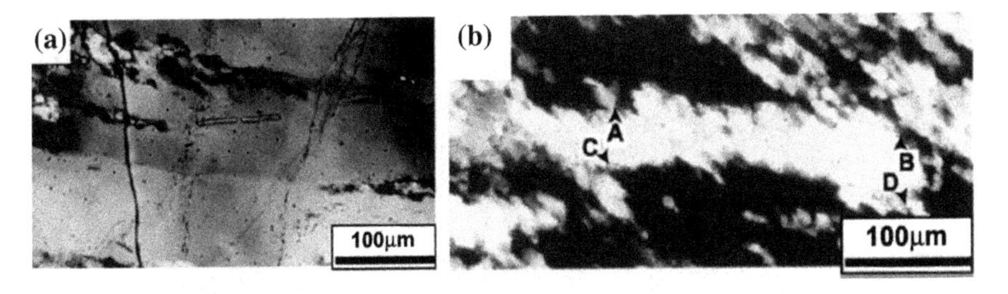

Fig. 6.86 Optical micrographs (crossed polarizers) showing examples of typical grain boundary shapes. **a** Smooth boundary developed between two B-grains. Misorientation angle is 7°. **b** Highly serrated grain boundary between W- and B-grains. Misorientation angle is 88°. *A–B* and *C–D* denote boundary segments used for the curvature analysis [68]. With kind permission of Elsevier

at the grain boundaries. This strengthening mechanism is used in high-temperature ceramics to provide safe use when creep deformation is a problem. Another method for improving creep resistance in materials is by the evolution of serrated grain boundaries. Serrated grain boundaries are effective in improving creep-strength properties. Serrated grain boundaries do not permit continued creep by GBS when stress is applied at high temperatures, unlike the case of non-serrated polycrystalline ceramics. Serrated boundaries develop in ceramics (as well in other materials) when grains do not lie on ordered planes. The effect of serration is equivalent to the 'self-locking' of the sliding process resulting from creep deformation. Thus, materials with irregular, serrated grain boundaries have improved resistance to creep-crack growth when compared to those with smooth grain boundaries. This is explained as a consequence of the difficulty in GBS and the increase in the path of grain-boundary diffusion. The strengthening mechanisms of serrated grain boundaries are principally the result of: (1) the inhibition of GBS; (2) the retardation of grain-boundary crack initiation, caused by the decrease in stress concentration at grain-boundary triple points, as a result of the decrease in sliding grain-boundary length and; (3) dynamic recovery at the serrated boundaries. In Fig. 6.86, both smooth and serrated grain boundaries in quartz may be seen. However, more constructive illustrations of serrated grain boundaries are shown in alloys in Fig. 6.87. In a recent publication [55], experimental data concerning the beneficial effect of serrated grain boundaries for reducing GBS were evaluated in a Nimonic 263 Ni-based alloy and the role of γ' precipitate was explored. Nimonic 263 is a wrought Ni-based superalloy used in gas-turbine combustion chambers and potentially applicable to outlet headers and steam lines of advanced coal-fired power plants.The precipitate, γ', has a volume fraction of ~ 10 % in the γ matrix. Grain-boundary serration occurs prior to the formation of $M_{23}C_6$ and without interaction with γ' particles, accompanied by the modification of its grain-boundary carbide characteristics. This occurs when a specimen is slow-cooled from the solution treatment temperature. The high-resolution observation of the lattice image of the serrated grain boundary suggests that the

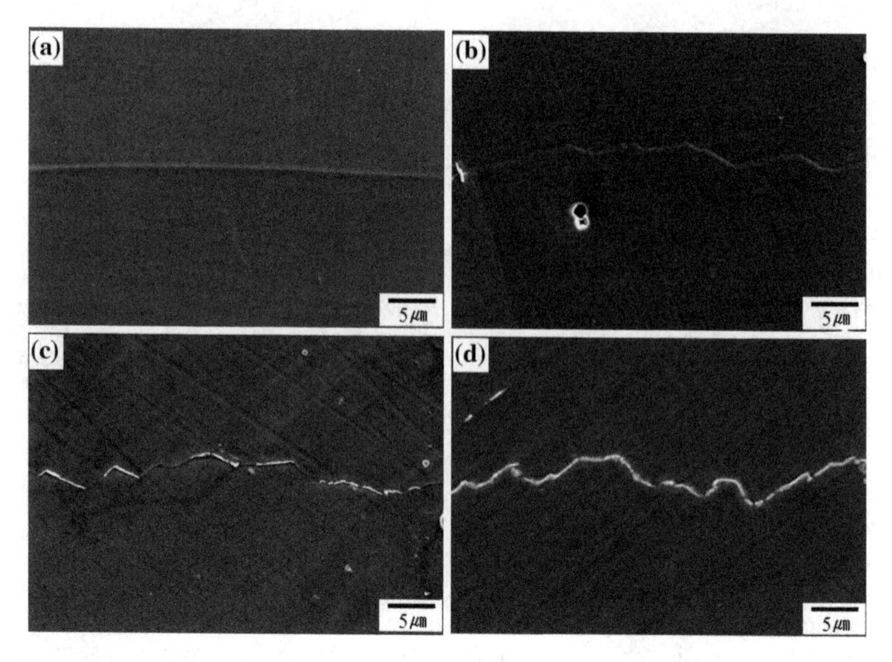

Fig. 6.87 The sequential development of grain boundary serration and subsequent carbide precipitation: **a** after solution-treatment (1150 °C/5 min/WQ), the grain boundary is flat **b** solution-treatment followed by slow cooling to 1000 °C at 10 °C/min, **c** solution treatment followed by slow cooling to 800 °C at 10 °C/min, and **d** solution-treatment followed by slow-cooling to 800 °C at 10 °C/min and aged for 8 h [55], pp. 638–642). With permission of Material Science Forum and the authors

grain boundaries tend to serrate when specific segments approach a {111} low-index plane at a boundary, in order to attain lower interfacial free energy. This finding implies that serration may be related to the local movement of grain boundaries intended to reduce the interfacial free energy.

The improvement with the serrated grain boundary and the creep life observed in the serrated sample are associated with a zigzag array of cavity formation in the serrated grain boundary, making it more difficult for the cavities to interlink and form an intergranular path for crack propagation than in the non-serrated sample. A lower rate of crack propagation along grain boundaries is expected in the serrated sample and, as a consequence, it may be inferred that the serrated sample is highly resistant to damage by cavity formation.

In addition to grain boundary serrations, GBS can also be reduced by tailoring the grain size. Figure 6.88 is an illustration of the grain-size effect in reducing GBS in an alumina-silicon carbide ceramic composite. Two grain sizes appear in this figure, containing 10 vol% SiC whiskers for the strengthening of the alumina.

In Fig. 6.88, creep strain and the creep-strain rate are shown versus time-quite similar to Fig. 6.1a and b. Analogous to curves Fig. 6.1a and b, the creep curves in Fig. 6.88a exhibit a primary stage and a well-developed second-stage creep over

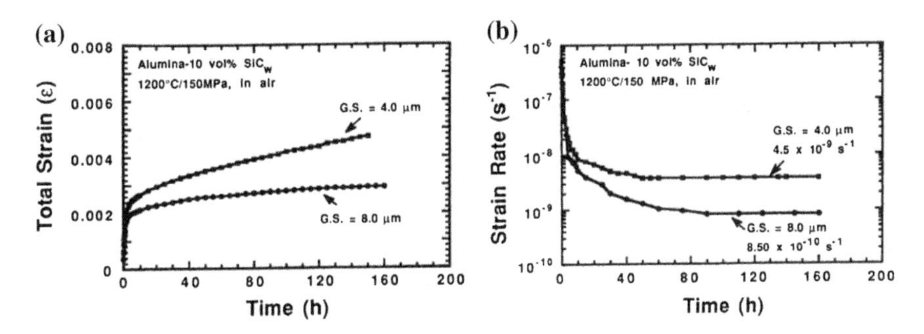

Fig. 6.88 **a** Strain and **b** strain rate versus time curves for alumina-SiC composites tested at 1200 °C and at 150 MPa [60]. With kind permission of John Wiley and Sons

time. The minimum creep rate was achieved after 140–160 h. From the stress dependence of the creep rate, a stress exponent of 2 was evaluated and is characteristic of the GBS mechanism. The creep resistance of alumina composites increases with an increase in matrix grain size and the creep rate (at constant applied stress) exhibits a grain size exponent of approximately 1. Further observations show that the prevalent site for creep cavity formation in coarse-grained alumina-SiC composites is at two-grain junctions, whereas, in the fine-grained ceramics, cavities form preferentially at three-grain junctions.

It has been indicated that the addition of a second phase distributed within the grain boundaries strengthens the ceramics and its creep properties (as in metals) by reducing GBS. This is illustrated in the case of Al_2O_3 composites with 20 vol% SiC particles at T = 1260 °C. Improvement in their creep resistance, compared to the creep behavior of monolithic Al_2O_3, is due to the pinning effect of the SiC particles at Al_2O_3 grain boundaries by reducing GBS during creep. Figure 6.89 shows that both composites, designated as ASS2 and ASL2, have uniform grain sizes and their SiC particles are not entrapped in the Al_2O_3 matrix. (In the SiC designation, S and L refer to grain sizes of 0.6 and 2.7 μm, respectively; the number 1 and 2 indicate 10 and 20 vol%, respectively). TEM micrographs of (a) ASL1 and (b) ASL2 in Fig. 6.90 shows the pinning effects of irregular and elongated SiC particles on Al_2O_3 grain boundaries (indicated by arrows). Figure 6.91 shows the stress dependencies of steady-state or minimum creep rates for monolithic Al_2O_3 and the composites with 20 vol% SiC particles at T = 1200 °C.

Linear fitting indicates that the stress exponent of monolithic Al_2O_3 is 1.45 and the stress exponents of ASS2 and ASL2 are 6.38 and 4.18, respectively. From Fig. 6.91, one may see that the strain rates of ASS2 and ASL2 are lower than those of monolithic Al_2O_3 in the low stress region. The ASL2 ceramic has a strain rate ~4–8 times lower than that of monolithic Al_2O_3. Since no dislocation motion was found in the crept specimens of the monolithic Al_2O_3 and its composites, the boundary creep mechanisms are predominant and the grain-size exponent, p, should vary from 1 to 3.

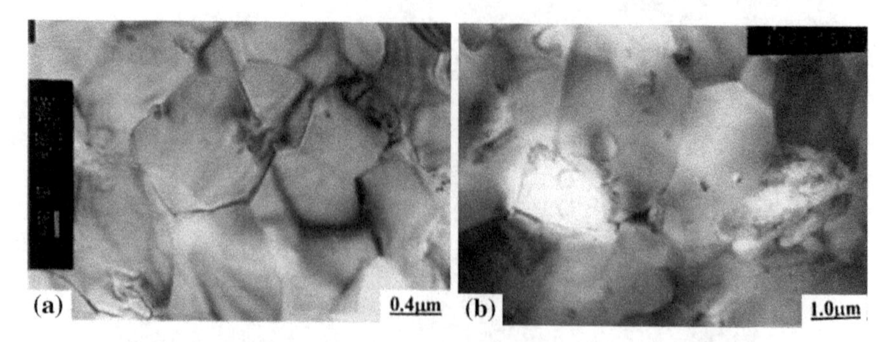

Fig. 6.89 TEM micrographs of **a** ASS2 and **b** ASL2, showing the equiaxed grain morphology of the composites with 20 vol% SiC particles [42]. With kind permission of Elsevier

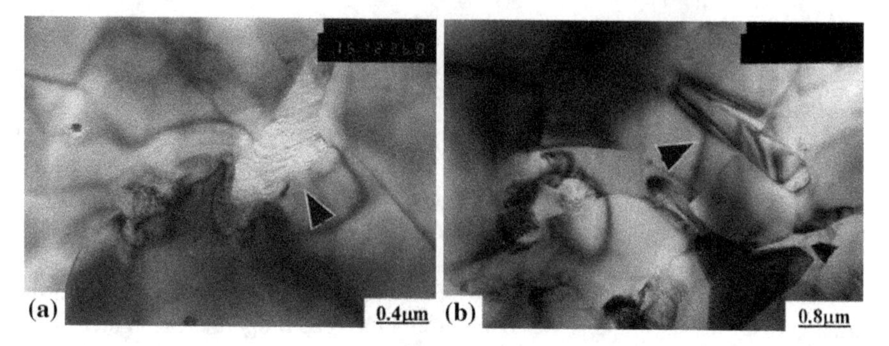

Fig. 6.90 TEM micrographs of **a** ASL1 and **b** ASL2, showing the pinning effect of irregular and elongated shaped SiC particles on Al_2O_3 grain boundaries (indicated by *arrows*) [42]. With kind permission of Elsevier

GBS, facilitated by boundary diffusion, is the predominant creep deformation mechanism in polycrystalline Al_2O_3 at temperatures below 1400 °C. TEM observations indicate that there are some cavities at triple-grain junctions or grain boundaries on the tensile sides of crept monolithic Al_2O_3 and its composite specimens. The shape of the Al_2O_3 matrix grain experienced no apparent change during the creep tests. Therefore, GBS, accommodated mainly by diffusion (though some remain unaccommodated), is the principal creep mechanism in the Al_2O_3 ceramics illustrated. The improvement in creep resistance seems to be due to the pinning of the SiC particles, which inhibit Al_2O_3 GBS during creep.

To summarize this section, note that GBS may account for 10–65 % of the total creep strain, depending on the alloy and the conditions of its use in service (temperature, load, etc.). In alumina, for example, experimental measurements of the offsets in marker lines at the grain boundaries reveal that the contribution of GBS to creep strain is 70 ± 6.2 % Chokshi [6]. Its contribution to creep strain increases with rising temperature and stress and with reduced grain size.

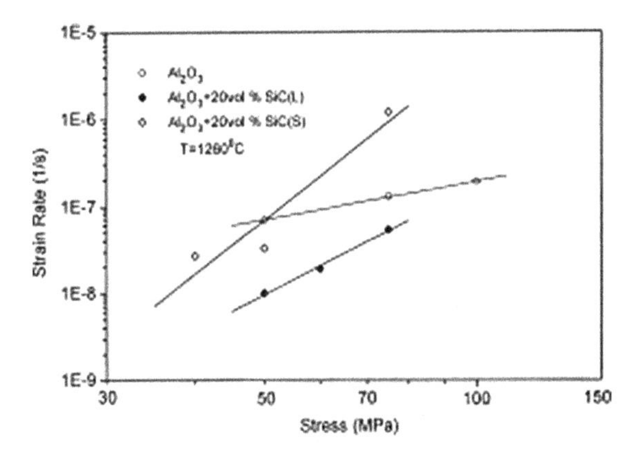

Fig. 6.91 Stress dependencies of steady-state or minimum creep rate for monolithic Al_2O_3 and the composites at T = 1200 °C. The stress exponent for strain rate is 1.45 for monolithic Al_2O_3, 6.38 and 4.18 for ASS2 and ASL2, respectively [42]. With kind permission of Elsevier

Above ~0.6 T_m, the grain-boundary region is thought to have lower shear strength than the grains themselves, probably due to the looser atomic packing at the grain boundaries. GBS may be reduced by introducing precipitates or grain-boundary serrations, which resist GBS and significantly reduce cavity formation of the types indicated above (which is a major factor in creep failure).

6.8 Creep Rupture in Ceramics

Creep rupture is actually related to creep deformation. Creep is a time-dependent deformation of a material while under an applied load that is below its yield strength. It most often occurs at elevated temperatures, but some materials creep at RT. Creep terminates in rupture, if steps are not taken to bring it to a halt. In essence, creep rupture tests are used to determine the elapsed time-to-failure. Generally, higher stresses are used for creep-rupture testing than in conventional creep tests and these tests are carried out until the specimen fractures. The objectives of the respective tests are to determine the minimum creep rate at stage II creep, on one hand, and to evaluate the time at which failure sets in, on the other. Such information is essential so that the proper ceramics will be selected to eliminate failure during service and to evaluate the time-period of safe use for high-temperature applications in which structural stability is essential. Various ceramic components operate at high temperatures and may experience creep. As in creep testing, stress-rupture testing involves the same testing elements (for example, a tensile specimen) and is performed under a constant load (or stress) at a constant temperature. Not surprisingly, creep failures may appear ductile or brittle, due to the nature of ceramics, but they are temperature dependent. Cavities, believed to be responsible for cracking by cavity coalescence, can be either r-type or w-type and either transgranular or intragranular. Figure 6.92 is a schematic illustration (based on Fig. 6.1a), showing where creep damage starts.

The location of creep damage coincides with the place where tertiary creep sets in and represents the minimum creep rate (also seen in Fig. 6.1b). This is one of the concepts regarding the time at which cavitation (either as microcavities or voids) develops. However, there are experimental indications (density measurements) that intergranular cavities may be observed before tertiary creep and are well-developed at the end of second-stage creep. Creep data for general design use are usually obtained under conditions of constant uniaxial loading and constant temperature. The results of tests are usually plotted as strain versus time-to-rupture. The experimental data indicate that, of the great variety of creep curves described by various laws, their shape is near-linear when the data are presented on a log-strain versus log-time basis. Time is expressed in hours. Very often, instead of using the rupture time, the time until reaching a steady-state or minimum creep is preferred, because then a much shorter period of testing time is needed to collect the creep data. The following discussion on creep structure will be exemplified by three ceramics of great technological interest: Al_2O_3, SiC and Si_3N_4.

6.8.1 Alumina

For high-temperature applications, major research efforts continue to be directed towards the development of a ceramic with improved toughness and damage tolerance, seeking to overcome the engineering design constraints imposed by the inherently brittle nature of monolithic ceramics. To that end, most of the monolithic ceramics are improved by various additives, among them ceramics, such as: ZrO_2, SiC, Si_3N_4, etc. Alumina is no exception. First, Fig. 6.93 illustrates the case of Al_2O_3 without any additives. The figure presents stress rupture data on a log–log plot, as is customary. (ARCO and AVCO are two manufacturers of alumina). Two distinct linear regions may be seen, indicating two mechanisms. At the two highest stresses, the data fit the known power relation, in which $t_r \sim \sigma^{-m}$ with m at about 2.5. At the lower stress, below 55 MPa, m is about 1.8. A transition region exists between these two regions, in which there is increased time-to-failure at decreased stress. Two types of tests were performed on the ARCO alumina, for flexural and tensile creep.

In Fig. 6.94, a combination of flexural and tensile tests was used; thus, two transitions in creep failure behavior are observed. At high stress (above about 175 MPa in flexure, performed by 4-point test), failure occurs very rapidly (52 h) and at low strain (51 %, not shown). Thus, failure occurs before steady-state creep is established. For a given stress, the scatter in time-to-failure is large (2 orders of magnitude). Thus, it is difficult to determine a stress exponent for stress rupture. However, it is quite high, about 40. At stresses below 175 MPa in flexure a transition occurs. The failure strain is much greater (12–16 % for AVCO; >18 % for ARCO) and the stress exponent is between 2 and 3.

The failure of ceramic polycrystals may generally be related to preexisting flaws (or in cases of high-temperature failure, to flaws generated during service).

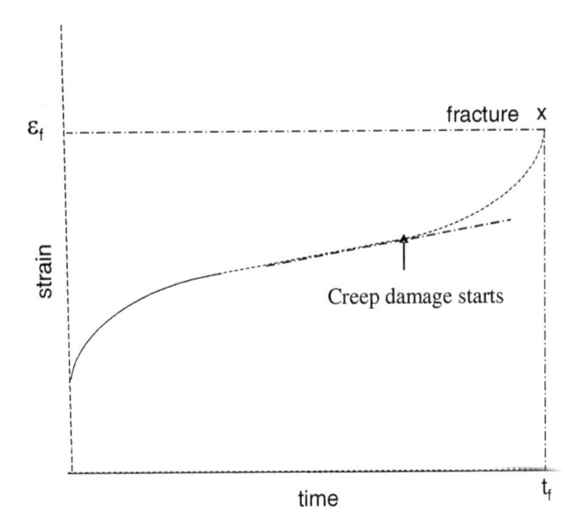

Fig. 6.92 A schematic creep curve (see Fig. 6.1a); ε_f and t_f are the strain and time to creep failure

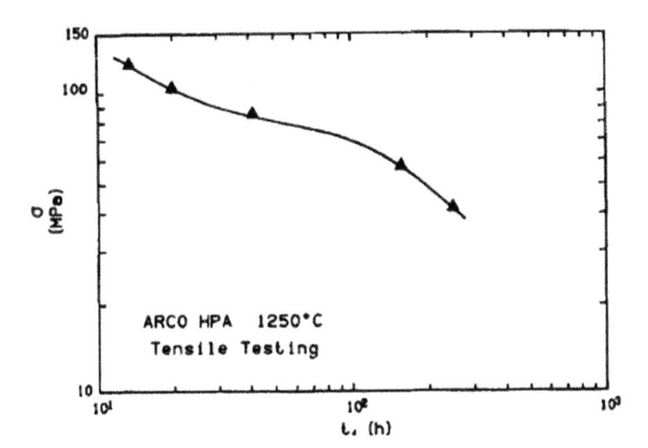

Fig. 6.93 Time to failure as a function of applied stress for ARCO alumina tested in tension. (ARCO is the manufacturer) [74]. With kind permission of John Wiley and Sons

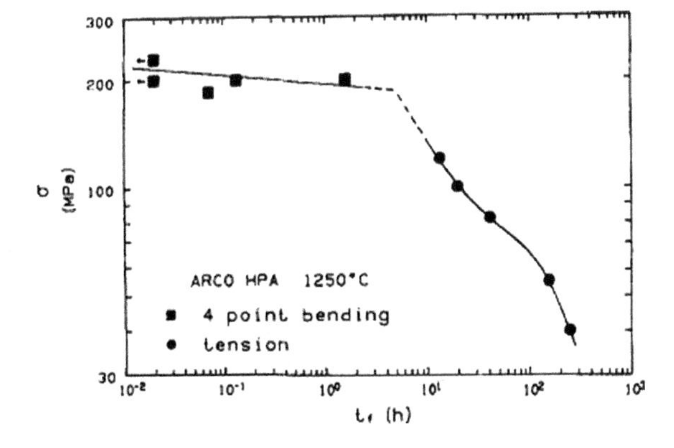

Fig. 6.94 Stress rupture data of Fig. 6.79 replotted including the high stress flexure data for ARCO alumina [74]. With kind permission of John Wiley and Sons

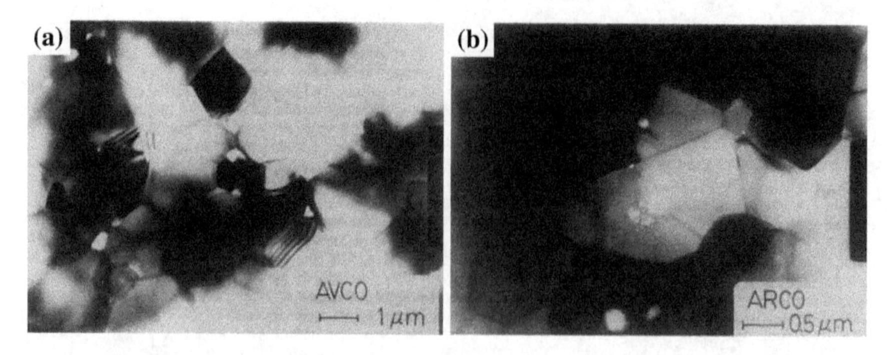

Fig. 6.95 Transmission electron micrograph of the as-received microstructure for the (**a**) AVCO and (**b**) ARCO materials [93]. With kind permission of John Wiley and Sons

In Fig. 6.95, the relatively very low porosity of each of the as-received materials is shown by TEM at a level of ~0.01 %, however their distribution is different. The AVCO alumina almost exclusively contains intergranular pores at three-grain junctions with a mean radius of 0.12 μm. The ARC0 alumina contains mostly intragranular porosity, typically in the form of clusters of very fine pores. Thermally etched, polished sections of as-received material do not reveal many microstructural defects. However, subsequent creep testing exposes defects in the form of porous regions, regions having large grain size and regions of chemical inhomogeneity. The defects are more numerous and more severe in the ARCO materials. Below a certain threshold, failure occurs because of the development of several microcracks, that grow simultaneously until one of them becomes critical. This has been evaluated by both tensile and flexure testing. In ARCO alumina in flexure, the samples develop relatively few microcracks. These are mostly nucleated at chemical inhomogeneities or are processing flaws (Fig. 6.96). They do not propagate easily through the material, but rather blunt or branch after propagating some distance (Fig. 6.96a). Other microcrack origins include large grains, either isolated (Fig. 6.96b) or in clusters (Fig. 6.96c).

For the maximum strains available in the bend rig (about 18 %), the larger cracks which develop never propagate to the point of failure. Figure 6.97 further illustrates the cracks in both types of alumina. These microcracks remain sharp as they grow (Fig. 6.97a), while cavities are flat or crack-like and cover a single-grain facet, in most cases (Fig. 6.97b). Failure occurs by the competitive growth of several microcracks, until one becomes critical.

The main difference between the two aluminas in this regime is the amount of microcracking and cavitation observed, which is much more extensive in the AVCO material. The higher density of microcracks in the AVCO alumina leads to the development of "shear" bands, which occur via microcrack coalescence during flexure deformation. High-magnification imaging of polished internal surfaces reveals a large density of cavities (Fig. 6.85) which, after coalescence, leads to the development of microcracks. These cavities are intergranular, having crack-like or

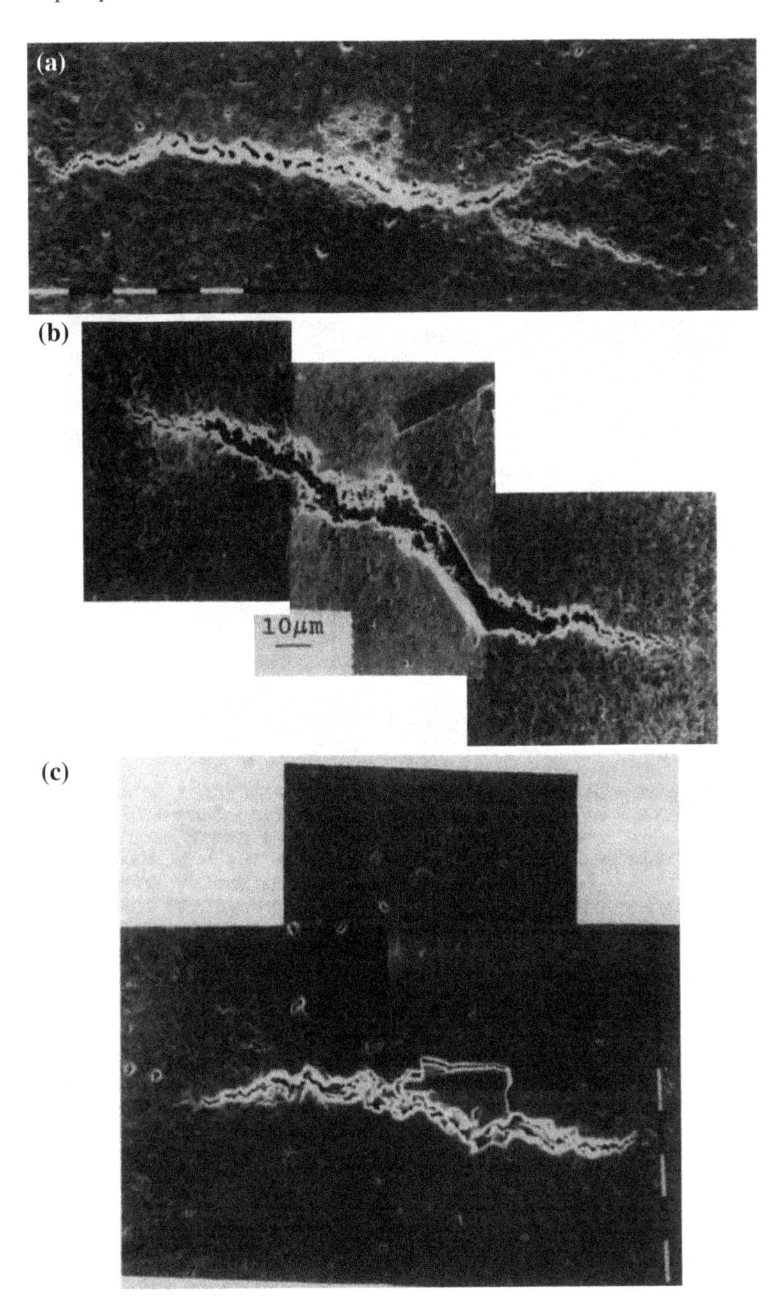

Fig. 6.96 Origins of microcracks include **a** chemical deposits sometimes leading to crack bifurcation (ARCO, 1250 °C, 140 MPa, flexure), **b** isolated large grains (ARCO, 1200 °C, 125 MPa, flexure), and **c** large grain clusters (ARCO 1250 °C, 170 MPa, flexure). Bars = 10 μm [93]. With kind permission of John Wiley and Sons

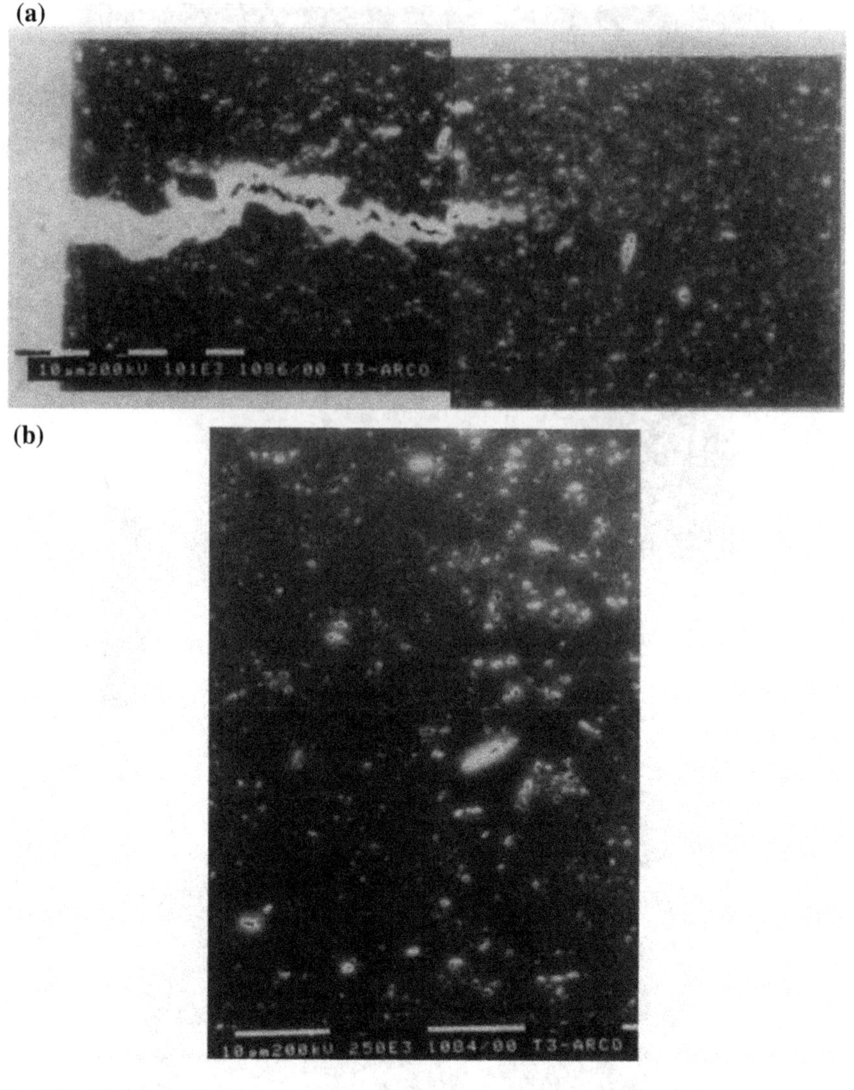

Fig. 6.97 High-magnification view of creep damage in an internal section of a tensile sample fractured of ARC0 at 82 MPa and 1250 °C: **a** near a microcrack and **b** typical general cavitation damage. Bars = 10 μm [93]. With kind permission of John Wiley and Sons

angular shapes, likely to be associated with surface diffusion and a GBS-induced cavity growth mechanism (Figs. 6.98 and 6.99).

Creep life is limited by the rate of crack propagation. Critical crack size development depends on the stress dependence of the failure time. The time-to-failure at stresses under tension below a certain limit (82 MPa for alumina) is a consequence of an increase in failure strain at the crack tip. Basically, failure is

Fig. 6.98 Microcracking at low stresses: **a** on the side of a tensile specimen and **b** an enlarged view showing the coalescence process. Tensile axis is vertical. ARCO, 40 MPa, 1250 °C [93]. With kind permission of John Wiley and Sons

(a)

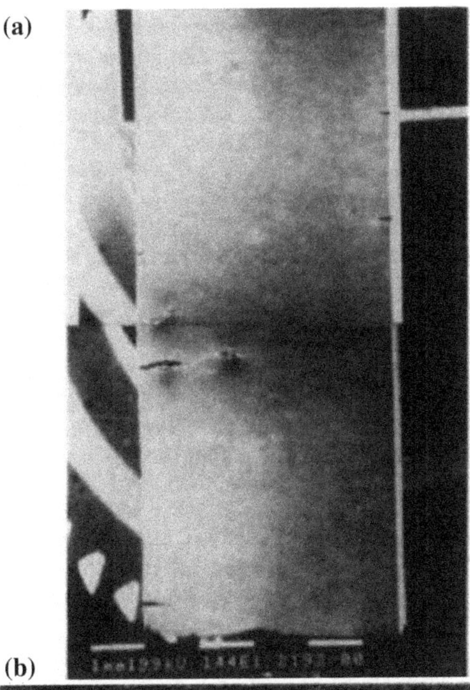

(b)

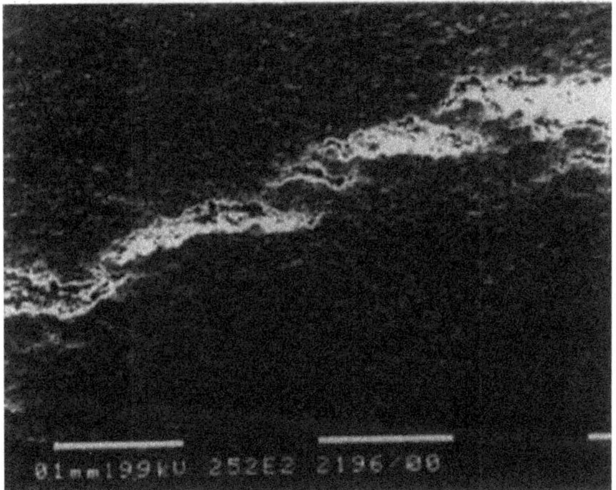

related to the presence of flaws, which, therefore, should be controlled by manufacturing in order to reduce creep rupture during high-temperature applications.

Developing appropriate ceramic–matrix composites [henceforth: CMCs] for possible aero-engine applications started more than a decade ago. CMCs continue to be developed for improved toughness, to overcome the inherently brittle nature of most of the monolithic ceramics. Many experiments have shown that long-term loading of CMCs (for thousands of hours) produce improved high-temperature properties in monolithic ceramics by causing the dispersion of ceramic whiskers or

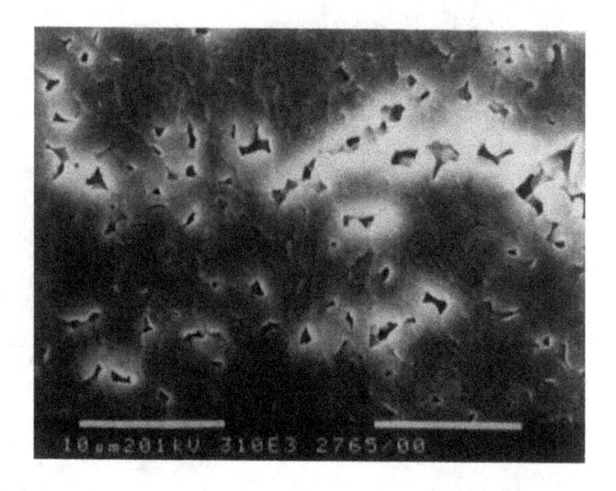

Fig. 6.99 General cavitation damage and the generation of a microcrack by cavity coalescence, in a tensile specimen of ARCO deformed at 40 MPa and 1250 °C. Tensile axis is vertical. Bars = 10 μm [93]. With kind permission of John Wiley and Sons

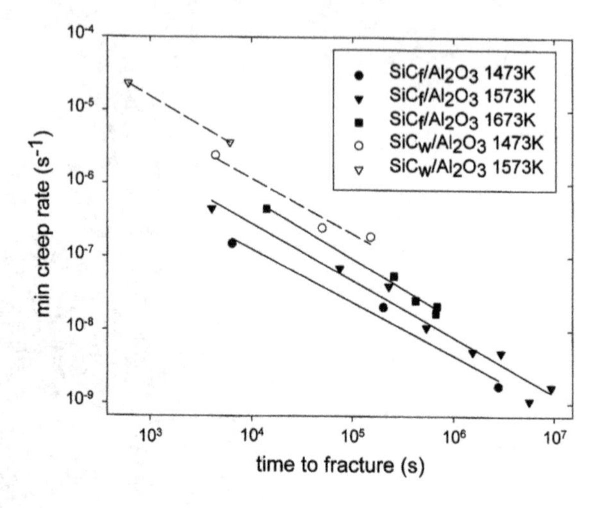

Fig. 6.100 The dependence of the time to fracture (t_f) on the minimum creep rate ($\dot{\varepsilon}$) for the 0/90° SiC/Al_2O_3 composite from 1473 to 1673 K, with data also included for a SiC/Al_2O_3 composite tested in tension at 1473 and 1573 K [94]. With kind permission of Elsevier

particles. Alumina, among others, has been strengthened by 10–33 vol% SiC whiskers, resulting in substantial improvements in toughness and strength, as well as significant increases in high-temperature creep and creep-fracture resistance. These improvements, following the dispersion of ceramic whiskers, rely on toughening mechanisms, including whisker bridging or pull-out and crack deflection. In Fig. 6.100, the minimum creep rate versus time-to-fracture is indicated at several temperature applications.

Creep and creep-rupture data may be presented as curves in different ways. One way is in terms of stress versus rupture time. Very often, instead of using the rupture time, the time until reaching a steady-state or minimum creep is preferred, because a much shorter period of time is required to collect creep-test data. Figure 6.100 is such a curve. Stress-rupture tests are used to determine the failure time, as mentioned above. The data are plotted as log–log curves. A straight line is

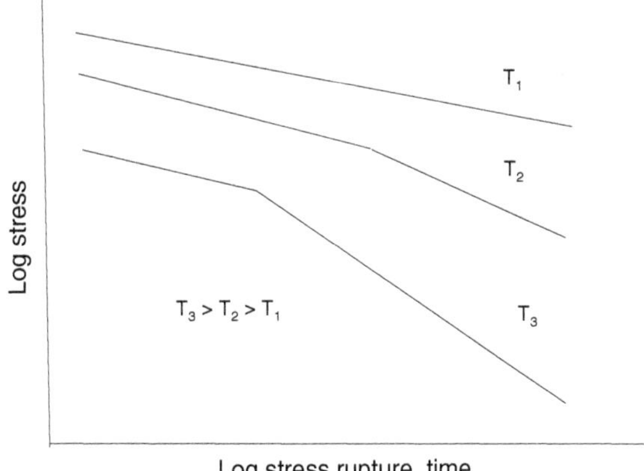

Fig. 6.101 A schematic illustration of logσ versus logt_r curves at three temperatures. At the higher temperatures, a change in slope occurs, indicating possible structural changes

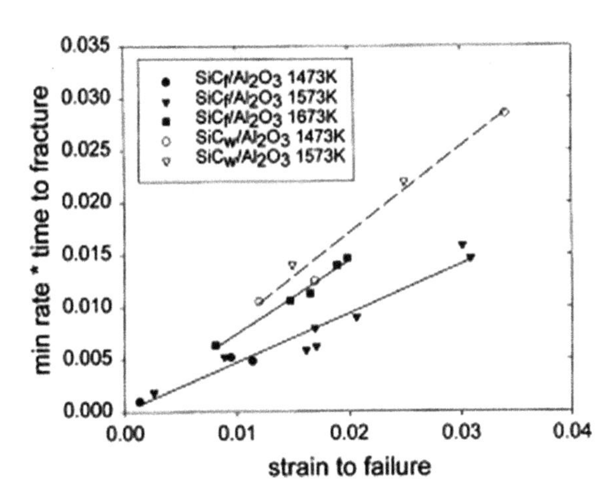

Fig. 6.102 The variation of the product of the minimum creep rate and the time to fracture ($\dot{\varepsilon}_m \cdot t_f$) with the total creep strain to failure (ε_f) for the 0/90° SiC_w/Al_2O_3 composite from 1473 to 1673 K, with data also included for a SiC_w/Al_2O_3 ceramic tested in tension at 1473 and 1573 K [94]. With kind permission of Elsevier

usually obtained for each temperature, if no structural changes occur. This information can then be used to extrapolate the time-to-failure for longer times. However, with increased temperature, structural changes are likely to occur, changing the creep resistance until rupture. It is important to be aware of such changes in material behavior. If structural changes do occur, then it is not safe to extrapolate the data of such curves to longer times. A schematic illustration (shown in Fig. 6.101) indicates a change in slope due to structural changes (Fig. 6.99).

The minimum rate x of time-to-failure versus strain-to-failure may be seen in Fig. 6.102, which is based on the relation:

$$t_f \propto \varepsilon_f / \dot{\varepsilon}_m \tag{6.85}$$

The microstructures of these ceramics are illustrated in Figs. 6.103 and 6.104. The longitudinal fibers transfer stress to the matrix, causing intergranular crack

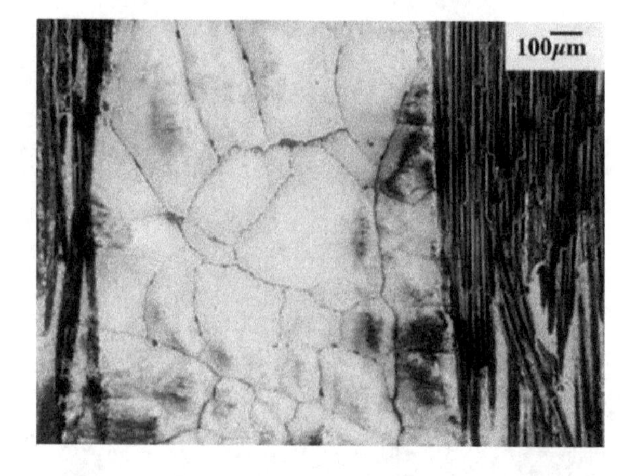

Fig. 6.103 Scanning electron micrograph showing crack development in the alumina matrix of the 0/90° SiC_f/Al_2O_3 composite [94]. With kind permission of Elsevier

Fig. 6.104 Scanning electron micrograph showing the zone characterized by oxidized fibres (*marked A*) and the final fibre pull-out zone on the creep fracture surface of a 0/90° SiC_f/Al_2O_3 crept to failure in air under a tensile stress of 79 MPa at 1573 K [94]. With kind permission of Elsevier

development (see Fig. 6.103). Cracking reduces matrix stiffness, reloading the fibers and inducing further creep. As crack growth occurs, the developing cracks become bridged by the longitudinal fibers; however, oxygen penetrates during the tests in air causing oxidation. Oxidation promotes failure of the crack-bridging fibers.

The regions showing oxidized fibers extend over substantial areas of the final fracture surfaces before the cracks grow to the length required to cause the sudden failures of the test-pieces by fiber pull-out (Fig. 6.104).

There is no difference in the high-temperature tension creep results between SiC-fiber- or SiC-whisker-reinforced alumina. However, the mechanism of creep seems to be different. In the case of SiC_w/Al_2O_3, creep occurs by GBS, resulting in cavitation leading to crack development and eventual failure. Yet, in SiC_f/Al_2O_3, tensile creep and creep fracture properties are determined by the strength of the fiber (Nicalon). In a case of fiber-reinforcement creep, cracking occurs in the porous matrix, developing to cavitation, which is dependent on the bridging by the fibers. This process, as expected, is time dependent. The use of continuous-fiber-

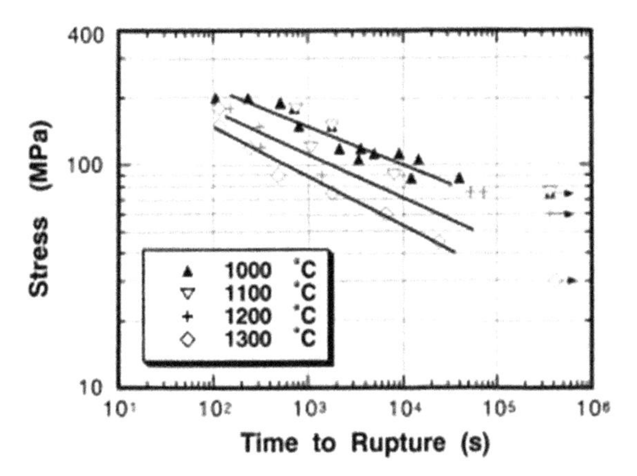

Fig. 6.105 Tensile stress versus time to rupture in argon at 1000, 1100, 1200 and 1300 °C [99]. With kind permission of Elsevier

reinforced composites for advanced CMC applications has great potential for long-term exposure at elevated temperatures.

6.8.2 SiC

An understanding of the creep behavior of ceramic materials is necessary in order to determine lifetime limits in applications where resistance to high temperatures is needed. Silicon carbide is one of the commercial ceramics that is used for various high-temperature structural applications.

To improve the performance of SiC and to increase its resistance against creep failure, generally various constituents are added to monolithic SiC ceramics. Additives in various shapes and sizes are usually added to SiC to achieve a better material for structural use and to extend its service lifetime. An evaluation of creep failure, commonly referred to as 'creep rupture' or 'stress rupture', is a critical step in evaluating the suitability of a certain ceramic for use in the desired application. The stress rupture and creep properties of a SiC matrix reinforced with SiC fiber (i.e., a SiC/SiC composite) has been evaluated by tests conducted in order to assess the propensity of SiC/SiC for high-temperature applications over an extended lifetime. In Fig. 6.105, plots of stress versus time-to-rupture are shown for several temperatures. As commonly done, these plots are on a log–log scale. Each curve can be fitted by means of an empirical relation, similar to the earlier exponential equation expressing the time-to-rupture, t_r, to a stress exponent for stress rupture as:

$$t_r = B\sigma^N \tag{6.86}$$

B is a constant and N is a stress exponent for stress rupture.

According to Eq. (6.86), N is 5.8 at 1000 °C, 4.1 at 1100 °C, 8.1 at 1200 °C and 4.2 at 1300 °C. Most of these results are similar to the stress exponents for

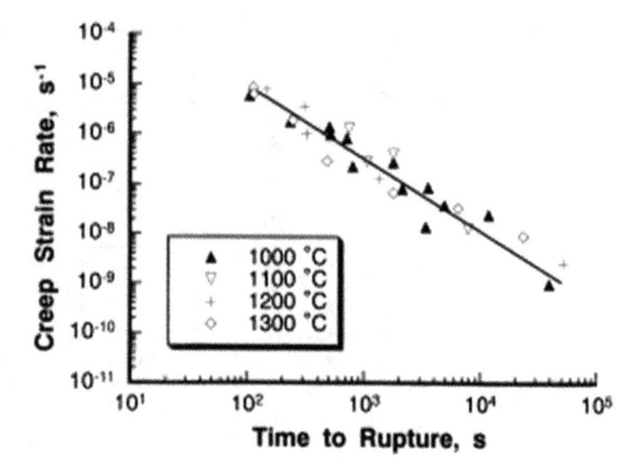

Fig. 6.106 Tensile minimum creep strain rate versus time to rupture at different stresses in argon at 1000, 1100, 1200 and 1300 °C [99]. With kind permission of Elsevier

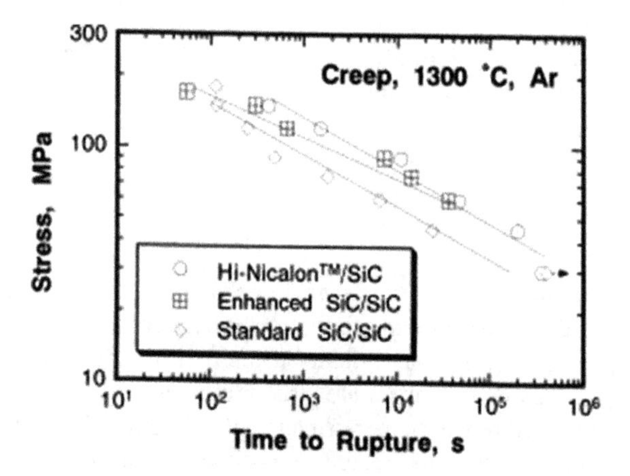

Fig. 6.107 Time to rupture versus stress in Hi-NicalonTM/SiC in air, enhanced SiC/SiC in air, and standard SiC/SiC in air and argon at 1300 °C [99]. With kind permission of Elsevier

creep at high stresses. The value at 1200 °C is unusually higher than others, since the data at 75 MPa are offset from others. Figure 6.106 illustrates the minimum creep-strain rate as a function of time-to-rupture at the same temperatures indicated in Fig. 6.105. Various reinforced SiCs were also tested by stress-rupture tests as illustrated in Figs. 6.107 and 6.108. In the illustrations, enhanced SiC/SiC refers to carbon-coated fiber reinforcement. There is some difference between these various fiber-reinforced SiCs, but both the Hi-Nicalon/SiC and the enhanced SiC/SiC are better than the standard SiC/SiC.

Note that in the ceramics community, flexural-creep testing using three- or four-point bending is common, since it provides an easy means for collecting creep data; this type of loading avoids the alignment and gripping problems of the test specimens often encountered in tensional-creep experiments.

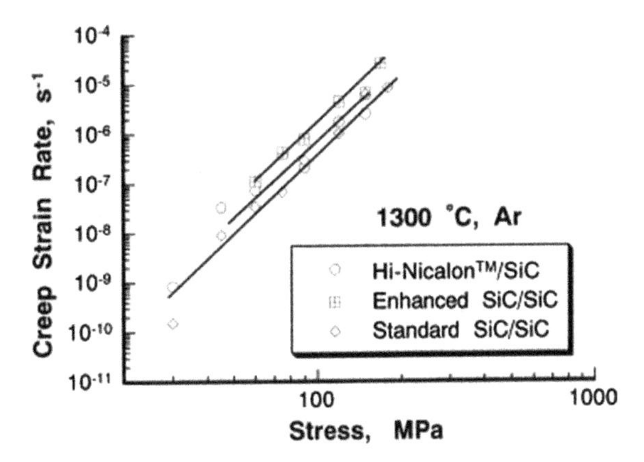

Fig. 6.108 Minimum creep strain rate as a function of stress in Hi-NicalonTM/SiC in air, enhanced SiC/SiC in air, and standard SiC/SiC in air and argon at 1300 °C [99]. With kind permission of Elsevier

6.8.3 Si_3N_4

The utilization of silicon nitride ceramics as components in advanced automotive, gas and turbine engines, among other potential applications, has received considerable attention over the last decade. Great advances have been made in the improvement of their mechanical properties at elevated temperatures, leading to their use as structural material for many engineering applications. It became necessary to also improve the reliability of monolithic and reinforced silicon nitrides by means of creep and creep-rupture testing. Below, some examples of monolithic Si_3N_4 are considered, followed by a comparison to Si_3N_4 reinforced by SiC whiskers. The addition of a reinforcing agent is one of the common methods for improving the performance of monolithic ceramics. Figure 6.109 is an illustration of creep rupture in Si_3N_4. As indicated in Fig. 6.101, a straight line is usually obtained for each temperature, if no structural changes occur. The dependence of creep-rupture lifetime on stress is shown in this figure. The dramatic change in the slope of the stress versus failure-time curves indicates that the kinetics of the growth mechanism change as stress decreases. The change in slope reflects a transition from stress-controlled to creep (strain)-controlled failure. Such behavior has been observed in other cases involving both polycrystalline alumina and silicon nitride. In the creep regime, the time-dependent evolution of damage in the form of cavities is thought to control failure. In this case, failure occurs when the strain related to the creep damage reaches a critical value. NT154 ceramic is a commercial material obtained by HIPing with low levels of yttria. A reasonable compromise in properties may be obtained via the fabrication of silicon nitride with an excellent combination of fracture strength and toughness. Engine components fabricated from this material have been successfully operated in gas-turbine engines at 1150 °C in a nitrogen atmosphere for 222 h. However, during

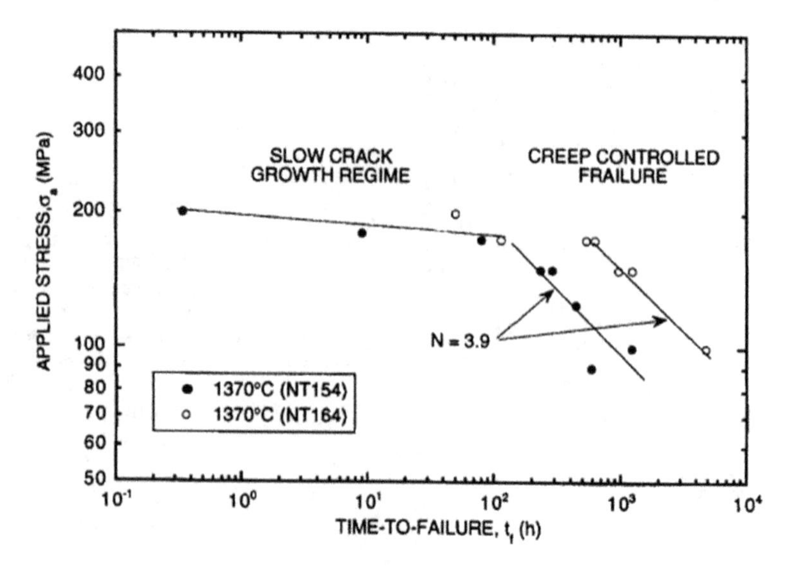

Fig. 6.109 Creep life dependence of stress. NT 154 has a transition from creep controlled failure to slow crack growth at 170 MPa. The *vertical arrows* indicate specimen failure [46]. With kind permission of John Wiley and Sons

long-term performance, cavities do develop. The performance of the newly-developed ceramic, NT164 silicon nitride, is compared in Fig. 6.109. Figure 6.110 shows the cavities observed through the gage section in both Si_3N_4 ceramics. TEM studies also indicate that the numbers of the multigrain-junction cavities in both the NT154 and NT164 were relatively low above the transition stress (170 MPa). However, for stresses less than 170 MPa, cavities are observed throughout the gage section in both materials (Fig. 6.96a–c). Consequently, as stress decreases below 170 MPa, the evolution of creep damage, in the form of cavities, controls specimen failure. An examination of re-fractured surfaces of the NT154 specimens, tested at stresses exceeding 170 MPa, revealed no evidence of grain-facet cavities. Thus, the creep-rupture data, for both the NT154 and NT164 Si_3N_4 ceramics, reveal a transition from stress-controlled to strain (creep)-controlled failure as the stress decreases below 170 MPa. This transition correlates well with the appearance of cavities at the lower stresses.

As mentioned earlier, the experimental data indicate that, of the great variety of creep curves described by various laws, their shapes are close to those of linear relations, when presented on a log-strain versus log-time basis. Time is expressed in hours. The best results are obtained for single-crystal components. Indeed, the usual practice in the industry is to fabricate single-crystal parts for certain elevated temperature applications. This is especially true in the high-pressure region in turbine engines, where the actual surface temperatures of the turbine blades can exceed ~940 °C; such high, homologous temperatures make creep and stress-rupture properties very important variables in the overall service lifetime of the

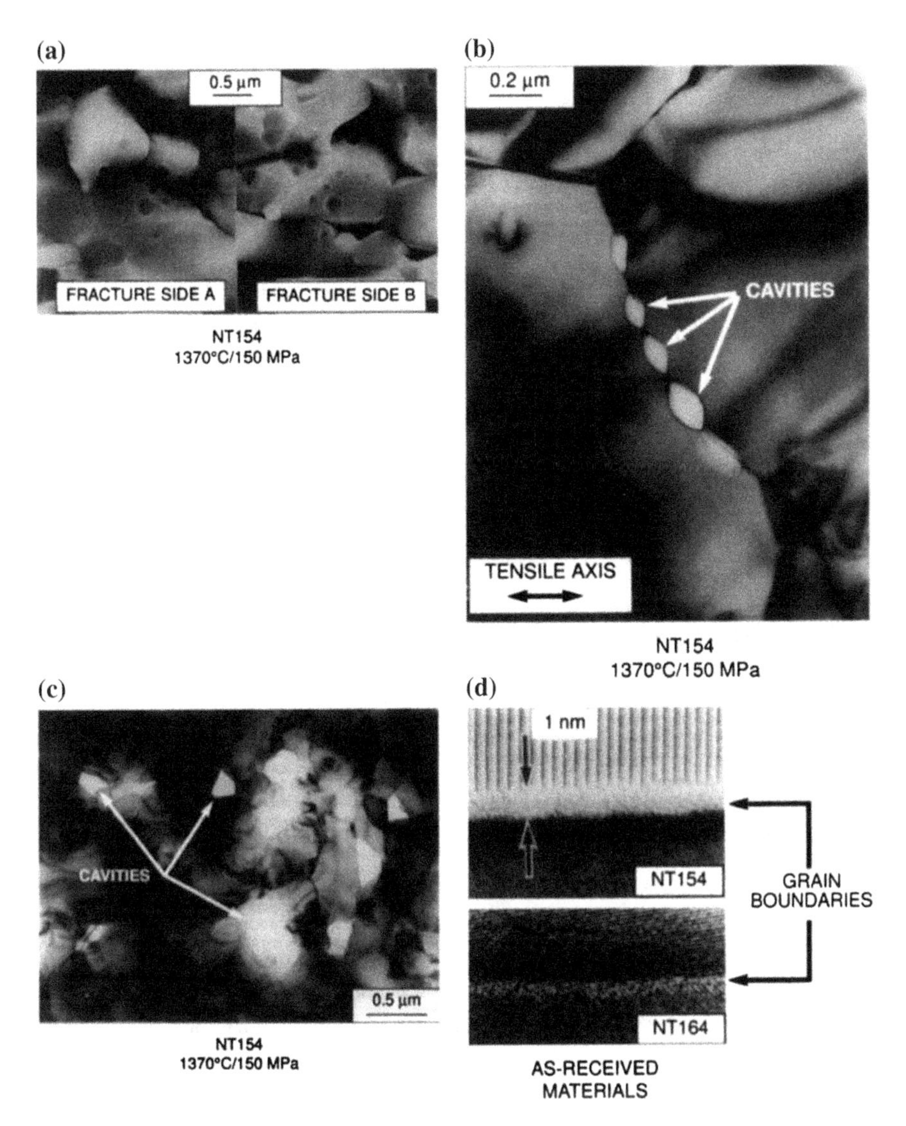

Fig. 6.110 The presence of lenticular cavities along two-grain junctions in the 00 was confirmed by **a** SEM of secondary fracture surface and **b** TEM of longitudinal section. **c** TEM also revealed extensive multigrain junction cavitation in the NT164. **d** The absence of two-grain junction cavitation in the NT164 was attributed to the nearly complete elimination of the intergranular phase [46]. With kind permission of John Wiley and Sons

blades. Nonetheless, the use of directionally-solidified ceramics is the next-best alternative for high-temperature use and is the present trend, mainly due to cost considerations. For example, directionally-solidified eutectic [henceforth: DSE] oxides present excellent resistance to oxidation at elevated temperatures, owing to

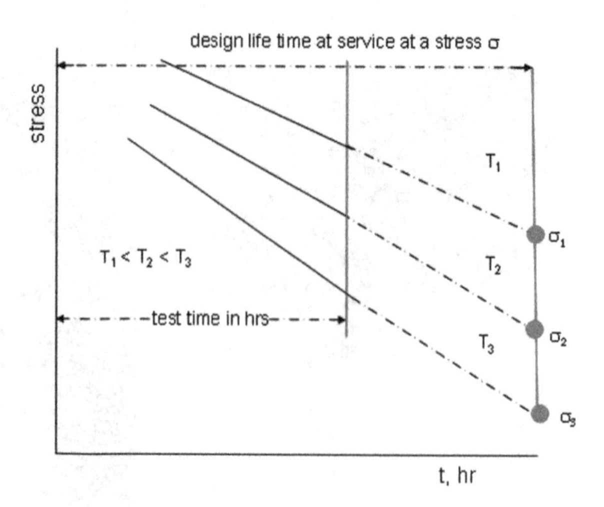

Fig. 6.111 A schematic extrapolation of creep-test data from the time indicated to the desired service lifetime; extrapolations for three temperatures at the desired, applied stresses

the inherent stability of the eutectic oxides and to the absence of impurities at the interfaces. For instance, no changes in weight or volume were detected in 6 × 6 × 6 mm^3 prismatic bars of Al_2O_3-YAG after 1000 h at 1700 °C in a laboratory atmosphere, whereas the shapes of Si_3N_4 and SiC specimens collapsed after 10–20 h under the same conditions due to oxidation [61].

Numerous experiments were performed on DSE ceramics for creep-resistant applications and the reader is referred to the technical literature on DSE ceramics and their high-temperature applications.

One of the relations used to express creep-rupture time, t_r, as a function of the activation energy of creep, Q_c, is:

$$t_r = t_0 \sigma^{-n} \exp(\frac{Q_c}{RT}) \tag{6.87}$$

where t_0 is a material constant and n is the stress exponent. As mentioned earlier, long-term creep-rupture time is predicted by extrapolation from short-term creep data. This is reasonably safe when no structural changes occur. This is not likely to be a meaningful method for curves, as in the case of Fig. 6.95 for Si_3N_4. A similar relation to Eq. 6.87 was used in the case of Si_3N_4 in the form of:

$$t_f = B_0 \left(\frac{\sigma}{\sigma_0}\right)^{-N} \exp\left(\frac{Q_f}{Rt}\right) \tag{6.87a}$$

where $t_f \equiv t_r$ is the creep rupture life and ***N***, Q_f, and ***B*** determine the stress and temperature dependencies of the dominant creep-rupture mechanism. Note that B_0/σ_0^N is equivalent to t_0 ($N = n$; $Q_c = Q_f$).

The schematic curves, illustrated in Fig. 6.111 for various temperatures may be used to evaluate the predetermined design lifetime of a material to prevent creep failure when no change in slope occurs.

6.8.4 Remarks on Creep Rupture: The Norton-Bailey Concept

Note the possible tests results from the schematic illustration in Fig. 6.111. First of all, no change in slope occurred, thus, allowing extrapolation from the test time to longer times. Secondly, the schematic illustration shows the test lines for three temperatures. At the highest temperature, the stress of the creep test is lowest and changes over time. Finally, the extrapolation of the test lines to some desired service lifetime indicates the stress that a material can sustain at the temperatures indicated (the point of intersection). Such creep tests help to predict the actual creep life of a material without failure throughout its designed lifetime. These tests are performed using the same creep strain, ε, for each temperature and the temperature level of the test is set higher than the highest anticipated service temperature.

Most creep analyses for long-term applications involve the determination of a steady-state or minimum creep rate and its examination as a function of applied stress and temperature. The steady-state or minimum, compressive creep rate ($d\varepsilon/dt_{min}$) may be related to the applied stress and temperature by an empirical Arrhenius power law or the familiar 'Norton-Bailey creep equation'. Often in creep testing, a minimal 1 % (of the expected lifetime) criterion is used. However, a 10 % criterion is preferable in order to obtain a meaningful prediction of the usable lifetime for creep applications. In critical applications, such as for turbine components, an ~25 % criterion is recommended, even preferred, so as to avoid failure by creep rupture.

Considering Eq. (6.87) and the graphs based on such equations, one may immediately recall the well-known 'Norton-Bailey relation', originally suggested for second-stage creep (or minimum creep rate) and used by many to predict the creep lifetime for engineering materials intended for high-temperature applications. A brief summary of the essence of the Norton-Bailey approach follows. The steady-state or minimum creep rate (dε/dtmin) relates to applied stress and temperature by an empirical Arrhenius power law, given as:

$$\left(\frac{d\varepsilon}{dt}\right)_{min} = A\sigma^n \exp\left(-\frac{Q}{RT}\right) \tag{6.88}$$

Multilinear regression may then be performed to determine the constants A, n and Q for each material of interest. Such an analysis implies that the same creep mechanism is acting as the dominant, rate-controlling mechanism at all the temperatures and stresses.

The constant thermal-creep behavior, ε_c, is a function of:

$$\varepsilon_c = \mathrm{f}(\sigma, \mathrm{T\ and\ t}) \tag{6.89}$$

It is usually assumed that this function may be separated into:

$$\varepsilon_c = \mathrm{f_1}(\sigma)\mathrm{f_1}(\mathrm{t})\mathrm{f_1}(\mathrm{T} \tag{6.89a}$$

Norton used what is called the 'power law' to describe stress dependence over time under constant stress (but many other expressions are also available). This power law function is given as:

$$f_1(\sigma) = A\sigma^n \tag{6.90}$$

The parameters, A and n, are material constants. Bailey suggested a time function, given as:

$$f_2(t) = Dt^m \tag{6.91}$$

(usually $1/3 \leq m \leq 1/2$) and:

$$f_3(T) = C\exp\left(-\frac{\Delta H}{RT}\right) \tag{6.92}$$

ΔH is activation energy in an Arrhenius-type relation. By combining these relations, Eq. (6.93) is obtained (which is about the same as Eq. 6.88):

$$\varepsilon_c = B\sigma^n t^m \exp\left(-\frac{\Delta H}{RT}\right) \tag{6.93}$$

At isothermal conditions, Eq. (6.93) is given as:

$$\varepsilon_c = B\sigma^n t^m \tag{6.94}$$

which is the 'Norton-Bailey law'. Equation (6.94) was developed for constant stress and is useful in describing all creep stages. Differentiating Eq. (6.94) for $n \gg 1$ and $m \leq 1$ yields the creep rate for varying stresses. Here, the strain rate is expressed as a function of stress and time (known as the 'time-hardening rule'). Another form (known as the 'strain-hardening rule') may be obtained by differentiating according to t; thus, Eq. (6.94) becomes:

$$\frac{d\varepsilon}{dt} = \dot{\varepsilon} = Bm\sigma^n t^{(m-1)} \tag{6.95}$$

It is possible to extract t from Eq. (6.94) as:

$$t = \left(\frac{\varepsilon_c}{B\sigma^n}\right)^{1/m} \tag{6.96}$$

and to insert it into Eq. (6.95) resulting in:

$$\dot{\varepsilon} = Bm\sigma^n\left(\frac{\varepsilon}{A\sigma^n}\right)^{\frac{(m-1)}{m}} \tag{6.95a}$$

which may also be written as:

$$\dot{\varepsilon} = B^{\frac{1}{m}} m(\sigma)^{\frac{n}{m}}(\varepsilon)^{\frac{(m-1)}{m}} \tag{6.97}$$

The parameters B, n and m depend on the material and the temperature, which can be determined by means of a uniaxial test.

Norton's power law and an Arrhenius-type equation are the most common expressions describing the stress and temperature dependence of the steady-state creep rate. Of the many relations suggested to describe creep-rupture life, the steady-state strain-rate relation was extended for the calculation of rupture strength and is used to predict service lifetimes. The Norton Bailey concept is used in regard to many solids, including ceramics (e.g., see Headrick et al. [50]).

To summarize this section, note that: (a) stress-rupture tests are used to determine the time necessary to produce failure; thus, this testing is always done until failure occurs; (b) the data are plotted as log–log plots (Figs. 6.100–6.109); (c) a straight line or the best fit curve is usually obtained for each temperature of interest; (d) this information may then be used to extrapolate time-to-failure for longer durations (a typical set of stress-rupture curves appear above) and; (e) changes in the slope of a stress-rupture line are due to structural changes in the material, which are significant changes in material behavior, because they may produce large errors.

6.9 The Prediction of LifeTime (Parametric Method)

There are parametric methods for determining the creep lifetime of materials. Such methods are based on evaluating the stress-rupture behavior. In essence, the results of short-duration, high-temperature tests are correlated with the performance of long-term tests at lower temperatures. The most popular parametric methods are: (a) Larson-Miller; (b) Manson-Haferd; (c) Orr-Sherby-Dorn and, (d) Monkman–Grant. Of these methods, the following is a discussion on the Larson-Miller and the Monkman–Grant methods to the evaluation of ceramic-material lifetimes.

6.9.1 The Larson-Miller Method

The Larson-Miller method is applicable to a variety of materials, including ceramics, and is most commonly used because of its simplicity. This relation is given as:

$$T(C + \log t) = P \tag{6.98}$$

T is given in degrees Rankin (i.e., °F + 460), t in hours and the constant C $\approx$ 20. The value of C seems to be applicable to many cases and materials, but deviation from this value has been observed and its value may be in the range of 15–30, depending on the material. Selecting the proper C value, which may be determined for a material of interest, can narrow the scatter by reducing the scatter problem, which is very common in ceramics experiments. Equation (6.98) is a stress-dependent, temperature-compensated rupture-life function. The Larson-Miller relation may be obtained as follows. Suppose that creep rate is adequately described by an Arrhenius-type equation, since creep is a thermally- activated process and that the minimum secondary-creep rate may be described by an Arrhenius equation (see Eqs. (6.3) or (6.4)). Thus:

$$\dot{\varepsilon} = A \exp -(\frac{Q}{kT}) \tag{6.99}$$

$$\ln \dot{\varepsilon} = \ln A - \frac{Q}{kT} \tag{6.99a}$$

by rearranging Eq. (6.99a) as:

$$\ln A - \ln \dot{\varepsilon} = \frac{Q}{kT} \tag{6.99b}$$

Assuming that the creep-strain-to-rupture, ε_r, is a constant over the temperature range of interest and, if the strain is predominantly in the steady-state creep regime, then the average creep rate for the specimen lifetime to rupture, namely, t_r, is given by:

$$\dot{\varepsilon} = \frac{\varepsilon_r}{t_r} \tag{6.100}$$

and Eq. (6.99b) may be written as:

$$\ln A - \ln \frac{\varepsilon_r}{t_r} = \ln A - \ln \varepsilon_r + \ln t_r = \frac{Q}{kT} \tag{6.101}$$

Write $\ln A - \ln \varepsilon_r = C$ and rearrange Eq. (6.101) to get:

$$T(C + \ln t_r) = \frac{Q}{k} = P \tag{6.102}$$

$P = T(C + \ln t_r)$ is the Larson-Miller parameter [henceforth: LMP] and is $f(\sigma)$. Under the assumption that activation energy is independent of applied stress, this equation may be used to relate the difference in rupture life to differences in temperature for a given stress. In many cases, C is indeed ~20, which is obtained from the intercept with the $\log t_r$ axis of a $\log t_r$ versus 1/T plott (Fig. 6.111). The slope of such a plot, namely Q/k (=P), is a function of stress, as seen in the schematic Fig. 6.111. For a graphic presentation, Eq. (6.102) may be written as:

$$\log t_r = 0.434 \frac{Q}{kT} - \log C \tag{6.102a}$$

As shown in Fig. 6.112, the intercept on $\log t_r$ is $-C$.

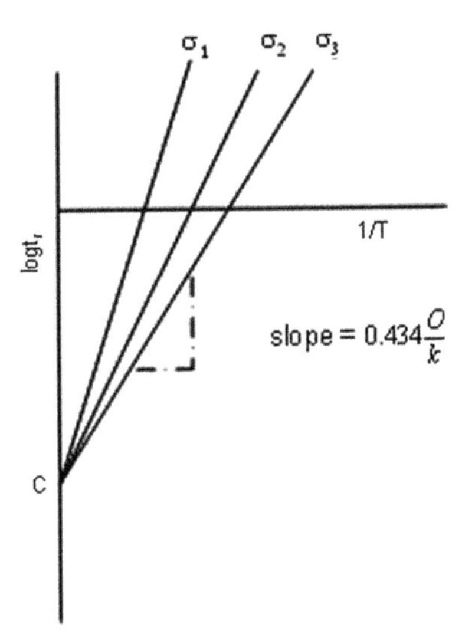

Fig. 6.112 A schematic plot according to Eq. (6.101) for various stresses

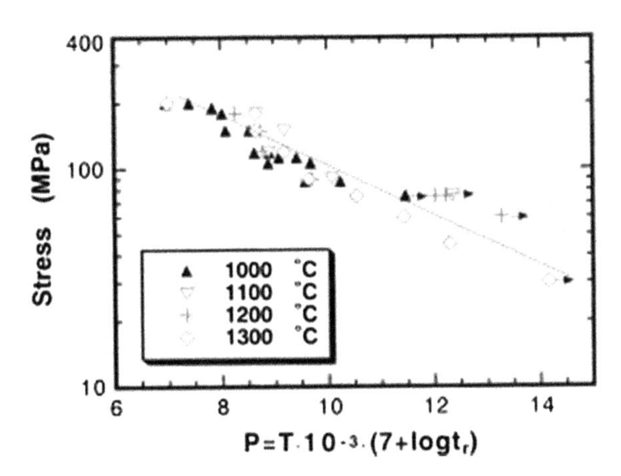

Fig. 6.113 Tensile stress versus Larson-Miller parameter at different stresses in argon at 1000, 1100, 1200 and 1300 °C [100]. With kind permission of Elsevier

The parametric method is commonly used to predict creep-rupture strength at longer times (often up to 10^5 h or even longer) based on tests performed for much shorter times. The relation between stress and the LMP is used to predict a probable time for the onset of creep-rupture failure. If one knows the temperature and stress at which a material is operating, the predicted time to creep-rupture failure for that set of conditions may be determined from a plot of LMP versus stress. The procedure using the LMP method requires a family of stress-rupture curves representing different test temperatures for a given material, which are then re-plotted on a revised temperature-compensated time axis, i.e., the LMP. The family of curves chosen is superimposed on a single master curve. Such a curve is illustrated in Fig. 6.113 for SiC-fiber/SiC composite.

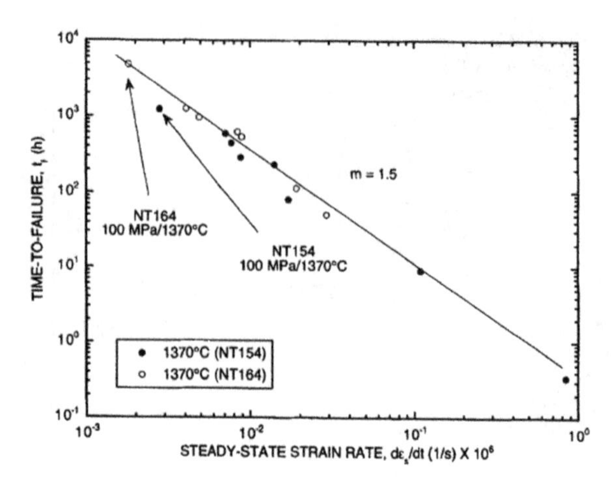

Fig. 6.114 Monkman-Grant plot shows that NT154 and NT164 follow the same linear relationship, indicating that the same mechanism is controlling creep rupture life for both materials [46]. With kind permission of John Wiley and Sons

The Larson-Miller parameter, P, in Eq. (6.102), is one of the useful parameters used for predicting creep life in metallic materials, but it is useful for ceramics as well. The LMP may be used to describe the stress-temperature-life relation in a SiC/SiC composite by means of the following expression:

$$P = T(C + \log t_r) \tag{6.102a}$$

which is Eq. 6.102 explicitly expressing P. It was found that data at different temperatures fall on the same line with the best fit when the constant C is between 5 and 10, when curves of log t_r versus 1/T are drawn. Figure 6.113 shows the relation of stress to the LMP with a C value of 7. Note that this value is different than the commonly-assumed one of ~20, which was found for various materials, particularly metallic.

6.9.2 The Monkman-Grant Method

The Monkman-Grant [henceforth: MG] relation between minimum creep rate (or steady-state strain rate) and time-to-fracture, t_f, is given as:

$$\dot{\varepsilon}_{mer}^{m} t_f = C \tag{6.103}$$

$$\log t_f = -m \log \dot{\varepsilon} + \log C \tag{6.103a}$$

where m and C are material constants. This relation holds for alloys and also for ceramics. Such a curve for Si_3N_4 NT154 and NT164 ceramics is indicated in Fig. 6.114. In addition, Fig. 6.106 above is actually a MG relation for the SiC fiber.

This relation is useful for industrial applications, when one knows the constants, m and C, of a material, since the above expression evaluates the fracture time on

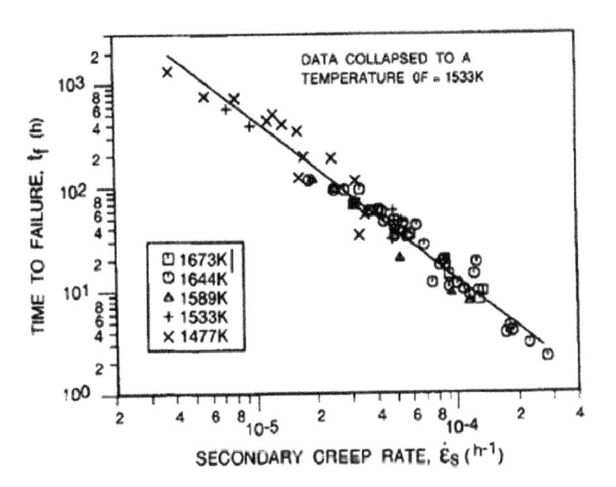

Fig. 6.115 Temperature dependence of the Monkman-Grant lines correlated with an additional temperature term Eq. (6.105) [62]. With kind permission of John Wiley and Sons

the basis of the minimum creep-rate data. There is no need for the long-term creep testing of high-temperature materials; generally, the time required to reach steady-state creep is much shorter than the time-to-fracture. This is particularly important regarding newly-developed materials for high-temperature applications. Figure 6.13 is a logarithmic plot of Eq. (6.103) for the nitrides (Si_3N_4) NT154 and NT164.

Originally, the MG relation was developed for alloys, but it also has the ability to predict the rupture life of ceramic materials. This MG relation was experimentally applied to various ceramics at various temperatures and stresses and an example of its use to predict their creep lifetime is shown for advanced silicon nitrides. Certain departures from the uniqueness of the MG relation, in both metals and ceramics, have been noted by previous investigators. Consequently, some improvements were suggested by Mamballykalathil et al. [62]. in order to achieve a modified MG relation for ceramics:

$$t_f = K(\dot{\varepsilon}_s)^{b_1} \tag{6.104}$$

Here, K and *b*, are constants. Note that this relation is equivalent to Eq. (6.103). The time-to-fracture (rupture), t_f, may be expressed as:

$$\ln(t_f) = b_0 + b_1 \ln(\dot{\varepsilon}) + b_2/T \tag{6.105}$$

In this relation, b_0 and b_2 are constants and T is the absolute temperature. Comparing this to Eq. (6.104), the first and the last terms of Eq. (6.105) would equal lnK in Eq. (6.104). The plot of this modified relation is illustrated in Fig. 6.115. The experimental observations relating to the creep testing of NT154 silicon nitride are presented in Fig. 6.115. The data at different temperatures sit reasonably well on one single line. In this figure, both the average line and the data had to be normalized to a particular temperature (1533 K) for all to be shown in one plot. Figure 6.116 shows the prediction from the above equation for the

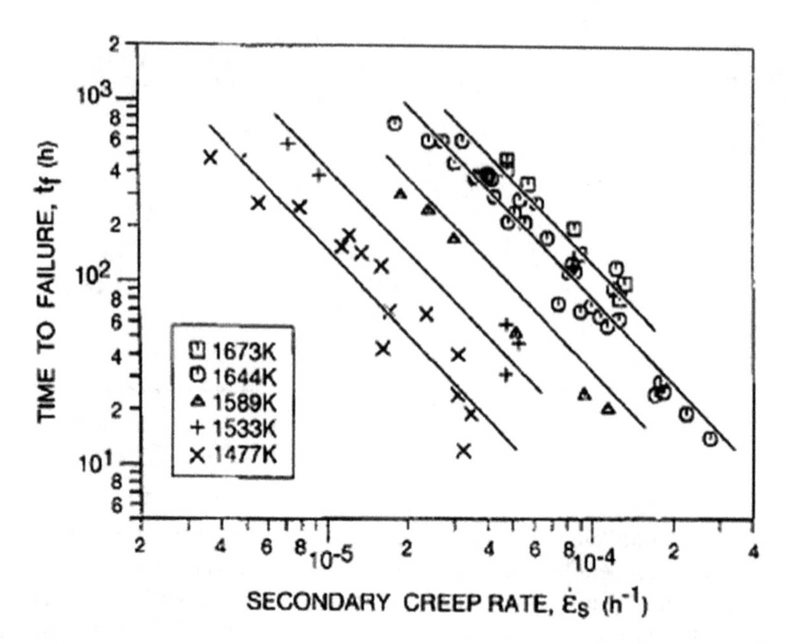

Fig. 6.116 Comparison of the average lines predicted by Eq. (6.105) with the unmodified data [62]. With kind permission of John Wiley and Sons

individual temperatures along with the unmodified data. The values of b_0, b_1 and b_2 were determined to be 15.87, −1.53 and −4.21 × 104, respectively. The negative sign of b_2 means that rupture life increases with increasing temperature for the same value of the creep rate. Furthermore, including the stress, σ which is related to 1/T at constant, $\dot{\varepsilon}$, and expressing this in Eq. (6.106) as:

$$\ln(\dot{\varepsilon}) = \ln(A) + n\ln(\sigma/E) - (Q/R)(1/T) \tag{6.106}$$

where, A, n, σ, E, Q, R are, respectively, a constant, stress exponent, applied stress, Young's modulus, activation energy and the universal gas constant. The stress-rupture data may be correlated with a stress term also which is seen in Eq. (6.107) and the curve plotted on the basis of this equation: (Fig. 6.117).

$$\ln(t_f) = c_0 + c_1\ln(\varepsilon_s) + c_2\ln(\sigma/E) \tag{6.107}$$

This revised approach improved the scatter in both Eqs. (6.100) and (6.102), as seen in the illustrations Fig. 6.116. Various parameters were also suggested for the extrapolation of time-to-rupture with varying success; certain ones have been used to predict the in-service lifetime of a component operating at high temperatures. Of these methods, the two most popular ones have been discussed in this section. The reader may turn to the professional literature in order to choose the most appropriate method for a given specific application.

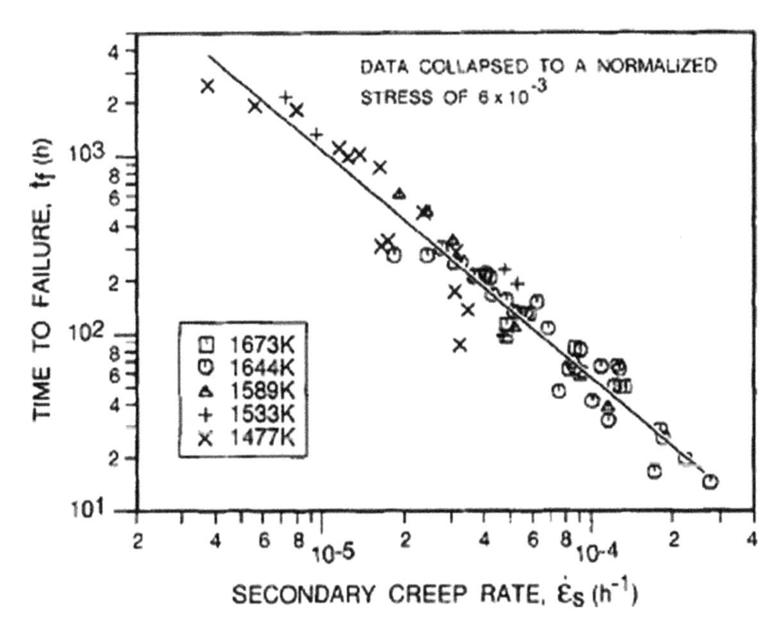

Fig. 6.117 Stratification of the Monkman-Grant lines can be correlated with an additional stress term (Eq. (6.107)) [62]. With kind permission of John Wiley and Sons

6.10 Concepts of Designing (Selecting) Creep-Resistant Ceramics

In this section, the presented design considerations and suggestions are general ones, applicable to all materials, regardless if they are ceramics, metals or even polymers; thus, no specific references are made to particular materials. General designs for universal, creep-resistant purposes, based on theoretical considerations, do not exist, despite the vast quantity of experimental results and the deep understanding of creep behavior in materials under various conditions. Primary design problems are complex, due to the many material parameters, the various stages of creep and the varying service conditions. Today, the decision to select appropriate materials for use or to design and develop new materials relies heavily on dependable, experimental observations and their interpretations according to physical principles. The following may be stated regarding creep, as a high-temperature deformation—it is associated with: (a) the presence of dislocations in materials and dislocation motion under stress; (b) temperature; (c) stress; (d) time; and (e) structure. Points (b)–(d) may be considered as environmental conditions, but, to these environmental factors, one should add the ambient atmosphere prevailing during the intended service. Each of these factors will now be discussed briefly.

(a) The role of dislocations. Dislocation slip is hindered by obstacles, such as: grain boundaries (in polycrystalline materials), precipitates or impurity particles, stress fields around solute atoms in solid solutions, strain fields of other dislocations and pile-ups. Various interpretations of dislocation motion under stress have been considered by a number of researchers in the field, based on the different stages of creep and the perceived contribution of dislocations to creep strain. It has been determined that dislocations can: (i) glide, leading to slip in their slip planes; (ii) climb, leading to subgrain formation; (iii) cross-slip; or (iv) slide at grain boundaries, considered by many as being induced by slip to satisfy strain compatibility at those grain boundaries. Actions (ii) and (iii) are ways of overcoming resistance by obstacles and allowing further strain to occur.

(b) The temperature. The creep resistance of a material depends on its melting point, T_m, and, at a given temperature, the higher the T_m of a material, the longer its lifetime will be. This is simply related to the rate of self-diffusion, which is slower in high-T_m materials. A specimen may be exposed to levels above or below $0.5T_m$, respectively. Low-temperature creep, at or below $0.5\ T_m$, is believed to be governed by non-diffusion-controlled mechanisms, whereas high-temperature creep (above $0.5\ T_m$) is diffusion-controlled. At and above $0.3\ T_m$, creep becomes a significant factor in materials, to be considered when designing creep-resistant materials. Creep resistance is an important material property in high-temperature design, but it is difficult to suggest a function for calculating this property. Therefore, parametric methods are used to achieve the desired long-term creep resistance of specific materials. Since climb is one of the mechanisms for overcoming obstacle resistance and produces creep strain in high-T_m materials, climb will not readily occur below $0.5\ T_m$.

(c) Stress. The role of stress is obvious, since it is responsible for creep deformation. However, one may talk about low-stress creep and high-stress creep. The effect of stress at constant temperatures depends on its level. Accordingly, when increasing stress from a low level to higher values, the creep curves pass through all the stages, from transient to tertiary creep. At a specific stress, all three stages of creep may be obtained.

(d) Time. Creep is a time-dependent deformation. Unlike brittle fracture, creep deformation does not occur suddenly upon the application of stress, since ceramics at high temperatures (usually at the creep temperature) are ductile. Instead, strain develops as a result of the application of long-term stress. At a given stress and temperature, the resulting creep strain depends on the length of time. The other parameters (σ, T) determine the rate of creep. Given the aforementioned parameters (σ, T and t), one may state that the rate of deformation is a material property.

(e) Structure. Generally, microstructural changes may occur in materials as a consequence of thermal effects in the absence of stress. Such changes are augmented when stress operates at high temperatures (creep conditions). Microstructural changes that may occur in a material under the combined long-term effects of acting stress and temperature are of prime concern when designing creep-resistant materials. Designers must be familiar with the structural properties and the possible changes that may occur in the wake of long-term exposure to the combined effects of stress and temperature.

Intentional structural changes are often induced in order to improve the duration of material lifetimes. Solid solutions have a dual effect, namely they strengthen materials (solid-solution strengthening) and the stress field around solute atoms acts as an obstacle blocking dislocation motion. Furthermore, precipitates obtained in material, as by precipitation hardening, also have a dual effect with regard to solid solutions. The size of these precipitates is significant: (a) for preventing dislocation bowing between the precipitates and (b) because, if they are very small, they may dissolve into the matrix, being unstable at the service temperature. Dispersion-hardened precipitates, such as oxides, nitrides, etc., are stable at high temperatures, generally insoluble in a matrix and are also stable at high temperatures; they are more effective material strengtheners for creep resistance.

Additional structural aspects to be considered are: (i) the crystallinity of materials; (ii) orientation, in the case of single crystals; (iii) the SF energy; and (iv) grain size. Concerning crystallinity: (i) The material may be single -crystal or polycrystalline. Single crystals, although not as strong as polycrystalline materials, due to the absence of boundaries, have been widely used for high-temperature applications (turbine blades, etc.), but cost considerations are an important factor in industry. Polycrystalline materials obtained by directional solidification are almost as good for creep resistance as single crystals. There is a tendency to replace single crystals, wherever possible, by directionally-solidified material. (ii) From a design point of view, it is desirable to use an orientation in which single-crystalline material has the best strength properties. (iii) High-SF-energy materials are inferior to those with low-SF-energy for creep-resistant applications. The uniting of partials is performed more readily in high-SF-energy materials, enabling climb and cross-slip and, thus, avoiding creep-resisting obstacles. Note that it is advantageous to use solutes for solution strengthening, which can also simultaneously reduce the SF energy. (iv) Grain size has a dual effect. On one hand, small-grained materials are stronger than large-grained materials, but, on the other hand, larger grains improve creep resistance. This is a consequence of vacancy formation and flow. In fine-grained materials with a large number of grain boundaries, climb may be rapid, because the grains are sources (not only sinks) of vacancies. For climb to occur readily, vacancies must be available. In large-grained materials with fewer boundaries, the number of vacancies (that might induce climb) is lower and, therefore, climb is slower with the consequent lower creep rate. As such, the effect of grain size on creep rests on the diffusion rate of the vacancies in polycrystalline materials. A compromise must be made between the strengthening effect of small grains and the reduction of the number of vacancies required for climb by the use of coarser grains.

Summing up the accepted rules for producing creep-resistant materials, the approach should be such that: (i) high-melting materials are preferential;

(ii) if one-phase materials are required, the matrix should be strengthened by effective solutes; (iii) low SF energy should be chosen (sometimes a solute which reduces the SF energy may also be used for solution strengthening; (iv) dispersed stable, non-soluble particles may be added to achieve a specific creep application; and (v) stable structures should be chosen for certain high-temperature applications, thus, short-duration tests can be extrapolated for long-term use in service.

References

1. ASM (1957) Creep and recovery. American Society of Metals, Metals Park
2. Bell RL, Langdon TG (1967) J Mater Sci 2:313
3. Chang HC, Grant NJ (1956) Trans AIME 206:544
4. Chen I-W, Xue LA (1990) J Am Ceram Soc 73:2585
5. Chen CF, Wiederhorn SM, Chuang T (1991) J Amer Ceram Soc 74:1658
6. Chokshi AH (1993) Mater Sci Eng A 166:119
7. Damask AC, Dienes GJ (1971) Point defects in metals. Gordon and Breach, New York
8. Dorn JE (1964) In: Mueller WM (ed) Energetics in metallurgical phenomena, vol 1. Gordon and Breach, London
9. Evans RW, Wilshire B (1985) Creep of metals and alloys. The Institute of Metals, London
10. Friedel J (1964) Dislocations. Addison-Wesley Publishing Company, Inc., Reading
11. Garofalo F (1965) Fundamentals of creep and creep-rupture in metals. The Macmillan Company, New York
12. Koumoto K, Takeda S, Pai CH, Sato T, Yanagida H (1989) J Am Ceram Soc 72:1985
13. Langdon TG (1993) Mater Sci Eng A166:67
14. McLean D (1962) Mechanical properties of metals. Wiley, New York
15. McLean D, Hale KF (1961) Structural processes in creep. Special Report No. 70, Iron and Steel Institute, London, p 19
16. Nabarro FRN (1948) Report on a conference on strength of solids. The Physical Society of London, p 75
17. Nabarro FRN, De Villers HL (1995) The physics of creep. Taylor & Francis, London
18. Norton FH (1929) The creep of steels at high temperatures. McGraw-Hill, New York
19. Poirier JP (1985) Creep in crystals. Cambridge University Press, Cambridge
20. Ruoff AL (1973) Materials science. Prentice-Hall Inc, Englewood CliffsJ
21. Singh D, Lorenzo-Martín M, Chen G, Gutiérrez-Mora F, Routbort JL (2007) J Eur Ceram Soc 27:3377
22. Steffens K, Wilhelm H (2000) Next engine generation: materials, surface technology, manufacturing processes. What Comes after 2000? MTU Aero Engines
23. Zener C (1948) Elasticity and anelasticity. University of Chicago Press, Chicago

Further References

24. da C. Andrade AN (1910) Proc Royal Soc Lond, A 84:1
25. da C. Andrade AN (1914) Proc Royal Soc Lond, A 90:329
26. da C. Andrade AN, Chalmers B (1932) Proc Royal Soc Lond, A 138:348
27. Armstrong WM, Irvine WR (1964) J Nucl Mater 12:261
28. Bilde-Sörensen JB (1972) J Am Ceram Soc 55:606
29. Blanchard CR, Chan KS (1993) J Am Ceram Soc 76:1651
30. Bretheau T, Castaing J, Rabier J, Veyssiere P (1979) Adv Phys 28:835

31. Brunner D, Taeri-Baghbadrani S, Sigle W, Rühle M (2001) J Am Ceram Soc 84:1161
32. Ch. Jou Z, Virkar AV (1990) J Am Ceram Soc 73:1928
33. Chan KS, Page KA (1992) J Mater Sci 27:1651
34. Chokshi (1990) J Mater Sci 25:3221
35. Clauer AH, Seltzer MS, Wilcox BA (1979) J Am Ceram Soc 62:85
36. Coble RL (1963) J Appl Phys 34:1679
37. Coble RL, Guerard YH (1963) J Am Ceram Soc 46:353
38. Cottrell AH (1952) J Mech Phys Solids 1:58
39. Cottrell AH, Aytekin V (1950) J Inst Met 77:389
40. de Arellano-López AR, Meléndez-Martínez JJ, Cruse TA, Koritala RE, Routbort JL, Goretta KC (2002) Acta Mater 50:4325
41. Chen D, Sixta ME, Zhang XF, de Jonghe, LC, Ritchie RO (2000) Acta Materialia 48:4599
42. Deng ZY, Shi JL, Zhang YF, Jiang DY, Guo JK (1998) J Eur Ceram Soc 18:501
43. Ding JL, Liu KC, More KL, Brinkman CR (1994) J Am Ceram Soc 867:77
44. Dobeš F (1998) J Mater Sci 33:2457
45. Doukhan N, Duclos R, Escaig B (1973) J De Phys 34:C9 379 (P. Dennison)
46. Ferber MK, Jenkins MG, Nolan TA, Yeckley RL (1994) J Am Ceram Soc 77:657
47. Gandhi C, Raj R (1981) Met Trans A12:515
48. Graham A, Walles KFA (1955) J Iron Steel Inst 179:105
49. Gordon RS, Terwillger GR (1972) J Am Ceram Soc 55:450
50. Headrick WL, Moore RE, Karakus M, Liang X (2003) Report DE-FC07-01ID14250, U S. Department of Energy, Final report for the period 1 Oct 2001–1 Oct 2003
51. Hemker KJ, Mills MJ, Nix WD (1991) Acta. metall. Matter 39:1901
52. Hensler JH, Cullen GV (1968) J Am Ceram Soc 51:557
53. Hensler JH, Cullen GV (1972) J Am Ceram Soc 55:450
54. Herring C (1950) J Appl Phys 21:437
55. Hong HU, Choi BG, Jeong HW, (2010) Kim S, Yoo YS, Jo CY. In: Chandra T, Ionescu M, Wanderka N, Reimers W (eds) Material science forum THERMEC, vol 2245, p 638
56. Hynes AP, Doremus RH, Ame J (1991) Ceram Soc 74:2469
57. Kašiarová M, Shollock B, Boccaccini A, Dusza J (2009) J Am Ceram Soc 92:439
58. Langdon TG (2006) J Mater Sci 41:597
59. Le Hazif R, Poirer JP (1975) Acta Met 23:865
60. Lin HT, Alexander KB, Becher PF (1996) J Am Ceram Soc 79:1530
61. Lorca JL, Orera VM (2006) Prog Mater Sci 51:711
62. Mamballykalathil NM, Ho TF, Wu DC, Jenkins MG, Ferber MK (1994) J Am Ceram 77:1235
63. Mariani E, Mecklenburgh J, Wheeler J, Prior DJ, Heidelbach F (2009) Acta Mater 57:1886
64. Martinez-Fernandez J, Jimenez-Melendo M, Dominguez-Rodriguez A, Heuer AH (1990) J Am Ceram Soc 73:2452
65. Mordike BL (2002) Mater Sci Eng A324:103
66. Mott NF (1953) Proc Royal Soc London, A 220:1
67. Nabarro FRN (2001) Mater Sci Eng A309:227
68. Nishikawa O, Saiki K, Wenk HR (2004) J. Struct Geol 26:127
69. Pelleg (2013) Mechanical properties of materials. Springer
70. Raj R (1981) Met Trans A12:1089
71. Rendtel A, Hübner H, Herrmann M, Schubert C (1998) J Am Ceram Soc 81:1109
72. Reynolds L, Burton B, Speight MV (1975) Acta Metall 23:573
73. Rieth M, Falkenstein A, Graf P, Heger S, Jantsch U, Klimiankou M, Materna-Morris E, Zimmermann H (2004) Inst. For Materialforschung Program Kernfusion, Forschungszentrum, Karlsruhe, in der Helmholtz-Gemeinschaft Wissenschaftliche Berichte, FZKA 7065
74. Robertson AG, Wilkinson DS, Cácerest CH (1991) J Am Ceram Soc 74:915
75. Sarfarazi M, Ghosh SK (1987) Eng Fract Mech 27:257
76. Schoeck G, Seeger A (1955) Report on the conference on defects in a crystalline solid. The Physical Society of London, p 340

77. Sherby OD, Lyton JL (1956) Trans AIME 206:928
78. Sherby OD, Orr RL, Dorn JE (1954) Trans AIME 200:71
79. Sherby OD, Lyton JL, Dorn JE (1957) Acta Met 5:219
80. Stroh AN (1955) Proc Royal Soc (Lond) 223A:548
81. Stroh AN (1957) Adv Phys 6:418
82. Suzuki M, Kimura T, Koike J, Maruyama K (2004) Mater Sci Eng A387:706
83. Tan JC, Tan MJ (2003) Mater Sci Eng A339:81
84. Terzic A, Pavlović Lj, Milutinović-Nikolić A (2006) Sci Sinter 38:255
85. Uchic MD, Chrzan DC, Nix WD (2001) Intermetallics 9:963
86. Valiev RZ, Yu Gertsman V, Kaibyshev OA (1986) Phys State Solid (a) 97:11
87. Wadsworth J, Ruano OA, Sherby O (2002) Met and Mater Trans 33A:219
88. Wang Z, Karato SI, Fujino K (1993) Phys Earth Planet Inter 79:299
89. Weertman J (1955) J Appl Phys 26:1213
90. Weertman J (1957) J Appl Phys 28(196):1185
91. Weertman J (1957) J Appl Phys 28(196):362
92. Weertman J (1968) Trans ASM 61:680
93. Wilkinson DS, Cácerest CH, Robertson AG (1991) J Am Ceram Soc 74:922
94. Wilshire B, Carreño F (1999) Mater Sci Eng A272:38
95. Wu X, Holmes JW (1993) Com Am Ceram Soc 76:2695
96. Wu MS, Zhou H (1996) Inter J Fract 78:165
97. da C. Andrade AN ASM, creep and recovery. American Society of Metals, p 176
98. Yoo HI, Wuensch BJ, Petuskey WT (2002) Solid State Ion 150:207
99. Zhu S, Mizuno M, Nagano Y, Kagawa Y, Kaya H (1997) Compos Sci Technol 57:1629
100. Zhu S, Mizuno M, Kagawac Y, Mutoh Y (1999) Compos Sci Technol. 59:833

Chapter 7
Cyclic Stress: Fatigue

Abstract Components in engineering applications operating under cyclic loads, commonly known as fatigue, may become unstable and cause catastrophic failure to occur unexpectedly because of structural instability. It is generally thought that over 80 % of all service failures are associated with fatigue. Therefore, operation of machines or their components under cyclic loads are of prime concern. To overcome the difficulty in predicting fatigue failure—because of a large spread of statistical results—it is essential to use many test specimens to reach a meaningful average value below which the probability for fatigue fracture is quite low. Fatigue-resistance evaluation is done by plotting applied stress against the number of cycles, usually referred to as the S–N (curve) relation. In some cases a horizontal line is observed in the plot known as the "knee" representing the endurance limit. At this level of stress or below it, ceramics have the ability to endure a large number of stress-cycles without failure. A favored location of failure initiation is the surface; therefore good surface finish (often by polishing) is recommended which significantly improves fatigue resistance. Introducing compressive stress by any of the following methods, namely, laser treatments, sand blasting or shot peening improve greatly the fatigue resistance. Regardless of the origin of stress when cycling is applied, fatigue damage may result. Thus stress cycling associated with temperature changes is of great concern because it can induce fatigue damage known as thermal fatigue with premature failure in components operating at elevated temperatures. Design to overcome fatigue failure and to increase resistance to cyclic deformation is essential. Environmental effects, among them corrosion, are important in design considerations. Corrosive environments may accelerate the growth of fatigue cracks, which initiate at the surface and, therefore, reduce overall fatigue performance.

7.1 Introduction

When materials are used for high temperature engineering applications under cyclic loads (fatigue), their structures become unstable and catastrophic failure usually occurs unexpectedly, which limits their use under those circumstances.

J. Pelleg, *Mechanical Properties of Ceramics*, Solid Mechanics and Its Applications 213, DOI: 10.1007/978-3-319-04492-7_7,

Note that the integrity of any machine element is, therefore, often limited by its response to the mechanical load; thus, its performance under cyclic loads is of prime concern when applying materials not only at high temperatures, but also at RT. In fact, it is generally thought that over 80 % of all service failures are associated with fatigue. As such, it is insufficient to consider ceramics merely for their toughness, which is usually of common interest to designers of ceramic parts, but it is mandatory to consider their eventual response during cyclic-load applications as well.

This subject has been discussed widely in the past mainly in regard to metallic materials, in which plastic deformation before failure is a key factor leading to fracture. Ceramics, which are mostly brittle at RT, fail almost completely without deformation, but are, nevertheless, exposed to fatigue fracture. It is useful to first consider this concept in metals in general before focusing our attention on ceramics.

The most common failure that occurs in materials, such as metals, happens due to fatigue. The simplest way of looking at fatigue is by considering a specimen which is being repeatedly stressed under tension and compression. It is not only repeatedly-applied tensile stresses that may cause. Fatigue failure, but any force which is acting in a reverse direction may ultimately produce such failure. fatigue failure may be induced by repeatedly loading a test piece and applying a force acting axially, torsionally or flexurally. The danger in fatigue failure is that it may occur without any warning at stress levels considerably below the yield stress. Over the years, much experience has been accumulated by exploring the possible reasons for this occurrence and tests have been suggested to evaluate the propensity for failure of machine elements which are exposed to pulsating or vibrational stresses. Nonetheless, the difficulty in predicting fatigue failure produces a wide spread of statistical results and often a deviation of ~50 % from the average value is observed. In contrast, other mechanical tests, such as for yield stress, do not deviate from an average value by more than ~2–3 %. This explains why many test specimens are used in fatigue experiments to reach a meaningful average value, below which the probability for fatigue fracture is quite low. Below, in this chapter, there are commonly-used tests having repeated stress with reverse loading as against the number of cycles in order to evaluate the endurance of a specimen. The results of such tests are plotted on S–N curves (stress vs. number of cycles). It is not only alternating stress that may cause failure, but also the duration of exposure, as in materials such as glass, that may fracture after long-term exposure to stress without undergoing any plastic deformation. The term for such behavior is 'static fatigue'. 'Thermal fatigue' is the term assigned to material failure caused by repeated changes in stress due to the rise and fall of thermal gradients for various reasons, involving restrictions in thermal expansion or contraction. In the following sections, various aspects related to fatigue are considered for the case of ceramics.

Recalling that most ceramics are brittle at RT and become ductile at elevated temperatures, one must consider the fatigue of both types. First, here is an introduction on the concept of S–N curves.

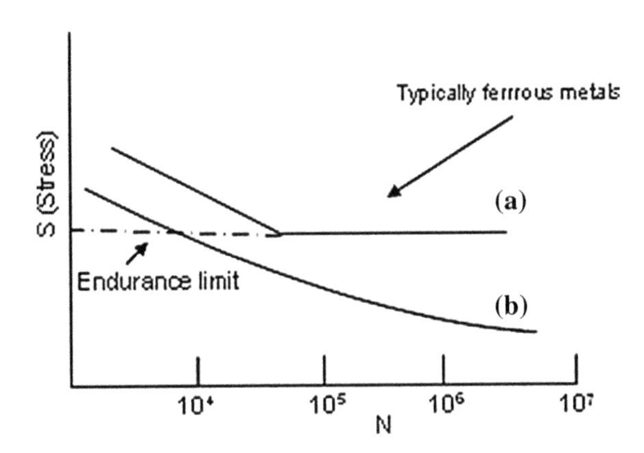

Fig. 7.1 S–N curves: **a** with a well-defined endurance limit, **b** without a definite fatigue limit

7.2 S–N Curves; Endurance Limit

One of the most frequently used tests for fatigue-resistance evaluation is the well-known plotting of stress versus the number of cycles, usually referred to as the 'S–N (curve) relation'. Various wave forms of cyclic stresses may be applied to a specimen to test its suitability to withstand prolonged strain. Machine elements are assessed to determine their practical endurance of industrial applications to which they may be exposed. Such tests focus on the nominal stress required to cause fatigue failure at some number of cycles. A logarithmic scale is almost always used for N, the number of cycles to failure. A schematic S–N plot is shown in Fig. 7.1. Note the horizontal line in plot (a), known as 'a knee', which represents 'the endurance limit'. As implied by its name, at this level of stress, the specimen is characterized by its ability to endure a large number of stress-cycles at the stress level of the horizontal line and below it without failure. In plot (b), no such horizontal line is observed and the curve continues to decrease, indicating that the stress must be reduced for the test specimen to be able to withstand a certain number of cycles.

Some materials, mostly metallic, indeed show a knee when tested under fatigue. The best known examples are alloys of steel and Ti. As illustrated, some ceramics may show a definite knee, signifying an endurance limit. Various wave forms of cyclic stresses may be applied to a specimen exposed to fatigue testing in order to evaluate S–N plots (see Fig. 7.1). The wave forms used for cyclic stress may be regular or irregular. The first of these (excluding **c**) is usually sinusoidal or repeated waves. On airplane wings, for example, the stress type is irregular and, therefore, difficult to analyze. Figure 7.2 shows the various wave forms of the cyclic stresses that may be applied to a specimen during fatigue testing. Some of the figures represent sinusoidal-type cycles and the last one is a repeated cycle, possibly representing a specimen exposed to repeated stress (Fig. 7.2).

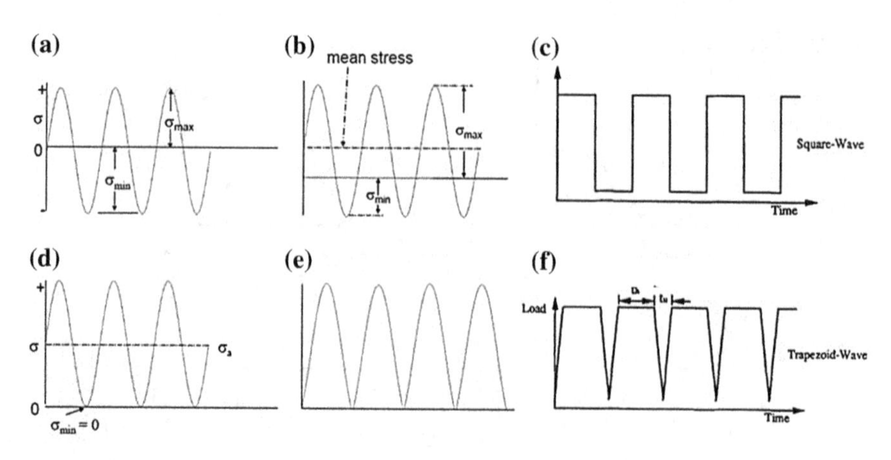

Fig. 7.2 Various forms of cyclic stresses: **a**, **b** and **d** are sinusoidally varying cycles **e** represents a repeating cycle, **c** a square cycle and **f** a trapezoid cycle

7.2.1 Brittle Ceramics

It has been thought for some time that ceramics are not candidates for fatigue failure, since they lack plasticity, especially in the vicinity of a crack tip, namely no crack blunting is likely. However, plenty of experimental data indicate that test specimens have failed under cyclic stress applications compared to static tensile testing. It was customary to compare cyclic test performance to tensile results. Here, S–N curves of several technically important ceramics are illustrated. In Figs. 7.3 and 7.4, the static and cyclic fatigue results are compared for polycrystalline alumina. Recall that the failure of a part under an ongoing static load is termed 'static fatigue'.

Figure 7.4 shows the cumulative probability plots as a function of cycles and time to failure. These results are from the sinusoidal tension-tension cyclic loading of specimens with controlled indentation flaws in biaxial flexure in water. The controlled Vickers indentation flaws at a load of 30 N were placed at the centers of the tensile faces of each specimen. These cyclic load tests were conducted under sinusoidal-tension tension on a servo-hydraulic fatigue testing machine. The minimum tensile stress in each series of tests was maintained at 20 MPa, but the maximum stress was adjusted to coincide with the constant stress levels applied in the static loading tests. At least five specimens were tested at each selected peak stress, at frequencies of 1 and 50 Hz, in water. In all the tests, the broken specimens were examined to verify that failure was initiated at the indentation site. Note that the purpose of indentation flaws is to act as notched fatigue tests. Observe in Fig. 7.3b that a knee exists in the plot, similar to the known fatigue tests of steel or Ti. Not all materials have a fatigue threshold and, for these unlimited materials, the test is usually terminated.

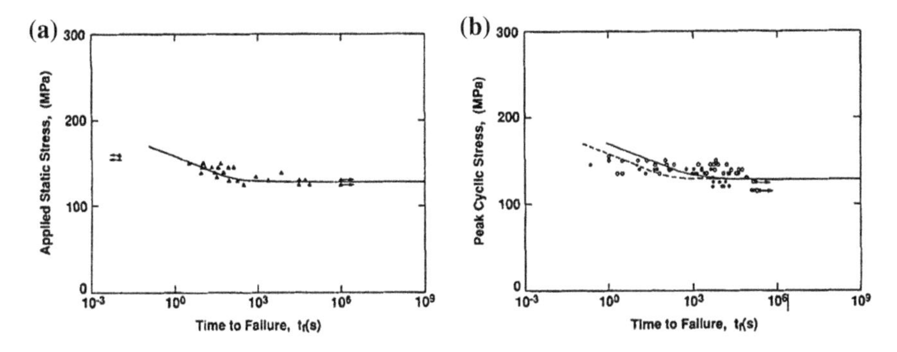

Fig. 7.3 **a** Static fatigue plot for polycrystalline alumina in water, for Vickers indentations at P = 30 N. *Data points* are results of individual tests. Arrows at *right* designate interrupted tests; at *left* breakages during ramp loading to maximum applied stress. *Solid curve* is theoretical prediction. **b** Cyclic fatigue plot for polycrystalline alumina in water, for Vickers indentations at P = 30 N. *Data points* are results of individual tests: *open symbols* are for 1 Hz, *closed symbols*, 50 Hz. *Solid curve* is prediction assuming only slow crack growth. Static fatigue curve from (**a**) is included for comparison [15]. With kind permission of John Wiley and sons

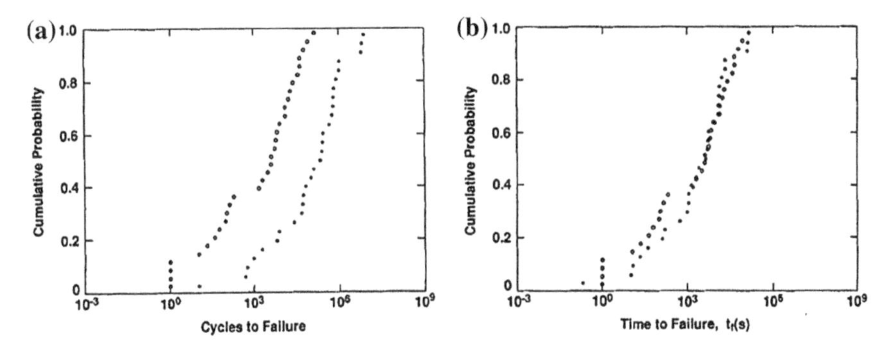

Fig. 7.4 Cumulative probability plots for **a** cycles and **b** time to failure for data in Fig. 7.3. *Open symbols* are for 1 Hz, *closed symbols*, 50 Hz [15]. With kind permission of John Wiley and sons

The time to failure may be computed numerically for any time-dependent applied stress function:

$$\sigma_a = \sigma_a(t)\ \sigma_a = \text{const} \quad (\text{static case}) \tag{7.1}$$

$$\sigma_a = \sigma_M + \sigma_0 \sin(2\pi \nu t) \quad (\text{cyclic}) \tag{7.2}$$

with v representing cyclic frequency and σ_M and σ_0, the mean and half-amplitude stresses, respectively.

In another illustration (Fig. 7.5), an S–N type curve is illustrated for Si_3N_4. Here, as in Fig. 7.3, static and cyclic fatigue are compared. Typical fractographs for cyclic and static fatigue specimens are shown in Fig. 7.6a, b, respectively. In

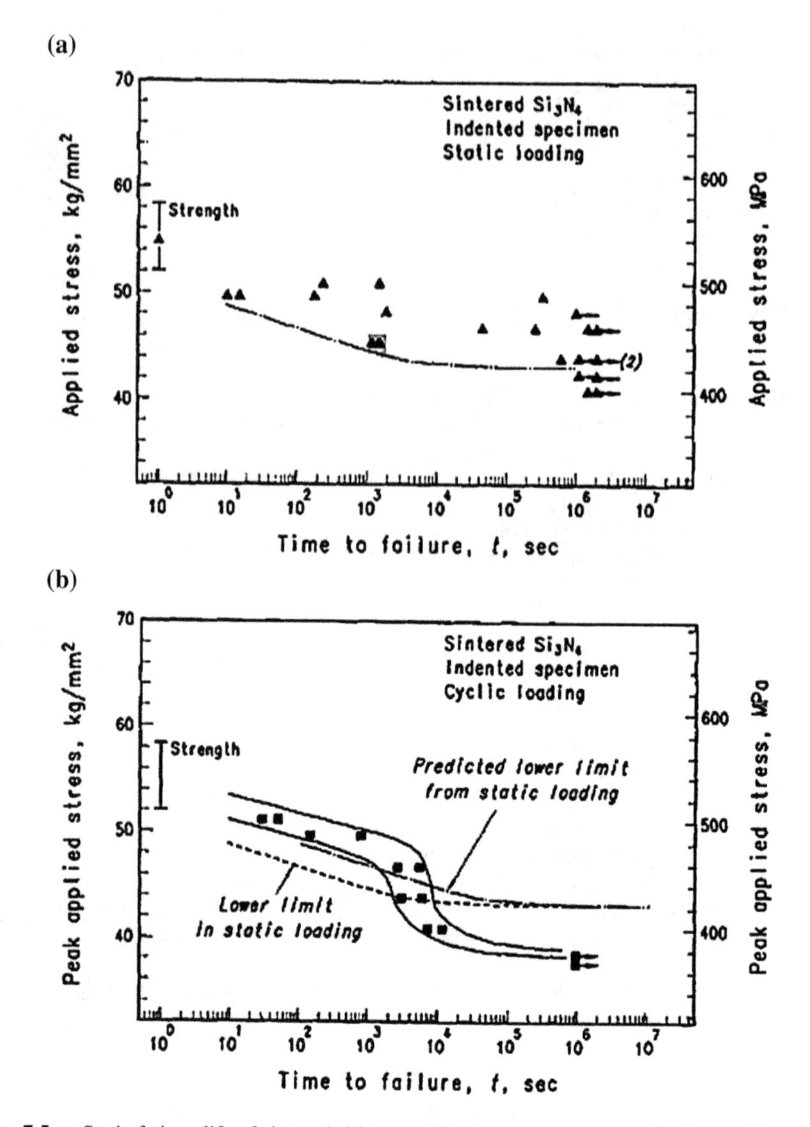

Fig. 7.5 **a** Static fatigue life of sintered silicon nitride at room temperature. **b** Cyclic fatigue life of sintered silicon nitride at room temperature [12]. With kind permission of John Wiley and sons

the case of cyclic fatigue specimens, several semicircular markings are observed, as well as many vague bow-like marks on the static fatigue specimens. The fracture surfaces of the flexural-strength specimens are similar to those of the cyclic fatigue specimens.

A high-magnification SEM fractograph of the portion between the two semicircular marks in the cyclic fatigue specimen is shown in Fig. 7.6c, after

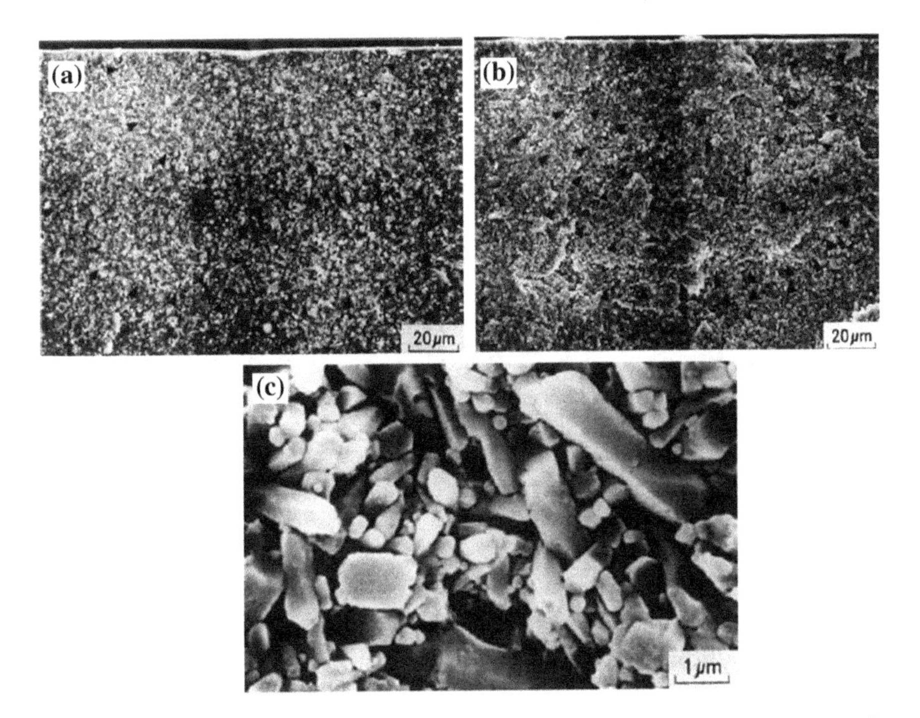

Fig. 7.6 SEM fractographs of fatigue specimen: **a** cyclic fatigue, **b** static fatigue, and **c** cyclic fatigue at high magnification [12]. With kind permission of John Wiley and sons

immersion in a solution of hydrochloric and hydrofluoric acids. Predominantly intergranular cracking, accompanied by partial transgranular cracking, are observed. The sintered silicon nitride shows considerable cyclic and static fatigue susceptibility in RT air.

7.2.2 Ductile Ceramics (RT and High Temperature)

Due to its excellent mechanical properties (i.e., bending strength and toughness) Y-TZP, is of special interest for acquiring fatigue information. In Fig. 7.7, fatigue information, in the form of S–N curves, is presented. Table 7.1 indicates the materials used for these fatigue tests. Cumulative survival plots (Kaplan–Meier) for all four materials for 1 million cycle data "as received" and "CoJet sandblasted" are displayed in Fig. 7.8. The beneficial effect of CoJet sandblasting is clearly shown for all four materials with a statistical significance of 95 % CI (confidence interval) for all zirconia materials except for the Zeno (ZW) (Wieland) ($p = 0.295$).

The microstructures of the ceramics used for the fatigue tests to evaluate the S–N plots are shown in Fig. 7.9.

(a)

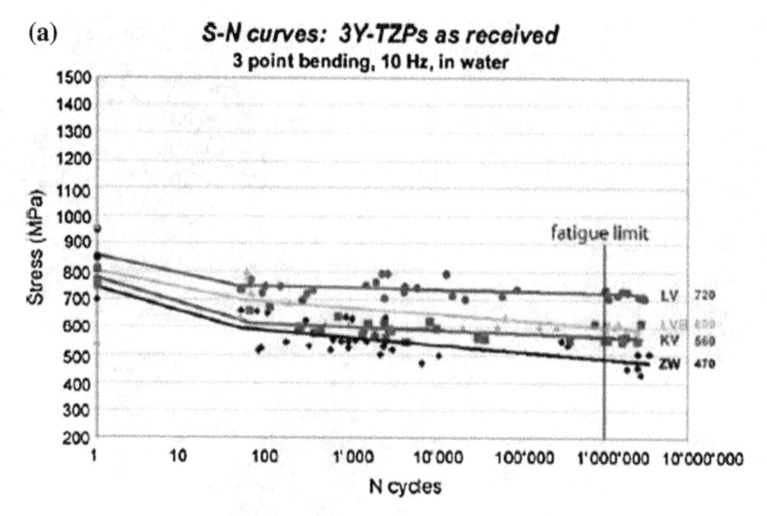

(b)

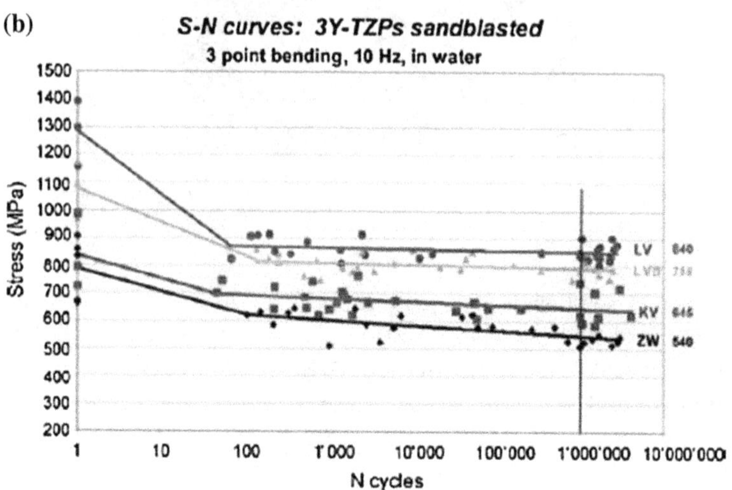

Fig. 7.7 **a** and **b** S–N fatigue data for Lava (LV) (3 M Espe), Lava colored (LVB) (3 M Espe), Everest ZS (KV) (KaVo), Zeno (ZW) (Wieland) "as received" and "after CoJet sandblasting". Fatigue limits at 10^6 cycles at 10 Hz in water were determined from trend lines using a least square linear regression fit. The fatigue limits "as received" were: LV = 720 MPa, LVB = 600 MPa, KV = 560 MPa, ZW = 470 MPa. The fatigue limits of "CoJet sandblasted" were: LV = 840 MPa, LVB = 788 MPa, KV = 645 MPa, ZW = 540 MPa. The initial strength for the "as received" was: LV = 854 ± 94 MPa, LVB = 749 ± 209 MPa (one outliner with major grinding flaw was responsible for dropping the mean initial strength), KV = 776 ± 31 MPa, ZW = 742 ± 39 MPa. The initial strength for the "after CoJet sandblasting" was: LVs = 1282 ± 119 MPa, LVBs = 1077 ± 102 MPa, KVs = 836 ± 135 MPa, ZWs = 787 ± 100 MPa [26]. With kind permission of Elsevier

Table 7.1 Y-TZP materials selected for fatigue testing [26] (with kind permission of Elsevier)

Brand	Materials	Manufacturer
LAVA(LV)	3Y-TZP	3M Espe, Seefeld, D
LAVA colored(LVB)	3Y-TZP	3M Espe, Seefeld, D
EVEREST ZS(KV)	3Y-TZP	KaVo Dental GmbH, Biberach, D
ZENO Zr(ZW)	3Y-TZP	Wieland, dental, Pforzheim, D

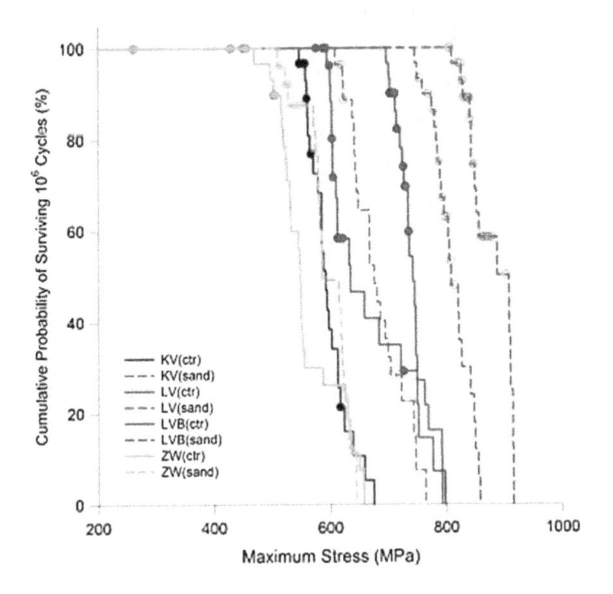

Fig. 7.8 Cumulative probability of survival at 1 million cycles for all four zirconia materials. *Dashed lines* denote the sandblasted groups. All four materials demonstrated a definite shift to the right as sandblasting improved their probability of survival. The median (50 % at 106 cycles) survival values for the "as received" and "CoJet sandblasted" groups were as follows: LV = 743 MPa, LVs = 908 MPa, LVB = 635 MPa, LVBs = 809 MPa, KV = 593 MPa, KVs = 676 MPa, ZW = 549 MPa, ZWs = 587 MPa. The increase for Zeno (ZW) (*green*—far left) was not statistically significant ($p = 0.295$) [26]. With kind permission of Elsevier

Surface properties affect the fatigue behavior of materials exposed to cyclic tests. In the case of ceramics, an accepted method for providing good surface properties for fatigue resistance is sandblasting. In Fig. 7.10, S–N curves of Lava zirconia are compared with and without sandblasting. On the right side, the curve representing the Kaplan–Meier survival plot from the same material at 1 million cycles indicates the significant improvement of the CoJet sandblasting treatment on its survival probability. A ~ 17 % increase in the fatigue limit (endurance) is

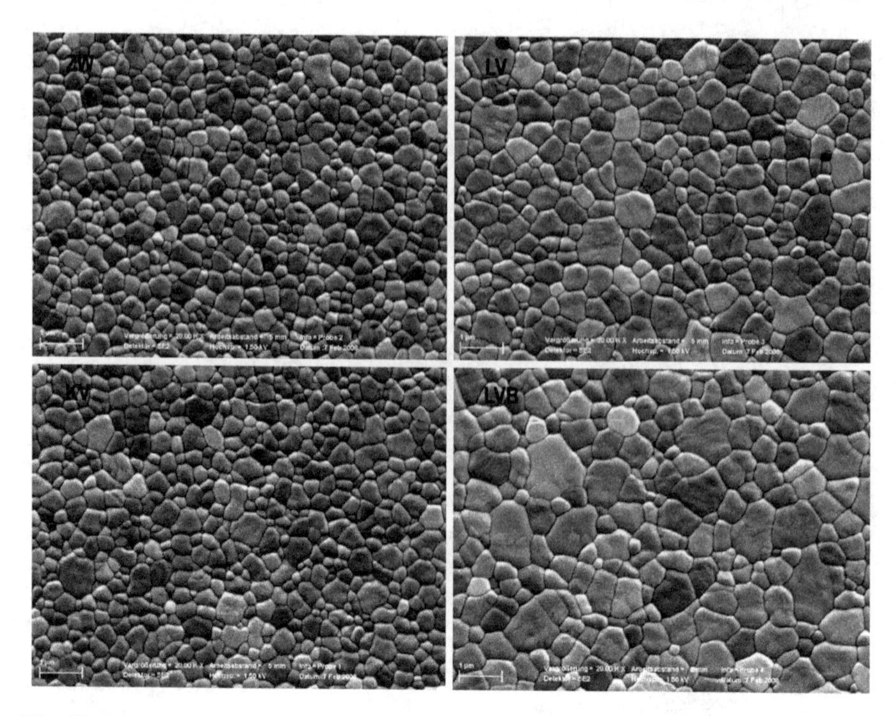

Fig. 7.9 Microstructure of the four Y-TZP ceramics (20,000×). The average grain size according to the 'Linear interceptive count method' was as follows: Zeno (ZW) (383 ± 47 nm); Everest (KV) (383 ± 47 nm); Lava (LV) 537 nm (±47); Lava colored (LVB) 643 nm (±61) [26]. With kind permission of Elsevier

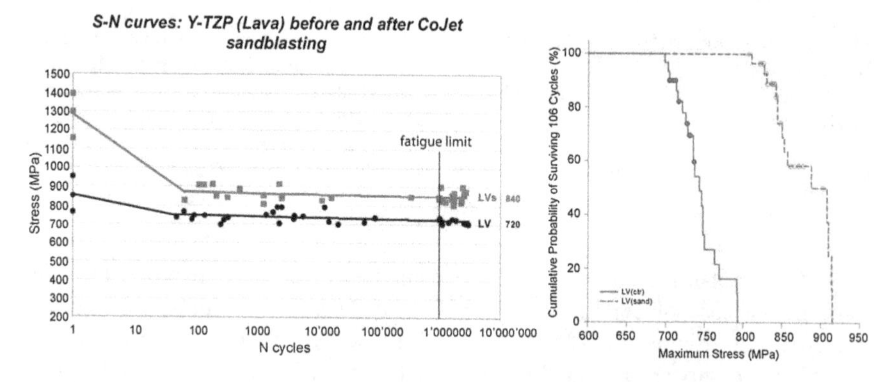

Fig. 7.10 On the left, *S–N* fatigue data (*left*) for Lava (LV) "as received" (*black*, *bottom*) and "CoJet sandblasted" (LVs) (*red*, *top*). Fatigue limits at 106 cycles at 10 Hz in water were 720 MPa "as received" and 840 MPa for "CoJet sandblasted" which represented an increase of 17 %. On the right, Kaplan–Meier cumulative probability of survival plot at 1 million cycles for Lava "as received"(*left*) and "CoJet sandblasted" (*right*). A clear shift to the right is seen when sandblasting, thus improving the survival probability [26]. With kind permission of Elsevier

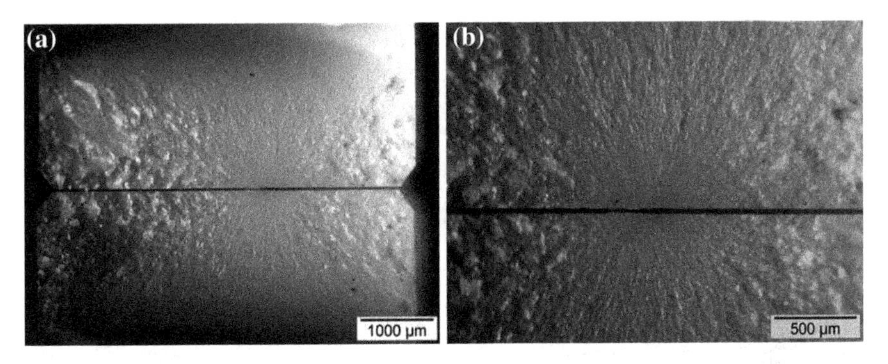

Fig. 7.11 Stereomicroscope views (32x and 57x magnification) of two broken halves of a Lava (3 M Espe) fatigue-failed zirconia specimen. **a** The mirrors (*smooth surfaces*) and **b** the *radiating hackle lines* surround the failure origin located on the tensile surface of this specimen [26]. With kind permission of Elsevier

seen in the figure. This proves the importance of surface finish. It is known that fatigue failure often starts at the surface, as may be seen also in Fig. 7.11.

The microstructures of the fractured test specimens were analyzed to determine the points of origin and means of initiation of each fatigue failure. Figure 7.11 shows two halves of a broken test piece which was fatigued. Processing flaws (fabrication, sintering and grinding) associated with fatigue failure are observed by SEM in Fig. 7.12. These flaws, which are the points of origin of the fatigue failure, act as stress raisers.

The above tests were performed under water (see, for example, Fig. 7.7), in order to more closely duplicate intraoral conditions. Recall that various zirconia ceramics are used for biomedical purposes (dental applications, etc.). Zirconia indeed suffers from low-temperature degradation in a humid environment due to alterations in its crystalline structure. Micrographs of a surface crack, initiated by a Vickers indent and grown under a cyclic fatigue load, are shown in Figs. 7.13 and 7.14. Under fatigue loading, the crack grew stably until it reached the size marked "f"; subsequent crack propagation was unstable, leading to catastrophic fracture. Note the striations seen clearly in Fig. 7.14.

Striations are known in the fatigue literature as 'beach marks' and are indicative of failure by crack growth, showing distinct crack-nucleation sites. In these figures, note the sites where fatigue-crack growth started, showing definite flows. R symbolizes the ratio, which is defined below in Sect. 7.3. Here, the possibility of the accelerated testing of specimens is considered by way of high-frequency, cyclic loading. Various results seem to indicate that materials with definite endurance limits do not necessarily show infinite-fatigue life when tested at cycles above 107. Pyttel [76] claims that: "The fatigue limit cannot be a general material property. The term 'fatigue limit' should not be used and it is better to substitute it by 'fatigue strength' at a definite number of cycles." As mentioned above and

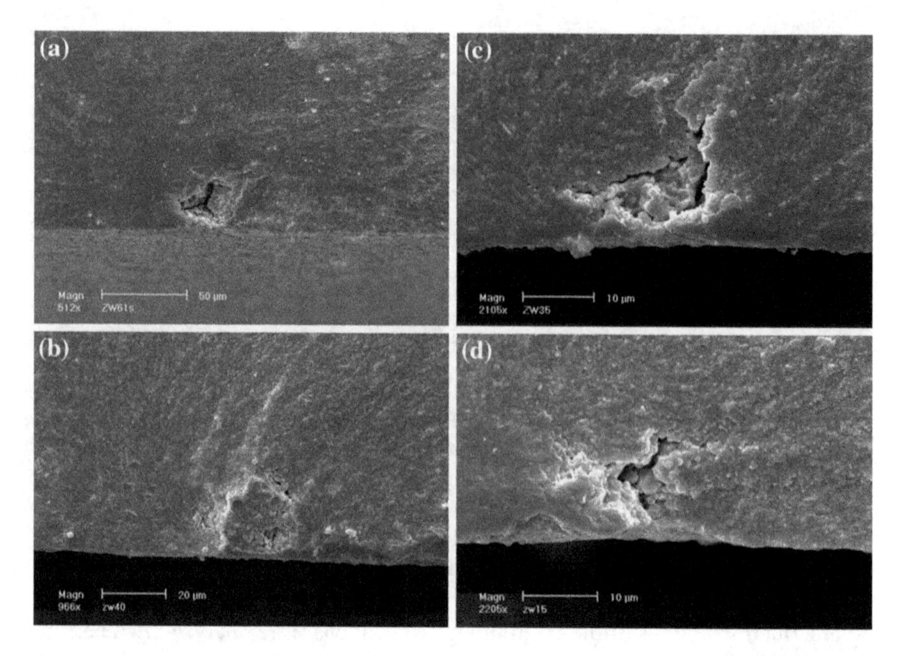

Fig. 7.12 Processing flaws for Zeno (ZW, Wieland) located at—or near the tensile surface, which acted as stress concentrators during fatigue testing. These flaws were incorporated into the ceramic during cold pressing of the blanks. The number of cycles and stress level for each image were as follows: **a** "sandblasted", 501, 550 cycles at 581 MPa; **b** "as received", 900 cycles at 635 MPa; **c** "as received", 51 cycles at 658 MPa; **d** "as received", 1577 cycles at 548 MPa [26]. With kind permission of Elsevier

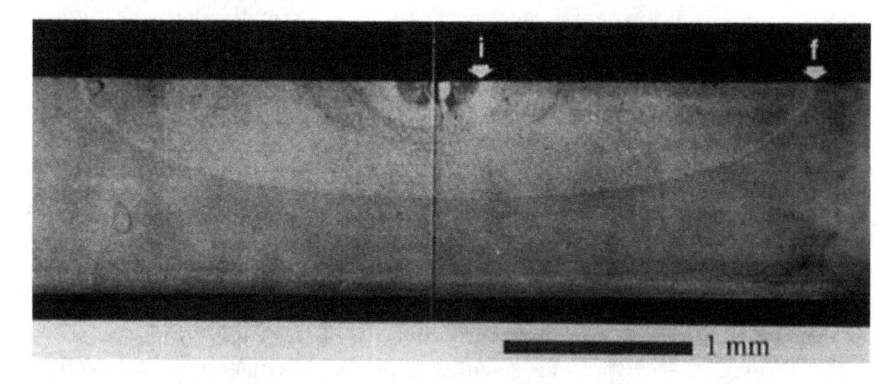

Fig. 7.13 Fatigue fracture surface of 3Y-TZP. Note the shape change, from initial indent crack indicated by "i" to the final fatigue crack indicated by "f" [18]. With kind permission of John Wiley and sons

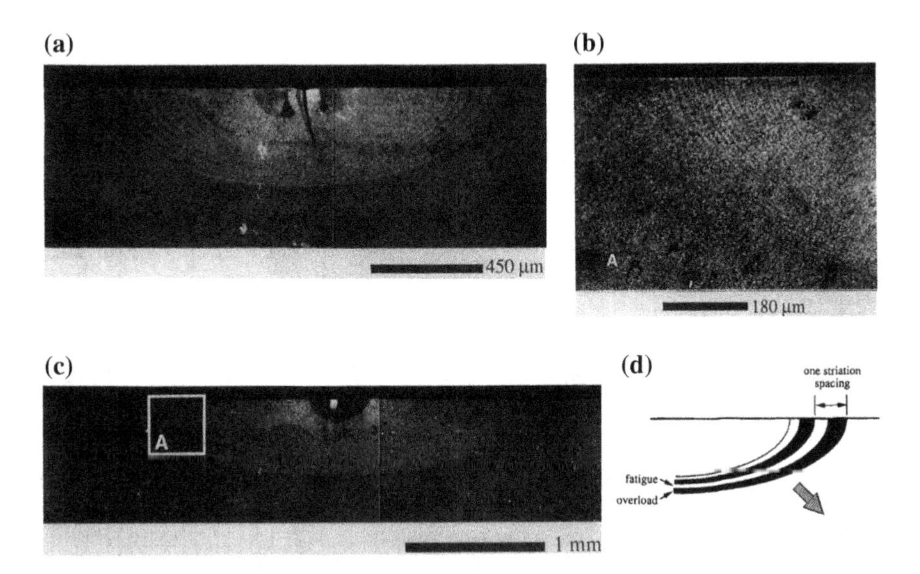

Fig. 7.14 Fatigue fracture surface of 3Y-TZP for **a** $R = 0.8$ and **b**, **c** $R = 0.01$. In both cases, fatigue striations appear during the last scores of cycles before catastrophic failure. **d** Schematic of fatigue striations in terms of two growth steps [18]. With kind permission of John Wiley and sons

indicated in Fig. 7.1, in many cases, no definite endurance limit exists and the curve continues to decrease with the number of cycles.

One of the lessons to be learned from the exemplary zirconia fatigue tests is that proper surface finish significantly improves the fatigue resistance of ceramics; thus, it is a recommended means of preparing specimens for exposure to cyclic stresses.

The following presents some definitions which are often used in fatigue testing.

7.3 Definitions

It is customary to use certain terms and concepts in fatigue studies and when analyzing fatigue tests, as follows. The safety factor, FS, is the ratio between allowed (σ_{al}) and applied (σ_{ap}) stress given as:

$$FS = \frac{\sigma_{al}}{\sigma_{ap}} \tag{7.3}$$

The mean stress is:

$$\sigma_{mean} = \frac{\sigma_{\max} + \sigma_{\min}}{2} \tag{7.4}$$

The ratio of load or stress is:

$$R = \frac{\sigma_{\min}}{\sigma_{\max}} \tag{7.5}$$

and the stress range is given by:

$$\sigma_r = \sigma_{\max} - \sigma_{\min} = \Delta\sigma \tag{7.6}$$

If $\sigma_{\max} = -\sigma_{\min}$, then the stress cycle is reversed. In such cases, the mean stress is zero.

The stress amplitude, σ_a, is one half the stress range and is given as:

$$\sigma_a = \frac{\Delta\sigma}{2} = \frac{\sigma_{\max} - \sigma_{\min}}{2} \tag{7.7}$$

In the above illustrations, the number of cycles does not exceed 10^7. In conventional fatigue tests, seldom more than 10^7 cycles are applied to check fatigue lives. However, in some industries, the required design lifetime of many components often exceeds 10^8 cycles. Note that materials with well-defined endurance limits do not necessarily show infinite fatigue lives when tested at cycles above 10^7. Time constraints usually prevent the performance of such extended tests.

7.4 The Stress Cycles

Cyclic loads may be applied in various forms, as mentioned earlier. These applied loads may be tensional, torsional or flexural. Any of them may induce fatigue failure. Rather than focusing on these various loading possibilities, this section will deal with the range of the cycles. A machine element may be exposed either to a large number of cycles or, often, to a small number of low cycle, inducing failure. The purpose of a designer is to obtain a long lifetime in service; therefore, testing must be aimed at obtaining information on the propensity of a material to endure a large number of cycles without failure.

7.4.1 Low-Cycle Fatigue Tests

Specimens may be tested during a relatively small number of cycles by repeated stress or strain until failure sets in. The number of cycles, determined arbitrarily, is in the range of 10^4–10^5 cycles, which may be considered the upper limit for low-cycle testing. Experience has taught an important lesson that structural materials may fail due to low-cycle fatigue, which imposes a serious structural problem. Performing tests at high cycles incurs time- and cost-related problems; therefore, low-cycle tests are often performed at high stresses. Low-cycle fatigue is strain-controlled and, frequently, the number of cycles does not exceed 10^3. The purpose is to evaluate fatigue life more precisely and to identify when fatigue cracks form.

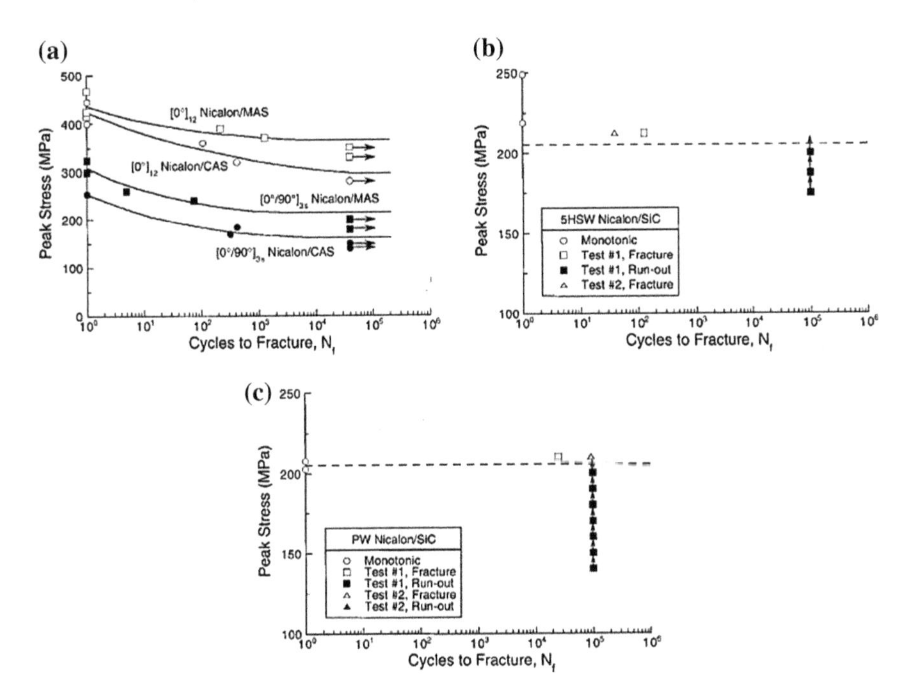

Fig. 7.15 Fatigue life of **a** the glass–ceramic-matrix composites and, **b** and **c**, the SiC-matrix composites. The *horizontal arrows* in **a** indicate run-out. In **b** and **c**, the *vertical arrows* indicate run-out to 10^5 cycles and subsequent cycling at the next highest stress amplitude and the *dashed horizontal lines* represent the apparent threshold [20]. With kind permission of Elsevier

In Fig. 7.15, an experimental investigation of damage and fracture in fiber-reinforced ceramic composites (CMC) under low-cycle fatigue is shown. Several different composites are indicated, each reinforced with ceramic-grade Nicalon2 fibers, but with varying fiber architectures and matrix materials.

The test specimens undergoing fatigue valuation at low cycles are: magnesium aluminosilicate [henceforth: MAS] glass–ceramic matrix; calcium aluminosilicate [henceforth: CAS] matrix and SiC matrix produced by chemical-vapor infiltration [henceforth: CVI]. In these experiments, it seems that the CMCs exhibit fatigue fracture at ambient temperature. Generally, fracture occurs only under conditions in which the peak stress is well above the matrix cracking limit. In this regime, multiple matrix cracks form during the first loading cycle, with attendant debonding and sliding occurring along the fiber/matrix interfaces. Additional sliding occurs during subsequent cycling, with the direction of slide reversing with each load reversal. The inference is that the mechanism giving rise to fatigue involves cyclic sliding and is not necessarily intrinsic to either the fibers or the matrix alone.

An interesting comparison exists between static and cyclic fatigues for Ce-TZP, in which, as is known, a t → m transformation takes place. Figure 7.16 shows a comparison of the measured results of static and cyclic fatigue tests for three Ce-TZP materials. The time-to-failure of the statically-loaded specimens is longer

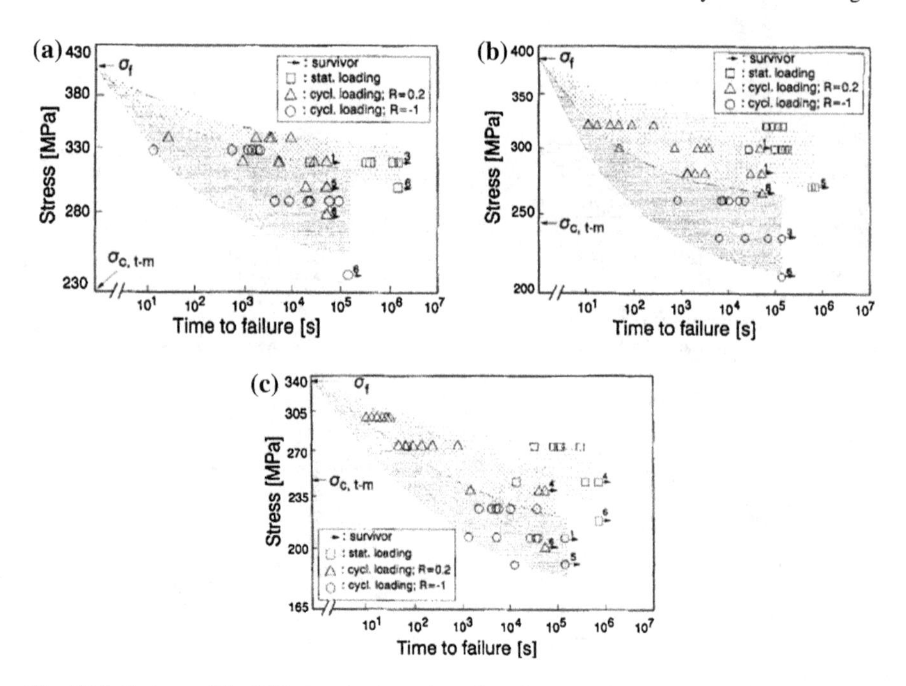

Fig. 7.16 Fatigue of Ce-TZP ceramics in static and cyclic (pulsating and reversed bending) tests: **a** Ce-TZP-11, **b** Ce-TZP-IV, and **c** Ce-TZP-V. The applied stress values are given as static stress s_s and maximum stress s_{max} for $R = 0.2$ and stress amplitude s_A for $R = -1$; σ_f and $\sigma_{c,\, t\rightarrow m}$ refer, to the average bending strength and the critical transformation stress, respectively. Data points in combination with figures indicate numbers of survivor specimens under identical loading conditions [8]

by 2–3 orders of magnitude at the same stress level than under cyclic loading ($R = 0.2$). The shorter time-to-failure is more pronounced under cyclic loading with increasing grain size. In addition, the fatigue limits with $R = -1$ are much lower than under static and cyclic loading with $R = 0.2$. Recall that R is the ratio given as $R = \frac{\sigma_{min}}{\sigma_{max}}$ (relation 7.5). These fatigue experiments and static fatigue tests were performed by four-point bending in air at 20 °C. The maximum duration of loading was fixed at 200 h. In these cyclic tests, the stress ratio is R = 0.2 for pulsating tests and R = −1 for reversed-bending tests. (For these types of cycles see Fig. 7.2). The test frequencies were 180 and 70 Hz for R = 0.2 and R = −1, respectively. The limited number of cycles was fixed at 10^7. The specimens and other data are listed in Table 7.2.

Ce-TZP ceramics are characterized by relatively low critical stresses for the stress-induced t–m transformation ($\sigma_{c,t} \rightarrow \sigma_f$). Pronounced transformation plasticity is observed, due to which the Ce-TZP ceramics are relatively flaw-tolerant and their strength is not controlled by the initial flaw size, but rather by the critical transformation stress, the zone size and the strain-hardening effect.

Table 7.2 Sintering parameters, average grain size, relative densities, Young's modulus and Vickers hardness of Ce-TZP materials [8]

Materials	Sintering parameters	Average grain size (μm)	Relative density (%TD)	Young's modulus (GPa)	Vickers hardness
Ce-TZP-I	1400 °C, 0.2 h	0.5	97.57	201	836
Ce-TZP-II	1400 °C, 2 h	1.0	99.83	202	849
Ce-TZP-III	1500 °C, 0.5 h	1.4	99.55	190	816
Ce-TZP-IV	1500 °C, 1 h	1.5	99.73	202	854
Ce-TZP-V	1600 °C, 1 h	2.7	99.36	199	780

7.4.2 High-Cycle Fatigue Tests

Today, various machines are available for high-cycle fatigue testing. A load-controlled servo-hydraulic test rig is one such machine commonly used in these tests, with frequencies of around 20–50 Hz. Resonant, magnetic machines are also in use as are rotating, bending machines. In such fatigue tests, a constant bending stress is applied to a round specimen, combined with the rotation of the sample around the bending-stress axis until its failure. The cycle range of these tests is 103–108. High-cycle fatigue strength may be described by stress-based parameters. High-cycle tests are usually performed until failure. The common procedure during these tests is to start testing specimens at high stress until their failure, which usually sets in after a relatively small number of cycles. Then, in tests done to the next specimens, the stress level is successively decreased to the point at which the test samples do not fail. When a few specimens do not fail under applied stress throughout all the cycles of the test, the data are noted and it is said that 'runout' has occurred. In most cases, this runout refers to at least 107 cycles. Thus, runout represents the highest stress of non-failure at the specified cycles. The point at which runout occurs is termed the 'endurance limit' (as mentioned earlier), which is not observed in all materials (also above). Therefore, these tests are usually terminated at ~108 cycles, and the materials are considered to be safe for practical use. Figure 7.17 illustrates a high-cycle test for M80 (80 % mullite/20 % alumina) and FGC (mullite/Al_2O_3 functionally graded ceramics).

7.5 Fatigue Lifetime

Many factors, in addition to stress, strain, amplitude, etc., influence the fatigue life of materials. Here, the effects of (a) stress-based and (b) strain-based evaluations of fatigue life will be considered for low- and high-cycle cases. In the past and even now, the stress-based approach is quite common.

(a) **The stress-based approach**. The stress-based evaluation of fatigue life clearly relates to the S–N curves observed in various metals. The applied-stress

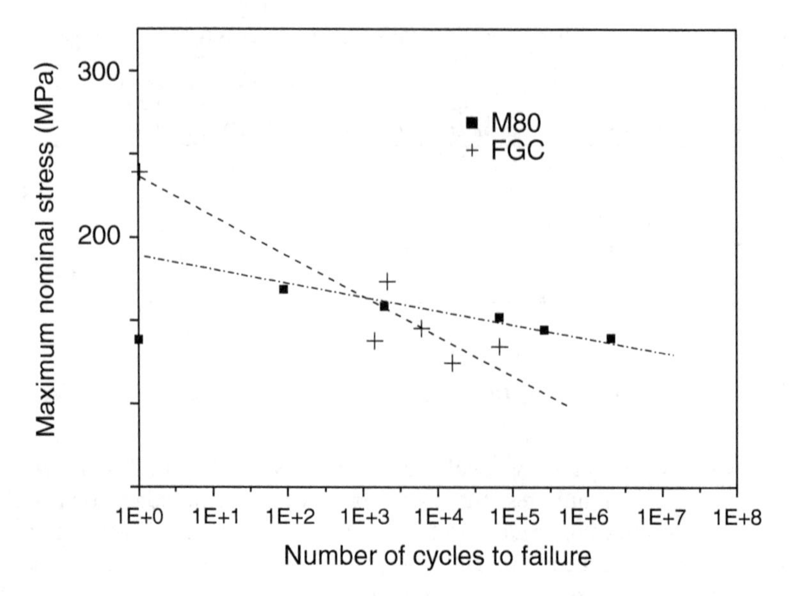

Fig. 7.17 Peak applied stress against the number of cycles to failure for monolithic mullite 80 % alumina, 20 % composites (M80) and FGC specimens [2]. With kind permission of John Wiley and Sons

range or rather the stress amplitude, σ_a, of the S–N curves is the significant factor determining a material's lifetime. Figure 7.18 is an illustration of the results of cyclic fatigue in alumina with different loading wave forms (see Fig. 7.2) at 1200 °C, where the maximum stress is plotted versus the number of cycles to failure. The arrow indicates that failure did not occur when the test was terminated. Each point represents a single specimen tested to failure. In Fig. 7.19a comparison is made between static and cyclic fatigues for the various wave forms used.

Fatigue lives were observed to increase with the decrease in applied stress, as expected for each cyclic wave form. No significant differences in cycle lifetimes between sine and square-wave forms under the same maximum applied load were observed. Note, however, that, when loading by a trapezoid cycle (trapezoid I), the number of cycles-to-failure is smaller than when loading by the other forms of waves. The number of cycles to failure is reduced by more than one order of magnitude. This comparison suggests that cyclic fatigue lifetimes are cycle shape or time-dependent. Static fatigue failure may be described by a power law and the correlation coefficient of the linear regression analysis is relatively high, having a value of $R^2 = 0.941$. The slope of the best-fitting curve is then used to draw the upper and lower limits of static fatigue life, which are shown as dashed lines in Fig. 7.19. Such data scatter is often observed in fatigue tests.

Many empirical relations have been suggested to predict fatigue lifetimes, one of which is the Manson-Coffin equation. This and other relations have been widely used for metals and alloys (J. Pelleg). The Manson-Coffin relation for lifetime

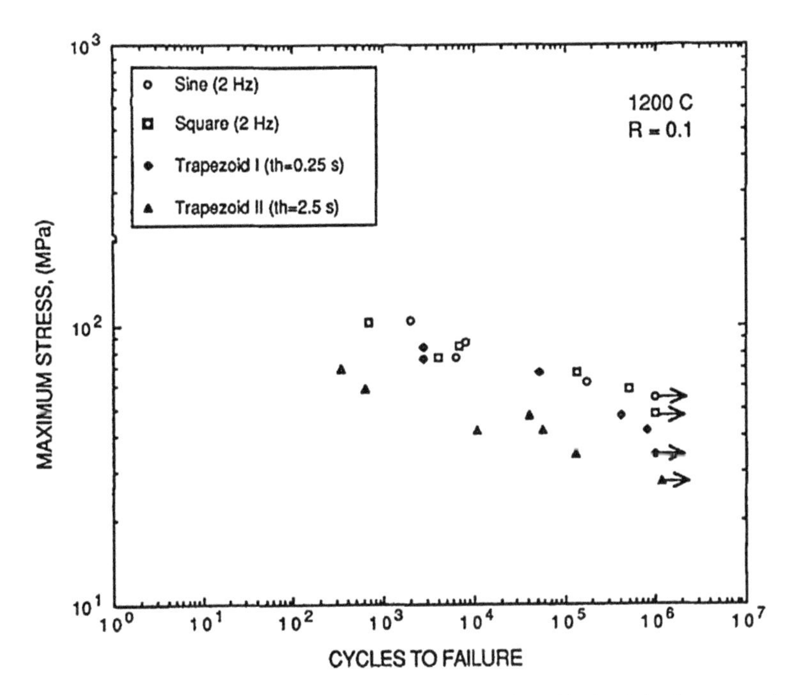

Fig. 7.18 Comparison of cyclic fatigue lifetime under various kinds of loading wave forms. (*Arrow* indicates that failure of specimen did not occur when the test was terminated) [14]. With kind permission of John Wiley and Sons

prediction has also been used for ceramic materials, to predict lifetimes by using low-cycle fatigue experiments [38]; for example, in SiC fiber reinforced SiC matrix (SiC/SiC). The Manson-Coffin relation is given as:

$$\Delta\varepsilon_p \cdot N_f^n = C \tag{7.8}$$

Here $\Delta\varepsilon_p$ is the plastic-strain range, N_f is the number of cycles-to-failure, n and C are constants. However, when no plastic deformation occurs, such as in SiC/SiC at RT, only the total strain range is relevant for controlling lifetimes. Thus, the total strain range should replace the plastic strain in Eq. (7.8). A log–log plot of Eq. (7.8), in terms of the strain range versus the number of cycle, allows for the evaluation of the exponent, n. The derived value of this exponent is 0.09, which is much smaller than the one characterizing metals. Figure 7.20 is a log–log plot of Eq. (7.8) in terms of the total strain range. The curve shows stress and strain controlled data. The cycles-to-failure during a low-cycle test at two ratios obtained under stress is shown in Fig. 7.21.

Almost no effect of the ratio is observed in SiC/SiC, as seen in the above figure, and the entire damage is caused by the applied tensile stress. In Fig. 7.22, cyclic softening may be observed where the strain amplitude, $\Delta\varepsilon/2$, is plotted against the

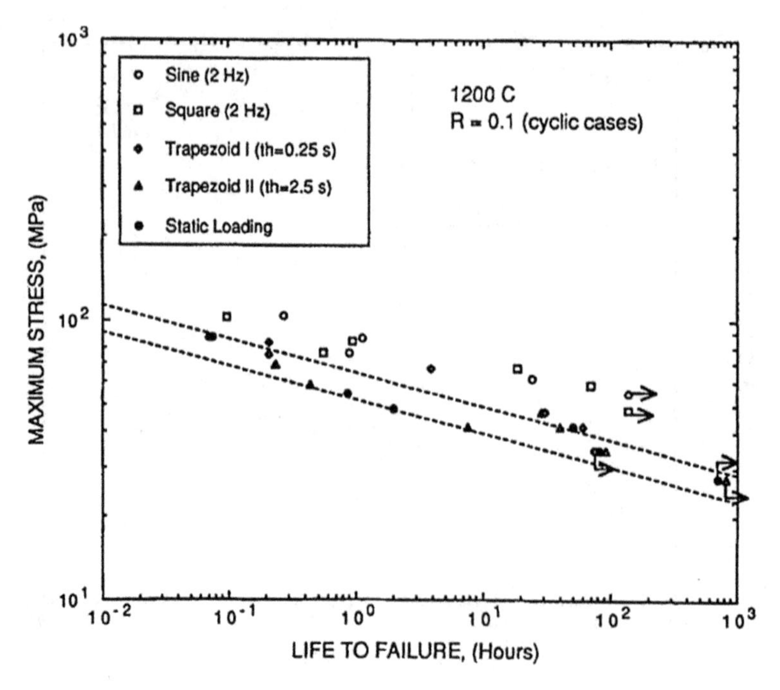

Fig. 7.19 Comparison of static and all cyclic fatigue lifetimes. (The *dashed lines* are the upper and lower bound of the static fatigue data) [14]. With kind permission of John Wiley and Sons

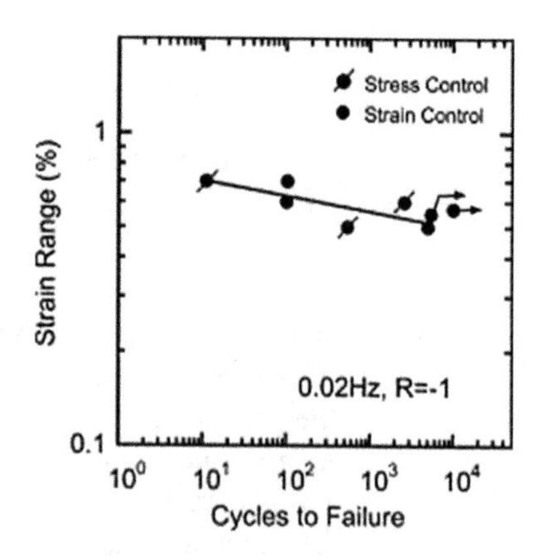

Fig. 7.20 Strain range as a function of cycles to failure for low-cycle fatigue [39]. With kind permission of Elsevier

number of cycles for various values of $\Delta\sigma/2$. Note that in Eq. (7.7) the applied half stress $\Delta\sigma/2$ equals $(\sigma_{\max} - \sigma_{\min})/2$, just as the strain amplitude equals $\Delta\varepsilon/2$.

The reason for the occurrence of strain softening seems to be associated with the initiation and propagation of cracks in the SiC matrix. The effect of frequency in stress-controlled fatigue in SiC/SiC, at a ratio of R = 0.1 is indicated in Fig. 7.23.

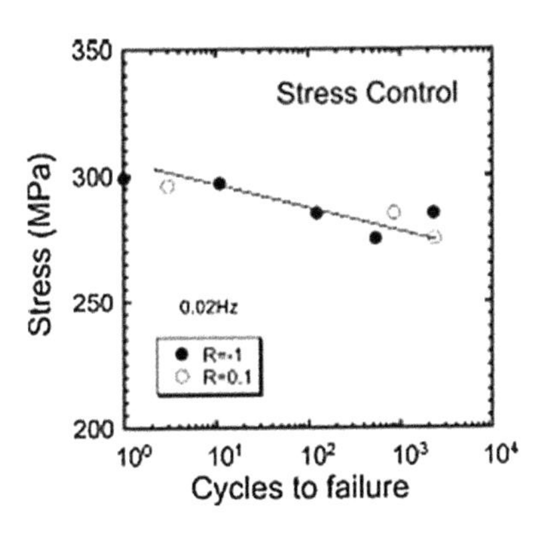

Fig 7.21 Maximum stress versus cycles to failure at different stress ratios [39]. With kind permission of Elsevier

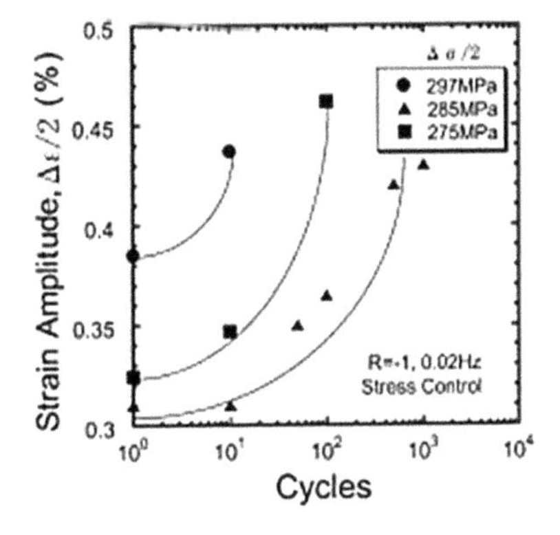

Fig. 7.22 Strain amplitude versus number of cycles for stress-controlled low cycle fatigue [39]. With kind permission of Elsevier

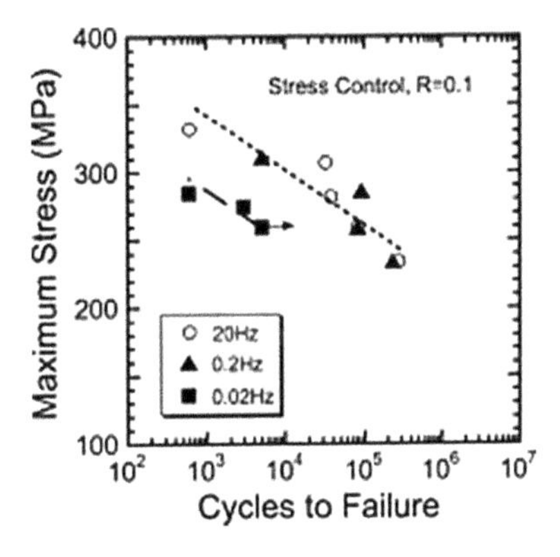

Fig. 7.23 Maximum stress versus cycles to failure at different frequencies [39]. With kind permission of Elsevier

A typical fracture surface of fibers with a mirror and hackle feature in SiC/SiC ceramics is shown in Fig. 7.24. The cyclic softening observed in SiC/SiC occurs either under stress control or strain control. Since this observation is crack initiation- and propagation-related, it is possible that similar phenomena may be more frequent after softening during fatigue deformation. The Manson-Coffin

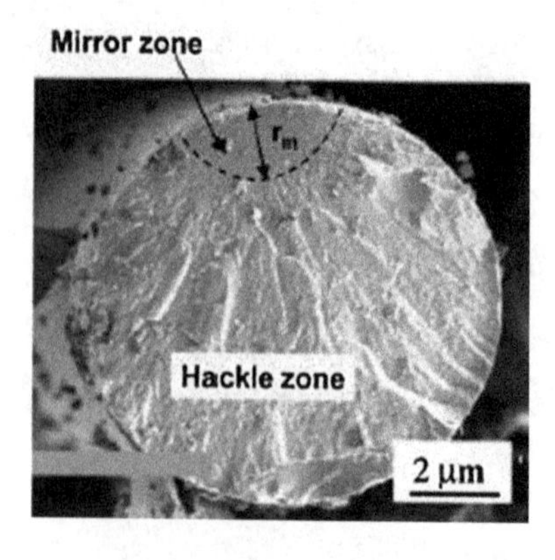

Fig. 7.24 Fracture surface of a fiber showing mirror and hackle zones [39]. With kind permission of Elsevier

equation is also applicable to the lifetime prediction of low-cycle fatigue in ceramic matrices, though this must be further empirically explored.

7.6 The Effect of Cracks: The Law of Paris

Most ceramics, being brittle at low temperatures and lacking crack-tip plasticity, cannot be blunt, thereby arresting fast crack propagation. The Paris Law describes crack growth from its subcritical dimension during fatigue stress to the stress intensity range, ΔK. The crack growth increment during fatigue, da/dN, is basically a function of the stress-intensity range and may be written as:

$$\frac{da}{dN} = C\Delta K^m \tag{7.9}$$

Here, "a" is the crack length, N, is the number of load cycles, C and m are experimentally determined scaling constants. ΔK is the difference between the maximum and minimum stress intensities for each cycle, namely $\Delta K = K_{max} - K_{min}$. Unlike metals, for which the exponent m of ΔK is in the range 2–4, for ceramics it can have values as high as $\sim 15 - 50$ (Dauskardt et al. [5]). Expressing Eq. (7.9) as a log–log plot of the crack-growth increment versus the difference in the stress intensities as:

$$\log \mathrm{da/dN} = \log \mathrm{C} + \mathrm{m}\Delta\mathrm{K} \tag{7.9a}$$

should give a straight line, as shown schematically in Fig. 7.25, with the slope providing m.

The crack intensity factor may be expressed in terms of stress, σ, as:

$$K = \sigma Y \sqrt{\pi a} \tag{7.10}$$

The uniform tensile stress, σ, is perpendicular to the crack plane and Y is a dimensionless parameter, depending on the geometry and the range of the stress-

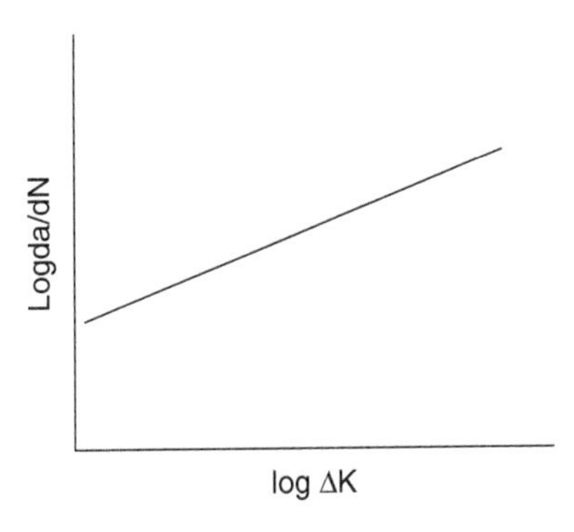

Fig. 7.25 Schematic plot of relation (7.9a). The slope is m

intensity factor. Thus, in terms of ΔK, Eq. (7.10) takes the form of:

$$\Delta K = \Delta\sigma Y\sqrt{\pi a} \tag{7.10a}$$

$\Delta\sigma$ is the range of the cyclic-stress amplitude. Expressing da/dN in terms of Eqs. (7.9), (7.10a) may be rewritten as:

$$\frac{da}{dN} = C\Delta K^m = C\left(\Delta\sigma Y\sqrt{\pi a}\right)^m \tag{7.11}$$

If it is assumed that Y is independent of the crack size (approximately good for small or short cracks) and taking the reciprocal value for the integration of Eq. (7.11), one can integrate the limits of the initial crack size, a_i, and the critical size, a_c, as follows:

$$\frac{dN}{da} = \frac{1}{C(\Delta\sigma Y\sqrt{\pi a})^m} \tag{7.11a}$$

$$\int_0^{N_f} dN = \int_{a_i}^{a_c} \frac{da}{C(\Delta\sigma Y\sqrt{\pi a})^m} = \frac{1}{C(\Delta\sigma Y\sqrt{\pi})^m}\int_{a_i}^{a_c} a^{-m/2}da \tag{7.12}$$

Integrating between the limits gives:

$$N_f = \frac{2\left(a_c^{\frac{2-m}{2}} - a_i^{\frac{2-m}{2}}\right)}{(2-m)C(\Delta\sigma Y\sqrt{\pi})^m} \tag{7.13}$$

The fatigue lifetime may be predicted by evaluating the respective parameters. C and m are experimentally measured. da/dN may be determined over a range of cycles and may be described in terms of a power-law function of the applied-stress intensity range, by choosing a value for m stated to be in the range of 15–50 for ceramics. These results potentially provide a means of analyzing experimental fatigue data and of obtaining some mechanisms of the fatigue process in materials.

Cyclic crack-growth experiments were performed in SiC whisker-reinforced Al_2O_3-SiC ceramic composites. The fatigue-crack propagation rate, da/dN, of long cracks (in excess of 3 mm) has indicated a power law-dependent function of the applied-stress intensity according to the Paris Law Eq. (7.9). The values of C and

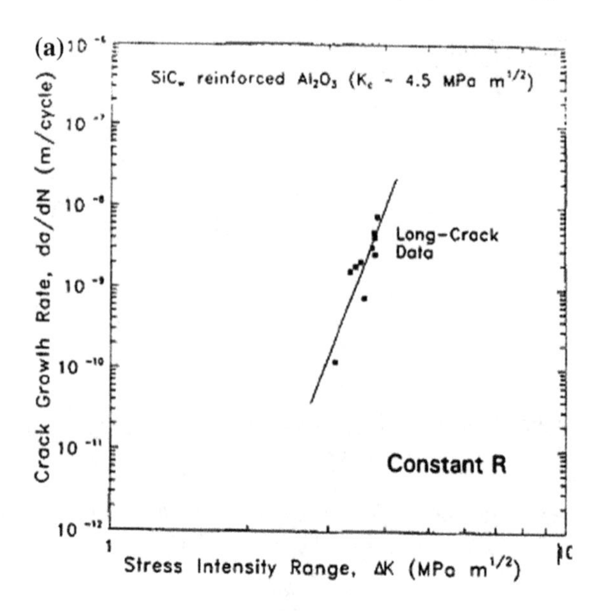

Fig. 7.26 **a** Cyclic fatigue-crack propagation rates, da/dN as a function of the applied stress-intensity range ΔK for SiC,-reinforced alumina. Data obtained for long cracks on C(T) samples in a room-air environment at 50 Hz with a constant load ratio ($R = K_{min}/K_{max}$) of 0.1 [5]. With kind permission of John Wiley and Sons

m are evaluated as above. The crack-growth rate versus the stress-intensity range, ΔK, is shown in Fig. 7.26.

The experimental technique is shown in Fig. 7.27. The value of K_{IC} is calculated from the highest load, where the elastic unloading compliance line deviates from linearity (Fig. 7.27a). The form of the cycle is also shown.

The crack-growth rates, da/dN, were determined for the range of -10^{-11} to 10^{-5} m/cycle under computer-controlled decreasing and increasing K conditions. The data arc presented in terms of the applied stress-intensity range ($\Delta K = K_{max} - K_{min}$) in the fatigue cycle. A series of microindentations (made using a 2-kg-load pyramidal indenter) were placed along the longitudinal axes of the surface to initiate multiple small cracks along the specimen's length (Fig. 7.28b).

Table 7.3 lists the relevant parameters determined for some ceramic materials. In the table ΔK_{TH} is the fatigue threshold, defined as the maximum value of ΔK at which growth rates did not exceed 10^{-10} m/cycle (ASTM E 647 procedure). Fig. 7.28 shows micrographs where a comparison is made between monotonic (monotonically increasing loads) and fatigue fractures. The lengths of selected microcracks, obtained by monitoring the top surface of the cantilever-beam specimens (Fig. 7.27b) at various maximum applied stress levels and plotted as a function of the number of stress cycles, are shown in Fig. 7.28a, b at load ratios R of 0.05 and −1, respectively (Fig. 7.29).

The morphology of a single microcrack after tension–compression cycling (maximum stress, $\sigma_{max} = 450$ MPa) is shown in the sequence in Fig. 7.30. Crack growth is predominantly intergranular. Evidence of crack bridging by both uncracked matrix ligaments (Fig. 7.30b) and SiC whiskers (Fig. 7.30c, d) in the wake of the crack tip is clearly apparent.

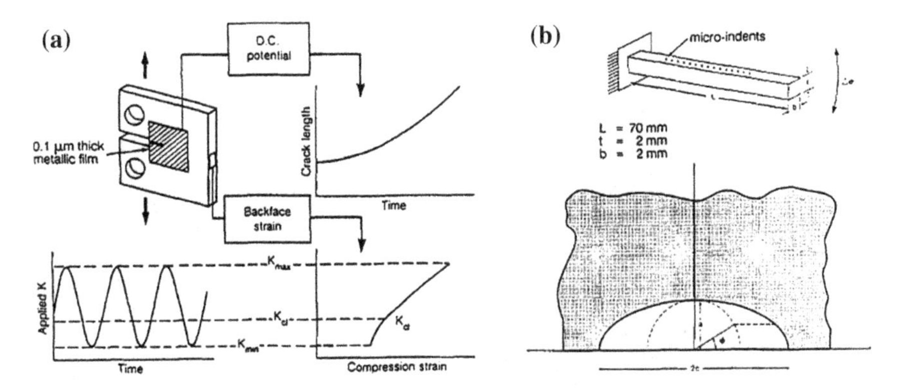

Fig. 7.27 Experimental techniques used to measure cyclic fatigue-crack growth rates showing **a** compact C(T) specimen and procedures used to monitor crack length and the stress intensity, K_{Cl}, at crack closure for long cracks, and **b** cantilever-beam specimen and semi-elliptical surface crack configuration for corresponding tests on small cracks [5]. With kind permission of John Wiley and Sons

Available data show that, in many ceramics, reduced lifetimes during cyclic fatigue stress/life (S/N) testing and significant cyclic-crack propagation occur at loads less than those required for environmentally-enhanced (static fatigue) crack growth during fracture mechanics testing.

Fatigue-crack growth in Al_2O_3-SiC ceramics is a mechanically-induced cyclic process. Growth rates (da/dN) may be described in terms of a Paris power-law function of the applied stress intensity range, ΔK with an exponent, m, on the order of 15. It was found that under constant amplitude loading, cyclic-crack growth shows evidence of crack closure, in addition to other crack-tip shielding mechanisms (crack deflection, uncracked ligament and whisker bridging).

7.7 Hysteresis

Various definitions are used for 'hysteresis', depending on the subject being considered. Thus, hysteresis may occur in ferromagnetic materials (magnetism), ferroelectric materials (ferroelectricity) or in materials under deformation (mechanical hysteresis). Considering mechanical properties, in general, and fatigue, in particular, for our purposes, one may define 'hysteresis' as being 'the directionality of strain when reversing an applied load. For example, in Fig. 7.31, hysteresis is illustrated for a metal, indicating a loading process from $-\sigma_{max}$ to $+\sigma_{max}$, forming a closed loop, known as an 'hysteresis loop'. In the illustration below, the hysteresis loop defines one single fatigue cycle. The area within the loop is the energy-per-unit-volume dissipated during a cycle.

In Fig. 7.32, a stress–strain response, appearing as hysteresis loops, is shown under the indicated conditions, namely for the steady states of: an homogeneous

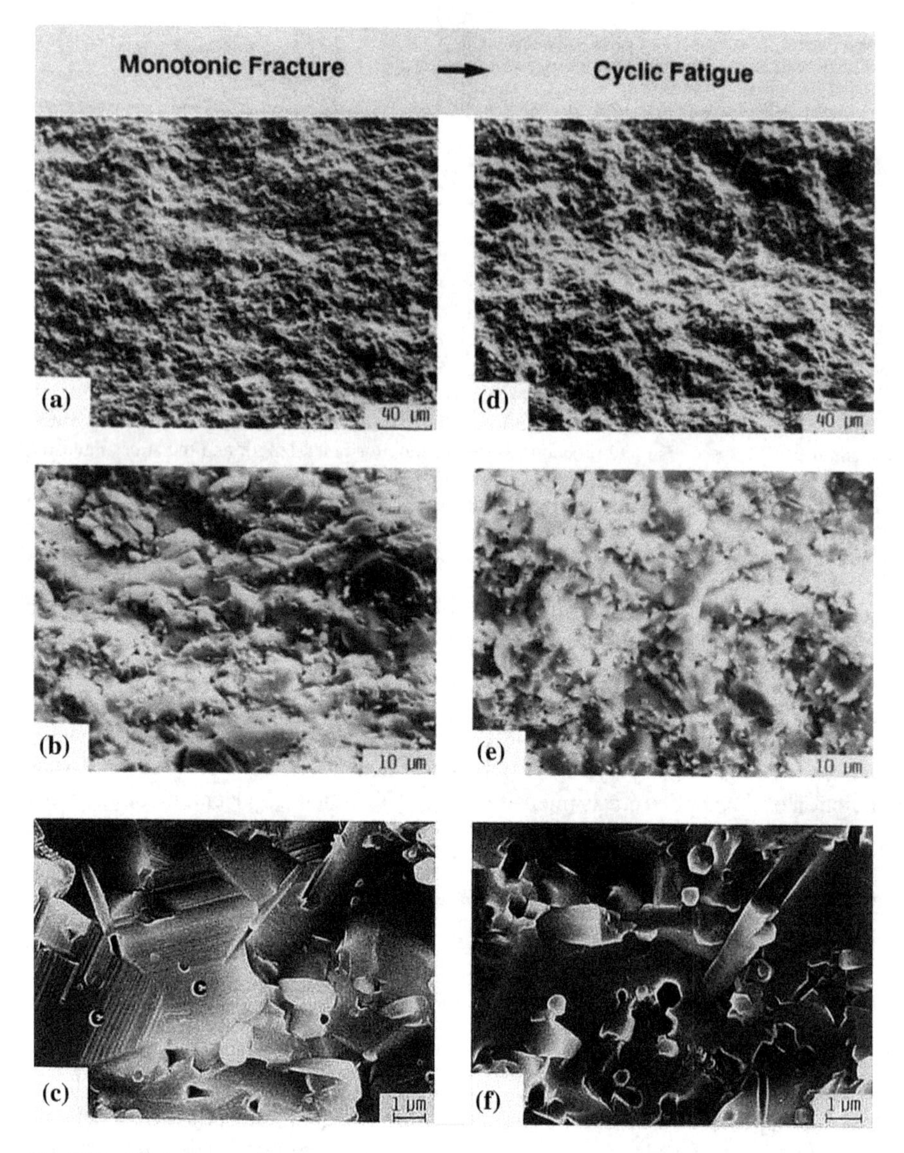

Fig. 7.28 SEM micrographs at increasing magnification of **a–c** monotonic fracture and **d–f** cyclic fatigue fracture in Al_2O_3-SiC, composite, showing the predominately transgranular nature of crack paths and regions of cleavage-like steps (indicated by letter C) formed under monotonic loading, compared to the rougher more intergranular fracture surfaces induced by cyclic loading. *Horizontal arrow* indicates direction of crack growth [5]. With kind permission of John Wiley and Sons

material, a cracked material under constant-strain control and a cracked material under steady-stress control.

In zirconia, two kinds of hysteresis loops develop. As mentioned previously, a tetragonal-to-monoclinic transformation takes place under the influence of stress,

Table 7.3 Values of C and m (in Eq. (7.1)) and the threshold, ΔK_{TH}, for some ceramic materials [5] (with kind permission of John Wiley and Sons)

	K_c (MPa m$^{1/2)}$	C (m/cycle (MPa m$^{1/2}$)$^{-m}$)	m	ΔK_{TM} (MPa m$^{1/2}$)	Ref.
Al_2O_3-SiC_w	4.5	1.12×10^{-17}	15	2.7	Present study
Alumina	~4.0		27–33	2.5–2.7	18, 20
Mg-PSZ (TS grade)	16.0	1.70×10^{-48}	42	7.7	4
(MS grade)	11.5	5.70×10^{-28}	24	5.2	4
(AF grade)	5.5	4.89×10^{-22}	24	3.0	1.4
(overaged)	2.9	2.00×10^{-14}	21	1.6	4
Silicon nitride	6.0	1.01×10^{-21}	12-18	2.0-4.3	21
3Y TZP	5.3	4.06×10^{-58}	21	2.4	10
Graphite/ pyrolytic C	~1.6	1.86×10^{-22}	19	~0.7	12

accompanied by microcracking. In Fig. 7.33, a transformation-induced hysteresis loop is illustrated.

When a material is subjected to cyclic loading, its stress–strain response may change with the number of applied cycles. If the maximum stress increases with the number of cycles, the material is said to 'cyclically harden'. If maximum stress decreases over the number of cycles, the material is said to 'cyclically soften'. If the maximum-stress level does not change, the material is said to be 'cyclically stable'. As seen in Fig. 7.33, the nature of these transformation-induced hysteresis loops is cyclically stable when the stress level is considered. However, the strain of these cycles upon unloading and under compression are different, possibly due to the asymmetric stress characteristic of phase transformation (the peak strain at compression point E is less than that at tension point B).

Microcracking-induced hysteresis may be seen in Fig. 7.34, where surface-crack friction and sliding occur. A comparison is made between such a case and one without crack friction and sliding.

Direct micrographic evidence for microcracking was observed by TEM for 3Y-TZP. Figure 7.35 is a TEM micrograph after fatigue. This microcrack is 0.1 μm. Calculations made on the basis of the estimated volume (10 μm^3) provided a crack density of 10^{-4}.

A tendency to strain softening with increased strain range has been observed and is illustrated in Fig. 7.36 by the fact that the hysteresis loop tilts progressively toward the strain axis as cycling proceeds. Also, the increase in the width of the loop is a sign of softening; in the case of Mg-PSZ, a relatively small loop width increase may be observed.

In the work of Liu and Chen [19], a series of hysteresis phenomena is described for 3YTZP and Mg-PSZ ceramics, as are the mechanisms controlling the evolved loop characteristics and the concept of cyclic softening. For more details on hysteresis please refer to the original work.

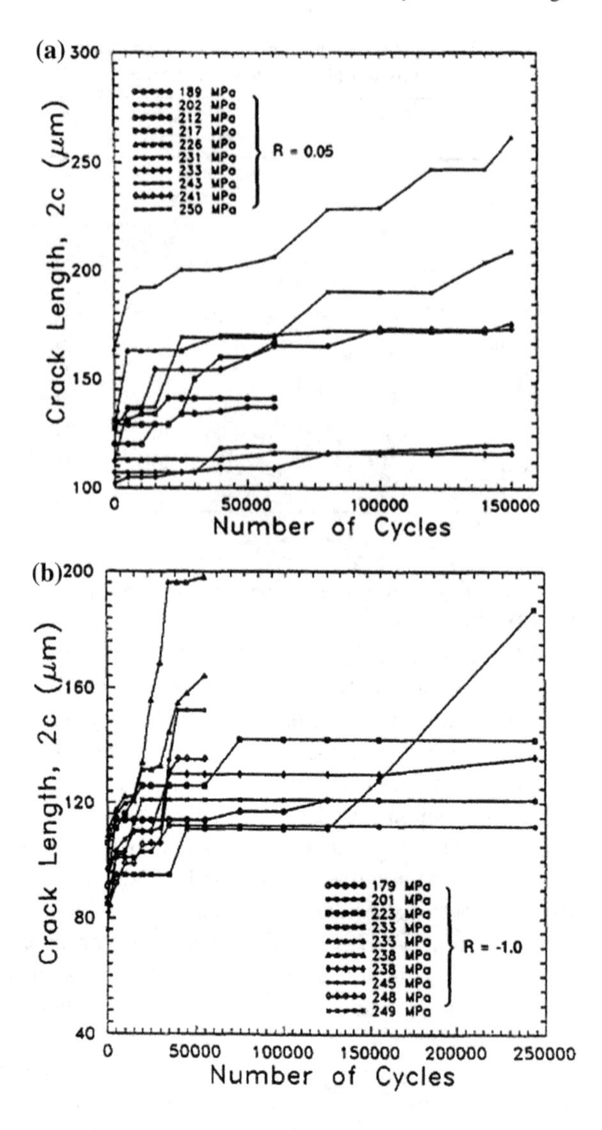

Fig. 7.29 Small-crack data in Al_2O_3-SiC, showing variation in surface crack length (2c) with number of cycles at **a** R = 0.05 and **b** R = −1.0 [5]. With kind permission of John Wiley and Sons

No discussion on hysteresis in ceramics is complete without an illustration of the case of fiber-reinforced ceramics. Most ceramics are strengthened by various means to induce certain improved mechanical (or physical) properties. Here, SiC is considered, after entering the spotlight thanks to its possible uses in aerospace structures, such as space shuttles, high-speed aircraft and in hot engines. Such ceramic composites, being resistance to both mechanical loads and thermal aggression, might potentially serve in many new high-technology applications. The expected improvement in mechanical load-bearing capacity is clearly related to the achievement of better toughness, basically because this property is not as good in bulk ceramics. The purpose of toughening the matrix is for crack-tip

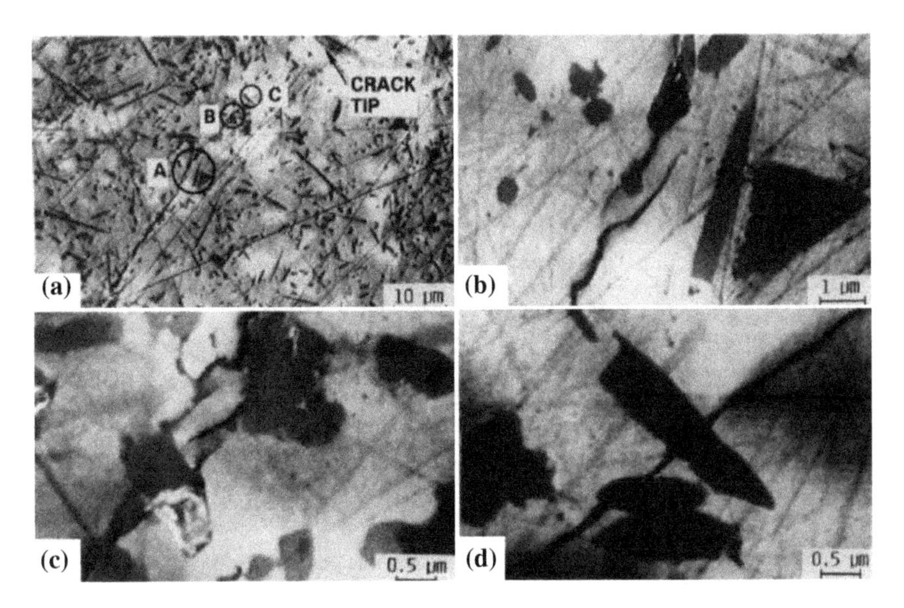

Fig. 7.30 Detailed SEM micrographs showing morphology of a single microcrack of surface length 167 μm in Al_2O_3-SiC_w showing **a** general morphology, **b** evidence of crack bridging by uncracked matrix ligaments, and **c**, **d** crack bridging by SiC [5]. With kind permission of John Wiley and Sons

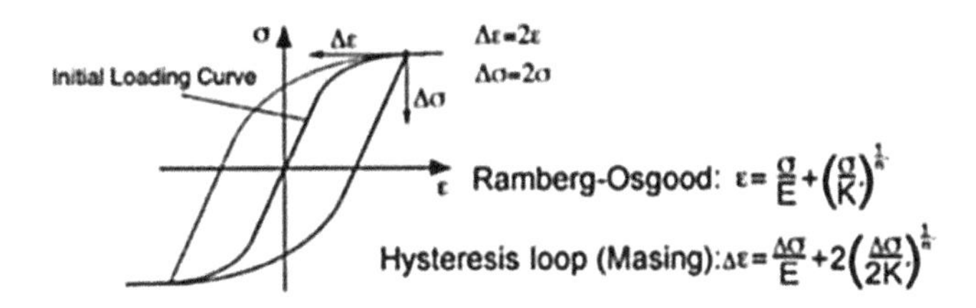

Fig. 7.31 A hysteresis loop with the Masing calculation of the curve [34]. With kind permission of Elsevier

shielding by bridging fibers to decrease fiber pull-out, which occurs when the specimen is broken. In aeronautic applications, fatigue behavior is extremely important. The illustrative figure (Fig. 7.37) represents a composite SiC matrix, in which bundles of Nicalon (SiC) fibers are embedded. Note that 2D refers to the bidirectional composite, while 1D refers to unidirectional loading.

Tests were carried out under tension–tension sinusoidal cycling, between zero and a controlled maximum stress, S (= σ), at a frequency of 1 Hz. The Nicalon itself has a failure strain of ~0.2–0.3 %. Figure 7.37 is based on a lifetime diagram (S–N relations) which is depicted in Fig. 7.38. Depending on the maximum stress, three stages may be observed: (a) samples broken during the first loading–scatter in ultimate tensile strength (O) is observed; (b) fracture occurred after a limited number

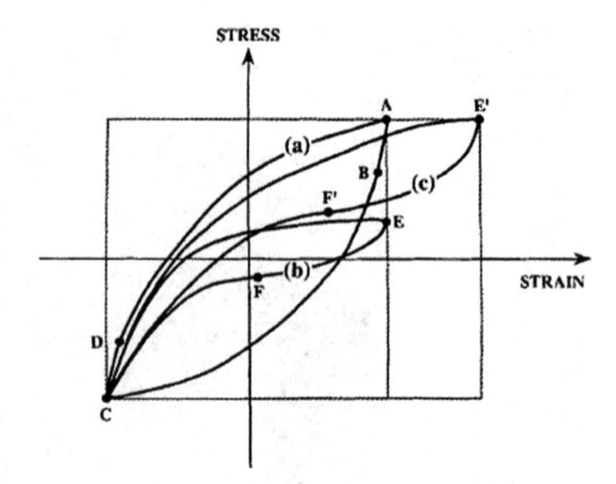

Fig. 7.32 Stress–strain response of **a** a dense, elastic-homogeneous plastic material, **b** a cracked material under constant strain-controlled condition, and **c** a cracked material under constant stress-controlled condition [19]. With kind permission of John Wiley and Sons

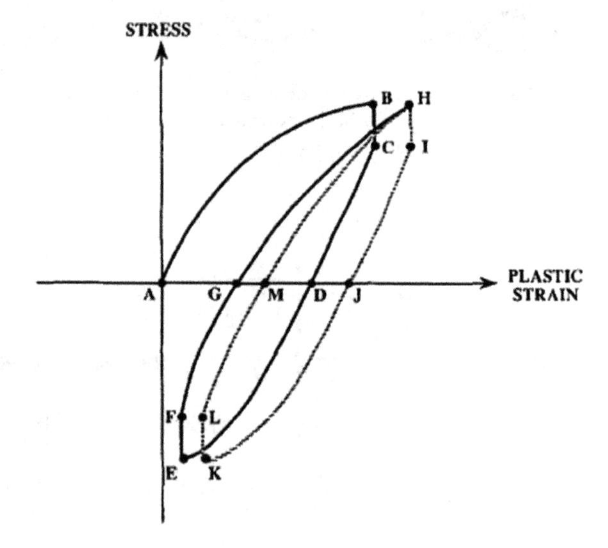

Fig. 7.33 Evolution of the transformation-induced hysteresis loops during stress cycling [19]. With kind permission of John Wiley and Sons

of cycles from 5 to 12000, in relation to the maximum fatigue stress (open square) and no fracture occurs after a given number of cycles, typically 106 (filled square), namely fatigue runout describes this case. One may see from Fig. 7.37a that the first stress/strain loading curve has a shape similar to that of a monotonic tensile test.

Furthermore, the maximum stress is beyond the linear line, which terminates at ~100 Mpa. Moreover, Fig. 7.37 shows that the areas of the hysteresis loops increase, which means that their main slope decreases. Figure 7.37b represents a theoretical hysteresis loop. Stress and strain measurements determine the changes in stiffness characteristics (tangent modulus) of the loop, namely E_{SL} at the start of

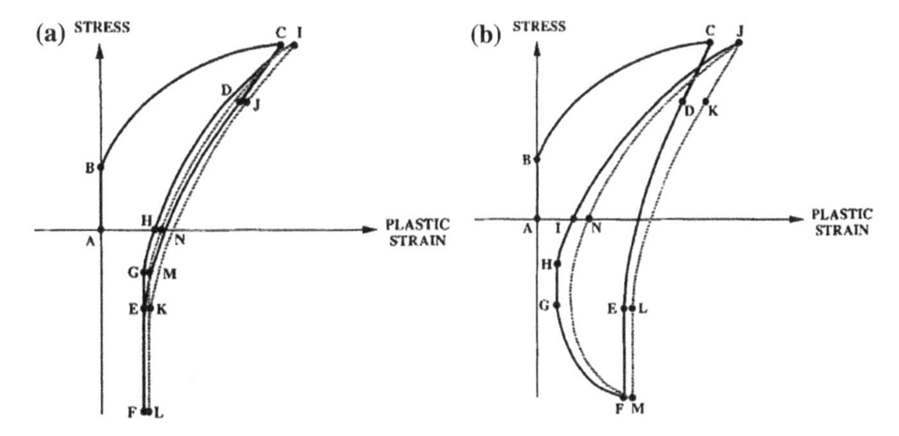

Fig. 7.34 Evolution of the microcracking-induced hysteresis loops during stress cycling **a** without crack surface friction and sliding, and **b** with crack surface friction and sliding [19]. With kind permission of John Wiley and Sons

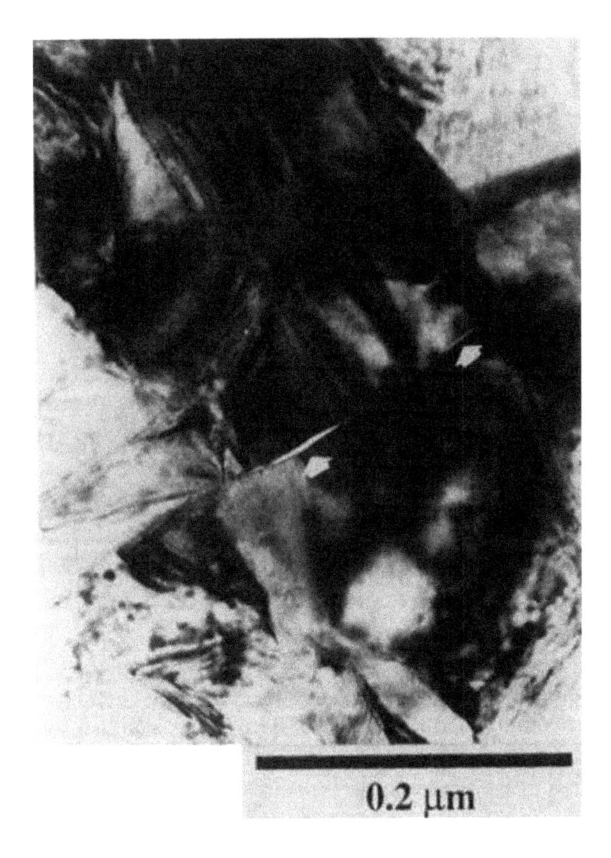

Fig. 7.35 TEM micrograph of 3Y-TZP after fatigue shows a lenticular microcrack along the grain boundary (as indicated by *arrows*) [19]. With kind permission of John Wiley and Sons

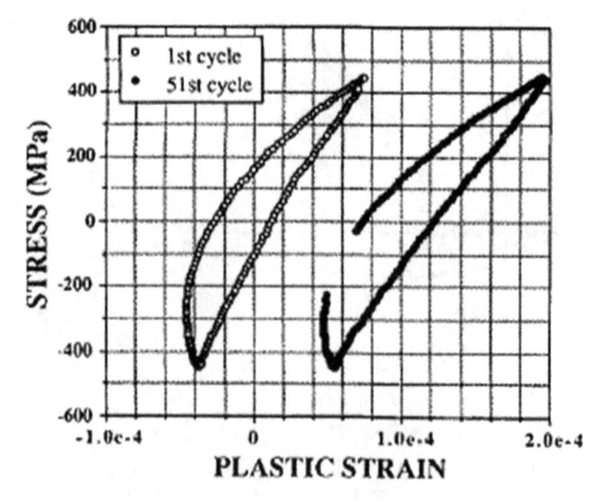

Fig. 7.36 1st and 51st hysteresis loops of Mg-PSZ at 0.25 Hz [19]. With kind permission of John Wiley and Sons

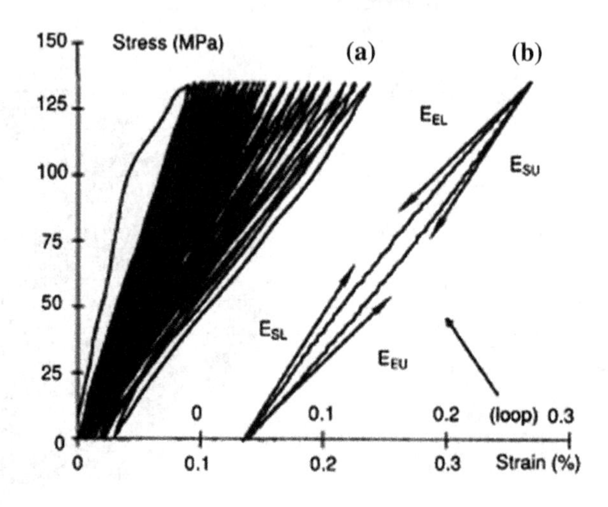

Fig. 7.37 2D SiC/SiC: fatigue in tension/tension. **a** Example of recorded stress/strain loops; **b** example of theoretical loop, definition of the tangent moduli [24]. With kind permission of Elsevier

loading, E_{EL} at the end of loading, E_{SU} at the start of unloading and E_{EU} at the end of unloading. Typical changes in these stiffnesses are shown in Fig. 7.39. E_{SL} and E_{EU} (at the bottom of the loop in Fig. 7.37b) are higher than the other two, owing to debris affecting normal crack closure. Thus, only E_{EL} and Esu (at the top of loop) are significant with respect to fatigue behavior.

Ceramics reinforced with continuous fibers exhibit delayed failure under a pulsating load (see Fig. 7.2c). Note that already during the first load cycle, the material exhibits multiple matrix cracks and also some fiber breaks. The higher the applied load, the higher the initial damage during the first cycle and the faster the instability condition sets in. The apparent fatigue limit is situated well beyond the proportional limit observed during tensile tests. For more mathematical analysis of the strengthening effect of fiber-reinforced ceramics and on decreased interfacial

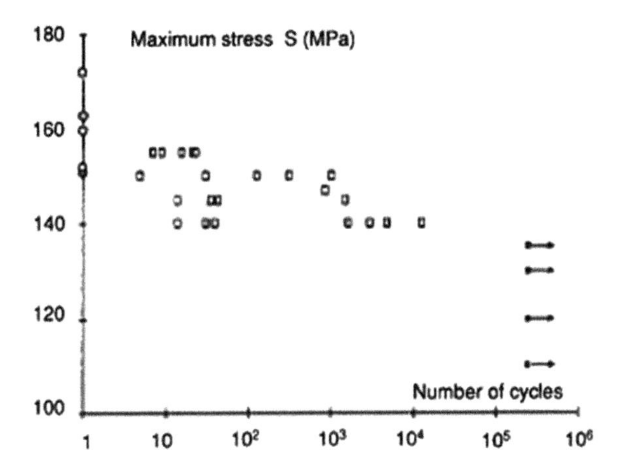

Fig. 7.38 Lifetime diagram for 2D SiC/SiC: fatigue in tension/tension. (O, failure at first backing; *Open square* fatigue failure; *Filled square* fatigue run-out) [24]. With kind permission of Elsevier

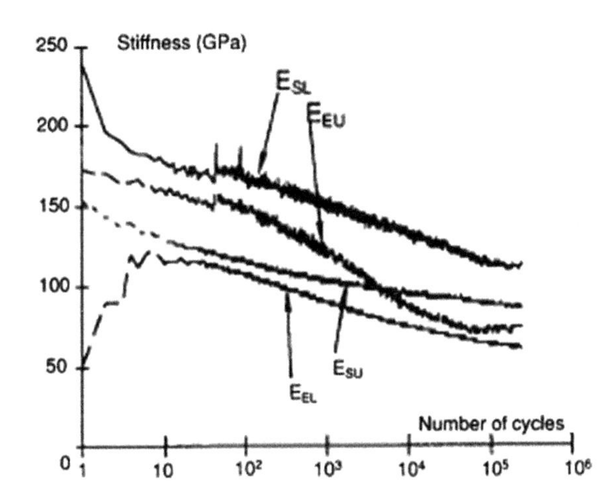

Fig. 7.39 2D SiC/SiC: fatigue in tension/tension (S = 135 MPa). Evolution of the tangent moduli as a function of number of cycles (no failure at 250000 cycles) [24]. With kind permission of Elsevier

stress and the critical instability, which determines failure as a consequence of the increase in broken fibers, one may consult the work of Rouby and Reynaud [24].

7.8 Cyclic Hardening (Softening)

The hardening ('work hardening') of ceramics occurs in a manner similar to monotonic (static) testing during deformation. Under cyclic stress, in addition to work hardening, the softening of materials ('work softening') is often also observed. It has been and is still customary to compare dynamic and static deformation results. There have also been attempts to present the fatigue behavior of materials in terms of ratios of dynamic-to-static test data, mainly the ratio of the

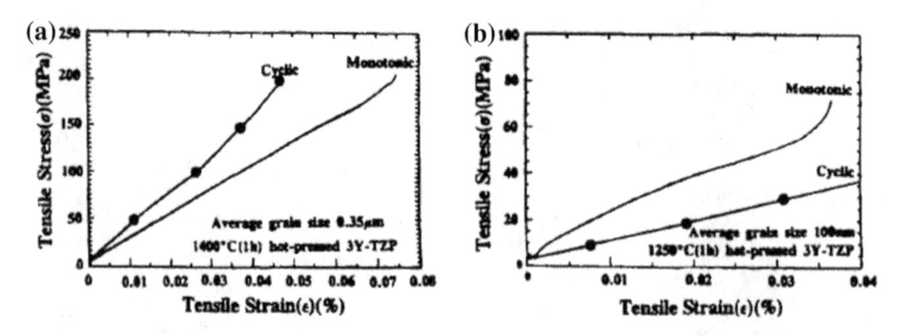

Fig. 7.40 Tensile stress–strain behavior, the cyclic curves are for a stress ratio, R of 0.1 and a frequency of 0.1 Hz at room temperature. **a** The submicron sized 3Y-TZP ceramics (0.35 μm) showing cyclic hardening, **b** the nanocrystalline 3Y-TZP ceramics (100 nm) showing cyclic softening [37]. With kind permission of John Wiley and Sons

fatigue limit to the tensile strength. K. J. Miller [65] states that attempting such a correlation is understandable, since the ability to predict fatigue behavior from a single monotonic test is highly desirable. However, he claims that correlating static and dynamic test results is dangerous and unwise and that static data should not be used by designers to assess fatigue behavior in materials, since cyclic softening, cyclic hardening or both may occur to some extent, depending on the strain range, strain rate, temperature, material composition, etc. Figure 7.40a indicates work hardening in a 3Y-TZP ceramic. In this case, this ceramic has submicron dimensions. The figure compares cyclic (fatigue) and monotonic tensile stress versus strain. One observes that the cyclic curve lies above the monotonic one, indicating that the hardening in cyclic deformation is greater than in monotonic straining. In the nanosized 3Y-TZP ceramic (100 nm), cyclic softening may be observed (see Fig. 7.40b). As expected in this ceramic, a t → m transformation may occur, just as a monoclinic phase has been observed by TEM in many micro-localities (illustrated in Fig. 7.41).

It is, therefore, reasonable to assume that the contribution of a cyclic-stress-induced phase transformation is associated with cyclic hardening. This cyclic hardening is an accumulative process, compared with hardening from static stress at the same stress level and so the strengthening is greater, as seen from the location of the curve representing cyclic hardening. The softening in the nanocrystalline 3Y-TZP (Fig. 7.40b) is associated with a large number of microcracks along the tetragonal grain boundaries. The formation of these microcracks is attributed to very large residual stresses resulting from martensitic transformations. Moreover, in the case of microcracking without transformation, the microcracks form along the tetragonal, rather than the monoclinic, grain boundaries. These microcracks coalesce and propagate to form a dominant crack under cyclic stress (see Fig. 7.41b).

SEM observations of the fatigue-fracture zone reveal that plastic deformation has occurred in some micro-areas after cyclic deformation at RT in the nanocrystalline (100 nm) 3Y-TZP, as seen in Fig. 7.42. This was affirmed by AFM, as

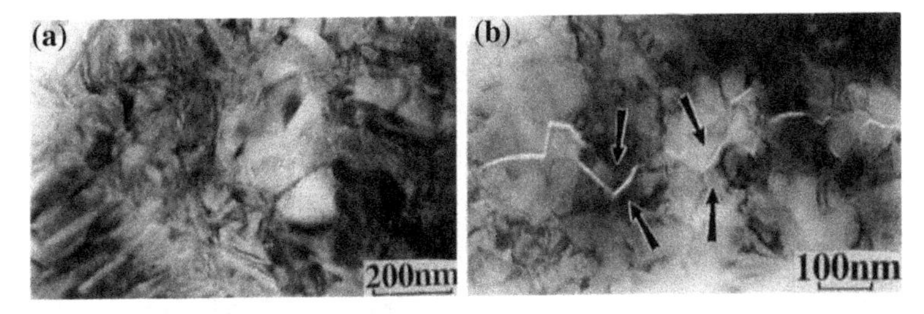

Fig. 7.41 TEM micrographs of microstructure **a** for the submicron material after cyclic hardening up to fracture (N > 500 cycles) showing that transformation occurred in many micro-localities without microcracking; **b** the nanocrystalline material after cyclic softening up to fracture (N > 500 cycles). The microcracks had coalesced and propagated into a main crack along the tetragonal grain boundary [37]. With kind permission of John Wiley and Sons

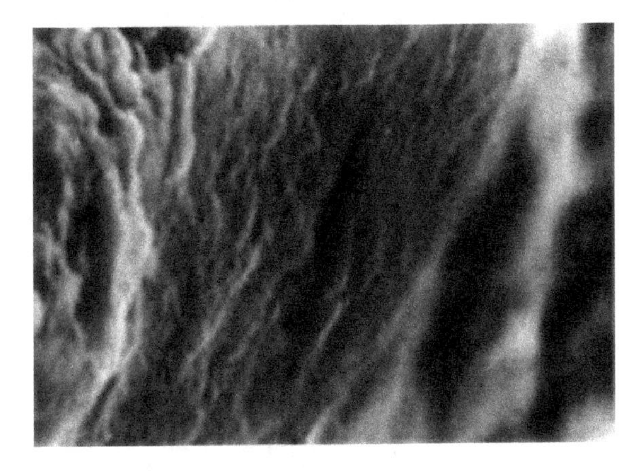

Fig. 7.42 A SEM micrograph showing the plastic deformation characteristics of a fatigue fracture at room temperature for the nanocrystalline 3Y-TZP ceramics (100 nm). The size of the micrograph is approximately 10 $\times$ 6 microns [37]. With kind permission of John Wiley and Sons

shown in Fig. 7.43. The isoaxial grains became banana-shaped, such that the long radius of the grains is 3–5 times longer than the short-grain dimension. This plastic deformation exceeded 100 %, which may be considered a superplastic deformation in the nanocrystalline 3Y-TZP (100 nm). The shapes of the isoaxial grains were not affected by fatigue after cyclic tension at RT in the submicron 3Y-TZP ceramic (0.35 μm).

It is interesting to observe the curved slip lines on the fractured surface of the nanocrystalline 3Y-TZP ceramic, where the role played by the dislocations in the plastic deformation is evident. Also note the appearance of the curved slip lines that resemble the microstructures observed in metals (the 'beach markings'; see, for example, Polakowski and Ripling). These lines represent hiatuses between working and rest periods during fatigue in metals (see Fig. 7.44).

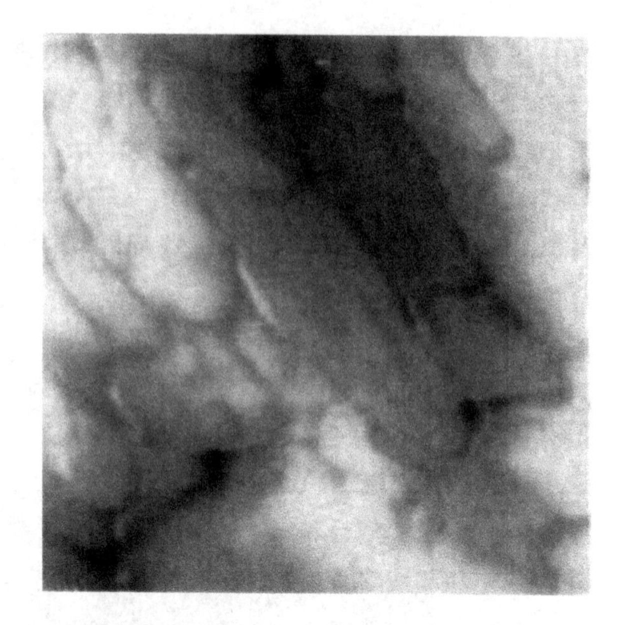

Fig. 7.43 An AFM micrograph illustrating superplastic deformation at room temperature for the nanocrystalline 3Y-TZP ceramic (100 nm). The grains have been tensioned into "banana" shapes, as viewed on the fatigue fracture surface. The size of the micrograph is approximately 600 × 600 nm [37]. With kind permission of John Wiley and Sons

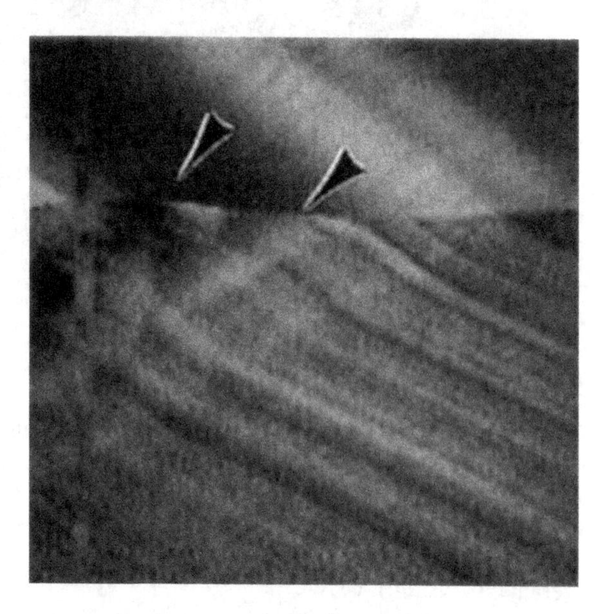

Fig. 7.44 An AFM micrograph of the curved slip lines and possibly a "young" extrusion (denoted by arrows) at room temperature for a nanocrystalline 3Y-TZP ceramic; a view of the side edge of the fatigue fracture. The size of the micrograph is approximately 350 × 350 nm [37]. With kind permission of John Wiley and Sons

The accumulated plastic strains during the cycles were continuously recorded after unloading for the nanocrystalline and the submicron-sized 3Y-TZP ceramics (see Fig. 7.45). This accumulated plastic strain saturated after 100 cycles or so, but was obviously different in nature and extent than that of metal. Such strain accumulation would eventually lead to failure. Under a 50 MPa maximum cyclic-tensile stress, very little strain accumulation was found under fully-cyclic loading

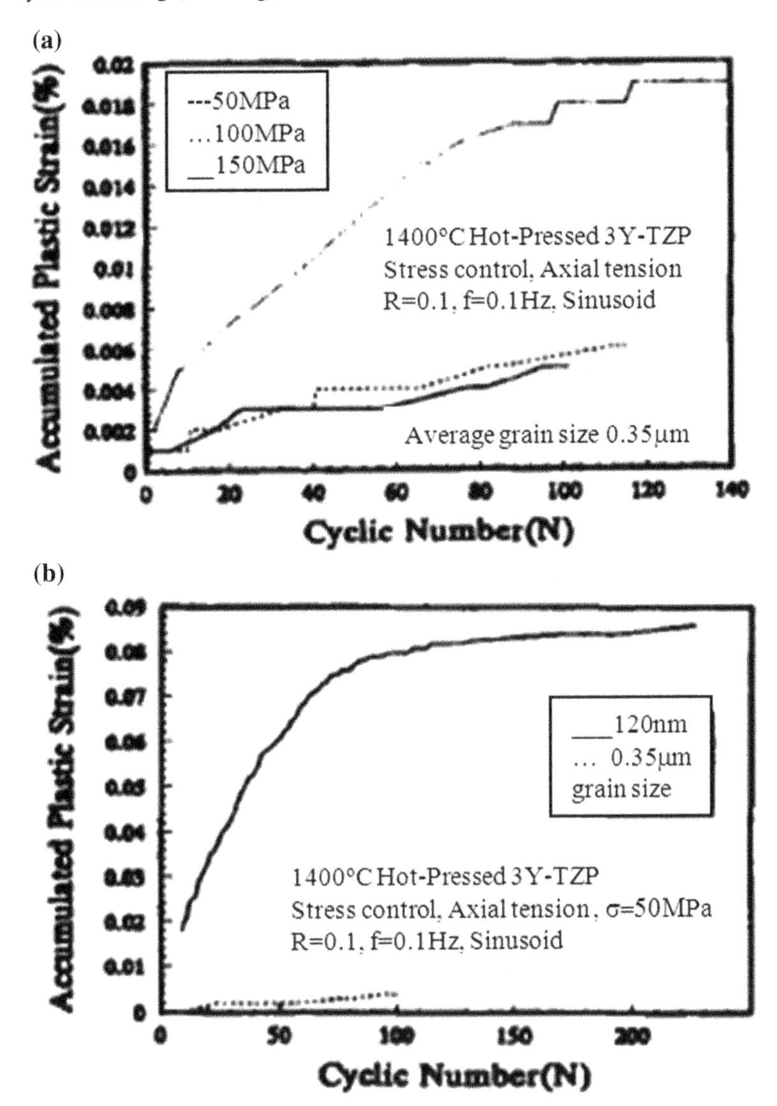

Fig. 7.45 Accumulated plastic strain versus cyclic number at the stress ratio, R of 0.1, and a frequency of 0.1 Hz (*sinusoidal*), and room temperature: **a** compared under different maximum tensile stresses for 0.35 μm sized 3Y-TZP ceramics; **b** under a tensile stress of 50 MPa, compared between 120 nm and 0.35 μm sized 3Y-TZP ceramics [37]. With kind permission of John Wiley and Sons

conditions for the submicron material (0.35 μm); but plastic strain was rapidly accumulated for the nanocrystalline material (100 nm) (see Fig. 7.45b). For the latter, the maximum accumulated plastic strain was close to 8.5×10^{-4} (after 230 cycles). The submicron material showed considerable accumulation until the stress increased to 150 MPa (see Fig. 7.45a). Even so, the maximum value of accumulated plastic strain for the submicron material under a 150 MPa stress was only 22 % of that of the nanocrystalline (100 nm) under a 50 MPa stress. Therefore, the nanocrystalline 3Y-TZP ceramic shows lower strength and a larger ability to accommodate plastic strain than the submicron material.

Thus, as indicated above, in the submicron-sized 3Y-TZP ceramic, the stress-induced cyclic hardening, due to transformation taking place, was higher than under static deformation. Nanocrystalline 3Y-TZP softened cyclically, due to the formation of a large number of microcracks. In the submicron structures, this observation basically reflects the effects of dislocations and dislocation–dislocation interactions. In the nanocrystalline 3Y-TZP ceramic, this greater ability to accommodate plastic strain is probably due to grain-boundary sliding, since in nanocrystalline structures dislocations cannot move, because slip distances are on an atomic scale (like the dimensions of dislocations themselves).

7.9 The Mean Stress and the Goodman Diagram

In Sect. 7.3 on definitions, the mean stress was defined in Eq. (7.4) as:

$$\sigma_{mean} = \frac{\sigma_{\max} + \sigma_{\min}}{2}$$

A large number of experiments on fatigue were performed with completely reversed cycles (indicated in Fig. 7.2a), where the mean stress is zero ($\sigma_{\max} = \sigma_{\min}$). Often in components exposed to fatigue, a pattern (as shown in Fig. 7.2b) is observed, resulting from the superposition of a static preload during the reversed cycle (see 7.2a); also note that the names of the cycle-patterns shown in Fig. 7.2 may have different nomenclature in the literature. This is also often stated differently, namely that the mean stress is $\sigma_{\mathrm{m}}(= \sigma_{\mathrm{mean}})$, represents a steady-state stress and the alternating stress is a variable stress. This stress cycle is asymmetrical, since the sum of $\sigma_{\max}$ and $\sigma_{\min} \neq 0$. Clearly, machine parts in service exposed to cyclic stresses may experience particular conditions in which $\sigma_{\min} \geq 0$ or $\sigma_{\max} \leq 0$.

To obtain an S–N curve, a very large number of tests are essential. In determining fatigue-stress levels using standard testing equipment, the test specimens are subject to alternating and reversed stress levels, as indicated in Fig. 7.2. Cyclic stress varies from σ_{a} (tensile) to σ_{a} (compressive). The mean stress equals 0. To overcome the difficulty posed by the large number of experiments, methods have been suggested, one of them by Goodman, known as the 'Goodman diagram', in which the alternate stress is plotted against the mean stress. The Goodman diagram is widely used for metallic materials, but its use has also been attempted for ceramics. Figure 7.46 is an illustration of a Goodman diagram for 3Y-TZP.

In the Fig. (7.46), the stress amplitude is plotted against the mean stress. The ratios (R) are indicated on the graph. The numerals appearing next to the symbols

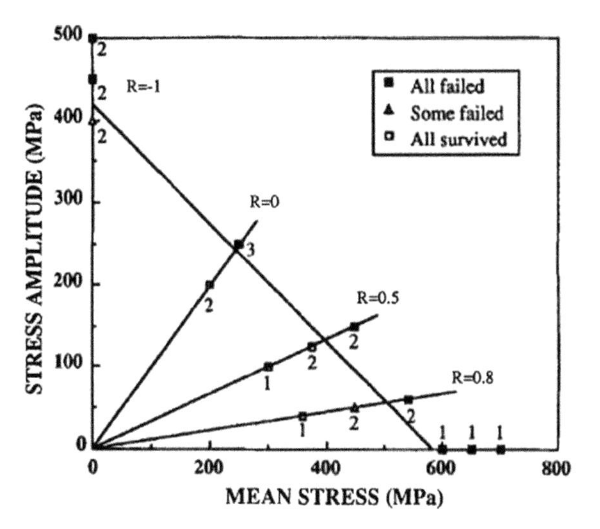

Fig. 7.46 Goodman diagram of 3Y-TZP for specimens tested for 10^4 cycles. Numbers next to the symbols indicate the number of specimens tested [17]. With kind permission of John Wiley and Sons

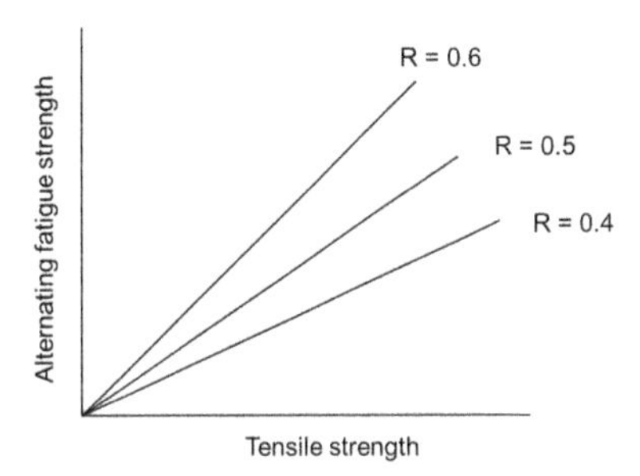

Fig. 7.47 A schematic plot of fatigue strength

indicate the number of specimens tested. The data fall into three categories: (a) all failed after 104 cycles; (b) all survived; and (c) some survived. With these data, a failure locus is drawn on the diagram (in Fig. 7.46) by means of the solid line having a negative slope, also expressed as:

$$\sigma_a = \sigma_e \left[1 - \left(\frac{\sigma_m}{\sigma_u} \right) \right] \tag{7.14}$$

where σ_a is the stress amplitude, σ_e is the fatigue strength at R $= -1$, σ_m is the mean stress and σ_u is the tensile strength. R $=$ 1 corresponds to a fully reversed cycle. Note that if fatigue life is controlled only by σ_{max}, then the failure locus should be a straight line with a slope of -1 (namely $\sigma_{max} = \sigma_m + \sigma_a$. For this equality, see the definitions in Sect. 7.4. However, when the stress amplitude is controlled, then the fatigue failure locus should be a horizontal line. These experimental results show that the slope is – 0.8, i.e., failure is mostly controlled by the maximum stress. Various combinations of stresses (such as mean stress or stress amplitude) are plotted against N_f or σ_a versus mean stress to get various values of N_f. In the above figure, stress amplitude (rather than alternating stress)

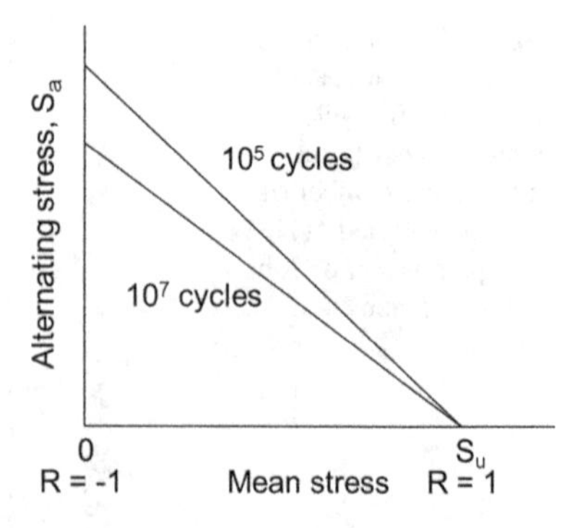

Fig. 7.48 A schematic plot of the alternating stress versus the mean stress during two cycles. S_u at R = 1 is the ultimate tensile strength

versus mean stress is shown. The input data for constructing a Goodman diagram may be an S–N curve.

One of the key limitations of the S–N curve is its inability to predict lifetimes at stress ratios different from those under which the curve was developed. To predict the lifetime of a certain component, a more useful presentation of fatigue life test data is the modified Goodman diagram.

It is worth restoring the Goodman diagram concept to appreciate its use. Essentially, the emphasis of Goodman's work was on tensile-mean stress with respect to fatigue life. The stress required to produce failures during a specified number of cycles is directly related to tensile strength, as indicated schematically in Fig. 7.47.

Tensile mean stresses reduce the fatigue life compared to the length observed in reversed loading. The original Goodman diagram is a graphic expression of this concept, shown schematically in Figs. 7.48 during two cycles.

The area below the curve (Figs. 7.46 and 7.48) indicates that the material should not fail at a given stress. The area above the curve represents the likelihood that fatigue failure will occur. The alternating stress is plotted on the ordinate and the mean stress on the abscissa. The allowable alternating stress, with no mean stress (0 mean stress), is the fatigue limit. The maximum mean stress, with zero alternating stress, is the ultimate tensile strength. A straight line is then drawn between the two points. Any combination of mean and alternating stress on this line will have the same fatigue life. Mathematically (in terms of the designation in the figure), this may be expressed as:

$$\frac{S_a}{S_{FL}} + \frac{S_{mean}}{S_u} = 1 \tag{7.15}$$

where S_{FL} is the fatigue strength. (In our earlier designation of Eq. (7.15), $S \equiv \sigma$). Equation (7.15) is the same as Eq. (7.14). The curves show the results of having compared the relationship of the tensile strength and the fatigue strength at the cycles indicated schematically. The stress required to produce failures in a specified number of cycles is directly related to the strength of the material.

Fatigue is a problem that may affect any movable part or component. Automobiles on roads, aircraft wings and fuselages, ships at sea, nuclear reactors, jet engines and land-based turbines are all subject to fatigue failure. There are three crucial factors that cause fatigue: (1) a maximum tensile stress of sufficiently high value; (2) a large enough variation or fluctuation in the applied stress; and (3) a sufficiently large number of cycles of applied stress. There are many types of fluctuating stresses as indicated in Fig. 7.2.

7.10 Load and Amplitude Effects on Crack in Fatigue

7.10.1 Introduction

Structural ceramic components, often working under cyclic conditions, may experience variations (overstressing or understressing) under common loading conditions. For example, metallic structures, such as airplane wings, are typical cases, in which common cyclic loading may be influenced by the absence of load constancy. Even ceramics which are increasingly used in a wide range of industries, including mining, aerospace, automotive, etc. are affected by the loading conditions. Therefore, designers should be concerned about the loading conditions when deciding to apply a ceramic component for some specific use. The aim of this section is to consider the possibilities of load variation, whether accidental or uncontrollable; it is important to understanding the consequences for fatigue life and crack propagation. Either compressive or tensile stress may be applied, but many researchers emphasize the importance of compressive stress on fatigue cracking. Cyclic loading may be performed under various conditions. Figure 7.2 illustrates some cycles and Eqs. (7.3)–(7.7) define certain important parameters that may be relevant to loading conditions. Of the various reports on loading conditions, this section discusses some principal, experimental results.

7.10.2 Loading Conditions and Effects

Here, examples of ceramics exposed to variable loading conditions will be illustrated. This test involves monitoring crack-growth rates at constant K_{max}, with: (i) the load cycled between K_{max} and K_{min} ($R = 0.1$) compared to being held constant at K_{max} (Fig. 7.49a) and; (ii) the value of K_{min} being varied (Fig. 7.49b). The specimens are of a MgO-PSZ ceramic, heat-treated to vary the fracture toughness, K, from ~3 to 16 MPa $m^{1/2}$ and tested both in inert and moist environments. Recall that cyclic fatigue-crack propagation, da/dN, is a function of the stress-intensity range, ΔK, and resembles metallic materials (where the growth rates may be fitted to a conventional Paris Law; see Sect. 7.6), namely:

$$da/dN = C(\Delta K)^m \tag{7.9}$$

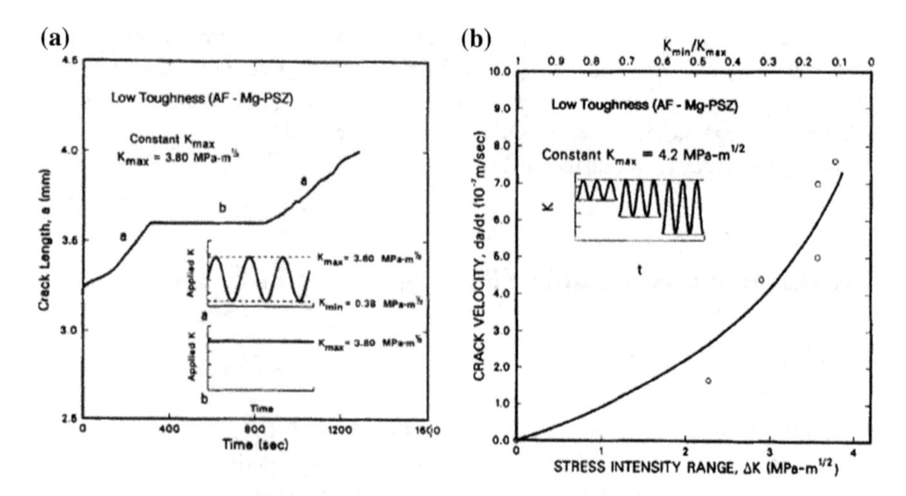

Fig. 7.49 Effect in low-toughness Mg-PSZ (AF) microstructure of **a** sustained and cyclic loading conditions on the crack velocity, da/dt, at constant K_{max}(= 3.8 MPa $m^{1/2}$), and **b** varying applied stress-intensity range ΔK on crack velocity at constant K_{max} (=4.2 MPa $m^{1/2}$) [17]. With kind permission of John Wiley and Sons

However, the exponent, m, in ceramics is considerably larger than reported for metals, i.e., in the 21–42 range (as opposed to 2–4 for metals) and having constant C scales inverse to the fracture toughness. ΔK is the stress-intensity range of the applied stress with a cyclic frequency varying between 1–50 Hz using a sinusoidal wave:

$$\Delta K = K_{max} - K_{min} \tag{7.10}$$

where K_{max} and K_{min} are the maximum and minimum stress intensities, respectively, in the fatigue cycle. Often, the stress-intensity factor related to the crack-opening stage and its blunting is expressed in terms of the following relations. The effective stress ratio is:

$$U = \frac{S_{max} - S_{op}}{S_{max} - S_{min}} \tag{7.11}$$

S is the stress and the subscript, op, stands for the crack-opening stress. Another factor to consider is that of stress-intensity, ΔK, which is given as:

$$\Delta K_{th} = \Delta K_{in} + \Delta K_{op} \tag{7.12}$$

$$\Delta K_{in} = K_{max} - K_{op} \text{ and } \Delta K_{op} = K_{op} - K_{min} \tag{7.13}$$

or

$$\Delta Kth = Kmax - Kmin \tag{7.12a}$$

ΔK_{th} is the threshold-intensity factor, ΔK_{in} is the intrinsic or basic stress intensity required for the extension of a crack opening and ΔK_{op} is an additional crack-opening stress intensity required to resist crack closure and maintain a fully opened

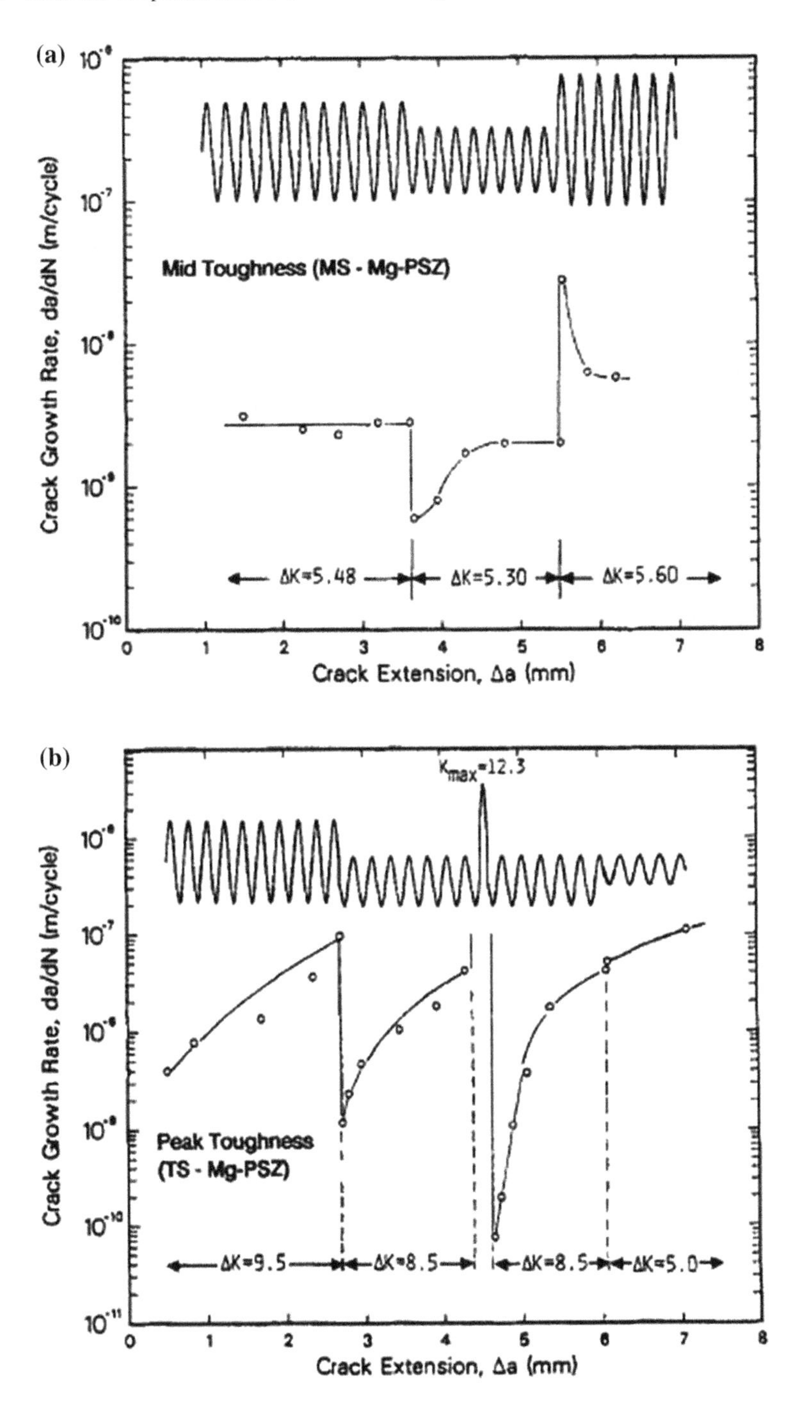
(a)
Mid Toughness (MS - Mg-PSZ)
ΔK=5.48
ΔK=5.30
ΔK=5.60
Crack Growth Rate, da/dN (m/cycle)
Crack Extension, Δa (mm)
(b)
Kmax=12.3
Peak Toughness
(TS - Mg-PSZ)
ΔK=9.5
ΔK=8.5
ΔK=8.5
ΔK=5.0
Crack Growth Rate, da/dN (m/cycle)
Crack Extension, Δa (mm)

◀ **Fig. 7.50** Transient fatigue-crack growth behavior in **a** mid-toughness (MS) and **b** peak-toughness (TS) Mg-PSZ due to variable-amplitude cyclic loads, showing immediate crack-growth retardations following high–low block overloads, immediate accelerations following low–high block overloads, and delayed retardation following a single tensile overload [6]. With kind permission of John Wiley and Sons

crack. The increased opening-stress intensity, caused by tensile load (overload) is an important factor in crack-growth retardation. Figure 7.49 above relates to constant-amplitude cyclic loading. To attain variable-cyclic loading, single and block overload sequences were applied during steady-state fatigue cracking (illustrated in Fig. 7.50). One may observe in Fig. 7.50a that crack advance, the crack-growth rate, remains approximately constant at ΔK (=5.48 MPa $m^{1/2}$). On reducing the cyclic loads, so that $\Delta K = 5.30$ MPa $m^{1/2}$ (high–low block overload), transient retardation is seen, followed by a gradual increase in growth rates, until a new steady-state is achieved. By subsequently increasing the cyclic loads, so that $\Delta K = 5.60$ MPa $m^{1/2}$ (low–high block overload), the growth rates show transient acceleration before decaying to the steady-state velocity. For more on overloads, under-loads and variable amplitude loads, one may read the Mechanical Properties of Metals (Pelleg). Similar crack-growth retardation, following a high–low block overload ($\Delta K = 9.5$–8.5 MPa $m^{1/2}$) is shown for peak-toughness Mg-PSZ in Fig. 7.50b. In addition, significant retardation may be seen following a single tensile overload to a K_{max} of 12.3 MPa $m^{1/2}$. Such results may be rationalized in terms of changes in crack-tip shielding from the transformation zone. Recall that zirconia undergoes a phase transformation. The tetragonal phase undergoes a stress-induced martensitic transformation to a monoclinic phase in the presence of a high-stress field near a crack-tip. The resulting dilatant transformation zone, in the wake of the crack, exerts compressive tractions on the crack surfaces and, hence, shields the crack-tip from the applied (far-field) stresses [51].

In Fig. 7.51a, fractographs are illustrated showing the transgranular nature of the crack path as observed via optical microscopy. The transgranular nature of crack paths is clearly evident. The grain boundaries are decorated by monoclinic zirconia phases. Frequent crack deflection is also seen. The degree of deflection appears to decrease progressively with decreasing toughness, as indicated by the reasonably flat crack path in Fig. 7.51b for an over-aged specimen also tested. SEM micrographs compare the appearance of the fracture surfaces in cyclic and monotonic (static) loading, which seem to be identical. Unlike metals, in the zirconia, fractured surface fatigue striations or crack arrest markings are not observed (Fig. 7.52) .

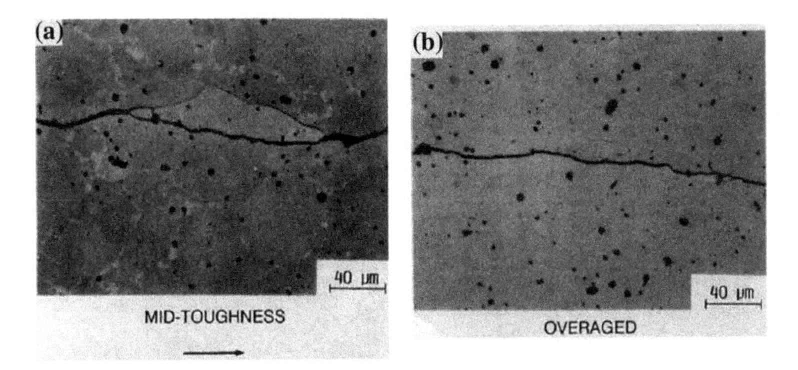

Fig. 7.51 Optical micrographs of the morphology of cyclic fatigue crack paths in Mg-PSZ, showing (**a**) an increasingly deflected crack path in the mid-toughened (MS) microstructure at $\Delta K \sim 6$ MPa $m^{1/2}$ compared to (**b**) an essentially linear crack path in an over-aged material at $\Delta K \sim 2$ MPa $m^{1/2}$. Note the transgranular fracture morphology and evidence of crack branching in the MS microstructure. *Arrow* indicates general direction of crack growth [6]. With kind permission of John Wiley and Sons

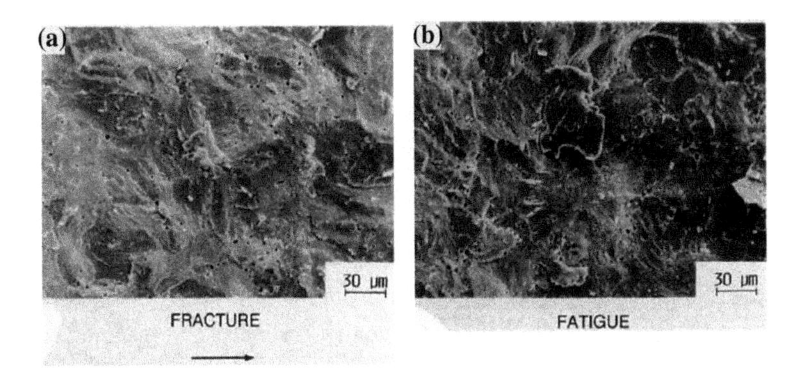

Fig. 7.52 Representative scanning electron micrographs of the nominally identical fracture-surface morphologies obtained in Mg-PSZ (MS grade) for **a** overload fracture under monotonic loads at $\Delta K \sim 11.5$ MPa $m^{1/2}$, and (b) fatigue fracture under cyclic loads at $\Delta K \sim 6$ MPa $m^{1/2}$. Note, in contrast to metals, the absence of striations or crack-arrest markings on the fatigue fracture surface. Arrow indicates general direction of crack growth [6]. With kind permission of John Wiley and Sons

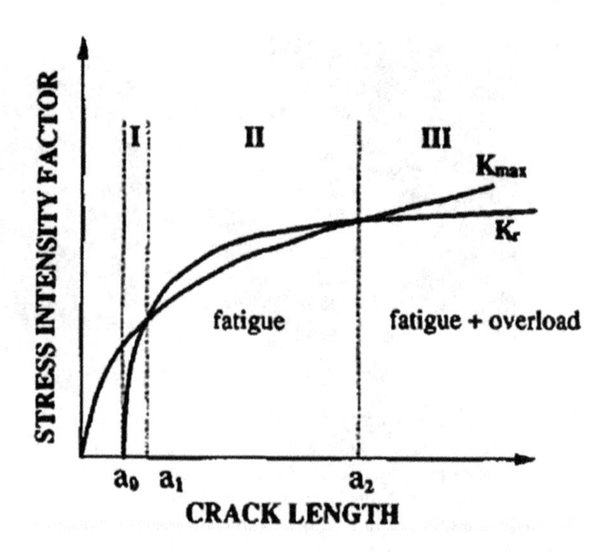

Fig. 7.53 Maximum stress intensity factor (K_{max}) and crack resistance (K_r) versus crack length: **I** growth due to overload, **II** growth due to fatigue, **III** growth due to overload and fatigue [18]. With kind permission of John Wiley and sons

In the Mg-PSZ, the transformation from the metastable tetragonal phase to the monoclinic phase absorbs part of the energy required for crack propagation with consequent increase in fracture toughness, manifested as crack-tip shielding.

7.11 Structural Observations in Fatigued Specimens

7.11.1 Striations

Common markings, found on the fractured surfaces of metals undergoing fatigue deformation, are the curved markings known as 'striations'. The striations characteristic of fatigue deformation in metals are crack-front markings, believed to be associated with the cycle applied during fatigue. Thus, it is thought that each striation (observed by examining fractured specimen surfaces after fatigue) is produced by one stress cycle. HRTEM and high-resolution SEM studies show that striations often have a saw-tooth profile, in which one side of the saw-tooth has a more jagged appearance, exhibiting more slip traces than the other side. The appearances and profiles of striations vary widely, depending on the metallic material. These structural features (striations), observed in fatigued machine parts, represent fatigue-progression marks, also known as 'beach marks'. Such fatigue striations have not, thus far, been widely reported in the ceramics literature, though they have been mentioned in some recent works. In Sect. 7.2.2, Fig. 7.13 is characterized by striations in 3Y-TZP and may also be seen in Sect. 7.3, Fig. 7.13 and, particularly, in Fig. 7.14b. Striations are considered in the work of Liu and Chen [18]. They state that fatigue striations result from alternate-overload fracture. The appearance of striations varies with the R ratio and is very sensitive to

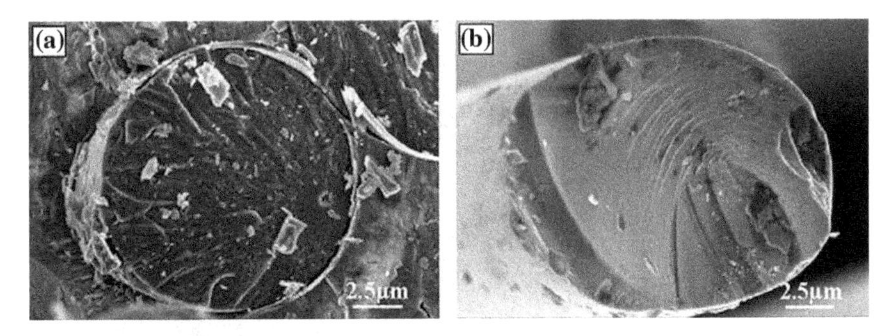

Fig. 7.54 Cross-section morphologies of the fibers in 3D SiC/SiC composite after. **a** *monotonic* tension test; **b** tension–tension fatigue test [31]. With kind permission of Elsevier

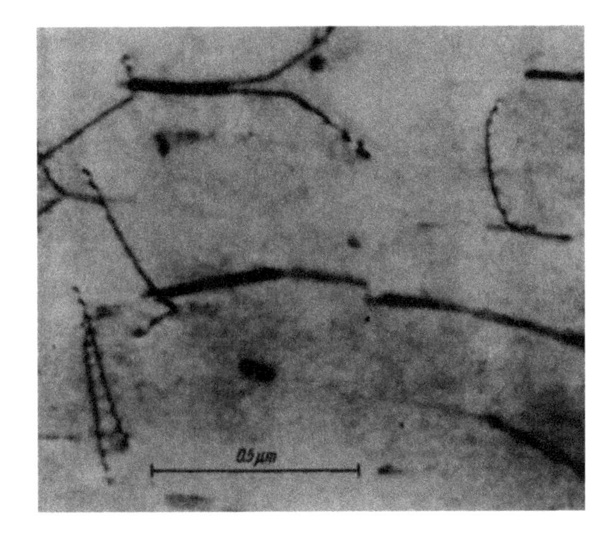

Fig. 7.55 Transmission electron micrograph showing unusually long pairs and elongated loops [27]. With kind permission of John Wiley and Sons

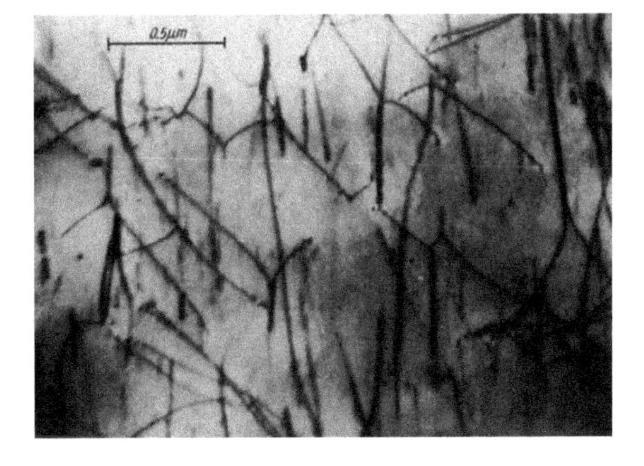

Fig. 7.56 Transmission electron micrograph showing unusually long dipoles in cyclically stressed MgO [27]. With kind permission of John Wiley and Sons

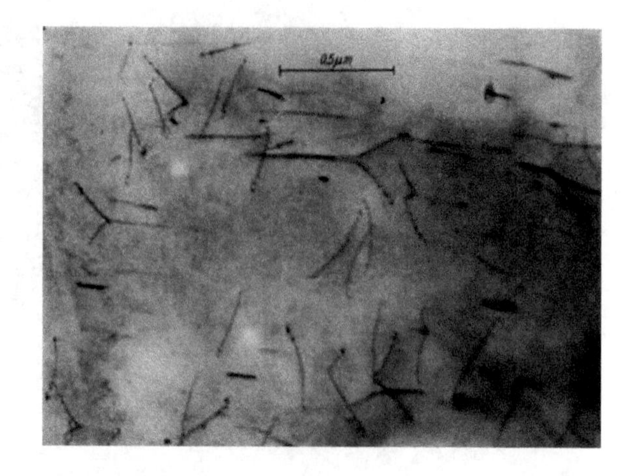

Fig. 7.57 Transmission electron micrograph showing long dipoles, elongated prismatic loops, and screw dislocations [27]. With kind permission of John Wiley and Sons

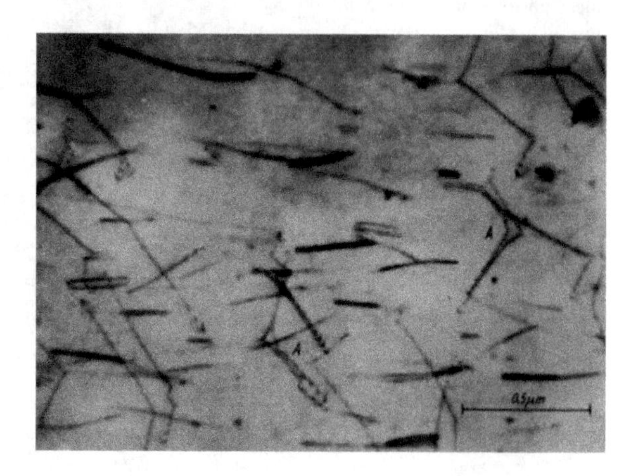

Fig. 7.58 Transmission electron micrograph showing inter-connected pairs. Dislocation pairs A are nearly in screw orientation [27]. With kind permission of John Wiley and Sons

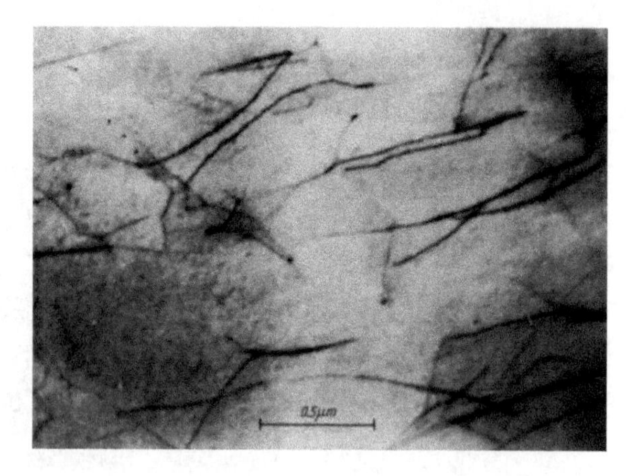

Fig. 7.59 Dislocation substructure in a specimen subjected to 1×10^6 cycles in fatigue [27]. With kind permission of John Wiley and Sons

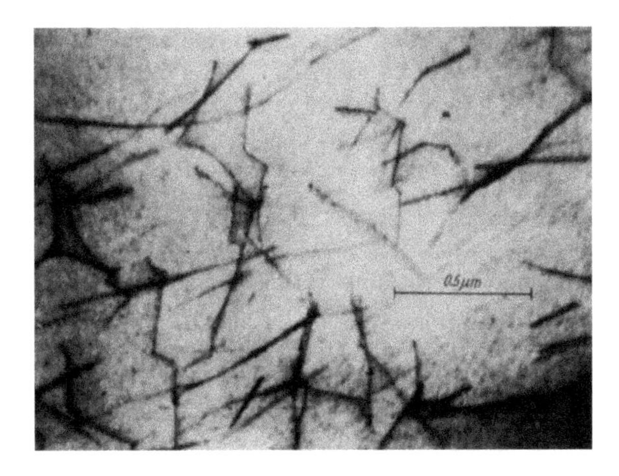

Fig. 7.60 Dislocation substructure in a specimen subjected to 4×10^6 cycles in fatigue [27]. With kind permission of John Wiley and Sons

the loading condition and the crack geometry. They also claim that the fatigue striations observed in 3Y-TZP have a different origin than those commonly observed in metals and polymers; they are actually the demarcations between fatigue fracture and overload fracture, which appear as alternating light and dark damage bands. When crack growth is entirely within the fatigue regime (as in regime II of Fig. 7.53), no striations should appear.

Additional striations were also observed in 3D SiC/SiC composite deformed by tension–tension fatigue, as shown in Fig. 7.54. Here, the fractured surface of monotonic tension and tension–tension fatigue results are compared. Note that the whole cross-section surface of the monotonic tension-tested specimens is coarse. On the cross-section surface of the tension–tension fatigue tested specimens, two different areas may be seen: one area composed of ringed striations, taking up about half of the whole section, while the other area is smooth.

7.11.2 Dislocation Structure in Fatigue

Examples of dislocation structures after fatigue deformation appear in a series of microstructures (Figs. 7.55–7.59) obtained by TEM. Subramanian [27] claims that these dislocation substructures are similar to that of unidirectionally-stressed MgO. The micrographs presented below are of single-crystal magnesia which underwent a large number of cycles (in the millions) of low strain amplitude. The maximum strain was about 0.1 % per cycle. The characteristics of the dislocation structure in MgO, having a rock-salt structure, is as follows:

(a) the dislocations are unusually long plus–minus pairs;
(b) there are elongated dislocation loops, wider than those observed in other types of RT tests, such as in the unidirectional straining;

(c) there are dislocation interactions leading to star-shaped pairs and interconnected pairs are formed;
(d) long jogs are observed in some dislocations;
(e) a large number of screw dislocations and a lower density of long edge-dislocation pairs and loops are visible.

It was further claimed, based on Figs. 7.59 and 7.60, that the dislocation substructure in fatigued single-crystal MgO specimens is independent of the number of cycles of repeated stress. Thus, in Fig. 7.59, the number of cycles is 1×10^6 and, in Fig. 7.60, the number of cycles is 4×10^6, but the dislocation substructures are similar in both these micrographs. The TEM data further indicate that the dislocation density in specimens after fatigue should be less than in unidirectionally-strained specimens and, thus, the dislocation density is independent of the cycles after a few initial testing cycles.

The existence of very long pairs after fatiguing is an outcome of screw dislocations that get pinned by long jogs at various points and then continue to bow out. The dislocation loops observed in these specimens are often quite wide and, in some loops, the separation between the edge components is as large as 280 Å. These pairs of dislocations are at 45° planes. From the observed separation between the glide planes, the maximum shear stress that might have been present in these regions, without driving the two dislocations past each other, was calculated to be on the order of G/1000 (G is the shear modulus).

Since MgO has a structure similar to that of NaCl crystal, its slip systems are also of the type $\{110\}$ $\langle 110 \rangle$. There are six such planes in a cubic structure. In a cubic structure, four of the six $\{110\}$ slip-plane projections onto the (001) slip planes leave traces along the $\langle 100 \rangle$ direction and lie at 45° to the surface; these are known as '45° slip planes'. Two have traces along the $\langle 110 \rangle$ direction, lie at 90° to the surface and are referred to as '90° slip planes'. The shortest Burgers vector for a perfect dislocation in the NaCl structure is a/2 $\langle 110 \rangle$. This is the operative slip direction. There is only one $\langle 110 \rangle$ slip direction in each $\{110\}$ slip plane. Therefore, there can be only one kind of mobile dislocation in each slip plane. The secondary slip systems are of the type $\{001\}$ $\langle 1\bar{1}0 \rangle$. Since the primary and secondary slip planes have a common Burgers vector, this is the cross-slip plane. No two primary slip planes have the same Burgers vector.

Experiments in rock-salt structure (among them MgO) indicate a correlation between fatigue failure and cross slip. Cyclic-stressing experiments of crystals that exhibit easy cross-slip and of those that do not show easy cross-slip suggest that the absence of easy cross-slip is the probable reason for the absence of fatigue failure. If the hypothesis that cross-slip is essential for fatigue failure is correct, a possibility for improving strength of certain materials exists, since crystals without mobile dislocations are very strong.

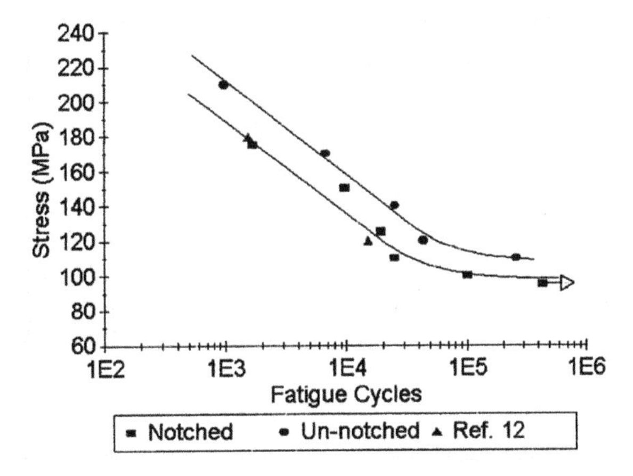

Fig. 7.61 S–N Curves [10]. (Ref. 12 is Headinger)

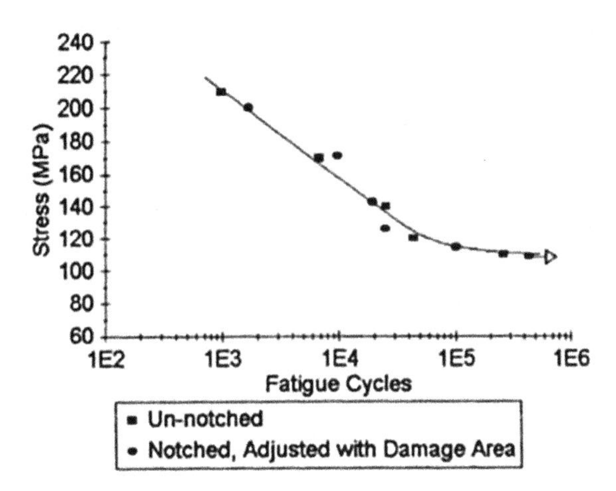

Fig. 7.62 S–N curves for un-notched and notched adjusted with damage area [10]. (Ref. 12 is Headinger)

7.12 Notch Effect

This section is of particular interest as a consequence of the fact that almost all ceramics have flaws, many in form of microcracks or even cracks on a larger scale. These are outcomes of the fabrication techniques; however, reducing such pores or cracks to a viable, minimal level is very costly. Many such cracks end at surfaces, where they act as stress raisers. Notches may also be considered as cracks, though the difference between notches and cracks is only a matter of size, not kind. Thus, the study of notched specimens has practical significance, because, in engineering components, fatigue starts mostly at the surface, at those places where stress-raisers are present in the form of 'micronotches' (i.e., traces of machining; artificial defects). In actual applications, such components are never perfectly smooth. Corners, fillets, holes, etc. all act as stress-raisers. Such geometries cause stress concentrations in the component and reduce the fatigue strength and life of a structure. The straightforward way to perceive the notch effect in fatigued specimens is by looking at the S–N curves. Figures 7.61 and 7.62 show experimental curves of 2D woven fabric-reinforced ceramic matrix composite (SiC/SiC), obtained in a tension–tension mode at 1100 °C.

Here, the load was applied as a 1.0 Hz triangular wave. (A triangular wave is clearly a non-sinusoidal wave form, but it is a periodic function). The minimum-to-maximum load ratio in all the cases was 0.1. The maximum applied-stress level is the reference for tests in which stress is defined as the applied tensile load per unit area. The area used in determining the stress level is the cross-sectional area at the gauge length and, in the case of the notched specimens, is reduced by the area of the hole. The notch is in the form of a central 3.06 mm diameter hole, D, in a 9.31 mm wide specimen, W, for a D/W ratio of 0.33. The S–N curves of the notched and un-notched fatigue tests are presented on a logarithmic scale of the x axis (fatigue cycles). Fatigue life is clearly a function of the stress level. In Fig. 7.61, a clear difference between the fatigue behavior of the notched and un-notched specimens is evident. The stress level of the notched specimens in the S–N curves is lower than that of the smooth specimens with a reduced endurance limit. The fatigue life of the un-notched specimens is longer by about a factor of four than that of the notched specimens. Due to the difference in applied-stress levels at a given fatigue life of the notched and un-notched specimens, a more appropriate comparison is fatigue (namely the applied-stress level) strength between these two. For this purpose, a fatigue-notch factor, β, is used as a ratio of the fatigue strength of the notched specimen (σ_{n}) to that of the un-notched specimen (σ_{un}) at a given fatigue life:

$$\beta = \frac{\sigma_n}{\sigma_{un}} \tag{7.16}$$

In Fig. 7.62, the adjusted notch-fatigue life fits within the scatter band of the S–N curve. In this case, the fatigue-stress level in the notched specimens was adjusted by using the damaged area, which extends about 0.4 mm into the specimen, and using the undamaged area beyond it. (The stress level is the load divided by the area and, to this end, an adjustment had to be made).

Equation (7.16) is often represented by a different symbol, K_f, the notch factor, relating the strength of smooth specimens to that of notched specimens, but having the same meaning as:

$$K_f = \frac{\text{fatigue strength of smooth specimen}}{\text{fatigue strength of notched specimen}}$$

Numerous experimental observations indicate that notches are among the key factors that determine the fatigue strength of structures and play an important role in the estimation of fatigue life and the strength of some structures exposed to fatigue deformation. Based on the definition of K_f, many expressions for the notch factor have been developed over the years, differing on the basis of their various assumptions regarding: notch geometry (e.g., sharpness); notch location (edge or centered); crack length and location; and material properties, such as tensile stress, etc. Traditional fatigue analysis of notched specimens is done using empirical approaches, though various models have been developed to consider the notch effect in regard to the strength and fatigue life of structures.

7.13 Failure Resulting from Cyclic Deformation (Fracture by Fatigue)

In almost all the sections discussed in this chapter, crack initiation and propagation are considered to be essential steps leading to failure. Again, fatigue cracks usually start from the surface of a component, while fatigue damage in ductile ceramics begins as shear cracks on crystallographic slip planes. Surfaces show slip planes as intrusions and extrusions. This is stage I of crack growth. After a transient period, stage II crack growth takes place in a direction normal to the applied stress. Finally, the crack becomes unstable and fracture occurs.

A separate portion of Chap. 8 (devoted to fracture) will consider this aspect of fatigue. Fracture by fatigue is a major part of all failures encountered in service, where components of structural systems are operating.

7.14 Effect of Some Process Variables

7.14.1 Surface Effects on Fatigue

Fatigue cracks generally initiate on specimen surfaces; therefore, their surface conditions are critical for good fatigue resistance of the materials. The surface roughness of a specimen has a great effect on its fatigue strength, since it is an effective stress-raiser. Surface roughness may originate from machining (very critical in metals); in order to reduce or eliminate its undesirable effect, surface-finishing or modifications of various kinds are usually a step in the manufacturing process. Surface roughness, such as scratches of various sizes, acts like micro-notches, producing similar effects on fatigue-strength properties as notches. Grinding must be such that those surface irregularities, introduced by machining, are removed to produce a smooth surface. The most effective surface-smoothing operation is polishing. Introducing compressive-residual stress on the specimen surface increases fatigue life. Sandblasting is also among the well-known techniques. There are certain factors that must be considered in order to ensure the best fatigue performance. Important material properties to be considered are: grain size, grain boundaries, temperature, size, etc.

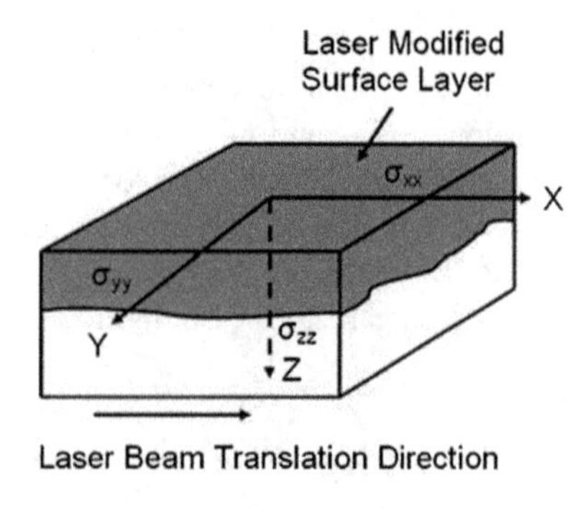

Fig. 7.63 Schematic of the coordinate system adopted in stress field evaluations [25]. With kind permission of Dr. Dahotre

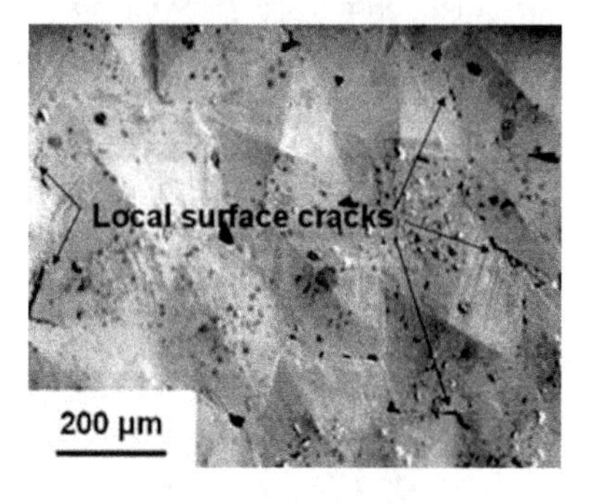

Fig. 7.64 Local surface cracks [25]. With kind permission of Dr. Dahotre

Many methods are available for the surface treatment of metals, such as the deposition of relatively thick coatings by one of the deposition techniques (e.g., chemical vapor deposition [henceforth: CVD], ion implantation, etc.). The techniques for modifying ceramic surfaces are more limited, due to their relative chemical inertness, the very strong bonds between the constituents comprising the ceramics and, last but not least, because of their high melting temperatures. The high melting temperature of ceramics means that surface modification processes involving the bulk diffusion of some desired constituent is an unlikely method for engineering ceramics, which require a chemical reaction or diffusion. However, non-equilibrium processes, involving surface bombardment by energetic atoms, are potentially viable methods for ceramic surface modification, such as treatments by plasma, ion beam or laser beam bombardment. Then, there are the other aforementioned physical methods, such as sandblasting, grinding and polishing. The examples provided below mostly relate to dental ceramics, where various surface treatments are necessary in order to obtain the most promising results for practical applications.

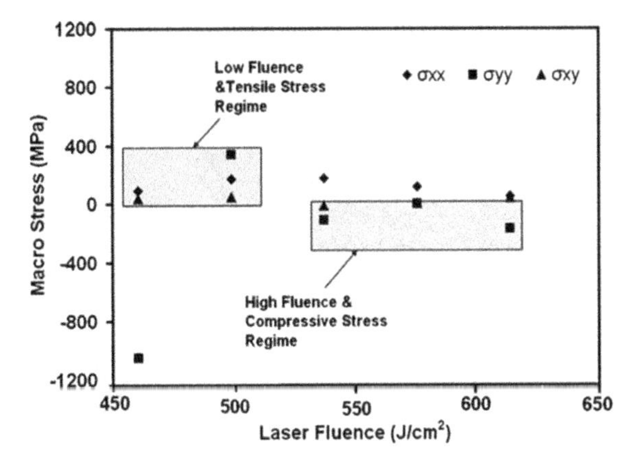

Fig. 7.65 Variation of macro stress with laser fluence [25]. With kind permission of Dr. Dahotre

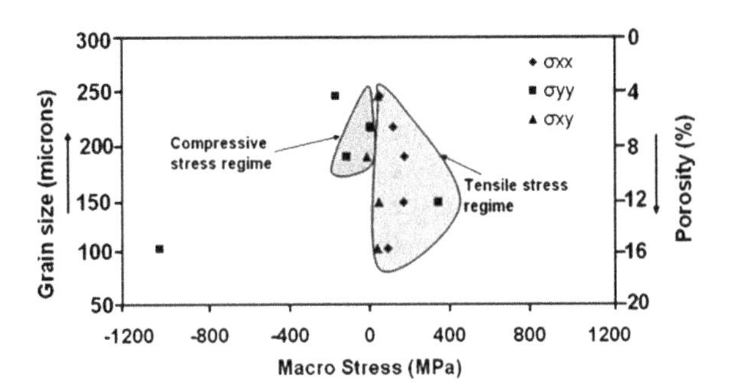

Fig. 7.66 Macro stress variation as a function of grain size and porosity [25]. With kind permission of Dr. Dahotre

7.14.2 Laser Treatments

The introduction of compressive stresses at the surface is known to improve material properties. Thus, compressive stresses improve the fatigue limit and wear resistance, while tensile stresses decrease fatigue strength and destroy the wear potential of surfaces. Laser applications for surface modification have become important processing tools for generating residual stresses after the cooling down following heating. These residual stresses form due to the great difference in temperature between the surface and the bulk of the laser-treated specimen. The high thermal stresses that are produced may modify the surface, but if the laser is applied incorrectly, surface cracks may develop, which (as stated above) may be the source of fatigue failure (when tensile-residual stress, rather than compressive-residual stress develops).

Here, laser fluences in the range of 459–611 J cm^{-2} were applied to modify alumina surfaces. The SEM micrographs show the surface features of the laser-treated alumina specimens. Finite-element simulations were carried out to predict

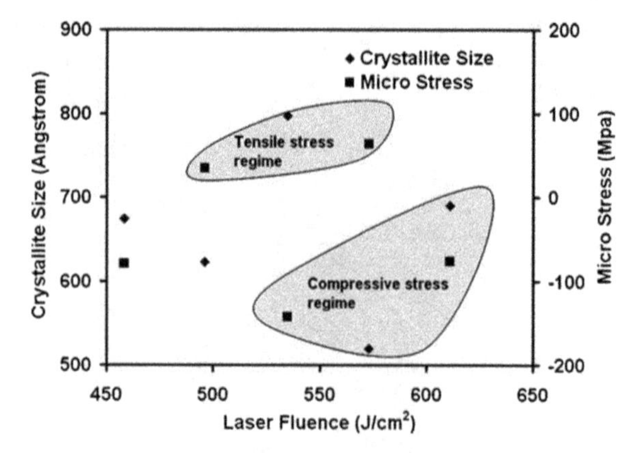

Fig. 7.67 Variation of crystallite size and micro stress with laser fluence [25]. With kind permission of Dr. Dahotre

the residual-stress state. In Fig. 7.63, a schematic illustration is shown, indicating the coordinate system used to evaluate the stress field.

The conventional signs in Fig. 7.63 are: the x axis is along the direction of the laser beam; the y axis is perpendicular to it; and both are along the surface; while the z axis is located along the depth of the specimen and perpendicular to the surface. The corresponding stresses acting along these directions are axial stress, σ_{xx}, lateral stress, σ_{yy}, normal stress, σ_{zz}, and shear stress, σ_{xy}. Isolated, localized stresses may be causing the local surface cracking, visible in the microstructure shown in Fig. 7.64.

The shear stress, σ_{xy}, values corresponding to all the laser fluences are negligible and also marginally vary as a function of laser fluence, indicating no significant plastic deformation within the area of analysis, as may be inferred from Fig. 7.65.

Note that the low laser fluence induces tensile stress, while high fluence stress promotes the favored compressive stress. Porosity and grain size are functions of laser fluence and the cooling rate and influence the macro stress. The influence of grain size and porosity on the variation in macro stress may be seen in Fig. 7.66.

Microscopic features, like grain size and porosity, are affected by laser fluence. Changing fluence will also affect the state of the stress. As observed in Fig. 7.66, stress values fall within the compressive regime for increased grain sizes, which is associated with decreasing cooling rates due to increasing laser fluences. In contrast, tensile stresses are favored by the generation of smaller grains, associated with high cooling rates as a consequence of decreasing laser fluences. Thus, laser fluence changes grain (crystallite) size and, consequently, the associated micro stresses will also vary. This may be seen in Fig. 7.67, where the variations in crystallite size and micro stress relative to fluence are indicated. For any particular stress regime, an increase in laser fluence, corresponding to lower cooling rates, leads to the formation of larger crystallites.

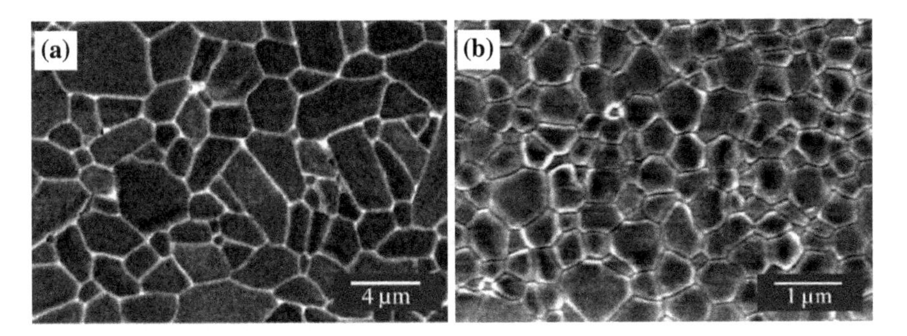

Fig. 7.68 Microstructures of **a** alumina (Al_2O_3) and **b** yttria-stabilized zirconia (Y TZP). SEM images. Surfaces are thermally etched [35]. With kind permission of Elsevier

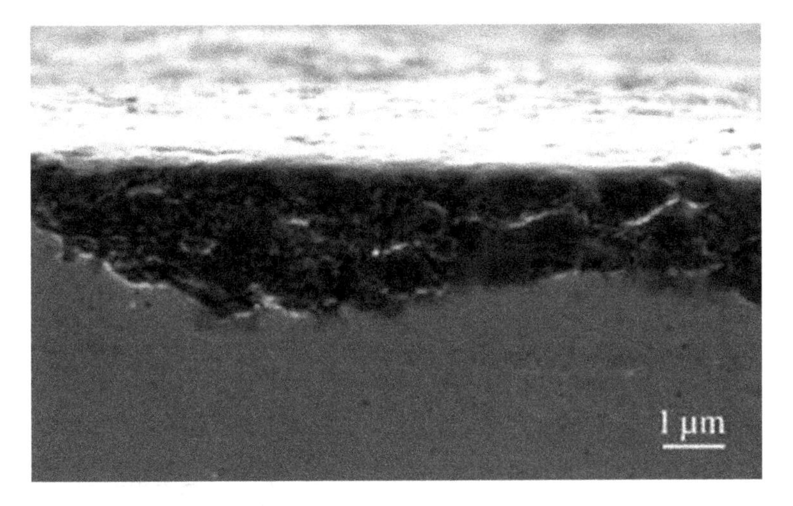

Fig. 7.69 Micrograph showing partial top and cross-section view of sandblast damage (by 50 μm Al_2O_3 particles) in Y-TZP [35]. With kind permission of Elsevier

Thus, in alumina ceramics, the surface residual stress sign (i.e., compressive or tensile) may be modified by laser, more specifically by the fluence applied, which changes the microstructure as a consequence of the cooling rate. Briefly, the macro stress values are in the compressive regime for increased grain size associated with the lower cooling rate due to increased fluence. Tensile macro stresses are generated with the formation of small grains, associated with high cooling rates, resulting from lower laser fluences. With systematic control of the laser processing conditions, control of grain size and porosity may be obtained for the generation of the stress pattern required for particular applications. The fatigue performance of a ceramic may be improved by inducing a compressive-stress pattern on the surface.

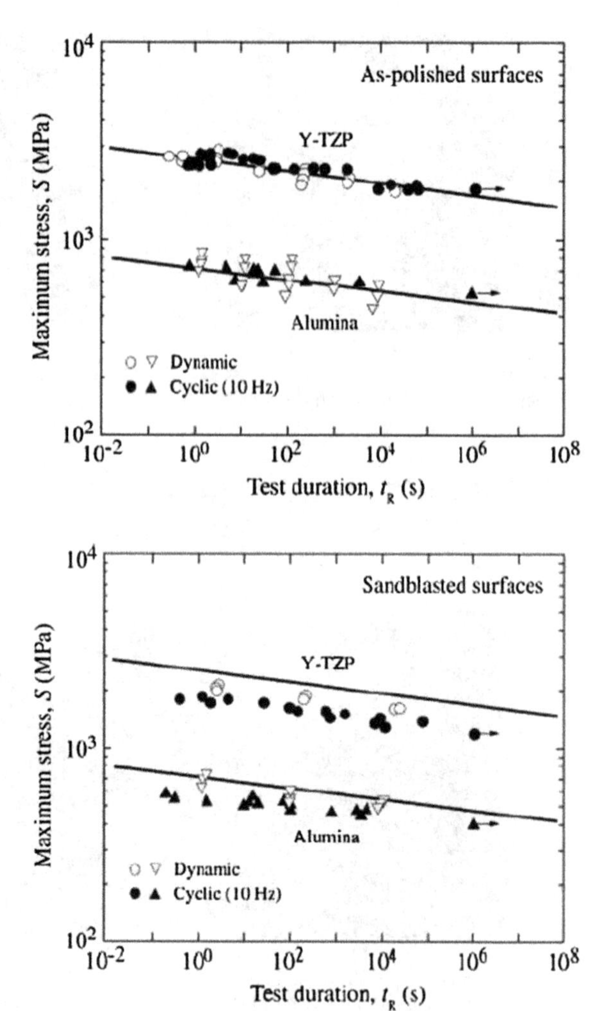

Fig. 7.70 Maximum tensile stress S in ceramic layer versus effective time to radial fracture t_R for as-polished Y-TZP and alumina plates bonded to polycarbonate substrates. Data represent individual tests at constant monotonic stressing rates (*unfilled symbols*) and in cyclic loading at 10 Hz (*filled symbols*). *Solid lines* are data fits in accordance with slow crack growth relations. *Arrows* indicate runouts [35]. With kind permission of Elsevier

Fig. 7.71 Same as Fig. 7.69, but for sandblasted Y-TZP and alumina plates. *Solid lines* are fits from Fig. 7.69 for as-polished surfaces [35]. With kind permission of Elsevier

7.14.3 Sandblasting

Y-TZP and alumina are attractive candidates for dental and other implant applications due to their good mechanical properties (for example, Kosmač et al. [59], Zhang et al. [35]). Their excellent mechanical properties, compared with others, is a consequence of a transformation mechanism, $ZrO_2(t) \rightarrow ZrO_2(m)$, which occurs as a diffusionless shear process. In dental clinical practice, an essential criterion for the selection of crown materials, besides aesthetics, is the resistance to fracture and deformation under long-term cyclic conditions. Surface conditions are important and their preparation may involve, grinding, sandblasting, polishing or any combination of these.

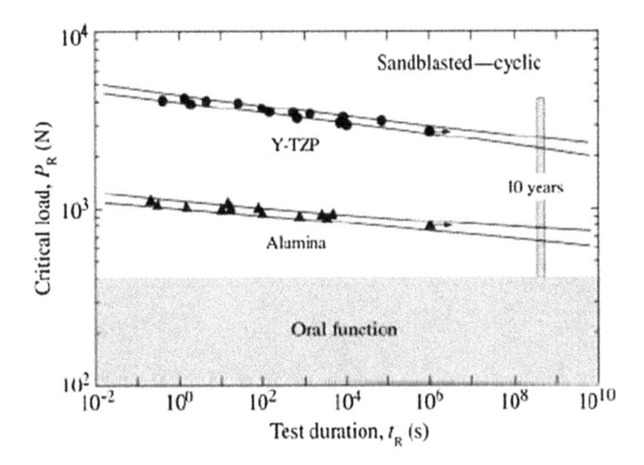

Fig. 7.72 Plots corresponding to cyclic fatigue data in Fig. 7.71 for sandblasted alumina and Y-TZP of thickness 1.5 mm, but in terms of critical loads instead of stress and for dentin-like substrate with intervening dental cement of thickness 100 μm, using Eqs. (7.17), (7.18), and (7.19) to convert the data. Ninety-five percent confidence bounds are used to evaluate uncertainties in sustainable loads at long lifetimes, $t_R = 10$ years. *Shaded band* indicates nominal oral function range [35]. With kind permission of Elsevier

The interior ceramic surface is an important factor governing the ultimate lifetime of ceramic-based prosthetic systems. Sandblasting of the interior surface is a common practice in all-ceramic crown restorations; the roughened surface enables a strong mechanical bond with resin-based dental cements. However, sandblasting introduces surface flaws and defects that may have an adverse effect on the strength of the crown. Therefore, the introduction of a compressive stress into the damaged layer (damage previously caused by the transformation in the zirconia) counteracts the probability of further damage. This countermeasure is dependent on the microstructure and the degree of the sandblast treatment. The effect of Hertzian contact with the spherical indenters stimulates the load experienced by a denture (i.e., the biting force). The experimental results below are based on monolithic ceramic layers, but are about the same as found in the ceramic cores of porcelain-veneered bilayers. Figure 7.68 illustrates the microstructures of alumina and Y-TZP. A cross-section SEM view of the sandblast damage in a bonded-interface specimen is shown for Y-TZP in Fig. 7.69. This figure reveals severe sandblasting damage extending ~4 μm below the surface. XRD analysis of the Y-TZP, before and after sandblasting, indicates only a small monoclinic phase content of ~4 vol% relative to a near-zero percent on as-polished surfaces. Furthermore, the Young modulus of the sandblasted surface is lower than in the as-polished specimens. Figure 7.70 plots the maximum stress, S, versus the effective time-to-radial-fracture, t_R, from dynamic fatigue and cyclic fatigue tests for polished Y-TZP and alumina bilayers (Fig. 7.71) are shown.

In Fig. 7.72, the critical load, rather than the stress, is plotted versus the test-duration time. Here, the cyclic-fatigue data from Fig. 7.71, for sandblasted alumina and Y-ZTP, have been converted to equivalent critical-load data. The relations used for this purpose are given by:

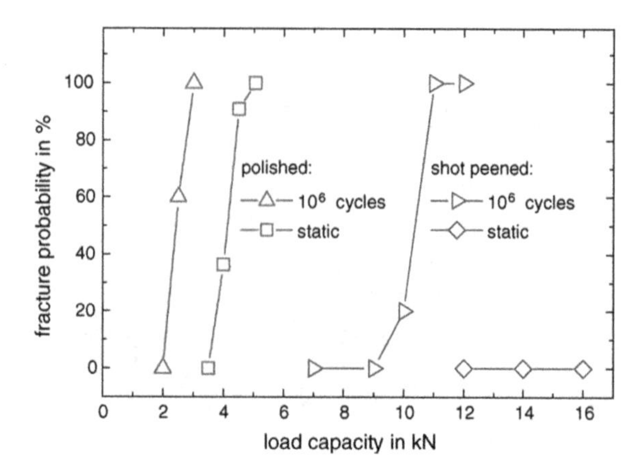

Fig. 7.73 Fracture probability of polished and shot peened samples (N3208), respectively, determined in static and cyclic ball-on-plate tests. Note that in case of the static tests of the shot peened samples the capacity of the test equipment was exceeded [23]. With kind permission of Elsevier

$$S = \left(\frac{P_R}{Bd^2}\right)\log\left(\frac{E_c}{E_s}\right) \tag{7.17}$$

$$E_* = E_i\left(\frac{E_s}{E_i}\right)^L \tag{7.18}$$

$$L = \exp\left\{-\left[\alpha + \beta\log\left(\frac{h}{d}\right)\right]^\gamma\right\} \tag{7.19}$$

In the above relations (see, for example, Zhang et al. [35], Kim et al. [58], Miranda et al. [66]), E_c and E_s are moduli of the ceramic plate and substrate, d is plate thickness and B = 1.35 is a dimensionless constant. In Eq. (7.18), E_s has been replaced by E_i, the cement modulus, h is the thickness between crown and dentin, and E_s, the actual substrate modulus, is replaced by an effective substrate modulus,. L = L(h/d) is an empirical Weibull function and $\alpha = 1.18$, $\beta = 0.33$, and $\gamma = 3.13$ were taken from Kim et al. [58]. The values used are d = 1.5 mm on thick dentin substrates with E_s = 16 cement interlayers of thickness h = 100 μm and modulus E_i = 5 GPa. The plots include 95 % confidence bounds to facilitate extrapolations to lifetimes at t_R = 10 years. Normal oral biting force (nominal range) is P_R = 0–400 N, as indicated in the shaded area in Fig. 7.72. These figures show that, for polished surfaces, no difference is observed between dynamic (constant load rate) and cyclic (sinusoidal loading) data. Sandblast damage introduced into ceramic undersurfaces causes a reduction in strength levels—10 % in single-cycle loading, but substantially more, 20–30 %, in cyclic loading at 10 Hz.

Sandblast treatment may generate superposed surface-compressive stresses by introducing open microcracks and inducing tetragonal to monoclinic phase transformations. Such stresses only serve to enhance strength. Kosmač et al. [59] was comparing the effects of various grinding processes, sandblasting or a combination of both on the surface conditions of Y-TZP ceramics by considering

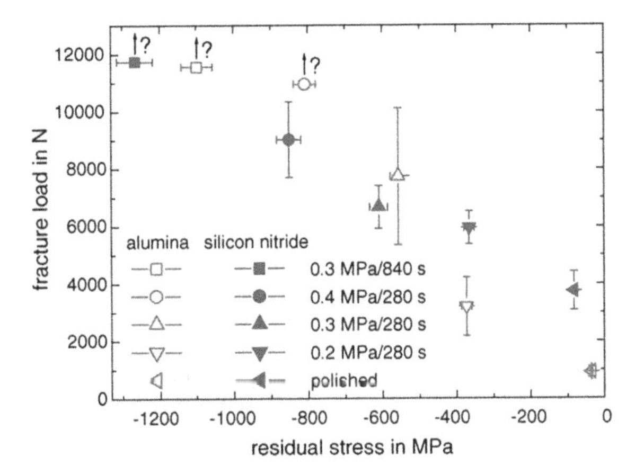

Fig. 7.74 Fracture load versus residual stress of silicon nitride (N3208) and alumina samples in polished and different shot peened condition [23]. With kind permission of Elsevier

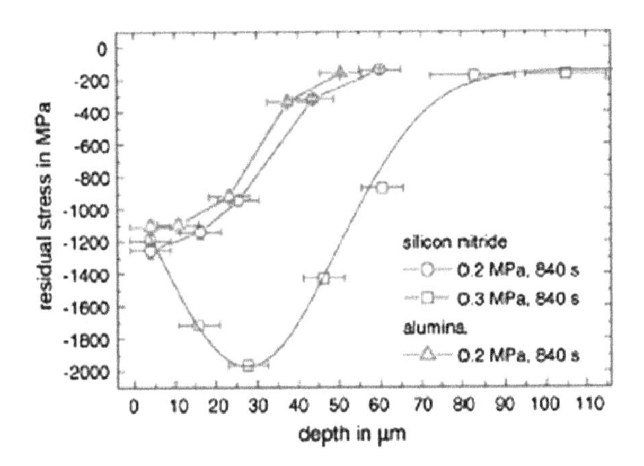

Fig. 7.75 Depth distribution of residual stresses of differently shot peened silicon nitride (N3208) and alumina samples [23]. With kind permission of Elsevier

flexural strength. In their Table 3, the mean flexural strength is listed and it is indicated that sandblasting provides the best results for both FG and CG Y-TZP. Wet and dry grinding yielded a lower flexural strength than the as-sintered value for both FG and CG ceramics. When a combination of grinding and sandblasting is such that the grinding precedes the sandblasting, flexural strength is much improved, but it is still somewhat lower than the outcome attained via a process of surface finishing by sandblasting alone. Sandblasting introduces a compressive-surface layer, which compensates for grinding-induced surface cracks. The highest amount of stress-induced monoclinic phase is observed in the sandblasted specimens (providing the stress needed for the transformation in zirconia).

In general, it is beneficial to attain the proper surface state for each particular application. The correct surface preparation of bio-materials is particularly important for orthodontic and other biomedical implant applications or their reconstructions.

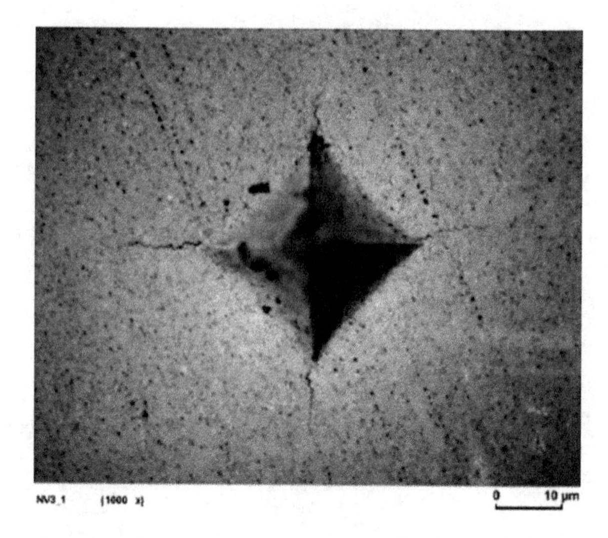

Fig. 7.76 Vickers indentation (9.81 N) in a polished silicon nitride sample [23]. With kind permission of Elsevier

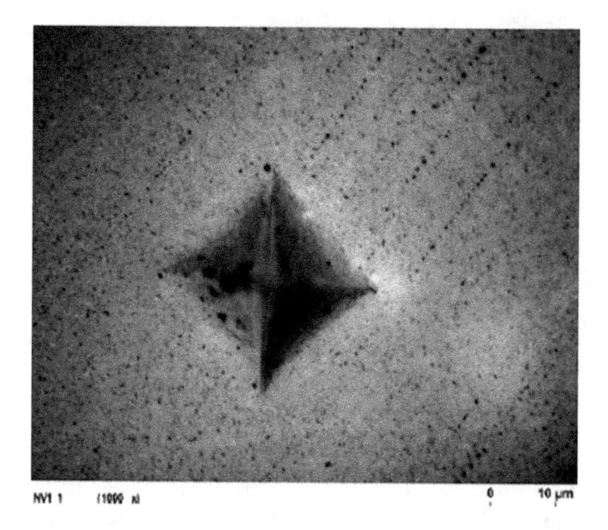

Fig. 7.77 Vickers indentation (9.81 N) in a shot peened (0.3 MPa pressure) silicon nitride sample. The sample was slightly repolished after shot peening to better show indentation-induced cracks [23]. With kind permission of Elsevier

7.14.4 Shot Peening

Commonly, shot peening [henceforth: SP] is not considered useful for the introduction of residual stress in ceramics, despite its wide use in modifying metal surfaces, due to its propensity for inducing non-healing cracks on ceramic surfaces or worse. However, in recent years, several reports have appeared in the literature indicating some beneficial effects of SP–if performed in the proper manner—for the introduction of the desirable compressive-residual stress crucial to fatigue resistance. The following two examples consider: (a) alumina and Si_3N_4 ceramics, and (b) Al_2O_3/SiC composites.

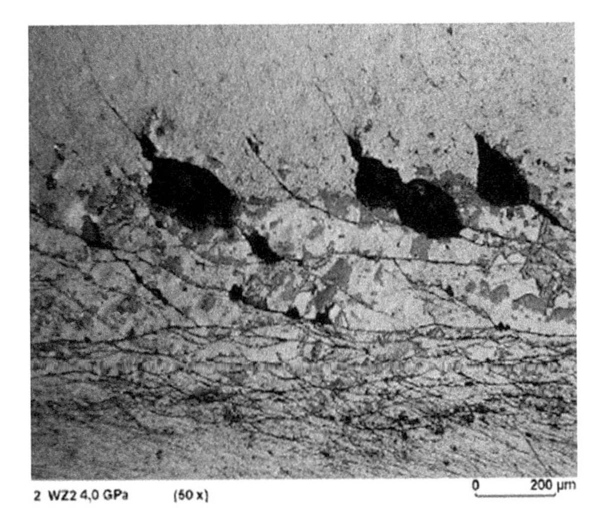

Fig. 7.78 Optical micrograph of a raceway after 55,000 cycles and a stepwise increase of the pressure up to 4.0 GPa. Interacting cracks and large-scale pitting in the polished sample [23]. With kind permission of Elsevier

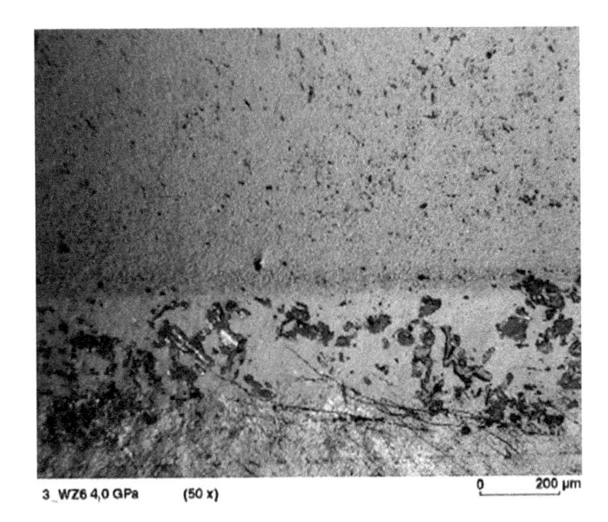

Fig. 7.79 Optical micrograph of a shot peened raceway after 55,000 cycles and a stepwise increase of the pressure up to 4.0 GPa. First small cracks are visible [23]. With kind permission of Elsevier

(a) Al_2O_3 and Si_3N_4

SP is a procedure commonly utilized in metals and alloys to increase the static and fatigue strengths. This strengthening effect is a result of the compressive-residual stress generated by SP, which prevents fatigue-crack propagation. Basically localized plastic deformation occurs in the near-surface regions, which strain-hardens (despite the fact that the bulk ceramic is brittle) with the associated increase in dislocation density and the consequent macroscopic compressive-residual stress generation. In the ceramics considered here, SP was applied by tungsten-carbide beads of 610–690 μm diameters at pressures of 0.2 up to 0.4 MPa. The peening time ranged from 280–840 s. The residual stresses and

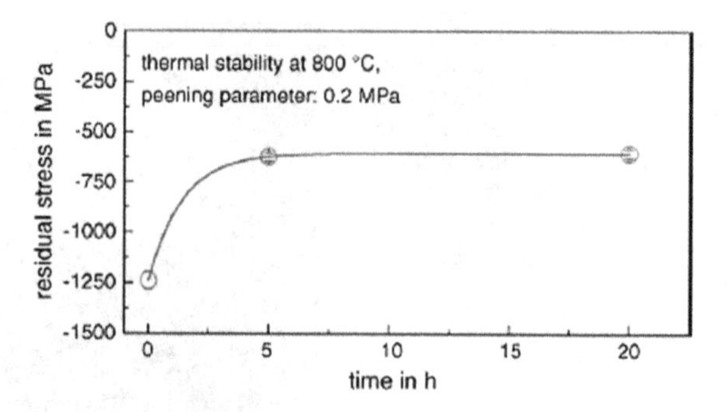

Fig. 7.80 Decline of shot peening induced residual stresses in silicon nitride N3208 at 800 °C [23]. With kind permission of Elsevier

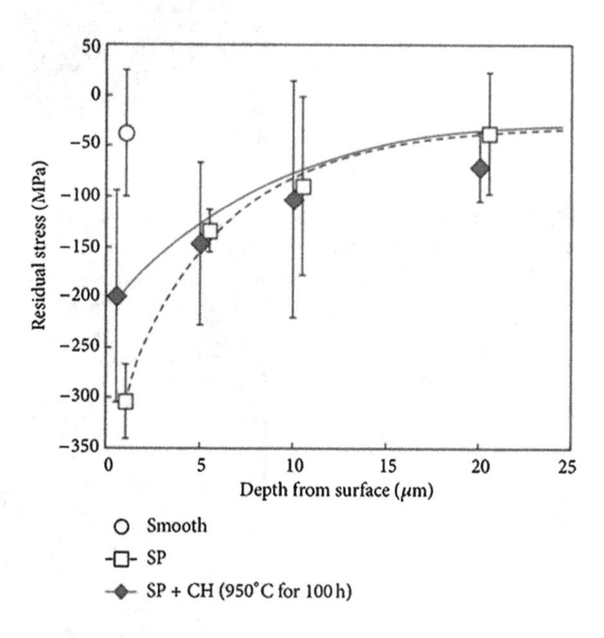

Fig. 7.81 Relationship between residual stress and depth from the surface [22]. Open access article permitting unrestricted use and reproduction

dislocation densities were determined by XRD. The full width at half maximum [henceforth: FWHM] was determined to characterize the dislocation density. The macroscopic residual stresses were derived from the peak shift using the $\sin^2\psi$-method [67]. Figure 7.73 shows the influence of SP on the cyclic-load capacity for silicon nitride N3208 using cyclic ball-on-plate tests. This figure shows the fracture probability as a function of the load for the SP and polished reference surfaces.

Observe in Fig. 7.73 that the static-load capacity of the polished reference samples is ~4 kN at a 50 % fracture probability, whereas the cyclic-load capacity of the SP silicon nitride samples is ~10.4 kN. The SP treatment increased the static-load capacity to more than 16 kN. Thus, the static load capacity is about 4 times higher. The capacity of the testing device was not high enough to introduce any cracks into the SP surfaces. From the no failure results of the static load up to

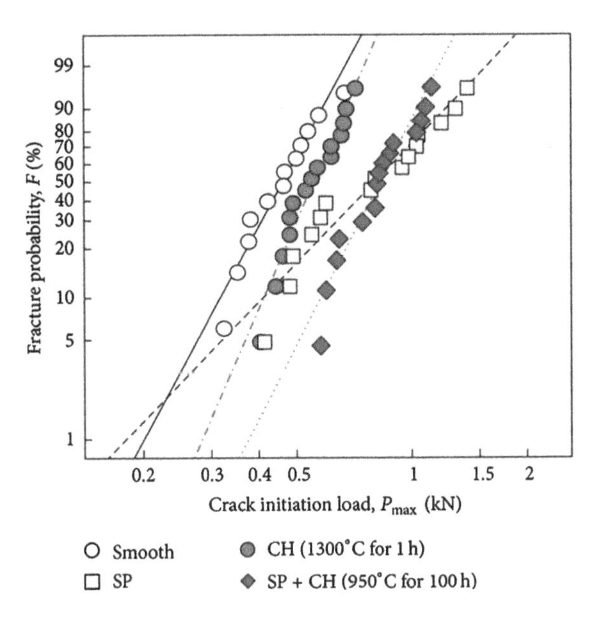

Fig. 7.82 Effects of shot peening and crack-healing on the Weibull distribution of crack initiation load [22]. Open access article permitting unrestricted use and reproduction

16 kN at least an increase by a factor of 4 may be seen. Figure 7.74 correlates the near-surface stress states and the load capacities of polished and differently SP alumina and silicon nitride samples.

SP enabled the creation of up to 1.3 GPa compressive-residual stresses near the surface. The depth distribution of this residual stress is shown in Fig. 7.75. The effect of high compressive-residual stresses may be seen from the indentation results made by the Vickers indenter in Figs. 7.76 and 7.77. Compare the two figures. The polished reference sample developed a typical crack formation at the edges of the indentation (Fig. 7.76), whereas, on the surface of the SP sample, only very small cracks may be detected (Fig. 7.77). Further indications of surface improvement (i.e., of increased fatigue resistance) may be obtained from a comparison of the following optical micrographs (Figs. 7.78 and 7.79).

Here, the damage in the polished reference sample, in the form of pitting and chipping after 45,000 cycles at a pressure of ~3 GPa (load 76 N), is compared with the damage pattern of the SP raceways after 55,000 cycles and up to 4.0 GPa pressure. The severe damage in the raceway of the reference sample, with numerous interacting cracks and large pitting areas, may be contrasted with the small, sporadic cracks in the SP raceway. This illustrates the strengthening effect of the SP process. The effect of SP on residual stress is SP-time-dependent and a decrease is obtained (as indicated in Fig. 7.80) showing a decay followed by a leveling off of the residual stress from a value of 1250 MPa at the surface to ~625 MPa. Note that tempering samples at 800 °C leads to a 50 % reduction of compressive-residual surface stresses within the first 5 h.

The above results indicate the importance of introducing compressive-residual stress by the SP of ceramics–which is a new development of the last decade. Even in brittle materials, like most ceramics, microplastic deformation is introduced

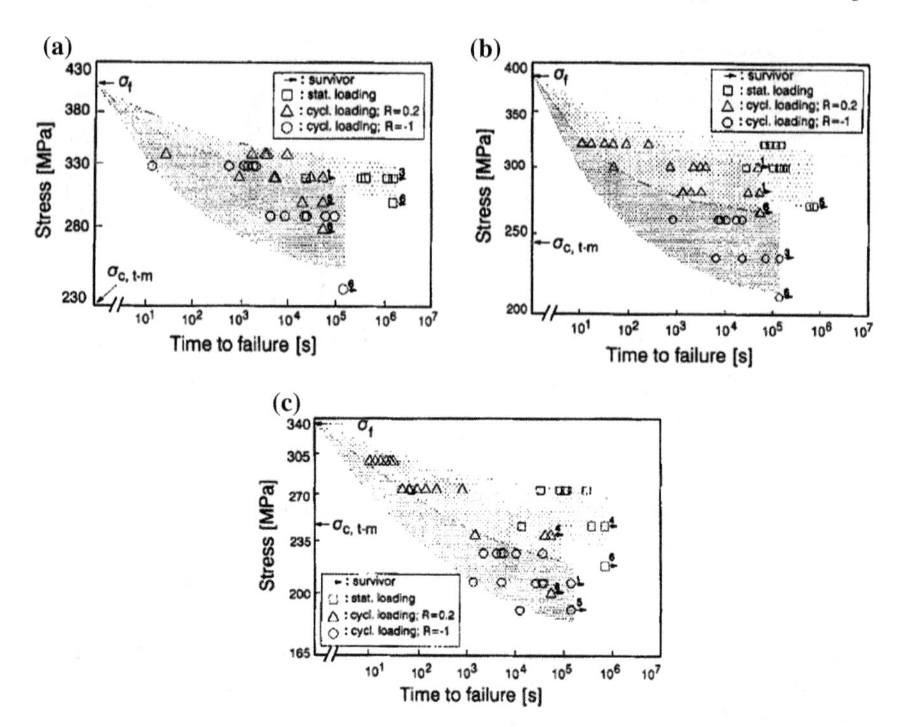

Fig. 7.83 Fatigue of Ce-TZP ceramics in static and cyclic (pulsating and reversed bending) tests: **a** Ce-TZP-II, **b** Ce-TZP-IV, and **c** Ce-TZP-V. The applied stress values are given as static stress σ_s, and maximum stress σ_{max} for R = 0.2 and stress amplitude σ_A for R = −1; σ_f and $\sigma_{c\ t\text{-}m}$ refer to the average bending strength and the critical transformation stress, respectively. Data points in combination with figures indicate numbers of survivor specimens under identical loading conditions [8]. With kind permission of John Wiley and Sons

at the SP surface, which strain hardens with consequent compressive-residual stress formation. The application, at elevated temperatures, of SP-treated ceramics is possible, due to their stability (Fig. 7.80). The successful increase in the near-surface strength of ceramics promises a greater use of static and cyclic applications in future.

(b) **Al_2O_3/SiC composites**

As indicated above, the compressive residual stress generated by SP prevents fatigue-crack propagation and the near-surface strength of ceramics can be improved. Thus, SP is a promising technique for increasing the strength of ceramics in the surface region for better fatigue resistance. Certain structural ceramics exhibit crack-healing [henceforth: CH] ability. Thus, if CH can be combined with SP, the surface strength and reliability of ceramics may be increased. As known, CH is a technique which heats a material before its use or application to some temperature, resulting in restoring its strength. In the present case of Al_2O_3/SiC composites, the specimens were CH in air at 1300 °C for 1 h.

Table 7.4 Sintering parameters, average grain size, relative densities, Young's modulus and Vickers hardness of Ce-TZP materials [40] (with kind permission of John Wiley and Sons)

Materials	Sintering parameters	Average grain size (μm)	Relative density (%TD)	Young's modulus (GPa)	Vickers hardness
Ce-TZP-I	1400 °C, 0.2 h	0.5	97.57	201	836
Ce-TZP-II	1400 °C, 2 h	1.0	99.83	202	849
Ce-TZP-III	1500 °C, 0.5 h	1.4	99.55	190	816
Ce-TZP-IV	1500 °C, 1 h	1.5	99.73	202	854
Ce-TZP-V	1600 °C, 1 h	2.7	99.36	199	780

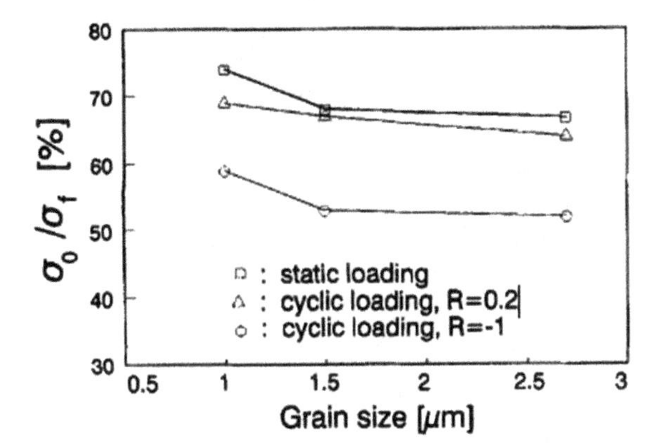

Fig. 7.84 Cyclic and static fatigue limits σ_0 related to strength σ_f as a function of the grain size of Ce-TZP [8]. With kind permission of John Wiley and Sons

SP was performed with ZrO_2 shots having a diameter of 180 μm. Figure 7.81 compares the residual stresses in smooth SP and SP + CH specimens. Note that SP + CH treatment gave the highest residual surface stress.

A maximum compressive-residual stress of 300 MPa is observed on the surface of the SP specimen, decreasing in the depth direction. In the case of the SP + CH specimen, the residual stress also decreased, but still exhibited a compressive-residual stress of 200 MPa. However, the induction of compressive-residual stress was accompanied by simultaneous CH. SP, in combination with CH, is a useful technique for improving the strength of ceramics, as may be seen in Fig. 7.82, which compares plots of the fracture probability against the crack-initiation load.

7.14.5 Grain Size

Recall the HP relation, which summarizes the effect of grain size on some static mechanical properties. Experimental results indicate that the resistance of materials to fatigue-crack initiation and propagation is significantly influenced by grain size. This applies to ceramics as well. It is widely recognized that, when all the other structural factors are kept approximately fixed, an increase in grain size will

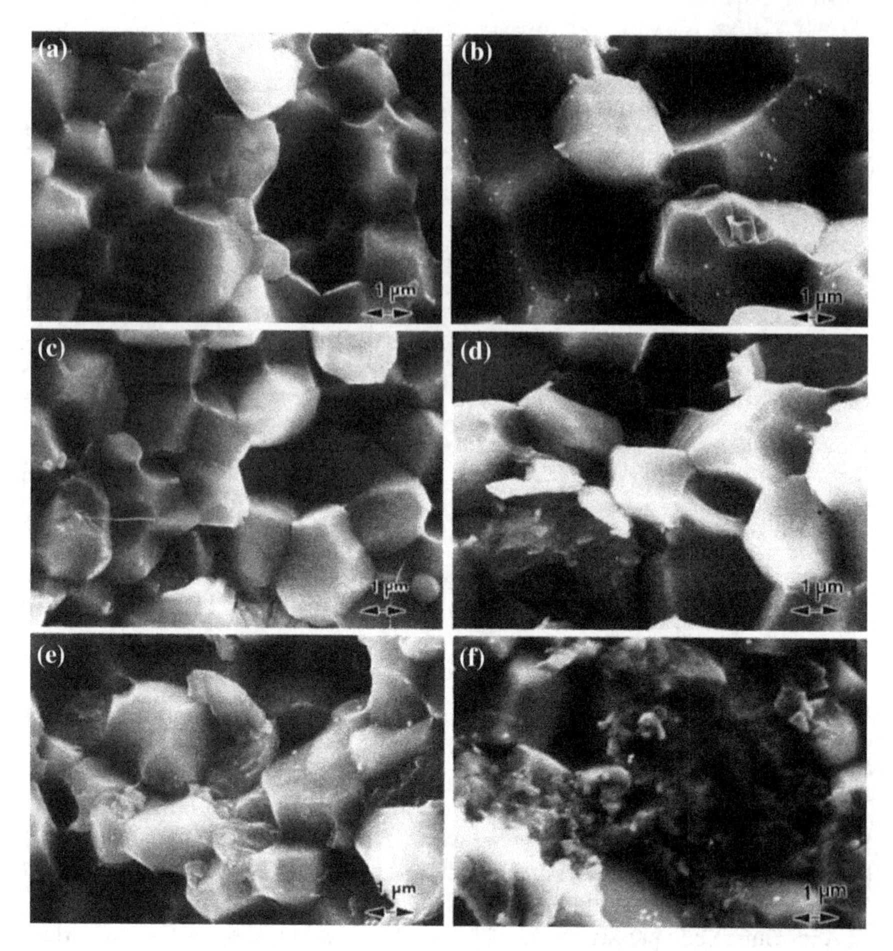

Fig. 7.85 Scanning electron micrographs of the fracture surfaces of **a**, **c**, and **e** Ce-TZP-IV and **b**, **d**, and **f** Ce-TZP-V as observed after **a** and **b** short-term strength and after cyclic fatigue tests with **c** and **d** R = 0.2 and **e** and **f** R = −1 [8]. With kind permission of John Wiley and Sons

generally result in a reduction in the fatigue-endurance limit. This observation parallels observations made of static deformation, namely that strength increases with decreasing grain size. An example of the effect of grain size appears below for CeO_2-stabilized tetragonal ZrO_2 (Ce-TZP). In Fig. 7.83, three grain sizes are presented to evaluate their effects on fatigue life.

The grain sizes are indicated in Table 7.4. Thus, in Fig. 7.83, the grain sizes are 1.0, 1.5 and 2.7 μm. Also note that, for the same time-to-failure (lifetime), the stress decreased with increasing grain size, regardless of the magnitude of R. Here, a comparison is made between static- and cyclic-fatigue tests. The time-to-failure of the statically-loaded specimens is longer by 2–3 orders of magnitude at the same stress level than under cyclic loading (R = 0.2), with the reduction of the time-to-failure being more pronounced under cyclic loading with increasing grain

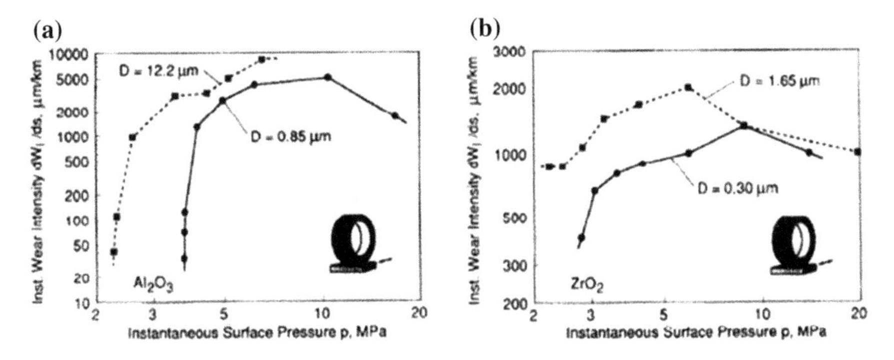

Fig. 7.86 Linear amount of wear of Al_2O_3 and ZrO_2 structures with two different average grain sizes plotted against the length of the wear path during **a** unidirectional and **b** reciprocating sliding contact [40]. With kind permission of Elsevier

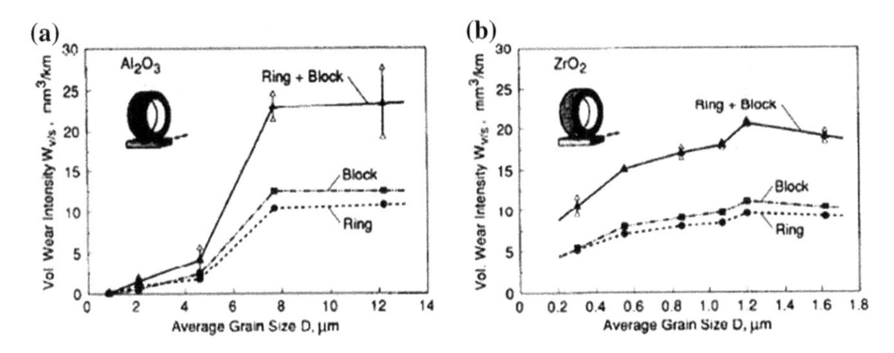

Fig. 7.87 Volumetric wear intensity of **a** Al_2O_3 and **b** ZrO_2 specimens during reciprocating sliding contact as a function of the average grain size of the structures [40]. With kind permission of Elsevier

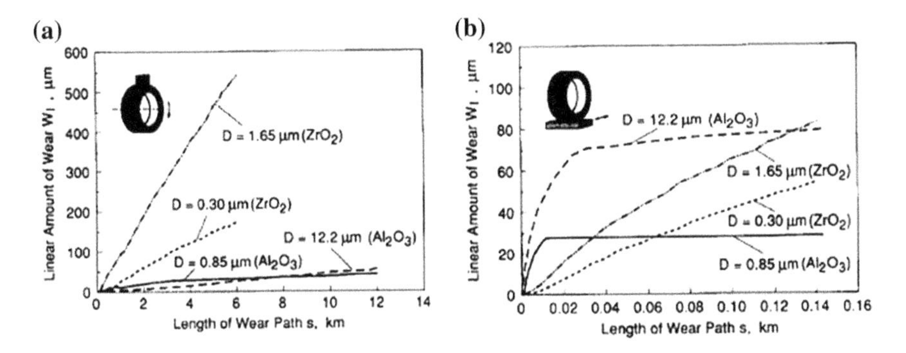

Fig. 7.88 Linear amount of wear of Al_2O_3 and ZrO_2 structures with two different average grain sizes plotted against the length of the wear path during **a** unidirectional and **b** reciprocating sliding contact [40]. With kind permission of Elsevier

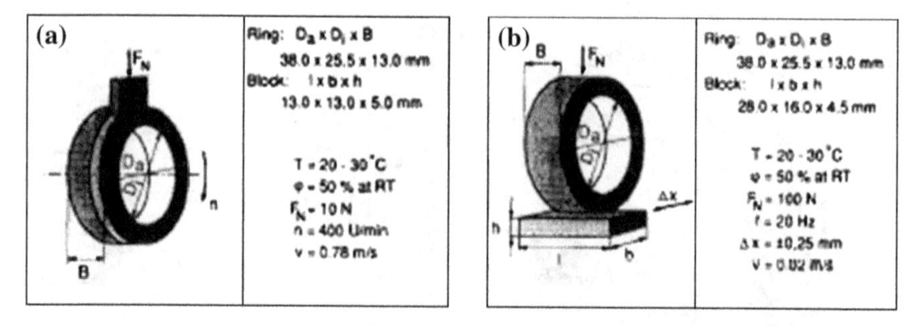

Fig. 7.89 Tribosystems and operating conditions for **a** unidirectional sliding and **b** reciprocating sliding wear [40]. With kind permission of Elsevier

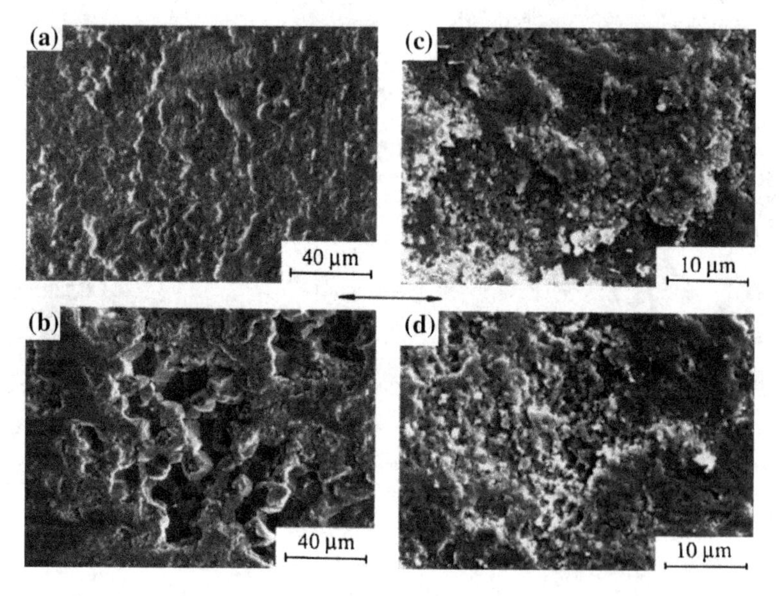

Fig. 7.90 Scanning electron micrographs of block specimens of Al_2O_3 with **a** D = 0.85 μm and **b** D = 12.2 μm, and ZrO, with **c** D = 0.55 μm and **d** D = 1.65 μm, worn in reciprocating sliding contact. *Arrow* indicates direction of sliding [40]. With kind permission of Elsevier

size. In addition, the fatigue limits, σ_0, under cyclic loading with R = −1 are much lower than under static and cyclic loading with R = 0.2. The static- and cyclic-fatigue limits are related to the mean value of the bending strength, σ_f, shown in Fig. 7.84, again indicating a decrease in the fatigue limits of the Ce-TZP ceramics under cyclic loading with R = −1 and their dependence on grain size.

Figure 7.85 compares the SEM micrographs of the areas fractured by the strength measurement with those of the specimens fractured under cyclic loading. Observe that the fatigue specimens have a higher degree of transcrystalline fracture in larger grain sizes. Recall that TZP undergoes a tetragonal to monolithic

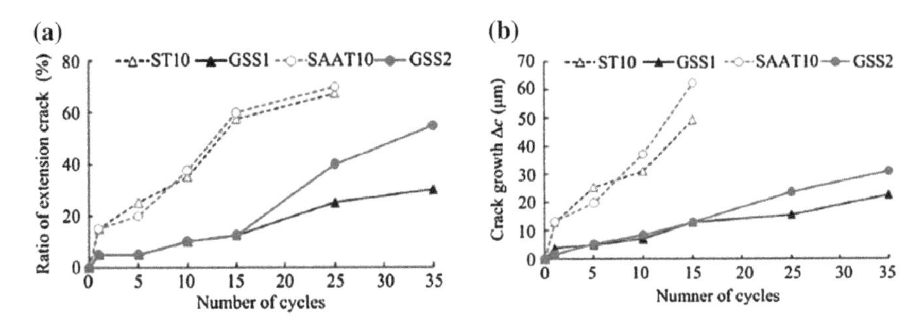

Fig. 7.91 Effect of the number of cycles on **a** ratio of extension crack and **b** crack growth (Δc) at ΔT = 400 °C [36]. With kind permission Elsevier

transformation under stress, which has occurred in the Ce-TZP. The enhanced degree of transcrystalline fracture in Fig. 7.85 indicates that damage to the structure, on account of the ZrO_2 t → m transformation within the grains, is crucial for crack initiation and propagation under cyclic loading. The Ce-TZP ceramics exhibit subcritical crack growth and an additional cyclic-fatigue effect, the latter being more pronounced in microstructures with larger grain sizes.

In the above illustration, grain-size variation and 'tetragonal toughening' (i.e., tetragonal to monolithic transformation) are acting simultaneously, although the effect of grain size is clear. In the following illustration, the wear resistance of alumina is considered; however, zirconia is also included for the purpose of comparison. Al_2O_3 and ZrO_2 ceramics are of interest for use as various wear components, such as bearing parts, cutting tools, valve seats and artificial hip joints.

These parts are exposed to reciprocating sliding wear (equivalent to cyclic-sliding deformation). Therefore, tests to check their resistance under cyclic (sliding) wear are also of great technological interest. Figure 7.86 compares the results of unidirectional and reciprocating tests of Al_2O_3 and ZrO_2 for the grain sizes indicated. In Fig. 7.87, the volumetric-wear intensities of Al_2O_3 and ZrO_2 ceramics are shown as a function of the average grain size. The wear intensity of Al_2O_3 increases strongly in microstructures containing average grain sizes, D, larger than ~4.5 μm, as seen in (a). The wear intensity of the fine grained (FG) structures is about two orders of magnitude smaller than that of the coarse grained (CG) structures. The volumetric-wear intensities of ZrO_2 increase more steadily, as seen in (b). The wear intensities of FG Al_2O_3 ceramics are substantially lower than those of the FG ZrO_2 ceramics. The linear amounts of wear for two grain sizes appear in Fig. 7.88 as a plot of wear versus the length of the wear path. Note that the sliding-wear tests were performed using a block-on-ring tribometer, shown in Fig. 7.89. This test was performed using unlubricated ceramic specimens. The normal force was 10 N and the sliding speed 0.78 m s^{-1}. The unlubricated oscillating-sliding test was performed at normal load, frequency of oscillation and stroke of 100 N, 20 Hz and 0.5 mm (±0.25 mm amplitude), respectively. These parameters were kept constant during each test and resulted in an average sliding speed of 0.02 m s^{-1}. These wear

Table 7.5 Composition (vol. %) of different composites [38] (with kind permission Elsevier)

Composites	Si_3N_4 (0.5 μm)	Si_3N_4 (0.02 μm)	Al_2O_3 (0.1 μm)	AIN (0.5 μm)	$TiC_{0.7}N_{0.3}$ (0.5 μm)	Al_2O_3 (0.5 μm)	Y_2O_3
SAAT10	53.25	17.75	10	5	10	0	4.0
ST10	61.50	20.50	0	0	10	3.2	4.8
ST15	57.75	19.25	0	0	15	3.2	4.8
ST20	54.00	18.00	0	0	20	3.2	4.8

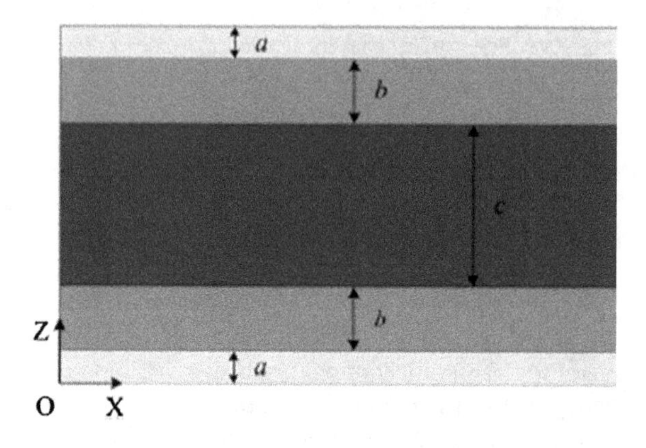

Fig. 7.92 Model of five-layered graded material with symmetrical structure [36]. With kind permission Elsevier

tests were run over 1.4×10^5 cycles or a total wear-path length of 0.14 km. SEM micrographs of worn Al_2O_3 and ZrO_2 surfaces are presented in Fig. 7.90. The layers show greater mechanical stability in FG than in CG Al_2O_3 and ZrO_2 ceramics. CG structures promote flaking of the layers, owing to cracking and delamination; this reduces the protection against wear to the material underneath. Intercrystalline fracture, followed by the spalling or fragmentation of individual grains, prevails on the surfaces of the CG structures of Al_2O_3 (Fig. 7.90b) and ZrO_2 (Fig. 7.90d). Relatively smooth surface areas are observed locally, owing to plastic deformation and shear fracture (Fig. 7.90b and d).

It must be emphasized again that the reciprocating-sliding test is equivalent to fatigue (or oscillating) deformation, which is the subject of this chapter.

7.15 Thermal Fatigue

Regardless of the origin of stress when cycling is applied, fatigue damage may result. Dangerous stress cycling is associated with thermal effects. Premature failure, resulting from cyclic stresses due to temperature changes, seems to be one way of approaching the problem of thermal fatigue in high-temperature structural components. This problem is of great concern in many fields in which structural components must operate at high service temperatures; if thermal gradients are

Table 7.6 Composite of each layer of five-layered FGM with symmetrical structure [38] (with kind permission Elsevier)

Specimen code	The 1st layer	The 2nd layer	The 3rd layer
GSS1	ST10	ST15	ST20
GSS2	SAAT10	ST15	ST20

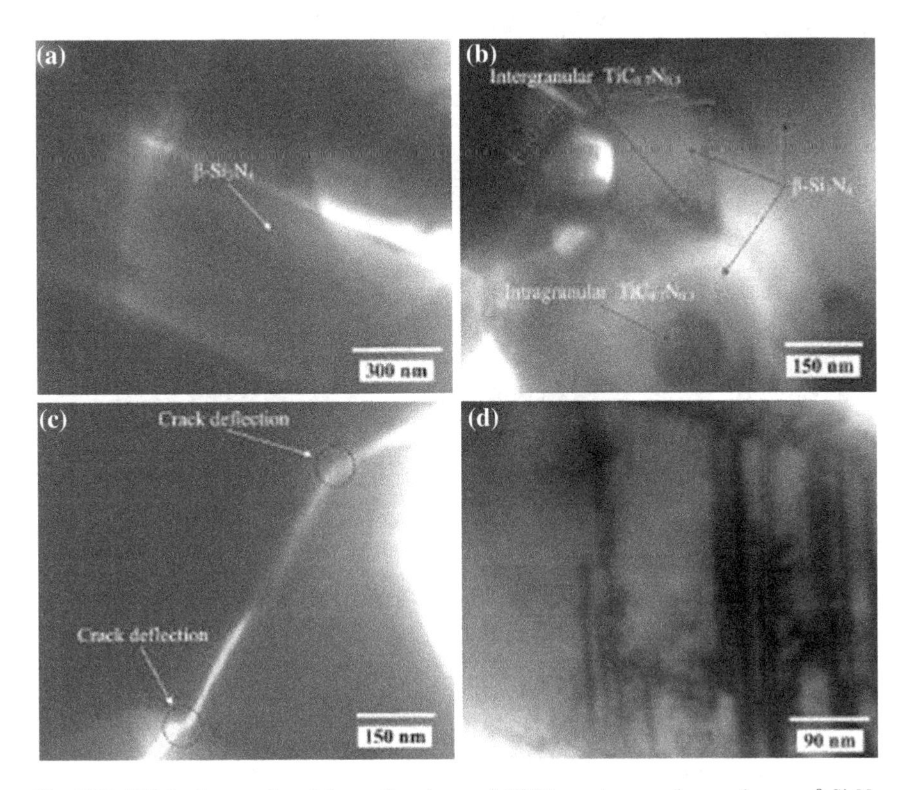

Fig. 7.93 TEM micrographs of the surface layer of GSS1 specimen, **a** long columnar β-Si_3N_4, **b** intragranular and intergranular $TiC_{0.7}N_{0.3}$ particles, **c** crack deflection and **d** twin structure [36]. With kind permission Elsevier

generated, the propensity for thermal-fatigue cracking increases. Thus, it is of great technological interest to strengthen those components that will be exposed to conditions in which thermal gradients are set up to inhibit thermal-fatigue damage.

The introduction of compressive stress and surface hardening are effective methods for preventing thermal fatigue. It is likely that their inhibiting effects are associated with both delayed-crack nucleation and crack growth, if cracks or pores are already present in the material (which is the actual case in most ceramics). Several methods for surface modification have been mentioned in the previous section, among them SP is a technique used to improve the fatigue properties by

Fig. 7.94 Fracture surface micrographs of the ceramic samples at ΔT = 600 °C, **a** ST10 and **b** the surface layer of GSS1 [36]. With kind permission Elsevier

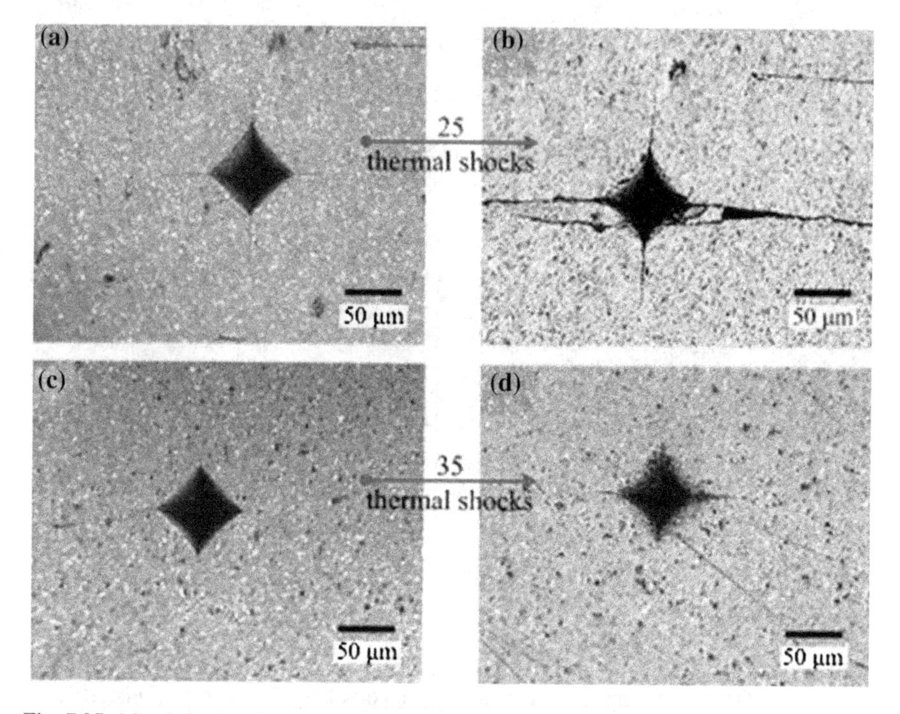

Fig. 7.95 Morphologies of the indentation cracks at ΔT = 400 °C after the different number of thermal shocks, **a** initial indentation cracks of ST10, **b** indentation cracks after 25 thermal shocks of ST10, **c** initial indentation cracks of GSS1 and **d** indentation cracks after 35 thermal shocks of GSS1 [36]. With kind permission Elsevier

introducing a compressive-stress pattern on the ceramic surface. An illustration of the effect of thermal cycles on crack growth is shown in Fig. 7.91. The symbols in Table 7.5 represent the composition of the ceramics. In this case, the major

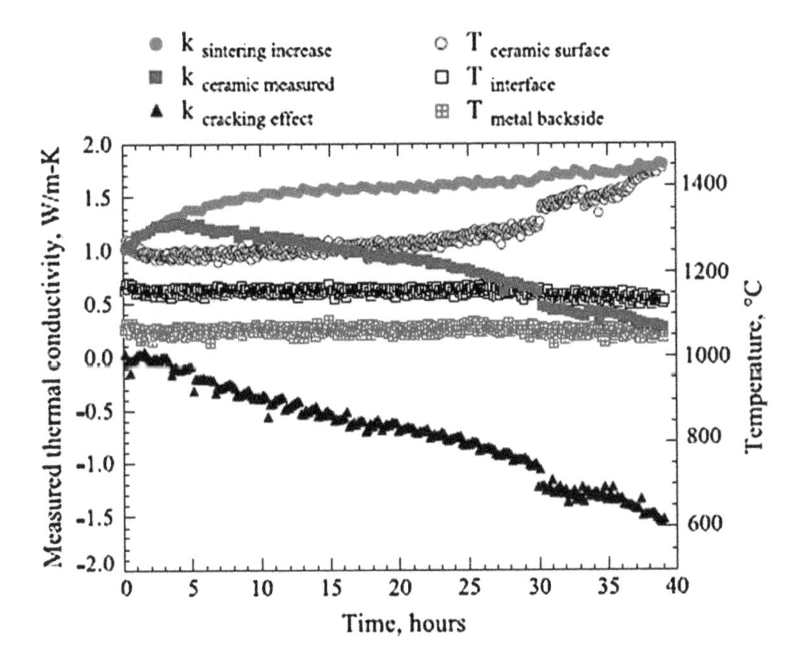

Fig. 7.96 Typical laser thermal fatigue test results of a 127-μm-thick, precracked TBC specimen showing the coating temperature and thermal conductivity changes as a function of cycle number under 10-min heating and 2-min cooling laser cycling. The ceramic surface temperature increases and the metal backside temperature slightly decreases as the delamination crack is initiated and propagated. The effective ceramic coating conductivity shows an initial increase due to the coating sintering, and then a decrease due to the crack propagation [38]. With kind permission of Elsevier

constituent in the HIP composite is Si_3N_4. GSS1 in Fig. 7.91 represents a functionally-graded ceramics. Figure 7.92 presents a model of five-layered, graded materials with symmetrical structures. The compositional distribution changes along the Z-axis. The thicknesses of the surface layer, the second layer and the third layer are: a, b and c, respectively. The thickness ratio is $e = a/b = b/c$ and a structural parameter is fixed at 0.3. Table 7.6 shows the composition of each letter (Fig. 7.93).

Note in Fig. 7.91b that the crack growth, Δc, of the graded ceramic material (GSS1) is much lower than that of homogeneous ceramic materials. These are all indications that the thermal-fatigue resistance of graded ceramics is higher than that of homogeneous ones. The fracture-surface micrographs of the ceramic samples at $\Delta T = 600$ °C are shown in Fig. 7.94. Many microcracks are visible on the fracture surface of ST10 (Fig. 7.94a), caused by thermal shock and causing a rapid drop in strength. Several microcracks are detectable on the fracture surface of the surface layer of the GSS1 (see Fig. 7.94b). The retained strength of the graded ceramic (GSS1) shows only a slight decrease after thermal shock; its cracks are fine and small (see Fig. 7.94b), producing only a minor reduction in strength. Further indications of the improved thermal-fatigue resistance may be obtained by means of indentation tests performed on homogeneous (ST10) and functionally-graded ceramics (GSS1) specimens. Notice in Fig. 7.95 that the GSS1 specimen has a much better resistance to indentation, even after 35 thermal cycles.

In Fig. 7.93, the TEM micrographs illustrate the surface layers of the GSS1 specimens. A long, columnar β-Si_3N4 grain is clearly visible in Fig. 7.93a. It may also be seen from Fig. 7.93b that the larger $TiC_{0.7}N_{0.3}$ particles are located at grain boundaries, while the smaller ones are trapped inside the β-Si_3N_4 grains.

Note in Fig. 7.91b that the crack growth, Δc, of the graded ceramic material (GSS1) is much lower than that of homogeneous ceramic materials. These are all indications that the thermal-fatigue resistance of graded ceramics is higher than that of homogeneous ones. The fracture-surface micrographs of the ceramic samples at $\Delta T = 600$ °C are shown in Fig. 7.94. Many microcracks are visible on the fracture surface of ST10 (Fig. 7.94 a), caused by thermal shock and causing a rapid drop in strength. Several microcracks are detectable on the fracture surface of the surface layer of the GSS1 (see Fig. 7.94b). The retained strength of the graded ceramic (GSS1) shows only a slight decrease after thermal shock; its cracks are fine and small (see Fig. 7.94b), producing only a minor reduction in strength. Further indications of the improved thermal-fatigue resistance may be obtained by means of indentation tests performed on homogeneous (ST10) and functionally-graded ceramics (GSS1) specimens. Notice in Fig. 7.95 that the GSS1 specimen has a much better resistance to indentation, even after 35 thermal cycles.

One of the reasons for the improved strength properties of the functionally-graded ceramics is due to the surface formation of compressive-residual stresses, which counteract some of the tensile stress generated during the thermal shock process.

There is growing interest in the potential uses of ceramic coatings as thermal barriers applied to components intended for in-service high-temperature exposure, e.g. in gas-turbine engines. Such applications rely on the understanding of ceramic coating behavior and ceramic failure modes under high temperatures, when high thermal gradients may exist under cyclic conditions. The lifetime properties of thermal barrier coatings [henceforth: TBC] are affected when cyclic deformation occurs in the presence of thermal gradients, associated with thermal fatigue. Specimens of ZrO_2-8 wt% Y_2O_3 are produced by air-plasma spraying for the use as TBC. Crack propagation, induced by laser-thermal fatigue, is shown in terms of thermal-conductivity tests in Fig. 7.96. This figure shows typical test results for a 127-μm-thick, precracked TBC specimen (with 1-mm diameter substrate hole) under laser-cyclic loading. The thermal conductivity of the ceramic coating may, thus, be determined as a function of the laser-cycle number. The effective thermal conductivity at any given cycle contains valuable information about the advancing delamination crack in the coating under laser-thermal cyclic loading.

The crack-propagation rate (i.e., delamination; da/dN) under laser-thermal cyclic loads may generally be expressed in terms of the Paris law, given below as:

$$\frac{da}{dN} = C(\Delta K)^m \tag{7.20}$$

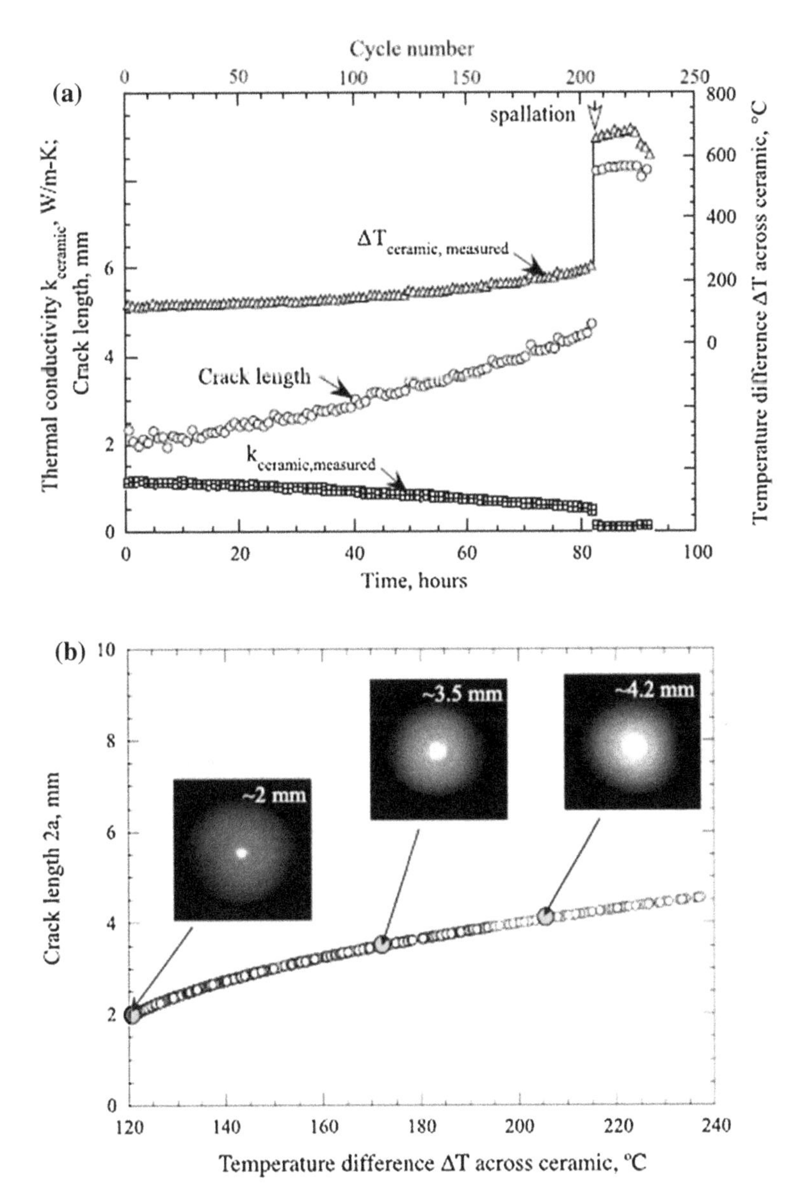

Fig. 7.97 The laser thermal fatigue test results of a 0.2-mm-thick TBC specimen with a 2-mm hole in the substrate when exposed to 20-min heating and 4-min cooling laser cycling. **a** A close relationship between the coating thermal conductivity and delamination crack length is demonstrated. **b** The coating propagation process has been monitored independently by a high sensitivity video camera, and expressed as a function of measured temperature difference across the coatings [38]. With kind permission of Elsevier

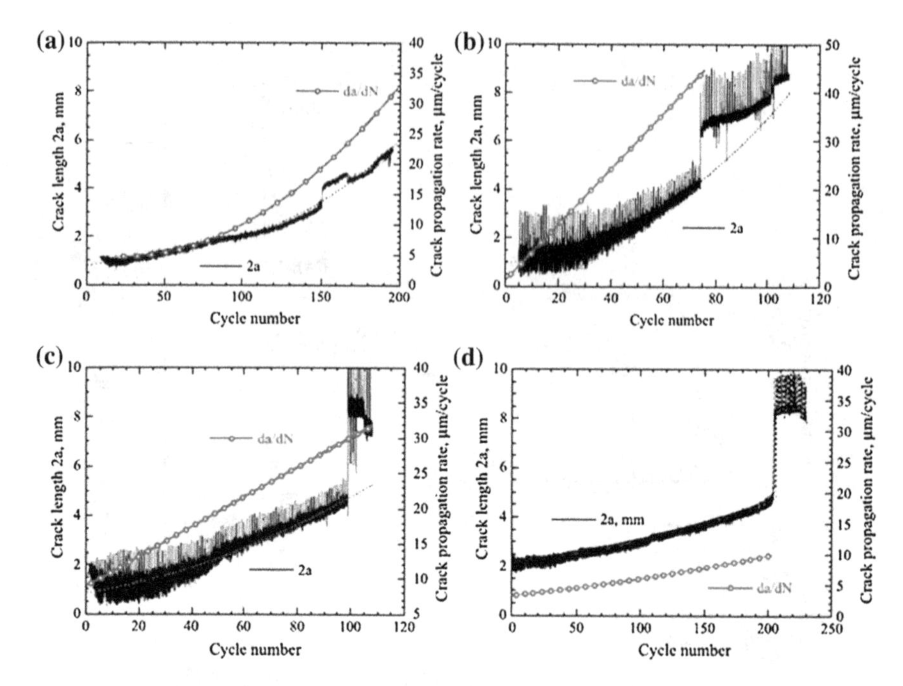

Fig. 7.98 Crack lengths and the corresponding crack propagation rates of laser tested TBC specimens as a function of cycle number. **a** 127-µm-thick coating, **b** 176-µm-thick coating, **c** 185-µm-thick coating, **d** 200-µm-thick coating [38]. With kind permission of Elsevier

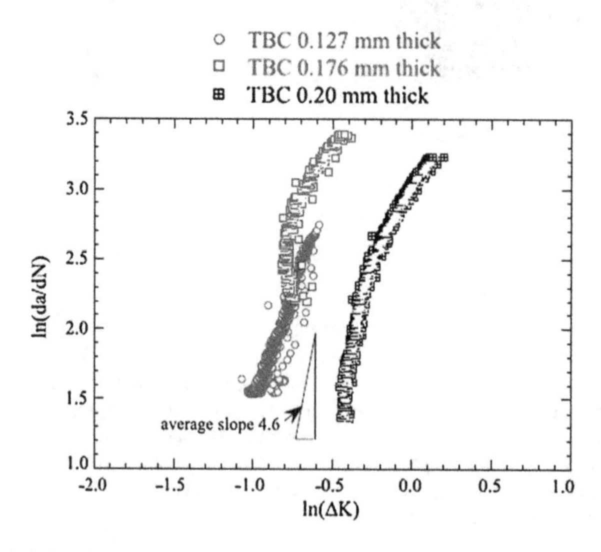

Fig. 7.99 The relationship between the delamination crack propagation rate da/dN and the laser thermal transient stress associated stress intensity factor amplitude ΔK [38]. With kind permission of Elsevier

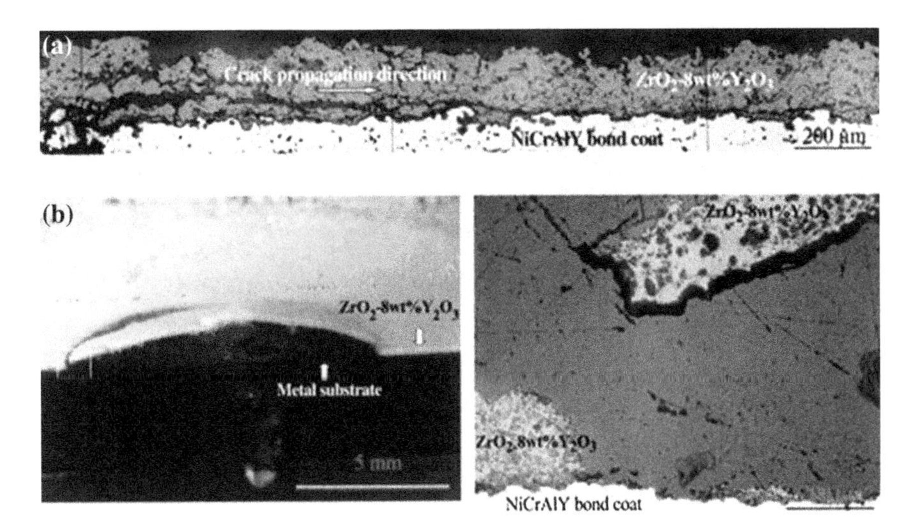

Fig. 7.100 Micrographs of laser thermal fatigue tested TBCs showing the coating delamination crack propagation and coating spallation. **a** Severe fatigue damages are observed near the early crack propagation wake surfaces with strong coating asperity/debris interactions and coating multiple delaminations under the laser thermal cyclic loading. The later crack paths show relatively smooth surfaces, which corresponds to the faster crack propagation regions under the increased crack propagation driving force. **b** Coating spallation morphology after the laser thermal fatigue test [38]. With kind permission of Elsevier

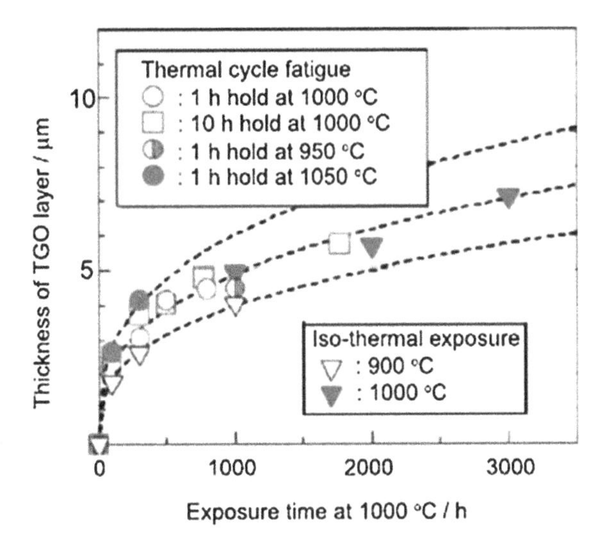

Fig. 7.101 Growth behaviour of the alumina type TGO during thermal cycle fatigue [32]. With kind permission of Professor Yamazaki for the editorial office

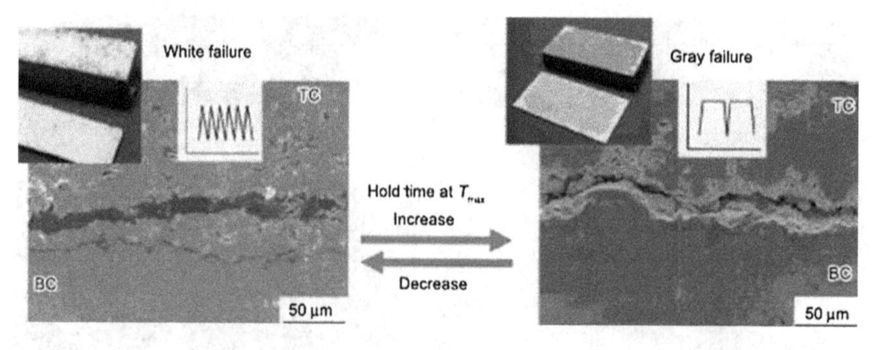

Fig. 7.102 Change of delamination failure morphologies [32]. With kind permission of Professor Yamazaki for the editorial office

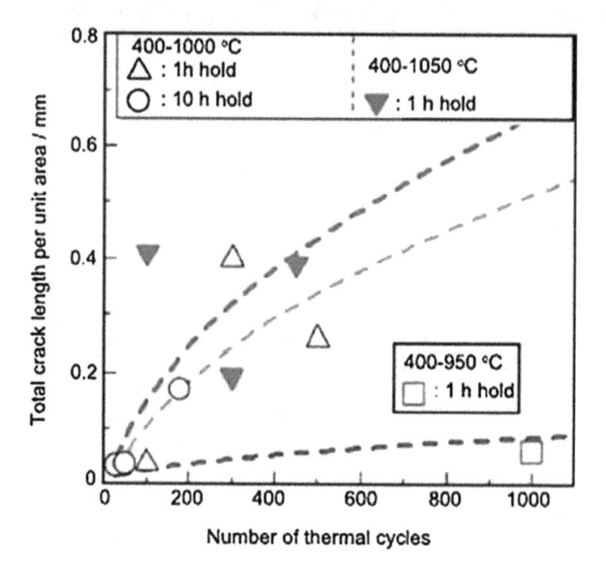

Fig. 7.103 Crack length density in top coat as a function of thermal cycles [32]. With kind permission of Professor Yamazaki for the editorial office

ΔK is the stress-intensity factor; da/dN is the crack-propagation rate; m is an appropriate exponent; and C is a constant:

$$\Delta K = K_{max} - K_{min} \text{ for } R > 0 \tag{7.21}$$

and:

$$\Delta K = K_{max} \text{ for } R \leq 0 \tag{7.22}$$

In this case, namely under laser-thermal cyclic loads, Eqs. (7.20–7.22) may be given as:

$$\begin{aligned} da/dN &= da/dN_{heating} + da/dN_{cooling} \\ &= C_1 \Delta K^m_{heating} + C_2 \Delta K^m_{cooling} \end{aligned} \quad (7.23)$$

$da/dN_{heating}$ and $da/dN_{cooling}$ are the crack propagation rates under laser-thermal transient heating and cooling, respectively; $\Delta K_{heating}$ and $\Delta K_{cooling}$ are the stress-intensity factor amplitudes associated with heating and cooling, respectively. Also note that Eq. (7.21), in terms of the measured thermal conductivity, $k_{measured}$, for a precracked TBC specimen, may be expressed as:

$$k_{measured} = k_{intact} + k_{cracking} \quad (7.24)$$

where k_{intact} is the thermal conductivity of non-cracked TBC specimens, which typically increases with increased time or cycle number due to the ceramic-sintering effect, while $k_{cracking}$ is the thermal-conductivity reduction due to a coating-cracking effect.

In Fig. 7.97, coating thermal conductivity and delamination-crack length are shown for a TBC specimen with a 2-mm hole in the substrate. The evolution of crack propagation and coating spallation behavior may be seen. In Fig. 7.98d, the effect of the cycle number on the crack-length increase is clearly visible, until spallation sets in at ~210 cycles. Here, the crack-length increase and the crack-propagation rates for different coating thicknesses, as a function of the cycle number may also be seen. Equation (7.23), on a logarithmic (ln) scale is illustrated in Fig. 7.99 for the thicknesses indicated.

In Fig. 7.99 the relationship between the delamination-crack propagation rate, da/dN, and the laser-thermal transient stress associated stress-intensity factor amplitude, ΔK, is illustrated during laser-thermal fatigue testing. The exponent, m, representing delamination-fatigue crack growth under a laser-thermal cycle driving force, ΔK, is about 5. Observe the micrographs of the laser-thermal fatigue-tested TBC specimens (Fig. 7.100). Here, (a) shows severe fatigue damage near the early crack-propagation wake surfaces with strong coating asperity/debris interactions and multiple coating delaminations under the laser-thermal cyclic loading. However, the later crack paths show relatively smooth surfaces, corresponding with faster crack-propagation regions under the increased crack-propagation driving force. A coating-spallation morphology, after the laser-thermal fatigue test, is shown in Fig. 7.100b.

7.16 Design for Fatigue

Regardless of whether a material is metallic or ceramic, design to overcome fatigue failure and to increase resistance to cyclic deformation is essential. However, for the time being, this task seems almost impossible. As indicated above, due to the basic nature of fatigue and of the parameters affecting it, it is difficult to have a perfect design. The necessary resistance to fatigue failure, required in order to overcome crack initiation and propagation, is clear. However, the translation of these aspects of fatigue into real designs of specific components is far from simple. It is virtually impossible to provide design prescriptions for each specific component and all the possible applications, due to the multitude structural applications and service environments and their various sizes (to mention only a few of the crucial, relevant factors). Nevertheless, here are a few of basic principles that may be applied to each individual case, as follows:

(a) A ceramic should be chosen so that the applied stress may be kept below the endurance limit, at which that material is expected to have an infinite lifetime. This design concept is also known as the 'infinite lifetime concept'. Often, a poorly-defined fatigue limit is not observed; in such ceramics, the use of an empirically evaluated lifetime at a specific stress is sound. A stress level of $\sim 10^7 - 10^8$ cycles is a reasonable design criterion.
(b) 'Safe life design' is a conservative approach. Parts are designed to operate for certain lifetimes, after which they are replaced by new parts. Thus, as per the designers' instructions, the empirically determined lifetime is only used with the proper safety factor. Clearly, in either (a) or (b), the random appearance of an unexpected stress, due to some irregular-stress cycle cannot always be foreseen.
(c) 'Damage-related design' requires the periodic, nondestructive inspection of crack formation and growth. By using one of the nondestructive-testing methods, it is possible to perform such inspections to see if existing cracks are nearing their critical-crack size. There are mathematical methods for making reasonably-accurate predictions of crack growth between inspections and for evaluating the time left for further safe use of that part following its inspection.

Despite these crucial design rules, a fatigue problem exists, since failure is usually sudden and often occurs at a stress level much lower than the ultimate-stress level. Frequent attempts have been made to relate static-deformation data to fatigue, but experiments have indicated little direct connection between the fatigue limit and yield strength, ductility and other static-deformation properties. In metals, a well-known relation exists between the ultimate tensile strength, σ_{UTS}, and fatigue strength. For steels, for example, Chapetti et al. [46] relate the fatigue strength, σ_{eR}, to σ_{UTS} by:

$$\sigma_{eR} = 0.5\sigma_{UTS} \tag{7.25}$$

indicating that fatigue fracture is related to surface-crack initiation. Another empirical relation for metallic materials is:

$$S_{FL} = 140 + 0.25\sigma_{UTS} \tag{7.26}$$

In this equation, S_{FL} is the fatigue limit. To-date, the author does not know if these relations hold for ceramic materials and whether similar relations exist between monotonic stress and fatigue strength for all ceramics.

Attempts have been made to relate fatigue properties to static parameters, based on the similarity between monotonic and fatigue mechanisms, which control cyclic straining and plastic flow. This is somewhat problematic, because most ceramics are brittle and may be observed in their plastic state only at elevated temperatures. It would be more practical to discover a relation between static-yield stress and fatigue strength, since the yield stress in brittle materials coincides more or less with the fracture strength.

Various degrees of stress concentration are usual features of fabricated-structural parts. Stress concentration acts at certain locations for fatigue-crack initiation. Residual stresses are common in machine parts, but only compressive stresses do not contribute to crack initiation or propagation. Residual-tensile stress or tensile components of applied loads are to be avoided, in so far as possible while designing for fatigue, because of their damaging natures. It has been observed that, all other things being equal, increasing specimen dimensions results in decreased fatigue strength. One explanation is that smaller-sized specimens have fewer microcracks than larger ones. Fatigue performance designers generally correct for the dimensions of the various parts.

To summarize the requirements for fatigue design, the following is a list of some of the many factors that influence fatigue life and must be taken into account during the design process:

(a) The ceramic material: Usually materials with high melting points are preferred, since, in general, physical and mechanical properties are related to the melting point via the cohesive properties. Ceramics should be free of inclusion porosity and other voids that interrupt material continuity;
(b) The conditions of material processing: The cyclic properties, like the static ones, are dependent on the processing of the machine elements;
(c) The surface conditions of a ceramic part: Parts should be scratch-free and polished specimens perform better than unpolished ones;
(d) The geometry of machine elements: Machine-element geometry is an essential design parameter–length, width, thickness and diameter produce the size effects of materials. Furthermore, the radius (size and sharpness) and the transition radius from location to location are of utmost importance, since they act as stress-raisers. Designers must account for the characteristic effects of different types of radii;
(e) The effect of the environment: Environmental effects are important design considerations. Corrosive environments are detrimental. Some structures are often located in corrosive environments; appropriate design steps should be taken to select the most resistant materials for parts exposed to corrosive environments. Corrosive environments may accelerate the growth of fatigue cracks, which initiate at the surface and, therefore, reduce overall fatigue performance;
(f) The effect of temperature: Temperature is a highly significant factor to be kept in mind. Creep and fatigue deformation may act in concert at high temperatures.

To conclude this chapter, it may be stated that fatigue damage is one of most frequent causes of the breakdown of structural elements in service. All the previous sections explained the failure of machine parts due to fatigue during service and the means for improving their resistance by realizing the various acting factors. Every structural material experiences the problem of fatigue life and any differences that exist are dictated by overall material properties. Clearly, it is beyond the possibility of one textbook to consider each of the many materials used for machine parts and all the particular applications of these parts. Therefore, one or two examples of certain materials sufficed to serve as representatives of the wide spectrum of applications for structural use exposed to fatigue deformation.

References

1. Anderson TL (1991) Fracture mechanics: fundamentals and applications. CRC Press, Boca Raton
2. Bartolomé JF, Moya JS, Requena J, Lorca J, Anglada M (1998) J Am Ceram Soc 81:1502
3. Beachem CD (1968) Microscopic fracture processes. In: Liebowitz H (ed) Fracture: an advanced treatise, microscopic and macroscopic fundamentals, vol 1. Academic Press, New York
4. Cottrell AH (1956) Dislocations and plastic flow in crystals. Oxford University Press, London
5. Dauskardt RH, James MR, Porter JR, Ritchie RO (1992) J Am Ceram Soc 75:759
6. Dauskardt RH, Marshall DB, Ritchie RO (1990) J Am Ceram Soc 73:893
7. Felbeck DK, Atkins AG (1996) Strength and fracture of engineering solids, 2nd edn. Prentice-Hall, Cambridge
8. Grathwohl G, Liu T (1991) J Am Ceram Soc 74:3028
9. Griffith AA (1924) In: Proceedings of the 1st international congress of applied mechanics (Delft), pp. 55–63
10. Groner DJ (1994) Thesis presented to the Faculty of the School of Engineering of the Air Force Institute of Technology, Air University. Approved for public release; distribution unlimited. Wright-Patterson Air Force Base, Ohio
11. Hertzberg RW (1996) Deformation and fracture mechanics of engineering materials, 4th edn. John Wiley and Sons, New York
12. Kawakubo T, Komeya K (1987) J Am Ceram Soc 70:400
13. Kitagawa H, Takahashi S (1976) In: Proceedings of the second international conference on mechanical behavior of materials. American Society for Metals, Metals Park, p. 627
14. Kuang C, Lin J, Socie DF (1991) J Am Ceram Soc 74:1511
15. Lathabai S, Mai YW, Lawn BR (1989) J Am Ceram Soc 72:1761
16. Li LS, Wang G (2008) Introduction to micromechanics and nanomechanics. World Scientific, Hackensack
17. Liu SY, Chen IW (1991) J Am Ceram Soc 74:1197
18. Liu SY, Chen IW (1991) J Am Ceram Soc 74:1206
19. Liu SY, Chen IW (1992) J Am Ceram Soc 75:1191
20. McNulty JC, Zok FW (1999) Compos Sci Technol 59:1597
21. Meyers MA, Chawla KK (2009) Mechanical behavior of materials, 2nd edn. Cambridge University Press, Cambridge
22. Oki T, Yamamoto H, Osada T, Takahashi K (2013) J Powder Technol 2013:1
23. Pfeiffer W, Frey T (2006) J Eur Ceram Soc 26:2639

24. Rouby D, Reynaud P (1993) Compos Sci Technol 48:109
25. Samant AN, Dahotre NB (2008) Appl Phys A 90:493
26. Scherrer SS, Cattani-Lorente M, Vittecoq E, de Mestral F, Griggs JA, Anselm Wiskott HW (2011) Dental Mater 27:e28
27. Subramanian KN (1968) Phys Status Solidi 98:9
28. Suresh S (1998) Fatigue of materials. Cambridge University Press, Cambridge
29. Tetelman AS (1969) Fundamental aspects of stress corrosion cracking. National Association of Corrosion Engineers, Houston, p 446
30. Wada H, Takimoto N, Kondo I, Murase K, Kennedy TC (2004) Estimation of dynamic fracture toughness from circumferentially notched round-bar specimens. Conference: 2004 SEM X international congress and exposition on experimental and applied mechanics
31. Wu S, Cheng L, Zhang J, Zhang L, Luan X, Mei H, Fang P (2006) Mater Sci Eng A 435–436:412
32. Yamazaki Y, S.-Ichiro Kuga, T. Ypshida (2011), Acta Metall Sin (Engl Lett) 24:109
33. Zener C (1948) The Macro-mechanism of Fracture, in Fracturing of Metals, American Society of Metals. Metals Park, Ohio, p 3
34. Zenner H, Renner F (2002) Int J Fatigue 24:1255
35. Zhang Y, Lawn BR, Rekow ED, Van P, Thompson J (2004) Biomed Mater Res, Part B: Appl Biomater 71B:381
36. Zheng G, Zhao J, Jia C, Tian X, Dong Y, Zhou Y (2012) Int J Refract Metals Hard Mater 35:55
37. Zheng Y, Yan D, Gao L (1996) Fatigue Fract Eng Mater Struct 19:1009
38. Zhu D, Choi SR, Miller RA (2004) Surf Coat Technol 188–189:146
39. Zhu S, Kaneko Y, Ochi Y, Ogasawara T, Ishikawa T (2004) Int J Fatigue 26:1069
40. Zum Gahr K-H, Bundschuh W, Zimmerlin B (1993) Wear 162–164:269

Further References

41. Armstrong RW (2010) Eng Fract Mech 77:1348
42. Beachem CD (1972) Metal Trans 3:437
43. Beevers CJ, Cooke RJ, Knott JF, Ritchie RO (1975) Met Sci 9:119
44. Birnbaum HK (1979) Hydrogen related failure mechanisms in metal. In: Foroulis ZA (ed) Environment-sensitiveve fracture of engineering materials, TMS-AIME, p. 326
45. Broek D (1973) Eng Fract Mech 5:55
46. Chapetti MD (2011) Int J Fatigue 33:833
47. Cottrell AH (1958) Theory of brittle fracture in steel and similar metals. Trans Metall Soc AIME 212:192
48. Cottrell AH (1963) Proceedings of the royal society of London. Ser A Math Phys Sci 276:1
49. Cuddy JK, Bassim MN (1990) Mater Sci Eng A125:43
50. de Luna S, Fernández-Sáez J, Párez-Castellanos JL, Navarro C (2000) Int J Press Vessels Pip 77:691
51. Evans AG, Cannon RM (1986) Acta merall 34:761
52. Eftis J, Liebowitz H (1972) Int J Fract Mech 8:383
53. Gardner RN, Pollock TC, Wilsdorf HGF (1977) Mater Sci Eng 29:169
54. Govila RK, Hull D (1968) Acta Metall 16:45
55. Griffith AA (1920) Philos Trans R Soc Lond A221:163
56. Hayne Shumate E Jr (2009) The radius of curvature in the prime vertical. ITEA J 30:159
57. Inglis CE (1913) Proceedings. Inst Naval Architects 55:219
58. Kim W-J, Wolfenstine J, Ruano OA, Frommeyer G, Sherby O (1992) Met Trans A 23A:527
59. Kosmač T, Oblak Č, Jevnikar P, Funduk N, Marion L Ceramics Department, JozefStefan Institute. Jamova 39, SI-1000 Ljubljana, Slovenia
60. Laird C, Smith GC (1962) Philos Mag 7:847

61. Levkovitch V, Sievert R, Svendsen B (2005) Int J Fract 136:207
62. Lynch SP (1988) Acta Metall 36:2639
63. Lynch SP (1989) Metallography 23:147
64. Mao WG, Zhou YC, Yang L, Yu XH (2006) Mechanics of Materials 38:1118
65. Miller LE, Smith GC (1970) J Iron Steel Inst 208:998
66. Miranda P, Pajares A, Guiberteau F, Deng Y, Lawn BR (2003) Acta Materialia 51:4347
67. Muller C, Wagner L (1992) Mater and Manufacturing Processes 7:423
68. Murakami Y, Beretta S (1999) Extremes 2:123
69. Murakami Y, Endo M (1994) Fatigue 16:163
70. Ohtani H, McMahon CJ Jr (1976) Acta Metall 23:377
71. Orowan E (1946) Trans Inst Eng Shipbuilders Scotland 89:165
72. Orowan E (1949) Rep Prog Phys 12:185
73. Paris PC, Gomez M, Anderson WE (1961) Trend Eng 13:1219
74. Polakowski NH, Ripling EJ (1966) Strength and Structure of Engineering Materials, Prentice-Hall, Inc. Englewood
75. Pressouyre GM, Dollet J, Vieillard-Baron B (1982) Mémoires et Etudes Scientifiques de la Revue de Metallurgie 79:161
76. Pyttel B, Schwerdt D, Berger CH (2010) Procedia Engineering 2:1327
77. Qing-fen Li, Li Li, Er-bao Liu, Dong Liu, Xiu-fang Cui (2005) Scripta Materialia 53:309
78. Quadrini E (1989) Mater Chem Phys 21:437
79. Ritchie RO, Knott JF, Rice JR (1973) J Mech Phys Solids 21:395
80. Sarfarazi M, Ghosh SK (1987) Eng Fract Mech 27:257
81. Shea MM, Stoloff NS (1973) Mater Sci Eng 12:245
82. Smith E (1968) Int J Fract Mech 4:131
83. Smith JF, Reynolds JH, Southworth HN (1980) Acta Metall 28:1555
84. Soderholm K-J (2010) Dental Materials 26:e63
85. Sohncke L (1869) Ann Phys Lpz. 137:177
86. Stroh AN (1955) Proc R Soc Lond 223A:548
87. Stroh AN (1957) Adv Phys 6:418
88. Teirlinck D, Zok F, Embury JD, Ashby MF (1988) Acta metall 36:1213
89. Troiano AR (1960) Trans ASM 52:54
90. Vosikovsky O (1979) Eng Fract Mech 11:595
91. Zhang XJ, Armstrong RW, Irwin GR (1989) Metallurgical Trans A 20A:2862

Chapter 8
Fracture

Abstract In ceramics it is essential to consider all kinds of fractures that a material might experience during its service life time as a consequence of deformation. Fracture propensity is critical in ceramics which does not show elongation (plasticity) because failure can set in at deformations which basically are elastic (brittle ceramics). It is important to understand the theories of fracture, and relate them to the theoretical strength of materials. Among the important theories one can mention Griffith's theory on fracture, Orowan's fracture theory, and the dislocation theory of brittle fracture including the Stroh model of fracture. One of the most important parameters regarding fracture is toughness. Fracture toughness is the property that describes the ability of a material containing a crack to resist fracture and is one of the most important properties of any material for design applications. Related to fracture toughness is the term R-curve, which refers to fracture toughness that increases as a crack grows. Prediction of the effect of existing flaws in ceramics on fracture strength is the R-curve. Fracture toughness is an indicator for failure in ceramics and the R-curve expresses ceramic crack resistance. Another way to characterize a ceramic is by the energy absorption concept which is related to its fracture toughness. The J-integral as a fracture criterion is used to express the energy absorbed during crack extension. Fracture may occur in ceramics under static load, time dependent and cyclic deformation. Toughness can be improved by changing the course of crack, by crack tip shielding, crack bridging and crack healing. In ceramics undergoing transformation, transformation-toughening can improve the toughness.

8.1 Introduction

This chapter presents fracture resulting from all the types of deformation considered in the previous chapters (static, time-dependent and cyclic), but first, the atomic bonds forming materials are discussed. 'Atomic cohesion' is the bond between atoms that holds them together to form an aggregate that does not

J. Pelleg, *Mechanical Properties of Ceramics*, Solid Mechanics and Its Applications 213, DOI: 10.1007/978-3-319-04492-7_8,

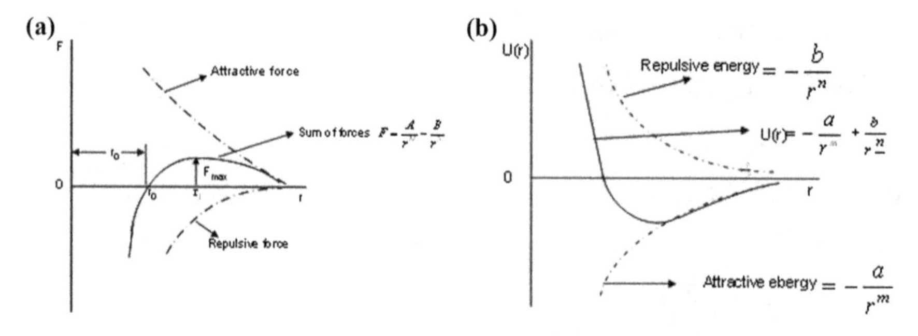

Fig. 8.1 The cohesion in a material, such as NaCl: **a** a plot of force versus distance and **b** the energy variation over distance. The *solid curves* are the sum of the attraction and the repulsion between two atoms

disintegrate under the normal conditions characteristic of that specific material. A brief look at the essentials of cohesion aids in the understanding of fracture, which occurs when a force of a certain magnitude is applied to the atomic bonds between atoms, causing the disintegration of a material. Those forces that hold the groups of atoms or molecules of a substance together are called 'bonds'. The formation of bonds between atoms is mainly due to their tendency to attain a minimum potential energy, thus reaching a stable state. In solid materials, it is usually assumed that two types of forces act between the atoms: (a) an attractive force, which keeps the atoms together, forcing them to form solids and (b) a repulsive force, which comes into play when a solid is compressed. Figure 8.1a illustrates the concept of cohesion based on the relation below:

In a material, such as NaCl, the overall cohesive energy may be expressed as:

$$F(r) = \frac{A}{r^M} - \frac{B}{r^N} \tag{8.1}$$

The first term is the attractive force, while the second term is the repulsive force. With decreasing distance between the atoms near the equilibrium position, the repulsive force (the second term in the relation) increases more rapidly than the first one and the exponent, N, must be greater than M. At the equilibrium distance, $r = r_0$ F(r) is zero, as seen in Fig. 8.1a. One can express Eq. (8.1) in terms of r_0 as:

$$\frac{A}{r_0^M} = \frac{B}{r_0^N} \tag{8.2}$$

The equilibrium distance is:

$$r_0 = \left(\frac{B}{A}\right)^{\frac{1}{N-M}} \tag{8.3}$$

Usually, the force is the negative derivative of the energy. Thus, integration of the force gives the cohesive energy as:

$$U(r) = \int F(r) = \int \left(\frac{A}{r^M} - \frac{B}{R^N}\right) dr \tag{8.4}$$

$$U(r) = -\frac{A}{r^{M-1}}\frac{1}{M-1} + \frac{B}{r^{N-1}}\frac{1}{N-1} + C \tag{8.5}$$

Equation (8.5) may be written as:

$$U(r) = -\frac{a}{r^m} + \frac{b}{r^n} + C \tag{8.6}$$

when U(r) = 0 and r goes to infinity, C = 0 and:

$$U(r) = -\frac{a}{r^m} + \frac{b}{r^n} \tag{8.7}$$

This relation is plotted in Fig. 8.1b, which gives the potential energy versus the distance.

Similarly, to the above, at r = r_0, the potential energy is a minimum and one can write:

$$\left(\frac{dU}{dr}\right)_{r=r_0} = 0 = \frac{am}{r_0^{m+1}} - \frac{bn}{r_0^{n+1}} \tag{8.8}$$

The cohesion above is for ionic solids; covalent and metallic bonds are different. In covalent bonds, electrons between atoms are shared, whereas in metallic solids, atoms of the same (or different) elements donate their valence electrons to form an electron gas throughout the space occupied by the atoms. Giving up their electrons to a common pool, known as an 'electron cloud' or 'electron gas', these atoms actually become positive (similar to positive ions). They are held together by forces similar to those of ionic bonds, but acting between ions and electrons. The electrostatic interaction between the positive ions and the electron gas holds metals together. Unlike other crystals, metals may be deformed without fracture, because the electron gas permits atoms to slide past one another, acting as a lubricant. In non-ductile materials, such as most ceramics, this is not possible and renders them brittle.

For further study of cohesive forces, refer to the literature. The main focus of this short introduction on cohesive energy and forces is on the shape of Fig. 8.1a, which is half a sinusoidal function (used to determine the theoretical cohesive strength of a crystal in Chap. 3, Sect. 3.3.3 via Eqs. (3.15–3.18). In Sect. 3.3, the critical amount of shear stress required to move adjacent atomic planes past one another was calculated, i.e., the energy per unit area involved in shearing two atomic layers from their equilibrium configuration. Calculations based on Fig. 3.1a can be most useful when fracturing a material. The critical shear stress calculated in Eq. (3.18) is ~G/6 (a more realistic value calculated for τ_{max} is ≈G/10–G/30). Even these refined values are ~2 orders of magnitude greater than the experimental values.

The calculation of the theoretical strength in Chap. 3 is based on the proper shear stress for ductile ceramics, some at RT, but the vast majority at high temperatures. For non-ductile materials, the normal stress, σ, rather than the shear stress, τ, is applicable. Applying a tensile stress normal to the planes, which separates two atomic planes, the theoretical cleavage stress may be evaluated in the same way as described in Sect. 3.3.3, but instead of considering the shear stress a normal stress is considered. One can derive an equation similar to Eqs. (3.17–3.18) in terms of the theoretical cleavage stress as:

$$\sigma_{max} = \sigma_0 = \frac{E}{2\pi} \cong \frac{E}{6} \tag{8.9}$$

with E being the Young's modulus. In polycrystalline ceramics, the values of E are in the range of 100–500 GPa. In alumina, for example, it is 460. A rough evaluation for alumina, according to Eq. (8.9), gives a theoretical strength of ~76.7 GPa. When using the refined value of E/10, the theoretical strength is ~46 GPa. This theoretical cleavage stress is about two orders larger than the experimentally observed strength.

The inevitable conclusion is that real crystals must contain defects, such as the dislocations suggested by Taylor, Orowan and Polanyi, which reduce mechanical strength or, more specifically, the resistance to slip when the applied stress reaches a critical value. Their postulate showed that shear is possible at much lower stresses than in a perfect crystal.

8.2 Fracture Types

Regardless of the kind of fracture (static, cyclic or creep), most types of failure by fracture are either brittle or ductile. The following figure illustrates the various types of fractures encountered in materials, some of which appear only in very ductile materials, especially in metals. For a complete picture of the possible types of fractured surfaces, even if they are not characteristics of ceramics, it is helpful to see them. Figure 8.2 schematically illustrates some types of fracture observed in materials, including metals such as steel or aluminum, and in very ductile and soft specimens (e.g., Pb, Au.). 'Fracture' means the separation of a body into two or more parts due to stress acting on it at temperatures below the melting point. 'Failure by fracture' consists of two steps: (a) crack formation and (b) crack propagation leading to complete fracture. In most ceramics, flaws in the form of pores or microcracks exist in the as-grown condition even without the application of a stress. The ability of a material to

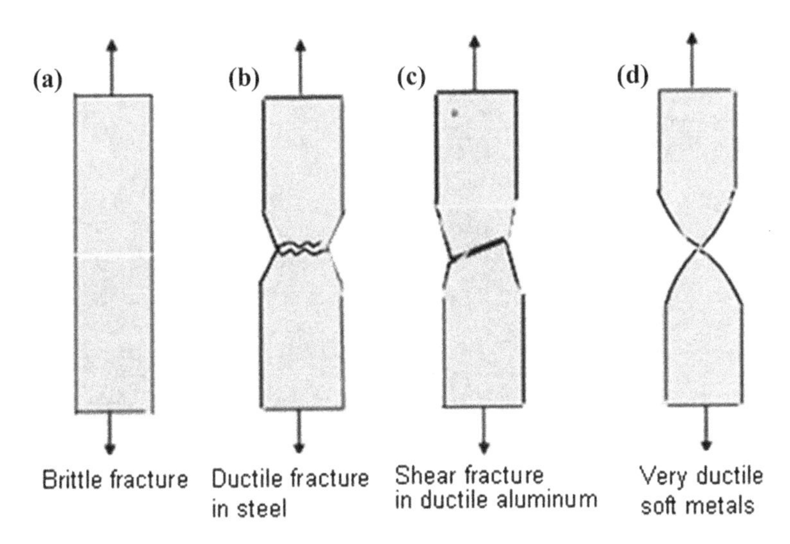

Fig. 8.2 A schematic illustration of some types of fractures observed in metals, such as steel or aluminum: **a** brittle fracture; **b** ductile fracture; **c** shear fracture; and **d** complete ductile fracture, also known as 'chisel point fracture'

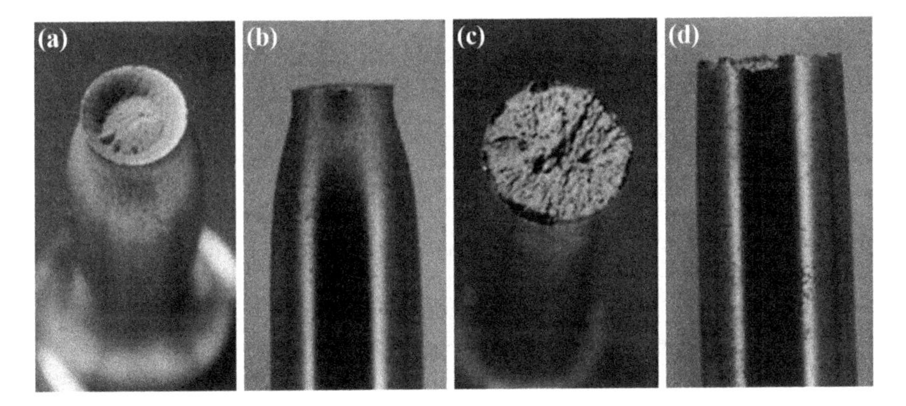

Fig. 8.3 Macroscopic appearance of ductile **a**, **b** and **c**, **d** brittle tensile fractures ([8], p. 102). With kind permission of ASM International. All rights reserved. http://www.asminternational.org

undergo complete or partial plastic deformation barely encompasses all the types of fractures that exist.

All the fracture types shown schematically in Fig. 8.2 may also be observed experimentally (see for example Pelleg). However, only two of these are illustrated in Fig. 8.3: the (a) classic cup-and-cone-type and (b) brittle fracture obtained by tensile testing to fracture. No necking occurs in brittle fracture and the percentage of elongation and reduction of the area are nearly zero. In this type of tensile fracture, the yield and tensile strengths are essentially identical.

Fig. 8.4 The river-pattern emitting from a pore on the fracture surface [56]. With kind permission of editorial office of IOP Science

8.3 Brittle Fracture

Since most ceramics are predominantly brittle at RT, it is important to mention some characteristic points:

(1) There is no gross, permanent deformation of the material;
(2) The surface of a brittle fracture tends to be perpendicular to the principal tensile stress, although other components of stress may be factors;
(3) Brittle fracture often occurs by cleavage fracture; the surface is characterized by flat normally grain-sized facets (in polycrystals);
(4) A river pattern consists of steps between parallel cleavages on parallel planes, that usually converge in the direction of the local crack propagation. This is mainly observed in metals, but also in ceramics (see Fig. 8.4);
(5) Characteristic crack advance markings frequently indicate where a fracture originated (fatigue fractures are a good example of this);
(6) The path that a crack follows depends on the material's structure. In polycrystalline materials, transgranular and intergranular cleavages are important. Cleavage clearly appears in SEM and other micrographs.

8.3.1 Theoretical Strength

Structural elements in brittle ceramics fail with little or no plastic deformation, often without warning. Since brittle fracture may lead to catastrophic results, it has been studied more intensively. Begin by assuming that a solid is perfect and that it breaks

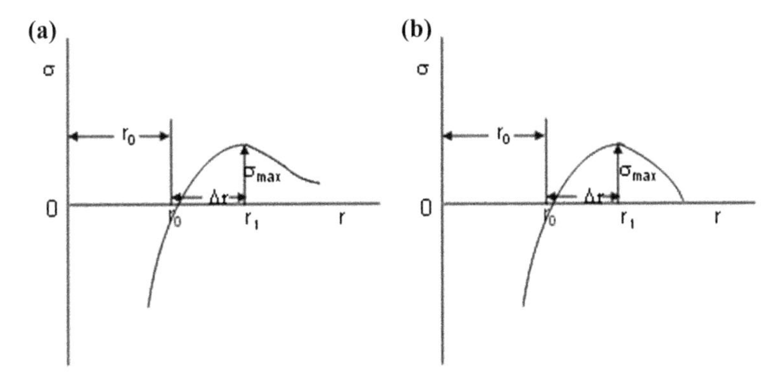

Fig. 8.5 The cohesive force (in terms of stress) as a function of the distance between atoms, representing (in this case) the stress needed to move an atom away from its neighbor: **a** the actual plot of stress versus distance and **b** presenting (**a**) as two equal halves

by separating along atomic planes. Differentiating energy versus distance results either in a force versus distance relation (as indicated in Fig. 8.1a or in a stress versus distance relation, because force is related to stress by the area. In Fig. 8.1a, r_0 represents the equilibrium separation between two atoms. To move an atom even a small distance by deformation requires a force (stress) which increases over the distance which the atom is being forced to move. A maximum force is required to bring an atom to distance r_1. Figure 8.5 is an illustration of the stress-distance variation.

One may now express the above statements in the following way. The sum of the areas equals:

$$\text{Sum of areas} = 2 \times (\Delta r \sigma_{max})/2 = \Delta r \sigma_{max} \tag{8.10}$$

This relation is a result of the assumption that each half-area may be approximated as a triangle. Thus, with two triangle-like areas, Eq. (8.10) is obtained. Now, $\Delta r = (r_1 - r_0)$ may be expressed in terms of strain, as:

$$\varepsilon = \frac{\Delta r}{r_0} \tag{8.11}$$

The energy is the stress multiplied by the distance, namely:

$$\text{energy} = \sigma_{max} 2(\Delta r) \tag{8.12}$$

Expressing Δr in terms of strain, taken from Eq. (8.11), and substituting into Eq. (8.12) results in:

$$\text{energy} = 2\sigma_{max} \varepsilon r_0 \tag{8.13}$$

Strain is given as:

$$\varepsilon = \frac{r_1 - r_0}{r_0} = \frac{\Delta r}{r_0} \tag{8.14}$$

Strain may also be expressed in terms of the Young's modulus, E ($\sigma = \varepsilon E$), to obtain:

$$\varepsilon = \frac{\sigma_{max}}{E} = \frac{\Delta r}{r_0} \tag{8.15}$$

Substituting this value of ε into Eq. (8.13), the energy may be expressed as:

$$energy = 2\sigma_{max}\frac{\sigma_{max}}{E}r_0 \tag{8.16}$$

Overcoming the theoretical cohesive strength of a material by fracturing creates two new surfaces and the surface energy of each is γ. The fracture energy is balanced by both energies of the two newly-formed surfaces. Thus, Eq. (8.16) may be written:

$$2\gamma = 2\sigma_{max}^2\frac{r_0}{E} \tag{8.17}$$

or

$$\sigma_{max} = \sqrt{\frac{\gamma E}{r_0}} \tag{8.18}$$

Equation (8.18) may be obtained more precisely by combining Eqs. (3.15b) and (3.17), as:

$$\tau = \tau_{max}\sin\frac{2\pi x}{a} \tag{8.19}$$

This is based on Fig. 3.27. The fracturing of a brittle solid requires work to create two new surfaces. Each of the two surfaces has a surface energy of γ. The fracture energy in a brittle solid, or rather the work done per unit area of surface, is the integral under the stress displacement curve and may be expressed

$$energy = \int_0^{a/2}\tau_{max}\sin\frac{2\pi x}{a}dx = \frac{\tau_{max}a}{\pi} \tag{8.20}$$

Equation (8.20) is equivalent to the formation of two surfaces or:

$$\tau_{\max} \frac{a}{\pi} = 2\gamma \cong \frac{\sigma_{\max} a}{\pi} \tag{8.21}$$

Note that instead of the shear stress, τ, the stress, σ, is used in line with the notation in Fig. 8.5 for cohesive stress. Thus, re-expressing Eq. (8.21), gives:

$$a = \frac{2\pi\gamma}{\sigma_{\max}} \tag{8.22}$$

Substituting for a from Eq. (8.21) into Eq. (3.17) and recalling that σ is used instead of τ, one obtains:

$$\sigma_{\max}^2 = \frac{E\gamma}{h} = \frac{E\gamma}{a} \tag{8.23}$$

In Eq. (8.23), the approximation from Sect. 3.3.3 was used (that h $\cong a$). Rewriting Eq. (8.23) by expressing $\sigma_{\max}$, one gets:

$$\sigma_{\max} = \sqrt{\frac{E\gamma}{a}} \tag{8.24}$$

Also note that in Fig. 8.5 the distance was given in terms of r_0, rather than a. Thus, by changing a to r, one goes from Eq. (8.24) to (8.18), in line with the notation in Fig. 8.5. Equations (8.18) or (8.24) provide a value (see 'theoretical strength' in Sect. 3.3.3) which is much higher than the strength actually observed. It has been suggested that, in reality, the calculated fracture stress needed to create two new surfaces does not apply except in cases of flawless, perfect brittle materials, such as whiskers (i.e., silica fibers), because various flaws, such as cracks, are usually present in engineering ceramics and are responsible for the lower-than-theoretical fracture strength. Giffith was among the first to explain the discrepancy between the theoretical and the actual strengths of materials (discussed later on).

8.3.2 Theories of Brittle Fracture

Brittle fracture in single crystals is related to the resolved normal stress on a cleavage plane. This concept is similar to that of slip in ductile materials, as formulated by Sohncke [51] (see a schematic illustration in Fig. 8.6.

Sohncke states that fracture occurs when a resolved normal stress reaches a critical value. Under tension, a fracture acquires low index planes in which the density of the atoms is high. These planes are known as 'cleavage planes' and the tensile stress is acting normal to them. Recall that Schmid's law for slip is

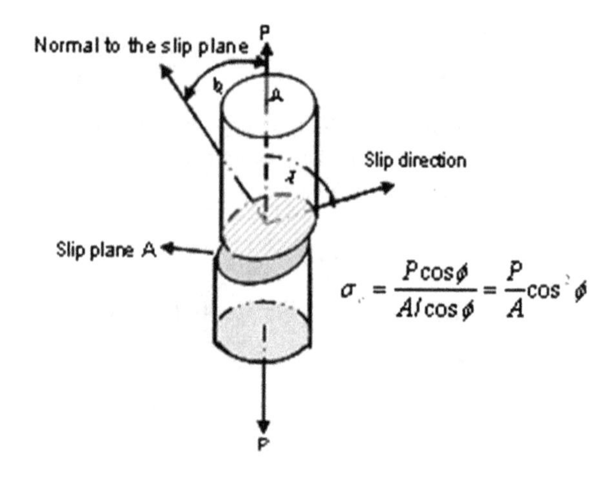

Fig. 8.6 A cleavage plane in single crystals related to normal stress, which is Pcosϕ

formulated in a similar manner (Sect. 4.6; Fig. 4.15b; relation 4.6), but it considers slip planes. The critical stress, σ_c, is given by:

$$\sigma_c = \frac{P\cos\phi}{A/\cos\phi} = \frac{P}{A}\cos^2\phi \tag{8.25}$$

as shown also in Fig. 8.6. Equation (8.25) was obtained as follows. The normal stress, which is the tensile stress acting on the cleavage plane (in the Fig. 8.6 it is the slip plane), is given in terms of the load, P, as Pcosϕ. The area on which normal stress is acting is A/cosϕ. Therefore, the load per area gives the stress (critical) as shown in Eq. (8.25). The difference between fracture strength and cohesive strength is a consequence of the inherent flaws in a material (mentioned above), which lowers fracture strength, as stated by Griffith [22].

8.3.3 Griffith's Theory on Fracture

The existence of flaws in the form of microcracks explains why the actual strength is lower than the theoretical strength. Crack formation requires energy to produce two new surfaces. Thus, Griffith's [22] approach to brittle failure is often called an 'energy balance theory'. Observe the schematic illustration in Fig. 8.7 characterizing a crack. Here, the crack is shown at the center, though it may be at the sides of the specimen (half-cracks).

In his model, Griffith [22] assumed that glass (a perfect, brittle material) contains small, flat (slot-like) cracks, which act as stress raisers. Together with Inglis' [29] theory of a pre-existing crack and its growth, Griffith [22] was able to show that theoretical, cohesive strength is reached locally at the crack tip. Crack growth is associated with strain-energy release during growth. (Griffith's [22] relation is derived below).

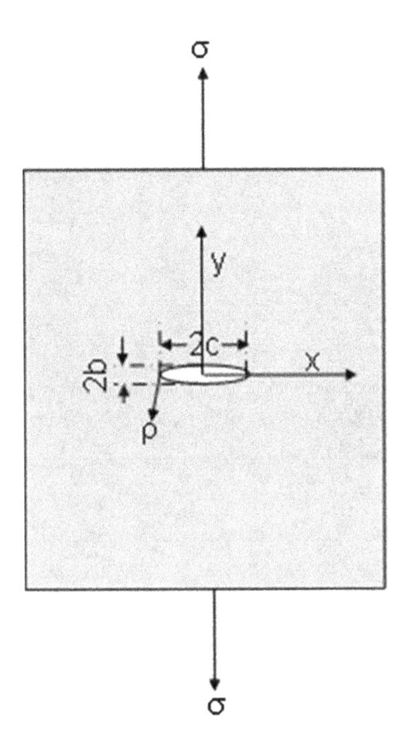

Fig. 8.7 The Griffith crack model for fracture. The flat, elliptical profile in the center of the plate represents a crack

Inglis' [29] description of the crack (Fig. 8.7 is a plate) assumes that the applied stress, σ_a, is magnified at the end of the ellipse along the x axis (the major axis). The expression of this ellipse in the configuration shown in Fig. 8.7 is:

$$\frac{x^2}{c^2} + \frac{y^2}{b^2} = 1 \tag{8.26}$$

The radius of curvature, ρ, is given (see, for example, Shumate, Jr. [50]) as:

$$\rho = \frac{b^2}{c} \tag{8.27}$$

The stress concentration at the edge of the crack (the end of the major axis), where c is half the major axis and b = half the minor axis (i.e., half the dimensions of the crack), is given as:

$$\sigma_{\max}(c,\ 0) = \sigma_a\left(1 + \frac{2c}{b}\right) = \sigma_a\left(1 + 2\left(\frac{c}{\rho}\right)^{1/2}\right) \tag{8.28}$$

The last term in Eq. (8.28) is a result of substituting for b from Eq. (8.27). When b ≪ c (e.g., a slot), Eq. (8.28) reduces to:

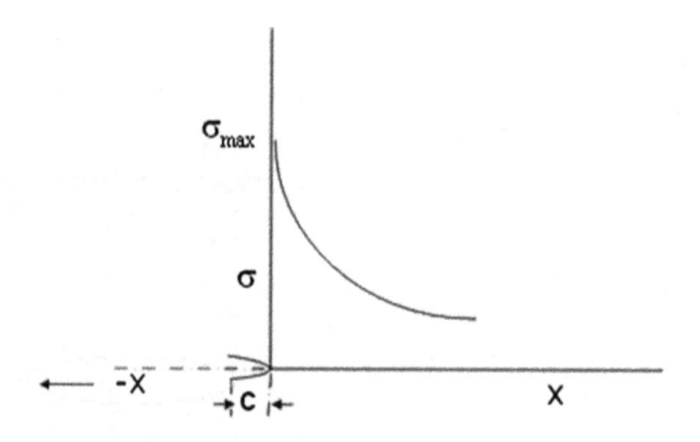

Fig. 8.8 Stress distribution away from the tip of a crack. Equation (8.29) relates $\sigma_{\max}$ to the crack by $\sigma_{\max} = \sigma_a 2\sqrt{c/\rho}$

$$\frac{\sigma_{\max}(c,0)}{\sigma_a} \approx \frac{2c}{b} = 2\left(\frac{c}{\rho}\right)^{1/2} \tag{8.29}$$

for circular holes, b = c. From Eq. (8.28), i.e., the central relation), one gets:

$$\frac{\sigma_{\max}(c,0)}{\sigma_a} \approx 3 \tag{8.30}$$

The stress distribution along x is shown in Fig. 8.8.

For a static-crack system, the total energy is the sum of three terms:

$$\mathrm{U} = (-\mathrm{W_L} + \mathrm{U_E}) + \mathrm{U_S} \tag{8.31}$$

U_E is the elastic energy, U_S is the surface energy and W_L is the mechanical energy of the body or, alternatively, the amount of work done by the applied loads. A decrease in strain energy results from the formation of a crack. Thermodynamic equilibrium is reached when:

$$\mathrm{dU/dc} = 0 \tag{8.32}$$

The mechanical energy of a body under constant, applied force during crack formation is:

$$\mathrm{W_L} = 2\mathrm{U_E} \quad \text{(at constant load)} \tag{8.33}$$

The Inglis' solution [29] for stress and strain fields around a sharp crack is:

$$U_E = \pi c^2 \frac{\sigma_L^2}{E} \quad \text{(plane stress for a thin plate)} \tag{8.34}$$

For a thick plate, the above should be modified to:

$$U_E = \pi(1 - \nu^2)c^2 \frac{\sigma_L^2}{E} \qquad \text{(plane stress for a thick plate)} \tag{8.35}$$

The surface energy of the crack in the plate is:

$$U_S = 4c\gamma \tag{8.36}$$

γ is the surface energy per unit area of the crack. In Eq. (8.36), the relation applies to a unit width of the crack. The total energy for the plane stress case is the sum of Eqs. (8.34) and (8.36), namely:

$$U = -\pi c^2 \frac{\sigma_L^2}{E} + 4c\gamma \tag{8.37}$$

The minus sign is a consequence of the fact that the growth of a crack releases strain energy. The tensile stress acts normally to the surface of the crack. Following Griffith's [22] concept by using the equilibrium condition in Eq. (8.32) for Eq. (8.37), an expression for the constant load and plane stress conditions is obtained as:

$$\sigma_L = \sqrt{\frac{2\gamma E}{\pi c}} \tag{8.38}$$

For constant load and plane strain conditions, one obtains:

$$\sigma_L = \sqrt{\frac{2\gamma E}{(1 - \nu^2)\pi c}} \tag{8.39}$$

The change in stress over distance in the vicinity of the crack tip in Fig. 8.7 is illustrated schematically in Fig. 8.8.

Despite the great impact that Griffith's theory had on the understanding of fracture in truly brittle materials, such as glass, some problems remain unresolved. Why do large cracks tend to propagate more easily than small cracks? Is this because dimensional change also modifies the radius of curvature? Where do cracks originate? Griffith's [22] theory (discussed above in regard to perfect, brittle materials, such as glass) is entirely satisfactory for amorphous materials, but cannot, in principle, be extended to metals or ductile materials, due to the different nature of plastic deformation in ductile materials.

8.3.4 *Orowan's Fracture Theory*

Orowan considered Griffith's [22] theory. If no plastic deformation occurs, the radius of curvature at the tip of a crack must be equal to the atomic radius, "a", representing the sharpest crack. By using the stress-concentration factor at the tip of an elliptical crack, Orowan arrived at a similar relation to Griffith's [22], in which the stress-concentration factor is defined as K_t, given as:

$$K_t = \frac{\sigma_{\max}}{\sigma_a} \tag{8.40}$$

from Eq. (8.28):

$$\sigma_{\max}(c,0) = \sigma_a\left(1 + \frac{2c}{b}\right) \tag{8.28}$$

and for b ≪ c (e.g., a slot), Eq. (8.28) reduces to Eq. (8.29) or is expressed in terms of K_t as:

$$K_t = \frac{\sigma_{\max}(c,0)}{\sigma_a} \approx \frac{2c}{b} \tag{8.41}$$

This relation is a consequence of b ≪ c and, thus, 1 can be neglected. When the radius of curvature ρ is of atomic dimensions (i.e., "a"), Eq. (8.27) may be written as Eq. (8.27a):

$$a = \frac{b^2}{c} \tag{8.27a}$$

Equation (8.41), in terms of "a" after substituting for b becomes:

$$K_t = \frac{2c}{\sqrt{ac}} \tag{8.42}$$

If it is assumed that the atomic radius is "a", then the fracture stress at the crack tip becomes equal to the theoretical stress (as calculated in Chap. 3, Sect. 3.3.3, Eq. (3.18)) and is approximately:

$$\tau_{\max} = \tau_0 \sim \frac{G}{2\pi} \sim \frac{G}{6} \tag{3.18}$$

or in terms of theoretical stress, $\sigma_T = \sigma_{\max}$:

$$\sigma_T = \frac{E}{10} - \frac{E}{20}$$

Then, with the value of ~E/10 (i.e., the theoretical stress) replacing $\sigma_{\max}$ (c, 0) in Eq. (8.41) and by using K_t from Eq. (8.42), one obtains for σ_a:

$$\sigma_a = \left(\frac{E}{20}\right)\sqrt{\frac{a}{c}} \tag{8.43}$$

One of the questions that Eq. (8.43) poses is whether a crack of the size obtained by the calculations in Eq. (8.43) really exists in glass. (For σ_T, an experimental strength of ~0.01–0.1 is often observed and with stress-concentration factors on the order of 10–100 crack sizes, 100–10,000 Å have been calculated). Whether such cracks are present in freshly-formed glass remains an unanswered question in the theories of both Griffith [22] and Orowan [41]. Since glass is easily damaged, mechanically or during handling, it is likely that, in most commercial glasses, surface defects will result. An indication of such a possibility is the well-known fact that hydrofluoric acid (HF) glass etching results in higher strength; such etching removes surface flaws and, therefore, increased strength is observed. Etching away the surface of the glass produces an effect known as the 'Joffé's effect'.

A definite size effect exists in brittle materials, such as glass, and their strength depends on the volume of the material. This volume-dependence is simply explained by the fact that the probability of finding proper-sized cracks increases with volume.

8.3.5 The Dislocation Theory of Brittle Fracture

8.3.5.1 Introduction

Although this is a discussion on brittle materials, such as ceramics (glass is a perfect, brittle material), several researchers have developed theories of fracture based on dislocation models. More specifically, the shear stress created by dislocation pile-ups at some obstacle, specifically grain boundaries in polycrystalline materials, reaches a sufficient value for crack formation. The following illustrates Stroh's [52] basic concept of microcrack formation, ultimately leading to the occurrence of fracture in brittle materials.

8.3.5.2 The Stroh Model of Fracture

Similarly to Zener's model [9] of microcrack formation at a pile up of edge dislocations, Stroh [52] developed a theory of fracture based on the concept of cracks initiated by the stress concentration of a dislocation pile-up. For brittle materials in which crack growth is not damped-out by plastic flow, Stroh calculated that the conditions for crack initiation may be given by:

$$\tau = \sqrt{\frac{3\pi G\gamma}{8(1-\nu)L}} \tag{8.44}$$

τ is resolved shear stress acting on the pile-up, γ is the surface energy of the cleavage plane and L is the distance of the pile-up. Another expression for shear stress, τ_s is given as:

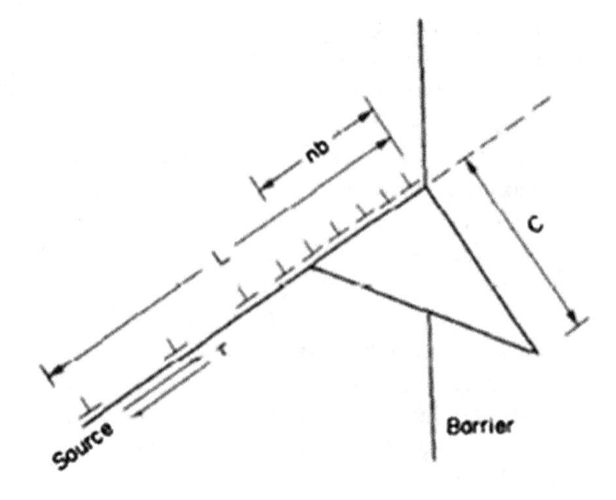

Fig. 8.9 Nucleation of a wedge crack due to piled-up dislocations on a slip plane [47]. With kind permission of Elsevier

$$\tau_s \approx \tau_i + \left(\frac{2\gamma_s}{nb}\right) \tag{8.45}$$

τ_i is the lattice-friction stress in the slip plane. In this case, the obstacle is grain boundary, d, which is taken into account by Stroh [52], as seen in Eq. (8.45a):

$$\tau_{eff} = \tau_y - \tau_i\sqrt{\frac{E\pi\gamma}{4(1-v^2)d}} \tag{8.45a}$$

Stress, τ_s, is created by the internal pile-up of n dislocations. τ_{eff} is an effective stress and τ_y is the yield stress. γ is the surface energy per unit area of the plane, as indicated earlier, and d/2 is the length of the dislocation pile-up. One illustration of Stroh's [52] concept is shown below:

In Fig. 8.9, one sees a 2D crack dislocation with a giant Burgers vector, nb, of length c. Stroh's [52] dislocation model for spontaneous microcrack formation, calculates the elastic energy associated with the wedge deformation of length, c, extending to a barrier by means of:

$$W_e = \frac{Gn^2b^2}{4\pi(1-v)}\ln\frac{4R}{c} + 2\gamma_s c \tag{8.46}$$

Some of these symbols represent their conventional usages. R is the bounding radius in the stress field. The surface energy term, $2\gamma_s c$, is added to obtain the total energy of the system. Differentiating Eq. (8.46), the critical length, c_{min}, may be found:

$$\frac{\partial W_s}{\partial c} = 0 \tag{8.46a}$$

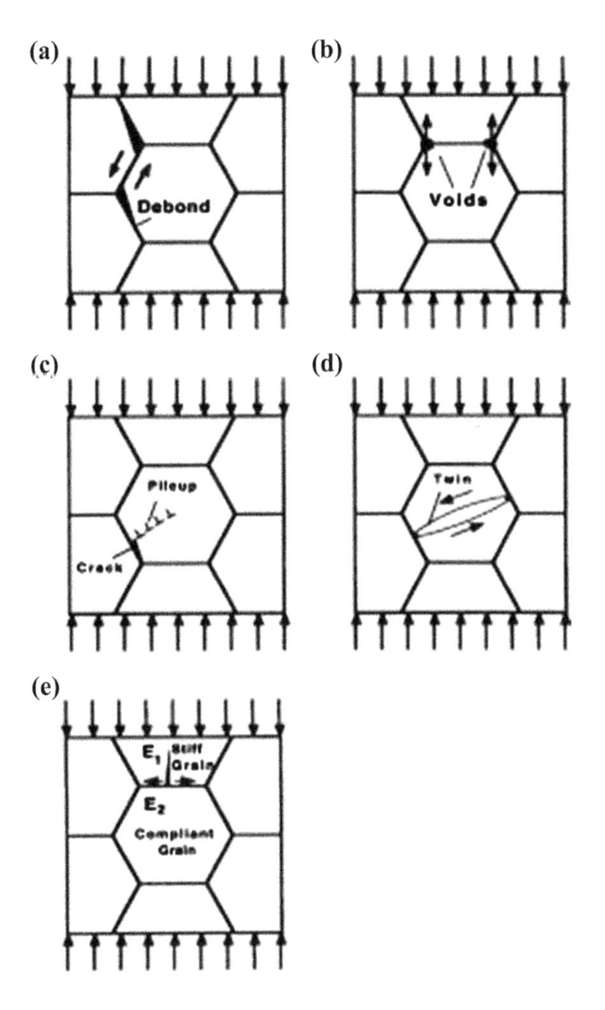

Fig. 8.10 Schematic overview of principal damage initiation mechanisms in SiC: **a** grain boundary debonds and voids; **b** foreign particles, such as inclusions; **c** dislocation pileups, leading to Zener-Stroh cracks; **d** twins and stacking faults; **e** dilatant crack produced by elastic anisotropy [49]. With kind permission of Elsevier

$$c_{\min} = G\frac{n^2b^2}{4\pi(1-v)}\frac{1}{2\gamma_s} \tag{8.47}$$

In polycrystalline solids, typical values [47] of b, G, v and γ_s are: b = 2×10^{-8} cm; G = 1012 dynes/cm^2; $v = 1/3$ and γ_s = 103 dynes/cm^2, respectively, which yields $c_{min} = 2.4 \times 10^{-8}$ for c_{min}. To obtain the c_{min} for ceramics, the specific values of G (or E) should be used for the surface energy and v, rather than the values indicated.

Probably, Stroh's [52] concept may be applied to semi-brittle materials, but its use for typically brittle materials is questionable. Thus, Low [39] questions the validity of Stroh's [52] concept for various reasons (see his work for a detailed discussion), among them the absence of observations of large pile-ups by etch pit

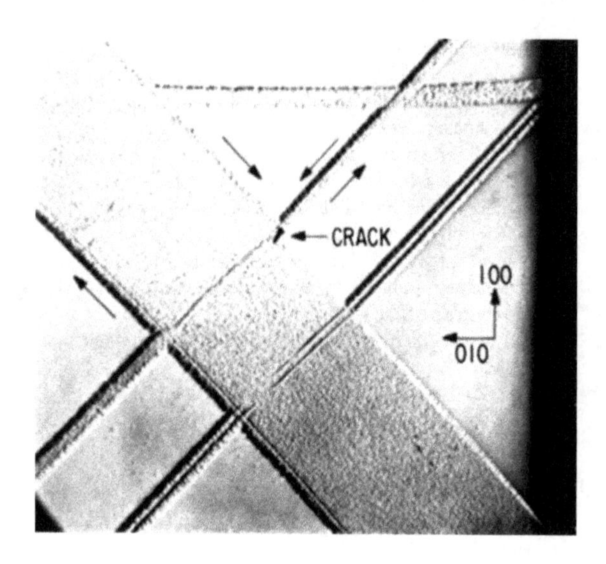

Fig. 8.11 Crack nucleation at slip band intersection in MgO crystal [39]. With kind permission of Elsevier

or TEM at the vicinity of the yield stress. Furthermore, Low [39] claims that theoretical shear strength must be reached in order to produce a pile-up large enough to exceed the theoretical cleavage strength. Nevertheless, many have applied the Stroh [52] concept to various materials, such as MgO (Johnston [31]) or SiC (e.g., Park [43], Shih [49]). Johnston's [31] Fig. 2 resembles Fig. 8.9, which represents the applied stress to crack initiation in MgO. Shih [49], in his publication on fracture criteria in ceramics, indicates that there are a number of microstructural mechanisms that have been identified as being the most likely sites for failure initiation during compressive loading. These are shown schematically in Fig. 8.10. Note the mechanism of pile-ups for fracture in (c), due to grain-boundary obstacles. Basically, Fig. 8.10c resembles Fig. 8.9. Thus, pile-ups are formed when they are stopped by an obstacle blocking their motion within their slip planes. A high shear stress is generated at the leading dislocations of the pile-ups, which may be related to microcrack nucleation and coalescence, for example at grain boundaries, resulting in intergranular fracture. Stroh [52] derived an expression for the minimum stress to nucleate at the end of a sliding interface:

$$\tau_{\min} = \left(\frac{12\mu\gamma_s}{\pi L}\right) \tag{8.48}$$

(In 8.48) $\mu(\equiv G)$ is the shear modulus, γ_s the appropriate surface energy and L is the length of the sliding interface. Low [39] reproduced the Johnston's [31] figure, illustrating crack nucleation in MgO at a slip-band intersection (shown in Fig. 8.11). Figure 8.11 was obtained by tension at RT. It indicates that a crack formed at a slip-band intersection and, once formed, could be prevented from propagating to produce complete fracture if it intersected the slip band. A crack can be nucleated at the intersection of two slip bands. Cracks of this type initially

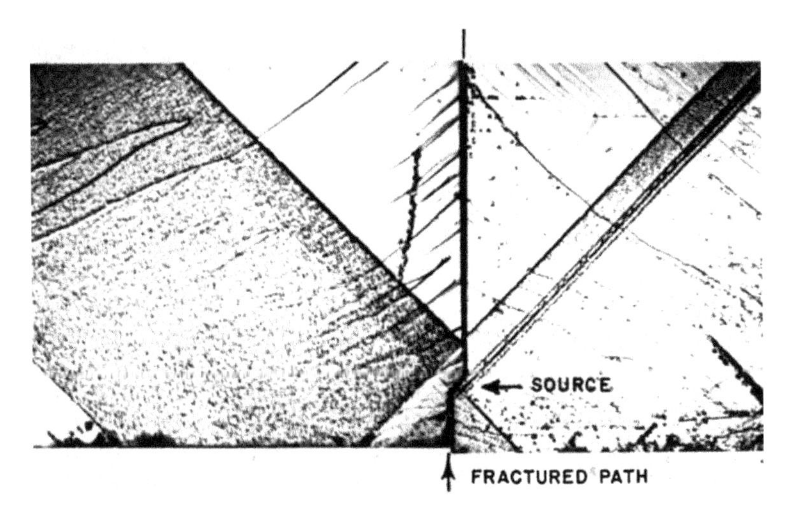

Fig. 8.12 Crack nucleated at (110) slip band intersection in MgO and propagated on (001) cleavage plane. x 225 [39]. With kind permission of Elsevier

Fig. 8.13 Microcracks in MgO stabilized by slip bands. x 150 [39]. With kind permission of Elsevier

propagate along a (110) plane (the slip plane), but eventually switch to the (100) plane normal to the tension axis to produce complete fracture, as shown in Fig. 8.12 at RT. If double slip is prevented and the crystal deformed in single slip

Fig. 8.14 Microcracks at slip bands blocked by grain boundary in MgO bicrystal [39]. With kind permission of Elsevier

only, then ductile behavior is observed and the crystals remain unbroken after elongations of up to 8 %. Examples of cracks stabilized by intersection with slip bands are seen in Fig. 8.13. This condition is produced by sprinkling the surface of the crystals with silicon carbide abrasive before testing, to produce many surface sources for slip bands. The most brittle condition is that in which only two intersecting slip bands form and a crack nucleates at their intersection, immediately propagating to complete fracture (Fig. 8.12). Low [39] indicates that Johnston et al. [31] also investigated the role of grain boundaries in crack nucleation by loading bicrystals of magnesium oxide in such a way that slip bands originating in the crystals impinged on the boundary.

The boundary misorientation was varied systematically and the probability of crack nucleation at the boundary, due to the blocking of the slip band, was found to depend on the degree of misorientation. For boundaries of 'medium' misorientation (5–20° tilt plus twist), screw bands penetrated the boundary, but edge bands were blocked and small cleavage cracks nucleated in the second grain in the manner envisaged by Zener [9] and Stroh [52]. These cracks may be seen at a number of points in Fig. 8.13. In the case of larger angle grain boundaries (tilt plus twist), cracks nucleated in the boundary itself, presumably because the cohesive strength of the large misorientation was less than the cleavage strength of the second grain. In all the cases, cracking was observed only at blocked edge dislocation bands. Screw dislocation bands either terminated at the boundary or penetrated into the adjacent grain without crack nucleation (see Fig. 8.14).

It may be summarized that Stroh [52] assumed that the dislocation sources in the grain adjacent to the pile-up are locked by Cottrell atmospheres and, thus,

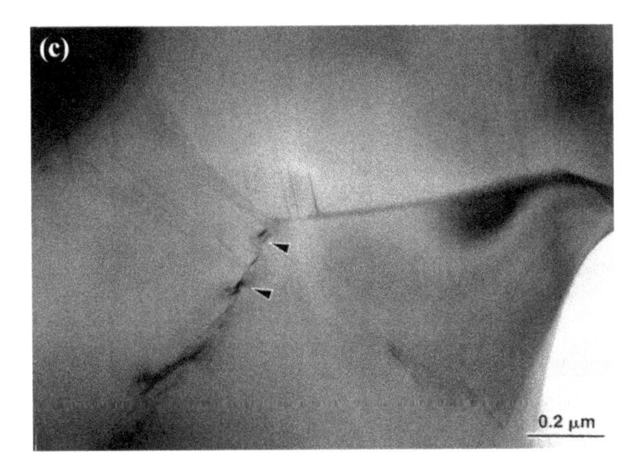

Fig. 8.15 TEM micrograph of grain boundary, indicated by *arrows*, showing **c** intergranular inclusion in isolated region [49]. With kind permission of Elsevier

prevented from relieving the stress concentration by slip. This mechanism operates only during yielding, when the density of the operative sources is low and the few sources that do operate may be expected to form a large pile-up in a short time. The probability of brittle fracture is the same as the probability that dislocations near a piled-up group will not be released by yielding in the next grain.

8.3.6 Fracture Toughness

8.3.6.1 Introduction

In the Sect. 8.2, one of the dislocation mechanisms was discussed–the pile-up concept–as the origin of brittle fracture due to the stress concentration occurring at the leading dislocation in the pile-up. However, in materials that is entirely brittle, as are most ceramics at RT and low temperatures, plastic deformation by dislocation motion does not occur or occurs to such a limited extent that cracks are sharp up to the atomic level. In order to understand the fracture behavior of ceramic materials, it is first necessary to understand the fracture mechanisms of materials that are entirely brittle. In such materials, the mechanism of fracture is associated with various flaws inherent or intentionally added to the ceramics. A list of most of the flaws that may induce brittle failure by crack formation is given below, as indicated in Fig. 8.10:

(a) Flaws formed by debonding along grain boundaries. These occur due to the presence of ellipsoidal defects, either in the form of foreign atoms or second phases. Tensile stresses are formed, triggering crack formation; the cracks then become initiating sites for intergranular fracture.

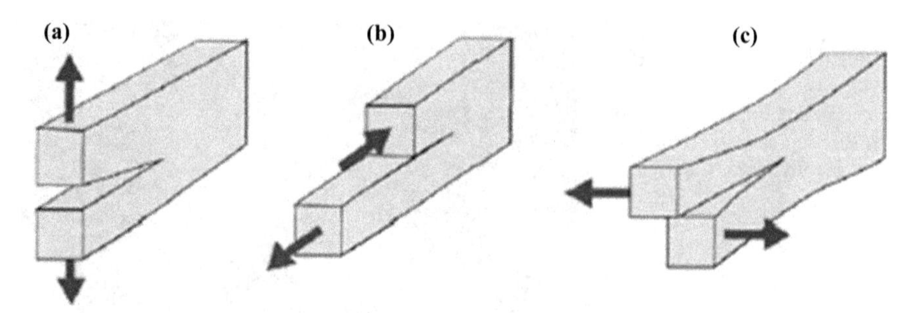

Fig. 8.16 Different failure modes: **a** Represents mode-I (tensile force), **b** represents mode-II (shear force) and **c** represents mode-III (torsional force) [6]. With kind permission of Elsevier

(b) Voids and various inclusions, which become initiation sites for failure in ceramics. Such voids are always present in conventional and imperfect processing techniques. Inclusions are often located at grain boundaries. Figure 8.15 shows the graphite inclusions of second-phase particles in SiC. Transgranular microcracks develop.
(c) Stress concentration development at piles-ups, according to the Zener-Stroh [9] model (see the Sect. 8.2). High stresses are generated at the leading dislocation in a pile-up, which may be dissipated by the formation of microcracks along the grain boundaries, resulting in intergranular fracture.
(d) Stress concentration due to twinning and stacking faults. The induced tension stress initiates and propagates a dilatant crack leading to transgranular fracture.
(e) Elastic anisotropy effect. Different grains have different stiffnesses due to elastic anisotropy. Under a compressive applied traction, internal tensile stresses, perpendicular to the loading direction, may develop. Such stresses are higher along the grain boundaries and inside the stiffest grains. It has been shown by Shih et al. [49] that high stress concentrations may be achieved at triple points subjected to compressive stresses, where transgranular microcracks develop.

This list of flaws shows that, in all cases at ambient temperatures, fracture is usually initiated via crack propagation by cracks present in the ceramics. The failure strength of ceramics is strongly dependent on the stress state and lateral confinement.

8.3.6.2 Microcrack Growth and Crack Propagation

For practical purposes, the 'fracture toughness' of a material is of interest; this is the property which describes the ability of a material containing a crack to resist brittle fracture and is one of the most important properties of any material for

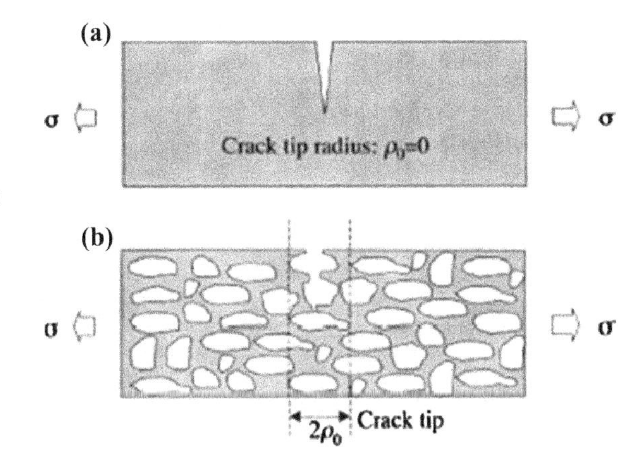

Fig. 8.17 Schematic representation of a crack propagated in **a** a dense ceramic material and **b** a porous ceramic material, showing a sharp crack-tip in the dense material and a blunt crack-tip in the porous material [15]. With kind permission of Elsevier

design applications. The linear-elastic fracture toughness of a material is determined by the stress-intensity factor, K_{Ic}. The subscript, Ic, denotes a mode-I crack opening under a normal tensile stress perpendicular to the crack.

The stress-to-fracture in conventional tests, such as tensile tests, is not a safe guide for the prevention of fracture resulting from crack propagation in structures used in service, since crack growth to catastrophic dimensions might often occur even below the tensile yield stress of the material. A critical-stress intensity factor, Kc, is a measure of fracture toughness. The fracture resistance of a material is known as its 'fracture toughness', which generally depends on: temperature, environment, loading rate, the composition of the material and its microstructure, together with the geometric effects of the tip. The existence of flaws, such as pores and microcracks, pose a special problem for the field of ceramic technology. The subject of fracture toughness is generally related to the amount of stress necessary to propagate a preexisting crack. Various defects commonly exist in materials and may develop to failure by fracture. The I subscript in K_{Ic} is used to denote the crack-opening mode, as indicated in the various types in Fig. 8.16.

The stress intensity factor is given in Eq. (8.49):

$$K_{Ic} = \sigma\sqrt{\pi a B} \tag{8.49}$$

As mentioned earlier, this stress-intensity factor is a function of loading, crack size and structural geometry. σ is the applied stress, a is the crack length and B is a dimensionless factor, which depends on specimen geometry. Thus, fracture toughness is the resistance of a material to failure from fracture initiated by a preexisting crack. The occurrence of cracks in structural components indicates a certain threat to their reliable operation, because these cracks can grow during service and reach critical sizes, leading to fracture. For a crack to grow, the stress at the crack tip must be greater than the strength of the material. The cohesive bond in the vicinity (front) of the crack is thus broken. However, this, by itself, is not a sufficient criterion for the growth of a crack to failure. An additional

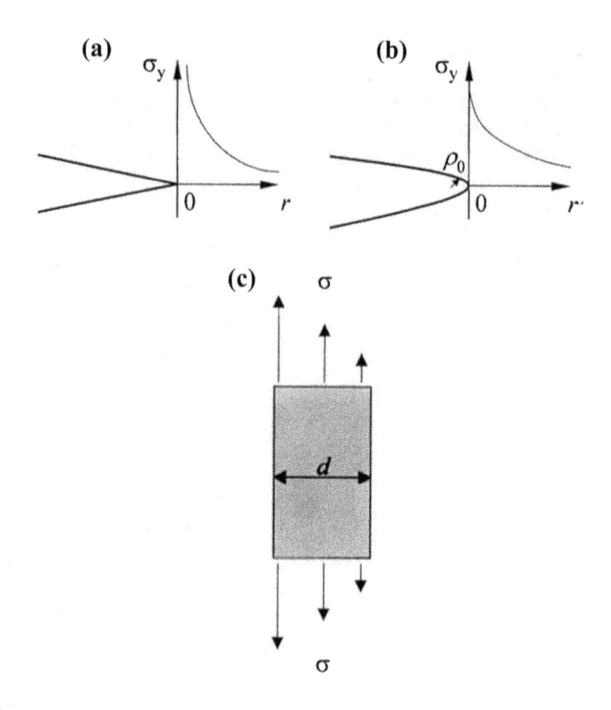

Fig. 8.18 Schematic representation of **a** a sharp crack, **b** a blunt crack with a root radius of ρ_0 and **c** a grain suffering a non-uniform tensile stress [15]. With kind permission of Elsevier

requirement is the release of energy that must occur when a crack grows to a size that forms two new crack surfaces. Both of these prerequisites are essential. On the one hand, if the force introduced into the system is of sufficient magnitude to break the bonds, but there is not enough energy to form new surfaces, the crack will not grow. On the other hand, if insufficient force is introduced into the system, such that the atomic bonds cannot be broken, even though there is enough energy to form new surfaces, the crack will also not grow. The geometry of a crack, or more precisely its sharpness, is a decisive factor in establishing the magnitude of the stress-concentration factor which develops at the crack tip. At the tip of a sharp crack, this factor requires only sufficient energy to form new surfaces, since the magnitude of the stress at the tip is of the order of the theoretical strength of the material (its cohesive stress). Therefore, a relatively small load will provide the necessary conditions for crack propagation. When the crack tip is dull (i.e., blunt), it is relatively easy to provide the energy for the formation of two new surfaces, but the stress is insufficient at the crack tip, because the stress concentration is much lower than that of a sharp crack. In this case, higher loads are needed to provide the required stress for crack growth. Thus, crack-tip blunting increases the fracture toughness of ceramics. Porous ceramics favor crack-tip blunting over the joining of pores and cracks (see Fig. 8.17).

Figure 8.18 schematically illustrates a sharp tip and a blunt one. From Eqs. (8.50) and (8.51), it may be seen that stress decreases with the distance away

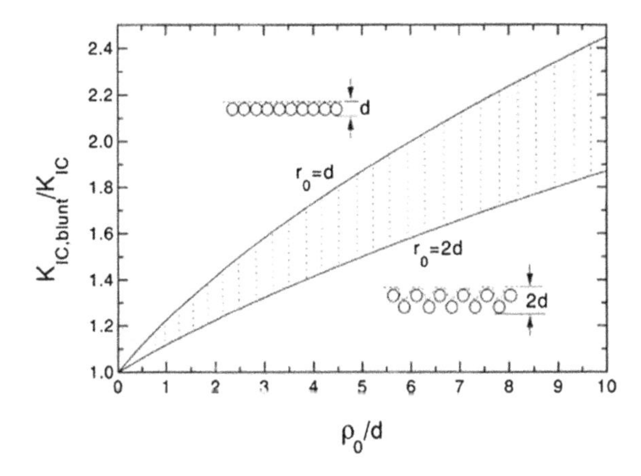

Fig. 8.19 Relationship between the toughness ratio of a blunt crack to a sharp crack and the ratio of the blunt crack-tip radius to the effective grain width, where the hatched region represents the results for different characteristic zone size. The *dashed lines* in the insets are the crack front of different grain arrangement [15]. With kind permission of Elsevier

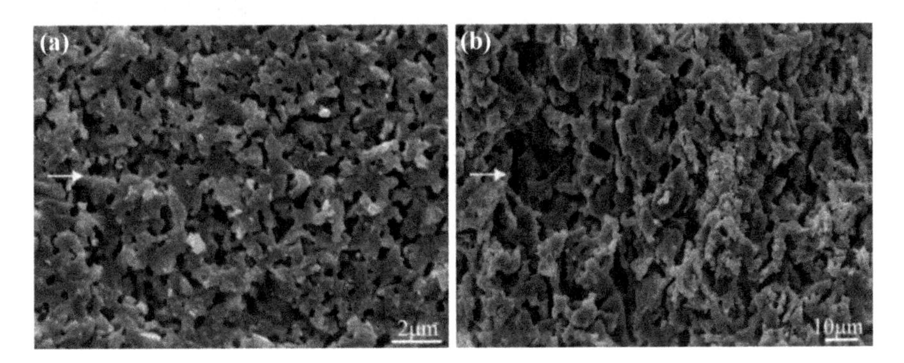

Fig. 8.20 SEM micrographs of grain and pore morphology in **a** FSC porous specimen and **b** CSC porous specimen, where the *arrows* represent crack propagation direction during strength and toughness tests [15]. With kind permission of Elsevier

from the crack tip. This implies that the stress applied to a grain at the crack tip is not a constant, as shown in Fig. 8.18c. However, the grain does not fracture immediately; even the stress at some positions exceeds its fracture strength.

$$\sigma_y = \frac{K_I}{\sqrt{2\pi r}} \qquad \text{for a sharp crack} \tag{8.50}$$

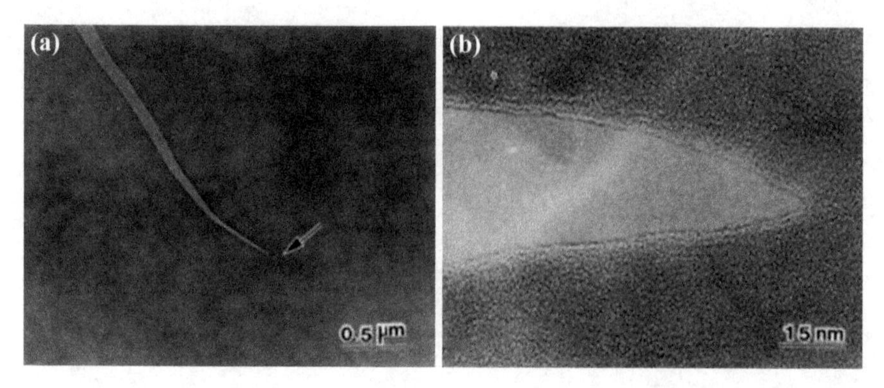

Fig. 8.21 Crack geometry of thin silica glass film at a low magnification and b high magnification. showing crack tip [12]. With kind permission of John Wiley and Sons

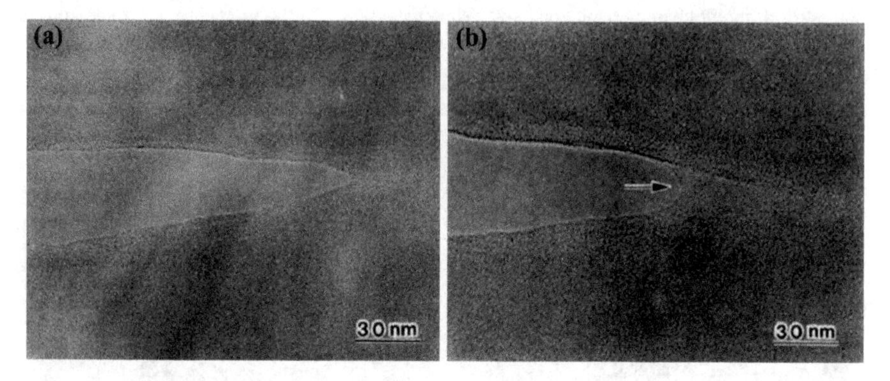

Fig. 8.22 Blunting of crack tip a immediately after crack introduction and b after film was soaked in water at 90 °C for 7 d [12]. With kind permission of John Wiley and Sons

and

$$\sigma_{y,\,blunt}\frac{K_{I,\,blunt}}{\sqrt{2\pi r}}\frac{1+\rho_0/r}{\left(1+\rho_0/2r\right)^{3/2}} \qquad \text{for a blunt crack} \tag{8.51}$$

Equation (8.50) is the same as Eq. (8.49). The variation of the toughness ratio of a blunt crack to a sharp crack and the ratio of the blunt crack-tip radius to the effective grain width are illustrated in Fig. 8.19. The grain and pore morphologies of the two types of porous SiC ceramics are shown in Fig. 8.20. Note that the specimen labeled FSC has fine pore structures, while the pores in the CSC specimen are large and have definite alignment. Both the FSC and CSC specimens exhibited trans-granular fractures, regardless of their different compositions. The crack-tip blunting reinforcement in porous ceramics, based on a grain fracture model, shows that crack

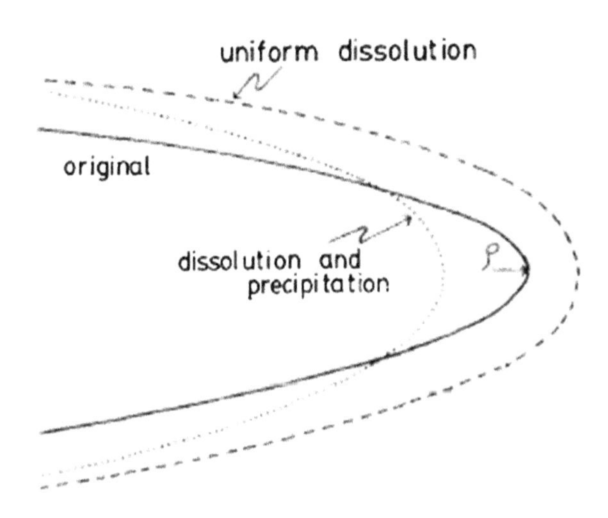

Fig. 8.23 Diagram of two crack blunting mechanisms [12]. With kind permission of John Wiley and Sons

tip blunting increases the fracture toughness of porous ceramics to an extent depending on the pore morphology. From the above, may infer that changing the tip geometry from a sharp to a blunted configuration strengthens ceramics. Crack-tip geometry may be directly observed by high-resolution-high-voltage electron microscopy (HR-HVEM). An illustration of the tip geometry of SiO_2 glass appears in Fig. 8.21, representing a silica glass thin film 20 nm thick. The strength of glass is strongly influenced by surface flaws or cracks. The shape of the cracks, especially their tip radii, are important parameters for analyzing fracture. A similar example is shown in Fig. 8.22 for a 40 nm thick glass film.

Figure 8.22a is a magnified view of a crack tip immediately after crack formation. The radius of curvature of the crack tip is 1.5 nm, the same as in Fig. 8.21b, indicating that tip geometry is independent of specimen thickness. This specimen was soaked in water at 90 °C for 7 d and the results are in Fig. 8.22b, which shows the position of the new crack tip, as indicated by the arrow. Note that crack blunting took place during soaking and that the radius of curvature of the crack tip increased from 1.5 to $\cong$5 nm. A diagram of the mechanisms involved is reproduced in Fig. 8.23.

One may also observe a variation in the crack geometry with aging after the glass was immersed in water. This suggests, as indicated in Fig. 8.23, that the mechanisms acting during the blunting process are dissolution and precipitation, rather than dissolution alone. An increase in the strength of the glass following blunting is obtained. Figure 8.23 represents those types of glass known to be very sensitive to external effects and exemplifies the facts that crack blunting improves the fracture resistance of ceramics while also illustrating the geometry of a blunted crack under HREM.

Of these three modes of cracking (shown in Fig. 8.16), mode-I (crack opening) is the most commonly discussed in the literature. Eftis and Leibowitz [16] presented opening-mode stresses in polar coordinates as:

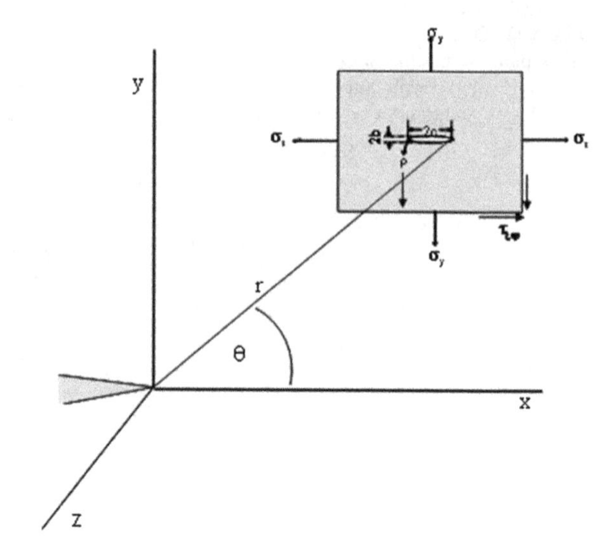

Fig. 8.24 Distribution of the stresses in the vicinity of a crack tip

$$
\begin{aligned}
\sigma_y &= \frac{K_{Ic}}{\sqrt{2\pi r}}\cos\frac{\theta}{2}\left(1+\sin\frac{\theta}{2}\sin\frac{3\theta}{2}\right)\\
\sigma_x &= \frac{K_{Ic}}{\sqrt{2\pi r}}\cos\frac{\theta}{2}\left(1-\sin\frac{\theta}{2}\sin\frac{3\theta}{2}\right)\\
\tau_{xy} &= \frac{K_{Ic}}{\sqrt{2\pi r}}\left(\sin\frac{\theta}{2}\cos\frac{\theta}{2}\cos\frac{3\theta}{2}\right)
\end{aligned}
\tag{8.52}
$$

$$
\begin{aligned}
\tau_{\mathrm{yz}} &= \tau_{\mathrm{xz}} = 0\\
\sigma_{\mathrm{z}} &= \nu(\sigma_{\mathrm{y}} + \sigma_{\mathrm{x}})\\
\sigma_z &= 0 \text{ for plane stress.}
\end{aligned}
$$

The stress distribution in the vicinity of a crack is shown in Fig. 8.24. Figure 8.7 is inset at the center of Fig. 8.24 and the origin of the coordinate system is located at the crack tip.

Note that the stress approaches infinity as the crack tip is approached, since the denominator factor, $(2\pi \mathrm{r})^{-1/2}$, approaches infinity with r going to zero. The literature contains expressions for K (and α) for a large number of crack and loading geometries and both numerical and experimental procedures exist for determining the stress-intensity factor in actual, specific geometries. From Eq. (8.49), it is possible to express critical stress as:

$$
\sigma_f = \frac{K_{Ic}}{\sqrt{\pi a B}} = \frac{K_{Ic}}{\alpha\sqrt{\pi a}} \tag{8.53}
$$

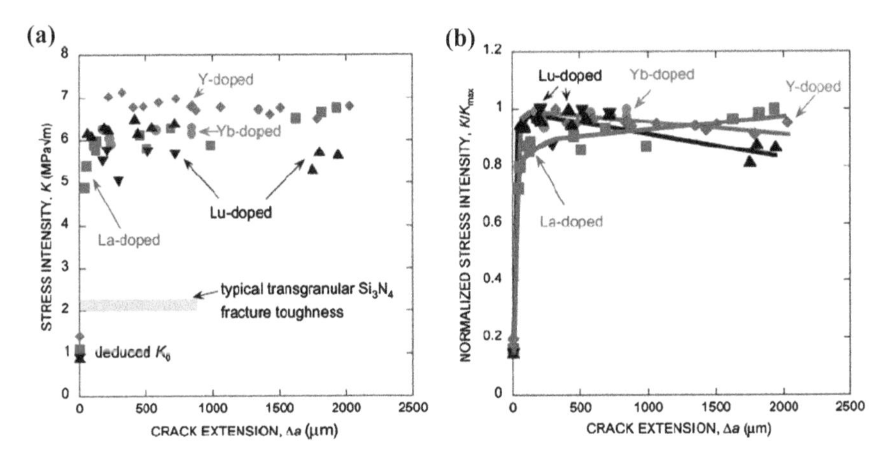

Fig. 8.25 R-curves were normalized to the peak toughness to clearly show the change in shape with sintering additive. Values of K_0 were deduced for two samples (La and Lu doped) from measured crack-opening profiles in R-curve samples, while K_0 for Y-doped material was taken from Kruzic et al. [35] Also shown are typical data for a Y_2O_3 doped Si_3N_4 that exhibits transgranular fracture. With kind permission of John Wiley and Sons

This is basically Eq. (8.50). In Eq. (8.49) above, α (= $B^{1/2}$) is a geometrical factor equal to 1 for edge cracks and generally of the order of unity in other situations. In design analysis, it is assumed that a material can withstand a stress up to a critical value of the stress-intensity factor. Beyond this critical value of K_{Ic}, cracks propagate rapidly.

8.3.6.3 R Curves

The term 'R-curve' refers to fracture toughness which increases as a crack grows– a desirable material property. Constructing R-curves is important for the prediction of the effect of existing flaws in ceramics on fracture strength. Fracture toughness is an indicator for failure in ceramics and the R-curve expresses ceramic crack resistance. The R-curve is crack-size dependent and, therefore, often the measurement of toughness properties, as a function of crack length, is required for constructing R-curves. Though mostly studied in metals and composites, investigations during the last decade have revealed similar properties of fracture toughness behavior in monolithic ceramics, such as Al_2O_3, and others, e.g., PSZ, α-SiAlON, Si_3N_4, etc. In Fig. 8.25, the R-curves for Si_3N_4, doped with various rare-earth [henceforth: RE] elements, are shown as stress intensity and normalized stress intensity with crack size change (crack extension). The curves provide information on extrinsic toughening by the edition of RE elements (RE)-MgO to Si_3N_4 ceramics. These materials tend to exhibit superior strength and fracture

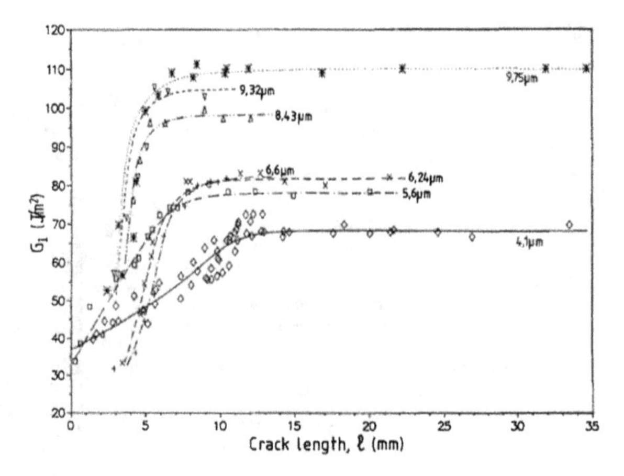

Fig. 8.26 R-curves of ADS96R with various grain sizes [55]. With kind permission of Elsevier. (ADS96R is a MgO-CaO-Al_2O_3 glass matrix)

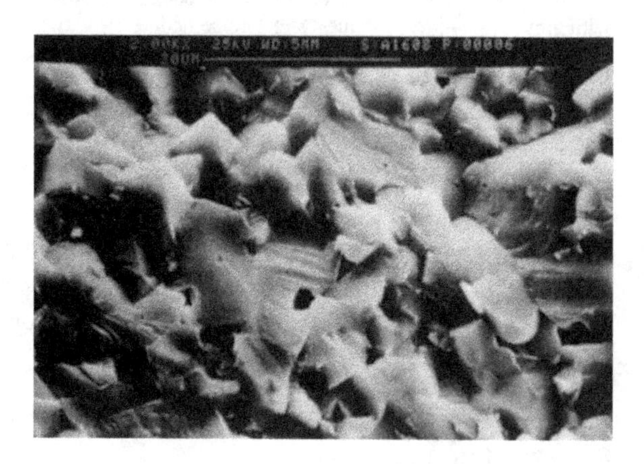

Fig. 8.27 SEM micrograph of the fracture surface of the material showing the predominantly transgranular fracture crack mode [55]. With kind permission of Elsevier. (ADS96R is a MgO-CaO-Al_2O_3 glass matrix)

toughness, properties which result in a wider range of potential commercial applications than those of only intrinsically toughened Si_3N_4.

Such additions give (i) sufficiently weak grain boundaries and (ii) bimodal grain distributions and/or high aspect ratio self-reinforcing grains. These structures fail predominantly intergranularly, which leads to extrinsic toughening via crack deflection and crack bridging. Crack deflection and crack bridging are considered to be basic mechanisms in the toughening process. Extrinsic toughening via grain-bridging involves intact grains that span the crack flanks and sustain part of the applied loads that would otherwise be experienced at the crack tip, thereby effectively toughening the material by lowering the near-tip stress intensity, K_{tip}, relative to the applied stress intensity, K_{app}. Thus:

$$K_{tip} = K_{app} - K_{br} \tag{8.54}$$

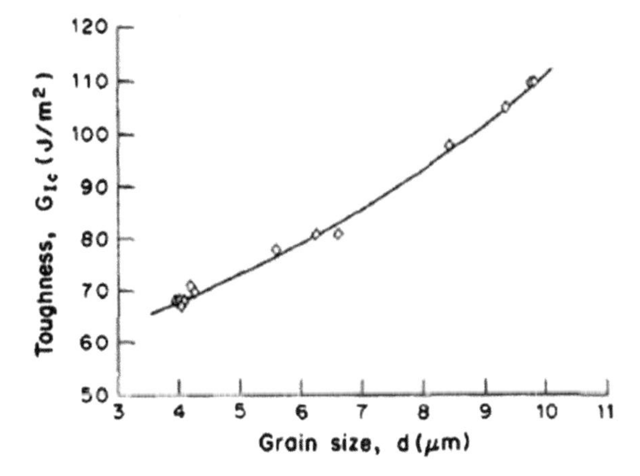

Fig. 8.28 The dependence of the limiting (steady-state) toughness, G_{Ic}, on the average grain size of the material [55]. With kind permission of Elsevier

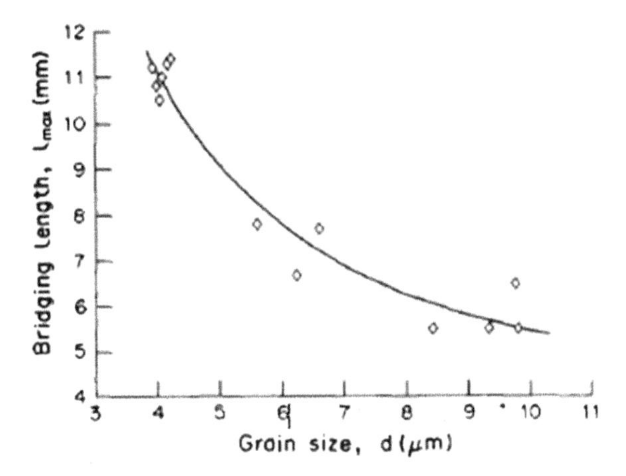

Fig. 8.29 The dependence of the bridging-zone length, l_{max}, on the grain size [55]. With kind permission of Elsevier

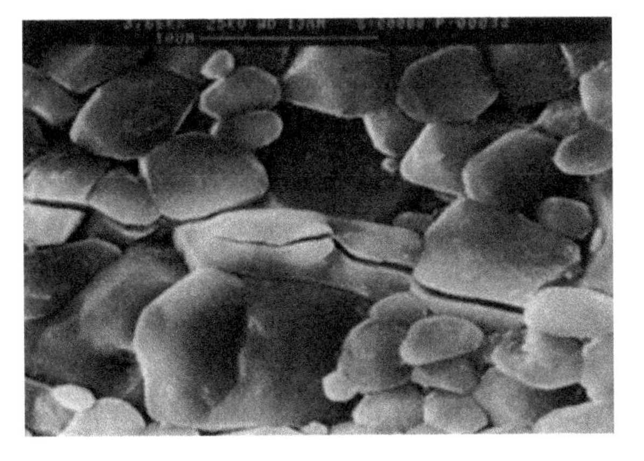

Fig. 8.30 An example of the beam-like ligaments bridging the cracks [55]. With kind permission of Elsevier

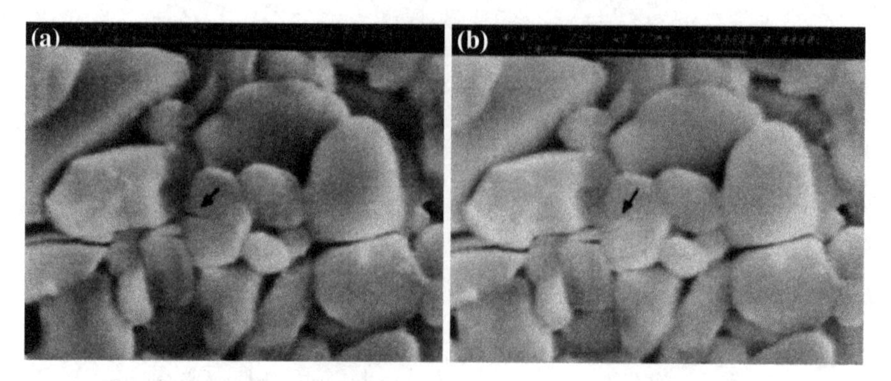

Fig. 8.31 A ligamentary microcrack [*arrowed* in (**a**)] is seen to close in (**b**) after propagation of the main crack, probably due to the influence of the residual thermal stresses in the material. Micrographs from video recording [55]. With kind permission of Elsevier

Fig. 8.32 The dependence of the change in toughness, ΔG_I, on the grain size [55]. With kind permission of Elsevier

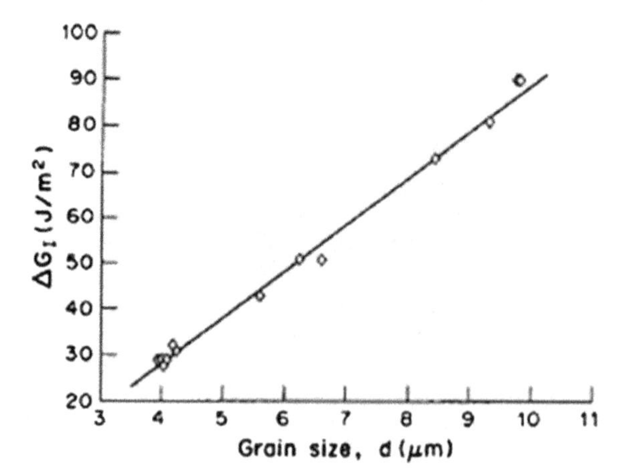

where K_{br} is the bridging-stress intensity. Grain bridging occurs as a result of intergranular fracture and, accordingly, the boundaries must be sufficiently weak to provide a preferential crack path. As a consequence of a reduction in boundary adhesion, the local intrinsic fracture resistance at the boundary should increase steady-state bridging, as $K_{br} \propto 1/K_{tip}$, by allowing for more high-angle deviations from the mode-I crack path (Fig. 8.16a), providing a more tortuous path, more bridging and less transgranular fracture. In consequence, higher toughness may be expected when lowering boundary adhesion in polycrystals.

The effect of grain size on R-curves is shown for Al_2O_3 in Fig. 8.26, obtained by instrumented in situ dynamic SEM, using the double torsion technique. The R-curve of the alumina glass is characterized by the energy dissipation requirement (during load-crack opening and closure), which provides data on whether the localized grain bridging is responsible for the toughening mechanism observed. In Fig. 8.26 of as-sintered and aged samples, the energy release, G_{Ic}, is plotted against crack extension, Δa. The fracture mode displayed in the experiments is predominantly transgranular, as seen in Fig. 8.27. The results from Fig. 8.26 show that increasing the grain size increases both the limiting (or 'steady-state') toughness and the toughening rate (the slope of the R-curve), while the bridging-zone length (measured by the crack extension required to reach a steady-state) decreases. These bridging-zone lengths are large: up to 12 mm or 3000 grain diameters for the as-sintered materials. The dependence of 'steady-state' (limiting) toughness and bridging-zone length on grain size are seen in Figs. 8.28 and 8.29. Notice in Fig. 8.30 that the cracks are largely transgranular with short intergranular steps. Grain bifurcation leads to beam-like ligaments which bridge the crack. The role of the visible ligamentary microcrack shown in Fig. 8.31 is to open and reclose after the advance of the main crack, possibly due to residual stresses in the material. Such elastic ligaments obviously exert a closing force, which has been suggested as the reason for the additional toughening. Furthermore, toughening depends on grain size, as may be seen in Fig. 8.32.

Al_2O_3 is one of the many ceramics displaying R-curve behavior characterized, as mentioned in the beginning of the section, by increasing resistance as a crack grows–a remarkable phenomenon, when considering the fact that cracks are the "root of all evil" in the nascent state of brittle fracture. Much experimental evidence indicates that this is caused by a crack-bridging mechanism, by grains which have not fractured. The geometry of these grains involves frictional forces, i.e., increased force is required to overcome frictional- and bridging- induced resistances, for the induction of fracture. In other words, toughening of the material has occurred.

Here, in Figs. 8.30 and 8.31, shown in connection with alumina toughening, toughness is indicated by the symbol G_{Ic} (following the authors' notation), however the notation found in most of the professional literature is K_{Ic}.

It seems that it is possible to design ceramics with increased toughness by inducing the frictional sliding of constrained grains, once a crack has been formed, by controlling grain morphology, interface strength and by inducing residual stress.

8.3.6.4 J-Integrals

The fracture of ceramics, like alumina, may also be characterized by the 'energy absorption' concept. Thus, toughness, K_{Ic}, (which is a linear elastic fracture mechanics parameter) and $G_{Ic,}$ the critical strain energy release were measured in ceramics as a fracture criterion. The strain energy concept (Irwin [30]) of linear elasticity may be generalized by defining a new parameter, J, which may be applied to nonlinear elastic behavior (for example to fracture). As a fracture criterion, it is applied to express the energy absorbed during crack extension. Notched beam specimens are used and the fracture surface energy, γ_{nbt}, and the work of fracture, γ_{wof}, are calculated from the linear elastic relations of K_{Ic} and ½ G_{Ic},

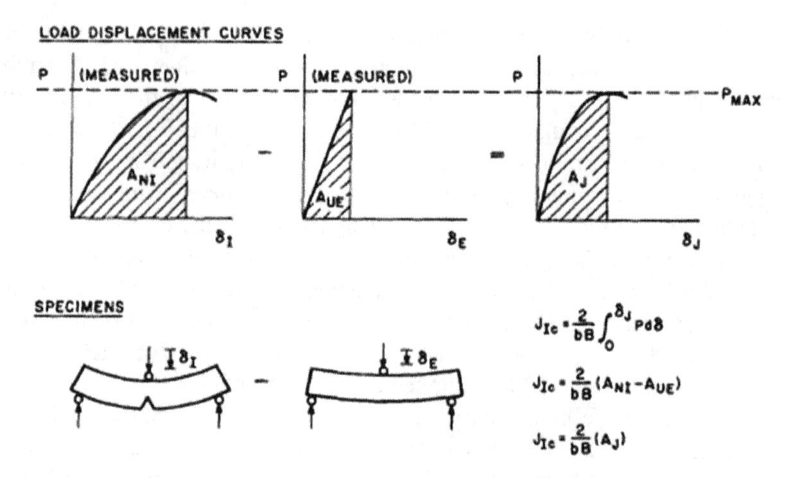

Fig. 8.33 Schematic of J measurement technique [27]. With kind permission of John Wiley and Sons

which are used as the fracture-surface energy for crack initiation. The range of typical values of γ_{nbt} is 5–20 J/m^2, while the energy absorption values, γ_{wof}, of a slowly extending crack often has a value of >100 J/m^2. It has been established that, for refractories, γ_{wof} is $\gg \gamma_{nbt}$. The ratio ($\gamma_{wof}/\gamma_{nbt}$) has been applied as an indication of the refractories' resistance to thermal shock damage which may be as high as 20, depending on the particular refractory of interest. This ratio can be controlled to some extent through chemistry, microstructure and processing.

The critical value of J (i.e., J_{Ic}) is about the same as K_{Ic} and G_{Ic}, as mentioned above for the fracture criterion. It is based on a crack-path independent energy-line integral that encloses the crack-tip region. In other words, it is a parameter describing the total energy of the crack-tip and the stress–strain field. One of the experimental methods for measuring J_{Ic} is to determine the energy difference between a cracked and an uncracked specimen loaded under the same conditions, when the load is at its maximum prior to catastrophic failure of the cracked specimen. Since, in general, brittle ceramics exhibit a distinguishable load maximum before fracture, the J-integral evaluation provides a reasonably reliable method for determining the energy dissipation during fracture initiation in materials such as ceramics.

As mentioned at the start of this section, the J-integral concept was originally applied to metals, in which the linear elasticity concept might include the plastic observations, i.e., plastic flow in the close vicinity of the tip. Reservations regarding the use of the J-integral in brittle materials, such as ceramics, are a consequence of the presence of microcracks or rather subcritical crack growth before fracture sets in, making it not strictly applicable. However, in ceramics, in which extensive inelastic processes are active in the crack-tip process zone, the use of J is not more restrictive than the use of the K_{Ic} criterion in linear elastic fracture mechanics.

An accepted method for J-integral measurements (as performed by Rice et al. [44]) is by a compliance method. This technique involves the subtraction of the load–displacement curve of an unnotched elastic (UE) specimen from that of a notched inelastic (NI) specimen for the same load of P_{max}. Figure 8.33 illustrates this technique of J-integral measurement.

Note that the measuring technique depends on parameters such as specimen type, dimensions and loading technique. From such measurements, J_{Ic} is calculated. The critical value, J_{Ic}, according to Eqs. (8.55) or (8.56) is calculated after performing the integral (represented by the area under the load curve vs. extension) at fracture:

$$J_{Ic} = \frac{2}{bB}\int_0^{\delta_J} P d\delta = \frac{2A_J}{bB} \tag{8.55}$$

or

$$J_{Ic} = \left(\frac{2}{bB}\right)(A_{NI} - A_{UE}) \tag{8.56}$$

In the above equations, A is the area, b is the uncracked ligament and B is the specimen thickness. The above was obtained according to Homney et al. [27]. following their approach as follows. The elastic property measurements were performed using a dynamic mechanical resonance method. The shear modulus, G, measurement was first performed by torsional resonance and then by flexural resonance to estimate the Young's modulus. E was calculated by estimating Poisson's ratio, ν, and then by using an iteration based on:

$$\nu = \left(\frac{E}{2G} - 1)\right) \tag{8.57}$$

(This appeared as Eq. (1.12a) in Chap. 1). Next, the fracture toughness, K_{Ic}, was measured on single-edge crack-like specimens with notches of one-half the specimen thickness. The specimen was tested in three-point flexure over a 6-in. span on a commercial testing machine at a crosshead speed of 0.05 in./min. The K_{Ic}'s were calculated from Brown and Strawley's [14] equation:

$$K_{Ic} = \sigma_f \sqrt{c} f\left(\frac{c}{d}\right) \tag{8.58}$$

where σ_f is the fracture stress, c the notch depth and f(c/d) is a polynomial describing the geometry of the test specimen. In the linear elastic case, K_{Ic} is related to G_{Ic} and γ_{nbt} through:

$$K_{Ic}^2 \frac{(1 - \nu^2)}{E} = G_{Ic} = 2\gamma_{nbt} \tag{8.59}$$

Table 8.1 Summary of Creep Tests for Hot-Pressed Silicon Nitride [40] (with kind permission of John Wiley and Sons)

Test temperature (c)	Stress (MPa)	Minimum strain rate (l/s)	Life (h)	Total strain (%)	Note
Monolith					
1200	250	5.0×10^{-8}	45	1.0	1
1200	250	4.4×10^{-8}	40	0.9	I
1200	250	4.1×10^{-8}	51	1.0	1
1200	200	1.8×10^{-8}	148	1.4	2
1200	150	8.2×10^{-9}	202	1.0	2
1200	150	7.5×10^{-9}	350	1.6	2
1200	100	1.4×10^{-9}			4
1200	70	5.0×10^{-10}	3280	0.8	2
1250	100	5.8×10^{-9}	360	1.3	2
1300	150	1.6×10^{-6}	2	1.1	3
1300	100	4.1×10^{-7}	12	1.4	3
1300	70	8.7×10^{-8}	25	1.4	3
1300	70	7.2×10^{-8}	19	1.2	3
1350	70	8.1×10^{-7}	4	2.7	3
1350	40	9.0×10^{-8}	55	3.4	3
Composite					
1200	250	4.0×10^{-8}	56	1.0	1
1200	200	1.6×10^{-8}	95	0.9	2
1200	200	1.5×10^{-8}	132	1.1	2
1200	200	1.4×10^{-8}	195	1.5	2
1200	150	4.8×10^{-9}	254	0.8	2
1200	100	8.0×10^{-10}			4
1200	100	6.4×10^{-10}			4
1250	100	3.1×10^{-9}	605	1.1	2
1300	100	2.5×10^{-7}	14	1.4	3
1350	70	8.2×10^{-7}	6	2.9	3
1350	40	1.2×10^{-7}	48	3.2	3

Notes 1. Fracture in a transient creep regime. 2. Fracture in steady state creep regime. 3. Fracture in a accelerated creep regime. 4. Interrupted before fracture

G_{Ic} is the critical strain energy release rate and γ_{nbt} is the fracture surface energy of crack initiation, often referred to as the 'notched-beam-test' fracture-surface energy in refractory literature. Also note that the stress intensity factor (Eq. 8.58) was given in Eq. (8.49) as:

$$K_{Ic} = \sigma\sqrt{\pi a B} \tag{8.49a}$$

Table 8.1 is a summary of all the creep tests performed on S_3N_4. Stress intensity factors for various geometries are given in Table 8.1 of Suh and Turner [7] (pp. 422–423), among them for edge cracks, though not strictly applicable to simple edge-crack-like specimens tested by the 3-point flexure method.

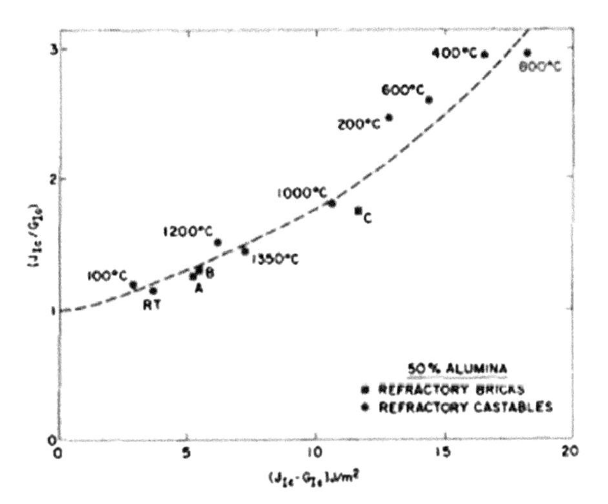

Fig. 8.34 Crack brittleness ratio, J_{Ic}/G_{Ic} versus (J_{Ic}-G_{Ic}) [27]. With kind permission of John Wiley and Sons

At a later stage, the work-of-fracture, $\gamma_{wof,}$ was evaluated in a specimen with a side notch having the geometry of a v-notch. These specimens were broken on a commercial testing machine over a 6-in. span at a crosshead speed of 0.002 in./min to yield fully stable fractures. The work-of-fracture was calculated from:

$$\gamma_{wof} = \int Fds/2A \tag{8.60}$$

$\int Fds$ is the area under the load-displacement curve and A is the area of the remaining v-section of the specimen after notching. Comparing J_{Ic} data with other fracture parameters, such as K_{Ic}, G_{Ic} and $2\gamma_{wof}$, is of interest. Thus, in a perfectly linear case (as indicated earlier), the critical value of J (i.e., J_{Ic}) is about the same as K_{Ic} and G_{Ic} for the fracture criterion and Eq. (8.59) may be rewritten as:

$$J_{Ic} = G_{Ic} = K_{Ic}^2\left(1 - v^2\right)/E \tag{8.61}$$

Like other measurements, the results of the fracture of 50 % alumina refractories indicate that the order observed is $G_{Ic} < J_{Ic} < 2\gamma_{wof}$. This trend appears to be consistent for all the ceramic materials which have been studied to date. Since J_{Ic} exceeds G_{Ic} for each refractory investigated, the magnitude of the difference (J_{Ir}-G_{Ic}) may indicate the refractory's propensity for exhibiting inelastic energy dissipation processes in the crack-tip zone during fracture initiation. Figure 8.34 relates the crack-brittleness ratio to the difference (J_{Ir}-G_{Ic}).

The other measure of energy dissipation by inelastic processes during fracture is $2\gamma_{wof}$. It is also a measure of the presence of energy-consuming inelastic processes, but of those occurring during slow crack motion through a complete fracture process. J_{Ic} also exceeds $2\gamma_{wof}$. In a manner similar to Fig. 8.34, a plot may be constructed relating the normalized work of fracture to the ($2\gamma_{wof} - G_{Ic}$) parameter shown in Fig. 8.35.

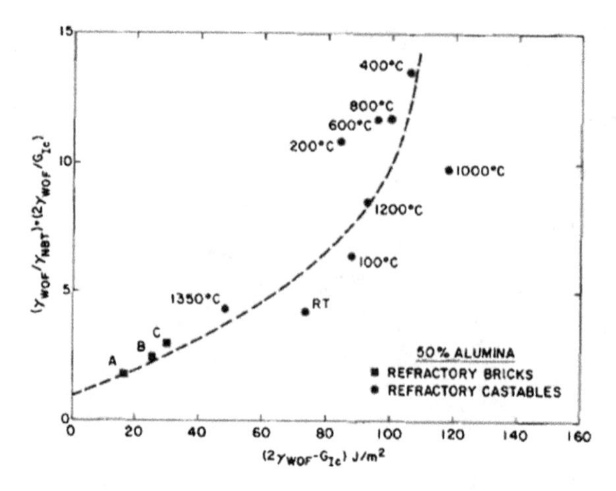

Fig. 8.35 Work-of-fracture normalized similar to the J_{Ic} fracture criterion [27]. With kind permission of John Wiley and Sons

Thus, the above discussion on the J-integral shows that its measurement in various refractories yields J_{Ic} values which exceed G_{Ic}. With regard to J_{Ic}, these measurements appear promising as a technique for quantitatively describing the inelastic energy consumption in the crack-tip process zone during fracture initiation in refractory ceramics.

8.3.7 Fatigue Fracture in Ceramics

Fatigue (non-static fatigue) is a cyclic deformation, such that damage in ceramics under cyclic loading is associated with its microstructure, specifically with the presence of flaws. Unlike in metals, which are designed for strength, in ceramics the major interest is toughness, which is very much affected by the presence of cracks and their growth. Ceramics are of interest for various structural applications in which cyclic loading is involved. Among these uses, ambient temperature is also being considered in particular for biomedical implants. However, in predominantly brittle ceramics, the focus has been on their use at elevated temperatures, such as in aviation (for fuselages) and for engine components, to mention a few. However, in general, applications of ceramics have been limited by their sensitivity to the presence of preexisting flaws in the form of cracks, pores and microcracks. Therefore, ceramics are expected to show marked susceptibility to premature failure under cyclic fatigue loading. Fracture is associated with fatigue-crack propagation and, therefore, the growth rate, da/dN, as a function of the stress intensity, K, is a subject of considerable interest. Figure 8.36 compares da/dN as a function of ΔK of several ceramic materials, intermetallics and metals.

Two aspects related to failure are observed in fatigue crack growth at the tips in materials. The first, which promotes crack advance, is the microstructural damage

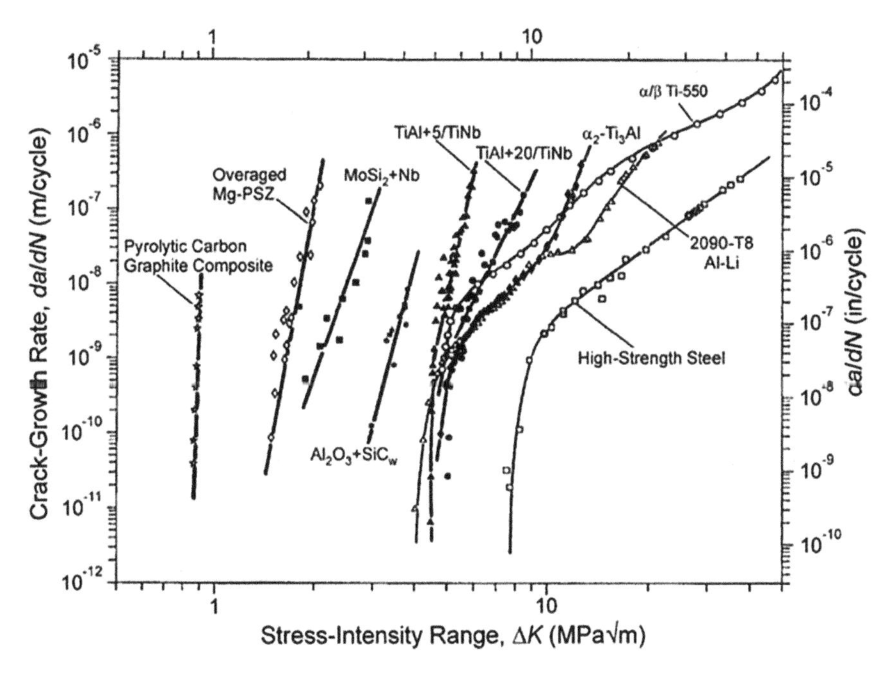

Fig. 8.36 Schematic variation of fatigue-crack propagation rate (da/dN) with applied stress intensity range (ΔK), for metals, intermetallics and ceramics [4]. With kind permission of Professor Ritchie

in front of the tip and the second, which impedes crack propagation at the tip, is a shielding mechanism implemented by crack bridging or crack deflection and fatigue crack closure. The crack-closure mechanism is related to ΔK. Recall that the crack-growth rate is related to ΔK by Eq. (7.9), which is Paris' relation, reproduced here as:

$$\frac{da}{dN} = C\Delta K^m \tag{7.9}$$

Essentially, creep blunting is associated with the amount of localized plastic deformation that occurs at the crack tip prior to crack propagation and represents the controlling feature in tip blunting. Such behavior cannot occur in commonly brittle material at RT due to its lack of plasticity. Therefore, ceramic materials may be toughened by using various crack-tip shielding mechanisms, whose effect is to impede crack extension by locally reducing near-tip intensity (or rather stress concentration) by modifying microstructural factors to promote toughening. Here, a list of the known methods (following Ritchie's [4] classification) is presented (Fig. 8.37).

One should note that transformation toughening is a possible method only in ceramics undergoing a transformation, for example in a zirconia-based ceramic in which tetragonal-monoclinic transformation might occur. One of the roles of

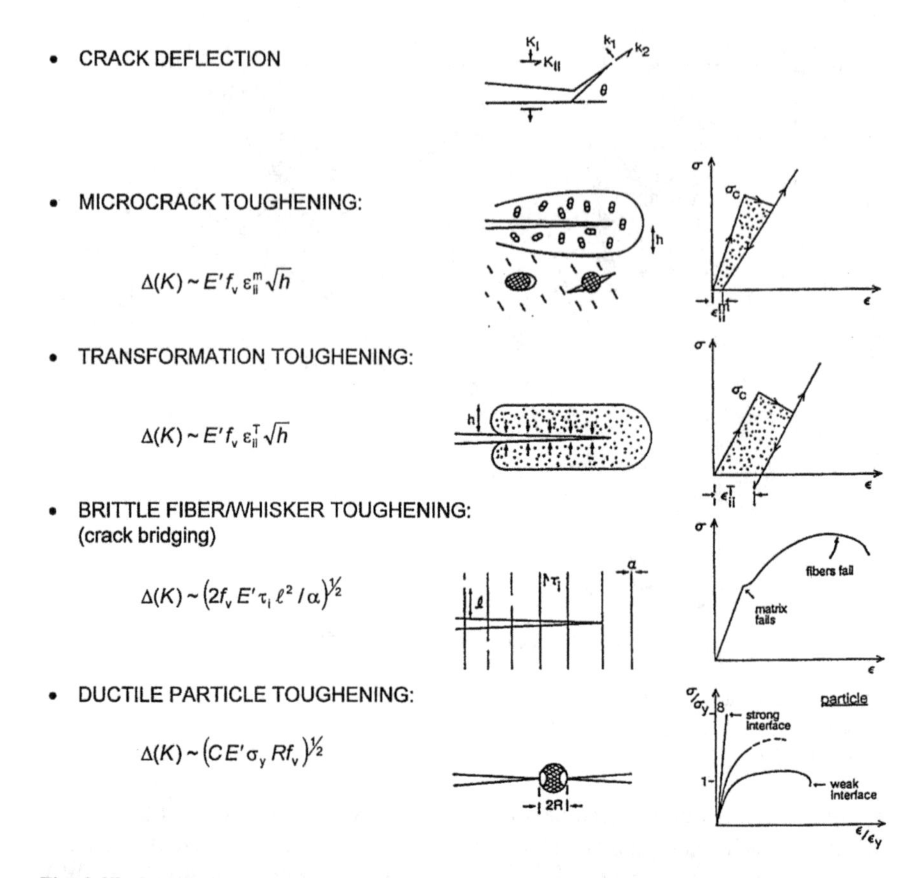

Fig. 8.37 Schematic illustration of the primary toughening mechanisms in ceramics and ceramic-matrix composites. Note that all mechanisms are extrinsic in nature and promote inelastic deformation which results in a nonlinear stress/strain relationship [4]. With kind permission of Professor Ritchie

particle additions is to induce crack deflection and, by so doing, the toughness of the ceramics is improved. An important aspect of toughening is by crack-tip shielding involving crack bridging by fibers (in fiber-enforced ceramic composites) or when a crack intersects a ductile phase (particle). The propagation of cracks eventually leads to failure. It is of great importance to extend the lifetimes of components operating under cyclic conditions. The fracture mode in cyclic loading is characterized by striation growth (which was believed to be a feature only in metallic fatigue). The fracture surfaces in ceramics were considered to be almost identical to those under monotonic loading (see Fig. 8.38) except that more debris is often present on fatigue-fractured surfaces. However, it turns out that striations are also found in ceramics, as seen, for example, in 3Y-TZP (Fig. 7.13, Sect. 7.2.2). Thus, striation cannot be considered a distinguishing fatigue feature

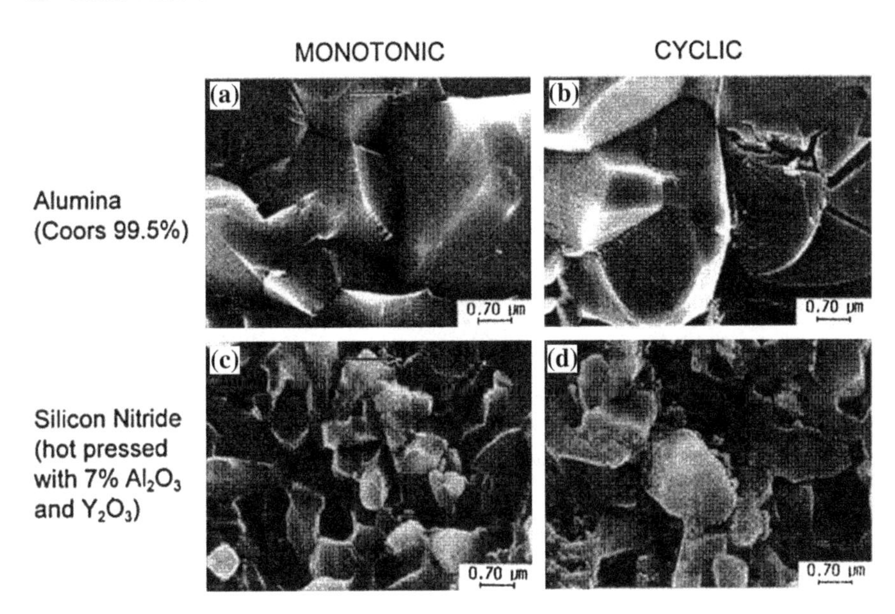

Fig. 8.38 Fractography of ceramic fatigue showing nominally identical fracture surfaces under, respectively, monotonic and cyclic loading in (**a**, **b**) alumina (Coors 99.5 %) and (**c**, **d**) silicon nitride (hot-pressed with 7 wt% $Al_2O_3 + Y_2O_3$). Note, however, the more debris and surface damage on the fatigue surfaces [4]. With kind permission of Professor Ritchie

of fractured surfaces characterizing only metals. Microstructure can have an important effect on resistance to fatigue fracture. Thus, the effect of grain size on the fatigue-crack growth rate is indicated in Fig. 8.39.

Regardless of the origin of stress when cycling is applied, fatigue damage may result. Dangerous stress cycling is associated with thermal effects. Premature failure, resulting from cyclic stresses due to temperature changes, seems to be one way of approaching the problem of thermal fatigue in high-temperature structural components. This problem is of great concern in many fields in which structural components must operate at high service temperatures; if thermal gradients are generated, the propensity for thermal-fatigue cracking increases. Thus, it is of great technological interest to strengthen those components that will be exposed to conditions in which thermal gradients are set up to inhibit thermal-fatigue damage.

Note that the grain size effect is evident in Fig. 8.39 and this effect is not emphasized or masked in b of this figure. Normalizing the growth rate in terms K_{max}/K_c, the microstructural effects (i.e., the grain-size effects) are essentially eliminated. Rather than using the Paris Eq. (7.9), as

$$\frac{da}{dN} = C\Delta K^m \tag{7.9}$$

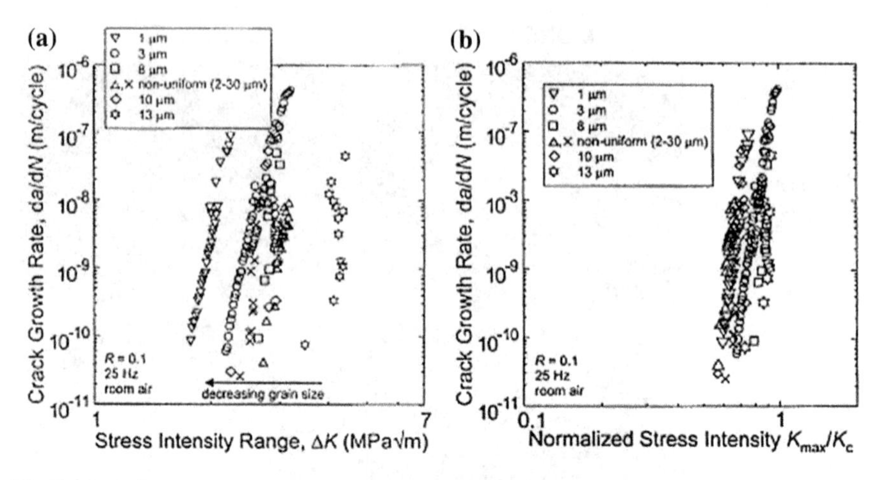

Fig. 8.39 Fatigue-crack growth rates, da/dN, as a function of **a** the applied stress-intensity range, ΔK, and **b** the maximum stress intensity normalized by the fracture toughness, K_{max}/K_c, for a range of polycrystalline aluminas [4]. With kind permission of Professor Ritchie

Liu and Chen [37] suggested:

$$\frac{da}{dn} = C'(K_{max})^n(\Delta K)^p \tag{7.9a}$$

where K_{max} appears in the equation in addition to ΔK. C′ is a constant and the exponents, n and p, are ~10 and 15 (in alumina reinforced with SiC whiskers) as compared to n = 0.5 and p = 3 for metal fatigue of a nickel-based superalloy. C′ is a constant equal to C (1 − R)n and (n + p) = m, the exponent of Paris' relation.

Ritchie [4] classified the fatigue mechanisms related to crack growth as being either 'intrinsic' or 'extrinsic' mechanisms, associated with the crack-tip characteristics. The intrinsic mechanism generally characterizes metals and their alloys, in which crack advance is a consequence of damage in the crack-tip region characteristics under cyclic loading. In the case of an extrinsic mechanism, crack-tip shielding behind the tip is degraded by accelerated crack growth during the unloading cycle. In ceramics, the cyclic process is predominantly extrinsic. One vital means of crack-tip shielding is bridging, which plays an important role in closing crack surfaces. Grain-bridging stress, p, may be expressed as a function of the distance behind the crack tip, X, and crack-opening displacement, 2u, in terms of the length of the bridging zone, L, and an exponent, k, as:

$$\begin{aligned} p(X) &= P_{\max}\left(1 - \frac{x}{L}\right)^k \\ p(u) &= P_{\max}\left(1 - \frac{X(u)}{L}\right)^k \end{aligned} \tag{8.62}$$

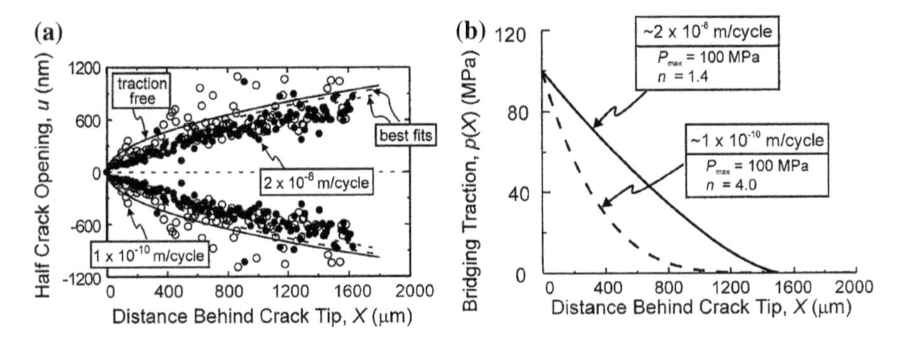

Fig. 8.40 Measured crack-opening profiles in an ABC-SiC ceramic for **a** Case I where the crack was grown near instability (at ~2 × 10^{-8} m/cycle at ~92 % K_c), and Case II where the crack was grown near threshold (at ~1 × 10^{-10} m/cycle at ~75 % K_c). In **b**, the best-fit bridging traction distributions are plotted for each case [4]. With kind permission of Professor Ritchie

The function p(X) describes the bridging stress distribution from a value of P_{max} at the crack tip (X = 0) and decreasing to zero at the end of the bridging zone, where X = L. At this location, 2u = $2u_f$, the critical crack opening for creep rupture. Bridging is a toughening mechanism. However, under cyclic loads, the repetitive opening and closing of the crack results in a decrease in the toughening capacity of the bridging zone due to the reduction in grain-bridging stress. This means that the accumulated damage at the grain/matrix interfaces causes premature damage by grain debonding. Measured crack-opening profiles in SiC ceramics may be seen in Fig. 8.40.

In Fig. 8.40a, SEM measurements were performed an in situ toughened ABC SiC of the opening profile for a crack approximating R-curve behavior and the plot is a function of distance, X, behind the crack-tip (see Sect. 8.3.6.3). In 10b, the best-fit profile is shown as determined by the following relation by Ritchie [4] (after Barenblatt):

$$u(x) = \frac{K_A}{E'}\left(\frac{8x}{\pi}\right)^{1/2} + \frac{2}{\pi E'}\int_0^L p(X')\ln\left|\frac{\sqrt{X'}+\sqrt{X}}{\sqrt{X''}-\sqrt{X}}\right| dX' \tag{8.63}$$

Here, the net crack opening profile, u(X), for a linear elastic crack under an applied far-field stress intensity, K_A, with a bridging traction distribution, p(X), of length, L, acting across the crack faces is expressed in terms of the elastic modulus, E′ (= E in plane stress of E/(1 − ν^2) in plane strain). X′ is the integral variable where the stress, p, acts. Also shown are the best-fit profiles determined from Eq. (8.63), shown as the dashed line, and the calculated opening profile for a traction-free track, indicated by the solid line. In Fig. 8.40b, the best fits, p(X) and u(X), were estimated by fitting the data. These data may further used to evaluate p(X) and p(u) in Eq. (8.62).

For further calculations to predict the bridging contribution of a crack that has already propagated some amount Δa or to predict other pertinent data on fatigue

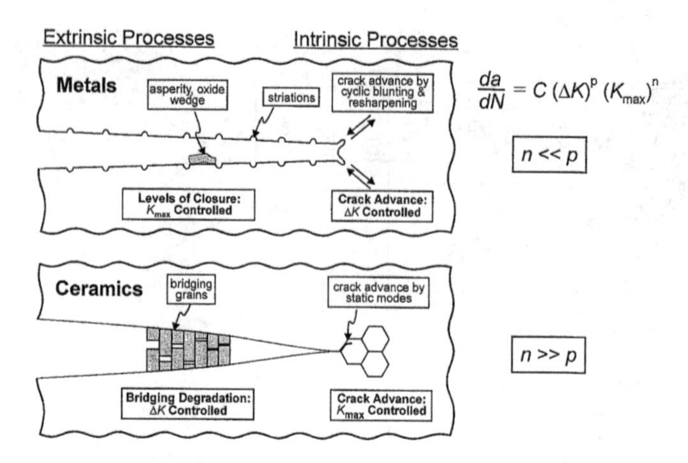

Fig. 8.41 Schematic illustrations of the intrinsic and extrinsic mechanisms involved in cyclic fatigue-crack growth in **a** metals and **b** ceramics, showing the relative dependencies of growth rates, da/dN, on the alternating, ΔK, and maximum, K_{max}, stress intensities [4]. With kind permission of Professor Ritchie

lifetimes, one is referred to Ritchie's [4] original work. This section is not complete without a schematic summary of the intrinsic and extrinsic processes leading to fracture by crack opening and propagation in metals and ceramics (Fig. 8.41).

8.3.8 Improving Toughness

There are various techniques that may be used to improve the toughness of ceramics, such as: (i) changing the course of a crack (deflection, bowing, branching, etc.); (ii) crack-tip shielding by bridging, transformation or causing plastic yielding in the tip vicinity; (iii) and crack healing. In (i) and (ii), the roles of second-phase particles, whiskers or fibers may explain the process occurring at the crack tip, though not necessarily, since the grain boundaries of polycrystalline ceramics may do just the same. In (iii), the major role of crack healing is temperature and duration at some temperature.

8.3.8.1 Changing the Course of a Crack

a) Crack deflection is one way to change the course of a crack, thus improving resistance to fracture. Toughening occurs due to deflection, whether particles or only grain boundaries are involved. A crack approaching some sort of microstructural inhomogeneity, such as a second-phase particle, may become tilted at some angle off its original plane. The deflection of a crack into a non-planar crack might be caused by the presence of residual strains. The origin of such residual

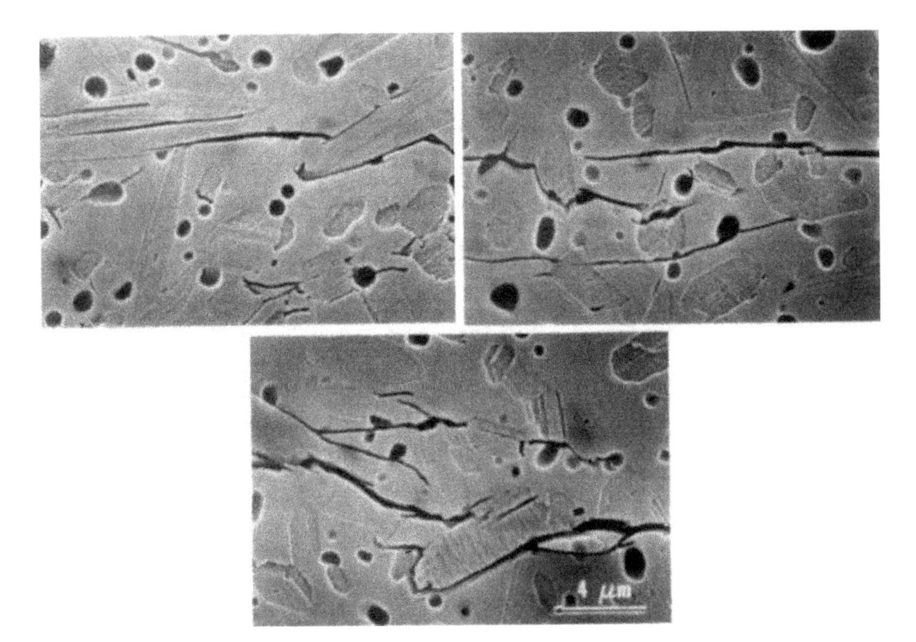

Fig. 8.42 Indentation crack profiles in lithium–alumin–silicate glass ceramic (heat treated 250 h at 850 °C, etched in 2.5 % HF for 5 s) indicating fracture of $Li_2Si_2O_3$ grains accompanying crack deflection. Also note microcrack [20]. With kind permission of Elsevier

strain, which may be compressive or tensile, may result from a mismatch of the thermal expansion during fabrication between the matrix and a second-phase particle. The sign of the residual strain (developed between the particle and the matrix) determines the direction of the deflection and also the position of the particle with respect to the movement of the crack. Crack deflections in lithium-alumino-silicate glass ceramics and Si_3N_4 composites are considered below. Figure 8.42 shows crack profiles in lithium-alumino-silicate glass. In this structure, residual strain develops due to the thermal expansion mismatch between a single crystalline phase of $Li_2Si_2O_3$ produced by heat treatment and the glass matrix. It is desirable to use an independent measure to specifically substantiate that crack deflection is associated with an increase in fracture toughness. Post-fracture information on the type of fracture and the origin of failure does not indicate the essential path and direction of the crack propagation. The crack-front tilt (or twist) is a significant piece of data for the unique determination of the effect of deflection on the toughening of the material under consideration. In Fig. 8.43, representative traces of end members of two Si_3N_4 specimens (samples A and G) are shown and the frequency distributions from smoothed histograms associated with these measurements appear in Fig. 8.44.

Since direct aspect ratio information could not be obtained for the Si_3N_4 systems, toughness correlations are, thus, based upon the measured mean deflection angles. As a reference required for comparison, the highest toughness at a median

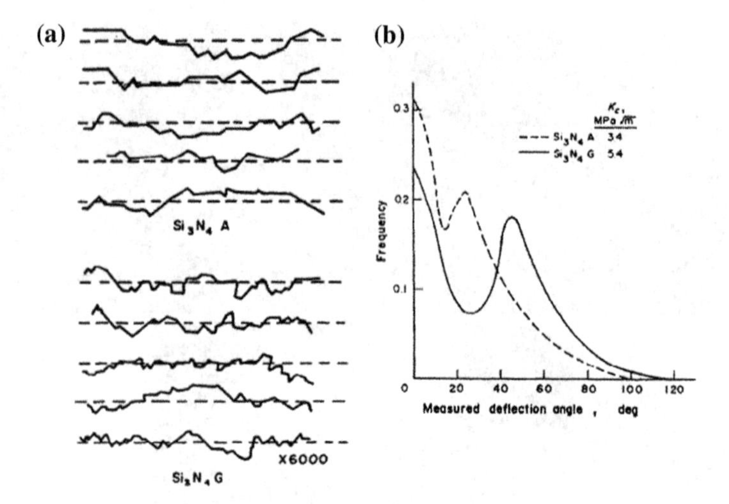

Fig. 8.43 **a** Crack deflection profiles for end member of the Si_3N_4 series traced from scanning electron micrographs. **b** Frequency distributions of measured detection of the angles for end members of the Si_3N_4 series [20]. With kind permission of Elsevier

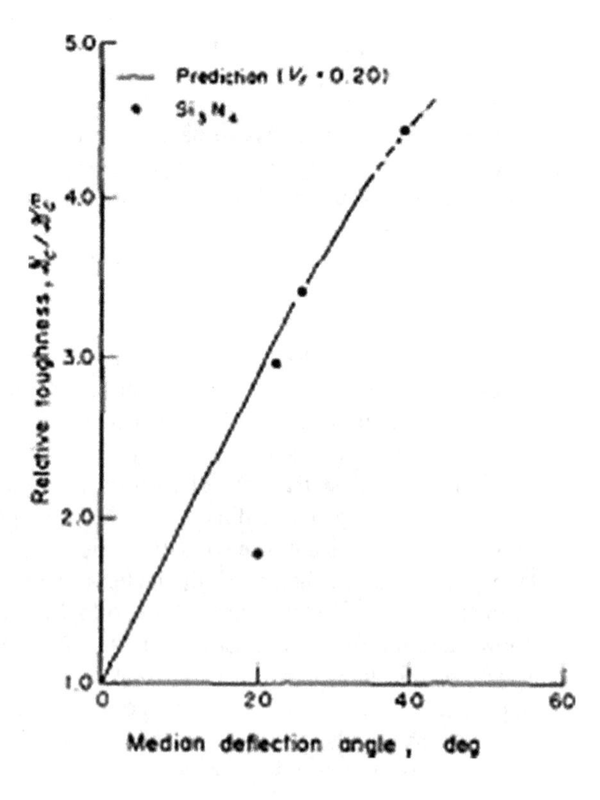

Fig. 8.44 Relative toughness-median deflection angle correlations for the hot-pressed Si_3N_4 series. (V_f in the figure is the volume fraction) [20]. With kind permission of Elsevier

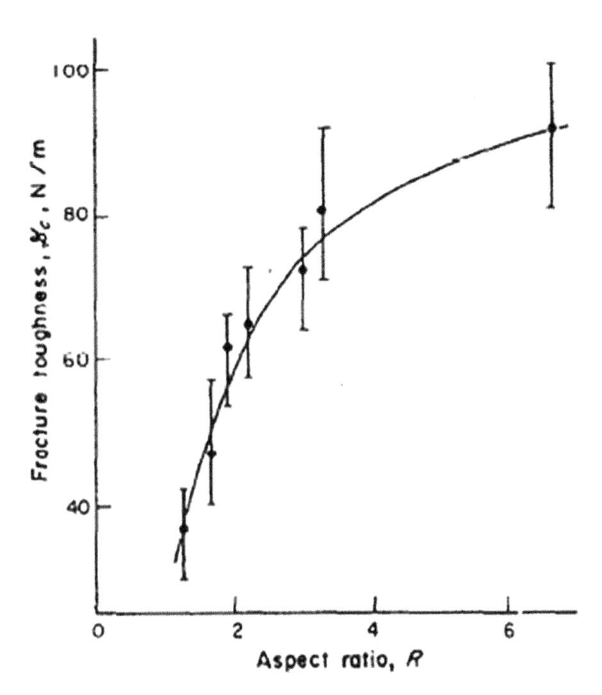

Fig. 8.45 Fracture toughness of a series of hot-pressed Si_3N_4 materials versus calculated aspect ratio [20]. With kind permission of Elsevier

deflection angle of 40° (composition G, median deflection angle of 40°) was selected. The prediction for rod-shaped particles with measured relative toughness and deflection angle is compared in Fig. 8.44. Only the lowest toughness (equi-axed grained material) deviates significantly from the prediction. From a series of hot-pressed silicon nitrides of rod-shaped grains having various aspect ratios, it was established that particle size does not toughen brittle ceramics due to crack deflection, but that its shape does have a strengthening influence. Clearly, an increase in the volume fraction also increases toughness. This observation is in line with the common concept of crack-deflection strengthening. The shape and size effects are shown in Figs. 8.45 and 8.46. Particle morphology effects may be seen in a series of hot-pressed silicon nitrides comprised of rod-shaped grains having various aspect ratios (Fig. 8.45; shape effect).

Lithium-alumina-silicate glass ceramics, containing $Li_2Si_2O_3$ lath-shaped crystals indicate the particle-size effects (Fig. 8.46). The most effective morphology for increasing toughness is the rod shape, followed by the disc and sphere shapes. Beyond particle morphology, additional features that affect the toughness of ceramics are volume fracture and the particle aspect ratio. Mathematical evaluation of the effect on toughening of rod-, sphere- and disc-shaped morphologies may be found in the work of Faber and Evans [20]. The relative toughness predictions for crack deflection for spherical, rod-like and disc-shaped particles are summarized in Figs. 8.47–8.49. From these toughness predictions, it is evident that an increase in toughness depends solely on particle shape and the volume fraction of the second phase. The most effective morphology for deflecting propagating cracks is a rod with

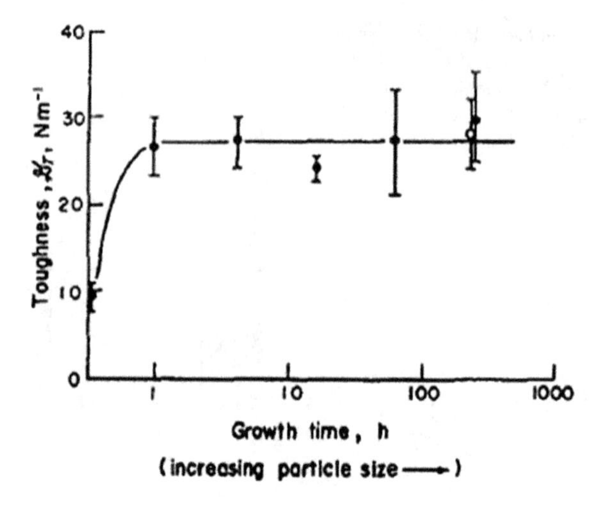

Fig. 8.46 Fracture toughness of a series of lithium-alumino-silicate glass ceramics plotted versus increasing growth time of the crystalline phase [20]. With kind permission of Elsevier

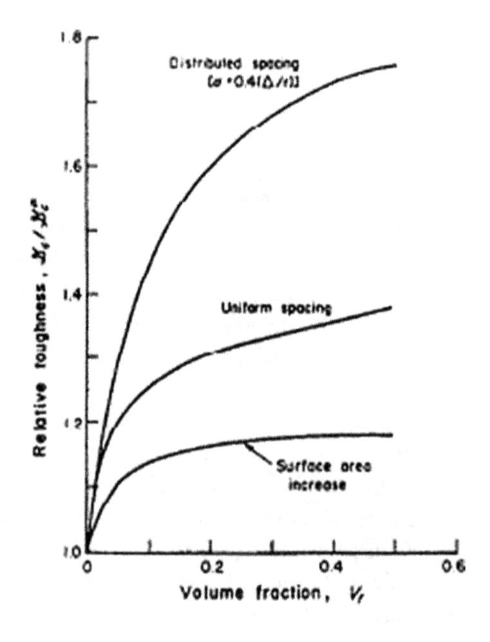

Fig. 8.47 Relative toughness predictions from crack deflection model for spherical particles including the effect of a distribution in interparticle spacing. Predictions are compared to the relative toughness due to surface area increase [20]. With kind permission of Elsevier

a high aspect ratio. Toughening results primarily from the twist of a crack front between particles, as indicated by the deflection profiles. Less effective for toughening are disc-shaped particles and spheres, respectively. These predictions provide the basis for the design of high toughness, two-phase ceramic materials. In addition, the ideal second phase should be present in amounts of 10–20 vol%. If the percentage is higher, the effect diminishes. Particles with high aspect ratios are most suitable for maximum toughening, especially particles with rod-shaped morphologies.

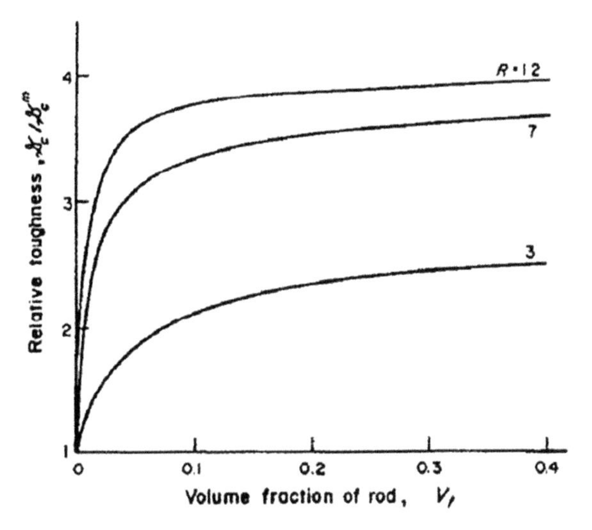

Fig. 8.48 Relative toughness predictions from crack deflection model for rod-shaped particles of three aspect ratios [20]. With kind permission of Elsevier

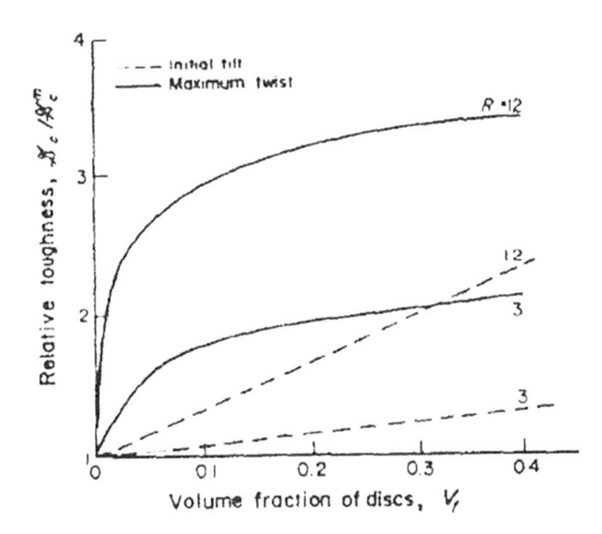

Fig. 8.49 Relative toughness predictions for initial tilt and for maximum twist based on crack deflection model for disc-shaped particles of two aspect ratios [20]. With kind permission of Elsevier

The experimental evidence correlates well with the guidelines developed for toughening brittle materials by crack deflection processes.

b) Crack bowing is relevant to an explanation of the difference between 'crack deflection' and 'crack bowing'. In the latter case, a nonlinear crack front develops, due to the resistance of a second-phase particle which happens to be in the path of an advancing crack.

Because of the resistance of the particles, the crack bows between them, causing the stress intensity, K, at the particle to increase, while, in the bowed segment of the crack, a softening effect occurs, i.e., K decreases along this segment. Bowing

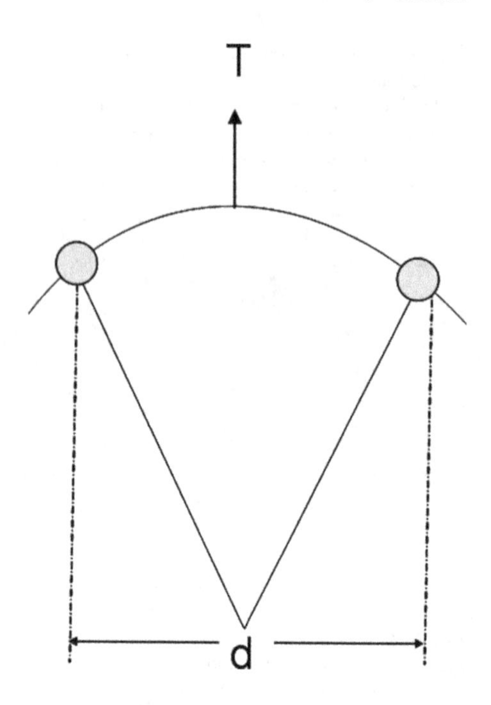

Fig. 8.50 Modified illustration of Fig. 3.45 from Sect. 3.38, Chap. 3

increases until the fracture toughness of the particle is overcome, thus enabling the crack to advance. As such, strong particles act as barriers to crack propagation, acting as pinning points with a distance, d, between them. The bowing out resembles the line tension, T (discussed in Sect. 3.3.8 in the chapter on dislocations). Figure 8.50 is a modified illustration of Fig. 3.45 (from Sect. 3.3.8, Chap. 3).

Lange [36] has analyzed the effect of the pinning points and expressed the fracture energy (i.e., the energy/unit area required to initiate fracture) as:

$$\gamma = \gamma_0 + \frac{T}{d} \tag{8.64}$$

where γ_o is the energy/unit area required to form a new fracture surface, T is the critical line energy/unit length (tension is the energy-per-unit length) of the crack front and d is the distance between pinning points. An estimate for the critical line energy is given by:

$$T = C\gamma_0 \tag{8.65}$$

where the value of C may be approximated by the crack size in the material. By substituting values for C and γ_0, one may evaluate the line tension. The line energy of a crack front in a glass composite (sodium borosilicate glass-Al_2O_3 composite), for example, was estimated as 15–60 ergs/cm. Often in the literature, the factor 2 precedes Eq. (8.64), to indicate the energy required to form two surfaces. The fracture energy data obtained from the work of the sodium borosilicate

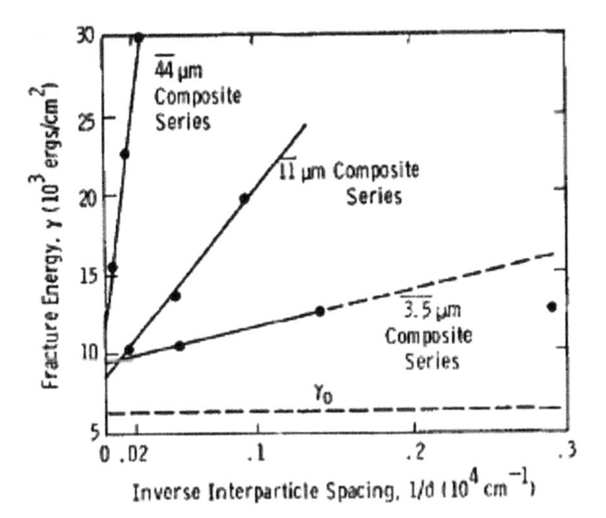

Fig. 8.51 Fracture energy of composite series versus average inverse interparticle spacing [39]. With kind permission of John Wiley and Sons

glass-Al_2O_3 composite are plotted as a function of the inverse average interparticle distance (l/d) in Fig. 8.51.

As may be seen, each of the series resulted in a linear plot. The slope of each plot equals the line energy of the crack front. The effect of the particle size may be incorporated into Eq. (8.64) as:

$$\gamma = \gamma_0 + F(D)\frac{T}{d} \tag{8.66}$$

F(D) is a dimensionless function of particle size D, where $0 \leq F(D) \leq 1$. When the pinning positions do not effect crack propagation, $F(D) = 0$ and, thus, the second term has no effect on the fracture energy. Unlike the case of deflection, no size effects were observed from the toughening of brittle materials, like ceramics, by the bowing process. The size of the pinning particles (and their fraction) determine the spacing, d, between the pinning points. Like dislocation pinning by obstacles and the concept of break-away stress (Pelleg), crack fronts are thought to break away from their pinning positions. The distance between the pinning positions decreases as the volume fraction of the particles increases; this, in turn, increases γ, as seen in Eq. (8.66). A schematic illustration of the break-away position of a crack front from its pinning sites appears in Fig. 8.52.

The calculated strength is compared with experimental strength data in Fig. 8.53, which is a plot of the flexural strength versus the volume fraction of Al_2O_3 dispersed in sodium borosilicate glass-Al_2O_3. We may conclude this section on crack bowing as follows. In contrast to deflection, there is a particle-size effect on toughness. A volume fraction increase increases toughness due to the decreasing distance between crack-anchoring particles, according to the second term in Eq. (8.66) or (8.64). This means that there is interaction between the crack front and the second phase (particle) dispersion.

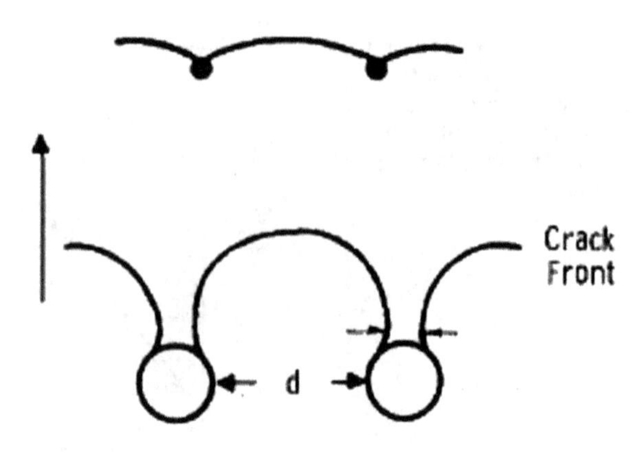

Fig. 8.52 Breakaway position of crack front from 2 pairs of pinning positions, each separated by distance d. Distance between 2 arms of crack front anterior to each pinning position (indicated by 2 *arrows*) is hypothesized to control breakaway position because of overlapping stress field. Large arrow is direction of crack propagation [36]. With kind permission of John Wiley and Sons

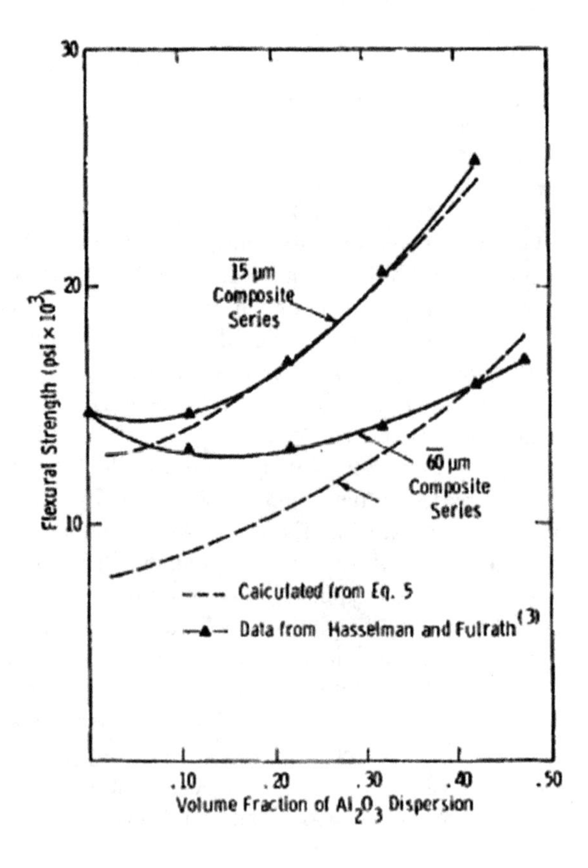

Fig. 8.53 Comparison of strength data of composite series reported in Hasselman and Fulrath with strength values calculated from Eq. (8.66) [36]. With kind permission of John Wiley and Sons

An increase in particle size decreases the distance between them and decreasing distance means an increase in fracture energy. Larger particle-size dispersions are more effective in interacting with crack fronts than smaller ones. Apparently, strengthening brittle materials by means of the dispersion of larger-sized particles is more effective than doing so by means of smaller ones, though there is a particle-size limit, since large particles tend to increase the size of the crack. Therefore, the best results demand a compromise when choosing the particle size for dispersal.

8.3.8.2 Crack-Tip Shielding

As indicated above, crack-tip shielding is a method for strengthening brittle materials by increasing their toughness. Several techniques are possible for inducing crack-tip shielding, among them bridging, transformation and causing plastic yielding in the tip vicinity. These techniques will be briefly considered here:

(a) **Bridging in Pure Ceramics (grain-bridging)**

In ceramics, flaws, in the form of cracks of various sizes, usually exist, unless very special techniques are applied to greatly reduce (though not completely eliminate) their volume fraction. An existing crack in a brittle-ceramic matrix propagates and stress concentrations exist in the vicinity of the crack tip. The microstructure effect on crack propagation and its influence on toughness are considered below. This was directly observed during Vickers-induced flaw experiments. The experimental set up of the test specimen and the schematic indentation tests are shown in Fig. 8.54. Reflected light micrographs are seen in Fig. 8.55. A microstructural observation of grain bridging in alumina appears in Fig. 8.56. Grain-localized 'bridges' across the crack interface, over large distances of several millimeters behind the tip, in nominally pure, CG alumina were observed. A major feature of this experimental procedure is its facility in monitoring the evolution of fracture during the application of stress. Accordingly, direct observations were made of the crack growth by optical microscopy, using the two loading configurations shown in Fig. 8.55a and b. Grain boundaries of the specimen (for observing bridging and additional features) were outlined by polishing with 0.3-μm Al_2O_3 powder. The observations indicated that, after the sudden propagation of the cracks to the edges, the fractured segments tended to remain intact and additional forces were needed for the separation of the pieces. This observation provided the clue for the idea that the advancing tip is retained by some forces acting across the interface.

Forces resisting further activity were observed to come from some distinct, local ligamentary entity. In Fig. 8.55, one may see that crack extension has occurred along the length of the specimen and in the areas labeled (a), (b) and (c). The progressive evolution through the loading sequence is visible. The areas of this figure are magnified in Fig. 8.56a, b, and c. In zone A of Fig. 8.55, the formation and rupture of a single ligamentary bridge through all six stages may be followed (better seen in the magnified illustration in Fig. 8.56).

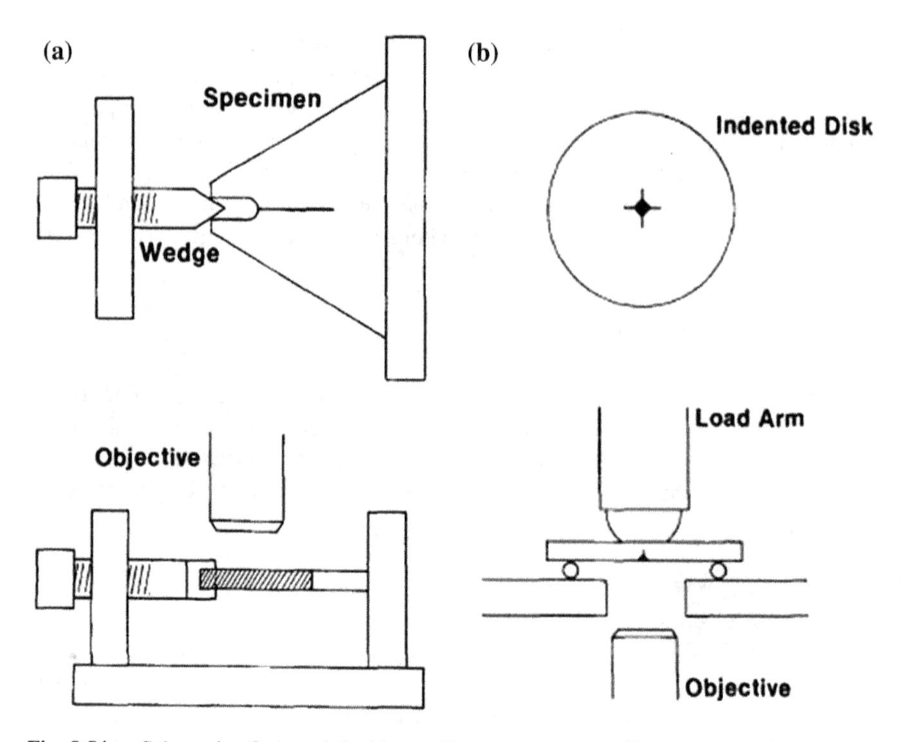

Fig. 8.54 **a** Schematic of tapered double cantilever beam test specimen used to observe crack growth during loading: **a** top view; **b** side view. Specimen cut from triangular slab, 12-mm edge length and 2-mm thickness to produce crack 7-mm long. Starter notch length 300 μm, radius 100 μm. Wedge angle 60°; (**b**) Schematic of indentation flaw test used to observe radial crack evolution to failure: (**a**) plan view, showing Vickers flaw on tensile surface; (**b**) side view, showing flexure system. Specimen dimensions 25-mm diameter by 2-mm thickness. Biaxial loading, 2-mm radius punch on 10-mm-radius (3-point) support [54]. With kind permission of John Wiley and Sons

More detailed information on the crack interface events are found in Fig. 8.57 and indicate physical contact restraints that persist even though that the crack interface is open to a large extent. Figure 8.58 presents a slightly more complex picture. Here, the grains in the centers of the fields of view have developed secondary microfractures in the base region of attachment to one of the crack walls. There is a strong element of transgranular failure associated with this microfracture process, particularly evident in Fig. 8.58a. Figure 8.59 illustrates a case in which a bridging grain has broken away from both walls and is presumably on the verge of detachment from the interface. Indeed, some minor fragments of material have already been thrown off as fracture debris, visible at lower left of the micrograph. These micrographs present bridging in pure alumina. There are indications in the literature that the interface-bridging mode may be far more widespread than hitherto suspected.

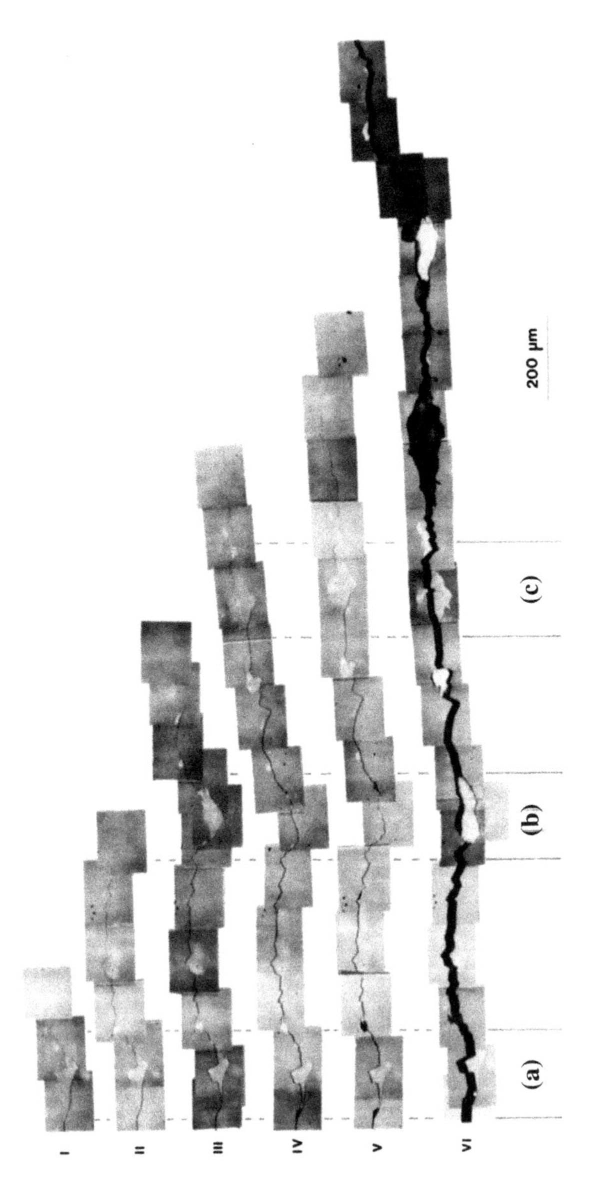

Fig. 8.55 Reflected light micrograph of crack evolution in tapered DCB specimen of alumina, shown at six stages of loading. Wedge remains inserted in notch (just out of field at *left*) in all stages [54]. With kind permission of John Wiley and Sons

(b) **Particle bridging**

In (a), grain-localized bridging was considered as one of the toughening mechanisms in ceramics, responsible for increased toughness. Bridging grains are wedged in the microstructure by internal compressive forces, which lead to an increase in fracture toughness as the crack grows. In this section, the effect of particles (second phase) on bridging is considered. The particles added during

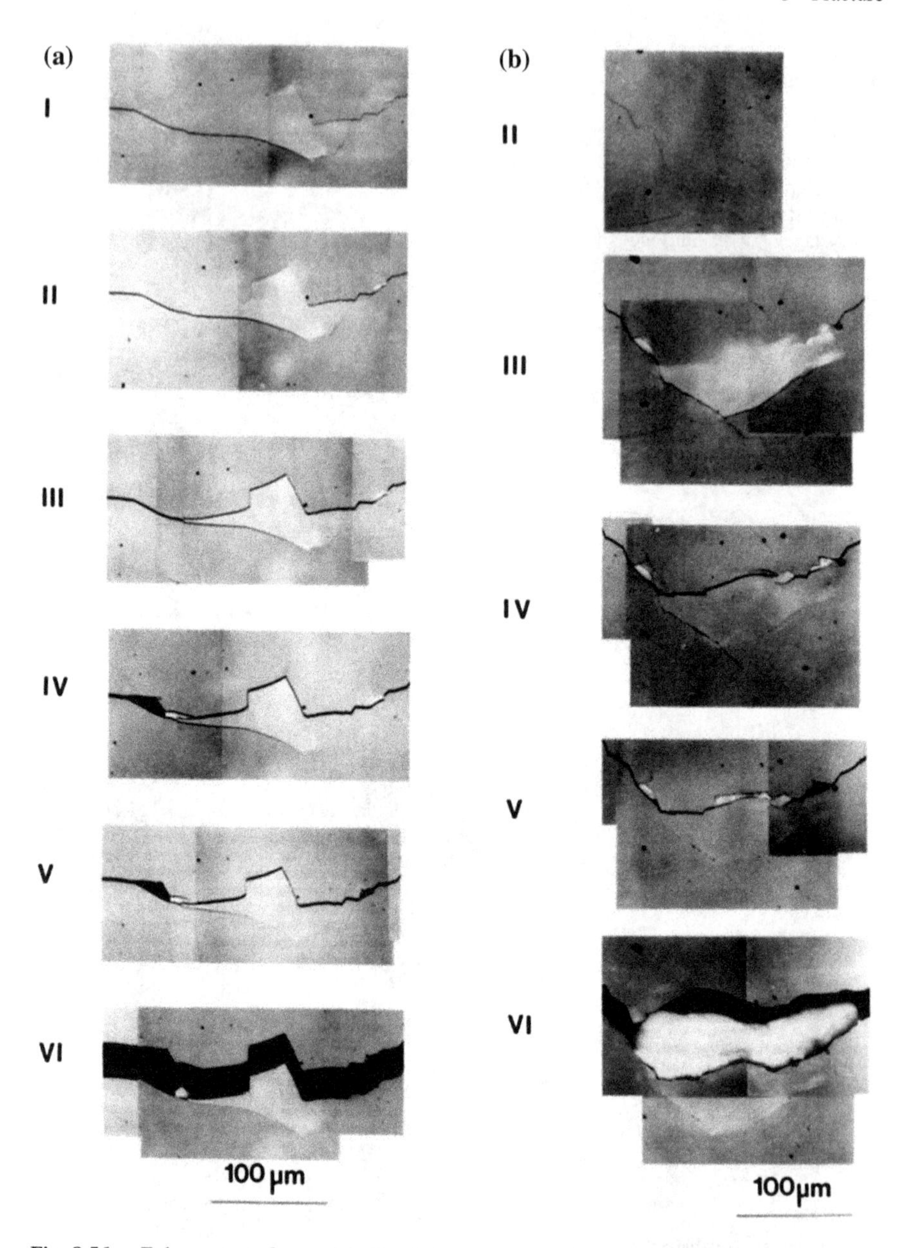

Fig. 8.56 **a** Enlargement of zone A in Fig. 8.55, showing evolution of a grain bridging site from inception to failure. Persistence of interface related secondary cracking is apparent through stage V. **b** Enlargement of zone B in Fig. 8.55. Note continually changing course of the local fracture path through the loading to failure. **c** Enlargement of zone C in Fig. 8.55 [54]. With kind permission of John Wiley and Sons

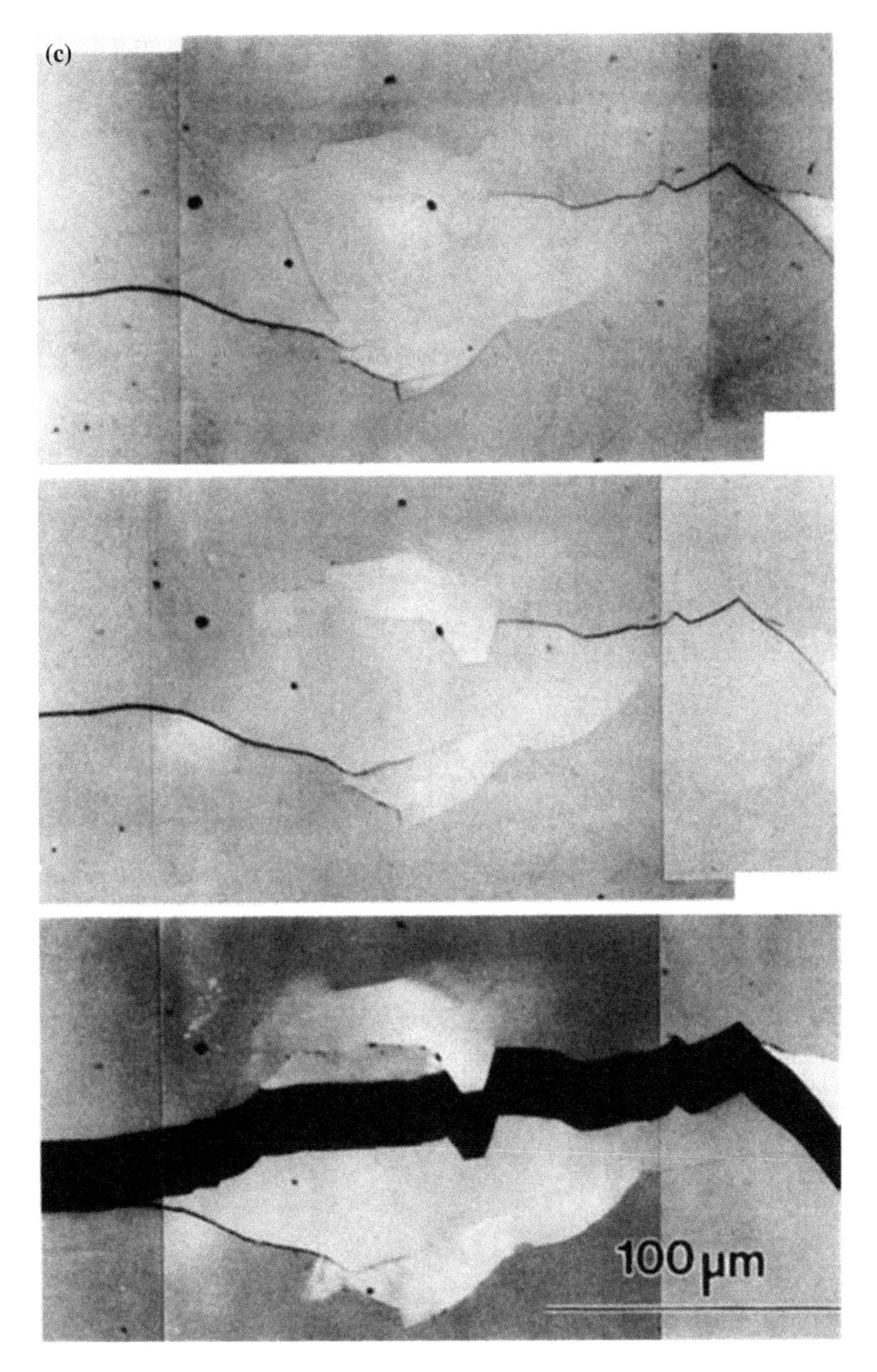

Fig. 8.56 (continued)

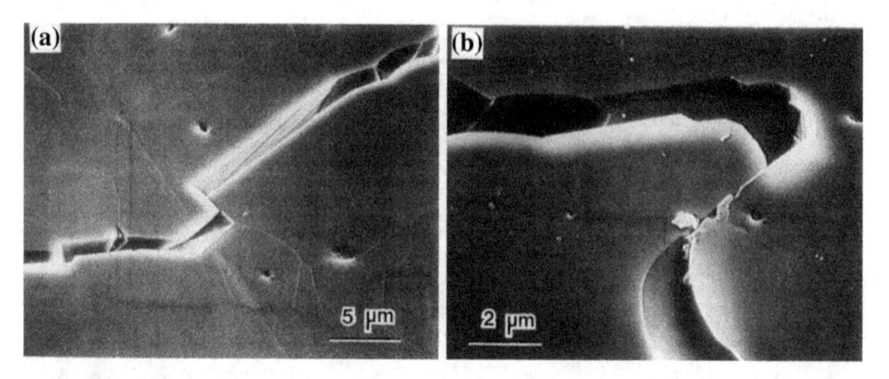

Fig. 8.57 Scanning electron micrographs of fractured-but-intact alumina disk, showing examples of apparent frictional interlocking at grain bridging sites [54]. With kind permission of John Wiley and Sons

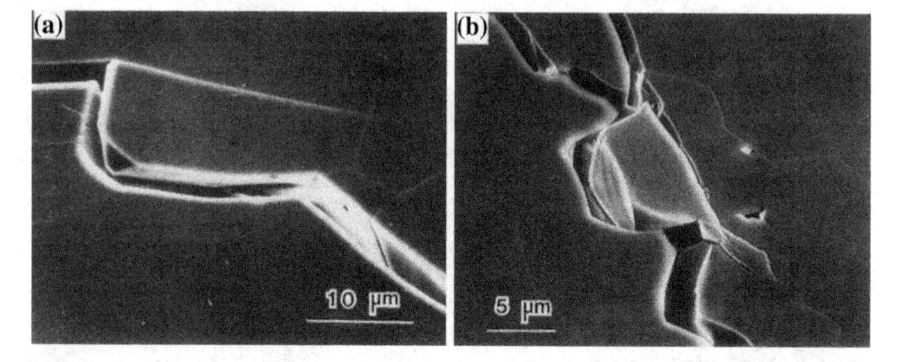

Fig. 8.58 Scanning electron micrographs, showing secondary micro-fracture about bridging grains indicating the intensity of interface traction forces [54]. With kind permission of John Wiley and Sons

Fig. 8.59 Scanning electron micrograph, showing detachment of bridging grain from fracture interface [54]. With kind permission of John Wiley and Sons

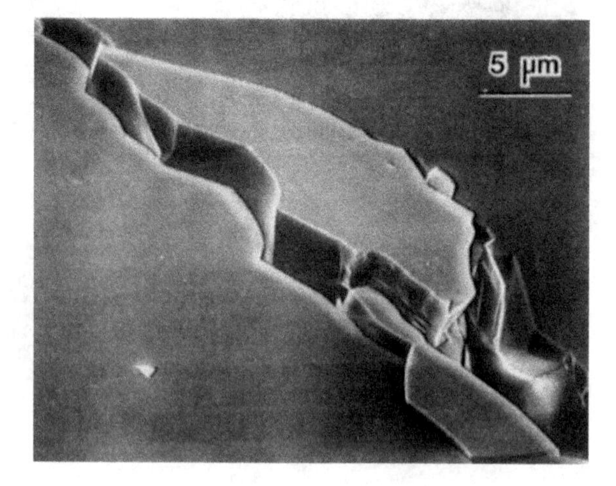

processing may be brittle or ductile. Nevertheless, since the formation of a ductile surrounding in the vicinity of a crack tip is considered a basic reason for the observed strengthening by bridging, the particles causing the toughening are considered to be ductile.

But before moving ahead, a few general words on crack toughening. In Fig. 8.60, a summary of extrinsic toughening mechanisms is listed (an extension of Fig. 8.41). A schematic illustration of crack-tip shielding, including ductile particle toughening, is shown in Fig. 8.61. Crack-tip shielding occurs in the vicinity of the crack tip, either in the region ahead of the crack tip (referred to as the 'frontal zone') or behind it (in the 'wake zone'), as indicated in Fig. 8.61. Toughening by the interaction of a propagating crack, involving crack-tip shielding may be expressed (according to F. J. Lino [38]) as follows: Overall toughness, T, is given by:

$$T = T_0 + T_\mu \tag{8.67}$$

and the crack resistance by:

$$R = R_0 + R_\mu \tag{8.68}$$

In the above relations, T_0 is intrinsic toughness, T_μ is the extrinsic toughness mechanism (crack-tip shielding), R_0 is the fracture resistance energy and R_μ is the crack resistance energy contribution. The critical condition for crack extension is then given by:

$$K_c = K_a = T_0 + T_\mu = T \tag{8.69}$$

or

$$G_c = G_a = R_0 + R_\mu = R \tag{8.70}$$

where K_c and K_a are the critical and applied stress intensity factors and G_c and G_a are the critical and applied mechanical-strain energy release rates.

Since crack-tip shielding events are irreversible by nature, it is expected that the toughness of ceramics will increase with the crack extension (T- or R-curve behaviors). This implies that the toughening terms, T_μ or R_μ, are functions of crack length, c. Therefore, Eqs. (8.69) and (8.70) may take the form:

$$T(c) = T_0 + T_\mu(c) \tag{8.69a}$$

or

$$R(c) = R_0 + R_\mu(c) \tag{8.70a}$$

In Fig. 8.62, plots of fracture strength, σ_f, and toughness versus crack size, c, are shown for materials exhibiting both non-R-curve and R-curve behaviors. Bridge formation and evolution is illustrated schematically in Fig. 8.63.

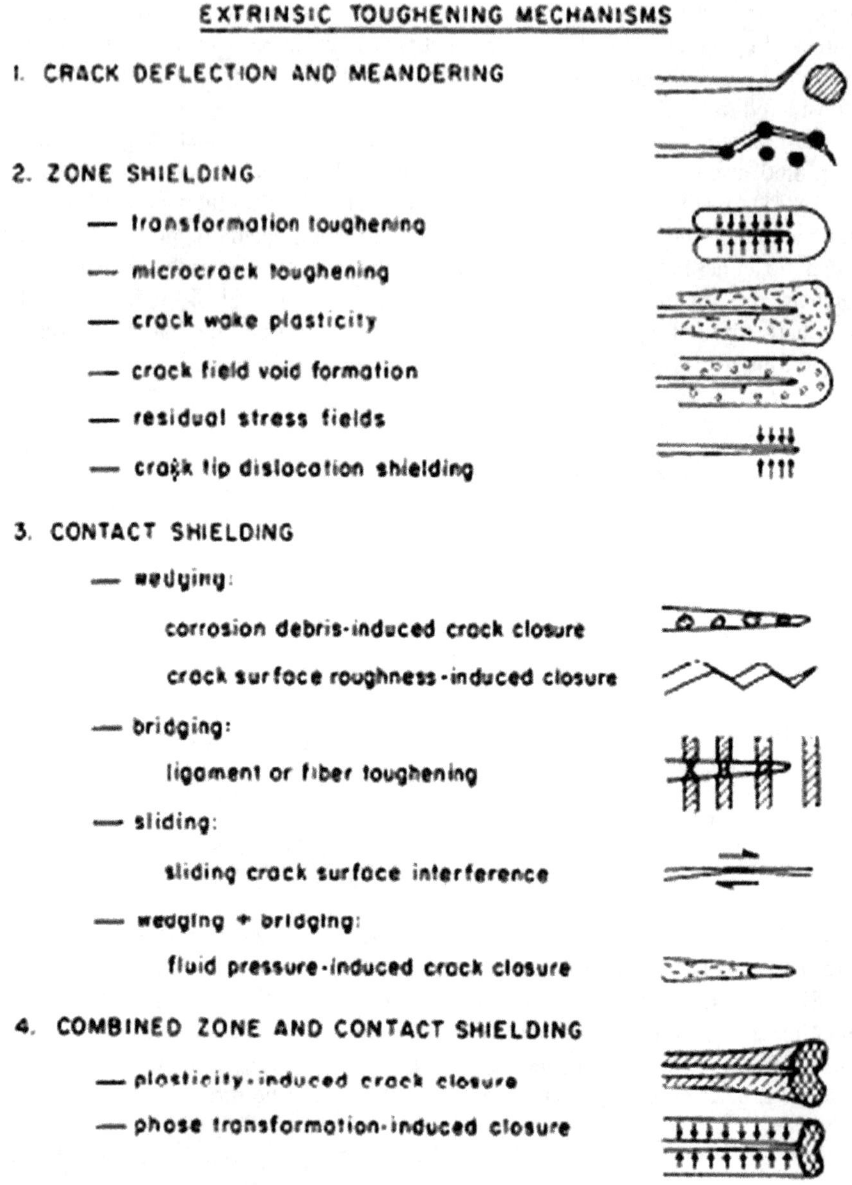

EXTRINSIC TOUGHENING MECHANISMS

1. CRACK DEFLECTION AND MEANDERING

2. ZONE SHIELDING

— transformation toughening

— microcrack toughening

— crack wake plasticity

— crack field void formation

— residual stress fields

— crack tip dislocation shielding

3. CONTACT SHIELDING

— wedging:

corrosion debris-induced crack closure

crack surface roughness-induced closure

— bridging:

ligament or fiber toughening

— sliding:

sliding crack surface interference

— wedging + bridging:

fluid pressure-induced crack closure

4. COMBINED ZONE AND CONTACT SHIELDING

— plasticity-induced crack closure

— phase transformation-induced closure

Fig. 8.60 Schematic representation of the classes and mechanisms of crack-tip shielding [46]. With kind permission of Professor Ritchie

A SEM micrograph of crack propagation, obtained in situ in an Al_2O_3-Al_2TiO_5 composite, is shown in Fig. 8.64. The advancing crack tip appears to be attracted to the alumina-aluminum titanate [henceforth: A-AT] interphase interfaces. This observation implies that the high residual stress (and possibly the elastic modulus

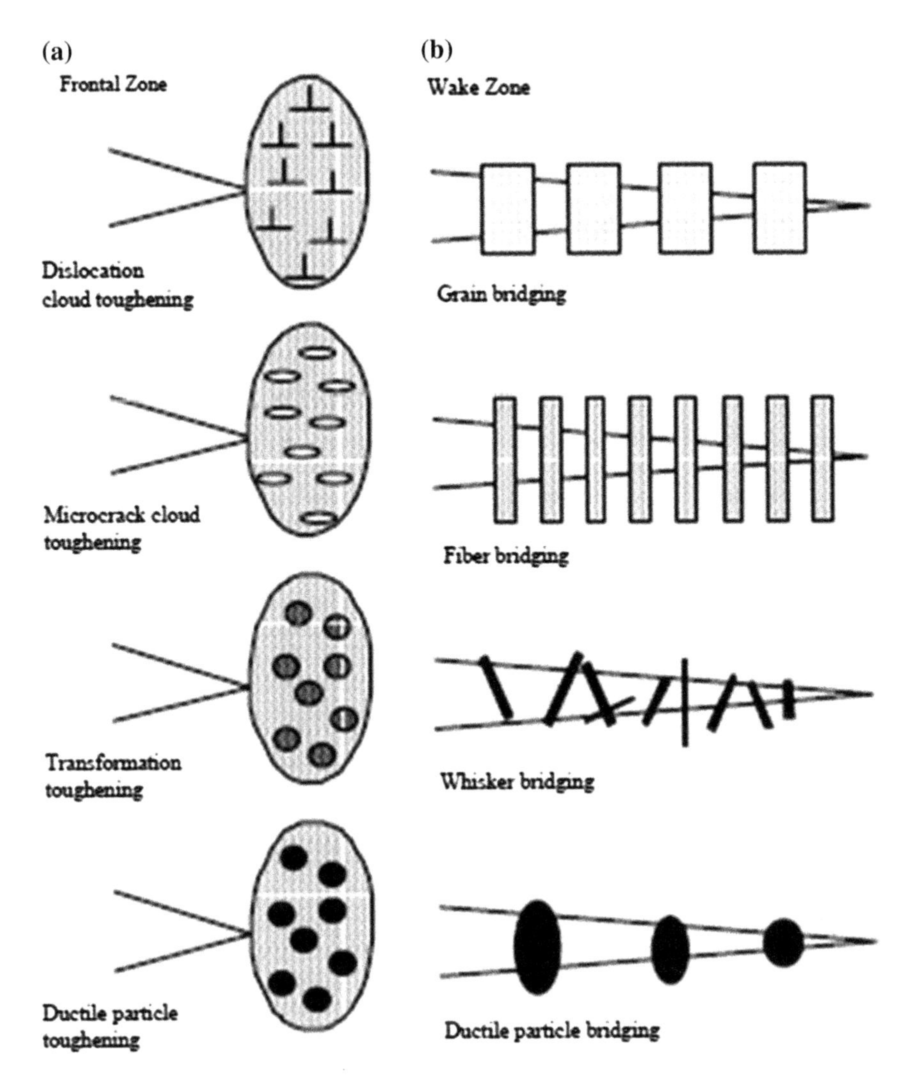

Fig. 8.61 Schematic illustrations of crack-tip shielding processes in ceramics (redrawn after Lawn) [21]. With kind permission of Dr. Jorge Lino

mismatch), due to the presence of AT, is instrumental in the formation of bridging elements in the A-AT composites. Here, AT is a second-phase involved in the bridging.

Now, back to the effect of ductile particles on bridging in brittle materials which cause toughening–a more specific schematic illustration of brittle ceramic toughening via crack bridging by ductile particles is illustrated in Figs. 8.65–8.67. Generally, it is assumed that the crack-bridging forces provided by the still unbroken particles improve the fracture toughness of the matrix, thus also of the composite aggregate. The toughening of the brittle matrix is due to the bridging action involving

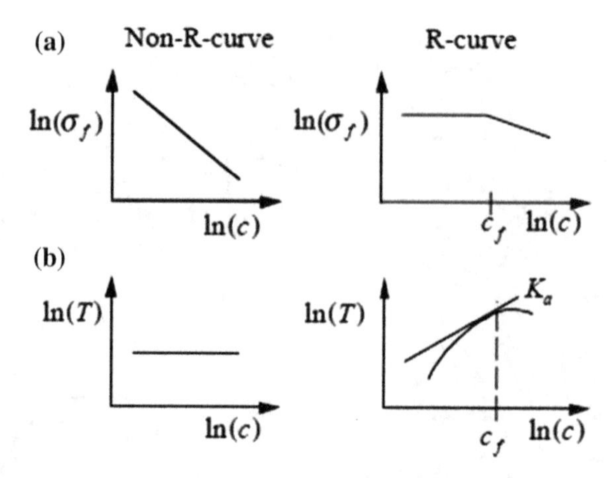

Fig. 8.62 Comparison of materials exhibiting non-R-curve and R-curve behavior; **a** for non-R curve materials, the fracture strength (σ_f) decreases with increasing flaw size. R-curve materials, however, exhibit a range of crack sizes over which the fracture strength is invariant, i.e., they are "flaw tolerant"; **b** for non-R-curve materials, the toughness (T) is a constant, independent of crack size. For R-curve materials, the toughness increases with crack size. c_f denotes the crack size below which the fracture stress is constant (redrawn after Harmer et al.) [21]. With kind permission of Dr. Jorge Lino

reinforcement of the crack surfaces; the toughening mechanism is indicated in Fig. 8.65. An existing flaw in a ceramic, in the form of a crack, will propagate and, because of the stress concentration around the particle, will encircle it while it is in the unbroken stage, becoming approximately a macroscopic planar crack perpendicular to the applied stress (Fig. 8.65a). In this configuration the inclusions act as bridges between the opposing faces of the crack, preventing excessive crack opening and, thereby, reducing the stress-intensity factor, K_I, at the crack tip (Fig. 8.65c).

For crack extension, a critical value of the intensity factor, K_{Ic}, is required which is greater than it would be in the absence of the particles; a higher applied stress is required for crack propagation. This indicates ceramic toughening. According to Fig. 8.65, complete crack-surface bridging is effective until the particle stretch in the center of the crack reaches a critical value, δ corresponding to the crack size a=ac as shown in Fig. 8.65b. For $a > a_c$ and $K_I = K_{Ic}$, the particles in the center portion of the crack rupture and the crack-surface bridging is only partial (see Fig. 8.65c). Note that the deflection of the crack toward the adjacent particle and the stress concentration around the particle in the plane perpendicular to the applied load are possible only if the stiffness (Young's modulus) of the particle is less than that of the matrix (Fig. 8.65a). If, however, the stiffness of the particle is larger than that of the matrix, the crack would generally be repelled by the particle and in the net ligament, between the inclusions (particles), there would stress reduction, rather than concentration (Fig. 8.65d). In such a case, the crack would remain entirely within the matrix and there would be no significant influence of the inclusions on the toughness of the material.

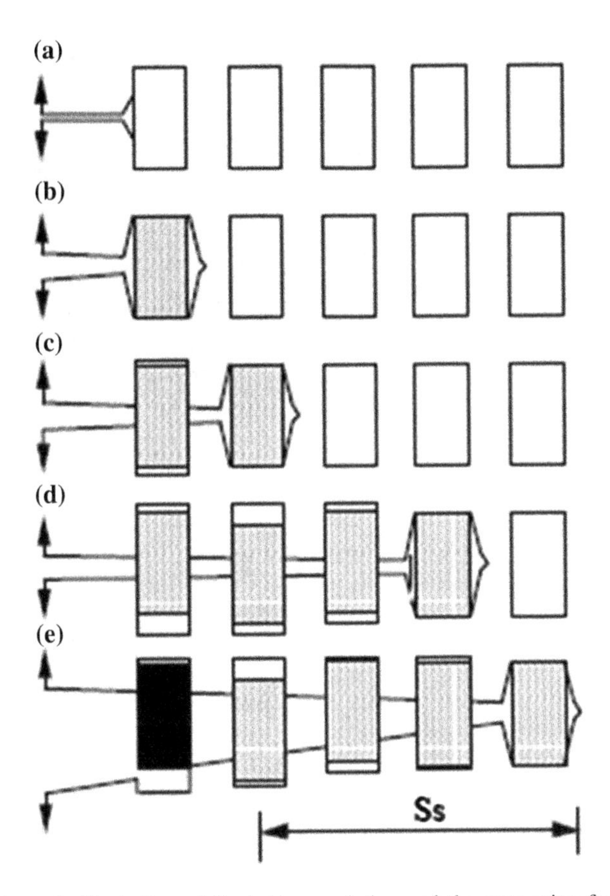

Fig. 8.63 Schematic illustration of the bridge evolution and the successive formation of the bridging zone. Open grains denote potential bridges; shaded grains represent active bridges and closed grains disengaged bridges. **a** crack deflection, **b** debonding, **c** grain pullout, **d** expansion of the bridging zone, and **e** formation of the steady-state (Ss) bridging zone (redrawn after Padture) [21]. With kind permission of Dr. Jorge Lino

Physical considerations require a certain amount of ductility of the particle (inclusion) in order to attain a noticeable improvement in toughness. A mathematical analysis indicates that the effective toughness of the composite ceramics is a function of δc. However, the yielding model for particles in the presence of cracks (shown in Figs. 8.66 and 8.67) indicates that, if the inclusion/matrix interface is very strong (due to high degree of geometrical constraint), the critical crack opening displacement, δc, corresponding to the inclusion rupture, is relatively small; therefore there is no significant increase in toughness. High values for δc are insufficient for producing very ductile inclusions (particles). Thus, Fig. 8.66 shows if the interface strength of an inclusion/matrix is already very low, then debonding of the inclusion occurs and so there is no crack-surface bridging action

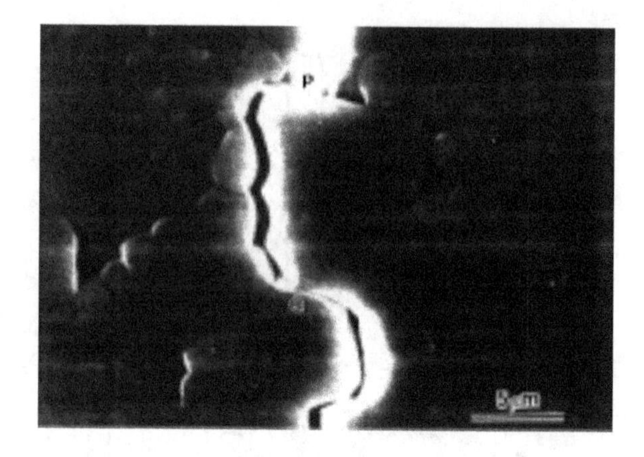

Fig. 8.64 Micrograph of a bridging site taken in situ in the SEM (secondary electrons) during crack propagation in an alumina/aluminium titanate. P and Q denote frictions points during grain pull-out (courtesy of Nitin Padture) [21]. With kind permission of Dr. Jorge Lino

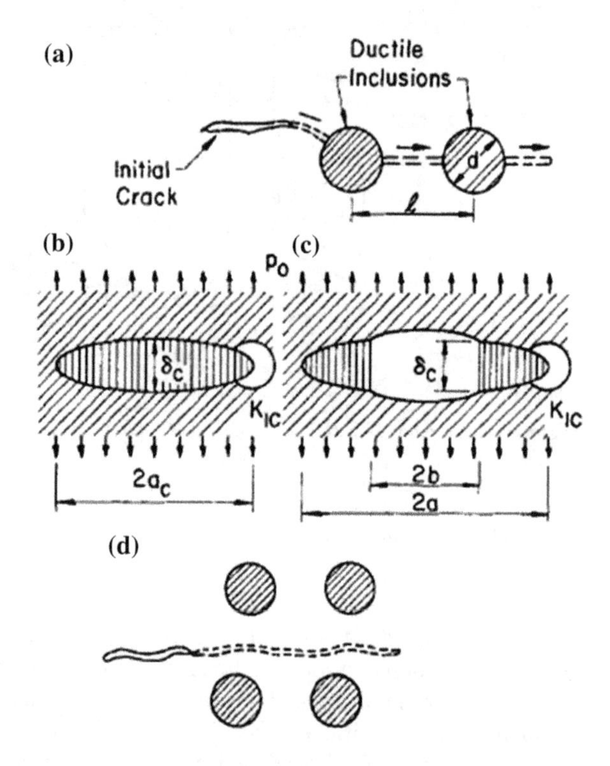

Fig. 8.65 Cracking model for a brittle matrix containing ductile inclusions [17]. With kind permission of John Wiley and Sons

and, consequently, no significant increase in toughness. Significant toughening is expected under the conditions described in Fig. 8.67.

The constraint is removed by some debonding, allowing for the necking of the ductile particles. The bridging force variations in these two cases are shown in Figs. 8.66b (very high interfacial bond) and 8.67b (particle necking). Mathematical

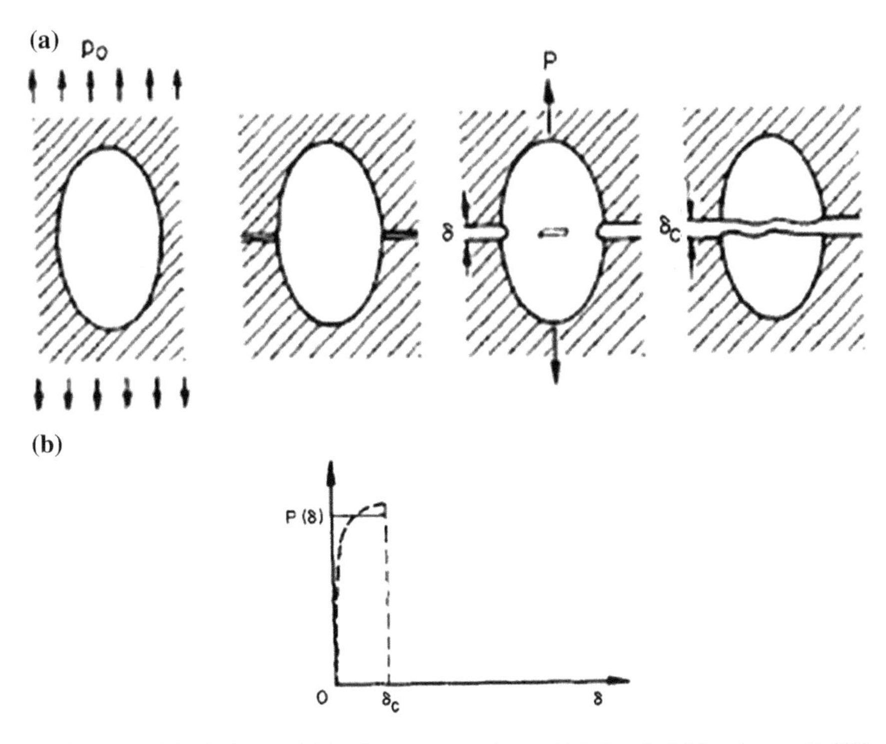

Fig. 8.66 Crack-inclusion model in the presence of very high interfacial bond strength [17]. With kind permission of John Wiley and Sons

analysis and a solution for a penny-shaped crack are provided by Erdogan and Joseph [17] and also an evaluation of the plane strain problem. Interested readers may consult their original work.

The contribution of particles, especially ductile ones, may be briefly summarized as follows. Crack-surface bridging by ductile particles is widely accepted as being one of the primary mechanisms explaining toughening in ceramics. In regard to crack propagation leading to fracture, the significant material parameters are: the fracture toughness of the brittle matrix; the yield strength of the ductile particles; and the degree of ductility or the elongation limit of the particles dispersed in the ceramics during processing. Several quantitative analyses have been performed to help with the design of ceramic materials having more reliability in service.

In this section, limits were imposed on the improvement of toughness in ceramics by the incorporation of brittle, discontinuous reinforcing phases. Observations indicate that such additions bridge cracks in the regions behind the crack tips. Whiskers act in the same manner as discontinuous reinforcing phases. Bridging models may be used to optimize toughening effects and to allow for the choice and modification of pertinent material characteristics, such as physical properties and microstructure.

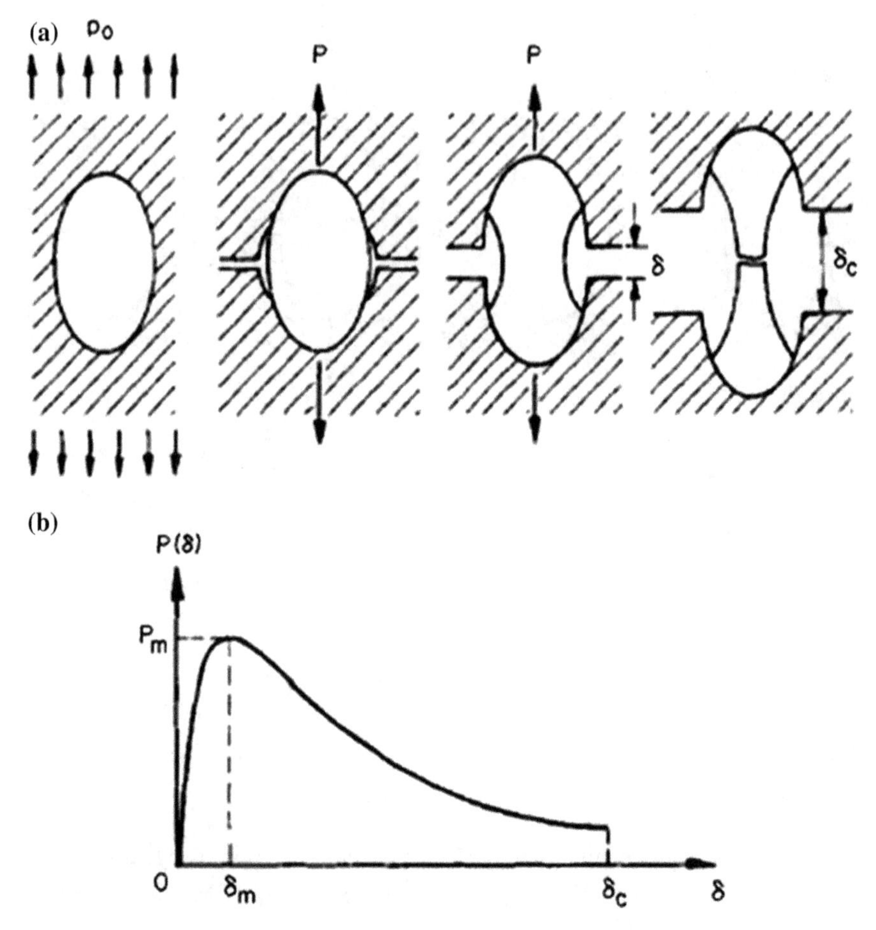

Fig. 8.67 Nonlinear crack surface bridging force model [17]. With kind permission of John Wiley and Sons

(c) **Transformation toughening at the crack tip**

In the literature, the most frequently cited ceramics undergoing transformation is zirconia, which has three allotropes. This section considers its tetragonal monoclinic transformation, which essentially is a Martenisitic one. Zirconia ceramics are highly attractive for a large number of applications, due to their well-researched physical and chemical properties. Among their beneficial mechanical features are: hardness, wear-resistance, a high elastic modulus and creep resistance, even at high temperatures. In addition, zirconia is known to be chemically inert. Despite its poor thermal conductivity and due to the aforementioned properties, zirconia is an important ceramic of great thermomechanical interest for technical use. Moreover, it features a transformation which induces toughening.

In Figs. 8.60 and 8.61, the list on crack-tip shielding includes transformation toughening, which will be briefly considered below.

The critical stress intensity factor (fracture toughness) of a two-phase zirconia system is:

$$K_{Ic} = K_0 + \Delta K_c \tag{8.71}$$

where K_{Ic} is the matrix toughness and ΔKc is the contribution from the shielding mechanisms. Among the toughening contributions to K_{Ic} are: (i) transformation toughening, ΔK_{cT}; (ii) microcrack toughening, ΔK_{cM}; (iii) and crack-deflection toughening, ΔK_{cD}. Earlier, only bridging in pure ceramics and particle bridging were considered. Other mechanisms are also listed in Figs. 8.60 and 8.61, but this section is only concerned with ΔK_{cT}–the transformation induced toughening. For a steady-state transforming zirconia tetragonal particle (as given in Eq. 8.71), ΔK_c becomes ΔK_{cT}, which is given (see, for example [18, 19, 23]) as:

$$\Delta K_{cT} = \frac{\eta E^* e^T V_f \sqrt{h}}{(1 - \nu)} \tag{8.72}$$

Here, η is a factor dependent on the zone shape at the crack tip in the range of 0.22–0.38 and dependent on the nature of the zone, whether it is hydrostatic or shear (Evans [19] was using 0.22). E* is the effective Young's modulus of the material, e^T is the dilatational strain (volume strain) change due to transformation, V_f volume fraction of the transforming particle, h is the width of the transformation zone and ν is Poisson's ratio. Typical examples of the three common forms of stress-induced transformation-toughening microstructures are shown in Fig. 8.68. As is known, the transformation t → m is sluggish when under applied stress at RT. Increasing temperature means the stabilization of the tetragonal phase and, therefore, strength and fracture toughness decrease with increasing temperature in transformation-toughened materials like ZrO_2. As a consequence of the reduced t → m transformation, the volume fraction, V_f, of the transforming particle and the transformation zone, h, decrease considerably, making ΔK_{cT} in Eq. (8.72) smaller.

In PSZ, the toughening is due to the transformation of the precipitates, which can take lenticular shapes. A common PSZ is Mg-PSZ, while the TZP's in use are Y-TZP and Ce-TZP. Also note that in TZP, transformation occurs within the grains (Fig. 8.68b), whereas in ZTA it occurs in the dispersed zirconia particles (Fig. 8.68c). Often the abbreviation DZC (i.e., dispersed zirconia ceramics) is applied to the stress-activated transformation in t-ZrO_2 particles dispersed in a ceramic matrix. In Fig. 8.68c, the ZTA is a commercially developed DZC system. The main objective of the fabrication of a transformation-toughened ceramic is the retention of t-ZrO_2, which transforms into m-ZrO_2 at or close to RT under the influence of an applied stress having a shear component. Control of the composition and thermal treatment are also important parameters in the fabrication of

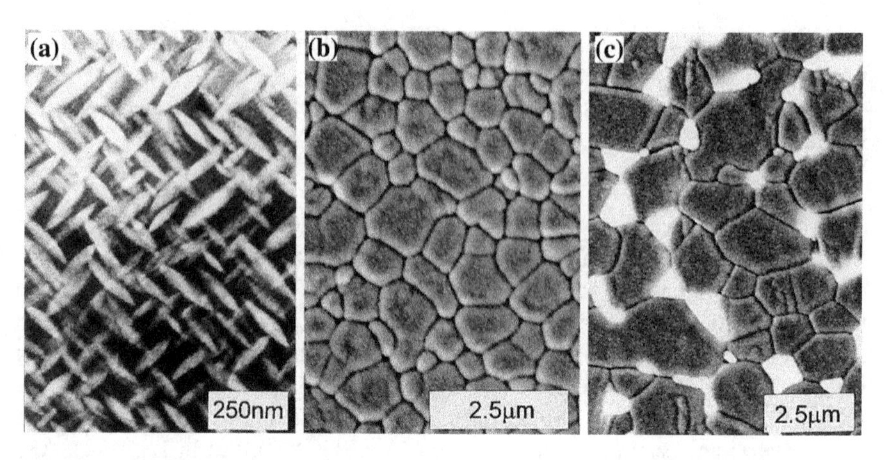

Fig. 8.68 Typical microstructures of the three common forms of TTZ alloy: **a** TEM micrograph of t precipitates in Mg-PSZ; and SEM micrographs of **b** Y-TZP and **c** ZTA. In **c**, the ZrO_2 grains are in bright contrast [24]. With kind permission of John Wiley and Sons

transformation-toughened ceramics and should be such that the tetragonal phase does not transform spontaneously into a monoclinic phase upon cooling. It is desirable for some stabilized tetragonal phase to remain (perhaps by using various stabilizing additives, like yttria), when stress-aided transformations occur. In order to determine the ΔK_{cT} for Eq. (8.72), the parameters, E*, e^T, V_f and h may be obtained from independent measurements before or after the sample has been fractured.

The crystallography of the t → m transformation was thoroughly discussed by Hannink et al. [24] and is available to those with further interest. Anyhow, two distinct orientation relationships have been established (indicated below) and the corresponding microstructure is shown in Fig. 8.69.

Orientation relationship 1	$(100)_m \| (010)_t$ *and* $[001]_m \| [100]_t$
Orientation relationship 2	$(001)_m \| (100)_t$ *and* $[100]_m \| [010]_t$

In pure ZrO_2, the t → m transformation has been widely investigated, although difficulties were encountered in retaining the tetragonal ZrO_2. Therefore, suitable stabilizing oxides, such as Y_2O_3 and CeO_2, are added to the zirconia; there is also a production requirement for fast cooling after the sintering and solution treatment temperatures to retain metastable t-ZrO_2 at RT. The grain structure obtained is equiaxed. Thus, FG TZP may be attained. In the absence of stress-induced transformation, the t → m transformation may occur athermally (similar to a Martensitic transformation) in specimens cooled below ambient temperature. The orientation relationships indicated above and the microstructure refer to CeO_2-stabilized TZP (i.e., 12CeTZP). MgO is another frequent additive to zirconia, which partially stabilizes it. Figure 8.70 is a TEM micrograph of Mg-PSZ. In Eq. (8.72), the volume-fraction and square-root parameters of the particles are

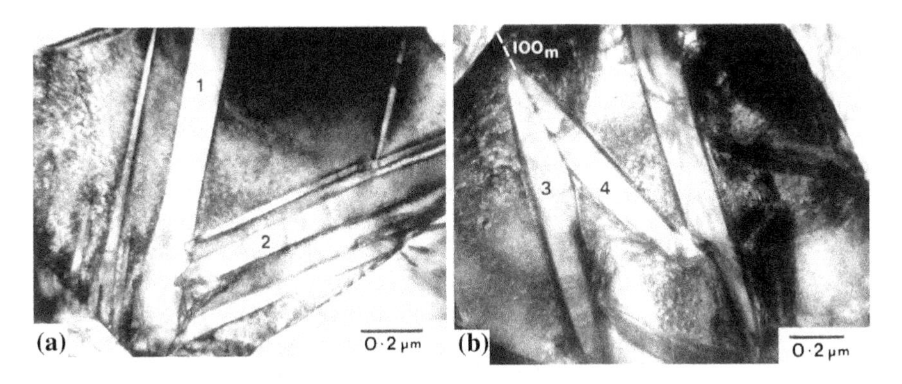

Fig. 8.69 TEM micrographs of partially-transformed t-ZrO_2 grains in CeO_2-stabilized polycrystalline TZP. Note that the m variants form in partially self-accommodating networks [24]. With kind permission of John Wiley and Sons

indicated. In Fig. 8.71, K_{cT} is plotted against $V_f x h^{1/2}$ for Ce-TZP, Y-TZP and Mg-PSZ. The straight lines obtained permit the determination of the slopes, which define values for constant η.

Other transformations, such as ferroelastic transformation and twin formation in a system may also induce toughening effects. The former discussion on stress-induced transformation was Martensitic, involving both dilation and shear components of the transformation strain. Twin transformation typically only has a shear component.

Finally, recall the effect of the transformation zone width and the overall strong particle-size dependence on the width as important parameters of toughening (Eq. 8.72). By measured parameters and knowing the toughening magnitude, the width, h, of a particle may be evaluated in accordance with Eq. (8.72a) or, with independent evaluations of the parameters (as indicated above), the value of ΔK_{cT} may be obtained:

$$h = \left[\frac{(\Delta K_{cT})(1 - v)}{\eta E^* V_f}\right]^2 \tag{8.72a}$$

Suffice it to say that crack shielding applies to various types of deformation, among them fatigue (i.e., cyclic deformation). Several toughening mechanisms are listed in Fig. 8.60, among them fiber- (or whisker-) induced toughening. However, having provided enough examples of toughening, the concept of crack healing will now be considered.

8.3.8.3 Crack Healing

The process of crack healing in engineering ceramics is of great technical interest, due to its potential relevance to saving money and extending the service time of

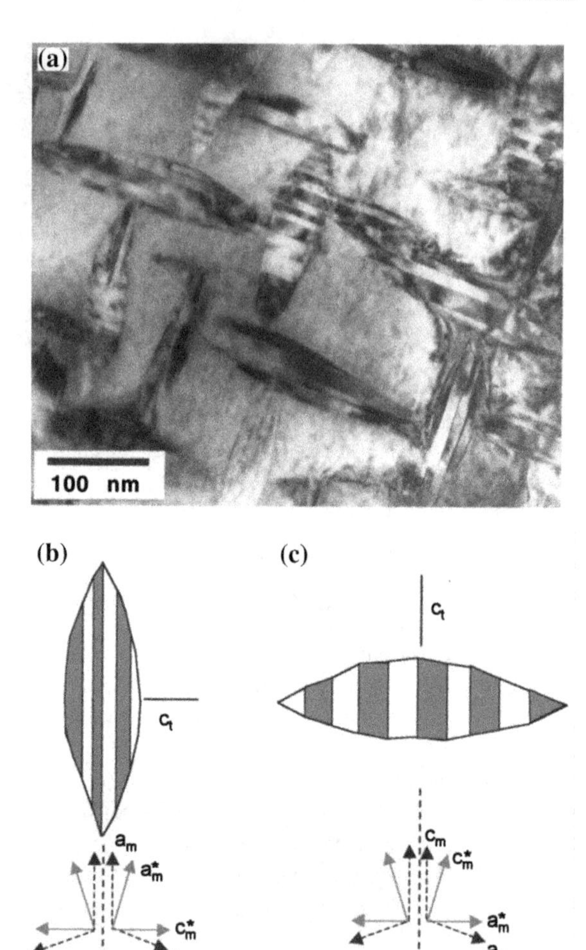

Fig. 8.70 **a** TEM micrograph of m phase particles in Mg-PSZ. **b** and **c** ORs between t and m lattices for the two substructures of parallel m variants. (OR—orientation relationship) [24]. With kind permission of John Wiley and Sons

ceramic manufactures. The concept basically relates to a self-healing process and one would anticipate that temperature and time-at-temperature are critical parameters for inducing self-healing. Ceramics under monotonic and cyclic deformation can be self-healed by various types of loading. The area of flaw and crack repair (recovering the initial properties of the ceramics) has recently become greatly significant in the field of engineering and activity has increased toward discovering more and more ceramic materials that are capable of self-healing. Due to the very good mechanical properties of Si_3N_4, its shock resistance, thermal performance and self-healing properties, this material has been chosen to exemplify the subject under consideration. In particular, Si_3N_4/SiC has very high crack-healing ability (Ando et al. [10]). Crack-healing behavior was mainly investigated under cyclic stress at 1100 and 1200 °C and the resultant cyclic-fatigue strength at the healing temperatures were recorded. The properties of this ceramic material

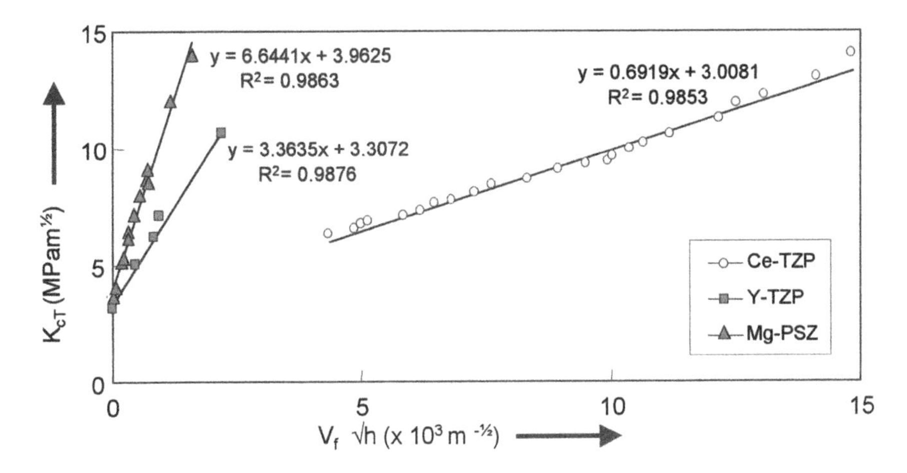

Fig. 8.71 Experimental data for fracture toughness as a function of $V_f \sqrt{h}$ for Mg-PSZ, Y-TZP, and Ce-TZP [24]. With kind permission of John Wiley and Sons

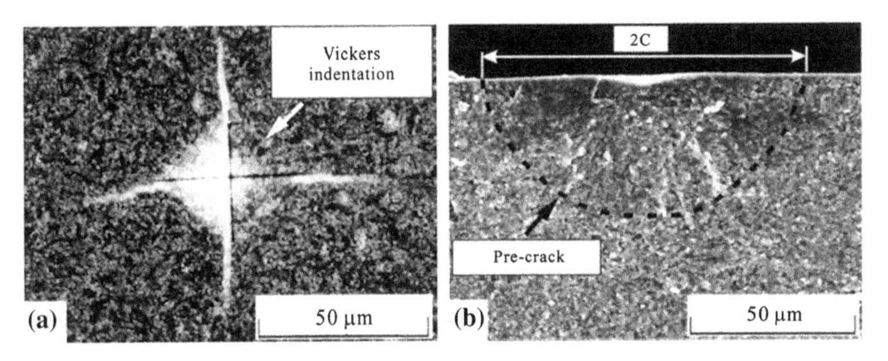

Fig. 8.72 SEM micrographs of **a** indentation crack and **b** fracture surface [10]. With kind permission of John Wiley and Sons

are: mean particle size = 0.2 μm, the volumetric content of α-Si_3N_4 was ~95 % and the rest was β-Si_3N_4. The SiC powder had a 0.27 μm mean particle size. Further details of the ceramic fabrication appear in Ando's paper. In Fig. 8.72, Vickers indentation cracks are shown. The specimen shown in Fig. 8.72 represents sintered ceramics. A semi-elliptical surface crack of 100 μm surface length was introduced at the center of the tension surface of the test pieces with a Vickers indenter using a load of about 20 N. The ratio of depth (a) to half-surface length (c) of a crack (aspect ratio) was a/c = 0.9.

The specimens used were short for two reasons: (a) the aim was to measure the bending strength of the crack-healed zone and not the matrix itself, (b) to reduce the strain energy so as to eliminate breakage of the test specimen into many pieces, which makes it difficult to identify the crack-initiation site. As mentioned at the

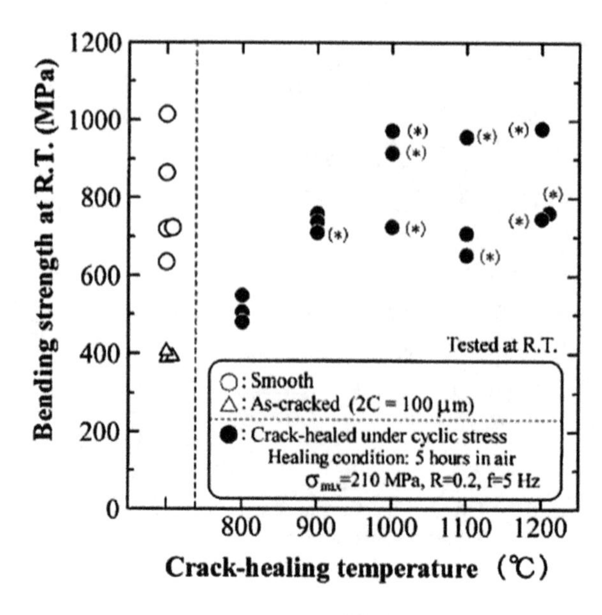

Fig. 8.73 Effect of crack-healing temperature on the bending strength of Si_3N_4/SiC at room temperature. Data marked with an asterisk indicate that fracture occurred outside of the crack-healed zone [10]. With kind permission of John Wiley and Sons

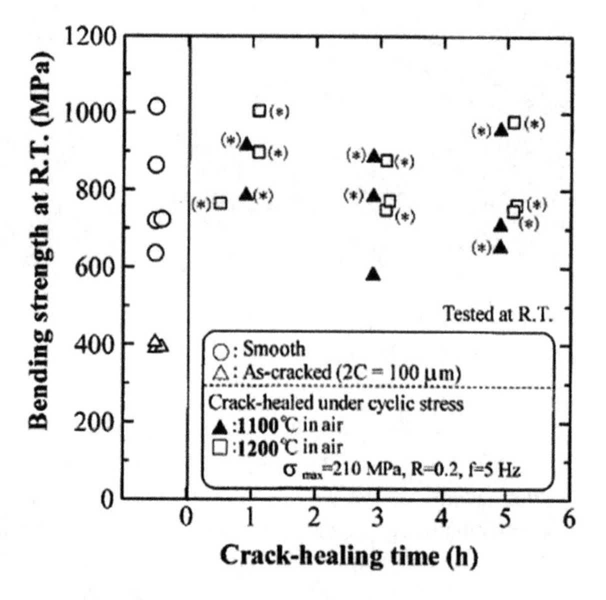

Fig. 8.74 Effect of crack-healing time at 1100 and 1200 °C on the bending strength of Si_3N_4/SiC at room temperature. Data marked with an asterisk indicate that fracture occurred outside of the crack-healed zone [10]. With kind permission of John Wiley and Sons

beginning of this section, temperature and time are critical parameters in the healing process. Thus, healing was performed at 1100 and 1200 °C under cyclic-bending stress, where the maximum bending stress (σ_{max}) was 210 MPa, the stress ratio (R) was 0.2 and the frequency was 5 Hz. The bending strength of as-cracked samples was ~400 MPa, as shown in Figs. 8.73 and 8.74 by the open triangles.

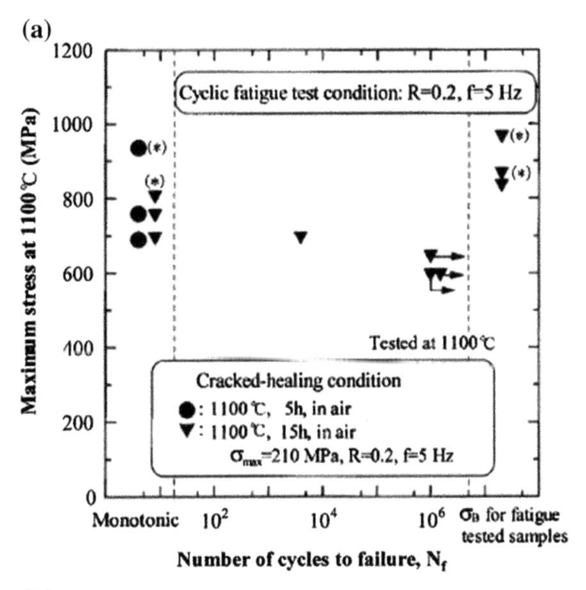

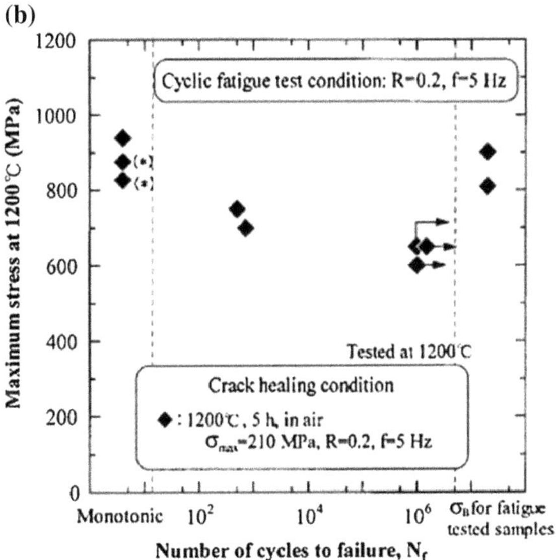

Fig. 8.75 Cyclic fatigue test results of Si_3N_4/SiC at **a** 1100 and **b** 1200 °C. Data marked with an *asterisk* indicate that fracture occurred outside of the crack-healed zone [10]. With kind permission of John Wiley and Sons

The applied stress of 210 MPa was ~53 % of the bending strength of the as-cracked samples and was higher than the RT cyclic-fatigue limit (~200 MPa) for a precracked sample. For performing the self-healing process, at first a cyclic-bending stress was applied and then the temperature increase was conducted at a rate of 10°/min. This way, unexpected crack healing, without applying a stress, was avoided. After crack healing, the monotonic bending tests were applied at RT and crack-healing temperatures of 1100 and 1200 °C. As indicated, cyclic tests were also performed at the crack-healing temperatures. Both the monotonic and

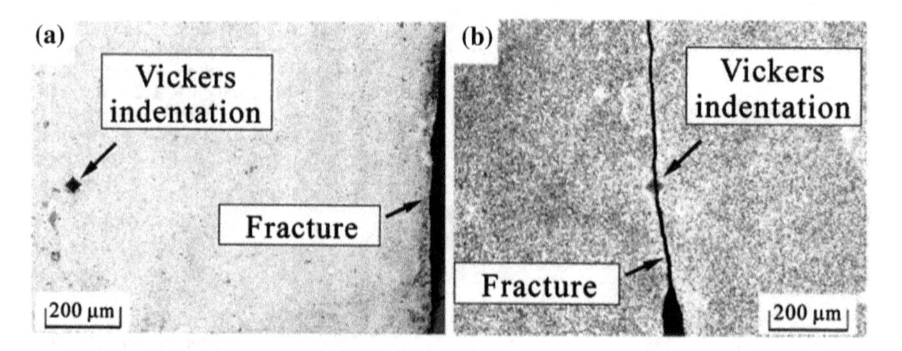

Fig. 8.76 Fracture patterns of crack-healed Si_3N_4/SiC: **a** Fracture occurred outside of the crack-healed zone. (Healing conditions: 1100 °C for 5 h in air, $\sigma_{max} = 210$ MPa, R = 0.2, f = 5 Hz, bending strength at room temperature 653 MPa.) **b** Specimen fractured across the crack-healed zone. (Healing conditions: 1100 °C for 15 h in air, $\sigma_{max} = 210$ MPa, R = 0.2, f = 5 Hz, bending strength at 1100 °C = 759 MPa) [10]. With kind permission of John Wiley and Sons

the cyclic-fatigue tests were conducted using a three-point loading system with a span of 16 mm. The crosshead speed during the monotonic test was 0.5 mm/min. Fracture initiation sites were identified by optical microscope and fracture surfaces were analyzed by SEM.

Note the healing temperatures and the durations of treatment under cyclic stress until crack-healing is attained in the specimens. Clearly, the crack-healing temperature (1000 °C) is a significant parameter of the process. Thus, at temperatures above the crack-healing temperature, e.g., at 1100 or 1200 °C, a surface crack can be completely healed even under cyclic stress. The crack-healed samples recovered their bending strength both at RT and at the crack-healing temperature. Most of the samples failed to heal outside the crack-healed zone (seen marked by an asterisk in Figs. 8.73 and 8.74). Below the crack-healing temperature, e.g. at 800 or 900 °C, the bending strength of the crack-healed zone was insufficient and most samples failed. Since these tests primarily exemplify cyclic stresses, it is of interest to know the numbers of cycles to failure, provided in Fig. 8.75.

Monotonic tests are also presented along the left margins of Fig. 8.75a, and b and symbols indicate the duration at the healing temperatures. The mean values of monotonic bending strength for 1100 and 1200 °C crack-healed samples at each crack-healing temperature are 775 and 881 MPa, respectively. These bending-strength results are comparable to the RT bending strength of smooth unnotched samples of ~800 MPa. The cyclic-fatigue tests were stopped at $N = 10^6$ cycles. Those samples that did not fracture during testing are marked by arrows (→). The applied stress at which a sample did not fracture (up to $N = 10^6$) is defined as the cyclic-fatigue limit (~650 MPa). This value is quite high compared to the value of a smooth specimen (~800 MPa). One might add (as seen from the figures) that healing times between 0.5 and 5 h show no significant difference in crack-healing behavior and a large percentage of the original bending strength can be recovered.

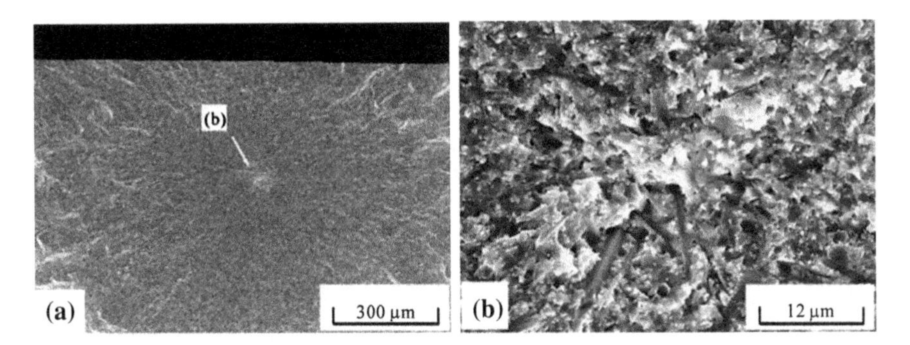

Fig. 8.77 **a** SEM micrographs of fracture surface of crack-healed Si_3N_4/SiC tested at room temperature. Fracture occurred outside the crack-healed zone. (Healing conditions: 1100 °C for 5 h in air, $\sigma_{max} = 210$ MPa, R = 0.2, f = 5 Hz, bending strength at room temperature = 653 MPa), **b** shows the detail of **a** [10]. With kind permission of John Wiley and Sons

To complete this section, illustrations of the crack patterns and SEM micrographs of the fractured surfaces are added (see Figs. 8.76 and 8.77).

Finally, in the results for Si_3N_4/SiC, the contribution of the crystallized additives, SiO_2 and $Y_2Si_2O_7$, to the healing of the precracked specimens should be mentioned. This should not be surprising, since various additives strengthen ceramics. It might be interesting to perform experiments in basic ceramics entirely without additives. In the literature, one may find much research on the healing of a large variety of engineering ceramics.

8.4 Creep Fracture in Ceramics

Unlike conventional deformation, where the response of a material to applied stress is almost instantaneous (particularly in brittle materials), in creep deformation, material deforms slowly under the influence of a permanent load. Deformation may occur when the material is exposed for long time at a relatively high stress, but which is below the yield stress. During creep, strain accumulates as a result of long-term stress. Therefore, creep is a time-dependent deformation. However, creep phenomena are more critical when the materials are exposed to high temperatures. The index of high temperatures is always related to the melting point of the specific material being tested. Generally, the higher the melting point of a material, the lower (but not zero) the likelihood of considerable creep deformation, but this depends on the deformation temperature. As such, modern technology is interested in high T_m materials for the potential manufacture of components or parts for high temperature applications. Creep always increases with temperature. Thus, a summary of the important parameters that determine the creep properties of specific materials is in order: (1) material properties (e.g., T_m); (2) duration of exposure under load; (3) temperature during exposure; and

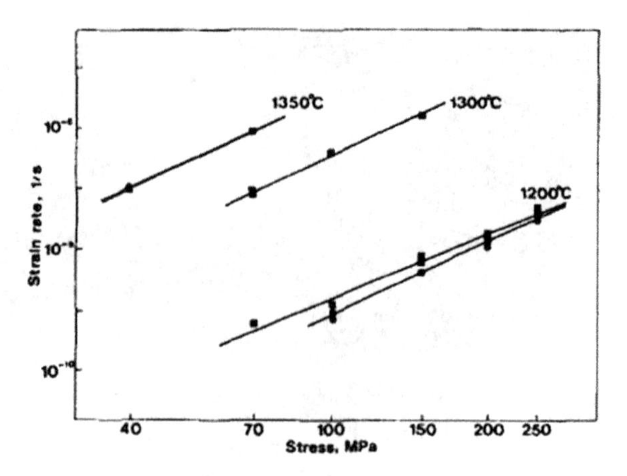

Fig. 8.78 Stress dependence of steady-state creep rate. Creep exponent is 3.2 for the monolithic material (filled square box) and 4.3 for the composite material (filled circle) at 1200 °C, 4.5 for the monolithic material at 1300 °C, and 4.0 for the composite material [40]. With kind permission of John Wiley and Sons

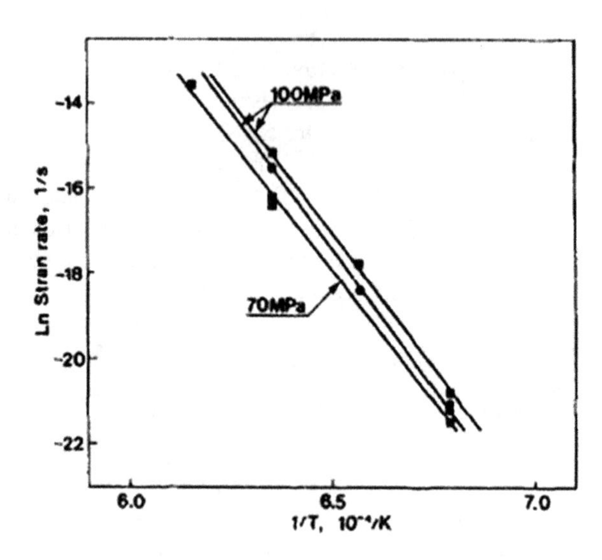

Fig. 8.79 Temperature dependence of the steady-state creep rate. The activation energy is 1065 kJ/mol for the monolithic material (filled square box) and 1190 kJ/mol for the composite material (filled circle) at 100 MPa, and 1032 kJ/mol for the monolithic material at 70 MPa [40]. With kind permission of John Wiley and Sons

(4) applied load. Depending on the magnitude of an applied stress and its duration, a deformation may become so large that a component can no longer perform its function and eventually fails. The effect of all these parameters on the creep rate was previously stated in Eq. (6.2) as:

$$\dot{\varepsilon} = f(\sigma, t, T) \tag{6.2}$$

In Chap. 6, creep was discussed in general, but a considerable portion of the chapter was devoted to creep fracture and creep rupture, so these will not be reviewed here. However, since various additives do influence the basic properties of materials (e.g., second phases, fibers, whiskers), to complete this discussion on creep phenomena in ceramics, an example follows of the effect of whiskers in a

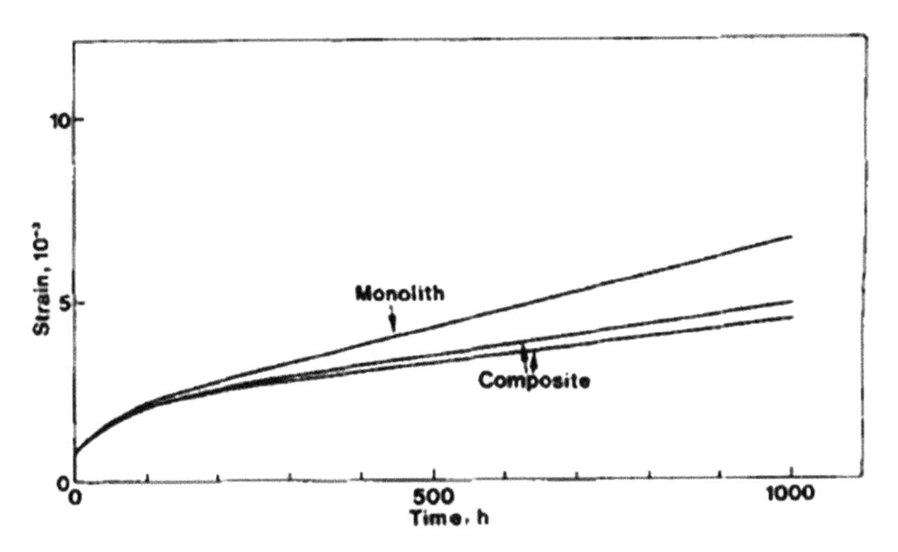

Fig. 8.80 Creep curves of whisker-free (*monolithic*) and whisker-reinforced (*composite*) hot-pressed silicon nitride at 1200 °C and 100 MPa. The tests were interrupted at 1000 h [40]. With kind permission of John Wiley and Sons

whisker-reinforced material on the overall creep properties leading to failure, compared with a monolithic ceramic (in this case, Si_3N_4). Tensile creep at high temperatures, such as 1200 and 1350 °C, is characterized by the formation of microcracks and their effect on creep resistance. Figure 8.78 shows the stress dependence of steady-state creep rates for a monolithic, whisker-strengthened ceramic.

In Chap. 6, one of the empirical relations was given as:

$$\dot{\varepsilon} = B\sigma^n \exp - \left(\frac{Q}{kT}\right) \tag{6.4}$$

The exponent in Eq. (6.4) is indicated in Fig. 8.78 for both monolithic and reinforced Si_3N_4 as 3.2 and 4.3, respectively. A natural logarithmic plot of strain rate versus 1/T allows for the determination of the activation energy according to Eq. (6.4). The plots for both the monolithic and the strengthened Si_3N_4 appear in Fig. 8.79. Figure 8.80 relates the creep strains of the monolithic- and whisker-strengthened silicon nitride. It is believed that this reinforcement suppressed the occurrence of cavity sliding and GBS. The time-to-failure by creep of the monolithic and reinforced materials are compared in Fig. 8.81. The time-to-failure follows the Monkman–Grant relationship discussed in Chap. 6, reproduced as:

$$\dot{\varepsilon}^m_{mer} t_f = C \tag{6.103}$$

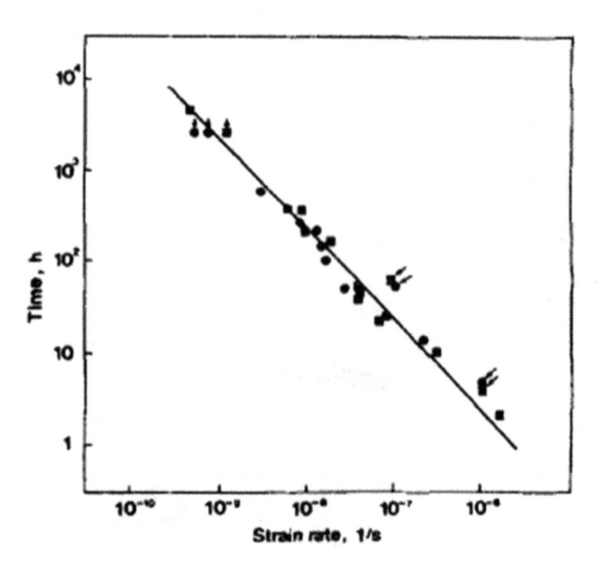

Fig. 8.81 Relationship between time-to -failure and steady-state creep rate for the monolithic (*filled square box*) and composite (*filled circle*) materials. The downward arrows indicate tests with large accelerated creep (tests at 1350 °C). The upward arrows indicate interrupted tests [40]. With kind permission of John Wiley and Sons

8.5 Ductile Fracture in Ceramics

Some ceramics (a very few) show a certain degree of ductile behavior. Single crystal $SiTiO_3$ is one of them. As is common in ductile alloys before fracture sets in, plastic deformation accompanied by 3-stage strain hardening is observed. Surprisingly, $SiTiO_3$ ceramics behave in a similar manner. Thus, dislocation mobility and dislocation interactions must be involved. Indeed, as shown below, this unique ceramic also presents three stages, of which the two first represent real hardening, whereas the third stage shows softening occurring before fracture. Figure 8.82 shows the three stages of the deformation curve in the ductile $SiTiO_3$. The heavy line represents the σ-ε curve and the thin line indicates the stages of the work-hardening rate. Note that if the zero stage (stage 0) is also included, it may be considered a 4-stage hardening curve. By the slope of the elastic region, E may be derived for $SiTiO_3$ as 228 GPa. The decreasing strain-hardening rate portion following the elastic region is stage 0. Unlike the case of metals, where stage I represents linear strain hardening, in $SiTiO_{3,}$ the stress remains low and almost unchanged up to ~7 % strain. In stage II, there is higher work hardening (up to ~17 % strain) followed by a sharp decrease in strain rate (stage III) until fracture sets in. Thus, Ductile ceramics (at least in $SiTiO_3$) somewhat resemble the deformation behavior of FCC metals. The designated red roman numerals represent samples retrieved from seven points corresponding to four stages of different deformation strains. Based on the observed deformation behavior, one might expect to observe slip lines, as traces of dislocation motion in this ceramic, and dislocation structures characteristic of the aforementioned stages of work hardening. Figure 8.83 is an illustration showing slip lines, cross slip and eventual fracture due to crack formation in $SiTiO_3$ ceramics.

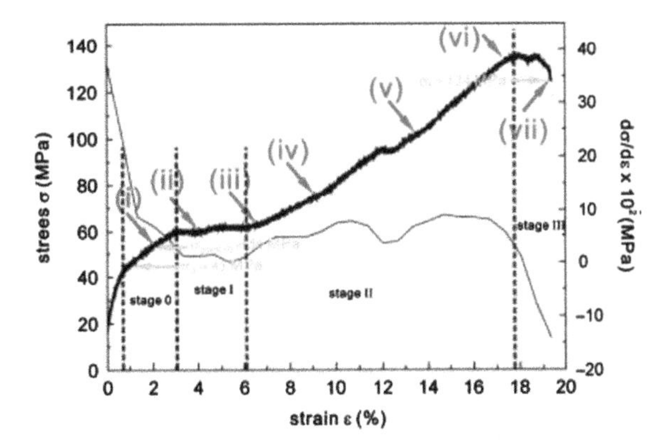

Fig. 8.82 A representative σ–ε curve (*thick curve*) and corresponding hardening rate (*thin curve*) with hardening stages indicated [57]. With kind permission of John Wiley and Sons

The dislocation structure at stage 0 may be seen in Fig. 8.84. Intersecting dislocations are visible in stage 0; the density of the dislocations is $\rho = 8.19 \times 10^{13}\ m^{-2}$ compared to $4.8 \times 10^{13}\ m^{-2}$ of as-received samples (shown in Fig. 8.84b). Four distinctive Burgers vectors of the same ⟨011⟩ family are determined using contrast analysis and the **g. b = 0** invisibility criteria. The dislocations shown in Fig. 8.84 are predominantly of the pure-edge type.

These results confirm that the four [011] $(0\bar{1}1)$, $[0\bar{1}1]$ (011), [101] $(\bar{1}01)$ and $[10\bar{1}]$ (101) slip systems for the six physically distinctive systems in $\langle 0\bar{1}1\rangle$ {011} are activated by a compressive stress (of ~56 GPa at point (i)) along [001]. In stage I, long straight dislocations are illustrated in Fig. 8.85. It has been suggested that, instead of being dislocation dipoles, they are collinear partial dislocations with b = ½$[0\bar{1}1]$ and ½[011], respectively. Kink pairs are formed along these dislocation lines. For stage II to operate, an increase in stress is required to activate the operation of the secondary slip system. In Fig. 8.86, the microstructure characterizing stage II is seen. Two additional slip systems, [101] $(\bar{1}01)$, and $[10\bar{1}]$ (101), are activated, indicating the multiple slip of the stage-II hardening. Cell and wall structures break up upon entering stage III, as seen in Fig. 8.87. As indicated earlier, stage III is characterized by softening. Clearly, some sort of dynamic recovery is involved in the process associated with the break-up of cell-and-wall structures. Dislocation annihilation and edge dipoles or mixed dislocation dipoles pinching-off into loops also occur.

As seen from the above discussion, fracture in ductile ceramics is preceded by dislocation interactions similar to those in metals and the motion of the dislocations resembles that found in ductile materials, leading to fracture usually in stage III. It is possible that in other ductile ceramics similar events also occur. It is usually thought that ceramic materials are brittle at ambient temperatures and that fracture sets in due to crack formation and propagation, which then induces further crack opening until fracture sets in.

In ceramics, such a mechanism is in complete contradiction to the aforementioned dislocation mechanism leading to failure. Other ductile ceramics may involve dislocations in their plastic deformation, for example MgO single

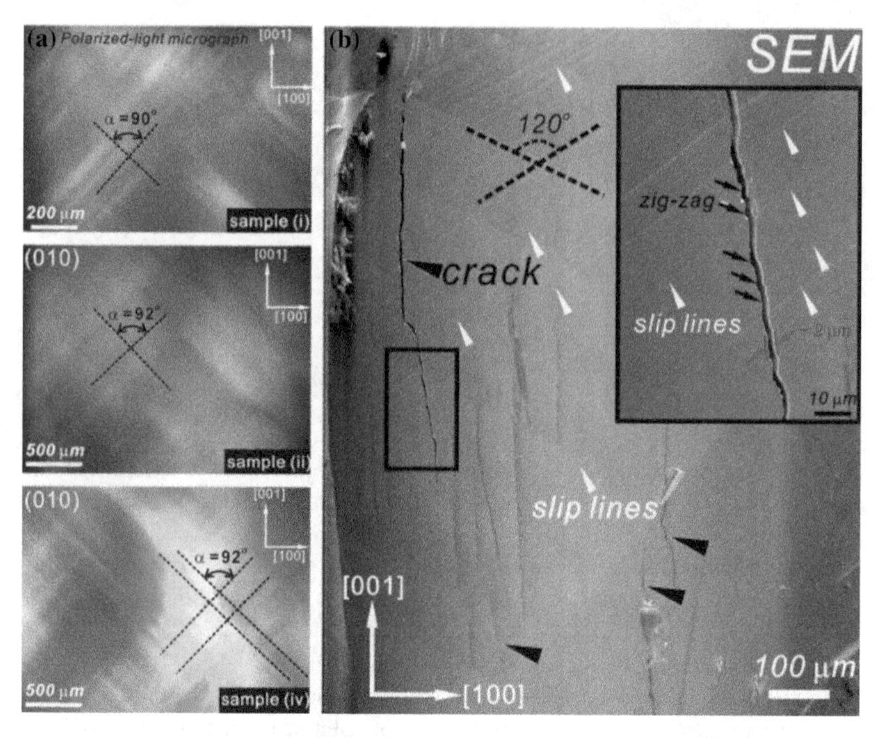

Fig. 8.83 **a** Two sets of *cross-cross slip lines* for samples (i) of stage 0, sample (ii) of stage I, and (iv) of stage II making an angle of α ≈90°, ≈92°, and ≈92° between each other, respectively, suggesting {101} is the primary slip planes (polarized light (OM), **b** those in sample (v) subtending an angle of $\alpha \approx 120°$, and cracks of ~800 μm in size found in this sample (SEM-SEI) [57]. With kind permission of John Wiley and Sons

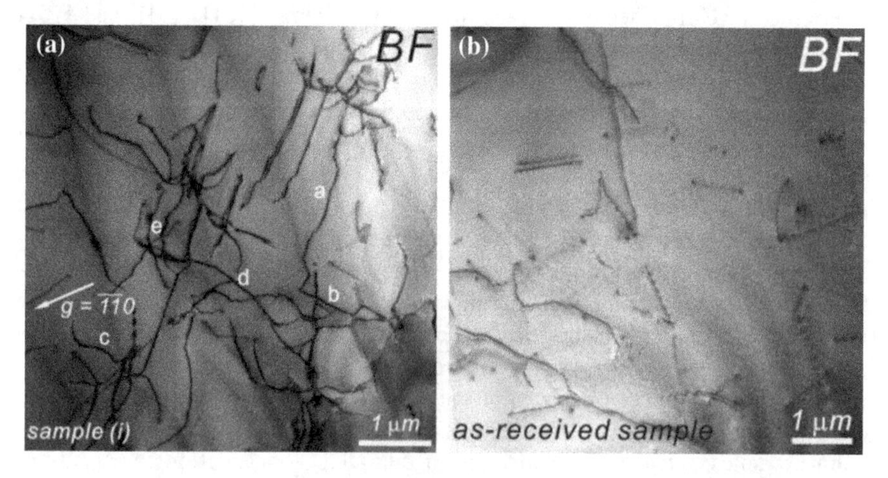

Fig. 8.84 Representative dislocation substructure observed in **a** sample (i) of stage 0 suggesting multiple slip, comparing to that of **b** as-received sample (BF image-TEM). (BF is bright field) [57]. With kind permission of John Wiley and Sons

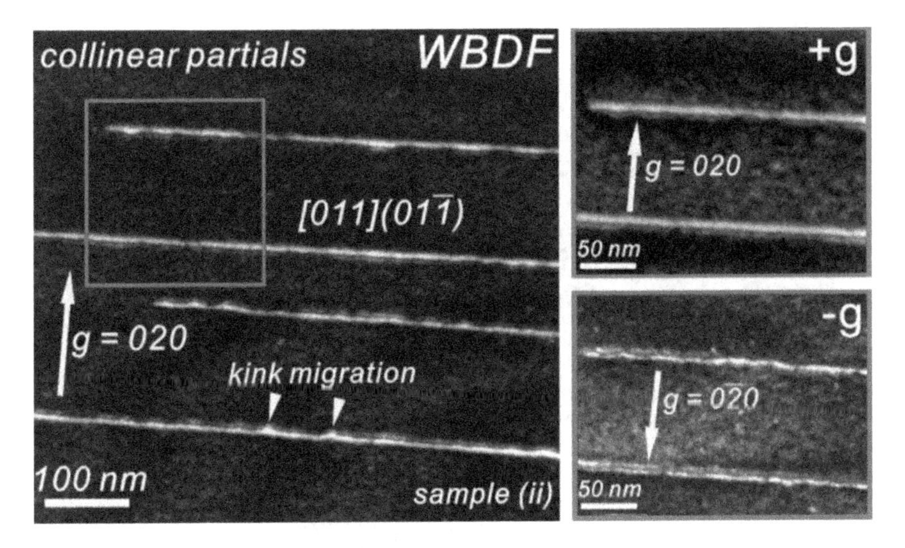

Fig. 8.85 Collinear partials from [011] (01$\bar{1}$) slip system activated in stage I showing kink migration, which are imaged with **+g** and **−g** vectors (WBDF image-TEM). (WBDF means weak beam dark field.) [57]. With kind permission of John Wiley and Sons

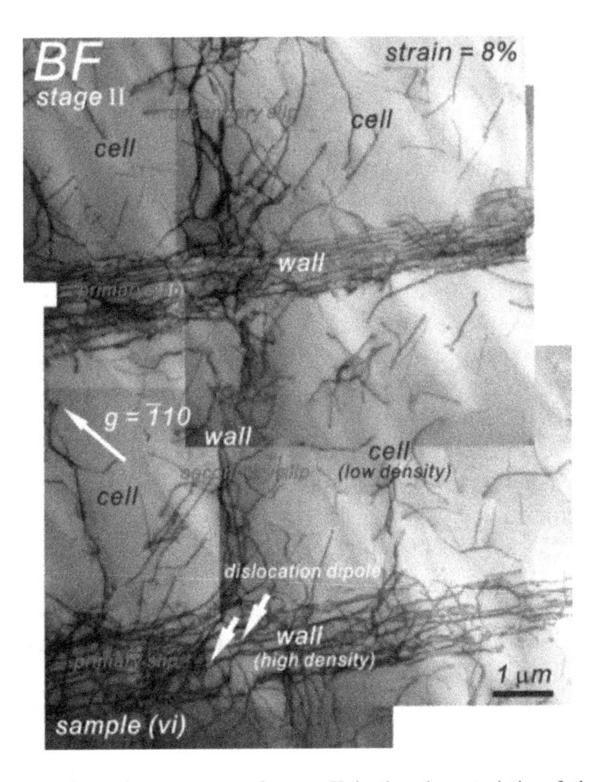

Fig. 8.86 Representative microstructure of stage II is the characteristic of the cell-and-wall structure of the composite model proposed for the plastic deformation of metals (BF image-TEM). (BF is bright field) [57]. With kind permission of John Wiley and Sons

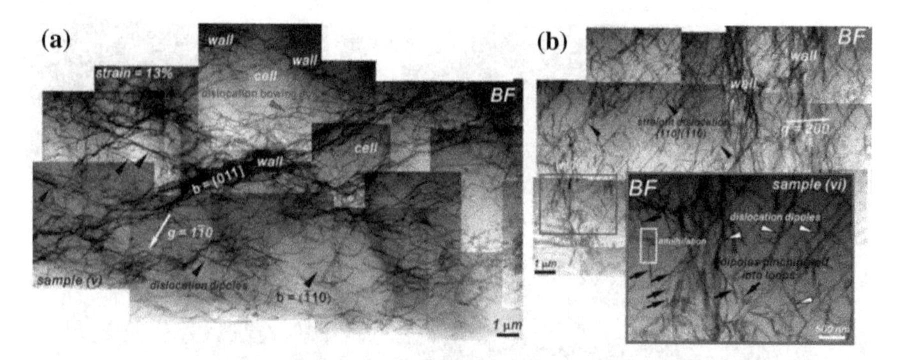

Fig. 8.87 Cell-and-wall structure is breaking up shown by sample retrieved **a** in the later stage of stage II, **b** just before entering stage III (WBDFimage-TEM) [57]. With kind permission of John Wiley and Sons

crystals [11, 28]. More research is needed to evaluate dislocation mechanisms and fracture in ductile ceramics. Such experimental work is of interest, because even the usually brittle ceramics become ductile at elevated temperatures.

8.6 Fracture in Superplastic Ceramics

Superplasticity has been discussed in detail in earlier chapters (Chaps. 2 and 5) and the conditions for obtaining such phenomena have been detailed. Many ceramics can attain superplasticity under certain conditions, among them Si_3N_4, SiC and ZrO_2 to mention a few. In this section, essential observations of fracture are discussed. But first, once again note the importance of superplasticity for technology; it is of prime importance in material-shaping processes (such as: rolling, forging and extrusion, some of which are only characteristic of the metal industry) since large deformations can be produced at relatively low loads, with relatively simple equipment and at relatively low cost.

One of the properties of superplasticity in alloys is a high strain rate. However, this is likely to induce premature failure in ceramics, due to the presence of various cracks (microcracks) and cavities (usually pores). Therefore, in ceramics using a slow strain rate is the acceptable practice. Certain composite ceramics, such as tetragonal zirconia, may be deformed even at strain rates up to 1.0 s^{-1}, resulting in superplasticity. This composite also exhibits a large tensile elongation, exceeding 1050 % at a strain rate of 0.4 s^{-1}. Another critical criterion for superplasticity is grain size. In particular, a small grain size is a prerequisite, since often grain boundary sliding (GBS) occurs during superplasticity. (Remember that GBS is considered to be the leading deformation mechanism). GBS without a change in the grain shape is accelerated by a small grain size. Grain growth during high-temperature deformation may occur, making the material unsuitable for extended superplastic forming. One may limit grain growth by adding a second phase, which pins grains, preventing growth. Also (as indicated in earlier chapters), the strain-rate-sensitivity exponent is an important parameter when observing superplastic behavior. The strain-rate-sensitivity parameter, m, appears in Eq. (2.21) from Chap. 2:

$$\sigma = K\left[\dot{\varepsilon}\exp\left(\frac{Q_c}{RT}\right)\right]^m \tag{2.21}$$

Observe that temperature is an important parameter and Q is the activation energy, also given in Eq. (2.20):

$$\dot{\varepsilon} = \frac{A\sigma^n}{d^p}\exp\left(-\frac{Q}{RT}\right) \tag{2.20}$$

Superplastic ceramics deform without necking and fail by intergranular cracks that propagate perpendicular to the applied tensile axis. Grain size has been considered as a critical factor in superplasticity. In Fig. 8.88, fracture strain is indicated as a function of flow stress in FG ceramics, while the strain-rate sensitivity, m, remains high. In Fig. 8.88, various ceramics are plotted, (the reader is referred to the original works):

The main observation here is that the tensile elongation in a number of ceramics exhibits the same trend, regardless of the test temperatures and strain rates employed. The true fracture strain, as a function of grain size, is specifically indicated for various ceramics in Fig. 8.89. Note that the tensile elongation of the FG ceramics indeed shows superplastic behavior, as expected in sufficiently small grain sizes. Among the superplastic materials, Y-TZP is a well-known ceramic, exhibiting optimal superplastic elongation-to-failure of ~700 % at 1823 K and a strain rate of $8.3 \times 10^{-5}\ s^{-1}$. However, a detailed microstructural investigation of the superplastically-deformed Y-TZP specimens reveals the occurrence of extensive concurrent grain growth and internal cavitation. For the best results, both should be avoided, as far as possible. The tendency toward cavity interlinkage in a direction perpendicular to the tensile axis is an important factor influencing the total elongation-to-failure, as observed in superplastic materials (Fig. 8.90).

Figure 8.90 indicates that the elevated-temperature superplastic-deformation process enhances concurrent grain growth. Since most of the data at a given strain rate were obtained from a specimen tested to failure, implying the same exposure durations, the results demonstrate the significance of superplastic strain for grain growth. Figure 8.91 illustrates the strain rate and temperature dependence of deformation-enhanced concurrent growth.

The data in Fig. 8.91a are shown at a fixed temperature of 1823 K and at a local true strain of 0.5, while those in Fig. 8.91b are shown at a fixed strain rate of $2.7 \times 10^{-5}\ s^{-1}$ and local true strain of 0.5. $\bar{L}_0$ refers to the grain size in the gripping region at zero local true strain. Thus, $\bar{L} - \bar{L}_0$ reflects the deformation-enhanced

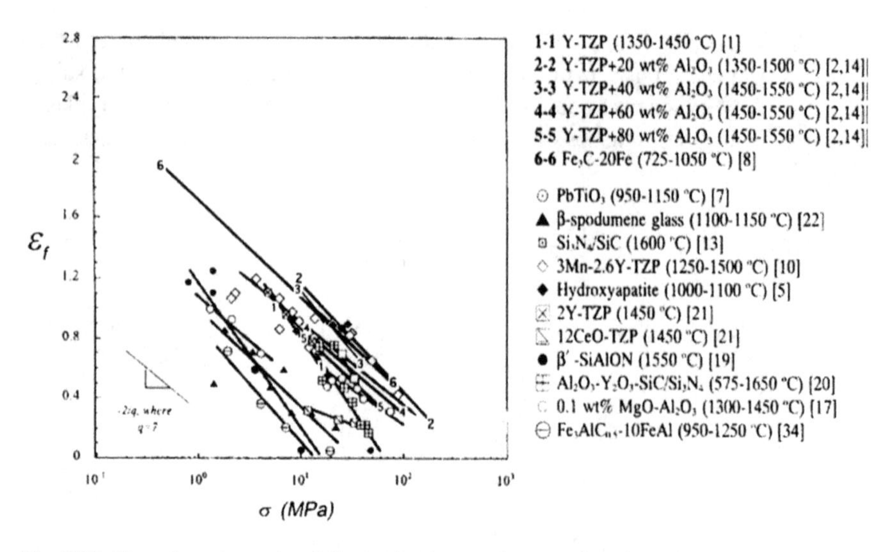

Fig. 8.88 True fracture strain of fine-grained ceramics as a function of flow stress, when m remains high [34]. With kind permission of Springer

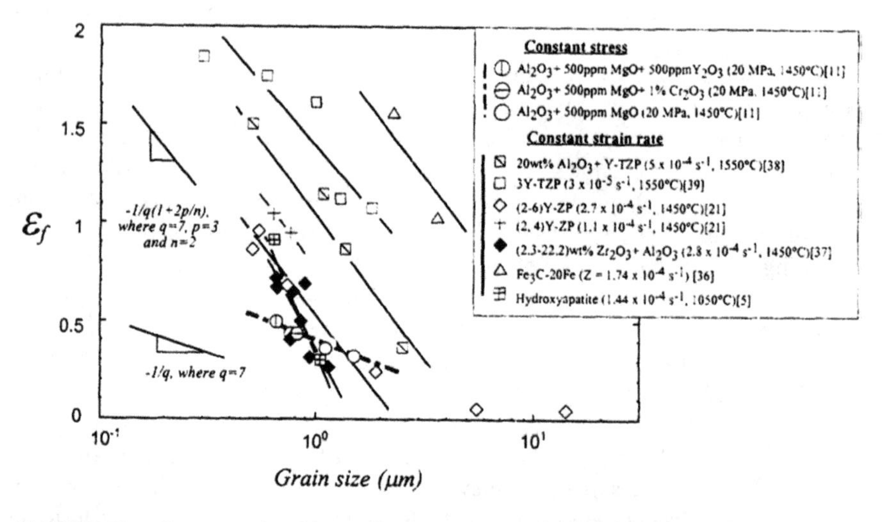

Fig. 8.89 True fracture strain of fine-grained ceramics as a function of grain size under constant strain rate conditions and constant flow stress condition [34]. With kind permission of Springer

concurrent grain growth at elevated temperatures. From these data, the strain-rate dependence of deformation-enhanced concurrent grain growth may be expressed as $(\bar{L} - \bar{L}_0) \propto \dot{\varepsilon}^{-0.6}$ and the temperature dependence as $(\bar{L} - \bar{L}_0) \propto \exp(-170000/RT)$.

Using the experimental results shown in Figs. 8.90 and 8.91, the kinetics of deformation-enhanced concurrent grain growth is:

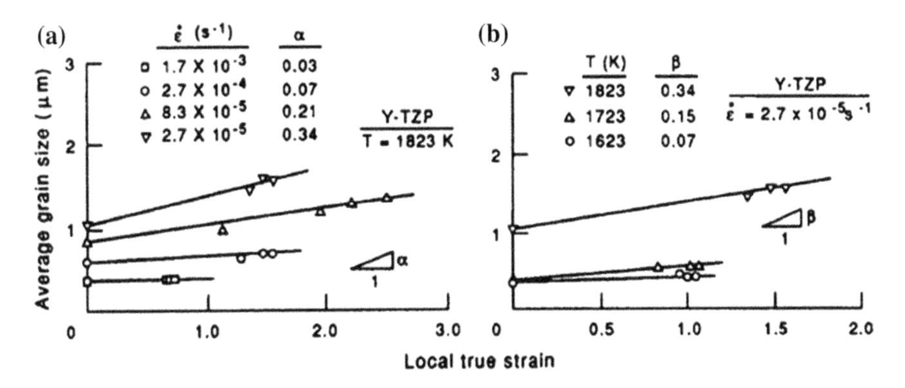

Fig. 8.90 Variation in average grain size with local true strain for **a** a fixed temperature of 1823 K and **b** a fixed strain rate of 2.7×10^{-5} s^{-1} [48]. With kind permission of Elsevier. pp. 3227–3236, 1991 0956-7151/91 \$3.00 + 0.00

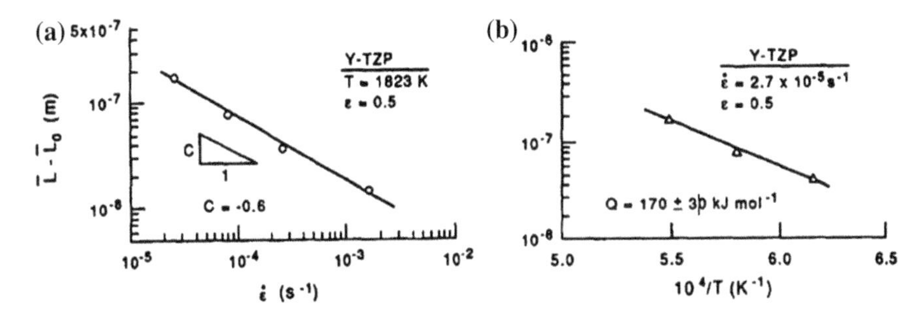

Fig. 8.91 Variation in deformation enhanced grain growth with **a** strain rate and **b** temperature [48]. With kind permission of Elsevier

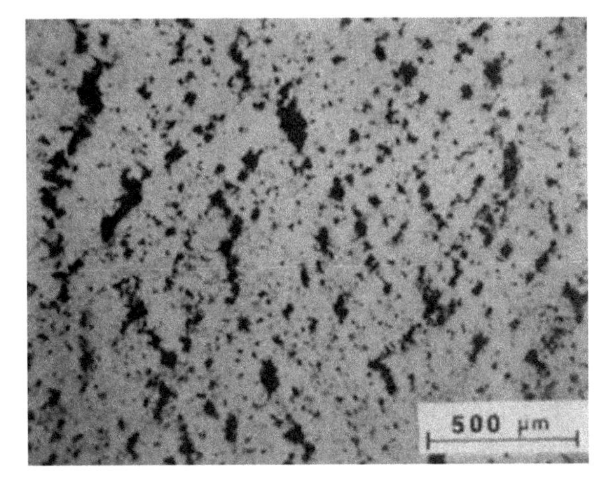

Fig. 8.92 Optical micrograph illustrating extensive cavity interlinkage perpendicular to the tensile axis in a specimen tested to an elongation to failure of ~150 % at 1723 K and a strain rate of 2.7×10^{-5} s^{-1}. The tensile axis is horizontal

$$\bar{L} - \bar{L}_0 = A\varepsilon\dot{\varepsilon}^{-0.6} \exp(-170000/RT).$$

Figure 8.92 is an optical micrograph of the cavities formed and interlinked during the deformation process. The microstructure reveals extensive concurrent cavitation, with levels of cavitation approaching ~30 % area fraction. The cavities appear to nucleate predominantly at triple points and they tend to have quasi-equilibrium spherical cap shapes. The reduction in elongation-to-failure at low strain rates is attributed to an increased tendency towards the interlinkage of cavities in a direction perpendicular to the tensile axis.

8.7 Concluding Remarks

For some time now, it was believed that ceramics could not show high ductility, due to their brittleness at low temperatures. However, certain brittle ceramics do become ductile, permitting very large plastic deformations at low temperatures, particularly if they are polycrystalline ceramics produced with a crystal size of only a few nm. More and more ceramic materials have been found to show superplasticity, provided that the proper experimental conditions are used. This is a potentially cost-saving deformation process, because large deformations can be achieved at relatively low temperatures. There are similarities and differences between superplastic metallic materials and ceramic ones. The similarities include the variation of strain rate with stress and grain size. A major difference is the possible presence of intergranular glassy phases in ceramics. The following conditions enhance superplastic ductility: (a) high strain-rate sensitivity; (b) limited concurrent grain growth; (c) reduced cavitation; and (d) hindrance of cavity interlinkage transverse to the loading (tensile) axis.

References

1. Griffith AA (1924) The phenomena of rupture and flows in solids. In: Proceedings of the 1st international congress of applied mechanics, Delft, pp 55–63
2. Kelly A, MacMillan NH (1986) Strong Solids, 3rd edn. Oxford University Press, Oxford
3. Lawn B (1993) Fracture of brittle solids, 2nd edn. Cambridge University Press, Cambridge
4. Ritchie RO (1999) Int J Fract 100:55
5. Schmid E, Boas W (1935) Kristalplastizitat. Springer Verlag, Berlin. English edition: Hughes FA et al. (1950) Plasticity in Crystals
6. Soderholm K-J (2010) Dent Mater 26:e63
7. Suh NP, Turner APL (1975) Elements of mechanical behavior of solids. Scripta Book Company, Washington, D. C, pp 422–423
8. Vander Voort GF (1987) ASM Handbook, vol 12. ASM International Fractography, USA, p 102
9. Zener C (1948) The macro-mechanism of fracture, in Fracturing of metals. Am Soc Metals 3(Metals Park, Ohio)

Further References

10. Ando K, Takahashi K, Nakayama S, Saito S (2002) J Am Ceram Soc 85:2268
11. Argona AS, Orowana E (1964) Phil Mag 9:1003
12. Bando Y, Ito S, Tomozawa M (1984) Commun Am Ceram Soc 67:C36
13. Barenblatt GI (1962) Adv Appl Mech 7:55
14. Brown WF Jr, Srawley JE (1976) Am Soc Test Mater 13–15 (Spec Tech Publ No:410)
15. Deng ZY, She J, Inagaki Y, Yang JF, Ohji T, Tanaka Y (2004) J Eur Ceram Soc 24
16. Eftis J, Liebowitz H (1972) Inter J Fract Mech 8:383
17. Erdogan F, Joseph PF (1989) J Am Ceram Soc 72:262
18. Evans AG, Cannon RM (1986###) Acta merall 34:761
19. Evans AG (1990###) J Am Ceram Soc 73:187
20. Faber KT, Evans AG (1983###) Acta metal 31:565, 577
21. Fernando Jorge Lino Alves, Faculdade de Engenharia, Departamento de Engenharia Mecânica e Gestão Industrial, Porto, Portugal
22. Griffith AA (1920) Phil Trans Roy Soc A221:163 (London)
23. Hannink RHJ, Swain MV (1994) Annu Rev Mater Sci 24:359
24. Hannink RHJ, Kelly PM, Muddle BC (2000) J Am Ceram Soc 83:461
25. Harmer MP, Chan H, Miller G (1992) J Am Ceram Soc 75:1715
26. Hasselman DPH, Fulrath RM (1966) J Am Ceram Soc 49:68
27. Homeny J, Darroudi T, Bradt RC (2008) J Am Ceram Soc 91:1986, 68:326 (1980)
28. Hulse CO, Pask JA (2006) J Am Ceram Soc 43:373
29. Inglis CE (1913) Institute of naval architects (Proceedings) 55, pp 219–241
30. Irwin GR (1957) J Appl Mech 24:361
31. Johnston TL, Stokes RJ, Li CH (1959) Phil Mag 4:1316
32. Johnston TL, Stokes RJ, Li CH (1962) Phil Mag 7:23
33. Johnston WG (1960) Phil Mag 5:407
34. Kim W-J (1995) Met Mater 1(2):117–124
35. Kruzic JJ, Satet RL, Hoffmann MJ, Cannon RM, Ritchie RO (2008) J Am Ceram Soc 91:1986
36. Lange FF (1971) J Amer Ceram Soc 54:614
37. Liu S-Y, Chen I-W (1991) J Am Ceram Soc 74:1197
38. Lino Alves FJ, Faculdade de Engenharia, Departamento de Engenharia Mecânica e Gestão Industrial, Porto, Portugal
39. Low JR Jr (1963–65) The fracture of metals. Prog Mater Sci 12:3
40. Ohji T, Yamauchi Y (1993) J Am Ceram Soc 76:3105
41. Orowan E (1949) Rep Prog Phys 12:185
42. Padture NP (1991) Crack resistance and strength properties of some alumina-based ceramics with tailored microstructures. Ph.D. Thesis, Lehigh University
43. Park S, Sun CT (1995) J Am Ceram Soc 78:1475
44. Rice JR, Paris PC, Merkle JG (1973) J Am Soc Test Mater 231–245 (Spec Tech Publ No:536)
45. Rice JR (1968) J Appl Mech 35:379
46. Ritchie RO, Yu W, Bucci RJ (1989) Eng Fract Mech 32:361
47. Sarfarazi M, Ghosh SK (1987) Eng Fract Mech 27:257
48. Schissler DJ, Chokshi AH, Nieh TG, Wadsworth J (1991) Acta metal Mater 39:3227
49. Shih CJ, Meyers MA, Nesterenko VF, Chen SJ (2000) Acta mater 48:2399
50. Shumate EH Jr (2009) The radius of curvature in the prime vertical. ITEA J 30:159
51. Sohncke L (1869) Ann Phys L p z 137:177
52. Stroh AN (1955) Proc Roy Soc (London) 223A:548
53. Stroh AN (1957) Adv Phys 6:418

54. Swanson PL, Fairbanks CJ, Lawn BR, Mai Y-W, Hockey BJ (1987) J Am Ceram Soc 70:279
55. Vekinis G, Ashby MF, Beamont PWR (1990) Acta metall mater 38:1151
56. Wang Z, Jiang Q, White GS, Richardson AK (1998) Smart Mater Struct 7:867
57. Yang K-H, Ho N-J, Lu H-Y (2011) J Am Ceram Soc 94:3104

Chapter 9
Mechanical Properties of Nanoscale Ceramics

Abstract The mechanism of deformation in nanosize ceramics occur either by dislocation motion or by grain boundary sliding depending on the size of the grains. In nanoceramics of grain size above ~ 100 nm the main deformation mechanism is by dislocation motion. At ultra-fine nano grain sizes below ~ 100 nm in the range <50 nm, the deformation mechanism is by grain boundary sliding. Dislocations cannot be accommodated conveniently in such nanosize materials and are prevented from motion and interactions. At levels in the hundreds of nanosized grains, a probable partial-dislocation mechanism may occur concurrently with other deformation mechanisms such as grain boundary sliding. For grain boundary sliding atomic mobility is essential, which results in a metal-like plasticity in nanoscale ceramics. One is interested in the behavior of nanoceramics under applied loads; therefore the various responses effecting static mechanical properties (tension–compression, hardness, etc.) time-dependent deformation (creep) and cyclic (fatigue) deformation are relevant. Making ceramics superplastic requires producing ultra-fine grains in the lower nanosize level, preferentially below 50 nm or even less. Various sophisticated techniques have been developed over the past decade or so, such that certain nanoceramics can now be produced with some measure of superplasticity. Superplastic materials may be thinned down, usually in a uniform manner, before breaking, without neck formation. The actual deformation mechanism is still under debate and may be material-dependent as well. Despite the various views on the exact mechanism responsible for the observed nano-behavior, it is clear from the experiments that nanoceramics may exhibit increased strength (hardness, for example), improved toughness, improved ductility and high resistance to fatigue. All these improved properties serve as safeguards against unexpected or premature fracture in service.

J. Pelleg, *Mechanical Properties of Ceramics*, Solid Mechanics and Its Applications 213, DOI: 10.1007/978-3-319-04492-7_9,

9.1 Introduction

Thus far, little attention has been paid to the effects of grain size. In recent years, however, more and more attention is being focused on 'nanoscale ceramics', i.e., ceramics with nanosized grains, perhaps because now a large number of ceramics may be made ductile to various degrees, even attaining superplasticity. The previous chapter indicated that ceramics may show some ductility at high temperatures, usually at temperatures $\geq 0.5\ T_m$. In such cases, dislocation mechanisms are involved, along with associated phenomena, such as work hardening, slip, etc., and dislocation mobility and interactions play key roles. Yet, in nanocrystalline ceramics, due to their miniscule dimensions, it is highly unlikely that dislocation-related deformation is involved in the various mechanical phenomena. The role of lattice dislocation slip, if at all present, is greatly diminished, due to the aforementioned nanoscale. However, one might expect that alternative deformation modes are operative, such as the sliding of nanosized grains and atomic mobility (a diffusional process) and responsible for nanometric ceramic ductility. Atomic mobility is essential for enabling GBS, which results in a metal-like plasticity in nanoscale ceramics. In any event, the actual dimensions of nanosized grains determine whether some sort of dislocation mechanism is involved, for example partial dislocations (since SFs are observed in experiments on nanoscale ceramics). At levels in the hundreds of nanosized grains, a probable partial-dislocation mechanism may occur concurrently with other deformation mechanisms. The actual deformation mechanism is still under debate and may be material-dependent as well. Experiments indicate that this alternative mechanism is the dominant mechanism, GBS, since the stress level required to nucleate dislocations is high and dislocation gliding occurs in nanoscale ceramics when the grain dimensions are at in the upper range (above ~ 100 nm). At the lower range of nanosized grains, dislocation gliding does not contribute to overall strain.

Here, the focus is on the various responses of nanoscale ceramics to applied loads, presenting the similar or different observations of certain aspects (tension–compression, hardness, cyclic- or time-dependent deformation) as was previously discussed regarding macroscale ceramics.

9.2 Static Properties

Does a very small grain size make materials stronger? Knowledge of small-sized materials is important for the understanding of their mechanical properties, which dictate their practical applications. At small dimensions, a significant departure from classical behavior is observed. The strength of a material increases either when its structure is small or, in our case, when nanometric crystals are involved. At small sizes, dislocation motion, if present, is restricted. Recall that when no dislocations are involved in deformation, high strength (at a level approaching the theoretical strength) is required to induce strain in the test specimen. Unlike conventionally-sized test specimens, in which ductility usually decreases with increased strength, nanocrystalline-sized specimens show high strength combined with good elongation. Moreover, such nano-specimens may reach high values of plasticity, which, in some ceramics, may lead to superplastic behavior before fracture. Here, some of the properties discussed in earlier chapters in regard to macroscale materials will be reviewed for nanoscale-material behaviors.

9.2.1 Stress–Strain Relation

Remember that since (1 nm $= 10^{-9}$ m $= 10$ Å) nanocrystalline materials are typically less than ~ 100 nm, their grain sizes are so small that a major part (or even all) of the microstructural volume consists of interfaces, mostly in the form of grain boundaries. A frequently presented illustration representing a typical nanostructure may be seen in Fig. 9.1.

Consequently, it is reasonable to expect that nanomaterials will exhibit mechanical properties different than those of coarse-grained ceramics, among them higher strength and hardness. High ductility is unique to nanocrystalline materials, along with strength (unlike conventional, coarse-grained materials).

Based on nanometric dimensions, often the various nanomaterials may be categorized as being: one-dimensional, 1-D (as layered structures, in which the nano size is in only one dimension); two-dimensional, 2-D (e.g., rod-shaped); and three-dimensional, 3-D (crystallized, like equiaxed grains). Nanocomposites are often used in ceramics in technological applications, so much research is devoted to them. The matrices of these nanocomposites may be in the micron range and their additives are nanoparticles (fibers, whiskers). Furthermore, various combinations are of interest, such as ceramics strengthened by either metallic or ceramic particles, or metallic matrices strengthened by nanosized ceramic particles. The most obvious advantage of nanocrystalline ceramics with homogeneous and dense microstructures is their improved mechanical properties, compared to conventional ceramics which under proper conditions is accompanied by good ductility. It may be stated that, while most coarse-grained ceramics are brittle, nanograined ceramics may exhibit significant ductility before failure. Such ductility is primarily contributed by their grain-boundary phase, but in grains at a certain ceramic phase some plastic deformation has been observed, which contributes to overall plastic strain.

In ceramics, compressive tests (rather than tensile tests) are used to obtain strength values and, thus, the stress–strain relation is expressed by plotting compressive stress against strain. Figure 9.2 shows the calculated compressive stress–strain curves for two porosities at various grain sizes. The purpose of these calculations is to predict the compressive yield strength of nanograined ceramics as the grain size decreases from a coarse-grained to a nanometric scale. The effects of porosity and the second elastic phases are also considered. The method of calculation (not presented here) and the theoretical model are given in Jiang and Weng's paper [10], which may be consulted. Note that the calculated variation of the compressive stress with grain size

Fig. 9.1 Reproduced from Fig. 1 of Jiang and Weng according to Schiøtz et al.; **a** Molecular dynamic simulation of grains and grain boundary in a nano-grained copper, showing the grain boundary has finite volume concentration [13]. With kind permission of Elsevier

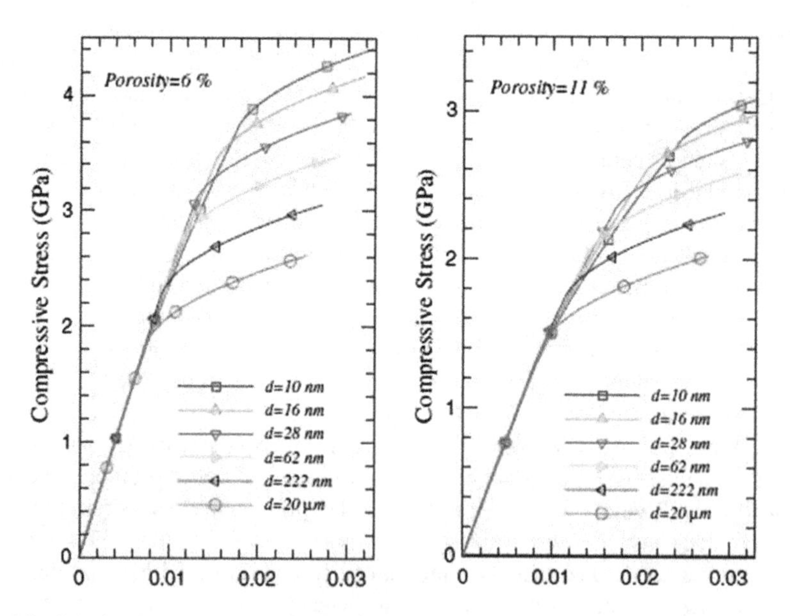

Fig. 9.2 Calculated compressive stress–strain relations as grain size decreases from 20 μm to 10 nm at low porosity levels: **a** c_{pore} ¼ 6 % and **b** c_{pore} ¼ 11 % [13]. With kind permission of Elsevier and Dr. Weng (Note that c_{pore} indicates volume concentration of pores)

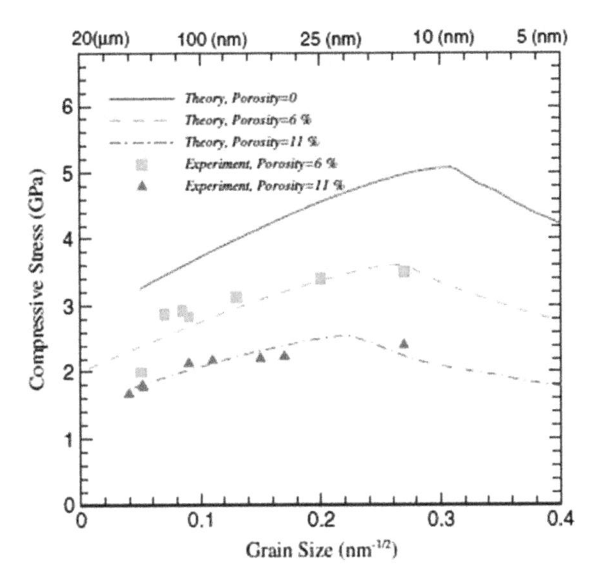

Fig. 9.3 Compressive yield stress as a function of grain size d at three levels of porosity (experimental data from [4, 13]). With kind permission of Elsevier and Dr. Weng

shows an inverse H–P relation, quite similar to the experimental shapes of the plots, as illustrated in Fig. 9.3.

Observe that, below some value of the grain size, a change in slope occurs. These experimental data are for TiO_2. More on the inverse H–P relation are considered below in the section on hardness. Usually, ceramic materials are composed of added second phases. A second phase may be elastic or show plasticity (e.g., metallic second phases). The second phase may influence compressive strength and, thus, the overall stress–strain relation in nanoceramic composites. In Fig. 9.4, the effect of an elastic phase, with an elastic modulus other than TiO_2, was artificially introduced into the system, having 6 % porosity and the calculated compressive strength for three grain sizes is seen. In this figure, three different cases are indicated: when the elastic modulus of the second phase is the same as that of TiO_2; when it is stiffer than that of TiO_2; and when it is softer. Differences may be seen in the compressive stresses at a given strain, e.g., at ~0.02 strain. The variation of the compressive stress with grain size, when a 10 % elastic phase is added, is indicated in Fig. 9.5. These curves are compared to the variations in stress with grain size without a second phase. Observe the thickness of the grain boundaries (see Fig. 9.1); one might think that this has an effect on the yield stress variation with grain size. The variations in yield stress with grain size for three boundary thicknesses are illustrated in Fig. 9.6.

The effect of the boundary thickness is small, except in the low nano range, below 100 nm. It should also be noted that both the compressive stress and the compressive yield stress decrease in the very low nano range, indicating an inverse H–P relation. Porosity reduces the compressive yield strength, but the shape of the curves is not affected.

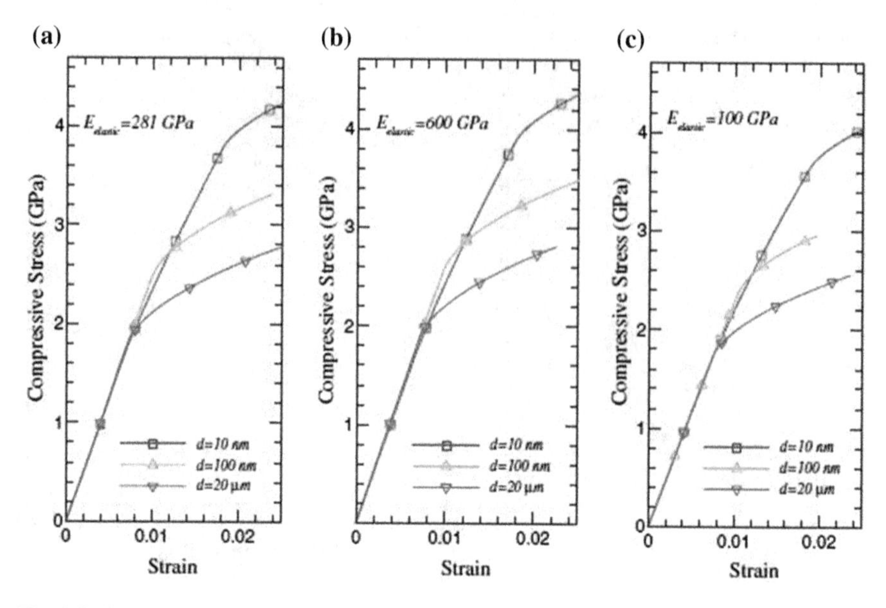

Fig. 9.4 Calculated compressive stress–strain relations at three different grain sizes when the elastic phase has: **a** the same, **b** stiffer, and **c** softer Young's modulus than TiO_2 [13]. With kind permission of Elsevier and Dr. Weng

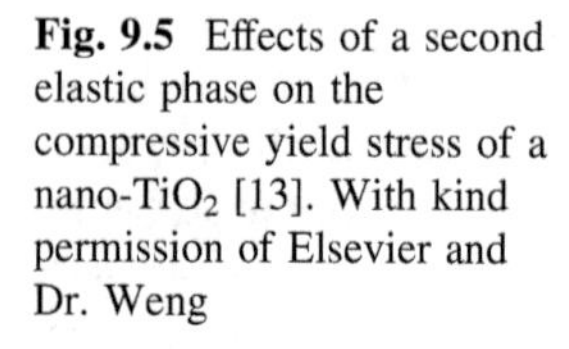
Fig. 9.5 Effects of a second elastic phase on the compressive yield stress of a nano-TiO_2 [13]. With kind permission of Elsevier and Dr. Weng

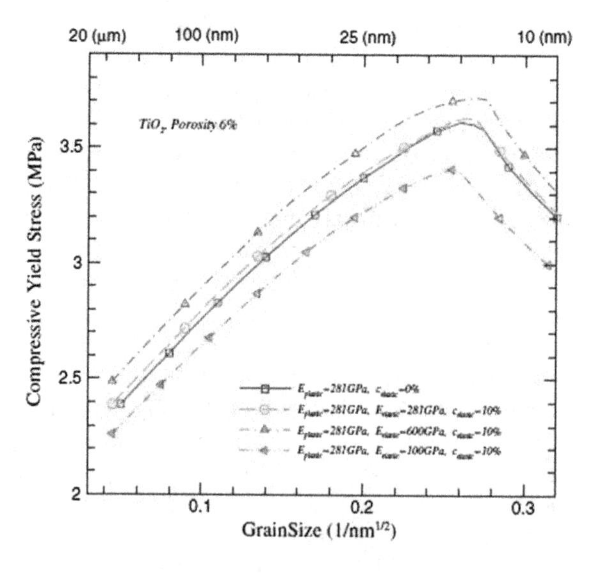

The variation of true stress with true strain is shown in Fig. 9.7 for nanosized silicon nitride ceramics; it does not indicate strain hardening and very large strains may be achieved, as seen from the flow stress.

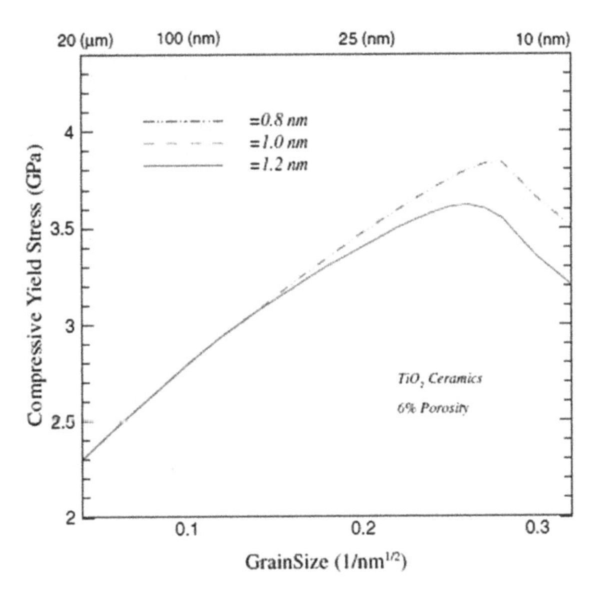

Fig. 9.6 Effect of the grain-boundary thickness on the compressive yield stress of TiO_2 [13]. With kind permission of Elsevier and Dr. Weng

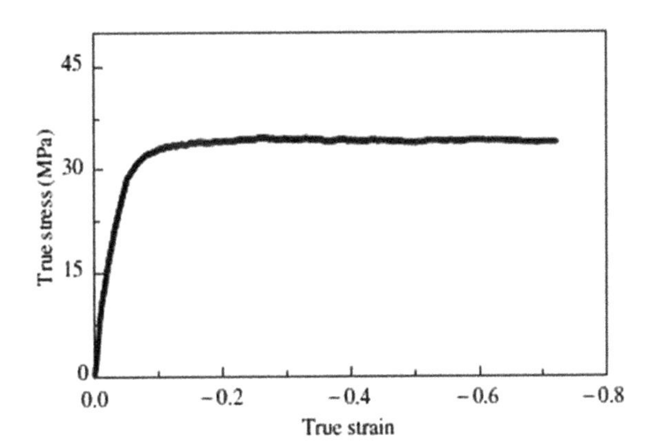

Fig. 9.7 True stress-true strain curve in compression obtained for silicon nitride nanoceramics [40]. With kind permission of John Wiley and Sons

This deformation was performed at a strain rate of 10^{-4} s^{-1} and 1500 °C, indicating that a very large strain may be achieved under a flow stress of ~34 MPa with no strain hardening. This is possible due to the stability of Si_3N_4 against grain growth. Such stability, that can prevent grain growth in certain ceramics, among them Si_3N_4, is crucial to superplastic behavior (discussed below). The fractured surface of one of the specimens (denoted by P2), sintered at 1600 °C, is shown in Fig. 9.8.

The method of fabrication has an important effect on the final results. Spark plasma sintering (SPS) produces homogeneous ceramics and, in the case of Si_3N_4, an average grain size of 70 nm is obtained. The results of such a fabrication technique for Si_3N_4 are high hardness and good high-temperature ductility.

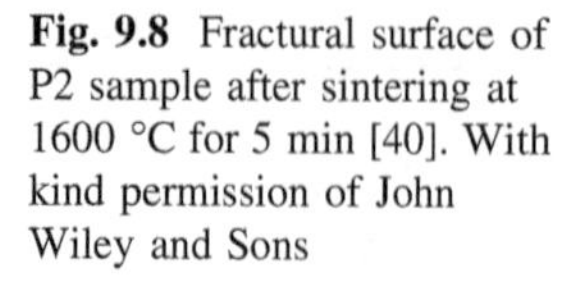

Fig. 9.8 Fractural surface of P2 sample after sintering at 1600 °C for 5 min [40]. With kind permission of John Wiley and Sons

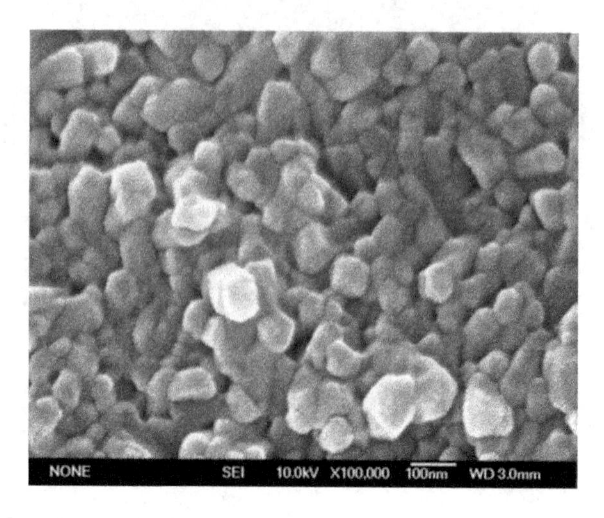

9.2.2 *Flexure (Bending) Stress*

It has been known for quite some time that the addition of nanosized ceramic particles into a ceramic matrix significantly enhances its strength properties. Here, bending stress is considered. Figure 9.9 illustrates bending stress variations for different temperatures in alumina strengthened by SiC.

The technique used for sintering the powders was SPS, a relatively new process that sinters ceramic powders rapidly to almost full density. The highest bending strength (980 MPa) was obtained when the sample was sintered at 1450 °C, which is much higher than that used for monolithic Al_2O_3 ceramics (350 MPa). The addition of nano-SiC particles improved the microstructure of the composites and enhanced the stability of the grain boundaries. The metallography of the nano-composite SiC–Al_2O_3 is shown in Figs. 9.10 and 9.11 by SEM and TEM, respectively.

The reinforcement of alumina may be accomplished by the use of other additives, such as boron nitride nanotubes (henceforth: BNNTs). Boron nitride has many excellent properties, such as low density, high thermal conductivity, stability and good mechanical performance. BNNTs are stable in air at 800 °C and even at higher temperatures. It is a good substitute for carbon nanotubes (henceforth: CNTs) for strengthening purposes. The strengthening capacity of BNNTs is a consequence of high tensile strength and a Young's modulus of ~30 and ~900 GPa. The average values of the bending strengths of BNNT/Al_2O_3 composites, as a function of BNNT content, are shown in Fig. 9.12. (The corresponding fracture toughness also appears below). The bending strength and the fracture toughness, as well, are greatly dependent on the amount of BNNTs; the bending strength at ~2 wt% has a highest value of ~532 MPa, a 67 % increase, compared to pure Al_2O_3 (~319 MPa).

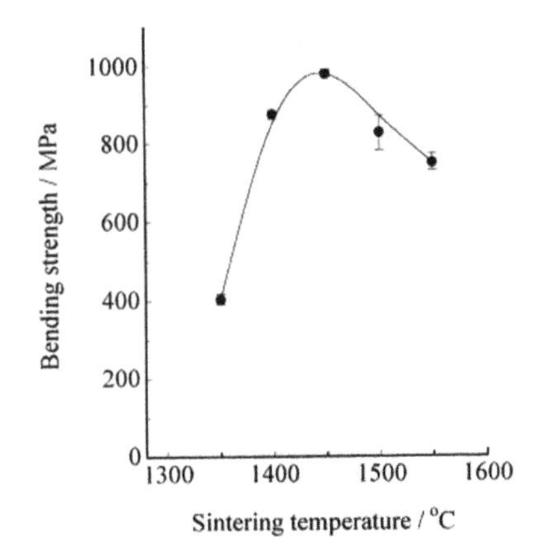

Fig. 9.9 Bending strength versus sintering temperature for 5 vol% $SiC–Al_2O_3$ superfast sintered by SPS [10]. With kind permission of Elsevier

Fig. 9.10 SEM micrograph of fracture surface of $SiC–Al_2O_3$ nanocomposites [10]. With kind permission of Elsevier

Fig. 9.11 TEM micrograph of $SiC–Al_2O_3$ nanocomposites [10]. With kind permission of Elsevier

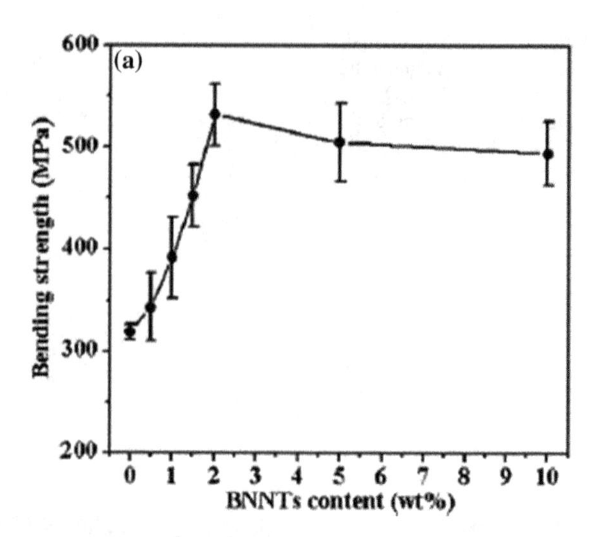

Fig. 9.12 Dependence of bending strength (**a**) on the amount of BNNTs in the composites [39]. With kind permission of Elsevier

9.2.3 Hardness

The strength of nanocomposite ceramics, in terms of hardness, may be exemplified for 1.5–3-mol% Y-TZP and Al_2O_3/Y-TZP nanocomposite ceramics with 1–5 wt% of alumina. Figure 9.13 shows TEM micrographs of 3-mol% Y-TZP produced by the hydrolysis of metal chlorides and a urea aqueous solution (150 °C for 10 h), washed, treated by microtip ultrasonication and calcined at 450–800 °C for 0.5–2 h. This figure has an aggregate size of 25 nm and shows that the primary crystallites of zirconia were bound together into the aggregate. The low sintering temperature (1150 °C) is a consequence of obtaining a uniform green body, virtually free of agglomerates, by using both the colloidal technique and ultrasonic dispersion. The addition of alumina to Y-TZP provides the ceramics with improved toughness. Also, small quantities of Al_2O are known to aid densification.

Figure 9.14 shows these specimens with and without alumina after sintering. The average grain size is about 110 nm. Square samples were used for the Vickers hardness indentation having an approximate 4 mm height and 12 mm side. The surfaces were scratch-free, as observed by optical microscopy. The grain sizes were determined by a linear analysis of SEM micrographs of the polished and etched surfaces. The hardness test results are illustrated in Fig. 9.15. As may be seen from Fig. 9.15, average hardness increased with increasing hold times. The hardness of the 1.25 wt% alumina–zirconia composites reached maximum values at an average grain size of 105 nm (24 h hold time) and at a relative density of 99.8 %. The density effect may be seen in Fig. 9.16. Furthermore, one can also see, from Fig. 9.15, that the average hardness of the 2.5 wt% alumina/zirconia composite reached a maximum value of 16.2 GPa at an average grain size of 94 nm (15 h duration) with a relative density of 99.2 %. The longer duration at

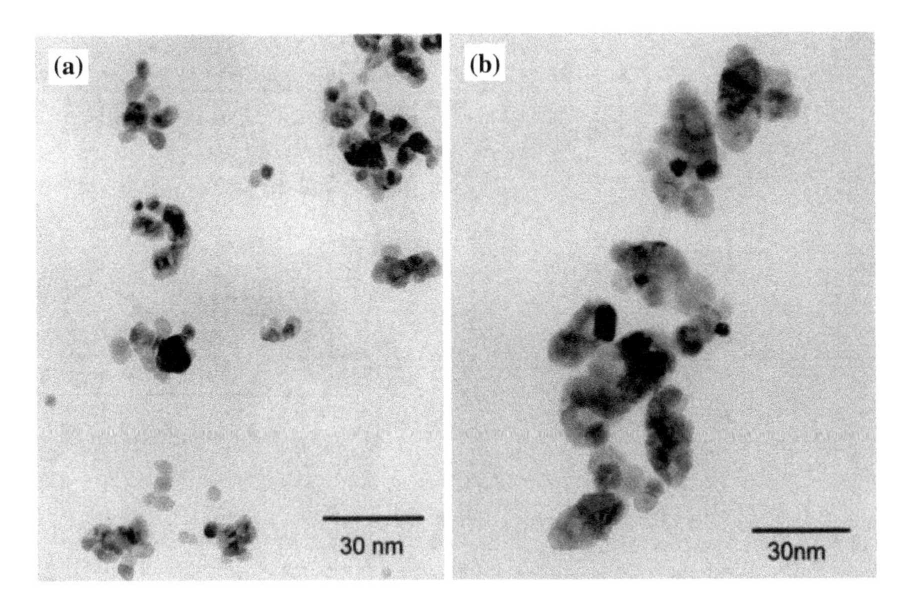

Fig. 9.13 **a** TEM micrograph of 3Y-TZP primary crystallite aggregation (urea hydrolysis at 150 °C for 10 h, calcinations at 450 °C for 1 h). **b** TEM micrograph of 3Y-TZP dense nanoaggregates [35]. With kind permission of John Wiley and Sons

Fig. 9.14 **a** SEM micrograph of 3Y-TZP ceramic sintered at 1150 °C for 30 h. **b** SEM micrograph of 2.5 wt% alumina/3Y-TZP ceramic sintered at 1150 °C for 20 h [35]. With kind permission of John Wiley and Sons

this temperature allowed the relative density to increase to full density; however, at the same time, average grain size also increases and the average hardness of such a ceramic gradually decreases (Figs. 9.15 and 9.16). These hardness indentations were performed by applying forces of 4.9 and 9.81 N for a dwell time of 15 s. For each sample, 10 indentations were made to obtain the average hardness and standard deviation.

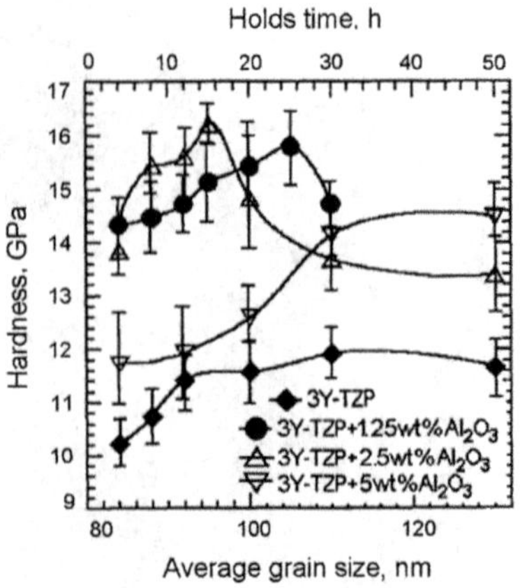

Fig. 9.15 Dependence of Vickers hardness on the hold time during sintering at 1150 °C [35]. With kind permission of John Wiley and Sons

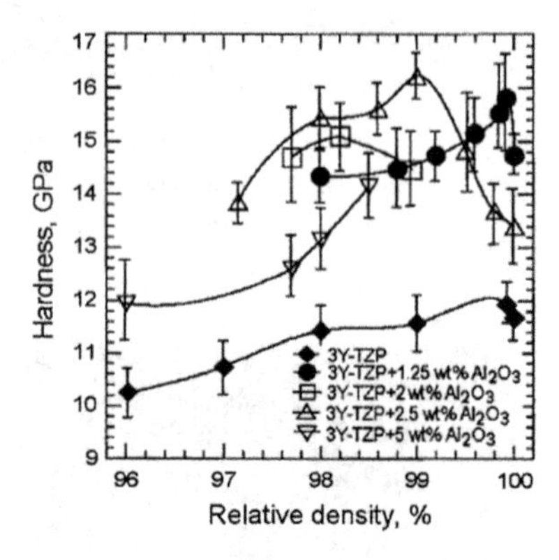

Fig. 9.16 Dependence of Vickers hardness on the relative densities of ceramic specimens [35]. With kind permission of John Wiley and Sons

The bending test and the microstructure are illustrated above for SiC–Al_2O_3. Hardness may be seen in this ceramic in Fig. 9.17 as a function of temperature. This plot closely follows the relative-density variations across sintering temperatures, as indicated in Fig. 9.18. The figure shows that an almost full density may be obtained at 1450 °C for SiC–Al_2O_3 powder by using the SPS technique, compared with the HIP sintering technique, requiring at least 1650 °C and 1 h soaking time. This observation indicates that SPS is a potential method for fabricating nano-SiC-oxide composites at much lower temperatures and within very short time periods.

HV measurement of bulk nano-twinned cubic boron nitride (henceforth: nt-cBN) samples with a standard square-pyramidal diamond indenter is shown in Fig. 9.19. An explanation of the connection between this figure and observations of the inverse H–P effect are found below. Reliable hardness values are best determined from the asymptotic hardness region. In Fig. 9.19, the variations in Vickers hardness are recorded by applying a series of loads. The asymptotic hardness value obtained at loads above 3 N was extremely high, 108 GPa, which

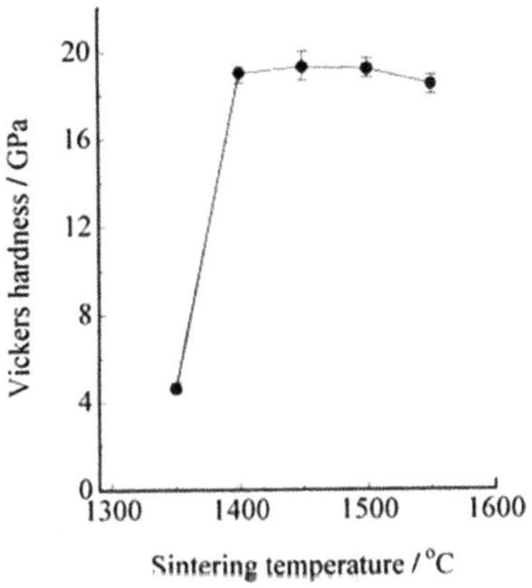

Fig. 9.17 Vickers hardness versus sintering temperature for 5 vol% SiC–Al_2O_3 superfast sintered by SPS [10]. With kind permission of Elsevier

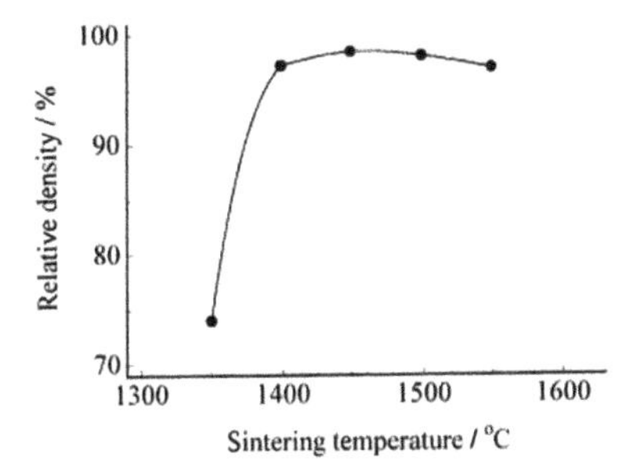

Fig. 9.18 Relative density versus sintering temperature for 5 vol% SiC–Al_2O_3 superfast sintered by SPS [10]. With kind permission of Elsevier

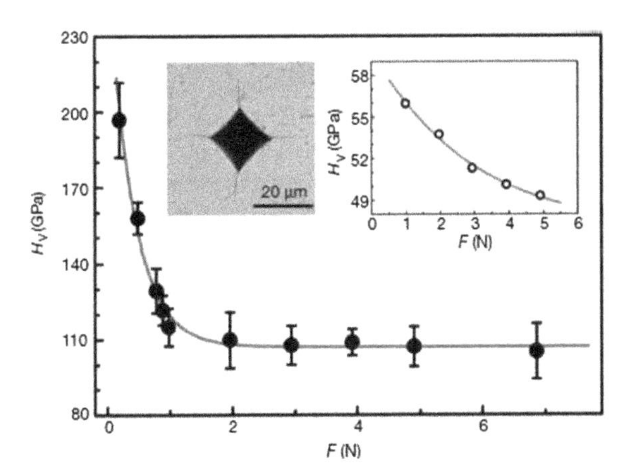

Fig. 9.19 The H_V of an nt-cBN bulk sample as a function of applied load (F). The H_V of the nt-cBN bulk decreases from ~196 GPa at 0.2 N to its asymptotic value, 108 GPa, beyond 3 N. Error bars indicate s.d. (n = 5). *Left inset*, an optical micrograph of the Vickers indentation with cracks produced at a load of 19.6 N. *Right inset*, the H_V-F curve of a 0.3-mm cBN single crystal. H_V does not reach its asymptotic value, and the crystal fractures when F exceeds 4.9 N [33]. With kind permission of the Nature Publishing Group

is the highest hardness reported so far for polycrystalline cBN and even exceeds that of synthetic diamonds.

Unlike macroscale materials, in which the mechanical properties depend mainly on the dislocation dynamics and their soundness (concentration of various flaws, such as cracks, etc.), in nanosized materials, grain size plays an especially important role, affecting behavior. Although grain size has some effect in macroscale materials, has less impact on certain properties, such as hardness. The H–P law, originally formulated regarding metals, states that in polycrystalline materials, hardness, yield stress and possibly tensile stress increase with decreasing grain size and varying linearly as a function of d^{-n}, where d is the grain size and $n > 0$ (see Sect. 4.93 and Eqs. (4.18) and (4.18a)). This increase in hardness (yield stress) is explained by the retardation of dislocation motion. In nanoscale materials, including ceramics, it is expected that a large increase should appear on the H–P plots, larger than in macroscale materials with decreasing grain size. Although this is observed at some stage, it occurs only in a very small nano range and, rather than observing increased hardness, a decrease (softening) is recorded. The term 'inverse H–P relation' is used to describe such plots.

9.2.3.1 Inverse H–P Effect

An inverse H–P phenomenon has been observed by researchers in various metals and also in some ceramics, e.g., in boron nitride. Attempts to explain this phenomenon have been made, though some claim that it is not a real characteristic (perhaps even a sort of artifact). Recent studies tend to indicate that the inverse H–P effect is real, true and apparently universal. However, the debate is now focused on its mechanism. Explanations do exist suggesting various mechanisms that may be responsible for this effect [7], even that of twinning (e.g., Tian et al. [32]). At these small nanosizes, dislocations cannot be the cause of this observed mechanical behavior, since such extended defects have certain dimensions that cannot be accommodated within the space of low range nanograins; dislocations are unlikely to reside in such miniscule structures. As stated above, in nanoscale materials, GBS is the likely mechanism affecting behavior. In Fig. 9.20, a Vickers indentation is shown for a nanocrystalline cBN, as a function of crystallite size. Here, ABNNC stands for 'aggregated boron nitride nanocomposite'.

The hardness value approaches that of single-crystal and polycrystalline diamonds and aggregated diamond nanorods. The other properties (unusually high fracture toughness and wear resistance) of this material are combinations with high thermal stability (above 1600 K in air), making this ceramic an exceptional superabrasive material. This unusually hard boron nitride was produced by high-pressure, high-temperature synthesis. Experimental observations and simulations suggest that, for many polycrystalline materials, there is an optimal grain size (usually in the range of several nanometers, i.e., nanocrystalline), which produces a significant (20–30 %) increase in the hardness of the material, compared with that of its coarse-grained counterpart. Usually, hardness measurements, for hard

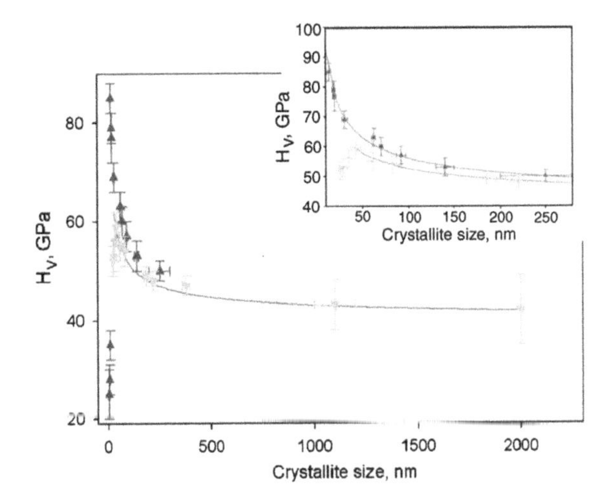

Fig. 9.20 (*Color online*) Vickers hardness (H_V) as a function of the crystallite size. *Triangles* correspond to the data for ABNNC; *inversed triangles*, c-BN [7]. With kind permission of AIP

and brittle materials, are in the asymptotic hardness region. Note in the H_V versus crystallite size curve that, with the decrease in grain size, the hardness values of both the ABNNC and the cBN increase. However, for polycrystalline cBN, hardness increases from 40 GPa and reaches a maximum of 59 GPa at a grain size of ~40 nm, while ABNNC hardness increases from 60 GPa, reaching its maximum of 85 GPa at 14 nm crystallite size. If the grain size decreases by a few more nanometers, hardness drops down to 25 GPa. Thus, the curve indicates that there is a transition during the hardening process due to the grain size (H–P effect) to softening (an inverse H–P effect) as grain size decreases into the range of a few nanometers. As indicated earlier, some researchers attribute this softening to GBS. Another approach relates it to quantum confinement. More specifically, in the observed inverse H–P, two factors should be noted (1) the nanosize effect, which restricts dislocation propagation through the material and (2) quantum confinement, which increases the hardness of individual crystallites. This concept is a consequence of ab initio calculations made by Tse [34], who suggested that the hardness of nanocrystals depends on the 'effective' band gap, which in turn is inversely proportional to the size of the crystallites, in other words, of the combined H–P and quantum confinement effects.

In sharp contrast to the above observation (that hardness decreases significantly in low-range nanosized cNB), Tian et al. [32] did not record an inverse H–P effect, despite the fact that their experiments were carried out down to the size of 3.8 nm. They consider twinned cBN to be involved in nanoscale hardening, and their results appear in Fig. 9.21. Furthermore, continuous hardening behavior with decreasing size or twin thickness is contrary to the findings for metals, in which both yield strength and hardness do show softening, as seen in the inset in Fig. 9.21 for Cu.

The continuous hardening behavior with decreasing microstructural sizes, down to 3.8 nm in cBN, may be explained as follows. For nano-twins with thicknesses

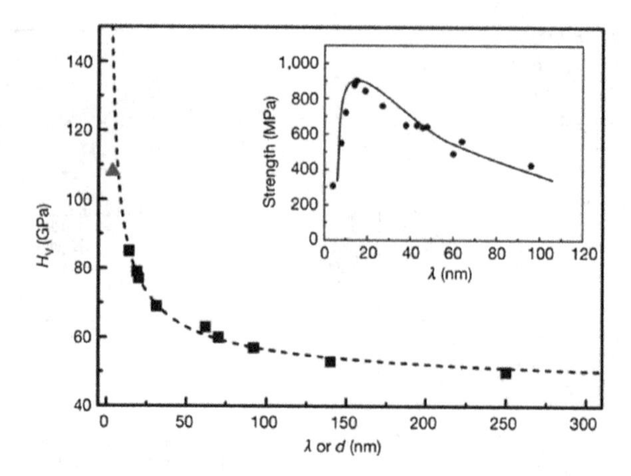

Fig. 9.21 H_V as a function of average grain size (d) or twin thickness (λ) for polycrystalline cBN bulk materials. Experimental data for nt-cBN bulk material (*triangle*) and for nt-cBN bulk materials[1] (*squares*) are shown. Using Eq. (9.1) to fit the experimental data, we obtained $H_V = 42.6 + 126d^{-1/2} + 130.7d^{-1}$. k (126 GPa $nm^{1/2}$) is taken from a previously reported value. The fitted C (130.7 ± 16.8 GPa nm) parameter characterizing the quantum confinement effect is in excellent agreement with the theoretical value[1] (136 GPa nm) from $211N_e^{1/3}\exp(-1291f_i)$ for cBN. This coincidence is not accidental and may provide proof of the existence of the quantum confinement effect in the synthesized nt-cBN. *Inset*, the yield strength as a function of λ for nt-Cu, in which the critical λ is about 15 nm (Ref. 15) [33]. With kind permission of the Nature Publishing Group. [In the legend, reference λ refers to Ref. [7] and 15 to Ref. [18]]

below 3.8 nm, the quantum confinement effect (which is inoperative in metals) becomes dominantly large, according to Eq. (9.1):

$$H_V = H_{HP} + H_{QC} = H_0 + kd^{-1/2} + Cd^{-1} \tag{9.1}$$

Here, $H_{HP} = (H_0 + kd^{-1/2})$ and $H_{QC} = (Cd^{-1})$ represent dislocation-related dislocation hardening based on the H–P effect and bandgap-related hardening based on the quantum confinement effect, as indicated above, following Tse's [34] calculations. H_0 is the single-crystal hardness and k is a material constant. C is a material-specific parameter equal to zero for metals and equal to $211N_e^{1/3}\exp(1.191f_i)$ for covalent materials, where N_e is the valence electron density and f_i is the Philips ionicity of the chemical bond.

Both the H–P and quantum confinement effects account for hardening in nanograin cBN (ng-cBN) with small grain sizes of 14 nm. This featured nanoscale geometry, as well as the very strong covalent B–N bonding, severely confines the nano twins {nt) migration of twin boundaries, (which is known to induce hardness softening in nt metals). These two hardening mechanisms remain valid when the twin thickness is significantly reduced to 3.8 nm for cBN. The continuously increasing hardening occurring with decreasing size is in contrast with the findings recorded for metals, in which both yield strength and hardness decrease at the low

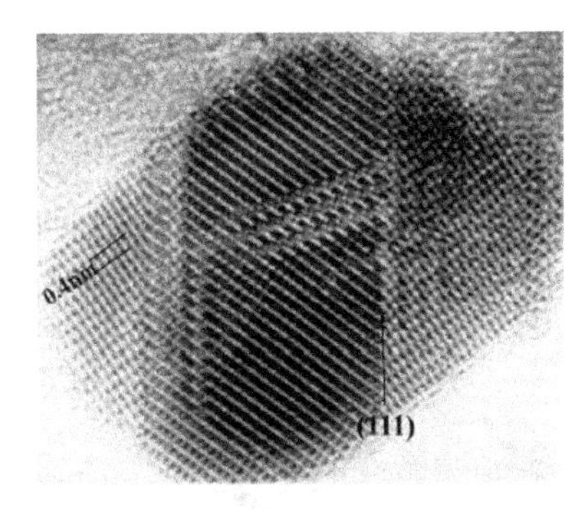

Fig. 9.22 High-resolution transmission electron microscopy image of a 10-nm SAG BaTiO3 nanocrystal with complex set of twins on (111) planes (prepared at 650 °C) [12]. With kind permission of Elsevier

range of the nanoscale. For polycrystalline cBN or diamonds, reducing the grain size has been the most widely used technique for enhancing hardness.

As previously mentioned, some explain the inverse H–P effect by twinning [7]. From the above examples, it may be assumed that, despite the claims of some researchers of the universality of the inverse H–P effect, its validity remains debatable. The above data are from experiments performed quite recently (2006 and 2012), so this debate is still unresolved.

9.2.4 Twinning

Twinning was discussed in Chap. 4, specifically in Sect. 4.7, and a schematic illustration of twin structure appears in Fig. 4.39. A high resolution transmission electron (HRTEM) image of a 10-nm stearic acid-gel (SAG) $BaTiO_3$ is also shown in Fig. 9.22, showing twin structure.

Twins, generated in SiC particles, were observed in a Al_2O_3/SiC composite [17], as shown in Fig. 9.23. These twins were observed in the larger SiC particles identified by EDS and diffraction patterns in Fig. 9.23c and d, respectively. The twin is another form of deformation when slip is difficult. Twins form rapidly during the crystal deformation. The stress needed for twin nucleation is higher than for its extension. The load or stress is known to fall suddenly, since twins are generated continuously and produce a jagged stress–strain curve (Pelleg, Chap. 3 on Plastic Deformation, Fig. 3.14). The appearance of the curve mentioned above has been observed in nanocomposite ceramics as well. For example, twinning in Al_2O_3/SiC (Fig. 9.23) produces a jagged (zigzag) curve during flexural-strength testing on the load–displacement plot [17]. Moreover, the existence of twins indicates that the material had a ductile character during fracture. The fracture energy of the Al_2O_3/SiC interface resulted from a thermal mismatch in the energy absorbed by each of the twins generated in the SiC particle. It was observed that both toughening and flexural strength increased. The contribution of twins to fracture toughness and flexural strength may be less than that of dislocations, but, nevertheless, it exists and influences overall strength.

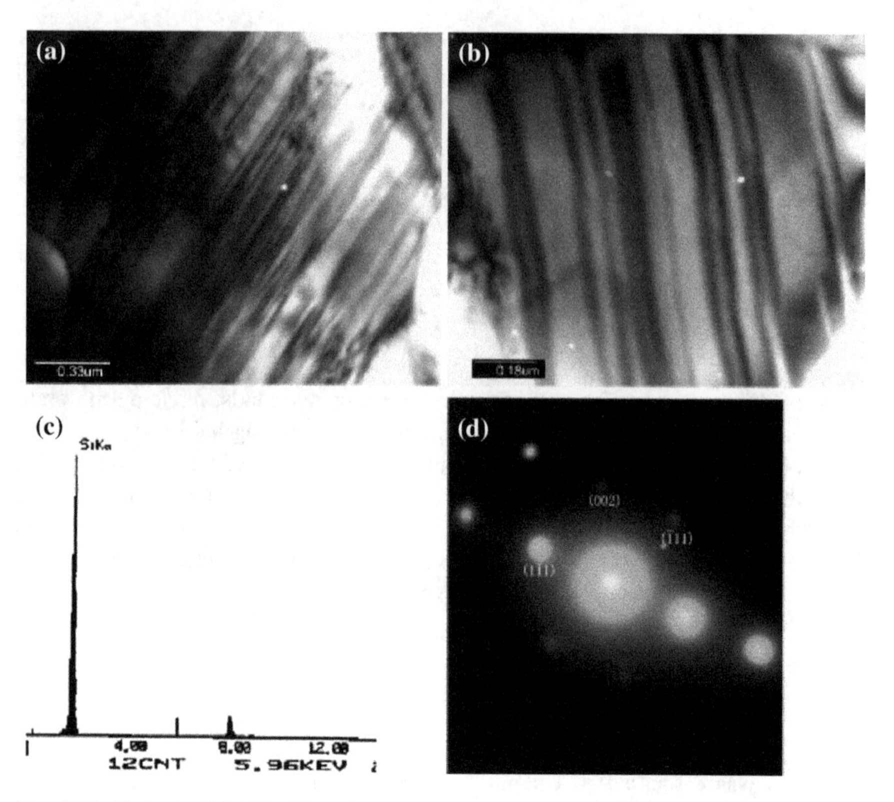

Fig. 9.23 Twins in SiC [17]. With kind permission of Elsevier

Some (Ovid'ko [24], Li [16]) claim that deformation involving nanotwins generated at grain boundaries and associated with partial dislocations induce toughening in the materials, metals and ceramics alike. In fact, Li [16] suggested that deformation twins (recall that they may be either deformation or annealing twins) are generated near cracks in nanoscale materials. Accordingly, deformation twins nucleate by stress-driven emissions of twinning dislocations from a grain boundary distant from the crack tip. During the process of deformation nucleation twins release high, local stresses near the crack-tip, enhancing fracture toughness. In Fig. 9.24, nanotwin growth in a deformed nanocrystalline specimen is shown.

Note that two types of twins were considered in Chap. 3 on plastic deformation [27], mechanical (deformation) and annealing (also known as growth or recrystallization) twins. These types are crystallographically related, but the nature of their formation is different. Both play an important role in the deformation of nanocrystalline materials. Twinning is a competing mechanism to slip, but is directly related to (stacking faults (SFs), which are themselves dislocation-related. Twinning shear stress has been expressed [37] as a function of source size and SF energy:

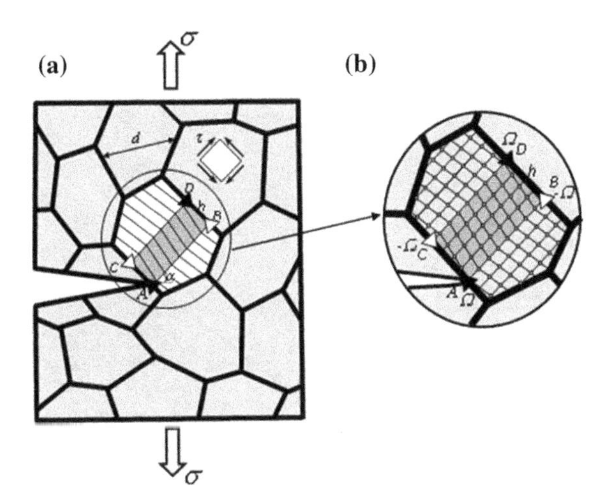

Fig. 9.24 Nanotwin growth in a deformed nanocrystalline specimen containing a pre-existent crack (a two-dimensional model). **a** General view. A deformed nanocrystalline specimen contains both a crack and a rectangular nanotwin ABCD. The nanotwin as a stress source represents a quadrupole of wedge disclinations located at its vortices. **b** The atomic structure of the grain contains the nanotwin ABCD (schematically) [31]. With kind permission of Dr. Skiba

$$\frac{\tau_T}{G_T} = \frac{\gamma}{Gb} + \frac{\alpha b}{l} \tag{9.2}$$

where τ_T is the shear stress, l is the size and γ is the SF energy. Equation (9.2) indicates that the twinning stress increases with γ and with decreased source size. Clearly, b is the Burgers vector of the partial dislocation (SF-related). The relations between normalized twinning stress and normalized SF for several metals are shown in Fig. 9.25.

To indicate that the Burgers vector is related to partial dislocation, it is occasionally written as b_p. Another parameter of importance is the grain size. For the following relations, grain size is taken into account in slip and twinning. Notice that these relations are similar, the difference being indicated by the subscripts for slip (S) and twinning (T); they are basically H–P equations:

$$\sigma_T = \sigma_{0T} + k_T d^{-1/2} \tag{9.3}$$

$$\sigma_S = \sigma_{0S} + k_S d^{-1/2} \tag{9.3.a}$$

The literature [20] indicates that $k_T \geq k_S$. It is clear from these equations that, in both cases, the respective stresses increase with decreasing size. For nanocrystalline materials, a critical radius has been suggested (Meyers et al. [20]) for twinning, expressed as:

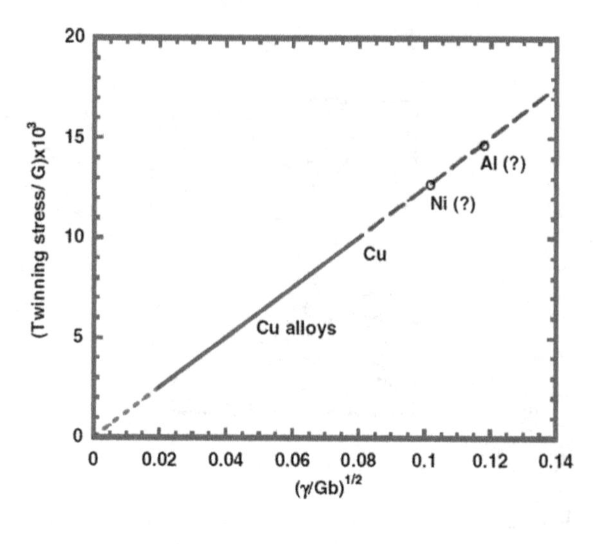

Fig. 9.25 Effect of stacking fault energy on twinning stress [20]. With kind permission of Elsevier

$$r_c = \frac{5\pi G\gamma_{TB}}{4\tau'_T} \tag{9.4}$$

where γ_{TB} is the twin-boundary energy (directly related to the SF energy) and τ'_T is the local (not global) twinning shear stress.

9.3 Time-Dependent Deformation (Creep)

Creep deformation was broadly discussed in Chap. 6 in general. Here, the specific character of creep is considered in nanoscale ceramics. The consolidation techniques of the raw materials and the production methods generally determine the mechanical, physical and other properties of a ceramic. This is particularly true in the case of nanoscale ceramics, since the distribution of the phases in composite ceramics, for example, and their location relative to the commonly found microcracks are decisive factors in obtaining strong, tough, fracture-resistant substances. By using an appropriate production method to eliminate glassy phases at grain boundaries, a strong, significantly-improved, creep-resistant, nanocomposite ceramic was obtained—a silicon nitride/silicon carbide nano–nano composite. The final sample size was a 19 mm diameter disk with a thickness of 3–4 mm. The electron microprobe analysis shown in Fig. 9.26 reveals that the initial Si–C–N powder had a nominal composition of $Si_{1.00}C_{1.55}N_{0.81}O_{0.17}$. The presence of oxygen in this composite is a result of surface oxidation, due to handling in air. The SEM observation shows that the mean particle size for this powder is about 1 μm. A TEM analysis reveals that, after decreasing the amount of additive, the grain size of the composites decreases monotonically and there is a transition from a micro–nano structure, to a nano–nano type structure. When the material is

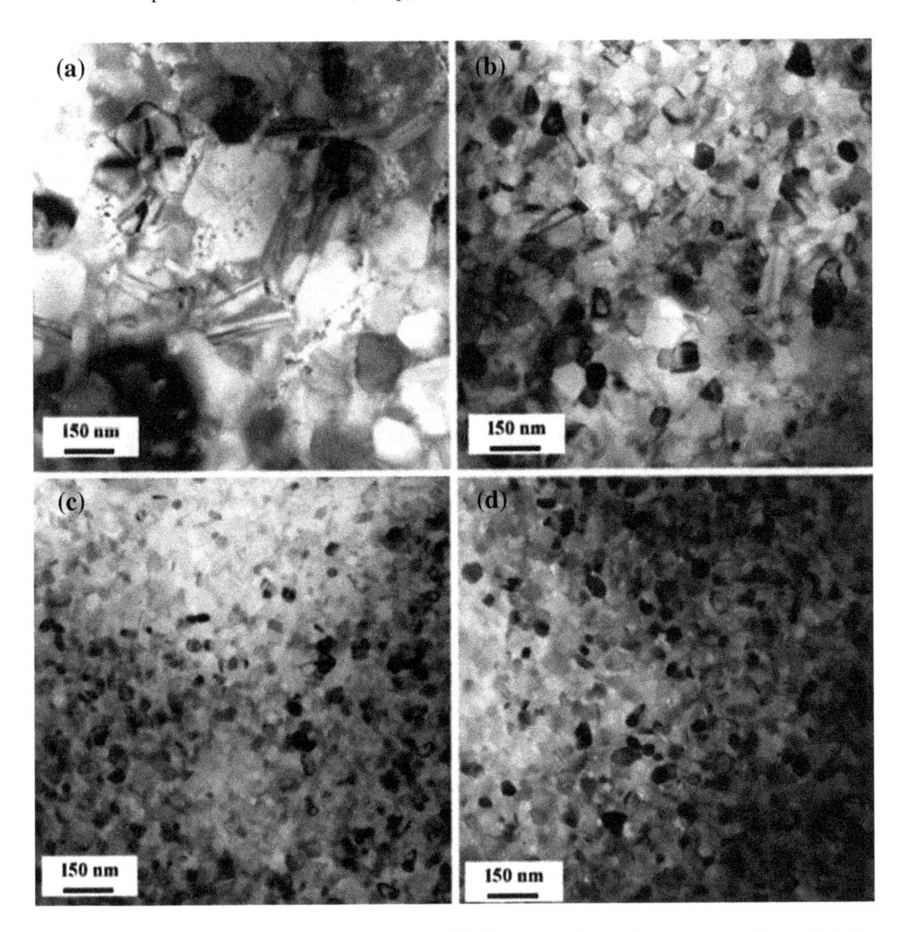

Fig. 9.26 Transmission electron microscopy (TEM) observations of nanocomposites of Si_3N_4–SiC **a** sintered with 8 wt% Y_2O_3 at 1600 °C for 10 min, micro–nano structure, **b** sintered with 3 wt% Y_2O_3 at 1600 °C for 10 min., nano–nano structure, **c** sintered with 1 wt% Y_2O_3 at 1600 °C for 10 min, nano–nano structure, **d** sintered without additive at 1600 °C for 30 min, nano–nano structure [38]. With kind permission of John Wiley and Sons

sintered without additives, 10 min at 1600 °C leads to a grain size of about 27 nm. After 30 min of sintering, the grain size reaches about 40 nm, as shown in Fig. 9.26d. In Fig. 9.27, an elemental analysis, by means of electron energy loss spectroscopy (EELS), indicates that the two phases in this material, namely, Si_3N_4 and SiC, were randomly mixed and had roughly equal grain size. The oxygen, which had been diffused from the surface into the material during the high temperature processes, caused the formation of some glassy material. HRTEM illustration in Fig. 9.28 indicates the grain-boundary region of specimens without additives, as shown in Fig. 9.26d. Only a small amount of the glassy phase, which is dependent on the amount of oxygen and its distribution, may be observed. In the ceramic under consideration, the oxygen was not homogeneously distributed in the grain-boundary

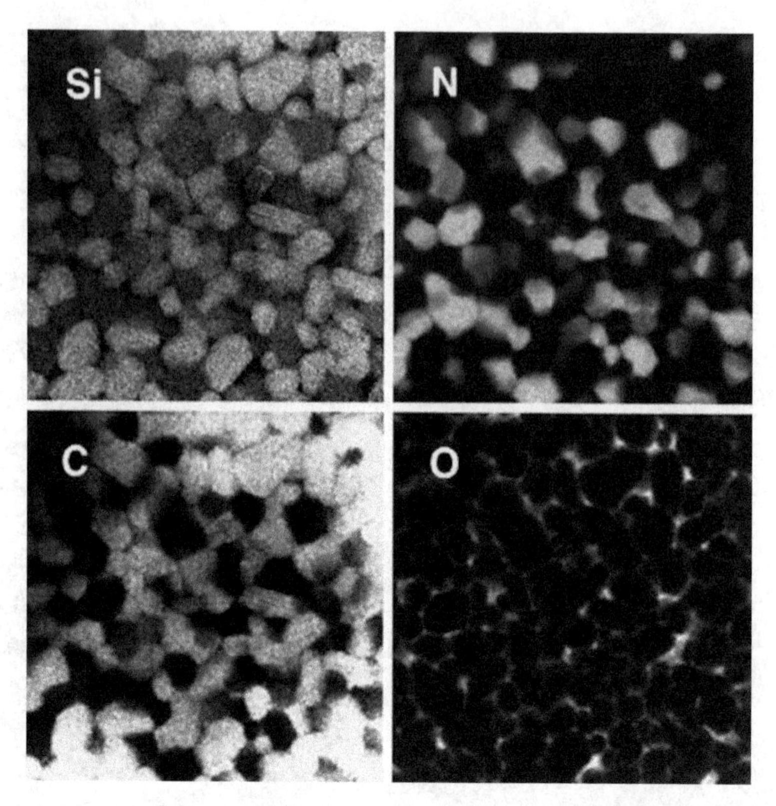

Fig. 9.27 Electron energy loss spectroscopy (EELS) analysis of the component elements in the Si_3N_4–SiC nanocomposite sintered at 1600 °C for 30 min without additive [38]. With kind permission of John Wiley and Sons

regions, some having more than others. Most of the glassy grain-boundary phase exists at multigrain junctions (e.g., see Fig. 9.28c). To avoid common complications, compression creep tests were conducted to examine the creep behavior of the nano–nano composites, rather than tensile tests, that are likely to induce cavitation. The steady-state creep of nano–nano composites at various temperature and stress levels is shown in Fig. 9.29 as a plot of strain as a function of time. A comparison between the creep properties of the nano–nano composite ceramics, silicon nitride/silicon carbide and microcrystalline silicon–nitride, based on data taken from the literature, is provided in Fig. 9.30. These nano–nano composites show extraordinarily high creep-resistance (corresponding to a low creep-strain rate). (For the references in Fig. 9.30, the reader is referred to the original work). As may be seen, the nanocomposite sintered without additives shows a creep rate as low as 6.3×10^{-11} 1/s at 50 MPa stress. The steady-state creep deformation of crystalline materials may be expressed by one of the empirical relations as:

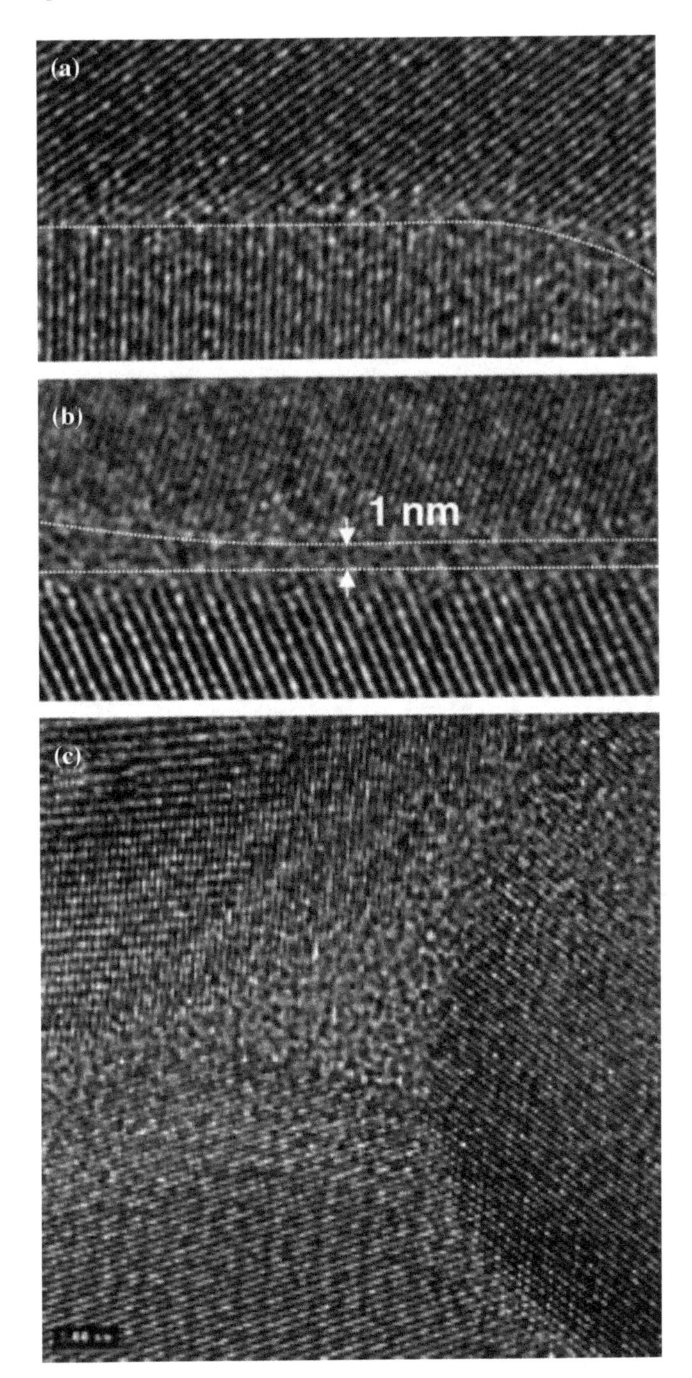

Fig. 9.28 High-resolution transmission electron microscopy (HRTEM) analysis of the grain boundary of the nano–nano composite (no additive, 1600 °C/30 min sintered). **a** Glassfree grain boundary, **b** grain boundary containing glassy layer, **c** triple junction [38]. With kind permission of John Wiley and Sons

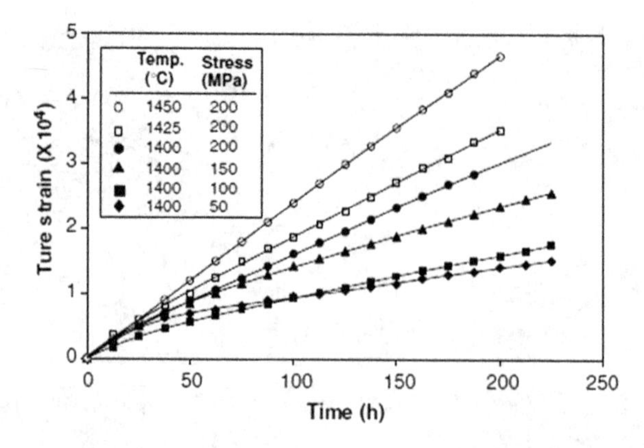

Fig. 9.29 Compression *creep strain–time curves* for one of the nano–nano composites (1 wt% Y_2O_3, 1600 °C/10 min sintered) [38]. With kind permission of John Wiley and Sons

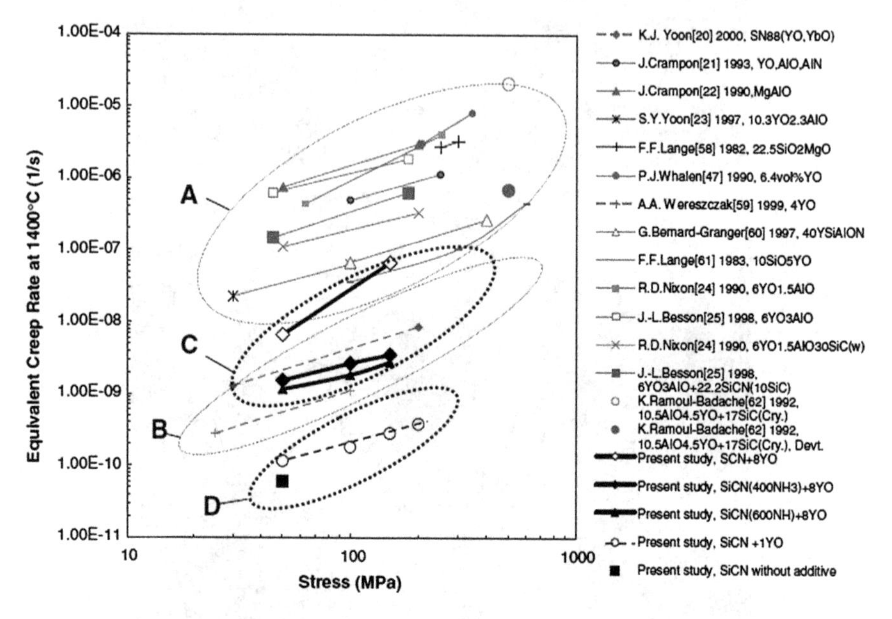

Fig. 9.30 Comparison of the compression creep property of nanocomposites with those of existing silicon–nitride ceramics (additive in weight percentage unless specified, molecular formula simplified for clarity. For instance, "6YO" in figure legend stands for "6 wt%Y_2O_3") [38]. With kind permission of John Wiley and Sons

$$\dot{\varepsilon} = A\frac{\sigma^n}{d^p}\exp\left(-\frac{Q}{RT}\right) \tag{9.5}$$

For similar empirical relations, Chap. 6 (Eq. 6.30) may be consulted. In Eq. (9.5), A is a constant, σ the applied stress, n the stress exponent, d the grain size, p the

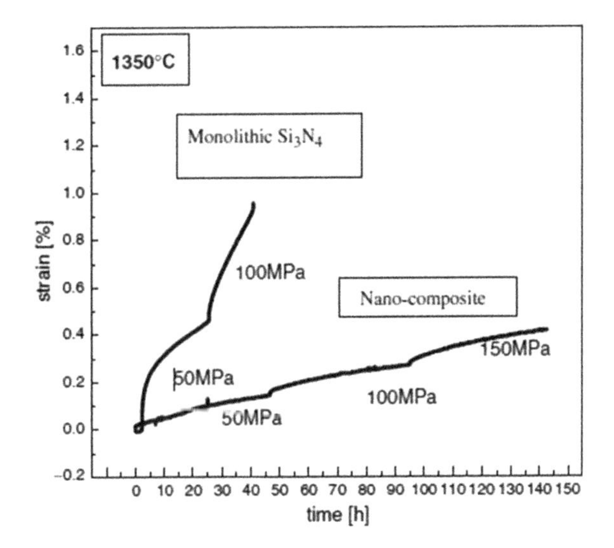

Fig. 9.31 Comparison of the creep deformation of monolithic silicon nitride and of the C-derived nanocomposite [9]. With kind permission of John Wiley and Sons

grain-size dependent exponent, Q the activation energy, R the gas constant and T is the absolute temperature. The stress exponent is often considered to indicate the creep mechanism. The experimentally-determined stress exponent, n, for various silicon–nitride system ceramics, may be as low as <1 or as high as >3, and the activation energy, Q, may be as low as 300 kJ/mol or as high as 1200 kJ/mol. The observed p is ~ 1–3, which means the strong dependence of the creep rate on the grain size. A large increase in creep rate (1–3 orders) with a corresponding decrease in grain size when going from the micron to the nano range is supposed to occur. However, the very large creep resistance found in nanocomposites strongly suggests a different acting creep mechanism than is found in relatively large, micron-sized materials. Another sign that the creep mechanism may be different in nano–nano composites is their low activation energy. This suggests that the resulting highly-dispersed distribution of oxygen at the interfaces may prevent the formation of the intergranular glassy phases that are effective as fast routes for mass transfer, leading to extraordinarily high creep resistance in such nanocomposites. Thus, Eq. (9.5) does not describe the increased creep resistance of Si_3N_4/SiC nanoceramic.

The Si_3N_4/SiC nanocomposite illustrated in Fig. 9.31 indicates the well-known fact that monolithic ceramics are weaker than nanocomposite ceramics. In this figure, the creep strain of monolithic Si_3N_4 is substantially greater than that of the nanocomposite.

Also in this case, Y_2O_3 was added to the nanocomposite. The creep tests were performed by four-point bending at temperatures of 1200 and 1450 °C within a stress range of 50–150 MPa. The creep rate was calculated from the slope of the ε versus t curve (Fig. 9.31) and steady-state creep was evaluated using Eq. (9.5), i.e., the Norton equation. An alternative explanation for the observed increase in creep resistance in the nanocomposite is that the SiC nanoparticles hinder the grain growth

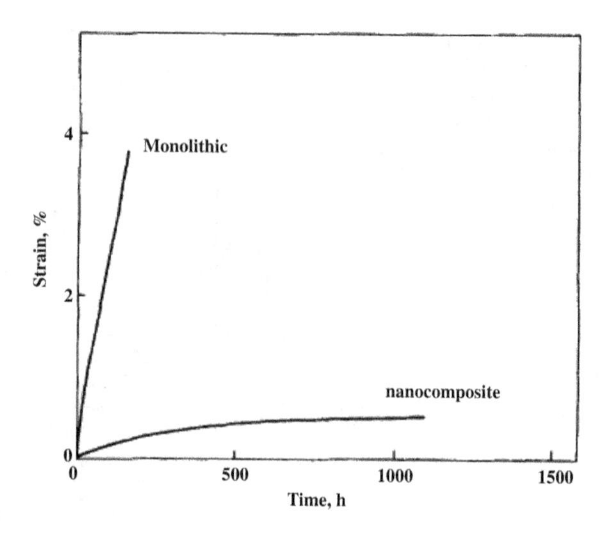

Fig. 9.32 *Tensile creep curves* of the monolith and nanocomposite at 1200 °C and 50 MPa. Slight accelerated creep and steady-state creep were present in the monolith, while they were little observed in the nanocomposite [23]. With kind permission of John Wiley and Sons

of Si_3N_4 by interlocking with the neighboring Si_3N_4 grains. In this manner, changes in the volume fraction and chemical composition at the grain-boundary phase occur, modifying the creep mechanisms, with consequent, visible improvement to the Si_3N_4/SiC nanocomposite. For a more general picture of improved creep resistance in nanoceramics (both monolithic and composite), additional examples are provided of Al_2O_3-based nanocomposites and of nano-zirconia. Figure 9.32 shows monolithic and nanocomposite creep curves of strain versus time.

The curve of the monolith consists of primary, steady-state and very small tertiary creep. The specimen lifetime was ~150 h and ~4 % of creep strain was obtained at fracture. A large number of microcracks were also identified by optical microscopy. Compared to the monolith, the nanocomposite exhibited excellent creep resistance. At 1200 °C and 50 MPa, its creep life was 1120 h, which is 10 times longer than that of the monolith. The creep strain at fracture was 0.5 %, which is eight times smaller than that of the monolith. Furthermore, the superior creep resistance of the nanocomposite was also obtained by flexure creep tests. Similar to tensile-creep curves, the strain of the nanocomposite tended to decrease over time, while the monolith exhibited steady-state creep and sometimes accelerated creep.

The strain rate, as a function of applied stress, is shown in Fig. 9.33 for the steady-state creep rates of both the monolith and the nanocomposite ceramics. It may be observed that the creep rate of the nanocomposite is about three orders of magnitude lower than that of the monolith under tension, and three to four orders of magnitude lower under flexure. One of the most characteristic changes in microstructures during creep is the rotation of the intergranular silicon-carbide particles, accompanied by GBS and small cavity formation around the particles. This may be seen in Fig. 9.34a.

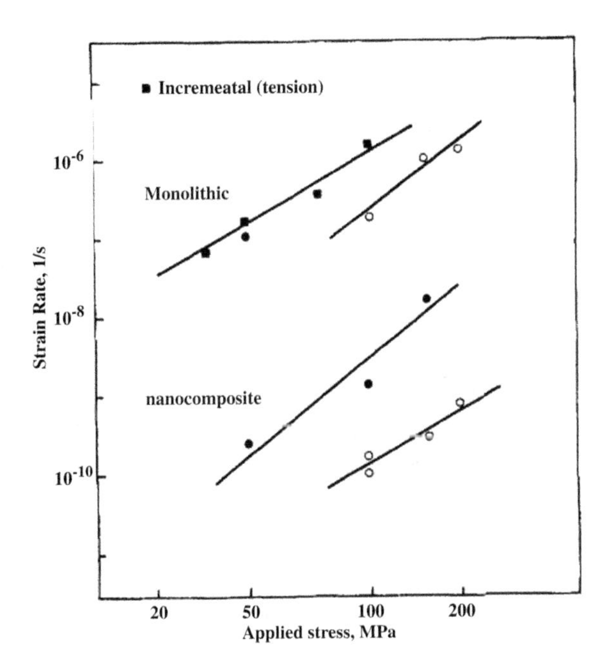

Fig. 9.33 Stress dependencies of steady-state or minimum creep rates in the tension (*closed symbol*) and the flexure (*open symbol*) for the monolith and the nanocomposite. The temperature is 1200 °C. The stress exponent for creep rate is 2.2 for the monolith and 3.1 for the nanocomposite in tension, and 2.9 for the monolith and 2.2 for the nanocomposite in flexure [23]. With kind permission of John Wiley and Sons

One may see that the interface between the upper and lower alumina grains is curved towards the intergranular silicon-carbide particle. Significant strain contrast contours have formed at the corners (top right and top left) of the particle and small cavities exist between the particle and the upper grain. Based on this microstructure, the following mechanism was suggested for GBS or its prevention. Before the sliding occurred, the intergranular particle was supposedly located on a straight alumina–alumina grain boundary. When creep occurred, the upper alumina grain slid downward over the lower grain. In order to maintain such sliding, the particle rotated counterclockwise and plunged into the lower grain. As byproducts of this process, strain contrast contours and small cavities were generated, and the grain boundary curved. This plunging increases the pinning effect of the particle and, consequently, improves creep resistance, resulting in transient creep. More evidence of the rotating and plunging of intergranular silicon-carbide particles and associated cavity formation is given in Fig. 9.34c. Figure 9.35 shows a trace of intergranular crack propagation. In this case, the crack proceeded along the alumina–alumina grain boundary, where small cavities formed around the particles, as a result of GBS.

The intergranular small cavities, formed during the plunging of the interfacial particles, weaken interfacial bonding and induce crack formation at the grain boundaries. The important role of the transgranular nanoparticles is to inhibit lattice diffusional creep. Grain boundary diffusion, however, is the most

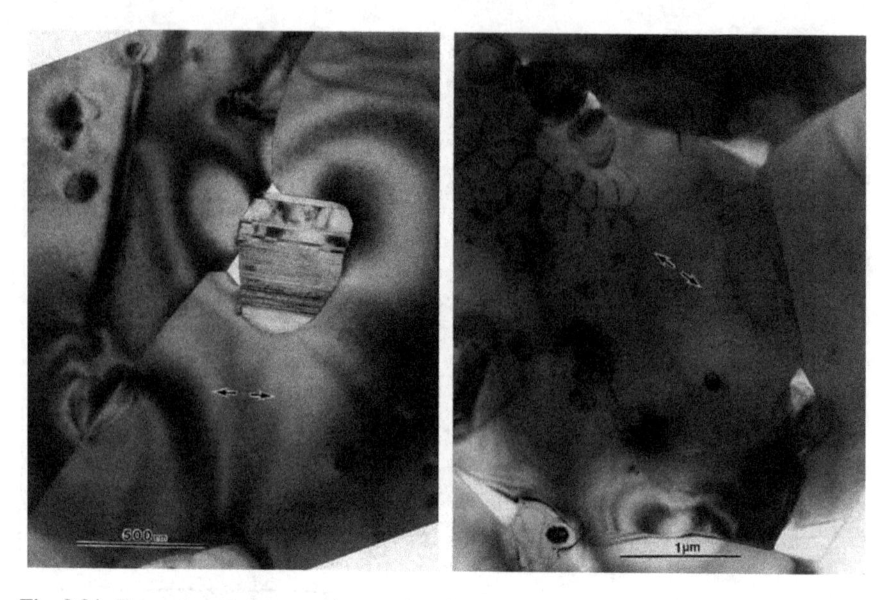

Fig. 9.34 Transmission electron micrographs of microstructures of the nanocomposite tested at 1300 °C and 50 MPa in tension, showing examples of rotating and plunging of intergranular silicon carbide particles and associated cavity formation. The stress direction is indicated by *arrows* [23]. With kind permission of John Wiley and Sons

predominant deformation mechanism of polycrystalline alumina at temperatures around 1200 °C. GBS requires diffusion processes, because of the role played by vacancies in sliding. The diffusion process at the particle matrix interface is significantly lower than at the matrix–matrix interface (which may explain the higher creep rate in monolithic alumina). As GBS proceeds, the intergranular silicon-carbide nanoparticle rotates and plunges into the alumina matrix, significantly increasing the creep resistance. This results in only one stage of creep, namely that of transient creep. Rapid crack formation at the grain boundaries during the final stages of creep deformation bring about fracture.

An additional example is a zirconia-based nanocomposite. In this case, various amounts of C nanotubes (CNTs) have been added to zirconia. Figure 9.36 compares the nanocomposite having various amounts of CNTs or none. The addition of CNTs to polycrystalline nanograined zirconia leads to a significant increase in creep resistance (and fracture toughness). These CNTs have a multiwalled C nanotubes [henceforth: MWCNTs] form. The creep resistance of the nanograined zirconia may be explained by the ability of CNTs to inhibit GBS at high temperature, thus improving creep resistance. MWCNTs substantially reduce grain growth and may also be able to pin grain boundaries and reduce their mobility during creep. Zirconia containing CNTs has a lower creep rate than monolithic zirconia. With higher amounts of MWCNTs, up to 5 %, after a transient

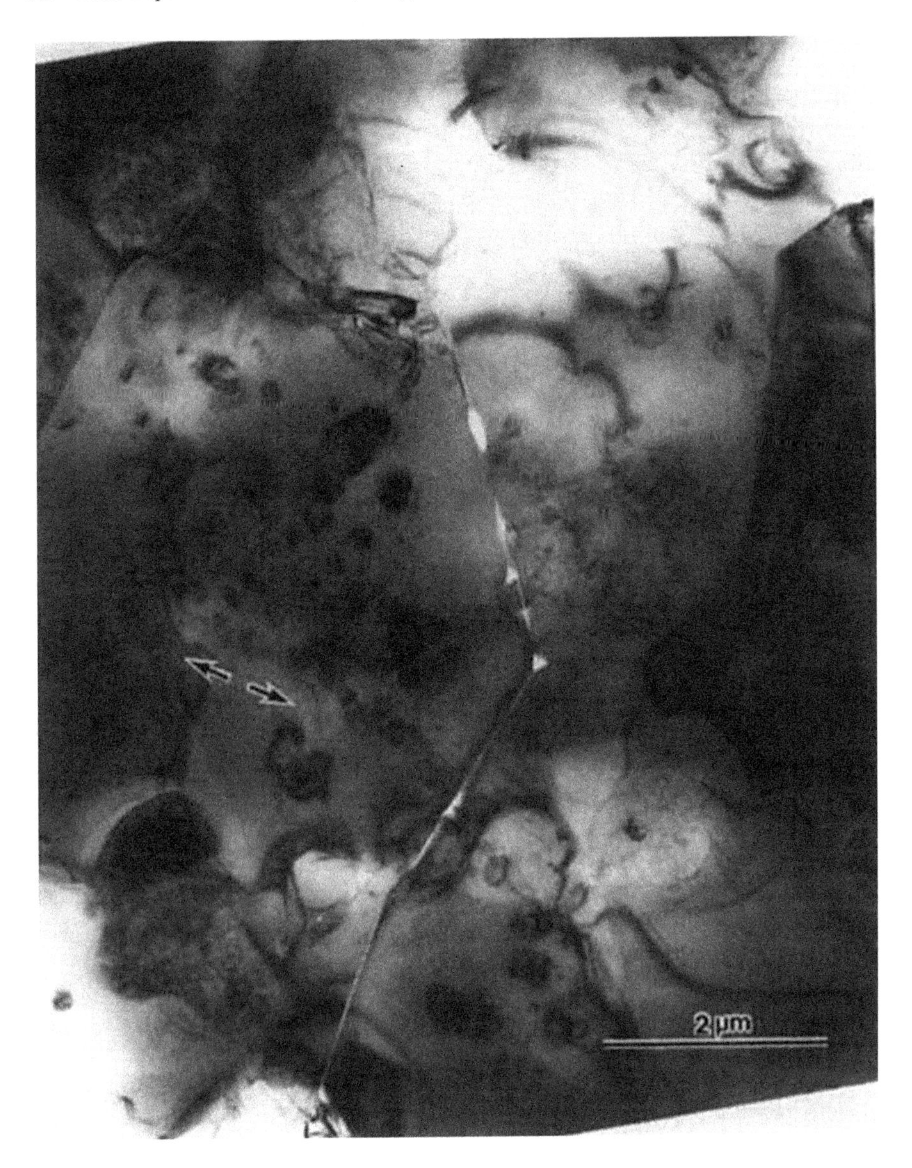

Fig. 9.35 Transmission electron micrograph of a trace of intergranular crack propagation. The sample was tested at 1200 °C and 100 MPa in tension. Note the transgranular-fractured nanoparticle. The stress direction is indicated by an *arrow* [23]. With kind permission of John Wiley and Sons

decreasing creep, the creep rate drops almost to zero (Fig. 9.36b). Other properties (hardness, fracture toughness) are also improved by additives. The most pronounced effect is associated with a 5 % CNT content.

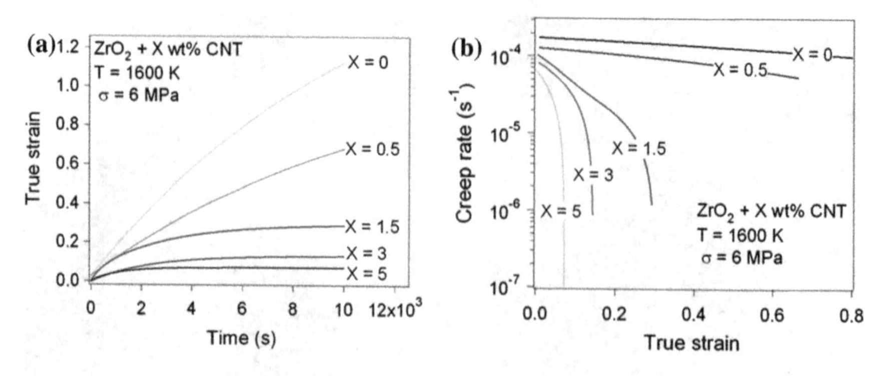

Fig. 9.36 Creep strain (**b**) recorded as a function of time (**a**) and creep rate as a function of creep strain measured during compression creep test for composites with different amounts of CNTs [19]. With kind permission of Elsevier

9.4 Dynamic Deformation in Nanoceramics

9.4.1 Introduction

In recent years, much effort is being devoted to the production of structures, among them ceramic and nanostructures, because they exhibit valuable, advanced mechanical properties for the construction of various components for functional applications. A major challenge in modern materials science is to produce homogeneous ceramics that show good static performance in applications involving dynamic deformation during service. Since most bulk ceramics are brittle at ambient conditions and many show ductile properties in the nanoscale range, the search for ceramics having superior properties, with special emphasis on their responses to dynamic deformation, is ongoing. Some hope for the improved performance of materials lies with nanoceramics. Clearly, one of the prerequisites for achieving good (or the best) static and dynamic properties is the production of ceramics without pores, microcracks or other flaws. This is a formidable task, but modern production techniques have been developed bringing the scientific and industrial communities several steps closer to that goal.

The fundamental behaviors of nanoscale materials may be completely different from those of bulk materials. Due to the very small sizes, especially at the low range of the nanoscale, surface and atomistic properties dominate behavior and one must be aware that bulk theories may not be completely applicable. Therefore, the study of nanoscale materials is a challenging route to the understanding of experimental observations.

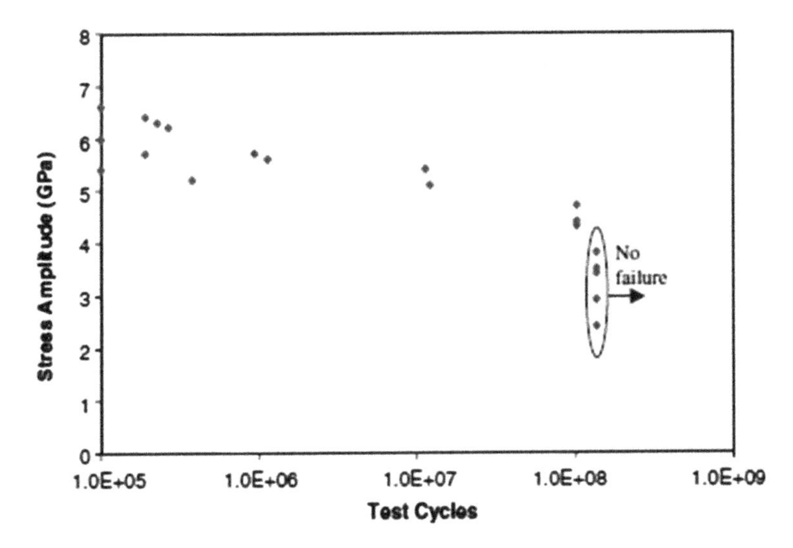

Fig. 9.37 Stress-life testing data for nano-scale tensile samples. The *circle* with a *horizontal arrow* indicates devices that did not fail under cyclic loading up to 10^8 cycles [6]. With kind permission of IOP Publishing and Ghodssi for the authors

9.4.2 Fatigue

This section begins with thin-film ceramics having nanosized dimensions, such as the low pressure chemical-vapor deposition [henceforth: LPCVD] of silicon nitride, tested via a nanoscale tensile test. The test devices are fabricated using a surface micromachining technique, in combination with deep reactive ion etching and ion milling. In situ fatigue measurements were performed on silicon nitride test structures using a 200 nm beam width inside a focused-ion-beam system, which is a recently developed experimental technique. The fatigue result is shown in Fig. 9.37.

By reducing the applied tensile stress to 3.8 GPa, these silicon nitride test structures can survive cyclic loadings up to 10^8 cycles.

When designing devices, such as nanoelectromechanical systems (NEMS), an understanding of the nanoscale mechanical properties of materials is essential, since they may incur thermal or mechanical stress during operation. Tensile tests may be performed within such systems by means of thin-film samples and the amplitude of the applied stress may be controlled by the frequency and amplitude of the input electrostatic energy. During such experiments, mechanical-amplifier actuators exhibit time-delayed failure and their resonant frequencies decrease monotonically over the test time, when the maximum operating stress exceeds

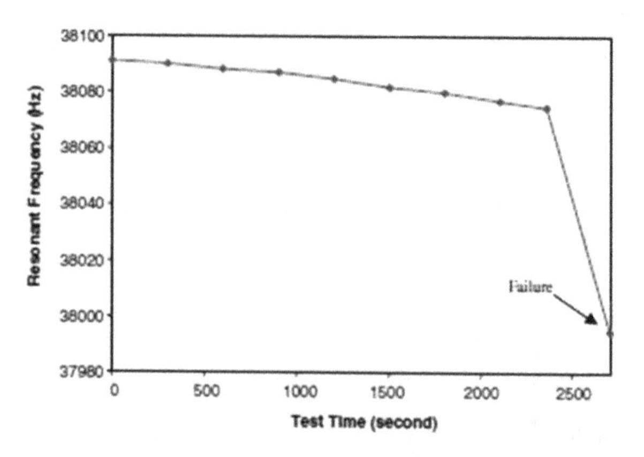

Fig. 9.38 Variation of resonant frequency with time for a mechanical-amplifier actuator during fatigue testing (test cycle: 10^8 cycles at stress amplitude 4.4 GPa) [6]. With kind permission of IOP Publishing and Ghodssi for the authors

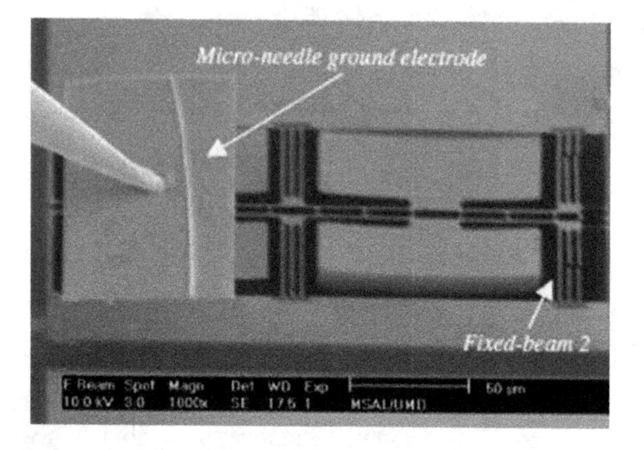

Fig. 9.39 Micrograph of a mechanical-amplifier actuator for fatigue tests with resonator 2 at its first resonant mode (f0 = 38.09 kHz). In this device, only one suspended beam is used as fixed-beam 2 [6]. With kind permission of IOP Publishing and Ghodssi for the authors

4.3 GPa. The variation in resonance frequency over time is shown in Fig. 9.38. The fatigue results appearing in Fig. 9.37 were obtained by means of a mechanical-amplifier actuator with resonator 2 at its first resonant mode. In this experiment, the input voltage to resonator 1 is 14.1 V_{RMS} and the resonant frequency of resonator 2 is found to be 38.09 kHz. The mechanical-amplifier actuator with resonator 2 at its first resonant mode is indicated in Fig. 9.39.

To summarize the results of this thin-film mechanical testing, note that the change in resonant frequencies indicates that tensile samples become more compliant during the test and that the failure of the tensile samples occurs as a result of the progressive accumulation of damage. Nonetheless, no fatigue failure of nanoscale LPCVD silicon nitride thin films is found up to 108 cycles during testing at stress amplitudes below 3.8 GPa with a load ratio of 0.48.

The benefits of using nanosized alumina-based composites, such as Al_2O_3/SiC follow. For instance, alumina is a commercial product having very good hardness, wear and corrosion resistance, and it responds very well to the sintering process; however, it has low bending strength and low fracture toughness. Even the

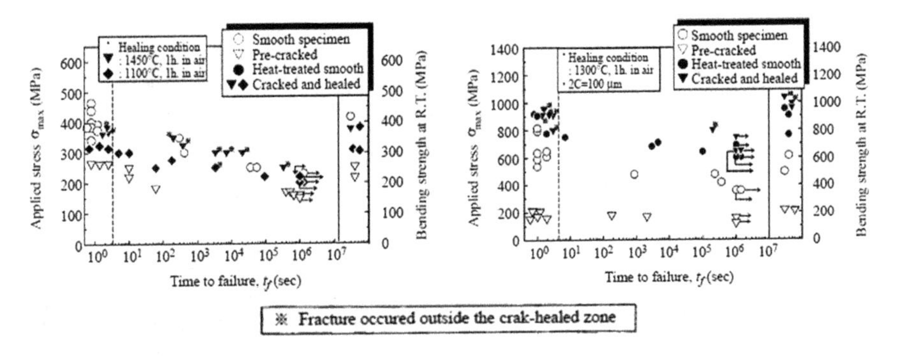

Fig. 9.40 **a** Relationship between applied stress and time to failure of monolithic Al_2O_3. **b** Relationship between applied stress and time to failure of Al_2O_3/SiC. With kind permission of Professor Ando for the authors of Ref. [1] and Professor of Ref. [2]

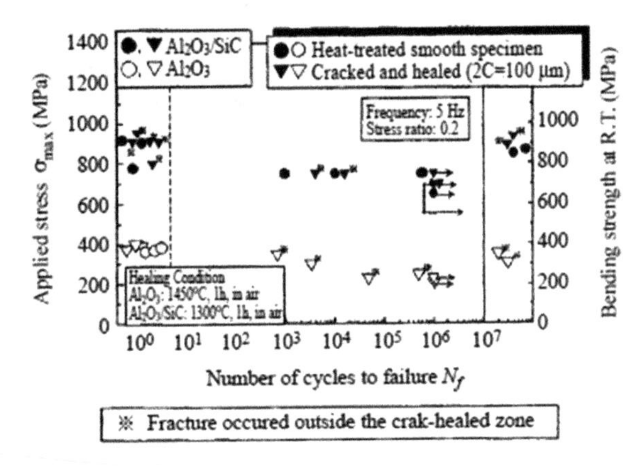

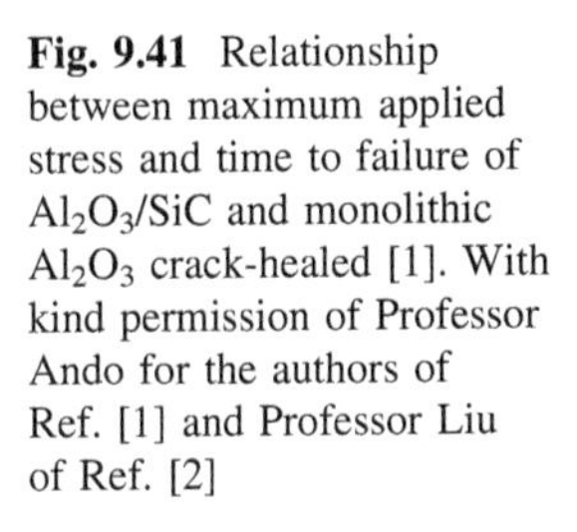
Fig. 9.41 Relationship between maximum applied stress and time to failure of Al_2O_3/SiC and monolithic Al_2O_3 crack-healed [1]. With kind permission of Professor Ando for the authors of Ref. [1] and Professor Liu of Ref. [2]

addition of ~5 % SiC improves the properties of the composite, giving it good crack healing ability. Applied stress, as a function of time (static fatigue), in monolithic Al_2O_3 and Al_2O_3/SiC composites are compared in Fig. 9.40.

Applied stress-to-failure in a plot against the number of cycles is shown in Fig. 9.41, comparing Al_2O_3 with the Al_2O_3/SiC composite. Here, the strengthening effect on fatigue in SiC may be seen. In the Figs. 9.40 and 9.41, in addition to the applied stress, the bending stress is also indicated. Note that all the specimens broke outside the crack-healed zone, supporting the importance of crack healing in ceramics. A fracture-surface photograph of the composite ceramic crack-healed specimen is shown in Fig. 9.42. The healing of cracks in ceramics is a means of improving fatigue resistance, not just the static mechanical properties (including static fatigue). In order to obtain this improvement, nanosized particles of SiC are dispersed in the alumina matrix. The alumina powder used for the alumina composite material has an average particle size of 0.5 μm and 99.99 % purity. In the Al_2O_3/SiC composite considered above, only the SiC are

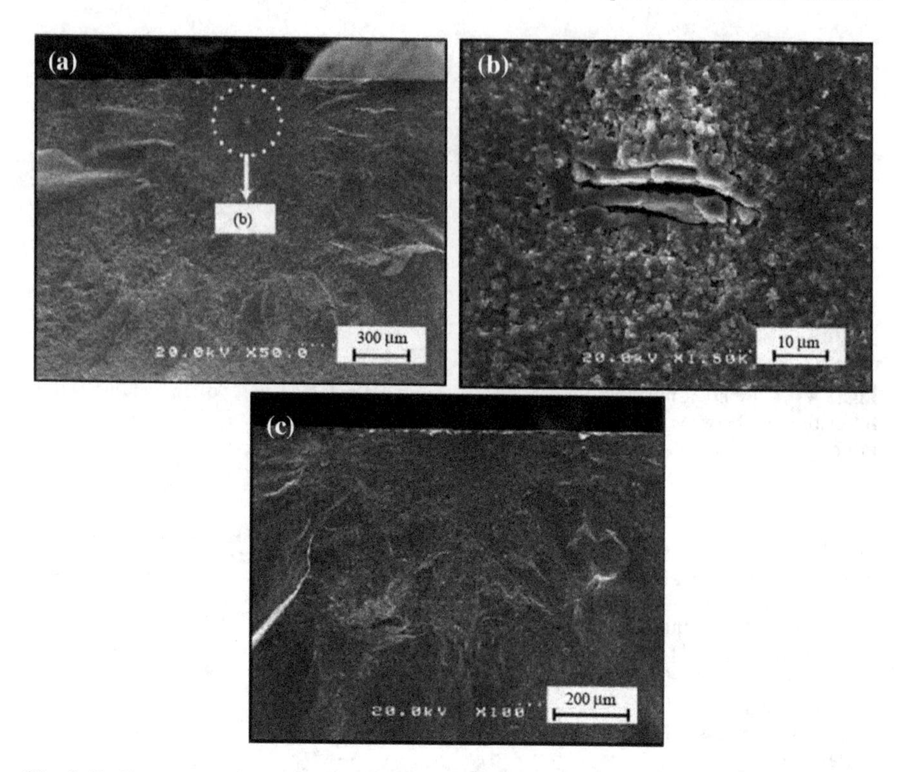

Fig. 9.42 Fracture surface of Al_2O_3/SiC crack-healed **a** Macro-SEM photograph of crack initiation site ($N_f = 3900$ cycle); **b** crack initiation site and flaw ($N_f = 3900$ cycles); **c** crack initiation site of $N_f = 15885$ cycles sample, no-flaw can be seen at the crack initiation site. With kind permission of Professor Ando for the authors of Ref. [1] and Professor Liu of Ref. [2]

nanoparticles, as seen in Fig. 9.43. Note that the healing of the base specimens and of the precracked specimens were performed at 1300 °C for 1 h. The crack-healing ability of the Al_2O_3/SiC composite material is better than that of monolithic alumina. In Fig. 9.42, one may observe that these base specimens broke in many places. The macroscopic fracture surface of the specimen ($N_f = 3900$ cycles) is shown in Fig. 9.42a. An enlarged photograph of the crack-initiation point appears in Fig. 9.42b, where there are large defects. Yet, in the fracture-surface photograph of the specimen ($N_f = 15885$ cycles) in Fig. 9.42c, the special defect cannot be recognized at the crack initiation point. One may conclude from the σ_{f0} result in Fig. 9.41 that the cyclic-fatigue strength of the base specimen is equal to or higher than the value of the crack-healed zone of the alumina composite material.

Recall that nanostructured materials are those materials whose structural elements (clusters, crystallites or molecules) have dimensions in the range of 1–100 nm. Finally, it may be said that the S–N response of nanocrystalline

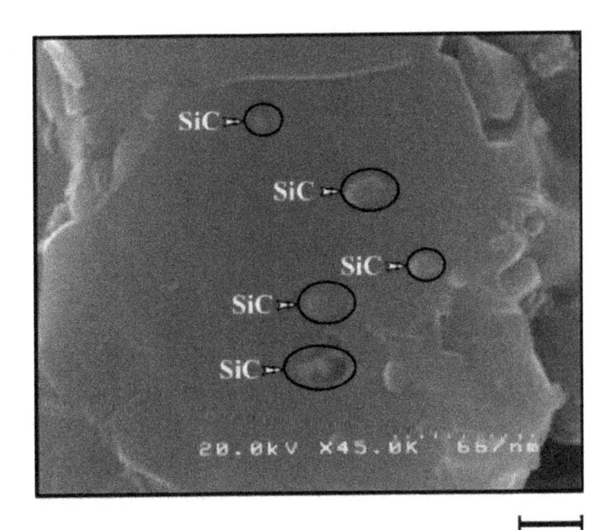

Fig. 9.43 Nano-size SiC particle located in alumina grain. With kind permission of Professor Ando for the authors of Ref. [1] and Professor Liu of Ref. [2]

materials to cyclic loading are superior to those of their corresponding conventionally-grain-sized ones, mainly because they have a higher yield and monotonic stresses.

9.5 Superplastic Observations in Nanoceramics

9.5.1 Introduction

In an earlier chapter, reference was made to a relation called the Mukherjee-Bird-Dorn equation for superplastic deformation (in metals), reproduced here as:

$$\dot{\varepsilon} = A\frac{D_0 Gb}{kT}\left(\frac{b}{d}\right)^p\left(\frac{\sigma}{G}\right)^n \exp\left(-\frac{Q}{RT}\right) \tag{9.6}$$

in which, as usual, G is the elastic shear modulus, b the Burger's vector, k Boltzmann's constant, with T representing the absolute temperature. The grain size is commonly denoted by d; p is the grain-dependent coefficient, while n is the stress exponent. Q and D_0 are the diffusion parameters, namely the activation energy and the frequency factor, respectively, and R is the gas constant. Also recall that the inverse of n is the strain-rate sensitivity, m. It has been widely considered that GBS is the dominant mechanism in superplastic flow. The values of n or m for GBS have been experimentally evaluated and are given as: n = 2 or m = 0.5. It is immediately obvious from Eq. (9.6) that, at constant stress and temperature, the high strain rate increases with decreasing grain size.

Various sophisticated techniques have been developed over the past decade or so, such that certain nanoceramics can now be produced with some measure of superplasticity. The present discussion will begin with static mechanical properties, such as the stress–strain relation and hardness.

9.5.2 *Static Properties Observed in Superplastic Ceramics*

9.5.2.1 Stress–Strain Relation

The stress–strain relation is used to determine the value of m, the strain-rate sensitivity exponent. The value of m is derived from the slope of such plots, indicating the existence of GBS, leading to superplastic behavior. Such plots are illustrated in Fig. 9.44b. The plotted material provided in Fig. 9.44 is a zirconia 3 mol% yttria–alumina–alumina–magnesia spinel nanoceramic composite. Here, PS-HEBM-SPS stands for specimens that were processed from plasma-sprayed [henceforth: PS] powders, undergoing high-energy ball-milling [henceforth: HEBM] (24 h) and spark plasma sintering (SPS).

A typical jump test is shown in Fig. 9.44 of flow stress versus strain. The stress temperature is 1450 °C. The slopes of each line in Fig. 9.44b yield for m $\sim$0.5, meaning that the stress exponent of the strain rate is $\sim$2; this indicates the superplastic behavior of the zirconia–alumina-spinel composite under the test conditions of temperature and strain rate. In order to determine the activation energy, a plot of strain rate versus the inverse absolute temperature must be made (as in Fig. 9.44c). The average activation energy of PS-HEBM-SPS is 945 kJ/mol, which is much higher than that of the composite processed from nanopowder mixtures (622 kJ/mol). This should represent GBS, if the concept of superplasticity is the dominant mechanism of deformation. Table 9.1 summarizes the strain rates and various temperatures of two and/or three specimens. PS-SPS appears in the Table 9.1 as PS-SPS and is listed under column C. For the purpose of comparison, the flow-stress results for nanopowder mixtures are also listed in Table 9.1 and are smaller than those processed from PS powders with/without HEBM.

It is interesting to see the relevant microstructures of these ceramics, shown in Fig. 9.45. SEM images of the fracture surfaces of the deformed (PS-HEBM-SPS) specimen (Fig. 9.45a, c and e) show three different kinds of microstructures: in a, equiaxed grains with rounded corners; in c, deformed dense agglomerates; and in e, non-deformed dense PS agglomerates. In a, evidence of GBS is visible, which is related to superplasticity. It represents deformed specimens processed from nanopowder mixtures; this structure was formed from the fine particles created by HEBM. Due to the short sintering time, no meaningful grain growth occurred. The shape of the particles in e were not affected by HEBM and, thus, kept their shapes during SPS and deformation. The strongly-bound grains inside the hard

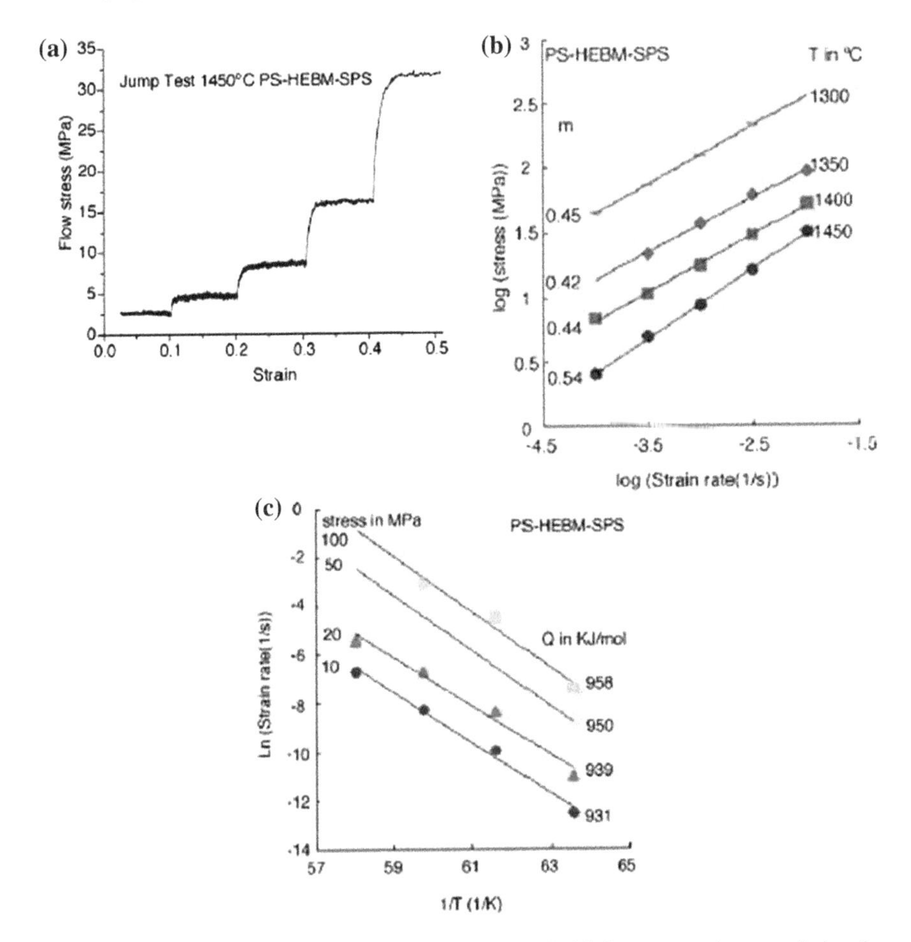

Fig. 9.44 **a** Jump test of PS-HEBM-SPS specimen at 1450 °C; **b** stress–strain *curves* in log–log scale to determine strain rate sensitivity m; **c** log(strain rate)-1/T *curves* to determine the apparent activation energy [41]. With kind permission of Elsevier

agglomerates did not contribute to GBS. The microstructures in b, d and f are of the deformed PS-SPS samples.

It is assumed that the difference in the superplastic behaviors, especially in their activation energies, implies different GBS energies. The grains in the specimens processed from nanopowder mixtures are more random in orientation and the grain boundaries are more prone to have high angles and high energy. As a result, GBS should be easier in nanopowder-based specimens than in PS powder-based ones which did not change during deformation. (In PS powder-based specimens, the primary grains inside the particles were formed by a nucleation and growth process from the metastable phase at elevated temperatures.)

Table 9.1 Flow stresses of the three ceramics composites *A*: nanopowder mixtures, *B*: PS-HEBM-SPS, *C*: PS-SpS [41] (with kind permission of Elsevier)

Strain rate (s^{-1})	Flow stress (MPa)									
	1300 °C		1350 °C		1400 °C			1450 °C		
	A	B	A	B	A	B	C	A	B	C
1×10^{-4}	–[a]	–[a]	–[a]	46	–[a]	7	118	–[a]	3	–[a]
3.16×10^{-4}	–[a]	74	–[a]	22	–[a]	11	170	–[a]	5	–[a]
1×10^{-3}	65	125	28	37	13	18	236	7	9	90
3.16×10^{-3}	127	216	56	61	24	30	–[b]	13	16	149
1×10^{-2}	227	–[c]	108	110	48	52	–[b]	24	32	197

Notes [a] not tested
[b] sample fractured during deformation
[c] beyond equipment load limit

9.5.2.2 Hardness in Superplastic Nanoceramics

Having already reviewed hardness in nanoceramics, the discussion here will focus specifically on the nanohardness observed in superplastic nanoceramics, again exemplified by TZP, since it shows superplasticity and is (among others) amenable to nanograined, bulk ceramic production with improved mechanical properties. TZP has attracted considerable attention due to these favorable qualities. Moreover, Y-TZP has good toughness and, by the addition of small quantities (3 mol%) of yttria, its fracture toughness (K_{Ic}) also increases. The variations in the Vickers hardness under true compressive stress and in relation to grain size also are illustrated in Figs. 9.46 and 9.47. Note that the grain size is 0.3 μm or 300 nm.

These plots show also fracture toughness (discussed below). One may see that the relatively insignificant increase in Vickers hardness as a function of the compressive strain is temperature-dependent and, at a somewhat higher temperature, it even decreases for both strain rates applied. The variation of the Vickers hardness with grain size may be seen in Fig. 9.47.

Note that these illustrations describe a high nano level (in the hundreds), although proper observations of ceramic toughness and hardness during superplastic deformation are best observed at a lower nano levels. The author believes that one of the reasons for the insignificant increase in the hardness level is to be expected. A common, if not general, observation in metals for example is that an increase in strength (or strength properties) is achieved at the expense of ductility. Superplastic deformation is an extreme ductile behavior and, thus, some loss of hardness, in certain cases, should not be surprising. According to the authors of these illustrations, the reason for the decrease in observed hardness is due to the formation of additional cracks; Fig. 9.48 makes their case clear.

An additional explanation may be added–the initial increase in hardness during superplastic deformation under compression is an outcome of the shrinkage of residual pores already contained within the specimen before deformation.

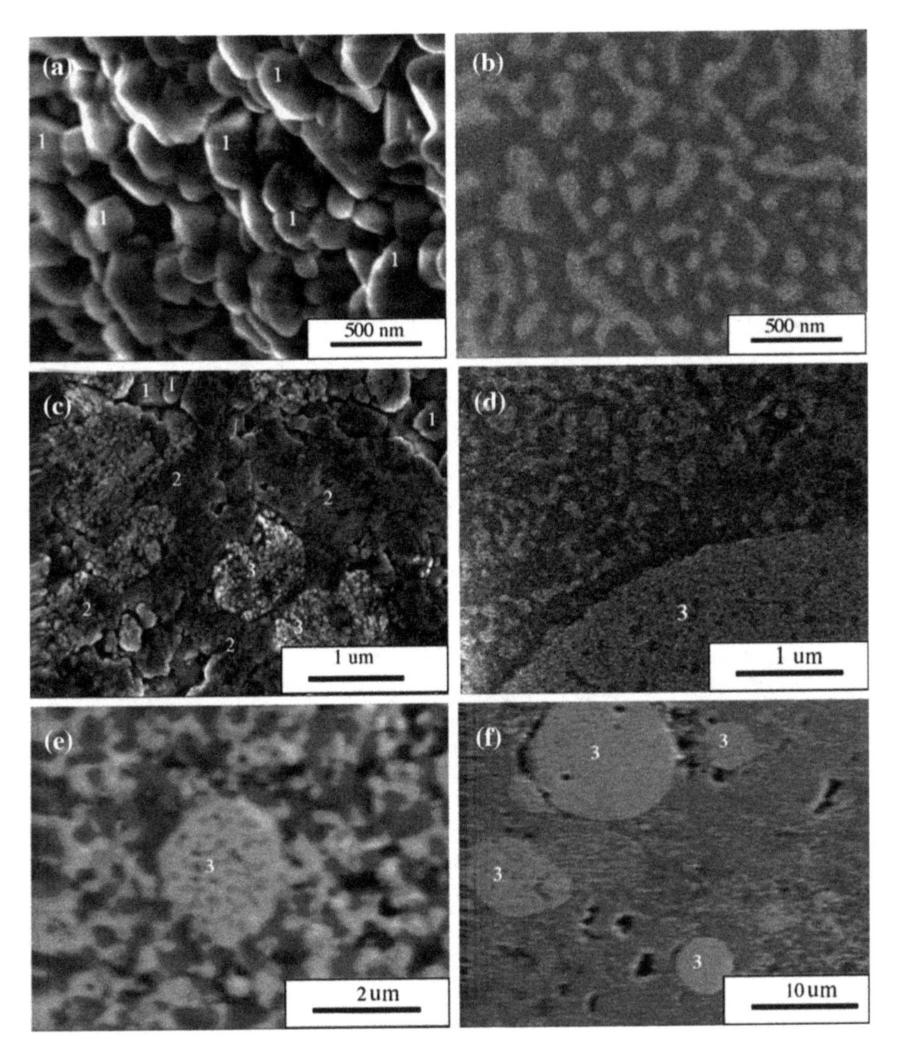

Fig. 9.45 SEM images of deformed PS-HEBM-SPS (**a**, **c**, **e**) and PS-SPS (**b**, **d**, **f**) at 1400 °C (*arrow 1* grains with round corners; *arrow 2* deformed hard agglomerates; *arrow 3* not deformed hard agglomerate in spherical shape) [41]. With kind permission of Elsevier

The action of the compressive stress shrinks those innate pores causing densification. The following decrease in hardness, after further compressive stress, is then due to a decrease in densification with the formation of new cavities by GBS.

In an earlier chapter, the effect of grain size on the strength of ceramics at ambient temperature was given as:

$$\sigma_b = Ad^{-n} \tag{9.7}$$

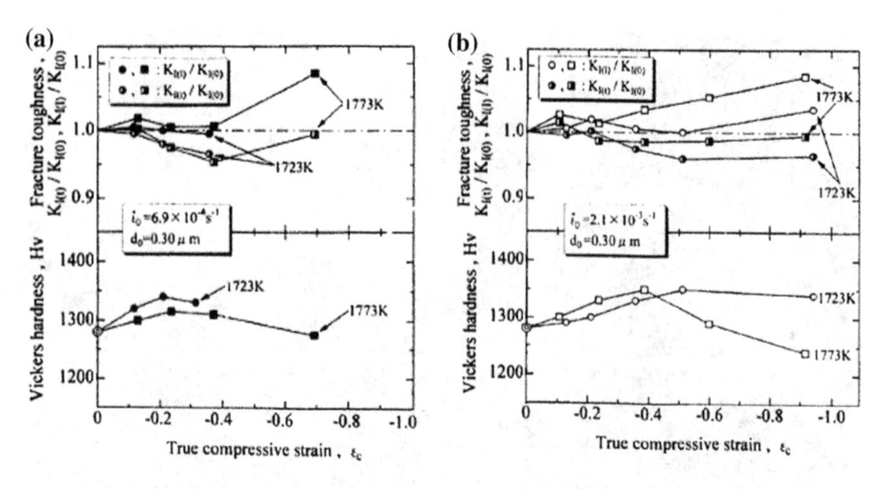

Fig. 9.46 Variations in micro-Vickers hardness and normalized fracture toughness with true compressive strain for specimen tested at strain rates of: **a** $6.9 \times 10^{-4}\ s^{-1}$; and **b** $2.1 \times 10p^{-3}\ s^{1}$ [21]. With kind permission of Elsevier

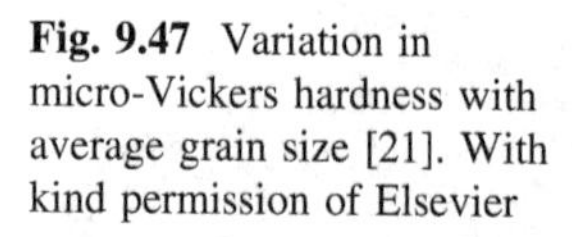

Fig. 9.47 Variation in micro-Vickers hardness with average grain size [21]. With kind permission of Elsevier

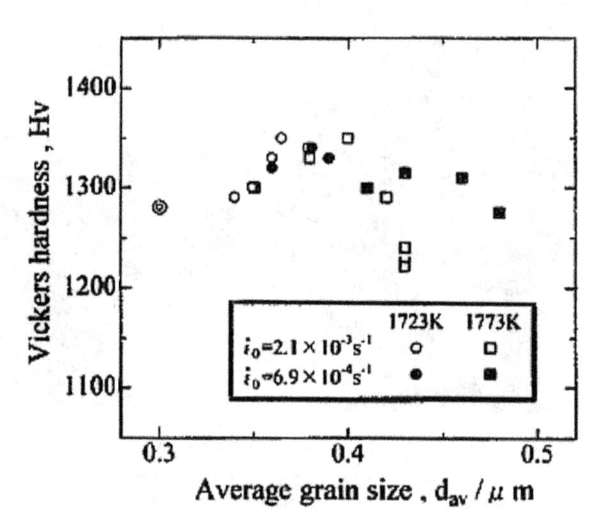

Here, the stress subscript indicates bending and, as before, n represents the grain-size exponent, while A is a constant. The value of n depends on the material having a usual value of 0.4–0.5. As seen from this relation, bending strength becomes smaller with increasing d (grain size). It is often mentioned in the literature that the bending strength Eq. (9.7) also holds for hardness, which also shows a similar variation, i.e., when bending stress increases so does hardness. Furthermore, bending strength depends on the pores present in the ceramics, often expressed as:

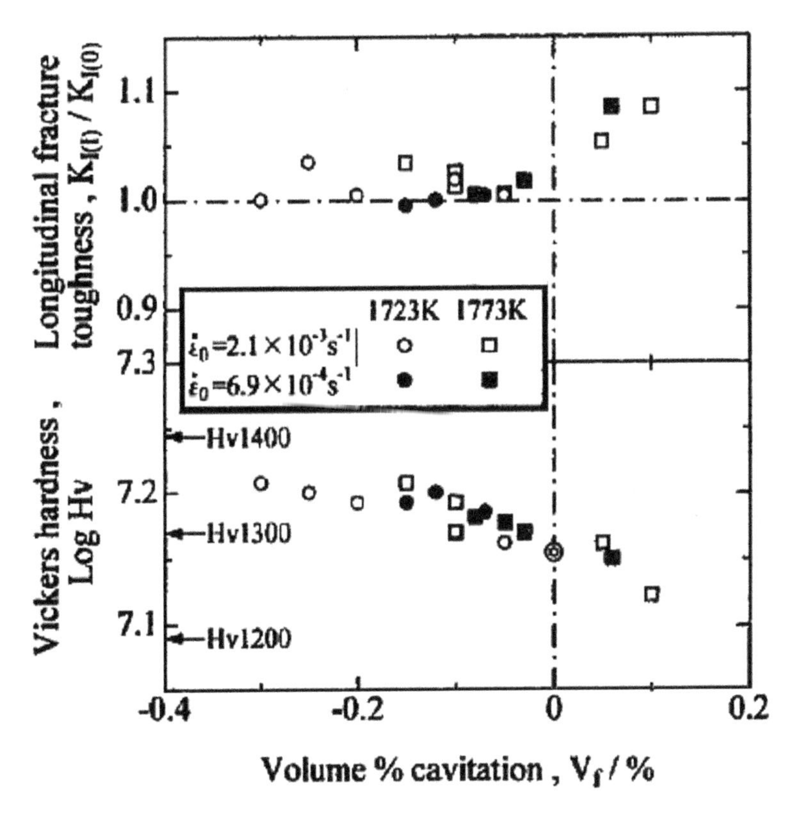

Fig. 9.48 Variations in micro-Vickers hardness and normalized longitudinal fracture toughness with volume fraction of cavities [21]. With kind permission of Elsevier

$$\sigma_b = B\,\exp(-bp) \tag{9.8}$$

Therefore, hardness may be expressed by a similar expression:

$$H_v = B'\,\exp(-b'V_{t^-}) \tag{9.9}$$

In the above relations, B, b, B′ and b′ are constants and p is the volume fraction of porosity. V_t– is similar to the volume fraction of porosity. Thus, the relation implies that if hardness increases so does the bending stress, and vice versa. However, the hardness of 3Y-TZP may be more sensitive to the total amount of cavitation and residual pores, namely to the apparent density than to the effect of grain growth, as in the case of alumina [26], in which all the strength properties are more sensitive to the total amount of porosity.

The above data, on the strength of superplastic materials, indicate that the plastic deformation of 3Y-TZP by the application of compressive stress (e.g., by forging) may be accomplished without too much difficulty and without significantly reducing its strength. Variations in strength properties, such as hardness and toughness, are relatively small and there may even be some improvement in fracture toughness. Compressive strain seems to be sufficient for the practical shaping of superplastic ceramics to desired dimensions.

9.6 Fracture

9.6.1 Introduction

Fracture, fracture types and the theoretical strength of materials were thoroughly discussed in Chap. 8. Now it is interesting to investigate whether the fracture patterns and resistance (in terms of fracture toughness) are the same or different in nanoceramics. Recall that nanoceramics are characterized by two conflicting effects regarding strength (and, hence, resistance to fracture). First of all, a large number of grain boundaries are expected to strengthen macro-materials. Consider that larger nanomaterials (in the 100–800 nm range), those that do have dislocations that are impeded in their motion through grain boundaries, are not unlike bulk macrosized specimens. Secondly, GBS is a contributor to strain during deformation, which basically reduces the strength properties of a material. Whether these two effects compensate each other or whether one of them is dominant should be considered. Nevertheless, it is known that the strength properties of nanomaterials are higher than found in macroscopic specimens and, therefore, their resistance to fracture is also greater.

9.6.2 Fracture Toughness

In continuation of the discussion on the mechanical properties of nanoceramics, this section is dedicated to fracture toughness. Nanoceramics are superior if low-temperature sintering densifies them to a desired maximum level. Since this greatly reduces grain growth in the starting nanoceramic powder, carrying the strength advantages into the manufactured nanoceramics. However, since high-temperature processing, such as sintering, is likely to provide a fully-dense ceramic composite, the mechanical properties, especially hardness and strength, may be greatly improved even for high-temperature applications. One of the reasons for improvement in high-temperature fabrication is the elimination, or rather reduction, of the pores. The purpose of a proper additive in a ceramic composite is to pin grain boundaries and prevent growth.

Below are several examples, parallel to those presented previously for regular ceramics, demonstrating the concept of fracture toughness for nanoceramics. First, take zirconia and zirconia–alumina nanoceramics, in which cracking occurs in a Palmqvist mode. In this case, Niihara et al. [22] express fracture toughness as:

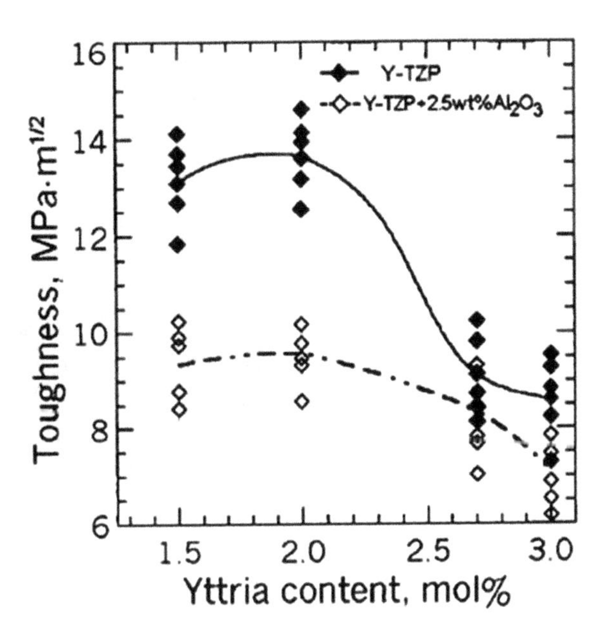

Fig. 9.49 Fracture toughness versus yttria-stabilizer content for fully tetragonal Y-TZP ceramics, and 2.5 wt% Al_2O_3/Y-TZP composites [35]. With kind permission of John Wiley and Sons

$$\left(\frac{K_{Ic}\phi}{Ha^{1/2}}\right)\left(\frac{H}{E\phi}\right)^{2/5} = 0.035\left(\frac{l}{a}\right)^{-1/2} \tag{9.10}$$

H is the Vickers hardness, E is the Young's modulus, 2a = d is the diagonal of the indentation, ϕ is the constrained factor and l is the crack length. As is clear from many hardness indentations on ceramics, Palmqvist cracks are observed at the ends of the diagonals of the indentations. The criteria for such cracks are given as:

$$0.25 \leq \frac{1}{a} \leq 2.5^{23} \tag{9.11}$$

Expressing fracture toughness according to the above relations yields:

$$K_{Ic} = 9.052 \times 10^{-3} H^{3/5} E^{2/5} d(l)^{-1/2} \tag{9.12}$$

A value of E = 210 GPa has been assumed for all of the ceramic samples, irrespective of their compositions. In addition, the crack lengths were measured immediately after the indentation was conducted to avoid slow crack growth after removing the load. The ceramic under consideration is YSZ. The above relations show the connection between hardness and fracture toughness. Fracture toughness versus the yttria-stabilizer content in fully tetragonal Y-TZP ceramics and 2.5 wt% Al_2O_3/Y-TZP composites are shown in Fig. 9.49. The average value of 8.62 MPa $m^{1/2}$ is indicated. The microstructure of the 2.5 % yttria content, represented by the dashed line in Fig. 9.49, is illustrated in Fig. 9.50.

Fig. 9.50 SEM micrograph of 2.5 wt% alumina/3Y-TZP ceramic sintered at 1150 °C for 20 h [35]. With kind permission of John Wiley and Sons

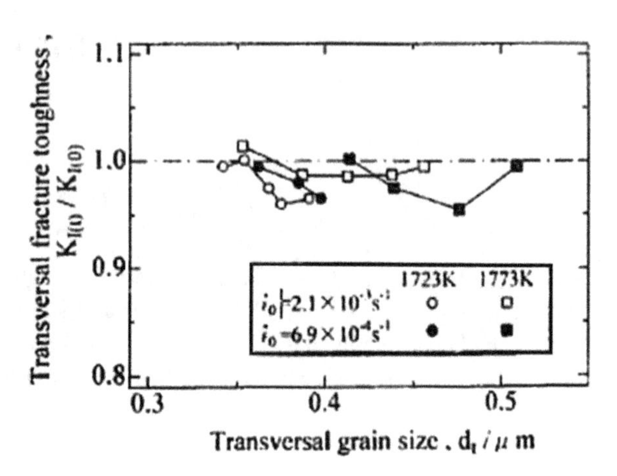

Fig. 9.51 Variation of normalized transversal fracture toughness with transversal grain size [21]. With kind permission of Elsevier

Grain size (which can grow) is one of the major factors that can affect fracture toughness. Other key factors are pores and grain boundary cavities. The change in the normalized transverse fracture toughness, $K_{I(t)}/K_{I(0)}$, with the transverse grain size, d_t, is shown in Fig. 9.51. Note that the $K_{I(t)}/KI_{(0)}$ value decreases initially with an increase in d_t, implying that fracture toughness decreases when individual grains grow. With a further increase in d_t, the $K_{I(t)}/K_{I(0)}$ value begins to rise after showing a minimum. The initial strain rates indicated in Fig. 9.51 are 6.9×10^{-4} and $2.1 \times 10^{-3}s^{-1}$ and the test temperatures are 1723 and 1773 K. The fracture toughness values are calculated [15] by Eq. (9.13), where E is clearly a Youngs modulus, C is the radius of the median crack, and the subscripts refer to longitudinal and transverse directions.

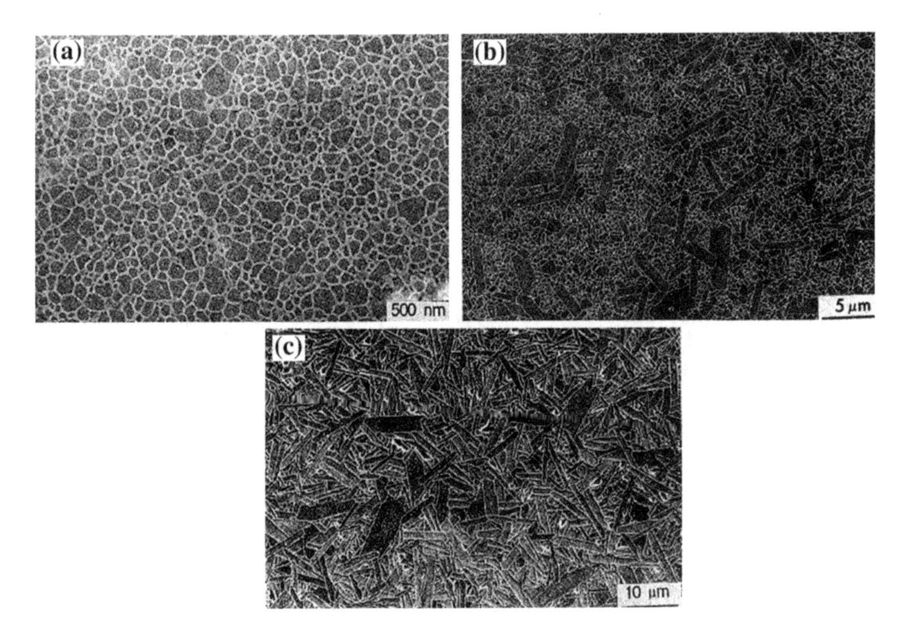

Fig. 9.52 SEM micrographs of hot-pressed and annealed materials: **a** SC0, **b** SC1, and **c** SC2 [14]. With kind permission of John Wiley and Sons

$$K_{I(t)} = 0.019\, E^{0.5} a_t^{-0.5} \left(\frac{C_t}{a_t}\right)^{-1.5} \tag{9.13}$$

Fracture toughness, calculated from C of the crack parallel to the compressive stress, is defined by longitudinal $K_{I(l)}$ and the normal, as the transverse fracture toughness, $K_{I(t)}$. As usual, P in these relations represents the load–in this case, the indentation load. Observe in Fig. 9.51 that the degradation of the mechanical properties (expressed as normalized fracture toughness) is quite small, if not negligible, which means that, under compressive stress, the plastic working of 3Y-TZP, such as forging, may be performed without too much difficulty.

The next example is β-SiC, to which 7 wt% Al_2O_3 is added. The microstructure of this HIPed and annealed powder is shown in Fig. 9.52. The average grain size of the FG SiC was 0.11 μm. SC0 is a FG ceramic with equiaxed 110 nm SiC grains. SC1 refers to large, elongated grains grown in a matrix of small grains. In SC2, the shapes of both the matrix and large grains are elongated. The frequencies of the grain distributions for all three types of composites are shown in Fig. 9.53.

Tables 9.2 and 9.3 list microstructural characteristics, fracture toughness, densification and annealing conditions, respectively. In Fig. 9.54, crack bridging by large SiC grains is indicated. Recall that crack bridging improves fracture resistance by acting as bridges between opposite faces of a crack and, during the

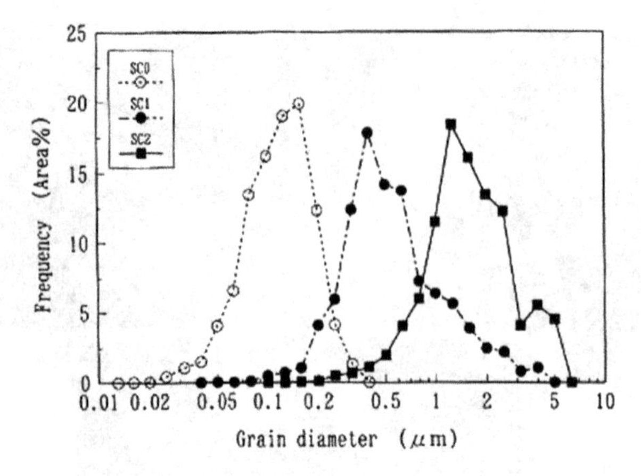

Fig. 9.53 Grain size distribution revealed by the relation between grain diameter and areal frequency [14]. With kind permission of John Wiley and Sons

Table 9.2 Microstructural characteristics and fracture toughness of hot-pressed and annealed materials [14] (with kind permission of John Wiley and Sons)

Materials	Densification and annealing conditions	Relative density (%)	Fracture toughness ($MPa{\cdot}m^{1/2}$)	Matrix grains		Large grains		
				d_{AVG} (μm)	R_{95} (μm)	Vol fraction (%)	d_{LG} (μm)	R_{LG} (μm)
SC0	Hot-pressed at 1750 °C for 15 min with 20 MPa	97.2	1.9	0.11	2.25			
SC1	SC0 is annealed at 1850 °C for 6 h	96.3	4.2	0.40	3.41	17.6	1.34	3.88
SC2	SC0 is annealed at 1850 °C for 12 h	95.4	6.1	1.28	5.25	10.1	3.59	3.26

Table 9.3 Polytypes in hot-pressed and annealed materials [14] (with kind permission of John Wiley and Sons)

Materials	Densification and annealing conditions	Composition (%)	
		3C	4H
SC0	Hot-pressed at 1750 °C for 15 min with 20 MPa	93	7
SC1	SCO is annealed at 1850 °C for 6 h	90	10
SC2	SCO is annealed at 1850 °C for 12 h	88	12

course of crack opening and propagation, the cracks deform plastically and fail by ductile rupture. Bridging prevents excessive crack opening. The variation of fracture toughness with indentation load appears in Fig. 9.55 for all three types of

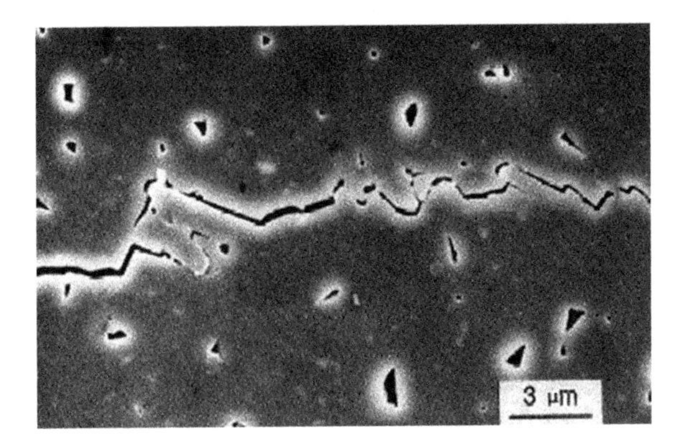

Fig. 9.54 SEM view of crack bridging by large SiC grains in SC2 [14]. With kind permission of John Wiley and Sons

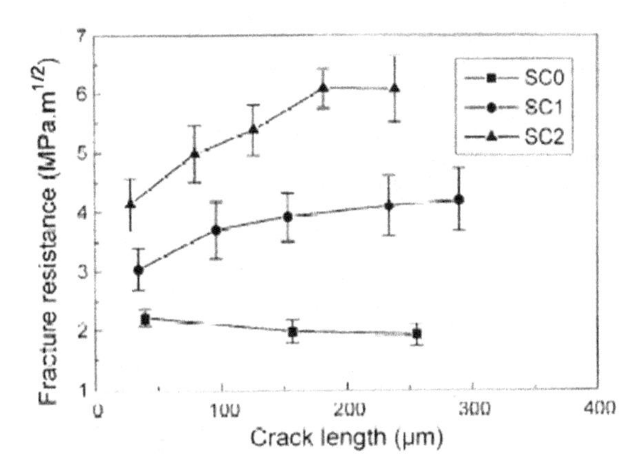

Fig. 9.55 Relation between fracture toughness and crack length for SC0, SC1 and SC2 [14]. With kind permission of John Wiley and Sons

Table 9.4 Nomenclature assigned for the nanoceramic composites developed [5] (with kind permission of John Wiley and Sons)

Sample designation	ZrO_2 matrix (mol% yttria)	Comments
TM2B	2	Mixed grade
TM2.25B	2.25	Mixed grade
TM2.5B	2.5	Mixed grade
T3B	3	Co-precipitated grade
T2B	2	Co-precipitated grade

B stands for ZrB_2

composites. As can be seen, there is an increase in fracture resistance with crack length in SC1 and SC2, but the toughness in SC0 is insensitive to indentation load. This suggests that annealing improves the fracture resistance of SiC, as seen from the data in Tables 9.2 and 9.3. Table 9.2 also shows that the fracture toughness of β-SiC increases as the result of the bridging by the larger, elongated grains.

For a more comprehensive picture of fracture and fracture resistance, here is yet another exemplary nanocomposite–zirconia reinforced with ZrB_2. The many

Table 9.5 Overall properties, i.e., relative density (RD), elastic modulus (E), hardness (H_{V10}) and fracture toughness (K_{Ic}) of the ZrO_2/ZrB_2 nanocomposites, SPSed at 1200 °C [5] (with kind permission of John Wiley and Sons)

Material designation	Relative density (% ρ_{th})	Elastic modules (E, GP_a)	Vickers hardness (Hv_{10}, GPa)
TM2B	95.4	250	12.3 ± 0.1
TM2.25B	97.8	260	1.31 ± 0.2
TM2.5B	98.4	257	13.7 ± 0.1
T3B	98.4	266	13.9 ± 0.3
T2B	98.5	261	13.9 ± 0.2

The standard deviation in the hardness data measured is also shown

Fig. 9.56 SEM fractographs of spark plasma-sintered ZrO_2–ZrB_2 nanoceramic composites, sintered at 1200 °C for a holding period of 5 min in vacuum: T3B grade (**a**), T2B grade (**b**), and TM2B grade (**c**). The presence of finer ZrO_2 grains (100–300 nm) and coarser ZrB_2 particulates (2–3 μm) can be distinguished. A model ceramic nanocomposite microstructure with nanosized matrix particles reinforced with microsized reinforcement particulates, as observed in the newly developed materials, is shown in (**d**) [5]. With kind permission of John Wiley and Sons

applications of zirconia-based ceramics have earned them their reputation as an important structural ceramic, thanks to their excellent combination of fracture toughness and strength. Tables 9.4 and 9.5 characterize some of these zirconia-based composites. Microstructures of these composites are shown in Fig. 9.56.

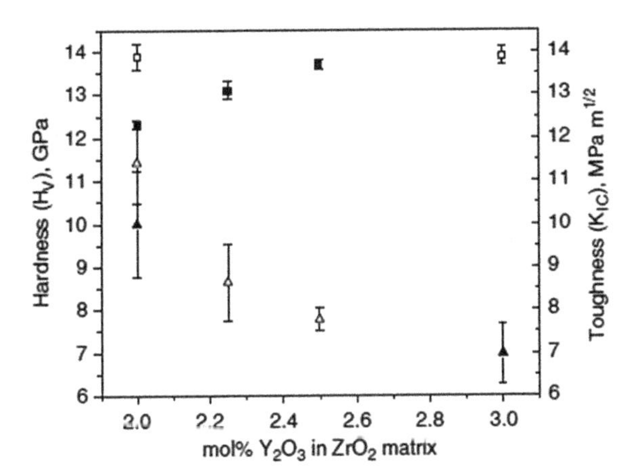

Fig. 9.57 Plot of hardness and toughness versus yttria stabilization ZrO_2 matrix for spark plasma-sintered (1200 °C, 5 min) nanocomposites. Different symbols represent the mechanical property measured with various composites based on the ZrO_2 matrix processed from either coprecipitated or mixed-grade ZrO_2 powders: *filled square*, H_{v10} of the mixed grades; *open square*, H_{v10} of the co-precipitated grades; m, KIc of the co-precipitated grades; *filled triangle* K_{Ic} of the coprecipitated grades (using the Palmqvist formulae); and *open triangle*, KIc of the mixed grades (using the Plamqvist formulae) [5]. With kind permission of John Wiley and Sons

Selected SEM fractographs of both mixed and co-precipitated ZrO_2 powder-based composites are shown in this figure.

Based on SEM images, the microstructure is characterized by the presence of coarser tabular elongated ZrB_2 particles (~2–3 mm) and equiaxed nano-ZrO_2 particles (~100–300 nm). A schematic illustration of the phases in the microstructure is also presented. The ZrO_2 grains in T3B are relatively finer (100–150 nm) than those in T2B and TM2.25B composites. The average ZrO_2 grain size in T2B is ~200–300 nm. The presence of a few coarse ZrO_2 grains of ~300 nm or more is found in TM2B. The variation of the hardness and the fracture toughness with the amount of Y_2O_3 in the ZrO_2 matrix is seen in Fig. 9.57.

There is almost no change in hardness with the addition of Y_2O_3; the values are between 12 and 14 GPa, as listed in Table 9.5. The absence of improvement of hardness, may be attributed to the presence of pores or microcracks, apparently formed during post-fabrication cooling or possibly developed due to indentation stresses. This may occur even after a 30 wt% ZrB_2 (a hard additive) was incorporated into the nanosized zirconia composite. Grain growth in ZrB_2, which may also have occurred, is another probable factor involved in the hardness values that were obtained. The toughness values were measured from an analysis of the indentation. The K_{Ic} of brittle materials, exhibiting radial-median cracks when $l/a > 2.5$, may be calculated (Anstis's model [3]) according to Eq. (9.14). In this relation, E, H and P have their usual meanings: elastic constant, hardness and indentation load, respectively, and c is the half-crack:

Table 9.6 Indentation data, i.e., average indent diagonal length (2*a*) and total crack length (2*c*), as well as the fracture toughness values for the SPSed ZrO_2/ZrB_2 nanocomposites [5] (with kind permission of John Wiley and Sons)

Material designation	Indent diagonal (2*a*) (μm)	Crack length (2*c*) (μm)	*l*/*a*	Indentation toughness (MPa m$^{1/2}$)
TM2B	122	158	0.3	11.4 ± 1.0
TM2.25B	118	185	0.6	8.7 ± 0.9
TM2.5B	115	197	0.7	7.8 ± 0.3
T3B	115	210	0.9	6.9 ± 0.7
T2B	114	166	0.4	10.0 ± 1.2

The crack length parameter (l) is defined as $l = c - a$. The standard deviation in the measured toughness data is also shown

$$K_{Ic} = \eta \left(\frac{E}{H}\right)^{1/2} \frac{P}{c^{3/2}} \tag{9.14}$$

η is an indenter-geometry-dependent, dimensionless constant. Another (Palmqvist [25]) equation may also be applied to the case of Palmqvist-type cracks, when $0.25 < l/a < 2.5$, given as:

$$K_{Ic} = \eta \left(\frac{E}{H}\right)^{2/5} \frac{P}{al^{1/2}} \tag{9.15}$$

The parameters a and c of Eqs. (9.14) and (9.15) are indicated in Table 9.6 as 2a the average indent diagonal length, 2c the crack length and $l = c - a$. Equations (9.14) and (9.15) associate hardness measurements with fracture toughness. The indentation tests used to evaluate fracture toughness are shown in Fig. 9.58. The indentation data, including the indentation diagonal, crack length and indentation toughness are listed in Table 9.6.

The basis of these composites, shown in Tables 9.4, 9.5, 9.6, is zirconia, which is known to transform into a monoclinic polymorph if not stabilized. As mentioned previously, prior transformation toughening by ZrO_2 $t \rightarrow m$ ZrO_2 may occur in sintered and/or annealed composite powders, depending on the stabilization of the zirconia. In the case of ZrO_2–ZrB_2 stabilized by yttria additives, the transformability of % t-ZrO_2, as a function of the amount of yttria in the zircomia matrix, is presented in Fig. 9.59. It may be seen from the graphs that transformability, the difference between the fracture and the polished surface m-ZrO_2 content (%), varies around 50–60 % for all the composites. The data in Fig. 9.59 indicate that toughness shows an almost linear correlation with transformability.The difference in the transformability between the two grades (the mixed and the coprecipitated grades) should be attributed to the use of two different ZrO_2 starting powders in the production of the composites (see Table 9.4). Toughness increased in both grades of composites (see Fig. 9.57) with decreasing yttria (to a level of 2 mol%) in the zirconia matrix; this correlated with the increase in measured t-ZrO_2 transformability. This is expected, since toughness is associated with the volume fraction of the transformable t-ZrO_2.

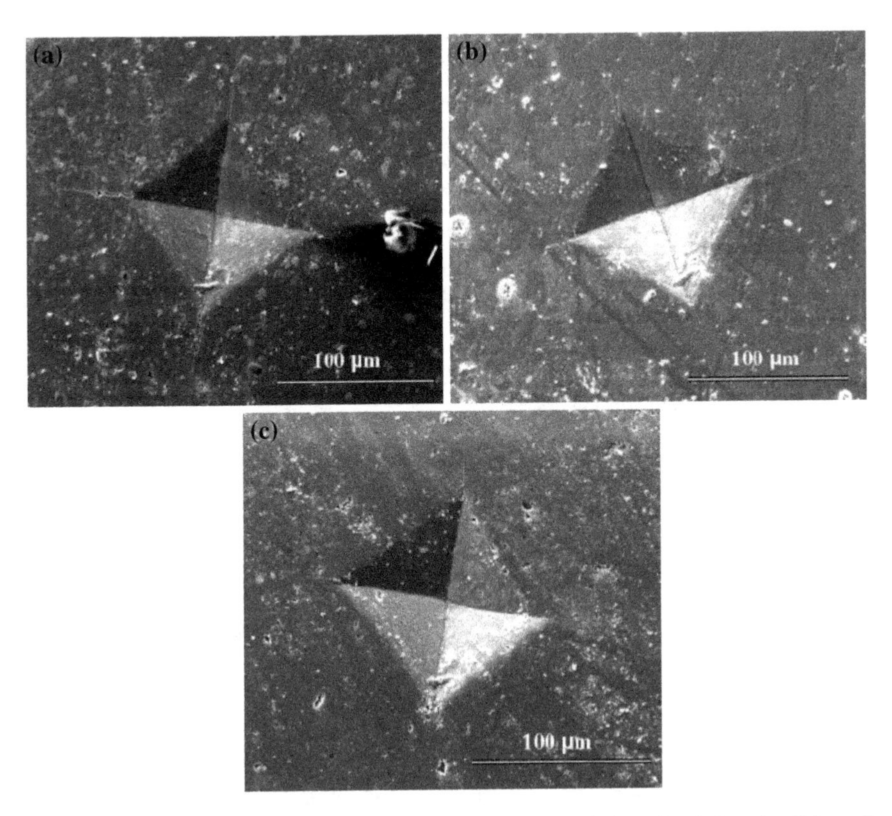

Fig. 9.58 SEM topography images of the Vickers indents and indentation-induced radial crack pattern in the SPS-processed zirconia nanoceramic composites: T3B, T2B, TM2B [5]. With kind permission of John Wiley and Sons

The t-ZrO_2 transformability is connected with the fact that a large amount of m-ZrO_2 was measured on the fractured surface, which is clearly associated with the transformable t-ZrO_2 in spark plasma sintered (SPSed) composites. The SPS processing route enables the retention of the finer t-ZrO_2 grains (100–300 nm) and the ZrO_2–ZrB_2 composite developed exhibits optimum hardness up to 14 GPa. The toughening mechanism for improving the fracture toughness of fine grained (FG) SPS composites is attributed to the contribution of transformation toughening.

Thus far, it has been stated that transformation toughening in zirconia-based composites may contribute to fracture toughness, in addition to the known grain-size effect. However, the paths of cracks should also be considered. If the path is obstructed and the crack is deflected from its route (known as tortuous path) by some hard particle, such as ZrB_2, toughening is expected. In Fig. 9.60, crack propagation may be observed.

Figure 9.60 reveals an increase in crack-path tortuosity due to crack deflection by hard and coarser ZrB_2 particles. A closer observation of Fig. 9.60b also reveals the

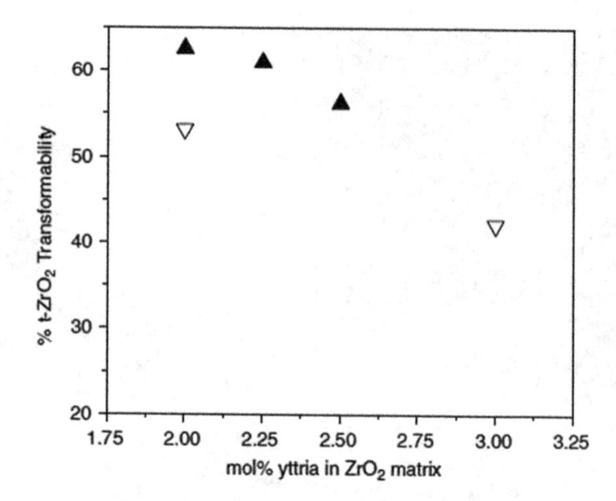

Fig. 9.59 Percentage transformability of the ZrO_2 matrix against the mol% of yttria content in that matrix at sintering temperatures of 1200 °C. The t-ZrO_2 transformability is defined as: (%m-ZrO_2 on the fractured surface—%m-ZrO_2 on the polished surface). The transformability data of various composite grades processed from the use of ZrO_2 powders are indicated by different symbols: *inversed open triangle*, co-precipitated grades; and *filled triangle*, mixed grades [5]. With kind permission of John Wiley and Sons

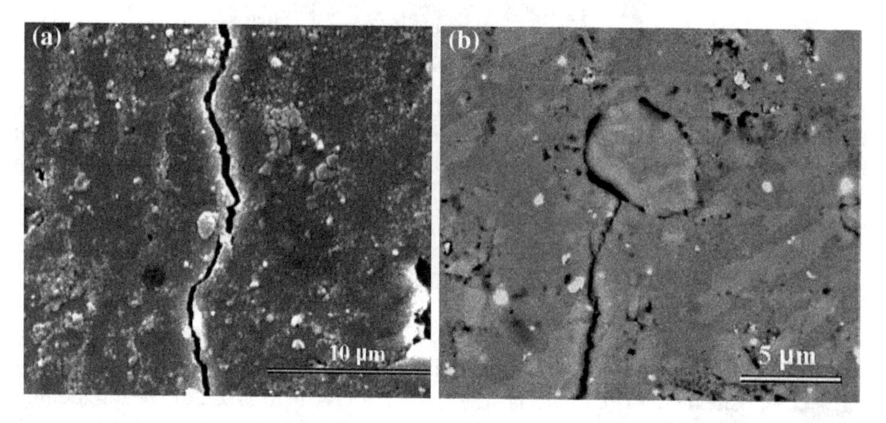

Fig. 9.60 Crack deflection by the hard second-phase ZrB_2 particles (**a**) and crack wake debonding of the coarser ZrB_2 particles at the interface (**b**) in the T2B nanocomposite, SPSed at 1200 °C for 5 min (heating rate: 600 K/min) [5]. With kind permission of John Wiley and Sons

crack-path debonding of the ZrB_2, as evident from the crack propagation around the ZrO_2/ZrB_2 interface. The ZrB_2 particles limit further crack propagation.

The experiments on composites shown in Tables 9.4, 9.5, 9.6 and Figs. 9.56, 9.57, 9.58, 9.59 and 9.60 raise the hope that in the future more experiments will be carried out on zirconia-based composites, exploring the possibilities and methods for utilizing transformation-induced toughening as an important contributor to increased toughness, improved fracture toughness, and enhanced resistance to fracture of these and other composites.

9.7 Epilogue

Almost all the aspects of mechanical behavior, which were discussed in the earlier chapters, have now been considered in regard to nanoceramics (in the submicron range). The mechanical properties of bulk macroceramics are governed by imperfections (defects), namely various point and line defects (dislocations). Point defects associated with properties such as creep engage in diffusion-governed-exchanges of such defects with atoms in their vicinity. This phenomenon also occurs in nanoceramics. Line defects also operate both in macro- and nanoceramics. However, a limit exists in nanoscale ceramics (below $\sim$100 nm), where dislocation motion is hindered due to spatial limitations. Then, a different deformation mechanism sets in–that of GBS. If dislocation plasticity is absent or impaired, the possibility of grain-boundary accommodation mechanisms, involving GBS, play a major role in deformation.

Strain rates associated with this process are proportional to the grain-boundary diffusivity coefficient and the grain diameter at some power. Generally, deformation by these mechanisms is observed in polycrystalline materials only at elevated temperatures. The hope is that higher, more effective grain-boundary diffusion coefficients may be realized in nanocrystalline ceramics, so that diffusion-based mechanisms may be induced and activated also at RT.

Although the observed mechanical properties in FG ceramics have been related to the GBS mechanism, the actual deformation mechanism is still under debate and may be material-dependent. A concept of plastic deformation has been suggested as a mechanism involving the non-local, homogeneous nucleation of nanoscale loops of partial dislocations; also unusual, nonlinear stress and grain-size dependence is assumed to facilitate nanocrystalline plasticity. However, the dominant mechanism is still GBS. The stress level required to nucleate a dislocation is much higher than usually encountered in experimental data. Therefore, dislocation gliding itself is not expected to contribute to total strain.

Naturally, different deformation mechanisms have different mechanical properties. Specifically, from available experimental evidence, nanoscale ceramics (and other materials as well) indeed show different mechanical properties. Higher strength is a common trait of nanoceramics, often found together with ductility; but the foremost characteristic of some nanoceramics is their ability to deform superplastically to elongations over 100 % and in certain cases to strains of hundreds of percents. Large strains, without necking, occur in superplasticity, if there are the necessary FGs (in the nano range). Sophisticated techniques, such as SPS, have been developed over the past two decades, enabling the manufacture of

FG nanoceramics from nanosized powders; the very small dimensions are prerequisite to the production of ceramics with reduced flaw content. Sometimes, it is also necessary to densify these powders by means of additives applied at elevated temperatures. However, low-temperature sintering and densification methods have been recently developed. Thus, there should not be any problems in obtaining nanoscale ceramics to facilitate the forming of materials into desired shapes. Forging, for example, might become cost-saving, because of the great plasticity of nanoceramics.

Because the difference in the deformation behavior in nanoscale materials from those of bulk macroscopic ones it is unlikely that their mechanical properties may be evaluated by means of direct extrapolation from tests performed on macroscopic specimens. Note, however, that the methods and equipment used in regular mechanical tests are generally not suitable for performing experiments on nanometric samples. Due to the sophisticated systems invented for directly studying nanoscale materials (including ceramics), much understanding of their behavior has been attained. Their mechanical properties, such as elasticity, plastic creep, fatigue and fracture can be investigated, as well as a variety of structures (not all of them included in this book), such as: nanofilms, nanowires, nanotubes and nanorods of various materials. The experimental research processes have led to the development of new test methods.

Despite the various views on the exact mechanism responsible for the observed nano-behavior, it is clear from the experiments that nanoceramics may exhibit increased strength (hardness, for example), improved toughness, improved ductility and high resistance to fatigue. All these improved properties serve as safeguards against unexpected or premature fracture in service. This attitude does not deny the existence of conflicting or contradictory results originating from experiments by various investigators in different laboratories. One has to remember that nanoscale materials are very sensitive to their initial microstructure, different starting powders for sintering and densification and different preparation procedures. Some contradictory experimental results did not prevent engineers from putting nanoscale materials into immediate use in industry, while keeping their eyes open and seeking reasons for the discrepancies.

References

1. Ando K, Kim B-S, Kodamama S, Liu S-P, Takahashi K, Saito S (2003) J Soc Mater Sci Jpn 52:1464
2. Ando K, Liu S-P (2004) J Chin Inst Eng 27:395
3. Anstis GR, Chantikul P, Lawn BR, Marshall DB (1981) J Am Ceram Soc 64:553
4. Averback RS, Höfler HJ, Hahn H, Logas JC (1992) Nanostruct Mater 1:173
5. Basu B, Venkateswaran T (2006) J Am Ceram Soc 89:2405
6. Chuang W-H, Fettig RK, Ghodssi R (2007) J Micromech Microeng 17:938
7. Dubrovinskaia N, Solozhenko VL, Dmitriev V, Oleksandr, Kurakevych O, Dubrovinsky L (2007) Appl Phys Lett 90:101912

8. Dubrovinskaia N, Solozhenko VL, Dmitriev V, Oleksandr, Kurakevych O, Tse LJS, Klug DD, Gao F (2006) Phys Rev B 73:140102
9. Dusza J, Kovalčík J, Hvizdoš P, Šajgalík P, Hnatko M, Reece MJ (2005) J Am Ceram Soc 88:1500
13. Gao L, Wang HZ, Hong JS, Miyamoto H, Miyamoto K, Nishikawa Y, Torre SDDL (1999) J Eur Ceram Soc 19:609
11. Hall EO (1951) Proc Phys Soc B 64:747
12. Jiang B, Peng JL, Bursilli LA, Ren TL, Zhang PL, Zhong WL (2000) Physica B 291:203
10. Jiang B, Weng GJ (2004) Int J Plast 20:2007
14. Kim Y-W, Mitomo M, Hirotsuru H (1995) J Am Ceram Soc 78:3145
15. Lawn BR, Evans AG, Marshall DB (1980) J Am Ceram Soc 63:574
16. Li JCM (1972) Surf Sci 31:12
17. Liu H, Huang C, Teng X, Wang H (2008) Mater Sci Eng A 487:258
18. Lu L, Chen X, Huang X, Lu K (2009) Science 323:607
19. Mazaheri M, Mari D, Hesabi ZR, Schaller R, Fantozzi G (2011) Composites Sci Technol 71:939
20. Meyers MA, Mishra A, Benson DJ (2006) Prog Mater Sci 51:427
21. Motohashi Y, Sekigami T, Sugeno N (1997) J Mater Process Technol 68:229
22. Niihara K, Morena R, Hasselman DPH (1982) J Mater Sci Lett 1:13
23. Ohji T, Nakahira A, Hirano T, Niihara K (1994) J Am Ceram Soc 77:2359
24. Ovid'ko IA, Scheinerman AG (2011) Rev Adv Mater Sci 29105
25. Palmqvist S (1962) Arch Eisenhuettenwes 33:629
26. Passmore EM, Spriggs RM, Vasilos T (1965) J Am Ceram Soc 48:1
27. Pelleg J (2013) Mechanical properties of materials. Springer, New York
28. Petch NJ (1953) J Iron Steel Inst 173:25
29. Petch NJ (1964) Acta Met 12:59
30. Shinoda Y (2012) Mechanisms of superplastic deformation of nanocrystalline silicon carbide ceramics, Army research laboratory, Aberdeen Proving Ground, MD 21005-5066, ARL-CR-702 August 2012 (Approved for public release; distribution is unlimited)
31. Skiba NV, Morozov NF, Ovid'ko IA (2013) Nanocon 10:16–18
32. Tian Y, Xu B, Yu D, Ma Y, Wang Y, Jiang Y, Hu W, Tang C, Gao Y, Luo K, Zhao Z, Wang L-M, Wen B, He J, Liu Z (2013) Nature 38(5):493
33. Tian Y, Xu B, Yu D, Ma Y, Wang Y, Jiang Y, Hu W, Tang C, Gao Y, Luo K, Zhao Z, Wang L-M, Wen B, He J, Liu Z (2013) Nature 493:385
34. Tse, S, Klug DD, Gao F (2006) Phys Rev B 73:140102
35. Vasylkiv O, Sakka Y, Skorokhod VV (2003) J Am Ceram Soc 86:299
36. Vasylkiv O, Sakka Y, Skorokhod VV (2003) J Am Ceram Soc 86:299
37. Venables JA (1961) Phil Mag 6:379
38. Wan J, Duan R-G, Gasch MJ, Mukherjee AK (2006) J Am Ceram Soc 89:274
39. Wang W-L, J-Q Bi, Wang S-R, Sun K-N, Du M, Long N-N, Bai Y-J (2011) J Eur Ceram Soc 31:2277
40. Xu X, Nishimura T, Hirosaki N, Xie R-J, Zhu Y, Yamamoto Y, Tanaka H (2005) J Am Ceram Soc 88:934
41. Zhou X, Hulbert DM, Kuntz JD, Sadangi RK, Shukla V, Kear BH, Mukherjee AK (2005) Mater Sci Eng A 394:353

Index

J. Pelleg, *Mechanical Properties of Ceramics*, Solid Mechanics and Its Applications 213, DOI: 10.1007/978-3-319-04492-7,

CPSIA information can be obtained
at www.ICGtesting.com
Printed in the USA
LVHW021923110819
627252LV00003B/9/P